LIFE *in the* FROZEN STATE

LIFE *in the* FROZEN STATE

Barry J. Fuller
Royal Free and University College London
London, U.K.

Nick Lane
Royal Free and University College London
London, U.K.

Erica E. Benson
University of Abertay Dundee
Dundee, U.K.

CRC PRESS

Boca Raton   London   New York   Washington, D.C.

Life in the Frozen State
Cover Images

1. Left-hand panel: Fjord scene – Iceland. (Courtesy E. Benson).

2. Right-hand panel: Top right – Cryomicroscope image of rat embryo during slow cooling. (Courtesy of L. McGann and J. Acker).

3. Second from top: Cryomicroscope image of initial extracellular ice formation in a suspension of the ciliated protozoan *Tetrahymena pyriformis.* (Courtesy of Dr. John Morris).

4. Third from top: Cryomicroscope image of the unicellular alga *Micrasterias americana* surrounded by extracellular ice; different ice crystals have different colors due to the optics used (differential interference contrast). In some cells intracellular ice has formed – evident as the cell interior appearing black. (Courtesy of Dr. John Morris).

5. Fourth from top: The unicellular alga *Micrasterias rotata* surrounded by extracellular ice, the cell interior has blackened due to the formation of intracellular ice. (Courtesy of Dr. John Morris).

Image 2 was provided by Professor Locksley McGann, Department of Laboratory Medicine and Pathology, University of Alberta, 8249-114 Street, Edmonton, Alberta T6G 2R8.

Images 3, 4 and 5 were provided courtesy of Dr. John Morris, Asymptote Ltd., St. John's Innovation Centre, Cambridge CB4 0WS, U.K. (www.asymptote.co.uk).

Library of Congress Cataloging-in-Publication Data

Life in the frozen state / edited by Barry J. Fuller, Nick Lane, Erica E. Benson.
    p. cm.
  Includes bibliographical references and index.
  ISBN 0-415-24700-4 (alk. paper)
   1. Cryobiology. I. Benson, Erica E. II. Fuller, Barry J. III. Lane, Nick.

QH324.9.C7L545 2004
571.4'645--dc22

2003069764

This book contains information obtained from authentic and highly regarded sources. Reprinted material is quoted with permission, and sources are indicated. A wide variety of references are listed. Reasonable efforts have been made to publish reliable data and information, but the authors and the publisher cannot assume responsibility for the validity of all materials or for the consequences of their use.

Visit the CRC Press Web site at www.crcpress.com

© 2004 by CRC Press LLC

No claim to original U.S. Government works
International Standard Book Number 0-415-24700-4
Library of Congress Card Number 2003069764
1 2 3 4 5 6 7 8 9 0

Foreword

The science of cryobiology pretty much began in the 1940s shortly after World War II. Before that time, research into subfreezing temperatures was largely the physicist's domain, with people like Bridgeman and Kammerling-Ones unravelling the physics and physical chemistry of ice. Biology was just emerging from its preoccupation with taxonomy. Histology and physiology extended no further than the limits of light microscopy and the early beginnings of modern biochemistry. Electron microscopy, newly born, was limited to transmission microscopy of nonhydrated materials. Neither sectioning nor scanning existed. At the 1953 meeting of the newly formed Electron Microscope Society, it was agreed that electron microscopy had no future in biology. Antibiotics were just being discovered and the "filterable viruses" were mysterious organisms known only through their effects. The term "genetic manipulation" meant selective breeding. Chemists relied heavily on massive German compendiums of known reactions. It had yet to be generally appreciated that molecules had physical shapes and specific charge distributions and that it was possible not only to understand the mechanism of chemical reactions but even to predict them. Detergents were just being developed. Computers used vacuum tubes and occupied warehouses.

In this context it is hardly surprising that biology and physics still had little in common. It was well into the 1950s before biophysics became a generally recognized specialty, and even at that time the number of true biophysicists was probably still in the double digits and populated primarily by physicists just discovering the frustrating and erratic complexity of living things.

It was in this environment, primitive compared to twenty-first-century science, that cryobiology struggled out of infancy. In the inclusive bibliography of Luyet's *Life and Death at Low Temperature* (Luyet and Gehenio, 1940), the first book devoted to low-temperature biology, there are only 97 references to biological freezing before 1900. Almost all of these were anecdotal observations of death or survival. Even in 1940, the many hypotheses regarding the nature of freezing injury were largely unsubstantiated by experimental evidence and even included such mysterious phenomena as the "direct action of cold."

There were a few first-class pioneers such as Sir William Hardy at the Low Temperature Institute at Cambridge University in England and Nord in Germany. Even as late as 1966, the foreword to the volume, *Cryobiology* (Meryman, 1966), states that, "the number of investigators primarily concerned with the fundamental nature of freezing injury can still be counted on the fingers of two hands."

Under these circumstances it was not surprising that early forays into low-temperature biology were largely descriptive and mostly lacked the rigor of physics. Luyet, commonly referred to as the Father of Cryobiology, focused almost totally on observation, description, and categorization, leaving most of the experimental work, which was largely empirical, to his students and coworkers. Most early cryobiologists were untrained in the physical sciences. Audrey Smith was a classical biologist. Ronald Greaves, second president of the Cryobiology Society and a pioneer in freeze-drying of living organisms, was a physician pathologist with a bent for engineering. I was a newly minted physician whom the Navy, during the aftermath of WWII, had assigned to mother the first U.S.-made electron microscope, the RCA EMB, despite the fact that my relevant experience was limited to the repair and maintenance of the Model A Ford. My tiny lab, staff of one, was also presented with the first, huge Pickels ultracentrifuge as well as an original Tiselius electrophoresis apparatus complete with giant water bath and a 15-foot optical path frequently obstructed by wandering cockroaches.

Luyet based most of his efforts at applied cryopreservation on the premise that ultra-rapid freezing and thawing could prevent freezing injury. The rates required, however, were and are, in the absence of glass-inducing solutes, still beyond reach. The critical development that put the nascent field of cryobiology on the map was the report (Polge et al., 1949) of the cryoprotection of fowl sperm by glycerol and, a year later, the confirmation by Audrey Smith (Smith, 1950) of the effectiveness of glycerol for red cell freezing. In the absence of insight into the nature of freezing injury, the discovery of glycerol was basically empirical. Equally empirical was our development of red cell preservation by spray-freezing onto liquid nitrogen (Meryman and Kafig, 1955), which resulted from the observation of intact erythrocytes viewed by electron microscopy, using a technique for replicating hydrated samples at liquid nitrogen temperature (Meryman, 1950). The cryoprotective properties of polymers was another empirical discovery, resulting from a survey of many possible additives by Arthur Rinfret and colleagues (Doebbler et al., 1966) at the Linde division of Union Carbide as part of their efforts to engineer the rapid freezing of red cells to practicality.

Jim Lovelock was probably the first serious cryobiologist with real biophysical credentials, and the discovery of dimethyl sulfoxide (Me_2SO) by Lovelock and Bishop (1959) was based on rational prediction. Peter Mazur brought physical and mathematical skills to the field and new, more rigorous standards to replace the largely anecdotal and uncontrolled experimental work of the earlier days, stimulating a burst of serious efforts to understand freezing injury.

Lovelock had originally proposed that denaturation by concentrated salts was responsible for cell damage during freezing (Lovelock, 1953). Mazur's two-factor hypothesis (Mazur, 1965), mechanical injury from intracellular ice during fast freezing and dehydration injury from extracellular ice during slow freezing, was a landmark event, although the precise mechanism of slow freezing injury remained unresolved. The long-held suspicion that living cells did not contain heterogeneous ice nuclei and could be deeply supercooled was finally confirmed in 1975 (Rasmussen et al., 1975). Mazur (1966) subsequently proposed that the moment of injury coincides with the temperature at which the radius of curvature of a growing extracellular ice crystal approximates the radius of pores in the membrane, seeding the intracellular solution with physically destructive ice. We (Meryman, 1968) proposed a compressive membrane stress from osmotic cell shrinkage that resulted in an undefined membrane breakdown, the "minimum volume" hypothesis. Bob Williams (Williams et al., 1975) demonstrated that Karkhov wheat cells, like many other plant cells, lost membrane material under hypertonic stress but achieved hardiness by recovering it on return to isotonicity.

The recent elegant studies of membrane dehydration by the Crowes (Crowe et al., 2001) have contributed greatly to an understanding of dehydration injury and tolerance—certainly one of the many factors at play during cell and tissue freezing. Cold denaturation of proteins was first predicted by Brandts (Brandts, 1964) and was subsequently confirmed by Tsonev and Hirsh (Tsonev and Hirsh, 2000). Although any or all of these concepts are probably, in one way or another, involved in freezing injury, and recognizing that the totality of the process is complex, it is still surprising that so little effort has been made over the last 40 years to construct a comprehensive picture of this important biological event.

So here we are at the birth of the twenty-first century. We have more sophisticated—and expensive—analytical devices. We no longer make our own apparatus (a lathe, a mill, and a soldering iron were the most important tools in my laboratory during the 1950s), and we are now wholly dependent on factory representatives to minister to the mysterious insides of our instruments. We spend more time on our computers and our multicolored, animated PowerPoint displays than we ever spent with pencil and graph paper. We spend more time writing grant proposals to cover escalating costs for overhead and instrumentation. And much of that time comes, unfortunately, at the expense of hands-on time at the bench, and in particular, from reading. I can recall spending wonderful days in the library following the trail of previous discoveries—the only way not to repeat the past. Today's literature searches are conducted within the limited scope of the Internet, and anything published before 1960 has become invisible. Even though much of the early literature may have been based on empirical (naive?) research, there are gold mines of information there, perhaps because it *was* more random, and I see now much of the early ground being replowed, often by equally empirical methods, albeit at far greater expense.

Another profound change over the years has been the increasing tendency for applied cryobiology to replace rather than supplement basic cryobiology. This is an understandable trend driven in part by the fact that useful applications tend to be more attractive to granting agencies, but also driven in part by the emergence of technology transfer as a desirable and profitable endpoint for research. Perhaps the most unhealthy side effect of this trend is the emergence of "intellectual property" as something to be held close to the chest, to remain secret, unpublished and unshared until patented, and even then to be leaked out slowly. The concept of science as a community of colleagues engaged in a public service (Greaves called it "fun and games for adults") has been eroded by the escalation in the cost of research and the emergence of industry as not only a major source of research funding but as the ultimate exploiter of the results, and we have little choice but to play the game.

There has long been a sentiment, though bias might be a better word, that only basic research is pure and that applied research is somehow second class, the caboose on the research train. However, useful applications are, biases to the contrary, an end result of research without which basic studies become an academic luxury. For cryobiology in particular, applications are everything, and it is no wonder that this has been the emphasis right from the beginning. So far, applied cryoprotection really has not progressed much beyond applications of glycerol, Me$_2$SO, and the polymers, empirically optimized for each specific use by the addition of other solutes and a choice of cooling and warming rates. Many of the articles in current issues of *Cryobiology* are no less empirical than Basil Luyet's freezing of chick embryo hearts and vinegar eels.

Classic research strategy says that an understanding of basic mechanisms must precede and underlie applied research and that only this can enable one to create a logical hypothesis. However, we do need to appreciate that basic cryobiology—in fact, all of biology—is an incredibly complex system. As Bob Williams once put it, "The answer to Nature's little secrets is always another secret." True understanding of the effects of cold, of dehydration, and of ice on biological elements will ultimately need to account for interactions right down to the molecular level, perhaps depending on microscopic and analytical techniques not yet developed. A complete exploitation of the many potential applications of cryobiology will require this depth of understanding. A more superficial picture of freezing injury could suggest tricky ways to plug membrane pores, increase membrane deformability, or prevent cold denaturation, but a comprehensive understanding of what really goes on will be no piece of cake.

The empirical approach has so far served cryobiology pretty well, but trying this and trying that to see what works and abandoning what doesn't, even based on good guesses, is a risky endeavor. The things that don't work will always be far more abundant than those that do, and sooner or later we have to worry about life expectancy.

There is one laboratory that has had spectacular success with the empirical approach. Mother Nature has the advantage of an extended life expectancy and has made good use of it. The number of her trials and errors defies imagination, and the countless successes are all here before us, potentially revealing not only the how but sometimes the why as well. In the absence of a complete understanding of events at the most basic level, it may be more fruitful at this time, in the interest of applications, to build on what nature has already discovered rather than to engage in independent trial and error without the advantage of a few billion years to do it in.

The most immediate example of the virtues of this approach is the discovery of the antifreeze proteins, which tells us that important new approaches to cryoprotection can arise without depending in any way on a theoretical understanding of freezing injury and that there is nothing holy about basic research when applications are the objective. The antifreeze proteins also present a unique opportunity for the induction of frost hardiness through gene transfer. Many of the other tolerance strategies found in both plants and animals may also be candidates for a similar end run around the need to fully understand the mechanisms involved, an understanding that may be many decades away.

Whatever the route taken, whether through better comprehension of basic phenomena or an exploitation of Nature's empiricism, most of cryobiology's real achievements lie ahead. However, as always in research, though frequently forgotten, the most important tools to this end will not be just new sophisticated techniques and instruments, but a determination to aggressively maintain an open mind, to avoid the tunnel vision of a favorite hypothesis, and to enjoy, as a consequence, the vital capacity to "stay in motion." I hope that this book will provide a valuable stepping stone in that direction.

Harry Meryman
Biomedical Research Institute, Rockville, MD

REFERENCES

Brandts, J.F. (1964) Thermodynamics of protein denaturation. I. The denaturation of Chymotrypsinogen, *J. Am. Chem. Soc.*, 86, 4291–4301.

Crowe, J.H., Crowe, L.M., Oliver, A.E., Tsvetkova, N., Wolkers, W., and Tablin, F. (2001) The trehalose myth revisited: Introduction to a symposium on stabilization of cells in the dry state, *Cryobiology*, 43, 89–105.

Doebbler, G.F., Sakaida, R.R., Cowley, C.W., and Rinfret, A.P. (1966) Cryogenic preservation of whole blood for transfusion: *In vitro* study of a process using rapid freezing, thawing and protection by polyvinylpyrrolidone, *Transfusion*, 6, 104–111.

Lovelock, J.E. (1953) The mechanism of the protective effect of glycerol against hemolysis by freezing and thawing, *Biochim. Biophys. Acta*, 11, 28–36.

Lovelock, J.E. and Bishop, M.W.H. (1959) Prevention of freezing damage to living cells by dimethyl sulphoxide, *Nature (Lond.)*, 183, 1394–1395.

Luyet, B.J. and Gehenio, P.M. (1940) *Life and Death at Low Temperatures*, Biodynamica, Normandy, MO.

Mazur, P. (1965) Causes of injury in frozen and thawed cells, *Fed. Proc.*, 24, S175–S182.

Mazur, P. (1966) Physical and chemical basis of injury in single-celled micro-organisms subjected to freezing and thawing, in Meryman, H.T., Ed., *Cryobiology*, Academic Press, London.

Meryman, H.T. (1950) Replication of frozen liquids by vacuum evaporation, *J. Appl. Phys.*, 21, 68.

Meryman, H.T., Ed. (1966) *Cryobiology*, Academic Press, London.

Meryman, H.T. (1968) A modified model for the mechanism of freezing injury in erythrocytes, *Nature*, 218, 333–336.

Meryman, H.T. and Kafig, E. (1955) The freezing and thawing of whole blood, *Proc. Soc. Exp. Biol. Med.*, 90, 587–589.

Polge, C., Smith, A.U., and Parkes, A.S. (1949) Revival of spermatozoa after vitrification and dehydration at low temperatures, *Nature (Lond)*, 164, 666.

Rasmussen, D.H., Macaulay, M.N., and Mackenzie, A.P. (1975) Supercooling and nucleation of ice in single cells, *Cryobiology*, 12, 328–339.

Smith, A.U. (1950) Prevention of hemolysis during freezing and thawing of red blood cells, *Lancet*, 2, 910–911.

Tsonev, L.I. and Hirsh, A.G. (2000) Fluorescence ratio intrinsic basis states analysis: A novel approach to monitor and analyze protein unfolding by fluorescence, *J. Biochem. Biophys. Methods*, 45, 1–21.

Williams, R.J., Hope, H.J., and Willemot, C. (1975) Membrane collapse as a cause of osmotic injury and its reversibility in a hardy wheat, *Cryobiology*, 12, 554–55.

About the Editors

Barry Fuller, a Professor in the Royal Free and University College Medical School, London, has worked in the area of cryobiology applied to medicine in the University of London for 30 years. He has edited two previous books and published some 200 papers and articles on various aspects of the topic. He holds visiting fellowships at the University of Wales College of Medicine, Northwick Park Institute for Medical Research, and the University of Luton. He is also a Professor in the UNESCO Chair of Cryobiology based at the Ukraine Academy of Sciences.

Nick Lane is an honorary senior research fellow at the Royal Free and University College Medical School, University of London. A biochemist by background, his doctoral and post-doctoral research focused on ischemia-reperfusion injury in hypothermically stored transplanted organs. He is the author of *Oxygen: The Molecule that Made the World,* published by OUP, and of numerous articles in international journals, including *Scientific American*, *New Scientist*, *The Lancet*, and the *British Medical Journal*.

Erica E. Benson, Reader in Plant Conservation at the University of Abertay Dundee, Scotland, has worked in the fields of conservation and environmental stress physiology for 20 years. Her main research interests are low-temperature biology, the cryoconservation of genetic resources, and fundamental studies of free-radical-mediated oxidative stress in biological systems.

Acknowledgments

It is with sadness that the Editors wish to acknowledge the recent death of Professor Alban Massip, an author and enthusiastic supporter of this book. Professor Massip was an internationally recognized expert in cryopreservation of reproductive cells in domestic species. He completed his Ph.D. at the School of Veterinary Medicine in Brussels on cryopreservation of mammalian oocytes in 1986, and went on to become Professor of Anatomy and Embryology of Domestic Species at Universite catholique de Louvain, Louvain-la-Neuve, Belgium. He published widely on his topic, and in 1999 he was awarded the distinguished European Embryo Transfer Pioneer Award given by the European Embryo Transfer Association. We offer our condolences to his family and colleagues.

On a personal level, we wish to thank the contributors to the book for their obvious hard work, willingness to meet deadlines in an increasingly time-starved world, and their cooperation among themselves to help mold the final product.

Last, but certainly not least, we wish to thank our spouses and partners (Mary Mitchell, Dr. Ana Hidalgo, and Dr. Keith Harding, respectively) for their tireless patience, encouragement, and advice throughout the project.

Barry Fuller
Nick Lane
Erica E. Benson
London and Dundee

Introduction

We began the task of compiling *Life in the Frozen State* in 1999. In embarking on what is the first major textbook on cryobiology in the genomic era, we felt it was important to bring together in one place a picture of the current understanding of "life in the frozen state" at the turn of the twenty-first century. The field now encompasses disciplines ranging from mathematically based biophysics, to the molecular biology of stress gene expression, to the infinitesimally slow metabolic adaptations of permafrost bacteria. It is therefore not just timely, but critical to the future coherence of the subject, to meld a cross-disciplinary platform of knowledge in cryobiology.

Our own interests in the ability of living organisms to withstand the transition into and out of the frozen state lie on a sliding scale. At one end is a pragmatic desire to harness low-temperature technologies to solve many practical problems in health care, conservation, and the biosciences; at the other end, we share an enthusiastic appreciation that this plasticity is an amazing feat of nature. The ability of life to survive the frozen state is at the very edge of our comprehension of animated processes. As we deepen our knowledge of each facet of this extraordinary transition, we can appreciate that further progress toward the practical goals of cryopreservation depends on applying insights gained from fundamental knowledge in each of these disciplines. This, in turn, depends on new collaborations, wider awareness, and thinking outside the traditional boxes.

The realization of this book, and the melding of its disparate contents, has been fuelled by our own very different professional backgrounds in medical cryobiology (Barry Fuller), stress physiology (Nick Lane), and plant genetic resources conservation (Erica Benson). From the start, we felt that to understand biological responses to freezing we needed to break down the classical subdivisions of research into "animals," "plants," "biochemistry," "ecology," and so forth, which are often not strictly applicable, and are rarely desirable. If we are to progress and apply our knowledge of life in the frozen state, we must seek to exploit both theoretical and experimental studies in the broadest sense. To do so, we have structured the book to bring together new collaborations of ideas, technologies, and applications. A necessary feature of this overarching scheme is a degree of overlap between chapters, which we have been careful to cross-reference, in the hope that this will encourage readers to step with confidence from familiar disciplines into more distant fields, and so to cross-fertilize. The areas of overlap largely concern what we currently know of the responses of cells, tissues, and organs to freezing across a remarkable, cross-kingdom diversity of species.

The term cryobiology was originally coined by Sir Alan Parkes, who defined it as the study of "frosty life." In keeping with this guiding spirit, we were especially keen to cover those life-forms that do encounter freezing (the ice transition), rather than those that survive at low temperatures *per se* (as in hibernation and chill tolerance). That said, it will become apparent in later chapters that under certain circumstances the ability to modulate the state of water in cells at temperatures above the traditional 0°C is of direct relevance to cryobiology, and so we have included chapters covering dry glasses. Regardless of the detailed content, however, one of the most exciting outputs of the compilation is the noisy "cross-talk" between evolving ideas on how to make cells tolerant to the freezing process. These ideas derive from comparative studies of *in vitro* freezing in the laboratory from and what has evolved *in vivo* in the "natural" cold laboratories of polar environments and high-altitude alpine habitats.

This progression in understanding has not been achieved by the smooth, step-by-step, planned exploitation of facts so beloved of politicians, administrators, and science fiction writers. Rather, it has come about by pulling together and ordering the divergent and disparate ideas from many different fields of pure and applied cryobiology—a process that we hope will allow the reader, from whatever field of cryobiology, to reevaluate his or her own ideas, and to apply new technologies to old problems. Such willingness to share new ideas and experiences, to discard old certainties and rediscover forgotten or misunderstood discoveries, is the reality of progress in scientific understanding.

This book is not a done deal, and we also wish to point out gaps and inconsistencies so that new ideas can be born of new thinking; this is essential to move the science of cryobiology forward. Indeed, such a mixture of theoretical knowledge, good scientific method, intuition, serendipity, and an ability to see clear patches of understanding through a fog of "accepted wisdom" is how cryobiology took its first hesitant steps. That was only 50 years ago, at a time when there were very clear distinctions between plant and animal biology. Despite our admonitions above, we have chosen to present the historical basis of cryobiology as two separate "animal" and "plant" chapters. It has only been with the benefit of hindsight that we can reevaluate the development of the subject in the context of the now obvious similarities in philosophy and experimentation that mark the development of both fields. However, to impose such similarities with hindsight is to risk losing the context of the pioneers themselves, and it is this context that continues to make their work relevant today. Ideas are born of context, and they sometimes flounder only because there were no robust means of testing them at the time. Today we are better armed with technology, and we can test many forgotten early ideas—but we will only grasp their relevance if we also grasp their context. For these reasons, we have chosen not to place the historical chapters at the start of the book, where they can be safely skipped by scientists with scant regard for history, but decided that they held more meaning if placed next to, and read with, the chapters that detail the modern developments flowing from such pioneering early investigations.

We have ordered the book into four themes. In *Theme 1: Fundamental Aspects*, Peter Mazur discusses his quantitative predictions on cellular responses to ice formation and debates the latest modifications to his theories. He presents these ideas so clearly that even those who harbor a deep fear of equations will gain a robust understanding of the biophysical basis of cryobiology from this chapter. In the next chapter, Kenneth Muldrew et al. cover their interpretation of these same phenomena, which are based on Mazur's early ideas but have digressed slightly in detail. We chose to include these chapters side by side, so that readers can gain a sense of the areas of accepted wisdom that have evolved, but also of the "shifting sands" of new hypotheses, experimentation, and technologies that are helping us to unravel and probe the parts of the cryobiological puzzle that have yet to be organized into a comprehensive and integrated picture.

Theme 2: Life and Death at Low Temperatures (the title being a nod to Basil Luyet's classic 1940 text) brings together a diverse collection of chapters by experts in the natural sciences. Despite strikingly different perspectives, they all study life at the extremes, in juxtaposition with ice, not just in the polar regions and the frozen tundras but also in more temperate climates, where freezing is a seasonal or even a nocturnal event.

Josef Elster and Erica Benson provide a comprehensive review of the polar terrestrial environment, the changes over geological time, and the constraints that these harsh conditions impose on life. Indeed, even in such conditions, few life-forms have adapted to truly survive ice formation in their living parts. This chapter focuses on polar algae as a model to describe how cold extremophilic life-forms can exist. In comparison, Monica Ponder et al. describe how microorganisms survive the permafrost of the frozen tundra under conditions where they have been trapped—frozen in the soil—for multiples of centuries and millennia, rather than weeks. These authors consider the fine balance between maintaining a low basal metabolism in extreme cold vs. retreating into a latent, anabiotic state, and the implications of each to cryobiology and related fields such as astrobiology.

Moving to more temperate environments, the low-temperature adaptations and responses of plants are explored. These (usually) sedentary organisms are (usually) fixed in their local environments and forced to adapt *in situ* to the fluctuating freezing temperatures that accompany winter.

In comparison, most (but not all) animal species employ "freeze-avoidance" by moving to habitats that exclude ice formation, even though they may be cold. Using plants as a model, the molecular basis of freezing tolerance is addressed by Roger Pearce, who describes the development and implications of the molecular technologies that are now being employed. This area of research has been most extensively applied to unravel the genetic basis of the changes that contribute to plant freeze-tolerance. These studies are predictably very important in crop plant agriculture, but they are also beginning to assume real significance in such diverse areas as medical cryobiology, the cryoconservation of genetic resources, and helping us to understand the incipient effect of climate change.

At the submolecular level, Erica Benson and David Bremner draw attention to the very real interactions of freezing with free radical-mediated oxidative stress. That free radicals and antioxidants may play a role in cryogenic processes has been guessed at and inferred over the past 30 years, but the role of oxidative processes in the frozen state is only now beginning to be illuminated, owing to developments in modern experimental methodology. Finally, Kenneth Storey and Janet Storey present in their chapter the intriguing molecular detail of freeze-tolerance in the wood frog, one of the select band of vertebrates where "life-force" and ice crystals can coexist in the same body without disastrous consequences. The parallels between genetic adaptations to freeze-tolerance in vertebrates and in plants, and the role of oxidative stress in both, raise many possibilities for enhancing freeze-tolerance in cryopreservation protocols, as we discuss in the final chapter of the book.

From all of these examples, it will be possible to see intermixing strands between fundamental understanding and natural adaptation, which (again with hindsight) appear as a composite and predictable weave. This was not, and is not, how the science of cryobiology "really" developed, but it charts a more meaningful course, one by which we can enhance our applications of "life in the frozen state" for great and far-reaching benefit in many different sectors.

In *Theme 3: Freezing and Banking of Living Resources*, we consider the growing applied interest in how to harness the natural ability of living cells to survive freezing. Cryopreservation of somatic and reproductive cells, tissues, and organs has a wide range of applications in biotechnology, biomedicine, agriculture, forestry, aquaculture, and biodiversity conservation. The capability of "freezing time" in cryogenic storage for months and years offers huge and compelling benefits, some of which necessitate the careful and sensitive consideration of ethical and regulatory issues.

As the book examines applied research, it will become evident that the *in vivo* characteristics required for cells and organisms to survive the two main components of cryoinjury (freezing and dehydration/desiccation) are complementary, but not completely superimposable, and that each has important implications for aiding the development of cryopreservation methodologies *in vitro*. In this context, Shu-hui Tan and Cor van Ingen discuss the application of freezing and drying (lyophylization) to fungi and bacteria, where strategies have been developed to counteract the stresses of freezing and drying. Drying after freezing has significant benefits if it can be achieved in a wider context, hence the interest in adapting the technology to medical cryobiology and to preserving reproductive cells from higher species (discussed in Theme 4).

Storage of plant cells in the frozen state is of great importance for the development of plant gene banks and culture collections and for the conservation of endangered and at-risk plants. The application of cryopreservation to plant-cell cultures in biotechnology is also very important. Akira Sakai describes the recent advances in plant cell cryobiology, in which new techniques such as vitrification (manipulation to limit true ice crystal growth) are coming to the fore. Allied with this chapter, Erica Benson provides one of the history chapters—it charts the progression, over 150 years, of early experiments that explored freezing processes in plants, many of which we now recognize as fundamental principles of freezing. The main subject of this chapter is, however, to provide a timeline of the development and application of cryopreservation in the context of helping to conserve the Earth's precious plant and algal diversity.

Subsequent chapters explore the applications of freezing for the preservation of gametes and embryos of animal species, both in agriculture and in the banking of germplasm for zoological diversity conservation. It is important to appreciate that many of the current theories and models for predicting freeze-survival in animal cells were pioneered by the early work on mammalian reproductive cells, and this historical resource is still being mined for inspiration today, as discussed earlier. This is why we here introduce the second chapter in the history of cryobiology, laid out in the review by Stanley Leibo. It is from this background that many of the ideas of application of cryoprotectant chemicals to abrogate damaging effects of freezing on cells were first formulated.

Proceeding from this historical perspective, Alban Massip et al. provide an overview of the use of cryopreservation in animal reproductive technologies. Cryogenic storage has made a very significant impact on animal husbandry in the last four decades of the twentieth century, and one that will undoubtedly expand in importance and range of species involved in the coming years. Amanda Pickard and William Holt provide a thoughtful discourse on the potential (only partially realized so far) and the limitations on the application of freezing to endangered and at-risk zoological species. The crucial question is: How can we protect and safeguard the future of natural diversity? The major challenge here is to enhance our basic understanding of the complex vista of species population genetics and reproduction. Similar, but possibly even more refractory, problems beset the cryopreservation of aquatic species, a field of potentially huge economic significance. The chapter by Tiantian Zhang examines current applications in this field: There remain significant problems in developing cryoconservation strategies for some life stages of certain species, and these areas will become important in terms of future cryogenic developments.

Theme 3 concludes with a chapter by Glyn Stacey, who sets out the presently accepted guidelines for the cryobanking of living resources. "Freezing time," in gene, cell, and organ banks must address critical issues related to quality control, the optimizing and standardizing of specialist technologies, and the implementation of safe cryopreservation protocols. Thus, the ability to cryopreserve "life in the frozen state" must carry with it the responsibility to create well-organized and safe cell, tissue, organ, and germplasm repositories. Just as the path to hell is paved with good intentions, so the path to failed cryopreservation is paved with good theory and poor attention to practical detail.

The major themes of the book conclude with *Theme 4, The Medical Applications of Cryobiology*. At this juncture the main thrust of the book has been on preserving "life in the frozen state." However, an alternative possibility relates to how we can use our preservation knowledge to divergent medical advantage by using cryogenics in surgical procedures that can kill harmful cells and tissues. Even a cursory reading of Theme 1 will have illustrated the fact that the destruction of life by freezing is an ever-present possibility when extremes of low temperatures are encountered. The chapter by Nathan Hoffman and John Bischof reviews where and how this destructive capability of ice is being gainfully employed in cryosurgery for the destruction of cancers and tumors. As in other fields of cryobiology, cryosurgery is currently benefiting from cross-fertilization (e.g., the use of antifreeze proteins to potentiate injury outside the ice-ball), but the molecular sequelae of sublethal injury are also set to feed back into cryopreservation protocols.

Continuing with our desire to embrace historical perspectives of cryobiology in the book, Andreas Sputtek and Rebekka Sputtek provide a chapter on the freeze preservation of blood cells, one of the earliest clinical applications of cryobiological principles, pioneered by James Lovelock and colleagues in the early 1950s. The Sputteks provide a thorough review of this important topic, highlighting from where we have progressed to where we are now. In this specialty, too, new concepts are being derived from our growing appreciation of the fundamentals (here it is the use of polymeric extracellular cryoprotectants), and are beginning to have an effect. There still remain significant challenges ahead for the application of low-temperature banking to some blood-derived cell types.

The expansion of assisted-reproductive technologies, outlined in Theme 3, is developed further in Theme 4, feeding into the fields of clinical reproductive medicine, where again a growing range of cells and tissues of therapeutic value are now candidates for freezing. These are discussed by Barry Fuller et al. In some areas (such as the cryogenic freezing of embryos), the techniques are widely applied, whereas in others (such as storage of unfertilized oocytes), recent advances have led to their introduction into clinical practice in as yet only a few places.

The concept of freezing larger, complex, and multicellular tissues presents one of the main challenges for medical cryobiology, and one that introduces novel problems concerned with the nucleation and location of ice crystals. In some respects, our knowledge of the destructive effects of ice formation in tissues (as discussed by Nathan Hoffman and John Bischof) can be exploited in organ preservation, but they do introduce complications when considering freeze-banking of tissues for medical use. Thus, in their chapter, Monica Wusteman and Charles Hunt describe how far we have come in the banking of complex tissues, and where developments in the near future might lie.

The final three chapters in Theme 4 examine biomedical applications of the principles of cryobiology to ice-free preservation. Some of these technologies will truly be seen as twenty-first-century medicine, but already some areas promise exciting new avenues for exploitation in the near future. Vitrification will increasingly play an important role in the development of the medical applications of cryobiology. In the chapter by Jason Acker et al., the authors set out their vision of applying novel technologies for the preservation of living cells "frozen in time" in the glassy state at ambient temperatures, rather than suspended in animation by extreme cold. Many of the problems of dehydration mirror those encountered during ice formation, but there are also new and different concerns. The goal of achieving cell preservation without the need for powerful cryogens (with the associated problems of cost and supply) appears to be at least in our sights, if not yet in our grasp. In the chapter by John Crowe et al., recent moves toward developing lyophylization (freeze-drying) of mammalian cells are set out—an approach with huge potential, but that has so far proved an intractable problem. In this case, a molecular understanding of mechanisms of stress-tolerance in natural organisms has pointed the way to new possibilities in the lyophilization of platelets, which are currently being evaluated clinically. In the final chapter of Theme 4, Michael Taylor et al. describe exciting developments in the preservation of complex tissues, which (as discussed above) suffer injurious damage from ice-crystal growth in a manner not experienced by cells in suspension. Central to all of these developments will be our ability to understand and control the different states of water. Thus, new ideas on achieving the glassy "vitreous" state at low temperatures, while controlling the growth of ice crystals by molecular means, will undoubtedly be an area of extreme importance when attempting to cryopreserve larger tissues and organs—situations that currently defy our best attempts.

To conclude the book as a whole, Nick Lane considers possible future directions in cryobiology. The discussion draws its inspiration in large part from other chapters in the book, and so ties together themes developed in more detail elsewhere. It is short, discursive, and unreferenced (except to other chapters). By drawing some strands of disparate thinking together, it is hoped to foster the spirit of cross-disciplinary thinking and collaboration. For good measure, the synthesis is used as a launchpad for a few thought-provoking, cross-disciplinary hypotheses that we hope will bear fruit in the years ahead.

In sum, we are extremely grateful to the many authors who have provided their time and knowledge in the compilation of this text. Any success from the venture is entirely down to their efforts; any failures or omissions must be laid at the editors' door. As mentioned at the outset, we have attempted to record the status of "life in the frozen state" at the start of the twenty-first century. Some areas have been dealt with only briefly. This is not to deny their importance, but merely to recognize that we could only present a limited number of subjects in the pages of a single volume to illuminate the current understanding of cryobiology. In other areas (such as the work on dehydration of cells to a glassy state at ambient temperatures), we have included topics that stretch the classical definition of freezing to the limits of scientific respectability. However, we believe that this represents a true (and exciting) futuristic picture of the current attempts to preserve cells suspended in time.

Which brings us, finally, to a word on the title—*Life in the Frozen State*. Although the freezing of water is not necessarily a part of dehydration and vitrification, these techniques do freeze life in time, if not in physical structure. Indeed, some might argue that by petrifying life's matrix in an essentially motionless state, vitrification freezes life in structure better than does ice. Be that as it may, we hope that our title will be taken in the metaphorical spirit that we intended: as a paean to the amazing feats of nature. We hope that you, the reader, will likewise be intrigued and excited by the realities and potentials afforded as we refine our understanding of "life in the frozen state."

Barry Fuller, Nick Lane, and Erica Benson

Contributors

Jason P. Acker
Department of Laboratory Medicine and
  Pathology
University of Alberta
Edmonton, Alberta
Canada

Erica E. Benson
Plant Conservation
Conservation and Environmental Chemistry
  Centre
School of Contemporary Science
University of Abertay Dundee
Dundee, Scotland

John C. Bischof
Departments of Biomedical Engineering,
  Urologic Surgery, and Mechanical
  Engineering
University of Minnesota
Minneapolis, Minnesota

Elisabeth Blesbois
Station de Recherches Avicoles
Institut National de la Recherche Agronomique
Nouzilly
France

David Bremner
Conservation and Environmental Chemistry
  Centre
School of Contemporary Science
University of Abertay Dundee
Dundee, Scotland

Kelvin G.M. Brockbank
Organ Recovery Systems
Charleston, South Carolina
U.S.A.

Tani Chen
Center for Engineering in Medicine and Surgical
  Services
Massachusetts General Hospital
Harvard Medical School
and
Shriners Hospitals for Children
Boston, Massachusetts
U.S.A.

John H. Crowe
Center for Biostabilization
University of California
Davis, California
U.S.A.

Lois M. Crowe
Center for Biostabilization
University of California
Davis, California
U.S.A.

Janet A.W. Elliott
Department of Chemical and Materials
  Engineering
University of Alberta
Edmonton, Alberta
Canada

Josef Elster
Institute of Botany
Academy of Sciences of the Czech Republic,
  Třeboň
and
Biological Sciences University of South
  Bohemia, České
Budéjovice
Czech Republic

Alex Fowler
Department of Mechanical Engineering
University of Massachusetts
Dartmouth, Massachusetts
U.S.A.

Barry Fuller
Department of Surgery
Royal Free and University College Medical
  School
University of London
London
U.K.

Nathan E. Hoffmann
Department of Biomedical Engineering
University of Minnesota
Minneapolis, Minnesota

William V. Holt
Institute of Zoology
Regent's Park
London
U.K.

Charles J. Hunt
U.K. Stem Cell Bank
National Institute for Biological Standards
  and Control
Potters Bar, Hertfordshire
U.K.

Nick Lane
Department of Surgery
Royal Free and University College Medical
  School
University of London
London
U.K.

S.P. Leibo
Department of Biological Sciences
Audubon Center for Research of Endangered
  Species
University of New Orleans
New Orleans, Louisianna
U.S.A.

Alban Massip (deceased)

Peter Mazur
Fundamental and Applied Cryobiology Group
Department of Biochemistry and Molecular and
  Cellular Biology
The University of Tennessee
Knoxville, Tennessee
U.S.A.

Locksley E. McGann
Department of Laboratory Medicine and
  Pathology
University of Alberta
Edmonton, Alberta
Canada

John McGrath
Center for Microbial Ecology
Center for Genomic and Evolutionary Studies
  on Microbial Life at Low Temperatures
Michigan State University
East Lansing, Michigan
U.S.A.
and
Department of Aerospace and Mechanical
  Engineering
University of Arizona
Tucson, Arizona
U.S.A.

Ken Muldrew
Department of Cell Biology
University of Calgary
Calgary, Alberta
Canada

Ann E. Oliver
Center for Biostabilization
University of California
Davis, California
U.S.A.

Sharon J. Paynter
Department of Obstetrics and Gynaecology
University of Wales College of Medicine
Cardiff, Wales

Roger S. Pearce
School of Biology
University of Newcastle upon Tyne
U.K.

Amanda R. Pickard
Frozen Embryo and Sperm Archive
Medical Research Council
Harwell, Oxfordshire
U.K.

Marcia Ponder
Center for Microbial Ecology
Center for Genomic and Evolutionary Studies
  on Microbial Life at Low Temperatures
Michigan State University
East Lansing, Michigan
U.S.A.

Akira Sakai
Hokkaido University (retired)
Asabucho, Kitaku
Sapporo
Japan

Ying C. Song
Organ Recovery Systems
Charleston, South Carolina
U.S.A.

Andreas Sputtek
Universitatsklinikum Hamburg-Eppendorf
Institut fur Transfusionsmedizin
Hamburg
Germany

Rebekka Sputtek
Universitatsklinikum Hamburg-Eppendorf
Institut fur Transfusionsmedizin
Hamburg
Germany

Glyn Stacey
Division of Cell Biology and Imaging
National Institute for Biological Standards and
  Control
South Mimms, Hertsfordshire
U.K.

Janet M. Storey
Institute of Biochemistry
College of Natural Sciences
Carleton University
Ottawa, Ontario
Canada

Kenneth B. Storey
Institute of Biochemistry
College of Natural Sciences
Carleton University
Ottawa, Ontario
Canada

Fern Tablin
Center for Biostabilization
University of California
Davis, California
U.S.A.

Shu-hui Tan
Centraalbureau voor Schimmelcultures
Utrecht, The Netherlands

Michael J. Taylor
Organ Recovery Systems
Charleston, South Carolina
U.S.A.

James Tiedje
Center for Microbial Ecology
Center for Genomic and Evolutionary Studies
  on Microbial Life at Low Temperatures
Michigan State University
East Lansing, Michigan
U.S.A.

Mehmet Toner
Center for Engineering in Medicine and Surgical
  Services
Massachusetts General Hospital
Harvard Medical School
Shriners Hospitals for Children
Boston, Massachusetts
U.S.A.

Nelly M. Tsvetkova
Center for Biostabilization
University of California
Davis, California
U.S.A.

Cor van Ingen
Netherlands Vaccine Institute
Bilthoven, The Netherlands

Tatiana Vishnivetskaya
Center for Microbial Ecology
Michigan State University
East Lansing, Michigan
U.S.A.
and
Institute for Physicochemical and Biological
  Problems in Soil Science
Russian Academy of Sciences
Moscow Region
Russia

Paul Watson
The Royal Veterinary College
University of London
London, U.K.

Willem Wolkers
Center for Biostabilization
University of California
Davis, California
U.S.A.

Monica Wusteman
Medical Cryobiology Unit
Department of Biology
University of York
York
U.K.

Tiantian Zhang
Luton Institute of Research in the Applied
  Natural Sciences
University of Luton
Luton
U.K.

Table of Contents

Theme 1

Fundamental Aspects

1 Principles of Cryobiology*

Peter Mazur

CONTENTS

---

* This chapter is dedicated to my deceased wife, Sara Jo.

0-415-24700-4/04/$0.00+$1.50
© 2004 by CRC Press LLC

1.1 INTRODUCTION

When cells or cellular systems are exposed to the ice formation that accompanies exposure to low subzero temperatures, they are subjected to profound changes in the physical state and chemical properties of their surrounding milieu, and the cells undergo major physical responses to these changes. Liquid water plays a quintessential role in the structure and function of living systems, and the most obvious change accompanying freezing is that the amount of this liquid is progressively reduced and eventually vanishes. The chapter begins by briefly reviewing the structure of both liquid water and ice, some of the ways that liquid water influences the structure of cellular macromolecules, and the nature of water on the surfaces of those macromolecules. The discussion is then extended to cells and their membranes.

Even if liquid water/macromolecule/cell structure interactions were unaffected by freezing, the mere reduction in the concentration of liquid water introduces large osmotic disequilibria to which the cells must respond. We first examine the magnitude of these disequilibria and the magnitude of the volume changes that cells undergo in response to them. We then review the rate at which these responses occur, rates that depend on the permeability of the cell to water and solutes. Finally, this first section relates the phenomenon of permeability to the structure of the cell membrane and to a recently discovered class of transmembrane proteins, the aquaporins.

This background material sets the stage for a discussion of cryobiological principles *per se.* The discussion begins by pointing out that plots of cell survival vs. cooling rate usually exhibit an inverted U—findings that indicate that an injury occurring at high cooling rates has a different genesis than that occurring during slow cooling. The discussion elaborates on the former by pointing out that rapid cooling sets up an osmotic disequilibrium between a cell and its environs and causes cell water to supercool by an amount proportional to the cooling rate and inversely proportional to the permeability of the cell to water. This supercooling cannot be maintained, and eventually supercooled cell water will freeze internally, usually with lethal consequences. The fact that a cell can supercool at all in the presence of external ice means that the plasma membrane is at least initially a barrier to the growth of ice from the exterior to the interior, yet intracellular freezing appears to be dependent on the presence of external ice, and this chapter reviews the three major theories that propose to account for the dependence.

Slowly cooled cells do not undergo intracellular freezing, yet they are generally killed if cooled too slowly and if cryoprotective agents of appropriate type and concentration are not present. Most of the ensuing discussion deals with the colligative theories of both the injury and the protection; that is, theories that propose that injury is associated either with the rise in solute concentration (especially electrolytes) that accompanies ice formation or with its mirror image, the reduction in the unfrozen fraction, and that propose that cryoprotection is related to the ability of certain permeating nontoxic nonelectrolytes like glycerol to reduce the concentrations of damaging solutes and to increase the magnitude of the unfrozen fraction at a given subzero temperature. Some researchers, however, believe that slow-freezing injury is a direct result of the dehydration of cell membranes and that cryoprotection results from the ability of cryoprotective agents to substitute for water or to otherwise protect against the injury that results from the removal or perturbation of the hydration layer on cell membranes.

Finally, freezing is only half the story. To function, a frozen cell must be returned to normal temperatures and must survive the accompanying warming and the thawing of the ice. In the case of slowly frozen cells, the events during the return to ambient temperature are mostly a mirror image of those that occurred during cooling. However, that is not the case with rapidly cooled cells. Commonly, rapidly cooled cells fare substantially better if the subsequent warming is rapid than if it is slow. That benefit is commonly explained in terms of the size of the ice crystals formed during the initial cooling and the effect of warming rate on that size. Higher cooling rates form smaller internal ice crystals, and small crystals appear to be less damaging than large ones. However, if subsequent warming is slow, those small crystals can enlarge to damaging size by the process of recrystallization. Another related possibility is that even at low cooling rates, not all the cell water freezes—a portion may be converted to a glass. If subsequent warming is too slow, this glassy water may devitrify (freeze), with lethal results. This partial glass formation is an introduction to purposeful attempts to induce vitrification of all the cell water—attempts that are being increasingly pursued in the cryopreservation of tissues and organs, systems that are damaged by extracellular as well as intracellular ice. Vitrification approaches are not covered in this chapter but are treated by Taylor et al. in Chapter 22.

1.2 UNDERLYING BIOPHYSICS AND CELL BIOLOGY

1.2.1 Water

Water plays a central role in cryobiology and, more important, a central role in biology. Aspects of the latter and their dependence on the unusual and unique properties of liquid water were elegantly described nearly a century ago by L. J. Henderson (1913). Its unusual properties can be seen in a comparison with those of its analog on the periodic table: hydrogen sulfide. For example, the boiling points are 100° and –60°C, respectively; the melting points are 0° and –86°C; the heats of vaporization are 10.7 and 4.5 kcal/mole; and the dielectric constants are 78 and 9 (Lide, 1992; Nemethy and Scheraga, 1964). These properties arise from the structure of the water molecule and its distribution of charge. Infrared and Raman spectroscopy have shown that the water molecule is shaped like an Australian boomerang or a winged seed from a maple tree. The oxygen is at the center of the "boomerang," and the two hydrogens are located at the tips. The distance between the hydrogen and the oxygen is 0.97 Å; the distance between the two hydrogens (i.e., the diameter of the molecule) is 1.54 Å (Robinson and Stokes, 1959, p. 2).

The oxygen atom has six electrons in its outer valence shell. Two of these combine with the single electron on the two hydrogen atoms to form the two O–H covalent bonds. These bonds (i.e., the arms of the boomerang) are at an angle of 105 degrees, which is close to the 109 degrees from the center of a tetrahedron to the vertices (Eisenberg and Kauzmann, 1969, p. 4; Robinson and Stokes, 1959, p. 2). The hydrogens bear a net positive charge, with these two positive charges directed toward two of the vertices of a tetrahedron. The remaining four electrons are referred to as "lone-pair" electrons. The two lone-pair electrons confer a net negative charge to the oxygen side of the water molecule, and these two charges are directed toward the other two vertices of the tetrahedron. Thus, the water molecule bears two net positive charges directed toward two vertices of a tetrahedron and two net negative charges directed toward the other two vertices. An important consequence of this is that the two positive charges on the hydrogen sides of a given water molecule are electrostatically attracted to the lone-pair electrons on the oxygen side of two other water molecules, and the two centers of negativity on the oxygen side of a given water molecule are electrostatically attracted to the protons of two other water molecules. In other words, a given water molecule can be attracted to four other water molecules located at the vertices of the tetrahedron. Two of these surrounding water molecules have their oxygen sides facing the central water molecule, and two of them have their hydrogens facing the central molecule.

The (predominantly) electrostatic attraction between the protons and the oxygens is referred to as the hydrogen bond. Hydrogen bonds are weak (6 to 8 kcal/mole; Eisenberg and Kauzmann, 1969, p. 139) compared, for example, with C–C covalent bonds (145 kcal/mole), but in biological aqueous systems, hydrogen bonds are so numerous as to play a vital role. It is these hydrogen bonds that are responsible for the abnormally high boiling and melting points of water and for its abnormally high heat capacity.

One other important attribute of the water molecule is that the centers of gravity of the two negative charges and the two positive charges do not coincide. This gives water a sizable dipole moment and makes it a polar molecule (Robinson and Stokes, 1959, p. 2). Furthermore, if the water molecule is free to rotate, it possesses a large dielectric constant (78). The large dielectric constant means that it can reduce the electric field between two charged layers or between two charged molecules like ions to a small fraction of what would be the case in a nonpolar solvent.

1.2.1.1 The Structure of Ice

The structure of ice (ordinary Ice I) represents the physical reality of what has just been discussed. That is to say, in ice, each central water molecule is surrounded by, and hydrogen-bonded to, four surrounding water molecules located as though at the vertices of a tetrahedron. The O–O distance is 2.76 Å (Eisenberg and Kauzmann, 1969, p. 71). This structure repeats indefinitely. The open structure of ice is responsible for its density, which is much lower than that of most other solids. The extensive hydrogen bonding is responsible for its high heat of fusion and heat of sublimation. The latter is 12.2 kcal/mole and represents mostly the heat necessary to completely break an average of two hydrogen bonds per molecule.

1.2.1.2 The Structure of Liquid Water

The structure of liquid water has been debated for at least 40 years and is still not resolved, but a few generalizations are possible. The structure is similar to ice in that each water molecule tends to be tetrahedrally bound by hydrogen bonds to four other water molecules. To be consistent with the 10% greater density of liquid water at the melting point, however, a liquid water molecule is considered to have ~4.4 nearest neighbors rather than four such neighbors (Robinson and Stokes, 1959, p. 4). It has a short-range structure (i.e., out to about three layers of neighbors) that can be seen by x-ray diffraction, but not the long-range order of ice. The lifetime of a given structure is estimated to be short (i.e., 10^{-11} sec), although it is relatively long compared with the temporal resolving power of the measuring techniques, which are ~10^{-13} sec for x-ray and dielectric relaxation.

Models of liquid water structure fall into two broad classes: continuum models and mixture models (Eisenberg and Kauzmann, 1969, p. 254–270; Frank, 1972). Mixture models hypothesize that liquid water consists of a mixture of water molecules forming one, two, three, and four hydrogen bonds and monomeric water forming no hydrogen bonds. Continuum models argue that these discrete classes do not exist, but that liquid water has a distorted hydrogen-bonded structure. The majority of investigators seem to favor the former. The mixture models are further divided into two groups. One model, the interstitial model, hypothesizes that monomeric water molecules reside inside ice-like cages formed by tetrahedrally H-bonded water. These would be analogous to clathrates or gas hydrates in which a nonpolar molecule like methane resides within an ice-like cage. The other mixture model has been picturesquely referred to as the "flickering-cluster" or "iceberg" theory. It was initially proposed by Frank and Wen (1957) and later elaborated on by Nemethy and Scheraga (1962). It depicts islands or clusters of tetrahedrally H-bonded water lying interspersed in a sea of unbonded water molecules. The lifetime of the clusters is estimated to be 10^{-10} to 10^{-11} sec. The argument is that if, as a result of local thermal or energy fluctuations, two water molecules form H-bonds, it becomes energetically favorable for them to form a third and a fourth H-bond. Conversely, if the local fluctuations result in the disruption of an H-bond, it becomes energetically favorable for the other H-bonds to break. Although there appear to be defects in the flickering cluster model (Frank, 1972, p. 533), one attractive feature, a feature that seems to have been ignored by those discussing water models, is that it is consistent with the mechanism proposed for the homogeneous nucleation of supercooled water. In that picture, supercooled water freezes spontaneously when critically sized icelike "embryos" form in the water as a consequence of energy fluctuations. The number of water molecules that constitute the critical size decreases with temperature, and it is only at about –40°C that the random formation of an "embryo" with that critical number becomes a probable event (Fletcher, 1962, p. 202).

1.2.1.3 Water around Ions

An attractive electrostatic force exists between the asymmetrical directional charges on a water molecule and ions or ionized groups. The difference between those forces and those in water–water interactions is that the orientations of the water molecules are unidirectional around ions. That is, in the first layer of water around a cation, water molecules will tend to be oriented with their negative oxygen sides toward the ion. Around an anion, they will be oriented with their positive protons toward the ion. The attractions can be sufficiently strong to immobilize the water, as manifested in a large decrease in the dielectric constant (Robinson and Stokes, 1959, pp. 15–17).

The forces are short range, affecting, in most ions, only the first layer of water. Small ions such as F^- or Li^+ are an exception in that the effect may extend to the second or third layers. Beyond that, normal water structure is restored. However, as the unidirectional water structure around ions is different from the alternating structure of water, there exists an intermediate region in which water structure is disrupted. Robinson and Stokes (1959, p. 15) point out that when the concentration of an electrolyte solution rises to ~1 M, the second or third water layers from a given ion fall into the sphere of influence of a neighboring ion. As Robinson and Stokes put it, "The farther from England, the nearer is France." The net result of these factors is that in most electrolyte solutions, the net water "structure-breaking" aspects outweigh those of "structure making."

1.2.2 WATER AND MACROMOLECULAR CONFORMATION

The function of macromolecules and complex cellular structures like cell membranes that contain macromolecules depends critically on their conformation. The conformation will be that which minimizes the free energy, and water plays a vital role in that minimization.

1.2.2.1 Maximizing H-Bonds

Hydrogen bonds are not restricted to water. They can form between many groups in macromolecules (Franks, 1973, p. 20) and between these groups and water. This includes the peptide bonds in proteins and the bases in DNA. It is a general tenet that macromolecules will assume a conformation that maximizes the number of internal hydrogen bonds consistent with maximizing the other forces involved. It is important to note, however, that internal hydrogen bonds by themselves do not confer net stability because they are no stronger and, indeed, may be slightly weaker than the hydrogen bonds between the groups and water (Klotz and Franzen, 1962). The most stable conformation will be that which maximizes internal hydrogen bonding, maximizes hydrogen bonding between surface groups on the macromolecule and water, and maximizes other stabilizing forces.

1.2.2.2 The Hydrophobic Interaction

One of the other important water-related stabilizing forces is the hydrophobic interaction. Nonpolar gases like methane are nearly insoluble in water. By definition, this means that the free energy for the transfer of such a gas from a nonpolar solvent to water is positive. Ordinarily, the heat of a reaction or enthalpy parallels the free energy; that is, a reaction that has a positive free energy will also have a positive enthalpy (absorb heat) of similar magnitude. Frank and Evans (1945) made the important discovery that this is not the case for reactions of the sort just cited. The enthalpy is negative. Free energy (F) and enthalpy (H) are related by the equation

$$\Delta F = \Delta H - T\Delta S,$$

where ΔS is the change in entropy and T is absolute temperature.

A positive value for ΔF when ΔH is negative arises only when ΔS is negative. A negative value for the change in entropy means an increase in order in the system, and Frank and Evans made the innovative suggestion that the increase was a consequence of the water becoming more ordered or "icelike." The conventional view is that the ordering reflects increased hydrogen bonding, but Scatena et al. (2001) dispute this. Their infrared vibrational measurements indicate that hydrogen bonds between water molecules at apolar surfaces are actually weakened or nonexistent. Nevertheless, interactions between water and those surfaces cause the water molecules to become oriented.

In 1959, Walter Kauzmann realized the implications of this decrease in entropy for protein structure. A number of the amino acids possess hydrocarbon-like nonpolar side chains; for example, the methyl group in alanine and the isopropyl group in valine. He proposed that this being the case, it ought to be energetically favorable if proteins folded in such a way as to reverse the Frank and Evans reactions; that is, if they assumed a conformation in which the maximum number of nonpolar side chains were removed from contact with water and buried in the interior of the protein. A consequence of this is that the water at the surface of the protein would become more disordered than would be the case if the nonpolar groups jutted into the surrounding water. He coined the term "hydrophobic interaction" for this reaction. In the ensuing 40 years, Kauzmann's hypothesis has received striking conformation from structural studies of native proteins. For example, the aquaporins, about which more will be said later, are protein complexes that span the plasma membranes of some cells. That portion of the complex that lies within the lipid bilayer portion of the membrane has been found to be rich in amino acids with nonpolar side chains (Verkman et al., 1996).

1.2.3 HYDRATION

The sizes of molecules can be estimated by a number of hydrodynamic methods such as the rate at which they move by diffusion and the rates at which they move when subjected to centrifugation or to an electric field (electrophoresis). When the computed sizes are compared to direct measurements of size by, for example, x-ray diffraction, the former is found to exceed the latter. The excess is mostly ascribed to water that is closely associated with the molecule in question. It is referred to as water of hydration. A number of measurements have been used to assess the physical properties of such water. These measurements include adsorption isotherms, dielectric constants, nuclear magnetic resonance, and calorimetry.

1.2.3.1 Adsorption Isotherms

Adsorption isotherms are generated by equilibrating solutions of macromolecules in air at various relative humidities and, after equilibration, measuring the amount of water remaining in the system by techniques such as oven drying. The water content is then plotted as a function of the relative humidity (RH). RH is defined as the ratio of the vapor pressure of water over the solution to the vapor pressure of pure water (i.e., p/p_o), which from Raoult's law equals the activity of the water (a_w). Figure 1.1 shows the adsorption isotherms for native and denatured hemoglobin (Eley and Leslie, 1966). The result is typical of that for a variety of proteins, for DNA, and for lipids. It shows that when the water content of the protein drops to 20% or below, the residual water possesses reduced activity. For example, when the protein has 0.2 g water per gram, the activity of that water is reduced to 0.8. A reduction in activity is equivalent to a reduction in chemical potential (μ_w) according to the relation $\mu_w - \mu^o_w = RT \ln a_w$, so that water with reduced activity is water with reduced chemical potential relative to that of pure water (R is the universal gas constant). In that sense, it is "bound." The binding appears to be an equilibrium state, not a kinetic barrier, for the graph is based on the weights of samples exposed to the desired RH for 24 h, and these weights were nearly attained after only 5 h. Note, however, that there is not a discrete fraction of bound

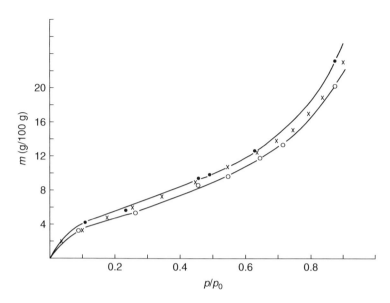

FIGURE 1.1 Adsorption isotherms for water on bovine hemoglobin at 25°C. (●, ×), native; (○), denatured. (From Eley and Leslie, 1966, by permission of the Royal Society of Chemistry.)

water but, rather, a continuum in which a progressively decreasing fraction is bound with progressively lower chemical potential.

1.2.3.2 Dielectric Measurements

Water exhibits its high dielectric constant (78) when the water molecules are able to orient themselves in an electric field. The value drops as the frequency of the alternating field increases and the molecules have insufficient time to change their orientation. It also drops if the water is immobilized by hydration forces. Figure 1.2 shows that as the water content of powdered bovine serum albumin is decreased, the dielectric constant (expressed as the permittivity) decreases until it reaches a breakpoint at a water content of 0.25 g per gram of protein (Rosen, 1963). At that point and below, the residual water is irrotationally bound.

1.2.3.3 Nuclear Magnetic Resonance

Protons spin, and in doing so, they generate a magnetic field at right angles to the spin. The spinning proton has a magnetic moment—a separation of north and south magnetic poles (analogous to dipole moment). If a molecule rich in protons is placed in a strong external magnetic field, H_0, it will precess like a gyroscope at a rate proportional to the magnitude of the external applied magnetic field.

   If a second variable magnetic field of magnitude H_1 (created by an alternating radio frequency [RF] current) is applied at a right angle, as its frequency approaches the precession rate of the proton, the proton will tip more and more until suddenly, when the frequency of H_1 equals the rate of precession, the proton abruptly flips; that is, exchanges north and south poles. In doing so, it absorbs energy, which can be detected. The value of H_0 (or H_1, depending on which is varied) at which resonance occurs depends on the group with which the proton is associated. For example, it is different in OH than in CH_3. This is referred to as a "chemical shift." Even on a given group, however, each proton is exposed not only to field H_0 but also to a local field H_{loc} from identical neighboring nuclei. These local fields cause the protons of given molecules to precess at slightly different rates depending on their neighbors, and thus the resonance condition varies slightly. In

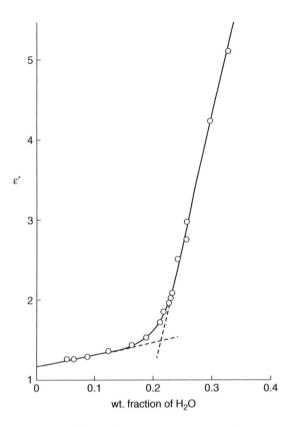

FIGURE 1.2 Permittivity (a measure of dielectric constant) of powdered bovine serum albumin as a function of its hydration expressed as grams of water per gram of dried protein plus water. The measurements were made at a frequency of 10^5 Hz. (From Rosen, 1963, by permission of the Royal Society of Chemistry.)

liquid water, these variations are so transient that all the protons are exposed to essentially the same $H_0 + H_{loc}$, and the resonance signal is a narrow peak at a fixed H_0. In a solid or in water protons with reduced mobility, however, the local variations in H_{loc} will persist for a long time relative to the RF frequency of H_1, and the resonance peak will be greatly broadened. In ice, it is broadened so much that it merges with the baseline and becomes invisible.

Kuntz et al. (1969) used nuclear magnetic resonance (NMR) to determine whether the water associated with proteins has different properties from bulk water. Ordinarily, the NMR signal from the bulk water in which the proteins are dissolved would swamp out the protein-associated water. Kuntz solved this problem by carrying out the NMR measurements at –35°C, at which temperature the bulk water was converted to ice and its signal rendered invisible by line broadening. When Kuntz did this, he found that the water proton signal from the residual liquid water in a variety of proteins was broadened relative to that of pure liquid water. He interpreted this to mean that the water of hydration was less mobile than that in free liquid water, although much more mobile than the protons in ice. From the area under the broadened signal, he was able to back calculate how much water this represented; he arrived at a value of ~0.4 g water per gram of protein. This quantity of liquid water with reduced mobility remained present down to –50°C, the minimum temperature studied.

1.2.3.4 Calorimetry

When water freezes, it releases 80 cal/g of heat. If one knows the total mass of water in a sample (by, say, oven drying), one can calculate the amount of heat that should be released by freezing. One can use calorimetry to measure the amount that actually freezes. One way to do this is to drop a frozen sample of a known mass of water into a known mass of water at room temperature in a calorimeter (method of mixtures). One then accurately measures the drop in temperature in the latter. From that drop and from the known masses, one can calculate the heat required to melt the sample and, from that, calculate the amount of ice that had to melt to absorb that amount of heat. Another method of calorimetry is to place a frozen sample of given mass at a known temperature into a calorimeter and to measure the electric energy required to warm and melt the sample. Privalov (1968) has made the latter measurements on a number of protein solutions and found that the heat released is less than expected, indicating that 0.25 to 0.3 g water per gram of proteins does not exhibit the expected latent heat of fusion. The usual interpretation is that that quantity of water is incapable of freezing; that is, it is water that is incapable of assuming the structure of ice.

Thus, all four methods discussed agree that there is water associated with proteins and other macromolecules that has reduced activity and that has reduced or nonexistent rotational mobility, and the methods converge on a similar value for the quantity of that water; that is, 0.25 to 0.4 g water per gram of protein. (A detailed comparison of estimates by these and other methods of water of hydration in various proteins was published by Privalov [1968].) Some authors (e.g., Berendsen, 1975, p. 329) have gone further to designate specific classes of this water and specific structures and to propose specific forces involved. Unfortunately, a discussion of these views is beyond the scope of this chapter (and beyond the abilities of the author). However, it is fair to say that hydrogen bonding must be one of the important forces involved.

1.2.4 WATER IN CELLS

Water composes some 60 to 85% of cells. Because cells contain complex mixtures of proteins, nucleic acids, lipids, and other solutes, it is not surprising that some of the water is hydrated or bound to these molecules. The properties have been assessed by similar methods to those used on purified components, and the amounts have been found to be comparable. For example, Sun (1999) has used differential scanning calorimetry (DSC) to measure the amount of water in acorns that can freeze. He dried the acorns to known water contents and then determined with DSC the extent to which that water released heat during freezing or absorbed heat during warming. As he reduced the water content of the acorns, the heat decreased proportionally until the water content reached 0.2 g per gram dry weight, at which point there was no heat released during subzero cooling of the sample and none was absorbed during warming. In other words, that quantity of water was incapable of freezing. Similar results were obtained by Wood and Rosenberg (1957) some 40 years earlier for yeast cells using method-of-mixtures calorimetry. About 10% of the water in the yeast cells, or ~0.25 g H_2O per gram dry weight, was incapable of freezing. Koga et al. (1966) conducted similar studies on yeast in which the hydration water in the cells was assessed by adsorption isotherms, dielectric constant measurements, and NMR. The adsorption isotherm that Koga obtained is similar in shape to that shown for hemoglobin in Figure 1.1. (The adsorption isotherm published by Sun [1999] for oak seeds is also similar.) The curves of water content vs. RH flattened out in cells with their water contents reduced to 5 to 10% of wet weight. The dielectric measurements showed a transition from mobile water molecules to irrotationally bound molecules when the residual water dropped below 11%. The NMR broadline measurements showed a decrease in the intensity of the narrow water signal that was linearly proportional to the cell water content until the water content was reduced to 8%. At that point, there was a sharp break, and the signal intensity remained constant as the water content was reduced toward zero. Similarly, Schreuders et al. (1996) have determined by DSC that 9% of the total water in eggs of the mosquito *Anopheles* does not freeze. Thus, as is the case with purified proteins, the results from these various methods and cell types converge in indicating that some 0.2 to 0.3 g water per gram dry weight or some 10% of the water in fully hydrated cells does not exhibit heat of crystallization or melting (i.e., presumably does not freeze), is irrotationally bound, or remains attached to cell components even when equilibrated with water vapor at an a_w of 0.8 or below.

These estimates of the water of hydration in cells, and particularly those from the calorimetric measurements, are of considerable cryobiological significance. Freezing is a form of water removal in that it transforms cellular liquid water into ice, either, as we shall see, within the cell or after flowing out of the cell and freezing externally. The calorimetric measurements indicate that cooling to low subzero temperatures will convert about 90% of the cell's water to ice by either route, but will not convert the remaining 10%. Expressed differently, freezing affects the free water in cells but tends not to affect the quantity that is bound by hydration forces.

Although freezing cannot "remove" the 10% residual water of hydration from cells, that fraction can be removed by freeze-drying or by air or vacuum drying from the liquid state. The dehydration in freeze-drying is a three-step process. First, the ~90% fraction represented by "free" water is removed by being transformed into ice. Second, the ice is sublimed away (under vacuum to speed up the process by increasing the mean free path of the water molecules). Third, once the ice has sublimed, the residual hydrated water is slowly removed by exposing the sample to surroundings in which the a_w is essentially zero. In air or vacuum drying from the liquid state, the dehydration is a continuous process involving no phase change. The sample effectively follows the adsorption isotherm. Because the rate of water removal is proportional to the chemical potential of the water, the cell water with the highest chemical potential is removed first (i.e., the "free" water), followed by water bound with progressively lower chemical potentials.

Are there different consequences to the removal of free water and the removal of water of hydration? The answer with respect to viability is yes. Using electrolyte leakage as the criterion, Sun (1999) assessed the damage to acorns as a function of the reduction in their water content. He found no damage until the water content was reduced below 0.4 g per gram dry weight, but below 0.3 g per gram dry weight, the damage increased sharply. The latter water content is close to the amount of nonfreezable water that Sun found by DSC in these seeds (0.2 g per gram dry weight). Further instances of this correlation will be given later in discussing dehydration theories of freezing injury.

It is perhaps not surprising that the removal of the water of hydration can have untoward effects, as interactions between water and macromolecules play a role in the conformation of those molecules. Water of hydration has another important role. In enzymatic reactions, enzyme and substrate have to come into intimate contact. If each is surrounded by a layer or layers of water of hydration, contact requires that the intervening water be removed. Cells are surrounded by and contain lipid bilayer membranes. Contact between two membranes is a first step in membrane fusion that can either be a normal process or one with pathological consequences. In either case, for contact to occur, the water molecules that hydrate the polar surface of each membrane have to be pushed aside. Leikin et al. (1993) and Wolfe and Bryant (1999) have reviewed the elegant studies in the past decade, particularly those by Parsegian and Rand and colleagues (e.g., Leikin et al., 1993), in which direct measurements have been made of the forces required to bring two hydrated membranes or two hydrated surfaces into contact. The typical result is that the required force increases markedly as the distance between the two surfaces decreases below some 25 Å. (At separations of ~2 Å, the force changes to an attraction as van der Waal forces become predominant.) The usual interpretation is that the repulsive force represents the resistance of the water of hydration to being pushed out of the way. As might be imagined, the situation is complex (and controversial). Israelachvili and Wennerstrom (1996), for example, have argued that the degree of repulsion depends on the degree to which the water molecules on the two approaching surfaces are in register or out of register.

It is clear that freeze-drying and air and vacuum drying affect water of hydration (they remove it) and that that removal can have adverse consequences (denaturation, fusion, etc.). An important question to which we shall return, however, is whether freezing *per se* affects the water of hydration of membrane surfaces and whether those effects have injurious consequences.

1.2.5 Osmotic Relations in Cells

Cryobiology is not only concerned with state of water in cells and the role of water in the structure and function of components but is also equally concerned with factors affecting the movement of water into and out of cells. There are two aspects to this concern. One is the factors that change the equilibrium amounts of water in cells; that is, the osmotic relations, which is the subject of this section. The other is the kinetics with which these equilibria are attained, which is the subject of the succeeding section.

1.2.5.1 Basic Osmotic Thermodynamic Equations

The fundamental principle of osmotic relations is that in the absence of metabolic forces, a cell will adjust the concentrations of water and solutes so that each is in chemical potential equilibrium with the concentrations outside the cell, provided the cell is permeable to that component. For water, this means that $\mu^i_w = \mu^e_w$. For water in a solution, $\mu_w = \mu^o_w + RT \ln a_w$, where a_w = vapor pressure of H_2O in a solution/vapor pressure (V.P.) pure H_2O = p/p^o and R is the universal gas constant, 0.082 lit atm/mole deg. In an ideal solution, $p/p^o = x_w$, the mole fraction of water in solution. If we know x_w, we can compute x_s and, from that, the molality m of the solute; that is, $\mu_w = \mu^o_w - v_w RTm$, where v_w = molar volume of water (18 cc/mole; the precise quantity to use is the concentration-dependent partial molal volume, but the differences between this volume and the molar volume are trivial).

In a real solution, $\mu_w = \mu^o_w - v_w RT[\Sigma]\phi\nu m$, where ν is the number of species into which solute dissociates (e.g., two for NaCl) and ϕ is the osmotic coefficient. (Values of ϕ for several cryobiologically relevant solutes are tabulated in Robinson and Stokes, 1959). In this equation, $RT[\Sigma]\phi\nu m$ is the osmotic pressure, Π, and $[\Sigma]\phi\nu m$ is the osmolality M. Hence, $\Pi = RTM$.

1.2.5.2 Osmotic Responses of Cells (Assuming Semipermeability)

First, a cell will adjust its water volume until the chemical potential of its water equals that in the medium; that is, until the osmolalities are equal, namely, until $M^i = M^e$.

However,

$$M^i = N^i/V^i, \tag{1.1}$$

where N^i = intracellular osmoles and V^i is volume of intracellular water.

Second, a cell in physiological medium is at its isotonic volume, and

$$M^i_{iso} = N^i/V^i_{iso} = M^e_{iso}. \tag{1.2}$$

Third, if we divide Equation 1.2 by Equation 1.1, we obtain

$$V^i/V^i_{iso} = V_{rel} = M^e_{iso}/M^e; \tag{1.3}$$

that is, the relative volume of cell water (V_{rel}) is inversely proportional to the external osmolality.

Fourth, if b is the volume of cell solids and nonosmotic water relative to the volume of the isotonic cell, then the volume of the cell, V_c, relative to its isotonic volume, will be

$$V_c = V_{rel} (1 - b) + b.$$

If we substitute Equation 1.3 for V_{rel}, we obtain

$$V_c = b + (1 - b)M^e_{iso}/M^e. \tag{1.4}$$

A cell that obeys Equation 1.4 is said to behave as an ideal osmometer; that is, a plot of V_c vs. $1/M^e$ will be linear with a slope of $(1 - b) M^e_{iso}$ and a volume b at infinite osmolality. This is

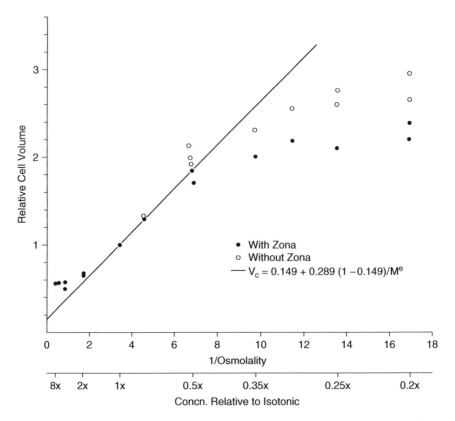

FIGURE 1.3 The volume of eight-cell mouse embryos relative to their isotonic volume as a function of the reciprocal osmolality of the external medium and as a function of whether the zona pellucida is present or not. The solid line represents ideal osmotic behavior as generated by the indicated equation. (From Mazur and Schneider, 1986, by permission of Humana Press.) Unlike these eight-cell embryos, mouse and bovine blastocysts showed no departure from ideal in the hypertonic range and much less departure in the hypotonic range.

referred to as a Boyle–van't Hoff plot (BVH). Many instances have been published of cells obeying the BVH relation, but some cells exhibit a linear response over part of the osmolality range and a nonlinear response over other parts. Both linear and nonlinear BVH responses are shown by eight-cell mouse embryos as illustrated in Figure 1.3 (Mazur and Schneider, 1986). The response is linear in solutions that are half to twice isotonic, but it departs markedly from linearity when the osmolality drops below ~50% of isotonic. A central premise of ideality is that the number of osmoles of solute in the cell remains constant with change in cell volume; that is, that $N^i = \phi n^i$ (a constant). That would become invalid if the membrane becomes leaky, causing n^i to change, or it could become erroneous if ϕ, the osmotic coefficient, changes with change in cell water volume and with solute concentration. The first scenario could occur under either strongly hyperosmotic or strongly hypoosmotic conditions, and it is probably the explanation for the departure from linearity noted in Figure 1.3 for dezonated eight-cell mouse embryos at extreme hypotonicities. The second source of nonlinearity, changes in ϕ, is more likely to be a factor under strongly hyperosmotic conditions, for the osmotic coefficients of proteins in particular can rise to high values in concentrated solutions.

Another cause of apparent nonideality under hypoosmotic conditions arises when the volume expansion called for to meet osmotic equilibrium is prevented or impeded by an external "shell." That is not a factor in most animal cells, as most lack a shell and as the plasma membrane is incapable of resisting more than the slightest pressure. However, mammalian embryos are surrounded by a shell-like structure, the zona pellucida, and the resistance of that shell to expansion of the embryo proper is probably the explanation for some of the nonideality shown in the curve

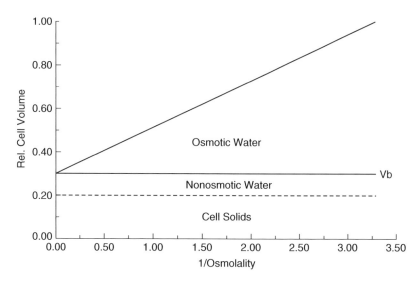

FIGURE 1.4 Schematic of relative cell volume as a function of reciprocal osmolality for a cell exhibiting an ideal osmotic response. The cell volume is considered to be composed of three regions: The volume occupied by cell solids, the volume occupied by water that does not contribute to the osmotic pressure (nonsolvent water), and the water that contributes to the osmotic pressure. The sum of the first two is referred to as V_b.

in Figure 1.3 labeled "with zona." An extreme example of resistance to osmotic expansion is the higher plant cell. Such cells normally exist in water or very dilute solutions, but they cannot expand to satisfy Equation 1.4 because of the presence of a rigid cell wall. How, then, do they keep internal and external water in chemical potential equilibrium? The answer is that the earlier equation ($\mu_w = \mu^o_w - v_w RT[\Sigma]\phi vm$) is incomplete. The fuller equation is

$$\mu_w = \mu^o_w - v_w RT[\Sigma]\phi vm + v_w P, \tag{1.5}$$

where P is the hydrostatic pressure. (Again, the exact expression for v_w is the partial molar volume.)

In other words, as the plant cell tries to expand, the protoplast encounters a hydrostatic pressure from the cell wall, and this increased hydrostatic pressure increases the chemical potential of the intracellular water (note the plus sign) until the increase balances the decrease caused by the osmolality term. Note that compensating for a 1 osmolal difference in osmotic pressure requires 24 atm of hydrostatic pressure at 20°C. One can also have intermediate cases in which the cell wall is elastic, not rigid, and exerts an increasing pressure on the protoplast as the latter attempts to expand in response to increasingly hypoosmotic conditions. That seems to be the situation in the yeast cell (Mazur, 1961).

1.2.5.3 The Meaning of *b* and Osmotic and Nonosmotic Water

The term *b* in Equation 1.4 is the extrapolated volume occupied by a cell when the external osmolality becomes infinite. It is composed of two elements: the solids in the cell and cell water that cannot act as solvent and therefore cannot contribute to the osmotic pressure.

These two components of *b* are illustrated schematically in Figure 1.4. The volume of solvent or osmotic water as a function of the external osmolality (i.e., the volumes > *b* in the example plotted) can be determined, for example, by electron spin resonance of a spin label like Tempone, which dissolves only in the water space (Du et al., 1993, 1994b) or by tritiated water (Armitage, 1986). The total water in the isotonic cell (the region above the dashed line) can be determined by techniques like oven drying. When Du et al. and Armitage plotted the measured water volumes of mouse and human sperm and of human platelets, respectively, as a function of the reciprocal of external osmolality, they obtained plots that were similar to that in the region above the dashed line in Figure 1.4. They were similar, both in being linear and in not extrapolating to zero cell water at infinite osmolality. The total (gravimetric) relative cell water in this example is 0.8 (i.e., 1.0 – 0.2). The measured osmotic water volume is 0.7 (i.e., 1.0 – 0.3). The ratio of osmotic water to total water is 0.7/0.8, or 0.875. This ratio is referred to as Ponder's *R* after the red cell physiologist Eric Ponder, and it is close to the value of *R* found by the above authors for the platelets and sperm. By subtraction, the volume of nonosmotic cell water is 12.5% of the total water in the isotonic cell (i.e., 0.1/0.8). Finally, the ratio of the volume of nonosmotic water to the volume of cell solids is 0.1/0.2, and the ratio of the mass of nonosmotic water to the mass of cell solids (assuming their density is 1.2) is 0.42 g H_2O per gram solids. Both values (12.5% and 0.42 g per gram solids) are similar to the values for unfreezable water and immobilized water found in cells and proteins by the other methods discussed above for determining water of hydration. This implies that nonosmotic water and water of hydration are roughly equivalent.

Values of Ponder's *R* < 1 are common but not universal (Dick, 1966, p. 61). Elford (1970), for example, found that dimethyl sulfoxide (Me_2SO) is accessible to the entire water space of guinea pig intestinal smooth muscle at 37°C (*R* = 1). However, interestingly, that was not the case at –7°C in the unfrozen state. At –7°C, 6.6 g of water per 100 g muscle were inaccessible to Me_2SO (*R* = 0.91), which translates to a nonsolvent water fraction of 0.36 g per gram of muscle protein, similar to that in the example shown in Figure 1.4.

I need to mention that some like Dick (1966, pp. 50–57) have argued than nonosmotic water is an artifactual consequence of the development of high osmotic coefficients in cell solutes (chiefly proteins) when they become highly concentrated in strongly hypertonic conditions. Mazur and Schneider (1986) and Pegg et al. (1987) discuss this in some detail and argue against it.

1.2.6 PERMEABILITY OF CELLS TO WATER AND SOLUTES

The osmotic relations just discussed describe the volume that cells attain when they have equilibrated in anisotonic media, but they say nothing about the kinetics of attaining that equilibrium. Those kinetics are of central importance in cryobiology (see also Chapter 2).

1.2.6.1 Permeability Equations for Water

The rate at which water leaves or enters a cell is proportional to the difference between the internal and external osmotic pressures; namely,

$$dV/dt = L_p A(\Pi^i - \Pi^e), \qquad (1.6)$$

where L_p is the hydraulic conductivity with units μ^3/μ^2/min-atm and A is the area of the cell surface. Π^e is the external osmotic pressure and is also the osmotic pressure within the cell after it has attained volume equilibrium with the external medium. This equation is equivalent to saying that water moves at a rate that is proportional to the difference in the chemical potentials of that water inside and outside the cell.

From the previous section, however, we have

$$\Pi = RTM = NRT/V,$$

where N is osmoles of solute and V is the volume of water. Therefore,

$$dV/dt = L_p ANRT(1/V - 1/V_{eq}), \qquad (1.7)$$

where V_{eq} is the cell water volume after equilibration.

If one knows b, the nonosmotic volume of the cell, the above equation can be expressed in terms of cell volumes rather than the volumes of cell water.

1.2.6.2 Methods of Measuring Permeability to Water and L_p

The procedure for estimating L_p is simple to state: Place cells in a hyperosmotic or hypoosmotic solution of a nonpermeating solute like NaCl or sucrose and measure the cells' volume as a function of time at a constant temperature. L_p is estimated by determining what value L_p results in a curve calculated from the whole-cell version of Equation 1.7 that most closely agrees with the experimental curve.

With large spherical cells, the rate of volume change can be followed most directly by using video or regular photomicrography to measure the changes in the cross-sectional area of the cells with time. With small or highly nonspherical cells, one has to use indirect methods such as an electronic sizer like the Coulter counter or light scattering. Because the latter two methods are indirect, the validity of the results they yield depends on the care of the calibrations used. If such measurements are made at different temperatures, one can calculate the temperature coefficient or activation energy for water permeation. One point about Coulter measurements is that the height of the voltage spike produced as a cell of given volume passes through a small orifice depends not only the cell volume but also on the electrical conductivity of the medium, and the electrical conductivity depends on the composition of the medium and its temperature.

Measured values of L_p are numerous. A number of older values have been tabulated by Dick (1966, p. 103) and Stein (1967, p. 110). More recent determinations are widely scattered in the literature. The values vary some 100-fold in different cells from ~0.1 μ^3/μ^2/min-atm to ~10 at room temperature. In general, the L_p for "typical" mammalian cells like white blood cells (Hempling, 1973; Hempling and White, 1984; Porche et al., 1986), pancreatic islet cells (Benson et al., 1998; Liu et al., 1995, 1997), and oocytes (Agca et al., 1998; Benson and Critser, 1994; Leibo, 1980; Marlow et al., 1994; Myers et al., 1987; Paynter et al., 1999) is about 0.2 to 0.5. However, the L_p for mammalian red cells (Shaáfi et al., 1967; Terwilliger and Solomon, 1981) and some mammalian sperm (Gilmore et al., 1995; Noiles et al., 1997) is some four- to 20-fold higher. I will have more to say shortly on the cause of the large difference. Commonly, L_p in the presence of cryoprotective solutes is roughly half the value of cells in saline (e.g., Hempling and White, 1984; Gilmore et al., 1995; Phelps et al., 1999; Rule et al., 1980;). Sometimes, however, as in oocytes, there is little or no difference (e.g., cf. Benson and Critser, 1994; Leibo, 1980; Paynter et al., 1997, 1999).

One should note that L_p as estimated is really a phenomenological constant, not a true permeability coefficient. What one really calculates from fitting Equation 1.7 to experimental data is the product $L_p \times A$, the surface area. For spherical cells, A is computed as the area of a sphere with the measured cross-sectional area. For a nonspherical cell, it may be computed as the area of an ellipsoid. In either case, A is assumed constant, and L_p is the adjustable parameter. (Some have assumed that A varies as the two-thirds power of V. However, Mazur [1990] argues that this rubber balloon-like behavior is not consistent with the inability of bilayer membranes to stretch or compress.) That is a reasonable estimate of the true surface area of red cells and sperm that have smooth plasma membranes, but it is an underestimate of the true surface area of cells like white cells, embryos, and tissue culture cells, the surfaces of which usually possess numerous folds, pleats, and microvilli. The true surface area of such cells has been estimated to be two (Knutton et al., 1976) to nine (Loo et al., 1996, p. 13,367) times the simple geometric surface area. In most cases in which it has been examined, the L_p calculated from water efflux is close to the value calculated from water influx (Armitage, 1986; Terwillger and Solomon, 1981). One striking exception is in zebra fish embryos, where L_p for efflux is manyfold higher than that for influx (Hagedorn and Kleinhans, 2001; Hagedurn et al., 2002).

1.2.6.3 Permeability Equations for Solutes

Statements made about chemical potentials apply to permeating solutes as well as to water; that is, if a cell is placed in a solution of a permeating solute in which $\mu_s^i \neq \mu_s^e$, then solute will move in or out of the cell to make the chemical potentials equal. For practical purposes, chemical potentials of solutes are proportional to their concentrations, so that after equilibration, the concentrations of the solute will be the same inside and outside the cell and the rate of that movement will be proportional to the differences in concentration; that is,

$$dn_s/dt = P_sA(m_s^c - m_s^i).$$

However, as $m_s^i = n_s^i/V^i$,

$$dn_s/dt = P_sA(m_s^c - n_s^i/V^i), \tag{1.8}$$

where P_s is the permeability coefficient for the solute (with units cm/min).

Note that n_s^i and V^i are both variables. The water content of the cell cannot remain constant during solute permeation because if n_s^i changes, the total osmolality of the cell changes, which changes μ_w^i so that $\mu_w^i \neq \mu_w^c$. As a consequence, for each increment of solute that enters or leaves, an increment of water has to move in or out of the cell to reestablish osmotic equilibrium.

The overall movement of water and permeating solute is thus determined by the simultaneous effects of the equations

$$dV/dt = L_pANRT(1/V^i - 1/V^i_{eq}); \tag{1.7}$$

$$dn_s/dt = P_sA(m_s^c - n_s^i/V^i). \tag{1.8}$$

L_p can be expressed in the same units as P_s (cm/min) by the relation

$$P_f = L_pRT/v_w,$$

where P_f is the filtration coefficient with units of cm/min (if L_p has the units cm/min-atm), R is 82 cm^3 atm/mole-deg, and v_w is the molar volume (more precisely, the partial molar volume) of water (18 cm^3/mole). When the comparison is made, P_f is nearly always found to be much greater than P_s (100- to 1000-fold). The consequence is that when a cell is placed in a hyperosmotic solution of a permeating solute, there is an initial abrupt shrinkage determined primarily by water being lost at a rate determined by P_f (or L_p; Equation 1.7), followed by a slower return to normal volume as the solute permeates at a rate determined by P_s (Equation 1.8). Plots of this volume response vs. time in solutions of permeating solutes are colloquially referred to as shrink/swell curves, and many have been published. One example is shown in Figure 1.5 (Paynter et al., 1999).

The volume to which the cell returns or asymptotes depends primarily on the concentration or osmolality of nonpermeating solutes in the test solution (e.g., NaCl or phosphate buffered saline [PBS])—not on the concentration of the permeating solute. If the osmolality of the nonpermeating salts in the medium is the same as that normally present in an isotonic solution, the cell water volume will return to the value in an isotonic cell. If the osmolality of salts is less (or more) than normal, the cell will equilibrate at a water volume that is greater (or less) than normal. If no nonpermeating solutes are present, the cell will expand infinitely (i.e., until it lyses).

Note that the above refers to cell water volumes—not cell volumes. Therein lies a subtle, but in some cases important, point. If the nonpermeating salts in, say, a glycerol/PBS solution are prepared so as to have the same molality (mole/kg water) or osmolality as in isotonic PBS alone, then the volume of cell water after the completion of the shrink/swell process will be equal to the volume of water originally in the cell in isotonic PBS, but the volume of the whole cell will be greater by the space occupied by the permeated glycerol. In 1 M glycerol that amounts to ~10%—not too significant. However, if, say, 5 M glycerol were present, the excess volume would be very

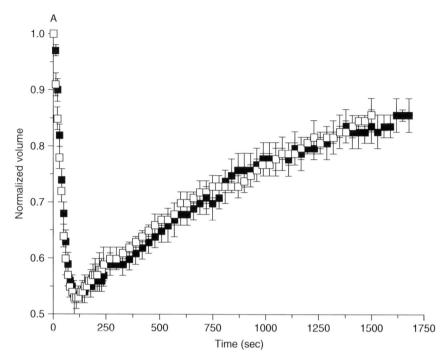

FIGURE 1.5 Volume relative to isotonic volume of mature mouse oocytes as a function of time in 1.5 *M* ethylene glycol (■) or dimethyl sulfoxide (Me$_2$SO) (□) in isotonic saline at 10°C. (From Paynter et al., 1999, by permission of Academic Press.)

significant. Indeed, Meryman and Douglas (1982) have shown that it is sufficient to cause red cells to exceed their critical volume and hemolyze. If, however, the nonpermeating solutes are maintained at the same molarity (moles/L solution) as in isotonic PBS, then the cells will equilibrate to the same total volume as existed initially, but with reduced water content. These facts were analyzed in detail by Pegg (1984).

1.2.6.4 Methods for Measuring Permeabilities

All the methods used to estimate L_p from the kinetics of shrinkage of cells in nonpermeating solutes are amenable to the calculation of solute permeabilities; for example, photomicrography and Coulter measurements. In addition, other methods become available because the reswelling process is so much slower than shrinkage. The most direct of these methods is the rate of uptake of an isotopically labeled solute. Jackowski et al. (1980), for instance, used ^{14}C glycerol to assess the permeation kinetics of glycerol in early mouse embryos and obtained values close to those calculated from shrink/swell curves. Mazur and Miller (1974) developed a time-to-lysis procedure for red cells based on the fact that if the concentration of nonpermeating solutes in, say, a glycerol/NaCl solution is sufficiently hypotonic, the cell will swell beyond its critical lytic volume. From knowledge of that critical volume and the time taken to attain it, one can use versions of Equations (1.7) and (1.8) to calculate P_s. Du et al. (1994a) have used an electron spin resonance technique to assess P_s in sperm. The method depends on the fact that the entry of a solute into a cell is accompanied by the entry of water and the entry of a spin label (Tempone) dissolved in that water. The signal from the external Tempone is eliminated by the presence of a compound to which the cell is impermeable. As a consequence, the magnitude of the observed signal is proportional to the amount of intracellular Tempone, which in turn is proportional to the amount of water and cryoprotective additive (CPA) that has entered the cell.

1.2.7 PERMEABILITY AND THE STRUCTURE OF THE CELL MEMBRANE

The above discussion of permeability has been phenomenological, but the movement of water and solute occurs through the plasma membrane, and the rates of movement are determined by its structure and composition. The basic structure originally proposed by Davson and Danielli (1952, Chapter 6) is a bimolecular lipid sheet. Nonpolar chains of the lipids face inward; polar head groups face into the aqueous cytoplasm and the external medium. That original concept was modified by Singer and Nicholson in 1972 to a structure they termed the fluid mosaic model. The chief difference is that it depicts intermittent proteins spanning the bilayer.

For several decades there was controversy as to whether water moves through this lipid bilayer by passive diffusion or whether it moves through small pores. Much of the research was done on human erythrocytes (red cells), which have a very high permeability to water (L_p = 5 μm/min.atm). In the late 1960s, Arthur Solomon's group found that water movement under osmotic flow was several-fold higher than water movement by diffusion with no osmotic flow, the latter being assessed using tritiated water (i.e., $P_f > P_d$). This inequality argues for the existence of pores because the movement of water by bulk flow is greater than the movement of single molecules of water by diffusion. The counterargument was that the diffusional value was artifactually low because unstirred water layers made the diffusional distance greater than the thickness of the membrane. However, that argument was refuted by Solomon's group (Shaàfi et al., 1967). In 1984, Maccy summarized evidence for the existence of water channels in red cells; to wit,

- P_f is much higher than that demonstrated for pure lipid bilayers
- $P_f/P_d \gg 1$
- E_a (activation energy) for water permeation is 4 to 6 kcal/mole rather than the 11 to 14 kcal/mol characteristic of permeation through lipid bilayers
- In the presence of mercurial sulfhydryl reagents (p-CMBS), the above values revert to those characteristic of lipid bilayers; that is, the pores close (interestingly, the closure was enthalpically unfavorable but was driven by large increase in entropy [hydrophobic interaction], indicating a structure-breaking effect on the water in the pores)
- p-CMBS did not affect the permeation of ions and small nonelectrolytes, indicating that they were transported by a different route

About 4 years later, a 28-kDA protein was discovered in the red cell membranes (and kidney proximal tubes) that was suspected to be related to the water channel. It was called CHIP28 (channel-forming integral protein of 28 kDA). In 1992, Preston et al. proved this to be the case. They injected transcribed human CHIP28 RNA into *Xenopus* eggs, and after a few hours they found CHIP28 to be expressed in the eggs. Coincident with that expression was an eightfold increase in the water permeability of the eggs, as measured by rate of swelling in hypotonic media, and a more than threefold decrease in the E_a of permeation from greater than 10 kcal/mole to less than 3 kcal/mole. All these changes were inhibited by mercurial compounds. Finally, they pointed out that the number of CHIP28 monomers in red cells is approximately equal to the number of water channels as estimated from measurements of P_f/P_d ratio in the red cell (~3 × 10^5/cell).

In the ensuing 9 years, nine additional water channel proteins have been found in a broad range of different mammalian cells and tissues (Edashegi et al., 2000; Sui et al., 2001; Verkman et al., 1996), and others have been found in insects and plants. As a consequence, the class has now been named aquaporins, and CHIP28 has been named aquaporin 1. Most aquaporins, similar to CHIP28, specifically allow the transport of only water and not solutes, although aquaporin 3 and perhaps aquaporins 7 and 9 allow the cotransport of water and small nonelectrolytes like glycols and urea (Edashige et al., 2000; Ishibashi et al., 1994; Zeuthen and Klaerke, 1999). The detailed molecular structure of CHIP28 has now been elucidated at 2.2-Å resolution by x-ray diffraction (Sui et al., 2001) and by molecular dynamic simulation (de Groot and Grubmüller, 2001). It consists of a tetramer with each monomer composed of six transmembrane helices. As expected from hydrophobic bonding theory, the transmembrane helices consist primarily of nonpolar residues. The pore forms in each monomer from the overlap of two interhelical loops. The regions of the pore near the extracellular medium and the cytoplasm are funnel-like, with outer diameters of ~14 Å, but the central portion narrows to a diameter of ~3 Å. (The constricted region is a major factor in their selectivity for water.) In other words, the pore diameter ranges from two to six times that of a water molecule.

Although aquaporins are being identified in increasing numbers of cell types, the cells possessing them still represent a distinct minority. The basis for this conclusion is that most cells do not meet the phenomenological requirements for them (e.g., high L_p and low activation energy for L_p). In the cells that lack them, the consensus is that water is transported by passive diffusion through the lipid bilayer (Finkelstein, 1987). That is also the consensus for the mechanism of permeation in most cells of most nonelectrolytic solutes of cryobiological concern. However, according to Stein (1986, pp. 96–100) the diffusion is not classical Stokesian, but involves a random walk through various size cavities that form transiently in the bilayer. (I will return to this matter later when discussing intracellular ice nucleation.)

The discussion so far has assumed that the permeation of water and solutes can be described by two phenomenological coefficients: L_p (or P_f) and P_s. In 1958, Kedem and Katchalsky published the important conclusion that <u>when solute and solvent permeate through a common pathway</u>, the two interact, and a third phenomenological coefficient (and irreversible thermodynamics) is required to adequately describe the flows. That coefficient is referred to as the reflection coefficient or sigma, and when it is used, it appears in both Equations (1.7) and (1.8). Sigma can assume values between 0 and 1. The former is applicable to a membrane that cannot distinguish between water and solute; the latter to a membrane that is impermeable to the solute. A number of cryobiologists have used the Kedem–Katchalsky formulation to analyze the volumetric responses of cells to solutions of permeating solutes in water. More accurately, they have used the shrink/swell response of cells to calculate all three phenomenological coefficients. This usage, however, ignores the above underlined caveat.

Because the preponderance of evidence is that water and cryobiologically relevant solutes in most cases do not permeate through common pathways (Macey, 1984), several authors have argued that the use of sigma and irreversible thermodynamics may be of little or no value and that in many cases it is conceptually erroneous (Kleinhans, 1998; Verkman, 2000; Verkman et al., 1996). (Aquaporin 3, which cotransports water and glycerol, might appear to be an exception; however, even there, Echevarria et al. [1996] have argued that on the basis of large differences in activation energies and differences in responses to inhibitors, the two compounds do not share the same pathway.) Moreover, Kleinhans has shown that the use of sigma is unnecessary. That is, he and others (Woods et al., 1999) have shown that shrink/swell curves computed from a two-parameter model (P_f and P_s) fit experimental data as accurately as those computed from a three-parameter model (P_f, P_s, sigma) and that the values of P_f and P_s derived from curve fitting are closely similar in the two cases. When solute and solvent do not interact, sigma has the value $1 - P_s v_s / P_f v_w$), where v_s and v_w are the molar (partial molar) volumes of solute and water (Kleinhans, 1998). For common cases in which P_s / P_f lies between 0.01 and 0.001, the noninteracting sigma has a value near unity (i.e., 0.96 and 0.996 when the solute is glycerol), and Macey (1984) and Verkman et al. (1996) state that the more carefully that sigma has been evaluated by independent methods in solutions of permeating solutes, the more the value in fact approaches unity.

1.3 CRYOBIOLOGY PRINCIPLES

1.3.1 THE INVERTED "U"

A major factor determining whether or not cells survive freezing to low subzero temperatures is the rate at which they are cooled. Commonly, plots of their survival vs. cooling rate take the form

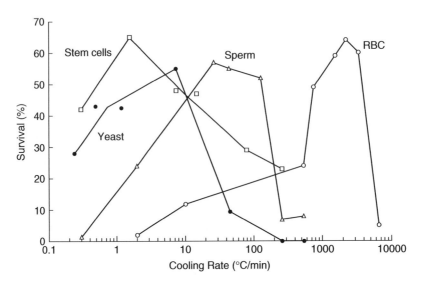

FIGURE 1.6 Survival of mouse marrow stem cells, yeast, mouse sperm, and human red cells as function of cooling rate. The data for the sperm are from Koshimoto and Mazur (2002) by permission of the Society for Reproduction. The other curves are from Mazur et al. (1970) by permission of the Ciba Foundation. The sources of the underlying data are given there.

of an inverted U, as exemplified in Figure 1.6. Maximum survival occurs at an intermediate rate. Fewer cells survive if the cooling rate is too low, and many fewer survive if the cooling rate is too high. Other examples abound; for example, human fibroblasts (Böhmer et al., 1973), mouse marrow stem cells (Leibo et al., 1970), hamster V79 tissue culture cells (Mazur et al., 1972), human lymphocytes (Scheiwe and Körber, 1983; Taylor et al., 1987), mammalian sperm (Henry et al., 1993; Koshimoto and Mazur, 2002; Woelders et al., 1997), yeast (Lepock et al., 1984), and *Chlamydomonas* (Morris, 1979).

An inverted U curve is most easily explained as the resultant of two opposing damaging factors. One tends to become damaging at high cooling rates, the other at low cooling rates. A first task of cryobiology is to explain what these two factors might be. A second task is to account for the fact that the optimum cooling rate can vary over a broad range in different cell types. In the examples shown in Figure 1.6, it is 1°C/min for mouse marrow stem cells and over 1000°C/min for human red cells. In other cell types (e.g., mouse embryos) the optimum can be below 1°C/min (Whittingham et al., 1972). Two other facts have to be explained. One is that the two limbs of the inverted U tend to behave differently with respect to the effects on survival of the rate at which the samples are warmed and thawed and with respect to the effects of the type and concentration of cryoprotective solute or additive present (CPA). In general, cells that have been cooled at supraoptimal rates tend to exhibit higher survivals when they are warmed rapidly than when they are warmed slowly, whereas cells that are cooled at suboptimal rates often respond oppositely or are relatively unaffected by the rate of warming. The effects of cryoprotectants also tend to be asymmetric: Cryoprotectants tend to protect slowly cooled cells and to do so in proportion to their concentration (Figure 1.7), but they tend to confer no protection to cells cooled at supraoptimal rates (Wellman and Pendyala, 1979) or even to cause damage to occur at lower supraoptimal rates than in the absence of CPA (Diller, 1979).

1.3.2 CAUSES OF LETHALITY AT HIGH COOLING RATES

The thermodynamic freezing point of most cells (the highest temperature at which ice can coexist with the protoplasmic solution) is about –0.5°C, but cells do not freeze even in the presence of

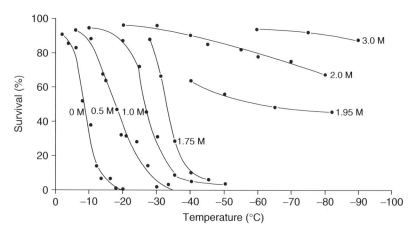

FIGURE 1.7 Survival (percentage unhemolyzed cells) of frozen-thawed human red cells a function of the concentration of glycerol in isotonic buffered saline and as a function of the temperature to which they were frozen. Cooling was slow (1.7°C/min); thawing was rapid. (From Souzu and Mazur, 1978, by permission of the *Biophysical Journal*).

external ice unless the temperature falls from 5° to 40°C below that temperature. By definition, water below its freezing point is supercooled or undercooled, and supercooled water has a higher vapor pressure, activity, or chemical potential at a given subzero temperature than that of ice or of water in a solution in equilibrium with ice. The consequence is that as long as the cell contents remain supercooled, the resulting vapor pressure or chemical potential difference will provide a driving force for intracellular water to leave the cell and freeze externally. In other words, the cell will tend to dehydrate during cooling. The rate and extent of that dehydration depends primarily on two variables. One is the inherent permeability of the cell to water; that is, the hydraulic conductivity, L_p. The other is the cooling rate. For a cell of given L_p, the slower it is cooled, the more it is able to lose sufficient water to remain in near chemical potential equilibrium with external ice and solution, and conversely, the faster it is cooled, the less it is able to dehydrate and the more its water will become supercooled as the temperature falls.

This qualitative description can be described quantitatively by four coupled equations. The first equation relates the rate of loss of cytoplasmic water to the difference in chemical potentials of intracellular and extracellular water expressed as a vapor pressure ratio; that is,

$$dV/dt = (L_p ART \ln p_e/p_i)/v_w, \tag{1.9}$$

where V is the volume of cell water, t is time, L_p is the permeability coefficient for water (hydraulic conductivity), A is the cell surface area, R the gas constant (μm^3 atm/deg mole), and v_w the molar volume of water. The ratio p_e/p_i is that for the external and internal vapor pressures of water. It is less than 1 because the intracellular water is supercooled and the vapor pressure of supercooled water is greater than that of ice or of water in a solution in equilibrium with ice. The change in this vapor pressure ratio with temperature can be calculated from a second differential equation derived from the Clausius-Clapeyron relation and Raoult's law:

$$d \ln (p_e/p_I)/dT = L_f/RT^2 - [n_2 v_w/(V + n_2 v_w)V]dV/dT. \tag{1.10}$$

Here, n_2 is osmoles of solute in the cell and L_f is the molar latent heat of fusion of ice.

Time and temperature are related by the cooling rate, which, if linear, is given by

$$dT/dt = B. \tag{1.11}$$

Finally, the hydraulic conductivity, L_p, decreases with falling temperature. If it is assumed to follow an Arrhenius relation, its value at a given absolute temperature, T, is given by

$$L_p = L_{pg} \exp \{-E_a/R' [(1/T) - (1/T_g)]\}, \tag{1.12}$$

where the subscript g is the value at a given reference temperature (commonly $22°$ or $0°C$) and R' is the gas constant, here expressed in the units cal/deg mole. E_a is the activation energy of L_p in cal/mole. R, R', L_f, and v_w are constants, the values of which are given in Mazur et al. (1984). The values of A, n_2, and L_{pg} are constant for a given cell but differ in different cells. L_p and E_a are adjustable parameters. Knowledge of L_{pg}, E_a, n_2, and A/V (the surface-to-volume ratio of the cell) permit one to compute the volume of cell water (and the extent of supercooling) vs. subzero temperature and cooling rate.

The results of such computations are shown in Figure 1.8 for mouse oocytes (Figure 1.8A) and mouse sperm (Figure 1.8B). Both plot the volume of cell water during freezing as a fraction of the volume of water in the unfrozen cell for a range of cooling rates. Both depict an equilibrium curve (*Eq*). It is the volume of water that a cell has to possess to remain in chemical potential equilibrium with external ice; that is, the volume of water in a cell that is cooled infinitesimally slowly. The curve is generated by the equation

$$V' = V/V_i = v_w M_i \times 10^{-15}/\exp[L_f/R(1/T - 1/273)] - 1, \tag{1.13}$$

where V' is the fractional water volume, V_i is the initial water volume, and M_i is the initial osmolality (Mazur et al., 1984).

In each case, the higher the cooling rate, the more the curves shift to the right of the equilibrium curve. The number of degrees the curve is shifted is the number of degrees the cell water is supercooled at given temperatures. The two plots differ primarily in the numerical values of the cooling rates that produce a given degree of dehydration. Thus, the ova are calculated to lose 75% of their water when cooled to $-20°$ at $2°C/min$. In sperm, the same extent of dehydration at $-20°C$ occurs at a calculated cooling rate of $\sim2000°C/min$. In human red cells it occurs at a cooling rate of $\sim10,000°C/min$. There several reasons for the ≥1000-fold variation. The very high cooling rate in red cells is a consequence of a very high L_p (12 times that of the ova), a low E_a of L_p (one third that of ova), a small volume, and a high surface-to-volume ratio as a result of their being biconcave disks rather than spheres. The high cooling rate required to produce comparable shrinkage in mouse sperm is a consequence of an L_p that is 2.5 times that of the ova, of small size, and of a very high surface-to-volume ratio.

Changes in L_p shift the dehydration curves left or right but do not change their shapes. Thus, if L_p for mouse ova were double the value of 0.2 μm/min atm used in Figure 1.8A, the curve for a cooling rate of $2°C/min$ would become the curve for $4°C/min$. Changes in E_a, in contrast, not only shift the curves but produce major changes in their shape. The dashed curve in Figure 1.8A shows that increasing the value of E_a from 14 to 17 kcal/mole slows the dehydration substantially at lower temperatures and eventually stops it. The reason, of course, is that the higher the E_a, the greater the drop in L_p with decreasing temperature.

The above equations were originally derived by Mazur (1963) and were modified by Mazur et al. (1984) to introduce an Arrhenius relation for L_p. A number of variants have been published by Fahy (1981), Toner et al. (1990), Karlsson et al. (1993), Liu et al. (2000), and others. One variant is to express the surface area of the cell as the two-thirds power of the cell volume, not as a constant as done here. The former assumes that the surface of the plasma membrane behaves like the surface of a rubber balloon as it shrinks in volume. As detailed elsewhere (Mazur, 1990) that is almost certainly not correct. The plasma membrane bilayer is incapable of no more that the slightest compression or expansion. However, in practice, whether A is assumed to be a constant or a variable

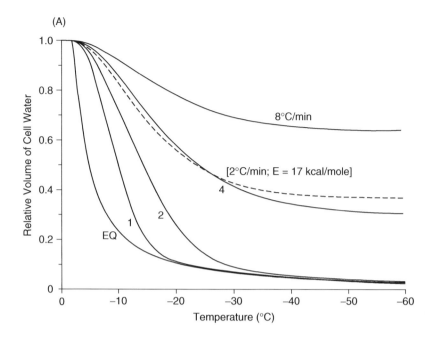

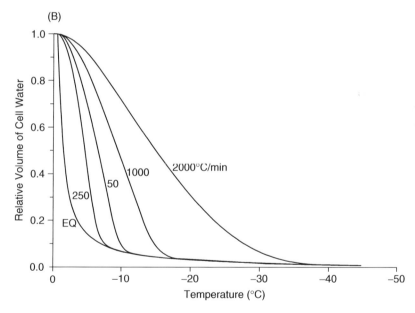

FIGURE 1.8 Computed kinetics of water loss from (A) mouse ova cooled at 1° to 8°C/min in 1 *M* dimethyl sulfoxide (Me₂SO), and (B) mouse sperm in 18% raffinose cooled at 250°–2000°C/min. The curves were calculated from Equations 1.9 to 1.12. Curve EQ is the water loss for ova and sperm subjected to equilibrium freezing; that is, cooled infinitely slowly. It was calculated from Equation 1.13. The solid curves in A were calculated using an activation energy (E_a) for L_p of 14 kcal/mole. The dashed curve shows the effect of increasing E_a to 17 kcal/mole. Plot A is from Mazur (1984), by permission of the American Physiological Society; Plot B is from Mazur and Koshimoto (2002), by permission of the Society for the Study of Reproduction. Values for the required parameters for the sperm are given there. Values for the required parameters for the ova are given in Mazur (1990).

has only a minor effect on the computed curves. The above equations are also predicated on both the solution and cytoplasm behaving as ideal solutions; that is, obeying Raoult's Law. That is a substantial simplification and approximation. Some like Fahy (1981) have attempted to increase the precision by more accurately defining the chemical potential of water in the external medium as a function of temperature and by incorporating coefficients obtained from phase diagrams to correct for nonideality of a specific ternary solution like glycerol/NaCl/water inside the cell. However, these refinements do not result in major changes in curves computed from the assumption of ideality. Furthermore, cell cytoplasm is not a ternary solution but a complex array of solutes, the aggregate osmotic coefficients of which are not known.

Another assumption is that water is the only component that moves across the cell membrane during freezing; that is, endogenous or added solutes do not. That is probably valid. As noted earlier, the permeability of cells to nearly all cryobiologically relevant solutes is 100 to 1000 times lower than that of water, and the activation energies for CPA permeation are generally equal to or higher than E_a for water permeability. Liu et al. (1997, 2000) have incorporated an equation to eliminate that assumption and allow the possibility of CPA permeation during freezing; however, their papers do not indicate how much permeation would be expected to occur. Pegg and Diaper (1988, p. 486) have argued on essentially thermodynamic grounds that little additional permeation should occur in cells that have previously been equilibrated with the CPA because there is little driving force for it to do so. The loss of cell water during equilibrium slow freezing causes the intracellular CPA concentration to remain close to the increasing extracellular concentration, thus eliminating the driving force. Other underlying assumptions were evaluated by Mazur (1963).

A more serious assumption underlying the equations is the appropriate value of L_p and its E_a at subzero temperatures. The curves in Figure 1.8 are predicated on the assumption that the E_a computed from measurements of L_p at 0°C and above continues to be applicable at subzero temperatures during freezing. There is indirect support for this assumption from instances that I will cite shortly that show that the probability of intracellular ice formation (IIF) vs. cooling rate computed from the equations agrees well with observed incidence of IIF. More persuasive would be direct cryomicroscope measurements of L_p from cell shrinkage kinetics at subzero temperatures in partly frozen solutions. Such measurements are difficult, but some have been made; three support the assumption. One by Levin (1979) on yeast, the second by Toner et al. (1990) on mouse oocytes, and the third by Devireddy et al. (1998) on human lymphocytes.

Some do not fully support the assumption, however, or are equivocal. Schwartz and Diller (1983) obtained rather similar values for L_p for yeast from shrinkage rates during freezing as did Levin, but somewhat lower values of E_a; consequently, extrapolated values of L_p at 20°C are different. Conversely, McCaa et al. (1991) obtained values for the E_a of monocytes at subzero temperatures that are similar to the value reported by Hempling (1973) for lymphocytes at >0°C, but McCaa et al. obtained a higher extrapolated value of L_p at 20°C. Aggarwal et al. (1988) observed the shrinkage kinetics of isolated human keratinocytes between –2° and –9°C in the presence of external ice, and from these observations, they computed the L_p at 0°C to be 0.035 μm/min.atm and an E_a of 10.7 kcal/mole. However, there are no estimates of L_p and E_a from above-zero measurements for comparison. Lin et al. (1989) and Pitt et al. (1991) have made such a comparison in permeabilized *Drosophila* eggs. They found that L_p at 0°C has about the same value (0.25 and 0.17 μm/min atm) whether calculated from above-zero shrinkage measurements or from cryo-microscope measurements between –2° and –9°C in the partly frozen state, but that the activation energy from the latter was fivefold higher than that from above-zero measurements (39 vs. 8 kcal/mole). When the ice nucleation temperature (see following section) is high (e.g., –7°C), that difference in E_a has little consequence on the computed cooling rate at which IIF will occur, and Lin et al. (1989) found relatively good agreement between the computed critical cooling rate for IIF (1 to 2°C/min) and that observed by Myers et al. (1989) (~1°C/min). If the nucleation temperature is significantly below –7°C, however, the use of the value of 8.1 kcal/mole will lead to serious overestimates of the critical cooling rate.

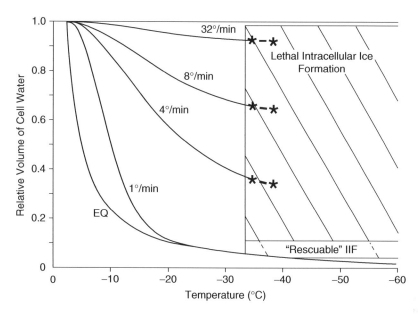

FIGURE 1.9 Kinetics of water loss from mouse ova as a function of the cooling rate during freezing in 1 *M* dimethyl sulfoxide (Me₂SO) or glycerol. The vertical solid line at –33°C is the median ice nucleation temperature observed by Rall et al. (1983) for eight-cell mouse embryos. It, and the zone labeled "Rescuable IIF," are discussed in the text. (From Mazur, 1990, by permission of Humana Press.)

Devireddy et al. (1998, 1999, 2000) have used an entirely different method to estimate L_p and its E_a at subzero temperatures; namely, DSC. With lymphocytes, the DSC values agree well with those they obtained from cryomicroscopy, and they agree well with Hempling's (1973) data at >0°C. However, in the case of human and mouse sperm, they have obtained values that are widely different from those extrapolated to subzero temperatures from volume measurements made at above-zero temperatures (Gilmore et al., 1995, 2000; Noiles et al., 1993, 1997; Phelps et al., 1999). The Devireddy et al. values of L_p for mouse sperm are 30-fold lower than the latter, and their values of E_a are twofold higher. The consequence is that computations based on their DSC values lead to the prediction that IIF will occur in mouse sperm cooled at 25 to 40°C/min, whereas computations based on extrapolations of above-zero estimates of L_p predict that IIF will only occur at cooling rates well above 1000°C/min (Mazur and Koshimoto, 2002). Experimentally, the drop in survival in the right-hand limb of the inverted U, presumably a manifestation of IIF, occurs at a cooling rate of about 250°C/min (Figure 1.6; Koshimoto and Mazur, 2002). The basis of the discrepancies is unknown.

1.3.2.1 The Ice Nucleation Temperature of Cells

The curves generated by solutions to the above equations or their variants permit an estimate of the extent to which the cell water will be supercooled at various subzero temperatures when cooled at various rates. The question is, What is the fate of that supercooled water? The answer is that at some temperature it must freeze *in situ*. More specifically, if cells contain supercooled water, at what temperature will that supercooled water nucleate to form ice? In the case of mouse embryos suspended in 1 to 2 *M* glycerol or Me₂SO and cooled at ~20°C/min (a rate that produces neither observable nor computed dehydration; Figure 1.8A), Rall et al. (1983) have determined the median nucleation temperature to be about –33°C. In Figure 1.9, that nucleation temperature has been superimposed on the dehydration curves of Figure 1.8A, and from that superimposition, one can draw some important inferences.

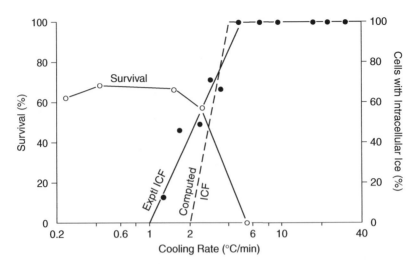

FIGURE 1.10 Comparison between the percentages of unfertilized mouse ova that underwent intracellular freezing as a function of cooling rate to –70°C (●) and the percentages that survived based on a fluorescence assay (○). (From Leibo et al., 1978, by permission of Academic Press.) The dashed line shows the percentage of cells that are computed to undergo intracellular freezing as a function of cooling rate. The calculations are based on solutions to 1.9 to Equation 1.13 assuming a nucleation temperature of about –30°C (Mazur et al., 1984).

The first such inference is that the curve of the water volume of oocytes cooled at 1°C/min merges with the equilibrium curve at temperatures well above the nucleation zone, whereas the curves for cooling rates of 4°C/min and higher lead to the prediction that their water contents will still be well above equilibrium at –33°C. An oocyte cooled at 1°C/min will contain no supercooled water by –22°C, whereas the water in an oocyte cooled 4°C/min will be supercooled 27°C at that temperature. A cooling rate of 2°C/min is borderline. In 1977, I proposed that a cell that enters the nucleation zone still containing 10% or more of its isotonic water volume and with that water supercooled 2°C or more will undergo IIF. The 10% value reflects the belief that ice formation would be difficult when protoplasmic macromolecules are concentrated more than 10-fold. The 2°C value simply reflects the fact that some supercooling is required for freezing to occur.

On the basis of these criteria, the dehydration curves in Figure 1.9 can be translated into estimates of the probability of IIF as a function of cooling rate. Those probabilities are given by the dashed curve in Figure 1.10. It shows a sharp transition from a probability of near zero for oocytes cooled at 2°C/min to a probability of ~1 for oocytes cooled at 4°C/min. The closed circles in that figure show Leibo et al.'s (1978) microscope observations on the percentage of oocytes undergoing IIF as a function of cooling rate. The agreement between computed and observed is good. The third curve, open circles, shows survival as a function of cooling rate. One sees that the cooling rates that are predicted to induce IIF and that are experimentally observed to do so also result in the death of the cells. As summarized by Mazur (1984) such agreement has been found in yeast, human red cells, human lymphocytes, hamster tissue culture cells, and plant protoplasts. From such agreement has come the important conclusion that IIF is a lethal event and that IIF is responsible for the drop in survival in cells cooled at supraoptimal rates; that is, the right-hand limb of the inverted U. (The zone in Figure 1.9 labeled "rescuable" IIF will be discussed later, in the section on warming and thawing.)

Somewhat counterintuitively, cells that contain CPAs tend to undergo IIF at lower cooling rates than those cells that do not (e.g., Diller, 1979). There are two reasons. First, CPAs generally reduce the L_p at all temperatures, perhaps because they increase viscosity. Second, as discussed by Myers et al. (1989, p. 481) and Karlsson et al. (1994, p. 4451), because they reduce the freezing point of the cytoplasm, they reduce the subzero temperature at which a given driving force for water efflux is attained, and at that lower temperature, the L_p is reduced.

In 1989, Pitt and Steponkus derived an alternative and more rational probabilistic model for the conditions under which IIF will or will not occur. The chief modification was to allow the value of the required amount of supercooling (what they refer to as supercooling tolerance, ΔT^*) to be an adjustable parameter rather than an arbitrary preassigned value. They then determined the value of ΔT^* that yielded curves of the probability of IIF vs. cooling rate that best fit experimental data for the percentage of cells undergoing IIF vs. cooling rate. For rye protoplasts, a value of ~1.5°C yielded the best fit. For permeabilized *Drosophila* eggs (Pitt et al., 1991), the best fit value of $\Delta T°$ was 4°C. Both these estimates are close to Mazur's (1977) arbitrary value of 2°C. In 1990, Toner et al. derived alternative mechanistic criteria for the likelihood of IIF as a function of the degree of cell dehydration and the extent of supercooling. These criteria were based on heterogeneous nucleation theory and will be discussed in more detail shortly. The estimates of the probability of IIF in mouse oocytes as a function of cooling rate on the basis of these mechanistic criteria agreed almost exactly with cryomicroscope observations.

The upper boundary of the nucleation zone in Figure 1.9 is set at –33°C on the basis of Rall et al.'s (1983) data on mouse embryos in 1 M Me$_2$SO, and it is drawn vertically. The vertical line assumes that the nucleation temperature is independent of cooling rate and is independent of the degree of dehydration of the cell. How valid are these assumptions? Is the temperature of approximately –30°C applicable to other cells, and is it independent of whether CPA is present and independent of the concentration and type of CPA? The answers to these questions are frequently no.

Observed median nucleation temperatures of single cells or cell complexes like embryos range from –10°C and above in hepatocytes (Harris et al., 1991), hamster tissue culture cells (Acker and McGann, 2000; Muldrew and McGann, 1994), permeabilized *Drosophila* eggs (Myers et al., 1989; Pitt et al., 1991), *Spirogyra* (Morris and McGrath, 1981), and nonacclimated rye protoplasts (Pitt and Steponkus, 1989), and to below –30°C or even below –40°C in mouse and bovine one-cell embryos (Leibo et al., 1978; Leibo, 1986) and oocytes (Ruffing et al., 1993), hamster oocytes (Shabana and McGrath, 1988), acclimated rye protoplasts (Pitt et al., 1989), and human stem cells (Hubel et al., 1999). Although some of this large range is characteristic of the cell type, more important determinants are the medium in which the cells are suspended and (sometimes) the cooling rate. The nucleation temperature generally drops sharply with increasing solute concentration in the suspending medium. Thus, Rall et al. (1983) reported that the median nucleation temperature of eight-cell mouse embryos fell progressively from –12°C for embryos frozen in PBS alone to –30° to –35°C for embryos frozen in PBS containing 1 to 2 M glycerol or Me$_2$SO. Only a small portion of this drop is the result of the suppression of the thermodynamic freezing point in the higher-osmolality solutions (e.g., from –0.6°C in PBS to –4.4°C in 2 M glycerol in PBS). The empirical observation is that the supercooling point of a solution is suppressed twice the thermodynamic freezing point depression (Rasmussen and MacKenzie, 1972; Rall et al., 1983; Myers et al., 1989). In bovine and hamster oocytes, the nucleation temperature similarly falls from –12°C in saline to –40° to –45°C in 1 M Me$_2$SO (Shabana and McGrath, 1988) or 1.5 M; for example, glycerol, and propane diol (Ruffing et al., 1993). Myers et al. (1989) made similar observations on permeabilized *Drosophila* eggs: The nucleation temperature dropped from –10°C for eggs suspended in isotonic saline to –25° to –30°C for eggs in saline plus 1 to 2 M ethylene glycol, Me$_2$SO, or propylene glycol. In ethylene glycol, they obtained about a 12°C drop for each molar increment of solute. Harris et al. (1991) reported analogous, but smaller, drops in the nucleation temperature of hepatocytes (i.e., from –8°C in saline to –17°C in 2 M Me$_2$SO).

The suppression of nucleation temperature by added solute appears independent of whether the solute is permeating or not. It was permeating in the above examples, but Toner et al. (1991) have reported similar suppressions for mouse oocytes suspended in hyperosmotic, nonpermeating saline; that is, from –12°C in isotonic saline to –32°C in 1.04 osmolal saline. Thus, the suppression of the nucleation temperature in higher-osmolality solutions appears to be dependent on the concentration of solutes in the external medium and not on whether the cells are osmotically dehydrated before freezing. In hyperosmotic permeating additives, the cells will be at normal volume when freezing is initiated. In hyperosmotic nonpermeating solutes, the cells will be osmotically shrunken at that point. Another line of evidence supporting the view that the suppression of the nucleation temperature is a consequence of the external solute concentration and not cell volume comes from experiments by Dowgert and Steponkus (1983) on ice nucleation in cold-acclimated (ACC) and non–cold acclimated (NA) rye protoplasts. ACC protoplasts suspended in 1.0 osmulal NaCl/CaCl$_2$ have the same volume as NA protoplasts suspended in 0.5 osmolal NaCl/CaCl$_2$. (This is because the former accumulate more solutes than the latter.) Yet the median nucleation temperature is much lower in the former (–42° vs. –14°C, respectively). If the two cells are suspended in the same osmolality medium, the difference in median nucleation temperature is substantially reduced in spite of the fact that the volumes of ACC cells in a given medium are substantially larger than those of NA cells. For example, when both cells were suspended in 0.5 osmolal saline, the nucleation temperatures with cooling at 10°C/min were –19° and –14°C in ACC and NA, respectively. When both were suspended in 1.0 osmolal saline, the nucleation temperatures were –42° and –28°C, respectively. (They interpret the residual difference to be a reflection of differences in the inherent stability of the two membranes.)

The effect of osmolality on nucleation rate tends to interact strongly with the cooling rate, and the effect of cooling rate on nucleation temperature is highly variable. When cells are frozen in 0 to 0.5 osmolal saline lacking CPA, the nucleation temperature is generally unaffected by the cooling rate, or is only weakly affected. That is the case for mouse oocytes cooled at rates ranging from 1° to 120°C/min (Toner et al., 1991, 1993a), hamster oocytes cooled at 4° to 40°C/min (Shabana and McGrath, 1988), hepatocytes cooled at 75° to 400°C/min (Harris et al., 1991), *Spirogyra* cooled at 3° to 20°C/min (Morris and McGrath, 1981), nonacclimated rye protoplasts cooled at 3° to 20°C/min (Dowgert and Steponkus, 1983; Pitt and Steponkus, 1989), and permeabilized *Drosophila* eggs cooled at 16° to 64°C/min (Myers et al., 1989). However, at lower cooling rates (4° and 16°C/min), the nucleation temperature of the eggs decreased from –10° to –17°C with an increasing rate.

The picture tends to be different for cells in higher concentrations of CPA or higher osmolalities of nonpermeating solutes. In some cases, the nucleation temperature is unaffected by cooling rate; for example, *Drosophila* eggs in 1 *M* ethylene glycol, or Me$_2$SO cooled at 1° to 16°C/min (Myers et al., 1989, Figures 4 and 6). In a few cases, it decreases with increasing cooling rate. Thus, Myers et al. (1989) observed a lowered nucleation temperature in permeabilized *Drosophila* eggs in propylene glycol as the cooling rate was increased from 1° to 16°C/min. (As mentioned, they observed a similar response in the absence of CPA.)

The more typical finding, however, is that in the presence of hyperosmotic solutes (permeating and nonpermeating), the nucleation temperature rises with increasing cooling rate. That is the observation of Myers et al. (1987), Leibo (1986), and Ruffing et al. (1993) on bovine oocytes and zygotes suspended in 1 to 1.5 *M* glycerol or ethylene glycol. They report large rises in the nucleation temperature (from ca. –30° to –40°C at a cooling rate of 2° to 4°C/min to –10° to –20°C at a cooling rate of 8° to 16°C/min). Shabana and McGrath (1988) found a 20°C increase in the nucleation temperature of hamster oocytes suspended in 1 *M* Me$_2$SO when the cooling rate was increased from 8° to 50°C/min. Leibo et al. (1978) also reported (much smaller) increases in the nucleation temperature of mouse oocytes in Me$_2$SO with increased cooling rate. The effect is not restricted to permeating solutes. Pitt and Steponkus (1989) found that the nucleation temperature of acclimated rye protoplasts in about half the tested 1.0 osmolal nonpermeating salts or sugars rose as the cooling rate was increased from 5° to 20°C/min. In some cases the rise was large (e.g., from –60° to –37°C). In other media, these authors and Dowgert and Steponkus (1983) found that cooling rate had no effect.

How can one explain the different relations between nucleation temperature and cooling rate? One possible explanation for the decrease in nucleation temperature with increased cooling rate in

Drosophila eggs is that some 2 to 10 min elapse between the onset of nucleation at a constant temperature and the visual manifestations of IIF (Myers et al., 1989, Figure 7; Pitt et al., 1991). If such eggs are observed while in the process of cooling, the higher the cooling rate, the more the temperature will drop during the time required for IIF to become visible. The opposite and more common effect, an increase in nucleation temperature with increased cooling rate, is more difficult to explain. One possible explanation is that rapidly cooled cells will be more hydrated than slowly cooled cells at any given temperature, and the more they are hydrated, the higher the temperature at which they can nucleate. However, this hypothesis seems inconsistent with the evidence mentioned above, that the suppression of the nucleation temperature has more to do with the osmolality of the medium than with the osmotic volume of the cells. Another possibility is that nucleation temperature is affected by the form of the external ice crystals and that higher cooling rates induce a more dendritic ice structure than lower cooling rates, especially in solutions containing CPA (cf. Figures 2 and 3 in Leibo et al., 1978).

Regardless of causation, the practical effect of an influence of cooling rate on nucleation temperature is that the vertical nucleation boundary drawn in Figure 1.9 should in some cases be drawn as slanting forward to lower temperatures with increasing cooling rate, or in other more common cases as slanting backward toward higher temperatures. If the slant is substantial, it will affect inferences made with respect to the cooling rate dependence of IIF.

There are still other factors influencing nucleation temperature. One is that the state of the cell membrane and its nucleation properties may depend on how effectively the cell is protected during freezing. The membranes of cells suspended in isotonic saline lacking CPA are not protected. Perhaps that is one reason why they tend to nucleate at substantially higher temperatures than cells those frozen in media containing CPA. Another factor is that the nucleation temperature of a given cell type can depend on the state of aggregation of that cell. It tends to be lower for single cells in suspension than for the same cell attached to glass or aggregated into multicellular spheres (Acker et al., 1999; Acker and McGann, 2000). A third interesting factor is the effect of antifreeze (thermal hysteresis) proteins (AFP). In two cases reported (cardiomyocytes; Mugnano et al., 1995) and hamster tissue culture cells (Larese et al., 1996), AFPs in mg/mL concentrations substantially raise the nucleation temperature, often to above –5°C. Those concentrations of AFP also change the ice crystal habit to more dendritic and, at the upper level of concentrations, to needle-like. This correlation supports the idea raised earlier that the nucleation temperature may be affected by the form of the external ice. (For further discussions on AFP and ice blocking agents, see Chapter 22.)

1.3.2.2 Mechanisms of Ice Nucleation and Factors Blocking Nucleation

Theories as to the mechanism of intracellular nucleation have to account for several experimental facts:

1. IIF above about –30°C requires the presence of extracellular ice; furthermore, it appears necessary that the ice be in close contact with the cells. Toner et al. (1991) demonstrated the former by cryomicroscopy of mouse oocytes; that is, no IIF occurred in oocytes in PBS supercooled to –20°C. Mugnano et al. (1995), Larese et al. (1996), and Köseoglu et al. (2002) observed the latter; that is, IIF in cardiomyocytes, hamster V79 cells, and starfish oocytes occurred only when external ice crystals came in direct contact with the cells. Similarly, Berger and Uhrik (1996) found that IIF occurred in 5- to 14-cell strands of cells from salivary glands of *Chironomus* only when the cells were encapsulated in external ice.

2. IIF does occur in the absence of external ice when temperatures fall to –30°C or below. That can be demonstrated in cases where the external water can be removed without major effects on the cells. Intact *Drosophila* and *Anopheles* eggs are two such cases. Their shells are so impermeable that the surrounding water can be removed by brief air-drying

without dehydrating the eggs. DSC (Myers et al., 1989; Schreuders et al., 1996) and differential thermal analysis (Mazur et al., 1992b) show that such surface-dried intact eggs do not undergo intraembryonic freezing until cooled to –28° to –32°C. Franks et al. (1983) studied ice nucleation temperatures in red cells, yeast, and three species of plant cells in the absence of external ice by a different approach; namely, by suspending cells in tiny emulsified droplets, only a small fraction of which contained external ice nucleators. The yeast and plant cells, like *Drosophila* eggs, froze at about –30°C. The red cells nucleated at about –39°C, only 1°C above the homogeneous nucleation temperature.

3. Cells and their surrounding aqueous medium can often be supercooled to –15° or even –20°C. If that is below the cell nucleation temperature, the induction of external ice immediately causes IIF. That was demonstrated clearly by Toner et al. (1991) in mouse oocytes in PBS. The median cell nucleation temperature is –12.8°C. Oocytes in supercooled media at –20°C are also unfrozen, but they froze instantly when the external medium was seeded.

4. However, if external ice forms above the cell nucleation temperature, cells will remain unfrozen even though their water is supercooled. That is, above the nucleation temperature, IIF does not occur even in the presence of external ice. Published data on nucleation temperatures and the factors affecting them were discussed earlier.

5. The nucleation temperature decreases substantially with an increase in the external concentration of solutes. The suppression is far greater than the depression of the thermodynamic freezing point. The suppression is independent of whether the solutes are nonpermeating (e.g., NaCl), or permeating (e.g., glycerol or Me_2SO). It therefore appears to be independent of whether the cells are or are not osmotically shrunken before the initiation of freezing.

As summarized above, there is persuasive direct evidence that the nucleation that occurs below about –30°C has a different genesis from that occurring at higher temperatures. The low-temperature nucleation is clearly a consequence of the action of weak intracellular heterogeneous nucleators that act independently of whether ice is present outside the cell. These endogenous nucleators are referred to as weak because they are acting only at temperatures about 10°C above the homogeneous nucleation temperature of water. There is also other, indirect, evidence that low-temperature nucleation is different from high-temperature nucleation. As mentioned, Pitt et al. (1991, Figure 5) showed that between –15° and –30°C, the formation of visible intracellular ice in *Drosophila* eggs requires about 10 min. At –34°C, however, IIF occurs almost instantly. Toner et al. (1991, Figures 3 and 12) show that plots of the percentage of rapidly cooled mouse oocytes that undergo IIF vs. temperature exhibit very different slopes when the nucleation temperature is above –30°C than when it is below –31°C. This, they conclude, is evidence for two different mechanisms.

Although the mechanism of low-temperature nucleation appears clear, the mechanism of high-temperature nucleation is more controversial. Findings 1, 3, and 4 demonstrate that it requires the presence of extracellular ice in close contact with the cells and that it requires that the cell water be supercooled at least some 2° to 4°C. These two requirements, however, are not sufficient. The cell must also be below its nucleation temperature. At temperatures above the nucleation temperature, some barrier prevents the passage of external ice into the supercooled cell interior. That barrier, in all likelihood, is the plasma membrane. Below the nucleation temperature, that barrier disappears or is breached.

There have been three major theories proposed to account for these facts. Theory number one is that nucleation occurs through preexisting pores in the cell membrane. In 1965 and 1966, I proposed that nucleation occurs as the result of the growth of external ice through preexisting Angstrom-sized pores in plasma membranes. Such growth cannot occur above a critical subzero temperature because the crystal would have to possess a radius of curvature approaching that of the radius of the pore, and ice crystals of such small radii, because of their higher surface energies, cannot exist unless the temperature is well below 0°C. As discussed in the above publications and in more detail by Acker et al. (2001), the suppression of the melting point of ice below that of a planar ice crystal in the same solution as a function of the radius of the pore is given by the Kelvin equation

$$\Delta T = 2vT^\circ \sigma_{SL}\cos \theta/aL_f,$$

where v is the molar volume of ice, T° is the melting point of a planar crystal, σ_{SL} is the interfacial tension between liquid and ice, a is the radius of the pore, and L_f is the latent heat of fusion. Cos θ is the contact angle—the angle made where the ice contacts the wall of the pore. Its value is determined by the interfacial energies between liquid and ice, between liquid and the wall of the pore, and between ice and the wall of the pore. The latter two values are not known, so a value of θ has to be assumed. To give an example of the magnitude of the Kelvin effect, if θ is assigned a value of 75°, the temperature would have to be about 10°C below the freezing point of the bulk solution for an ice crystal to grow through a pore of 12 Å radius (Acker et al., 2001).

The concept of Angstrom-sized pores in membranes was not mere speculation. As mentioned earlier, the work of Solomon's group on differences between the osmotic and diffusional water permeability of human red cells had yielded strong inferential evidence on the existence of such pores (Shaàfi et al., 1967) and that inference received striking confirmation with the discovery of aquaporins, some characteristics of which I discussed earlier. However, there are two problems with proposing nucleation via the passage of ice through aquaporin channels. One is that although they are being found in an increasing variety of cell types, the plasma membranes possessing them still constitute a distinct minority. In cells that lack them, the consensus is that water that enters and leaves cells does so by a diffusional mechanism (Finkelstein, 1987; Macey, 1984). It is difficult to conceive of how an organized ice crystal, even one of small radius, could possibly "diffuse" through a cell membrane (but see following). The other problem is that even where cell membranes possess aquaporins, the center constriction in their channels would seem to preclude the passage of even an extremely small organized ice crystal. As mentioned earlier, the pores in aquaporin 1 are hourglass in shape, with outer and inner regions of 14 Å in diameter joined by a constricted region 3 Å in diameter. Although one could conceive of crystalline ice growing into the 14-Å outer region, it is difficult to conceive of it traversing a constriction that is only twice the diameter of a water molecule.

This assumes that the pore dimensions remained unaffected by attempts by external ice to grow through them. That may not be so. Consider the situation where external ice has grown into the outside 14-Å funnel. Below some temperature described by the Kelvin equation, the chemical potential of that ice will drop below that of the supercooled water in the constricted region, in the inner funnel, and in the cytoplasm. As a result of the chemical potential difference, water will move from the inner regions through the pore constriction until it comes in contact with the ice in the outer funnel, at which point it will freeze. When it freezes, it will produce an increase in the dimensions of the ice in the outer funnel. That growth may generate forces that enlarge the constricted area of a pore until its diameter becomes sufficient to permit the passage of crystalline ice into the cell interior. It should be noted that this enlargement would only have to occur in a single pore out of the 100,000 or so present for nucleation to occur. If the distortion is restricted to a small number of pores, it, *per se*, may be an innocuous event.

There is good evidence that ice can grow through pores of some 16 to 30 Å in diameter that exist in the contiguous membranes connecting adjoining cells of many tissues. Chambers and Hale (1932) were among the first to demonstrate that IIF proceeds in a sequential fashion from one cell to another in a tissue (onion epidermis) with brief pauses. Berger and Uhrik (1996) made similar observations in the linear array of cells isolated from the salivary gland of the insect *Chironomas*. Cell-to-cell propagation occurred at approximately –12° to –14°C. They proposed that it occurred through the channels in gap junctions—channels that have a radius of 10 to 15 Å in insect tissues (smaller in mammalian cells). This conclusion was based on the abolishment of sequential freezing in glands treated with dinitrophenol or heptanol, compounds that block gap junctions. That hypothesis has been strengthened by the studies of Acker et al. (2001) and Irimia and Karlsson (2002). The former used low-Ca^{++} media to suppress the formation of gap junctions in MDCK cells; the latter used the specific inhibitor 18β-glycyrrhetinic acid for the same purpose in pairs of HepG2 cells. In both cases, the inhibitors suppressed (but did not totally eliminate) cell–cell ice propagation. In addition, Acker et al. (2001) found that IIF in confluent hamster V79 cells (which do not form gap junctions) does not proceed sequentially from cell to cell, but occurs randomly and at a 5°C lower temperature. In cells systems with gap junctions, the temperatures at which sequential cell–cell ice propagation occurs fall in the range predicted by the Kelvin equation for the suppression of the melting point of ice attempting to grow through pores of the diameters of these channels.

What about ice propagation into the majority of cells where water moves by diffusion through the lipid bilayer and not through aquaporins? Three points. First, the presence or absence of aquaporins may well not be absolute. Mouse oocytes and embryos are generally considered not to contain aquaporins, but Edashige et al. (2000) have found that they do, in fact, express a number of them, including aquaporins 3 and 7 (but not aquaporin 1), albeit at low levels, In theory, it would only take one or a few aquaporin channels to permit ice nucleation by the mechanism just suggested. Second, even cells that lack aquaporins almost certainly contain other channels that mediate ion transport. Although these channels differ in structure and properties from aquaporins (e.g., Dutzler et al., 2002), they too could be subject to physical distortion by external ice. Third, the current theory of diffusional permeability is that cavities in membrane bilayers of various sizes are forming and disappearing spontaneously with time and that solute molecules move or diffuse when a cavity of sufficient size forms (Stein, 1986, pp. 96–100). That process could conceivably also apply to the movement of an ice crystal of small radius of curvature through a membrane.

Theory number two is plasma membrane-catalyzed nucleation. In a series of papers, Toner and colleagues (Toner et al., 1990, 1993b) have proposed a mechanism for the higher-temperature nucleation in which external ice plays an indirect—not a direct—role. Toner et al. propose that external ice induces structural changes in the plasma membrane that cause the inner surface of the membrane to become an effective heterogeneous nucleator of the supercooled cytoplasm. They refer to this as surface catalyzed nucleation. Heterogeneous nucleation involves kinetic factors and thermodynamic factors. Important in the former are the viscosity, temperature, cell surface area, solid–liquid interfacial energy, and contact angle between nucleator and ice. Important factors in the latter are the equilibrium freezing point as well as the contact angle and solid–liquid interfacial energy. The values of some of the variables were obtained from the literature. Others were obtained by adjusting their values to obtain a best fit between the probability of ice formation vs. temperature from heterogeneous nucleation theory and experimental observations on the percentage of rapidly cooled mouse oocytes undergoing IIF vs. temperature. (By cooling oocytes at a high rate, Toner et al. precluded water transport and cell dehydration and its effect on ice nucleation.) The agreement between the theoretical predictions and experimental observations was excellent (Toner et al., 1990, Figure 6; Toner et al., 1993b, Figures 4 and 6).

The main difficulty with this otherwise elegant treatise is that it can only provide speculation as to how the inner surface of the plasma membrane can become the heterogeneous nucleator under the influence of external ice. The authors argue that it does so by somehow affecting the value of the contact angle θ. One wonders whether the analysis might not be equally applicable if external ice itself were the heterogeneous nucleator; that is, some blend with the pore theory discussed above.

I noted above their observation that the nucleation temperature was sharply reduced in oocytes frozen in 735 to 1000 mosm saline and that the shape of the curve of percentage cells with IIF vs. temperature changed. They interpret these two observations to mean that the cell shrinkage at the higher osmolalities suppressed and eventually eliminated the surface catalyzed nucleation, leaving only the intracellular nucleators acting by what they refer to as volume catalyzed nucleation (which

I refer to as low-temperature nucleation). However, as I indicated in the discussion of nucleation temperature, this suppression of nucleation temperature appears to be a consequence of increased concentration of solutes in the external medium rather than a consequence of an osmotic reduction in the volume of the cells. One line of evidence is that the extent of the suppression is about the same for cells suspended in solutions of hyperosmotic permeating solutes (Rall et al., 1984) as for those in hyperosmotic saline of comparable osmolality (Toner et al., 1990). In the former, the cells are at normal isotonic volume before freezing; in the latter case, they are shrunken before freezing. One simple explanation for the suppression of nucleation temperature in high-osmolality media, an explanation suggested by both Dowgert and Steponkus (1983) and Rall et al. (1984), is that the higher the osmolality of the external medium, the greater is the unfrozen fraction at any temperature and the greater is the unfrozen fraction, the lower the probability that a cell plasma membrane will come in direct contact with ice. As mentioned, such direct contact appears to be required for nucleation.

Theory number three is the osmotic rupture hypothesis. The predicate in the above two hypotheses is that nucleation leads to IIF and that IIF leads to membrane and cytoplasmic damage and lethality. Some researchers, however, have proposed the opposite sequence; namely, some event or events during cooling produce lesions in the membrane, and IIF is a consequence of the presence of these lesions—not the cause. In other words, they propose that IIF is a consequence of injury and not the cause. Dowgert and Steponkus (1983) and Steponkus et al. (1983) ascribe to this view based on phenomena they observed at plant protoplast cell surfaces 0.03 to 1 sec before IIF. The phenomena include outward flow of intracellular solution and fluttering of the plasma membrane. The challenge to this alternative picture is to explain why damage above −30°C requires the presence of external ice, why the probability of IIF increases sharply above a critical cooling rate, and why the quantitative analyses discussed above are able quite successfully in many cases to predict that critical cooling rate.

The progressive formation of external ice has two consequences. One is that it increases the concentration of solutes in the unfrozen channels in the extracellular milieu—channels in which the cells lie. The other is that it decreases the chemical potential of water in that unfrozen fraction and thereby establishes the driving force for the efflux of supercooled water out of the cells. It is difficult to see why the first—the increase in external solute concentration—should become increasingly damaging to membranes with increased cooling rate and become abruptly so above a critical cooling rate. Indeed, as will be discussed later, the opposite is generally hypothesized; that is, the increase in external solutes is considered to be a major factor in the slow-freezing injury that occurs at cooling rates below those that produce IIF. The second concept, however, is possible. The faster cells are cooled, the more they will be supercooled, and the more they are supercooled, the greater will be the driving force for the efflux of water and the higher will be the maximum instantaneous water flux. Pitt et al. (1992) proposed that the extent and kinetics of IIF are related to the extent and duration of supercooling, but Muldrew and McGann (1990, 1994) have gone farther by proposing that IIF is related to the rate of efflux of water, and they have proposed a mechanism that they refer to as the "osmotic rupture hypothesis" of intracellular freezing. Their proposal, in brief, is that as water molecules move through a membrane either by diffusion or by osmotic flow through pores, they impart a frictional force on the bilayer from collision with the lipid molecules. The magnitude of that force will be proportional to the velocity of the water molecules. If the frictional force exceeds the tensile strength of the membrane, it will rupture. If it ruptures, it will allow the passage of external ice and the subsequent seeding of the remaining internal supercooled water. Muldrew and McGann's 1994 paper goes on to experimentally measure the flux in hamster tissue culture cells under various freezing conditions that do or do not produce IIF, to estimate the magnitude of the frictional force or pressure imparted as a function of flux rate, to evaluate whether that force is sufficient to rupture membranes, and to determine whether freezing conditions that yield that critical force are also the conditions that lead to IIF. From literature values for some of the parameters, they estimate that the critical flux rate for the rupture of hamster V79 membranes lies between 17 and 17,000 μm^3 of water/sec. From cryomicroscopy, they find that IIF begins to occur in cells held at –5°C, and their measurements of L_p indicate that the maximum flux rate at –4°C is 38 μm^3/sec, just above the bottom of the critical range. Analogous analyses on liposomes and mouse oocytes yielded similar results. If they fitted the parameters for membrane failure (chiefly the critical pressure) to the computed flux rates vs. cooling rate and temperature that yielded IIF, they obtained values of that critical pressure that fall at the lower end of observed values. Furthermore, with estimates of the critical pressure in hand, their equations permitted them to predict the percentage of cells or liposomes that are subjected to critical pressures and hence would undergo IIF as a function of cooling rate. The predicted percentages of IIF agreed closely with the observed values.

In spite of the excellent phenomenological agreement, this hypothesis faces difficulties, most of which the authors themselves raise. (For further discussions on the osmotic rupture hypothesis, see Chapter 2.) First, they conclude that the critical flux rate for hamster fibroblasts at –4°C is slightly above 38 μm^3/sec. The external concentration of NaCl or other solutes at –4°C is 2.2 osmolal. Yet, they themselves report (1990) that some 90% of these cells survive the water efflux that accompanies abrupt immersion in ~3 osmolal Me_2SO at 0°C. At 20°C, the L_p and shrinkage rate in the same osmolality of NaCl or other impermeant solute will be four to 10 times higher than at 0°C, but abrupt exposure to such osmolalities in determining BVH response or in L_p determinations is generally innocuous. (Gervais and colleagues have shown that yeast cells are killed by abrupt subjection at room temperature to much higher concentrations of glycerol [~39 molal] solutions that produce extremely high osmotic pressures [100 MPa or 1000 atm]. However, exposures to half that concentration were not damaging, nor was exposure at 5° or 11°C [Beney et al., 2001].) One way to meet that objection is to assume, which Muldrew and McGann do, a strong negative temperature coefficient of membrane brittleness; that is, assume that cell membranes at 20°C can withstand much higher fluxes than those at –4°C. In two examples analyzed (1994), they estimate the temperature coefficients of the critical flux rate to be 18.5 kcal/mole and 5 kcal/mole.

Second, as noted earlier, the greater the L_p of a cell (with other factors constant), the higher the cooling rate required to produce IIF. An increased ability to withstand cooling at higher rates translates to increased ability to withstand the higher maximum water fluxes that accompany higher cooling rates. To be explicable in terms of their hypothesis, the critical flux rate for membrane damage and consequent IIF would have to be higher in cells with larger L_p than in cells with smaller L_p. The problem is there is no nonspeculative basis for such a correlation.

Third, although the establishment of an osmotic efflux requires the presence of external ice, it does not require direct contact between the ice and the cells. IIF, however, does appear to require that direct contact (Berger and Uhrik, 1996; Mugnano et al., 1995; Köseoglu et al., 2001; Larese et al., 1996; Toner et al., 1991).

Fourth, if high water effluxes during rapid cooling cause membrane damage, then high influxes during rapid warming ought also to cause damage. Yet, as we shall discuss and as the authors themselves point out, rapid warming is usually beneficial. One possible resolution is that membrane lesions produced during the rapid thawing of cells that have not undergone IIF during cooling are reversible because ice does not grow through the defect to exacerbate the damage.

Fifth, another challenge to the osmotic rupture hypothesis is that cells that undergo limited IIF can often survive. Acker et al. (2001) report that to be the case in confluent monolayers of MDCK cells. They find that the gap junction channels through which the ice is propagating are neither transiently nor permanently enlarged by the propagation. Furthermore, as will be discussed, cells containing limited ice can often be rescued by rapid warming, presumably because rapid warming minimizes the recrystallization of ice. For the osmotic rupture theory to resolve the latter problem, one has to postulate that the flux-induced defect that results in IIF is reversible (presumably by resealing) if the ice crystals passing through it do not increase in size by recrystallization during warming.

It may turn out that the resolution of the conflicts among the three nucleation theories will be something like the following: The presence of external ice in direct contact with membranes causes the flickering cavities in bilayers or preexisting pores to become enlarged in rapidly cooled cells. The enlargement may be the result of pressures produced by high efflux rates of water, but it is more likely caused by forces exerted by external ice itself as it grows by accretion from supercooled water moving from the cell interior. These perturbations are not in themselves damaging and are therefore reversible. However, if the extracellular ice is able to grow through the membrane and cause the cytoplasm to freeze, and if the cytoplasmic and intramembranous ice crystals grow by recrystallization during slow warming, the result is irreversible damage. (For further discussion on intracellular ice formation, see also Chapter 2.)

1.3.3 SLOW-FREEZING INJURY

Although cooling rates low enough to prevent intracellular freezing are generally necessary for survival, they are insufficient. As we have seen from the left-hand limb of the inverted U, slow freezing itself can be injurious. For the purposes of the ensuing discussion, I will define a slow cooling rate as one that allows sufficient water to leave the cell to keep the remaining cell water in near chemical potential equilibrium with extracellular water and ice throughout cooling; that is, conditions under which $\mu^i_w = \mu^e_w$. Because the only ice that forms under these conditions is extracellular, any observed injury must be a consequence of the direct action of that ice, of an increase in the proportion of ice to unfrozen extracellular solution, or of increases in the concentration of solutes in the external solution brought about by the conversion of water into ice. With respect to the last, as the cells are assumed to remain in chemical potential equilibrium with the external ice and solution, the cells will dehydrate and the concentration of their endogenous solutes will rise. Injury could be a consequence of solute concentration either outside or inside the cell. With respect to the physical route by which cells seek to attain equilibrium with the external medium, I make one further assumption; namely, that equilibrium is maintained solely by water efflux during freezing and influx during warming, and not by the movement of added permeating CPAs or by the development of leaks to other extracellular or intracellular solutes (chiefly electrolytes) to which cells are normally quite impermeable. As already discussed, the basis for the assumption about CPAs is that the permeability of cells to them is nearly always at least 100 to 1000 times lower than their permeability to water, and furthermore, that there is little driving force for their entry. With respect to electrolytes and the like, the assumption is that if leaks develop, they are a pathological consequence of the equilibration process and not a normal alternative route of equilibration.

1.3.3.1 Slow-Freezing Injury from the Concentration of Electrolytes during Freezing

As ice progressively forms in the medium, it progressively withdraws pure water from the solution and causes the concentration of extracellular solutes to rise to very high values. If the cells are frozen in isotonic NaCl or PBS, the solutes that are concentrating are electrolytes. Because by our definition of slow cooling the cell contents are in equilibrium with the outside solution, their solutes are also concentrating. The quantitative relation is

$$M^c = \phi v m^c = \phi n_2/V^e = \Delta T/1.86 = M^i, \tag{1.14}$$

where M^c and M^i are the external and internal osmolality, ϕ is the osmotic coefficient, v is number of species into which the solutes dissociate, m^c is the molality, n_2 is moles of solute, V^e is the volume of extracellular water, and ΔT is number of degrees below 0°C. The value 1.86 is the molal freezing-point depression constant for water. As the right-hand side of the equation indicates, the extracellular and intracellular concentrations rise to high levels as the temperature falls; for example, 5.4 osmolal or ~2.7 molal by −10°C

Lovelock (1953a) proposed that it was this concentration of solutes (specifically electrolytes) that is responsible for slow freezing damage. The experimental basis for his conclusion was that when he calculated the concentrations of NaCl in partly frozen solutions at which human red cells began to hemolyze, using a version of Equation 1.14, he obtained closely similar extents of hemolysis in suspensions that were exposed to those same concentrations of NaCl in the unfrozen state at 0°C. He and others knew at that time that the damage to cells during freezing was suppressed by the presence of a CPA such as glycerol. In red cells, the protective effect of increasing glycerol concentrations is dramatic (Figure 1.7). Lovelock proposed (1953b) that the protection arises because the CPA solutes colligatively suppress the concentration of the damaging electrolytes at a given subzero temperature. The basis of this suppression is the following.

In partly frozen solutions, M^c is independent both of the nature of the solutes and of their total concentration before freezing. At constant pressure, it is dependent only on temperature. For a solution containing a single solute, this is also roughly true of the molality m^c. (It is only roughly true because ϕ changes somewhat with concentration.) As a consequence, the total osmolal concentration of solutes in the unfrozen portion of a solution at a given temperature is not influenced by the addition of solutes like glycerol. For example, the unfrozen portions of both an isotonic salt solution and an isotonic solution containing 1 M glycerol (1.4 osmolal total) will have the same osmolal concentration at −10°C; namely, 5.4 osmolal.

The presence of additive, however, does reduce the concentrations of salt at the given temperature according to the relation

$$M_{NaCl} = M^0_{NaCl}/(1 + R),\qquad\qquad (1.15)$$

where M_{NaCl} and M^0_{NaCl} are the osmolal concentrations of sodium chloride in the partly frozen solution in the presence and absence of additive and R is the mole ratio of additive to NaCl before freezing.

The changes in osmotic coefficients with temperature and concentration make it difficult to solve Equations (1.14) and (1.15) accurately, but accurate determinations of the composition and relative amounts of the concentrated liquid and ice can be made from phase diagrams. Detailed diagrams of the ternary system glycerol-NaCl-H_2O, for example, have been published (Shepard et al., 1976), and from these it is possible to determine the exact NaCl concentration at any temperature. Such derivative phase diagrams show, for example, that the concentration of NaCl at −10°C will be reduced from 2.7 molal in the absence of glycerol to 0.9 molal and 0.6 molal if 0.5 M and 1.0 M glycerol are added to the 0.15 M NaCl solution, respectively. (Mazur, 1984, Figure 8). It is important to emphasize that the ability of glycerol to reduce the concentration of NaCl in the unfrozen portion of the solution has nothing to do with the chemical nature of glycerol. It is a colligative effect that will be manifested by adding any other solute to NaCl. The benefits of glycerol (and other CPAs like Me_2SO) derive from the fact that they are highly soluble, and so can concentrate during freezing without falling out of solution, and the fact that they are relatively nontoxic.

Lovelock (1953b) showed to a first approximation that hemolysis of human red cells began to occur during freezing at whatever subzero temperature the concentration of NaCl in a series of glycerol/NaCl/water solutions reached a mole fraction of 0.014 (~0.8 molal, depending on the glycerol concentration). The effect of increasing the concentration of glycerol was to decrease the temperature at which that NaCl concentration was reached. We obtained similar results as shown in Figure 1.11A. In both 0.5 and 1.0 M glycerol, 50% of red cells hemolyze when the NaCl concentration has risen to 1.6 molal, but that concentration is attained at −17° and −29°C, respectively. In the absence of glycerol, that concentration is attained at −6°C. According to his hypothesis, that is the reason that the survival curves in Figure 1.7 move to progressively lower temperatures as the glycerol concentration increases from 0 to 1.75 M. If the concentration of glycerol gets high enough, the critical concentration of NaCl is never attained at any temperature. That presumably

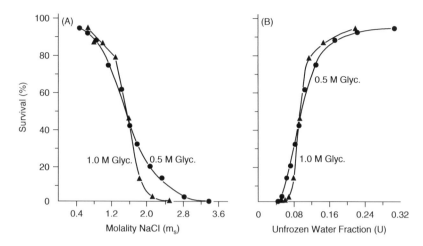

FIGURE 1.11 Survival of human red cells (percentage unhemolyzed) (A) as a function of the molality of NaCl to which they are exposed after being frozen at 1.7°C/min to various subzero temperatures while suspended in solutions of 0.5 or 1.0 M glycerol in isotonic saline, and (B) as a function of the fraction of water that remains unfrozen at those temperatures. Thawing was rapid. (From Mazur, 1984, by permission of the American Physiological Society.)

is why survivals in Figure 1.7 remain high down to –80°C when the red cells are frozen in 2 M glycerol or higher.

Lovelock's basic finding that the hemolysis produced by freezing red cells in saline to a given temperature and then thawing them can be mimicked by exposing the cells to the same concentrations of NaCl at 0°C and returning them to isotonic has been confirmed by Pegg and Diaper (1988, 1991) and Mazur and Cole (1989). Fahy and Karow (1977) have pointed out, however, that Lovelock's conclusion that hemolysis occurs at a given concentration of NaCl during freezing regardless of the glycerol concentration applies only if based on very low levels of hemolysis. When based on higher percentages of hemolysis, the damage occurs at a lower concentration of NaCl in the presence of glycerol than in its absence. For example, the mole fraction of NaCl yielding 60% hemolysis drops from 0.04 in NaCl alone to 0.025 for red cells frozen in saline containing 1 M glycerol. Expressed differently, although glycerol protects frozen-thawed red cells from hemolysis in proportion to its concentration, its protection is less than would be predicted from the Lovelock hypothesis. Rall et al. (1978) have confirmed that finding, as have Pegg and Diaper (1988).

Although Lovelock's basic hypothesis continues to have broad acceptance 50 years later, there is only limited direct evidence to support its specific conclusion that damage from slow freezing is caused by the attainment of a damaging concentration of extracellular or intracellular electrolytes. For the extracellular medium, that translates to a damaging concentration of NaCl. One piece of direct evidence comes from the study of Santarius and Giersch (1983). They showed that when spinach thylakoid membranes were slowly frozen to –20° to –25°C while suspended in media containing various mole ratios of nonpermeating sugars to NaCl, the greater the ratio, the lower was the concentration of NaCl in the unfrozen portion of the medium and the higher was the survival. A second example is the reports by Stacheki et al. (1998) and Stacheki and Willadsen (2000) that the survival of slowly frozen mouse oocytes is substantially enhanced by the substitution of choline for NaCl. A second problem is that a finding of equivalent effects of attaining high NaCl concentrations by freezing and by exposure at ≥0°C is not universal. Watson and Duncan (1988), for example, reported that the latter is considerably more damaging than the former to ram sperm. A third problem is that of distinguishing extracellular events from intracellular. By our definition of slow freezing as equilibrium freezing, water flows out of the cell osmotically and freezes externally. One consequence is that the intracellular solutes also concentrate. A second consequence is that substances added to the external medium will have no ability to suppress that concentration of internal solutes unless they have been able to permeate the cell before the onset of freezing. This has led to the general conclusion that to be an effective cryoprotectant, an added solute must permeate the cell. For example, Kasai et al. (1981) showed that the permeating CPAs glycerol, ethylene glycol, and Me$_2$SO protect slowly frozen mouse morulae, whereas the nonpermeating solute sucrose confers no protection. Similarly, Fuller and Bernard (1986) showed from shrink/swell osmotic response curves that the protection of slowly frozen two-cell mouse embryos by glycerol was dependent on the extent to which it had permeated. McGann (1979) found no protection when hamster V79 cells were slowly frozen after an exposure of less than 5 min to 5% glycerol at 0°C or when they were frozen in 5% hydroxyethyl starch. Permeation in both cases would be minimal or zero. Eroglu et al. (2000) have published an elegant example of the need for permeation. Mammalian fibroblasts, like most or all cells, are impermeable to trehalose. However, using a genetically engineered mutant of *Staphylococcus*-hemolysin to create reversible pores in the plasma membranes of the fibroblasts, Eroglu et al. were able to load the cells with 0.2 to 0.4 *M* trehalose. Survival after freezing was 80%, whereas in the absence of poration, it was only ~10%.

Although there are many examples substantiating a need for CPA permeation, there are also some striking exceptions. One exception is mouse spermatozoa, which are protected by the nonpermeating trisaccharide raffinose and the disaccharide sucrose in the complete absence of a permeating cryoprotectant (Koshimoto et al., 2000). The permeating CPA glycerol is not only unnecessary, its presence is detrimental. Another is bovine red cells. They will survive slow freezing in sucrose as effectively as in glycerol and they will survive freezing in glycerol when freezing is initiated before the glycerol has had time to permeate (Mazur et al., 1974). A third example is that of Santarius and Giersch (1983), just cited. They found that a variety of sugars protect thylakoid membranes even though (presumably) they do not permeate. In the case of these exceptions, one has to conclude that the additional protection the sugars confer on the external surface of the cell is more important than added protection to the cell interior.

A fourth problem is that there are well-documented cases of precisely the opposite effects of nonpermeating sugars. For example, when plant protoplasts are frozen in solutions of the nonpermeating solute sorbitol that lack electrolytes (Steponkus and Gorden Kamm, 1985), they exhibit a temperature dependence of freezing damage very similar to that observed in red cells frozen in isotonic saline (Figure 1.7). Similarly, the inactivation of ram sperm at room temperature as a function of the osmolality of external nonpermeating solutes is identical whether the external solute is NaCl or sucrose (Watson and Duncan, 1988). The unavoidable conclusion from such results is that the damage is either a direct consequence of the osmotic dehydration of the cells or a consequence of the increase in the concentration of the internal solutes. Steponkus and colleagues, and to a certain extent Watson and Duncan (1988), have chosen to emphasize the former interpretation. I shall return to this matter shortly.

A fifth problem is that the rise in solute concentration during freezing is a consequence of a decrease in the fraction of the extracellular or intracellular solution that remains unfrozen at any temperature. Freezing essentially involves the progressive removal of water from the solution and its deposition as pure ice. The unfrozen fraction at any temperature can be calculated from the same phase diagrams used to calculate the increase in NaCl concentration. This permits one to plot cell survival as a function of unfrozen fraction. When that plot is made for red cells initially suspended in 0.5 and 1 *M* glycerol in isotonic saline, for example, the result (Figure 1.11B) is a mirror image of the plot of survival vs. the concentration of NaCl in Figure 1.11A. As a consequence, based on this alone, one can just as readily attribute cause and effect to the former as to the latter.

1.3.3.2 Partial Separation of NaCl Concentration and Unfrozen Fraction

There is a way, however, to experimentally separate the solute concentration attained at a given temperature from the fraction that is unfrozen at that temperature. We noted in connection with

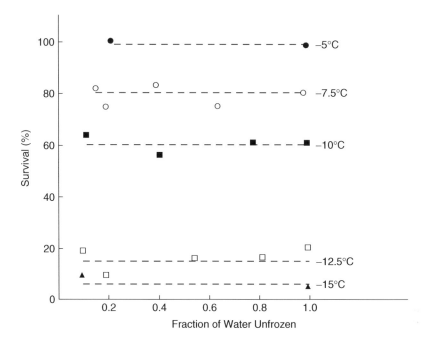

FIGURE 1.12 Survival of rye protoplasts as a function of the fraction of unfrozen water (U). (From Steponkus and Gordon-Kamm, 1985, by permission of *Cryo-Letters*.)

Equation 1.14 that the total concentration of solute in a partly frozen solution depends on temperature alone and is therefore independent of the starting solute concentration. This also applies to a ternary solution like glycerol-NaCl-H_2O for a given weight or mole ratio of glycerol to NaCl in the initial solution. However, this is not the case with the unfrozen fraction. Its magnitude is dependent on both the starting solute concentration and the temperature. It is possible then, by varying the starting solute concentrations (holding the initial glycerol/NaCl ratio constant), to vary the fraction unfrozen while maintaining a constant solute concentration at a given temperature in that unfrozen fraction. Conversely, one can hold the unfrozen fraction constant and vary the solute concentration by varying both starting concentration and final temperature. Our group applied this approach to human red cells in a series of studies (Mazur and Cole, 1985, 1989; Mazur and Rigopoulos, 1983; Mazur et al., 1981); that is, red cells were slowly frozen under conditions that produced exposures to various unfrozen fractions (U) at constant NaCl concentration (m_s) and others that produced various values of m_s at constant U. To do this, we prepared solutions with a fixed molar ratio of glycerol to NaCl in which the concentrations of glycerol were 0.38, 0.5, 1.0, 1.5, and 2.0 M and the concentrations of NaCl were 0.75, 1, 2, 3, and 4× isotonic, respectively. The red cells were then frozen slowly to specific subzero temperatures ranging from –5° to –25°C. The ternary phase diagram permitted us to calculate the concentration of NaCl at each temperature (independent of starting concentration) and the value of U for each temperature and starting concentration. For example, if the five solutions are frozen to –10.7°C, m_s will remain constant at 1.0, but U will vary from 0.11 to 0.70. Conversely, if the 1, 2, 3, and 4× solutions are frozen to –5.1°, –10.7°, –17.6°, and –25.4°C, U will remain constant at 0.3, but m_s will vary from 0.5 to 2.3 molal. The final step was to plot survival of the red cells as a function of U for the various values of m_s.

If survival is controlled solely by m_s, one would expect a series of parallel horizontal lines as has been found by Steponkus and Gorden-Kamm (1985) for plant protoplasts (Figure 1.12). However, as shown in Figure 1.13, that was not the finding for red cells. At low U (5 to 15% unfrozen), survival depends almost solely on U and is independent of m_s. At higher U, survivals

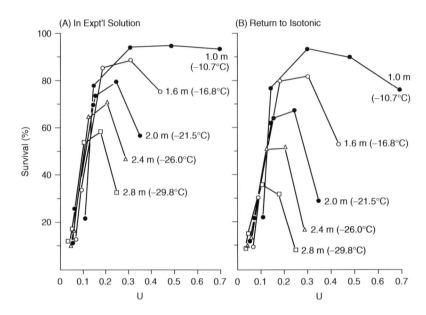

FIGURE 1.13 Survival as a function of the unfrozen fraction of water of a 2% suspension of human red cells frozen at 0.6°C/min to indicated temperatures in several solutions of glycerol/NaCl with a constant weight ratio (5.42) of the two solutes but with varying total concentrations. The salt concentration (m_s) in the unfrozen fraction (U) depends only on temperature. U was varied independent of m_s by appropriate choice of the initial total solute concentration. (A) Survival of thawed cells still in the solutions in which they were frozen. (B) Survival after the cells were returned to near isotonic NaCl. (From Mazur and Cole, 1989, by permission of Academic Press.)

becomes dependent on both m_s and U and decrease with increase in both. The left-hand panel shows the hemolysis of the thawed cells still in the experimental solutions. The right-hand panel shows the survivals after returning the cells to isotonic saline. The effect of the return is to exaggerate the turn-down in survival at high U. I and my colleagues proposed the following physical explanation of deleterious effects of low unfrozen fractions; namely, that the cells, which lie in unfrozen channels between growing ice crystals, are damaged when the size of these channels diminishes. The smaller the channels, the greater become the contacts among the increasingly crowded cells and the greater becomes the likelihood of close contact between the cells and the encroaching ice. Mazur and Cole (1989) refer to these effects as rheological. At higher values of U, the rheological effects disappear and injury becomes ascribable to the detrimental effects of exposure to hypertonic solutions and subsequent dilution (so-called posthypertonic hemolysis; Zade-Oppen, 1968).

Pegg and Diaper (1988, 1989) have confirmed the essentials of these experimental findings for human red cells, and Schneider and Mazur (1987) and Watson and Duncan (1988) have shown them to be applicable to slowly frozen mouse embryos and ram spermatozoa. However, Pegg and Diaper argue for an alternative explanation. They point out that red cells in glycerol solutions with differing starting tonicities of NaCl undergo differing volume excursions before and during freezing. These volume excursions are illustrated in Panels I and II, respectively, of Figure 1.14. All the red cells begin at isotonic volume (a relative volume of 1.0). In Panel I, they initially shrink to variable extents in proportion to the starting concentration of NaCl + glycerol. Then as the glycerol permeates, they reswell to an equilibrium volume that is determined almost exclusively by the impermeant species, NaCl. Those initially suspended in 1× tonicity, return to their normal volume; those suspended in 0.75× solutions return to a volume of about 120% above isotonic; and those that are suspended in 2×, 3×, and 4× solutions return to volumes that are well below isotonic. Survival is near 100% at this point. Panel II illustrates the subsequent shrinkage during slow

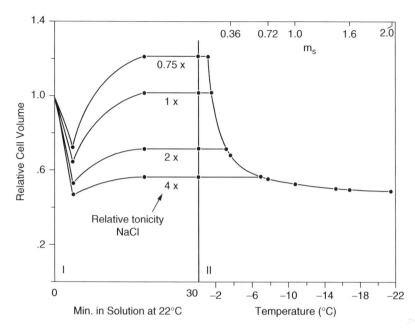

FIGURE 1.14 Computed relative volume changes in human red cells during (I) equilibration in four glycerol-NaCl-water solutions and (II) freezing to indicated temperatures. The four test solutions depicted contained 0.75, 1, 2, 3 and 4× the isotonic concentration of NaCl and corresponding concentrations of glycerol such that the weight ratio of glycerol to NaCl remained constant at 5.42. The curve for 3× isotonic saline is omitted for clarity. Note that the cell volumes attained in Panel II are dependent only on temperature and m_s and are independent of the starting solute concentrations. (From Mazur and Cole, 1989, by permission of Academic Press.)

(equilibrium) freezing. Note that the cell volumes follow a single curve. Once the cells have cooled to below the freezing point of the most concentrated solution (4×, –7.0°C), all have the same volume relative to their initial isotonic volume. Also note from the dual abscissa on Panel II that, regardless of the tonicity of the starting solution, all the cells see the same concentrations of NaCl during freezing, concentrations that are determined solely by temperature. As described in Figure 1.13, however, the unfrozen fractions at given temperatures differ widely and the survivals differ widely. The low values of U that produced high hemolysis (i.e., U < 0.15), were obtained in the 0.75× and 1× solutions. Pegg and Diaper point out that these lower tonicities also produce the greater volume decrease during freezing. Furthermore, the cells in 0.75× solutions have swelled above isotonic volume before the initiation of freezing. They argue that it is this greater volume excursion during freezing combined in the 0.75× solution with the somewhat swollen state before the onset of freezing that is responsible for the increased hemolysis, and not the lower unfrozen fraction. Survival as a function of the volume reduction during freezing is shown in Figure 1.15 both before and after returning the thawed cells to isotonic. Here, the right-hand side of each plot (large reductions in cell volume) correspond to the low values of U in Figure 1.13. One important difference between the two plots is that in plots of survival vs. U, the survivals at low U are nearly independent of m_s, whereas in plots of survival vs. ΔV, survival depends on both ΔV and m_s throughout. If the latter is the correct depiction of cause and effect, then the superimposition of the survival curves for various m_s at low U in Figure 1.13 has to be fortuitous.

The chief basis for Pegg and Diaper's assertion that hemolysis is a consequence of large decreases in cell volume during slow freezing rather than low values of the unfrozen fraction is that they obtain rather similar survivals when red cells are exposed by sequential dialysis at room temperature to the same concentrations of NaCl and glycerol as those experienced by cells subjected to the several solutions and then frozen to different temperatures and thawed; that is, the treatments

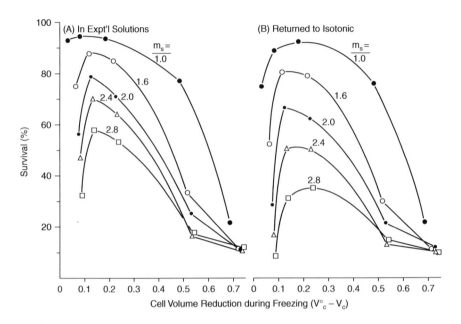

FIGURE 1.15 A replot of the data in Figure 1.13 showing survival of the slowly frozen red cells as a function of the relative reduction in cell volume that occurs during freezing. Large reductions in cell volume correspond to low values of U in Figure 1.13. (A) Survival of thawed cells still in the solutions in which they were frozen. (B) Survival after cells were returned to near isotonic conditions. (From Mazur and Cole, 1989, by permission of Academic Press.)

shown in Panels I and II of Figure 1.14. Because the dialysis experiments were conducted at room temperature, no ice was present, and the unfrozen fraction was 1 in all cases. Mazur and Cole (1989) have argued, however, that although no ice is present, the dialysis also results in decreases in the volumes of the liquid surrounding the cells within the dialysis sac—decreases that mimic those occurring during freezing. The result is that both dialysis and freezing cause an increase in cell concentration and the possibility of cell–cell interactions.

Survival in the above experiments was assessed after thawing, and it is quite possible that the dilution accompanying thawing contributes importantly to damage. The volume decreases described in Panel II of Figure 1.14 during freezing are reversed during slow thawing. At the completion of thawing, the cells return to the same volumes they had at the end of Panel I. Again, those cells that had been originally suspended in the lower tonicities (0.75× and 1×) undergo the larger volume excursions during thawing. The final step in the process is to return the cells to their original isotonic volume by transferring them to isotonic saline. The effects of such a return on survival vs. U and survival vs. ΔV are shown in panels B of Figure 1.13 and Figure 1.15. The return to isotonic does not affect the overall pattern, but it does exaggerate the damaging effects of freezing to high U (right side of Figure 1.13B) or of freezing involving small-volume excursions (left side of Figure 1.15B). In other words, the return to isotonic amplifies the detrimental effects of posthypertonic hemolysis.

Although freezing and thawing and its mimicry by dialysis produces similar survivals of red cells, differences appear when the thawed or dialyzed cells are returned to isotonic saline (Pegg and Diaper, 1991); namely, those suspended in 0.6× or 1× NaCl are substantially more damaged by freezing and thawing than are those subjected to room-temperature dialysis. Furthermore, although cells subjected to freezing and thawing and return to isotonicity show sizable effects of the initial tonicity of NaCl (damage decreased in the sequence 0.6× > 4× > 1× > 2×; Pegg and Diaper, 1991, Figure 2B), those subjected to dialysis mimicry show little or no effect of initial tonicity. The survival curves for cells in 0.6×, 1×, and 4× NaCl are superimposible, and the curve for those in a 2× solution was closely similar (Pegg and Diaper, 1991, Figure 3B). Tenchini et al.

(1980) have reported analogous effects of initial NaCl tonicity on frozen human embryonic epithelial cells. Those slowly frozen to –20°C in 0.36 M NaCl (2.6×) showed a six- to sevenfold higher survival after thawing than those frozen in 0.14 M NaCl.

These results lead to an important conceptual point made by Mazur and Cole (1989). Consider two groups of cells that both start at isotonic volume in isotonic saline. One group is then suspended in a solution made with 0.75× saline; the other in a solution of 3× saline. Both are then frozen to –10.7°C, at which temperature they both become exposed to the same concentration of NaCl (1.0 m) and they both shrink to the same extent (50% of isotonic; Panel II, Figure 1.14). They are then thawed, and both are finally returned to isotonic saline and isotonic volume. Yet in spite of the identity of the initial state, the state after freezing, and the final state, the survivals differ more than fourfold (20% and 90%, respectively). The large differences in survival must be dependent on differences in the paths taken between these states or must depend on some other state characteristic at –10.7°C. We have attributed the differences in survival in the two instances to differences in the unfrozen fractions to which the cells are subjected at the minimum freezing temperature (0.1 and 0.5, respectively, in the example cited). Pegg and Diaper have attributed it to differences in the volume shrinkage during freezing experienced by the cells in the two cases (70% and 8%, respectively, in the example cited). With respect to the volume excursions described in Panels I and II of Figure 1.14, the difference between the 0.75× and 3× solutions is that in the former, all the shrinkage occurs during freezing and all the restoration in volume occurs during thawing, whereas in the 3× solution nearly all the shrinkage occurs in Panel I before freezing, and nearly all the reexpansion to normal volume occurs when the cells are returned to isotonic after thawing.

It is reasonable to argue that differences in when, where, and at what temperature cell volume excursions occur could influence the survival of frozen-thawed cells. It is more difficult to make that argument for cells that are dialyzed to mimic the freezing and thawing process. Because all the exposures in the dialysis experiments are at room temperature, it is reasonable to ask, as did Mazur and Cole (1989), how the cell can distinguish between the volume excursions that occur primarily in Panel I of Figure 1.14 (the 3× and 4× solutions) and those that occur entirely in Panel II (0.75× and 1× solutions), especially in dialysis, where both are occurring at room temperature. The more recent experiments of Pegg and Diaper (1991) suggest that red cells in fact do not make the distinction when they are returned to isotonic concentrations and volumes. The survivals after dialysis mimicry are the same or nearly the same regardless of whether the cells are initially suspended in 0.6×, 1×, 2×, 3×, or 4× saline solutions. (Glycerol was not present in Pegg and Diaper's 1991 experiments; as a consequence, the cells did not go through the shrink/swell excursions depicted in Panel I of Figure 1.14, but shrank to and remained at the equilibrium volumes shown at the right-hand side of that panel.)

The above has been an attempt to extract the salient points of complex experiments, analyses, and interpretations by the two groups, but even though "simplified," it may have obscured the central question, which is, Can slow-freezing injury of cells be ascribed solely to the concentration of solutes produced by external ice formation and the osmotic consequences thereof (Lovelock and Pegg), or is slow-freezing injury also a consequence of direct or indirect physical forces exerted on the cells by the growing ice or a consequence of increased cell–cell interactions resulting from cells being crowded into progressively smaller unfrozen channels (my view)? I would like to summarize some other evidence that supports the latter.

First, Nei (1981), Pegg (1981), Scheiwe et al. (1982), Pegg and Diaper (1983), Mazur and Cole (1985), and De Loecker et al. (1998) have shown that the hemolysis of slowly frozen red cells increases substantially when the cells are frozen at high hematocrit (>40%). De Loecker et al. (1998) found a similar effect in slowly frozen hepatocytes except that in that case the injury increased linearly with increased cytocrit. Kruuv (1986) found the survival of V79 hamster tissue culture cells after slow freezing in 0.45 M Me₂SO as suspensions of individual cells to be two to three times higher than that of cells frozen as aggregates of less than or equal to 1000 cells (≤150 µm diameter). Wells et al. (1979) reported that the survival of human stem cells after slow freezing at

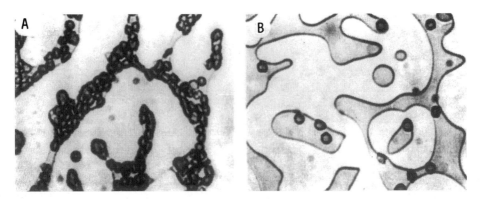

FIGURE 1.16 Photomicrographs of human red cells after slow freezing in isotonic saline to –4°C. The hematocrits of the cell suspensions were ~20% (left) and ~2% (right). The cells lie in unfrozen channels between the ice crystals. (From Nei, 1981, by permission of Academic Press.)

a concentration of 200×10^6/mL was only one sixth the survival of those frozen at concentrations of $10–145 \times 10^6$/mL. They did not specify the cytocrits, but the critical factor appears to be the volume fraction occupied by the cells rather than the numbers of cells per unit volume. For instance, Watson and Duncan (1988) found that a fourfold variation in the latter had no effect on the motilities of ram sperm that were slowly frozen at low cytocrit (estimated 2 to 8%). This adverse effect of cell packing is consistent with the view that cell–cell interactions are an important factor in slow-freezing injury, for the higher the initial cytocrit, the more the cells will be crowded together in the shrinking unfrozen channels (Figure 1.16). Mazur and Cole (1986, 1988) have simulated the packing effect by subjecting a pellet of centrifuged red cells to osmotic dehydration in an unfrozen solution of 5.5 molal glycerol/1.6 molal NaCl at –5 to –8°C, the concentration they would be exposed to after freezing to –16°C when initially suspended in 0.5 m glycerol/0.15 m NaCl. Only 20% survived. In contrast, when the cells were subjected at subzero temperatures to the concentrated glycerol/NaCl solution while in suspension and not centrifuged, or when unshrunken cells were pelleted by centrifugation, 95% survived. In other words, damage resulted from the shrinking of cells in close contact. Some authors like Crowe et al. (1990) have speculated that dehydrated cells placed in close contact have an increased possibility of damaging membrane fusional events. Fujikawa (1981) and Fujikawa and Miura (1986) have evidence from freeze-cleaving electron microscopy that the plasma membranes of slowly frozen red cells or fungal hyphae in fruiting bodies possess patches that are free of intramembrane particles or areas that show particle aggregation. These areas could be sites of potential fusion. Anchordoguy et al. (1987) have demonstrated that fusion occurs in unprotected slowly frozen liposomes.

Second, two lines of evidence indicate that the survival of slowly frozen cells can be substantially influenced by the form of the extracellular crystals. One line of evidence stems from directional solidification techniques that permit one to independently vary the two components of cooling rate—thermal gradient and ice crystal growth velocity—while holding the cooling rate constant. Variations in these components affect the form of the ice crystals (type of dendrites and interdendritic spacing), and Beckmann et al. (1990) and Hubel et al. (1992) have shown that they also affect the survival of human lymphocytes and lymphoblasts that are frozen at the same cooling rate and that are therefore subject to the same sequence of changes in chemical potentials of water and solutes in the medium.

A second line of evidence stems from the response of red cells frozen in solutions containing antifreeze proteons (AFPs). These AFPs in milligrams per milliliter concentrations induce major changes in the form of the external ice crystals, converting them from "cellular" or dendritic to needle-like spicules or to bipyrimidal crystals. Ishiguro and Rubinsky (1994) and Carpenter and Hansen (1992) have reported that these AFP-induced changes in ice crystal habit correlate with greatly increased hemolysis. This is in spite of the fact that their concentrations are too low and their molecular weights too high to influence the chemical potentials of water and solutes in the medium.

Although these two lines of evidence indicate that cell–ice interactions can influence survival, the nature of the interactions remains conjectural. Olien and Smith (1977) proposed that damage to barley cells was a consequence of adhesion between membranes and ice, and Tondorf et al. (1987) have measured strong adhesions between liposomes and ice; however, Hendl et al. (1987) find no such adhesions in dezonated mouse oocytes. Nevertheless, even in the absence of adhesion, growing ice crystals could generate shearing and other rheological forces that are damaging. What is clear is that the external ice crystals can exert force, for as I shall discuss shortly, cells become distorted in their presence.

Third, the most direct way to distinguish between the Lovelock–Pegg hypothesis and the unfrozen fraction hypothesis is to compare the survival of cells subjected to freezing and thawing with that of cells subjected to the same changes in chemical potential at the same temperatures and at the same rates but in the complete absence of ice (that differs from the Pegg and Diaper dialysis experiments with respect to identity of temperature and rates). Because such experiments are difficult to execute, there have been very few of them. One such study involved smooth muscle, and the authors (Taylor and Pegg, 1983) found that contractility after extracellular freezing to −21°C was only one third of that following exposure to the same concentration of solutes at −21°C in the complete absence of ice. However, the fact that extracellular ice physically disrupts a multicellular tissue does not mean that it is physically damaging to single cells in suspension. (For further discussions on models of slow-feezing injury, see Chapter 2.)

1.3.3.3 Dehydration Theories of Slow-Freezing Injury

Several groups have proposed that slow-freezing injury is a consequence of excessive cell dehydration. These hypotheses differ from that of Pegg and Diaper's just discussed, which proposes that damage in red cells is a consequence of the magnitude of the decrease in cell volume during freezing and not the final level of dehydration. In 1968 and 1970, Meryman proposed that the injury is a consequence of damage that develops when cells are unable to shrink osmotically below a certain level in an attempt to achieve osmotic equilibrium. He referred to this as the "minimum volume" hypothesis of slow-freezing damage. As summarized below and as discussed by Pegg and Diaper (1988), there are a number of arguments against this hypothesis, especially with respect to red cells. First, careful measurements show that there is no minimum volume for red cells as the hypertonicity is increased up to the point where nearly all free water is osmotically removed. Second, as we have discussed in reference to Panel II of Figure 1.14, all the cells frozen in the several glycerol/NaCl solutions to temperatures below the freezing points of the solutions shrink to the same volume, yet their survivals differ widely depending on either the volume decrement during freezing (Pegg and Diaper's view) or the magnitude of the unfrozen fraction (my view). Third, as discussed in connection with Lovelock's hypothesis, some cells survive slow freezing well even when suspended in a permeating solute like glycerol for too short a time and too low a temperature to permit permeation or when suspended in nonpermeating sugars like sucrose or raffinose. If the CPA cannot or does not permeate, the extent of cell dehydration at a given temperature will be as extensive as in the complete absence of CPA. Fourth, substitution of other salts for sodium chloride at equal osmolality can substantially affect the survival of frozen-thawed cells even though all should produce the same degree of cell dehydration at given subzero temperatures. Sodium iodide increases the damage of red cells; sodium acetate decreases it (Farrant and Woolgar, 1970). The substitution of choline for NaCl substantially increases the survival of frozen-thawed mouse oocytes (Stachecki and Willadsen, 2000).

Steponkus also believes that slow-freezing injury is a consequence of plasma membrane destabilization brought about by cell dehydration, at least in plant protoplasts, and he and Lynch (1989) have suggested highly detailed mechanistic explanations of that injury. Cold-acclimated (ACC) and nonacclimated (NA) rye protoplasts respond very differently to slow freezing. NA protoplasts exhibit two very different forms of slow-freezing injury depending on the degree of dehydration attained. Down to –5°C, at which point ~80% of the cell water is lost, injury on thawing is caused by "expansion-induced" lysis. The dehydration causes the NA protoplasts to lose a portion of their plasma membrane by blebbing into the cell interior. The protoplasts are osmotically responsive during the initial stages of thawing, but lysis occurs during the latter stages because the diminished membrane does not have sufficient area to permit the protoplast to return to full isotonic volume. If the NA protoplasts are subjected to further freezing and dehydration (e.g., –10°C and 90% water loss), they become completely osmotically unresponsive during thawing. ACC protoplasts behave dramatically differently in response to freeze-induced dehydration. They extrude membrane material to the exterior, but the material remains attached and is reincorporated into the main membrane during thawing. As a consequence, normal reexpansion can occur. The authors attribute these differences to differences in lipid composition, which in turn lead to different responses to the stresses on membranes resulting from the dehydration.

The chief source of the stress is proposed to derive from forces established when the dehydration attempts to force membranes close together and that approach is opposed by the high energies required to remove the intervening water of hydration (hydration forces). Damage is a consequence of the response of the membranes to these stresses. The concept has been elaborated on in considerable detail by Bryant and Wolfe (1992) and Wolfe and Bryant (1999). They consider a situation in which two or more bilayer membranes separated by intervening water are exposed to bulk outside water, the chemical potential of which is progressively dropping with temperature as a result of ice formation in the bulk phase. The assumption is that ice cannot form in the interlamellar water; as a consequence, its chemical potential is transiently higher than that of the external ice (it is transiently supercooled). The standard way to reduce the difference in chemical potential would be for the interlamellar water to leave that region and freeze in the bulk phase. However, that efflux of water would cause the two membranes to approach each other, and that approach will be increasingly repulsed by the hydration forces, the value of which increases exponentially to very high values as the separation decreases to less than ~2 nm. If the chemical potential of the interlamellar water cannot be reduced to that of the bulk ice by efflux, it has to be reduced by another route. That other route is the establishment of a negative hydrostatic pressure (i.e., suction) on the lamellar water (Equation 1.5). Equilibrium is established when the suction force is equal to the repulsive force. (Indeed, they point out that one way to measure the repulsive force as a function of separation distance is to calculate the suction forces developed by freezing or by the introduction of rather large solutes in the bulk phase and to determine by x-ray diffraction the resulting separation.)

Although the forces are balanced normal to the plane of the membrane, they are not balanced parallel to the membranes. As a consequence, the suction force acts to compress the membrane in that plane. The compression leads to other responses that may be deleterious. These responses may involve gel–liquid crystal transitions, they may involve the demixing of lipid components in the membrane, and they may involve the formation of inverse hexagonal phases (Hexagonal II) in the membrane in which tubes of water are surrounded by lipid rather than the normal reverse situation. Which of the responses occur and to what degree depends importantly on the lipid composition of the membrane. Furthermore, the forces generated by the dehydration are also affected by the chemical properties of the solutes present in the medium and by whether or not the solutes can enter the interlamellar water. The picture presented is elegant, but it raises questions.

First, the models and associated physical measurements and thermodynamics apply to two or more bilayer membranes that are brought in close apposition by the dehydration that accompanies freezing, unless an isolated single membrane forms folds or pleats during shrinkage. Closely apposed dual membranes could arise if the plasma membrane of a cell makes close contact with the membranes of its intracellular organelles, or if two cells approach each other closely. If the latter, the hypothesis would predict that cells slowly frozen at low cytocrit should be quite resistant as the probability of cell-to-cell contact would be remote (Figure 1.16). That is clearly not the case with red cells frozen in lower concentrations of glycerol (Figure 1.7), although, as we have seen, injury can be exacerbated in cells frozen at high cytocrits.

Second, nearly all the experimental evidence cited by the above authors for the connection between freeze-induced dehydration, membrane stacking, and phospholipid transformations comes from studies on plant protoplasts and model systems. To date, however, these protoplasts are unique in exhibiting no effect of variations in unfrozen fraction or volume excursions (Figure 1.12). One instance of possibly similar phenomena in animal cells is Fujikawa's (1981) electron microscope study on human red cells. He found that the plasma membranes of red cells frozen at rates below those inducing IIF exhibited patches that were free of the usual intramembranous particles (IMPs), presumably a consequence of lateral demixing. When such slowly frozen cells were subsequently vacuum dried and sectioned, he observed multilamellar structures in the membranes that seemed on the basis of size to correspond to those patches. Whether these multilamellar structures were present after freezing but before the drying is unknown. It is also unclear whether these IMP-free patches represent damaging events. The membranes from cells frozen at 3°C/min exhibited very few (but large) patches, yet they were 100% hemolyzed. The membranes from cells frozen at 1800°C/min exhibited 200 times as many (small) patches, yet 50% of the cells survived.

Even if applicable to plant protoplasts, the Steponkus/Wolfe hypothesis does not seem applicable to slow-freezing injury in a number of other cell types. As already noted, yeast cells and cell of *Escherichia coli* survive slow freezing to –60°C and below, by which temperature they have lost ~90% of their total water as a result of efflux to the external ice, a degree of dehydration that Steponkus and Lynch (1989) refer to as severe (p. 26) and damaging to the higher plant cell protoplasts on which they base their hypothesis. Finally, I have already summarized evidence indicating that the survival of slowly frozen red cells does not depend primarily on the final level of cell dehydration attained

Third, Bryant and Wolfe (1992) indicate that a consequence of the repulsive force generated as membranes closely approach each other during dehydration is that cell shrinkage at high hypertonicities ought to be less than that predicted by a linear BVH plot; that is, the measured volumes at high osmolalities should fall above the line. However, there are a number of published instances in which the BVH relationship remains linear up to osmolalities of 8× isotonic, and one instance in which that holds to 12× isotonic. Examples of the former are platelets (Armitage, 1986), mouse embryo blastocysts and bovine morulae and blastocysts (Mazur and Schneider, 1986), and corneal epithelium (Pegg et al., 1987). Figure 1.3 for mouse eight-cell embryos shows some departure, but that could represent solute leakage.) The latter (linearity to 12× isotonic) was found for human red cells using isotopic techniques (Farrant and Woolgar, 1972; Wiest and Steponkus, 1979). Cells that shrink in an ideal osmotic fashion in solutions that are 8× the isotonic osmolality will have lost 88% of their osmotically available water; those that shrink ideally in a 12× solution will have lost 92% of their osmotic water. These values are well within the range in which Steponkus and colleagues argue for a dominant role of repulsive hydration forces, but in the instances cited, if these forces are operating, their effects are not manifested in resistance to shrinkage.

Fourth, central to the Steponkus/Wolfe hypothesis appears to be the supposition that the water of hydration on membranes is water of high activity. That does not seem consistent with adsorption isotherms, which show that macromolecules retain significant quantities of water when equilibrated with low a_w atmospheres, and it does not seem consistent with dielectric and NMR studies, some of which I reviewed earlier, which show that at least one layer of the water of hydration is irrotationally bound and that additional layers exhibit reduced mobility, presumably as a result of hydrogen bonding with the underlying surface. Nor does it seem consistent with calorimetric studies that show that approximately that same amount of water (~0.25 g H_2O/g dry wt) does not freeze (or at least does not exhibit latent heat of fusion). Wolfe and Bryant (1999, p. 106) argue that ice does not form in the transiently supercooled interlamellar water because the Kelvin effect, discussed above in relation to nucleation through pores, reduces the temperature at which it would be stable.

The alternative interpretation is that the water of hydration does not freeze because its activity is already reduced by the binding forces of hydration. It is easier to see how the Steponkus/Wolfe picture might apply to the damage resulting from freeze-drying, which does remove water of hydration, than it is to see how it applies to the damage resulting from freezing *per se*, which seems not to perturb that water. Crowe et al. (1990) draw a similar sharp distinction between the degree of dehydration produced by freezing and that achieved by freeze-drying or other forms of desiccation, the types of water removed in the two cases (freezable and hydrated), and the consequences of removing these two types of water to the properties of cell constituents like lipid bilayers and to viability. Wesley-Smith et al. (2001) have found that embryos from plant seeds dried to water contents of 0.2 to 0.35 g H_2O per gram dry weight survive rapid freezing to −196°C. Sun (1999) determined calorimetrically that oak seeds dried to a water content of 0.3 g per gram dry weight survived rapid freezing in LN_2 better than those dried to higher or lower water contents. Those dried to lower water contents are damaged by the desiccation. Those dried to higher water contents are presumably killed by IIF. (DSC indicated that no ice formation is measurable in seeds dried to below 0.3 g water per gram dry weight.) O'Dell and Crowe (1979) obtained similar results for the relation between the water content of nematodes and the percentages that survived rapid freezing. Those with water contents less than 0.3 g H_2O per gram dry weight survived; 75% of those with a water content of 0.4 g per gram dry weight were killed. Fully hydrated yeast and *E. coli* survive slow freezing to −60°C, at which temperature about 0.23 g of their water per gram of solids is unfrozen (Souzu et al., 1961; Wood and Rosenberg, 1957), but survival drops when their water content is reduced below that value by air-drying or freeze-drying (Mazur, 1968, p. 360).

Bronshteyn and Steponkus (1993) have disputed the view that unfreezable water is water of low activity and have disputed DSC estimates of its magnitude. With respect to the latter, they claim that if DSC baseline values are assessed with high precision, one sees that significant amounts of ice begin to melt above approximately −50°C even in a pure ice/water sample. They argue that the heat absorbed in that low-temperature melting is not included in other DSC studies and that this omission leads to an underestimate of the total amount of frozen water. They ascribe the low-temperature melting to a Kelvin suppression of the melting point of small ice crystals in thin films, but that seems highly unlikely or impossible thermodynamically. Were ice to melt at, say, −40°C, the resulting pure liquid water would be supercooled 40°C relative to large ice crystals with little curvature and would rapidly refreeze on the surface of those larger crystals. Indeed, this may be one of the mechanisms involved in the recrystallization of ice, the process by which larger ice crystals grow at the expense of smaller ones in a polycrystalline sample. Furthermore, Wilson et al. (1999), using three different DSC instruments, have failed to find any evidence of melting of pure ice below −3°C except possibly a tiny endotherm in one case at −15°C. Moreover, other calorimetric procedures such as the "method of mixtures" used by Wood and Rosenberg (1957) measure the total amount of heat involved in the warming of a frozen system from, say, −60°C through melting, and they lead to similar percentages of unfrozen water as those determined by DSC. Bronshteyn and Steponkus's other argument is that the barrier to the freezing of the unfrozen water is kinetic, not thermodynamic. They present DSC evidence that the measured amount of unfrozen water in a multilamellar liposome system decreases over a 30-min time period at subzero temperatures. However, the decrease amounts to only about 10% and appears to be asymptotic. Furthermore, it is difficult to see how the reduced mobility of water as assessed by NMR or its immobilization as assessed by dielectric measurements would represent a metastable situation; yet, the fractions of water exhibiting reduced or zero mobility agree reasonably well with the fraction that does not freeze.

Fifth, Steponkus and colleagues have chosen to relate the experimental manifestations of injury to dehydration *per se*, and in particular, to the dehydration of the plasma membrane, but their experiments cannot exclude the alternative hypothesis that the effects derive from the concentration of intracellular solutes that is an unavoidable consequence of the dehydration. There have been numerous theories that emphasize the concentration aspects such as Levitt's (1966) disulfide hypothesis. The strengths and weaknesses of some of these have been reviewed by others (e.g., Crowe et al., 1990), but as they are subject to the same confounding, I do not think it profitable to review them.

Fujikawa and Miura (1986) have published an interesting electron microscope study on the response of fungal hyphae to slow freezing that may represent one example of a sort of synthesis of these several conflicting theories. They found that the fungal fruit bodies became increasingly damaged as slow cooling progressed from $-5°$ and $-20°C$. Concomitantly, the hyphae became increasingly dehydrated, they became increasingly deformed, and increasing percentages of their plasma membranes display aggregated IMPs. IMP aggregation occurred only in regions in which the deformation caused the inner surfaces of the plasma membranes in a given hyphum to come in contact. A 5.4-osmolal sorbitol solution has the same osmolality as that experienced by the hyphae at $-10°C$, but the hyphae in pieces of fruit body exposed to that solution at $22°C$ did not distort, and their IMPs did not aggregate. In other words, slow freezing appears to be damaging primarily because the external ice caused hyphal deformation (a direct physical effect) and not because it caused osmotic dehydration (a solution effect). The combination led to plasma membrane contact and changes in IMP pattern. Perhaps these changes in pattern were a prelude to deleterious membrane fusion. Mouse oocytes represent another example in which slow freezing produces extensive deformation of dehydrated cells (Leibo et al., 1978), whereas osmotic shrinkage of cells in hyperosmotic unfrozen media does not (Jackowski et al., 1980). The same is true of barley tissue culture cells (Olien and Smith, 1977).

1.3.3.4   Mode of Action of Cryoprotectants

Both the Lovelock/Pegg and the Mazur hypotheses of slow-freezing injury ascribe the protective effects of low-molecular-weight cryoprotectants such as glycerol to their nonspecific colligative ability to reduce the concentration of damaging solutes, increase the unfrozen fraction, or reduce volume excursions during freezing and thawing. The major support for the colligative theory comes from detailed studies like those of Santarius and Giersch (1983), who showed with spinach thylakoid membranes that survival after slow freezing in media containing a variety of sugars and NaCl depended on the concentration of NaCl that was attained in the partly frozen state and was independent of the chemical nature of the sugar (although at very low NaCl concentrations, sugar-specific effects were noted).

However, there is an opposing view that ascribes the protection from CPA to solute-specific interactions with the phospholipids of bilayers that lead to the stabilization of the plasma membrane. Major proponents of that view are Crowe and colleagues (e.g., Anchordoguy et al., 1987). Their experiments with liposomes indicate that the sugars sucrose and trehalose are especially protective on a molar basis and are considerably more effective than glycerol, which in turn is considerably more effective than Me_2SO. Although the issue is important mechanistically, it is not currently resolvable, in spite of an enormous amount of data on the effects of various additives on various cells and model systems. Each concept has its difficulties. For example, the colligative school has difficulty explaining why high–molecular weight compounds such as polyvinylpyrrolidone and hydroxyethyl starch can protect some cells; for example, Hamster V79 (Mazur et al., 1972) and red cells (Scheiwe et al., 1982). The specific interaction school has difficulty not only with that finding but with explaining why glycerol and Me_2SO are often equivalently protective on a molar basis in mouse embryos, for example. Part of the problem is that for a rigorous test of the colligative hypothesis, one needs phase diagrams of the CPA/NaCl/water system under study, but these are only available for a handful of systems like glycerol/NaCl/water. (For further discussions on cryoprotectant action, see also Chapters 2 and 21.)

Still another possibility is raised by the interesting finding of Ishiguro and Rubinsky (1994) that human red cells slowly frozen in saline are pushed into unfrozen channels between the growing ice crystals, whereas those frozen in 2.7 M glycerol/saline appear to be encapsulated within the

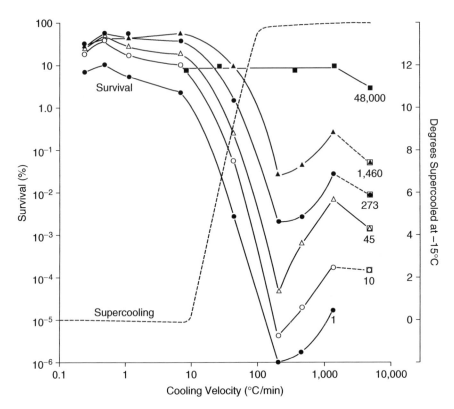

FIGURE 1.17 Effect of cooling velocity to –70°C on the survival of cells of the yeast *Saccaharomyces cerevisiae* subsequently cooled to –196°C and warmed at the rates indicated on each curve. The dotted curve labeled "Supercooling" and the right-hand ordinate show the calculated extent to which the intracellular water is supercooled at –15°C in cells cooled at the indicated rates. (From Mazur and Schmidt, 1968, by permission of Academic Press.)

ice. The latter survive; the former do not. (Interestingly [and perhaps significantly], this glycerol concentration lies in the range shown in Figure 1.7 in which striking protection is manifested.) Cells frozen in 2.7 *M* glycerol/saline containing antifreeze proteins behave like those frozen in saline alone; that is, they are pushed between the spicular ice crystals and do not survive.

1.4   WARMING AND THAWING

Frozen cells have to be warmed and thawed before they can be used or their viability assessed, and the effects of rate of warming can be as profound as the effects of cooling rate. The two factors interact strongly; that is, the effects of warming rate depend importantly on the prior rate of cooling. When cells have been cooled at rates that lie in the right-hand side of the inverted U, survivals tend to be substantially higher when subsequent warming is rapid than when it is slow. Yeast represents a striking example (Figure 1.17). Other instances have been reported in *Neurospora* spores (Wellman and Pendyala, 1979), rat lymphocytes (Taylor et al., 1987), and mouse morulae (Kasai et al., 1980). Because such cooling rates are computed or shown to induce IIF, the conclusion can be rephrased to say that when cooling rates are high enough to induce IIF, rapid warming tends to be beneficial. That has generally been explained on the basis of its ability to inhibit the recrystallization of the internal ice or to prevent the devitrification of glassy water. Recrystallization is the conversion of small ice crystals to large ones. It occurs as a consequence of the fact that small ice crystals possess higher surface free energies than larger crystals. The argument is that because crystal size decreases with increasing cooling rate (Van Venrooij et al., 1975), the crystals that form in cells at high cooling rates will tend to be small, and because they are small, they will have high surface energies. Because they have high surface energies they will tend to grow by recrystallization given sufficient time at appropriate temperatures. Slow warming provides more time at gradually increasing temperatures than does rapid warming. The final step in the argument is that large intracellular ice crystals are more damaging than small ones either because of size *per se* or because of forces arising from the growth process. Recrystallization in cells and foods was reviewed in considerable detail by Fennema et al. (1973), and there have been several reports experimentally demonstrating a correlation between visible recrystallization of intracellular ice and cell death during the slow warming of rapidly cooled cells; namely, in yeast, higher plant cells, ascites tumour cells, hamster tissue culture cells, and mouse embryos (reviewed in Mazur, 1984). Although the temperatures for the two events do not always correlate exactly (Rall et al., 1984), I suggested in my 1984 review that the discrepancies are not sufficient to negate the hypothesis.

The effects of warming rate on slowly cooled cells are more complex and more difficult to interpret. Again, I define slowly cooled cells as those cooled slowly enough to preclude IIF; that is, those depicted by the left-hand limbs of the inverted U curves in Figure 1.6. The three possibilities are that warming rate is without effect, that slow warming is more damaging than rapid, and that rapid warming of slowly frozen cells is more damaging than slow warming. There are examples in all three groups. Examples of the first group are human stem cells frozen slowly in Me_2SO (Wells et al., 1979); eight-cell mouse embryo frozen slowly in glycerol to –50°C or below (Miyamoto and Ishibashi, 1983; Rall and Polge, 1984); one-cell mouse embryos frozen slowly in propanediol to –35°C or below (Van den Abbeel et al., 1994); rat morulae in Me_2SO, ethylene glycol, or glycerol (Kasai et al., 1982); mouse lymphoma cells frozen slowly in glycerol or Me_2SO to –100°C (Akhtar et al., 1979); and rat lymphocytes frozen slowly in Me_2SO (Taylor et al., 1987). (The CPA concentrations in these studies were 1 to 2 M [mostly 1.3 to 1.6 M.])

Examples of the second group are equally common, as illustrated by the left side of Figure 1.17 for yeast. In this group, slowly cooled cells, like rapidly cooled ones, are more damaged with slow warming than with rapid warming, but usually much less so. Such a pattern is also seen in mammalian sperm (Fiser et al., 1986, 1993; Henry et al., 1993; Koshimoto and Mazur, 2002). It is probably not fruitful to try to assign specific physical/chemical causes to this second type of response, as there are too many possibilities. Most of what I discussed with respect to injury from slow cooling may also apply to slow thawing. The cell volume excursions during freezing are to a first approximation reversed during warming. Solutes that have concentrated during freezing become diluted during thawing, and the dilution may set the stage for posthypertonic hemolysis or lysis. Furthermore, injurious events that have been activated during freezing may only be expressed during thawing. That seems to be the case with respect to the release of some of the hemoglobin in red cells (Gupta, 1975). In general, if the development of a damaging event is time-dependent, one might expect the sequence of slow cooling/slow warming to be more damaging than that of slow cooling/rapid warming, simply because of an increase in the total exposure time at subzero temperatures.

The third and opposite type of warming rate dependence is that slowly cooled cells are more damaged by subsequent rapid warming than by slow warming. Although less common than the other two groups of responses, instances of this third pattern have been reported in mouse embryos frozen in 1 or 1.5 M Me_2SO (but not 1.5 M glycerol or propanediol; Rall and Polge, 1984; Van den Abbeel et al., 1994; Whittingham et al., 1972, 1979), human red blood cells (Mazur, 1984, Figure 18), and in higher plant cells (Levitt, 1966). One suggested explanation for the increased sensitivity to rapid warming is that it induces osmotic shock. There is evidence that additional additive may be driven into cells during slow freezing, a process referred to as solute loading (Fuller and De Loecker, 1985; Griffiths et al., 1979). In such cases, there may not be sufficient time for the excess solute to diffuse back out when thawing is rapid, so the cells swell and lyse as the medium becomes abruptly diluted by the melting of extracellular ice. Van den Abbeel et al.'s (1994) experiments are consistent with this view.

When they added 0.1 *M* sucrose to the 1.5 *M* Me$_2$SO as an impermeable osmotic buffer, the detrimental effect of rapid thawing disappeared. However, there are complications with this interpretation. Pegg and Diaper (1988) agree that solute loading can occur during freezing and thawing, but they maintain that there is no driving force for permeating solutes to enter during freezing; rather, the solutes that enter are ones to which the cells are normally impermeable, such as cations or sucrose. Furthermore, they believe that the inward leak develops during the initial stages of thawing and not during freezing, and that the membranes subsequently reseal to trap excess solute.

One potential cause of injury during the slow warming of slowly cooled cells that is not self-evident is the devitrification of extracellular or intracellular glass formed even when the initial freezing is slow. In elegant studies Rall et al. (1984) and Rall and Polge (1984) showed that the warming rate dependence of the survival of slowly frozen (0.5°C/min) eight-cell mouse embryos depended on the temperature at which the slow cooling was interrupted by very rapid cooling to –150°C or below. When the transition from slow to rapid cooling was made below –55°C, slow warming was innocuous, but when the transition was effected at –30° to –40°C, slow warming was much more damaging than rapid warming. A combination of cryomicroscopy and DSC provided the explanation. Although the cells dehydrate during the initial slow freezing, they still contain "freezable" water at –40°C. That dehydration, however, has caused the external and internal solutes to concentrate with a resulting large increase in the viscosity of the solutions. Because of the high viscosity, the residual freezable water at –40°C does not crystallize during the subsequent rapid cool to –150°C or below, but undergoes vitrification to form a metastable glass. When subsequent warming is slow, the glass devitrifies (freezes), the resulting ice recrystallizes, and the embryos are killed. In contrast, when the subsequent warming is rapid, there is insufficient time for devitrification, and the embryos survive. When the slow cooling is continued to –55°C or below, most of the remaining freezable water leaves the cells, so that when rapid cooling is then initiated, the remaining unfreezable water forms a stable glass that does not undergo devitrification even during subsequent slow warming. Van den Abbeel et al. (1994) have obtained very similar results for one-cell mouse embryos frozen in propanediol, except that the critical transition temperature occurs at about –30° to –35°C. When the shift from slow to very rapid cooling is made above –30°C, the embryo survival is far higher when subsequent warming is rapid than when it is slow. When the shift is made below –40°C, warming rate is without effect. Karlsson (2001) has recently published a detailed thermo-dynamic model of these sorts of events. One counterintuitive conclusion from the analysis is that with certain combinations of slow cooling and slow warming, cells that contain residual vitrified solution after the dehydration that accompanies the initial slow cooling will actually continue to shrink during the initial stages of slow warming. Karlsson argues that the additional shrinkage can be beneficial in lessening the chances of devitrification. However, if the dehydration during slow freezing is damaging, any additional shrinkage during the beginning of slow warming may become even more damaging.

1.5 SUMMARY AND CONCLUSIONS

This review has been concerned with analyzing the main elements affecting the survival of cells subjected to ice formation during their journey to low subzero temperatures. Although many of the elements are clear, others are not. With respect to intracellular freezing, an important unresolved question is the mechanism by which the plasma membrane mediates the nucleation of the intracellular milieu by extracellular ice. With respect to slow freezing, there continue to be a number of unresolved questions. Does external ice play merely a neutral role, exerting its effects solely by modifying the composition of the outside solution, reducing the chemical potential of the external water, and inducing cell dehydration? Or does it play a more direct role through mechanical effects on the cells and through enhancing cell–cell interactions? Is the protection from CPAs primarily physico-chemical and colligative, or do they exert chemically specific effects on cell membranes and on the properties of the external and internal solutions?

A number of topics highly relevant to cryopreservation but not directly involved in the freezing of individual cells have been omitted or only incidentally touched on. One of these is the osmotic events involved in the introduction and removal of CPAs. If these steps are carried out in overly large increments or too rapidly, the osmotic damage can be considerable. As nicely illustrated by Gao et al. (1995) and Katkov (2000), one can use values of L_p and P_s to develop procedures that are innocuous. A second omitted topic is chill injury in the absence of ice formation. In some cases such as *Drosophila* eggs (Mazur et al., 1992a, 1992b) and pig morulae and bovine morulae derived from *in vitro* matured oocytes (Pollard and Leibo, 1994), it can be so severe as to preclude cryopreservation by slow equilibrium freezing. A third topic touched on only incidentally here but studied in detail by many others is that the cryopreservation of tissues and organs introduces problems that are not present in the individual cells that compose those multicellular systems. In those systems, it is clear that extracellular ice or intercellular ice (ice growing between the cells) is mechanically destructive. Furthermore, as organs are composed of a variety of cell types, a preservation procedure optimized for one type may be suboptimal for others (Karow and Pegg, 1981).

A fifth topic, also touched on only incidentally, is vitrification. It is a potential solution both to severe chill injury and to the detrimental mechanical effects of external ice. In the former, it permits the use of cooling rates that are high enough to outrun chilling injury without introducing lethal intracellular ice, and it has permitted the cryopreservation of, for example, *Drosophila* eggs (Mazur et al., 1992a, 1992b; Steponkus et al., 1990). For the latter, it can avoid mechanical damage from extracellular ice by preventing the formation of that ice. The concept of the utility of vitrification was pioneered by Luyet and colleagues, who published extensively on the conditions under which it would and would not occur (e.g., Luyet and Rasmussen, 1967; MacKenzie and Luyet, 1967). Many others have since continued to study both the physics of the vitrification of biological solutions (e.g., Angell and Senapti, 1987; Boutron, 1987; MacFarlane and Forsyth, 1987) and its biological applicability (e.g., Fahy, 1987; Rall, 1987). (For more detailed discussions on vitrification and current applications of this technique in cells, see Chapters 10, 12, and 18; in Chapter 22, Taylor et al. describe the extension of vitrification to complex tissues.)

Finally, although I have tried to extract the commonality of cryobiological events, there are a number of specific cell types that present specific challenges and problems both to cryopreservation and to understanding. In some cases, such as insect and fish eggs and to a certain extent mammalian oocytes, the problem is poor permeability to water and cryoprotective solutes. Successful cryopreservation demands high permeability to both. In some cases, most notably mammalian sperm, the experimental responses to freezing variables are not always in accord with those inferred from physical–chemical analysis. Presumably, the same laws of physics apply to sperm as to other cells, so that the discrepancies must be caused by unidentified cell biological factors or by the use of incorrect parameters in the models. In some cases, difficulties in cryopreservation may stem from the complex concatenation of conflicting variables. Thus, cooling rates that are low enough to avoid IIF may be so slow as to induce damage from solution effects or chilling. The use of higher CPA concentrations to minimize solution effects may introduce toxicity or exacerbate osmotic damage. Toxicity may be reduced at lower temperatures, but lower temperatures slow the permeation and further exacerbate osmotic damage. Damage from external ice can be prevented by vitrification, but the induction of the vitrified state requires high CPA concentrations that exacerbate both toxicity and osmotic problems. These incompatibilities may not be challenges to our understanding, but in some cases they remain challenges to achieving successful cryopreservation.

REFERENCES

Acker, J.P., Elliot, J.A.W., and McGann, L.E. (2001) Intracellular ice propagation: experimental evidence for ice growth through membrane pores. *Biophysical J.*, 81, 1389–1397.

Acker, J.P., Larese, A., Yang, H., Petrenko, A., and McGann, L.E. (1999) Intracellular ice formation is affected by cell interactions, *Cryobiology*, 38, 363–371.

Acker, J.P. and McGann, L.E. (2000) Cell-cell contact affects membrane integrity after intracellular freezing, *Cryobiology*, 40, 54–63.

Agca, Y., Liu, J., McGrath, J.J., Peter, A.T., Critser, E.S., and Critser, J.K. (1998) Membrane permeability characteristics of metaphase II mouse oocytes at various temperatures in the presence of Me₂SO, *Cryobiology*, 36, 287–300.

Aggarwal, S.J., Diller, K.R., and Baxter, C.R. (1988) Hydraulic permeability and activation energy of human keratinocytes at subzero temperatures, *Cryobiology*, 25, 203–211.

Akhtar, T., Pegg, D.E., and Foreman, J. (1979) The effect of cooling and warming rates on the survival of cryopreserved L-cells, *Cryobiology*, 16, 424–429.

Anchordoguy, T.J., Rudolph, A.S., Carpenter, J.F., and Crowe, J.H. (1987) Modes of interaction of cryoprotectants with membrane phospholipids during freezing, *Cryobiology*, 24, 324–331.

Angell, C.A. and Senapti, H. (1987) Crystallization and vitrification in cryoprotected aqueous systems, in *The Biophysics of Organ Cryopreservation*, Pegg, D.E. and A.M. Karow, Jr., Eds., Plenum, New York, pp. 147–172.

Armitage, W.J. (1986) Effect of solute concentration on intracellular water volume and hydraulic conductivity of human blood platelets, *J. Physiol.*, 374, 375–385.

Beckmann, J., Korber, C., Rau, G., Hubel, A., and Cravalho, E.G. (1990) Redefining cooling rate in terms of ice front velocity and thermal gradient: first evidence of relevance to freezing injury of lymphocytes, *Cryobiology*, 27, 279–287.

Beney, L., Marachel, P. A., and Gervais, P. (2001). Coupling effects of osmotic pressure and temperature on the viability of *Saccharomyces cerevisiae*. *Appl. Microbiol. Biotech.* 56, 513–516.

Benson, C.T. and Critser, J.K. (1994) Variation of water permeability (L_p) and its activation energy (Ea) among unfertilized golden hamster and ICR murine oocytes, *Cryobiology*, 31, 215–223.

Benson, C.T., Liu, C., Gao, D.Y., Critser, E.S., Benson, J.D., and Critser, J.K. (1998) Hydraulic conductivity (L_p) and its activation energy (Ea), cryoprotectant agent permeability (Ps) and its Ea, and reflection coefficients (σ) for golden hamster individual pancreatic islet cell membranes, *Cryobiology*, 37, 290–299.

Berendsen, J.C. (1975) Specific interactions of water with biopolymers, in *Water—A Comprehensive Treatise*, Vol. 5. Franks, F., Ed., Plenum Press, New York, pp. 293–330.

Berger, W.K. and Uhrik, B. (1996) Freeze-induced shrinkage of individual cells and cell-to-cell propogation of intracellular ice in cell chains from salivary glands, *Experientia*, 52, 843–850.

Böhmer, H.V., Wöhler, W., Wendel, U., Passarge, E., and Rüdiger, H.W. (1973) Studies on optimal cooling rate for freezing human diploid fibroblasts, *Exp. Cell Res.*, 79, 496–498.

Boutron, P. (1987) Non-equilibrium formation of ice in aqueous solutions: efficiency of polyalcohol solutions for vitrification, in *The Biophysics of Organ Cryopreservation*, Pegg, D.E. and A.M. Karow, Jr., Eds., Plenum, New York, pp. 201–236.

Bronshteyn, V. and Steponkus, P.L. (1993) Calorimetric studies of freeze-induced dehydration of phospholipids, *Biophys. J.*, 65, 1853–1865.

Bryant, G. and Wolfe, J. (1992) Interfacial forces in cryobiology and anhydrobiology, *Cryo-Letters*, 13, 23–36.

Carpenter, J.F. and Hansen, T.N. (1992) Antifreeze protein modulates cell survival during cryopreservation: Mediation through influence on ice crystal growth, *Proc. Natl. Acad. Sci. USA*, 89, 8953–8957.

Chambers, R. and Hale, H.P. (1932) The formation of ice in protoplasm, *Proc. Royal Soc. B (Lond.)*, 110, 336–352.

Crowe, J.H., Carpenter, J.F., Crowe, L.M., and Anchordoguy, T.J. (1990) Are freezing and dehydration similar stress vectors? A comparison of modes of interaction of stabilizing solutes with biomolecules, *Cryobiology*, 27, 219–231.

Davson, H. and Danielli, J.F. (1952) *The Permeability of Natural Membranes*, Cambridge: Cambridge University Press.

de Groot, B.L. and Grubmüller, H. (2001) Water permeation across biological membranes: mechanism and dynamics of Aquaporin-1 and GlpF, *Science*, 294, 2353–2357.

De Loecker, W., Koptelov, V.A., Grishenko, V.I., and De Loecker, P. (1998) Effects of cell concentration on viability and metabolic activity during cryopreservation, *Cryobiology*, 37, 103–109.

Devireddy, R.V., Raha, D., and Bischof, J.C. (1998) Measurement of water transport during freezing in cell suspensions using a differential scanning calorimeter, *Cryobiology*, 36, 124–155.

Devireddy, R.V., Swanlund, D.J., Roberts, K.P., and Bischof, J.C. (1999) Subzero water permeability parameters of mouse spermatozoa in the presence of extracellular ice and cryoprotective agents, *Biol. Reprod.*, 61, 764–775.

Devireddy, R.V., Swanlund, D.J., Roberts, K.P., Pryor, J.L., and Bischof, J.C. (2000) The effect of extracellular ice and cryoprotective agents on the water permeability parameters of human sperm plasma membranes during freezing, *Hum. Reprod.*, 15, 1125–1135.

Dick, D.A.T. (1966) *Cell Water*, Butterworths, Washington, D.C.

Diller, K.R. (1979) Intracellular freezing in glycerolized red cells, *Cryobiology*, 16, 125–131.

Dowgert, M.F. and Steponkus, P.L. (1983) Effect of cold acclimation on intracellular ice formation in isolated protoplasts, *Plant Physiol.*, 72, 978–988.

Du, J., Kleinhans, F.W., Mazur, P., and Critser, J.K. (1993) Osmotic behavior of human spermatozoa study by EPR, *Cryo-Letters*, 14, 285–294.

Du, J., Kleinhans, F.W., Mazur, P., and Critser, J.K. (1994a) Human spermatozoa glycerol permeability and activation energy determined by electron paramagnetic resonance, *Biochim. Biophys. Acta*, 1194, 1–11.

Du, J., Tao, J., Kleinhans, F.W., Mazur, P., and Critser, J.K. (1994b) Water volume and osmotic behavior of mouse spermatozoa determined by electron paramagnetic resonance, *J. Reprod. Fertil.*, 101, 37–42.

Dutzler, R., Campbell, E.B., Cadene, M., Chait, B.T., and MacKinnon, R. (2002) X-ray structure of a ClC chloride channel at 3.0Å reveals the molecular basis of anion selectivity, *Nature*, 415, 287–294.

Echevarria, M., Winhager, E.E., and Frindt, G. (1996) Selectivity of the renal collecting duct water channel Aquaporin-3, *J. Biol. Chem.*, 271, 25079–25082.

Edashige, K., Sakamoto, M., and Kasai, M. (2000) Expression of mRNAs of the aquaporin family in mouse oocytes and embryos, *Cryobiology*, 40, 171–175.

Eisenberg, D. and Kauzmann, W. (1969) *The Structure and Properties of Water*, Oxford University Press, Oxford.

Eley, D.D. and Leslie, R.B. (1966) Kinetics of adsorption of water vapor and electrical conduction in bovine plasma albumin, *Trans. Faraday Soc.*, 62, 1002–1014.

Elford, B.C. (1970) Non-solvent water in muscle, *Nature*, 227, 282–283.

Eroglu, A., Russo, M.J., Bieganski, R., Fowler, A., Cheley, S., Bayley, H., and Toner, M. (2000) Intracellular trehalose improves the survival of cryopreserved mammalian cells, *Nat. Biotechnol.*, 18, 163–171.

Fahy, G.M. (1981) Simplified calculation of cell water content during freezing and thawing in nonideal solutions of cryoprotective agents and its possible application to the study of solution effects injury, *Cryobiology*, 18, 473–482.

Fahy, G.M. (1987) Biological effects of vitrification and devitrification, in *The Biophysics of Organ Cryopreservation*, Pegg, D.E. and A.M. Karow, Jr., Eds., Plenum, New York, pp. 265–297.

Fahy, G.M. and Karow, A.M., Jr. (1977) Ultrastructure-function correlation between electrolyte toxicity and cryoinjury in the slowly frozen, cryoprotected rat heart, *Cryobiology*, 14, 418–427.

Farrant, J. and Woolgar, A.E. (1970) Possible relationships between the physical properties of solutions and cell damage during freezing, in *The Frozen Cell*, Wolstenholme, G.E.W. and M. O'Connor, Eds., J.A. Churchill, London, pp. 97–114.

Farrant, J. and Woolgar, A.E. (1972) Human red cells under hypertonic conditions: a model system for investigating freezing damage 1. Sodium chloride, *Cryobiology*, 9, 9–15.

Fennema, O.W., Powrie, W.D., and Marth, E.H. (1973) *Low-Temperature Preservation of Foods and Living Matter*, Marcel Dekker, New York.

Finkelstein, A. (1987) *Water Movement through Lipid Bilayers, Pores, and Plasma Membranes—Theory and Reality*, Wiley-Interscience, New York.

Fiser, P.S., Fairfull, R.W., Hansen, C., Panich, P.L., Shrestha, J.N.B., and Underhill, L. (1993) The effect of warming velocity on motility and acrosomal integrity of boar sperm as influenced by the rate of freezing and glycerol level, *Mol. Reprod. Develop.*, 34, 190–195.

Fiser, P.S., Fairfull, R.W., and Marcus, G.J. (1986) The effect of thawing velocity on survival and acrosomal integrity of ram spermatozoa frozen at optimal and suboptimal rates in straws, *Cryobiology*, 23, 141–149.

Fletcher, N.H. (1962) *The Physics of Rainclouds*, Cambridge, Cambridge University Press.

Frank, H.S. (1972) Structural models, in *Water—A Comprehensive Treatise Vol 1: The Physics and Physical Chemistry of Water*, Franks, F., Ed., Plenum, New York, pp. 515–543.

Frank, H.S. and Evans., M. (1945) Free volume and entropy in condensed systems. III. Entropy in binary
    liquid mixtures; partial molal entropy in dilute solutions; structure and thermodynamics in aqueous
    solutions, *J. Chem. Physics*, 13, 507–532.
Frank, H.S. and Wen, W.-Y. (1957) Ion-solvent interaction. Structural aspects of ion-solvent interaction in
    aqueous solutions: A suggested picture of water structure, *Trans. Faraday Soc.*, 24, 133–140.
Franks, F. (1973) The solvent properties of water, in *Water—A Comprehensive Treatise Vol 2: Water in
    Crystalline Hydrates; Aqueous Solutions of Simple Nonelectrolytes*, Franks, F., Ed. Plenum, New
    York, pp. 1–54.
Franks, F., Mathias, S.F., Galfre, P., Webster, S.D., and Brown, D. (1983) Ice nucleation and freezing in
    undercooled cells, *Cryobiology*, 20, 298–309.
Fujikawa, S. (1981) The effect of various cooling rates on the membrane ultrastructure of frozen human
    erythrocytes and its relation to the extent of haemolysis after thawing, *J. Cell Sci.*, 49, 369–382.
Fujikawa, S. and Miura, K. (1986) Plasma membrane ultrastructural changes caused by mechanical stress in
    the formation of extracellular ice as a primary cause of injury in fruit-bodies of basidionmycetes
    (*Lyophyllum ulmarium*) (Fr.) Kuhner, *Cryobiology*, 23, 371–382.
Fuller, B.J. and Bernard, A. (1986) The relationship between intracellular glycerol permeation and survival
    following cryopreservation of the *in vitro* fertilized 2-cell murine embryo, *Cryo-Letters*, 7, 254–259.
Fuller, B.J. and De Loecker, W. (1985) Changes in the permeability characteristics of isolated hepatocytes
    during slow freezing, *Cryo-Letters*, 6, 361–370.
Gao, D.Y., Liu, J., Liu, C., McGann, L.E., Watson, P.F., Kleinhans, F.W., Mazur, P., Critser, E.S., and Critser,
    J.K. (1995) Prevention of osmotic injury to human spermatozoa during addition and removal of
    glycerol, *Hum. Reprod.*, 10, 1109–1122.
Gilmore, J.A., Liu, J., Woods, E.J., Peter, A.T., and Critser, J.K. (2000) Cryoprotective agent and temperature
    effects on human sperm membrane permeabilities: convergence of theoretical and empirical
    approaches for optimal cryopreservation methods, *Hum. Reprod.*, 15, 335–343.
Gilmore, J.A., McGann, L.E., Liu, J., Gao, D.Y., Peter, A.T., Kleinhans, F.W., and Critser, J.K. (1995) Effect
    of cryoprotectant solutes on water permeability of human spermatozoa, *Biol. Reprod.*, 53, 985–995.
Griffiths, J.B., Cox, C.S., Beadle, D.J., Hunt, C.J., and Reid, D.S. (1979) Changes in cell size during the
    cooling, warming and post-thawing periods of the freeze-thaw cycle, *Cryobiology*, 16, 141–151.
Gupta, K.C. (1975) The mechanism of cryohemolysis: by direct observation with the cryomicroscope and the
    electron microscope, *Cryobiology*, 12, 417–426.
Hagedorn, M. and Kleinhans, F.W. (2001) Asymmetry in membrane permeability in zebrafish embryos.
    Cryobiology, 43, 339 (abstract).
Hagedorn, M., Lance, S.L., Fonesca, D.M., Kleinhans, F.W., Artimov, D., Fleischer, R., Hoque, A.T.M.S., and
    Pukazhenthi, B.S. (2002) Altering fish embryos with aquaporin-3: an essential step toward successful
    cryopreservation, *Biol. Reprod.*, 67, 961–966.
Harris, C.L., Toner, M., Hubel, A., Cravalho, E.G., Yarmush, M.L., and Tompkins, R.G. (1991) Cryopreser-
    vation of isolated hepatocytes: intracellular ice formation under various chemical and physical con-
    ditions, *Cryobiology*, 28, 436–444.
Hempling, H.G. (1973) Heats of activation of the exosmotic flow of water across the membrane of leucocytes
    and leukemic cells, *J. Cell. Physiol.*, 81, 1–10.
Hempling, H.G. and White, S. (1984) Permeability of cultured megakaryocytopoietic cells of the rat to dimethyl
    sulfoxide, *Cryobiology*, 21, 133–143.
Henderson, L.J. (1913) *The Fitness of the Environment*, Beacon Press, Boston.
Hendl, A., McGrath, J.J., and Olien, C.R. (1987) On the adhesive interaction between ice and mouse oocytes,
    Cryo-Letters, 8, 334–345.
Henry, M., Noiles, E.E., Gao, D., Mazur, P., and Critser, J.K. (1993) Cryopreservation of human spermatozoa.
    IV. the effects of cooling rate,and warming rate on the maintenance of motility, plasma membrane
    integrity, and mitochondrial function, *Fertil. Steril.*, 60, 911–918.
Hubel, A., Cravalho, E.G., Nunner, B., and Korber, C. (1992) Survival of directionally solidified B-lympho-
    blasts under various crystal growth conditions, *Cryobiology*, 29, 183–198.
Hubel, A., Norman, J., and Darr, T.B. (1999) Cryobiophysical characteristics of genetically modified hemato-
    poetic progenitor cells, *Cryobiology*, 38, 140–153.
Irimia, D. and Karlsson, J.O.M. (2002). Kinetics and mechanism of intercellular ice propagation in a micro-
    patterned tissue construct, *Biophysical J.* 82, 1858–1868.

Ishibashi, K., Sasaki, S., Fushimi, K., Uchida, S., Kuwahara, M., Saito, H., Furukawa, T., Nakajima, K., Yamaguchi, Y., Gogobori, T., and Marumo, F. (1994) Molecular cloning and expression of a member of the aquaporin family with permeability to glycerol and urea in addition to water expressed at the basolateral membrane of kidney collecting duct cells, *Proc. Natl. Acad. Sci. USA*, 91, 6269–6273.

Ishiguro, H. and Rubinsky, B. (1994) Mechanical interactions between ice crystals and red blood cells during directional solidification, *Cryobiology*, 31, 483–500.

Israelachvili, J. and Wennerstrom, H. (1996) Role of hydration and water structure in biological and colloidal interactions, *Nature*, 379, 219–225.

Jackowski, S., Leibo, S.P., and Mazur, P. (1980) Glycerol permeabilities of fertilized and unfertilized mouse ova, *J. Exp. Zool.*, 212, 329–341.

Karlsson, J. (2001) A theoretical model of intracellular devitrfication, *Cryobiology*, 42, 154–169.

Karlsson, J.O.M., Cravalho, E.G., and Toner, M. (1993) Intracellular ice formation: Causes and consequences, *Cryo-Letters*, 14, 323–334.

Karlsson, J.O.M., Cravalho, E.G., and Toner, M. (1994) A model for diffusion-limited ice growth inside biological cells during freezing, *J. Appl. Phys.*, 75, 4442–4455.

Karow, A., Jr. and Pegg, D.E., Eds. (1981) *Organ Preservation for Transplantation*, Marcel Dekker, New York.

Kasai, M., Niwa, K., and Iritani, A. (1980) Survival of mouse embryos frozen and thawed rapidly, *J. Reprod. Fertil.*, 59, 51–56.

Kasai, M., Niwa, K., and Iritani, A. (1981) Effects of various cryoprotective agents on the survival of unfrozen and frozen mouse embryos, *J. Reprod. Fertil.*, 63, 175–180.

Kasai, M., Niwa, K., and Iritani, A. (1982) Survival of rat embryos after freezing, *J. Reprod. Fertil.*, 66, 367–370.

Katkov, I.I. (2000) A two-parameter model of cell membrane permeability for multisolute systems, *Cryobiology*, 40, 64–83.

Kauzmann, W., (1959) *Some factors in the interpretation of protein denaturation, Adv. Protein Chem.*, 14, 1–63.

Kedem, O. and Katchalsky, A. (1958) Thermodynamic analysis of the permeability of biological membranes to non-electrolytes, *Biochem. Biophys. Acta*, 27, 229–246.

Kleinhans, F.W. (1998) Membrane permeability modelling: Kedem-Katchalsky vs a two-parameter formalism, *Cryobiology*, 37, 271–289.

Klotz, I.M. and Franzen, J.S. (1962) Hydrogen bonds between model peptide groups in solution, *J. Am. Chem. Soc.*, 84, 3461–3466.

Knutton, S., Jackson, D., Graham, J.M., Micklem, K.J., and Pasternak, C.A. (1976) Microvilli and cell swelling, *Nature*, 262, 52–54.

Koga, S., Eshigo, A., and Nunomura, K. (1966) Physical properties of cell water in partially dried *Saccharomyces cerevisiae*, *Biophysical J.*, 6, 665–674.

Köseoglu, M., Eroglu, A., Toner, M., and Sadler, K.C. (2001) Starfish oocytes form intracellular ice at unusually high temperatures, *Cryobiology*, 43, 248–259.

Koshimoto, C., Gamliel, E., and Mazur, P. (2000) Effect of osmolality and oxygen tension on the survival of mouse sperm frozen to various temperatures in various concentrations of glycerol and raffinose, *Cryobiology*, 41, 204–231.

Koshimoto, C. and Mazur, P. (2002) Effects of cooling and warming rate to and from −70°C and effects of cooling from −70 to −196°C on the motility of mouse spermatozoa, *Biol. Reprod.*, 66, 1477–1484.

Kruuv, J. (1986) Effects of pre- and post-thaw cell-to-cell contact and trypsin on survival of freeze-thaw damaged mammalian cells, *Cryobiology*, 23, 126–133.

Kuntz, I.D., Brassfield, T.S., Law, G.D., and Purcell, G.V. (1969) Hydration of macromolecules, *Science*, 163, 1329–1331.

Larese, A., Acker, J., Muldrew, K., Yang, H., and McGann, L. (1996) Antifreeze proteins induce intracellular nucleation, *Cryo-Letters*, 17, 175–182.

Leibo, S.P. (1980) Water permeability and its activation energy of fertilized and unfertilized mouse ova, *J. Membrane Biol.*, 53, 179–188.

Leibo, S.P. (1986) Cryobiology: preservation of mammalian embryos, in *Genetic Engineering of Animals*, Evans, J.W. and Hollaender, A., Eds., Plenum, New York, pp. 251–272.

Leibo, S.P., Farrant, J., Mazur, P., Hanna, M.G., Jr., and Smith, L.H. (1970) Effects of freezing on marrow-stem cell suspensions: Interactions of cooling and warming rates in the presence of PVP, sucrose, or glycerol, *Cryobiology*, 6, 315–332.

Leibo, S.P., McGrath, J.J., and Cravalho, E.G. (1978) Microscopic observations of intracellular ice formation in unfertilized mouse ova as a function of cooling rate, *Cryobiology*, 15, 257–271.

Leikin, S., Parsegian, V.A., Rau, D.C., and Rand, R.P. (1993) Hydration forces, *Ann. Rev. Physical Chem.*, 44, 369–395.

Lepock, J.R., Keith, A.D., and Kruuv, J. (1984) Permeability changes in yeast after freeze-thaw damage; comparison to reproductive survival, *Cryo-Letters*, 5, 277–280.

Levin, R.L. (1979) Water permeability of yeast cells at sub-zero temperatures, *J. Membrane Bio.*, 46, 91–124.

Levitt, J. (1966) Winter hardiness in plants, in *Cryobiology*, Meryman, H.T., Ed., Academic Press, London, pp. 495–563.

Lide, D.R., Ed. (1992) *CRC Handbook of Chemistry and Physics*, CRC Press, Boca Raton, FL.

Lin, T.-T., Pitt, R.E., and Steponkus, P.L. (1989) Osmometric behavior of *Drosophila melanogaster* embryos, *Cryobiology*, 26, 453–471.

Liu, C., Benson, C.T., Gao, D., Haag, B.W., McGann, L.E., and Critser, J.K. (1995) Water permeability and its activation energy for individual hamster pancreatic islet cells, *Cryobiology*, 32, 493–502.

Liu, J., Woods, E.J., Agca, Y., Critser, E.S., and Critser, J.K. (2000) Cryobiology of rat embryos II: a theoretical model for the development of interrupted slow freezing procedures, *Biol. Reprod.*, 63, 1303–1312.

Liu, J., Ziegler, M.A.J., Lakey, J.R.T., Woods, E.J., and Critser, J.K. (1997) The determination of membrane permeability coefficients of canine pancreatic islet cells and their application to islet cryopreservation, *Cryobiology*, 35, 1–13.

Loo, D.D.F., Zeuthen, T., Chandy, G., and Wright, E.M. (1996) Transport of water by the Na+/glucose cotransporter, *Proc. Natl. Acad. Sci. USA*, 93, 13367–13370.

Lovelock, J.E. (1953a) The haemolysis of human red cells by freezing and thawing, *Biochem. Biophys. Acta*, 10, 414–426.

Lovelock, J.E. (1953b) The mechanism of the protective effect of glycerol against haemolysis by freezing and thawing, *Biochem. Biophys. Acta*, 11, 28–36.

Luyet, B. and Rasmussen, D. (1967) Studies by differential thermal analysis of the temperatures of instability in rapidly cooled solutions of polyvinylpyrrolidone, *Biodynamica*, 10, 137–147.

Macey, R.I. (1984) Transport of water and urea in red blood cells, *Am. J. Physiol.*, 246 (*Cell Physiol.* 15), C195–C203.

MacFarlane, D.R. and Forsyth, M. (1987) Devitirication and recrystallization of glass forming aqueous solutions, in *The Biophysics of Organ Preservation*, Pegg, D.E. and A.M., Karow, Jr., Eds., Plenum, New York, pp. 237–263.

MacKenzie, A.P. and Luyet, B. (1967) Electron microscope study of recrystallization in rapidly frozen gelatin gels, *Biodynamica*, 10, 95–122.

Marlow, D.C., McGrath, J.J., Sauer, H.J., and Fuller, B.J. (1994) Permeability of frozen and non-frozen mouse oocytes to dimethylsulfoxide, Advances in Heat and Mass Transfer in Biological Systems ASME, 288, 71–79.

Mazur, P. (1961) Manifestations of injury in yeast cells exposed to subzero temperatures, II. Changes in specific gravity and in the concentration and quantity of cell solids, *J. Bacteriol.*, 82, 673–684.

Mazur, P. (1963) Kinetics of water loss from cells at subzero temperatures and the likelihood of intracellular freezing, *J. Gen. Physiol.*, 47, 347–369.

Mazur, P. (1965) The role of cell membranes in the freezing of yeast and other single cells, *Ann. N.Y. Acad. Sci.*, 125, 658–676.

Mazur, P. (1966) Physical and chemical basis of injury in single-celled micro-organisms subjected to freezing and thawing, in *Cryobiology*, Meryman, H.T., Ed., Academic Press, London, pp. 213–215.

Mazur, P. (1968) Survival of fungi after freezing and drying, in *The Fungi*, Ainsworth, G.C. and A.S., Sussman, Eds., Vol. 3, Academic Press, New York, pp. 325–394.

Mazur, P. (1977) The role of intracellular freezing in the death of cells cooled at supraoptimal rates, *Cryobiology*, 14, 251–272.

Mazur, P. (1984) Freezing of living cells: Mechanisms and implications, *Am. J. Physiol.*, 247 (*Cell Physiol.*, 16), C125–142.

Mazur, P. (1990) Equilibrium, quasi-equilibrium, and non-equilibrium freezing of mammalian embryos, *Cell Biophys.*, 17, 53–92.

Mazur, P. and Cole, K.W. (1985) Influence of cell concentration on the contribution of unfrozen fractions and salt concentration to the survival of slowly frozen human erythrocytes, *Cryobiology*, 22, 509–536.

Mazur, P. and Cole, K.W. (1986) Responses of packed and suspended human red cells to hyperosmotic glycerol-NaCl solution at subzero temperatures in the absence of ice formation, *Cryobiology*, 23, 574 (abstract).

Mazur, P. and Cole, K.W. (1988) Contact between shrunken red cells as a factor in freezing injury, *Cryobiology*, 25, 510–511 (abstract).

Mazur, P. and Cole, K.W. (1989) Roles of unfrozen fraction, salt concentration, and changes in cell volume in the survival of frozen human erythrocytes, *Cryobiology*, 26, 1–29.

Mazur, P., Cole, K.W., Hall, J., Schreuders, P.D., and Mahowald, A.P. (1992a) Cryobiological preservation of *Drosophila* embryos, *Science*, 258, 1932–1935.

Mazur, P. and Koshimoto., C. (2002) Is intracellular ice formation the cause of death of mouse sperm frozen at high cooling rates? *Biol. Reprod.*, 66, 1485–1490.

Mazur, P., Leibo, S.P., and Chu, E.H.Y. (1972) A two-factor hypothesis of freezing injury—evidence from Chinese hamster tissue culture cells., *Exp. Cell Res.*, 71, 345–355.

Mazur, P., Leibo, S.P., Farrant, J., Chu, E.H.Y., M.G. Hanna, Jr., and Smith, L.H. (1970) Interactions of cooling rate, warming rate, and protective additive on the survival of frozen mammalian cells, in *The Frozen Cell*, Wolstenholme, G.E.W. and M., O'Connor, Eds., J & A Churchill, London, pp. 69–88.

Mazur, P. and Miller, R.H. (1974) Permeability of the bovine red cell to glycerol in hyperosmotic solutions at various temperatures, *J. Membrane Biol.*, 15, 107–136.

Mazur, P., Miller, R.H., and Leibo, S.P. (1974) Survival of frozen-thawed bovine red cells as a function of the permeation of glycerol and sucrose, *J. Membrane Biol.*, 15, 137–158.

Mazur, P., Rall, W.F., and Leibo, S.P. (1984) Kinetics of water loss and the likelihood of intracellular freezing in mouse ova: influence of the method of calculating the temperature dependence of water permeability, *Cell Biophys.*, 6, 197–214.

Mazur, P., Rall, W.F., and Rigopoulos, N. (1981) Relative contributions of the fraction of unfrozen water and of salt concentration to the survival of slowly frozen human erythrocytes, *Biophysical J.*, 36, 653–675.

Mazur, P. and Rigopoulos, N. (1983) Contributions of unfrozen fraction and of salt concentration to the survival of slowly frozen human erythrocytes: Influence of warming rate, *Cryobiology*, 20, 274–289.

Mazur, P. and Schmidt, J.J. (1968) Interactions of cooling velocity, temperature, and warming velocity on the survival of frozen and thawed yeast, *Cryobiology*, 5, 1–17.

Mazur, P. and Schneider, U. (1986) Osmotic response of preimplantation mouse and bovine embryos and their cryobiological implications, *Cell Biophys.*, 8, 259–284.

Mazur, P., Schneider, U., and Mahowald, A.P. (1992b) Characteristics and kinetics of subzero chilling injury in *Drosophila* embryos, *Cryobiology*, 29, 39–68.

McCaa, C., Diller, K.R., Aggarwal, S.J., and Takahashi, T. (1991) Cryomicroscopic determination of the membrane osmotic properties of human monocytes at subfreezing temperatures, *Cryobiology*, 28, 391–399.

McGann, L.E. (1979) Optimal temperature ranges for control of cooling rate, *Cryobiology*, 16, 211–216.

Meryman, H.T. (1968) Modified model for the mechanism of freezing injury in erythrocytes, *Nature*, 218, 333–336.

Meryman, H.T. (1970) The exceeding of a minimum tolerable cell volume in hypertonic suspension as a cause of freezing injury, in *The Frozen Cell*, Wolstenholme, G.E.W. and M. O'Connor, Eds., J & A Churchill, London, pp. 51–64.

Meryman, H.T. and Douglas, M. (1982) Isotonicity in the presence of penetrating cryoprotectants, *Cryobiology*, 19, 565–569.

Miyamoto, H. and Ishibashi, T. (1983) Survival of mouse embryos frozen-thawed slowly or rapidly in the presence of various cryoprotectants, *J. Exp. Zool.*, 226, 123–127.

Morris, G.J. (1979) The cryopreservation of *Chlamydomonas*, *Cryobiology*, 16, 401–410.

Morris, G.J. and McGrath, J.J. (1981) Intracellular ice nucleation and gas bubble formation in Spirogyra, *Cryo-Letters*, 2, 341–352.

Mugnano, J.A., Wang, T., Layne, J.R. Jr., DeVries, A.L., and Lee, R.E. (1995) Antifreeze glycoproteins promote intracellular freezing of cardiomyocytes at high subzero temperatures, *Am. J. Physiol.*, 269 (*Regulatory Integ. Comp. Physiol.*, 38), R474–R479.

Muldrew, K. and McGann, L.E. (1990) Mechanisms of intracellular ice formation, *Biophysical J.*, 57, 525–532.

Muldrew, K. and McGann, L.E. (1994) The osmotic rupture hypothesis of intracellular freezing injury, *Biophysical J.*, 66, 532–541.

Myers, S.P., Lin, T.-T., Pitt, R.E., and Steponkus, P.L. (1987) Cryobehavior of immature bovine oocytes, *Cryo-Letters*, 8, 260–275.

Myers, S.P., Pitt, R.E., Lynch, D.V., and Steponkus, P.L. (1989) Characterization of intracellular ice formation in *Drosophila melanogaster* embryos, *Cryobiology*, 26, 472–484.

Nei, T. (1981) Mechanism of freezing injury to erythrocytes: effect of initial cell concentration on the post thaw hemolysis, *Cryobiology*, 18, 229–237.

Nemethy, G. and Scheraga, H.A. (1962) Structure of water and hydrophobic bonding in proteins. I. A model for the thermodynamic properties of liquid water, *J. Chem. Phys.*, 36, 3382–3400.

Nemethy, G. and Scheraga, H.A. (1964) Structure of water and hydrophobic bonding in proteins. IV. The thermodynamic properties of liquid deuterium oxide, *J. Chem. Phys.*, 41, 680–689.

Noiles, E.E., Mazur, P., Kleinhans, F.W., and Critser, J.K. (1993) Determination of the water permeability coefficient for human spermatozoa and its activation energy, *Biol. Reprod.*, 48, 99–109.

Noiles, E.E., Thompson, K.A., and Storey, B.T. (1997) Water permeability, L_p, of the mouse sperm plasma membrane and its activation energy are strongly dependent on interaction of the plasma membrane with the sperm cytoskeleton, *Cryobiology*, 35, 79–92.

O'Dell, S.J. and Crowe, J.H. (1979) Freezing in nematodes: the effects of variable water contents, *Cryobiology*, 16, 534–541.

Olien, C.R. and Smith, M.N. (1977) Ice adhesions in relation to freeze stress, *Plant Physiol.*, 60, 499–503.

Paynter, S.J., Fuller, B.J., and Shaw, R.W. (1997) Temperature dependence of mature mouse oocyte membrane permeabilities in the presence of cryoprotectant, *Cryobiology*, 34, 122–130.

Paynter, S.J., Fuller, B.J., and Shaw, R.W. (1999) Temperature dependence of Kedem-Katchalsky membrane transport coefficients for mature mouse oocytes in the presence of ethylene glycol, *Cryobiology*, 39, 169–176.

Pegg, D.E. (1981) The effect of cell concentration on the recovery of human erythrocytes after freezing and thawing in the presence of glycerol, *Cryobiology*, 18, 221–228.

Pegg, D.E. (1984) Red cell volume in glycerol/sodium chloride/water mixtures, *Cryobiology*, 21, 234–239.

Pegg, D.E. and Diaper, M.P. (1983) The packing effect in erythrocyte freezing, *Cryo-Letters*, 4, 129–136.

Pegg, D.E. and Diaper, M.P. (1988) On the mechanism of injury to slowly frozen erythrocytes, *Biophysical J.*, 54, 471–488.

Pegg, D.E. and Diaper, M.P. (1989) The "unfrozen fraction" hypothesis of freezing injury to human erythrocytes: a critical examination of the evidence, *Cryobiology*, 26, 30–43.

Pegg, D.E. and Diaper, M.P. (1991) The effect of initial tonicity on freeze-thaw injury to human red cells suspended in solutions of NaCl, *Cryobiology*, 28, 18–35.

Pegg, D.E., Hunt, C.J., and Fong, L.P. (1987) Osmotic properties of the rabbit-corneal endothelium and their relevance to cryopreservation, *Cell Biophys.*, 10, 169–189.

Phelps, M.J., Liu, J., Benson, J.D., Willoughby, C.E., Gilmore, J.A., and Critser, J.K. (1999) Effects of percoll separation, cryoprotective agents, and temperature on plasma membrane permeability characteristics of murine spermatozoa and their relevance to cryopreservation, *Biol. Reprod.*, 61, 1031–1041.

Pitt, R.E., Chandrasekaran, M., and Parks, J.E. (1992) Performance of a kinetic model for intracellular ice formation based on the extent of supercooling, *Cryobiology*, 29, 359–373.

Pitt, R.E., Myers, S.P., Lin, T.-T., and Steponkus, P.L. (1991) Subfreezing volumetric behavior and stochastic modeling of intracellular ice formation in *Drosophila melanogaster* embryos, *Cryobiology*, 28, 72–86.

Pitt, R.E. and Steponkus, P.L. (1989) Quantitative analysis of the probability of intracellular ice formation during freezing of isolated protoplasts, *Cryobiology*, 26, 44–63.

Pollard, J.W. and Leibo, S.P. (1994) Chilling sensitivity of mammalian embryos, *Theriogenology*, 41, 101–106.

Porche, A.M., Korber, C., English, S., Hartmann, U., and Rau, G. (1986) Determination of the permeability of human lymphocytes with a microscope diffusion chamber, *Cryobiology*, 23, 302–316.

Preston, G.M., Carroll, T.P., Guggino, W.B., and Agre, P. (1992) Appearance of water channels in *Xenopus* oocytes expressing red cell CHIP28 protein, *Science*, 256, 385–387.

Privalov, P.L. (1968) Water and its role in biological systems, *Biophysics* (USSR) [English translation], 13, 189–208.

Rall, W.F. (1987) Factors affecting survival of mouse embryos cryopreserved by vitrification., *Cryobiology*, 24, 387–402.

Rall, W.F., Mazur, P., and McGrath, J.J. (1983) Depression of the ice-nucleation temperature of rapidly cooled mouse embryos by glycerol and dimethyl sulfoxide, *Biophys. J.*, 41, 1–12.

Rall, W.F., Mazur, P., and Souzu, H. (1978) Physical-chemical basis of the protection of slowly frozen human erythrocytes by glycerol, *Biophys. J.*, 23, 101–120.

Rall, W.F. and Polge, C. (1984) Effect of warming rate on mouse embryos frozen and thawed in glycerol, *J. Reprod. Fertil.*, 70, 285–292.

Rall, W.F., Reid, D.S., and Polge, C. (1984) Analysis of slow-freezing injury of mouse embryos by cryomicroscopical and physiochemical methods, *Cryobiology*, 21, 106–121.

Rasmussen, D. and MacKenzie, A.P. (1972) Effects of solute on the ice-solution interfacial free energy; calculation from measured homogeneous nucleation temperatures, in *Water Structure at the Water–Polymer Interface,* Jellinek, H.H.G., Ed., Plenum, New York, pp. 126–145.

Robinson, R.A. and Stokes, R.H. (1959) *Electrolyte Solutions*, Academic Press, New York.

Rosen, D. (1963) Dielectric properties of protein powders with adsorbed water, *Trans. Faraday Soc.*, 59, 2178–2191.

Ruffing, N.A., Steponkus, P.L., Pitt, R.E., and Parks, J.E. (1993) Osmometric behavior, hydraulic conductivity, and incidence of intracellular ice formation in bovine oocytes at different developmental stages, *Cryobiology*, 562–580.

Rule, G.S., Law, P., Kruuv, J., and Lepock, J.R. (1980) Water permeability of mammalian cells as a function of temperature in the presence of dimethylsulfoxide: correlation with the state of the membrane lipids, *J. Cellular Physiol.*, 103, 407–416.

Santarius, K.A. and Giersch, C. (1983) Cryopreservation of spinach chloroplast membranes by low-molecular weight carbohydrates, *Cryobiology*, 20, 90–99.

Scatena, L.F., Brown, M.G., and Richmond, G.L. (2001) Water at hydrophobic surfaces: weak hydrogen bonding and strong orientation effects, *Science,* 292, 908–912.

Scheiwe, M.W. and Körber, C. (1983) Basic investigations on the freezing of human lymphocytes, *Cryobiology*, 20, 257–273.

Scheiwe, M.W., Nick, H.E., and Körber, C. (1982) An experimental study on the freezing of red blood cells with and without hydroxyethyl starch, *Cryobiology*, 19, 461–477.

Schneider, U., and Mazur, P. (1987) Relative influence of unfrozen fraction and salt concentration on the survival of slowly frozen 8-cell mouse embryos, *Cryobiology*, 24, 17–41.

Schreuders, P.D., Smith, E.D., Cole, K.W., Valencia, M.-P., Laughinghouse, A., and Mazur, P. (1996) The characterization of intraembryonic freezing in *Anopheles gambiae* embryos, *Cryobiology*, 33, 487–501.

Schwartz, G.J. and Diller, K.R. (1983) Osmotic response of individual cells during freezing II. Membrane permeability analysis, *Cryobiology*, 20, 542–552.

Sha'afi, R.I., Rich, G.T., Sidel, V.W., Bossert, W., and Solomon, A.K. (1967) The effect of the unstirred layer on human red cell water permeability, *J. Gen. Physiol.*, 50, 1377–1399.

Shabana, M. and McGrath, J.J. (1988) Cryomicroscope investigation and thermodydamic modeling of the freezing of unfertilized hamster ova, *Cryobiology*, 25, 338–354.

Shepard, M.L., Goldston, C.S., and Cocks, F.H. (1976) The H2O-NaCl-glycerol phase diagram and its application in cryobiology, *Cryobiology*, 13, 9–23.

Singer, S.J. and Nicholson, G.L. (1972) The fluid mosaic model of the structure of cell membranes, *Science*, 175, 720–731.

Souzu, H. and Mazur, P. (1978) Temperature dependence of the survival of human erythrocytes frozen slowly in various concentrations of glycerol, *Biophys. J.*, 23, 89–100.

Souzu, H., Nei, T., and Bito, M. (1961) Water of microorganisms and its freezing, with special reference to the relation between water content and viability of yeast and coli cells, *Low Temp. Sci., Ser. B*, 19, 49–57.

Stachecki, J.J., Cohen, J., and Willadsen, S. (1998) Detrimental effects of sodium during mouse oocyte cryopreservation, *Biol. Reprod.*, 59, 393–400.

Stachecki, J.J. and Willadsen, S.M. (2000) Cryopreservation of mouse oocytes using a medium with low sodium content: effect of plunge temperature, *Cryobiology*, 40, 4–12.

Stein, W.D. (1967) *The Movement of Molecules across Cell Membranes*, Academic Press, New York.

Stein, W.D. (1986) *Transport and Diffusion across Cell Membranes*, Academic Press, New York.

Steponkus, P.L., Dowgert, M.F., and Gordon-Kamm, W.J. (1983) Destabilization of the plasma membrane of isolated plant protoplasts during a freeze-thaw cycle: the influence of cold acclimation, *Cryobiology*, 20, 448–465.

Steponkus, P.L. and Gordon-Kamm, W.J. (1985) Cryoinjury of isolated protoplasts: a consequence of dehydration or the fraction of the suspending medium that is frozen? *Cryo-Letters*, 6, 217–226.

Steponkus, P.L. and Lynch, D.V. (1989) Freeze/thaw induced destabilization of the plasma membrane and the effects of cold acclimation, *J. Bioenergetics Biomembranes*, 21, 21–41.

Steponkus, P.L., Myers, S.P., Lynch, D.V., Gardner, L., Bronshteyn, V., Leibo, S.P., Rall, W.F., Pitt, R.E., Lin, T.-T., and MacIntyre, R.J. (1990) Cryopreservation of *Drosophila melanogaster* embryos, *Nature*, 345, 170–172.

Sui, H., Han, B.-G., Lee, J.K., Wallan, P., and Jap, B.K. (2001) Structural basis of water-specific transport through the AQP1 water channel, *Nature*, 414, 872–878.

Sun, W.Q. (1999) State and phase transition behaviors of *Quercus rubra* seed axes and cotyledonary tissues: relevance to the desiccation sensitivity and cryopreservation of recalcitrant seeds, *Cryobiology*, 38, 372–385.

Taylor, M.J., Bank, H.L., and Benton, M.J. (1987) Selective destruction of leucocytes by freezing as a potential means of modulating tissue immunogenicity: membrane integrity of lymphocytes and macrophages, *Cryobiology*, 24, 91–102.

Taylor, M.J. and Pegg, D.E. (1983) The effect of ice formation on the function of smooth muscle tissue stored at −21 or −60°C, *Cryobiology*, 20, 36–40.

Tenchini, M.L., Bolognani, L., and Carli, L.D. (1980) Effect of hypertonicity on survival of unprotected human cultured cells following freezing and thawing, *Cryobiology*, 17, 120–124.

Terwilliger, T.C. and Solomon, A.K. (1981) Osmotic water permeability of human red cells, *J. Gen. Physiol.*, 77, 549–570.

Tondorf, S., McGrath, J.J., and Olien, C.R. (1987) On the adhesive interaction between ice and cell-size liposomes, *Cryo-Letters*, 8, 322–333.

Toner, M., Cravalho, E.G., and Karel, M. (1990) Thermodynamics and kinetics of intracellular ice formation during freezing of biological cells, *J. Appl. Phys.*, 67, 1582–1593.

Toner, M., Cravalho, E.G., and Karel, M. (1993a) Cellular response of mouse oocytes to freezing stress: prediction of intracellular ice formation, *J. Biomechanical Eng.*, 115, 169–174.

Toner, M., Cravalho, E.G., Karel, M., and Armanti, D.R. (1991) Cryomicroscopic analysis of intracellular ice formation during freezing of mouse oocytes with cryoaddtives, *Cryobiology*, 28, 55–71.

Toner, M., Cravalho, E.G., Stacheki, J., Fitzgerald, T., and Tomkins, R.G. (1993b) Nonequilibrium freezing of one-cell mouse embryos. Membrane integrity and developmental potential, *Biophysical J.*, 64, 1908–1920.

Van Den Abbeel, E., Van Der Elst, J., and Van Steirteghem, A.C. (1994) The effect of temperature at which slow cooling is terminated and of thawing rate on the survival of one-cell mouse embryos frozen in dimethyl sulfoxide or 1,2-propanediol solutions, *Cryobiology*, 31, 423–433.

Van Venrooij, G.E.P.M., Aertsen, A.M.H.J., Hax, W.M.A., Ververgaert, P.H.J.T., Verhoeven, J.J., and Vorst, H.A.V.D. (1975) Freeze-etching: freezing velocity and crystal size at different locations in samples, *Cryobiology*, 12, 46–61.

Verkman, A.S. (2000) Water permeability measurement in living cells and complex tissues, *J. Membrane Biology*, 173, 73–87.

Verkman, A.S., Hoek, A.N.V., Ma, T., Frigeri, A., Skach, W.R., Mitra, A., Tamarappoo, B.K., and Farinas, J. (1996) Water transport across mammalian cell membranes, *Am. J. Physiol.*, 270 (*Cell Physiol.* 39), C12–C30.

Watson, P.F. and Duncan, A.E. (1988) Effect of salt concentration and unfrozen water fraction on the viability of slowly frozen ram spermatozoa, *Cryobiology*, 25, 131–142.

Wellman, A.M. and Pendyala, L. (1979) Permeability changes in membranes of *Neurospora crassa* after freezing and thawing, *Cryobiology*, 16, 184–195.

Wells, J.R., Sullivan, A., and Cline, M.J. (1979) A technique for the separation and cryopreservation of myeloid stem cells from human bone marrow, *Cryobiology*, 16, 201–210.

Wesley-Smith, J., Walters, C., Pammenter, N.W., and Berjak, P. (2001) Interactions among water content, rapid (nonequilibrium) cooling to −196°C and survival of embryonic axes of *Aesculus hippocastanum* L. seeds, *Cryobiology*, 42, 196–206.

Whittingham, D.G., Leibo, S.P., and Mazur, P. (1972) Survival of mouse embryos frozen to −196 and −269 C., *Science*, 178, 411–414.

Whittingham, D.G., Wood, M., Farrant, J., Lee, H., and Halsey, J.A. (1979) Survival of frozen mouse embryos after rapid thawing from −196°C, *J. Reprod. Fertil.*, 36, 11–21.

Wiest, S.C. and Steponkus, P.L. (1979) The osmometric behavior of human erythrocytes, *Cryobiology*, 16, 101–104.

Wilson, P.W., Arthur, J.W., and Haymet, A.D.J. (1999) Ice premelting during differential scanning calorimetry, *Biophys. J.*, 77, 2850–2855.

Woelders, H., Matthijs, A., and Engel, B. (1997) Effects of trehalose and sucrose, osmolality of the freezing medium, and cooling rate on viability and intactness of bull sperm after freezing and thawing, *Cryobiology*, 35, 93–106.

Wolfe, J. and Bryant, G. (1999) Freezing, drying, and/or vitrification of membrane-solute-water systems, *Cryobiology*, 39, 103–129.

Wood, T.H. and Rosenberg, A.M. (1957) Freezing in yeast cells, *Biochem. Biophys. Acta*, 25, 78–87.

Woods, E.J., Liu, J., Gilmore, J.A., Reid, T.J., Gao, D.Y., and Critser, J.K. (1999) Determination of human platelet membrane permeability coefficients using Kedem-Katchalsky formalism: estimates from two- vs three-parameter fits, *Cryobiology*, 38, 200–208.

Zade-Oppen, A.M.M. (1968) Posthypertonic hemolysis in sodium chloride systems, *Acta Physic. Scand.*, 73, 341–364.

Zeuthen, T. and Klaerke, D.A. (1999) Tranport of water and glycerol in Aquaporin 3 is gated by H+, *Proc. Natl. Acad. Sci. USA*, 274, 21631–21636.

2 The Water to Ice Transition: Implications for Living Cells

Ken Muldrew, Jason P. Acker, Janet A.W. Elliott, and Locksley E. McGann

CONTENTS

2.1 INTRODUCTION

When liquid water is cooled, a temperature is reached at which the solid phase, ice, becomes stable. That is to say that an ice crystal, when added to the cold water, will not melt. If the liquid is below that fusion temperature, the ice crystal will, in fact, grow; water molecules will leave the jostling, random walk of the liquid phase and join the fixed lattice of the ice crystal. When the pure liquid is cooled below the fusion temperature without any external ice being added, however, the solid phase does not necessarily appear. There is a temperature range in which the liquid is metastable; the presence of a seed crystal (or other nucleating enhancement) is required to initiate the phase transition. In this region there are clusters of water molecules joining together to form networks that extend the correlation length between water molecules beyond the scale of a few molecules. As these networks grow, the number of possible paths for continued growth increases exponentially, but the correlation length also decays exponentially because of collisions that break the network. Only the latter phenomenon is dependent on temperature, so there is a critical temperature at which the multiplicity of growth paths can overwhelm the collision-based fractures. Below this temperature a nucleation event can occur whereby the length scale of a particular network exceeds the length scale of the destructive collisions that break these extended clusters. The growing network becomes an ice crystal.

All living organisms primarily consist of water, so their response to freezing is necessarily dramatic. Though humans suffer profound injury when any part of their body freezes, we see an enormous number of organisms that can withstand the harshest winter climate, and therefore we know that life can withstand these conditions. Our investigation of this phenomenon is not merely out of curiosity, either, as there is an enormous practical benefit to being able to store living tissues at low temperatures. Life exists far from equilibrium and will therefore decay quickly without an expensive energy throughput to maintain itself. A steady state that avoids decay can be established, with respect to the timescale that is important for living organisms, by lowering the temperature. The clocks that determine the timescale for life are biochemical reactions, and the rate at which these reactions proceed is controlled by an energy barrier. Though the energy balance favors the forward reaction, the reactants must overcome an intermediate energy barrier before the reaction proceeds. The simplest way to slow the rate with which these reactions occur, and thereby slow biological time, is to lower the free energy of the reactants by lowering the temperature. Indeed, it would be as easy as that were it not for the troublesome tendency of water to form ice when it is cooled. We will look at some of the problems that biological systems encounter that are caused by the water to ice transition. The topic is far too broad to cover in this chapter, and thus we will concentrate on our own investigations that have centered around the direct observation of the freezing process under a microscope, giving only a brief background to set this work in context and to introduce the conceptual tools that have been used to design experiments.

2.2 WATER TO ICE

2.2.1 AQUEOUS SOLUTIONS

Solution thermodynamics describes a variety of processes that occur when a cell suspension is subjected to lowering temperatures. Chemical potential is the thermodynamic property of a solution that governs many of these processes. The chemical potential of a solution is, in general, a function of pressure, temperature, and concentration. If a solution is assumed to be thermodynamically ideal and dilute, then the chemical potential at a given temperature and pressure depends only on the concentration of solute, not on the type of solute. For many biologically relevant solutions, however, both the concentration and types of solutes are important.

At equilibrium, the chemical potential of ice is equal to the chemical potential of the solution with which the ice is in contact. This gives us an equation for the freezing point as a function of solution concentration (freezing point depression). For an ideal, dilute solution the freezing point depression is given by

$$\Delta T = \left(\frac{RT_m^2}{\Delta H_{fus}} \right) X_B \tag{2.1}$$

where ΔT is the freezing point depression, T_m is the melting point of the pure solvent, ΔH_{fus} is the enthalpy of fusion of the solvent, and X_B is the mole fraction of the solute. For a solution that is not ideal and dilute, the relationship between equilibrium temperature and mole fraction will not be linear. As an aqueous solution is cooled, ice will begin to form when the extracellular solution reaches its freezing point (if a nucleation event occurs). As ice forms, the pure crystal excludes

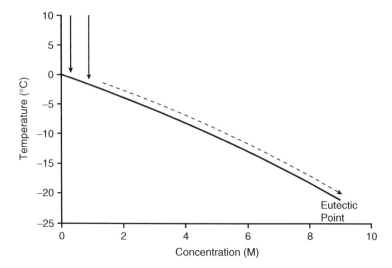

FIGURE 2.1 The liquidus curve for the NaCl–water phase diagram is shown. The arrows indicate starting concentrations of isotonic (0.3 *M*) and 3× isotonic (0.9 *M*). If the temperature of these solutions is dropped, starting at room temperature, the composition of the solution remains the same until the liquidus curve is reached. Any further lowering of the temperature will follow this curve until the eutectic is reached at −21.2°C.

solutes that concentrate in the residual liquid, further lowering the freezing point. The dotted line in Figure 2.1 shows the path of a simple physiological solution (0.15 *M* NaCl) during cooling and warming. Starting at a temperature above freezing, the solution concentration remains constant as the temperature is lowered. Freezing can occur when the sample temperature goes below the melting point of the solution (−0.55°C). As the temperature is further lowered, more ice is formed and the concentration of NaCl in the liquid phase increases. In the presence of ice, the phase diagram in Figure 2.1 shows that the composition of the liquid phase is dependent only on temperature, not on the concentration in the initial sample. Ice continues to form with decreasing temperature until the sample reaches the eutectic point at which the entire system solidifies and there is no further change in composition. During warming, the sample retraces the path, diluting the liquid phase as ice melts. The unfrozen solution in this sample would therefore develop highly concentrated solutions (up to 9 *M*) at subzero temperatures. The nonlinear curve shown in Figure 2.1 indicates that a sodium chloride solution cooled to its eutectic cannot be considered ideal and dilute.

2.2.2 OSMOTIC RESPONSE OF CELLS

At equilibrium, the chemical potential of the intracellular solution is equal to the chemical potential of the extracellular solution. Because the cell membrane is effectively semipermeable (impermeant to many of the solutes that determine the chemical potential but permeant to water and some added cryoprotectants), water will move in or out of the cell in response to changes in solution concentration outside the cell. With freezing, the extracellular solution becomes more concentrated and water therefore leaves the cell until the gradient in chemical potential is neutralized, establishing "osmotic" equilibrium across the plasma membrane. This new equilibrium results from the concentration of the intracellular solutes. The equilibrium cell volume, *V*, is a function of the extracellular solute concentration and has been traditionally given by the Boyle–van't Hoff equation:

$$V = \pi_c^0 \left(V_o - b \right) \frac{1}{\pi_e} + b , \qquad (2.2)$$

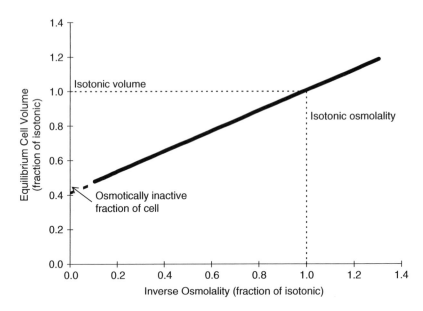

FIGURE 2.2 The Boyle–van't Hoff plot for cell volumes as a function of inverse osmolality. The graph uses an osmotically inactive fraction of 0.41, a value reported for bovine chondrocytes (From McGann et al., 1988. With permission.).

where π_e is the osmolality of the extracellular solution, b is the osmotically inactive volume of the cell and V_o and π_e^0 are the cell volume and the extracellular osmolality, respectively, for the isotonic condition. Note that the Boyle–van't Hoff equation is derived by assuming that the cell contents form a thermodynamically ideal, dilute solution; it is not valid for solutions that are not ideal and dilute. In addition, osmolality in the equation should not be replaced by osmotic pressure if the equation is to be used at a variety of temperatures. The Boyle–van't Hoff plot in Figure 2.2 shows the relative cell volume as a function of inverse osmolality. Extrapolation of this line to infinite osmolality (to zero on the abscissa) has been used to estimate the osmotically inactive fraction of the cell volume. If the cell contents were ideal, dilute solutions, the value of this parameter would represent the proportion of the total cell volume that does not participate in osmotic activity. In real cells, the solutions are not ideal and thus the physical interpretation may be misleading.

When cells are exposed to an increased concentration of an effectively impermeant solute, as occurs when ice forms in a physiological solution, the cell will shrink to a volume defined by the Boyle–van't Hoff plot in Figure 2.2. At the temperatures at which ice forms, the cells do not have the metabolic capacity to alter their intracellular osmotic environment, and thus the cells are essentially ideal osmometers.

The rate of movement of water across a membrane is limited by the permeability properties of the membrane. If there is a chemical potential gradient across the membrane, the rate of change of cell volume will be determined by this gradient and by the permeability of the membrane. Figure 2.3 shows the cell volume as a function of time after exposure to, and dilution from, a solution containing an impermeant solute. The rate at which the cell loses water is a function of the magnitude of the osmotic gradient and the permeability of the plasma membrane (the area of the membrane and its conductivity to water) and is given by:

$$\frac{dV_W}{dt} = L_p ART \left(\pi_i - \pi_e \right), \tag{2.3}$$

where V_W is the water volume of the cell, t is time, L_p is the hydraulic conductivity, A is the membrane surface area, R is the gas constant, T is the absolute temperature, and π_i is the intracellular osmolality.

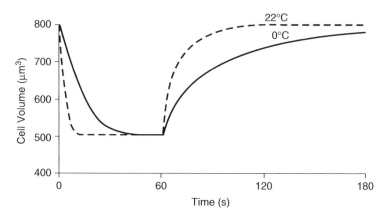

FIGURE 2.3 Cell volume changes caused by water movement are shown for hamster fibroblast cells placed into a 3× isotonic salt solution and then returned to isotonic. The dashed line shows the response at 22°C, and the solid line is at 0°C.

Water movement through the membrane occurs through aqueous pores created by hollow protein cylinders that are embedded in the membrane as well as directly by diffusion through the lipid bilayer (Finkelstein, 1987). There are a variety of water transport pores, known as aquaporins, and the type and quantity of these pores in a given cell type can significantly alter the water permeability (Verkman et al., 1996). Even in cells with aquaporins, though, water will continue to move through the bilayer by a solubility/diffusion mechanism; this permeability will be determined by the lipid composition of the bilayer. In both of these cases permeability will be temperature dependent because of the temperature dependence of water self-diffusion, for both aquaporins and solubility/diffusion, and because of the fluidity of the bilayer lipids in the case of solubility/diffusion (Elmoazzen et al., 2002). Because both processes are caused by diffusion of water, the rate will be proportional to the osmotic gradient across the plasma membrane (but limited by the permeability of the membrane, a property that is different for each cell type). The temperature dependence of the kinetics of water loss is demonstrated in Figure 2.3 by the different curves at 22° and 0°C, though the equilibrium volume depends solely on the solution osmolality. The difference between the two temperatures arises from the explicit temperature dependence of Equation 2.3 and the temperature dependence of L_p (described by an Arrhenius relation). On dilution back to isotonic conditions, the cell volume simply returns to its isotonic value according to the same equation.

In the presence of a permeant solute (such as Me$_2$SO, glycerol, propylene glycol, or many other cryoprotectants) there is both water and solute movement, and there may be an interaction between these fluxes in the membrane. In a situation in which there is an osmolality difference caused by a solute that cannot permeate the membrane as well as a solute that can permeate the membrane, we need to account for the flow of this penetrating solute as well. The equations to describe such coupled transport have been derived from nonequilibrium thermodynamics by Kedem and Katchalsky (1958). We will introduce a simplified form of the K-K equations given by Johnson and Wilson (1967) so that the interested reader can reproduce the simulations described elsewhere in the chapter.

Johnson and Wilson wrote the water flux equation in terms of both solutes as

$$\frac{dV_W}{dt} = L_p ART \left[\left([I]_i - [I]_e \right) + \sigma \left([S]_i - [S]_e \right) \right], \tag{2.4}$$

where $[I]_i$ is the intracellular concentration of impermeant solute and $[I]_e$ is the extracellular concentration of impermeant solute, $[S]_i$ is the concentration of permeant solute inside the cell, and $[S]_e$ is the

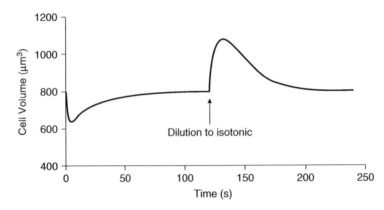

FIGURE 2.4 Cell volume changes caused by water and solute movement are shown for hamster fibroblast cells placed into a 1 M Me$_2$SO solution and then diluted out to isotonic at 22°C.

concentration outside the cell (all concentrations are expressed as osmolalities). The reflection coefficient, σ (on the interval 0–1), was introduced to account for the interaction of solute and solvent.

Johnson and Wilson gave the solute flux equation as

$$\frac{dS}{dt} = P_s A\left([S]_e - [S]_i\right) + (1-\sigma)\overline{[S]}\frac{dV_w}{dt}, \qquad (2.5)$$

where S is the number of moles of permeant solute, P_s is the solute permeability of the membrane, and $[S]$ is the average concentration of permeant solute within the membrane:

$$\overline{[S]} = \frac{[S]_e + [S]_i}{2}. \qquad (2.6)$$

Because there are a number of assumptions made in using Equation 2.5, it should be recognized that P_s will, in general, be a function of permeant solute concentration; the importance of this concentration dependence needs to be checked for particular circumstances.

These two Equations (2.4 and 2.5) are often referred to as the coupled transport equations and can be used to describe the osmotic behavior of a cell with a permeant cryoprotectant present. They are based on the more general coupled transport equations derived from nonequilibrium thermodynamics (Kedem and Katchalsky, 1958).

The advantage of the thermodynamic approach is that the parameters, L_p, P_s, and σ, are not tied to a physical model of membrane transport. For cryobiology, in which extrapolation of osmotic behavior is the dominant concern, rather than membrane biophysics, it is preferable to use the most general formalism. In addition, there are some inconsistencies in the model involving the definition of σ. Thus the physical interpretation of the parameters used in the above model, although pedagogically helpful, should be used with caution unless careful experiments are performed to distinguish physical mechanisms. The interested reader is directed to the review by Kleinhans (1998), in which these issues are discussed, and a two-parameter formalism (i.e., without σ) is reintroduced that may be sufficient for the demands of cryobiological osmotic modeling.

Figure 2.4 shows the typical transient volume changes that will occur in a cell during addition and dilution of a permeant cryoprotectant. The magnitude and duration of the transient volume excursions are dependent on the osmotic properties of the cells, on the concentrations of solutes, and on the temperature. Not only may these transients exceed tolerable limits of cell volume and cell surface increments, particularly during dilution but they also expose cells to high osmotic

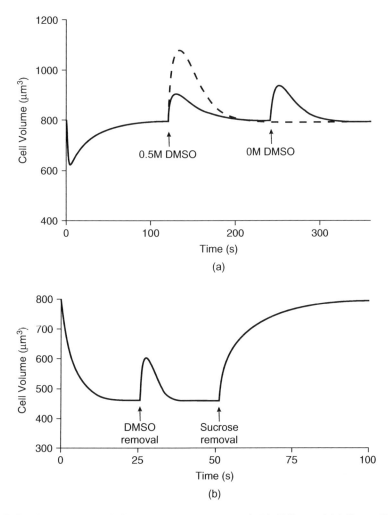

FIGURE 2.5 Cell volume changes during cryoprotectant removal. (a) Cells are initially equilibrated in a 1 *M* Me$_2$SO solution, then transferred into 0.5 *M* Me$_2$SO at 120 sec and later into 0 *M* Me$_2$SO at 240 sec. The dashed line shows the volume change if the cells are placed directly into a 0 *M* Me$_2$SO solution at 120 sec. (b) Cryoprotectant removal in a sucrose solution. Initially the cells are in equilibrium in a 1 *M* Me$_2$SO solution and are placed into a solution with 1 *M* sucrose and 1 *M* Me$_2$SO. At 26 sec, the cells are placed into a solution with 1 *M* sucrose and no Me$_2$SO. At 52 sec, the cells are placed into an isotonic solution with neither sucrose nor Me$_2$SO.

gradients across the membrane and high water flux through the membrane, which may also be damaging (Muldrew and McGann, 1990). Strategies to reduce osmotic stresses during removal of cryoprotectants include stepwise dilution and sucrose dilution. In stepwise dilution, the concentration of permeant solute is progressively reduced, allowing enough time between steps for the cell to return to an equilibrium volume to avoid exceeding the osmotic tolerance of the cells. Figure 2.5a compares the dilution of a 1 *M* Me$_2$SO solution in a single step vs. two steps; the maximum cell volume is substantially lower in the two-step removal. Alternatively, a sucrose dilution (Figure 2.5b) is often used for rapid removal of the cryoprotectant (important if the cryoprotectant exhibits some toxicity) while remaining within the osmotic tolerance of the cells. Sucrose, an impermeant solute, is added to the suspending solution to counteract the water influx that occurs during removal of the permeant solutes from the extracellular solution. The presence of the impermeant solute in the extracellular solution maintains volume excursions and osmotic stresses within tolerable limits during efflux of the permeant solute from the cells. The impermeant solute is then removed in a subsequent single step and the cell returns to its isotonic volume.

2.2.3 MEASUREMENT OF OSMOTIC RESPONSES

Determination of osmotic properties requires measurement of the equilibrium and nonequilibrium changes in cell volume as a function of the osmolality of the suspending solution. Fitting experimental data to theoretical predictions yields estimates of osmotic permeability coefficients of the plasma membrane to water and various solutes. Since the early 1950s there have been a number of mathematical models developed that allow researchers to determine the osmotic parameters of many cell types (Davson and Danielli, 1952; Kedem and Katchalsky, 1958; Johnson and Wilson, 1967; Kleinhans, 1998; Mazur, 1963a). The classic approach has been to apply the methods of Kedem and Katchalsky (1958) who followed Staverman (1948) and Kirkwood (1954) in their use of irreversible thermodynamics. This approach to determine the osmotic parameters of biological membranes has since been revised and applied to the conditions occurring during cryopreservation (Kleinhans, 1998; Mazur, 1963a).

Over the last 50 years there have been a number of techniques developed to measure changes in cell volume that occur during exposure to hypo- and hypertonic solutions. These techniques include stopped-flow spectrophotometers (Boroske et al., 1981; Terwilliger and Solomon, 1981), diffusion and perfusion chambers (McGrath, 1985; Walcerz and Diller, 1991; Woods et al., 1997), and electronic particle counters (Armitage and Juss, 1996; Gilmore et al., 1998; McGann et al., 1982). The spectrophotometric method was developed exclusively for the determination of red blood cell volumes during exposure to hypo- and hypertonic solutions. The linear relationship between the volume of a cell and the intensity of light scattered by the cell suspension is used as a basis for monitoring cell volume. Although this technique was widely used, there were a number of inherent problems that limited its widespread application. The indirect nature of the measurement of the average cell sizes coupled with the requirement for large sample volumes and the considerable shear stresses produced during the procedure adversely affected the use of stopped-flow spectrophotometers. Microscopic diffusion chambers proved to be much more versatile for the determination of cell volume changes.

Diffusion chambers have been widely used for the measurement of cell volume changes since the early 1940s. Many innovative designs have been documented that allow cell suspensions to be viewed using a standard light microscope while the extracellular media is rapidly changed (Aggarwal et al., 1984; Gao et al., 1996; McGrath, 1985; Walcerz and Diller, 1991; Woods et al., 1997). Changes in the physical dimensions of individual cells (typically cross-sectional area) are used to calculate the change in cell volume. By observing and measuring this response, calculations can be made to determine cellular osmotic parameters and critical osmotic limits, though the technique cannot be used to measure solute permeabilities.

Electronic particle counters have also been used to determine cell membrane permeability characteristics (Acker et al., 1999b; Armitage and Juss, 1996; Buckhold et al., 1965; Gao et al., 1998; Gilmore et al., 1998). On the basis of the principle developed by Coulter, cells traversing an aperture in an electronic particle counter displace a volume of conducting fluid proportional to the volume of the cell (Adams et al., 1967; Gregg and Steidley, 1965; Hurley, 1970; Kubitschek, 1958). This displacement results in an electronic pulse that allows for the determination of cell concentration (pulse count) and cell volume (pulse height). Interfacing an electronic particle counter with a computer allows one to track the sequence of pulses and to perform detailed studies on the dynamic osmotic response (McGann et al., 1982). Electronic particle counters provide a rapid, reproducible means to collect the equilibrium and kinetic cell volume data needed to perform osmotic modeling and to extract cell membrane permeability parameters. Although diffusion chambers and electronic particle counters have inherent advantages and disadvantages, both techniques are useful in studies on the osmotic response of cells (Acker et al., 1999b).

2.2.4 CELLULAR OSMOTIC BEHAVIOR AT LOW TEMPERATURES

During cooling of cells in the presence of ice, there are two kinetic processes—the growth of ice and the loss of water from the cell both occur with characteristic rates. These are the primary determinants of osmotic responses during freezing for isolated cells in bulk solution. The rate at which cells lose water during freezing is strongly dependent on the particular cooling profile. If cooling proceeds at a constant rate, then a slow cooling rate will allow the cell to remain close to osmotic equilibrium by losing water at the same rate that water is being lost from the extracellular solution (to form ice). A rapid cooling rate will result in water being converted to ice in the extracellular solution at a faster rate than the cell can lose water (because of the permeability barrier); thus the osmolality gradient across the plasma membrane continues to increase as cooling proceeds.

The differing osmotic responses at different cooling rates can be simulated using the cell permeability model. In bulk samples, the rate of freezing must be considered because the latent heat of fusion is often liberated faster than the cooling apparatus can remove the heat. This is mainly a problem at the time of nucleation after supercooling below the freezing point, when the majority of the water turns to ice. For our simulation, we will assume that the latent heat is completely removed and thus the composition of the extracellular solution is given by the phase diagram. At low rates of cooling, the cells are able to lose water through the membrane fast enough to maintain chemical equilibrium across the cell membrane; therefore, they become progressively dehydrated during exposure to solutions of increasing concentrations as the temperature is lowered (Mazur, 1963a); an example is shown in Figure 2.6a. At higher rates of cooling, the rate of ice growth is faster than cellular water loss, and the cytoplasm becomes increasingly supercooled. The amount of supercooling is defined as the difference in temperature between the melting point of a solution and its actual temperature. The probability of nucleation in a solution increases as the amount of supercooling increases. Figure 2.6b shows the amount of supercooling in cells as a function of temperature during cooling at different rates. Increasing the cooling rate increases the amount of supercooling, which correlates with an increase in the likelihood of intracellular freezing.

Cellular osmotic responses have significant consequences for the survival of cells following a freeze–thaw cycle. Cells are damaged at rapid and at slow cooling rates, so an optimal cooling rate exists between these two in which cell survival is maximal (Mazur, 1984). Because "rapid" and "slow" cooling are relative terms, related to the water permeability of a given cell type, it is clear that the optimal cooling rate will also be specific to a given cell type. The current understanding of the cellular response to freezing is that there are two competing mechanisms of injury; one that occurs during slow cooling but that can be avoided by increasing the cooling rate, and another that occurs during rapid cooling but that can be avoided by decreasing the cooling rate (Mazur et al., 1972). Slow-cooling injury has been shown to accumulate with exposure to high concentrations of solutes (though it manifests itself on dilution to isotonic; Lovelock, 1953b) and is often referred to as "solution-effects" injury, whereas rapid cooling injury is associated with the formation of intracellular ice.

That rapid cooling injury should be intimately related to the water permeability of a particular cell type follows from the fact that the cytoplasm has to be supercooled before intracellular ice can form. That solution-effects injury should also be related to a cell's water permeability may be related to the temperature at which a cell is exposed to concentrated solutes; the reduced rate of injury at low temperatures may prevent any accumulation of damage from occurring. Although this explanation provides an understanding of why solution-effects injury would not occur at rapid cooling rates, it does not explain the curious relationship between the cooling rates at which this injury declines and the water permeability of the cell. Indeed, over several orders of magnitude of water permeabilities, the optimal cooling rate for a particular cell is directly related to the cell's water permeability (Mazur, 1984). If the two types of injury were truly independent, then we would expect to see some cells with a broad plateau at the optimum, whereas other cells would show

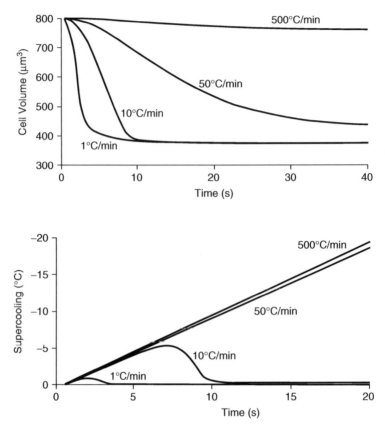

FIGURE 2.6 (a) Cell volume changes are shown for various constant cooling rates. (b) Supercooling of the cytoplasm is shown for the same cooling rates (note that the timescale is changed for clarity).

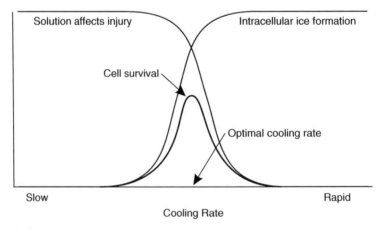

FIGURE 2.7 Overlapping injury mechanisms that occur at slow cooling rates (solution effects injury) and at rapid cooling rates (intracellular ice formation) are shown with the resulting cell survival vs. cooling rate curve.

overlapping injuries with no survival at any cooling rate. Instead we always see an optimal cooling rate that falls away sharply on both sides (Figure 2.7). Thus, both rapid- and slow-cooling injury appear to be related to the presence of an osmotic pressure gradient across the plasma membrane, although other factors such as temperature are clearly important. Nevertheless, this indicates that a reduction in the osmotic pressure gradient should result in less injury at any given cooling rate.

2.2.5 TWO-STEP AND NONLINEAR PROTOCOLS

During cryopreservation, an optimal cooling rate is normally selected to avoid intracellular ice formation while minimizing the deleterious effects of concentrated solutes. Constant cooling and warming rates are therefore often mistakenly considered as fundamental variables in cryobiology. An alternate strategy to minimize damage is the "two-step method" for cooling (Farrant et al., 1977; McGann and Farrant, 1976a, 1976b), in which samples are cooled rapidly to a constant subzero temperature and held at that temperature to allow sufficient osmotic water efflux to avoid intracellular freezing on the subsequent rapid cooling to a low storage temperature ($-196°C$). This approach allows adequate osmotic dehydration at an intermediate temperature (Farrant et al., 1977), which reduces the probability of intracellular nucleation. A subsequent rapid cooling step then minimizes damage caused by exposure to the concentrated solutes at lower temperatures.

2.3 CRYOPROTECTIVE ADDITIVES

Many chemicals have now been identified as having a cryoprotective action; by adding these cryoprotective additives (CPAs) to a cell suspension, the survival following freezing and thawing can be substantially increased. CPAs are broadly classed as penetrating (for chemicals that will diffuse through the plasma membrane and equilibrate in the cytoplasm) and nonpenetrating (chemicals that do not enter the cytoplasm). Penetrating CPAs are small, nonionic molecules that have a high solubility in water at low temperatures and low cellular toxicity. Their activity can be understood by their effect on the freezing point. Most important, the presence of penetrating CPAs lowers the concentration of the salts that are normally found in physiological solutions for a given temperature (below the freezing point, when ice formation causes the concentration of these salts). They do this by lowering the amount of ice present for a given temperature (Figure 2.8a) as well as by acting as a secondary solvent for salt (Pegg, 1984). Thus, a cell that is being frozen in a simple saline solution will see a higher sodium chloride concentration at $-10 °C$ than will the same cell in a saline solution with 1 M dimethyl sulfoxide at the same temperature (Figure 2.8b). By lowering the temperature at which the cell is exposed to the increased extracellular solute concentration, the magnitude of injury, and the kinetics at which damage accumulates, is reduced. Penetrating CPAs are able to greatly mitigate slow-cooling injury but provide little protection against rapid-cooling injury.

Nonpenetrating CPAs are generally long-chain polymers that are soluble in water and that have large osmotic coefficients (they increase the osmolality far in excess of their molar concentration). They are thought to act by dehydrating the cell before freezing, thereby reducing the amount of water that the cell needs to lose to remain close to osmotic equilibrium during freezing. The cytoplasm does not supercool to the same extent, and therefore intracellular ice becomes less likely at a given cooling rate. These compounds provide little protection from slow-cooling injury.

The inability of nonpenetrating CPAs to provide protection against solution-effects injury seems to indicate that the intracellular concentration of solutes is at least as important as the extracellular concentration for this type of injury. This indicates that the aforementioned relationship between solution-effects injury and a cell's water permeability is caused by the intracellular solute concentration. During rapid cooling, the intracellular solute concentration remains low because the cell cannot lose water fast enough to remain in equilibrium with the extracellular solution. This delays the effects of having a highly concentrated intracellular solution to lower temperatures (where it has a less damaging effect). Thus it appears that solution-effects injury may be dependent on the solute concentration of both the intracellular and extracellular solutions as well as on the integrated exposure to these solutions as a function of temperature.

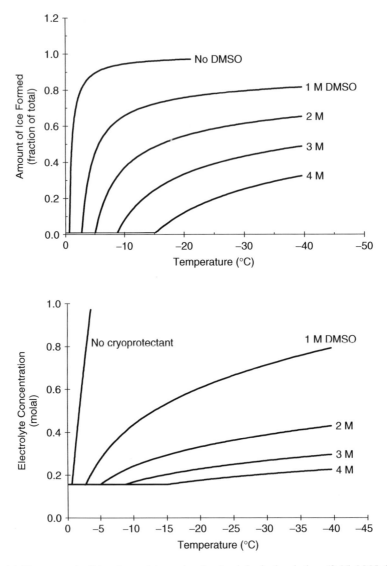

FIGURE 2.8 (a) The amount of ice formed in a simple physiological solution (0.15 *M* NaCl) containing different concentrations of Me₂SO. (b) The concentration of electrolyte in a simple physiological solution (0.15 *M* NaCl) containing different concentrations of Me₂SO. (Data were calculated from Pegg et al., 1987.)

Studies of natural systems that survive extreme environmental stress have shown that one of the adaptive mechanisms used is the overproduction and accumulation of sugars (Crowe and Crowe, 2000). Recent work has shown that these "natural cryoprotectants" are capable of stabilizing and preserving the biological activity of proteins, viruses and bacteria (see Chapter 21, this volume). High concentrations of mono- and disaccharides can act to protect biological structures during dehydration through the formation of a stable glassy matrix (Buitink et al., 1998; Crowe et al., 1998; Wolfe and Bryant, 1999) or through binding to sites previously stabilized by water (Crowe et al., 1993a, 1993b; Gaber et al., 1986). In the absence of more traditional cryoprotectants (e.g., Me₂SO, glycerol), sugars have been shown to be effective protectants in mammalian cell cryopreservation (Eroglu et al., 2000). Although it has been shown that sugars need to be present on both sides of the plasma membrane to be maximally effective (Chen et al., 2001; Eroglu et al., 2000, 2002), the mechanisms by which these "penetrating" agents function to protect cells during cryopreservation are relatively unknown. Further work in this area is warranted.

2.4 MECHANISMS OF CRYOINJURY

2.4.1 SLOW-COOLING INJURY

During slow cooling, the cell is able to maintain osmotic equilibrium with the extracellular solution through dehydration. Although osmotic equilibrium is maintained and intracellular ice is avoided, damage can still occur. There are two basic mechanisms by which damage is thought to occur during slow cooling: solute toxicity (Lovelock, 1953b; Mazur et al., 1972) and physical changes to the cell induced by excessive cell shrinkage under an osmotic stress (Lovelock, 1953a; Meryman, 1970, 1974; Steponkus and Wiest, 1978). During slow cooling, the increase in concentration of the extracellular and intracellular solutes occurs because of the formation of external ice and the resulting efflux of water from the cells, respectively (Mazur, 1963a). Although the molecular mechanisms by which concentrated solutes damage a cell are unknown, it is foreseeable that alterations in the chemical equilibrium of the cell could result in numerous biophysical and biochemical changes that could lead to cell death. One of the mechanism proposed by Lovelock was a lyotropic effect on the plasma membrane by highly concentrated salt (Lovelock, 1953b); little progress has been made in this area since then.

The second mechanism by which slow cooling is thought to injure cells is by the shrinkage of the cell that results from exposure to hypertonic solutions. Work with red blood cells showed that cells remained intact during exposure to high concentrations of solutes, but then lysed when they were diluted back to isotonic (Lovelock, 1953a). The so-called posthypertonic lysis was thought to be a result of salt loading caused by a limitation on the minimum size that a cell could attain through osmotic shrinkage. Once the cell had reached its minimum volume (the membrane simply could not collapse any further about the physical structure of the cell interior), it could no longer lose water to maintain osmotic equilibrium. The gradient in chemical potential could only be decreased by the movement of salt from the extracellular solution into the cytoplasm. On thawing, the cytoplasm would have a higher than isotonic concentration of salt and would draw water in to reach equilibrium. If the swelling-induced expansion of the cell surpassed the yield strength of the membrane, then the cell would lyse (Zade-Oppen, 1968). Later it was shown that there was indeed an uptake of sodium during hypertonic exposure, but there was no minimum volume that could be surpassed osmotically; the previous result had been an artifact of the measurement technique (Farrant and Woolgar, 1972).

A variant on the minimal volume hypothesis suggested that the osmotic pressure gradient that developed once the minimum volume had been reached led to a mechanical stress on the cell (caused by an induced hydrostatic pressure gradient) that caused injury (Meryman, 1974), though it should be noted that this hypothesis requires a minimum volume below which the cell could not shrink osmotically (something that seems not to occur). Alternatively, it has been proposed that the surface area of the cell is reduced as a result of a loss in membrane material during slow cooling. Shrinkage of the cell leads to membrane fusion events, triggered by the dehydration of the membrane itself, whereby some of the plasma membrane is internalized as vesicles. This reduction in the surface area of the plasma membrane reduces the magnitude of expansion that the cell can tolerate during dilution to isotonic (Steponkus and Wiest, 1978).

Another, more recent hypothesis (Muldrew et al., 2000a) suggests an alternative mechanism for salt loading during hypertonic exposure in which cytoplasmic proteins with salt bridges are brought into solution through the interaction of dissolved ions with fixed charges on the proteins. This sinking of ions by cytoplasmic proteins is the putative mechanism behind the experimentally observed uptake of sodium chloride during hypertonic exposure (Farrant and Woolgar, 1972). On thawing (or dilution to isotonic), the excess ions are released back into solution in the cytoplasm, and the cell may swell beyond its elastic limit.

2.4.2 INTRACELLULAR ICE FORMATION

Intracellular ice formation (IIF) occurs when a cell is unable to maintain equilibrium with the external environment. During rapid cooling, the formation of ice and the concentration of extracellular solutes occur too quickly for the cell to respond by exosmosis. Thus, the cytoplasm becomes increasingly supercooled, and there is a corresponding increase in the probability of intracellular freezing. Supercooling, however, is not the only prerequisite condition for IIF. Because the cell membrane serves as an effective barrier to ice growth (Chambers and Hale, 1932; Luyet and Gibbs, 1937; Mazur, 1965) and the cytoplasm contains few effective nucleators (Franks et al., 1983; Rasmussen et al., 1975), the mechanism by which the supercooled cytoplasm becomes nucleated has been the subject of much debate.

There is evidence to indicate that extracellular ice and the plasma membrane are involved in the initiation of IIF. Many cells in an isotonic solution will freeze intracellularly between –5° and –15°C (reviewed in Mazur, 1965). However, IIF will only occur under these conditions when extracellular ice is present. Attempts to observe ice formation in a supercooled cytoplasm in the absence of extracellular ice have so far been unsuccessful. In all cases, the cell cytoplasm reaches temperatures very close to the homogeneous nucleation temperature before nucleation occurs (Franks et al., 1983; Rasmussen et al., 1975). Furthermore, investigations with liposomes, membranes with no intracellular components to act as nucleation centers, demonstrated IIF at temperatures above –10°C (Callow and McGrath, 1985).

The cell membrane has been shown to be an effective barrier to ice growth (Mazur, 1965). If extracellular ice is an important element in the nucleation of ice in the cytoplasm, then the plasma membrane must be involved. Work with nonacclimated and acclimated plant protoplasts has shown that compositional alterations of the plasma membrane during cold acclimation change the conditions under which IIF occurs (Dowgert, 1983; Pitt and Steponkus, 1989; Steponkus, 1984). Similarly, work with hydrophilic antifreeze proteins has suggested that by promoting a closer interaction of ice with the plasma membrane, the incidence of IIF can be affected (Ishiguro and Rubinsky, 1994; Larese et al., 1996; Mugnano et al., 1995). In fact, extracellular ice appears to grow right through the plasma membrane when antifreeze proteins are present (Larese et al., 1996; Mugnano et al., 1995). This may be a result of the creation by the amphipathic nature of the antifreeze proteins of a hydrophobic surface on a growing ice crystal that allows the ice to penetrate the hydrophobic region of the membrane.

At present, there are three dominant hypotheses that attempt to explain the mechanism by which extracellular ice interacts with the plasma membrane to initiate intracellular ice formation. The protein-pore theory of Mazur (1965) was motivated by the observations that supercooled cells could freeze internally well above the homogenous nucleation temperature and that the plasma membrane was an effective inhibitor of ice only above a certain temperature. Mazur hypothesized that external ice could seed the supercooled cytoplasm by growing through aqueous pores in the membrane. For this to occur, the tip radius of the growing ice crystal must approximate the radius of the pores in the plasma membrane. The Kelvin equation relates temperature to the smallest stable radius of an ice dendrite through the freezing-point depression because of the curvature of the tip (Acker et al., 2001; Elliott, 2001; Mazur, 1965). For any given pore size, therefore, there is a corresponding temperature below which ice should be able to grow through the pore. The mechanism of injury is thought to be an enlarging of the membrane pore by the process of recrystallization during warming (a reduction in the ratio of surface area to volume by the loss of ice from regions of high curvature and accretion at regions of low curvature). In support of this mechanism of IIF, recent studies have suggested that ice growth through stable proteinaceous membrane pores is responsible for the intercellular propagation of intracellular ice that is observed in confluent monolayers (Acker and McGann, 1998; Acker et al., 2001) and tissues (Berger and Uhrik, 1996), although as we will discuss later, this type of intracellular ice does not necessarily correlate with membrane failure on thawing.

Disruption of the plasma membrane has been proposed as an alternative mechanism by which extracellular ice can nucleate the cytoplasm. Working with unfertilized eggs of the sea urchin *Hemicentrotus pulcherrimus*, Asahina proposed in 1962 that the cause of intracellular ice formation was damage to the plasma membrane, concluding that membrane damage precedes IIF. Similarly,

Steponkus and Dowgert (1984) were able to directly observe ruptures in the membranes of plant protoplasts immediately before the formation of intracellular ice. They later suggested that the disruption occurs as a result of the development of electrical transients created by charge separation at the interface of the growing ice and the aqueous solution (Steponkus et al., 1984). Motivated by the fact that cells can be damaged at 0°C in the absence of ice because of an osmotic pressure gradient remarkably similar to those produced during freezing, Muldrew and McGann proposed that membrane damage may be a result of a critical osmotic pressure gradient across the membrane (Muldrew and McGann, 1990). They noted that the occurrence of IIF correlated with the peak water flux across the plasma membrane and mistakenly postulated that the movement of water could create an outward force on the membrane that could lead to rupture (Muldrew and McGann, 1994). A refined version of this hypothesis will be developed in a later section. In all variants of this basic mechanism, there is damage to the plasma membrane before IIF, and there is a trans-membrane ice crystal following IIF; therefore, recrystallization on warming could serve to enhance the preexisting lesion.

Intracellular ice that results from an intracellular nucleation event has been hypothesized, where the membrane of a cell can behave as a nucleation site for internal ice when acted on by extracellular ice. Though the detailed mechanism by which this might occur has yet to be explored, the thermodynamics of the process have been found to agree with experimental data. Toner initially proposed the idea of surface-catalyzed heterogeneous nucleation in 1990 (Toner et al., 1990). The theory attributes the formation of intracellular ice to the ability of external ice to interact with, and alter, the structure of the cell membrane. The nature of this interaction could be chemical, electrical, mechanical, ionic, or thermal, but it appears to make the plasma membrane an effective nucleator. The external ice can therefore induce the formation of intracellular ice without physically disrupting the integrity of the plasma membrane (Toner et al., 1990). A second form of intracellular nucleation, volume catalyzed IIF, was also proposed to account for IIF events that occur at low temperatures when the cell has lost much of its cytoplasmic water (Toner et al., 1990). The mechanism by which intra-cellular ice leads to failure of the plasma membrane on thawing is presently lacking in this model.

The three mechanisms proposed for the formation of intracellular ice each assume a different role for the plasma membrane. In the protein-pore theory, the cell membrane is an effective inhibitor of external ice only above the temperature at which the ice crystals are too large a size to propagate through permanent transmembrane pores. The membrane failure hypothesis requires that the integrity of the membrane be disrupted and postulates that cell damage precedes the initiation of IIF. Finally, the surface-catalyzed nucleation theory suggests that the barrier properties of the cell membrane do not have to be compromised for the initiation of internal ice. Although each of these theories proposes an alternative means by which ice can enter the cell, each one supports the assertion that it is the extracellular ice interacting with the plasma membrane that is responsible for the formation of intracellular ice.

Although the mechanism of IIF is still controversial, the ability to predict the probability of IIF for a given freezing protocol has been attained using phenomenological curve fitting techniques. This use of mathematical models that can predict the occurrence of IIF has been motivated by the potential utility of these models in the design of cryopreservation protocols and their ability to further the investigation of the mechanisms of IIF. Numerous phenomenological models (Mazur, 1984; Pitt and Steponkus, 1989; Pitt et al., 1991, 1992) have been developed that use statistics and an understanding of the conditions surrounding the cell just before freezing to predict the likelihood of intracellular freezing. Mechanistic hypotheses have also been developed into mathematical models by fitting parameters to IIF data rather than determining their values from more fundamental principles (Muldrew and McGann, 1994; Toner et al., 1990). These two approaches, phenomeno-logical and mechanistic, although fundamentally different in design, have been shown to give relatively similar end results (Karlsson et al., 1993). Each model is able to predict with some degree of certainty the degree of intracellular ice formation in a cellular system under well-defined conditions.

2.5 INVESTIGATION OF THE MECHANISMS OF CRYOINJURY

2.5.1 INTRODUCTION

The nature of cellular injury caused by freezing and thawing is undoubtedly complex. Many of the investigations aimed at understanding this injury can be said to fall into the study of first-order mechanisms of injury. This type of injury refers to the catastrophic events that lead to cell death. It is highly likely that there are second-order effects that occur in cells that represent sublethal injury (i.e., the cell is able to repair the damage once metabolism is restored). These second-order effects may be quite independent of the first-order effects that have been studied thus far, adding multiple injuries to the "two factors" that are normally considered. Indeed, the possibility of overlapping modes of first-order damage is also a reasonable proposition. The common behavior of the wide variety of cell types and experimental arrangements might only reflect a common proximal cause (e.g., an osmotic pressure gradient), whereas independent molecular mechanisms could be the detailed causes leading to a common endpoint (e.g., cell lysis). Nevertheless, it is important to identify the primary modes of injury before studying the secondary modes because we need to understand this injury before we can properly run controlled experiments in which we separate out the overlapping effects.

Our own program of investigation into the mechanisms of cryoinjury has exploited two fundamental tools to feed the processes of building theories, generating hypotheses, experimentally testing the hypotheses, and then returning full circle and refining the theories. We are still going around this loop; certainly we are ahead of where we started, but still with many interesting questions and uncertainties remaining. This section will attempt to give an overview of our current thinking along with some of the developmental work that led us to this position. Our primary conceptual tool in this progression has been the mathematical modeling of the osmotic responses of cells. By simulating cellular responses on a computer, we can visualize our idealized model of what the cells are experiencing during freezing and thawing under arbitrary conditions. This tool has allowed us to refine vague notions about how cells might be injured into precise hypotheses that can be tested. Our primary experimental tool has been the cryomicroscope. The computer-controlled cryostage allows an incredibly diverse range of thermal protocols to be generated, and it allows direct visual observation of the cells during the process. Individual cells can be followed during freezing and thawing, and after, allowing direct testing of various cellular parameters. Through the use of these two tools we are beginning to develop a theoretical understanding of first-order cryoinjury, and we have collected a substantial body of experimental work that puts limits on what alternative theories will have to account for. The following descriptions will largely focus on our own developments, leaving different perspectives to be advanced in other chapters.

2.5.2 OSMOTIC PORATION AND INTRACELLULAR ICE FORMATION

Hydrophilic pores have been postulated to form spontaneously in the lipid bilayer through thermal fluctuations (Bordi et al., 1998; Glaser et al., 1988; Paula et al., 1996). These pores are small, short lived, and widely spaced, though they are stable enough to substantially raise the permeability of cell membranes to small ions (Bordi et al., 1998; Paula et al., 1996). In fact, the permeability of cell membranes to ions (e.g., sodium) that is predicted from considering the transport of ions in the absence of pores is at least three orders of magnitude lower than the measured value (Paula et al., 1996). If the membrane is subjected to a high potential difference, then these thermal pores can expand through a process of reversible electric breakdown to become relatively stable (Abidor et al., 1979; Glaser et al., 1988). Similarly, the presence of thermal pores during periods of large osmotic water flux will allow the water molecules to preferentially pass through the pore. These water molecules will have collisions with the edge of the pore that tend to enlarge the pore, allowing more water to flow through it if the driving force is sufficient. This positive feedback loop will proceed as long as there is a local osmotic pressure gradient to drive water movement. A corollary of this hypothesis is that cells that are subject to high osmotic stresses would require dedicated water channels (such as aquaporins) to avoid the potential injury of osmotic poration. This corollary is supported by the high concentration of aquaporins in red blood cells, the cells of the renal collecting ducts, and other cells that are subject to osmotic stresses (Verkman et al., 1996).

Data from osmotic pulse experiments (a technique for loading impermeable solutes into a cell) seem to indicate that osmotic stresses can, in fact, lead to the presence of reasonably long-lived aqueous pores in the membrane (Franco et al., 1986). We have pursued the hypothesis that pores generated by osmotic stresses are responsible for intracellular ice formation during rapid cooling and its concomitant injury. Simply put, the pores form spontaneously because of thermal fluctuations and are then expanded by the interaction of water with the edge of the pore during periods of high osmotically driven water movement. The pore will collapse once the water flux diminishes, but if it opens wide enough when extracellular ice is present (i.e., during freezing), then the supercooled cytoplasm can be seeded by ice growth through the pore. The minimum size of pore that will lead to IIF is given by the relationship between the minimum radius of an ice crystal and the temperature (given by the Kelvin equation; Elliott, 2001). The formation of intracellular ice is not specifically tied to the lysis of the cell on thawing. If the pore through which the ice crystal grows is below the critical size that causes cellular lysis, then the pore will simply reseal when the ice melts. A complicating factor is that ice crystals will undergo recrystallization during warming (the ratio of surface area to volume is reduced). If the transmembrane ice crystal expands during warming, the pore will become enlarged. Rapid warming will tend to lead to resealing (if the pore is below the critical size), whereas slow warming will lead to cell lysis.

When isolated cells are cooled on a cryostage to a temperature just below the freezing point and ice formation is seeded, the ice can be observed to grow around the cells confining them to unfrozen channels between ice dendrites. Unless some insult has been imposed on the cells, it is a universal observation that the ice never grows through the plasma membrane into the cytoplasm, even in the absence of cryoprotective compounds. For this reason, the increased solution concentration in the unfrozen channels imposes an osmotic gradient across the cell membrane and the cell loses water until the osmotic pressure of the cytoplasm equals that of the unfrozen solution. If pores are present in the membrane when extracellular ice is present, then their size must be smaller than the minimum radius of curvature of an ice crystal at that temperature for the membrane to retain this barrier property.

Because the hypothesis of osmotic poration predicts that these pores are not linked to ice growth, except by the osmotic conditions generated by the formation of ice, we postulated that they could be formed through osmotic stresses at temperatures above the freezing point. If large, stable pores could be produced, then we hypothesized the cell membrane would lose its barrier properties to ice growth just below the freezing point. To test this experimentally, we used dimethyl sulfoxide (Me_2SO) to produce an osmotic gradient across the cell membrane at 0°C. Because cells are relatively permeant to Me_2SO, the osmotic gradient developed was transient, and thus we could add it in one dose for a maximal osmotic pressure gradient or in small increments to keep the osmotic gradient low while still developing the same concentration of Me_2SO in the cells. The Me_2SO was then diluted and the cells placed on a cryostage, and ice formation was seeded just below the freezing point. The hypothesis was supported, as the proportion of cells that lost their barrier properties was a function of the originally imposed osmotic pressure gradient (Table 2.1; Muldrew and McGann, 1990).

These data gave us a powerful tool to investigate whether these pores were the cause of intracellular ice, as we could correlate the temperature dependence of IIF and osmotic poration in the absence of ice. If the correlation existed, then we would have additional confidence that the same mechanism was responsible for both phenomena.

Our osmotic modeling work that had first led us to suspect that IIF was caused by excessive osmotic pressure gradients suggested a line of inquiry using cryomicroscopy that would help to answer this question. Cryomicroscopy would allow us to actually measure some of the physical

TABLE 2.1
**Membrane Damage Caused by the Osmotic Stress
of Rapid Addition of a Permeant Cryoprotectant at
0°C; Toxicity Effects Are Shown to be Negligible by
the Low Rate of Injury with Slow Addition**

Me_2SO (M)	Rapid-Addition Rupture (%)	Slow-Addition Rupture (%)
3	11	2
4	47	0
5	88	2

parameters that had been implicated in the literature on IIF with direct visualization of the formation of IIF in individual cells. In addition, we could also use this data to see whether the osmotic pressure gradients associated with IIF correlated with osmotic poration in the absence of ice.

We chose to combine the modeling of cellular responses to osmotic stresses that we had developed with cryomicroscopic observation of IIF so that we could estimate the physico-chemical environment during freezing and thawing. The two experimental protocols that we focused on were based on freezing cells at constant cooling rates and holding cells at constant temperatures (subfreezing) while seeding ice. The convection cryostage was ideal for these experiments, as it was possible to supercool the sample without ice forming in the extracellular compartment. For the constant cooling experiments, ice had to be seeded just below the freezing point, and then the cells had to be held at that temperature to bring them into osmotic equilibrium (so that our simulations would give reasonable estimates of the conditions). For both of these experimental protocols we used cells that were equilibrated in several different concentrations of cryoprotectant. This helped to prevent any artifacts that might have been caused by cellular injury from exposure to the freezing environment (the solution-effects injury), and it also gave altered conditions for the formation of IIF so that we could look for mechanisms.

The isothermal technique showed, first of all, that IIF was not likely to be caused by electrical transients at the ice interface, as suggested by Steponkus et al. (1984). The magnitude of charge separation at an ice interface is proportional to the interface velocity, but when we measured this, we found that the formation of IIF occurred at lower velocities as the concentration of cryoprotectant was increased (Muldrew and McGann, 1990). With the buffering action of the cryoprotectant, we expected either the opposite trend or an independence from cryoprotectant.

The comparison between the isothermal experiment and the constant cooling experiment led us to initially discount the hypothesis, first postulated by Mazur, that extracellular ice was able to grow through membrane-bound proteinaceous aqueous pores (Mazur, 1960, 1965, 1966). Although this theory is very similar to the osmotic poration theory, the size of the pores is independent of water flux (and is also very small). If the temperature and composition of the solution determine the minimum radius of a growing ice crystal, then there should be a characteristic IIF temperature for a given solution. When we compared the temperature associated with 50% IIF, for each solution, between the isothermal and the constant cooling experiments, we found that this temperature was about a factor of two lower in the constant cooling experiment in each case (Muldrew and McGann, 1990). This conclusion is being reevaluated in light of our recent work on innocuous intracellular ice formation (see following).

This technique was unable to discount intracellular heterogeneous nucleation, as hypothesized by Levitt (Levitt and Scarth, 1936) and others (Pitt and Steponkus, 1989; Toner et al., 1990), as being responsible for IIF. In fact the isothermal experiment supported the hypothesis. The amount of supercooling that correlated with IIF increased as the temperature of IIF decreased (because of increasing cryoprotectant concentration; Muldrew and McGann, 1990). If the likelihood of nucleation gets smaller as solute is added to a solution, then we would expect the degree of supercooling required for IIF to similarly increase.

These experiments were also consistent with the hypothesis of osmotic poration being responsible for IIF. The osmotic pressure gradient that correlated with IIF decreased as the temperature of IIF decreased (again because of increasing cryoprotectant concentration; Muldrew and McGann, 1990). This is expected because of the relationship between the critical pore size and temperature. We were able to extrapolate the relationship between the osmotic pressure gradient and the temperatures that correlated with 50% IIF to 0°C to compare them with the osmotic pressure gradient that gave 50% osmotic poration from rapid addition of Me$_2$SO. The values matched exactly (Muldrew and McGann, 1990), giving us confidence in this interpretation. We further tested the osmotic poration hypothesis by looking for the presence of membrane pores immediately after intracellular ice formation. By performing an isothermal experiment in which 50% of the cells underwent IIF, then warming them to just above the freezing point (so that both intracellular and extracellular ice would melt but no metabolic repair could occur), and then cooling to just below the freezing point and seeding ice formation, we were able to correlate IIF with the loss of membrane barrier properties for individual cells. We found that although 7% of cells that did not form IIF had lost their barrier properties (perhaps because of the fact that no cryoprotectant was used), 94% of the cells that had formed IIF were also found to have a pore in the membrane (indicating that some resealing of osmotic pores probably occurred).

Although this evidence strongly supported osmotic poration as the mechanism by which IIF occurred, it also weakly supported a mechanism based on intracellular nucleation. To further test the hypothesis, we developed a physical model of how the water flux might lead to a loss of membrane integrity. The theory that we developed was based on the frictional drag of water moving through the membrane under an osmotic stress (through a solubility/diffusion mechanism rather than through pores). We mistakenly generalized the frictional force to an outward pressure on the membrane (because the membrane moves inward during water efflux, the actual hydrostatic pressure must also be inward) and postulated that this pressure could exceed the yield strength of the membrane (Muldrew and McGann, 1994). When we calculated the magnitude of the "pressure" that correlated to IIF, it matched exactly the hydrostatic pressure required to rupture a lipid membrane. This unlikely coincidence, coupled with the strong experimental support for an osmotically induced membrane defect, led to our inability to see the obvious flaw.

Fortunately, the reduction of the theory to a practical implementation was carried out using statistical arguments and phenomenological parameters and was thus not tied to the interpretation of a water pressure. The equation describing the probability of IIF was based on a critical water flux, which we still hypothesize as the driving force for pore expansion. Unfortunately, our error served to divert attention from this mechanism for IIF despite the experimental support for the general hypothesis of osmotic poration. The phenomenological equation also was found to work very well for describing experimental data involving IIF, as we were able to use it to successfully describe IIF in bovine chondrocytes (Muldrew and McGann, 1994). We measured the permeability parameters for isolated chondrocytes and then performed both constant cooling and isothermal experiments on the cryostage (with various concentrations of cryoprotectant as before). The osmotic response of the cells for both freezing protocols was simulated, and the parameters describing IIF were fit to the experimental data using the simulated water flux. The figures below show that the equation was able to reasonably describe IIF under a wide variety of conditions (it should be noted that the parameters could be adjusted to provide an almost exact fit for either the data in Figure 2.9a or the data in Figure 2.9b; the discrepancies in the fit come from using the same parameters to simulate both experiments).

To further uncouple the effects of temperature (and hence the likelihood of forming IIF through proteinaceous pores) from the effects of water flux, we developed an experimental method that

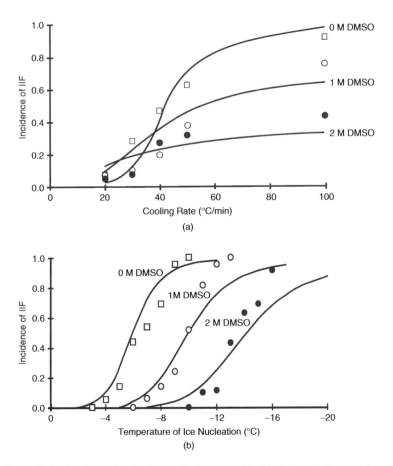

FIGURE 2.9 Intracellular ice formation is shown for hamster fibroblasts in solutions with no Me$_2$SO, 1 M Me$_2$SO, and 2 M Me$_2$SO. (a) IIF is shown as a function of cooling rate. (b) IIF is shown as a function of holding temperature (the cell suspension is supercooled at the holding temperature and then ice formation is initiated while holding the sample at a constant temperature). The solid lines represent the predictions of the osmotic rupture equation using a single set of parameters for the cell type.

would introduce a lag between the lowest temperature reached and the peak water flux. Even though this would not separate the relative magnitudes of these peak values, by separating them temporally we hoped to be able to correlate IIF with one or the other. The technique relied on cycling the temperature of the cryostage sinusoidally such that the cells would be alternately subjected to hypertonic and "hypotonic" conditions (relative to the instantaneous state of the cytoplasm). The temperature remained below the freezing points of the solutions, so that ice was always present, but the amount of ice was varied so that the osmotic pressure of the unfrozen solution would vary. Figure 2.10 shows the thermal protocol as well as the water flux and the degree of supercooling for a given condition.

Figure 2.10 also shows that the peak water flux occurs just past 180°C into the cycle whereas the minimum temperature occurs at 270°C. The results of the experiment showed that not only did IIF correlate most strongly with the point in the cycle corresponding to the peak water flux, but the magnitude of supercooling of the cytoplasm that corresponded to IIF was inordinately small (Muldrew and McGann, 1994). Indeed, IIF was observed with as little as 2°C of supercooling—a magnitude that would require an astonishingly efficient nucleation site for heterogeneous nucleation to occur. Furthermore, once IIF was observed to occur in a cell, it was also observed in that cell on all subsequent cycles, with the magnitude of water flux (or supercooling) required becoming

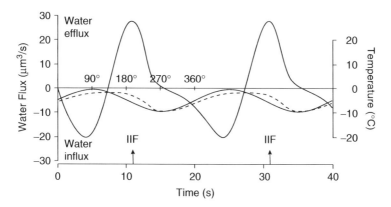

FIGURE 2.10 Water flux is shown during sinusoidal thermal cycling for hamster fibroblasts. The temperature is cycled between –0.5° and –9.5°C. Intracellular ice forms at the point in the cycle corresponding to the peak water flux as well as the peak supercooling. The dashed line shows the freezing point of the cytoplasm; thus, supercooling is given by the difference between this line and the temperature line. In this case, though maximum supercooling is only 3°C, 25% of the cells undergo IIF by the fifth cycle.

less and less with each subsequent cycle (Muldrew and McGann, 1994). In fact, there were some instances in which the hole in the cell membrane was so enlarged by the fifth cycle that we could actually see the ice crystal growing into the cell (the radius of the "pore" was about one-quarter the radius of the cell).

The hypothesis that osmotic poration is the mechanism by which IIF forms remains a viable alternative to nucleation-based theories. It also provides a mechanism of injury (unlike the nucleation theories) and a rationale for avoiding that injury. The substantial base of experimental evidence in support of osmotic poration remains as solid today as ever and should be considered on its own rather than in the context of a previous mechanistic model that was incorrect. The broad range of applicability (explaining poration phenomena above freezing, the type of injury associated with IIF, and the phenomenon of IIF itself) and the very useful extension of the hypothesis toward finding ways to achieve rapid cooling without the lethal injury associated with IIF (discussed below) must surely be taken as strong support for continuing to investigate this phenomenon.

2.5.3 INNOCUOUS INTRACELLULAR ICE FORMATION

A long-held tenet in cryobiology is that intracellular ice is lethal to cells. Because it has been shown that intracellular ice formation in cells in suspension occurs during rapid freezing (Chambers and Hale, 1932; Luyet and Gibbs, 1937), and that rapid freezing causes cell death (Luyet and Gehenio, 1940; Mazur, 1966), it has been assumed that intracellular ice causes cell death (Diller et al., 1972; Karlsson et al., 1993; Mazur, 1966, 1977, 1984; Muldrew and McGann, 1990). However, the degree to which intracellular ice is damaging to cells and the mechanisms by which this injury occurs have been largely speculative. The most widely held view is that IIF damage occurs as a result of mechanical damage caused by a surface-area-to-volume redistribution of the ice crystal (Karlsson, 2001; Mazur et al., 1972; Muldrew and McGann, 1990). Recrystallization of the ice during slow warming will manifest itself as a net increase in the size of an intracellular ice crystal. The observation that cells could be "rescued" from intracellular ice formation as a result of rapid-warming techniques (Fowler and Toner, 1998; Mazur et al., 1972) has provided support for this mechanism of damage. Recrystallization, however, is not the only means by which IIF can be lethal. Various nonmechanical mechanisms have also been proposed including solution effects and thermal shock (Farrant and Morris, 1973), osmotic injury (Farrant et al., 1977), protein denaturation (Levitt, 1962), and gas bubble formation (Ashwood-Smith et al., 1988; Morris and McGrath, 1981; Steponkus and Dowgert, 1981). Although the presence of intracellular ice has been thought to result in irreversible damage to cells frozen rapidly, conclusive evidence to support this claim has been difficult to obtain.

Intracellular ice may not be in itself lethal. The fact that rapidly cooled cells can survive if rapidly warmed suggests that the intracellular ice *per se* is not damaging. Instead, the amount of ice, the size of the ice crystals (Shimada and Asahina, 1975), the location of the ice formed (Bischof and Rubinsky, 1993; Farrant et al., 1977; Hunt, 1984), or the mechanism of formation (Acker and McGann, 2000, 2001, 2002; Acker et al., 2001) have been identified as conditions that could lead to cellular damage. There have been numerous studies that suggest that innocuous intracellular ice formation is possible if the amount of ice formed can be controlled (Mazur, 1990; Rall et al., 1980). Minimizing the size of the ice crystals formed during rapid cooling and warming to the point that the sample remains transparent has been shown to be successful at preserving tumor cell function (Asahina et al., 1970; Sherman, 1962). It is thought that there is a critical size of ice crystal that the cell can tolerate before damage to internal organelles occurs. Shimada has suggested that this critical size is 0.05 μm (Shimada and Asahina, 1975).

Historical attempts to obtain vitrified tissues using rapid freezing provide an excellent example of the success that researchers have had in minimizing the size of intracellular ice crystals and enhancing postthaw survival (Keeley et al., 1952; Luyet and Hodapp, 1938; Mider and Morton, 1939). Although the initial assumption was that ice formation had been prevented by rapid freezing (Keeley et al., 1952; Mider and Morton, 1939), it was subsequently determined using x-ray diffraction that these cells did indeed form intracellular ice but that the crystals were too small to be resolved using light microscopic techniques (Luyet and Rapatz, 1958). Even today, there are frequent reports of tissues being "vitrified," using rapid-freezing techniques, with superior survival (Day et al., 1999; de Graaf and Koster, 2001; Zieger et al., 1997). Because complete vitrification of bulk systems using traditional freezing protocols is unlikely (Fahy et al., 1990; Meryman, 1957), these results may be the product of "partial vitrification," or the presence of innocuous intracellular ice.

Although it is generally accepted that intracellular ice formation inevitably results in lethal damage to cells in suspension and single attached cells (Acker and McGann, 2000, 2002; Diller et al., 1972; Diller, 1975; Karlsson et al., 1993; Mazur, 1966, 1977; Muldrew and McGann, 1990), the relationship between cell damage and intracellular ice formation has proven to be difficult to experimentally verify (Karlsson et al., 1993; Acker and McGann, 2001). It has not yet been conclusively established whether intracellular ice formation is the cause of (Levitt, 1962; Farrant et al., 1977; Mazur, 1966; Mazur and Koshimoto, 2002; Mazur et al., 1972) or results from (Dowgert, 1983; Muldrew and McGann, 1994; Toner et al., 1990) damage to cellular components. Identifying the temporal sequence of events involved in intracellular ice formation, the role of biology in IIF, and the IIF-related sites of cell injury are critical outstanding issues in the field of cryobiology. Over the last 15 years, we have made some progress toward understanding these issues.

There is an innate relationship between intracellular ice formation and the cell plasma membrane (Diller et al., 1972; Karlsson et al., 1993; Mazur, 1965; Muldrew and McGann, 1990; Toner et al., 1990, 1993). The limited permeability of the membrane to water and solutes results in cytoplasmic supercooling during rapid cooling. Because the membrane is a barrier to ice propagation (Chambers and Hale, 1932; Luyet and Gibbs, 1937), and there are few efficient nucleators in a cell (Franks et al., 1983; Rasmussen et al., 1975), nucleation of ice within the supercooled cytoplasm at high subzero temperatures is thought to involve extracellular ice. One of the consistent elements in the study of IIF has been the observation following warming that there is significant damage to the plasma membrane. Although it is not known whether this damage to the plasma membrane occurs during freezing (Asahina, 1962; Muldrew and McGann, 1994; Steponkus et al., 1983) or thawing (Mazur, 1963b, 1965; Fowler and Toner, 1998), or whether IIF is the cause of, or results from, this damage, the fact is that damage to the plasma membrane is almost always observed after IIF (Chambers and Hale, 1932; Dowgert, 1983; Fujikawa, 1980; Mazur, 1965, 1966, 1984; Muldrew and McGann, 1990; Steponkus et al., 1983; Steponkus and Dowgert, 1984). Because a damaged plasma membrane and subsequent loss of semipermeability are strong indicators of lethal injury

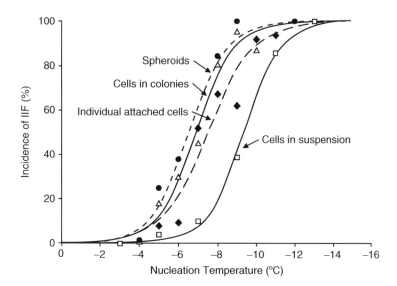

FIGURE 2.11 Cumulative incidence of intracellular ice formation of hamster fibroblasts as a function of temperature of extracellular ice nucleation for four tissue models: cells in suspension (□); individual cells attached to glass (♦); cells in monolayers (△); and cells in a spheroid (●). Lines are logistic curves fitted to the data using a least squares method. (Reprinted from Acker et al., 1999a. With permission.)

to a cell (Acker et al., 1999a; Dankberg and Persidsky, 1976; Mazur, 1965, 1966; Yang et al., 1995), intracellular ice formation has always been correlated with cell death.

In 1997, while studying the effects of cell–cell and cell–matrix interactions on IIF, we were able to dissociate intracellular ice formation and membrane damage. We observed that a majority of cells in confluent monolayers that form intracellular ice at low temperatures did not display a damaged plasma membrane following thawing (Acker and McGann, 2000). Prior work had already shown an increased probability of intracellular ice formation in confluent monolayers and cell spheroids (Acker and McGann, 1998; Acker et al., 1999a), which indicated that cell–cell contact was an important mediator of IIF (Figure 2.11). By directly observing and measuring the diffusion of a membrane impermeable fluorescent stain, ethidium bromide, into frozen cells and correlating the results with the formation of intracellular ice and the postthaw integrity of the plasma membrane, we were able to examine the time course of IIF-related membrane damage (Acker and McGann, 2000, 2001). We found that in monolayers, cells that form intracellular ice before any of their nearest neighbors have a higher probability of having a disrupted cell membrane than cells that freeze next to an already frozen cell. We now had a condition in which intracellular ice formation could occur without damaging the cell membrane, allowing us to begin to understand the causal relationship between IIF and membrane damage.

From this work, it became clear that IIF-related membrane damage was dependent on the location of the frozen cell within a monolayer and that cell–cell contact was an important determinant of both the incidence and kinetics of IIF and membrane damage. Using the cryomicroscope to observe the pattern in which IIF occurred in confluent monolayers of different cell lines gave us important information on the mechanism by which intracellular ice forms in these systems (Figure 2.12). We found that with fibroblast monolayers, the nucleation of ice in one cell was followed, after a brief delay, by the freezing of an adjacent cell. This resulted in well-defined clusters of cells with intracellular ice when the monolayers were frozen at constant subzero temperatures. This effect was amplified in epithelial monolayers in which a wave-like propagation of intracellular ice occurred in the samples, resulting in all cells forming intracellular ice at relatively high subzero temperatures. Using a simple statistical method to test the degree of randomness of IIF in the confluent monolayers, we were able to conclude that the formation of ice in a cell that

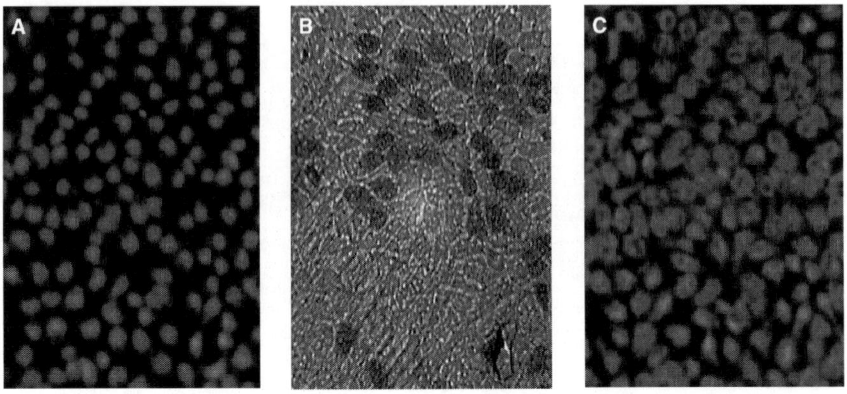

FIGURE 2.12 Pattern of intracellular ice formation in confluent cell monolayers. Photomicrographs of a hamster fibroblasts stained with the fluorescent nucleic acid stain SYTO 13 before freezing on a convection cryomicroscope. (a) Before freezing at $-7°C$; (b) brightfield photomicrograph of cells frozen at $-7°C$ showing characteristic darkening of the cytoplasm; (c) fluorescent photomicrograph of frozen cell at $-7°C$.

is part of a monolayer increases the likelihood of intracellular freezing in adjacent cells (Acker and McGann, 1998).

As the propagation of intracellular ice between adjacent cells in monolayers and tissue systems had been previously documented in the literature (Asahina, 1956; Berger and Uhrik, 1996; Brown, 1980; Brown and Reuter, 1974; Chambers and Hale, 1932; Levitt, 1966; Luyet and Gibbs, 1937; McLeester et al., 1969; Molisch, 1982; Stuckey and Curtis, 1938; Tsuruta et al., 1998), we became interested in understanding why this phenomenon occurred and how it affected postthaw cell viability. In 1992, while working with single strands of salivary tissue, Berger and Uhrik demonstrated that the induction of ice between cells could be inhibited using a chemical agent that uncoupled the cells (Berger and Uhrik, 1992, 1996). They concluded that intercellular junctions (gap junctions) were responsible for the propagation of ice between adjacent cells. Critical to their hypothesis was the assumption that ice could grow through small-diameter pores. To test this idea, we examined the theoretical relationship between the equilibrium ice crystal radius and the temperature of extracellular ice nucleation and compared it with experimental data on the temperature of intercellular ice propagation (Acker et al., 2001). By examining cell lines with and without gap junctions, we found a temperature that resulted in a significant deviation in the incidence and pattern of IIF that we attributed to gap junction-facilitated intercellular ice propagation. This temperature agreed (at least qualitatively) with our theoretical predictions. This work provided strong evidence to support the concept that intracellular ice can propagate between adjacent cells via gap junctions. This hypothesis has recently been tested and verified in a much more rigorous manner by Irimia and Karlsson (2002), using micropatterned substrates to control cell–cell interactions.

It is important to note that nonrandom intercellular ice propagation has also been observed in cell lines that do not form pores between cells (Acker and McGann, 1998; Acker et al., 2001; Irimia and Karlsson, 2002,), and therefore, ice propagation via gap junctions cannot be the sole mechanism for ice propagation between adjacent cells. Recent studies have suggested that the induction of ice between adjacent cells can occur as a result of surface catalyzed nucleation in which ice in one cell, using the adhesion between cells, results in the nucleation of ice in adjacent cells (Acker et al., 1999a; Tsuruta et al., 1998). This hypothesis awaits theoretical and experimental validation.

With a better understanding of how intracellular ice formation occurs in confluent monolayers and the role of cell–cell contact on membrane integrity, it became clear to us that intracellular ice formation in confluent monolayers was markedly different from IIF in cell suspensions. As much of our understanding on IIF had come from work with cell suspensions, we believed that studies with confluent monolayers would enhance our ability to more precisely define the relationship between IIF and cell injury. Specifically, we were interested in testing the long-standing tenets that intracellular ice is lethal and that intracellular ice formation should be avoided during cryopreservation.

We knew that intracellular ice could propagate between adjacent cells and that the majority of the cells that formed intracellular ice as a result of intercellular ice propagation had intact plasma membranes. Because we now had a means to form intracellular ice in cells and thaw them without incurring membrane damage, we could now dissociate IIF from membrane damage and focus on the important question of whether intracellular ice is lethal. We induced all of the cells in confluent monolayers to form intracellular ice by freezing at defined subzero temperatures. We then thawed the cells and assessed the postthaw survival using three different indices: membrane integrity, metabolic activity, and clonogenic function. We found that the postthaw survival of confluent monolayers was dependent more on the presence of an intact plasma membrane than on the presence of intracellular ice (Acker and McGann, 2002). As approximately 80% of the cells that had intracellular ice survived freezing and thawing, this implied that the presence of ice inside cells was not inherently lethal. Intracellular ice itself can be innocuous. It would appear from this preliminary work that the mechanism of intracellular ice formation—the mechanism that results in membrane damage—is the lethal event.

The most surprising results from our work on intracellular ice came when we compared cells frozen to –40°C in 10% Me_2SO to those cells that formed innocuous intracellular ice. By inducing intracellular ice formation in confluent monolayers in the absence of any chemical cryoprotectant, we obtained high cell viability following freezing and thawing (Acker and McGann, 2002). Although the recovery was not as great as that obtained using a standard cryopreservation protocol with 10% Me_2SO, these data would indicate that intracellular ice is not only innocuous but can have a cryoprotective effect. We followed this work with an examination of the effect of innocuous intracellular ice on cell survival following slow cooling (Acker and McGann, 2003). We found that the survival of confluent cell monolayers cooled to –40°C at 1°C/min was highest when the incidence of intracellular ice was close to 100%. Together, these studies support the concept that conditions exist where intracellular ice may confer cryoprotection, a strategy that has been overlooked in our development of protocols for the cryopreservation of cells and tissues.

Recent work has focused on the mechanism by which innocuous intracellular ice functions to protect cells from cryoinjury. We propose that if cryoinjury occurring during cooling results from excessive changes in cell volume during freezing and thawing, then preventing osmotic volume changes would protect cells from this injury. Once intracellular ice forms, there is no longer a driving force for the efflux of water across the plasma membrane during cooling, as osmotic equilibrium is maintained by the formation of more intracellular ice. Innocuous intracellular ice will therefore eliminate osmotic cell shrinkage during cooling. This proposed theory is consistent with our understanding of osmotic poration.

The concepts arising from our work clearly indicate a need to reassess our understanding of intracellular ice formation and the role of intracellular ice in biological systems. We contend that it is not the intracellular ice that is lethal to cells, but the mechanism of formation, and that innocuous intracellular ice is a potentially important means to protect cells from freezing injury. The dramatic differences we have observed in the effect of intracellular ice indicates the need for a better understanding of the differences between the low-temperature response of cells in suspension and the constituent cells of tissues.

2.6   ICE GROWTH IN TISSUES

Although techniques for the successful cryopreservation of isolated cells have been available for over 50 years, there has been little success in scaling these techniques to cryopreserve organized tissues or organs. The ability to model cellular responses to freezing and thawing has certainly not yet delivered on its promise of facilitating the cryopreservation of tissues. This failure has led us to question the underlying assumptions that are currently being used to develop cryopreservation techniques. The basic method comes from consideration of the two types of cellular cryoinjury mentioned earlier; cooling too quickly leads to intracellular freezing (and cell death), whereas cooling too slowly leads to solution-effects injury (and cell death). The strategy is to cool as rapidly as possible while avoiding intracellular ice so as to minimize the time of exposure to concentrated solutes. Also, chemical cryoprotective additives are added to reduce the concentration of solutes at a given temperature. Under this premise, one need only know the phase diagram for the extracellular solution and the permeability parameters for a particular cell type (to predict the cooling rates that will allow the cell to remain in osmotic equilibrium, as well as the rates that will cause sufficient supercooling of the cytoplasm so that IIF occurs) to plan a cryopreservation protocol. When scaling this program up to tissues, where cells are embedded in an organized three-dimensional structure, it is assumed that the physical and chemical conditions produced by freezing are the same as in bulk solution and that the cellular responses will also be the same.

One of the motivating factors for pursuing this line of inquiry is the curious case of articular cartilage. Articular cartilage is the tissue that is located at the ends of the long bones, distributing load between the bones and providing a near-frictionless surface for articulation. The transplantation of bone is second only to blood in terms of the frequency with which it is used in modern medical practice. Unfortunately, all attempts to store bone with articular cartilage for transplantation have failed to keep the cartilage viable (and the body has no way to repair or regenerate this tissue). Recent work that attempted to scale cryopreservation techniques for isolated chondrocytes (the cells that reside in and maintain the articular cartilage) to the intact tissue met with very poor results (Muldrew et al., 1994, 1996). Follow-up work led to the hypothesis that the structure of the tissue was affecting ice growth and thereby negating all the assumptions that went into designing the cryopreservation protocol (Muldrew et al., 2000b, 2001a, 2001b).

Articular cartilage basically consists of an extracellular matrix with cells embedded within it. The matrix is about 80% water but has a highly organized network of collagenous fibers that constrain the so-called ground substance (a collection of enormous molecules with a high degree of exposed negative charges). Whereas the fibers give the cartilage tensile strength, the ground substance draws water into the matrix and restricts its movement in compression. Nutrients and waste products of the cells must diffuse through the ground substance as well. The structure can be considered as a capillary-porous network in which the water channels exist in small pores that are connected through tortuous paths.

Figure 2.13 is a micrograph of the structure of frozen articular cartilage. The outer surface of the cartilage is visible at the top of the image, whereas the surface on the right side was created with a scalpel blade before freezing. The large holes inside the matrix are lacunae, where the chondrocytes were situated. The two distinct regions visible in the figure (the dense region adjacent to a surface and the spongy region of the interior) are caused by fundamentally different ice morphologies in these regions (the sample was freeze-substituted; thus, vacancies in the image show where ice crystals were). Near the surface of the tissue, ice appears to have grown into the matrix as a single crystal, leaving the tissue without apparent disruptions. Although cartilage is a porous medium, the pore size is on the order of 5 nm (Maroudas, 1970); therefore, we do not expect to see any aqueous vacancies in the tissue except for the lacunae at this magnification. The matrix of the interior region, however, is porous and open, suggesting polycrystalline ice formation and mechanical disruption of the matrix architecture. The growth of ice in this region is of a completely different nature from that near the surface. Since the recovery of cells exactly correlates with these two different ice morphologies (live cells in the outer dense region and dead cells in the spongy interior; Muldrew et al., 2000b), understanding ice growth in this tissue appears to be crucial to understanding cryoinjury.

If this phenomenon were limited to articular cartilage, it would have an important but specialized application to cryobiology. However, all mammalian tissues have connective tissue as part of their

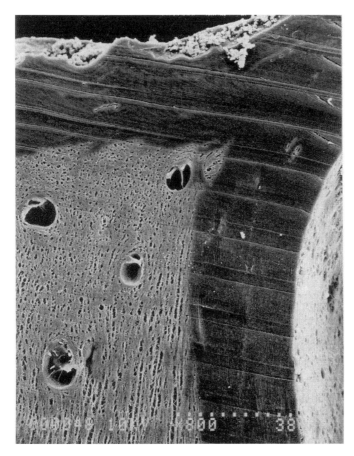

FIGURE 2.13 Scanning electron micrograph showing a cartilage section following freeze substitution (the tissue was cooled at 1°C/min and then freeze substituted without thawing). The top surface is the articular surface and the surface on the right is a scalpel cut that was administered before freezing. The vacant holes are chondrocyte lacunae. The spongy interior is thought to be to the result of mechanical damage to the matrix caused by ice lenses. Bar, 38 μm.

architecture (connective tissue refers to the extracellular matrix, usually collagen based; articular cartilage is a rather specialized form where almost the entire tissue is extracellular matrix). Thus, an understanding of ice growth in connective tissue is relevant to the whole field of medical cryobiology. Before pursuing the specific case of articular cartilage further, we will revisit the process of ice growth in general.

The structure of ice is such that it does not allow the inclusion of impurities, except within defects in the crystal structure. Thus, when an ice nucleus begins to grow, any solutes that are present in the liquid will be excluded from this growing ice front. If the rate of crystal growth is faster than the rate at which diffusion of the particular solutes can carry them away from the ice front, then a concentration gradient will very quickly form in the liquid that surrounds the ice crystal. The concentrated solute in the liquid phase, just in advance of the ice front, will then lower the freezing point of the solution (Figure 2.14). When a sufficient amount of ice has formed, then the solution at the interface will have a freezing point equal to the temperature of the interface; at this point, if the ice interface is planar (and stable), then continued growth will be limited by diffusion of the solute away from the crystal. If this situation arises when the solution well away from the ice crystal (in which the solute is not concentrated) is supercooled (at a temperature below the melting point), then we have constitutional supercooling of the solution in this region. Eventually

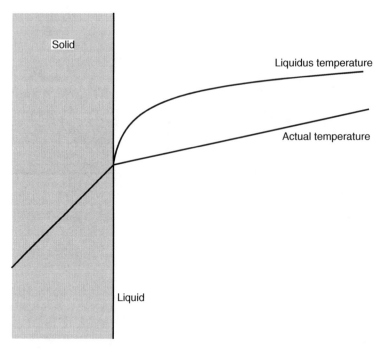

FIGURE 2.14 Constitutional supercooling occurs in advance of the ice front because of a high concentration of excluded solutes immediately adjacent to the interface.

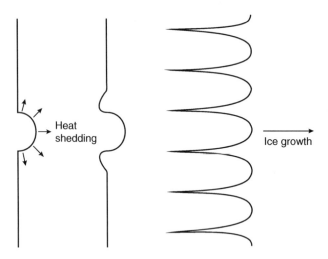

FIGURE 2.15 Ice growing into supercooled solution has an unstable interface because of the advantage that a protruding region has for shedding its latent heat into a colder region of liquid.

diffusion will ensure that the system goes to equilibrium; however, an unstable situation is created when this occurs.

The growth of ice in bulk solution is only planar for the case when heat is removed through the solid phase (i.e., ice is growing on the surface of a heat sink). When nucleation occurs in the solution (such that the crystal is completely surrounded by liquid, so-called equiaxed growth), then the interface is unstable because of the release of heat from crystallization. The heat is created at the crystal surface and must either be shed through the crystal or through the liquid (Figure 2.15).

FIGURE 2.16 Growth of ice into supercooled solution proceeds as dendritic growth. The branches of ice have smaller branches growing off them in turn. Solute is encapsulated between the dendrites.

The removal of this heat occurs by conduction and can only occur through the liquid if it is supercooled. If the latent heat of fusion is conducted away through the ice (e.g., if the crystal is in connection with a heat sink), then the growing crystal will remain essentially smooth, as any part of the interface that grows beyond the planar front will not be able to lose its heat as quickly as the ice on either side of it; thus, these instabilities quickly die out. If the heat is conducted away through the liquid, however, then there are several aspects in which crystal growth is altered. For instance, growth occurs preferentially along the a-axis compared with growth along the c-axis: This occurs because of the rise in temperature of the liquid surrounding the crystal. As the molecules become more energetic, they are less likely to join a planar surface where they can only hydrogen bond with a single neighbor. On an a-axis face, they can bond with at least two neighbors, losing more of their kinetic energy to potential energy of bond formation. Thus, we see a symmetry about a hexagonal disk in ice crystals growing under such circumstances.

The existence of constitutional supercooling in advance of a growing ice crystal results in an unstable situation. Because there is a gradient in the degree of supercooling, which is maximal a little way out in front of the interface, a planar ice front will be susceptible to small perturbations. If a local region of the interface advances just slightly ahead of the plane, then its growth rate will increase as well. This is because of the fact that it will now be able to shed its latent heat of fusion to a greater volume of liquid, as well as to liquid that is more supercooled. Such an instability will grow through the supercooled region until the supercooling is reduced to the level at which the rate of growth is limited by the conduction of latent heat once again. In fact, an entire planar interface will form an array of these "cells" when constitutional supercooling occurs.

Further to the conduction of latent heat, there is also the destabilizing effect of solute exclusion. The (ice) cells will exclude solute to the sides as well as in front; therefore, the regions between cells will contain concentrated pockets of solute. If the conditions leading to cellular growth are particularly pronounced, then the cells may turn to dendrites—protuberances that start to grow side branches (Figure 2.16). The sides of the cells become plane fronts in themselves—also subject to the same destabilizing effects of heat conduction and solute exclusion. Because the ice crystal is built on a hexagonal symmetry, these side branches will follow that symmetry.

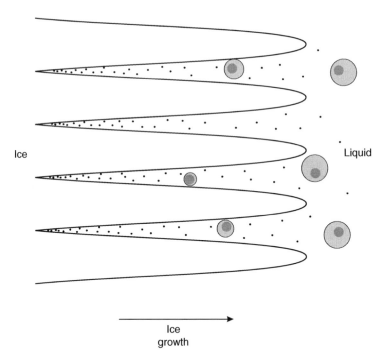

FIGURE 2.17 Dendritic ice growth encapsulates solute and biological cells between the dendrites; the unfrozen solution becomes progressively more concentrated causing the cells to lose water.

Once dendritic breakdown (the formation of growing dendrites from a planar interface) occurs, then the excluded solute will be encapsulated in channels that are sandwiched between the growing dendrites. If the freezing is initiated by a single nucleation event, then these channels will all be connected. Any (biological) cells that are present in the solution will be encapsulated in these channels along with the solute that was originally present in the solution (Figure 2.17).

Ice growth in bulk physiological solution proceeds dendritically, allowing the ice crystal to extend throughout the solution and encapsulate solute in unfrozen channels. No region of the solution is shielded from the ice crystal; thus, significant supercooling does not occur unless the sample is small and the cooling very efficient. In contrast, however, ice growth in a capillary-porous medium can proceed very differently, as the solid regions of the medium impose additional constraints on ice morphology. Ice growth in a narrow capillary proceeds as a confined crystal with a hemispherical interface with liquid water (Figure 2.18). The radius of the capillary, as well as the contact angle between ice and the capillary wall, imposes a curvature on the interface. The temperature at which a crystal of a given radius will be thermodynamically stable is given by the Kelvin equation, which shows that as capillaries get to the submicron size, the curvature effects can be appreciable. It is also possible for a capillary-porous medium to have capillaries that are too small to allow cellular or dendritic ice growth, and all solute will be excluded at the ice front. The rate of ice growth will then be limited by the diffusion of solutes away from the ice interface (diffusion also has to proceed through the same channels), as the solutes cannot be encapsulated between dendrites.

Ice growing into such a medium from one side could appear macroscopically planar in cross-section because of the growth of many ice dendrites through the tortuous paths defined by the medium. Though a real capillary-porous medium will have a complicated topology (it is not simply connected) and nonuniform porosity, we will consider the simpler model in which the aqueous channels are parallel and separate from each other (Figure 2.19).

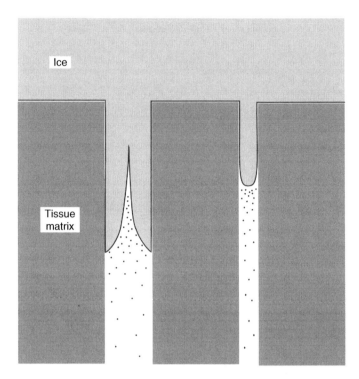

FIGURE 2.18 Ice dendrites have a minimum radius for a given temperature; therefore, capillaries that are below some critical radius will not allow the excluded solutes to be encapsulated in the unfrozen channels between dendrites. Ice growth in capillaries below this radius will then be rate-limited by diffusion of the solutes away from the interface because diffusion will have essentially one dimension, rather than two.

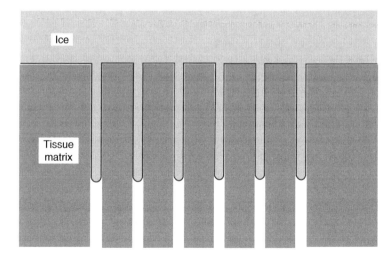

FIGURE 2.19 Schematic of a capillary-porous medium in which ice growth appears to be planar macroscopically because of growth along porous channels from the outer surface. The rate of ice growth in these channels (and hence through the medium) is diffusion limited rather than temperature limited.

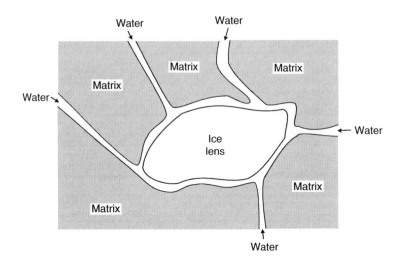

FIGURE 2.20 When an ice crystal cannot grow through the aqueous channels adjacent to it because of the necessity of a highly curved interface in the channel, water is drawn through the channels to the ice crystal. Equiaxed growth in this situation forms an ice lens that can produce substantial mechanical forces.

In a porous medium with heterogeneous interstitial spaces, ice will form initially at the surfaces (if ice is present externally) or in the larger aqueous cavities during freezing. For example, if ice is growing into a capillary-porous tissue, then the growth of ice will be diffusion-limited. Continued cooling of this tissue will lead to substantial constitutional supercooling of the interior, making nucleation of ice likely. Once an ice crystal forms in a cavity in such a tissue, if it is connected to water sources through capillaries that are small enough so that freezing cannot occur within them, then water will flow to the ice crystal, leaving part of the sample in a desiccated state. The ice crystal in the cavity continues to grow, generating mechanical forces that are responsible for enlarging the cavity. In soils, where this phenomenon occurs frequently, the ice cavities are usually shaped like a convex lens and are thus called "ice lenses" (Figure 2.20).

Because all biological tissues contain an extracellular matrix that consists of a macromolecular framework with a porous aqueous component, considerations of ice growth through capillary porous media are relevant to the freezing of biological tissues. Musculoskeletal tissues, in which extracellular matrix is a primary structural component, are likely to show effects that dominate the cryobiological behavior of the tissue. Even in more cellular tissues, the presence of basement membranes and blood vessels (with tight junctions between the endothelial cells and a connective tissue basal lamina) could severely affect the growth of ice and the redistribution of water during freezing. The presence of ice on one side of an extracellular matrix boundary will preferentially draw water from the other side if the temperature is too high for ice to grow through the porous structure of the matrix. For example, ice is very likely to grow through the vascular channels of a tissue without impedance, but it may not be able to grow into the interstitial compartment. As the temperature is dropped, the interstitial space will become increasingly dehydrated and the ice crystal in the vasculature will continue to expand, analogous to the aforementioned ice lens in soil. Both mechanical and osmotic stresses could disrupt the integrity of the tissue architecture to an extent where the tissue will not be able to function on thawing. Cryoinjury caused by tissue structure (and cryodestruction of the tissue) has to be considered, in addition to cellular cryoinjury, when organized tissues are frozen.

Cells, too, may have additional modes of cryoinjury when ice growth is altered by extracellular matrix. If ice growth is rate-limited by the diffusion of solutes away from the ice front, the consequence of further lowering of the temperature will be an increased supercooling of the interior of the tissue and, eventually, spontaneous nucleation at many sites within the matrix and the

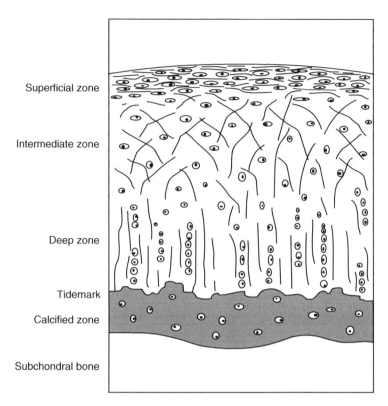

FIGURE 2.21 A cross-sectional view of articular cartilage showing the various morphological zones. Cell survival was restricted to the top portion of the superficial zone and most of the deep zone following cryopreservation. (Reprinted from Muldrew et al., 1994. With permission.)

formation of ice lenses. Because ice lenses draw water to them from every direction, they can expand against a resisting force and create mechanical effects. Cells adjacent to an ice lens may be disrupted by mechanical forces resulting from ice growth. As a further consequence of ice lens formation, the distribution of water is altered. Although water and solute would be drawn through the capillaries to an ice lens, only pure water will join the crystal. Because the crystal can enlarge without forming a network of aqueous channels (as occurs with dendritic ice growth), on warming, the potential exists for cells to be subject to transient hypotonic stresses. Thus, the formation of ice lenses could produce mechanical forces and osmotic stresses that lead to cellular injury and the disruption of tissue architecture, acting in parallel with the conventional mechanisms of cryo-injury.

We have used articular cartilage as a model to study ice growth in biological tissues because of its relative simplicity. The tissue has neither nerves nor blood vessels, and only a single cell type (the chondrocyte), yet there is still a great deal of complex biology that occurs in the tissue. The cells in adult articular cartilage are immobile, and they are not active in the cell cycle (Figure 2.21). They do, however, monitor the integrity of the extracellular matrix in which they are embedded (the functional component of the tissue—it distributes loads between bones and allows smooth articulation of the joints) and remodel it to suit the biomechanical environment. This process involves the trafficking of enzymes and structural molecules extracellularly to effect the remodeling process; as yet, these processes are but poorly understood. Nevertheless, from a structural point of view, we will consider the tissue to be composed of cells embedded in a homogeneous capillary-porous medium.

Our initial attempts to cryopreserve articular cartilage were based on a rationale that treated the cartilage matrix as bulk solution (i.e., we assumed that ice would grow dendritically throughout the matrix). Working with osteochondral dowels (cylinders of cartilage and bone of about 1 cm diameter and 1 cm length), and using Me_2SO as a cryoprotectant, we were able to achieve a 40% cell survival (assessed by measuring the integrity of the plasma membrane) following storage in liquid nitrogen (Muldrew et al., 1994). The pattern of recovery was far from random, however, with a thin zone of recovery adjacent to the articular surface and a larger zone of recovery adjacent to the cartilage-bone interface; there were no surviving cells in the middle of the tissue when viewed in cross section (Muldrew et al., 1994). When cryopreserved osteochondral dowels were transplanted into sheep and left for up to 12 months, the pattern of recovery seen just after thawing was identical to that found following implantation in a live host, although cells without an intact membrane before transplantation had been resorbed in the live animal, leaving an acellular zone in the interior of the matrix (Hurtig et al., 1998; Schachar et al., 1999).

The two hypotheses that could explain this differential recovery were that the cells of the different zones had different osmotic behaviors or that the structure of the matrix was imposing different physicochemical conditions in the interior as compared with those found near the surfaces. To test these ideas, we made scalpel cuts in the cartilage, extending well into the interior zone where the cells had been killed previously. These cuts opened up new surfaces in the cartilage matrix, thereby putting some of the interior cells adjacent to a surface. After freezing and thawing, the cells adjacent to the cut surface showed identical recovery to the cells adjacent to the articular surface (Muldrew et al., 2000b), providing evidence that the pattern of cryoinjury in articular cartilage was caused by the physico-chemical environment in the matrix rather than by a difference in cell types.

Figure 2.13 shows the ice morphology (using freeze-substitution and scanning electron microscopy [SEM]) in the cartilage matrix of osteochondral dowels that had been cut before being frozen. In this study we found a remarkable correlation between the presence of vacancies in the tissue (left behind by ice crystals) and the zone of cell death (Muldrew et al., 2000b). The interior of the cartilage matrix (where the cells had died) had a spongy appearance with large vacancies on the order of 1 μm across, at least 100 times the average pore size in normal cartilage. Ice growth in this region had clearly caused a mechanical disruption to the tissue structure, whereas the region adjacent to the surfaces (where the cells survived) had no visible vacancies, as would be expected if the pore size remained on the order of 10 nm (Muldrew et al., 2000b). The boundary between these two regions, as with the boundary between live and dead cells, was a sharp line, exactly parallel to the outer surfaces.

We hypothesized that ice was growing into the cartilage matrix in a planar fashion rather than dendritically, at least macroscopically. On a microscopic level, the ice would grow through the aqueous channels as dendrites, but within each channel, the curvature imposed on the ice interface would be too high to allow more than one dendrite to form in a channel. The effect of such a mode of ice growth would be to impose a diffusion-limited growth rate on each ice dendrite, and hence on the macroscopic ice front as a whole. The rate that ice could grow into the matrix from the surface would then be limited by diffusion of solutes away from the interface (toward the interior of the cartilage), even in the presence of a steep temperature gradient. A consequence is that the interior of the tissue would become constitutionally supercooled, making nucleation events more likely. Any ice crystal that formed in the interior of the matrix because of a nucleation event would thus become an ice lens, drawing water from the nearby matrix and expanding the porosity of the matrix. This would explain the spongy morphology of the matrix seen with freeze substitution and also give a plausible reason for cell death. The chondrocytes in this region may have been injured by mechanical effects of ice lenses, or the redistribution of water that accompanies the formation of ice lenses may have imposed an osmotic stress that led to cell injury.

To test this hypothesis, we decided to look at the rate of cooling. If the growth of ice into the matrix was truly diffusion-limited, then we could allow ice growth to proceed further into the tissue—and reduce the likelihood of interior nucleation events—by slowing the cooling rate. Because this would also have the effect of increasing the solution-effects injury to the cells that were encapsulated in the ice growing into the tissue, we attempted to keep the temperature high, where solution-effects injury accumulates slowly, and to then proceed with freezing at a much faster rate once lower temperatures were reached. To simplify the procedure, we used holding times at various subzero temperatures rather than various cooling rates; the samples were held at –4°C for 60 min, –8°C for 30 min, and –40°C for 10 min and then plunged into liquid nitrogen. Using this method, we were able to see the overall cell recovery almost double, compared with the 1°C/min protocol, and the regions of cell recovery extend much further into the matrix, though there remained a zone of dead cells in the middle (Muldrew et al., 2001b).

Although this appeared to be a substantial improvement over previous techniques for cryo-preserving articular cartilage, when we transplanted this tissue into sheep we found that it was far from optimal. At 3 months following transplantation, the number of live cells was dramatically lower than we had observed immediately after thawing (Muldrew et al., 2001a). This was a surprise because our previous experience had shown that a membrane integrity assay had given a reliable indication of cell death caused by cryoinjury in articular cartilage (because chondrocytes are immobile, we could compare the results of the membrane integrity assay with the cells that remained in the host following transplantation). As a further surprise, the chondrocytes that remained after 3 months in the host had reentered the cell cycle, and by 12 months there were chondrocyte clusters, clonal in origin, that consisted of over 100 cells in some cases (Muldrew et al., 2001a). We hypothesized that the extended holding period at high subfreezing temperatures, although facilitating ice growth into the tissue, was responsible for the loss of cellular processes that extended into the cartilage matrix (because of hypertonic shrinkage). Because the normal chondrocyte phenotype is dependent on these connections, their loss may have induced the cells to either initiate apoptosis or to reenter the cell cycle (both of which actions are seen in late-stage osteoarthritis when cell-matrix connections are lost because of matrix degradation).

Although the successful cryopreservation of articular cartilage remains an elusive goal, there are lessons that may be applied to tissue cryopreservation in general. The most important are that the tissue structure can alter the growth of ice, which can lead to osmotic conditions that are far from the phase diagram; that ice lenses can cause mechanical and osmotic stresses; and that even novel modes of cryoinjury may occur (because of altered cell-matrix interactions), when compared with the cryopreservation of cells in bulk solution. Any attempt to optimize cell recovery within a tissue using mathematical models should not assume that the cells are on the liquidus curve of the phase diagram or that simple mechanisms of cellular injury are dominant. Articular cartilage is unique in that it has almost no capacity to repair any injury (and thus cryopreservation must be almost perfect for successful transplantation), but it has been precisely this property that has shown us that the problem is much more complex than the mere cryopreservation of cells. Although empirical methods have been very successful in developing cryopreservation techniques for different cells, a sound understanding of ice growth, and the cellular responses to that growth, will be more likely to lead to success with tissues.

2.7 MELTING OF ICE

During rewarming at temperatures above the eutectic temperature, as ice melts, the concentration of solutes in the residual phase decreases. For cells equilibrated by dehydration during slow cooling, there is a progressive rehydration during warming (Figure 2.22). Most practical protocols for cryopreservation use the highest cooling rate that avoids intracellular freezing to minimize cryo-injury. These cells are, therefore, not in osmotic equilibrium during cooling, so there will be further dehydration occurring during the warming phase (Figure 2.22) as cells respond to the lower extracellular chemical potential. However, as more ice melts, diluting the extracellular solution, the water flux is reversed, and the cell becomes rehydrated.

For cells containing intracellular ice, recrystallization is a thermodynamic phenomenon that occurs during thawing and that is thought to add further to cryoinjury. Small ice crystals will have a higher internal pressure than larger ice crystals because of their increased interfacial curvature

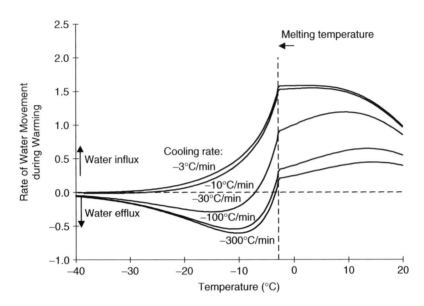

FIGURE 2.22 The water movement across the plasma membrane of cells during warming at 100°C/min after cooling at different rates to –40°C in the presence of 1 M Me$_2$SO.

and the existence of ice–solution interfacial tension. Because of the dependence of chemical potential on pressure, large ice crystals may grow while small ice crystals are melting at a given temperature. These growing ice crystals may cause damage to cells (Mazur, 1984).

2.8 CONCLUSIONS AND FUTURE DIRECTIONS

The ability to cryopreserve many cellular systems with high postthaw viability has led to many important developments in agriculture, animal husbandry, and medicine. Concomitant increases in understanding the behavior of cells at the ice–water interface have refined these processes to the point at which many cellular systems are routinely preserved with high recovery. Despite these advances, new challenges for cellular preservation are emerging as the requirements for even higher cell survival is becoming important, as in the areas of stem cell preservation, where a limited number of valuable cells—natural or engineered—are available for specific applications. There also are still some important cell types for which effective methods for cryopreservation are not available, such as sperm cells and oocytes from several species and some blood cells (platelets and granulocytes). The detailed mechanisms of injury need to be explored and understood, both to provide a rational basis for extrapolating cryopreservation techniques and for discovering novel techniques that may provide unforeseen benefits.

The cellular studies have provided a starting point for modeling low-temperature responses of the cells within a tissue matrix, but the extrapolation is not trivial. At the very least, tissues are three-dimensional structures, so heat and mass transfer within the tissue are important factors. Tissues generally contain several cell types, with specific characteristics that affect low-temperature responses. Cell–cell and cell–matrix adhesions strongly influence low-temperature responses, and these interactions appear to be the factors limiting cryopreservation of tissues. Therefore, the limits to our understanding on low-temperature responses of biological systems are most evident at the ice–water interface in tissue systems. Like natural tissues, engineered tissues are constructed in a variety of configurations—encapsulated beads, monolayers on flat or spherical surfaces, embedded sheets, and three-dimensional structures with different morphologies.

Cryopreservation is often the only method for preserving physiological structure, viability, and function in living tissues for extended periods of time, so considerable research efforts around the world are now exploring methods for cryopreservation of tissue systems, motivated by the fact that, in medicine, tissue transplantation is increasingly being used in the treatment of a variety of disorders, and engineered tissues are emerging for many biomedical applications. Our ability to preserve these natural or engineered constructs for distribution has lagged significantly behind their development and use. The pending use of engineered tissues (corneal, skin, liver) for toxicology testing has increased the imperative for effective methods for routine preservation. Current indications are that low-temperature responses of engineered tissues are similar to those of natural tissues.

Experience over the last few decades has demonstrated that low-temperature responses of tissues are too complex for purely empirical development of effective cryopreservation protocols, so conceptual developments at the ice–water interface for tissues will require increased understanding to guide empirical experimentation.

REFERENCES

Abidor, I.G., Arakelyan, V.B., Chernomordik, L.V., Chizmadzhev, Y.A., Pastushenko, V.F., and Tarasevich, M.R. (1979) Electric breakdown of bilayer lipid membranes I. The main experimental facts and their qualitative discussion, *J. Electroanal. Chem.*, 104, 37–52.

Acker, J.P., Elliott, J.A.W., and McGann, L.E. (2001) Intercellular ice propagation: Experimental evidence for ice growth through membrane pores, *Biophys. J.*, 81, 1389–1397.

Acker, J.P., Larese, A., Yang, H., Petrenko, A., and McGann, L.E. (1999a) Intracellular ice formation is affected by cell interactions, *Cryobiology*, 38, 363–371.

Acker, J.P. and McGann, L.E. (1998) The role of cell-cell contact on intracellular ice formation, *Cryo-Letters*, 19, 367–374.

Acker, J.P. and McGann, L.E. (2000) Cell–cell contact affects membrane integrity after intracellular freezing, *Cryobiology*, 40, 54–63.

Acker, J.P. and McGann, L.E. (2001) Membrane damage occurs during the formation of intracellular ice, *Cryo-Letters*, 22, 241–254.

Acker, J.P. and McGann, L.E. (2002) Innocuous intracellular ice improves survival of frozen cells, *Cell Transplant.*, 11, 563–571.

Acker, J.P. and McGann, L.E. (2003) Protective effect of intracellular ice during freezing? *Cryobiology*, 46, 197–202.

Acker, J.P., Pasch, J., Heschel, I., Rau, G., and McGann, L.E. (1999b) Comparison of optical measurement and electrical measurement techniques for the study of osmotic responses of cell suspensions, *Cryo-Letters*, 20, 315–324.

Adams, R.B., Voelker, W.H., and Gregg, E.C. (1967) Electrical counting and sizing of mammalian cells in suspension: An experimental evaluation, *Phys. Med. Biol.*, 12, 79–92.

Aggarwal, S.J., Diller, K.R., and Baxter, C.R. (1984) Membrane water permeability of isolated skin cells at subzero temperatures, *Cryo-Letters*, 5, 17–26.

Armitage, W.J. and Juss, B.K. (1996) Osmotic response of mammalian cells: Effects of permeating cryo-protectants on nonsolvent volume, *J. Cell Physiol*, 168, 532–538.

Asahina, E. (1956) The freezing process of plant cell, *Contrib. Inst. Low Temp. Sci. Ser. B*, 10, 83–126.

Asahina, E. (1962) Frost injury in living cells, *Nature*, 196, 445–446.

Asahina, E., Shimada, K., and Hisada, Y. (1970) A stable state of frozen protoplasm with invisible intracellular ice crystals obtained by rapid cooling, *Exp. Cell Res.*, 59, 349–358.

Ashwood-Smith, M.J., Morris, G.W., Fowler, R., Appleton, T.C., and Ashorn, R. (1988) Physical factors are involved in the destruction of embryos and oocytes during freezing and thawing procedures, *Hum. Reprod.*, 3, 795–802.

Berger, W.K. and Uhrik, B. (1992) Dehydration and intracellular ice formation during freezing in single cells and cell strands from salivary glands, *Cryobiology*, 29, 715–716.

Berger, W.K. and Uhrik, B. (1996) Freeze-induced shrinkage of individual cells and cell-to-cell propagation of intracellular ice in cell chains from salivary glands, *Experientia*, 52, 843–850.

Bischof, J.C. and Rubinsky, B. (1993) Large ice crystals in the nucleus of rapidly frozen liver cells, *Cryobiology*, 30, 597–603.

Bordi, F., Cametti, C., and Naglieri, A. (1998) Ionic transport in lipid bilayer membranes, *Biophys. J.*, 74, 1358–1370.

Boroske, E., Elwenspoek, M., and Helfrich, W. (1981) Osmotic shrinkage of giant egg-lecithin vesicles, *Biophys. J.*, 34, 95–109.

Brown, M.S. (1980) Freezing of nonwoody plant tissues. IV. Nucleation sites for freezing and refreezing of onion bulb epidermal cells, *Cryobiology*, 17, 184–186.

Brown, M.S. and Reuter, F.W. (1974) Freezing of nonwoody plant tissues. III. Videotaped micrography and the correlation between individual cellular freezing events and temperature changes in the surrounding tissue, *Cryobiology*, 11, 185–191.

Buckhold, B., Adams, R.B., and Gregg, E.C. (1965) Osmotic adaptation of mouse lymphoblasts, *Biochim. Biophys. Acta*, 102, 600–608.

Buitink, J., Claessens, M.M.A.E., Hemminga, M.A., and Hoekstra, F.A. (1998) Influence of water content and temperature on molecular mobility and intracellular glasses in seeds and pollen, *Plant Physiol.*, 118, 531–541.

Callow, R.A. and McGrath, J.J. (1985) Thermodynamic modeling and cryomicroscopy of cell-size, unilamellar and paucilamellar liposomes, *Cryobiology*, 22, 251–267.

Chambers, R. and Hale, H.P. (1932) The formation of ice in protoplasm, *Proc. R. Soc.*, 110, 336–352.

Chen, T., Acker, J.P., Eroglu, A., Cheley, S., Bayley, H., Fowler, A., and Toner, M. (2001) Beneficial effect of intracellular trehalose on the membrane integrity of dried mammalian cells, *Cryobiology*, 43, 168–181.

Crowe, J.H., Carpenter, J.F., and Crowe, L.M. (1998) The role of vitrification in anhydrobiosis, *Annu. Rev. Physiol.*, 60, 73–103.

Crowe, J.H. and Crowe, L.M. (2000) Preservation of mammalian cells: Learning nature's tricks, *Nature Biotechnol.*, 18, 145–146.

Crowe, J.H., Crowe, L.M., and Carpenter, J.F. (1993a) Preserving dry biomaterials: The water replacement hypothesis, Part 1, *Biopharm*, 4, 28–33.

Crowe, J.H., Crowe, L.M., and Carpenter, J.F. (1993b) Preserving dry biomaterials: The water replacement hypothesis, Part 2, *Biopharm*, 5, 40–43.

Dankberg, F. and Persidsky, M.D. (1976) A test of granulocyte membrane integrity and phagocytic function, *Cryobiology*, 13, 430–432.

Davson, H. and Danielli, J.F. (1952) *The Permeability of Natural Membranes*, The University Press, Cambridge, U.K.

Day, S.H., Nicoll-Griffith, D.A., and Silva, J.M. (1999) Cryopreservation of rat and human liver slices by rapid freezing, *Cryobiology*, 38, 154–159.

de Graaf, I.A.M. and Koster, H.J. (2001) Water crystallization within rat precision-cut liver slices in relation to their viability, *Cryobiology*, 43, 224–237.

Diller, K.R. (1975) Intracellular freezing: Effect of extracellular supercooling, *Cryobiology*, 12, 480–485.

Diller, K.R., Cravalho, E.G., and Huggins, C.E. (1972) Intracellular freezing in biomaterials, *Cryobiology*, 9, 429–440.

Dowgert, M.F. (1983) Effect of cold acclimation on intracellular ice formation in isolated protoplasts, *Plant Physiol.*, 72, 978–988.

Elliott, J.A.W. (2001) On the complete Kelvin equation, *Chem. Eng. Educ.*, 35, 274–278.

Elmoazzen, H.Y., Elliott, J.A.W., and McGann, L.E. (2002) The effect of temperature on membrane hydraulic conductivity, *Cryobiology*, 45, 68–79.

Eroglu, A., Russo, M.J., Bieganski, R., Fowler, A., Cheley, S., Bayley, H., and Toner, M. (2000) Intracellular trehalose improves the survival of cryopreserved mammalian cells, *Nat. Biotechnol.*, 18, 163–167.

Eroglu, A., Toner, M., and Toth, T.L. (2002) Beneficial effect of microinjected trehalose on the cryosurvival of human oocytes, *Fertil Steril*, 77, 152–158.

Fahy, G.M., Saur, J., and Williams, R.J. (1990) Physical problems with the vitrification of large biological systems, *Cryobiology*, 27, 492–510.

Farrant, J. and Morris, G.J. (1973) Thermal shock and dilution shock as the causes of freezing injury, *Cryobiology*, 10, 134–140.

Farrant, J., Walter, C.A., Lee, H., and McGann, L.E. (1977) Use of two-step cooling procedures to examine factors influencing cell survival following freezing and thawing, *Cryobiology*, 14, 273–286.

Farrant, J. and Woolgar, A.E. (1972) Human red cells under hypertonic conditions: A model system for investigating freezing damage, *Cryobiology*, 9, 9–15.

Finkelstein, A. (1987) *Water Movement Through Lipid Bilayers, Pores and Plasma Membrane*, Wiley, New York.

Fowler, R. and Toner, M. (1998) Prevention of hemolysis in rapidly frozen erythrocytes by using a laser pulse, *Ann. N.Y. Acad. Sci.*, 858, 245–252.

Franco, R.S., Barker, R., Novick, S., Weiner, M., and Martelo, O.J. (1986) Effect of inositol hexaphosphate on the transient behavior of red cells following a DMSO-induced osmotic pulse, *J. Cell. Physiol.*, 129, 221–229.

Franks, F., Mathias, S.F., Galfre, P., Webster, S.D., and Brown, D. (1983) Ice nucleation and freezing in undercooled cells, *Cryobiology*, 20, 298–309.

Fujikawa, S. (1980) Freeze-fracture and etching studies on membrane damage on human erythrocytes caused by formation of intracellular ice, *Cryobiology*, 17, 351–362.

Gaber, B.P., Chandrasekhar, I., and Pattabiraman, N. (1986) The interaction of trehalose with the phospholipid bilayer: A molecular modeling study, in *Membranes, Metabolism and Dry Organisms*, Leopold, A.C., Ed., Cornell University Press, Ithaca, NY, pp. 231–241.

Gao, D.Y., Benson, C.T., Liu, C., McGrath, J.J., Critser, E.S., and Critser, J.K. (1996) Development of a novel microperfusion chamber for determination of cell membrane transport properties, *Biophys. J.*, 71, 443–450.

Gao, D.Y., Chang, Q., Liu, C., Farris, K., Harvey, K., McGann, L.E., English, D., Jansen, J., and Critser, J.K. (1998) Fundamental cryobiology of human hematopoietic progenitor cells I: Osmotic characteristics and volume distribution, *Cryobiology*, 36, 40–48.

Gilmore, J.A., McGann, L.E., Ashworth, E., Acker, J.P., Raath, C., Bush, M., and Critser, J.K. (1998) Fundamental cryobiology of a selected African mammalian spermatozoa and its role in biodiversity preservation through development of genome resource banking, *Anim. Reprod. Sci.*, 53, 277–297.

Glaser, R., Leikin, S., Chernomordik, L.V., Pastushenko, V.F., and Sokirko, A. (1988) Reversible electrical breakdown of lipid bilayers: Formation and evolution of pores, *Biochim. Biophys. Acta*, 940, 275–287.

Gregg, E.C. and Steidley, K.D. (1965) Electrical counting and sizing of mammalian cells in suspension, *Biophys. J.*, 5, 393–405.

Hunt, C.J. (1984) Studies on cellular structure and ice location in frozen organs and tissues: The use of freeze-substitution and related techniques, *Cryobiology*, 21, 385–402.

Hurley, J. (1970) Sizing particles with a coulter counter, *Biophys. J.*, 10, 74–79.

Hurtig, M., Novak, K., McPherson, R., McFadden, S., McGann, L.E., Muldrew, K., and Schachar, N.S. (1998) Osteochondral dowel transplantation for the repair of focal defects in the knee: An outcome study using an ovine model, *Vet. Sur.*, 27, 5–16.

Irimia, D. and Karlsson, J.O.M. (2002) Kinetics and mechanism of intercellular ice propagation in a micro-patterned tissue construct, *Biophys. J.*, 82, 1858–1868.

Ishiguro, H. and Rubinsky, B. (1994) Mechanical interactions between ice crystals and red blood cells during directional solidification, *Cryobiology*, 31, 483–500.

Johnson, J. and Wilson, T. (1967) Osmotic volume changes induced by a permable solute, *J. Theor. Biol.*, 17, 304–311.

Karlsson, J.O.M. (2001) A theoretical model of intracellular devitrification, *Cryobiology*, 42, 154–169.

Karlsson, J.O.M., Cravalho, E.G., and Toner, M. (1993) Intracellular ice formation: Causes and consequences, *Cryo-Letters*, 14, 323–334.

Kedem, O. and Katchalsky, A. (1958) Thermodynamic analysis of the permeability of biological membranes to non-electrolytes, *Biochim. Biophys. Acta*, 27, 229–246.

Keeley, R.L.A., Gomez, A.C., and Brown, I.W. (1952) An experimental study of the effects of freezing, partial dehydration and ultra-rapid cooling on the survival of dog skin grafts, *Plast. Reconstr. Sur.*, 9, 330–344.

Kirkwood, J.G. (1954) Transport of ions through biological membranes from the standpoint of irreversible thermodynamics, in *Ion Transport Across Membranes*, Clarke, H.T., Ed., Academic Press, New York, pp. 119–127.

Kleinhans, F.W. (1998) Membrane permeability modeling: Kedem-Katchalsky vs. a two parameter formalism, *Cryobiology*.

Kubitschek, H.E. (1958) Electronic counting and sizing of bacteria, *Nature*, 182, 234–235.

Larese, A., Acker, J.P., Muldrew, K., Yang, H., and McGann, L.E. (1996) Antifreeze proteins induce intracellular nucleation, *Cryo-Letters*, 17, 175–182.

Levitt, J. (1962) A sulfhydryl-disulfide hypothesis of frost injury and resistance in plants, *J. Theor. Biol.*, 3, 355–391.

Levitt, J. (1966) Winter hardiness in plants, in *Cryobiology*, Meryman, H.T., Ed., Academic Press, London, pp. 495–563.

Levitt, J. and Scarth, G.W. (1936) Frost hardening studies with living cells. II. Permeability in relation to frost resistance and the seasonal cycle, *Can. J. Res.*, 14, 285–305.

Lovelock, J.E. (1953a) The haemolysis of human red blood cells by freezing and thawing, *Biochim. Biophys. Acta*, 10, 414–426.

Lovelock, J.E. (1953b) The mechanism of the protective action of glycerol against haemolysis by freezing and thawing, *Biochim. Biophys. Acta*, 11, 28–36.

Luyet, B.J. and Gehenio, P.M. (1940) The mechanism of injury and death by low temperature, *Biodynamica*, 3, 33–99.

Luyet, B.J. and Gibbs, M.C. (1937) On the mechanism of congelation and of death in the rapid freezing of epidermal plant cells, *Biodynamica*, 1–18.

Luyet, B.J. and Hodapp, E. (1938) Revival of frog's spermatozoa vitrified in liquid air, *Proc. Soc. Exp. Biol.*, 39, 433–444.

Luyet, B.J. and Rapatz, G. (1958) Patterns of ice formation in some aqueous solutions, *Biodynamica*, 8, 1–68.

Maroudas, A. (1970) Distribution and diffusion of solutes in articular cartilage, *Biophys. J.*, 10, 365–379.

Mazur, P. (1960) Physical factors implicated in the death of microorganisms at subzero temperatures, *Ann. N.Y. Acad. Sci.*, 85, 610–629.

Mazur, P. (1963a) Kinetics of water loss from cells at subzero temperatures and the likelihood of intracellular freezing, *J. Gen. Physiol.*, 47, 347–369.

Mazur, P. (1963b) Studies on rapidly frozen suspensions of yeast cells by differential thermal analysis and conductometry, *Biophys. J.*, 3, 353.

Mazur, P. (1965) The role of cell membranes in the freezing of yeast and other single cells, *Ann. N.Y. Acad. Sci.*, 125, 658–676.

Mazur, P. (1966) Physical and chemical basis of injury in single-celled micro-organisms subjected to freezing and thawing, in *Cryobiology*, Meryman, H.T., Ed., Academic Press, London, pp. 213–315.

Mazur, P. (1977) The role of intracellular freezing in the death of cells cooled at supraoptimal rates, *Cryobiology*, 14, 251–272.

Mazur, P. (1984) Freezing of living cells: Mechanisms and implications, *Am. J. Physiol.*, 247, C125–C142.

Mazur, P. (1990) Equilibrium, quasi-equilibrium, and nonequilibrium freezing of mammalian embryos, *Cell Biophys.*, 17, 53–92.

Mazur, P. and Koshimoto, C. (2002) Is intracellular ice formation the cause of death of mouse sperm frozen at high cooling rates? *Biol. Reprod.*, 66, 1485–1490.

Mazur, P., Leibo, S.P., and Chu, E.H.Y. (1972) A two-factor hypothesis of freezing injury—evidence from Chinese hamster tissue-culture cells, *Exp. Cell Res.*, 71, 345–355.

McGann, L.E. and Farrant, J. (1976a) Survival of tissue culture cells frozen by a two-step procedure to –196°C. I. Holding temperature and time, *Cryobiology*, 13, 261–268.

McGann, L.E. and Farrant, J. (1976b) Survival of tissue culture cells frozen by a two-step procedure to –196°C. II. Warming rate and concentration of dimethyl sulphoxide, *Cryobiology*, 13, 269–273.

McGann, L.E., Stevenson, M., Muldrew, K., and Schachar, N.S. (1988) Kinetics of osmotic water movement in chondrocytes isolated from articular cartilage and applications to cryopreservation, *J. Orthopaed. Res.*, 6, 109–115.

McGann, L.E., Turner, A.R., and Turc, J.M. (1982) Microcomputer interface for rapid measurement of average volume using an electronic particle counter, *Med. Biol. Eng. Comput.*, 20, 117–120.

McGrath, J.J. (1985) A microscope diffusion chamber for the determination of the equilibrium and nonequilibrium osmotic response of individual cells, *J. Microscopy*, 139, 249–263.

McLeester, R.C., Weiser, C.J., and Hall, T.C. (1969) Multiple freezing points as a test for viability of plant stems in the determination of frost hardiness, *Plant Physiol.*, 44, 37–44.

Meryman, H.T. (1957) Physical limitations of the rapid freezing method, *Proc. R. Soc.*, 147, 452–459.

Meryman, H.T. (1970) The exceeding of a minimum tolerable cell volume in hypertonic suspension as a cause of freezing injury, in *The Frozen Cell*, Wolstenholme, G.E. and O'Connor, M., Eds., Churchill, London, pp. 51–64.

Meryman, H.T. (1974) Freezing injury and its prevention in living cells, *Annu. Rev. Biophys.*, 3, 341–363.

Mider, G.B. and Morton, J.J. (1939) The effect of freezing *in vitro* on some transplantable mammalian tumors and on normal rat skin, *Am. J. Cancer*, 35, 502–509.

Molisch, H. (1982) Investigation into the freezing of plants, *Cryo-Letters*, 3, 331–390.

Morris, G.J. and McGrath, J.J. (1981) Intracellular ice nucleation and gas bubble formation in spirogyra, *Cryo-Letters*, 2, 341–352.

Mugnano, J.A., Wang, T., Layne, J.R., DeVries, A.L., and Lee, R.E. (1995) Antifreeze glycoproteins promote intracellular freezing of rat cardiomyocytes at high subzero temperatures, *Am. J. Physiol.*, 269, R474–R479.

Muldrew, K., Acker, J.P., and Wan, R. (2000a) Investigations into quantitative post-hypertonic lysis theory using cultured fibroblasts, *Cryobiology*, 41, 337.

Muldrew, K., Chung, M., Novak, K., Schachar, N.S., Rattner, J.B., and Matyas, J. (2001a) Chondrocyte regeneration in adult ovine articular cartilage following cryoinjury and long-term transplantation, *Osteoarthritis Cartilage*, 9, 432–439.

Muldrew, K., Hurtig, M., Novak, K., Schachar, N.S., and McGann, L.E. (1994) Localization of freezing injury in articular cartilage, *Cryobiology*, 31, 31–38.

Muldrew, K. and McGann, L.E. (1990) Mechanisms of intracellular ice formation, *Biophys. J.*, 57, 525–532.

Muldrew, K. and McGann, L.E. (1994) The osmotic rupture hypothesis of intracellular freezing injury, *Biophys. J.*, 66, 532–541.

Muldrew, K., Novak, K., Studholme, C., Wohl, G., Zernicke, R., Schachar, N.S., and McGann, L.E. (2001b) Transplantation of articular cartilage following a step-cooling cryopreservation protocol, *Cryobiology*, 43, 260–267.

Muldrew, K., Novak, K., Yang, H., Zernicke, R., Schachar, N.S., and McGann, L.E. (2000b) Cryobiology of articular cartilage: Ice morphology and recovery of chondrocytes, *Cryobiology*, 40, 102–109.

Muldrew, K., Schachar, N.S., and McGann, L.E. (1996) Permeation kinetics of dimethyl sulfoxide in articular cartilage, *Cryo-Letters*, 17, 331–340.

Paula, S., Volkov, A.G., Van Hoek, A.N., Haines, T.H., and Deamer, D.W. (1996) Permeation of protons, potassium ions, and small polar molecules through phospholipid bilayers as a function of membrane thickness, *Biophys. J.*, 70, 339–348.

Pegg, D.E. (1984) Red cell volume in glycerol/sodium chloride/water mixtures, *Cryobiology*, 21, 234–239.

Pegg, D.E., Hunt, C.J., and Fong, L.P. (1987) Osmotic properties of the rabbit corneal endothelium and their relevance to cryopreservation, *Cell Biophys.*, 10, 169–189.

Pitt, R.E., Chandrasekaran, M., and Parks, J.E. (1992) Performance of a kinetic model for intracellular ice formation based on the extent of supercooling, *Cryobiology*, 29, 359–373.

Pitt, R.E., Myers, S.P., Lin, T., and Steponkus, P.L. (1991) Subfreezing volumetric behavior and stochastic modeling of intracellular ice formation in *Drosophila melanogaster* embryos, *Cryobiology*, 28, 72–86.

Pitt, R.E. and Steponkus, P.L. (1989) Quantitative analysis of the probability of intracellular ice formation during freezing of isolated protoplasts, *Cryobiology*, 26, 44–63.

Rall, W.F., Reid, D.S., and Farrant, J. (1980) Innocuous biological freezing during warming, *Nature*, 286, 511–514.

Rasmussen, D.H., Macaulay, M.N., and MacKenzie, A.P. (1975) Supercooling and nucleation of ice in single cells, *Cryobiology*, 12, 328–339.

Schachar, N.S., Novak, K., Hurtig, M., Muldrew, K., McPherson, R., Wohl, G., Zernicke, R., and McGann, L.E. (1999) Transplantation of cryopreserved osteochondral dowel allografts for repair of focal articular defects in an ovine model, *J. Orthopaedic Res.*, 17, 909–920.

Sherman, J.K. (1962) Survival of higher animal cells after the formation and dissolution of intracellular ice, *Anatomical Rec.*, 144, 171–177.

Shimada, K. and Asahina, E. (1975) Visualization of intracellular ice crystals formed in very rapidly frozen cells at −27°C, *Cryobiology*, 12, 209–218.

Staverman, A.J. (1948) Non-equilibrium thermodynamics of membrane processes, *Trans. Faraday Soc.*, 48, 176–185.

Steponkus, P.L. (1984) Role of the plasma membrane in freezing injury and cold acclimation, *Annu. Rev. Plant Physiol.*, 35, 543–584.

Steponkus, P.L. and Dowgert, M.F. (1981) Gas bubble formation during intracellular ice formation, *Cryo-Letters*, 2, 42–47.

Steponkus, P.L. and Dowgert, M.F. (1984) Phenomenology of intracellular ice nucleation in isolated protoplasts, *Plant Physiol.*, 67, S58.

Steponkus, P.L., Dowgert, M.F., and Gordon-Kamm, W.J. (1983) Destabilization of the plasma membrane of isolated plant protoplasts during a freeze-thaw cycle: The influence of cold acclimation, *Cryobiology*, 20, 448–465.

Steponkus, P.L., Stout, D., Wolfe, J., and Lovelace, R. (1984) Freeze-induced electrical transients and cryo-injury, *Cryo-Letters*, 5, 343–348.

Steponkus, P.L. and Wiest, S.C. (1978) Plasma membrane alterations following cold acclimation and freezing, in *Plant Cold Hardiness and Freeze Stress—Mechanisms and Crop Implications*, Li, P.H. and Sakai, A., Eds., Academic Press, New York, pp. 75–91.

Stuckey, I.H. and Curtis, O.F. (1938) Ice formation and the death of plant cells by freezing, *Plant Physiol.*, 13, 815–833.

Terwilliger, T.C. and Solomon, A.K. (1981) Osmotic water permeability of human red cells, *J. Gen. Physiol.*, 77, 549–570.

Toner, M., Cravalho, E.G., and Karel, M. (1990) Thermodynamics and kinetics of intracellular ice formation during freezing of biological cells, *J. Appl. Phys.*, 67, 1582–1593.

Toner, M., Cravalho, E.G., Stachecki, J., Fitzgerald, T., Tompkins, R.G., Yarmush, M.L., and Armant, D.R. (1993) Nonequilibrium freezing of one-cell mouse embryos: Membrane integrity and developmental potential, *Biophys. J.*, 64, 1908–1921.

Tsuruta, T., Ishimoto, Y., and Masuoka, T. (1998) Effects of glycerol on intracellular ice formation and dehydration of onion epidermis, *Ann. N.Y. Acad. Sci.*, 858, 217–226.

Verkman, A.S., Van Hoek, A.N., Ma, T., Frigeri, A., Skach, W.R., Mitra, A., Tamarappoo, B.K., and Farinas, J. (1996) Water transport across mammalian cell membranes, *Am. J. Physiol.*, 270, C12–C30.

Walcerz, D.B. and Diller, K.R. (1991) Quantitative light microscopy of combined perfusion and freezing processes, *J. Microscopy*, 161, 297–311.

Wolfe, J. and Bryant, G. (1999) Freezing, drying, and/or vitrification of membrane–solute–water systems, *Cryobiology*, 39, 103–129.

Woods, E.J., Zieger, M.A.J., Lakey, J.R.T., Liu, J., and Critser, J.K. (1997) Osmotic characteristics of isolated human and canine pancreatic islets, *Cryobiology*, 35, 106–113.

Yang, H., Chen, A., Muldrew, K., Novak, K., Zernicke, R., Schachar, N.S., and McGann, L.E. (1995) *In situ* assessment of cell viability in tissues, *Cryobiology*, 32, 582–583.

Zade-Oppen, A.M.M. (1968) Posthypertonic hemolysis in sodium chloride systems, *Acta Physiol. Scand.*, 73, 341–364.

Zieger, M.A.J., Tredget, E.E., Sykes, B.D., and McGann, L.E. (1997) Injury and protection in split-thickness skin after very rapid cooling and warming, *Cryobiology*, 35, 53–69.

Theme 2

Life and Death at Low Temperatures

3 Life in the Polar Terrestrial Environment with a Focus on Algae and Cyanobacteria*

Josef Elster and Erica E. Benson

CONTENTS

---

*Dedicated to Dr. Jan Kvĕt on the occasion of his 70th birthday and to Professor William Block, with thanks for encouraging our collaborative research

0-415-24700-4/04/$0.00+$1.50
© 2004 by CRC Press LLC

3.1 INTRODUCTION

This review pertains to the coldest regions on Earth, highlighting the importance of applying a fundamental knowledge of *in vivo* "life in the frozen state." An overview of the geographical, ecological, and physiological properties of the polar terrestrial environment and polar biodiversity is presented. Emphasis is placed on the central importance of low temperature status and water availability and the way in which these factors affect terrestrial polar habitats and define the life therein. The first part provides a general account of polar environmental research and biodiversity. Exemplar studies of selected permanent residents of Arctic and Antarctic terrestrial ecosystems (algae, higher plants, insects, and arthropods) provide an excellent opportunity for applied low-temperature biologists to explore extreme survival strategies and adaptive responses, thus offering unique insights into how life tolerates and endures the "frozen state" *in vivo*. The second part specifically focuses on physiological and biochemical adaptive responses in selected, terrestrial polar biota, and particularly algae and cyanobacteria. The chapter concludes by discussing the present and future potential uses of polar research in applied cryobiology.

3.1.1 SETTING THE SCENE: THE ARCTIC AND ANTARCTIC REGIONS

Polar regions may be defined as those areas that lie within the Arctic and Antarctic circles. Fogg (1998) confines these to parallels of latitude at 66°33' north and south, respectively, corresponding to the angle between the axis of rotation of the earth and the plane of orbit round the sun. They are vast areas, covering 16.5% of the Earth's surface and comprising 84 million km^2. The Polar regions are of outstanding international scientific and environmental significance, and increasingly studies of these areas are making major contributions to understanding global change—in terms of the Earth's ancient evolutionary history as well as with respect to contemporary climate change issues. Studies of Antarctica's ozone hole in particular assist the global monitoring of pollution effects on the Earth's outer atmosphere. The need to conserve, manage, and protect the polar regions is exemplified especially by the Protocol on Environmental Protection and the Antarctic Treaty, adopted in 1991 (see http://www.antarctica.ac.uk).

The seasonal and diurnal variations of polar terrestrial environments represent a series of water availability gradients ranging from aquatic and semiaquatic to dry habitats. These patterns differ in periodicity, amplitude, synchronicity, and regularity, and they initiate a number of different ecological and physiological acclimation and adaptation responses. However, to understand the basis of survival responses in polar regions, it is essential to appreciate that these environments are subjected not only to extremes but also to rapid fluctuations across these extremes. For example, the speed at which water states (liquid to solid ice/snow) and low temperatures can change is one of the most important ecological and physiological factors in polar regions. On a physiological level, changes in temperature and water status provoke a series of adaptive responses comprising resistance and tolerance to cold, freezing, drought, desiccation, and salinity stress. All these injurious components support, at the environmental level, the two-factor theory (see Chapter 1) of freezing injury (ice and dehydration). Interestingly, most of these ecological and physiological acclimation/adaptation strategies can also be found at temperate latitudes (cold season periodicity, mountainous, and Alpine regions), but it is considered in this chapter that the physical and chemical conditions that are uniquely imposed by polar environments have selected for a particularly specific and resilient set of adaptive biological characters.

The severe biological constraints of polar habitats include diurnal and seasonal fluctuations as well as the instabilities that arise from geological events, as over time these can affect polar organisms by accelerating evolutionary processes. In this sense, the polar terrestrial environment represents a unique natural history laboratory documenting the evolution of life (Vincent, 2000). Studies of those organisms (extremophiles) that survive and even thrive in extreme polar terrestrial habitats can provide unique insights into the mechanisms of low-temperature adaptation and tolerance.

3.1.2 Why Algae and Cyanobacteria?

These photosynthetic organisms have a distinct and important status in Arctic and Antarctic ecology, exemplified by their key role as primary producers. Terrestrial prokaryotic (cyanobacteria) and eukaryotic microalgae have therefore been selected as "model organisms," particularly as the lead author is a polar phycologist and is able to consider the broader, environmental context of *in situ* "life in the frozen state." Freshwater and soil cyanobacteria and microalgae are the main "engines" of primary production: they are widespread in all polar terrestrial environments and frequently produce visible biomass (Figure 3.1; Vincent, 1988). Algae and cyanobacteria posses an oxyphototrophic type of photosynthesis and are well adapted to the extreme natural conditions of polar terrestrial habitats, within which there is an absence of severe grazing pressure from animals and a lack of strong competition from a wide range of other species (Hawes, 1989; Howard-Williams et al., 1989; Vincent and Howard-Williams, 1986). Algae and cyanobacteria can therefore provide a valuable and informative "model" for the study of adaptive strategies required to survive in one of the Earth's coldest and most extreme environments.

3.2 PHYSICAL AND BIOLOGICAL ATTRIBUTES OF THE POLAR TERRESTRIAL ENVIRONMENT

The north and south polar regions differ considerably with respect to their physical and biological attributes and also in terms of the ability of researchers to access and monitor these respective regions. This section will overview the differences and similarities between the two areas. It will also provide a critique of the problems and solutions encountered in undertaking temperature-related polar research in such remote and extreme places.

3.2.1 Comparing the Physical Environments of the Arctic and Antarctica

Because of their different geological histories, the Arctic and Antarctic regions differ remarkably in their geology, energy transport, and balance. This is largely the result of the fact that the main part of the Arctic is covered by the Arctic Ocean, which is connected to temperate oceans, which critically affects the transport and exchange of heat energy in the region. Air circulation, influenced

FIGURE 3.1 Dry and frozen biomass morphologies of perennial, prokaryotic cyanobacteria (A, B, C) and annual eukaryotic algae (D, E, F) photographed in various localities of the Antarctica and the Arctic during which temporary summer or autumn temperatures are greater than 0°C. (A) Wet wall covered by black cyanobacteria biomass (arrow) from the northern part of Sverdrup Pass, central Ellesmere Island, 79°N, Canada. This cyanobacteria community is composed mainly of *Aphanothece, Gloeocapsa, Chroococcus* species (Chroocoocales), and *Dichothrix, Scytonema, Calothrix* species (Nostocales). (B) *Nostoc* sp. gelatinous dark brown rosette from Signy Island, 60°S maritime Antarctica photographed in dry and frozen stage (arrow) at the end of summer season. (C) Dry and frozen *Phormidium* sp. dark brown mats (arrow) from King George Island, (62°S) maritime Antarctica. (D) Dry and frozen biomass with snow flakes of *Prasiola crispa* a macroscopic green alga (arrow) producing foliose irregular, crisped monostromatic blades aggregated into dense dark-green carpets on hydroterrestrial soils in and adjacent to bird colonies photographed at Signy Island, 60°S maritime Antarctica. (E) Dry and frozen community of filamentous algae (*Klebsormidium rivulare, Klebsormidium crenulatum* and *Ulothrix mucosa*) black-green (arrow) and or white-grey (dotted arrow) in color during summer temperature falls and glacial stream dry up periods in Sverdrup Pass, central Ellesmere Island, 79°N, Canada (after Elster et al., 1997). (F) Dry and frozen white-gray mats (arrow) of green algae (Conjugathophyceae) *Zygnema* cf. *leiospermum* and *Spirogyra groenlandica* covered temporal pools in area of deglaciated moraine in Sverdrup Pass, central Ellesmere Island, 79°N, Canada.

by oceanic currents, provides another major input of heat into the Arctic. The warm air moving northward interchanges with cold polar air in cyclones associated with low pressure, thereby providing heat energy input to the Arctic region. The south polar region is a huge area, and the glaciated Antarctic continent is located exactly at its center, the presence of which strongly influences the main transport of thermal energy through atmospheric circulation. The westerly winds produce a succession of cyclones, which show a regular procession of air masses that spiral around the Antarctic continent. In the sea there is a northward transport; however, the wind-driven Antarctic Circumpolar Current deflects this. Detailed information concerning the polar physical environments and their development in geological time can be found elsewhere (Fogg, 1998; Huntington et al., 2001; Walton, 1987).

3.2.2 WHY SO COLD?

The severity and complexity of polar environment is a consequence of the Earth's geometry. The axis of the Earth's rotation is not at right angles to the plane in which the orbit revolves around the sun. Because of this asymmetry, the North and South Poles, instead of having a daily alternation of diurnal cycles, have nightless summer months, followed by sunless winters. The frigid climate of the polar regions is the most important and hostile feature, but it is not necessarily a result of reduced solar radiation. Because of the 24 h of radiation received around midsummer, the sum of energy reaching the Earth's surface at the poles is higher than that on the equator at same time. However, because of the low angular height of the sun and the entire lack of radiation in winter, as well as reflection losses, the overall sum of energy per unit surface delivered during the year is low. Reflection losses and thermal energy acquired by the absorption of solar radiation mainly depends on surface characteristics. The ratio of reflected to incident radiation is higher in areas covered by ice or snow than that of the snow-free terrestrial landscape or open water, and thermal energy absorption of the areas covered by ice or snow is therefore negligible. Both the extent to which ice and snow cover the landscape and its geographical characteristics are the most dominant factors influencing the transport and balance of radiated energy in polar environments; however, these properties have changed over geological time. Changes are also directed by the earth's rotation and the position of landmasses, which consequentially affect the circulation of the atmospheric and oceanic currents, thereby affecting heat balance in the polar regions (Fogg, 1998; Huntington et al., 2001; Walton, 1987).

3.2.3 HOW LONG HAVE THE POLAR REGIONS BEEN COLD?

The present distribution of continents results from their drifting (Walton, 1987). About 220 million years ago, the northern part of Pangea fragmented and separated into the Siberian coast, Greenland, and the North American archipelago, enclosing an area of sea containing the North Pole. The movement of landmasses into the vicinity had a profound effect on climate. The heat storage capacity of the sea was reduced, and winter cooling of the land brought a fall in mean annual temperature leading to extensive glaciation, which began about 3 million years ago (Fogg, 1998). About 180 to 200 million years ago, the southern part of Pangea, known as Gondwana, drifted south to give rise to landmass pieces that would become South America, South Africa, India, Australia, and Antarctica. Antarctica thus established its position in the vicinity of the South Pole, and during that time, the gap between Antarctica and its neighboring landmasses widened as they drifted north. Until around 25 million years ago, when the "gap" opened sufficiently for the Circumpolar Current to become established, the southern continent was isolated. This had a profound effect on the Antarctic climate, and from that time on the ice caps began to expand (Walton, 1987; Fogg, 1998).

3.2.4 TEMPERATURE GRADIENTS, BIODIVERSITY, AND THE DISTRIBUTION OF LIFE
IN POLAR REGIONS

In terms of vegetation, soil, and climate the polar regions have been subject to various classification schemes. In the Arctic, the most common schemes are Low (Subarctic), Mid- (True), and High Arctic (Polar Desert; Aleksandrova, 1988). In Antarctica the most frequently used terminologies refer to Continental Antarctica, Maritime Antarctica, and the Subantarctic Islands (Holdgate, 1970).

These zones have been distinguished in various ways, none of which is entirely satisfactory. The scheme of Longton (1988) is the one most commonly used for comparisons of the Arctic and Antarctica; these are zones designated as Mild, Cool, and Cold in the Arctic, and Mild, Cool, Cold, and Frigid in the Antarctic. Continental Antarctica, along with neighboring seas, is recognized as the Frigid Zone (such a severe environment is not present in the Arctic) and is frequently compared with the environment on Mars. Passing through these zones equatorward from the poles, a general increase in productivity and species diversity can be found (Rosenzweig, 1995). The standard explanation for this is that fewer and fewer plants can survive as the environment becomes increasingly severe. However, the transitions are not smooth or definite because of local features of climate, geology, and topography (Fogg, 1998). This pattern is broadly true for terrestrial environments; however, the species richness and productivity of the marine polar ecosystem is quite different. Recent data from lists of large marine species covering broad geographical ranges strongly indicate that there is a gradient of increasing species richness from the Arctic to the tropic. In contrast, the Southern Ocean has high species richness, and in the southern hemisphere there is no clear evidence of a cline. A continuous variation in form within members of a species and their populations (as well as an increase in biodiversity) results from the gradual changes or transitions over the polar to tropical range (Gray, 2001). There are two types of natural selection that can occur in the polar regions (Dunbar, 1977; Elster, 1999), and these lead to different types of ecosystem stability, discussed below.

3.2.4.1 Marine, Nonoscillatory Steady State

This habitat has a poor input of energy, as polar oceans are frequently covered by ice, which restricts solar input, and they also have an impoverished mineral nutrient content. However, because they present a stable environment, evolutionary time is extended and genetic diversity is high. This system is sensitive to unpredictable, externally introduced perturbations (climate change and pollutants).

3.2.4.2 Terrestrial, Frequent Oscillation

Organisms in this environment have developed adaptive abilities to respond to perturbations and return to status quo. The system has a poorer species diversity, reflecting a high production/biomass ratio, and the species are less limited by low solar and nutrient energy economies. As they lack competition and grazing pressures, their growth is punctuated by episodes of explosive growth and development.

In nature, a series of transient ecological states can also exist, and it has been suggested (Dunbar, 1977) that successful maintenance of an ecosystem under conditions of considerable oscillation depends on geographical scale. The main focus of this chapter will be given to polar terrestrial and, thus, frequently oscillating environments. It is highly probable that those low-temperature terrestrial environments (polar or mountainous), which demonstrate oscillating environmental conditions, existed in very ancient geological time, even before Pangea or Godwana were respectively fragmented. Indeed, it is probable that low-temperature, terrestrial environments dated from a historical phase when the first organisms started to expand from the sea to take residence in the terrestrial environment.

The diversity of polar terrestrial nonvascular plants, especially oxyphototrophic microorganisms, is generally much higher than the diversity of vascular plants (Broady, 1996; Elster, 2002; Lewis-Smith, 1984; Longton, 1988; Vincent, 1988). A similar situation can be observed for nonphotosynthetic microbial life. Thus, bacteria, fungi, and protozoa constitute the major part of polar biodiversity, encoding the genetic information of a vast number of species (Beyer and Bölter, 2002; Friedmann, 1993; Vincent, 1988; Wynn-Williams, 1996; Zöcker et al., 2002).

In the polar terrestrial environment, the classical explanations for an equatorward gradient in productivity and species diversity include increased environmental severity and instability, occurring simultaneously with differences in acclimation and adaptation responses, dispersal mechanisms, and a release from competition. Very few resistant biotas can survive on land in the extreme conditions of the polar regions. Because of this, these terrestrial ecosystems are usually very simple with respect to the structure and complexity of their trophic levels, food webs, and polar food chains. For example, in polar desert soils, groups of algae will be the only primary producers (comprising 20 to 50 species), and two species of microscopic animals will consume the algae. In addition, there may be many (50 to 100) species of bacteria and fungi that use this biomass energy and nutrients.

3.2.5 TEMPERATURE MONITORING IN THE POLAR REGIONS: LIMITATIONS, PROBLEMS, AND SOLUTIONS

Physical parameters of the polar terrestrial environment, particularly low temperatures, clearly affect the distribution of prokaryotic cyanobacteria and eukaryotic microalgae. Surprisingly, despite the importance of such parameters, there is a paucity of information related to the continuous monitoring of temperatures in polar terrestrial habitats. Only a limited number of field-based ecological studies (Caulson et al., 1995; Friedmann et al., 1987; McKay et al., 1993) measure diurnal, seasonal, and annual fluctuations (within and between years) in temperature on a truly comprehensive basis. Such measurements are important for understanding the life cycle and physiological adaptations of the organisms that live in these extreme and fluctuating environments. There are several key reasons why temperature-defined ecological data are limited for terrestrial polar environments.

Practical and technical limitations related to inaccessibility include that the remoteness of locations greatly restricts long-term monitoring and continuous field surveillance in polar regions. To date, measurements mostly involve "classical" meteorological air temperature data acquisition (at weather stations). These air temperatures are measured about 1.5 to 2 m above the ground surfaces, where there are almost no or very low microbial populations. The lowest temperature ever measured in Antarctica was –89.3°C, recorded at the Russian Vostok station. Other lowest temperatures ever recorded in the Asian Arctic at Oimekon (–62.3°C); in Europe at Ust Shchugor it was –46.4°C. These places are both in Russia. Finally, in North America the lowest temperature was measured at Snag, Yukon, Canada (–56.3°C), and at Northice, Greenland (–60.2°C; see http://www.ncdc.noaa.gov/oa/climate/globalextremes.html). These low-temperature minima were measured at various altitudes and also latitudes, as thermal profiles are influenced not only by altitude but also by elevation. The general rule of thumb used by meteorologists is that temperature decreases approximately 0.82°C per 100 m in the troposphere. This is an average value and varies from season to season. Thus, the mountain tops or nunataks are, in terms of their thermal properties, the most extreme habitats in the polar terrestrial environment. In addition to the latter, extremely low temperatures can also be considered in atmospheric supercooled cloud droplets, where actively growing bacteria have been detected (Sattler et al., 2001). Amazingly, living microorganisms have been detected even in the stratosphere (Imshenetsky et al., 1978).

There are a limited number of examples of polar temperature measurements being recorded throughout the whole year in terrestrial polar habitats. The development and use of automatic data-loggers for continuous measurement of ecological parameters, including temperature, has allowed the monitoring of various types of terrestrial habitats (Caulson et al., 1995; Friedmann et al., 1987; McKay et al., 1993), and as a result, data are now becoming increasingly available. However, the difficulties encountered are further compounded by the fact that polar habitats possess highly diverse properties with respect to their thermal behaviors. Thus, the complexity and heterogeneity of different substrates and states (solid, liquid, and gaseous) make it very difficult to accurately ascertain temperature changes and fluctuations. Different components can moderate the incidence of solar radiation reflection, and they have very disparate temperature conductivities and absorption properties. The polar terrestrial environment also comprises a complicated mosaic of fluctuating microenvironmental conditions (wind, sun, shadows, aspect, cold air, drainage from ice) that can significantly influence the thermal properties of specific habitats. These factors intercalate the effects of water status, thermal state (liquid/solid) and longevity, geological substrate (density, granularity, presence or absence of gaseous substances), and vegetation cover. Thus, it can be concluded that the Arctic and Antarctica are the most extensive low-temperature regions in the world, and they experience a wide range of temperatures with different thermal regimes. For example, even the most extreme localities in high/low latitudes can reach extremes of from about +20° to +25°C to about –50° to ≤ 60°C. By comparison, the lowest known recorded temperature for the universe is approximately –272°C, which is about 1°C greater than absolute zero (1K; http://uni-data.ucar.edu/staff/blynds/tmp.html).

3.2.6 EXAMPLES OF STABLE AND UNSTABLE TEMPERATURE ENVIRONMENTS

Two different (stable and unstable) ecosystems can be considered with respect to temperature classification in the polar terrestrial environment.

3.2.6.1 Stable Low-Temperature Environments

Stable low-temperature environments comprise terrestrial ecosystems of frozen ground (permafrost) and ice bodies (i.e., icecaps, glaciers), including melting surfaces and subglacial systems manifested as either frozen permafrost or unfrozen subglacial lakes and soil. Temporary snowfields, in which meltwater percolates through upper layers of snow, are also representative of this type of ecosystem, even though this habitat is temporary. Stable low-temperature systems will now be considered in more detail.

3.2.6.1.1 Glacial Ice

Atmospheric circulation in the polar regions provides an air-mass exchange with neighboring latitudes. Microorganisms can thus be transported to these areas on terrestrial dust and in precipitation; as a result they become embedded in ice formed from falling snow. Direct sampling has repeatedly demonstrated the presence of microorganisms characteristic of the microflora of more northern and southern latitudes (e.g., Elster et al., 2003; Marshall, 1996; Wynn-Williams, 1991). Microorganisms have accumulated in the ice sheets over many thousands of years, and certain ones remain viable for long periods at low temperatures (Abyzov, 1993; Abyzov et al., 1998). For example, isotope studies of the Vostok ice core showed that the lake beneath 3750 m of the Antarctica ice sheet started to be formed about 420,000 years ago (Jouzel et al., 1999) and that it contains viable microorganisms (Karl et al., 1999). Ice sheet temperature depends on ice thickness, geothermal conditions of the basal ground, latitude, altitude, and local climate. The mean surface and basal temperatures of the high Arctic (e.g., the Aggasiz ice cap, Ellesmere Island) are approximately –22.9° and –18.3°C, respectively; those for the Greenland ice cap –21.2° and –13.1°C. In comparison, the Antarctica station of Vostok is at –55.5°C, and the Byrd station is at –28.0° and –1.6°C (Abyzov et al., 1998; Koerner, 1989). Because of the subglacial lake under the Vostok station, the basal temperature has not been directly measured yet. During the last decade, great attention has been given to the microbiological analyses of the Vostok ice core and its deeper ice part, derived from an ice subglacial lake (Christner et al., 2001; Karl et al., 1999; Priscu et al., 1999). It can thus be concluded that the Vostok ice core contains a wide diversity of microorganisms; bacteria prevail, although yeasts, fungi, and cyanobacteria and microalgae can also occur. However, microbial ice core analyses have resulted in a cautionary debate, as they provide more questions than answers, and future examinations must avoid problems associated with contamination (Vincent, 1999).

3.2.6.1.2 Permafrost

Permafrost is frozen ground comprising different (ranging from dry desert soils to water-saturated soils) water profiles existing in the solid frozen state (see also Chapter 4). It may be regarded as the most stable polar environment for microorganisms, and studies of their distribution reveal their numbers to be only one magnitude lower than that of surface soils (Gilichinsky et al., 1993; Vishnivetskaya et al., 2000). From the soil surface down a few hundred meters, temperatures vary from above zero to between –10° and –12°C in the Arctic and from –25° to –30°C in the Antarctic (Vorobyova et al., 1997). It has been shown by Gilichinsky et al. (1995) that significant numbers of viable microorganisms, representing various ecological and morphological groups, have been preserved under permafrost conditions for thousands and sometimes millions of years. In northeastern Siberia, the oldest cells discovered date back 2 to 3 million years (Gilichinskĭ et al., 1995). This Russian team (Gilichinskĭ) isolated viable cyanobacteria and algae and have produced a collection of living cultivable strains dated from almost 2 million years ago; other microorganisms may be even older. However, the occurrence of photosynthetic microorganisms in permafrost is numerically much lower than that of bacteria. Nonetheless, viable ancient phototrophs (cyanobacteria and green algae) have preserved their photosynthetic apparatus in deep darkness and were found and isolated from Arctic permafrost soils of Pliocene–Pleistocene age (Vishnivetskaya et al., 2001).

3.2.6.1.3 Subglacial Systems

Unfrozen lakes and soils warmed by geothermal energy are a common feature in both polar regions. Over 70 lakes have been identified beneath the Antarctic ice sheet (Ellis-Evans and Wynn-Williams, 1996; Jean-Baptiste et al., 2001). Although none of the water from these lakes has been directly sampled, analyses of lake ice frozen at the underside of the ice sheet above Lake Vostok have been undertaken. Findings from these studies support the proposition that small numbers of viable microbes live in this unique and extreme ecosystem. All subglacial lakes are subject to high pressure, low temperatures (about –3°C), and permanent darkness (Siegert et al., 2001). Similar interdisciplinary research studies focusing on the subglacial ecosystem are presently ongoing in the Arctic Svalbard archipelago (Elster et al., 2002; Glasser and Hambrey, 2001).

3.2.6.1.4 Glacial Melting Surfaces

Meltwaters on glaciers contain a variety of aquatic biota, particularly within the habitat formed by cryoconite holes. These cryoconite holes occur in the ablation regions of glaciers, where ice loss occurs because of melting, wind erosion, evaporation, and calving (Wharton et al., 1985). The holes fill with meltwater in summer, and they freeze in winter (Gerdel and Drouet, 1960). Their basal layer is covered by dark sediment, and they can coalesce into bigger holes or become connected by meltwater channels (Wharton et al., 1985). In extreme cases, this type of growth can result in broad, shallow depressions holding a large amount of sediment. Summer water temperatures are usually remarkably uniform, from –0.1° to +0.1°C, and they rank among the coldest polar freshwater ecosystems (Säwström et al., 2002). As solar radiation decreases at the end of summer, ice forms at the water surface and grows downward. During the freezing process, solutes are rejected because they impede the formation of a regular hexagonal crystal structure. The phenomenon of "freeze-out" is responsible for solute concentration until the temperature drop stabilizes or the solutes precipitate out and the remaining water freezes (Lock, 1990). This process is less effective with faster freezing rates: Because solutes do not have a chance to fractionate, they become trapped in bubbles or brine pockets (Mueller et al., 2001). The holes freeze solid after reaching a minimum brine temperature; they are most typically 10 to 60 cm deep and provide a suitable habitat for microbial colonization and growth. The communities in these habitats are complex microbial consortia of bacteria, cyanobacteria, eukaryotic algae, and protists, and the holes may also contain microinvertebrates (Mueller et al., 2001; Säwström et al., 2002).

3.2.6.1.5 Temporary Snowfields

Where meltwater percolates through the upper layer of temporary snowfields, their surfaces may be colored by various microbial communities, of which photosynthetic cyanobacteria and eukaryotic algae are the most frequently represented. This type of ecosystem functions in a cryophilic stable state, but only for a particular period of time (e.g., several days to ≥3 months). The rest of the time it behaves as a hydro-terrestrial or edaphic environment, and as the snow melts, part of the soil flora survives in shallow freshwater ecosystems. A wide spectrum of cyanobacteria and algae has also been found in the melted snows of the Arctic and Antarctica (Kol, 1972; Ling, 1996; Müller et al., 1998). Leya et al. (2001) recorded snowfield temperatures in Kongsfjorden, Ny-Ålesund, Svalbard, over 2 years, where cryophilic communities regularly occurred. During the melting period, snow and wet soils present stable temperatures of around –0.6°C for the month of June and half of July. As summer progresses, however, the soil surface dries, becoming exposed on occasions; the temperature rises to +22°C but usually ranges between +3° and +10°C.

3.2.6.2 Unstable Low-Temperature Environments

Historically, three ecological categories, as presented by Vincent (1988), Vincent and James (1996), Broady (1996), and Elster (2002), make up unstable low-temperature environments: terrestrial, hydro-terrestrial, and lacustrine (glacial lakes in which water column stratification permanently or temporarily occurs).

The lacustrine environment has fewer temperature oscillations and could therefore be considered a different ecological category. Alternatively, it could also be included in the low-temperature stable ecosystems (see above). Water gradient and status reflect the local climate and determine the properties of the polar terrestrial ecosystem. There are great differences in heat exchange between air and ground temperatures because of the presence or absence of water (dry soil, wet soil, or shallow wetlands). Because of its large heat storage capacity, water in either liquid or solid form acts as an important buffer to temperature change. Local climatic conditions of unstable terrestrial ecosystems are the principal external factors controlling functionality. Near the ground surface (microclimate) there is a zone of energy exchange between the substratum and the atmosphere above. The microclimate and environment surrounding individual organisms are of the highest significance. This very limited zone, comprising just a few centimeters above and below the ground surface, forms the environment for life existing in bare landscapes. For the organisms living there, these microhabitats are a "climatic paradise" or "haven" in what are otherwise cold and frozen ecosystems.

Polar unstable terrestrial ecosystems also differ substantially from those of temperate regions because of the presence of permafrost. The thickness of the active layer of the soil surface changes periodically (seasonally and diurnally), causing temperature fluctuations. It is in this layer that the most active biological processes occur, and this layer above the permafrost is dependent on solar heat balance. Permafrost limits and delineates the subsurface system from the active layer, and in doing so restricts the migration of liquid water to the deeper frozen soil. In the case of active layers, which are in milder localities, they frequently become saturated with water, and in addition to wet soil, various shallow, temporary freshwater ecosystems (lotic and lentic wetlands) arise. The lotic freshwater ecosystem pertains to ecological communities living in flowing rivers or streams, and the lentic system designates ecological communities living in still water. Favorable conditions can exist for those organisms that are able to acclimate or adapt to these unstable types of environment. However, such habitats are affected by seasonal and diurnal variation and by environmental factors that can impose severe conditions. Organisms living there have to overcome a series of very severe environmental restrictions. It has been shown that water status (liquid or frozen extracellularly), desiccation, and the availability of "free water" in biotic tissues are crucial factors influencing the ability of life to exist in the polar terrestrial unstable ecosystem.

The distribution of organisms in active layer "active life zones" capable of supporting growth, reproduction, and development in polar regions is influenced by a complex interaction between temperature and water availability (influencing the changing of water to ice and vice versa). In the polar terrestrial environment, water is available in liquid form for only a short period (as snow melt, summer rain, or snowfall); it is also made available as air vapor absorbed from the air (humidity). In nearly all polar terrestrial environments, microorganisms are invariably encountered both on and beneath the surface of soil and rocks. These microorganisms range from communities on periodically wetted soils and rocks to two clearly distinguished terrestrial categories comprising soil (edaphic) and aerophitic (lithophitic) communities. They play especially important ecological roles in polar desert–semidesert and in young periglacial ecosystems. The edaphic environment of wetted soils and rocks overlap in many cases with the hydro-terrestrial environment, which will be discussed later.

3.2.6.2.1   The Soil Environment

Polar substrata (ground–bedrock) vary greatly in their physical and chemical characteristics (Tedrow, 1991; Walton, 1984). Soil substratum is derived from a range of parent rock types in which concentrations of salts differ widely. The organic content is usually extremely low, but soil can sometimes contain higher levels if it is water saturated (e.g., peat bog, marsh, moss carpets). These conditions support layers of high biomass, some of which have accumulated over hundreds and even thousands of years. Soils vary in texture, pH, and nutrient content and can differ markedly in their moisture status, ranging from extremely arid to water saturated. Cold and aridity slow down chemical weathering and the rate of biological activity. Water saturation, in contrast, results in the leaching of salts, exclusion of oxygen, and accumulation of organic matter. Of course, water saturation and status also decelerates temperature change; this occurs either during the summer or in prolonged transient periods in spring and autumn. However, these soil factors can change remarkably over short distances and with time and altitudinal/latitudinal gradients. The soil factors also affect the composition and abundance of soil polar algal communities (Broady, 1996; Elster et al., 1999; Longton, 1988). It has been proposed that cyanobacteria and algae predominate in the soils of the colder and more arid coastal and slope provinces of continental Antarctica. In contrast, in milder and moister maritime areas of Antarctica, bryophytes and lichens are more common (Broady, 1996; Elster et al., 1999; Lewis-Smith, 1984). Soil temperature may be greater than 20°C during summer (Davey, 1991; Walton, 1982) and fall to less than –10° to –20°C. This may be even lower in the Ross Desert of Antarctica (McMurdo Dry Valleys), where surface ground temperatures can fall to from –42.7° to –47.2°C (Friedmann et al., 1987; McKay et al., 1993) during the winter, although minimum temperatures are dependent on the depth of snow cover and on amelioration by plant litter (Caulson et al., 1995; Davey, 1991; Leya et al., 2001). So, for example, Worland and Block (2003) recorded a typical soil surface temperature (including winters) for four field polar sites in the Arctic and the Antarctic. Minimum temperatures followed approximately the same pattern, ranging from –20° to –4°C, with a highest maximum surface temperature of 22.5°C and a low of 11.2°C. Snow cover and timing of snow fall acts as an efficient ground insulator, and a continuous snow cover throughout the winter helps many biotas remain active in the subnivean (beneath snow cover) environment (Caulson et al., 1995). By contrast, biotas overwintering above the snow or in sites without snow experience low temperatures and repeated freeze/thaw cycles.

3.2.6.2.2   Lithophytic and Aerophytic Communities on or within
Rock Substrata

Communities of cyanobacteria, algae, filamentous fungi, yeasts, and heterotrophic bacteria can also occur on or within rock substrata elevated above the ground (soil). Algae and fungi can be associated as lichens, and they can cover or grow within wood, bone, and other natural or artificial materials found in polar regions. Rock surface offers some hospitality to life, and those organisms that are able to take advantage of such surfaces are termed epilithic. Endolithic (growing beneath the rock surface) microbial life-forms predominate within rock outcrops, and cyanobacteria, green algae, and members of the Xanthophyceae are the most dominant oxyphototrophic microorganisms of lithophytic habitats (Friedmann, 1980; Golubic et al., 1981). In extreme desert areas, lithophytic cyanobacteria and algae are the only photoautotrophic organisms (Cameron and Black, 1967). Endolithic communities are the most thoroughly investigated terrestrial algal communities in Antarctica and are located in regions ranging from southern Victoria Land to Signy Island in maritime

Antarctica (Broady, 1981; Feingold et al., 1990; Friedmann and Weed, 1987; Friedmann et al., 1988). Differences in the millimeter range pertain to the nanoclimate and control the environment in which communities can exist. In the most extreme climates, the absolute limit for life is reached in terrestrial lithophytic communities (McKay et al., 1993). In the Antarctic, desert temperatures are far below the biological optima, and these organisms exist near the limit of their physiological potential. Biological processes take place in rocks only at temperatures greater than –10°C, whereas the lower temperature ranges dominate yearly temperature averages.

3.2.6.2.3 Hydro-Terrestrial

Liquid water is available for most of the entire period of summer in hydro-terrestrial environments (wetlands). A common distinction between hydro-terrestrial (wetland) and lacustrine (lake) areas in the polar regions is that wetlands freeze solid during winter, whereas most lakes do not (Hawes et al., 1992). This inevitability is a strong habitat-defining characteristic, which places considerable stress on resident organisms (Hawes et al., 1992; Vincent et al., 1993). Under the category of the hydro-terrestrial environment, the environment of glacial melting surfaces and temporary snowfields can also be taken in consideration. Hydro-terrestrial habitats can occur up to highest and lowest possible latitudes and altitudes as long as water is available for some time during the year. Algal distribution, abundance, and species diversity vary in accordance with habitat (microhabitat) characteristics. In localities with a steady moisture and nutrient supply, abundance and species diversity of algae is relatively high. However, as the severity of living conditions increases (mainly changes in desiccation–rehydration and subsequent changes in salinity), algal abundance and species diversity decreases. A wide spectrum of ecological studies have recorded seasonal, diurnal, and year-round temperature fluctuations and changes in water state transitions in all types of polar hydro-terrestrial habitats (Elster and Komárek, 2003; Hawes, 1989; Hawes and Brazier, 1991; Hawes and Howard-Williams, 1998; Howard-Williams et al., 1986; Sheath and Müller, 1997; Vincent and Howard-Williams, 1989). Seasonal water temperatures range usually from 0 to 8 to 10°C. Diurnal air temperature can fall below 0°C, resulting in superficial freezing, but wetland bed temperatures can fall below freezing only occasionally. These distinct diurnal spikes appear mainly at the beginning and at the end of a summer season. Larger wetlands tend to be warmer, more temperature stable, and frozen during the winter. However, wetlands covered by snow frequently show a minimum temperature of only –4°C (whereas air temperatures can fall lower). Accumulation of snow and ice offers a soil environment very effective insulation. In contrast, sites without snow and ice freeze solid, and temperatures can fall down to very low temperatures (in the continental Antarctica locations, up to ≥ –40°C). Schmidt et al. (1991) studied the limnological properties of Antarctic ponds (at Cape Evans, Ross Island) during winter freezing. The bottom waters became increasingly saline as freezing and water temperatures decreased to less than 0°C. In colder periods of the year (June), the remaining bottom water overlaying the sediments had conductivities greater than 150 mS/cm and a temperature of –14°C.

3.2.6.2.4 Lacustrine

Lakes provide some of the most favorable habitats for microbial growth in the polar regions (Kalff and Welch, 1974; Vincent, 1988). Many are capped by thick ice, and the water beneath remains unfrozen, despite winter temperatures dropping far below 0°C. The ice cover has a pervasive influence on physical, chemical, and biological properties (Canfield and Green, 1985; Ellis-Evans, 1981a, 1981b; Wharton et al., 1987). Ice protects the underlying water from wind so that the water column is stabilized and mixing is restricted to the ice-free period. On this basis, polar lakes have been defined as having surface temperatures always less than 4°C, with a short ice-free period accompanying horizontal water mixing. Thick ice cover overlying the lake occurs mainly in cold continental regions and persists even during summer. Extreme and unusual salinities can influence the solute content of lakes, which can be concentrated several times by freezing and evaporation. Trophic status and temperature are also important factors in polar lakes, many of which contain high concentrations of dissolved solids. Water layers can be stabilized by salt gradients, which enhance the absorption of solar radiation; as a result, temperatures can rise well above ambient. The most dramatic example of this occurring is at Lake Vanda in Antarctica, Wright Valley, Southern Victoria land. Here temperatures rise from 0°C immediately beneath the permanent ice cap to 24°C at the bottom of the lake. This inverse distribution of heat is stabilized by the increasing salinity gradient with depth, from about 0.1% near the surface to 123% at the bottom (Vincent, 1987). In contrast, a different affect is observed in the same region as represented by Don Juan Pond (about 300×100 m, and only about 10 cm deep). In the winter, the bottom of the water has a nearly saturated solution of calcium chloride that remains unfrozen at temperatures which are less than or equal to –50°C (Vincent, 1987). These factors cause remarkable differences in the composition and abundance of microbial floras (largely cyanobacteria, phytoflagellates, and chlorophytes) that exist in Arctic and Antarctic lacustrine environments (Vincent, 1997; Vincent and Hobbie, 2000; Vincent et al., 1998).

3.3 LIFE STRATEGIES AND STRESS FACTORS IN THE TERRESTRIAL POLAR ENVIRONMENT

Vincent (2000) reviews the present knowledge of the evolutionary origins of Antarctica's microbiota and life strategies, and because of the similarities in physical environment, this can be applied to both polar regions, but it is important to take into account Antarctica's geographical isolation. Current debate concerns to what extent polar microbial flora is genetically different from the rest of the globe's microbial genepool and the effect that severe ecological constraints have on influencing the direction and speed of evolution in the polar regions. Life cycle strategies and stress factors will undoubtedly influence both evolutionary trends.

3.3.1 LIFE STRATEGIES

It is accepted that the global transport of microbial genomes over distant or local ranges is possible, through atmospheric circulation, ocean currents, and via vectors (birds, fish, marine, terrestrial mammals, humans). These vectors and transportation mechanisms can permanently displace prokaryotic and eukaryotic microbial genomes (Broady, 1996; Elster et al., 2003; Marschall, 1996; Vincent, 2000). However, our present knowledge of microbial dispersal is limited because relatively little research has been performed to date (Broady, 1996; Broady and Smith, 1994; Elster et al., 2003; Marschall, 1996; Marshall and Chalmers, 1997; Schlicting et al., 1978, Vincent, 2000). We are also restricted by the fact that only a very small range of individual organisms respond to cultivation after their dispersion and deposition in polar environments. This is because cells and spores reduce their metabolic activities during airborne dispersal. In addition, their introduction into colder polar habitats induces so-called cryptic dormant states that allow long-term persistence sustained by a much-diminished metabolic state. Under these dormant conditions, they can be considered to represent only a "potential component" of polar genetic diversity. This is because this special proportion of the microbial biota is not capable of reproducing while dormant and will not be exposed to the polar evolutionary pressures and genetic recombination, as would reproductive representatives of polar biodiversity. To date, an understanding of those factors that are responsible for changing the metabolic activity of introduced biota from normal to dormant states and vice versa is limited. Similarly, it is not known how long these polar organisms persist in a dormant or resting state. This is particularly the case for photosynthetic species, for which information is either absent or very limited. However, it can be expected that once microbial viable cells or resting spores arrive from either distant or local regions at a potential habitat (via physical or vector transport), the limiting conditions will exert strong selection pressures. On the basis of different habitat conditions, different microbial life strategies have been proposed as operating in polar environments (Vincent, 2000; Vincent and Quesada, 1997), discussed below.

Specialist organisms occupy narrow niches within extreme but stable environments, and their life strategies are more common to the nonoscillating low-temperature marine ecosystem. However, specialist psychrophiles prefer to inhabit an environment in which the optimal temperature for growth is at or >15°C. These psychrophiles have only a minor ecological role in polar terrestrial and hydro-terrestrial habitats of the active layer, and this is defined within a wide range of temperature fluctuations. In contrast, habitats such as those of deep lacustrine (lake); glacial ice; permafrost; and marginal, cryophilic habitats (e.g., microbiota growing in melting snow fields and cryoconite holes), which have stable temperatures, contain a much higher percentage of specialized genotypes (Vincent, 1988).

A more generalist genotype is one that grows suboptimally but survives because it is able to tolerate environmental extremes. Many polar terrestrial and hydro-terrestrial microbial biotas show a limited metabolic "tuning" to their ambient environment, with the manifestation of optimal growth conditions that are outside the environmental range of their polar habitats (Vincent, 2000). For example, mat-forming cyanobacteria are frequently the dominant biomass in the various shallow wetlands (hydro-terrestrial habitats) of polar environments. Growth assays of 27 high-latitude strains showed that they were all psychrotrophic (microorganisms growing at $\Sigma5°C$ but that have a temperature optimum for growth of >15°C) but had a mean temperature optimum for growth significantly below that for strains from temperate latitudes (Tang et al., 1997). These observations imply an environmental selection for adaptive genotypes and imply the potential for evolutionary divergence from temperate latitude microbiota (Vincent, 2000). Generalist genotypes also occur that occupy broad niches that support periods of optimal and suboptimal growth and acclimation. Again, this strategy can be found in various terrestrial and hydro-terrestrial habitats in which microbial biotas seem to be more likely cosmopolitan species that can grow in many parts of the world (Vincent, 2000). It is thus evident that there is not a sharp distinction between the different strategies and that the polar terrestrial environment has the capacity to modulate the physiological, metabolic, and genetic properties of life over long periods of time. The evolution of organisms through these processes is exhibited by, and expressed through, different life strategies, and these strategies are frequently engaged as other responses and adaptations, as seen in nonpolar latitudes and ecosystems.

3.3.2 Extremes and Fluctuations in Environmental Parameters: Impacts on Microbial Life

Fluctuations in extremes of low temperatures or water status have a wide physiological range of influence. For this reason, prokaryotic and eukaryotic oxyphototrophic microorganisms from polar marine ecosystems are frequently used as a model for investigating cold acclimation, adaptation, and survival (Kirst and Wiencke, 1995). In the case of the unstable polar terrestrial ecosystem, Vincent (1988) and Friedmann (1993) have overviewed low-temperature and water status fluctuations with respect to the microbial ecology and ecophysiology of the Antarctica ecosystem. To date, the ecosystem of terrestrial ice bodies has not received detailed consideration, and the freshwater ice bodies overlying the polar terrestrial ecosystem lack the habitat (brine channels) in which microorganisms could live. However, as more studies ensue, it is notable that microbial cells/spores introduced into Antarctica are transported together with air mass circulation, and they are later deposited and stored on glacial surfaces. Because little is known about "storage" of microbial life in continental icecaps, it is becoming increasingly important that intensive biological and interdisciplinary research is performed in both polar regions. Such a study promises to expand an exciting and potentially fruitful (in terms of acquiring both fundamental and applied knowledge) area of polar environmental biology research.

Low temperatures and associated changes in water state (particularly at freezing point) have a dramatic effect on chemical and physical characteristics of all potential habitats throughout the polar regions. Ice formation physically changes the environment, and it can cause biologically significant chemical changes (in salinity, pH, conductivity, and gas content). Water availability, and thus freezing and melt cycles, are also a dominant influence on the cellular environment of most terrestrial life forms.

3.3.3 How Does Polar Life Cope with Environmental Extremes?

Polar organisms have developed a wide range of adaptive strategies that allow them to avoid, or at least minimize, the injurious effects of extreme and fluctuating environmental conditions. In addition, a wide spectrum of ecological and physiological studies have shown that in terms of environmental priorities, demands for moisture (water in a liquid form) precede demands for nutrients, which in turn, precede demands for high temperatures (Elster, 2002; Kennedy, 1993; Lewis-Smith, 1996, 1997; Svoboda and Henry, 1987; Vincent, 1988). The three main strategies for coping with living in a polar habitat are avoidance, protection, and forming partnerships with other organisms.

3.3.3.1 Avoidance

In polar terrestrial cyanobacterial and algal communities, a diverse range of ecological and physiological life strategies and behaviors manifest the ability to avoid low temperatures and fluctuations in water status. Prokaryotic cyanobactcria and eukaryotic algae are, together with bacteria, lichens, and a selected group of mosses, poikilohydric organisms that are able to tolerate desiccation to different capacities. This is because that they do not contain vacuoles, which in higher eukaryotic algae and plants are the cellular component responsible for water control. The external environment directly manages the metabolic activity of poikilohydric organisms by affecting the presence or absence of water in either liquid or vapor forms. In ecological terminology these organisms are called "ecological opportunists." Thus, poikilohydric organisms have great ecological advantage in the severe polar terrestrial environment.

Shelter strategies present another type of avoidance mechanism against low temperatures and water status fluctuations. This strategy can be compared with those used in life in more stable environments such as lakes, wetland bottoms, and substrata. Shelter strategies physically protect oxyphototrophic microorganisms against stress factors caused by dynamic temperature and water gradients. Poikilohydricity and shelter strategies are frequently intercalated, and when combined with cellular and physiological modifications (cell motility, development of complex life cycles, multilayered cell walls, sheet and mucilage production, life associations with other organisms), they afford considerable advantages. These avoidance regimes have thus been developed because of considerable evolutionary pressures.

3.3.3.1.1 Motility and Avoidance

Some polar prokaryotic cyanobacteria and eukaryotic algae are motile at vegetative or reproductive cell stages. Mobility can facilitate avoidance of the most stressful conditions and propel the organism to a more favorable environment (Castenholz et al., 1991; Spauldin et al., 1994; Wiedner and Nixdorf, 1998). Motile cyanobacteria and algae can also react to stimuli generated by a wide spectrum of environmental conditions and chemical factors (pH, temperature, viscosity). Among their different motility types, flagella are most frequently responsible for movement in eukaryotic unicellular algae. In contrast, filamentous prokaryotic cyanobacteria (Oscillatoriales) move using straight-lined gliding. The actual mechanism of this type of motility is not known, but it is combined with mucilage production. The existence of several different types of motility in Oscillatoriales has been described (Castenholz, 1982; Van Liere and Walsby, 1982), and their properties have very important ecological effects. Because of their motility and mucilage production, polar terrestrial and hydro-terrestrial cyanobacteria are important in soil formation and aggregation (Bailey et al., 1973). Poikilohydricity and motility are associated with the oldest evolutionary and simplest forms of phototrophic life on the planet. In the polar terrestrial environment they can therefore be considered as the Earth's most simple, and in evolutionary terms oldest and most primitive, form of oxyphototrophic microorganism. Of course, there are a few exceptions, (e.g., the foliose macroscopic green algae *Prasiola*) associated in a cosmopolitan genus proliferating in very different, and sometimes even atmophytic, habitats of the Antarctica and the Arctic (Broady, 1989; Hamilton and Edlund, 1994; Kováčik and Pereira, 2001).

3.3.3.1.2 Complex Life Cycles and Avoidance

Complex life cycles are a major avoidance strategy and usually involve the development of resting (dormant), vegetative, and reproductive stages that change during an organisms' life cycle to accommodate seasonally incurred environmental fluctuations. A good example includes the snow algae—inhabitants of melting temporary snowfields for which about 100 species have been described (Akiyama, 1977; Kol, 1969, 1971). More recent studies (Ling and Seppelt, 1993, 1998; Müller et al., 1998) revised these classifications, showing that for some, the original descriptions were wrong because they showed single vegetative, resting, and reproductive stages. At present, it is clear that some snow algae have developed complicated life cycles to protect themselves against the severe and changeable environmental conditions that occur in melting snow. This is a very important adaptive trait because of the seasonal instability that occurs within the polar terrestrial environment, concurrent with the ability to produce resting, dormant stages through anhydrobiosis. The environmental physiology of these stages has, however, received infrequent attention in the field of polar algal ecology. Undertaking studies in this area is important, as dormant stages in cyanobacteria and algae facilitate their transportation around the world and across the polar regions. This has implications in terms of influencing global microbial geneflow and the distribution of genomes across long distances.

Dormant stages accumulate high concentrations of soluble carbohydrates that substitute for water molecules during dehydration. This stabilizes the structure and functions of macromolecules, membranes, and cellular organization (Crowe et al., 1984). The presence of protective sugars in cells enables the vitrification of cytoplasm on drying and supports the formation of a high-viscosity, metastable glassy state (Bruni and Leopold, 1991). This preserves cell viability during dry frozen storage through immobilizing cellular constituents and suppressing deleterious chemical or biochemical reactions that threaten survival (Sun and Leopold, 1994a, 1994b).

3.3.3.2 Protection Strategies

The extracellular production of protective compounds and structures such as multilayered cell walls, sheets, and mucilage is a very common phenomenon in polar terrestrial cyanobacteria and algae, and it protects them against fluctuations in water status. The observation that many polar algae and cyanobacteria have a layer of mucilage and thick protective sheet, which retards water loss from cells, has been widely reported (Bawley, 1979; Whitton, 1987). The multilayered envelopes surrounding these cells frequently contain sporopollenin or sporopollenin-like compounds. The sporopollenins are a group of resistant biopolymers considered to be products of the oxidative polymerisation of carotenoids or of their esters (Wiermann and Gubatz, 1992). The vegetative or dormant cells covered by these multilayered compounds are widely distributed in nature, including the polar regions, and they contribute to the strengthening of the cell walls of cyanobacteria and algae, plant pollen, and fungal spores. Studies (Caiola et al., 1996) using the coccal cyanobacterium *Chroococcidiopsis* sp. Chroococcales showed that simple organisms are able to modify the composition of their cell envelopes by altering the proportion of acid and beta-linked polysaccharides, lipids, and proteins. Moreover, single cells of *Chroococcidiopsis* sp. scattered in desiccated cultures might be regarded as resting forms, allowing survival under prolonged and severe dry conditions, so there may be a trade-off between adopting structures able to withstand short-term water loss and those useful for prolonged survival. Such a phenomenon is observed in cyanobacteria and algae in which dense mucilage-adhesive mats and multilayered cell walls and sheets are produced. They are also observed in periphyton ecology comparisons of continental and maritime Antarctica stream ecosystems. Vincent and Howard-Williams (1986) and Vincent et al. (1993) showed that the stream periphyton communities of the McMurdo Sound region of Antarctica are composed of dense, cohesive, and frequently multilayered mats. These may be adaptive structures that accommodate the short summer season and frequent freeze–thaw cycles of the habitats. In contrast, for the relatively milder localities of the maritime Antarctica stream ecosystem (South Shetlands, King George Island), the longer summer season and reduced-fluctuation freeze–thaw cycles do not require such protective responses. In these cases the stream periphyton communities are composed of free-floating mats and clusters of unicellular diatoms (Elster and Komárek, 2003; Kawecka and Olech, 1993).

3.3.3.3 Life-Form Associations

Life-form associations (symbiosis and mutualism) may also offer protection against low temperatures and water status fluctuations. They are traditionally defined as the living together of two or more unlike organisms to the benefit of both organisms. However, this definition has been modified to include the acquisition and active maintenance of one genome by another or the acquisition of novel metabolic capacity and structure (Douglas, 1994). These definitions avoid the oversimplifications inherent in the term "mutual benefit" and indirectly promote the linkage of symbiosis to horizontal gene transfer mechanisms. The life associations that commonly occur in the polar terrestrial environment, with the exception of physical protection, also have physiological and metabolic advantages. The most frequent and also the most ecologically important association in polar terrestrial environment is the association of cyanobacteria/algae with fungi in lichens. Asco-mycetous lichens, especially those associated with green-algal photobionts, dominate the terrestrial vegetation and are among the largest primary producers of polar terrestrial environment (Barrett and Thomson, 1975; Kappen, 1993). With respect to water management, the algal and fungal association (lichen symbiosis) is one of the most successful mutualistic life strategies found in the polar regions.

Prokaryotic and eukaryotic oxyphototrophic microorganisms have very different life strategies with respect to their susceptibility to low temperatures and freezing. Prokaryotic (cyanobacteria) microorganisms are particularly well adapted to the severe and changeable conditions involving cycles of desiccation, rehydration, salinity, and freeze–thaw episodes. This gives them a great ecological advantage, and importantly, it allows them to be perennial (Davey, 1989; Hawes, 1989; Hawes et al., 1992; Vincent and Howard-Williams, 1986). The cyanobacteria may have lower rates of photosynthesis, but their biomass production is higher than is their subsequent decomposition. This leads to an accumulation of cyanobacteria biomass in particular polar habitats. Eukaryotic oxyphototrophic microorganisms (microalgae) have higher rates of photosynthesis (Davey, 1989; Hawes, 1989) and lower resistances to freeze–thaw cycles. This predetermines their annual character (Davey, 1989; Elster and Svoboda, 1996; Elster et al., 1997; Hawes, 1989; Hawes et al., 1993, 1997). The retention of year-to-year persistence is provided by only a very small number of highly resistant cells (Hawes et al., 1992) that are able to survive winter conditions and extremes. This is a very simplified summary, and it is evident that features responsible for resistance against low temperature, cryoinjuries, desiccation, and salinity stress are taxonomically specific at the species and ecoform level. In addition, low temperature and cryoinjury stresses act in concert with other complex ecological factors (salinity, desiccation, rehydration, level of irradiation), and changes in single factors can operate in a synergistically positive or negative manner. The timing and duration of unfavorable conditions and the intensity of episodes of stress are also important factors that play a detrimental role in the extent of cell damage. Differences in life strategies of prokaryotic and eukaryotic oxyphototrophic microorganisms and their ecological behaviors in polar terrestrial environment are varied; some are documented in Figure 3.1.

Cyanobacteria (Figure 3.1A) frequently have multilayered cell walls or sheets with rich muci-lage. This wet wall occurs on the top of hill and is wetted only temporarily during spring melts or during summer rains. Most of season is dry with quick diurnal and seasonal temperature fluctuations. Fluctuations in temperature are followed by enormous changes in irradiation, salinity, pH, and other ecological factors. The Nostoc (Figure 3.1B) produces macroscopic colonies rich in mucilage surrounded by multilayered surfaces. Hawes et al. (1992) showed that the Nostoc-dominated mats rapidly recover from freezing and desiccation and that full mat recovery (photosynthesis and respiration) occurs within 10 min of an episode. Recovery of acetylene reduction activity (nitrogen fixation) is slower and requires about 24 h. Nostoc maintained high internal sugar-phosphate concentrations under a wide range of environmental conditions, which is probably added protection against freezing and desiccation damage. Nostoc is a widespread genus in various ephemeral freshwater habitats and frequently occurs on well-wetted localities (seepages, moss, and vascular plants carpets) of polar terrestrial habitats. *Phormidium* species (Figure 3.1C), envelop themselves in sheets and mucilage and produce rich skinlike mats, located at the base of various wetlands and on water-saturated soils. Davey (1989) undertook an Antarctic study showing that the survival of *Phormidium* sp. after freezing and desiccation is dependent on prevailing light conditions and the presence or absence of free water. High light intensities during freezing and desiccation negatively influenced survival of *Phormidium* sp. Photosynthesis completely ceased following five freeze–thaw cycles without an excess of water but showed no impairment in the presence of water. *Phormidium* sp. mats had low initial water content because of freezing and desiccation processes, but water loss is slow because of the presence of mucilage. Under some circumstances, *Prasiola crispa* (Figure 3.1D) produces lichen symbiotic associations with *Mastodia tesselata* (Huskies et al., 1997). In the same study previously described for *Phormidium* sp., Davey (1989) examined the freezing and desiccation survival strategy of the Antarctic *Prasiola crispa*. This depends on light conditions and on the absence or presence of free water. At low irradiances, *Prasiola crispa* did not show a decline in rate of photosynthesis during five daily freeze–thaw cycles at low irradiances. However, at high irradiance the decline of photosynthetic activity was recorded. *Prasiola crispa* is also sensitive to desiccation and freeze–thaw fluctuations in the absence of free water. Communities of filamentous algae demonstrate black-green parts of mats sheltered against direct irradiation (Figure 3.1D; Elster et al., 1997). These communities contained viable cells, contrasted with white-grey parts of the mats, which were dead (J. Elster, unpublished data). In the case of green algal mats (Figure 3.1E), the biomass of dead cells only can be observed (J. Elster, unpublished data). Indigenous Inuit peoples call such a structure "Martian paper," and it can persist in the field as dry, dead biomass for several years. Hawes (1990) studied the effects of cryoinjuries on the Antarctic alga *Zygnema* sp. During repeated overnight exposures to temperatures of down to –4°C, the alga's photosynthetic capacity was maintained. An increase in the duration of exposure or a decrease in temperature resulted in loss of photosynthetic capacity and in leakage of solutes and cell death. This alga is well suited to growth during the Antarctic summer, where diurnal freeze-thaw cycles rarely fall below –4°C and, in addition, these temperature changes are slow. Slow temperature changes (0.5°C/min) avoid ice nucleation, and in such populations a viability of 92% is reached. Hawes (1990) also showed that in a *Zygnema* sp. community frozen to a temperature of –4°C for 10 d, photosynthesis was undetectable, and viability decreased gradually to 40% after 16 d.

Clearly the ability of polar organisms to adapt to and survive in some of the most inhospitable parts of the planet has important consequences for other aspects of cryobiological research. The potential for interfacing polar and applied cryogenic research forms the final part of this chapter.

3.4 EXPLOITATION OF POLAR RESEARCH IN APPLIED CRYOBIOLOGY

As the rate of chemical reactions decreases in accordance with the Arrhenius equation, organisms affected by chilling and freezing will, within species and life strategy-specific limits, respond to a temperature reduction at the biomolecular level. All temperatures greater than or equal to –40°C, the temperature of homogeneous ice nucleation (and as long as liquid water is available) biochemical processes will continue to occur, albeit at low rates at temperatures less than 0°C. Tropical and, in some cases, temperate species, lethally succumb to critical limits of chilling and freezing at relatively elevated low temperatures when they are applied *in vivo*. Remarkably, however, it is possible to apply ultralow temperatures in the cryoconservation of many tropical as well as temperate plant species (see Chapters 9 and 10). This has been achieved using cryoprotective methodologies developed through the application of empirical, ecological, and physiological knowledge. Importantly, valuable insights into the basis of low-temperature adaptations can be gained through the pursuit of studies in comparative environmental stress physiology. When applied to diverse organisms adapted to different ecosystems, this approach may help elucidate the generic stress responses associated with freezing injury and cryotolerance.

Mankind's interest in preserving "life in the frozen state" spans the temperature limits and optima that polar organisms are naturally able to withstand. These pertain to studies (Fogg, 1998) of polar psychrophiles (organisms that grow at or <0°C) and psychrotrophs (organisms that grow at 0°C but that have optimum growth temperatures >15°C and upper limits at 40°C). There now exist exciting opportunities to use the outputs of polar research to help elucidate the generic adaptive mechanisms involved in freezing tolerance; such an approach has great potential in facilitating the wider application of low-temperature biology. Importantly, the knowledge gained may be used to help understand low temperature effects in nonpolar organisms. For example, understanding the basis of natural, *in vivo* low-temperature tolerances and associated stresses (desiccation, dehydration) has exciting implications (e.g., the use of ice-nucleating proteins) for medical cryobiology (cryosurgery and transplant organ storage). Furthermore, the application of polar knowledge may help in the development of improved and novel ways to cryopreserve both native biodiversity and the genetic resources of agricultural and forestry species, domesticated animals, and endangered species from tropical, temperate, and extreme ecosystems. This chapter will therefore appraise the potential applications of polar biology in applied research, specifically, acquiring a fundamental knowledge of polar low-temperature tolerances that can be used in medical, environmental, and stress physiology research; formulating the means to mitigate against cryoinjury; and developing improved and novel methods of cryoprotection and cryopreservation.

3.4.1 POLAR ECOLOGY: PROVIDING INSIGHTS INTO THE GENERIC BASIS OF *IN VIVO* AND *IN VITRO* FREEZING STRESSES

Convey (2000) identified seven interrelated stress factors that can affect polar organisms:

1. Temperature
2. Water
3. Nutrient status
4. Light availability
5. Freeze–thaw events
6. Length of growing season
7. Unpredictability

These factors have been discussed previously in this chapter in relation to their ecological implications; now their effects at the cellular, biochemical, and molecular levels will be considered. Importantly, many of the subcellular responses to these factors are similar to those that will affect organisms and their component parts exposed to cryogenic manipulations. Indeed, these factors involve the exposure of organisms and their component parts to extreme and episodic fluctuations in stress; for example, during low-temperature storage, cryosurgery, *in vitro* cryopreservation, and organ preservation.

Changes in water status may be considered one of the main generic stresses across both polar and nonpolar groups, and there is a very close relationship between the mechanisms of desiccation and cold tolerance (Convey, 2000). Clearly there are parallels between the generic environmental stresses that affect both cold extremophiles and nonpolar organisms. Thus, despite the fact that biodiversity exists in different and fluctuating environments, it is markedly apparent that representatives of the various Kingdoms possess common stress responses with respect to *in vitro* cryogenic freezing. This is especially so if low-temperature effects are considered at the subcellular level. For "life in the frozen state," the two-factor (colligative/dehydration stress and ice formation) hypothesis of cryoinjury (see Chapter 1) can be universally applied across different physiological levels (cells, tissues, and organs) from all taxonomic groupings. This has lead to the development of the most common cryoprotectant additives and the strategies that are now so widely applied to conserve biodiversity (animal, plant, fungal, and bacterial) across all taxa (see Chapters 8, 10, 11, and 15). Membrane damage, metabolic uncoupling, and free radical–mediated oxidative stress therefore represent generic stress responses (see Chapters 1, 6, and 20) in cryopreserved systems. An understanding of the common basis of cryoinjury can help develop novel and improved approaches to cryogenic storage in widely different fields of application. Within the field of applied (medical, veterinary, aquaculture, microbial and plant) cryobiological research there occurs significant evidence of multidisciplinary interaction and the exchange of ideologies. The following sections will therefore offer exemplars as to how terrestrial polar ecology research (specifically pertaining to polar invertebrates and higher plants) may aid applied low-temperature research.

3.4.2 SURVIVAL AT SUBZERO TEMPERATURES: POLAR TERRESTRIAL ANIMALS AND PLANTS

Surprisingly, there are, to date, few examples that present the importance of polar ecology in facilitating the development of cryoconservation and low-temperature storage methodologies in nonpolar organisms. However, such an approach has the potential to provide an exciting platform on which to develop new and innovative means of low-temperature conservation. For example, the discovery of glycerol's cryoprotective properties for avian spermatozoa was rapidly applied thereafter to plant, microbial, and animal cells (see Chapters 9 and 11 for historical reviews). Recent studies (Block, 2002; Dumet et al., 2000) and a European Consortium (http://www.cobra.ac.uk) do, however, promise exciting developments in the application of polar biology in cryopreservation research.

Some of the most important groups of polar terrestrial organisms that show adaptive responses to extreme cold are the various groups of arthropods and other invertebrates that inhabit polar environments. Fjellberg (1994) reports 34 species of Arctic Collembola in the Alexandra Fiord of Ellesmere Island (Canadian High Arctic) that inhabit varied and extreme habitats with considerable variations in water status. In the sub-Antarctic region in particular, there is a rich diversity of these invertebrate life-forms, and species such as the Antarctic springtail can be found in large numbers in the maritime Antarctic (Worland and Block, 2001), yet they must survive long periods at deep subzero temperatures during overwintering. These organisms provide a novel basis on which to investigate the comparative effects of freezing on different types of organisms (Block, 2002). It is beyond the scope of the present chapter to discuss these organisms in detail, but it is pertinent to draw attention to some aspects of the known strategies employed by polar invertebrates to adapt to the low-temperature environment (for more detailed discussion about insect freeze tolerance, see Bale, 2002; Lee and Denlinger, 1991).

Polar invertebrates show adaptive or "cold hardiness" changes, in response to seasonal variations, which encompass morphological, biochemical, behavioral, and ecological changes (Block, 1990). In general, invertebrates can survive exposure to very low temperatures by one of two routes: freeze avoidance or freeze tolerance (Block, 1990; Lee, 1991; Storey and Storey, 1992; Zachariassen, 1985). Those that choose the latter option have been identified as having evolved a very interesting battery of responses to survive freezing, which responses in some respects have been mimicked by the empirical experiments undertaken in the laboratory to cryopreserve living cells (see Chapter 1). In the case of freeze avoidance, seasonal adaptations have evolved to inhibit ice formation in body tissues. This can be achieved by an increase in the supercooling point of the body fluids (haemolymph) by increased production of solutes (such as glycerol and sorbitol) and sugars (such as trehalose and sucrose) to such an extent that they produce a colligative depression of freezing point (Lee, 1991; Zachariassen, 1985). This is also often combined with strategies to inhibit organic ice nucleators by removing them from the body (e.g., emptying of gut contents) and with synthesis of antifreeze proteins (or thermal hysteresis proteins, termed THPs), which can mask the growing surfaces of small ice crystals (Block, 1990; Duman, 2001; Chapter 7). The combined effects may decrease supercooling points in Antarctic arthropods to below –30°C (Block, 1990). For the most part, freeze-tolerant invertebrates employ similar strategies (accumulation of natural cryoprotectants such as glycerol to allay damage from dehydration during freezing), with protein agents, THPs, and ice nucleating proteins (INPs). It may seem counterintuitive to evolve mechanisms for nucleating ice while avoiding the lethal effects of freezing, but for the major categories of freeze-tolerant invertebrates, the control of the site and timing of ice crystal growth is an essential part of the scheme. Via INP action, ice formation may be limited to multiple extracellular compartments in the body, and by nucleating ice at high subzero temperatures, slow dehydrative freezing may take place (over many hours or days), which reduces body water content by more than 60% (Storey and Storey, 1992). There is some recent evidence that in certain selected species of invertebrates (e.g., the Antarctic nematode *Panagrolaimus davidi*), intracellular freezing may be tolerated (Wharton and Block, 1997). The nature and location of the intracellular ice remain to be fully investigated, but there was no clear evidence of existence of a glassy or vitreous state when these organisms were studied by calorimetric techniques; development of vitreous changes has been suggested previously as one mechanism of survival in insect freeze tolerance (Waslyk et al., 1988).

Applying acclimation treatments to help overcome low-temperature stress is an important aspect of applied cryobiology (see Chapters 5 and 10). However, this response has also been noted (Worland et al., 2000) in the larvae of the Diptera species *H. borealis*, obtained from Arctic habitats. This organism overwinters in the larval form in the Arctic, and it can be subjected to temperatures below –15°C, yet the larvae freeze at around –7°C. Interestingly, if the larvae were cold acclimated (at 5°C for 1 week followed by gradual reductions in temperature at 1° to 2°C/h to terminal transfer temperatures from 0° to –20°C and thence directly to –60°C), up to 80% were able to survive at –60°C. However, of those that were exposed to acclimating temperatures between –4° and –15°C, only very few (3%) were capable of pupation. Those cooled to –20°C had pupation levels of 44%, and 4% emerged as adults. As is the case with plant studies (see Chapter 10), the acclimation of *H. borealis* may be linked with the accumulation of cryoprotective sugars and polyols (trehalose, glycerol, sorbitol). Interestingly, larval fructose levels increased within the range –2° to –15°C. The accumulation of sugars and polyols remained at relatively low levels and was probably not sufficient to afford full cryoprotection. It was thus hypothesized that their presence restricted cell volume reduction during the formation of extracellular ice (Worland et al., 2000). It was also suggested that the main protective strategy was caused by the preconditioned hypometabolic state that the larvae developed before the onset of winter. This occurs as the larvae enter a dormant phase at a critical body mass. They maintain this state until they experience a low-temperature episode (<15°C) followed by a warm period (5°C).

The similarities between cold acclimation, sugar accumulation, dormancy, and life cycle strategies of polar invertebrates and temperate plants (e.g., see Chapter 5) are fascinating. A comparative study of such evidence also supports the application and development of common (cross-Kingdom) *in vitro* cryoprotection strategies for organisms from different taxa. Block (2002) has taken this approach by assessing the interactions of principle freezing phenomena in four very different biological systems exposed to freezing: (1) the freeze-intolerant Antarctic springtail *Cryptopygus antarcticus*, (2) the freeze-tolerant larvae of *H. borealis* from Arctic habitats, (3) the freeze-intolerant High Arctic springtail *Onychiurus arcticus*, and (4) cryopreserved meristems of *Ribes ciliatum* (a wild currant species)—a genotype originally collected from a volcanic region of Mexico. Block's

(2002) case study focused on examining the role of water in the different organisms. The main factors identified were water status, the presence or absence of ice nucleating agents, and time scales of desiccation. Prevention of ice nucleation and lowering the water content by osmotic means were important cryoprotective strategies in the invertebrates. *C. antarcticus* was desiccated by the atmospheric removal of water through the cuticle to the atmosphere, which has similarities to the application of evaporative desiccation in plant cryoprotection (see Dumet et al., 2000; Chapter 10). In the case of *C. antarcticus*, the number of ice nucleators and their activities increased with lowered ambient temperatures. However, for the plant system, cryoprotection is achieved via the osmotic and evaporative removal of water to a critical point at which the tissues survive and water forms stable glasses on exposure to cryogenic temperatures. In contrast, *H. borealis* larvae do not desiccate, but conserve body water in a form that is osmotically inactive. *O. arcticus* is, by comparison, dehydration tolerant and can survive 40% loss of water to allow supercooling and thus ensure wintering.

The impacts of freezing stress in polar organisms cannot, however, be considered in terms of physical parameters alone. There exist highly complex interrelationships between the geographical distribution of different polar organisms, their life histories, and their trophic status, all of which can contribute to freezing tolerance. For example, parasitism is considered just as important a factor as climate in the life-cycled development of the High Canadian Arctic insect *Gynaephora groenlandica* (Kukal and Kevan, 1994). Similarly, Bale et al. (1999) assessed insect thermal tolerance (under laboratory conditions) of three beetles sampled from South Georgia, sub-Antarctic. They were able to correlate cold hardiness and heat tolerance (using supercooling points and upper and lower lethal temperatures) with the altitudinal and latitudinal distributions of the species *in vivo*. When considering the application of *in vivo* studies, it is therefore also important to be aware of the interactive effects of both biological and ecological parameters on chilling freezing and desiccation responses.

3.4.3 POLAR PHYTOBIOTA: ECOPHYSIOLOGICAL INSIGHTS INTO CRYOTOLERANCE

Polar plant life (and algae and lichens) has to cope with incredible extremes and limiting factors that restrict photosynthesis, growth, and reproduction. Short summers and low inputs of thermal and light energy are characteristic of polar ecosystems, and they have a major effect on phytobiota. Even though there is a long photoperiod in the summer, the radiation received is of low radiant flux densities and can be influenced by snow and ice cover. Comparative studies of temperate and polar bryophytes and lichens (Convey, 2000) indicate that polar organisms show relatively high optimum photosynthetic temperatures and shallow response curves; this permits a significant net primary production at temperatures around zero. The possible physiological basis of subzero photosynthesis is also considered to equate to tolerance to low water potentials and desiccation. This is the particularly the case for cryptogams that can only become physiologically active when moist (see Lee, 1998, for a review). The Arctic lichen *Cetraria nivalis* was found to be photosynthetic at $-5°C$ and at low photon flux densities of less than 150 $-mol\ m^{-2}\ sec^{-1}$ (Kappen et al., 1995). This study revealed that at a subarctic site in Sweden, *C. nivalis* was able to accumulate 200 mmol CO_2 kg dry wt^{-1} over 6 d while growing under melting snow.

Even though water is in abundance in the polar regions, it is locked in ice. Low precipitation and high transpiration rates caused by wind and evaporative desiccation exacerbate water loss in terrestrial polar phytobiota. Poikilohydrous polar autotrophs (Ennos and Sheffield, 2000) may also reach a state of anhydrobiosis at which metabolic activity ceases and maintain this state until more favorable conditions ensue. This has been discussed earlier in this chapter. The importance of microgeography and inclination in polar survival is also demonstrated (Montiel, 2000) by Antarctica's vascular plants *Colobanthus quietensis* (cushion pearlwort) and *Deschampsia antarctica* (Hairgrass). These inhabit moist and sheltered north-facing slopes. Polar soils are frequently poor in nutrients, and this can also be affected by wind erosion, lack of microbial activity, the absence of legumes, and low phosphates. As the rate of soil formation is very slow, polar substrates tend to be poor and unstructured and have low water-retention capacities (Fitter and Hay, 1995).

The mechanical effects of ice, freezing/thawing, and melt cycles cause soil disruption, which uproots plants and impairs the ability of seedlings to establish their root systems. Reproductive problems are also important to polar plants, as the absence of insects, limited vegetation cover, and low densities of animals that can aid pollination and seed dispersal limit the potential for cross-pollination. Reproductive adaptations (Lee, 1998) in Arctic plants include the production of pre-formed flower buds that maximize the time to flowering and seed set. Some plants are heliotropic (following the sun); this helps pollination by attracting basking insects and raises the temperature of the gynaecium, promoting fertilization of embryo development. A further interesting adaptation is pseudovivipary, in which bulbils or miniplantlets are produced instead of seeds. These structures are relatively large and ready to undertake active growth and development. This ensures that plants are established more rapidly (than is the case for a germination cycle) in short-growing seasons.

The effects of all the above limiting factors are further compounded by interactions and synergism. Convey (2000) explains that water and light availability will be intimately related to meltwater patterns; for example, snow cover can provide an insulation layer, as has been discussed above in the case of algae. High light intensities and temperatures associated with melts necessitate the development by plants of mechanisms to overcome photoinhibition (see Chapter 6). These include saturation at low light intensities, reversible photo-inhibition, the accumulation of screening pigments (these can also be linked to ultraviolet-B radiation), avoidance of water loss, and enhancement of cuticular waxes (Convey, 2000). Dark anthocyanin pigments can also accumulate in some plants (Fitter and Hay, 1995). These increase the absorbance of radiant heat and have a key role in raising the temperature of the plant. Flower warming can accelerate reproduction and may have an effect on seed weight. Most polar plants adopt cushion or dwarfing growth patterns, closely spaced prostrate plants, and the clustering of insulating, dead foliage aground young, developing shoots. However, these are all energy- and resource-consuming tolerance strategies and influence other components (growth and reproduction) of polar flora life cycles. Growth limitation is thus a significant strategy, ensuring that there is sufficient energy to allow reproduction and development (Fitter and Hay, 1995). Life cycle adaptations that allow overwintering and dormancy are an important component of polar plant life strategies. Surviving very severe and long winters means that the growing and reproductive season is very short (sometimes only a few weeks), and it can be highly irregular. It is therefore critical that growth resumes at the earliest possible time, and for this reason polar plants often start to grow as soon as the temperature allows. They do not, therefore, generally display the highly programmed, innate dormancy that is found in temperate plant species.

3.5 BIOCHEMICAL CRYOPROTECTION STRATEGIES IN POLAR ORGANISMS: POTENTIAL FOR DEPLOYMENT IN APPLIED CRYOBIOLOGY

So far, this chapter has primarily examined ecophysiological adaptations to the terrestrial polar environment, focusing on algae as the main exemplar. A brief summary of polar plant and invertebrate adaptive physiology has also been included, providing an antecedent for considering the wider applications of polar biology research. At the whole-organism level, strategies for protection against freezing injury equate to either freeze-avoidance or freeze-tolerance, depending on the capacity to survive or to avoid extracellular ice formation. However, it is also important to assess the cellular and biochemical basis of polar "cryotolerance" *in vivo*, as this has great potential for aiding the discovery, development, and application of cryoprotective strategies that may be implemented in applied cryobiology. Many organisms cannot avoid exposure to subzero temperatures—polar organisms being a case in extremis—so they have to use a range of biochemical strategies that permit them to suppress the temperature at which ice freezes.

In vivo there exist three main biochemical cryoprotection strategies and they can alter the formation and pattern of ice crystallization as well as conferring protection through undercooling. They are:

- Manipulation of cellular osmotic status via the synthesis of sugars
- Sugar alcohols and polyols
- Manipulation of ice nucleation
- Production of antifreeze proteins (AFPs)

3.5.1 CARBOHYDRATES AND CRYOPROTECTION

A major biochemical approach to freeze avoidance involves the ability of organisms to change their intracellular solute composition, especially with respect to sugars and sugar alcohols. This results in the supercooling of water; *in vivo* freezing is therefore circumvented. This is a very important adaptation for surviving "the frozen state" in nature, but it also has great relevance to cryogenic manipulations (cryopreservation) *in vitro*.

Biochemical studies of polar organisms collected from remote polar regions, and especially Antarctica, are difficult, as collection and transport may compromise the biochemical integrity of the tissues (Johnstone et al., 2002). However, a comparative study of the importance of sugars in the cryoprotection of polar organisms has been documented by Montiel (2000). This comprises a detailed analytical investigation of carbohydrate-based cryoprotection in a wide range of polar biota. Biochemical analyses were performed using high-pressure liquid chromatography to detect, at the cellular level, seasonal variations in soluble carbohydrate compositions of Antarctica species from Signy Island. These species were selected from a wide range of taxa comprising invertebrates (*Alaskozetes antarcticus, Gamazellus racovitzai, C. antarcticus, Eretmoptera murphyi* and *Pseudo-boeckella* spp.), lichens (*Umbilicaria antarctica, Caloplaca regalis, Himantormia lugubris*), mosses (*Andreaea regularis/gainii, Polytrichum alpestre, Sanionia uncinata*), an alga (*P. crispa*), and two vascular plants (*C. quitensis, D. antarctica*) from terrestrial, maritime habitats. The study profiled environmental effects over air temperature ranges of –2° to <10°C in the spring/summer and –30° to 0°C in the winter. Total soluble carbohydrate content in these organisms could be correlated with low-temperature responses with a significant increase in most winter samples, as compared to those taken in the spring and summer. In the case of the phototrophic species, polyols were the main component in the lichens and the alga. These carbohydrates were absent in the vascular plants and mosses, and in these cases, glucose, sucrose, and fructose levels were higher. In the cryptogams that displayed seasonal changes in concentrations of solutes, carbohydrates were higher in the spring. Trehalose was a major component in winter samples in some of the invertebrates and the lichens. Montiel (2000) suggests that the production of this sugar may be correlated with a partial dehydration step that allows low-temperature acclimatization, and that this may also correspond with an overwintering dehydration state of the organism. Montiel also reports that there is strong ecophysiological evidence that there is a widespread occurrence of trehalose across several Antarctica biota (nematodes, tardigrades, fungi, lichens, arthropods). This certainly exemplifies the importance of this protective sugar in *in vivo*, polar cryoprotection strategies. It is considered to have a key role in thermal and dehydration tolerance (also see Chapter 21). In the case of the poikilohydric cryptogams, Montiel (2000) postulated that the relatively high concentrations of carbohydrates act as osmolytes in the winter period and as physiological buffering agents in spring and summer. In the case of overwintering invertebrates, soluble carbohydrates increase over the winter, and therefore reduce intracellular water status.

The manipulation of cell solute composition and viscosity is a key component of cryoprotective strategies applied to cryoconserve cells, tissues, and organs and low and ultralow temperatures. This can be achieved through the application of polyols, sugars, dimethyl sulfoxide (Me$_2$SO), and evaporative desiccation. Cells must achieve a solute composition at which, on exposure to ultralow temperatures (liquid nitrogen, –196°C), the cells avoid ice formation completely and an amorphous, vitrified glass is created. This is essentially cryogenic storage in the absence of ice (see Chapters 9 and 10), although in some cases the cryoprotectant can form ice while the cells and tissues are vitrified.

Alternatively, colligative cryoprotectants can be applied that allow supercooling to a critical point at which external (extracellular) ice is formed before intracellular ice. On the formation of extracellular ice, a vapor pressure deficit is created between the inside and the outside of the cell. As a result, intracellular water moves out of the cell, and in doing so, the water available for ice formation in the cell is reduced. The aim of the applied cryobiologist is therefore to manipulate this process (by optimizing cryoprotectant treatments, cooling rates, terminal transfer temperatures, and ice nucleation events) to achieve an intracellular water status such that the solute composition becomes concentrated enough (following the removal of water by the water vapor deficit gradient) to support vitirification. Penetrating colligative cryoprotectants play an essential role in this process as they mitigate damaging solution effects. The process of vitrification has been extensively used as a cryoprotective strategy across all applied cryobiological disciplines. It is also considered to have a very important role in plant freezing tolerance *in vivo* (see Chapter 10). Changes in cellular osmotic status can also be invoked *in vitro* to assist the survival of organisms and their component parts after cryogenic storage. This occurs through the process of cold acclimation (see Chapter 5), which can be applied as part of a cryoprotective strategy by simulating cold acclimation for *in vitro* systems (Chang et al., 2000; Dumet et al., 2000). Thus, cold acclimation has been simulated as a method of enhancing the recovery of plants cells from cryogenic storage. The application of soluble protective agents sugars, carbohydrates, polyols, and glycerol as cryoprotectants in germplasm cryoconservation and medical cryogenic storage could, therefore, be considered to simulate, in part, the natural cryoprotective responses of polar organisms; although, of course, trehalose and other soluble sugars are also involved in the adaptive responses of freeze- and desiccation-tolerant organisms from nonpolar regions. Broad-spectrum deployment of natural cryoprotectants in applied cryobiology is especially exemplified by the use of trehalose and sucrose. Trehalose has been similarly applied (Bhandal et al., 1985) for the cryopreservation of suspension cultures of plant cells (*Daucus carota, Nicotiana tabacum*, and *Nicotiana plumbaginifolia*). Pretreatment of (5 to 10% w/v for 24 h before freezing) cells with trehalose as the sole cryoprotectant (20 to 40% w/v trehalose on ice for 1 h) resulted in approximately 50 to 75% cell viability. Trehalose has been used as a cryoprotectant for the cryopreservation of the highly storage-recalcitrant and thermophillic bacterium *Lactobacillus bulgaricus* (De Antoni et al., 1989). At a concentration of 0.3 M, trehalose was able to maintain the viability and acidification functionality in the organism when stored at –60°C.

The application of sucrose in plant cryoconservation (as a pregrowth additive, cryoprotectant in vitrification solutions, and osmoticum in encapsulation/dehydration) demonstrates the extensive use of the sugar in plant cryogenic storage (see Chapter 9). However, applications of sugars in cryoprotection protocols are also significant in medical and animal cryobiology. Kravchenko and Sampson (1998) demonstrated that a cryoprotectant medium comprising 0.25 M sucrose, 1% (w/v) bovine serum albumen (BSA), and a natural plant antioxidant silibor (derived from *Silybum marianum*) preserved whole rat liver under hypothermic conditions (0°C) for 24 h and maintained good biochemical activity. Rudolph and Crowe (1995) examined the effectiveness of trehalose and proline in the preservation of membrane structure and function in frozen muscles derived from *Homarus americanus* (lobster). Their studies showed that both compounds retained the functionality of the tissue, and inhibited the membrane mixing on freezing, of phospholipid vesicles. Glycerol and Me$_2$SO were less efficacious with respect to the preservation of membrane stability.

3.5.2 APPLICATIONS OF ICE NUCLEATORS

Intracellular ice nucleators (templates on which ice crystals are first initiated) can prove highly problematic, yet they also offer important applications in environmental, agricultural, and food technologies (Lindow, 1987; Lee et al., 1995; Li and Lee, 1995). Ice-nucleating bacteria are very potent initiators of ice formation and are found in both temperate (Costanzo and Lee, 1996) and polar (Obata et al., 1999) habitats. For example, plants that are susceptible to frost may be injured by the presence of ice-nucleating epiphytic bacteria (e.g., *Pseudomonas syringae*). In nature these become associated with plant leaves, and they nucleate ice at relatively high temperatures. (−2° to −5°C). Plants that have a bacterial flora capable of ice nucleation are therefore far more sensitive to freezing stress. Orser et al. (1985) discovered the ice-nucleating gene from *P. syringae*. Strains of bacteria that lack the gene have been isolated from wild populations and used to infect plants such that they out-compete the ice-nucleating bacteria, reducing the possibility of frost damage. However, far more controversial (Miller, 1994) is the release of genetically manipulated bacteria containing an ice-minus gene *in vivo*, the rationale being that these bacteria will compete with wild strains that do contain the ice-nucleating gene and will enhance freezing tolerance in crops.

Studies of polar ice-nucleating capacities in different organisms offer considerable potential for the applied sector. The control of ice nucleation and ice growth in polar organisms can be a very important survival strategy (Storey and Storey, 1992). In the case of polar microinvertebrates, ice-nucleating organisms or other templates can be present in large numbers in the external and internal body floras. If they are not eliminated from the body, this can pose a major problem. Some polar organisms may therefore purge any potential ice-nucleating agents (food particles, proteins) from their bodies on a seasonal basis, such that in the winter months moderating diet and body fluids reduces the potential for ice nucleation to occur (Storey and Storey, 1992). This purge, coupled to a programmed desiccation response in which body fluids are reduced (as is the freezable water content), obviates the potential for lethal nucleation events to occur. The action of ice-nucleating proteins and microorganisms *in vivo* and *in vitro* can also be beneficial if extracellular nucleation allows the slow dehydration of cells through a controlled water vapor gradient (Storey and Storey, 1992; see also Chapter 7).

Interestingly, a strain (IN-74) of a novel ice-nucleating bacterium, *Pseudomonas antarctica*, has been identified as a potential freeze-texturing agent of food. The organism was originally isolated from Ross Island, Antarctica, and it produces ice-nucleating activity at temperatures of −15° to −22°C (Obata et al., 1999). Ice-nucleating proteins are also found in the blood of freeze-tolerant vertebrates, and they have very similar characteristics to those of tardigrades, molluscs, and insects. It has been proposed (Constanzo and Lee, 1996; Obata et al., 1999) that the selective expression of ice-nucleating proteins may be a critical factor in vertebrate freeze-tolerance. If these proteins fail, some organisms are able to use ice-nucleating bacteria that are harbored in the gut or on the skin.

Ice-nucleation activity (potentially assigned to clusters of bacteria and noncharacterized templates and particles) has been assessed using a droplet-freezing method in 19 cold-adapted species of phytobiota sampled from South Georgia and the maritime Antarctic (Worland et al., 1996). The main aim of the investigation was to develop a robust method to quantify the effect of different variables (ice-nucleator particle size, cooling rate, and sample preparation) on nucleating activity. Interestingly, this comparative study revealed a range of ice-nucleating capacities from 257,000 to 16,220 nuclei/gram for the respective successional order of activity: lichens > mosses > flowering plants. Clearly the deployment of such information related to polar organisms may offer useful insights into the wider applied use of ice nucleators (Lee et al., 1995; Li and Lee, 1995).

3.5.3 ANTIFREEZE PROTEINS: A DIVERSITY OF APPLICATIONS

Although ice nucleators are involved with the initial formation of ice, AFPs are associated with postnucleation modification of ice crystal growth and structure. This is very important in nature and in applied research (Griffith and Ewart, 1995), as the size and form of ice crystals can have a major effect on the potential injurious effects of freezing. AFPs become adsorbed onto ice crystals, and they actually modify crystal growth. At critically high concentrations the AFPs cause a thermal change to the solution in which they are associated. This is termed "thermal hysteresis," a phenomena that causes the depression of the freezing point of a solution without affecting the melting temperature. Importantly, at low concentrations, AFPs can also inhibit the recrystallization of ice (occurring at, or fluctuating near, 0°C). AFPs do not act colligatively, but because of their ability to prevent the growth of ice (through adsorption and changing their tertiary structure), they are up to 500 times more effective in lowering freezing temperature than other known solutes (Fletcher et al., 1999). This is an important capability in nature, as it means that organisms, such as polar fish, that are in possession of AFPs are able to moderate their cryoprotection mechanism without changing their osmotic status.

In nature, AFPs are found in plants, insects, and various polar and subpolar fishes, of which four classes have been characterized (Fletcher et al., 1999). About 30 angiosperm species demonstrate AFP activity after cold acclimation, and some of these species are commercially important crops (rye, wheat, barley, oat, a range of *Brassica* spp., potato, and carrot). Antifreeze glycoproteins (AFGPs) originate in Antarctic *Notothenioidei* teleost and northern cods; other fish types I–III are found in flounder, shorthorn and longhorn sculpin, sea ravens, smelts, and herring (Fletcher et al., 1999). *Notothenioidei* teleost fish are the dominant fish fauna of the coastal regions of the Antarctic Ocean. They are considered to be ecologically successful because they have adapted their blood composition to include antifreeze glycoproteins that have an equilibrium freezing point of –0.7 to –1.0°C. Chen et al. (1997a, 1997b) have characterized the antifreeze protein of the *Notothenioid* Antarctic fish taxon. The AFGP gene is thought to have evolved from pancreatic trypsinogen over 5 to 15 million years ago (coincident with the freezing of the southern Ocean). However, a remarkable discovery has also been made in that two groups of polar fish (the Antarctic *Notothenioid* teleost and the Arctic cod) originating from opposite poles of the earth have both evolved similar AFGPs. This research (Chen et al., 1997a, 1997b) demonstrates a unique direct link between the evolution of protein, the diversification of an animal via covergent evolution and environmental change in the polar regions. Although the Antarctic fish genes originate from trypsinogen, the AFGP of Arctic cod shares no sequence homology with this enzyme, showing that it has a different progenitor. However, the evolution of antifreeze proteins in these animals is suggested to be a major factor in allowing the polar fishes to colonize different levels of the cold Arctic and Antarctic waters.

Many applications of AFPs have been identified (Fletcher et al., 1999; Griffith and Ewart, 1995) in commercial, medical cryobiology and in agricultural sectors:

- Improving food textures (e.g., of frozen foods)
- Stabilizing the shelf life of frozen foods
- Cryogenic storage of plant and animal germplasm
- Enhancement of the success of surgical cryoablation
- Medical cryobiology (transplant organ and tissue storage)
- Crop protection against freezing injury
- Improvement of the characteristics of transgenic fish
- Application of transgenic technologies to introduce antifreeze genes into animals and plants

In terms of cryopreservation, the application of AFPs can be contentious, as some reports indicate that they can enhance survival, whereas others indicate that clear beneficial effects are equivocal. Comparing and contrasting the use of AFPs in cryopreservation with those in cryosurgery may offer some explanation as to their confounding deleterious and positive effects. In the case of cryosurgical applications, it is essential that all the tumor tissue is destroyed during cryoablation (Chapter 16). This can be critically dependent on cooling rates, and it is important to avoid vitrification processes that may confer cell survival (El-Sakhs et al., 1998). Pham et al. (1999) performed an *in vivo* study of AFP adjuvant cryosurgery using subcutaneous metastatic prostate tumors developed in nude mice. AFP TYPE I was applied in the tumor before freezing at a rate of 10 mg/L. This treatment was found to enhance the cryodestruction of the tissues under thermal conditions that would normally enhance cell survival. In a different application, Naidenko (1997) applied AFP1 at a rate of 0.1 to 0.2 mg/mL to oocytes, embryos, and larvae of *Crassostrea gigas* (oyster) and found that AFPI reduced cryoinjury in frozen oocytes.

Clearly there is an interesting disparity in the application of AFPs in causing or ameliorating cryoinjury. Pham et al. (1999) make the important comment that it is the way in which the AFPs modify the structure of ice crystals that determines their effects (see also Chapters 2 and 22). Studies have shown that at concentrations greater than 5 mg/L, AFPs can cause frozen cell destruction; it is proposed (Pham et al., 1999) that the effect is caused by the formation of needle-like ice crystals. These can be associated with the severe mechanical disruption of cellular and connective tissues. The damaging effect of the AFPs occurs as the temperature gradient is propagated and the formation of needle ice spicules thus concentrates the AFPs. Incipient concentration effects are also considered by Ekins et al. (1996), who applied AFP I and III to rat liver slices with a view to stabilizing their biochemical and functional activities. Pham et al. (1999) make a very important comment in noting that many researchers studying the cryoprotective effects of AFPS during cryopreservation may mistakenly apply them at too a high concentration, which promotes the production of spicular ice structures that cause mechanical cell damage and is not related to the cryoprotective effects of the proteins, which should be examined at much lower concentrations. Ekins et al. (1996) concur with this observation and show that at high concentrations (1 mg/mL), the AFPs applied to a rat liver system may act as ice nucleators, whereas at lower concentrations (0.2 mg/mL), AFPI and AFP III may be beneficial.

Further examples are demonstrated by Wang et al. (1994) such that AFGPs derived from the Antarctic nototheniod fish failed to protect cardiac explants in rats during hypothermic and freezing preservation. Similarly, O'Neil et al. (1998) applied AFGP to cryopreserved vitrified mature mouse oocytes. The AFGP was derived from the blood of the Antarctic nototheniod fish. Supplementation with 1 mg/mL of AFGP produced poor rates of survival, and these were highly variable, depending on the temperature of cryoprotectant application. Lowering the temperature of exposure to the treatment did, however, improve viability. Thus the protective and deleterious effects of AFPs may be critically concentration dependent. Ramlov et al. (1996) also consider the role of THP proteins in the freezing-tolerant Antarctic nematode *P. davidi*, and this study may have some relevance in helping to understand the action of AFPs *in vitro*. The possibility is presented that the role of THPs in insects is not solely to prevent ice recrystalization but also to prevent the migration of still-liquid domains present in the frozen tissue. These are operational because of the freeze concentration of body fluids moving along differential temperature gradients in the organism (Knight et al., 1995; Ramlov et al., 1996). Thus the migration of liquid domains could be harmful and the effects will be related to the size. The larger the organism, and presumably in medical cryobiological applications the larger the tissue, the more likely that differential temperature gradients will become significant on thawing/rewarming.

To summarize, clearly the application of AFPs/THPs in applied low-temperature research necessitates a greater understanding of the different modalities (cryoprotective and cryodestructive) of their complex functions. The size and shape of ice crystals formed in association with AFPs appears to be critically affected by the concentration at which they are applied. Furthermore, size may also have a critical effect on the movement of water (associated with AFPs) within different domains of tissues. In the future it may be beneficial to combine multidisciplinary knowledge and expertise (Block, 2002; Knight et al., 1995; Ramlov et al., 1996) of *in vivo*, *in vitro*, medical, and ecological aspects of AFPs with a view to enhancing our understanding of natural freeze-tolerance adaptations. Cross-disciplinary studies of AFPs in cryodestruction (cryosurgery) and cryoprotection (cryopreservation) may be particularly useful.

3.6 FUTURE DIRECTIONS OF POLAR CRYOBIOLOGICAL RESEARCH

This chapter has compared the ecological and biological attributes of the Arctic and Antarctic terrestrial environments. To conclude, a summary (Figure 3.2) is presented that redefines the environment of the different polar regions with respect to biotic and abiotic parameters affected by low temperatures and water availability. Furthermore, it simultaneously introduces the different

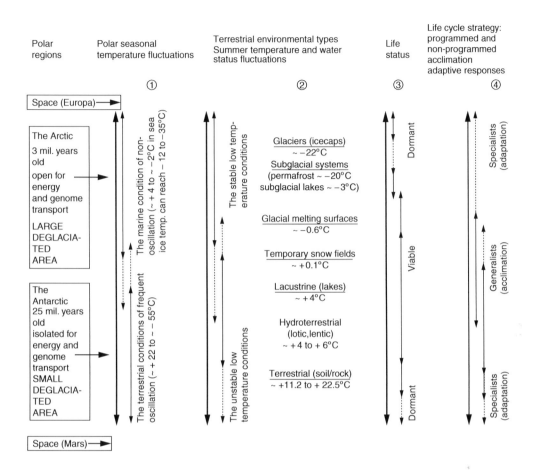

FIGURE 3.2 This scheme summarizes low-temperature and water status parameters in the polar regions and simultaneously profiles the adaptive responses of microorganisms, using cyanobacteria and algae as exemplars. In succession, from left (1) to right (4), the scheme depicts key differences between the Antarctica and Arctic regions related to: (1) Polar sea and terrestrial environment and their seasonal fluctuations; (2) summer temperature fluctuation inside of the terrestrial environment types; (3) life status; (4) life cycle strategy, incorporating programmed and nonprogrammed acclimation/responses and adaptations. For comparative purposes, the temperature ranges estimated for space research programs (e.g., relative to Mars and Jupiter's Moon, Europa) are indicated. Perpendicular thick arrows ↕ depict the main range of each primary parameter (Polar Region and 1–4); perpendicular thin arrows ↕ delineate the fluctuations and parameter ranges within each parameter; periodic arrows (----) delineate interfacing margins across and within each parameter range.

levels of polar adaptive responses using cyanobacteria and algae as exemplars. The north and south polar regions are remarkably different from each other in their geological histories, biodiversity, energy balance, and transport, yet they share the common challenges for the life that exists there—the fluctuating extremes of cold and desiccation stress. The Arctic is a geologically young and open area in which rapid glacial and periglacial processes occur. There is a relatively (to Antarctica) higher level of biodiversity and a greater capacity for geographical genome transfers to take place, aided by biological vectors, land mass continuity, and air and oceanic currents. In contrast, (see Figure 3.2), a thick ice sheet covers most of the Antarctic continent, which has been geographically isolated for a very long period of time. As a consequence, glacial and periglacial processes are much slower, and genome exchange and movement, and consequentially biodiversity, are more limited in the south polar region of Antarctica.

The different types of polar environment within a range of seasonal temperature fluctuations are shown in Figure 3.2, indicating the differentials between the terrestrial and marine environments. Seasonal temperatures of terrestrial environments vary greatly, with a maximum range of oscillation of ~50 to ~80°C. In contrast, the maritime system is more stable, with maximum seasonal fluctuations of ~2 to ~6°C. Figure 3.2 indicates the margins of terrestrial and marine environments and their boundary lines, where temperatures can range from ~10 to ~40°C. These boundaries have been estimated from previous sources presented in this chapter and from Deming and Huston (2000), Deming and Baross (2001), and Deming (2002). Extremes of, and fluctuations in, temperature and desiccation parameters (Figure 3.2) in Arctic and Antarctic regions have resulted in many polar organisms possessing highly specialized adaptive responses (see Figure 3.2). As such, they are described as "extremophiles," and interest in them is rapidly increasing, as gaining insights into the basis of their adaptive responses has great potential for applications in other fields; for example, as indicative markers of climate change and in biotechnology and conservation, through the deployment of extremophile protectants as cryoprotective (e.g., trehalose, antifreeze proteins) and cryosurgical adjuvants (ice-nucleating proteins). This chapter will therefore conclude by highlighting two future directions for the application of polar research.

3.6.1 POLAR CRYOBIOLOGY AND SPACE BIOLOGY

Polar biology, and especially extremophile research, provides an interesting platform from which to consider space biology. Recently this has been exemplified by the topical, and indeed controversial, issue of "Snowball Earth," which speculates on the involvement of ice in the development of early planetary life (Walker, 2003). Space and polar studies frequently interface with research undertaken in some of the Earth's most extreme environments, and most especially Antarctica (Wynn-Williams and Edwards, 2000a, 2000b). Polar temperature classifications together with their constraints are highlighted in Figure 3.2, and for comparative purposes, estimates of extraterrestrial temperatures are given. Temperature and the availability and state of water (as ice) may well dictate the most likely contenders within our planetary system for harboring extraterrestrial microbial life (Deming and Huston, 2000; DesMarais and Walter, 1999). One such candidate is Mars, as within its present polar ice caps, water may have been present in the past (Baker, 2001). That water and life existed on Mars previously, or that it exists now or indeed has the potential to exist in the future, is open for debate (Mullins, 2003). Another candidate for extraterrestrial life is Jupiter's moon, Europa, which is suggested to be home to a vast ocean below the ice (Chyba and Phillips, 2002). It is evident that Mars shows some environmental (temperature and water status fluctuations) similarities with extremes of Earth's terrestrial polar environment and Europa with Earth's marine polar environment. However, if the existence of life on Mars and Europa, or indeed elsewhere in our solar system, is to be validated, temperature will most likely not be the most important limiting factor. From terrestrial experiences of microbiology, extraterrestrial life would probably not survive the temperature fluctuations exceeding the limits of those found in the Earth's polar regions. The complexity and severity of other environmental factors (radiation, absence of atmosphere) in extraterrestrial environments will probably be more important factors limiting "Martian" life. Nevertheless, there still remains a high motivation to document biodiversity at high latitudes, as the polar marine and terrestrial environments provide a unique natural laboratory (see Foreword, this volume) for studying the evolution of life with respect to survival in extremis, either on, or (speculatively) off, our planet.

3.6.2 INTEGRATING POLAR BIOLOGY WITH APPLIED MEDICAL AND CONSERVATION CRYOBIOLOGY

Modulation of low temperatures and water status are central factors impacting all aspects of "life in the frozen state." In this chapter, emphasis has been placed on the *in vivo* responses of polar organisms to these parameters. Prokaryotic cyanobacteria and eukaryotic microalgae have been used as model organisms, selected because of their distinct and important status as primary producers in polar ecosystems. This review does, however, comment on the adaptations of some key higher taxa that are also resident in the polar terrestrial environment, allowing the extrapolation of knowledge gained from polar ecology to aid applied cryobiology. The algae and cyanobacteria represent an evolutionary point in the lineage of microorganisms, preceded by ancient and simpler microbial life forms (e.g., archaea, eubacteria, bacteria) and succeeded by complex higher plants. Discourse is also given to the invertebrates because they represent one of the most important groups of permanently resident, polar terrestrial animals, and their study offers unique perspectives of adaptive responses that may be applied to higher taxa.

Thus, our concluding scheme (Figure 3.2) demonstrates the different terrestrial and oceanic environments as affected by changes in seasonal temperature fluctuations, water state, and life responses depicted in terrestrial stable environments (glacial ice, subglacial systems, glacial melting surface). Also, this is the case to a lesser extent in snowfields and lacustrine ecosystems, for which summer temperatures and water state fluctuations are minimal or strongly restricted. In contrast, in unstable environments, partially hydroterrestrial and mainly in the terrestrial ecosystems, fluctuations in temperature and water availability are common and span wide ranges. Summer temperatures and water fluctuations of the stable and unstable low-temperature polar terrestrial environment are indicated in Figure 3.2. In certain terrestrial habitats (desert soil and rock, frozen glacial surfaces), liquid-state water is available for only a very limited time, and mist vapor is only sufficient to support microbial community development (Nienow and Friedmann, 1993). Similar studies of other frozen systems such as the Antarctic snow, deep glacial and lake ice matrices, and the Siberian permafrost show that there is adequate energy and space, for example, in the veins between ice crystals that remain liquid-filled at temperatures at –5° to –20°C. This type of microniche is capable of supporting bacterial life in the frozen state (Carpenter et al., 2000; Price, 2000; Rivkina et al., 2000).

Figure 3.2 demonstrates the distribution of the life strategies and their genetically programmed (e.g., seasonally induced dormancy) and nonprogrammed (responses to incidental stresses) adaptations and acclimation processes. Vincent (2000) and Elster (1999, 2002) have designated generalist (mesophile) genotypes as the most common and—ecologically—most important life strategy in polar terrestrial environments. These genotypes survive such severe ecological conditions because of their tolerance through acclimation to environmental extremes. Acclimation is described here as a response at the physiological or metabolic level that is not genetically programmed (e.g., such as seasonal dormancy) and that is expressed during sporadic environmental extremes.

The growth and development of mesophiles occurs suboptimally, meaning that most physiological and metabolic processes occur at much slower rates than those of optimal conditions. However, in some special niches of the polar terrestrial environment, optimal conditions (during short episodes of better weather) can occur irregularly and for a limited time. This offers the opportunity for growth and development under optimal conditions that are comparable to midlatitudes. Regular, generalist mesophiles are influenced by the geographical features of defined localities, and genotype exchanges probably occur between the polar terrestrial environment and midlatitudes. In our scheme (Figure 3.2), generalist genotypes are the most common, and specialist genotypes occupy only narrow niches of specific habitats that are at extreme low temperatures.

Figure 3.2 highlights the ecological, physiological, and genetic behavior of microorganisms in the polar terrestrial environment. It is evident that the stable and unstable components offer a wide spectrum of (micro-) habitat types for relatively long periods of time, in which various microorganisms including cyanobacteria and algae can grow and reproduce. Cyanobacteria and algae have a wide spectrum of ecological avoidance strategies, such as the avoidance of fluctuations (in time and space), in low temperatures and low water availabilities. However, in habitats in which appropriate temperature and simultaneous liquid water are time-limited, and in which there is frequent oscillation of these conditions, microorganisms produce dormant cells. During this process, cells significantly reduce their metabolic activity, meaning they can persist for long periods of time. In our spectrum of the polar terrestrial environments, dormant forms of microbial life can be found mainly at the upper and higher margins (Figure 3.2). In unstable terrestrial habitats, microbial life is found only in places (soil, on or within rock substrata) in which temperatures are low and water, existing in a liquid form, is available only for very short time. The endolithic and soil communities of the polar desert ecosystem represent these habitats.

One of the striking aspects of comparing *in vivo* (polar) and *in vitro* (cryoconservation and cryodestruction/cryosurgery) life at the very limits of cold endurance is that there exist so many parallels and similarities in their response phenomena. For example, the polar terrestrial environment is not only a place of extremes (constrained by low temperatures and water availability) but it also exposes the life that lives there to frequent and long-term successions of environmental fluctuations. Such rapid changes are also encountered during cryopreservation and cryodestructive processes, which may well occur at far greater rates (°C seconds/hours) and amplitudes than those experienced in nature. In contrast, temporal *in vivo* fluctuations in temperature and water stress are caused by instabilities precipitated by diurnal, seasonal, and geological changes and by, more recently, climate change. These changes have promoted and possibly accelerated polar evolutionary processes, leading to a diverse array of complex, physiological acclimation and adaptation responses. Cryobiology practitioners in conservation and medical sectors make good use of simulating nature's survival responses to aid cryoconservation and cryodestruction. A good example is in the application of AFPs from teleost fish of polar and northern temperate regions applied in cryosurgery (Chapter 16) to aid the destructive capacity of ablation treatments.

Perhaps one of the most exciting possibilities for future applications of Arctic and Antarctic research concerns studies of the longevity of organisms "trapped in polar ice." The work of Gilichinsky et al. (1995) may have particularly important implications for polar, space, and applied cryobiologists (see also Chapters 9 and 16). Viable microorganisms representing various ecological and morphological groups have been "cryo"-preserved under permafrost for thousands and sometimes millions of years; cells discovered in northeastern Siberia are the oldest and date back 2 to 3 million years. These cells provide phenomenal examples of longevity "on ice" and demonstrate the importance of really understanding the profound constraints that the natural polar environment imparts, not only physically but also biologically and, perhaps most importantly, ecologically. The application of "ecological " knowledge in applied cryobiology is yet to be fully realized, yet the benefits of pioneering Antarctic and Arctic environmental research in an applied context are enormous. Opportunities suggested in this volume range from the development of "natural" cryoprotective strategies to help cryopreserve algal, plant, animal, and human cells to applications in cryosurgery. Polar research promises exciting opportunities that have the potential to support future and emergent applications of cryobiology that affect all aspects of "life in the frozen state."

ACKNOWLEDGMENTS

We gratefully acknowledge the financial support of EU Commission Grant Quality of Life and Living Resources QLRT-2000-01645 (COBRA). In addition, the project experiences from the following grants have been used in this review: Natural Environment Research Council (LSF-82/2002), The Grant Agency of the Ministry of Education of the Czech Republic (KONTAKT–ME 576), the French–Czech cooperative research program BARRANDE (No. 99054), and The Grant Agency of the Ministry of Education of the Czech Republic (MSM 1231 00004). The first author gives a special appreciation to R. J. Delmas (Laboratoire de Glaciologie et Géophysique de l'Environnement du CNRS, St Martin d'Hères, France) and Professor P. Prosek (leader of the Czech Antarctic Research Group, Masaryk University in Brno) for their support of the polar research. We are also grateful to D. Svehlova for technical assistance. Both authors are very grateful to Barry Fuller and Nick Lane for supporting information and critical comments.

REFERENCES

Abyzov, S.S. (1993) Microorganisms in the Antarctic ice, in *Antarctic Microbiology,* Friedmann, E.I., Ed., Wiley-Liss, New York, pp. 265–295.

Abyzov, S.S., Mitskevich, I.N., Poglazova, M.N., Barkov, N.I., Lipenkov, V.Y., Bobin, N.E., Koudryashov, B.B., and Pashkevich, V.M. (1998) Antarctic ice sheet as a model in search of life on other planets, *Adv. Space Res.*, 22(3), 363–368.

Akiyama, M. (1977) Some ecological and taxonomical observation on the coloured snow algae found in Rumpa and Skarvsnes, Antarctica, *Mem. Natl Polar Res,* 11, 27–34.

Aleksandrova, V.D. (1988) *The Arctic and Antarctic: Their Division into Geobotanical Areas*, Cambridge University Press, Cambridge, U.K.

Bailey, D., Mazurak, A.P., and Rosowski, J.R. (1973) Aggregation of soil particles by algae, *J. Phycol.*, 9, 99–101.

Baker, V.R. (2001) Water and the Martian landscape, *Nature,* 412, 228–236.

Bale, J.S. (2002) Insects and low temperatures: from molecular biology to distribution and abundance, *Philos. Trans. R. Soc. Ser. B*, 357, 849–861.

Bale, J.S., Worland, M.R., and Block, W. (1999) Comparative thermal tolerance of herbivorous and predatory beetles on South Georgia, Sub-Antarctic, *Cryo-Letters,* 20, 83–88.

Barrett, P.E. and Thomson, J.W. (1975) Lichens from a high-arctic coastal lowland, Devon Island, N.W.T, *Bryologist*, 78, 160–167.

Bawley, J. D.(1979) Physiological aspects of desiccation tolerance, *Annu. Rev. Plant Physiol.*, 30, 195–238.

Beyer, L. and Bölter, M., Eds. (2002) *Geoecology of Antarctic Ice-Free Coastal Landscapes,* Ecological Studies, Vol. 154, Springer, Berlin, pp. 427.

Bhandal, I.S., Hauptman, R.M., and Widholm, J.M. (1985) Trehalose as a cryoprotectant for the freeze preservation of carrot and tobacco cells, *Plant Physiol.,* 78, 430–432.

Block, W. (1990) Cold tolerance of insects and other arthropods, *Philos. Trans. R. Soc. Ser. B,* 326, 613–631.

Block, W. (2002) Interactions of water, ice nucleators and desiccation in invertebrate cold survival, *Eur. J. Entomol.*, 99, 259–266.

Broady, P.A. (1981) Ecological and taxonomical observations on subaerial epilithic algae from Princess Elizabeth Land and Mac. Robertson Land, Antarctica, *Br. Phycol. J.*, 16, 257–266.

Broady, P.A. (1989) The distribution of *Prasiola calophylla* (Carmich.) Menegh. (Chlorophyta) in Antarctic freshwater and terrestrial habitats, *Antarctic Sci.*, 1, 109–118.

Broady, P.A. (1996) Diversity, distribution and dispersal of Antarctic algae, *Biodiv. Conserv.*, 5, 1307–1335.

Broady, P.A. and Smith, R.A. (1994) A preliminary investigation of the diversity, survivability and dispersal of algae into Antarctica by human activity, *Proc. Natl. Inst. Polar Res. Symp. Polar Biol.* 7, 185–197.

Bruni, F. and Leopold, A.C. (1991) Glassy state in soybean seeds: relevance to anhydrous biology, *Plant Physiol.*, 96, 660–663.

Caiola, M.G., Billi, D., and Friedmann, E.I. (1996) Effect of desiccation on envelopes of the cyanobacterium *Chroococcidiopsis* sp. (Chroococcales), *Eur. J. Phycol.*, 31, 97–105.

Cameron, R.E. and Black, G.B. (1967) Desert soil algae survival at extremely low temperatures, *Cryogenic Technol.*, 3, 151–156.

Canfield, D.E. and Green, W.J. (1985) The cycling of nutrients in a closed-basin Antarctic lake: Lake Vanda, *Biochemistry*, 1, 233–256.

Carpenter, E.J., Kin, S., and Capone, DG. (2000) Bacterial activity in South Pole snow. *Appl. Environ. Microbiol.,* 66, 4514–4517.

Castenholz, R.W. (1982) Motility and taxes, in *The Biology of Cyanobacteria*, N.B. Carr and B.A. Witton, Eds., *Botanical Monographs*, Vol. 19, Blackwell Scientific Publishers, Oxford, pp. 9–45.

Castenholz, R.W., Jørgensen, B.B., D'Amelio, E., and Bauld, J. (1991) Photosynthetic and behavioral versatility of the cyanobacterium *Oscillatoria boryana* in a sulfide-rich microbial mat, *FEMS Microbial Ecol.*, 86, 43–58.

Caulson, S.J., Hodkinson, I.D., Strathdee, A.T., Block, W., Webb. N.R., Bale, J.S., and Worland, M.R. (1995) Thermal environments of Arctic soil organisms during winter, *Arctic Alpine Res.*, 27(4), 364–370.

Chang, Y., Barker, R.E., and Reed, B.M. (2000) Cold acclimation improves recovery of cryopreserved grass (*Zoysia* and *Lolium* sp.*), Cryo-Letters*, 21, 107–116.

Chen, L., DeVries, A.L., and Cheng, Chi-Hing, C. (1997a) Convergent evolution of antifreeze glycoproteins in Antarctic notothenioid fish and Arctic cod, *Proc. Natl. Acad. Sci.*, 94, 3817–3822.

Chen, L., DeVries, A.L., and Cheng, Chi-Hing, C. (1997b) Evolution of antifreeze glycoprotein gene from a trypsinogen gene in Antarctic notothenioid fish, *Proc. Natl. Acad. Sci.*, 94, 3811–3816.

Christner, B.C., Mosley-Thompson, E., Thompson, G.L., and Reeve, J.N. (2001) Isolation of bacteria and 16S rDNAs from Lake Vostok, *Environ. Microbiol.*, 3, 1–9.

Chyba, C.F. and Phillips, C.B. (2002) Europa as an abode of life. *Orig. Life Evol. Biosph.* 32, 47–68.

Constanzo, J.P. and Lee, R.E. (1996) Mini-review: ice nucleation in freeze-tolerant vertebrates, *Cryo-Letters*, 17, 111–118.

Convey, P. (2000) How does cold constrain life cycles of terrestrial plants and animals? *Cryo-Letters*, 21, 73–82.

Crowe, J.H., Crowe, L.M., and Chapman, D. (1984) Preservation of membranes in anhydrobiotic organisms: the role of trehalose, *Science,* 223, 701–703.

Davey, M.C. (1989) The effects of freezing and desiccation on photosynthesis and survival of terrestrial Antarctic algae and cyanobacteria, *Polar Biol.*, 10, 29–36.

Davey, M.C. (1991) The seasonal periodicity of algae on Antarctic fellfield soils, *Holoarct. Ecol.*, 14, 112–120.

De Antoni, G.L., Peres, P., Abraham, A., and Anon, M.C. (1989) Trehalose, a cryoprotectant for *Lactobaccillus bulgaricus*, *Cryobiology*, 26, 149–153.

Deming, J.W. (2002) Psychrophiles and polar regions, *Curr. Opin. Microbiol.*, 5(3), 301–309.

Deming, J.W. and Baross, J.A. (2001) Search and discovery of microbial enzymes from thermally extreme environments in the ocean, in *Enzymes in the Environment*, Dick, R.P. and Burns, R.G., Eds., Marcel Dekker, New York, pp. 327–362.

Deming, J.W. and Huston, A.L. (2000) An oceanographic perspective on microbial life at low temperatures with implications for polar ecology, biotechnology and astrobiology, in *Cellular Origins and Life in Extreme Habitats*, Seckbach, J., Ed., Kluwer, Dordrecht, pp. 149–160.

DesMarais, D.J. and Walter, M.R. (1999) Astrobiology: exploring the origins, evolution, and distribution of life in the universe. *Annu. Rev. Ecol. Syst.* 30, 397–420.

Douglas, A.E. (1994) *Symbiotic Interactions*, Oxford, Oxford University Press.

Duman, J. (2001) Antifreeze and ice nucleator proteins in terrestrial arthropods, *Ann. Rev. Physiol.* 63, 327–357.

Dumet, D. Block, W., Worland, M.R., Reed, B.M., and Benson, E.E. (2000) Profiling cryopreservation protocols for *Ribes ciliatum* using differential scanning calorimetry, *Cryo-Letters*, 21, 367–378.

Dunbar, M.J. (1977) The evolution of polar ecosystems, in *Adaptations within Antarctic Ecosystems, Proceedings of Third SCAR Symposium on Antarctic Biology,* Llano, G.L., Ed., Washington, DC, Smithsonian Institution, pp. 1063–1076.

Ekins, S, Murray, G.I., and Hawksworth, G.M. (1996) Ultrastructural and metabolic effects after vitrification of precision-cut rat liver slices with antifreeze proteins, *Cryo-Letters*, 17, 157–164.

Ellis-Evans, J.C. (1981a) Freshwater microbiology in the Antarctic: II Microbial numbers and activity in nutrient-enriched Heywood Lake, Signy Island, *Br. Antarctic Surv. Bull.*, 54, 105–121.

Ellis-Evans, J.C. (1981b) Freshwater microbiology in the Antarctic: I. numbers and activity in oligotrophic Moss Lake, Signy Island, *Br. Antarctic Surv. Bull.*, 54, 85–104.

Ellis-Evans, J.C. and Wynn-Williams, D. (1996) A great lake under the ice, *Nature*, 381, 644–646.

El-Sakhs, S., Shimi, S., Benson, E.E., Newman, L., and Cuschieri, A (1998) Physical observations on rapid freezing of tumour cells to –40°C using differential scanning calorimetry, *Cryo-Letters*, 19, 159–170.

Elster, J. (1999) Algal versatility in various extreme environments, in *Enigmatic Microorganisms and Life in Extreme Environments*, Seckbach, J., Ed., Kluwer Academic, pp. 215–227.

Elster, J. (2002) Ecological classification of terrestrial algae communities of polar environment, in *GeoEcology of Antartic Ice-Free Coastal Landscapes, Ecological Studies*, Vol. 154, Beyer L. and Bölter M., Eds., Springer, Berlin, pp. 303–326.

Elster, J., Delmas, R.J., Petit, J., and Kaštovská, K. (2003) Composition of microbial communities of aerosol, snow and ice samples from remote glaciated areas (Antarctica, Alps, Andes), *Arctic Antarctic Alpine Res.* (In preparation).

Elster, J., Kaštovská, K., Brynychová, K, Lukešová, A., Stibal, M., and Kanda, H. (2002) Diversity of cyanobacteria and eukaryotic microalgae in subglacial soil (Ny-Ålesund, Svalbard) (extended abstract), *Proceedings of the Sixth Ny-Ålesund Workshop*, Norsk Polarinstitutt, Internrapport Nr. 10, Trømso, Norway.

Elster, J. and Komárek, O. (2003) Ecology of periphyton in a meltwater stream ecosystem in the maritime Antarctic, *Antarctic Sci.,* 15(2), 189–201.

Elster, J., Lukesova, A., Svoboda, J., Kopecky, J., and Kanda, H. (1999) Diversity and abundance of soil algae in the polar desert, Sverdrup pass, central Ellesmere Island, *Polar Rec.,* 35(194), 231–254.

Elster, J. and Svoboda, J. (1996) Algal seasonality and abundance in, and along glacial stream, Sverdrup Pass 79°N, Central Ellesmere Island, Canada, *Mem. Natl. Inst. Polar Res.,* 51, 99–118.

Elster, J., Svoboda, J., Komárek, J., and Marvan, P. (1997) Algal and cyanoprocaryote communities in a glacial stream, Sverdrup Pass, 79°N, Central Ellesmere Island, Canada, *Arch. Hydrobiol./Suppl. Algolog. Stud.,* 85, 57–93.

Ennos, R. and Sheffield, E. (2000) *Plant Life*, Blackwell Science, Oxford.

Feingold, L., Singer, M. A., Federle, T.W., and Vestal, J.R. (1990) Composition and thermal properties of membrane lipids in cryptoendolithic lichen microbiota from Antarctica, *Appl. Environ. Microbiol.,* 56, 1191–1194.

Fitter, A.H. and Hay, R.K.M. (1995) *Environmental Physiology of Plants*, 2nd ed. Academic Press, London.

Fjellberg, A. (1994) Habitat selection and biogeography of springtails (Collembola) from Alexandra Fiord, Ellesmere Island, in *Ecology of a Polar Oasis, Alexandra Fiord, Ellesmere Island, Canada*, Svoboda, J. and Freedman, B., Eds., Captus University Publications, Toronto, pp. 227–229.

Fletcher, G.L., Goddard, S.V., and Wu, Y. (1999) Antifreeze proteins and their genes: From basic research to business opportunity, *Chemtech,* 30, 17–28.

Fogg, G.E. (1998) *The Biology of Polar Habitats*, Oxford University Press, Oxford.

Friedmann, E. I. (1980) Endolithic microbial life in hot and cold deserts, *Orig. Life,* 10, 233–245.

Friedmann, E.I., Ed., (1993) *Antarctic Microbiology*, Wiley-Liss, New York.

Friedmann, E.I. and Weed, R. (1987) Microbial trace-fossil formation, biogenous, and abiotic weathering in the Antarctic cold desert, *Science,* 236, 703–705.

Friedmann, E.I., Hua, M., and Ocampo-Friedmann, R. (1988) Cryptoendolithic lichen and cyanobacterial communities of the Ross Desert, Antarctica, *Polarforschung,* 58, 251–259.

Friedmann, E.I., McKay, C.P., and Nienow, J.A. (1987) The cryptoendolithic microbial environment in the Ross Desert of Antarctica: satellite-transmitted continuous nanoclimate data, 1984 to 1986, *Polar Biol.,* 7, 273–287.

Gerdel, R.W. and Drouet, F. (1960) The cryoconite of the Thule area, Greenland, *Trans. Am. Microsc. Soc.,* 79, 256–272.

Gilichinsky, D.A., Soina, V.S., and Petrova, M.A. (1993) Cryoprotective properties of water in the Earth cryollitosphere and its role in exobiology, *Orig. Life Evol. Biosph.,* 23, 65–75.

Gilichinsky, D.A., Wagener, S., and Visnivetskaya, T.A. (1995) Permafrost microbiology, *Permafrost Periglacial Processes,* 6, 281–291.

Glasser, N.F. and Hambrey, M.J. (2001) Styles of sedimentation beneath Svalbard valley glaciers under changing dynamic and thermal regimes, *J. Geol. Soc.,* 158, 697–707.

Golubic, S., Friedmann, E.I., and Schneider, J. (1981) Lithobiontic ecological niche, with special reference to microorganisms, *J. Sedim. Petrol.,* 51, 475–478.

Gray, J.S. (2001) Marine Diversity: The paradigms in patterns of species richness examined, *Sci. Mar.,* 65, 41–56

Griffith, M. and Ewart, K.V. (1995) Antifreeze proteins and their potential use in frozen foods, *Biotechnol. Adv.,* 13, 375–402.

Hamilton, P.B. and Edlund, S.A. (1994) Occurrence of *Prasiola fluviatis* (Chlorophyta) on Ellesmere Island in the Canadian Arctic, *J. Phycolol.,* 30, 217–221.

Hawes, I. (1989) Filamentous green algae in freshwater streams on Signy Island, Antarctica, *Hydrobiologia,* 172, 1–18.

Hawes, I. (1990) Effects of freezing and thawing on a species of *Zygnema* (Chlorophyta) from the Antarctic, *Phycologia,* 29, 326–331.

Hawes, I. and Brazier, P. (1991) Freshwater stream ecosystems of Janes Ross Island, Antarctica, *Antarctic Sci.,* 3(3), 265–271.

Hawes, I. and Howard-Williams, C. (1998) Primary production processes in streams of the McMurdo Dry valleys, Antarctica, in *Ecosystem Dynamics in a Polar Desert, Antarctic Research Series*, Vol. 72, Priscu, J.C., Ed., American Geophysical Union, Washington, DC, pp. 129–140.

Hawes, I., Howard-Williams, C., and Vincent, W. F. (1992) Desiccation and recovery of antarctic cyanobacterial mats, *Polar Biol.*, 12, 587–594.

Hawes, I., Howard-Williams, C., and Pridmore, R.D. (1993) Environmental control of microbial communities in the ponds of the McMurdo Ice Shelf, Antarctica, *Arch. Hydrobiol.*, 127, 271–287.

Hawes, I., Howard-Williams, C., Schwarz, A.M., and Downes, M. (1997) Environment and microbial communities in a tidal lagoon at Bratina Island, McMurdo Ice Shelf, Antarctica, in *Antarctic Communities: Species, Structure and Survival*, Battaglia, B., Valencia, J., and Walton, D. H. W., Eds., Cambridge University Press, Cambridge, pp. 170–177.

Holdgate, M.W. (1970) *Antarctic Ecology*, Academic Press, London.

Howard-Williams, C., Pridmore, R., Downes, M.T., and Vincent, W.F. (1989) Microbial biomass, photosynthesis and chlorophyll *a* related pigments in the ponds of the McMurdo Ice Shelf, Antarctica, *Antarctic Sci.*, 2, 125–131.

Howard-Williams, C., Vincent, L., Broady, P.A., and Vincent, W.F. (1986) Antarctic stream ecosystems: variability in environmental properties and algal community structure, *Int. Rev. gesampten Hydrobiol.*, 71, 511–544.

Huntington, H. et al. (2001) Conservation of Arctic flora and fauna: Status and conservation, Helsinki, *Edita*, 272.

Huskies, A.H.L., Gremmen, N.J.M., and Francke, J.W. (1997) The delicate stability of lichen symbiosis: comparative studies on the photosynthesis of the lichen *Mastodia tesselata* and its free-living phycobiont, the alga *Prasiola crispa*, in *Antarctic Communities: Species, Structure and Survival*, Battaglia, B., Valencia, J., and Walton, D.W.H., Eds., Cambridge University Press, Cambridge, pp. 234–240.

Imshenetsky, A.A., Lysenko, S.V., Kazakov, G.A. (1978) Upper boundary to the biosphere, *Appl. Environ. Microbiol.*, 35, 1–5.

Jean-Baptiste, P., Petit, J.-R., Lipenkov, V. Ya., Raynaud, D., and Barkov, N. I. (2001) Constrains on hydrothermal processes and water exchange in Lake Vostok from helium isotopes, *Nature*, 411, 460–462.

Johnstone, C., Block, W., Benson, E.E., Day, J.G., Staines, H., and Illian, J.B. (2002) Assessing methods for collecting and transferring algae from Signy Island, maritime Antarctic to the United Kingdom, *Polar Biol.*, 25(70), 553–556.

Jouzel, J., Petit, J.R., Souchez, R., Barkov, N.I., Lipenko, V.Y., Raynaud, D., Stievenard, M., Vassiliev, N.I., Verbeke, V., and Vimeux, F. (1999) More than 200 meters of lake ice above subglacial lake Vostok, Antarctica, *Science*, 286, 2138–2141.

Kalff, J. and Welch, H.E. (1974) Phytoplankton production in Char Lake, a natural polar lake, and Meretta lake, a polluted polar lake, Cornwallis Island, Northwest Territories, *J. Fish. Res. Board Can.*, 31, 621–636.

Kappen, L. (1993) Lichens in the Antarctic region, in *Antarctic Microbiology*, Friedmann, E. I., Ed., Wiley-Liss, New York, pp. 433–490.

Kappen, L., Sommerkorn, and Schroeter, B. (1995) Carbon acquisition and water relations in lichens in Polar regions-potentials and limitations, *Lichenologist*, 27, 531–545.

Karl, D.M., Bird, D.F., Björkman, K., Houlihan, T., Shackelford, R., and Tupas, L. (1999) Microorganisms in the accreted ice of Lake Vostok, Antarctica, *Science*, 286, 2144–2147.

Kawecka, B. and Olech, M. (1993) Diatom communities in the Vanishing and Ornithologist Creek, King George Island, South Shetlands, Antarctica, *Hydrobiologia*, 269/270, 327–333.

Kennedy, A D. (1993) Water as a limiting factor in the Antarctic terrestrial environment: a biogeographical synthesis, *Arctic Alpine Res.*, 25, 308–315.

Kirst, G.O. and Wiencke, C. (1995) Ecophysiology of polar algae, *J. Phycol.*, 31, 181–199.

Knight, C.A., Wen, D., and Laurensen, R.A. (1995) Non-equilibrium antifreeze peptides and the recrystallization of ice, *Cryobiology*, 25, 5–60.

Koerner, R.M. (1989) Ice core evidence for extensive melting of the Greenland ice sheet in the last interglacial, *Science*, 244, 9649–9668.

Kol, E. (1969) The red snow of Greenland II., *Acta Bot. Acad. Sci. Hung.*, 15, 281–289.

Kol, E. (1971) Green snow and ice from the Antarctica, *Ann. Hist.-Nat. Mus. Hung.*, 63, 51–56.

Kol, E. (1972) Snow algae from Signy Island (South Orkney Islands, Antarctica), *Ann. Hist. Nat. Mus. Hung.*, 64, 63–70.

Kováčik, L. and Pereira, A.B. (2001) Green alga *Prasiola crispa* and its lichenized form *Mastodia tesselata* in Antarctic environment: general aspects, in *Proceedings of International Conference—Algae and Extreme Environments—Ecology and Physiology, Nova Hedvigia*, Elster, J., Seckbach, J., Vincent W., and Lhotský O., Eds., *Nova Hedvigia*, 123, 465–478.

Kravchenko, S.I. and Sampson, V. (1998) Efficiency of the sucrose-containing solution on the cold preservation of whole liver, *Cryo-Letters*, 19, 231–236.

Kukal, O. and Kevan, P.G. (1994) The influence of parasitism on the life history of the high-Arctic insect *Gynaephora groenlandica* (Wöcke) (Lepidoptera; Lymantriidae), in *Ecology of a Polar Oasis, Alexandra Fiord, Ellesmere Island, Canada*, Svoboda, J. and Freedman, B., Eds., Captus University Publications, Toronto, pp. 231–239.

Lee, J.A (1998) Arctic plants: Adaptations and environmental change, in *Physiological Plant Ecology*, Press, M.C., Scholes, J.D., and Barker, M.G., Eds., Blackwell Science, Oxford, pp. 313–330.

Lee, R. (1991) Principles of insect low temperature tolerance in *Insects at Low Temperatures*, Lee R., and Denlinger, D., Eds., Chapman & Hall, New York, pp. 17–46.

Lee, R. and Denlinger, D., Eds., (1991) *Insects at Low Temperatures*, Chapman & Hall, New York.

Lee, R.E. Jr, Warren, G.J., and Gusta, L.V. (1995) *Biological Ice Nucleation and Its Applications,* American Phytopathological Society Press, St. Paul, MN.

Lewis-Smith, R.I. (1984) Terrestrial plant biology of the sub-Antarctic and Antarctic, in *Antarctic Ecology*, Vol. 1, Laws R.M., Ed., Academic Press, London, pp. 63–162.

Lewis-Smith, R.I. (1996) Terrestrial and freshwater biotic components of the western Antarctic Peninsula, in *Foundations for cological Research West of the Antarctic Peninsula. Antarctic Research Series*, Vol. 70, Ross, R., Hofmann, E., and Quetin, L., Eds., American Geophysical Union, Washington, DC, pp. 15–59.

Lewis-Smith, R.I. (1997) Oases as centres of high plant diversity and dispersal in Antarctica, in *Ecosystem Processes in Antarctic Ice-free Landscapes,* Howard-Williams, L. and Hawes, I., Eds., Balkema, Rotterdam, pp. 119–128.

Leya, T., Müller, T., Ling, H.U., and Fuhr, G. (2001) Psychrophilic microalgae from north-west Spitsbergen, Svalbard: Their taxonomy, ecology and preliminary studies on their cold adaptation using single cell electrorotation, in *Proceedings of International Conference—Algae and Extreme Environments—Ecology and Physiology*, Elster, J., Seckbach, J., Vincent W., and Lhotský O., Eds., *Nova Hedvigia*, Vol. 123, 551–570.

Li, J.K. and Lee, T.C. (1995) Bacterial ice nucleation and its potential application in food industry, *Trends Food Sci. Technol.*, 6, 259–265.

Ling, H.U. (1996) Snow algae of the Windmill Islands region, Antarctica, in *Biogeography of Freshwater Algae*, Kristiansen J., Ed., *Hydrobiologia,* Vol. 336, 99–106.

Ling, H.U. and Seppelt, R.D. (1993) Snow algae of the Windmill Islands, continental Antarctica. 2. *Chloromonas rubroleosa* sp. nov. (Volvocales, Chlorophyta) an alga of red snow, *Eur. J. Phycol.*, 28, 77–84.

Ling, H.U. and Seppelt, R.D. (1998) Snow algae of the Windmill Islands, continental Antarctica. 3. *Chloromonas polyptera* (Volvocales, Chlorophyta), *Polar Biol.*, 20, 320–324.

Lock, G.S.H. (1990) The growth and decay of ice, in *Studies in Polar Research*, Lock, G.S.H., Ed., Cambridge, Cambridge University Press, Cambridge, U.K.

Longton, R.E. (1988) *Biology of polar bryophytes and lichens*, Cambridge University Press.

Marshall, W.A. (1996) Biological particles over Antarctica, *Nature*, 383, 680.

Marshall, W.A. and Chalmers, M. O. (1997) Airborne dispersal of Antarctic terrestrial algae and cyanobacteria, *Ecography*, 20, 585–594.

McKay, C.P., Nienow, J.A., Meyer, M.A., and Friedmann, E. I. (1993) Continuous nanoclimate data (1985–1988) from the Ross Desert (McMurdo Dry Valleys) cryptoendolithic microbial ecosystem. Antarctic meteorology and climatology: studies based on automatic weather stations, *Antarctic Res.*, 61, 201–207.

Miller, H.I (1994) A need to reinvent biotechnology regulation at the EPA, *Science*, 266, 1815–1817.

Monteil, P.O (2000) Soluble carbohydrates (trehalose in particular) and cryoprotection in polar biota, *Cryo-Letters,* 21, 83–990.

Mueller, D.R., Vincent, W.F., Pollard, W.H., and Fritsen, C.H. (2001) Glacial cryoconite ecosystems: A bipolar comparison of algal communities and habitats, in *Proceedings of International Conference—Algae and Extreme Environments—Ecology and Physiology*, Elster, J., Seckbach, J., Vincent, W., and Lhotský O., Eds., *Nova Hedvigia*, Vol. 123, 173–197.

Müller, T., Bleiβ, W., Martin, C-D., Rogaschewski, S., and Fuhr, G. (1998) Snow algae from northwest Svalbard: their identification, distribution, pigment and nutrient content, *Polar Biol*, 20, 14–32.

Mullins, J. (2003) Sink or swim, *New Sci.*, 2396, 28–31.

Naidenko, T. (1997) Cryopreservation of *Crassostrea gigas* oocytes, embryos and larvae using antioxidant echinochrome A and antifeeze protein AFP1, *Cryo-Letters,* 18, 375–382.

Nienow, J.A. and Friedmann, E.I. (1993) Terrestrial lithophytic (rock) communities, in *Antarctic Microbiology,* Friedmann, E.I., Ed., Wiley-Liss, New York, pp. 343–412.

O'Neil, L.O., Paynter, S.J., Fuller, B.J. Shaw, R.W., and DeVries, A.L. (1998) Vitrification of mature mouse oocytes in a 6 M Me$_2$SO solution supplemented with antifreeze glycoproteins: The effect of temperature, *Cryobiology,* 37, 59–66.

Obata, H., Muryoi, N., Kawahara, H., Yamade, K., and Nishikawa, J. (1999) Identification of a novel ice-nucleating bacterium of Antarctic origin and its ice nucleation properties, *Cryobiology,* 38, 131–139.

Orser, C.S., Staskawicz, B.J., Panapoulos, N.J., Dahlbeck, J., and Lindow, S.E. (1985) Cloning and expression of bacterial ice nucleation genes in *Escherichia coli, J. Bacteriol.,* 164, 359–366.

Pham, L., Dahiya, R., and Rubinsky, B. (1999) An *in vivo* study of antifreeze protein adjuvant cryosurgery, *Cryobiology,* 38, 169–175.

Price, P.B. (2000) A habitat for psychrophiles in deep Antarctic ice, *Proc. Natl. Acad. Sci. USA,* 97, 1247–1251.

Priscu, J.C., Adams, E.E., Lyons, W.B.,Voytek, M.A., Mogk, D.W., Brown, R.L., McKay, C.P., Takacz, C.D., Welch, K.A.,Wolf, F.C., Kirshtein, J.D., and Acci, R. (1999) Geomicrobiology of subglacial ice above lake Vostok, Antarctica, *Science,* 286, 2141–2144.

Ramlov, H., Wharton, D.A., and Wilson, P.W. (1996) Recrystalization in a freezing tolerant Antarctic nematode, *Panagrolaimus davidi* and an Alpine weta, *Hemideina maori* (Orthoptera; Stenopelmatidae), *Cryobiology,* 33, 607–613.

Rivkina, E.M., Friedmann, E.I., McKay, C.P., and Gilichinsky, D.A. (2000) Metabolic activity of permafrost bacteria below the freezing point. *Appl Environ Microbiol,* 66, 3230–3233.

Rosenzweig, M. L. (1995) *Species Diversity in Space and Time,* Cambridge, Cambridge University Press.

Rudolph, A.S. and Crowe, J.H. (1995) Membrane stabilization during freezing: The role of two natural cryoprotectants trehalose and proline, *Cryobiology,* 22, 367–377.

Sattler, B., Puxbaum, H., and Psenner, R. (2001) Bacterial growth in supercooled cloud droplets, *Geophys. Res. Lett.,* 28(2), 239–242.

Säwström, C., Mumford, P., Marshall, W., Hodson, A., and Laybourn-Parry, J. (2002) The microbial communities and primary productivity of cryoconite holes in an Arctic glacier (Svalbard 79°N), *Polar Biol.,* 25, 591–596.

Schlicting, H.E., Speziale, B.J., and Zink, R.M. (1978) Dispersal of algae and protozoa by Antarctic flying birds, *Antarctic J. US,* 13(4), 147–149.

Schmidt, S., Moskal, W., De Mora, S.J., Howard-Williams, C., and Vincent, W.F. (1991) Limnological properties of Antarctic ponds during winter freezing, *Antarctic Sci.,* 3, 379–388.

Sheath, R.G. and Müller, K.M. (1997) Distribution of stream macroalgae in four high Arctic drainage basins, *Arctic,* 50(4), 355–364.

Siegert, M.J., Ellis-Evans, J.C., Tranter, M., Mayer, C., Petit, J.-R., Salamatin, A., and Priscu, J. (2001) Physical, chemical and biological processes in lake Vostok and other Antarctic subglacial lakes, *Nature,* 414, 603–609.

Spauldin, S.A., McKnight, D.M., Smith, L.R., and Dufford, R. (1994) Phytoplankton population dynamics in perennially ice-covered lake Fryxell, Antarctica, *J. Plankton Res.,* 16(5), 527–541.

Storey, K. and Storey, J. (1992) Biochemical adaptations for winter survival in insects, in *Advances in Low Temperature Biology,* Vol. 1, Steponkus, P., Ed., JAI Press, London, pp. 101–140.

Sun, W.Q. and Leopold, A.C. (1994a) The glassy state and seed storage stability: a viability equation analysis, *Ann. Bot.,* 74, 601–604.

Sun, W.Q. and Leopold, A.C. (1994b) The role of sugar, vitrification and membrane phase transition in seed desiccation tolerance, *Physiol. Plant.,* 90, 621–628.

Svoboda, J. and Henry, G.H.R. (1987) Succession in marginal arctic environments, *Arctic Alpine Res.,* 4, 373–384.

Tang, E.P.Y., Tremblay, R., and Vincent, W.F. (1997) Cyanobacterial dominance of polar freshwater ecosystems: Are high-latitude mat-formers adapted to low temperature? *J. Phycol.,* 33, 171–181.

Tedrow, J.C.F. (1991) Pedogenic linkage between the cold desert of Antarctica and the polar desert of the high Arctic, Contributions to *Antarctic Res.* II, Vol. 53, 1–17.

Van Liere, L. and Walsby, A.E. (1982) Interactions of cyanobacteria with light, in *The Biology of Cyanobacteria,* Vol. 19. *Botanical Monograph,* Carr, N.G. and Whitton, B.A., Eds., Blackwell Scientific Publishers, Oxford, pp. 9–45.

Vincent, W.F. (1987) Antarctic limnology, in *Inland Waters of New Zealand*, Viner, A.B., Ed., SIPC, New Zealand, pp. 379–412.

Vincent, W.F. (1988) *Microbial Ecosystem of Antarctica*, Cambridge University Press.

Vincent, W.F. (1997) Polar desert ecosystems in a changing climate: a North-South perspective, in *Ecosystem Processes in Antarctica Ice-Free Landscapes*, Lyons, W.B., Howard-Williams, C., and Hawes, I., Eds., A.A. Balkema, Rotterdam, pp. 3–14.

Vincent, W.F. (1999) Ice life on a hidden lake, *Science*, 286, 2094–2095.

Vincent, W.F. (2000) Evolutionary origins of Antarctic microbiota: Invasion, selection and endemism, *Antarctic Sci.*, 12(3), 374–385.

Vincent, W.F. and Hobbie, J.E. (2000) Ecology of Arctic lakes and rivers, in *The Arctic: Environment, People, Policies*, Nuttall, M., Pocar, S., and Callaghan, T., Eds., Gordon and Breach, London, U.K., pp. 197–231.

Vincent, W.F., Howard-Williams, C., and Broady, P.A. (1993) Microbial communities and processes in Antarctic flowing waters, in *Antarctic Microbiology*, Friedmann, E.I., Ed., Wiley-Liss, New York, pp. 543–569.

Vincent, W.F. and Howard-Williams, C. (1986) Antarctic stream ecosystem: physiological ecology of a blue-green algal epilithon, *Freshwater Biol.*, 16, 219–233.

Vincent, W.F. and Howard-Williams, C. (1989) Microbial communities in southern Victoria Land streams: II. The effect of low temperature, *Hydrobiologia*, 172, 39–49.

Vincent, W.F. and James, M.R. (1996) Biodiversity in extreme aquatic environments: lakes, ponds and streams of the Ross Sea Sector, Antarctica, *Biodiv. Conserv.*, 5, 1451–1471.

Vincent, W.F., Howard-Williams, C., and Broady, P.A. (1993) Microbial communities and processes in Antarctic flowing waters, in *Antarctic Microbiology*, Friedmann, E.I., Ed., Wiley-Liss, New York, pp. 543–569.

Vincent, W.F., Pienitz, R., and Laurion, I. (1998) Arctic and Antarctic lakes as optimal indicators of global change, *Ann. Glaciol.*, 27, 691–696.

Vincent, W.F. and Quesada, A. (1997) Microbial niches in the polar environment and escape from UV radiation in non-marine habitats, in *Antarctic Communities: Species, Structure and Survival,* Battaglia, B., Valencia, J., and Walton, D.W.H., Eds., Cambridge University Press, Cambridge, pp. 388–395.

Vishnivetskaya, T., Kathariou, S., McGrath, J., Gilichinsky, D., and Tiedje, J.M. (2000) Low-temperature recovery strategies for the isolation of bacteria from ancient permafrost sediments, *Extremophiles*, 4, 165–173.

Vishnivetskaya, T.A., Erokhina, L.G., Spirina, E.V., Shatilovich, A.V., Vorobyova, E.A., and Gilichinski, D.A. (2001) Ancient viable phototrophs within the permafrost, in *Proceedings of International conference—Algae and Extreme Environments—Ecology and Physiology*, Elster, J., Seckbach, J., Vincent, W., and Lhotský O., Eds., *Nova Hedvigia*, Vol. 123, 427–441.

Vorobyova, E., Soina, V., Gorlenko, M., Minkovskaya, N., Zalinova, N., Mamukelashvili, A., Gilichinsky, D., Rivkina, E., and Vishnivetskaya, T. (1997) The deep cold biosphere: facts and hypothesis, *FEMS Microbiol. Rev.*, 20, 277–290.

Walker, G. (2003) *Snowball Earth*, Bloomsbury, London.

Walton, D.W.H. (1982) The Signy Island terrestrial reference sites. XV. Microclimate monitoring, *Br. Antarctic Surv. Bull.*, 55, 11–126.

Walton, D.W.H. (1984) The terrestrial environment, in *Antarctic Ecology*, Kerry, K.R. and Hempel, G., Eds., Springer, Berlin, pp. 51–60.

Walton, D.W.H., Ed., (1987) *Antarctic Science,* Cambridge University Press, Cambridge, U.K.

Wang, T., Zhu, Q., Yang, X., Layne, J.R., and DeVries, A.L. (1994) Antifreeze glycoproteins from Antarctic notothenioid fishes fail to protect the rat cardiac explant during hypothermic and freezing preservation, *Cryobiology*, 31, 185–192.

Waslyk J., Ties, A., and Baust, J. (1988) Partial glass formation: a novel mechanism of insect cryopreservation, *Cryobiology,* 25, 451–458.

Wharton, D. and Block, W. (1997) Differential scanning calorimetry studies on an Antarctic nematode (*Panagrolaimus davidi*) which survives intracellular freezing, *Cryobiology* 34, 114–121.

Wharton, R.A., McKay, C.P., Mancinelli, R. L., and Simmons, G. M. (1987) Perennial N_2 supersaturation in an Antarctic lake, *Nature*, 325, 343–345.

Wharton, R.A., McKay, C.P., Simmons, G.M., and Parker, B.C. (1985) Cryoconite holes on glaciers, *Bioscience*, 35, 499–503.

Whitton, B.A. (1987) Survival and dormancy of blue-green algae, in *Survival and Dormancy of Microorganisms*, Henis, Y., Ed., Wiley, New York, pp. 109–167.

Wiedner, C. and Nixdorf, B. (1998) Success of chrysophytes, cryptophytes and dinoflagellates over bluegreens (cyanobacteria) during an extreme winter (1995/96) in eutrophic shallow lakes, *Hydrobiologia*, 369/370, 229–235.

Wiermann, R. and Gubatz, S. (1992) Pollen wall and sporopollenin, *Int. Rev. Cytol.*, 140, 35–72

Worland, M.R. and Block, W. (2003) Desiccation stress at sub-zero temperatures in polar terrestrial arthropods, *J. Insect Physiol.*, 49(3), 193–203.

Worland, R. and Block, W. (2001) Experimental studies of ice nucleation in an Antarctic springtail (Collembola, Isotomidae), *Cryobiology*, 42, 170–181.

Worland R., Block, W., and Grubor-Lajsic, G. (2000) Survival of *Heleomyza borealis* (Diptera, Heleomyzidae) larvae down to –60°C, *Physiol. Entomol.*, 25, 1–5.

Worland, M.R., Block, W., and Oldale, H (1996) Ice nucleation activity in biological materials with examples from Antarctic plants, *Cryo-Letters*, 17, 31–38.

Wynn-Williams, D.D. (1991) Aerobiology and colonization in Antarctica—The BIOTAS Programme, *Grana*, 30, 380–393.

Wynn-Williams, D.D. (1996) Antarctic microbial diversity: the basis of polar ecosystem processes, *Biodiversity Conserv.*, 5, 1271–1293.

Wynn-Williams D.D. and Edwards H.G.M, (2000a) Antarctic ecosystems as models for extraterrestrial surface habitats. *Planet. Space Sci.*, 48, 1065–1075.

Wynn-Williams, D.D. and Edwards H.G.M. (2000b) Proximal analysis of regolith habitats and protective biomolecules *in situ* by laser Raman spectroscopy: Overview of terrestrial Antarctic habitats and mars analogs. *Icarus* 144, 486–503.

Zachariassen, K.A. (1985) Physiology of cold tolerance in insects, *Physiol. Rev.*, 65, 799–832.

Zöcker, C., Huntington, H.P., and Chernov, Y. (2002) North-South trends in terrestrial species diversity, in *Conservation of Arctic Flora and Fauna: Status and Conservation*, Huntington, H. et al., Eds., Helsinki, Edita, pp. 48–49.

4 Microbial Life in Permafrost: Extended Times in Extreme Conditions

Monica Ponder, Tatiana Vishnivetskaya,
John McGrath, and James Tiedje

CONTENTS

4.1 INTRODUCTION

The permafrost microbial community has been characterized as a "community of survivors" (Friedmann, 1994) based on the continued viability of many of their members for hundreds to millions of years in the frozen state. Despite the extreme conditions of low temperature, low nutrients, and low water activity associated with this environment, a large diversity of viable microorganisms exists in even the deepest layers of permafrost. This continually frozen soil matrix contain members of the bacteria, archaea, and some green algae and yeast that will grow on appropriate media. Other eukaryotes have not been cultivated from permafrost layers older than 10,000 years. These microbes may have responses to the stress conditions that prolong their survival under the stable conditions of permafrost. Known response reactions include physiological changes to membrane composition, which maintain fluidity; changes in protein production; a reduction in cell size; and the production of internal osmolytes.

Dispute remains about whether microbes simply persist in the long term or whether they actively survive in nongrowth-promoting conditions. A particularly controversial topic is whether sustained or intermittent metabolic activity occurs within the permafrost. Evidence presented below supports both intermittent metabolic activity and the existence of a sustained anabiotic state. The remarkable ability to survive in a continuously frozen matrix of permafrost for millions of years makes the permafrost community unique for several practical and scientific reasons. For example, the permafrost microbes may harbor novel enzymes of biotechnological importance. They also beg deep questions about the nature of life itself: How long can cells survive in continually frozen conditions? What is the lowest rate of metabolic activity necessary to retain viability? Is life likely to exist on the cooler outer planets?

The purpose of this chapter is to highlight the nature of the physical environment while primarily focusing on the microbial community and the metabolic state of these microbes *in situ*.

4.2 PERMAFROST PHYSICAL CHARACTERISTICS

The permafrost environment is composed of soil, bedrock, sands, and sediments that are exposed to temperatures of 0°C or below for a period of at least 2 years. Over 20% of the Earth's land surfaces are subjected to these permanently cold conditions, including 85% of Alaska, 55% of Russia and Canada, 20% of China, and the majority of Antarctica (Pewe, 1995).

The active layer of permafrost is the surface layer that freezes in the winter and thaws in the summer. The depth of the active layer is dependent on moisture content and can vary from several feet in well-drained soils to less than 1 ft in bog environments (Gilichinsky, 1993; Pewe, 1995).

The thickness of the permafrost is governed by the balance between its formation and its loss to the atmosphere because of heat flow from the earth's core through the permafrost layers. Regional thickness varies from a few meters in sub-Arctic regions to hundreds of meters in the northern Arctic (Gilichinsky et al., 1992). Permafrost is often much thicker in Antarctic regions because of their longer periods of subzero temperatures. Topography and vegetation also contribute to differing local variations in permafrost thickness. Vegetation and snow cover serve to insulate, by preventing heat from leaving the ground, and thereby reduce the permafrost thickness (Pewe, 1995; Spirina and Federov-Davydov, 1998).

Permafrost soil represents an extreme environment caused by low temperatures (–10 to –20°C [Arctic] to –65°C [Antarctic], low water content (1 to 7%), low carbon availability, and long-term exposure to gamma radiation. The composition of permafrost varies greatly in ice, organic carbon content, and physical characteristics. It is possible for permafrost to contain no moisture, and thus no ice, in dry, arid regions, such as the Ross desert of Antarctica. In other regions, the high salinity of permafrost may inhibit ice formation or growth well below freezing temperatures. However, the majority of permafrost in the Arctic and Antarctic regions is firmly cemented by ice. Ice content varies according to the physical and ionic content of the soil. Sandy, well-drained soils often contain between 15 and 25% ice, whereas loam soils vary between 30 and 60% ice (Gilichinsky, 1993). Sands of the Antarctic Dry valley contain between 25 and 40% ice, most likely because of the ability of the coarse grains to cement (Wynn-Williams, 1989).

The concentration of organic carbon varies greatly in localized areas of permafrost. Many sites within northeastern Siberia are oligotrophic, with total organic compositions between 0.85 and 1% (Matsumoto et al., 1995); others contain higher total organic compositions (Gilichinsky et al., 1992) as a result of the presence of detritus layers. Late Pliocene and Pleistocene Siberian soils often

TABLE 4.1
Diversity of Organisms Obtained from Permafrost Soils

	Number isolated (cells/gm soil)	Oldest sample (years)	Depth (meters)	Reference
Archaea	10^5–10^7	2 million	43	Rivkina et al., 1998; Tiedje et al., 1998
Eukaryotes	10^3–10^5			
Algae	10^3–10^5	3 million		Vishnivetskaya et al., 2001
Fungi and yeasts	10^3	3 million	78	Dmitrev et al., 1997
Protists	10^3	Modern	.3	Vishniac, 1993
Bacteria	10^7–10^9	3 million	43	Vishnivetskaya et al., 2000

contain peat, detritus layers, semidecayed plants, and humic-rich areas, which can cause large variations in organic carbon concentrations (Gilichinsky et al., 1992). High total organic compositions contents and a low C/N ratio are common in the soils of East Antarctica (Beyer et al., 2000).

Beneath the active layer lies the permafrost table. This is a physical barrier, restricting water percolation and solute penetration because of the presence of ice-filled pores. The absence of continuous water-filled horizons reduces the possibility of penetration of modern microorganisms into the ancient frozen layers. The remaining water remains in an unfrozen state because of the concentration of organic solutes and mineral particles. The amount of unfrozen water present in permafrost decreases with temperature (Ostroumov, 1992). In Arctic permafrost soils that experience temperatures between $-10°$ and $-12°C$, the amount of unfrozen water is estimated to be between 3 and 5% (Gilichinsky et al., 1995). In Antarctic soils in which temperatures are commonly below $-25°C$, unfrozen water is often undetected (Ostroumov, 1992). The ice-filled soil horizons prevent microbial movement because the thickness of the unfrozen water film (5 to 75 Å) is insufficient for the passage of 0.5 to 1.0-micron microorganisms (Gilichinsky et al., 1993). The presence of polygonal ice wedges, unique $^{16}O:^{18}O$ ratios, and the presence of pollen grains only from tundra plants in the soil horizons indicate that these soils have remained continually frozen for hundreds of thousands to millions of years (Gilichinsky et al., 1992; Vasilchuk and Vasilchuk, 1997). Soil age is determined by radiocarbon dating of the extracted pollen and other organic matter. Permafrost provides a unique opportunity to examine the microbial diversity present in ancient soils because of its continually frozen state.

4.3 PERMAFROST DIVERSITY

A vast diversity of microorganisms has been isolated from all layers of permafrost—some as old as 3 to 5 million years (see Table 3.1). Microbial diversity and numbers seem to be related to the age of the soil and are, therefore, dependent on the length of time frozen.

It might be construed that such harsh conditions would limit life to a very low diversity and cause the small number of remaining cells to exist only in a resting state. However, this is not the case. Omelansky isolated the first Siberian permafrost microorganisms in 1911, while excavating a mammoth carcass. This discovery was followed by isolation of viable microorganisms from permafrost soils all over the world. Microbes were soon characterized from snow and ice in Antarctica, Canadian permafrost subsoil, and Alaskan permafrost (Boyd and Boyd, 1964; James and Sutherland, 1942; McLean, 1918). Four different genera of bacteria were isolated from 20,000 to 70,000-year-old Alaskan permafrost cores (Becker and Volkmann, 1961). This was followed shortly by the first quantitative study of Alaskan permafrost. The active layer showed the presence of 55,000 viable bacteria per grams of soil, whereas the permafrost (8 to 15 ft in depth) revealed the presence of 5 to 130 cells/gm soil when incubated at 22°C on nutrient broth (Boyd and Boyd,

1964). These early studies also highlighted the differences in numbers of viable fungi (10^3/g soil) and bacteria (10^4/g soil) isolated from the active permafrost layer when incubated at 24°C (Kjoller and Odum, 1971). The first viable microflora older than the late Pleistocene (1 to 2 million years) were isolated in Antarctica by Cameron and Morrelli (1974).

These initial studies included few controls to ensure that contaminant microorganisms were not introduced into the samples. Early attempts to control the introduction of outside microorganisms relied on the surface sterilization of the core with a flame and then drilling into the center of the core with flame-sterilized drill bits, though some speculation remains about possible contamination by the drilling fluid (Cameron and Morrelli, 1974). At present, standardized methods adapted from temperate subsurface sampling processes minimize contamination risks (Beeman and Suflita, 1990). Samples are obtained by slow, rotary drilling without the use of fluid to prevent melting and contamination by surface organisms. In addition, the outside of the drill is coated with an indicator microorganism, *Serratia marcescens*. Internal segments of the cores are obtained by aseptically removing the outermost 1 cm under frozen conditions. These internal segments are then tested for the presence of the indicator strain (Khlebnikova et al., 1990).

4.3.1 BACTERIA

Bacteria dominate in all layers of the permafrost. Their diversity and numbers decrease with soil age, but large numbers (10^8 cells/g) of viable bacteria were obtained from the oldest samples (3 million years; Vishnivetskaya et al., 2000; Vorobyova et al., 1997). The majority of studies of the psychrotolerant permafrost community focus on aerobic systems where up to 10^8 cells/g can be isolated (Vishnivetskaya et al., 2000; Vorobyova et al., 1997). A small number of anaerobic studies reveal that permafrost samples also contain a diverse population of bacteria, accounting for 10^2 to 10^6 cell/g (Rivkina et al., 1998). Psychrophilic bacteria are surprisingly absent from permafrost communities, with the majority of isolates possessing psychrotolerant properties (Vishiniac, 1993).

A discrepancy exists between the number of bacteria that can be cultured and those present in the permafrost soils. Arctic permafrost sediment often contain between 10^7 and 10^9 cells per gram dry weight (dw), as determined by acridine orange direct microscopy counts, wheras Antarctic permafrost contains between 10^7 and 10^8 cells/g dw. However, only a small fraction of these bacteria (10^2 to 10^8 cfu /g dw) are viable by selected culturing methods (Khlebnikova et al., 1990; Vorobyova et al., 1997). Studies using olgiotrophic media such as peptone yeast glucose vitamin media or 1/10 tryptic soy broth obtained greater numbers of isolates, but a smaller diversity of morphotypes, than studies using rich media (Siebert and Hirsch, 1988; Vishnivetskaya et al., 2000). However, incubation at cold temperatures was shown to increase both the microbial diversity and the numbers of bacteria isolated from Arctic permafrost (Vishnivetskaya et al., 2000).

Characterization of the viable and culturable permafrost community has revealed the presence of equal numbers of Gram-positive and Gram-negative bacterial isolates in some soils and a preponderance of nonspore-forming Gram-positive bacteria in other permafrost soil samples (Vishnivetskaya et al., 2000). Previous studies by Russian scientists indicate that Gram-positive bacteria such as actinomycetes (85%), and cocci (5%), including the genera *Arthrobacter*, *Rhodococcus*, *Micrococcus*, *Deinococcus*, *Brevibacterium*, and *Streptomyces*, were predominant in Siberian permafrost (Zvyagintsev et al., 1990). Gram-negative rods represented about 12% of aerobic isolates, with the majority of isolates being members of the *Pseudomonas* and *Flavobacterium* genera (Vorobyova et al., 1997; Zvyagintsev et al., 1990). The majority of aerobic permafrost-bacterial isolates are nonspore-forming bacteria, especially in the most ancient permafrost soils, and therefore this classical persistence structure does not appear to be important for microbial survival in permafrost.

In late Pliocene (1.8 to 3 million years old) permafrost, the mesophilic actinomycetes and related high-GC Gram-positive bacteria were still found to be the predominant (55 to 75%) culturable isolates. However, the total number of cells isolated decreased with age of the soil

$(6 \times 10^4 – 1.4 \times 10^5$ cell/g). Classification based on culturable, morphological, and chemotaxonomical characteristics placed the isolates within the genera *Arthrobacter, Kocuria, Aureobacterium, Gordona, Nocardia, Rhodococcus, Mycobacterium, Nocardioides,* and *Streptomyces.* In the aforementioned study, a number of bacteria isolated differed from previously described genera in their cell wall composition, indicating that the bacteria may belong to a new species or genera (Karasev et al., 1998). Bacteria of the genus *Bacillus* were often isolated from old permafrost samples (Kuz'min et al., 1996).

Only a few studies have examined the presence of anaerobes in permafrost despite their presence in large numbers (10^7 to 10^8 total cells/g; Vorobyova et al., 1997). Denitrifiers, sulfate reducers, and Fe(III) reducers were isolated by viable plate counts and quantified by presence of end products. The largest numbers of anaerobes were detected in the younger soils (Rivkina et al., 1998). The anaerobic bacteria were identified as *Propionibacterium* species (sp.) (Karasev et al., 1998), *Desulfotomaculum* sp., and nitrifying and denitrifying bacteria of the genera *Nitrosospira, Nitrosovibrio,* and *Nitrobacter* (Rivkina et al., 1998).

Phylogenetic studies of Siberian permafrost isolates reveal a vast diversity of microorganisms belonging to the *Actinobacteria,* and the *Firmicutes* commonly referred to as the high-G+C and the low-G+C of the Gram-positive group, and responding *Proteobacteria* phylogenetic groups. (Note that G+C content is a means of distinguishing between different bacterial species.) *Arthrobacter* and a member of the Micrococcus subclass were the most common isolates independent of soil age within the *Actinobacteria. Exiguobacterium* represented the *Firmicutes* group throughout the soil column, whereas *Planococcus* was isolated only from the younger soils. Members of the α-*Proteobacteria, Sphingomonas,* were found in soils up to 3 million years old, whereas members of the Flavobacterium–Bacteroides–Cytophaga group and γ-*Proteobacteria, Flavobacterium* and *Psychrobacter,* were present only in the younger soils (Vishnivetskaya et al., in preparation). This age relationship is currently unexplained, but it may suggest that those genera present in ancient soils possess cold acclimation properties not present in the strains of younger origin. Shi et al. (1997) examined 29 viable permafrost isolates. Among them, 16 (55%) were Gram-negative and 13 (45%) were Gram-positive. Phylogenetic analysis revealed that the isolates fell into four categories: high-GC Gram-positive bacteria, β-*Proteobacteria,* γ-*Proteobacteria,* and low-GC Gram-positive bacteria. Most high-GC Gram-positive bacteria and β-*Proteobacteria,* and all γ-*Proteobacteria,* came from samples with an estimated age of 1.8 to 3.0 million years. Most low-GC Gram-positive bacteria came from samples with an estimated age of 5000 to 8000 years (Shi et al., 1997).

Recent studies have shown that the number of viable but nonculturable microorganisms present in permafrost is considerable. Extraction of genomic DNA from the active layer followed by observations of restriction fragment-length polymorphism (RFLP) patterns indicate that the Gram-negative bacteria are most often amplified (60.5%) compared with the Gram-positive bacteria that often dominate in viable diversity studies. The *Proteobacteria* dominate the Gram-negative isolates with clones belonging to the α (20.9%), δ (25.6%), β (9.3%), and γ (4.7%). Overall, 70% of clones were 5 to 15% different from the strains in current databases, whereas 7% differed more than 20% (Zhou et al., 1997). This would seem to indicate that the physiology and function of the dominant members of the community are unknown.

The reports about isolation of viable bacteria from Antarctic permafrost are numerous and are thoroughly reviewed by Vishinac (1993) in *Antarctic Microbiology.* Bacteria are able to survive in Antarctic permafrost at levels of nearly 10^5 cell/g, with the genera *Bacillus, Arthrobacter,* and *Streptomyces* predominating (Vorobyova et al., 1997).

4.3.2 CYANOBACTERIA AND ALGAE

Viable green algae and cyanobacteria were isolated from numerous permafrost samples, which varied in lithology genesis and depth. Permafrost samples were obtained from the Kolyma-lowland, Siberia, and Antarctic dry valleys (Victoria Land). The frequency of discovery of viable algae has

TABLE 4.2
Phylogenetic Group vs. Age (in Thousands of Years) of Permafrost Sediment

Phylo Group	Total (no.)	Modern	2–8	20–30	60–100	200–600	600–2000	2000–3000
Flavobacteria/Cytophaga/ Bacteroides group	6	1		4			1	
Fibrobacter group	7	7						
Proteobacteria	43	26	1	5		1		10
(alphas)	(11)	(9)				(1)		(1)
(betas)	(10)	(4)	(1)					(5)
(gammas)	(11)	(2)		(5)				(4)
(deltas)	(11)	(11)						
Bacillus/Clostridia group	20	2	8	5	1	1		3
(Bacillus)	(11)	(1)	(8)					(2)
(*Exiguobacterium*)	(4)	(1)		(1)		(1)		(1)
(*Planococcus*)	(5)			(4)	(1)			
Actinomycetes	32	1		13		4		14
(*Arthrobacter*)	(24)			(12)		(1)		(11)
(*Microbacteriaceae*)	(6)			(1)		(3)		(2)
(*Rhodococcus*)	(2)	(1)						(1)

Note. Compiled from phylogenetic diversity literature of Shi et al., 1997; Vishnivetskaya et al., in preparation; and Zhou et al., 1997.

been lower than bacteria, and decreases with increasing permafrost age (see Table 4.1). In contrast, the presence of viable algae did not depend on the depth of the soil core when samples of identical age were examined. Green algae and cyanobacteria were isolated more often from fine lake–swamp or lake–alluvium loams and sandy loams. No viable algae were isolated from coarse marine and channel-filled sands. The biodiversity of green algae and cyanobacteria was highest in the Holocene sediments (10,000 years old). Their occurrence in the Arctic early Pleistocene and late Pliocene, and in any Antarctic frozen sediments, can be characterized as sporadic. Statistically, green algae species were found more frequently and with greater species diversity than cyanobacteria. This may reflect the ability of the green algae to survive in the ancient tundra environments. Numerous studies to date have shown that green algae are dominant in the modern tundra environments in quantity and in biodiversity (Getsen, 1990, 1994; Zenova and Shtina, 1990).

Unicellular green algae *Pseudococcomyxa* and *Chlorella* were often isolated from different ages of Arctic permafrost samples. Nonmotile globular cells of *Chlorella* were more abundant among green algae isolates. Other *Chlorella* were dominant in the permafrost samples. Viable green algae from the Arctic permafrost were represented by *Chlorella* sp., *Chlorella vulgaris*, *Chlorella sacchorophilla*, *Pseudococcomyxa* sp., *Mychonastes* sp., *Chlorococcum* sp., *Chodatia* sp., *Chodatia tetrallontoidea*, *Stichococcus* sp., and *Scotiellopsis* sp. Unicellular green algae of *Mychonastes* sp. and *Pedinomonas* sp. were the only algae discovered in the Antarctic permafrost. Green algae isolates were unicellular coccoidal, elliptical, and rodlike. All cyanobacteria were filamentous in morphology and belonged to the Oscillatoriales and Nostocales orders. Among them *Nosctoc*, *Anabaena*, *Phormidium*, and two different species of *Oscillatoria* were identified. *Nostoc* and Oscillatoria, with straight, narrow trichomes, were most often isolated form Arctic permafrost. The Antarctic permafrost samples studied did not contain viable cyanobacteria (Vishnivetskaya et al., 2001).

Viable algal isolates possessed low growth rates, with doubling times of 10 to 14 d. Green algae strains were better able to grow at 27°, 20°, and 4°C, but cyanobacteria possessed good growth only at 25°C. Cyanobacteria *Nostoc* and *Anabaena* were capable of nitrogen fixation and chromatic adaptation (Erokhina et al., 1999; Vishnivetskaya et al., 2001).

It is amazing that green algae and cyanobacteria survive in deep permafrost sediments devoid of light and below the freezing point for thousands to millions of years. These algae may exist in some readily reversible dormant state from which they may easily regain photoautotrophic capabilities when light and water again become available.

4.3.3 ARCHAEA

Archaea are present in great numbers in permafrost. Methanogens are present between 10^5 and 10^7 total counts per gram (Rivkina et al., 1998). Phylogenetic diversity of Archaea from Siberian permafrost aged from 100,000 to 2 million years has been studied by RFLP distribution analysis. The archaeal diversity was demonstrated to be significantly lower than that of bacteria found in surface tundra soils from the same region. The majority of archaeal amplicons belonged to a novel lineage that was deeply rooted in the Crenarchaeota. Three amplicons were closely related to *Methanosacina* sp., contained in the Euryarchaeota (Tiedje et al., 1998). Since this study, 12 additional strains of methanogenic archaea have been isolated from permanently cold environments, including permafrost (Nozhevnikova et al., 2001).

4.3.4 EUKARYOTIC MICROORGANISMS

Fungi are found predominantly in the upper strata of permafrost, whereas yeast can be isolated from all layers of permafrost. Fungi have yet to be isolated from samples older than 10,000 years in age (Zvyagintsev et al., 1990; Gilichinsky et al., 1992). A complete overview of early taxonomic studies of fungi and yeasts isolated from Antarctic permafrost is provided elsewhere (Vishiniac, 1993; see also Chapter 8).

Electron microscopic investigations of deep permafrost samples show a decrease in the stability of eukaryotes under cryopreserved conditions. Samples older than 10,000 years contain low numbers of eukaryotic cells, all showing damaged internal structures but intact cell walls (Soina et al., 1995).

Recently, yeasts have been shown to comprise 20 to 25% of the aerobic heterotrophs in some 2 to 3-million-year-old Siberian permafrost soils. The large number of yeast (9700 cfu/g) and the absence of yeast from adjacent layers indicates that these yeast are neither contaminants nor percolated through from upper layers (Dmitriev et al., 1997). Large numbers of yeasts (17,000 to 70,000 cells/g) are more commonly encountered in soils that contain large amounts of undecomposed plant material (Spirina et al., 1998). Yeast abundance increases in coastal Antarctica because of increased moisture contents and plant cover, and particularly around human inhabited areas, such as McMurdo station (Atlas et al., 1978). A small number of studies have revealed yeasts to be the predominant microbe in Antartic soil; however, there seems to be no correlation between any of the physical characteristics of the soils and the microbial diversity (Atlas et al., 1978; Horowitz et al., 1972). The yeasts isolated from permafrost are generally obligate psychrophiles, unlike the viable bacteria, which tend to be psychrotrophic (Atlas et al., 1978).

Despite their smaller numbers throughout the permafrost column, the presence of eukaryotic microorganisms indicates their ability to survive extreme conditions for extended periods of time.

4.4 MECHANISMS OF SURVIVAL

Throughout the scientific community, there remains much dispute about what the existence of microbes in the permafrost indicates about the potential for long-term survival in nongrowth-promoting conditions. This section seeks to express some of the current theories about cell preservation and the possibilities of *in situ* metabolic activity.

4.4.1 PHYSICAL CHARACTERISTICS OF PERMAFROST

The soil composition itself can affect survival of bacteria at low temperatures. Soil particles decrease the depth of penetration of thermal oscillations and therefore help to buffer the microbes from rapid

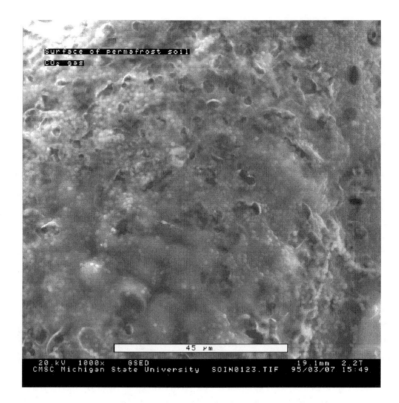

FIGURE 4.1 Microbes apparently encased in biofilm in frozen state in permafrost soil. Environmental scanning electron microscope. (Courtesy of Dr. Vera Soina.)

temperature changes. Soils with larger pore spaces, such as sand, have been shown to contain less unfrozen water and to yield fewer viable bacteria compared with peat-containing soils (Gilichinsky et al., 1995). Recent studies reveal a correlation between bacterial numbers and the total organic carbon and clay content of soil (Beyer et al., 2000). Increased content of organic carbon and clay will provide nutrients and have greater water holding capacity, preventing desiccation of the cells.

Unfrozen water surrounding the cells acts as a nutrient medium because as ice forms, it concentrates solutes. In addition, this unfrozen film acts as a cryoprotectant by preventing invasion from extracellular ice crystals. The amount of unfrozen bound water surrounding microbes was shown to decrease with temperature (Gilichinsky et al., 1995).

In addition to the ameliorating effect of unfrozen water and organic composition, the physical structure of soil particles may increase microbial survival. This is particularly apparent when one compares the high number of cells isolated from frozen soil to the low numbers from pure ice. Recent studies of Lake Vostok accretion ice indicate that the number of viable microbes increases when dust particles are present in the ice cores (Karl et al., 1999). This is further supported by evidence in cryopreserved soils modeled to resemble permafrost environments, which shows that the ability of cells to adhere to soil particles increases cell survival (Sidyakina et al., 1992). Scanning environmental microscopy further reveals the tight association of bacteria with the surrounding Siberian permafrost (Soina, unpublished observation; see Figure 4.1).

4.4.2 PHYSIOLOGICAL ADAPTATION

Physiological responses to extreme conditions, including low temperature, desiccation, and low nutrients, have been well described (see also Chapters 3, 5, and 8). Many cellular responses to a single stress are also protective against other types of stress. The low water activity of the permafrost environment may induce physiological changes that allow cells to survive radiation exposure, as seen in *Deinococcus radiodurans* (Venkateswaran et al., 2000). The response of permafrost-entrapped microbes to one or more of these conditions may prolong their survival.

4.4.2.1 Membrane Adaptation

Biological membranes are composed of multiple types of phospholipids and proteins that are transitioned between a highly ordered state (crystalline gel) and a fluid state (liquid crystalline). A reduction in temperature often leads to an ordering of the membrane system, resulting in a solid-state (gel) membrane. To maintain membrane fluidity, microorganisms induce changes in membrane composition at low temperatures (Russell, 1990). These membrane changes allow normal functions, such as transport, to continue by maintaining necessary interactions with membrane proteins. In general, a decrease in temperature brings about a subsequent decrease in saturation of fatty acids, acyl chain length, and proportion of cyclic fatty acids. Altogether, these changes function to lower the gel–liquid crystalline transition temperature (Morris and Clarke, 1987).

Membrane unsaturation is accomplished via changes in the synthesis of new lipids, as seen in *Escherichia coli*, and also by alteration of existing fatty acids, as seen in *Bacillus subtilis* and *Synechocystis*. *E. coli* has a 30-second response time before production of unsaturated fatty acids occurs after a temperature downshift (Ingraham and Marr, 1996), performed by regulating the enzyme β-keto-acyl ACP synthase. The alteration of existing membranes may be a more rapid, efficient response to cold temperature stress. The desaturase genes responsible for introducing the double bonds between specific C moietes allow for rapid cellular response to cold temperatures and are constitutively expressed to allow for immediate response to low-temperature stress (Aguilar et al., 1999). A two-component signal transduction system, composed of a sensor kinase and response regulator, is responsible for the cold induction of *des* genes in *B. subtilis* and *Synechocystis* (Aguilar et al., 1999; Suzuki et al., 2000). The inactivation of two sensor kinase domains decreases the amount of low-temperature induction of genes encoding δ6 and δ3 desaturases (Suzuki et al., 2001). The response regulator was identified in *B. subtilis* and shown to be involved in control of gene expression. It is also inactivated by the presence of unsaturated fatty acids; however, the exact mechanism is unknown (Aguilar et al., 2001).

An important consequence of membrane adaptation is the increased efficiency of nutrient uptake at cold temperatures in strains of permanently cold origin. Glucose use/uptake increases at 0°C in a cold-adapted psychrotroph, whereas a psychrotroph exposed to temperature fluctuations shows a decrease in uptake at 0°C (Ellis-Evans and Wynn-Williams, 1985). This suggests that fluctuations in temperature favor a different fatty acid composition of the membrane than the cold-acclimated membrane.

A temperature minimum does exist at which simple membrane changes no longer maintain the spatial organization to allow transport molecule interactions, effectively allowing the cell to starve even at high nutrient concentrations (Nedwell, 1999). Membrane adaptation may function to exclude ice crystals and allow continued nutrient transport while maintaining cellular integrity. However, the enzymatic and structural function of proteins not associated with the cell membrane must be maintained, necessitating other molecular mechanisms of cold adaptation.

4.4.2.2 Production of Cold-Induced Proteins

The ability to maintain structure, transport, storage, and enzymatic function at cold temperatures is necessary for continued cellular activity. Low temperature and nutrient conditions have been shown to induce high levels of expression of three types of proteins: cold-shock, antifreeze, and ice nucleation (Chong et al., 2000; Nemecek-Marshall et al., 1993; Sun et al., 1995).

Cold-shock proteins control many different types of functions from transcription and translation to general metabolism and recombination. A complete review of cold-shock proteins was made by

Phadtare et al. (2000). It is believed that their primary role is to halt protein synthesis until the cells have acclimated to their new environment. The signals necessary to restart protein synthesis are unknown (Russell, 1990).

Ice nucleation proteins are proteins whose structure mimics an ice crystal, providing a lattice for crystallization of water (Zachariassen and Hammel, 1976). This allows for crystallization of extracellular water at higher temperatures, releasing a small amount of heat, which may aid in delaying internal damage until the temperature is increased. The small extracellular crystals may be too small to penetrate the membrane and initiate freezing (Mazur, 1977; see also Chapter 1). In addition to the surface-catalyzed theories of intracellular ice formation, ice can form inside cells as osmotic pressure rises during extracellular freezing, which may lead to the rupturing of the membrane (Muldrew and McGann, 1994; Toner and Cravalho, 1990). Harmful intracellular ice formation is believed to be averted with ice-nucleation-active proteins because the small size of the formed crystals prevents membrane damage—the primary cause of cell death in freezing systems. Eventually, these small, thermodynamically unstable crystals have a tendency to reform into large, damaging structures, indicating that these proteins are unlikely to offer long-term survival benefits (Mazur, 1984; see also Chapter 1). Recently, the discovery of a protein that is capable of both ice nucleation and antifreeze activity has been described and may present a solution to this dilemma (Xu et al., 1998).

Antifreeze proteins have long been known to prevent the freezing intracellular water in fish by impeding the addition of water molecules to the existing crystal (DeVries and Wohlschlag, 1969; see also Chapters 2 and 22). Such thermal hysteresis activity can result in lowering of the freezing temperatures of water by as much as 9 to 18°C. Conversely, the freezing of extracellular water can also protect some terrestrial arthropods (Duman, 2001; see also Chapter 3) and invertebrates (see Chapter 7), presumably by enabling the intracellular environment to vitrify while maintaining ice in "safe" extracellular spaces. The ice-nucleation domain of the dual-action protein is believed to cause the formation of extracellular ice, whereas the antifreeze domain maintains these crystals at a nondamaging size (Xu et al., 1998). The ability of some plant- and animal-derived antifreeze proteins to inhibit microbial ice-nucleation activity indicates that antifreeze proteins may lower the supercooling point of water within the cells. However, no direct evidence supports this theory (Duman et al., 1991; Griffith and Ewart, 1995).

4.4.2.3 Compatible Solute Production

Ice formation increases solute concentration by decreasing the amount of free water available for biological use. This produces an environment with low water activity similar to those seen in desiccation and salt-stressed environments. Halophilic and halotolerant microorganisms have developed mechanisms to allow their continued growth in low water activities and high salinity. The mechanism excludes environmental solutes from the cell while accumulating other solutes (compatible solutes) that control osmotic balance and that are compatible with the organism's metabolism (Brown, 1990).

Compatible solutes reduce the normal cell shrinkage associated with desiccation (McGrath et al., 1994). Compatible solutes fall into five classes of compounds: sugars, polyols, betaines, ectoines, and some amino acids (Csonka, 1989; Galinski and Herzog, 1990; deSantos and Galinski, 1998; D'Souza-Ault et al., 1993; Frings et al., 1993; Glaasker et al., 1996; Ko et al., 1994; Larsen et al., 1987; Ruffert et al., 1997; Severin et al., 1992; Smith and Smith, 1989; Welsh, 2000). Studies of psychrotrophs and psychrophiles have revealed the presence of compatible solutes such as glycine betaine, glutamate, and proline (Russell, 1990). Proline has been shown to depress the freezing point of water in some plants (Aspinall and Paleg, 1981; see also Chapter 5). In addition, Gould and Measures have shown a correlation between the concentration of proline and growth at low water activities, such as those seen at freezing temperatures (Measures, 1975). It is believed that compatible solutes such as proline intercalate between membrane phospholipids, helping to stabilize membranes by maintaining hydrogen bonding (Rudolph et al., 1986).

TABLE 4.3
Listing of Common Compatible Solutes Accumulating Bacteria and Conditions under which Synthesis or Uptake Increases

Compatible Solute and Bacteria Known to Accumulate	Conditions	References
N-acetylglutaminylglutamine amide		
Pseudomonas aeruginosa	Osmotic	18
Ectoine		
Brevibacterium	Osmotic	24
Halomonas	Osmotic	72
Halovibrio	Osmotic	72
Marinococcus	Osmotic	72
Pseudomonas halophila	Osmotic	72
Vibrio costicola	Osmotic	72
Glutamate		
Arthrobacter	Cold	25
Escherichia coli	Osmotic	48
Exiguobacterium	Cold	25
Vibrio∗	Osmotic	14
Klebsiella∗	Osmotic	14
Lactobacillus	Osmotic	33
Glycine betaine		
Chromatium	Osmotic	95
Corynebacterium	Osmotic	70
Ectothiorhodospira	Osmotic	72
Enterobacteriaceae family	Osmotic	14
Listeria monocytogenes	Osmotic, cold	45
Rhizobium meliloti	Osmotic	76
Pseudomonas aeruginosa	Osmotic	14
Streptomyces griselous	Osmotic	72
Thiobacillus ferrooxidans	Osmotic	72
Proline		
Salmonella∗	Osmotic	72
Coreynebacterium	Osmotic	70
E. coli	Osmotic, cold	38
Exiguobacterium	Cold	25
Klebsiella	Osmotic	14
Lactobacillus	Osmotic	14
Pseudomonas aeurginosa	Osmotic, cold	14
Staphylococcus aureus	Osmotic	14
Serratia marcescens	Osmotic	14
Thiobacillus ferroxidans	Osmotic	14
Trehalose		
E. coli∗	Osmotic, cold	25, 38
Ectothiorhodospira halochris	Osmotic	24
Chlorobium	Osmotic	95

∗ Indicates that organism produces compound for use as compatible solute.

Studies of selected permafrost isolates indicate the dominance of glutamate and proline as compatible solutes, produced in high concentrations when exposed to low temperatures (Galinski, unpublished data). However, this has not been the case for one *Arthrobacter* sp. isolated from 600,000-year-old permafrost core (Mindock et al., 2001). Instead, in this case, the concentration of compatible solutes actually decreased with lowered temperatures. This curious finding might be explained by the idea that the remaining water in the interior of the cell does not behave like bulk water but is, instead, ordered along the surface of the membrane, producing a high enthalpic potential, which prevents the reordering of molecules and, thus, ice formation. Solutes are excluded from the ordered regions, as their inclusion would result in a decrease in the concentration of compatible solutes necessary to maintain homeostasis. The combination of ordered peripheral water molecules and centralized solute accumulation inhibits freezing (Mindock et al., 2001). It remains to be seen whether compatible solutes will be detected in other permafrost bacterial isolates.

ABC-transporter proteins mediate uptake of most compatible solutes. The only well-defined osmotic responsive ABC transporter is OpuA, responsible for glycine betaine uptake in *Lactococcus lactis*. The uptake of glycine betaine by OpuA, is dependent on the interactions of the lipid and transporter interactions (van der Heide and Poolman, 2000). As the outside osmolarity increases, the proteoliposomes react by decreasing their surface-to-volume ratio, therefore reducing the electrostatic interactions between the lipid headgroups and the protein, resulting in increased glycine betaine uptake (van der Heide et al., 2001). Similar systems are believed to exist in the ion-linked transporters ProP and BetP, involved in proline and betaine uptake, respectively (van der Heide et al., 2001).

4.4.2.4 Alternative Physiological Adaptations to Permafrost Conditions

It seems unlikely that membrane adaptation, compatible solute production, or cold-induced proteins are alone responsible for adaptation to permafrost conditions. Compatible solutes may act to stabilize proteins and membranes by substituting for hydrogen bonds until the membrane composition has changed. Selective dehydration may also be used to expel intracellular water and, therefore, reduce the chance of frozen intracellular water, as seen in spores to promote heat resistance (Gould and Measures, 1977). A decrease in cell size and ability to adhere to the surrounding microenvironment could greatly facilitate survival. A decrease in cell volume 14-fold is seen for one permafrost isolate when grown at 4°C compared with 24°C (Mindock et al., 2001). The reduction in metabolism required for cell survival at low temperatures is facilitated by cell dehydration, caused by a combination of freezing and nutrient limitation. Theoretically, the small size should decrease the probability of intracellular ice formation, but modeling studies indicate that permafrost cells are small enough to dehydrate rather than form intracellular ice at the freezing rates likely to be experienced in the environment (McGrath et al., 1994) (Figure 4.2). The diminished size of the cells would require a lower rate of basal metabolism, which could allow the cell to survive given the diminishing levels of energy caused by a decrease in active transport (Herbert, 1986). Decreased active transport of solutes has been shown to occur with decreasing temperature through the use of oxidation of radiolabeled substrates (Herbert, 1986).

The production of capsular polysaccharides is common in permafrost bacteria (Soina et al., 1995; Vorobyova et al., 1996). The capsules decrease in thickness at cold temperatures. The benefits of a decrease in capsule thickness are twofold: it conserves the energy-rich components while at the same time decreasing the diffusion barrier encountered by nutrients and other small molecules. Capsules also play a role in adherence. The survival rates of nonmotile bacteria have been shown to increase after a rapid freeze, compared with liquid environments in which no soil particles are present for adherence (Sidyakina et al., 1992). Regardless of thickness, the presence of capsular layers seen in some permafrost isolates could provide an extra barrier similar to the membrane to hinder penetration of extracellular ice (Mazur, 1977; Zvyagintsev et al., 1985).

The production of pigment proteins is a known consequence of general stress responses in microorganisms. Therefore, it is of no surprise that the majority of permafrost isolates are highly pigmented (Siebert and Hirsch, 1988; Vishnivetskaya et al., 2000). Antarctic red-pigmented isolates are more tolerant of alkaline conditions and intense ultraviolet light exposure than their nonpigmented relatives (Siebert and Hirsch, 1988). Recent studies have shown that ancient Siberian

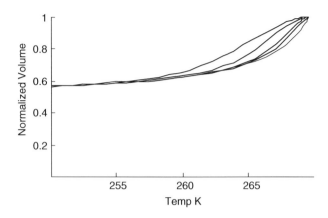

FIGURE 4.2 This plot shows predicted microbial plasmolysis during freezing at various cooling rates. The vertical axis represents the volume of the cytoplasm normalized with respect to the cytoplasm volume before freezing. The results show that the cytoplasm shrinks in response to external freezing. The right-most curve is for an extremely slow freezing rate (0.01°C/min), indistinguishable from equilibrium freezing at infinitely slow rates. The faster the cooling rate, the more displaced the curves are to the left in this figure. The cooling rates here are 5°, 10°, 25°, and 50°C/min, which would be high rates of cooling for most cases in the environment. The fact that the cell retains more of its initial volume for freezing at faster rates corresponds to the fact that more water remains "trapped" within the cell, making it more likely to freeze internally. Overall, these predictions indicate that permafrost microbes are not likely to form intracellular ice. Instead, they will respond to freezing in their environment by exosmosis. This will create a dehydrated, but unfrozen, intracellular state. It is assumed that the intracellular concentration of compatible solutes is 2.0 osmolal. The initial cell diameter is taken to be 1.0 μm, and it is assumed that 50% of the intracellular volume is incapable of participating in osmosis. Membrane water permeability and its temperature dependence are taken as representative values for yeast published in the literature.

permafrost isolates decrease in pigment intensity while maintaining rapid growth rates with decreased temperature and nutrient level (Vishnivetskaya et al., in preparation). The cause of this response is unknown. It seems likely that the cells are limiting metabolism to allow expression of only a few necessary proteins as the energy available to the cell decreases.

4.4.3 METABOLIC ACTIVITY

The fact that many of these microorganisms are viable and culturable leads to the question of what adaptations they possess to allow their survival in such harsh conditions. Bacteria might survive either by a very slow rate of metabolism, or by existing in a state of anabiosis. Both theories have supporting and contradictory evidence, and presumably both are true of some microorganisms.

4.4.3.1 Anabiosis Theory

The theory of anabiosis, or dormancy, is popular at least partly because little evidence exists for metabolic activity within the permafrost. The presence of intact membranes and autoregulatory factors that keep the cell in a resting state reinforce the anabiosis theory (Soina et al., 1995). Phenol lipids of acylresorcine, which have been shown to accompany dormancy of both spore- and nonspore formers in response to starvation, have been detected in permafrost isolates (Vorobyova et al., 1996).

Many studies have shown the presence of some enzyme-driven reactions but were unable to determine whether the cells producing the enzymes were still viable. Early studies failed to detect the presence of radiolabeled carbon dioxide in Antarctic soil after the soil had been amended with a substrate, allowing the authors to conclude that metabolic activity was absent (Horowitz et al.,

1972). However, in surface soils some radiolabeled glucose was assimilated even though a very small number of bacteria were cultivated (Horowitz et al., 1972). This activity was attributed to a combination of factors including the presence of nonculturable microorganisms, the presence of inorganic catalysts, and the presence of dead, albeit still enzymatically active, cells (Horowitz et al., 1972).

Activity at cold temperatures has been characterized for several enzymes. The best studied are isocitrate dehydrogenase and catalase (Vorobyova et al., 1996). Isocitrate dehydrogenase is cold labile and is used as a measure of the cryoprotective properties of other proteins (Koda et al., 2000). Within the permafrost, ionizing radiation produces hydroxyl radicals on contact with superoxides and peroxide. The continued activity of catalase could be beneficial for organisms within the permafrost as the role of catalase is to protect cells from harmful hydrogen peroxide by dismutation to water and oxygen, eliminating a substrate required for formation of more harmful hydroxyl radicals.

4.4.3.2 Low Rate of *in Situ* Metabolic Activity Theory

The major obstacles to metabolic activity in permafrost are low temperature, low water activity and radiation exposure. For a number of years, only circumstantial evidence of metabolic activity was available. The presence of large numbers of viable microorganisms isolated from ancient permafrost soils indicated that DNA- and membrane-repair mechanisms must exist to protect the cells from long-term exposure to free radicals produced by the γ-radiation emitted from the potassium-40 in surrounding rock (Friedmann, 1994). In addition, the presence of sufficient unfrozen water for a small rate of biological activity has also been detected (Rivkina et al., 2000; Zvyagintsev et al., 1990). By combining studies of bound water concentrations, the thin surface layer of water that surrounds the microbes, and measures of membrane lipid incorporation at decreasing temperatures, Rivinka et al. (2000) illustrated that the availability of free water diminishes with decreased temperature (well below Siberian permafrost in situ temperatures). This creates a diffusion barrier, effectively starving the cells (Rivkina et al., 2000).

A number of recent studies have shown the presence of metabolic activity at cold temperatures in snow, ice, and soil. Slow growth of microorganisms has been detected in Siberian permafrost at –8° to –10°C over an 18-month period (Gilichinsky et al., 1993). In addition, doubling rates of 1 d at 5°C, 20 d at –10°C, and 160 d at –20°C were measured by membrane lipid incorporation (Rivkina et al., 2000).

Low rates of protein and DNA synthesis were indicated by biological incorporation of [^{3}H] leucine (proteins) and methyl [^{3}H] thymidine (DNA) between –12° and –17°C in arctic snow (Carpenter et al., 2000). Furthermore, radioisotopic studies of Lake Vostok accretion ice indicate the presence of metabolic activity, with only a small portion resulting from macromolecular synthesis (Karl et al., 1999). These findings indicate that the microbes maintain only enough basal metabolism to allow repairs to membranes and DNA molecules but do not actively divide under the oligotrophic, freezing environment.

Amino acid racemization studies of aspartic acid extracted from 35,000-year-old permafrost indicate ongoing repair processes. Racemization studies examine the frequency of D-aspartic acid, which is formed after cell death when the hydrogen of the chiral α-C is detached and rejoined in the opposite orientation. In the presence of metabolically active organisms this D form is repaired to the biologically active L form of aspartic acid. The estimated ratio of D-aspartic acid, resulting from dead, inactive cells, to L-aspartic acid was smaller than predicted by mathematical models, allowing the authors to conclude that some metabolic activity was occurring to produce the diminished levels of D-aspartic acid (Brinton et al., 2002).

Metabolic activity of anaerobes, including nitrifying microorganisms, has been shown in permafrost frozen for up to 2 million years by measurement of metabolic end products. The youngest soils were shown to contain the highest amounts of anaerobic metabolic end products (ferrous iron, sulfide, and methane), and these decreased with the age of the soil (Rivkina et al., 1998). As the age of the soil increases, the concentration of the substrate, ammonium, increases, whereas the products of nitrification are found in smaller amounts (Janssen and Bock, 1994). However, it is uncertain whether these end products were formed before freezing and then simply trapped in the soil layers after freezing because of the low gas diffusion seen in permafrost.

4.5   IMPORTANCE OF THE PERMAFROST ENVIRONMENT

Low temperatures provide a stable environment for sustaining life in the permafrost. The permafrost is therefore likely to be a depository for ancient biomarkers such as biogenic gases, polyaromatic hydrocarbons, biominerals, biological pigments, lipids, enzymes, proteins, nucleotides, RNA and DNA fragments, molecules, microfossils, and viable cells (Vorobyova et al., 1997). The presence of potentially deadly human viruses also exists in permafrost (Reid et al., 1999). This makes the permafrost important for future advances in cryobiology, palaeontology, and exobiology.

The continually frozen nature of permafrost aids the study of microbial evolution because closely related microorganisms have been isolated from soils of different periods of geologic time. Discovery of similar species using phylogenetic methods may allow for better representations of the evolutionary clock in cold environments, as determined by the mutation rates of particular organisms in that environment (Gilichinsky et al., 1992).

Permafrost is believed to extend over many of the planets of our solar system, including the majority of the planet Mars. The low temperature, low water containing, and continuous exposures to radiation of Arctic and Antarctic permafrost are very similar to conditions on Mars (Malin and Edgett, 2000). Therefore, studies of the long-term preservation of microbiota in Earth's permafrost can provide a benchmark from which searches for extraterrestrial life can be launched (Soina et al., 1995). The age of the Earth's permafrost is much younger than the 3-billion-year-old estimate for the Martian surface, but it may still be used as a model to examine the long-term preservation of cells that cannot be modeled accurately.

Permafrost might also provide a better understanding of the types of microbes likely to survive on meteors or spacecraft, which may be able to regain their activity when conditions are again agreeable. This is the basis of the theory of panspermia, in which life is argued to survive as spores in space, and underpins the nascent science of astrobiology. The low temperatures, low levels of nutrients, and radiation exposure found in the permafrost is similar to that of space, at least within meteors or other debris (Mastrapa et al., 2001).

The remarkable ability to survive in a continuously frozen matrix of permafrost for millions of years makes the permafrost community unique, not only for its implications for the discovery of novel enzymes of biotechnological importance but also to begin to answer basic science questions. How long can a cell survive in continually frozen conditions? What is the lowest rate of metabolic activity necessary to retain viability? And does life exist outside the Earth?

REFERENCES

Aguilar, P.S., Hernandez-Arriaga, A.M., Cybulski, L.A., Erazo A.C., and Mendoza, D. (2001) Molecular basis of thermosensing: a two component signal transduction thermometer in *Bacillus subtilis, EMBO J.* 20, 1681–1691.

Aguilar, P.S., Lopez P., and de Mendoza D. (1999) Transcriptional control of the low temperature inducible *des* gene, encoding the δ5 desaturase of *Bacillus subtilis*, *J. Bacteriol.*, 181, 7028–7033.

Aspinall, D. and Paleg, J.G. (1981) Proline accumulation: Physiological traits, in *The Physiology and Biochemistry of Drought Resistance in Plants*, Aspinall D. and Paleg, J.G., Eds., Academic Press, New York, pp. 206–240.

Atlas, R., DiMenna, M., and Cameron, R.E. (1978) Ecological investigations of yeasts in Antarctic soils, *Antarctic Res. Ser.* 30, 27–34.

Becker, R.E. and Volkmann, C.M. (1961) A preliminary report on the bacteriology of permafrost in the Fairbanks area. *Proc. Alaskan Sci. Conference*, Alaska, Vol. 12, p. 188.

Beeman, R. and Suflita, J.M. (1990) Evaluation of deep subsurface sampling procedures using serendipitous microbial contaminants as tracer organisms, *Geomicrobiology*, 7, 223–233.

Beyer, L., Bolter, M., and Seppelt, R.D. (2000) Nutrient and thermal regime, microbial biomass, and vegetation of Antarctic soils in the Windmill Islands region of east Antarctica (Wilkes land), *Arctic Antarctic Alpine Res.* 32, 30–39.

Boyd, W.L. and Boyd, J.W. (1964) The presence of bacteria in permafrost of the Alaskan arctic. *Can. J. Microbiol.*, 10, 917–919.

Brinton, K.L.F., Tsapin, A.I., Gilichinskii, D.A., and McDonald, G. (2002) Aspartic acid racemization and age-depth relationships for organic carbon in Siberian permafrost, *Astrobiology*, 2, 77–82.

Brown, A.D. (1990) *Microbial Water Stress Physiology: Principles and Perspectives*, Wiley, New York.

Cameron, R.E. and Morrelli, F.A. (1974) Viable microorganisms from ancient Ross island and Taylor valley drill core, *Antarctic J. USA*, 9, 113–116.

Carpenter, E.J., Lin S., and Capone D. (2000) Bacterial activity in South Pole snow, *Appl. Environ. Microbiol.*, 66, 4514–4517.

Chong, B., Kim, J., Lubman, D., Tiedje, J., and Kathariou, S. (2000) Use of non-porous reversed-phase high-performance liquid chromatography for protein profiling and isolation of proteins induced by temperature variations for Siberian permafrost bacteria with identification by matrix-assisted laser desorption/ionization time-of-flight mass spectrometry and capillary electrophoresis-electrospray ionization mass spectrometry, *J. Chromatog. B*, 748, 167–177.

Csonka, L. (1989). Physiological and genetic responses of bacteria to osmotic stress, *Microbiol. Rev.*, 53, 121–147.

deSantos, H. and Galinski, E.A. (1998) An overview of the role and diversity of compatible solutes in bacteria and archaea, in *Biotechnology of Extremophiles*, Antranikan, G., Ed., Springer, New York, pp. 117–149.

DeVries, A.L. and Wohlschlag, D.E. (1969) Freezing resistance in some Antarctic fishes, *Science*, 163, 1073–1075.

Dmitriev, V.V., Gilichinskii, D.A., Faizutdinova, R.N., Shershunov, I.N., Golubev, V.I., and Duda, V.I. (1997) Detection of viable yeast in 3-million-year-old permafrost soils of Siberia, *Microbiology*, 66, 546–550.

D'Souza-Ault, M.R., Smith, L.T., and Smith, G.M. (1993) Roles of n-acetylglutaminylglutamine amide and glycine betaine in adaptation of *Pseudomonas aeruginosa* to osmotic stress, *Appl. Environ. Microbiol.*, 59, 473–478.

Duman, J.G. (2001) Antifreeze and ice nucleation proteins in terrestrial arthropods, *Annu. Rev. Physiol.*, 63, 327–357.

Duman, J.G., Xu, L., Neven, T.G., Tursman, D., and Wu, D.W. (1991) Hemolymph proteins involved in insect sub-zero temperature tolerance: ice nucleators and antifreeze proteins, in *Insects at Low Temperature*, Lee, R.E. and Denlinger, D.L., Eds., Chapman & Hall, New York, pp. 94–127.

Ellis-Evans, J.C. and Wynn-Williams, D.D. (1985) The interaction of soil and lake microflora at Signy island, in *Antarctic Nutrient Cycles*, Siegfried, W.R., Condy, P.R., and Laws, R.M., Eds., Springer, Berlin, pp. 662–668.

Erokhina, L.G., Spirina, E., and Gilichinsky, D.A. (1999) Accumulation of phycobiliproteins in cells of ancient cyanobacteria from arctic permafrost as dependent on the nitrogen source for growth, *Microbiologiya* 68, 628–631.

Friedmann, E.I. (1994) Permafrost as microbial habitat, in *Viable Microorganisms in Permafrost*, Gilichinsky D.A., Ed., Russian Academy of Sciences, Pushchino, Russia, pp. 21–26.

Frings, E., Kunte, H.J., and Galinski, E.A. (1993) Compatible solutes in representatives of the genera *Brevibacterium* and *Corynebacterium*: Occurence of tetrahydropyrimidines and glutamine, *FEMS Microbiol. Lett.*, 109, 25–32.

Galinski, E.A. and Herzog, R.M. (1990) The role of trehalose as a substitute for nitrogen-containing compatible solutes (*Ectothiorhodospira halochloris*), *Arch. Microbiol.*, 153, 607–613.

Getsen, M. (1990) Algae as a constitution base for life of high latitude ecosystems, *Bot. J.*, 75, 1641–47.

Getsen, M. (1994) Blue-green and desmidea algae in algal flora of the small reservoirs, in *Structural and Functional Organization of Phytocenoses on the High North*, Gecen M. and Nazarova, S., Eds., Syktyvkar Press, Syrtyvkar, Russia, pp. 61–71.

Gilichinsky, D. (1993) Viable microorganisms in permafrost: The spectrum of possible applications to investigations in science for cold regions, Fourth International Symposium on Thermal Engineering and Science for Cold Regions, U.S. Army Cold Regions Research and Engineering Laboratory, Hanover, NH, US Army Corps of Engineers, 28 September–1 October, 1993.

Gilichinsky, D., Vorobyova, E., Erokhina, L.G., Fyordorov-Dayvdov, D.G., and Chaikovskaya, N.R. (1992) Long term preservation of microbial ecosystems in permafrost, *Adv. Space Res.*, 12, 255–263.

Gilichinsky, D., Wagener, S., and Vishnivetskaya, T. (1995) Permafrost microbiology, *Permafrost Periglacial Processes* 6, 281–291.

Gilichinsky, D.A., Soina, V.S., and Petrova, M.A. (1993) Cryoprotective properties of water in the earth cryolithosphere and its role in exobiology, *Orig. Life Evol. Biosph.* 23, 65–75.

Glaasker, E., Konings, W.N., and Poolman, B. (1996) Osmotic regulation of intracellular solute pools in *Lactobacillus plantarum*, *J. Bacteriol.*, 178, 575–582.

Gould, G.W. and Measures, J.C. (1977) Water relations in single cells, *Philos. Trans. R. Soc. Lond. B*, 278, 151–166.

Griffith, M. and Ewart, K.V. (1995). Antifreeze proteins and their potential uses in frozen foods, *Biotechnol. Adv.*, 13, 375–402.

Herbert, R.A. (1986) The ecology and physiology of psychrophilic microorganisms, in *Microbes in Extreme Environments*, Herbert, R.A. and Codd, G.A., Eds., Academic Press, London, pp. 1–24.

Horowitz, N.H., Cameron, R.E., and Hubbard, J.S. (1972) Microbiology of the dry valleys of Antarctica, *Science* 176, 242–245.

Ingraham, J.L. and Marr, A.G. (1996) Effect of temperature, pressure, pH, and osmotic stress on growth, in *Escherichia coli and Salmonella: Cellular and Molecular Biology*, Neidhart, F., Ed., American Society for Microbiology Press, Washington, DC, pp. 1570–1578.

James, N. and Sutherland, M.L. (1942) Are there living bacteria in permanently frozen subsoil, *Can. J. Res. C*, 20, 228–235.

Janssen, H. and Bock, E. (1994) Profiles of ammonium, nitrite and nitrate in the permafrost soils, in *Viable Microorganisms in Permafrost*, Gilichinsky, D., Ed., Puschino Research Center, Puschino, Russia, pp. 27–36.

Karasev, S., Gourina, L., Adanin, D., Gilichinskii, D.A., and Evtoushenko L. (1998) Viable actinobacteria from the ancient Siberian permafrost, *Earth Cryosph.* 2, 68-74.

Karl, D.M., Bird, D.F,. Bjorkman, K., Houlihan, T., Shackelford, R., and Tupas, L. (1999) Microorganisms in the accreted ice of Lake Vostok, Antarctica, *Science,* 286, 2144–2147.

Khlebnikova, G.M., Gilichinsky, D., Federov-Davydov, D.G., and Vorobyova, E.A. (1990) Quantitative evaluation of the microorganisms in permafrost deposits and buried soils, *Microbiologiya,* 59, 148–155.

Kjoller, A. and Odum, S. (1971) Evidence for longevity of seeds and microorganisms in permafrost, *Arctic,* 24, 230–233.

Ko, R., Smith, L.T., and Smith, G.M. (1994) Glycine betaine confers enhanced osmotolerance and cryotolerance on *Listeria monocytogenes*, *J. Bacteriol.*, 176, 426–431.

Koda, N., Aoki, M., Kawahara, H., Yamade, K., and Obata, H. (2000) Characterization and properties of intracellular proteins after cold acclimation of the ice-nucleating bacterium *Pantoea agglomerans* (*Erwinia herbicola*) ifo12686, *Cryobiology*, 41, 195–203.

Kuz'min, N.P., Duda, V.I., Shershunov, I.N., Tomashevskiy, A., Yu, W., Lisenko, A.M., and Gilichinskii, D.A. (1996) Bacillus from ancient Siberian permafrost, in *Microbial Biodiversity: Conditions, Survival Strategy, Ecological Problems*, Institute of Ecology and Genetics of Microorganisms, Chereshnev, p. 41.

Larsen, P.I., Sydnes, L.K., Landfald, B., and Strom, A.R. (1987) Osmoregulation in *Escherichia coli* by accumulation of organic osmolytes: betaines, glutamic acid, and trehalose, *Arch. Microbiol.*, 147, 1–7.

Malin, M.C. and Edgett, K.S. (2000) Evidence for recent groundwater seepage and surface runoff on Mars, *Science,* 288, 2330–2335.

Mastrapa, R.M.E., Glanzberg, H., Head, J.N., Melosh, H.J., and Nicholson, W.L. (2001) Survival of bacteria exposed to extreme acceleration: implications for panspermia, *Earth Planet. Sci. Lett.*, 189, 1–8.

Matsumoto, G.I., Friedmann, E.I., and Gilichinsky, D. (1995) Geochemical characteristics of organic compounds in a permafrost sediment core sample from northeast Siberia, Russia, NIPR Symposium of Antarctic Geosciences.

Mazur, P. (1977) The role of intracellular freezing in the death of cells cooled at supraoptimal rates, *Cryobiology*, 14, 251–272.

Mazur, P. (1984) Freezing of living cells: Mechanisms and implications, *Am. J. Physiol.*, 247, C125–C142.

McGrath, J., Wagener, S., and Gilichinsky D. (1994) Cryobiological studies of ancient microorganisms isolated from Siberian permafrost, in *Viable Microorganisms in Permafrost*, Gilichinsky, D., Ed., Puschino Press, Puschino, Russia, pp. 48–67.

McLean, A.L. (1918) Bacteria of ice and snow in Antarctica, *Nature*, 102, 35.

Measures, J.C. (1975) Role of amino acids in osmoregulation of non-halophic bacteria, *Nature*, 257, 398–400.

Mindock, C., Petrova, M.A., and Hollingsworth, R.I. (2001) Re-evaluation of osmotic effects as a general adaptive strategy for bacteria in sub-freezing conditions, *Biophys. Chem.*, 89, 13–24.

Morris, G.J. and Clarke, A. (1987) Cells at low temperatures, in *The Effects of Low Temperatures on Biological Systems*, Morris, G.J. and Grout, B.W.W., Eds., Edward Arnold, Baltimore, MD, pp. 71–129.

Muldrew, K. and McGann, L.E. (1994) The osmotic rupture hypothesis of intracellular freezing injury, *Biophys. J.*, 66, 532–541.

Nedwell, D.B. (1999) Effect of low temperature on microbial growth: lowered affinity for substrates limits growth at low temperatures, *FEMS Microb. Ecol.*, 30, 101–111.

Nemecek-Marshall, M., LaDuca, R., and Fall, R. (1993) High level expression of ice nuclei in a *Pseudomonas syringae* strain is induced by nutrient limitation and low temperature, *J. Bacteriol.*, 175, 4062–4070.

Nozhevnikova, A.N., Simankova, M.V. Parshina S.N., and Kotsyurbenko O.R. (2001) Temperature characteristics of methanogenic archaea and acetogenic bacteria isolated from cold environments, *Water Sci. Technol.*, 44, 41–48.

Omelyansky, V.L. (1911) Bacteriological investigation of the Sanga mammoth and surrounding soil, *Arkhiv Biol. Nauk*, 16, 335–340 (in Russian).

Ostroumov, V.E. (1992) Unfrozen water types in frozen soils, Joint Russian-American Seminar on Cryopedology and Global Change, Gilichinsky, D., Ed., Puschino, Russia, Russian Academy of Sciences, 248–257.

Pewe, T. (1995) Permafrost, *Encyclopaedia Britannica*, 15th ed., Vol. 20, pp. 752–759.

Phadtare, S., Yamanaka, K., and Inouye, M. (2000) The cold shock response, in *Bacterial Stress Responses*, Storaz, G. and Hengge-Aronis, R., Eds., American Society for Microbiology Press, Washington, DC, pp. 33–47.

Reid, A., Fanning, T.G., Hultin, J.V., and Taubenberger, J.K. (1999). Origin and evolution of the 1918 "Spanish" influenza virus hemagglutinin gene, *Proc. Natl. Acad. Sci. USA*, 96, 1651–1656.

Rivkina, E., Friedmann, E.I., McKay, C.P., and Gilichinsky, D. (2000) Metabolic activity of permafrost bacteria below the freezing point, *Appl. Environ. Micro.*, 66, 3230–3233.

Rivkina, E., Gilichinsky, D., Wagener, S., Tiedje, J., and McGrath, J. (1998) Biogeochemical activity of anaerobic microorganisms from buried permafrost sediments, *Geomicrobiology*, 15, 187–193.

Rudolph, A.S., Crowe, J.H., and Crowe, L.M. (1986) Effects of three stabilizing agents—proline, betaine, and trehalose on membrane phospholipids, *Arch. Biochem. Biophys.*, 245, 134–143.

Ruffert, S., Lambert, C., Peter, H., Wendisch, V.F., and Kramer, R. (1997) Efflux of compatible solutes in *Corynebacterium glutamicum* mediated by osmoregulated channel activity, *Eur. J. Biochem.*, 247, 572–580.

Russell, N.J. (1990) Cold adaptation of microorganisms, *Philos. Trans. R. Soc. Lond. B*, 326, 595–611.

Severin, J., Wohlfarth, A., and Galinski, E.A. (1992) The predominant role of recently discovered tetrahydropyrimidines for the osmoadaptation of halophilic eubacteria, *J. Gen. Microbiol.*, 138, 1629–1638.

Shi, T., Reeves, R.H., Gilichinsky, D.A., and Friedmann, E.I. (1997) Characterization of viable bacteria from Siberian permafrost by 16s rDNA sequencing, *Microb. Ecol.*, 33, 169–179.

Sidyakina, T.M., Lozitskaya, N.D., Dobrovolskaya, T.G., and Kalakoutskii, L.V. (1992) Cryopreservation of various types of soil bacteria and mixtures thereof, *Cryobiology*, 29, 274–280.

Siebert, J. and Hirsch, P. (1988) Characterization of 15 selected coccal bacteria isolated from Antarctic rock and soil samples from the McMurdo-Dry valleys (South-Victoria land), *Polar Biol.*, 9, 37–44.

Smith, L.T. and Smith, G.M. (1989) An osmoregulated dipeptide in stressed *Rhizobium meliloti*, *J. Bacteriol.*, 171, 4714–4717.

Soina, V.S., Vorobyova, E.A., Zvyagintsev, D.G., and Gilichinskii, D.A. (1995) Preservation of cell structures in permafrost: a model for exobiology, *Adv. Space Res.*, 15, 3237–3242.

Spirina, E. and Federov-Davydov, D.G. (1998) Microbiological characterization of cryogenic soils in the Kolymskaya lowland, *Eurasian Soil Sci.*, 31, 1331–1344.

Sun, X., Griffith, M., Pasternak, J.J., and Glick, B. (1995) Low temperature growth, freezing survival, and production of antifreeze protein by the plant growth promoting rhizobacterium *Pseudomonas putida* gr12-2, *Can. J. Microbiol.*, 41, 776–784.

Suzuki, I., Kanesaki, Y., Mikami, K., Kaneshisa, M., and Murata, N. (2001) Cold-regulated genes under control of the cold sensor hik33 in Synechocystis, *Mol. Microbiol.*, 40, 235–244.

Suzuki, I., Las, D., Kanesaki, Y., Mikami, K., and Murata, N. (2000) The pathway for perception and transduction of low-temperature signals in synechocystis, *EMBO J.*, 19, 1327–1334.

Tiedje, J., Petrova, M.A., and Moyer, C. (1998) Phylogenetic diversity of archaea from ancient Siberian permafrost, 8th International Symposium on Microbial Ecology, Halifax, Nova Scotia, Atlantic Canada Society for Microbial Ecology, August 9–14.

Toner, M. and Cravalho, E. (1990) Thermodynamics and kinetics of intracellular ice formation during freezing of biological cells, *J. Appl. Phys.*, 67, 1582–1593.

van der Heide, T. and Poolman, B. (2000) Osmoregulated ABC-transport system of *Lactococcus lactis* senses water stress via changes in the physical state of the membrane, *Proc. Natl. Acad. Sci. USA*, 97, 7102–7106.

van der Heide, T., Stuart, M., and Poolman, B. (2001) On the osmotic signal and osmosensing mechanism of an ABC transport system for glycine betaine, *EMBO J.*, 20, 7022–7032.

Vasilchuk, Y.K. and Vasilchuk, A.C. (1997) Radiocarbon dating and oxygen isotope variations in late Pleistocene syngenetic ice-wedges, northern Siberia, *Permafrost Periglacial Processes*, 8, 335–345.

Venkateswaran, A., McFarlan, S.C., Ghosal, D., Minton, K.W., Vasilenko, A., Makarova, K., Wackett, L.P., Daly, M.J., et al. (2000). Physiologic determinants of radiation resistance in *Deinococcus radiodurans*, *Appl. Environ. Microbiol.*, 66, 2620–2626.

Vishiniac, H.S. (1993) The microbiology of Antarctic soils, in *Antarctic Microbiology*, Friedmann, E.I., Ed., Wiley-Liss, New York, pp. 297–341.

Vishnivetskaya, T., Erokhina, L.G., Spirina, E., and Shatilovich, A. (in press) Viable phototrophs: Cyanobacteria and green algae from permafrost darkness, in *Life in Ancient Ice*, Castello, J.D. and Rogers, S., Eds., Princeton University Press, Princeton, NJ.

Vishnivetskaya, T., Kathariou, S., McGrath, J., and Tiedje, J.M. (2000) Low temperature recovery strategies for the isolation of bacteria from ancient permafrost sediments, *Extremophiles*, 4, 165–173.

Vishnivetskaya, T., Petrova, M., Urbance J., Ponder, M., and Tiedje, J. (in preparation) Phylogenetic diversity of bacteria inside the arctic permafrost.

Vorobyova, E., Soina, V.S., and Mulukin, A.L. (1996) Microorganisms and enzyme activity in permafrost after removal of long term cold stress, *Adv. Space Res.*, 18, 12103–12108.

Vorobyova, E.A., Soina, V.S., Gorlenko, M., Minkovskaya, N., Natalia, Z., Mamukelashvili, A., Gilichinsky, D.A., Rivkina, E., and Vishnivetskaya, T. (1997) The deep cold biosphere: facts and hypothesis, *FEMS Microbiol. Rev.*, 20, 277–290.

Welsh, D.T. (2000) Ecological significance of compatible solute accumulation by microorganisms: From single cells to global climate, *FEMS Microbiol. Rev.*, 24, 263–290.

Wynn-Williams, D. (1989) Ecological aspects of Antarctic microbiology, in *Advances in Microbial Ecology*, Marshall, K.C., Ed., Plenum Press, New York, pp. 71–146.

Xu, H., Griffith, M., Patten, C., and Glick, B. (1998) Isolation and characterization of an antifreeze protein with ice nucleation activity from the plant growth promoting rhizobacterium *Pseudomonas putidia* gr12-2, *Can. J. Microbiol.*, 44, 64–73.

Zachariassen, K.E. and Hammel, H.T. (1976) Nucleating agents in the haemolymph of insects tolerant to freezing, *Nature,* 262, 285–287.

Zenova, G.M. and Shtina, E.A. (1990) *Soil Algae*, Moscow State University Press, Moscow.

Zhou, J., Davey, M.E., Figueras, J., Rivkina, E., Gilichinsky, D.A., and Tiedje, J.M. (1997) Phylogenetic diversity of a bacterial community determined from Siberian tundra soil DNA, *Microbiology*, 143, 3913–3919.

Zvyagintsev, D.G., Gilichinsky, D., Blagodatsky, S.A., Vorobyova, E., Khlebnikova, G.M., Arkhangelov, A.A., and Kudryavtseva, N.N. (1985) Survival time of microorganisms in permanently frozen sedimentary rocks and buried soils, *Microbiology*, 54, 131–136.

Zvyagintsev, D.G., Gilichinsky, D., Khlebnikova, G.M., Fedorov-Davydov, D.G., and Kudryavtseva, N.N. (1990) Comparative characteristics of microbial cenoses from permafrost rocks of different ages and genesis, *Microbiology*, 59, 332–338.

5 Adaptation of Higher Plants to Freezing

Roger S. Pearce

CONTENTS

5.1   INTRODUCTION

Light frost devastates the economic value of citrus and cauliflower, yet dwarf willow and tussock grass survive sub-artic and sub-Antarctic winters. Evolution has molded them to their different environments. Molecular analysis reveals the underlying factors, information that can be exploited to help crops cope with a changing and uncertain climate, to explain the distribution of wild plants, and to reveal cryopreservation strategies of general value.

Plant adaptation* to freezing is the possession of genes conferring freezing resistance (tolerance or avoidance; Levitt, 1980). These genes may be constitutively expressed, conferring permanent elements of resistance, or they may be expressed in response to an environmental signal. Acclimation* comprises those processes at the molecular, biochemical, and physiological levels, triggered

---

* This term may be used differently in the microbial and animal literature.

171

by that environmental signal, that confer enhanced resistance. The emphasis of most recent research has been on analysis of acclimation at the molecular level, often employing the model species *Arabidopsis thaliana* or cereals (Pearce, 1999; Thomashow, 1999; Xin and Browse, 2000). This has given considerable insight into the functional and regulatory genes and processes involved. Simultaneously, advances have continued in understanding plant freezing itself (Pearce, 2001), highlighting a need to understand how events at the molecular and at the physical and physiological levels might be connected. This chapter discusses the principles and the individual genes that are emerging as a result of molecular analysis, against a background of wider knowledge of plant adaptation to cold and freezing.

5.2 PLANT FREEZING

Plants can use both tolerance and avoidance strategies for surviving frost (Sakai and Larcher, 1987). The most common freezing pattern in plants is growth of extracellular ice, which, for example, occurs in the leaves of most plants and in the bark of woody species. Most of this ice grows at the expense of water drawn from within the nonfrozen cells. By freezing extracellularly, the cells avoid internal freezing, but at the expense of suffering partial or extensive dehydration, which may be tolerated if not too severe but that is itself a potentially lethal stress.

Supercooling of the cell contents (see Chapter 1) would avoid this dehydration. However, if the supercooling limit of cells is exceeded, then intracellular freezing occurs, killing the cells. Slight-to-moderate levels of supercooling occur in leaves of some plants such as the olive (Sakai and Larcher, 1987). Deep supercooling, from about –15°C down to a limit at about –40°C, occurs in many temperate woody plants such as apple and maple (Burke et al., 1976). Deep supercooling is limited to only some organs or tissues. For example, in apple the cells of the xylem parenchyma deep supercool but the bark freezes extracellularly (Ashworth et al., 1988), and in forsythia and maple, cells in flowerbud organs supercool but the bud scales and nearby stems, again, freeze extracellularly (Ashworth, 1990; Ishikawa et al., 1997).

More unusual phenomena may account for extreme freezing tolerance, below the –40°C limit to deep supercooling. In the xylem parenchyma of red osier dogwood, ice forms between the cell wall and the plasma membrane—outside the protoplast, and without damaging it (Ashworth, 1996). Vitrification (see Chapter 10) occurs in cells of twigs of some woody plants such as poplar (Hirsh et al., 1985), which could explain why twigs of several woody species can survive the temperature of liquid nitrogen (Sakai, 1960).

Could different freezing patterns be a chance result of different structural features, or are there factors in plants whose role is to control the pattern of freezing? Barriers to the spread of ice are common in woody plants, though often the nature or mode of functioning of the barrier is unclear (Wisniewski and Fuller, 1999). However, in forsythia and peach the growth of xylem into buds in the spring removes a barrier to ice (Ashworth et al., 1992), indicating that in this case, timing of differentiation controls supercooling of the bud organs. There are also simpler structures and specific macromolecules that appear to have a specific role in controlling nucleation of freezing or growth of ice.

5.2.1 NUCLEATION OF EXTRACELLULAR FREEZING

Freezing may be nucleated either homogeneously or heterogeneously. Homogeneous nucleation involves the spontaneous formation of an ice nucleus of sufficient stability to grow, whereas heterogeneous nucleation occurs when some other substance helps to form or stabilize the ice nucleus (see Chapters 10 and 22). Homogeneous nucleation is relatively unlikely except at temperatures near –40°C or in volumes of water far larger than are likely even in large trees (Pearce, 2001). This influences the sites at which freezing can be initiated. If heterogeneous nucleators are not present within a cell, then intracellular freezing is very unlikely at temperatures above –40°C. As a consequence, in plants in nature, ice usually, but not exclusively, first forms extracellularly.

Surface moisture is essential for nucleation of freezing of many herbaceous species including crops (Fuller and Wisniewski, 1998). This indicates that ice must first have formed on the surface of the plant and then grown into it. Surface freezing can be initiated by snow or sleet precipitation. Alternatively, surface moisture can be nucleated by ice-nucleation-active bacteria (Gurian-Sherman and Lindow, 1993; Wolber, 1993) or by organic and inorganic debris.

Ice can grow into leaves of herbaceous plants through stomata (Wisniewski and Fuller, 1999). However, freezing often first occurs at night, when the stomata of most plants are closed. The leaves of many species carry hydathodes; for example, grasses. These structures include a permanently open pore, and it is evident that ice, at night, can grow into grass leaves through these pores (Pearce and Fuller, 2001).

In woody plants, despite the presence of stomata, lenticels, and possible surface lesions, growth of surface ice into the plant appears much less important than spread of freezing through the plant from intrinsic (within the plant) nucleation events (Pearce, 2001; Wisniewski et al., 1997). Although intrinsic nucleation is frequent in woody plants (Sakai and Larcher, 1987), it also occurs in some herbaceous plants such as *Veronica* (Kaku, 1973). Where it does occur, it is often a constitutive feature, present at all times of year. Purification of the factor responsible is difficult. Attempts to isolate intrinsic nucleators results in loss of activity, indicating that a structural component may be essential for them to function fully (Griffith and Antikainen, 1996). Furthermore, they need not be abundant, as freezing is often initiated at only one or a few places in the plant. This is because when ice first forms at one site, it then rapidly grows through the shoot or through the whole plant (Fuller and Wisniewski, 1998; Pearce and Fuller, 2001; Kitaura, 1967; Wisniewski et al., 1997).

Nevertheless, several intrinsic nucleators have been partly characterized. Mucilage and carbohydrates, respectively, appear to be the nucleators in *Opuntia* and in giant high-altitude African species of *Lobelia* (Goldstein and Noble, 1991; Krog et al., 1979). Interestingly, cell wall arabinoxylans may have an opposite—antifreeze—effect in rye and barley crowns and seeds (Kindel et al., 1989; Olien, 1965; Olien and Smith, 1977; Williams, 1992). The nucleator in peach also appears to be neither proteinaceous nor lipidic in composition (Ashworth et al., 1985; Gross et al., 1988). In contrast, nucleators from rye are proteinaceous, though they also include phospholipid and carbohydrate components (Brush et al., 1994). Thus, plant intrinsic ice-nucleators are not all of one kind.

5.2.2 ANTIFREEZE PROTEINS

Antifreeze proteins (AFPs) occur in the apoplast* (in cell walls and between cells) of rye and other freezing-tolerant cereals (Hon et al., 1994; Pihakaski-Maunsbach et al., 1996, 2001) and are also found in expressed sap from a variety of vascular and nonvascular plants (Duman and Olsen, 1993; Urrutia et al., 1992). However, extracts from spruce, fir, and hemlock do not contain AFPs (Duman and Olsen, 1993), indicating that AFPs are not a universal component of freezing-tolerant plants. The apoplastic AFPs are cold inducible, as too are many of those found in exudates. Most plant AFPs are distinct from those found in insects and fish (which themselves are diverse), as they generally have different amino acid compositions and cause a lesser thermal hysteresis (0.1 to 0.7°C; Griffith and Antikainen, 1996). However, a carrot AFP is exceptional in having structural similarity to fish and insect AFPs (Worrall et al., 1998).

Apoplastic extracts of cold-acclimated winter rye leaves contain six polypeptides that modify ice growth (Hon et al., 1994). These correspond to three classes of PR (pathogenesis-related) proteins: two endochitinases, two β-1,3-endoglucanases, and two thaumatin-like proteins (Griffith and Antikainen, 1996). The native form of the rye AFPs is oligomeric, with each oligomer including proteins from more than one class of PR-like AFPs (Yu and Griffith, 1999). Proteins of the same

---

* The apoplast is the space outside the symplast: cell walls, intercellular spaces, and any dead cells such as xylem vessels. The symplast is the system of living protoplasts connected through plasmodesmata.

PR classes that are normally expressed in response to attack by pathogens do not have antifreeze properties, indicating that antifreeze activity is not a chance feature of the normal PR proteins.

Roles suggested for the apoplastic AFPs include controlling the sites of ice formation, controlling the rate at which ice grows, or inhibiting recrystallization (Griffith and Antikainen, 1996). One suggestion is that they may help prevent penetration of cell walls by ice, thus helping protect the protoplast from freezing. At present there is no direct evidence for this suggestion, and in fact there is no direct visualization of ice penetrating cell walls in either freezing-susceptible or freezing-tolerant species frozen under natural or natural-like conditions. The average pore size in plant cell walls is so small that it would effectively prevent ice penetration (Ashworth and Ables, 1984). However, what matters is the size of the largest pore present, though this would need to be of the order of 100 nm for penetration to occur (Ashworth and Ables, 1984).

There is no indication that the initial rate of growth of ice in plant organs is controlled by the plant. When freezing first occurs in leaves of freezing-tolerant grasses, the ice formed grows as rapidly or more rapidly through the organ (at up to 4 cm/s) than one would expect from the physical literature (Pearce and Fuller, 2001). The second phase of ice growth in leaves, drawing water from the cells, is much slower than the rate of initial growth of ice; however, this is so in species without as well as those with known apoplastic AFPs. In contract, ice growth between organs is often much slower than within organs, though this could be because of anatomical barriers (Luxova, 1986; Zámećník et al., 1994).

Recrystallization of extracellular ice could build-up massive ice crystals at some locations, especially during prolonged freezing stress. If the growth of these masses tears the structure of an organ, then even though the cells may not be directly damaged themselves, normal physiology after thawing could in some cases be affected. Also, the lesions may become sites of entry for pathogens. Some locations in a plant can tolerate ice masses, often because the organ affected is dispensable, such as bud scales (Ashworth, 1990), or because the void created by the ice can be tolerated, such as in the core of some stems. Clearly, factors that favor or disfavor recrystallization may have a powerful role in controlling the locations at which such masses form. Plant apoplastic AFPs strongly inhibit recrystallization (Figure 5.1) and are effective at low concentrations: below 10 μg mL^{-1} for a *Lolium perenne* AFP and at 25 μg total protein mL^{-1} for a rye apoplastic extract (Griffith and Antikainen, 1996; Sidebottom et al., 2000). Thus, there is a strong possibility that a function of some AFPs *in vivo* is to inhibit recrystallization.

In contrast, AFPs from fish are capable, in certain concentrations and subcellular environments, of exerting effects different from, even counter to, antifreeze. Thus they can nucleate ice formation, as well as inhibiting it, and can both stabilize and destabilize cell membranes (Wang, 2000). Moreover, a plant β-1,3-endoglucanase protects isolated thylakoid membranes from freezing-induced damage (Hincha et al., 1997a). Thus, it is possible that plant AFPs are a multifunctional group.

5.3 EXTRACELLULAR-FREEZING-INDUCED DAMAGE

Freezing can damage plants by a variety of mechanisms, some disrupting gross structure or interfering with water transport (Pearce, 2001; Sakai and Larcher, 1987) and others acting at the cellular or subcellular level (see Chapter 6). The most widespread mechanism appears to be extracellular-freezing-induced cellular dehydration. It is the main cause of death in herbaceous plants, which have been the most extensively and intensively studied objects of molecular analysis of their adaptation to freezing (see Section 5.4).

The vapor pressure of ice is lower than that of unfrozen solution (see Chapters 1 and 20). As a consequence, once extracellular ice forms, it can grow by drawing water from cells until the vapor pressure of extracellular ice and subcellular compartments are equal, thus dehydrating the cell contents. The vapor pressure of ice falls as temperature falls, drawing more water from the cells and dehydrating them further. Hence, cellular dehydration becomes progressively greater as

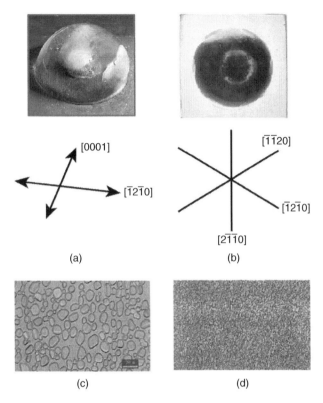

FIGURE 5.1 Binding of an antifreeze protein (AFP) from *Lolium perenne* to ice reduces ice recrystallization. A single ice crystal hemisphere grown in a dilute solution of the AFP was surface-etched by evaporation at subzero temperature. In general, where an AFP is absent the surface is mirror-smooth, whereas where it has become incorporated into the ice, the surface becomes etched, appearing similar to finely ground glass. When ice crystals are grown in the absence of AFPs, no such etched areas occur. In the present case, evaporation produced an etched pattern with sixfold symmetry, indicating that the AFP had bound to the ice, and did so specifically on the primary prism plane: (a) three elongated etched patches positioned on the primary prism plane; (b) planes symmetrically arranged around the crystal's c-axis (see Chapter 2 for explanation of ice crystal axes and principles of interactions with AFPs). The AFP had a high specific activity in an ice-recrystallization test: ice crystals were evident in a 30% sucrose solution held for 60 min at –6°C (c), but their growth was inhibited when <10 μg mL^{-1} was included in the solution (d). Scale bar 50 μm. (Reprinted by permission from *Nature*, Sidebottom et al., copyright 2000, Macmillan Publishers Ltd.)

temperature falls (Gusta et al., 1975; Pearce, 1988), down to a limit set by vitrification. Cooling in the field is slow enough for the vapor pressure of the cellular contents and extracellular ice to be at or close to equilibrium (Mazur, 1969). In some species cell walls partially resist the collapse in cellular volume, creating a divergence from equilibrium and reducing the extent of dehydration; however, substantial cellular dehydration still occurs (Zhu and Beck, 1991).

Many species show remarkable tolerance of this freezing-induced cellular dehydration. A simple equation describes the relationship between osmolarity of a solution in equilibrium with ice and the subzero temperature: osmolarity = degrees below 0°C/1.86 (Pitt, 1990). This equation can be used to get an idea of the cellular dehydration exerted by extracellular freezing in equilibrium with cellular contents. The contents of active plant cells generally have an osmolarity of below 1. From the above equation, at –18.6°C the intracellular osmolarity in equilibrium with extracellular ice would be 10. This increase in osmolarity in the cell will be caused by loss of water to the growing extracellular ice. Thus, at –18.6°C, which, for example, acclimated rye can tolerate, organelles, macromolecules, and solutes will be interacting with about one tenth of the water present in the nonstressed cell, and the water loss will also have resulted in a large volumetric reduction in the cell (Pearce, 1988).

Membrane structure is damaged when freezing-induced dehydration exceeds the dehydration tolerance of a cell (Levitt, 1980; Pearce and Willison, 1985a, 1985b; Steponkus, 1984; Steponkus et al., 1993). This damage is absent from supercooled cells (Pearce and Willison, 1985b), yet similar membrane damage is induced by desiccation at room temperature, indicating that the damage caused by freezing is caused by the desiccation involved, rather than the temperature (Pearce, 1985). The plasma membrane is often the membrane most vulnerable to this kind of damage; other membranes are also affected, but often at a lower temperatures (Pearce and Willison, 1985b). However, in some cases, tonoplast damage limits survival (Murai and Yoshida, 1998; Stout et al., 1980; Zhang and Willison, 1992).

Normal cellular function depends on maintaining an intact fluid-lamellar structure in lipoprotein membranes (Williams, 1990). Low temperatures or very low water contents can induce pure polar membrane lipids to undergo a transition from a fluid phase (L_α) to a gel phase (L_β). The latter would leak solutes and could also more broadly disrupt membrane organization. However, low hydration can, instead, induce pure polar membrane lipids to undergo a transition from the L_α phase to a nonlamellar phase, which is also inimitable to normal membrane structure and function. Formation of the normal fluid-lamellar structure of membranes depends on the presence of polar lipids such as phosphatidylcholine that have a cylindrical form at moderate temperatures and at normal levels of hydration. However, plasma membrane, for instance, also contains lipids such as phosphatidylethanolamine (PE) that have an inverted cone shape. Numbers of molecules of this shape together form inverted micelles (such as the hexagonal$_{II}$—hex$_{II}$—phase) rather than lamellae. Hence, the retention of the lamellar state in normal plasma membrane, for instance, despite the presence of PE, indicates that PE is normally dispersed in the membrane. This may be achieved by direct interaction with membrane-embedded proteins, stabilizing both the PE and the protein in the membrane (Williams, 1990).

Freeze–fracture methods have been used to detect local changes in membrane structure that could result from stress. Aparticulate domains are areas of membrane free of the particles representing membrane-embedded proteins. Aparticulate domains indicate phase separation in the plane of the membrane; that is, into particulate and aparticulate domains, at least. The aparticulate domains may or may not also differ in lipid phase (such as L_β) or in lipid composition, depending on what causes them. Abundant nonlamellar structures can easily be detected; thus, stacks of tubules with a diameter of about 8 nm would indicate the hex$_{II}$ phase. The term "lipidic particle" covers a number of interconvertible nonbilayer structures appearing as a variety of pits or pimples surrounded by bilayer (Verkleij, 1984). These structures locally involve much less of the membrane in a conversion to the nonbilayer phase than does hex$_{II}$, but they can create holes, create additional lamellae, facilitate flow of lipids from one lamella to another, and facilitate vesiculation.

The mechanism of damage in leaf cells does indeed involve the formation of aparticulate areas surrounded by apparently normal particulate areas, indicating phase separation in the plane of the membrane. The aparticulate areas are associated with underlying lamellae similarly showing aparticulate areas. This appearance was first described by Pearce and Willison (1985b) in both nonacclimated (Figure 5.2) and acclimated leaves given a freezing stress, and was later named "fracture jump lesion" by Steponkus et al. (1993). Aparticulate domains and fracture jump lesions are associated with lipidic particles in both freezing-stressed and dehydration-stressed leaves. They are also centers of membrane disorganization (Pearce, 1985; Pearce and Willison, 1985b; Figure 5.2). Clearly, a causal sequence is possible: stress – phase separation – lipidic particles – membrane disorganization – solute leakage and cell death (Pearce and Willison, 1985b). We speculated that phase separation could be a consequence of disruption to normal interactions between the cytoskeleton and membrane-embedded proteins, altering their distribution. The localized loss of proteins might then have freed inverted cone-shaped lipids to combine to form lipidic particles, leading to rearrangement and disorganization of membranes.

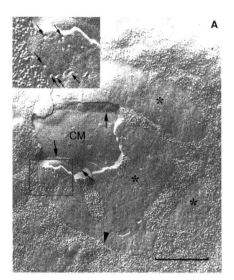
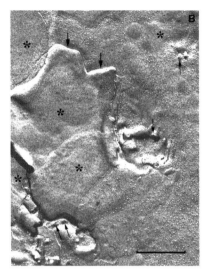

FIGURE 5.2 Disruption of plasma membrane structure by extracellular freezing, as shown by freeze–fracture electron microscopy. This technique reveals the inner (hydrophobic) plane of any lipoprotein membrane, which appears as surface carrying particles that are caused by membrane-embedded proteins. In these micrographs, the predominant surface is the inner protoplasmic face (the P face) of the plasma membrane. This has the protoplasm behind it. In normal (nonstressed) plasma membrane (not shown), this surface would carry an even scattering of the membrane-embedded protein particles. The micrographs show plasma membrane from extracellularly frozen tissue. It is clear that freezing had altered the normal plasma membrane organization, creating areas of membrane that contain no membrane-embedded protein particles (asterisks; the arrow head in [A] shows the margin between example particulate and aparticulate areas). Furthermore, these protein-free areas of membrane are associated with other particle-free membranes, both to the inside (A) and outside (B) of the plasma membrane, symptomatic of membrane reorganization. (A) Fracture-jump lesion where the fracture plane has jumped from the plasma membrane to an adjacent cytoplasmic membrane (CM). The cytoplasmic membrane revealed is, like the adjacent plasma membrane areas, mainly free of embedded protein particles. The boxed area is enlarged to show pits and grooves (small arrows) in the particle-free area of the plasma membrane, indicating "lipidic particles"; that is, nonlamellar structures that could be regions of connection between the plasma membrane and adjacent lamellae. Also note that in the main figure the protein particles present in the cytoplasmic membrane are located in bands (large arrows) that continue the bands of particles seen in the plasma membrane. This may indicate that freezing created the particle distribution in both membranes through an effect on intervening cytosol. (B) Fracture jump lesion where particle-free membranous lamellae lie on the outer side of the plasma membrane (e.g., asterisk top left; the plane of the plasma membrane itself is confirmed by the presence of plasmodesmatal openings—small arrows). An extensive particle-free area (left of center) clearly consists of a multilamellar fold (large arrows indicate the rounded edge of the fold). To create the lamellae and fold, membranous material must have been redeployed from the plane of the plasma membrane. Experimental details: Wheat plants were grown in a nonacclimating environment; the leaf sheath was excised, exposed to a freezing stress of –8°C, freeze-fixed, and freeze-fractured, and a replica of the fracture surface was examined by transmission electron microscopy. Scale = 0.5 μm. (Reprinted from Pearce and Willison, 1985b.)

Unlike Pearce and Willison (1985b), Gordon-Kamm and Steponkus (1984) found that freezing-stressed nonacclimated cells contained abundant hex$_{II}$ instead of fracture jump lesions. Thus, they thought that fracture jump lesions were confined to stressed acclimated leaves, indicating that the mechanisms of damage in acclimated and nonacclimated leaves may be fundamentally different (Steponkus et al., 1993), whereas Pearce and Willison (1985b) thought they were in principle the same.

Steponkus et al. (1993) used a cooling rate of –48°C/h (0.8°C/s), which is an order of magnitude faster than is usual in nature, but then allowed 30 min equilibration at the nadir temperature. They assumed, incorrectly, that 30 min equilibration would also be required with the much slower cooling rates used by Pearce and Willison (1985b; –1.5°C/h down to –2°C and rising to 7°C/h at lower temperatures), who they therefore suggested had not allowed sufficient dehydration to occur. The cooling treatment used by Pearce and Willison (1985b) had, in fact, caused severe dehydration in the nonacclimated leaves (Pearce and Willison, 1985a). The more probable explanation for the discrepancy in results with nonacclimated leaves is that the abundant hex$_{II}$ found by Steponkus et al. (1993) was an artifact of their unrealistically fast cooling rate. Replacing 48° with 4°C/h in the treatment protocol leads to replacement of hex$_{II}$ by fracture jump lesions (Pihakaski-Maunsbach and Kukkonen, 1997).

In the formation of hex$_{II}$, most of the membrane lipid loses its lamellar structure, whereas with formation of fracture jump lesions or vesiculation, most of the lipid remains lamellar but is reorganized into vesicles and stacks of lamellae. However, this difference deflects attention from an important similarity. Lipidic particles would be involved in the formation of abundant hex$_{II}$-phase lipid (Verkleij, 1984) just as much as in the formation of vesicles and multilamellar stacks. Clearly, in either case, steps in acclimation that would stabilize normal membrane structures by preventing formation of lipidic particles should be protective.

There is much less idea of exactly how freezing kills cells with no ability to acclimate. It was widely thought that intracellular freezing occurred in freezing-sensitive plants and that it was this that killed them. However, recent experiments show that in leaves of tomato, tobacco, cucumber, and phaseolus bean, freezing is extracellular and that it is thus the cellular desiccation that kills the cells (Ashworth and Pearce, 2002). Unfortunately, it is not known whether the cells are killed in the same way as in the freeze-adapted species, by dehydration-induced destabilization of membrane bilayers, or whether there is some other form of dehydration-induced damage in freezing-sensitive species. Interestingly, freezing in maize was more complex: the mesophyll froze extracellularly, whereas the epidermis and vascular bundle sheath froze intracellularly (Ashworth and Pearce, 2002). Thus, no single strategy for improving freezing tolerance would suit all species.

5.4 ACCLIMATION TO COLD AND FREEZING

Acclimation to cold and freezing is initiated by exposure to low positive temperatures. Thus, with respect to freezing, it is "anticipatory," occurring before the stress and not solely in response to it. However, exposure to freezing can further increase the level of freezing tolerance and associated gene expression (Pearce et al., 1996). In many woody plants, declining day length is also a factor (Sakai and Larcher, 1987).

Acclimation to cold is not an all-or-nothing phenomenon. For example, acclimation of barley only occurs when temperatures fall below some upper limit, the amount of freezing tolerance and gene expression then being progressively higher in plants grown in progressively lower temperatures (Pearce et al., 1996). Figure 5.3 shows data from barley crowns: The changes in level of freezing tolerance and in the expression of three gene sequences followed this pattern of progressively higher expression with lower temperature.

This apparently simple system masks complexity. Two of the cold-expressed barley genes, blt101 and blt14, had the same upper temperature limit for expression—between a night temperature of +9° and +10°C—and had similar relative responses of expression to the lower temperatures (Figure 5.3). This would not necessarily have been expected, as their expression is controlled by different mechanisms—transcriptionally and posttranscriptionally, respectively (Dunn et al., 1994). The third gene, blt4 (which is a nonspecific lipid transfer protein gene; Section 5.4.2.2), is constitutively expressed at modest levels in control plants. However, expression again increased with fall in temperature, following approximately the same pattern as the other two genes (Figure 5.3). Again, this would not necessarily be expected, as the genes are expressed in different tissues: blt4 in the epidermis, and blt14 and blt101 around the vascular transition zone (Pearce et al., 1998). It follows that there is an integrated response to cold acting through more than one regulatory pathway (Section 5.6).

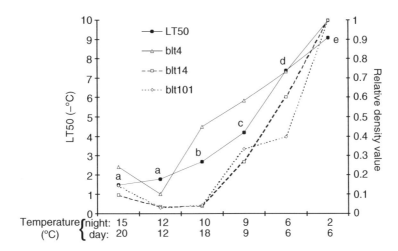

FIGURE 5.3 Acclimation to cold is not all or nothing: It increases progressively with fall in temperature below an upper limit. Barley plants were grown in a control environment for 21 d and then transferred for 7 d to the temperature environment shown. Crowns were then assessed for freezing tolerance and for mRNA expression of the cold-response genes *blt4*, *blt14*, and *blt101*. The left *y*-axis shows freezing tolerance as LT_{50}, and the right *y*-axis shows the level of mRNA as relative density values (calculated from densitometric analysis of Northern blots). The *x*-axis shows the day and night temperatures (using a 10-h photoperiod). The letters by the LT_{50} data points are for comparison of temperature effects on LT_{50}: where the same letters occur there is no difference; where the letters are different the LT_{50} values are different at the 5% significance level. The gene expression data are single observations based on extracts from 15 crowns; the same pattern of gene expression was obtained after 14 d acclimation and was also found when the experiment was repeated with a different preacclimation environment. Drawn using data from Pearce et al. (1996) with permission of Blackwell Publishing.

The physiological response is also complex: Acclimation to freezing is accompanied by acclimation to other stresses and by other changes. Plants in the field during winter will be exposed to other winter stresses as well as to freezing. Examples are wind, which can exacerbate damage caused by freezing and also causes damage of its own, and waterlogging, which reduces oxygen availability to roots. In addition, cold in interaction with day-length can induce developmental changes. Finally, the cold that induces acclimation also greatly reduces the plant's rate of metabolism and growth, and consequently the response to cold is partly homeostatic and compensatory. This component of the response is important for plant competition in the field but may only indirectly influence cell survival of stress. Thus, not all the physiological processes and not all the genes expressed in the field in autumn or in a cold environment in the laboratory will be involved in conferring freezing tolerance; many will be related to these other adaptations to winter.

5.4.1 OUTLINE OF PHYSIOLOGY AND BIOCHEMISTRY OF ACCLIMATION

Freezing-tolerant plants are of two broad types: those that have the potential to continue to grow (slowly) in milder periods during winter and those that become innately dormant. The latter include woody species with the greatest known tolerance of freezing. However, more is understood about acclimation in plants that remain potentially active.

The growth rate and metabolism of plants that remain potentially active is initially greatly reduced by cold but then partially recovers. At the physiological and biochemical level, this partial recovery is reflected in an increase in respiration, photosynthesis, and the protein synthetic machinery (Guy, 1990). This general upregulation of primary metabolism presumably is partly helpful for sustaining some growth in the cold, but it may also be needed to support the processes underlying increases in stress tolerance.

Other changes during acclimation are broadly similar in both winter-dormant and winter-active species. Cells accumulate solutes, particularly soluble carbohydrates but also others such as free amino acids (Levitt, 1980; Sakai and Larcher, 1987). The exact solutes accumulated vary; for example, among sugars, it is only the raffinose-family of oligosaccharides whose accumulation in a number of woody species correlates with frost tolerance (Stushnoff et al., 1993), whereas in overwintering grasses it is only sucrose and fructans that accumulate (Pollock and Jones, 1979). There are several possible advantages of solute accumulation during acclimation (see Chapter 10). It would undoubtedly have a colligative effect, reducing the amount of water lost from cells during extracellular-freezing-induced dehydration. Solutes such as sucrose, trehalose, and fructans can also stabilize macromolecules and membranes by noncolligative mechanisms (Crowe et al., 1992; Hincha et al., 2000; Strauss and Hauser, 1986). In addition, solutes form a reserve that can be mobilized to support rapid growth when the environment becomes warmer (Pollock and Jones, 1979) or to support recovery from sublethal injury (Eagles et al., 1993).

The imino acid proline is commonly accumulated during acclimation to cold and other desiccating stresses. Selection *in vitro* of hydroxyproline-resistant lines of wheat and barley gave variants that overexpressed proline and had enhanced freezing tolerance of up to 3°C (Dörffling et al., 1993). The correlation between these two characteristics was inherited over three generations, thus indicating a causal connection between proline accumulation and enhanced freezing tolerance (Dörffling et al., 1997). Further evidence that proline accumulation improves freezing tolerance was provided by transformation of *A. thaliana* to express an antisense sequence to the proline dehydrogenase gene sequence (Nanjo et al., 1999). This led to a reduced level of the corresponding enzyme, whose role is to catabolize proline. Hence proline accumulated, which caused an increase in freezing tolerance.

Membrane properties are a central factor in cold and freezing tolerance, as stress may cause transitions from L_α to L_β or to nonlamellar phases, disrupting normal membrane structure and function (Section 5.3). However, unsaturation has opposite effects on the two types of phase transition. In model membranes containing only one or two polar lipids, a greater unsaturation of lipid acyl groups causes the transition from L_α to L_β to occur at a lower temperature. This would be relevant to plant tolerance of low positive temperatures. In freezing stress, tolerance of low hydration levels is also important (Section 5.3). In this case, greater unsaturation of membrane lipid species could defer the L_α to L_β phase transition to a lower hydration state. In contrast, greater unsaturation, by making lipid species take a more inverted conical molecular shape, increases the probability of transition from the L_α phase to a nonlamellar phase.

Acclimation involves significant changes in membrane lipid composition. Plants grown in lower temperatures increase the unsaturation of their polar membrane lipids, but there is not a simple relationship between unsaturation, membrane fluidity, and adaptation or acclimation to cold. This is partly because the unsaturation of the acyl group at each position in different classes of polar lipid has to be considered and also because other membrane components, such as sterols and proteins, can also influence membrane phase transitions. However, adaptation to chill (low nonfreezing temperature) is consistently associated with a high level of unsaturation of phosphatidylglycerols in thylakoid membranes (Murata, 1983). Genetic manipulation to alter phosphatidylglycerol unsaturation altered chill tolerance, proving its importance for tolerance of this stress (Murata et al., 1992).

The unsaturation of polar lipids is also important for freezing tolerance. Acclimation of rye to freezing caused an increase in the content of phospholipids (PLs) in the plasma membrane of leaf cells and also caused changes in proportions of cerebrosides, free sterols, sterol glucosides, and acylated sterol glucosides (Lynch and Steponkus, 1987). However, relative proportions of PE and other phospholipids did not change appreciably, indicating that head group size, which can affect phase transitions, was not important. However, the proportions of diunsaturated PLs were higher in the acclimated plants (Lynch and Steponkus, 1987). To test the importance of this greater unsaturation, liposomes were made from plasma membrane PLs from leaves of acclimated rye plants and fused with the plasma membrane of rye leaf protoplasts from nonacclimated plants. This changed the lipid composition of the plasma membrane in a controlled way and conferred an increase in freezing tolerance. This effect could be reproduced by using liposomes comprised of diunsaturated phosphatidylcholines, thus demonstrating a role for unsaturation of plasma membrane PLs in acclimation to freezing (Steponkus et al., 1988).

A low level of oxidative stress is a constitutive phenomenon in organisms including plants, but cold, in common with many other stresses, can increase this level (Prasad, 1997). A number of the factors contributing to protection against oxidative stress are expressed at higher levels or with increased activity in cold-grown plants, including the potentially protective enzymes ascorbate peroxidase, glutathione reductase, and catalase, and protective substances such as ascorbate and reduced glutathione. Many of the reports of these changes are from species with little or only moderate chill tolerance, such as maize and tomato, when grown at chill-stress temperatures (Hull et al., 1997; Jahnke et al., 1991; Prasad, 1997; Prasad et al., 1994; Walker and McKersie, 1993). However, evidently it is also a factor in cold-acclimation of freezing-tolerant wheat (Bridger et al., 1994; Kocsy et al., 2000). Transgenic experiments confirm the practical importance of protection against oxidative stress in adaptation to stressful field environments. A Mn-superoxide dismutase from *Nicotiana plumbaginifolia* was constitutively expressed in alfalfa and targeted to chloroplasts or mitochondria (McKersie et al., 1993). This resulted in enhanced vigor and survival after exposure to sublethal freezing or water deficit in controlled environment tests (McKersie et al., 1993). The performances of transformants and control lines were also compared in the field in Southern Ontario, Canada, over 3 years. The transformants had significantly higher survival and yields: 26 to 66% (in different transgenic lines) of the transformed plants survived, compared with 17% of the untransformed plants (McKersie et al., 1996). However, tetrazolium staining showed that this improved survival in the field was not caused by protection at the primary site of freezing injury (McKersie et al., 1999). Overall, although oxidative stress is probably a factor in stress caused by low positive temperatures, this is less clear for freezing stress itself.

5.4.2 GENES* EXPRESSED DURING ACCLIMATION

Cold will have a pervasive effect on a plant, impinging on every aspect of its physiology and biochemistry, affecting constitutive processes such as primary metabolism as well as inducing changes that are protective or developmental. As a consequence, one would expect that expression of many genes will be affected, varying from small adjustments in levels of expression of some constitutively expressed genes, to induction of high levels of expression of genes specialized for roles in the cold. This expectation of a widespread effect has proved to be correct.

The outline of the physiology of acclimation, above, justifies more specific expectations. Some genes will have a role in relation to the upregulation of primary metabolism to sustain both constitutive processes, such as energy supply and some growth, and acclimation-specific processes, such as solute accumulation and adjustment of membrane composition. Others should have protective roles, particularly in relation to membrane stability but also more generally to stabilize macromolecules. Some genes would be involved in protection against more intense levels of constitutive stresses such as oxidative stress and damage to protein structure. There may, in addition, be expression of genes that address supracellular aspects of stress. These would include control of freezing (Section 5.2) and reducing gross tissue water loss. Also, acclimation must involve integration of the response, within cells and between organs, and therefore genes involved in a sophisticated signaling and regulatory system should be found. Finally, the extensive nature of the molecular response indicates that the new and unexpected will be involved.

---

* When discussing genes and proteins, the convention is used that names in italics refer to genes (base sequences), whereas names not in italics refer to the protein coded for.

Guy et al. made the first report in 1985 of a gene having cold-upregulated expression. By 1998 there was a count of 51 widely different types of gene (classified into types on the basis of sequence similarity) reported in a variety of plants during cold acclimation, and usually each type included several to many individual genes (Pearce, 1999).

Now microarray technology is accelerating discovery. It has verified principles, for example, that both defense and metabolism are involved in acclimation to cold, and has added new understanding by quantifying the great extent and general character of the response to cold (Fowler and Thomashow, 2002). Thus, the expression of 4% of *A. thaliana* genes changed by > threefold during cold acclimation mostly upregulated, but about one third were downregulated. In 35% of cold-responsive genes, the change in expression was long term; in 65% it was only transient, including many transcription factors, signaling, and communication genes, possibly indicating extensive homeostatic processes. Among the cold-responsive genes coding for transcription factors, expressed long term or transiently, many were not previously associated with cold acclimation, verifying its complexity. Finally, microarray data added 46 to the list of cold-associated genes of a novel type yet to be understood.

5.4.2.1 Genes Involved in Metabolism, Structure, and Regulation

The likely subcellular function of a number of cold-induced genes is indicated by their strong sequence similarity to genes of already known biochemical function. These include genes of types that are otherwise constitutively required, such as forms of elongation factor 1α, which is essential for protein synthesis (Berberich et al., 1995; Dunn et al., 1993; Rorat et al., 1997), sucrose phosphate synthase (Guy et al., 1992; Reimholz et al., 1997), β-tubulin (Chu et al., 1993), and xyloglucan endotransglycosylase, which can modify cell wall properties (Polinsky and Braam, 1996; Xu et al., 1996a). In these cases the cold-induced genes are forms of genes that are also required by the plant when it is not under stress, but presumably by having one or more extra genes of the type expressed under stress, some advantage is gained (Section 5.4.2.4). Thus, a form of alcohol dehydrogenase is cold induced (Jarillo et al., 1993), and presumably, this could help overcome oxygen-deficiency-induced energy supply problems resulting from waterlogging or ice encasement.

Some of the cold-upregulated genes are directly involved in stress-resistance, such as ascorbate peroxidase (Zhang et al., 1997), contributing protection against oxidative stress; cyclophilin, which helps to repair proteins (Droual et al., 1997); and galactinol synthase, contributing to solute accumulation (Liu et al., 1997). Yet others are known to have regulatory functions, such as protein kinases (Holoppa and Walker-Simmons, 1995; Hong et al., 1997; Monroy and Dhindsa, 1995) and CBF/DREB1 transcription factors (Sections 5.4.2.6 and 5.6.2.2). Others also may have a regulatory function: 14-3-3, a possible kinase regulator (Jarillo et al., 1994), and RNA-binding proteins (Carpenter et al., 1994; Dunn et al., 1996; Molina et al., 1997; Section 5.6.2.2).

Functionally important changes occur in plasma membrane PL unsaturation during acclimation (Section 5.4.1). The first double bond is introduced into acyl groups by Δ^9-desaturases, and this has a greater lowering effect on the L_α-L_β transition temperature than introduction of subsequent double bonds. *A. thaliana* contains two temperature-responsive Δ^9-desaturase genes: expression of *ADS2* is upregulated by cold ($10°C$), whereas expression of *ADS1* is downregulated (Fukuchi-Mizutani et al., 1998). An unusual feature is that both are more like animal, yeast, and cyanobacterial Δ^9-desaturases than like plant plastid Δ^9-desaturases and, consequently, were suggested to have a unique role in the complex plant metabolic system for lipid desaturation. The third double bond is introduced by ω^3-desaturases. Expression of several ω^3-desaturases is affected by temperature. In *A. thaliana*, the expression of the relevant gene, *FAD8*, is upregulated below $20°C$; Gibson et al. (1994) point out that its function could be a downregulation in relation to high-temperature stress rather than an upregulation in anticipation of chill temperatures. In contrast, in maize, which is neither freezing tolerant nor strongly chill tolerant, expression of Zm*FAD8* is upregulated at $5°C$ (Berberich et al., 1998). Interestingly, the amount of ω^3-desaturase in wheat roots is under temperature-dependent translational control (Horiguchi et al., 2000).

Desaturation can be controlled by temperature in other ways. Increased availability of oxygen, alone, at low temperatures can promote higher rates of desaturation (Harris and James, 1969; Rebeille et al., 1980). Speculatively, temperature-dependent activation of preexisting enzyme molecules might also contribute to desaturation, as it does with a Δ^9-desaturase in carp (Tiku et al., 1996).

5.4.2.2   Genes Coding for Likely Protective Proteins

There are a large number of genes that seem likely to have protective roles at a variety of organizational levels within the plant. The most commonly reported of these are LEA (late-embryogenesis-abundant) and LEA-like proteins, nonspecific lipid transfer proteins, and proteins with antifreeze activity (Section 5.2.2). Lectins could also be protective (Hincha et al., 1997b).

5.4.2.2.1   LEA Proteins

LEA proteins were first described from developing seeds, where they are accumulated during the period of rapid desiccation. However, a number of LEA proteins are also expressed in vegetative tissues in response to cold and dehydrating stresses.

LEA proteins are highly hydrophilic and are boiling-stable. The amino acid sequence comprises repeated amino acid sequence motifs separated by much less strongly conserved sequences. Their tolerance to denaturing conditions indicates that their role in stress may be to stabilize macromolecules in a low-water environment. On the basis of features of their primary sequences, several classes of LEA protein have been identified, and the protein structures inferred from these different sequences indicate possible protective functions (Dure, 1993a).

The most frequently reported type of cold-induced genes codes for group 2 (D-11 family) LEA proteins, also called dehydrins (Pearce, 1999, cites 20 references covering cold-induced dehydrins from 10 species, including woody as well as herbaceous plants). Their characteristic sequence motif is a 15-residue sequence rich in lysine (called the K-segment; Campbell and Close, 1997), which might form an alpha helix. However, in solution the interactions within the polypeptides are labile and the dehydrin molecules probably are in equilibrium between conformational states having preferentially extended substructures (Lisse et al., 1996).

Direct evidence of a protective function is limited. The cold-induced spinach leaf dehydrin, COR85, protects lactate dehydrogenase against freezing-induced denaturation *in vitro* (Kazuoka and Oeda, 1994); however, there appear to be no *in vitro* tests of membrane protection by dehydrins. Severe desiccation, by weakening the hydrophilic interactions between water and the surface of proteins and membranes, thus indirectly destabilizes the important hydrophobic interactions within the macromolecules. It has been speculated that dehydrins could stabilize these hydrophobic interactions in one of two ways (Close, 1996). They might interact directly with membrane or protein surfaces, through their K-segments, and thus, as hydrophilic molecules themselves, increase hydrophilic interactions with the surface. Alternatively, they could act without direct interaction with the surface: as hydrophilic molecules, they would interact strongly with the remaining aqueous phase, and thus would be excluded from the hydrophobic interior of proteins and membranes, thus stabilizing these proteins and membranes, as the exclusion of water normally does (Close, 1996).

Other LEA and LEA-like proteins are also important. Group 3 LEAs are suggested, when dimerized, to sequester ions (Dure, 1993b). This could be an important function in the cytosol of a dehydrated cell, as dehydration would concentrate ions to levels where they could be toxic. A group 3 (D-7 family) LEA protein, HVA1, is cold-induced in barley (Sutton et al., 1992). Overexpression of HVA1 increased the salt and drought tolerance of rice (Xu et al., 1996b). There are a number of cold-induced LEA-like proteins, too, such as cor15, which has similar sequence features to group 3 LEAs. *In vitro* cor15 can protect lactate dehydrogenase from freezing-induced denaturation (Lin and Thomashow, 1992), and *in vivo* overexpression of *cor15* confers a small increase in freezing tolerance, possibly through stabilizing chloroplast membranes (Section 5.4.2.6).

5.4.2.2.2 Nonspecific Lipid Transfer Proteins

Nonspecific lipid transfer proteins (nsLTPs) have the ability to bind certain lipids and thus to transfer them across an aqueous environment. NsLTPs are expressed in nonstressed plants, in which they probably have a constitutive role (Kader, 1996). They are also strongly expressed, above their constitutive level, in response to stresses such as cold, salt, and drought (Dunn et al., 1991; Hughes et al., 1992; Plant et al., 1997; White et al., 1994).

NsLTPs are extracellular. Their gene sequences encode polypeptides that include a leader sequence targeting the protein to the secretary pathway. In nonstressed plants the mature protein is present in the cuticle (Thoma et al., 1993, 1994) or wax layer (Pyee et al., 1994) or is present perivascularly (Sossountzov et al., 1991). Cold-upregulated nsLTP mRNAs are accumulated in the epidermis (Pearce et al., 1998), indicating a likely epidermal location and function for the mature protein.

A number of experiments indicate two possible functions for nsLTPs: in pathology or in cuticle formation (Kader, 1996). Inclusion of the protein in microbial culture media inhibits the growth of microbes, including plant pathogens (Molina et al., 1993). Barley plants upregulate expression of nsLTP mRNAs in response to pathogen attack; however, the response is not pathogen-specific (Molina and García-Olmedo, 1993), indicating a nonspecific role in cold-induced pathogen resistance. Carrot culture nsLTPs bind potential cutin monomers, possibly indicating a role in cuticle formation (Meijer et al., 1993). Lipid polymers are known to accumulate in the cuticle during acclimation (Griffith et al., 1985), but the role of nsLTPs in this process has not been experimentally defined. Both suggested functions are consistent with the location at which nsLTPs accumulate. Moreover, both could be relevant to the whole-plant survival of winter, as on the one hand crops such as cereals can be attacked by snow-mold, and on the other, wind and winter sun increase water loss from the shoot, exacerbating the dehydration stress caused by freezing.

A further function for some nsLTPs is also possible. A protein isolated from cabbage protects isolated thylakoids against freezing-induced damage. It was originally called "cryoprotectin" but is structurally a nsLTP (Hincha et al., 2001). However, it does not have lipid transfer properties, whereas a closely related nsLTP also from cabbage does have lipid transfer properties but is not cryoprotective, indicating that different members of the nsLTP multigene family (Section 5.4.2.4) may be specialized for different biochemical functions.

5.4.2.3 Genes of Unknown Biochemical Function

There are many cold-induced genes of this kind (Pearce, 1999). An overall picture is hard to provide, but the following illustrates the range of features of the proteins coded for. Some code for hydrophilic proteins, including ones that are also boiling-stable but have not been classified as LEA proteins. Nevertheless, the proteins coded for might be protective in a similar way to LEA proteins. This includes *rd29a* (Horvath et al., 1993; Luo et al., 1991; Nordin et al., 1991) and *kin1* and *kin2* (Hajela et al., 1990; Kurkela and Borg-Franck, 1992; Kurkela and Franck, 1990). The latter code for proteins rich in alanine. This could have indicated antifreeze activity, but when a similar protein, BN28 from oil seed rape, was tested, it was found not to have antifreeze activity (Boothe et al., 1997). In contrast to these genes coding for hydrophilic proteins, a number of genes code for much more hydrophobic proteins. An example is *blt101* from barley and similar genes from grasses, strawberry, and *A. thaliana* (Goddard et al., 1993; Medina et al., 2001; Ndong et al., 1997). These code for small polypeptides with two major hydrophobic regions. It is likely that the protein would form a V-shaped molecule, with two membrane-spanning regions and little of the amino acid sequence lying outside the membrane. One could speculate on a role in membrane stabilization, but no hard evidence addresses this possibility. Other genes code for proteins that may be targeted to the nucleus (*cas15* in alfalfa, Monroy et al., 1993a; *src1* in soybean, Takahashi and Shimosaka, 1997). Yet another has sequence similarity to a conserved breast-cancer-related gene from humans (*BnC24* in oil seed rape, SaezVasquez et al., 1993). Altogether, whether the hints of function are reliable or not, the structural diversity of these genes must indicate a diversity of subcellular functions and confirms that our knowledge of functional areas of acclimation to cold and freezing will need significant effort to complete.

5.4.2.4 Multigene Families

Multigene families are probably central to phenotypic plasticity (Smith, 1990) and thus are a key factor in adaptation to cold and freezing. Many cold-induced genes are members of small multigene families; that is, they belong to a group of genes coding for closely similar proteins. It is thought there could be several advantages to this. The existence of more than one copy of a gene allows evolution to diversify biochemical function, which may have occurred with AFPs and nsLTPs (Sections 5.2.2 and 5.4.2.2). This would create structural variation within the small multigene family. The different members could have different physiological functions; for example, being expressed in different environments or different tissues or organs. Existing as separate genes, the regulation of their expression could be fine-tuned to their physiological role. Another possible advantage is that by having more genes, the expression of the totality of all forms could increase much more rapidly in response to a signal.

Barley provides examples where these advantages appear to apply. Eleven barley dehydrins differ from each other markedly in size and sequence features, consistent with diversification of biochemical function, and different ones are expressed in response to drought or cold or both, indicating physiological specialization (Choi et al., 1999). The latter is also apparent in three nonspecific lipid transfer protein genes analyzed in barley. All were strongly expressed in response to cold; one was also strongly expressed in response to drought, whereas the other two were, respectively, moderately and only weakly expressed in response to drought (White et al., 1994). Another form of functional specialization involves differential expression in different organs, as indicated by the different genes of types *rlt14* and *blt14*, in rye and in barley, respectively (Phillips et al., 1997; Zhang et al., 1993). However, in all these examples, the exact advantage of having the different structures or expression patterns is not yet clear.

5.4.2.5 Apoplastic Proteins

Many of the proteins expressed during cold acclimation are outside the protoplast. Several lines of evidence indicate this. Electrophoresis of protein extracts from the apoplast of leaves of cold acclimated cereals shows the accumulation of specific polypeptides during acclimation (Antikainen and Griffith, 1997; Hon et al., 1994). Polypeptides coded for by a number of cold-upregulated mRNAs include leader sequences probably directing the polypeptide to the secretory pathway. The most intensively studied AFPs are extracellular, and so too are the nonspecific lipid transfer proteins, as discussed earlier (Sections 5.2.2 and 5.4.2.2, respectively). The role of some cold-expressed proteins may be to directly modify cell wall structure, for example, a xyloglucan endotransglycosylase from *A. thaliana* (Xu et al., 1996a). Other apoplastic proteins include many of unknown function such as MSACIA, which is glycine-rich and accumulates in cell walls during acclimation of alfalfa (Ferullo et al., 1997). The gene *BnPRP* codes for a protein whose structure indicates that it could link cell wall and plasma membrane (Goodwin et al., 1996). Overall, though the biological roles are only partly understood, change in the apoplastic complement of proteins appears to be a significant event during acclimation, presumably reflecting the importance of this compartment in acclimation.

5.4.2.6 Evidence of Physiological Function of Cold-Expressed Genes

As discussed above, the precise biochemical function of cold-upregulated genes may or may not be evident. In either case, the physiological function, for example, the role in relation to acclimation to freezing rather than to some other aspect of physiology, remains to be proved. Experiments that test for correlation between gene expression and freezing tolerance can be only a partial, and potentially misleading, guide to physiological function. Transgenic experiments provide more powerful tests.

5.4.2.6.1 Transgenic Tests of Function of Cold-Upregulated Genes

The only strict test of the role of a gene in acclimation is by testing the effect of downregulating its expression in the species from which the gene came. If reduced expression leads to reduced acclimation, a role in acclimation is indicated. However, a lack of effect is more difficult to interpret if, as is usually the case, the gene is a member of a multigene family, as other members of the multigene family may take over the function of the downregulated member. Although this problem could be overcome, downregulation experiments do not seem to have been tried.

The effect of upregulation of expression on freezing tolerance has been tested. If this enhances freezing tolerance, it simultaneously indicates that this may be the physiological role of the gene and that conferring this effect on other species might be tried. However, negative effects are again hard to interpret (and, perhaps for this reason, are not reported). At present, we have little idea of how many of the cold-upregulated genes act additively and how many act synergistically. Expression of a single gene having a synergistic role in acclimation, but in the absence of upregulation of expression of any other genes, may have no effect on the level of freezing tolerance. Even for genes that act additively, the effect of an individual gene could be too small to detect.

The most marked transgenic evidence for a role in acclimation to freezing is for the DREB (also called CBF)[4] transcription factor genes cloned from *A. thaliana*. These transcription factors are expressed in response to cold, drought, or salt stresses and bind to, and thus activate, the promoter of many stress-upregulated genes (Section 5.6). Overexpression of DREB1A and DREB1B (CBF3 and CBF1, respectively)* confers constitutive freezing tolerance on *A. thaliana*, as is shown in Figure 5.4. Significantly, the level of tolerance achieved can be equivalent to the acclimated freezing tolerance of the wild-type (Jaglo-Ottoson et al., 1998). Overexpression of these transcription factors also induces expression of biochemical changes usually associated with acclimation: expression of cold-upregulated genes, accumulation of proline, and accumulation of soluble sugars (Gilmour et al., 2000; Jaglo-Ottoson et al., 1998; Kasuga et al., 1999). It also confers enhanced tolerance to drought (Figure 5.4), consistent with the expected importance of dehydration tolerance in tolerance of freezing. The clear-cut results for overexpression of these genes is probably a result of their key position in the cold signal transduction pathway (Section 5.6), where they induce expression of a large number of the separate functional components of acclimation, so that the overall effect on freezing tolerance is large.

The *A. thaliana* gene *cor15a* codes for a small chloroplast-located polypeptide that has some sequence similarity, particularly in imperfectly repeated motifs, to *lea* genes (Lin and Thomashow, 1992). Its mRNA transcripts accumulate in *A. thaliana* during cold acclimation. Constitutive expression of *cor15a* in the chloroplast of transformed *A. thaliana* plants resulted in greater freezing tolerance of protoplasts isolated from the leaves and tested *in vitro* (Artus et al., 1996). Chloroplasts tested *in situ* in the control and the transgenic plants were 1° to 2°C more tolerant in the latter, as indicated by amelioration of freezing-induced reduction in quantum yield of photosystem II. In protoplasts, improved tolerance of up to 1°C was found only over the temperature range –5 to –8°C, whereas over the temperature range –2 to –4°C, the freezing tolerance was lower in the transformants than in the wild-type (Artus et al., 1996). Thus the experiments indicate both a positive and negative effect of expression of this gene on freezing tolerance. This result may be confirmation that an accurate test of function cannot be made by overexpression of a single functional gene.

In a comparison of wheat chromosome substitution lines, antifreeze activity of apoplastic extracts did not correlate with freezing tolerance in a laboratory test but did positively correlate

---

* There are two sets of nomenclatures for the members of this small multigene family, differing between the two originating laboratories, based on the differing response element names, DRE-binding factor (DREB) or CRT-binding factor (CBF). DREB1A, DREB1B, DREB1C, which are identical to CBF3, CBF1, and CBF2, respectively, are similar in sequence and are cold inducible. DREB2A and DREB2B are different from DREB1 sequences. However, all the DREB sequences contain a similar core that codes for an AP2 domain, also found in a large number of other transcription factors having other functions. Additional members of the DREB1 and DREB2 types (up to DREB1F and DREB2H, respectively) have now been found in the arabidopsis genome (Sakuma et al., 2002).

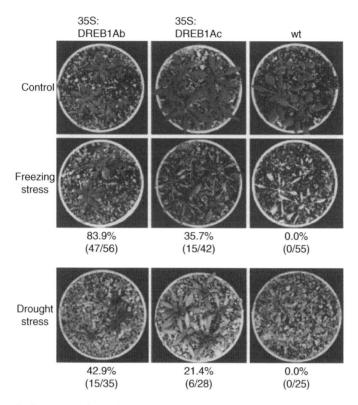

FIGURE 5.4 Constitutive expression of the cold-response transcription factor DREB1 confers greatly increased freezing tolerance on *Arabidopsis thaliana* and also increases drought tolerance in the absence of any cold or drought-induced acclimation. Percentage values indicate percentage of plants surviving the treatment; numbers in brackets indicate the number surviving out of the number tested. 35S:DREB1Ab and 35S:DREB1Ac indicate different lines transformed with the same construct to constitutively express DREB1A (see Liu et al. 1998, for data showing transgene expression). wt indicates wild-type (untransformed) plants. Control: 3-week-old plants grown in continuous light at 22°C. Freezing-stress: plants grown the same as the controls then exposed to –6°C for 2 d and returned to the control environment for 5 d. Drought stress: grown the same as the controls except water withheld in the second and third weeks. (Reproduced from Liu et al. (1998), copyrighted by the American Society of Plant Biologists and used with permission.)

with winter field survival (Chun et al., 1998). These correlative data indicate that plant AFPs may have a role in plant overwintering, but the precise role is unclear. However, infusion of a fish AFP into leaves lowered the leaf freezing temperature (Cutler et al., 1989). As a consequence, a number of attempts have been made to modify freezing behavior of plants by overexpressing AFPs. An AFP from carrot has been expressed in tobacco, conferring apoplastic antifreeze activity, but no effect on survival was reported (Worrall et al., 1998). Fish AFPs have also been expressed in plants, so far without any positive effect on survival that could be attributed to altered freezing behavior (Kenward et al., 1999; Wallis et al., 1997).

5.4.2.6.2 Correlative Evidence

Comparisons were made between levels of mRNAs or proteins coded for by particular cold-expressed genes and the level of freezing tolerance, using either contrasting plant types or contrasting environments to provide different levels of acclimation. Nearly all of these experiments showed a positive relationship between levels of expression and acclimation, including expression of dehydrins (five subtypes in a variety of species) and of five gene-types of unknown biochemical function (six subtypes again in a variety of species; summarized in Pearce, 1999). Such correlative tests provide weak evidence that any particular one of these genes influences acclimation. In contrast, the number of tests giving a positive correlation is impressive, and it is hard to imagine that plants would invest resources in such a range of gene products if many of them did not have a function during acclimation. This thus may be taken as evidence that, in general, cold-upregulated genes will have a role in acclimation to cold or winter. However, because different physiological aspects of acclimation to winter may themselves be correlated, correlation between gene expression and freezing tolerance is not necessarily indicative of a role in acclimation to freezing itself.

Correlations can also be sought between expression in different plant parts and their different freezing tolerances. A cereal plant can survive loss of its leaf blades or roots, but the survival of the crown is essential for plant survival. Within the crown, the vascular transition zone is one of the most susceptible parts (Tanino and McKersie, 1985). WCS120, a wheat dehydrin, is most strongly expressed in tissue surrounding the vascular strands in the vascular transition zone (Houde et al., 1995). The expression of two transcripts of unknown biochemical function, *blt101* and *blt14*, in the crown of acclimated barley was also strongest in these tissues (Pearce et al., 1998). These two studies link expression of the genes tested to acclimation of one of the most vulnerable parts of the shoot.

5.5 GENETIC ANALYSIS OF ADAPTATION TO FREEZING

Genes expressed during acclimation do not necessarily contribute to freezing tolerance; they may relate to other stresses or to development (Section 5.4). The possibility that these genes contribute to freezing tolerance may be tested by transgenic experiments, but so far few unambiguous results have been obtained (Section 5.4.2.6). An alternative, more traditional, approach to proving physiological function is to start with the phenotype rather than the gene. This can be done by isolating and studying freezing-tolerance mutants. Alternatively, one can analyze mapping populations for freeze-adaptation-related quantitative trait loci (QTL). The former can identify genes with a major role in the physiology and biochemistry of tolerance, whereas the latter can eventually lead to identification of genes that explain the evolution of tolerance. These approaches also have the advantage that they can identify not only the genes involved in acclimation but also any that are expressed constitutively. Interestingly, several of the mutations appear to be in regulatory genes.

5.5.1 MUTANTS

A. thaliana increases in frost hardiness by 6 degrees during acclimation. Seven *sfr* (susceptible-to-freezing) mutants were isolated that were less freezing tolerant than the wild-type by 1 to 2 degrees, indicating that the underlying genes had major roles in freezing tolerance of this species (McKown et al., 1996; Warren et al., 1996). The mutants were also tested for expression of typical features of *A. thaliana* acclimation: anthocyanin accumulation, sucrose and glucose accumulation, accumulation of three specific proteins, and fatty acyl composition of lipids. Four mutations affected one or more of these, including one (*sfr4*) that reduced both sucrose and glucose levels and prevented the changes in fatty acid composition. At the other extreme, three mutations affected none of these. When grown in the cold, most of the mutants expressed known cold-response genes. However, the *sfr6* mutant was different, being deficient in expression of several cold-response genes. This indicated that it could be a mutant in a gene that affects acclimation through controlling gene expression (Knight et al., 1999).

Even when not grown in the cold, plants have 1 to several degrees of freezing tolerance. A freezing-sensitive *A. thaliana* mutant, *frs1*, isolated by Llorente et al. (2000), had reduced levels of both constitutive and acclimated freezing tolerance. The phenotype was "wilty," and the plants were low, compared to the wild-type, in content of the plant desiccation-stress-related hormone, abscisic acid (ABA). Exogenous supply of ABA caused recovery of the wild-type phenotype. Complementation tests showed the mutant was an allele of the *ABA3* locus. The results clearly indicated a role for ABA in regulating both constitutive and acclimated levels of freezing tolerance (Llorente et al., 2000).

Mutants that have acquired an enhanced constitutive level of freezing tolerance are also informative. *A. thaliana* plants that carry a mutation at the eskimo 1 locus (*esk1*) accumulate high levels of proline, a protective solute, even when grown in a warm environment (Xin and Browse, 1998). Interestingly, the *esk1* mutants had become constitutively more freezing tolerant than the wild-type, yet had not acquired constitutively increased levels of expression of cold-response gene transcripts, other than those of genes involved in proline synthesis (Xin and Browse, 1998).

Strategies can be used to ensure recovery of mutations in the regulatory system. A transgenic line was made that expressed a reporter gene (luciferase) under the control of the stress-responsive *rd29a* promoter (Section 5.6.2.2). Mutant lines that had a modified luciferase expression were likely to contain mutations in the system regulating stress-gene expression. Large numbers of mutants were obtained, with enhanced or reduced expression of luciferase in response to one or more of cold, osmotic, or ABA signals, clearly indicating that parts of the respective signal transduction pathways were separate but that parts converged (Ishitani et al., 1997). Some of these mutations have been analyzed in detail. *Hos1* and *hos2* are negative regulators of the cold response; that is, the mutants had enhanced expression of the reporter gene. However, they differed in other respects: *hos1* also affected vernalization but did not affect the cold-induced level of freezing tolerance, whereas *hos2* did not affect vernalization but did reduce freezing tolerance (Ishitani et al., 1998; Lee et al., 1999, 2001).

5.5.2 QUANTITATIVE TRAIT LOCI

Freezing tolerance is a quantitative trait; that is, the intensity of freezing tolerance is a continuous variable. QTL are chromosome regions containing one or more genes that explain part of the quantitative trait. Their discovery involves creation and analysis of mapping populations. These are populations obtained by crossing inbred parents that differ in the genetic control of the trait of interest. Marker loci are established throughout the genome (Lister and Dean, 1993). The linkage of traits to these markers is used to statistically determine the location of the QTL, and the proportion of the quantitative trait that each QTL explains is calculated (Kearsey and Pooni, 1996).

Physiological and biochemical studies of acclimation and the numerous cold-responsive genes discussed in Section 5.4 attest the complexity of cold acclimation. Nevertheless, it would be possible for major differences in cold adaptation between species or cultivars to depend on allelic differences in a small subset of the genes involved. In cereals, in which winter hardiness has long been regarded as under complex genetic control, one chromosomal region can have a large effect. A quantitative trait locus on chromosome 7 of barley accounted for 37 to 68% of the variance in cold tolerance in the mapping population used (Hayes et al., 1993; Pan et al., 1994), indicating that there are gene(s) in this region of the chromosome with important role(s) in freezing tolerance. This same region contains a cluster of other winter-related genes, including ones controlling vernalization and sugar accumulation, and some dehydrin genes (Close, 1996; Pan et al., 1994). A different experimental approach has been used in wheat, using monosomic lines and substitution analysis. This showed that, although 10 of the 21 chromosomes are involved in control of frost tolerance and winter hardiness, chromosome 5A (which corresponds to chromosome 7 in barley) and chromosome 5D each carry a freezing-resistance gene with comparatively large allelic effects. As in barley, on chromosome 5A there are nearby genes controlling vernalization and cold-induced carbohydrate accumulation (Galiba et al., 1995, 1997; Roberts, 1990; Sutka, 1994).

The position of any QTL is determined with a 95% confidence interval covering a chromosomal region of tens of centimorgans. Regions of such a size contain very many individual genes, on the order of 10^3. In species that have a heavily mapped genome, it is often possible to identify genes within the QTL region that are candidates to explain the QTL. However, at present there will be many genes present that are of unknown biochemical type (roughly 30% of the predicted gene products coded for by the *A. thaliana* genome could not be assigned to functional categories; The Arabidopsis Genome Initiative, 2000). Thus, potential candidates may be overlooked. Moreover, more than one gene in a region may contribute to explaining the QTL. However, QTL, once identified, can be exploited by using marker-assisted selection or, potentially, by transferring parts of the region to other plants by cloning and transformation.

5.6 CONTROL OF ACCLIMATION

For plants to acclimate to a changed environment they must first detect a signal from their environment, then transduce this signal and transmit it to the components that then function to protect and maintain the plant. Transduction and transmission of the signal is through one or more signal transduction pathways (STPs). The functional system controlled is itself complex (Section 5.4). Hence, it is not surprising that signal transduction is also complex. Nevertheless, key steps in cold signaling are known.

5.6.1 COLD- AND WINTER-SENSING

How plants, and many other organisms, sense cold is unknown. A possible model is available in the cyanobacterium *Synechocystis*, in which cold is sensed through its effect on membrane fluidity. The membrane lipids of *Synechocystis* undergo a change in fluidity on cooling, and histidine protein kinases embedded in the plasma membrane are thought to sense this change and activate the STP (Nishida and Murata, 1996; Suzuki et al., 2000; Vigh et al., 1993). Mutant analysis showed there was a second, as yet unidentified, sensor (Suzuki et al., 2001). The temperature range involved in cold acclimation of *Synechocystis* involves a control temperature of 34°C and cold acclimation at 22°C. These temperatures are much higher than are required to acclimate temperate plants (below 10°C in barley; Figure 5.3). Thus it is not clear whether the cold sensors in *Synechocystis* would be relevant to acclimation to freezing in higher plants. However, a role for membrane fluidity is indicated by experiments with *Brassica napus* cell cultures. These show that membrane rigidification promotes acclimation, whereas fluidization prevents it (Sangwan et al., 2001). Interestingly, the experiments also showed that the cytoskeleton, which is disassembled by cold and which interacts with the cell membranes, may also have a role, promoting acclimation when destabilized and inhibiting it when rigidified.

There is evidence for other possibilities. The redox status of photosystem II is affected by cold, and this appears to control acclimation (Gray et al., 1997). However, this could be only part of the story, as the crowns of some plants, such as cereals, appear to be sensors of cold (Peacock, 1975; Watts, 1972), yet are nongreen. However, crowns and other sink organs could respond to signals from leaves. Thus, another hypothesis to test is that soluble carbohydrate supply to sink organs has a role. Cold and freezing cause a rapid rise in soluble carbohydrate content in the apoplast of source and sink organs (Livingston and Henson, 1998). In a barley cell culture, high soluble carbohydrate level in the medium was essential for acquisition of freezing tolerance and for upregulation of expression of cold-responsive mRNAs, indicating a regulatory role for sugars (Tabaei-Aghdaei et al., 2003).

The environmental factor sensed need not be cold itself, or not only cold: environmental or physiological correlants of cold could also be effective signals. Winter can expose plants to water stress. Experimentally, too, plants exposed to cold shock can experience water stress (Markhart et al., 1979), though this is slight in *Arabidopsis* (Lång et al., 1994). In barley, slight wilting is evident, and relative water content is reduced and stomatal resistance increased during several days of exposure to cold (Roberts and Pearce, unpublished data). Cold acclimation can involve changes in levels of the hormone ABA (Chandler and Robertson, 1994) and, conversely, ABA and drought can cause cold acclimation (Guy, 1990; Veisz et al., 1996). A mutation in ABA synthesis reduces the acclimatory response to cold (Xiong et al., 2001). Analysis of cold acclimation and of cold-induced gene expression in ABA-deficient and ABA-insensitive *A. thaliana* mutants indicates there are parallel STPs, one involving ABA, the other not (Gilmour and Thomashow, 1991; Gosti et al., 1995; Nordin et al., 1991). Furthermore, exposure to freezing after cold acclimation enhances the level of freezing tolerance and of gene expression (Pearce et al., 1998), indicating that the dehydrating effect of freezing might have a signaling role. Thus, a water-stress signal could be additional to or integral to the cold signal.

5.6.2 COLD SIGNAL TRANSDUCTION PATHWAYS

Three areas of research have been crucial in achieving current understanding of the plant STP. Some researchers have tested for involvement of factors already known to be involved in other STPs, including transient rises in cytosolic calcium and cold-induced expression of protein kinases. Others have used the promoter region of known cold-response genes as bait to bind and identify cold-response transcription factors. Third, mutant analysis has indicated converging and diverging regulatory factors (Section 5.1). Most progress has been made in two stages of the STP: surrounding a cold-induced calcium signal, an early stage in the cold STP, and surrounding the control of amounts of cold-specific mRNA, a late stage in the cold STP. The exact mechanisms connecting these stages are not yet clear.

5.6.2.1 Early Events

Cold shock causes a transient rise in the cytosolic calcium level (Knight et al., 1991, 1996). A variety of calcium influx transporters operate in plants (Bush, 1995), and mechanically operated channels are markedly affected by cold (Ding and Pickard, 1993). The cold-shock-induced transient rise in cytosolic calcium results partly from cold-induced opening of plasma membrane calcium channels, but influx from the vacuole also occurs (Knight et al., 1996; Lewis et al., 1997; Monroy and Dhindsa, 1995; Monroy et al., 1993b). The role of calcium could be partly homeostatic, helping integrate the adjustment of metabolism to stress, but it appears also to be essential for acclimation. Inhibitors of calcium influx from the apoplast prevent both cold acclimation and cold-induced expression of genes (Monroy et al., 1993b; Polinsky and Braam, 1996), and experimentally induced calcium influx in the absence of cold triggers the expression of cold-responsive genes (Monroy and Dhindsa, 1995).

The calcium signal would be expected to alter protein phosphorylation. The evidence supports this, as cold acclimation causes phosphorylation of a number of proteins (Monroy and Dhindsa, 1995; Monroy et al., 1993b) and several cold-responsive protein kinases have been identified (Holoppa and Walker-Simmons, 1995; Hong et al., 1997; Monroy and Dhindsa, 1995). Mitogen-activated protein (MAP) kinases have been implicated in stress STPs in a variety of organisms. In *A. thaliana*, two are transiently activated by cold stimuli (Ichimura et al., 2000).

5.6.2.2 Immediate Controls of mRNA Levels

Acclimation involves increased levels of expression of many genes, evident as an increase in amount of the corresponding mRNA. It is reasonable to think that understanding how this is controlled would provide considerable insight into how acclimation as a whole is controlled. Experiments in both *A. thaliana* and barley show that the transcript amounts of only about half of the cold-responsive genes are controlled by an increased rate of transcription. The remainder are controlled posttranscriptionally by a reduced rate of transcript turn over (Dunn et al., 1994; Hajela et al., 1990; Phillips et al., 1997). In the case of transcriptional control, promoter features are responsible, whereas for posttranscriptional control, the mechanism that controls transcript-specific mRNA breakdown would involve protein(s) able to bind to mRNAs. Several cold-upregulated sequences code for possible RNA-binding proteins. The nucleic acid-binding capacity of one, BLT801, was experimentally confirmed (Dunn et al., 1996). Interestingly, in view of the early calcium signal, BLT801 can be phosphorylated.

Transcriptional control in eukaryotes is exerted by a promoter region of the DNA, often running from the transcription start site to approximately 500 bases upstream, though much more distant regions of the DNA sequence can be involved. Transcription factors control transcription of a particular gene sequence by recognizing and binding to specific short sequences in the promoter that are known as response elements. This positions the transcription factors to interact with the other components of the transcription complex and thus influence local transcriptional activity.

The promoter of the *A. thaliana* gene *rd29a* has been analyzed in depth (Yamaguchi-Shinozaki and Shinozaki, 1984). This promoter contains response elements also found in many other cold-response promoters, and thus its study can identify regulatory elements of general importance in controlling cold acclimation. The cold-response element usually present is also responsive to drought and thus has been called the drought-response element (DRE; Yamaguchi-Shinozaki and Shinozaki, 1994), the low-temperature-response element (LTRE; Hughes and Dunn, 1996), and the C-repeat/drought-response element (CRT/DRE; Stockinger et al., 1997). This response element contains the core base sequence CCGAC found in cold-responsive promoters from both broad-leaved plants and cereals, indicating the likely universality of this regulatory element in higher plants. However, variants of the DRE sequence do occur, such as CCGAA, which is functional in barley (Dunn et al., 1998).

Other response elements may also be involved. Cold-response promoters often contain ABA response elements (ABRE), and these may contribute to the response to cold (Hughes and Dunn, 1996; Figure 9.5). MYB-response elements are also commonly present, binding drought-induced MYB transcription factors (Urao et al., 1993). In addition, the promoter of the cold-specific barley gene *blt101.1* contains a functional low-temperature response element with no close sequence similarity to any of the above (Brown et al., 2001). There is also evidence of negative regulatory factors in barley cold-responsive promoters and evidence that cold may conformationally alter the promoter sequence, possibly making it more accessible to regulatory factors (Brown et al., 2001; Dunn et al., 1998). Finally, the structures as well as compositions of cold-responsive promoters differ widely between different genes and between different species. Thus, the typical cold-responsive promoter probably has a relatively complex structure, conferring complex response properties, and it is possible that all these variations would contribute to species differences in adaptation.

The transcription factors that bind to the DREs have been studied in detail. They comprise a small multigene family in which the individuals are called CBF[4] (CRT-binding-factor: Stockinger et al., 1997) or DREB[4] (DRE-binding-factor: Shinwari et al., 1998). The importance of these transcription factors for acclimation to freezing has been proved by transgenic experiments (Section 5.4.2.6). CBF/DREB transcription factors are most completely described for the model plant, *A. thaliana*, but it is clear that genes coding for CBF-like proteins are of wide occurrence, including in cereals (Jaglo et al., 2001).

CBF/DREB1-like sequences are also expressed in tomato (Jaglo et al., 2001), which has no ability to acclimate to freezing. Rice, which lacks tolerance of temperatures much below 15°C, also produces CBF/DREB1-like proteins (Ito et al., 2002). Clearly, the CBF/DREB1 transcription factors are not specific to acclimation to freezing.

All plants have some ability to respond to a fall in temperature. Interestingly, rice, which has no freezing tolerance, possesses a dehydrin gene of a type expressed in freezing-tolerant cereals, but rice does not express it. Thus, what distinguishes freezing-tolerant from freezing-sensitive plants is probably not possession of the relevant functional genes but, rather, whether or not they are expressed in response to a cold signal (Danyluke et al., 1994). However, the regulatory difference responsible apparently does not reside in the possession or expression of CBF/DREB-like transcription factor genes, as shown in tomato and rice (Ito et al., 2002; Jaglo et al., 2001). Possibly, the difference is in undiscovered fine details of the regulatory mechanisms. However, speculatively, the CBF/DREB1 transcription factors may elicit as much of a response to a fall in temperature as a particular species is capable of, regardless of whether the plant is native to a tropical, temperate, or arctic environment. If so, species with markedly different adaptation to temperature may be using very similar regulatory factors and pathways to achieve their different responses to a fall in temperature.

Mutant analysis has added several important facts to this picture. The negative regulator *hos1* appears to be upstream of CBF/DREB1, as CBF/DREB1 expression is enhanced in the mutant (Lee et al., 2001). In contrast, *sfr6* is a positive regulator on a separate cold-STP that converges downstream of CBF expression, possibly interacting with CBF, as the mutant lacks expression of several cold-response genes including *rd29a* but does not lack expression of CBF itself (Knight et al., 1999). In contrast to both these, *esk1* is on a cold-STP that appears to be entirely separate, as the mutation has no effect on expression of cold-upregulated genes except those of proline synthesis (Xin and Browse, 1998). Thus, the complexity evident in the promoters is echoed in the whole interacting network of STPs, commensurate with the sophistication of control needed to fine-tune the plant's responses to its often stressful and constantly changing field environment, in which cold is only one factor.

5.7 CONCLUSION

There are several patterns of freezing and mechanisms of damage in plants. The most common type of freezing is extracellular. This reduces the water content of cells, killing the cells when the dehydration is sufficient to damage membrane structure. Plant antifreeze proteins do not appear to have the role of preventing freezing. Instead, they may act to control recrystallization, which could reduce structural damage to tissues or organs but would not avoid cellular dehydration.

The extent of the molecular response to cold is large, involving expression of a great diversity of genes of both known and unknown subcellular functions. This may include expression of some structurally similar proteins with different biochemical or physiological functions. A number of the genes expressed, such as those coding for LEA and LEA-like proteins, are probably directly protective. Others are involved in synthesizing protective solutes or adjusting membrane compositions to make the membrane less susceptible to stress. However, only part of the molecular response to cold relates to freezing tolerance in this direct way; other parts may address other winter stresses or adjustment of constitutive functions.

The regulatory system is complex, involving parallel and converging signal transduction pathways. Influx of calcium into the cytosol and phosphorylation of proteins are early steps in signaling. Later steps include transcriptional control of mRNA levels, where CBF/DREB1 and other transcription factors interact with promoters. However, amounts of only some cold-specific mRNAs are controlled in this way, as about half are controlled by mRNA turnover.

Freezing-tolerant and freezing-susceptible species appear to posses the same gene types and common regulatory components; however, differences in gene expression are apparent. A more detailed understanding of the molecular differences explaining their different physiology is needed. This can be achieved by QTL analysis combined with direct comparative molecular analysis focused on key regulators and on representative functional genes.

The existing molecular knowledge can be exploited. Overexpression of members of the CBF/DREB small multigene family of transcription factors enables control of freezing tolerance, but probably only in species with a natural capacity to acclimate to freezing. In species that cannot acclimate to freezing, overexpression of functional components of acclimation may help confer tolerance, such as proline accumulation, changes in lipid unsaturation, and accumulation of LEA and LEA-like proteins. Individually, these components may confer some tolerance on sensitive species, but they are inadequate alone to confer the levels of tolerance achieved by the most freezing-tolerant species. It is possible that simultaneous expression of several such genes might confer considerably higher levels of tolerance on nonadapted species. This remains to be shown.

Exploitation outside plants is also possible. It would be interesting to try inclusion of LEAs in cryopreservation protocols and to extend this to test the effect of their expression in a nonplant model organism. As with plants themselves, it is probable that a combination of factors would be most effectively protective. One could suggest high solute content, as at present, in combination with expression of LEAs and adjustment in membrane lipid composition. Cold-response gene discovery continues in plants, and these genes will offer unexpected possibilities.

REFERENCES

The Arabidopsis Genome Initiative. (2000) Analysis of the genome sequence of the flowering plant *Arabidopsis thaliana*, *Nature*, 408, 796–815.

Antikainen, M. and Griffith, M. (1997) Antifreeze protein accumulation in freezing-tolerant cereals, *Physiol. Plant.*, 99, 423–432.

Artus, N.N., Uemura, M., Steponkus, P.L., Gilmour, S.J., Lin, C.T., and Thomashow, M.F. (1996) Constitutive expression of the cold-regulated *Arabidopsis thaliana cor15a* gene affects both chloroplast and protoplast freezing tolerance, *Proc. Natl. Acad. Sci. USA*, 93, 13404–13409.

Ashworth, E.N. (1990) The formation and distribution of ice within forsythia flower buds, *Plant Physiol.*, 92, 718–725.

Ashworth, E.N. (1996) Responses of bark and wood cells to freezing, *Adv. Low Temp. Biol.*, 3, 65–106.

Ashworth, E.N. and Abeles, F.A. (1984) Freezing behaviour of water in small pores and the possible role in the freezing of plant tissues, *Plant Physiol.*, 76, 201–204.

Ashworth, E.N., Davis, G.A., and Anderson, J.A. (1985) Factors affecting ice nucleation in plant tissues, *Plant Physiol.*, 79, 1033–1037.

Ashworth, E.N., Echlin, P., Pearce, R.S., and Hayes, T.L. (1988) Ice formation and tissue response in apple twigs, *Plant Cell Environ.*, 11, 703–710.

Ashworth, E.N. and Pearce, R.S. (2002) Extracellular freezing in leaves of freezing-sensitive species, *Planta*, 214, 798–805.

Ashworth, E.N., Wilard, T.J., and Malone, S.R. (1992) The relationship between vascular differentiation and the distribution of ice within *Forsythia* flower buds, *Plant Cell Environ.*, 15, 607–612.

Berberich, T., Harda, M., Sugawara, K., Kodama, H., Iba, K., and Kusano, T. (1998) Two maize genes encoding ω-3 fatty acid desaturase and their differential expression to temperature, *Plant Mol. Biol.*, 36, 297–306.

Berberich, T., Suguwara, K., Harada, M., and Kusano, T. (1995) Molecular cloning, characterisation and expression of an elongation factor 1-alpha gene in maize, *Plant Mol. Biol.*, 29, 611–615.

Boothe J.G., Sonnichsen, F.D., deBeus, F.D., and Johnson-Flanagan, A.M. (1997) Purification, characterisation, and structural analysis of a plant low-temperature-induced protein, *Plant Physiol.*, 113, 367–376.

Bridger, G.M., Yang, W., Falk, D.E., and McKersie, B.D. (1994) Cold-acclimation increases tolerance of activated oxygen in winter cereals, *J. Plant Physiol.*, 144, 235–240.

Brown, A.P.C., Dunn, M.A., Goddard, N.J., and Hughes, M.A. (2001) Identification of a novel low-temperature-response element in the promoter of the barley (*Hordeum vulgare* L) gene *blt101.1*, *Planta*, 213, 770–780.

Brush, R.A., Griffith, M., and Mlynarz, A. (1994) Characterisation and quantification of intrinsic ice nucleators in winter rye (*Secale cereale*) leaves, *Plant Physiol.*, 104, 725–735.

Burke, M.J., Gusta, L.V., Quamme, H.A., Weiser, C.J., and Li, P.H. (1976) Freezing and injury in plants, *Annu. Rev. Plant Physiol.*, 27, 507–528.

Bush, D.S. (1995) Calcium regulation in plant cells and its role in signalling, *Annu. Rev. Plant Physiol. Plant Mol. Biol.*, 46, 95–122.

Campbell, S.A. and Close, T.J. (1997) Dehydrins: genes, proteins, and associations with phenotypic traits, *New Phytol.*, 137, 61–74.

Carpenter, C.D., Krebs, J.A., and Simon, A.E. (1994) Genes encoding glycine-rich *Arabidopsis thaliana* proteins with RNA-binding motifs are influenced by cold treatment and an endogenous circadian rhythm, *Plant Physiol.*, 104, 1015–1025.

Chandler, P.M. and Robertson, M. (1994) Gene expression regulated by abscisic acid and its relation to stress tolerance, *Annu. Rev. Plant Physiol. Plant Mol. Biol.*, 45, 113–141.

Choi, D.W., Zhu, B., and Close, T.J. (1999) The barley (*Hordeum vulgaris* L.) dehydrin multigene family: sequences, allele types, chromosome assignments, and expression characteristics of 11 Dhn genes of cv. Dicktoo, *Theor. Appl. Gen.*, 98, 1234–1247.

Chu, B., Snusted, P., and Carter, J.V. (1993) Alterations of β-tubulin gene expression during low-temperature exposure in leaves of *Arabidopsis thaliana*, *Plant Physiol.*, 103, 371–377.

Chun, J.U., Yu, X.M., and Griffith, M. (1998) Genetic studies of antifreeze proteins and their correlation with winter survival in wheat, *Euphytica* 102, 219–226.

Close, T.J. (1996) Dehydrins: Emergence of a biochemical role of a family of plant dehydration proteins, *Physiol. Plant.*, 97, 795–803.

Crowe, J.H., Hoekstra, F.A., and Crowe, L.M. (1992) Anhydrobiosis, *Annu. Rev. Physiol.*, 54, 570–599.

Cutler, A.J., Saleem, M., Kendall, E., Gusta, L.V., Georges, F., and Fletcher, G.L. (1989) Winter flounder antifreeze proteins improves the cold hardiness of plant tissues, *J. Plant Physiol.*, 135, 351–354.

Danyluke, J., Houde, M., Rassart, E., and Sarhan, F. (1994) Differential expression of a gene encoding an acidic dehydrin in chilling sensitive and freezing tolerant gramineae species, *FEBS Lett.*, 344, 20–24.

Ding, J.P. and Pickard, B.G. (1993) Modulation of mechanosensitive calcium-selective cation channels by temperature, *Plant J.*, 3, 713–720.

Dörffling, K., Dörffling, H., and Lesselich, G. (1993) *In-vitro*-selection and regeneration of hydroxyproline-resistant lines of winter-wheat with increased proline content and increased frost tolerance, *J. Plant Physiol.*, 142, 222–225.

Dörffling, K., Dörffling, H., Lesselich, G., Luck, E., Zimmermann, C., Melz, G., and Jurgens, H.U. (1997) Heritable improvement of frost tolerance in winter wheat by *in vitro* selection of hydroxyproline-resistant proline overproducing mutants, *Euphytica*, 93, 1–10.

Droual, A.M., Maaroufi, H., Creche, J., Chenieux, J.C., Rideau, M., and Hamdi, S. (1997) Changes in the accumulation of cytosolic cyclophilin transcripts in cultured periwinkle cells following hormonal and stress treatments, *J. Plant Physiol.*, 151, 142–150.

Duman, J.G. and Olsen, T.M. (1993) Thermal hysteresis protein activity in bacteria, fungi, and phylogenetically diverse plants, *Cryobiology*, 30, 322–328.

Dunn, M.A., Brown, K., Lightolers, R., and Hughes, M.A. (1996) A low-temperature-responsive gene from barley encodes a protein with single stranded nucleic acid binding activity which is phosphorylated *in vitro*, *Plant Mol. Biol.*, 30, 947–959.

Dunn, M.A., Goddard, N.J., Zhang, L., Pearce, R.S., and Hughes, M.A. (1994) Low-temperature-responsive barley genes have different control mechanisms, *Plant Mol. Biol.*, 24, 879–888.

Dunn, M.A., Hughes, M.A., Zhang, L., Pearce, R.S., Quigley, A.S., and Jack, P.L. (1991) Nucleotide sequence and molecular analysis of the low-temperature induced cereal gene, *blt4*, *Mol. Gen. Genet.*, 229, 389–394.

Dunn, M.A., Morris, A., Jack, P.L., and Hughes, M.A. (1993) A low-temperature-responsive translation elongation factor 1α from barley (*Hordeum vulgare* L.), *Plant Mol. Biol.*, 23, 221–225.

Dunn, M.A., White, A.J., Vural, S., and Hughes, M.A. (1998) Identification of promoter elements in a low-temperature-responsive gene (*blt4.9*) from barley (*Hordeum vulgare* L.), *Plant. Mol. Biol.*, 38, 551–564.

Dure, L. III (1993a) Structural motifs in LEA proteins of higher plants, in *Response of Plants to Cellular Dehydration During Environmental Stress*, Close T.J. and Bray, E.A., Eds., American Society of Plant Physiologists, Rockville, MD, pp. 91–103.

Dure, L. III (1993b) A repeating 11-mer amino acid motif and plant desiccation, *Plant J.*, 3, 363–369.

Eagles, C.F., Williams, J., and Louis, D.V. (1993) Recovery after freezing in *Avena sativa* L., *Lolium perenne* L. and *L. multiflorum* LAM, *New Phytol.*, 123, 477–483.

Ferullo, J.-M., Vezina, L.-P., Rai,l J., Laberge, S., Nadeau, P., and Castonguay, Y. (1997) Differential accumulation of two glycine-rich proteins during cold-acclimation of alfalfa, *Plant Mol. Biol.*, 33, 625–633.

Fowler, S. and Thomashow, M.F. (2002) Arabidopsis transcriptome profiling indicates that multiple regulatory pathways are activated during cold acclimation in addition to the CBF cold response pathway, *Plant Cell,* 14, 1675–1690.

Fukuchi-Mizutani, M., Tasaka, Y., Tanaka, Y., Ashikari, T., Kusumi, T., and Murata, N. (1998) Characterization of Delta 9 acyl-lipid desaturase homologues from *Arabidopsis thaliana*, *Plant Cell Physiol.*, 39, 247–253.

Fuller, M.P. and Wisniewski, M. (1998) The use of infrared thermal imaging in the study of ice nucleation and freezing of plants, *J. Thermal Biol.*, 23, 81–89.

Galiba, G., Kerepesi, I., Snape, J.W., and Sutka, J. (1997) Location of a gene regulating cold-induced carbohydrate production on chromosome 5A of wheat, *Theor. Appl. Genet.*, 95, 265–270.

Galiba, G., Quarrie, S.A., Sutka, J., Morgounov, A., and Snape, J.W. (1995) RFLP mapping of the vernalization (Vrn1) and frost-resistance (Fr1) genes on chromosome 5A of wheat, *Theor. Appl. Genet.*, 90, 1174–1179.

Gibson, S., Arondel, V., Iba, K., and Somerville, C. (1994) Cloning of a temperature-regulated gene encoding a chloroplast ω-3 desaturase from *Arabidopsis thaliana*, *Plant Physiol.*, 106, 1615–1621.

Gilmour, S.J., Sebolt, A.M., Salazar, M.P., Everard, J.D., and Thomashow, M.F. (2000) Overexpression of the Arabidopsis CBF3 transcriptional activator mimics multiple biochemical changes associated with cold acclimation, *Plant Physiol.*, 124, 1854–1865.

Gilmour, S.J. and Thomashow, M.F. (1991) Cold acclimation and cold-regulated gene expression in ABA mutants of *Arabidopsis thaliana*, *Plant Mol. Biol.*, 17, 1233–1240.

Goddard, N.J., Dunn, M.A., Zhang, L., White, A.J., Jack, P.L., and Hughes, M.A. (1993) Molecular analysis and spatial expression pattern of a low-temperature-specific barley gene, *blt101*, *Plant Mol. Biol.*, 23, 871–897.

Goldstein, G. and Noble, P.S. (1991) Changes in osmotic pressure and mucilage during low-temperature acclimation of *Opuntia ficus-indica*, *Plant Physiol.*, 97, 954–961.

Goodwin, W., Pallas, J.A., and Jenkins, G.I. (1996) Transcripts of a gene encoding a putative cell-wall plasma membrane linker protein are specifically cold-induced in *Brassica napus*, *Plant Mol. Biol.*, 31, 771–781.

Gordon-Kamm, W.J. and Steponkus P.L. (1984) Lamellar-to-hexagonal$_{II}$ phase transitions in the plasma membrane of isolated protoplasts after freeze-induced dehydration, *Proc. Natl. Acad. Sci. USA*, 81, 6373–6377.

Gosti, F., Bertauche, N., Vartanian, N., and Giraudat, J. (1995) Abscisic acid-dependent and -independent regulation of gene expression by progressive drought in *Arabidopsis thaliana*, *Mol. Gen. Genet.*, 246, 10–18.

Gray, G.R., Chauvin, L-P., Sarhan, F., and Huner, N.P.A. (1997) Cold acclimation and freezing tolerance, *Plant Physiol.*, 114, 467–474.

Griffith, M., and Antikiaien, M. (1996) Extracellular ice formation in freezing-tolerant plants, *Adv. Low-Temp. Biol.*, 3, 107–139

Griffith, M., Huner, N.P.A., Espelie, K.E., and Kolattukudy, P.E. (1985) Lipid polymers accumulate in the epidermis and mestome sheath cell walls during low temperature development of winter rye leaves, *Protoplasma*, 125, 53–64.

Gross, D.C., Proebsting Jr., E.L., and Macrindle-Zimmerman, H. (1988) Development, distribution, and characteristics of intrinsic, nonbacterial ice nuclei in *Prunus* wood, *Plant Physiol.*, 88, 915–922.

Gurian-Sherman, D. and Lindow, S.E. (1993) Bacterial ice nucleation: Significance and molecular basis, *FASEB J.*, 7, 1338–1343.

Gusta, L.V., Burke, M.J., and Kapoor, A. (1975) Determination of unfrozen water in winter cereals at subfreezing temperatures, *Plant Physiol.*, 56, 707–709.

Guy, C.L. (1990) Cold acclimation and freezing stress tolerance: role of protein metabolism, *Annu. Rev. Plant Physiol. Plant Mol. Biol.*, 41, 187–223.

Guy, C.L., Huber, J.L.A., and Huber, S.C. (1992) Sucrose phosphate synthase and sucrose accumulation at low temperature, *Plant Physiol.*, 100, 502–508.

Guy, C.L., Niemi, K.J., and Bramble, R. (1985) Altered gene expression during cold acclimation of spinach, *Proc. Nat. Acad. Sci. USA*, 82, 3673–3677.

Hajela, R.K., Horvath, D.P., Gilmour, S.J., and Thomashow, M.F. (1990) Molecular cloning and expression of *cor* (Cold Regulated) genes in *Arabidopsis thaliana*, *Plant Physiol.*, 93, 1246–1252.

Harris, P. and James, A.T. (1969) The effect of low temperatures on fatty acid biosynthesis in plants, *Biochem. J.*, 112, 325–330.

Hayes, P.M., Blake, T., Chen, T.H.H., Tragoonrung, S., Chen, F., Pan, A., and Liu, B. (1993) Quantitative trait loci on barley (*Hordeum vulgare* L) chromosome 7 associated with components of winter hardiness, *Genome*, 36, 66–71.

Hincha, D.K., Hellwege, E.M., Heyer, A.G., and Crowe, J.H. (2000) Plant fructans stabilize phosphatidylcholine liposomes during freeze-drying, *Eur. J. Biochem.*, 267, 535–540.

Hincha, D.K., Meins, F., and Schmitt, J.M. (1997a) Beta-1,3-glucanase is cryoprotective *in vitro* and is accumulated in leaves during cold acclimation, *Plant Physiol.*, 114, 1077–1083.

Hincha, D.K., Neukamm, B., Sror, H.A.M., Sieg, F., Weckwarth, W., Ruckels, M., Lullien-Pellerin, V., Schroder, W., and Schmitt, J.M. (2001) Cabbage cryoprotectin is a member of the nonspecific plant lipid transfer protein gene family, *Plant Physiol.*, 125, 835–846.

Hincha, D.K., Pfüller, U., and Schmitt, J.M. (1997b) The concentration of cryoprotective lectins in mistletoe (*Viscum album* L.) leaves is correlated with leaf frost hardiness, *Planta* 203, 140–144.

Hirsh, A.G., Williams, R.J., and Meryman, H.T. (1985) A novel method of natural cryoprotection—intracellular glass-formation in deeply frozen *Populus*, *Plant Physiol.*, 79, 41–56.

Holoppa, L.D. and Walker-Simmons, M.K. (1995) The wheat abscisic acid-responsive protein-kinase messenger-RNA, PKABA1, is upregulated by dehydration, cold temperature, and osmotic-stress, *Plant Physiol.*, 108, 1203–1210.

Hon, W.-C., Griffith, M., Chong, P., and Yang, D.S.C. (1994) Extraction and isolation of antifreeze proteins from winter rye (*Secale cereale* L.) leaves, *Plant Physiol.*, 104, 971–980.

Hong, S.W., Jon, J.H., Kwak, J.M., and Nam, H.G. (1997) Identification of a receptor-like protein kinase gene rapidly induced by abscisic acid, dehydration, high salt, and cold treatments in *Arabidopsis thaliana*, *Plant Physiol.*, 113, 1203–1212.

Horiguchi, G., Fuse, T., Kawakami, N., Kodama, H., and Iba, K. (2000) Temperature-dependent translational regulation of the ER omega-3 fatty acid desaturase gene in wheat root tips, *Plant J.*, 24, 805–813.

Horvath, D., McLarney, B.K., and Thomashow, M.F. (1993) Regulation of *Arabidopsis thaliana* L. (Heyn) *cor78* in response to low temperature, *Plant Physiol.*, 103, 1047–1053.

Houde, M., Daniel, C., Lachapelle, M., Allard, F., Laliberté, S., and Sarhan, F. (1995) Immunolocalisation of freezing-tolerance-associated proteins in the cytoplasm and nucleoplasm of wheat crown tissues, *Plant J.*, 8, 583–593.

Hughes, M.A. and Dunn, M.A. (1996) The molecular biology of plant acclimation to low temperature, *J. Exp. Bot.*, 47, 291–305.

Hughes, M.A., Dunn. M.A., Pearce, R.S., White, A.J., and Zhang, L. (1992) An abscisic acid-responsive, low temperature barley gene has homology with a maize phospholipid transfer protein, *Plant Cell Envir.*, 15, 861–865.

Hull, M.R., Long, S.P., and Jahnke, L.S. (1997) Instantaneous and developmental effects of low temperature on the catalytic properties of antioxidant enzymes in two *Zea* species, *Aust. J. Plant Physiol.*, 24, 337–343.

Ichimura, K., Mizoguchi, T., Yoshida, R., Yuasa, T., and Shinozaki, K. (2000) Various abiotic stresses rapidly activate Arabidopsis MAP kinases ATMPK4 and ATMPK, *Plant J.*, 24, 655–665.

Ishikawa, M., Price, W.S., Ide, H., and Arata, Y. (1997) Visualization of freezing behaviors in leaf and flower buds of full-moon maple by nuclear magnetic resonance microscopy, *Plant Physiol.*, 115, 1515–1524.

Ishitani, M., Xiong, L.M., Lee, H.J., Stevenson, B., and Zhu, J.K. (1998) HOS1, a genetic locus involved in cold-responsive gene expression in Arabidopsis, *Plant Cell*, 10, 1151–1161.

Ishitani, M., Xiong, L., Stevenson, B., and Zhu, J.-K. (1997) Genetic analysis of osmotic stress signal transduction in *Arabidopsis*: interactions and convergence of abscisic acid-dependent and abscisic acid-independent pathways, *Plant Cell*, 9, 1935–1949.

Ito, Y., Katsura, K., Shinozaki, K., and Yamaguchi-Shinozaki, K. (2002) Functional analysis of rice OsDREB genes by using transgenic rice, *Plant Cell Physiol.*, 43, S105.

Jaglo, K.R., Kleff, S., Amundsen, K.L., Zhang, X., Haake, V., Zhang, J.Z., Deits, T., and Thomashow, M.F. (2001) Components of the *Arabidopsis* C-repeat/dehydration-responsive element binding factor cold-response pathway are conserved in *Brassica napus* and other plant species, *Plant Physiol.*, 127, 910–917.

Jaglo-Ottosen, K.R., Gilmour, S.J., Zarka, D.G., Schabenberger, O., and Thomashow, M.F. (1998) *Arabidopsis CBF1* overexpression induces *COR* genes and enhances freezing tolerance, *Science*, 280, 104–106.

Jahnke, L.S., Hull, M.R., and Long, S.P. (1991) Chilling stress and oxygen metabolizing enzymes in *Zea mays* and *Zea diploperennis*, *Plant Cell Environ.*, 14, 97–104.

Jarillo, J.A., Capel, J., Leyva, A., Martinez-Zapater, J.M., and Salinas, J. (1994) Two related low-temperature-inducible genes of *Arabidopsis* encode proteins showing high homology to 14-3-3 proteins, a family of putative kinase regulators, *Plant Mol. Biol.*, 25, 693–704.

Jarillo, J.A., Leyva, A., Salinas, J., and Martinez-Zapater, J.M. (1993) Low temperature induces the accumulation of alcohol dehydrogenase mRNA in *Arabidopsis thaliana*, a chilling-tolerant plant, *Plant Physiol.*, 101, 833–837.

Kader, J.-C. (1996) Lipid-transfer proteins in plants, *Annu. Rev. Plant Physiol. Plant Mol. Biol.*, 47, 627–654.

Kaku, S. (1973) High ice nucleating ability in plant leaves, *Plant and Cell Physiol.*, 14, 1035–1038.

Kasuga, M., Liu, Q., Miura, S., Yamaguchi-Shinozaki, K., and Shinozaki, K. (1999) Improving plant drought, salt, and freezing tolerance by gene transfer of a single stress-inducible transcription factor, *Nat. Biotechnol.*, 17, 287–291.

Kazuoka, T. and Oeda, K. (1994) Purification and characterisation of COR85-oligomeric complex from cold-acclimated spinach, *Plant Cell Physiol.*, 35, 601–611.

Kenward, K.D., Brandle, J., McPherson, J., and Davies, P.L. (1999) Type II fish antifreeze protein accumulation in transgenic tobacco does not confer frost resistance, *Transgenic Res.*, 8, 105–177.

Kearsey, M.J. and Pooni, H.S. (1996) *The Genetical Analysis of Quantitative Traits,* Chapman & Hall, London.

Kindel, P.K., Liao, S.-Y., Liske, M.R., and Olien, C.R. (1989) Arabinoxylans from rye and wheat seed that interact with ice, *Carbohydrate Res.*, 187, 173–185.

Kitaura, K. (1967) Freezing and injury of mulberry trees by late spring frost, *Bull. Sericult. Exp. Stn. (Tokyo)* 22, 202–323.

Knight, M.R., Campbell, A.K., Smith, S.M., and Trewavas, A.J. (1991) Transgenic plant aequorin reports the effects of touch, cold-shock and elicitors on cytoplasmic calcium, *Nature*, 352, 524–526.

Knight, H., Trewavas, A.J., and Knight, M.R. (1996) Cold calcium signalling in *Arabidopsis* involves two cellular pools and a change in calcium signature after acclimation, *Plant Cell*, 8, 489–503.

Knight, H., Veale, E.L., Warren, G.J., and Knight, M.R. (1999) The sfr6 mutation in arabidopsis suppresses low-temperature induction of genes dependent on the CRT DRE sequence motif, *Plant Cell*, 11, 875–886.

Kocsy, G., Szalai, G., Vaguifalvi, A., Stehli, L., Orosz, G., and Galiba, G. (2000) Genetic study of glutathione accumulation during cold hardening in wheat, *Planta*, 210, 295–301.

Krog, J.O., Zachariassen, K.E., Larsen, B., and Smidsrød, O. (1979) Thermal buffering in Afro-alpine plants due to nucleating agent-induced water freezing, *Nature*, 282, 300–301.

Kurkela, S. and Borg-Franck, M. (1992) Structure and expression of *kin2*, one of two cold- and ABA-induced genes of *Arabidopsis thaliana*, *Plant Mol. Biol.*, 19, 689–692.

Kurkela, S. and Franck, M. (1990) Cloning and characterization of a cold- and ABA-inducible *Arabidopsis* gene, *Plant Mol. Biol.*, 15, 137–144.

Lång, V., Mantyla, E., Welin, B., Sundberg, B., and Palva, E.T. (1994) Alterations in water status, endogenous abscisic-acid content and expression of RAB18 gene during the development of freezing tolerance in *Arabidopsis thaliana*, *Plant Physiol.*, 104, 1341–1349.

Lee, H.J., Xiong, L.M., Gong, Z.Z., Ishitani, M., Stevenson, B., and Zhu, J.K. (2001) The Arabidopsis HOS1 gene negatively regulates cold signal transduction and encodes a RING finger protein that displays cold-regulated nucleo-cytoplasmic partitioning, *Genes Dev.*, 15, 912–924.

Lee, H., Xiong, L.M., Ishitani, M., Stevenson, B., and Zhu, J.K. (1999) Cold-regulated gene expression and freezing tolerance in an *Arabidopsis thaliana* mutant, *Plant J.*, 17, 301–308.

Levitt, J. (1980) *Responses of Plants to Environmental Stresses*. Vol. 1. Academic Press, New York.

Lewis, B.D., Karlin-Neumann, G., Davis, R.W., and Spalding, E.P. (1997) Ca^{2+}–activated anion channels and membrane depolarizations induced by blue light and cold in *Arabidopsis* seedlings, *Plant Physiol.*, 114, 1327–1334.

Lin, C. and Thomashow, M.F. (1992) DNA sequence analysis of a complementary DNA for cold-regulated *Arabidopsis* gene *cor15* and characterization of the COR15 polypeptide, *Plant Physiol.*, 99, 519–525.

Lisse, T., Bartels, D., Kalbitzer, H.R., and Jaenicke, R. (1996) The recombinant dehydrin-like desiccation stress protein from the resurrection plant Craterostigma plantagineum displays no defined three-dimensional structure in its native state, *Biol. Chem.*, 377, 555–561.

Lister, C. and Dean, C. (1993) Recombinant inbred lines for mapping rflp and phenotypic markers in *Arabidopsis thaliana*, *Plant J.*, 4, 745–750.

Liu, J.-J., Galvez, A.F., Krenz, D.C., and de Lumen, B.O. (1997) Galactinol synthase (GS), a key enzyme in biosynthesis of raffinose family oligosaccharides (RFO): Activation of enzyme activity and induction of gene expression by cold and desiccation, *Plant Physiol.*, 114, 600.

Liu, Q., Kasuga, M., Sakuma, Y., Abe, H., Miura, S., Yamaguchi-Shinozaki, K., and Shinozaki, K (1998) Two transcription factors, DREB1 and DREB2, with an EREBP/AP2 DNA binding domain separate two cellular signal transduction pathways in drought- and low-temperature-responsive gene expression, respectively, in Arabidopsis, *Plant Cell* 10, 1391–1406.

Livingston, D.P. and Henson, C.A. (1998) Apoplastic sugars, fructans, fructan exohydrolase, and invertase in winter oat: responses to second-phase cold hardening, *Plant Physiol.*, 116, 403–408.

Llorente, F., Oliveros, J.C., Martinez-Zapater, J.M., and Salinas, J. (2000) A freezing-sensitive mutant of Arabidopsis, *frs1*, is a new aba3 allele, *Planta*, 211, 648–655.

Luo, M., Lin, L.H., Hill, R.D., and Mohapatra, S.S. (1991) Primary structure of an environmental stress and abscisic acid-inducible alfalfa protein, *Plant Mol. Biol.*, 17, 1267–1269.

Luxova, M. (1986) The hydraulic safety zone at the base of barley roots, *Planta*, 169, 465–470.

Lynch D.V. and Steponkus P.L. (1987) Plasma membrane lipid alterations associated with cold acclimation of winter rye seedlings (*Secale cereale* L. Cv Puma), *Plant Physiol.*, 83, 761–767.

Markhart, A.H., Fiscus, E.L., Naylor, A.W., and Kramer, P.J. (1979) Effect of temperature on water and ion transport in soybean and broccoli systems, *Plant Physiol.*, 64, 83–87.

Mazur, P. (1969) Freezing injury in plants, *Annu. Rev. Plant Physiol.*, 20, 419–448.

McKersie, B.D., Bowley, S.R., Harjanto, E., and LePrince, O. (1996) Water-deficit tolerance and field performance of transgenic alfalfa over-expressing superoxide dismutase, *Plant Physiol.*, 111, 1177–1181.

McKersie, B.D., Bowley, S.R., and Jones, K.S. (1999) Winter survival of transgenic alfalfa overexpressing superoxide dismustase, *Plant Physiol.*, 119, 839–847.

McKersie, B.D., Chen, Y.R., Debeus, M., Bowley, S.R., Bowler, C., Inze, D., Dhalluin, K., and Botterman, J. (1993) Superoxide-dismutase enhances tolerance of freezing stress in transgenic alfalfa (*Medicago sativa* L.), *Plant Physiol.*, 103, 1155–1163.

McKown, R., Kuroki, G., and Warren, G. (1996) Cold responses of *Arabidopsis* mutants impaired in freezing tolerance, *J. Exp. Bot.*, 47, 1919–1925.

Medina, J., Catala, R., and Salinas, J. (2001) Developmental and stress regulation of RCI2A and RCI2B, two cold-inducible genes of arabidopsis encoding highly conserved hydrophobic proteins, *Plant Physiol.*, 125, 1655–1666

Meijer, E.A., de Vries, S.C., Sterk, P., Gadella, D.W.J., Wirtz, K.W.A., and Hendriks, T. (1993) Characterization of the nonspecific lipid transfer protein—EP2 from carrot (*Daucus carota* L.), *Mol. Cell. Biochem.*, 123, 159–166.

Molina, A. and García-Olmedo, F. (1993) Developmental and pathogen-induced expression of three barley genes encoding lipid transfer proteins, *Plant J.*, 4, 983–991.

Molina, A., Mena, M., Carbonera, P., and Garcia-Olmedo, F. (1997) Differential expression of pathogen-responsive genes encoding two types of glycine-rich proteins in barley, *Plant Mol. Biol.*, 33, 803–810.

Molina, A., Segura, A., and García-Olmedo, F. (1993) Lipid transfer proteins (nsLTPs) from barley and maize leaves are potent inhibitors of bacterial and fungal plant pathogens, *FEBS Lett.*, 316, 119–122.

Monroy, A.F., Castinguay, Y., Laberge, S., Sarhan, F., Vazina, L.P., and Dhindsa, R.S. (1993a) A new cold-induced alfalfa gene is associated with enhanced hardening at subzero temperature, *Plant Physiol.*, 102, 873–879.

Monroy, A.F. and Dhindsa R.S. (1995) Low-temperature signal-transduction—induction of cold acclimation-specific genes of alfalfa by calcium at 25-degrees-C, *Plant Cell*, 7, 321–331.

Monroy, A.F., Sarhan, F., and Dhindsa, R.S. (1993b) Cold-induced changes in freezing tolerance, protein-phosphorylation, and gene-expression—evidence for a role of calcium, *Plant Physiol.*, 102, 1227–1235.

Murai, M. and Yoshida, S. (1998) Vacuolar membrane lesions induced by a freeze-thaw cycle in protoplasts isolated from deacclimated tubers of Jerusalem artichoke (*Helianthus tuberosus* L.), *Plant Cell Physiol.*, 39, 87–96.

Murata, N. (1983) Molecular species composition of phosphatidylglycerols from chilling-sensitive and chilling-resistant plants, *Plant Cell Physiol.*, 24, 81–86.

Murata, N., Ishizaki-Nishizawa, Q., Higashi, S., Hayashi, H., Tasaka, Y., and Nishida, I. (1992) Genetically engineered alteration in the chilling sensitivity of plants, *Nature* 356, 710–713.

Nanjo, T., Kobayashi, M., Yoshiba, Y., Kakubari, Y., Yamaguchi-Shinozaki, K., and Shinozaki, K. (1999) Antisense suppression of proline degradation improves tolerance to freezing and salinity in *Arabidopsis thaliana*, *FEBS Lett.*, 461, 205–210.

Ndong, C., Ouellet, F., Houde, M., and Sarhan, F. (1997) Gene expression during cold acclimation in strawberry, *Plant Cell Physiol.*, 38, 863–870.

Nishida, I. and Murata, N. (1996) Chilling sensitivity in plants and cyanobacteria—the critical contribution of membrane-lipids, *Annu. Rev. Plant Physiol. Plant Mol. Biol.*, 47, 541–568.

Nordin, K., Heino, P., and Palva, E.T. (1991) Separate signal pathways regulate the expression of a low-temperature-induced gene in *Arabidopsis thaliana* (L.) Heynh, *Plant Mol. Biol.*, 16, 1061–1071.

Olien, C.R. (1965) Interference of cereal polymers and related compounds with freezing, *Cryobiology*, 2, 47–54.

Olien, C.R. and Smith, M.N. (1977) Ice adhesions in relation to freeze stress, *Plant Physiol.*, 60, 499–503.

Pan, A., Hayes, P.M., Chen, F., Chen, T.H.H., Blake, T., Wright, S., Karsai, I., and Bedö, Z. (1994) Genetic analysis of the components of winter hardiness in barley (*Hordeum vulgare* L), *Theor. Appl. Genet.*, 89, 900–910.

Peacock, J.M. (1975) Temperature and leaf growth of *Lolium perenne*. II. The site of temperature perception, *Ann. Appl. Ecol.*, 12, 115–123.

Pearce, R.S. (1985) A freeze-fracture study of membranes of rapidly drought-stressed leaf bases of wheat, *J. Exp. Bot.*, 36, 1209–1221.

Pearce, R.S. (1988) Extracellular ice and cell shape in frost-stressed cereal leaves: A low temperature scanning electron microscopy study, *Planta*, 175, 313–324.

Pearce, R.S. (1999) Molecular analysis of acclimation to cold, *Plant Growth Reg.*, 29, 47–76.

Pearce, R.S. (2001) Plant freezing and damage, *Ann Bot.*, 87, 417–424.

Pearce, R.S., Dunn, M.A., Rixon, J.E., Harrison, P., and Hughes, M.A. (1996) Expression of cold-induced genes and frost hardiness in the crown meristem of young barley (*Hordeum vulgare* L. cv. Igri) plants grown in different environments, *Plant Cell Environ.*, 12, 275–290.

Pearce, R.S. and Fuller, M.P. (2001) Freezing of barley (*Hordeum*) studied by infrared video thermography, *Plant Physiol.*, 125, 227–240.

Pearce, R.S., Houlston, C.E., Atherton, K.A., Rixon, J.E., Harrison, P., Hughes, M.A., and Dunn, M.A. (1998) Localization of expression of three cold-induced genes (*blt101, blt4.9, blt14*) in different tissues of the crown and developing leaves of cold-acclimated cultivated barley (*Hordeum vulgare* L. cv. Igri), *Plant Physiol.*, 117, 787–795.

Pearce, R.S. and Willison, J.H.M. (1985a) Wheat tissues freeze-etched during exposure to extracellular freezing: distribution of ice, *Planta*, 163, 295–303.

Pearce, R.S. and Willison, J.H.M. (1985b) A freeze-etch study of the effect of extracellular freezing on the cellular membranes of wheat, *Planta*, 163, 304–316.

Phillips, J.R., Dunn, M.A., and Hughes, M.A. (1997) mRNA stability and localisation of the low-temperature-responsive barley gene family *blt14*, *Plant Mol. Biol.*, 33, 1013–1023.

Pihakaski-Maunsbach, K., Griffiths, M., Antikainen, M., and Maunsbach, A.B. (1996) Immunogold localization of glucanase-like antifreeze protein in cold acclimated winter rye, *Protoplasma*, 191, 115–125.

Pihakaski-Maunsbach, K. and Kukkonen, M. (1997) Ultrastructural changes induced by sub-zero temperatures in the plasma membrane of protoplasts from winter rye, *Physiol. Plant.*, 100, 333–340.

Pihakaski-Maunsbach, K., Moffatt, B., Testillano, P., Risueno, M., Yeh, S.S., Griffith, M., and Maunsbach, A.B. (2001) Genes encoding chitinase-antifreeze proteins are regulated by cold and expressed by all cell types in winter rye shoots, *Physiol. Plant.*, 13, 359–371.

Pitt, R.E. (1990) Cryobiological implications of different methods of calculating the chemical potential of water in partially frozen suspending media, *Cryo-Letters*, 11, 227–240.

Plant, A.L., Cohen, A., Moses, M.S., and Bray, E.A. (1997) Nucleotide sequence and spatial expression pattern of a drought- and abscisic acid-induced gene in tomato, *Plant Physiol.*, 97, 900–906.

Polinsky, D.H. and Braam, J. (1996) Cold-shock regulation of the *Arabidopsis TCH* genes and the effects of modulating intracellular calcium levels, *Plant Physiol.*, 111, 1271–1279.

Pollock, C.J. and Jones, T. (1979) Seasonal patterns of fructan metabolism in forage grasses, *New Phytol.*, 83, 8–15.

Prasad, T.K. (1997) Role of catalase in inducing chilling tolerance in pre-emergent maize seedlings, *Plant Physiol.*, 114, 1369–1376.

Prasad, T.K., Anderson, M.D., Martin, B.A., and Stewart, C.R. (1994) Evidence for chilling-induced oxidative stress in maize seedlings and a regulatory role for hydrogen peroxide, *Plant Cell*, 6, 65–74.

Pyee, J., Yu, H., and Kolattukudy, P.E. (1994) Identification of a lipid transfer protein as the major protein in the surface wax of broccoli (*Brassica oleracea*) leaves, *Arch. Biochem. Biophys.*, 311, 460–468.

Rebeille, F., Bligny, R., and Douce, R. (1980) Oxygen and temperature effects on the fatty acid composition of sycamore cells (*Acer pseudoplatanus* L.), *Biochem. Biophys. Acta* 620, 1–9.

Reimholz, R., Geiger, M., Haake, V., Deiting, U., Krause, K.P., Sonnewald, U., and Stitt, M. (1997) Potato plants contain multiple forms of sucrose phosphate synthase, which differ in their tissue distributions, their levels during development, and their responses to low temperature, *Plant Cell Envir.*, 20, 291–305.

Roberts, D.W.A. (1990) Identification of loci on chromosome 5A of wheat involved in control of cold hardiness, vernalization, leaf length, rosette growth habit, and height of hardened plants, *Genome*, 33, 247–259.

Rorat, T., Irzkowski, W., and Grygorowicz, W.J. (1997) Identification and expression of novel cold induced genes in potato (*Solanum sogarandinum*), *Plant Sci.*, 124, 69–78.

SaezVasquez, J., Raynal, M., Mezabasso, L., and Delseny, M. (1993) Two related, low-temperature-induced genes from *Brassica napus* are homologous to the human tumor *bbc1* (breast basic conserved) gene, *Plant Mol. Biol.*, 23, 1211–1221.

Sakai, A. (1960) Survival of twigs of woody plants at −196°C, *Nature*, 185, 393–394.

Sakai, A. and Larcher, W. (1987) *Frost Survival of Plants*. Springer, Berlin.

Sakuma, Y., Liu, Q., Dubouzet, J.G., Abe, H., Shinozaki, K., and Yamaguchi-Shinozaki, K. (2002) DNA-binding specificity of the ERF/AP2 domain of *Arabidopsis* DREBs, transcription factors involved in dehydration- and cold-inducible gene expression, *Biochem. Biophy. Res. Commun.*, 290, 998–1009.

Sangwan, V., Foulds, I., Singh, J., and Dhindsa, R.S. (2001) Cold-activation of *Brassica napus* BN115 promoter is mediated by structural changes in membranes and cytoskeleton, and requires Ca2+ influx, *Plant J.*, 27, 1–12.

Sidebottom, C., Buckley, S., Pudney, P., Twiggs, S., Jarman, C., Holt, C., Telford, J., McArthur, A., Worrall, D., Hubbard, R., and Lillford, P. (2000) Heat-stable antifreeze protein from grass, *Nature*, 406, 256.

Shinwari, Z.K., Nakashima, K., Miura, S., Kasuga, M., Seki, M., Yamaguchi-Shinozaki, K., and Shinozaki, K. (1998) An Arabidopsis gene family encoding DRE/CRT binding proteins involved in low-temperature-responsive gene expression, *Biochem. Biophys. Res. Commun.*, 250, 161–170.

Smith, H. (1990) Signal perception, differential expression within multigene families and the molecular basis of phenotypic plasticity, *Plant Cell Environ.*, 13, 585–594.

Sossountzov, L., Riz-Avila, L., Vignols, F., Jolliot, A., Arondel, V., Tcheng, F., Grosbois, M., Guerbette, F., Miginiac, E., Delseny, M., Puigdomenèch, M., and Kader, J.-C. (1991) Spatial and temporal expression of a maize lipid transfer protein gene, *Plant Cell*, 3, 923–933.

Steponkus, P.L. (1984) Role of the plasma membrane in freezing injury and cold acclimation, *Ann. Rev. Plant Physiol. Plant Mol. Biol.*, 35, 543–584

Steponkus, P.L., Uemura, M., Balsamo, R.A., Arvinte, T., and Lynch, D.V. (1988) Transformation of the cryobehavior of rye protoplasts by modification of the plasma membrane lipid composition, *Proc. Natl. Acad. Sci. USA*, 85, 9026–9030.

Steponkus, P.L., Uemura, M., Webb, M.S. (1993) A contrast of the cryostability of the plasma membrane of winter rye and spring oats—two species that widely differ in their freezing tolerance and plasma membrane lipid composition, *Adv. Low Temp. Biol.*, 2, 211–312.

Stockinger, E.J., Gilmour, E.J., and Thomashow, M.F. (1997) *Arabidopsis thaliana CBF1* encodes an AP2 domain-containing transcription activator that binds to the C-repeat/DRE, a cis-acting DNA regulatory element that stimulates transcription in response to low temperature and water deficit, *Proc. Natl. Acad. Sci. USA*, 94, 1035–1040.

Stout, D.G., Majak, W., and Reaney, M. (1980) *In vivo* detection of membrane injury in freezing temperatures, *Plant Physiol.*, 66, 74–77.

Strauss, G. and Hauser, H. (1986) Stabilization of lipid bilayer vesicles by sucrose during freezing, *Proc. Natl. Acad. Sci. USA*, 83, 2422–2426.

Stushnoff, C., Remmele, R.L., Essensee, V., and McNeil, M. (1993) Low temperature induced biochemical mechanisms: implications for cold acclimation and de-acclimation, in *Interacting Stresses on Plants in a Changing Climate*; NATO ASI Series I: Global Environmental Change, Vol. 16, Jackson, M.B. and Black, C.R., Eds., Springer, Berlin, pp. 647–657.

Sutka, J. (1994) Genetic-control of frost tolerance in wheat (*Triticum aestivum* L), *Euphytica* 77, 277–282.

Sutton, F., Ding, X., and Kenefick, D.G. (1992) Group 3 LEA gene *HVA1* regulation by cold acclimation and deacclimation in two barley cultivars with varying freeze resistance, *Plant Physiol.*, 99, 338–340.

Suzuki, I., Kanesaki, Y., Mikami, K., Kanehisa, M., and Murata, N. (2001) Cold-regulated genes under control of the cold sensor Hik33 in *Synechocystis*, *Mol. Microbiol.*, 40, 235–244.

Suzuki, I., Los, D.A., and Murata, N. (2000) Perception and transduction of low-temperature signals to induce desaturation of fatty acids, *Biochem. Soc. Trans.*, 28, 628–630.

Tabaei-Aghdaei, S.R., Pearce, R.S., and Harrison, P. (2003) Sugars regulate cold-induced gene expression and freezing-tolerance in barley cell cultures, *J. Ex. Bot.*, 54, 1565–1575.

Takahashi, R. and Shimosaka, E. (1997) cDNA sequence analysis and expression of two cold-regulated genes in soybean, *Plant Sci.*, 123, 93–104.

Tanino, K.K. and McKersie, B.D. (1985) Injury within the crown of winter wheat seedlings after freezing and icing stress, *Can. J. Bot.*, 63, 432–436.

Thoma, S., Hecht, U., Kippers, A., Botella, J., De Vries, S., and Somerville, C. (1994) Tissue-specific expression of a gene encoding a cell wall-localized lipid transfer protein from *Arabidopsis*, *Plant Physiol.*, 105, 35–45.

Thoma, S., Kaneko, Y., and Somerville, C. (1993) A non-specific lipid transfer protein from *Arabidopsis* is a cell wall protein, *Plant J.*, 3, 427–436.

Thomashow, M.F. (1999) Plant cold acclimation: Freezing tolerance genes and regulatory mechanisms, *Annu. Rev. Plant Physiol. Plant Mol. Biol.*, 50, 571–599.

Tiku, P.E., Gracey, A.Y., Macartney, A.I., Beynon, R.J., and Cossins, A.R. (1996) Cold-induced expression of delta(9) desaturase in carp by transcriptional and post-translational mechanisms, *Science*, 271, 815–818.

Urao, T., Yamaguchi-Shinozaki, K., Urao, S., and Shinozaki, K. (1993) An arabidopsis *myb* homolog is induced by dehydration stress and its gene product binds to the conserved MYB recognition sequence, *Plant Cell*, 5, 1529–1539.

Urrutia, M.E., Duman, J.G., and Knight, C.A. (1992) Plant thermal hysteresis proteins, *Biochem. Biophys. Acta*, 1121, 199–206.

Veisz, O., Galiba, G., and Sutka, J. (1996) Effect of abscisic acid on the cold hardiness of wheat seedlings, *J. Plant Physiol.*, 149, 439–443.

Verkleij, A.J. (1984) Lipidic intramembranous particles, *Biochim. Biophys. Acta*, 779, 43–63.

Vigh, L., Los, D., Horvath, I., and Murata, N. (1993) The primary signal in the biological perception of temperature: Pd-catalyzed hydrogenation of membrane lipids stimulated the expression of the *desA* gene in *Synechocystis* PCC6803, *Proc. Natl. Acad. Sci. USA*, 90, 9090–9094.

Walker, M.A. and McKersie, B.D. (1993) Role of the ascorbate-glutathione antioxidant system in chilling resistance of tomato, *J. Plant Physiol.*, 141, 234–239.

Wallis, J.G., Wang, H.Y., and Guerra, D.J. (1997) Expression of a synthetic antifreeze protein in potato reduces electrolyte release at freezing temperatures, *Plant Mol. Biol.*, 35, 323–330.

Wang, J.H. (2000) A comprehensive evaluation of the effects and mechanisms of antifreeze proteins during low-temperature preservation, *Cryobiology* 41, 1–9.

Watts, W.R. (1972) Leaf extension of *Zea mays*. II. Leaf extension in response to independent variation of the temperature of the apical meristem, of the air around the leaves and the root zone, *J. Exp. Bot.*, 23, 713–721.

Warren, G., McKown, R., Marin, A., and Teutonica, R. (1996) Isolation of mutations affecting the development of freezing tolerance in *Arabidopsis thaliana* (L) Heynh, *Plant Physiol.*, 111, 1011–1019.

White, A.J., Dunn, A.M., Brown, K., and Hughes, M.A. (1994) Comparative analysis of genomic sequence and expression of a lipid transfer protein gene family in winter barley, *J. Exp. Bot.*, 281, 1885–1892.

Williams, R.J. (1992) Anomalous behaviour of ice in solutions of ice-binding arabinoxylans, *Thermochim. Acta*, 212, 105–113.

Williams, W.P. (1990) Cold-induced lipid phase transitions, *Phil. Trans. R. Soc. Lond. Series B—Biol. Sci.*, 326, 555–570.

Wisniewski, M. and Fuller, M. (1999) Ice nucleation and deep supercooling in plants: new insights using infrared thermography, in *Cold-Adapted Organisms—Ecology, Physiology, Enzymology and Molecular Biology*, Margesin R. and Schinner, F., Eds., Springer, Berlin, pp. 105–118.

Wisniewski, M., Lindow, S.E., and Ashworth, E.N. (1997) Observation of ice nucleation and propagation in plants using infrared video thermography, *Plant Physiol.*, 113, 327–334.

Wolber, P.K. (1993) Bacterial ice nucleation, *Adv. Micro. Physiol.*, 34, 203–237.

Worrall, D., Elias, L., Ashford, D., Smallwood, M., Sidebottom, C., Lillford, P., Telford, J., Holt, C., and Bowles, D. (1998) A carrot leucine-rich-repeat protein that inhibits ice recrystallization, *Science*, 282, 115–117.

Xin, Z.G. and Browse, J. (1998) *eskimo1* mutants of Arabidopsis are constitutively freezing-tolerant, *Proc. Nat. Acad. Sci.*, *USA*, 95, 7799–7804.

Xin, Z.G. and Browse, J. (2000) Cold comfort farm: the acclimation of plants to freezing temperatures, *Plant Cell Envir.*, 23, 893–902.

Xiong, L.M., Ishitani, M., Lee, H., and Zhu, J.K. (2001) The Arabidopsis LOS5/ABA3 locus encodes a molybdenum cofactor sulfurase and modulates cold stress- and osmotic stress-responsive gene expression, *Plant Cell*, 13, 2063–2083.

Xu, W., Cambell, P., Vargheese, A.K., and Braam, J. (1996a) The *Arabidopsis* XET-related gene family—environmental regulation and hormonal regulation of expression, *Plant J.*, 9, 879–889.

Xu, D.P., Duan, X.L., Wang, B.Y., Hong, B.M., Ho, T.H.D., and Wu, R. (1996b) Expression of a late embryogenesis abundant protein gene, *HVA1*, from barley confers tolerance to water deficit and salt stress in transgenic rice, *Plant Physiol.*, 110, 249–257.

Yamaguchi-Shinozaki, K. and Shinozaki, K. (1994) A novel *cis*-acting element in an *Arabidopsis* gene is involved in responsiveness to drought, low-temperature, or high-salt stress, *Plant Cell*, 6, 251–264.

Yu, X.M. and Griffith, M. (1999) Antifreeze proteins in winter rye leaves form oligomeric complexes, *Plant Physiol.*, 119, 1361–1369.

Záмećník, J., Bieblovaa, J. and Grospietch, M. (1994) Safety zone as a barrier to root-shoot ice propagation, *Plant and Soil*, 167, 149–155.

Zhang, L., Dunn, M.A., Pearce, R.S., and Hughes, M.A. (1993) Analysis of organ specificity of a low temperature responsive gene family in rye (*Secale cereale* L), *J. Exp. Bot.*, 44, 1787–1793.

Zhang, H., Wang, J., Nickel, U., Allen, R.D., and Goodman, H.M. (1997) Cloning and expression of an *Arabidopsis* gene encoding a putative peroxisomal ascorbate peroxidase, *Plant Mol. Biol.*, 34, 967–971.

Zhang, M.I.N. and Willison, J.H.M. (1992) Electrical impedance analysis in plant tissues: the effect of freeze-thaw injury on the electrical properties of potato tuber and carrot root tissues, *Can. J. Plant Sci.*, 72, 545–553.

Zhu, J.-J., and Beck, E. (1991) Water relations of *Pachysandra* leaves during freezing and thawing, *Plant Physiol.*, 97, 1146–1153.

6 Oxidative Stress in the Frozen Plant: A Free Radical Point of View

Erica E. Benson and David Bremner

CONTENTS

6.1 INTRODUCTION

The metabolism, development, and growth of photosynthetic organisms are influenced by environmental temperature. Thus, plant and algal evolution has largely been dictated by the ability of aerobic, photoautotrophic organisms to survive and reproduce in specific temperature zones and environmental niches (see Chapters 3 and 5). Survival in temperate, tropical, and extreme environmental conditions has important consequences for the exploitation of economically important plants cultivated outside their evolutionary center of origin. Survival of plants and algae (see Chapter 3) at low temperatures is influenced by their complex biochemical responses to environmental cues, which are manifest at molecular, genetic, and physiological levels (see Chapter 5). If we are to understand the basis of oxy-phototropic stress adaptations to "life in the frozen state," it is important to appraise the complex relationships between low-temperature stress, primary metabolism, and oxidative stress.

Oxidative, metabolic perturbations are encountered by all aerobic organisms, and they will consequentially alter the redox state of cells, eventually leading to generic stress responses mediated by reactive oxygen species (ROS). For this reason, free radical-mediated oxidative stress is an important, but little understood, factor in chilling and subzero cryogenic injury. Photosynthetic and photo/heterotrophic (i.e., photosynthetic organisms maintained under "artificial" *in vitro* heterotrophic conditions) organisms exposed to freezing are particularly at risk from free radicals. Low temperatures can disturb primary metabolic pathways, disrupt membranes, and compromise the efficient functioning of enzyme-driven antioxidant protection systems.

Oxygen is essential to all aerobic life (Lane, 2002), and it has a central role in primary metabolism, electron transfer reactions, and substrate level oxidations. Aerobes are therefore constantly at risk from free radical–mediated oxidative stress and, as a result, have developed complex antioxidant systems that protect against damaging radicals and their toxic reaction products. Plants are especially vulnerable because O_2 is a terminal electron acceptor in plant respiration and is produced by the water-splitting reactions of photosynthesis. Moreover, plant secondary metabolism comprises many reactions in which O_2 and ROS participate in pathogen defense. These "chemical" protection systems have evolved, in part, as a result of a sedentary response to stress and life-cycle adaptations. Oxidative cell signaling (Guan et al., 2000) may also account for the astonishing capacity that plants have to react to the different environmental cues that trigger stress responses and life-cycle changes (e.g., senescence, rejuvenation, flowering, and dormancy). Although ROS have an essential role in plants, oxidative pathways can become disrupted at low temperatures. Thus, the aim of this chapter is to provide a perspective of the chemical and biochemical basis of cryoinjury, explore the evidence for the occurrence of oxidative stress in the frozen state, and consider the value of this knowledge in applied, cryogenic plant research.

Knowledge derived from studies of other biological systems will also be discussed, as this offers multidisciplinary insights into the importance of oxidative stress in cryoinjury. Past and contemporary theories pertaining to understanding the biochemical basis of freezing stress will be reviewed and the work of early pioneers (e.g., Heber, 1968; Levitt, 1956, 1962, 1966; Tappel, 1966) revisited. These first advances in low-temperature biology are highly relevant today, particularly if reevaluated in relation to molecular studies (see Chapter 5).

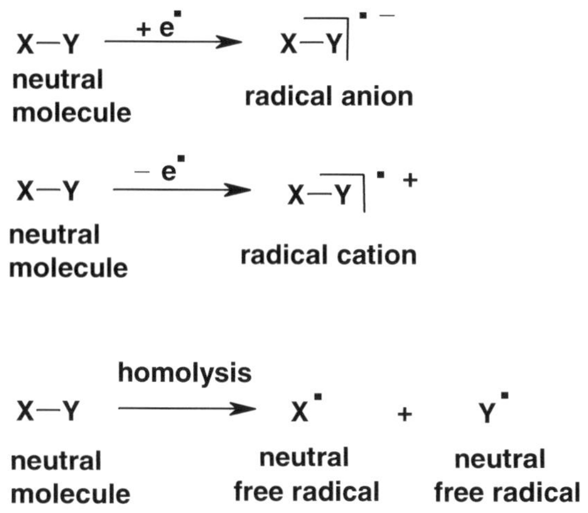

$$X-Y \xrightarrow{\;+e^{\bullet}\;} X\overline{-Y}\rceil^{\bullet\,-}$$

neutral
molecule   radical anion

$$X-Y \xrightarrow{\;-e^{\bullet}\;} X\overline{-Y}\rceil^{\bullet\,+}$$

neutral
molecule   radical cation

homolysis

$$X-Y \xrightarrow{\qquad} X^{\bullet} \quad + \quad Y^{\bullet}$$

neutral   neutral   neutral
molecule   free radical   free radical

SCHEME 6.1 Formation of radicals.

6.2 PRINCIPLES OF FREE RADICAL CHEMISTRY AND BIOCHEMISTRY

Most chemical and biochemical reactions proceed either by the breaking/forming of covalent bonds or with the involvement of ions. Little consideration was given before 1950s of a third possibility, in which molecules with unpaired electrons might exist, and to the idea that this type of reaction mechanism might be important in biological systems. We now know that a myriad of free radical reactions are crucial to the survival and, indeed, death of all living things. This section will concentrate on providing a background to free radical chemistry/biochemistry in general and for reactions at low temperature in particular so that the later discussion of events occurring in the frozen state may be better understood.

6.2.1 Free Radicals and ROS

A radical can be defined as any species that contains one or more unpaired electrons. However, there is considerable debate about what "free" means in relation to radicals, and it is best to consider such radicals as those that, when generated, will participate immediately in new chemical reactions. It is also important to note that stable radicals can also be formed by their interaction with radical trapping agents, and these are often used in electron paramagnetic resonance (EPR) studies. Free radical reactions involving biological molecules will be considered in more detail later. The presence of unpaired electrons means that free radicals are highly reactive (usually indiscriminately, though stabilities vary) and paramagnetic (a magnetic property arising from the presence of an unpaired electron), which accounts for the importance of free radicals in chemistry and biology.

Free radicals can be formed in a number of different ways, which can be represented by the addition of an electron to a neutral molecule, the loss of an electron from a neutral molecule, or cleavage of a bond (two electrons) between atoms such that each species receives an electron. This latter process is termed homolysis (Scheme 6.1).

It is generally not appreciated that oxygen gas (O_2), having two unpaired electrons with the same spin quantum number (parallel spins), is classed as a radical species, though not a "free radical." As a consequence, normal ground state diatomic oxygen is said to be in its triplet state ($^3O_2$), and although O_2 can act as an oxidizing agent, it tends to do so by accepting electrons only one at a time, and this, paradoxically and fortunately, renders it much less reactive than might be

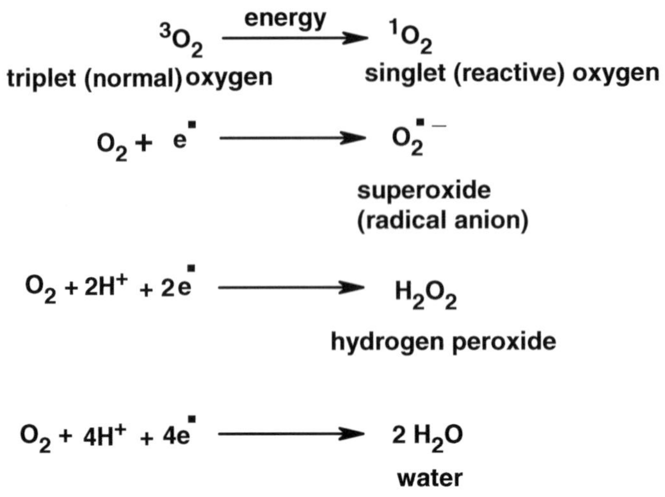

$$^3O_2 \xrightarrow{\text{energy}} {}^1O_2$$

triplet (normal) oxygen   singlet (reactive) oxygen

$$O_2 + e^{\bullet} \longrightarrow O_2^{\bullet -}$$

superoxide
(radical anion)

$$O_2 + 2H^+ + 2e^{\bullet} \longrightarrow H_2O_2$$

hydrogen peroxide

$$O_2 + 4H^+ + 4e^{\bullet} \longrightarrow 2\,H_2O$$

water

SCHEME 6.2 Addition of energy, electrons, and protons to oxygen.

expected for a true free radical. Different forms of oxygen species can be considered to be derived from $^3O_2$ by the addition of energy or electrons and can result in radicals or nonradicals (Scheme 6.2).

If energy, usually in the form of ultraviolet light, is imparted to "normal" diatomic oxygen ($^3O_2$), it is converted into the much more reactive singlet oxygen ($^1O_2$). Addition of one electron to $^3O_2$ leads to a radical anion, which is usually referred to as the superoxide radical (O_2^-) or superoxide (Asada, 1999). Although superoxide is indeed a radical, it is normally less reactive than O_2, but in its protonated state, it may become more reactive (De Grey, 2002). Certainly, superoxide does have significant involvement in plant stress, as will be seen later. When two electrons, along with two protons, are added to O_2, the resulting species is hydrogen peroxide, and if four electrons and four protons are added, then the final product is water. Arguably, the most important, and indeed, the most oxidizing species, is the hydroxyl radical ($\cdot OH$), which has a standard reduction potential of +2.31 V (oxidizing), which compares with –0.33 V (reducing) for oxygen going to superoxide—shown in Scheme 6.2 (Buettner, 1993). Biologically, the most common source of hydroxyl radicals is the Fenton reaction. Although the precise mechanism and the role of the metals involved is by no means delineated, it is common to consider that the hydroxyl radical is generated from H_2O_2 and some form of ferrous (Fe^{2+}) ion, as shown in Scheme 6.3.

Hydrogen peroxide is produced in plant systems via a number of pathways, and consequently, hydroxyl radicals may be readily generated by Fenton or related chemistry. Some enzymes such as those involved in xanthine and P450 hydroxylation give rise to hydrogen peroxide directly, but most are generated by the superoxide dismutase (SOD) group of enzymes (Fridovich, 1995). An overview of the generation of the most important reactive oxygen species is given in Scheme 6.4.

6.2.2 Free Radical–Mediated Secondary Oxidative Reactions

The generation of ROS, and the hydroxyl radical in particular, is only the beginning of a series of events that can lead to free radical–mediated secondary oxidative reactions. As discussed above, it

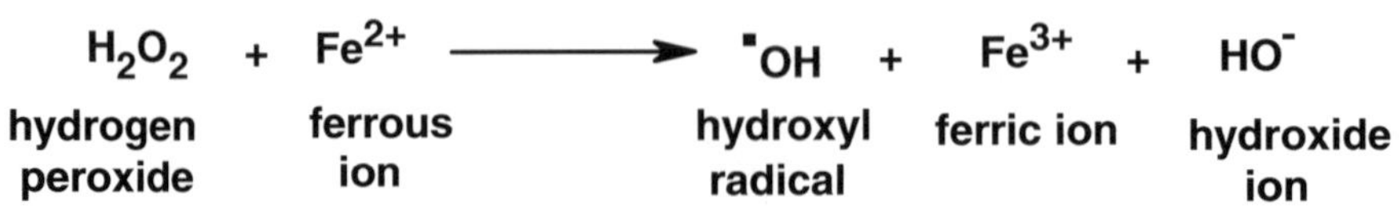

$$H_2O_2 + Fe^{2+} \longrightarrow {}^{\bullet}OH + Fe^{3+} + HO^-$$

| hydrogen peroxide | ferrous ion | hydroxyl radical | ferric ion | hydroxide ion |

SCHEME 6.3 The Fenton Reaction.

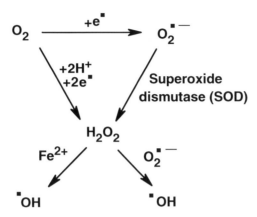

SCHEME 6.4 Generation of reactive oxygen species (ROS) from O_2.

is known that the hydroxyl radical is the most oxidizing and damaging of the ROS, and its reactions are exemplified by interactions with DNA, lipids, and proteins. Halliwell and Gutteridge (1999) give an excellent, extensive discussion on the reactions and the consequences of hydroxyl radicals interacting with biomolecules, so what follows is a brief overview.

6.2.2.1 Reaction of DNA with Free Radicals

Surprisingly, DNA or RNA bases do not react in significant amounts with superoxide, nor with hydrogen peroxide. However, hydroxyl radicals do oxidize, or hydroxylate, sugars, purines, and pyrimidines indiscriminately, and overall, the process is usually referred to as "oxidative damage to DNA." The resulting products are implicated in mutagenesis, carcinogenesis, and aging, and DNA damage mediated by free radicals will have important consequences for genetic stability (Breen and Murphy, 1995; Dizdaroglu, 1992, 1998; Halliwell and Gutteridge, 1999). Such damage can be repaired via base excision, but major problems arise when the system is overwhelmed by excess hydroxyl radicals (Friedberg et al., 1995). The number of possible reactions is enormous, but perhaps the most ubiquitous reaction occurs when ·OH reacts with guanine at the C-8 position. This adduct can be reduced, oxidized, or ring opened, depending on the conditions (Scheme 6.5). Such products have been used as measures of stress in mammals (Halliwell and Arouma, 1991) and in plants (Floyd et al., 1989). There is increasing evidence *in vivo* that DNA damage contributes to the age-related development of cancer (Cadenas and Davies, 2000; Marnet, 2000).

6.2.2.2 Lipids and Lipid Peroxidation

In a complex series of reactions, hydroxyl radicals begin the process of lipid peroxidation where any activated CH_2 groups of polyunsaturated fatty acids are attacked to generate lipid free radicals,

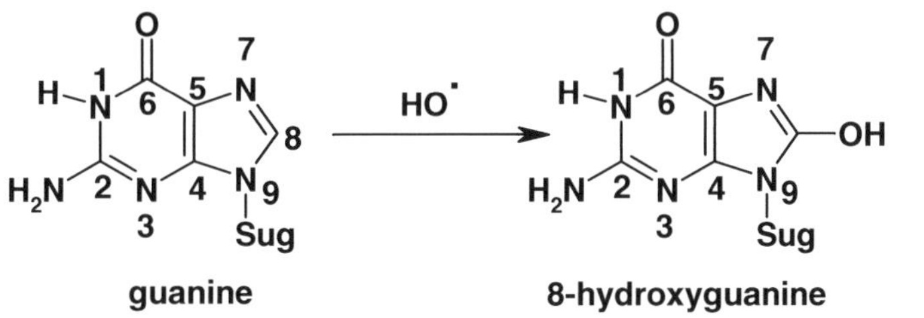

guanine   8-hydroxyguanine

SCHEME 6.5 Reaction of hydroxyl radical with guanine.

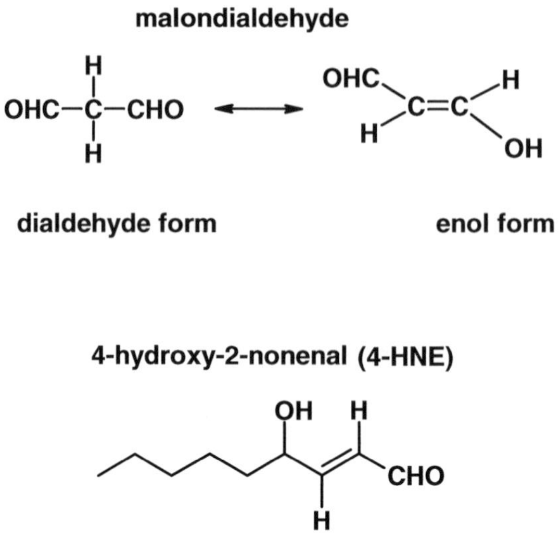

malondialdehyde

$$OHC-\overset{\overset{\displaystyle H}{|}}{\underset{\underset{\displaystyle H}{|}}{C}}-CHO \quad \longleftrightarrow \quad \overset{\displaystyle OHC}{\underset{\displaystyle H}{}}C=C\overset{\displaystyle H}{\underset{\displaystyle OH}{}}$$

dialdehyde form   **enol form**

4-hydroxy-2-nonenal (4-HNE)

SCHEME 6.6 Structures of malondialdehyde and 4-HNE.

which react with oxygen to form lipid hydroperoxides. Under biological conditions, these peroxides are reduced to the corresponding hydroxy acids, the most common of which are hydroxyoctadec-adienoic acids, which are often used as markers of lipid peroxidation. In some diseases the amount of hydroxyoctadecadienoic acids present in low-density lipoproteins is up to 100 times higher than in healthy individuals (Spiteller, 1998). Another fate of lipid hydroperoxides is decomposition and rearrangement into secondary reaction products. Some compounds, such as ethane and ethylene, are relatively innocuous, whereas others like malondialdehyde and 4-hydroxy-2-nonenal (Scheme 6.6) are considered to be highly toxic and mutagenic (Esterbauer et al., 1988, 1991).

Malondialdehyde (MDA) can react with DNA bases, particularly guanine, to give mutagenic lesions, but the evidence of MDA mutagenicity *in vivo* is by no means convincing (Benamira et al., 1995), and polymeric forms of MDA may actually account for the observed effects (Riggins and Marnet, 2001). Hydroxynonenal (HNE) has been shown to inhibit cell growth and is genotoxic and chemotactic (Muller et al., 1996). Esterbauer et al. (1988, 1991) have speculated that HNE was the major toxic component that caused the "Spanish cooking oil syndrome."

6.2.2.3 Proteins and Free Radicals

Although direct interaction of hydroxyl radicals with proteins is possible, it is the secondary products of lipid peroxidation that really cause problems. MDA, having two aldehyde functions present, can readily react with proteins containing lysine residues, and cross links are formed between protein chains (Scheme 6.7)

4-HNE, being an unsaturated hydroxyaldehyde, can undergo an array of reactions to form adducts with DNA, proteins, phospholipids, and amines (giving Schiff's bases) and also participates in Michael reactions with GSH (glutathione; Scheme 6.8)

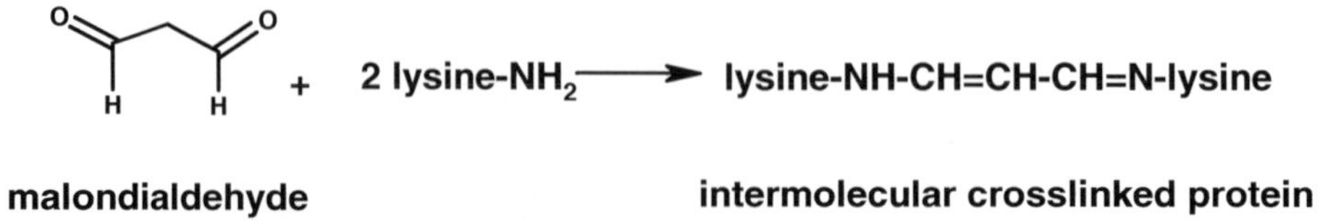

+ 2 lysine-NH$_2$ ⟶ lysine-NH-CH=CH-CH=N-lysine

malondialdehyde   **intermolecular crosslinked protein**

SCHEME 6.7 Formation of two Schiff's base bonds between proteins and malondialdehyde.

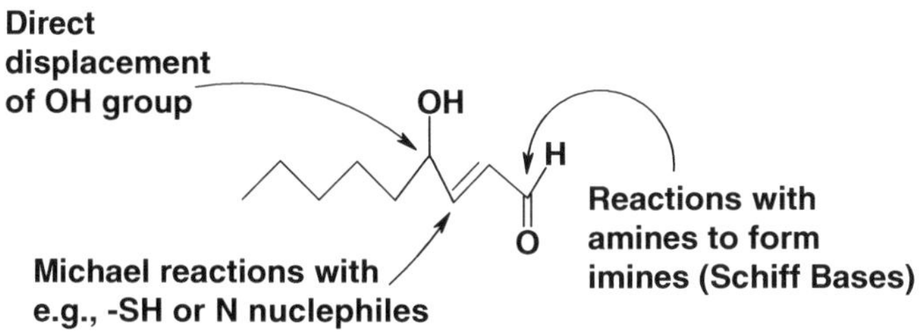

Direct displacement of OH group

OH

H

O

Reactions with amines to form imines (Schiff Bases)

Michael reactions with e.g., -SH or N nuclephiles

SCHEME 6.8 Potential reactions of 4-HNE.

For example, Esterbauer et al. (1991) reported that 4-HNE reacts with sulfhydryl groups of proteins, cysteine, and glutathione to yield stable thioether derivatives, and Uchida and Stadtman (1992) showed that in studies on insulin, which does not contain any sulfhydryl groups, histidine is the only amino acid that is modified on exposure to HNE. Polyclonal antibodies have been used to further characterize the location and type of reaction of HNE with proteins (Hartley et al., 1997). All of these potential reactions will cause alterations to the structure and function of proteins and enzymes, and this will have a profound effect on cell metabolism and stability.

6.2.3 Free Radical Chemistry at Low and Ultra-Low Temperatures

The reactions described above are those commonly found to occur in mammals at around 37°C and in plants between 5° and 30°C. However, it is now timely to consider the effects of low temperatures and cryogenic temperatures on fundamental free radical chemistry. This is because free radicals may have very important implications for "life in the cryogenic frozen state." Cryopreservation is increasingly being applied for the long-term maintenance of living cells, tissues, and organs, and there has been a recent progression toward using different modes of cryogenic storage. For example, storage in vapor as compared to the liquid phase of liquid nitrogen and storage in the vitrified as opposed to the crystalline (frozen) state. The latter comprises a system in which there remains a proportion of unfrozen water associated with either or both the cells/tissues and the extracellular and cryoprotectant medium. It may, therefore, be important to consider the comparative stability of the oxidative chemistry of cells held under such different cryogenic conditions and to pose the following questions:

1. Will all chemical reactions, including radical reactions, cease at low temperatures, or will free radical–mediated stress manifest itself even at −196°C?
2. Will free radical chemistry (as opposed to active, metabolically driven reactions) have significant consequences over longer-term periods of cryogenic (vitrified and frozen) storage?
3. Will free radical chemistry (as opposed to active, metabolically driven reactions) be influenced by the phase (vapor or liquid) of liquid nitrogen cryogenic storage?
4. Will water phase and moisture content status of the cryogenically stored cells influence the chemistry of free radical reactions and hence the long-term stability of preserved material?

Surprisingly, very little work has been carried out on nonbiological free radical reactions at low temperatures, probably because there is little desire for synthetic chemists to study reactions at such low temperatures, as they are most interested in reactivity at ambient temperature. Interestingly, superoxide radicals trapped using the spin-trap reagent 5-diethoxyphosphoryl-5-methyl-1-pyrroline N-oxide can be successfully recorded using EPR at −196°C (77K), and the signal intensity remains unaltered for up to 7 d, thus indicating that if the radicals are generated at this temperature, they may exist for a considerable amount of time (Dambrova et al., 2000). Studies of low temperatures on free radical generation in model nonbiological systems can offer valuable insights and help in the study of free radical damage in the context of cryobiology. Thus, Symons (1982) explored the generation of free radicals in irradiated liquid and frozen water using EPR spectroscopy. Such analytical approaches are very useful in the study of free radical processes at ultra-low temperatures, as unstable primary radical products can be "trapped" within the crystalline lattice and accumulated at concentrations that are detectable. In the case of Symons's (1982) studies, H· and ·OH were the only radical species considered to be of significance in irradiated low-temperature solids. Secondary processes also involved the reaction of H· and ·OH to form H_2O_2, but this reaction did not occur at $-269°C$ (4K). Only ·OH radicals were detected in frozen water irradiated at $-196°C$ (77K). In aqueous organic systems, a phase separation can be seen in which ·OH radicals accumulate in the ice phase and organic radicals accumulate in the organic phase (Symons, 1982). The formation of glasses (vitrification) at low temperatures is a survival strategy in plants, and achieving the glassy state is a crucial cryoprotective strategy. It may, therefore, also be important to consider the chemical stability of vitrified cells, especially in relation to the formation of ·OH radicals (see also Chapter 20).

6.3 METABOLISM, REDOX COUPLING, CELLULAR INTEGRITY: EFFECT OF LOW TEMPERATURES

Exposure of plants to low-temperature stress can comprise several sequential stress-inducing components: chilling, freezing (or vitrification), thawing, and recovery. The ability to survive different phases of low-temperature treatments and their associated stresses (desiccation, dehydration) depend on tolerance, sensitivity, and natural life-cycle adaptations. Many temperate plants are able to acclimate (see Chapters 5 and 10) after exposure to low positive temperatures, enabling the plant to survive further low-temperature episodes. The ability to cold acclimate and enter dormancy cycles forms an important survival strategy in temperate plants, allowing overwintering. The natural ability to acclimate to low-temperature stress has also been used as a cryoprotective strategy in the cryogenic storage of plant germplasm. However, exposure of nonacclimating species to chilling can also potentially effect the ability of cold-sensitive species to succumb to subsequent and more extreme episodes of chilling and freezing. The process of cold acclimation involves several cell-signaling pathways, but the mechanistic and molecular bases of these pathways are complex (Thomashow, 1999; Xin and Browse, 2000). Identifying freeze-specific genes from genes involved in other cold/dehydration stress responses is difficult (see Chapter 5), and assigning an oxidative stress component to freeze tolerance is, to date, still open for debate. Importantly, increased levels of antioxidants have been associated with freezing tolerance (Xin and Browse, 2000).

Ouellet (2002) reviews the induction of cold-tolerance gene expression in acclimated plants and cautions that as freezing stress has a symptomology common to drought and osmotic and salinity stress, it is difficult to ascertain the molecular signaling basis of freezing-specific stress. Mechanisms by which plant cells actually perceive low temperatures are not well understood, and Ouellet (2002) presents a signal detection and transduction pathway based on the interaction of a putative cold receptor coupled with light (chloroplast-based) perception, with the intracellular signal being mediated by Ca^{2+} and protein phosphorylation. Importantly, metabolic shifts are highlighted as being of importance in the induction of freezing tolerance, although free radical signaling processes *per se* have not been considered a major component of freeze-tolerance acquisition. However, Pearce (Chapter 5) discusses the importance of antioxidant gene expression and molecular signaling in cold acclimation and freezing. Thus, oxidative stress can certainly increase in plants challenged by cold treatments (and possibly freezing), and this has been evidenced by transgenic plant experiments. To complement the molecular study of Pearce (Chapter 5), this chapter will explore the potential effects of chilling and freezing in low-temperature metabolic pathology and under oxidative stress.

Loss of physical compartmentalization and metabolic uncoupling has a catastrophic effect on both primary metabolism and antioxidant defenses. When this loss occurs, as can be the case during freezing (Singh and Miller, 1984), metabolic disruption will undoubtedly lead to the production of toxic free radicals and their reaction products. Maintenance of metabolic control (e.g., by rate-controlling enzymes in primary pathways), redox coupling, and cellular integrity (membrane stability) are the first lines of defense against oxidative stress, together with the stringent scavenging of free radicals formed at metabolic sites involving O_2 and electron transfer systems.

Pioneering researchers (see Tappell, 1966, for a review) demonstrated that mitochondrial enzymes and ATP synthesis become uncoupled by freeze–thaw cycles. A number of enzymes (e.g., pyruvate phosphate dikinase, phosphoenol pyruvate carboxylase, ribulose bisphosphate carboxylase/oxygenase, phosphofructokinase, fructose-1,6,-bisphosphate, glyceraldehyde-3-phosphate dehydrogenase, pyruvate kinase, and NADP reductase) that are key to the regulation of primary metabolic pathways have been found to be chilling/freezing labile in cold-sensitive plants. Similarly, the functionality of antioxidant enzymes (i.e., catalase, glutathione reductase, and ascrobate peroxidase) is compromised at freezing and chilling temperatures (Guy, 1990).

Cold stress, can therefore, disrupt metabolism and disturb redox balance, and as a result, the potential exists for the formation of free radicals, which then attack key macromolecules such as polyunsaturated fatty acids (see Section 6.2). Secondary oxidative damage is promoted by lipid peroxidation, which interferes with membrane-based cell signaling. Under extreme stress, membrane destruction ensues with the complete and irreversible loss of cellular integrity. Plants therefore require highly complex biochemical and metabolic control mechanisms, which modulate the oxidative pathways of primary metabolism (photosynthesis and respiration). Metabolic control must also be fully integrated with antioxidant protection processes, most of which comprise enzymatic components. It thus follows that as low temperatures can severely compromise enzyme kinetics, plants exposed to chilling and freezing injury must have the capacity to both regulate and protect their oxidative processes in the photosynthetic and respiratory pathways.

6.3.1 REACTIVE OXYGEN SPECIES: PHOTOINHIBITION, PHOTOPROTECTION, AND PHOTOOXIDATION

Photosynthesis is the primary metabolic pathway in plants and comprises the "light" and the "dark" reactions. The former involves the capture of light energy by photosynthetic pigments and the transfer of this energy through an electron transport chain that generates reducing power (NADPH) and energy (ATP). These are used in the "dark reactions" of the Calvin cycle to "fix CO_2." The light reactions are highly susceptible to oxidative stress because of the generation of O_2 from the "splitting of water" (formed at photosystem II [PSII]) and the propensity to form excited electronic states in reaction centers situated in close proximity to electron transport chains. Photosystem I (PSI) and PSII are vulnerable to chilling stress, as these components of the photosynthetic unit contain the reaction center chlorophylls linked to light-harvesting antennae. Low-temperature stress can potentially exacerbate the production of free radicals and deleterious excited electronic states in the light reactions of photosynthesis. Moreover, as episodes of chilling may precede exposure to subzero freezing temperatures, it is particularly important to consider the combined effects of both chilling and freezing on the maintenance of photosynthetic integrity. Photoprotection mechanisms in oxygenic organisms form a complex set of integrated processes (see Niyogi, 1999, for a review):

1. Adjustment of the chorophyll antenna
2. Photoprotection by carotene and xanthophylls pigments
3. Thermal dissipation

4. Modulation of CO_2 fixation
5. Photorespiration
6. Repair and synthesis of pigment–protein complexes
7. Oxidative cycling of pigments
8. Antioxidant systems
9. PSI cyclic electron transfer
10. Water–water cycling
11. Quenching of excited reaction center complexes by carotenoids

Photoinhibition in this broad context can be considered to occur when primary photosynthetic metabolism is compromised by the uncoupling of the dark and light reactions, leading to the propagation of photooxidative pathways and free radicals (e.g., Lea and Leegood, 1999). Osmond et al. (1999) offer a more explicit definition of photoinhibition and photoprotection based on PSII photochemistry and chlorophyll fluorescence criteria that differentiate them *in vivo*. Photoprotection involves a light-dependent, rapid, and reversible decline in the efficiency of primary photochemistry associated with the antenna complexes of PSII. The fluorescence characteristics of the process indicate that excitation energy is dispersed in a thermal process rather than via transfer to a PSII reaction center in which energy is dissipated via photochemistry. Photoinactivation, as defined by Osmond et al. (1999), is defined as a fluorescence-specific, light-dependent reversible decline in the primary PSII photochemistry, which results in a reduction in the number of functional reaction centers. Photoprotection is thought to reduce photoinactivation. The replacement of nonfunctional PSII reaction centers involves the complex regulation of chloroplast protein synthesis and repair, with the D1 protein being the most sensitive in the P680 reaction centre. This protein is responsible for binding the redox components in photosynthetic charge separation and plastoquinone reduction, and is stabilizes the water-spilling complex (Kyle et al., 1984, as discussed by Osmond et al., 1999). It has been termed the "suicide polypeptide" because of its rapid turnover in photosynthetic destruction–repair cycles.

PSII centers that have lost their functionality may have a photoprotective role, as they can still dissipate excitation energy as heat and preferentially protect functional PSII complexes. In terms of low-temperature stress, it is thought that this defensive mechanism is of considerable importance, as protective enzymatic reactions will be impaired by low-temperature rate-limiting factors (Öttander et al., 1993, as discussed by Osmond et al., 1999). There is evidence for photoinhibition occurring in cold-challenged plants, manifest as a light-dependent decrease in the quantum yield of PSII photochemistry (Hurry et al., 1992). Osmond et al. (1999) report the unpublished findings of J.J.G. Egerton, who examined seasonal changes in the chlorophyll fluorescence of *Eucalyptus pauciflora* located in mountainous subalpine woodlands of southeast Australia. Unhardened plants experienced severe frosts in autumn and had fluorescence profiles typical of temporary photoinactivation. Cold-hardening of plants can reduce susceptibility to low-temperature-induced photoinhibition, as observed for rye by Öquist and Huner (1993). Hurry et al. (1992) examined the effects of long-term photoinhibition on the growth and photosynthesis of cold-hardened spring and winter wheat and concluded that in the case of cold-tolerant plants, photoinhibition PSII provides a mechanism for stable downregulation of photochemistry. This matches the overall demand for ATP and reducing power in the dark reactions.

The water–water cycle in chloroplasts has been proposed by Asada (1999) as a means of scavenging active O_2 species and dissipating excess photons. PSI has a primary function in the photoreduction of $NADP^+$ in the cyclic and noncyclic flow of electrons in the photosystem electron transport chain. In the case of PSI, Asada (1999) proposes that two molecules of O_2 are reduced by electrons from PSII to produce superoxide anion radicals (O_2^-). These are then removed at PSI by superoxide dismutases (Cu/Zn primarily) and the H_2O_2 so produced is reduced to water by ascorbate via the catalysis of ascorbate peroxidase. Oxidized ascorbate molecules (monohydroascorbate radicals and dehydroascorbate) are then reduced to ascorbate using two electrons from PSII.

Concomitantly, one molecule of O_2 is reduced to two molecules of water in PSI (using four electrons derived from two molecules of H_2O_2 in PSII). Thus, photoreduction of O_2 in PSI can generate O_2^- as a major product, and in the intact chloroplast O_2^- and H_2O_2 are rapidly scavenged by a complex array of antioxidants including CuZn superoxide dismutase and the ascorbate-monodehydroascorbate-glutathione recycling enzymes. These are located at the site of O_2 generation, and they prevent inactivation of the PSI. Thus, the whole process is dependent on the maintenance of chloroplast structural integrity and redox state and the control of the production of ROS through stringent metabolic coupling and antioxidant protection. The reaction rates for photoreducing-reduced species of O_2 to water is several orders of magnitude higher than that of O_2, which ensures that the water–water cycle (Asada, 1999) rapidly scavenges O_2^- and H_2O_2 before they have the chance to interact with and damage other molecules. Avoidance of stress scenarios that lead to the production of ·OH is particularly important (see Section 6.2), as disturbance of the water–water cycle can lead to the formation of ·OH from H_2O_2 via transition metal cation-mediated Fenton chemistry (see Scheme 6.3). Thus, the overall function of the water–water cycle is to operate in concert with other enzyme-based antioxidant systems to protect plants from photoinhibition and to dissipate excess photon energy. Light energy can become harmful if the CO_2 assimilation mechanisms are compromised by environmental challenges associated with stress-inducing fluctuations in temperature, water status, and light.

Fryer et al. (1998) reported that leaves of maize (a C4 plant, see Section 6.3.2.1) enhanced their water–water cycle activity in response to chilling. Similarly, the enzymes of the water–water cycle in *Zea* spp. are inactivated by chilling (Jahnke et al, 1991), and it is possible that low-temperature-induced photoinhibition of PSI can occur if antioxidant enzymes (e.g., SOD and ascorbate peroxidase) are inhibited (Tjus et al., 1998). Conversely, chilling-tolerant genotypes of maize contained higher activities of enzymes associated with the water–water cycle than a cold-intolerant genotype (Pinhero et al., 1997), and during cold acclimation, scavenging enzymes are induced (Fryer et al., 1998; O'Kane et al., 1996). In combination, these studies provide strong evidence for the involvement of antioxidants, ROS, and the water–water cycle of PSI in plant cold-stress tolerance and sensitivity. Thus, low temperatures can exacerbate photoinhibition, as rates of electron transport processes will be compromised by the low-temperature reduction of Rubisco and Calvin cycle enzyme activities (see Section 6.3.2.2). Moreover, it could be postulated that risks of photoinhibition may be greater if freezing stress exacerbated by colligative dehydration injuries compromises rate-limiting enzyme repair mechanisms.

6.3.2 ADAPTIVE MODULATION OF CARBON FIXATION PATHWAYS: IMPLICATIONS FOR LOW-TEMPERATURE RESEARCH?

Plants have developed a wide range of protective mechanisms that enable them to modulate (in tandem) the light and dark processes involved in photosynthesis. This supports survival in temperate, tropical, extreme, and fluctuating environments. Temperature, light, CO_2 and O_2, and water status are the main factors that signal adaptive changes, and these can be defined by specific carbon fixation processes.

6.3.2.1 C3 Metabolism and Photorespiration

Different modes of photosynthesis are defined as to whether the first stable product of carbon fixation is a C3 or a C4 carbon unit. The C3 pathway involves the Calvin cycle (see Uno et al., 2001, for a textbook account) in which atmospheric CO_2 is converted to sugars using the reducing power (NAPH) and energy (ATP) generated by the light reactions of photosynthesis. The enzyme responsible for the initial fixation of atmospheric CO_2 is ribulose-1,5-bisphosphate carboxylase/oxygenase (Rubisco), which "adds" CO_2 to the five-carbon sugar, ribulose-1,5-bisphosphate. The first stable products of the catalysis are two molecules of 3-phosphoglycerate, and this mode of photosynthesis is called C3 metabolism. If CO_2 levels are limited, Rubisco preferentially acts as an oxygenase (a process termed photorespiration) as opposed to a carboxylase. In photorespiration, a molecule of 3-phosphoglycerate and a 2-C molecule, phosphoglycolate, are the first stable products that are formed. There is no net carbon input into the Calvin cycle during the process of photorespiration, and this pathway is thought, by some, to be wasteful (Uno et al., 2001), as it results in a reduced fixation of CO_2. Note that the concentration at which plants show no net carbon fixation is termed the CO_2 compensation point and can be influenced by photorespiration and be considered a relict adaptation to ancient levels of atmospheric O_2 and CO_2, different to those that exist today (see Lane, 2002; Uno et al., 2001). However, the fact that photorespiration occurs in "contemporary" C3 plants exposed to high light intensities and elevated temperatures is important. C3 plants can, from time to time, be exposed to conditions that promote desiccation stress and, hence, reduce the levels of CO_2 resulting from stomatal closure. Importantly, photorespiration can still operate at the compensation point by recycling CO_2, and it is thought to have a major part in the amelioration of damage caused by the oxidative effects of "excess light energy" absorbed by the light reactions of photosynthesis (Lea and Leegood, 1999; Niyogi, 1999). Thus, photorespiration is thought to indirectly dissipate light excitation energy that cannot be used efficiently through compromised Calvin cycle enzymes and the carboxylase functionality of Rubisco. When leaves are exposed to bright illumination at low CO_2 levels, the reduction of oxygen by uncoupled electron transfer processes may occur, leading to the generation of potentially toxic free radical species (see Section 6.3.1). These radicals attack the photosynthetic membranes and cause the destruction of the pigment–protein complexes (Section 6.3.1). Photorespiration removes O_2 under conditions when the intercellular CO_2 concentration is low and there exists the possibility of O_2 photoreduction. Physiologically, this can occur when stomata close as a response to water stress and desiccation (Lea and Leegood, 1999).

6.3.2.2 C4 and Crassulacean Acid Metabolism Plants

Although C3 plants are able to cope with the problems associated with metabolic uncoupling and photoinhibition through the photorespiratory pathway and the adaptive alteration of the light reactions (Niyogi, 1999), this is not the case for those species that live in environments that are continually compromised by limiting combinations of high temperatures and levels of irradiance and reduced water and CO_2 availabilities. Under these conditions, photorespiration would severely compromise the carbon assimilation capacity. So, typically, plants that inhabit desert, tropical, aquatic, and epiphytic microenvironments have evolved different modes of photosynthesis that are not constrained by photorespiration. These adaptations involve C4 photosynthesis (Ku et al., 1996) and Crassulacean acid metabolism (CAM; Cushman and Bohnert, 1997) using carbon fixation pathways that initially capture CO_2 in the form of a stable C4 organic acid. C4 and CAM plants have specially adapted anatomies, physiologies, and biochemical pathways. CAM plants are usually found in desert regions, and they are able to fix CO_2 in the night by the temporal, nocturnal regulation of stomatal apertures and the "switching on" in the dark of a CO_2 concentrating mechanism or "pump." This "fixes CO_2" using the enzyme phosphoenolpyruvate carboxylase (PEPC) and the substrate phosphoenolpyruvate-producing oxaloacetate, a C4 carbon fixation product, which is then converted to malate. During the day, the stomata close but the "chemically trapped" CO_2 is released from malate and refixed by Rubisco. The light reactions generate ATP and NADPH during the day, and the Calvin cycle becomes operational. In the case of C4 plants, photosynthesis involves two different cell types: the mesophyll and bundle sheath cells (termed Kranz Leaf Anatomy). Initially, CO_2 is fixed by the PEPC pump and converted to oxaloacetate, which is rapidly reduced to malate in the mesophyll cell chloroplasts. C4 compounds are then transferred to the bundle sheath cells, where they are decarboxylated to release CO_2 that is refixed by Rubisco. The bundle sheath cells are "airtight" and allow the CO_2 to concentrate to a level that ensures the operation of the carboxylase functionality of Rubisco.

6.3.3 Modulation of ROS in Respiration

Mitochondria are especially vulnerable to the production of reactive oxygen species, as O_2 is the terminal electron acceptor in aerobes and the production of O_2^-, and H_2O_2 is therefore unavoidable. For some time (e.g., see Araki, 1977) deleterious structural changes to mitochondria have been considered a key factor in cryoinjury. Tsvetkov et al. (1985b) performed *in vitro* studies on isolated freeze-dried and frozen/thawed rat mitochondria and demonstrated an inhibition in ATPase activity, which could be partially prevented by the application of trehalose. In a similar study, Tsvetkov et al. (1985a) used $K_3Fe(CN)_6$, an artificial electron acceptor impermeable to the mitochondrial membrane, to study the relationships between succinate dehydrogenase, mitochondrial membrane integrity, and freezing stress. The data indicate a nonspecific perturbation in membrane integrity of intact mitochondria exposed to freezing/thawing. The extent of the injury was dependent on the rates of freezing and thawing. A direct link between freezing injury and the perturbation of mitochondrial electron transport chain activity (as loss of cytochrome c) has been substantiated in isolated rat mitochondria by Petrenko and Subbota (1986). However, the modulation and protection of plant respiratory function at low temperatures are likely to be different.

Unlike animal mitochondria, plants possess a bifurcated respiratory electron–transfer chain (Maxwell et al., 1999) that diverges from the main respiratory route at ubiquinone. Electron transfer through the alternative oxidase (AOX) pathway bypasses two out of the three sites of the cytochrome chain at which electron transport is coupled to ATP synthesis. The AOX is a single enzyme, located on the inner mitochondrial membrane; it is induced when the cytochrome c component of the pathway is impaired. It has been proposed that one of the functions of AOX pathway is to limit mitochondrial ROS formation when imbalances occur between upstream respiratory carbon metabolism and downstream electron transport (Vanlerberghe and McIntosh, 1997). Møller (2001) proposes that the AOX maintains the electron transport chain in a relatively oxidized state. In addition, the NADPH produced by NADP-isocitrate dehydrogenase and nonproton pumping transhydrogenase activities maintains plant mitochondrial antioxidants in the reduced state. The AOX has also been implicated in low-temperature plant adaptations. For example, temperatures fluctuate dramatically in arctic regions, especially during the growing season. Arctic plants frequently have high respiratory rates as compared with temperate plants. It has been suggested that the AOX pathway is used to generate heat in some Arctic plant species (Billings and Mooney, 1968; Breeze and Elston, 1978; McNulty and Cummins, 1987). Furthermore, the maintenance of efficient respiratory metabolic coupling will be critical to survival at arctic temperatures. Studies undertaken (McNulty and Cummins, 1987) on the arctic perennial herb, *Saxifraga cernua*, showed high levels of dark respiration attributed to AOX at lower temperatures, and it was also concluded that the alternative pathway was not as low-temperature sensitive as the normal cytochrome pathway. *In vitro* studies performed by Yoshida and Tagawa (1979) compared mitochondrial function in chilling-sensitive and chilling-resistant calli of *Cornus stolonifera* and *Sambucus sieboldiana*, respectively. Diversion of electron flow to the AOX pathway was an immediate response in *C. stolonifera*, but this was not observed in the stress-tolerant plant system. It was concluded that alterations in the regulation of respiratory electron transport chains comprise one of the most immediate responses to low-temperature exposure. This was further substantiated in a similar study (Yoshida and Niki, 1979) that concomitantly profiled changes in cell permeability and respiratory activity in *C. stolonifera* callus. Limited or no change in membrane permeability was observed in the first 24 h following chilling treatment. In contrast, a depression in respiratory activity was detected 12 h after chilling, and the researchers proposed that lesions to cell membranes might not be the primary injurious event in chilling injury.

The aforementioned studies provide interesting points for future speculation as to the causal factors in cryoinjury. It is well known that membrane damage (Levitt, 1980; Meryman and Williams, 1985; Steponkus, 1985) is the principal causal factor in low-temperature injury. However, it is also plausible that low-temperature impairment of primary oxidative metabolic pathways (respiration and photosynthesis) will also initiate a cascade of damaging oxidative reactions that eventually lead to the production of toxic free radicals and, hence, membrane damage.

6.3.4 Disturbed Primary Metabolism: A Key Causal Factor in Plant Low-Temperature Injury?

As is the case with other organisms, membrane damage is a primary cause of low and ultra-low temperature injury in plants (Levitt, 1980; Meryman and Williams, 1985; Singh and Miller, 1984; Steponkus, 1985; Fleck et al., 1997). However, low-temperature-induced perturbation of primary metabolic pathways and their temperature-sensitive enzymatic components may also be a major component of plant freezing stress. This may be particularly the case if chilling occurs before an episode of freezing in nonacclimated plants. The early research findings of Tappel (1966), Levitt (1956), and Heber (1968) provide compelling evidence for this, especially in the light of more recent insights into metabolic and enzyme regulation (see Asada, 1999; Maxwell et al., 1999; Møller, 2001). It is first important to consider the relationship between membrane damage and the uncoupling of primary metabolism *in vivo*. Even small environmental changes, such as a modest decrease in temperature, can provide a "metabolic cue" for plants to alter the flow of electrons in their primary electron transport systems (Asada, 1999; Maxwell et al., 1999; Møller, 2001), thereby altering the capacity for forming ROS and free radical species. Studies have correlated cold acclimation with ATP accumulation (Sobczyk and Kacperska-Palacx, 1978; Sobczyk and Marszalek, 1985) and cold sensitivity with a decrease in ATP levels and an inhibition of Ca^{2+}-dependent ATPase (Stewart and Guin, 1969). Indeed, Sobczyk and Marszalek (1985) postulated that cold-resistant cells that have been cold acclimated may acquire new characteristics at their phosphorylation sites that are protected against inactivation by freezing.

To gain a greater understanding of the role and interrelationship of different "types" of cryoinjury (e.g., physical, biochemical, molecular) in plants is not straightforward, and to do so it will be necessary to unravel the association between cause and effect. The following points may be useful to achieve this end:

1. Metabolic coupling has a direct role in the control of ROS scavenging in chloroplasts and mitochondria, and this is critically dependent on the maintenance of cell, organelle, and membrane integrity.
2. Plants have evolved highly complex regulatory processes (e.g., water–water cycling, photorespiration, C4 and CAM and AOX pathways) for the protection of their photosynthetic and respiratory electron transport systems during stress episodes. These processes are dependent on the cellular compartmentalization of specific components of the metabolic pathways. Such adaptations must be considered in the context of different environmental systems (e.g., tropical, polar, temperate).
3. The processes comprising the second point function in part to limit the production of ROS and free radicals.
4. Regulatory components of the second and third points are intimately integrated with enzyme antioxidant systems located in the mitochondria and chloroplasts. These systems provide a "dual protection system" in which electron transfers involving ROS production are "backed up" by antioxidant protection systems located at the site of ROS production.
5. Leakage of ROS and free radicals from the chloroplast and mitochondrial electron transport systems can potentially exacerbate other freeze-induced membrane injuries (e.g., those caused by physical factors such as ice formation) by attacking electron rich domains of membranes, which contain polyunsaturated fatty acids.
6. Primary metabolism will be impaired at low temperature, and this will result in the loss of reducing equivalents and will compromise the stability of the ATP/ADP pool.

7. Free radical–mediated lipid peroxidation arising from the first and fourth points will compromise antioxidant protection with the further exacerbation of freeze-induced membrane damage through secondary oxidative stress (lipid peroxidation).

8. Events in the fifth and seventh points will impair cell signaling processes, and this will have consequences for the molecular processes involved in cold acclimation and essential repair processes.

Understanding the effects of freezing on primary plant metabolism is also complicated by the fact that evidence (Vanlerberghe and McIntosh, 1997) indicates that protection from ROS in stressed photosynthetic and respiratory pathways requires integrated control. Therefore, protection mechanisms are most likely to be operational and coordinated across both primary (photosynthetic and respiratory) metabolic routes. For example, photorespiration may have a part in moderating reducing power in the light via the action of ribulose bisphosphate oxygenase, which indirectly modulates energy, reducing power, and attenuates carbon imbalances that arise between the dark and light reactions of photosynthesis. Interestingly, AOX also functions in CAM and C4 plants (Agostino et al., 1996; Rustin and Queiroz-Claret, 1985), and it may have an indirect role in preventing the formation of ROS in photosynthesis. This is highly likely, as CO_2 assimilation will influence the supply of carbon skeletons, reducing equivalents and ATP for respiration. For example, acclimation of *Arabidopsis* leaves that were developing at low temperatures enhanced activities of Calvin cycle enzymes as well as those in the sucrose biosynthetic pathway (Strand et al., 1999). The same study also demonstrated that leaves developing at 5°C had an increased cytoplasmic volume and a decreased vacuolar volume, indicating that this may be an important mechanism to increase key primary metabolic enzyme levels in cold-acclimated leaves. In terms of the C4 and CAM photosynthetic pathways, the effects of low temperatures are limited, as the species that have these adaptations generally occur in tropical and arid regions.

Understanding C4/CAM adaptations to low temperatures does have agronomic significance, as exposure of key C4 crops such as maize and sorghum to reduced temperatures in the early growing season in temperate climates can cause a reduction in photosynthetic performance. Fryer et al. (1998) examined the effects of chilling on maize carbon metabolism, electron transport, and free radical scavengers. Antioxidant enzymes activities were raised during chilling, indicating the production of activated oxygen species; however, the functionality of the C4 pathway was unclear because of analytical constraints. Maize plants exposed to 14°, 18°, and 20°C (Kingston-Smith et al., 1999) were evaluated for their Rubisco and PEPC activities in relation to antioxidant and H_2O_2 content. Plants grown at the lowest temperature had less chlorophyll and had reduced Rubisco and PEPC activities. The glutathione pool was similar across the different temperatures, and foliar H_2O_2 and ascorbate content were increased at the lowest temperature (14°C). Plants grown at this temperature had increased rates of CO_2 fixation and decreased quantum efficiencies for PSII in the light, but photoinhibition was not in evidence. Growth at the suboptimal temperature of 14°C decreased the abundance of the D1 repair protein.

Understanding the basis of low-temperature stress/freezing responses in tropical species is becoming more and more important in terms of enhancing the potential of growing tropical crops in temperate regions. Perhaps more important, however, is the application of cryogenic storage to tropical and aquatic plant and algal germplasm (see Chapter 9). Thus, the broadest consideration of cold effects on C3, C4 and CAM metabolism in tropical plant germplasm exposed to cryogenic treatments may, in the future, become more important. This is because fundamental research is imperative for the development of cryoconservation methods for recalcitrant tropical plant germplasm.

6.4 ANTIOXIDANT PROTECTION AT LOW TEMPERATURES

Antioxidants are required to prevent the "leakage" of damaging ROS, and in higher plants and most algae they comprise SODs, catalase and "guaiacol"-peroxidases, α-tocopherol (vitamin E),

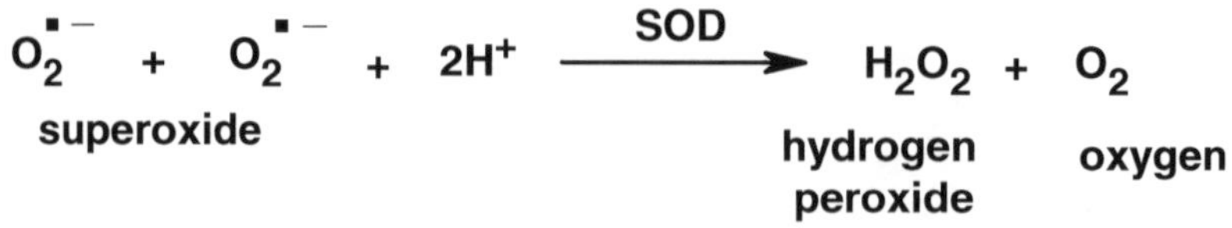

$$O_2^{\bullet -} + O_2^{\bullet -} + 2H^+ \xrightarrow{\text{SOD}} H_2O_2 + O_2$$

superoxide   hydrogen   oxygen
                                                              peroxide

SCHEME 6.9 Conversion of superoxide to hydrogen peroxide and oxygen.

ascorbate peroxidase, mono or dehydroascorbate reductases, glutathoine peroxidase, and glutathione reductase (Elstner and Osswald, 1994). Our understanding of plant antioxidant protection has increased considerably during the last decade, largely because of the application of transgenic technologies (see Chapter 5). As chilling, freezing, and desiccation stress seriously impair crop plant productivity, genetic manipulation techniques have been used to alter the expression of antioxidant enzyme genes in the context of low-temperature-induced oxidative stress.

6.4.1 SUPEROXIDE DISMUTASE

SOD converts superoxide radical anion into hydrogen peroxide and oxygen (Scheme 6.9). Three types of SODs are classified depending on their metal cofactor and subcellular location (Van Camp et al., 1997): manganese (MnSOD), present in mitochondria; iron (FeSOD), present in chloroplasts; and copper-zinc (Cu/ZnSOD), present in chloroplasts and the cytosol. SOD activities have also been associated with peroxisomes.

Considerable emphasis has been placed on the application of transgenic techniques to enhance SOD activities in crop plants exposed to chilling and freezing temperatures. Tobacco plants have been genetically manipulated to constitutively overproduce SOD, and the performance of the plants indicates that this antioxidant is a potentially rate-limiting factor in the prevention of oxidative damage in stressed plants (Van Camp et al., 1997). SOD enhances tolerance to freezing stress in transgenic alfalfa plants, which are overexpressing SOD genes (McKersie et al., 1993, 1996). A subsequent field trial investigation (McKersie et al., 1999) explored the possibility that enhanced tolerance to oxidative stress increased winter survival in alfalfa. The crop plant was transformed with Mn-SOD (in both chloroplast and mitochondrial genomes), and activities of the enzyme were elevated in the majority of the primary transgenic plants. Interestingly, McKersie at al. (1999) noted that the cytosolic and plastid forms of Cu/Zn-SOD had lower activities in the chloroplast SOD transgenic plants as compared to the nontransformed controls. Field trials of these plants showed that survival and yield was greater in most of the transgenic plants as compared to controls, but the researchers cautioned that glasshouse screening of these plants was not an effective indicator of tolerance. Freezing injury in leaf blades was monitored on the basis of electrolyte leakage (as an indicator of membrane damage), and the tolerant genotypes were found to be only 1°C more freezing resistant. McKersie et al. (1999) concluded that as there was only a limited level of enhanced tolerance, the trait was not associated with changes in the primary site of freezing injury.

The involvement of oxidative stress in freeze–thaw injury has also been investigated in mutant yeast cells, providing important corollaries for researchers studying freezing injury in other eukaryotes. Park et al. (1998) used mutant yeast lines, defective in a range of antioxidants, and found that only those affecting SOD (Cu/Zn SOD) encoded by *SOD1* and Mn SOD encoded by *SOD2* showed decreased freeze–thaw tolerance (at –20°C); concluding that O_2^- radicals are produced as a result of freezing stress. This was confirmed by treating the *SOD1* with O_2^- scavengers or by freezing in the absence of O_2. These researchers therefore concluded that oxidative stress is a major component of cryoinjury in yeast cells, which indicates that $O_2^{\bullet -}$ is produced in the cytoplasm and occurs as a result of an oxidative burst caused by the leakage of electrons from the mitochondrial electron transport chain and the subsequent reduction of O_2. Thomas et al. (1999) examined the role of Fe SOD in low-temperature protection in the cyanobacterium, *Synnechococcus*, using a strain that was

$$2\ H_2O_2 \xrightarrow{\textit{Catalase}} 2\ H_2O\ +\ O_2$$

hydrogen   **water   oxygen**
peroxide

SCHEME 6.10 Action of catalase on hydrogen peroxide.

$$\textbf{substrate-H}_2 + \textbf{H}_2\textbf{O}_2 \xrightarrow{\textit{peroxidase}} \textbf{substrate} + 2\ \textbf{H}_2\textbf{O}$$

SCHEME 6.11 Decomposition of hydrogen peroxide by peroxidase.

deficient in the enzyme. A number of chilling regimes (0 to 17°C) were applied to the organism to test the efficacy of SOD in cold tolerance. Results indicated that SODs may have a role in chilling resistance but that in this system, SOD is inactivated during severe chilling stress. Thomas et al. (1999) concluded that at low temperatures it might be the nonenzymatic antioxidants, such as β-carotene and α-tocopherol that confer the best protection, as antioxidant enzymes lose their activities during chilling.

6.4.2 CATALASE, PEROXIDASES, AND H_2O_2 REMOVAL

H_2O_2 is potentially extremely cytotoxic because of its capacity to form ·OH through Fenton chemistry (see Section 6.2). Removal by catalase, a tetrameric iron porphyrin (haem) protein is, therefore, a vital component of cellular antioxidant protection (Scheme 6.10).

Plants have multiple catalase isozymes (McClung, 1997), and individual catalase mRNAs respond differently to temperature stress. Transient accumulation of H_2O_2 during cold acclimation in maize induces antioxidant enzymes, including the Cat3 isomer, and this has been correlated with protection against low-temperature stress (Anderson et al., 1995; Prasad et al., 1994). Haem containing proteins that catalyze the H_2O_2-dependent oxidation of substrates (Scheme 6.11), including phenolics (e.g., guaiacol), are termed peroxidases (Elstner and Osswald, 1994).

H_2O_2 is produced via the reaction of SOD, and ascorbic acid reacts with H_2O_2 in a reaction catalyzed by ascorbate peroxidase to form monodehydroascorbate. Ascorbate is then regenerated from monodehydroascorbate through an enzymatic route (monodehydroascorbate reductase) or via spontaneous transformation of monodehydroascorbate into DHA (dehydrascorbate). Ascorbic acid is regenerated from DHA in a catalytic reaction by dehydrascorbate reductase coupled to a GSH–GSSG (oxidized glutathione) cycle in which reduced GSH is oxidized to GSSG. GSH is then regenerated by glutathione reductase (Scheme 6.12; adapted from Sharma and Davis, 1997; Halliwell, 1982).

The ability of plants to control H_2O_2 levels is a critical factor in ameliorating plant stress (Asada, 1992). Gillham and Dodge (1986) considered the spatial and cellular distribution of H_2O_2 scavengers, which they found to be primarily located in the chloroplast. They suggested that plastid antioxidants confer important adaptive responses when the photosynthetic apparatus is stressed (including at low temperatures). This provides further evidence that disturbances in primary metabolism (see Section 6.3) and the differential responses of antioxidants (Thomas et al., 1999) should be considered more fully in low-temperature plant-stress physiology. This may be especially pertinent if the "frozen plant" has been previously compromised by an episode of chilling injury. For example, Omran (1980) monitored activities of catalase, peroxidase, and indoleacetic acid oxidase and peroxide in chilled cucumber seedlings and found that catalase activity increased, peroxidase activity remained the same, indoleacetic acid oxidase activities increased, and peroxide levels increased. Differential responses may have important consequences in the ability to overcome

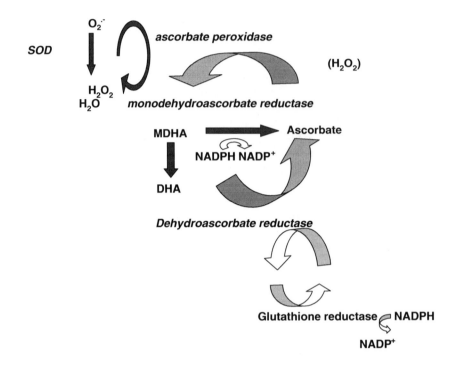

SCHEME 6.12 Ascorbate-glutathione (Halliwell–Asada) antioxidant pathway (adapted after Asada and Taka-hashi, 1987; Foyer and Halliwell, 1976).

sequential low-temperature stress. It is imperative to consider the interrelationships between Fenton chemistry (see Section 6.2), the fate of H_2O_2-removal systems, and H_2O_2 when interpreting experimental findings. Changes in the activities of H_2O_2 scavengers and H_2O_2 may not necessarily imply that H_2O_2 is not produced or "accumulated" in the cell. Absence of a clear inverse relationship (MacRae and Ferguson, 1995) between peroxide scavenging and accumulation in plants must be carefully interpreted, and it is prudent to take into consideration that H_2O_2 will participate in Fenton chemistry. Thus, reactions occurring "downstream" of H_2O_2 production should be taken into account in the design of experiments investigating the role of H_2O_2 and low-temperature injury.

H_2O_2 and H_2O_2-scavenging capacity will also have a major effect on the stability of chloroplast membranes and the exacerbation of low-temperature injury by high light intensities. This is because of the fact that catalase has a haem group moiety and the enzyme is known to be light sensitive under both *in vitro* and *in vivo* conditions (Cheng et al., 1981). Volk and Feierabend (1989) presented some interesting interpretations on observing low-temperature injury and the photoinactivation of catalase in rye leaves. They postulated that in their system, the decline in catalase activity was caused by photoinactivation, and low temperature alone did not induce this effect. Furthermore, the apparent loss of catalase was thought to be caused by the blocking of its resynthesis by low temperatures. Cryoinjury may well compromise the repair of light-sensitive components of the photosynthetic apparatus and prevent the induction of adaptive reactions in antioxidant enzyme synthesis as well as activity. This concurs with the well-established observation that photoinhibition at low temperatures also inactivates PSII (Öquist, 1983). Thus, photooxidation requires the constant *de novo* synthesis of proteins that are used to repair damaged sites in the photosynthetic electron transport chain and light-sensitive protective proteins such as catalase. At low temperatures, the

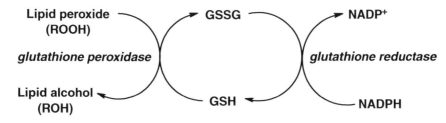

SCHEME 6.13 Glutathione protection mechanisms.

turnover of these repair processes will be inhibited or, indeed, cease (Feierabend et al., 1992), and this most likely accounts for the fact that low-temperature injury exacerbates photooxidation and *vice versa*. Streb and Feierabend (2000) investigated the significance of electron sinks and antioxidants in cold- and non–cold hardened rye leaves with respect to photo-oxidative stress. Ascorbate, glutathione, glutathione reductase, and SOD contents increased in non–cold hardened leaves to levels observed in hardened leaves when precursor substrates of L-galactonic acid-γ-lactone and 2-oxothiazolidine-4-carboxylate were applied. However, reduced GSH was rapidly removed from the non–cold acclimated tissues when an inhibitor of its biosynthesis (buthionine sulfoximine) was applied. Importantly, increased antioxidant contents did not establish resistance to low-temperature-induced photo-inactivation of PSII and catalase in non–cold hardened leaves. However, the resistance of cold-hardened leaves to low-temperature-induced photo-inactivation of PSII and catalase was dependent on low-temperature repair and active carbon fixation. Anderson et al. (1995) discovered that catalase3 was elevated in acclimated seedlings of maize, indicating that this response offers the first level of protection against H_2O_2 generated by mitochondria. It may be appropriate to consider the role of cold acclimation in enhancing antioxidant protection in relation to low-temperature injury and to relate this to photorepair processes, H_2O_2, and electron transfer mechanisms in more detail. A study of the potential occurrence of downstream Fenton chemistry and the production of the ·OH radical may be especially interesting.

6.4.3 VITAMIN E, ASCORBATE, AND GLUTATHIONE

Vitamin E (α-tocopherol) is lipophilic, and when embedded into membranes, it is in close proximity to potentially vulnerable sites of free radical damage. α-Tocopherol is thus preferentially oxidized during free radical attack as compared to polyunsaturated fatty acids. The resulting quinone is reduced by an ascorbate cycling mechanism to regenerate α-tocopherol. Glutathione is a low–molecular weight tripeptide sulfur compound that contains the thiol group (S–H) in the cysteine component (GSH). It protects oxygen-sensitive enzymes and proteins from oxidative degradation of their sulfhydryl groups. GSH reductase catalyzes the recycling of the protective molecule (see Scheme 6.12). GSH can also detoxify peroxidized membranes through the recycling (see Scheme 6.13) action of glutathione peroxidase and glutathione reductase (for reviews, see Alscher et al., 1997; Asada, 1992; Benson, 1990).

The peroxidized membrane is also repaired though the action of phospholipases, which preferentially excise lipid peroxides, making them more accessible to GSH peroxidase. The hydroxy fatty acid produced is reacylated and reinserted into the damaged membrane (Benson, 1990; Elstner and Osswald, 1994; Van Kuik et al., 1987). Many plant secondary metabolites are phenolic, and they can participate in stress metabolism as either pro- or antioxidants. The balance between these two functions may be regulated enzymatically through phenoloxidases and peroxidases (Asada, 1992; Elstner and Osswald, 1994).

The possibility that plant antioxidant defenses implicating GSH and ascorbate protection systems in freezing and low-temperature stress tolerance has been considered for several decades. Levitt (1962) developed the sulfhydryl-disulfide hypothesis of frost injury and resistance in plants, which proposes:

1. "Frost injury is due to an unfolding and therefore a denaturation of the protoplasmic proteins. This results from the formation of intermolecular S-S bonds induced by the close approach of the protein molecules due to frost dehydration."
2. "Frost resistance is a resistance towards S-H oxidation and S-H ↔ S-S interchange and therefore to these intermolecular S-S bonds" (Levitt, 1962, p. 358).

Levitt's exciting hypothesis was presented at a time when freezing injury was mainly considered in terms of physical phenomena associated with intra- and extracellular ice formation and included the colligative properties of the cell. However, the S–H/S–S concepts of cryoinjury were an important advance, as they considered both the chemical and physical basis of freezing injury. Thus, ice formation and dehydration was proposed in the context of causing profound (i.e., biochemically significant) changes to the chemical structures of essential macromolecules such as proteins exposed to low temperatures. In the context of this chapter, it is therefore appropriate to reconsider Levitt's pioneering research, which was the first to ascertain that decreases in SH were a manifestation of freezing injury in plants. This was particularly the case for non–frost tolerant genotypes and GSH decline was most likely caused by the formation of S–S. We now know that to keep ROS at low steady-state levels, stressed plants cells must operate a complex and integrated antioxidant system that involves both enzymes and antioxidant metabolites, including GSH, the GSH-GSSG recycling system (see above), and glutathione peroxidase. Thus, current studies entirely concur with Levitt's original hypothesis. Levitt (1962) also described four phases in frost injury:

1. The moment of freezing
2. While frozen (after freezing equilibrium has been reached)
3. The moment of thawing
4. Postthawing

Freezing *in vivo* and under certain *in vitro* circumstances (e.g., cryopreservation and cold acclimation) is usually preceded by a chilling phase, even if of a short duration. It is therefore essential to consider the role of oxidative stress at low temperatures and the effect that this will have on the antioxidant status of the plant before and just after exposure to freezing. These may result in two different scenarios depending on the type and level of stress and the ability of the plant to overcome the stress. Cold acclimation (see Chapters 5 and 10) is an important survival response allowing a plant to undertake metabolic changes that will enhance the ability to survive subsequent episodes of low-temperature exposure. Cold acclimation has been used effectively as a precryogenic treatment (see Chapters 9 and 10). However, plants that are not able to cold-acclimate or that have been exposed to chilling (e.g., leading to cold stress as opposed to a positive and protective stress acclimation response) and subsequent freezing will be compromised. Under these circumstances, chilling will most certainly influence antioxidant enzyme activities, and this will affect tolerance and sensitivity limits on subsequent exposure to freezing. Anderson et al. (1995) found that glutathione reductase izoenzyme profiles changed in cold-acclimated mesocotyls of maize and that levels of ascorbate, GSH, and lignin (phenolics) were also altered, indicating that they were implicated in cold-tolerance responses. Prasad (1996) investigated chilling injury and tolerance in maize seedlings exposed to chilling stress with and without acclimation treatments. Acclimated seedlings had elevated levels (35 to 120%) of glutathione reductase (as well as catalase and guaiacol peroxidase). Importantly, nonacclimated seedlings had higher levels of oxidized proteins and lipids (35 to 65%) as compared with the acclimated seedlings. These data were presented by Prasad (1996) as correlative evidence indicating that in nonacclimated seedlings, low-temperature injury is, in part, caused by the formation of ROS and that in acclimated seedlings, chilling tolerance is the result of enhanced antioxidant status.

Clearly there is now an accumulation of both historical and contemporary evidence to indicate that antioxidant defense is an important component of low-temperature tolerance and freezing stress in plants. Future investigations will continue to apply transgenic technologies (see CHapter 5) to gain a greater understanding of the mechanisms involved. However, the careful interpretation of findings from these studies must also take into account the need to characterize the production and activities of ROS and their reaction products as well as antioxidant enzymes *per se*.

6.5   OXIDATIVE STRESS IN THE FROZEN STATE: THE EVIDENCE

6.5.1   METHODS OF INVESTIGATION

The complexity and rapidity of free radical reactions (see Section 6.2) pose an analytical problem because to gain a thorough insight into oxidative stress, it is important to monitor primary and secondary oxidative processes and antioxidant status (see Section 6.4). Evidence for free radical damage in plants exposed to low temperatures can be gained by direct and indirect monitoring of "ROS markers" and also by studying antioxidant status (Benson, 1990; Harding and Benson, 1995). Because free radicals are highly reactive, the only direct means by which they can be measured is by using EPR spectroscopy, usually in combination with spin traps. The latter are organic molecules that react with free radicals to produce a longer-lived radical that can be more easily detected by spectroscopy. Alternatively, indirect detection using "marker" compounds that are formed as a result of free radical reactions can be applied. For example, ·OH readily reacts with dimethyl sulphoxide to produce the methyl radical and, subsequently, methane, which can be quantified by volatile headspace sampling and GC (Benson and Harding, 1995). Membrane lipids are a primary target for free radical attack, and oxidized fatty acid reaction products are frequently used as "markers" of oxidative stress (Benson, 1990). These products include conjugated dienes, lipid peroxides, aldehyde breakdown products, and volatile hydrocarbons. The aldehydic reaction products of lipid peroxides can cross-link with protein groups to form Schiff's bases, which are also measured as "markers" of free radical damage. Selection of the appropriate free radicals and "reaction product markers" in low-temperature oxidative stress studies can be aided by taking into consideration the physical and chemical factors involved in the generation of radical species.

6.5.2   EVIDENCE FOR FREE RADICAL FORMATION

The presence of O_2 in biological systems exposed to freezing can have deleterious effects, and the production of free radicals (confirmed using EPR) is implicated. In an early study on the effect of lyophilization on microbial cultures, Heckly and Quay (1983) found that the best survival rates were obtained by exclusion of air (oxygen). The production of free radicals was measured by EPR, and the authors concluded that the presence of water had a *"profound effect"* on free radical production, and indeed, if free radicals were generated in the presence of high levels of moisture, they rapidly decomposed. The biochemical effects of freezing in O_2 were examined in *Escherichia coli*, by Swartz (1971a), and oxygen-dependent freezing damage was characterized by the production of free radicals concomitant with a leakage of amino acids from frozen/thawed cells. An interesting point for genetic stability studies was that freezing also caused oxygen-dependent single-strand breaks in *E. coli* DNA. In a subsequent study, Swartz (1971b) applied the radioprotectant, β-mercaptoethylamine to a repair-deficient strain of *E. coli* exposed to freezing in the presence of oxygen. EPR spectroscopy showed that β-mercaptoethylamine participates in free radical reactions when *E. coli* cells were frozen and that these reactions included single-electron transfers with the S-centers of β-mercaptoethylamine and offered protection against O_2-freezing damage, mitigating the repair-deficiency effects.

The reactivity of hydroxyl radicals generated by gamma irradiation in phosphate buffered saline solutions containing golden-hamster cells has been studied at –196° and –162°C (77 and 111K)

$$CH_3SOCH_3 + OH \longrightarrow CH_3SOOH + {}^{\bullet}CH_3 \; (\longrightarrow CH_4)$$

dimethyl   hydroxyl   methane   methyl   methane
sulphoxide radical   sulphinic acid radical

SCHEME 6.14 Methane production from Me₂SO.

using EPR (Yoshimura et al., 1992). The authors suggest that because OH radicals do not diffuse in frozen cells at –196°C, they will not therefore react with protein or DNA at this temperature. When the previously irradiated samples were warmed up to –162°C (111K), the OH radicals decayed rapidly, whereas the amounts of organic radicals remained constant at this temperature. The results indicate that hydroxyl radicals do not react with organic substances during the warming of cells and that OH radicals are not the main reactive species responsible for the biological effects in gamma-irradiated frozen cells. The lack of movement of hydroxyl radicals when generated at –196°C (77K) is further confirmed by gamma-irradiation of frozen aqueous DNA solutions with or without a spin-trapping agent (Oshima et al., 1996). After thawing the solutions, EPR indicated that reactive hydroxyl radicals were produced in the hydration layer of gamma-irradiated DNA and were scavenged by the spin trap but that unreactive hydroxyl radicals were produced in the free water layer of gamma-irradiated DNA. It was concluded that hydroxyl radicals generated in the hydration layer of gamma-irradiated DNA did not induce strand breaks but induced base alterations by addition of ·OH to guanosine, in particular. The importance of the water content status of samples was further illustrated by the gamma-irradiation of soybean paste with differing moisture contents (Lee et al., 2001). Samples were kept in liquid nitrogen at –196°C (77K) during irradiation and subsequent EPR measurements. The EPR spectra, although rather complex, indicated that there was a direct correlation between water content and radical production.

The possibility that hydroxyl radicals are produced by plants exposed to cryopreservation (at –196°C) has been explored in higher plants and algae. Benson and Withers (1987) developed a novel nondestructive approach to study oxidative stress in cryopreserved *Daucus carota* cells by using dimethyl sulphoxide as a hydroxyl radical probe (Scheme 6.14).

Methane (the stable volatile reaction product of ·OH and Me₂SO) is used as a "marker" for using headspace volatile sampling and gas chromatography. Elevated levels of methane were detected immediately after thawing cryopreserved carrot cultures, and fluctuations in methane production could be correlated with postthaw stress and recovery responses. Using the same approach, Fleck et al. (2000) also demonstrated that ·OH activity was significantly enhanced when the freeze-recalcitrant algae, *Euglena gracilis*, was exposed to low temperatures (from –10° to –60°C) and liquid nitrogen treatments. The application of the potent iron chelating agent desferri-oxamine (an inhibitor of Fenton chemistry; see Section 6.2) reduced ·OH production and enhanced postfreeze recovery. Using a similar approach to study the formation of ·OH in stems of winter wheat exposed to freezing temperatures, Okuda et al. (1994) monitored the formation of meth-ansulphinic acid (MSA) formed from Me₂SO in stems exposed to a minimum temperature of –20°C. MSA was not detected in stems incubated at 28°C but was detectable at –5°C; importantly, MSA formation was observed very rapidly, within 1 min of cold treatment, and increased with decrease in temperature. Samples exposed to –20°C produced larger amounts (>0.3 µmol.g.f.wt⁻¹[grams fresh weight]) of MSA. Samples of wheat were exposed to cold treatments for up to 7 h, and these stress episodes were concomitant with increased production of MSA. Okuda et al. (1994) concluded that there is a close relationship between the formation of ·OH, plant stress and death at freezing temperatures. Plants exposed to freezing temperatures may also be susceptible to photooxidation and to the generation of ROS formed by the interaction of excited photosynthetic pigments, light, and O_2, during freezing stress. Evidence for this possibility was provided by Benson and Norhona-Dutra (1988), who demonstrated the production of singlet oxygen ($^1O_2$) by cryopreserved *Brassica napus* shoot cultures.

6.5.3 Evidence for Secondary Oxidative Stress

Free radicals and activated oxygen species (e.g., H_2O_2 and $^1O_2$) will initiate secondary oxidation pathways, and of these, lipid peroxidation is probably one of the most important. The short- and long-term consequences of lipid peroxidation are considerable, ranging from the disruption of cell membranes to molecular and genetic damage (Benson, 1990, 2000; Benson et al., 1997). In this respect, it is particularly pertinent to evaluate research arising from studies of different organisms (not only plants) and, indeed, of model systems because, at the biochemical level, lipid peroxidation processes may be very similar for even very diverse organisms, so long as they possess unsaturated fatty acids.

Whiteley et al. (1992) studied lipid peroxidation, measured by malondialdehyde production, as a cause of deterioration of liver tissue. At $-20°C$, lipid peroxidation increased, but this was not observed at $-196°C$. To determine the effect of temperature on production of hydroperoxides, Fennema and Sung (1980) examined the slow freezing for 4 minutes to subzero temperatures of -5, -10, $-15°C$ before transfer to $78.5°C$ and rapid freezing by direct plunging into dry ice (at $-78.5°C$) of samples containing linolenic acid, buffer, and lipoxygenase. It was found that the accumulation of oxidation products was temperature dependent, with lower yields of hydroperoxides being produced at lower temperatures. The authors speculated that the lower yield could be attributed to "greater reversible denaturation of lipoxygenase" as the subfreezing temperature was lowered or to progressive increases in resistance to diffusion of substrate and reaction products. In this work, no mention was made of the concentration of available oxygen or, indeed, moisture content.

Karl et al. (1982) performed a study on heart muscle and identified O_2-dependent cryoinjuries, which caused an increase in lipid peroxidation products that were decreased in the presence of seleno-methionine. As desiccation and dehydration stress are important components of cryoinjury, it is also essential to consider their role in oxidative stress. An EPR study by Leprince et al. (1995), although not involving exposure to subzero temperatures, did demonstrate that O_2 exacerbated free radical production in orthodox seeds exposed to desiccation. This is an important observation, as in the case of cryopreserved plant germplasm, desiccation tolerance is frequently a prerequisite for freezing tolerance (for reviews, see Benson, 1990; Dumet and Benson, 2000; Dumet et al., 2000). Apgar and Hultin (1982) monitored malondialdehyde (as a thiobarbituric acid reaction product) and lipid peroxide formation in the microsomal fractions of fish muscle tissue at freezing temperatures. Rates of lipid peroxidation were dependent on freezing temperature (overall, they decreased with decreasing temperature) and solute composition. These researchers indicate that the availability of oxygen to participate in reactions in the frozen state may be limited and that this will influence the rate of oxidative reactions. They also noted that the effects of low-temperature exposure in the presence and absence of ice might well be different. Oxidative reactions at above-freezing temperatures and at below-freezing temperatures, but in the presence of solvents that prevented freezing at less than $0°C$, were examined. Apgar and Hultin (1982) discuss the possibility that the process of ice formation has an accelerating effect on oxidative reactions, concluding that this may be because of the concentration of reactants. It may be construed that ice could have a potentially injurious colligative consequence with respect to the enhancement of oxidative stress. The process by which secondary oxidative reactions proceed during freezing can be the result of both nonenzymatic and enzymatic reactions, and the colligative behavior of the solute/solvent system may thus be a significant factor in how prooxidative enzyme reactions proceed. However, this is clearly highly complex, as each component of the oxidative system (e.g., enzymatic and nonenzymatic processes) will differentially respond to freezing, chilling, and colligative factors. Fennema and Sung (1980) examined lipoxygenase-catalyzed oxidation of linolenic acid at subfreezing temperatures and found that the accumulation of the oxidative reaction products was temperature dependent. Reaction completeness (the degree of lipid peroxidation) decreased with decreasing subfreezing temperatures (range $-5°$ to $-15°C$), concurring with previous studies on other oxidative enzymes (Bengtsson and Bosund, 1966). That prooxidative enzyme activity (e.g., lipoxygenases, lipases, peroxidases) is reduced at freezing temperatures is, of course, a positive finding for proponents of the low-temperature storage of viable cells. The actual mechanism by which this inhibition occurs will be enzyme dependent and influenced by physico-chemical factors such as the water activity of tissues before freezing. Reversible enzyme damage may also occur during freezing; irreversible damage was ruled out, as lipoxygenase activity was found to be intact after a freeze/thaw cycle (Fennema and Sung, 1980). This finding does, however, highlight the need to evaluate postthaw oxidative stress in low-temperature storage, as rewarming may well incur considerable secondary oxidative damage. The relationship between cell viscosity, freezing, and hydration status is most frequently investigated with respect to the physical attributes and stabilities of glasses; however, the relationship between the frozen and vitrified state will also have an important effect on oxidative processes. Cell viscosity increases on freezing, and as previously considered above (Symons, 1982), it is possible that the glassy state in water–organic solute systems may be a critical factor in determining both the extent and localization (e.g., aqueous or organic phase distributions) of secondary oxidative reactions. Therefore, it would be interesting to investigate the stability of vitrified and nonvitrified (frozen) systems with respect to the formation of free radicals and the progression of secondary oxidative reactions.

The study of lipid oxidation processes in the frozen food industry has been ongoing for a number of decades and also provides some interesting evidence as to the propensity for plant tissues to undergo lipid peroxidation reactions at subzero temperatures. Duden (1985) reports on the enzymatic lipid degradation in deep-frozen leafy vegetables, showing that in parsley, 70% of monogalactosyl diglycerides are lost after 3 months of storage at –18°C and that similar losses were observed for phospholipids after only 3 d. At –32°C, losses were almost 20% after 30 d of storage. However, studies of viable frozen plant tissues, organs, and whole plants provide information that is of more direct relevance to applied research such as cryopreservation. Comparative studies of freezing injury (as quantitative changes in lipid content) in acclimated and nonacclimated *Citrus* spp. were performed by Nordby and Yelenosky (1985), who showed that following freeze–thaw stress (to –6.7°C), total lipid and triacylglycerol fatty acid profiles of cold-hardened hybrids were similar to profiles of the hardy *Poncirus trifoliate*. In contrast, the less hardy *Citrus sinensis* lost 22% of leaf fatty acids during freezing and 13% during thawing. Linolenic acid (a key target for free radical attack) accounted for 98% of the total fatty acid decrease. Importantly, triacylglycerol (which is rich in linolenic acid) increased by 12% in the cold-hardened hybrid, and three other triacylglycerols (also rich in linolenic acid) increased during the freeze–thaw regimes. Nordby and Yelenksy (1985) postulated that in the case of freeze-tolerant *Citrus* genotypes, levels of highly unsaturated triacylglycerols increase during hardening and freeze–thawing as an adaptive response to maintaining the fluidity of membranes during freezing. Clearly, biochemical approaches have an important role in enhancing our understanding of the structural and molecular changes that occur *in situ* in plant plasma membranes exposed to freeze/thaw cycles. That secondary free radical and lipid peroxidation reactions are important components of freezing stress is highly probable. Studies of oxidative stress in plant and algal germplasm stored at low (–20°C) and ultra-low (–196°C) temperatures during cryopreservation have demonstrated the formation of free radicals, volatile hydrocarbons' breakdown of lipid peroxides, and aldehydic lipid peroxidation products (Benson, 1990; Benson and Withers, 1987; Benson et al., 1992, 1995; Fleck et al., 1999, 2000; Magill et al., 1994). As these studies comprise a diverse range of species representing both tropical and temperate plants and algal cells, the evidence is compelling that both primary and secondary oxidative stress are important components of plant cryoinjury.

6.6 FREE RADICALS AND CRYOPRESERVED PLANT GERMPLASM

One of the most important applications of plant low-temperature biology concerns the development of storage procedures for the cryopreservation (at –196°C) of plant genetic resources (see Chapter 10). Cryogenic storage of plant germplasm in the "frozen state" offers the possibility of slowing down and, indeed, stopping the aging process. That plant germplasm is actually maintained in the frozen state is, of course, dependent on whether or not it is vitrified. However, whatever the physical state (solid or glassy) of the germplasm conserved, the total inhibition of chemical and biological reactions is required to stop aging processes. Denham Harman (1956) first proposed that free radicals have a central role in animal aging, and as this elegant theory progressed (Harman, 1988, 1991), it soon became evident that stress, diet, metabolic rate, and disease were key factors in promoting aging though free radical damage. By extrapolating this theory, it follows that cryogenic storage must obviate biochemical (e.g., metabolically driven), chemical and free radical reactions. The latter are especially pertinent, as these are the only reactions likely to occur at ultra-low, subzero temperatures (see Section 6.2.1 and Symons, 1982).

When studying oxidative stress in cryopreserved plant germplasm, it is also important to take into consideration the two key factors of dehydration and ice formation in freezing injury, as both may have profound effects on the ability of a system to participate in free radical chemistry. Indeed, water activity can change, enhance, and suppress free radical reactions (e.g., Heckly and Quay, 1983). As desiccation tolerance (Dumet and Benson, 2000) is an important prerequisite for the successful cryogenic storage of many types of plant germplasm, it is important to consider free radical damage in relation to dehydration/desiccation stress as well as freezing.

6.6.1 LESSONS FROM SEED STORAGE STUDIES

Seed storage (usually at –20°C) is still the main, and indeed preferred, method of conserving plant germplasm and free radical processes in stored and desiccated seeds, and has been examined in some detail (Benson, 1990; Hendry, 1993; Leprince et al., 1993; Senaratna et al., 1985). For researchers interested in oxidative stress and cryogenic storage, much can be gained from exploring seed storage literature, as evidence that seeds stored at about –20°C are susceptible to oxidative stress is quite considerable (Hendry, 1993). However, the data are often conflicting and are mainly correlative because of the physiological complexity of studying seed storage parameters. Prestorage status of seeds can greatly influence the oxidative stress profiles obtained with respect to seed hydration, age, storage behavior (recalcitrant, intermediate, orthodox), and biochemical composition (e.g., lipid content). Free radical stress in seeds has mainly considered storage criteria (viability and germination) and, in some cases, the application of accelerated aging treatments. Hendry (1993) makes the very important point that this experimental approach poses a major problem in ascertaining the definitive role of free radicals in the deterioration of seeds actually held in storage. He considered that the key to this relates to understanding the relationship between damage and viability. In the case of viable seeds, the evidence for free radical involvement in storage stress is indeed most persuasive; however, this is not so clear-cut for seeds that have already lost viability and that have been dead, in storage, for considerable periods. For these systems, the "fingerprints" of oxidative stress will have changed over time and will largely reflect chemical changes postmortem, rather than the free radical processes incurred when the germplasm was still viable and that may have contributed to the actual loss of viability. Thus, when considering the importance of oxidative stress in any type of plant germplasm stored in the "frozen state," it will be very important to take into consideration the following factors:

1. Primary oxidative metabolism is the main driving force for free radical formation in biological systems.
2. Storage at low temperatures will have a differential effect on biological and chemical reactions.
3. Oxidative profiles of cryopreserved tissues must be very carefully interpreted with respect to viability, moisture and antioxidant contents, aging, and postmortem status.

6.6.2 CRYOPROTECTION, CRYOINJURY, AND FREE RADICALS

Plant cryopreservation now involves a range of different cryoprotective strategies (see Chapter 10) based on either traditional controlled-rate freezing or vitrification. Traditional methods involve the application of penetrating colligative cryoprotectants, the precise control of cooling rate, ice nucleation, and freeze-induced cellular dehydration. Cryoprotectant mode of action is largely colligative and is achieved by penetrating cryoprotectants such as Me$_2$SO, sometimes used in combination with nonpenetrating cryoprotectants (e.g., polyols, sugars), which offer additional protection by removing freezable water from the cells through osmotic dehydration.

In the case of cryopreservation protocols based on vitrification, the glassy state can be achieved by concentrating cell solutes through osmotic dehydration and evaporative desiccation or the loading of highly concentrated penetrating cryoprotectants such as Me$_2$SO. Both approaches have the same overall effect of increasing cellular viscosity, such that on cryogenic storage, ice formation is inhibited and an amorphous glassy state is created. Although plant cryopreservation protocols have now been categorized as either "traditional" or "vitrified," it may be more appropriate to view them, respectively, as cryopreservation in the presence or absence of ice. In the case of traditional cryopreservation, the formation of extracellular ice is critical in creating a vapor pressure gradient between the frozen extracellular and the unfrozen intracellular compartments. Unfrozen water thus moves to the outside of the cell, and in doing so, the solutes become concentrated and the cell viscosity is increased. It is therefore highly likely that, even in the case of "traditional" cryopreservation, the majority of the surviving cells do, in fact, vitrify (see Chapter 10) when placed in cryogenic storage, even though the extracellular components exist in a frozen state.

Fundamental knowledge of the physical mode of action of cryoprotectants has provided valuable insights that have been used to advance the field of medical cryobiology and to aid storage protocol development. The possibility that cryoprotective additives offer "biochemical" as well as "physical" protection at ultra-low temperatures has also been explored. However, despite the major advances in the application of cryopreservation to plant germplasm, our present understanding of the modes of action of the different cryoprotective strategies is still very limited. A fundamental knowledge (physical and chemical) of cryoprotectant mode of action is, practically, very important when studying the involvement of oxidative stress in the "frozen plant." It is also highly pertinent when considering the long-term study of genetic stability in plants recovered from cryogenic storage. Primary and secondary oxidative stresses are both implicated (see Section 6.2.2) in genetic and molecular damage (Benson, 1990; Harding, 1999), and understanding of the role of oxidative stress in cryoinjury may offer insights into the future development of both "traditional" and contemporary cryoconservation protocols (e.g., see Fleck et al., 2000).

6.6.2.1 Radioprotectants, Cryoprotectants, and Free Radical Scavengers

Following the serendipitous discovery that glycerol has cryoprotective properties (Polge et al., 1949), a number of researchers screened other radioprotective compounds such as dimethyl sulphoxide (Me$_2$SO) for cryoprotectant activity (Ashwood-Smith, 1975; Miller and Cornwell, 1978). Thus, it is not too surprising that many radioprotectants (glycerol, Me$_2$SO, alcohols, glycols, polyols, sugars, tetramethylureas, and homologues) also possess excellent cryoprotective properties. A main factor linking radioprotection with cryoprotection is the ability of compounds to scavenge ·OH. If hydroxyl radicals are the most likely ROS to be produced at low and ultra-low temperatures (see Section 6.3.1), it must follow that the ·OH free radical scavenging capacities of cryoprotectants will make a significant contribution to their protective efficacy as well as to their colligative and osmotic properties. In this respect, Me$_2$SO is especially interesting, as it is one of the most potent cryoprotective and radioprotective agents ever discovered. Another property of Me$_2$SO worthy of note is its capacity to reversibly substitute for water in the hydration sheath of polysaccharides, proteins, and nucleic acids, thereby altering their structure (Barnett, 1972; Chang and Simon, 1968; Rammler and Zaffaroni, 1967). It is thus possible that by changing the conformation of cellular macromolecules, Me$_2$SO renders them less amenable to radiation-induced injury. It would be most interesting to determine whether this property also confers a protective effect on cryopreserved systems.

Kaul (1970) examined the efficacy of Me_2SO radioprotection in the seeds of several cereal species and speculated that the biochemical mode of action of Me_2SO may be to the result of protection against the radio-sensitizing effects of $O_2 \cdot^-$ as well as of O_2 alone, conferring preferential protection of essential biological molecules, as Me_2SO is such a potent $\cdot OH$ scavenger.

Me_2SO may also repair unstable intermediates through the donation of hydrogen atoms, and the presence of the S atom could assist radical scavenging and prevent free radicals from first attacking essential macromolecules that contain a sulfur atom (Kaul, 1970). Dickinson et al. (1967) demonstrated that Me_2SO protects tightly coupled mitochondria from freezing damage, and this highlights the potency of the cryoprotectant, as primary damage to the electron transfer chain has the potential to generate free radicals. Evidence for the role of Me_2SO as a free radical scavenger in plant cryoprotection relates to the use of Me_2SO as a "chemical probe" to detect $\cdot OH$ produced by cryopreserved algal and plant cells (Benson and Withers, 1987; Fleck et al., 2000) formed as a result of cryoinjury. The finding is further supported by the fact that the application of desferrioxamine (a potent inhibitor of Fenton chemistry) to cryopreserved plant and algal cells enhances poststorage recovery and limits the production of $\cdot OH$ (Benson et al., 1995; Fleck et al., 2000). Other related antioxidants have been applied in clinical freezing research. Thus, selenium compounds, known to be important in activating the glutathione peroxidase system in animals, have also been investigated as possible antioxidant, cryoprotective agents. Seleno-L-methionine, which is a component of the glutathione antioxidant protection system involved in the prevention of lipid peroxidation damage, was found to confer some freezing protection on rat heart muscle (Matthes et al., 1981). Further studies (Armitage et al., 1981) of seleno-D-L-methionine, improved the cryoprotective action of ethanediol in whole rabbit hearts cooled to $-15°C$. It was concluded that selenium increased the tolerance of myocardial cells to freezing. An alternative approach to reducing free radical damage has been studied by Mazur et al. (2000). Because free radical damage in mouse sperm is proportional to oxygen concentrations, they used an *E. coli* membrane preparation, Oxyrase, to reduce the oxygen to less than 3% of atmospheric levels. Experiments showed that if mouse sperm are exposed to 0.8 *M* glycerol, 0.17 *M* raffinose, and 3.5% Oxyrase and then cooled to $-75°C$, 50% motility relative to untreated controls was obtained. In the absence of Oxyrase, motility was at 31% for frozen/thawed sperm samples. However, the authors note that the Oxyrase treatment appeared to protect the sperm against centrifugation and osmotic shock as opposed to freeze/thaw cycles *per se*.

In conclusion, it would appear that to protect against free radical stress caused by reduced temperatures it is best to

1. Minimize the amount of water present
2. Limit the levels of oxygen
3. Include a free radical scavenger in the cryoprotectant mixture or use a cryoprotectant that also offers free radical scavenging (e.g., Me_2SO or glycerol)
4. Have an antioxidant or iron chelator present to reduce the levels of metal cations such as Fe^{2+}

Perhaps all four factors should be considered to maximize the chances of recovering living cells after exposure to low and ultra-low freezing temperatures. Moreover, such an approach would also help limit the potentially damaging effects of free radicals on the functionality and long-term genetic and biochemical stability of cryopreserved cells.

6.6.2.2 Oxidative Stress in Vitrified, Cryopreserved Plant Germplasm

The more recent development of vitrification-based cryopreservation protocols has had a major effect on increasing our ability to cryopreserve plant germplasm (see Chapters 9 and 10). In this context, vitrification is considered as cryogenic storage in the absence of ice, such that both the biological and cryoprotectant component are vitrified. Note that vitrification is also considered of major importance in medical cell, tissue, and organ cryogenic storage (see Chapter 22). Vitrification has enabled the cryogenic storage of a wider range of previously recalcitrant (especially tropical) plant germplasm. Furthermore, the approach is simple and cost-effective and circumvents the need to purchase expensive programmable freezing equipment.

In terms of the long-term stability of cryopreserved cells and tissues, vitrification-based cryogenic storage presents an interesting discussion point with respect to the stability of vitrified germplasm as compared to germplasm cryopreserved in the frozen state (see Section 6.3). Taylor and colleagues (Chapter 22) emphasize the fact that vitrification is the solidification of a liquid without crystalline structure. They caution that as cooling proceeds, viscosity increases to a point at which translational molecular motion is essentially (as opposed to definitively) halted when the solution becomes a glass. Consideration is given to the fact that a glass is thermodynamically metastable, which in the long-term timescale of low-temperature preservation my well be of practical significance.

The application of vitrification in plant and algal cryoprotection is a relatively recent development, and to the authors' knowledge, there has hitherto been no study performed that has examined (including comparisons with cryopreservation in the frozen state) the long-term effects of vitrified cryogenic storage on stability. It would be most interesting to undertake such an investigation by assessing stability at the physical, chemical, and molecular levels. However, it is important to note that viable germplasm stored using "traditional" freezing protocols will most likely comprise a biological vitrified component and a nonvitrified (or partially frozen) phase.

It is possible that the free radical chemistry of vitrified biological systems may be quite different to that which occurs in a system where ice is present. Moreover, as the glassy state is physically metastable, it may be important to assess the effect that devitrification and glass relaxation could have on the oxidative chemistry of vitrified germplasm (see Chapter 22). Importantly, the chemical treatments and desiccating manipulations that lead to the establishment of a vitrified state in cryopreserved germplasm also comply with those conditions that would certainly limit, if not stop, free radical chemistry. These are the application of high concentrations of cryoprotectants (e.g., Me_2SO, polyols, glycerol) and sugars (especially sucrose), which have also been discovered to be potent free radical (especially $\cdot OH$) scavengers in other applications (see Section 6.6.2.1). Also, the reduction of water activity in vitrified germplasm (through the application of evaporative and osmotic dehydration/desiccation treatments) concurs with the diminution of free radical chemistry. Symons (1982) postulated that for glasses (comprising aqueous organic systems) irradiated at low temperatures, the organic compound acts as an electron-acceptor and the "hole" is trapped by the solvent as $\cdot OH$. Interestingly, he also noted that compounds frequently used as cryoprotectants (e.g., aqueous alcohols, especially glycols, glycerol, and sugars) formed good glasses on cooling and that their irradiation gave high yields of trapped electrons, their centers being either $\cdot OH$ or RO. Detailed EPR studies performed on lyophilized microbial cultures could also offer some clues as to the potential for free radical reactions occurring in highly desiccated, vitrified, cryopreserved germplasm. Moisture had a significant effect on the formation of free radicals from lyophilized cells (Heckly and Quay, 1983) and on the production of free radicals from macromolecules (Ruuge and Blyumenfeld, 1965). Thus, as water content increases, so do the levels of intensity of EPR signals assigned to free radical activities. However, this is a dynamic process, and it is highly dependent on moisture content. At a critically high humidity level, free radical signals become unstable and disappear. Heckly and Quay (1983) performed a study of the effects of various carbohydrates on the generation of free radicals from propyl gallate and histidine under continuous drying. Sucrose appeared to confer a protective "antioxidant" affect, as it was the most efficient carbohydrate for limiting free radical production. These researchers also studied the effects of sucrose and propyl gallate on the protection of lyophilized bacteria and found that at high sucrose levels (6 to 12%) survival was about 100%, which was concomitant with a reduced free radical activity. These studies are most interesting when viewed in the context of plant cryopreservation research, as sucrose is generally found to be highly effective in conferring cryotolerance to plant germplasm (Dumet and Benson, 2000).

6.7 TO WHAT EXTENT DO OXIDATIVE PROCESSES OCCUR IN THE FROZEN AND CRYOPRESERVED STATE?

Ensuring the long-term stability of cryopreserved germplasm is an important aspect of low temperature (e.g., seeds held at $-20°C$) and cryogenic storage research (Harding, 1999). However, to date, our understanding of the fate and reactivity of free radicals formed (or "trapped") in viable, frozen/vitrified biological systems is limited. Of course the propensity for reactions to proceed at low temperatures will also depend on terminal freezing temperature and hydration status (Fennema and Sung, 1980). Hydrated tissues held at high negative temperatures (e.g., $-0°C$) do indeed appear to support some free radical activity (Whiteley et al., 1992; Okuda et al., 1994).

There is still, however, a paucity of information related to the study of "biologically relevant" redox and free radical processes at ultralow cryogenic temperatures. This is, in part, because of the fact that monitoring free radicals, their reaction products, and antioxidants in the frozen milieu is technically very difficult. Prooxidative chemistry comprising electron transfers, free radicals, and enzymatic reactions (e.g., lipoxygenase/$O_2\cdot^-$) cannot be easily detected in the frozen state *per se*. Cells and tissues are normally returned to a positive temperature for assaying (unless EPR spin traps are used in a cryogenic system). This confounds data interpretation related to those events that may actually occur in storage. Consideration must also be given to the effects of the oxidative processes that occur before (chilling) at the point of freezing/vitrification and during thawing/devitrification. Viable tissues recovered from low and ultra-low (cryopreserved) temperatures will thus be influenced by their prestorage "metabolic history." The status of cell signaling mechanisms and redox states (especially of the chloroplast/mitochondrial electron transfer chains) on freezing is particularly important with respect to the subsequent manifestation of free radical injury during and after retrieval from storage. Thus, incipient, deleterious metabolic reactions may be "fixed" at the point of freezing and "held in suspension" until their effects are manifest on thawing and devitrification. This scenario (also see Benson, 1990) has received little attention in plant cells, tissues, and organs maintained under low-temperature storage but has received detailed consideration in the study of free radical damage associated with cold ischemia and reperfusion injury in mammalian organs preserved at subzero temperatures (Fuller et al., 1988). Thus, when a blood supply (and hence oxygen) is returned to a transplant organ that has been held in hypothermia, there exists the potential to exacerbate oxidative stress. Damaging radical reactions can arise from multiple pathways (Fuller et al., 1988) and can ensue for hours after return to normal temperatures. Antiradical agents (e.g., desferrioxamine) applied on reperfusion help to alleviate these effects, and indeed a similar approach has been pioneered in plant cryopreservation research (see Benson and Withers, 1987; Benson et al., 1995; Fleck et al., 2000). Thus, unraveling the complexity of the pro- and antioxidant metabolic processes that occur before, during, and after low-temperature storage offers practical potential; for example, by enhancing survival through the application of agents that limit the negative, prooxidative effects of cold ischemia and reperfusion injury. Interestingly, a similar scenario can be considered for plant germplasm that has been dehydrated/desiccated and hermetically sealed (e.g., in seed banks) or stored under liquid nitrogen storage and subsequently returned to a hydrated, aerated, and metabolically active state.

The information presented thus far (Sections 6.5 to 6.6) does offer compelling evidence that free radical–mediated stress is associated with chilling, freezing, and thawing. However, importantly, cryogenic liquid nitrogen temperatures (−196°C) must be expected to constrain thermodynamic and molecular mobility (e.g., related to enzyme-substrate proximities) to a level that will inhibit those oxidative processes (enzymatic, electron transport, and chemical) that "drive" both the formation of free radicals and secondary oxidative reactions. However, as the length of time that viable cells, tissue, and organs are held in cryogenic storage increases, it may be cautionary to include fundamental studies of "free radicals in the frozen state" in future research objectives. A starting point for this may be to revisit some of the more pertinent EPR and radiation literature (Oshima et al., 1996; Symons, 1982), as well as constructing technically stringent experimental designs that examine oxidative processes in the frozen state *per se*.

6.8 CONCLUSIONS AND FUTURE STUDIES

Investigations of free radical chemistry and biology in the "frozen state" still remain very limited despite the fact that there is now considerable evidence to indicate that low temperatures can have a profound effect on primary oxidative metabolism and antioxidant protection in both plants and animals. This chapter has examined the role of free radicals in plant-low temperature biology on the premise that applied cryobiological research will benefit from an increased knowledge of generic (oxidative) stress. In conclusion, it appears that free radical–mediated oxidative stress and antioxidant status are influenced by low and ultra-low temperatures and that the potential exists (e.g., from EPR studies of model systems) for free radical chemical reactions to occur even at very low subzero temperatures, albeit the fact that clearer evidence is still required to substantiate whether this is the case for viable biological material held under cryogenic storage. If free radicals can participate in reactions at ultra-low temperatures, it is essential to critically evaluate whether they do, indeed, pose a significant risk. However, this may be most unlikely, and it is more probable that oxidative stress becomes more significant on return to normal temperatures.

Thus, the possibility that chemical (as opposed to metabolically driven) free radical reactions occur in long-term cryogenically stored germplasm has yet to be explored. Theoretically this may be more important with respect to vitrified germplasm held in long-term storage (as base collections) at temperatures at, or near, the glass transition temperature. Oxidative stress in germplasm may also be influenced by the transient changes in temperature that inevitably occur when samples are serially withdrawn from cryotank inventory systems over prolonged periods. Maintaining a critically low temperature that ensures glass stability and the stabilization of incipient oxidative (especially free radical) reactions becomes very important in the long-term storage of germplasm. Thus, the importance of developing robust inventories and handling procedures for the periodic retrieval of cryovials from large-scale storage tanks (particularly those held in the vapor phase) cannot be overstated.

Importantly, the limited studies that have to date examined cryopreserved plant germplasm have not indicated problematic consequences with respect to genetic stability (Harding, 1999), which is most encouraging. However, the possibility that the physical stability of metastable glasses may affect biochemical and molecular integrity does call for further consideration with respect to the following:

1. Material in cryogenic storage probably comprises mixtures of frozen or vitrified states, each of which will have different physical stabilities, particularly if there is a risk of devitrification occuring during handling procedures.
2. Vitrification requires cryopreservation in a highly desiccated/dehydrated state, and this may have important consequences for recalcitrant, metabolically active germplasm.
3. There is an increasing propensity (e.g., for medical safety legislation and economic reasons) to use vapor-phase as opposed to liquid-phase liquid nitrogen storage.

Although the final issue is not presently a factor in plant cryopreservation, it is most likely (e.g., for economic reasons) that plant cryobiologists will increasingly use vapor-phase storage in the future. If this is applied to vitrified germplasm, then it will be important to consider the long-term effects of vapor-phase cryogenic storage on glass stability (see Chapters 20 and 22); for example, in relation to the consequences that this will have on the physical stability of different water states, and how this may influence submolecular, free radical chemistry.

As cryogenic storage methodologies change and diversify, it may well be prudent to ascertain whether different types of cryoprotection, storage, handling, retrieval (from cryotanks) and recovery procedures differentially influence stability. For example, vitrification is increasingly used as the method of choice for plant cryopreservation (see Chapter 9). This approach uses highly concentrated cryoprotectants (sucrose, Me_2SO, glycerol), which also have excellent free radical scavenging properties. Thus, the future development of cryopreservation methods (particularly for metabolically active, recalcitrant germplasm) may benefit from a more informed and fundamental understanding of the role of cryoprotectants in protecting oxidative metabolism and ameliorating free radical–induced cryoinjury.

ACKNOWLEDGMENTS

The comments of Nick Lane and Keith Harding were gratefully received. We gratefully acknowledge the collaborative inputs and support of those colleagues and postgraduate students who have been involved in the research presented in this chapter, which was compiled in accordance to the fulfillment of European Commission Research Grant deliverables. The authors gratefully acknowledge the financial support of EU Commission Grants Quality of Life and Living Resources QLRT-2000-01645 (COBRA) and QLRT-2001-01279 (CRYMCEPT).

REFERENCES

Agostino, A., Heldt, H.W., and Hatch, M.D. (1996) Mitochondrial respiration in relation photosynthetic C4 acid decarboxylation in C4 species, *Aust. J. Plant Physiol.*, 23, 1–7.

Alscher, R.G., Donahue, J.L., and Cramer, C.L. (1997) Reactive oxygen species and antioxidants: relationships in green cells, *Physiol. Plant.*, 100, 224–233.

Anderson, M.D., Prasad, T.K., and Stewart, C.R. (1995) Changes in the isozyme profiles of catalase, peroxidase and glutathione reductase during acclimation to chilling in mesocotyls of maize seedlings, *Plant Physiol.*, 109, 1247–1257.

Apgar, M.E. and Hultin, H.O. (1982) Lipid peroxidation in fish muscle microsomes in the frozen state, *Cryobiology*, 19, 154–162.

Araki, T. (1977) Freezing injury in mitochondrial membranes. Susceptible components in the oxidation systems of frozen and thawed liver mitochondria, *Cryobiology*, 14, 144–150.

Armitage, W.J., Matthes, G., and Pegg, D.E. (1981) Seleno-DL-methionine reduces freezing injury in hearts protected with ethanediol, *Cryobiology*, 18, 370–377.

Asada, K. (1992) Ascorbate peroxidase—a hydrogen peroxide-scavenging enzyme in plants, *Physiol. Plant.*, 85, 235–241.

Asada, K. (1999) The water-water cycle in chloroplasts: scavenging of active oxygen species and dissipation of excess photons, *Annu. Rev. Plant Physiol. Plant Mol. Biol.*, 50, 601–639.

Asada, K. and Takahashi, M. (1987) Production and scavenging of active oxygen in photosynthesis, in *Photoinhibition*, Kyle, D.J., Osmond, C.B., and Arntzen, C.J., Eds., Elsevier, Amsterdam, pp. 227–287.

Ashwood-Smith, M.J. (1975) Current concepts concerning the radioprotective and cryoprotective properties of dimethyl sulphoxide in cellular systems, *Ann. N.Y. Acad. Sci.*, 243, 246–256.

Barnett, B.M. (1972) Radioprotective effects of dimethyl sulfoxide in *Drosophila melanogaster*, *Radiat. Res.*, 51, 134–141.

Benamira, M., Johnson, K., Chaudhary A., Bruner K., Tibbits C., and Marnett L.J. (1995) Induction of mutations by replication of MDA-modified M13 DNA in *E. coli*: determination of the extent of DNA modification, genetic requirements for mutagenesis and types of mutation induced, *Carcinogenesis*, 16, 93–99.

Bengtsson, B. and Bosund, I.I. (1966) I. Lipid hydrolysis in unbalanced frozen peas (*Pisum sativum*), *J. Food Sci.*, 32, 474–481.

Benson, E.E. (1990) *Free Radical Damage in Stored Plant Germplasm*, IPGRI, Rome.

Benson, E.E. (2000) Do free radicals have a role in plant tissue culture recalcitrance? *In vitro Plant Cell Dev. Biol.*, 36, 163–170.

Benson E.E., Lynch P.T., and Jones, J. (1992) The detection of lipid peroxidation products in cryoprotected and frozen rice cells: consequences for post-thaw survival, *Plant Sci.*, 85, 107–114.

Benson, E.E., Lynch, P.T., and Jones, J. (1995) The effects of the chelating agent desferrioxamine on the post-thaw recovery of rice cells: a novel approach to cryopreservation, *Plant Sci*, 110, 249–258.

Benson E.E., Magill, W.J., and Bremner D.H. (1997) Free radical processes in plant tissue cultures: implications for plant biotechnology programmes, *Phyton*, 37, 31–38.

Benson, E.E. and Noronha-Dutra, A.A. (1988) Chemiluminescence in cryopreserved plant tissue cultures: the possible role of singlet oxygen in cryoinjury, *Cryo-Letters*. 9, 120–131.

Benson, E.E. and Withers, L.A. (1987) Gas chromatographic analysis of volatile hydrocarbon production by cryopreserved plant cultures: a non-destructive method for assessing stability, *Cryo-Letters*, 8, 35–36.

Billings, W.D. and Mooney, H.A. (1968) The ecology of arctic and alpine plants, *Biological Rev.*, 43, 481–529.

Breen, A.P. and Murphy, J.A. (1995) Reactions of oxyl radicals with DNA, *Free Radical Biol. Med.*, 18, 1033–1077.

Breeze, V. and Elston, J. (1978) Some effects of temperature and substrate upon respiration and carbon balance of field beans, *Ann. Bot.*, 42, 863–876.

Buettner, G.R. (1993) The pecking order of free radicals and antioxidants: lipid peroxidation, α-tocopherol and ascorbate, *Arch. Biochem. Biophys.*, 300, 535–543.

Cadenas, E. and Davies, K.J. (2000) Mitochondrial free radical generation, oxidative stress and aging, *Free Radical Biol. Med.*, 29, 222–230.

Chang, C. and Simon, E. (1968) The effect of Me_2SO on cellular systems, *Proc. Soc. Exp. Biol. Med.*, 128, 60–66.

Cheng, L., Kellog, E.W., and Packer, L. (1981) Photoinactivation of catalase, *Photochem. Photobiol.*, 34, 125–129.

Cushman, J.C. and Bohnert, H.J. (1997) Molecular genetics of crassulacean acid metabolism, *Plant Physiol.*, 113, 667–676.

Dambrova M., Baumane L., Kalvinsh I., and Wikberg J.E.S. (2000) Improved method for EPR detection of DEPMPO-superoxide radicals by liquid nitrogen freezing, *Biochem. Biophys. Res. Comm.*, 275, 895–898.

De Grey, A.D. (2002) HO2*: The forgotten radical, *DNA Cell Biol.*, 21, 251–257.

Dickinson, D., Misch, M.J., and Drury, R.E. (1967) Dimethyl sulfoxide protects tightly coupled mitochondria from freezing damage, *Science*, 156, 1738–1739.

Dizdaroglu, M. (1992) Oxidative damage to DNA in mammalian chromatin, *Mutation Res.*, 275, 331–342

Dizdaroglu, M. (1998) Mechanisms of free radical damage to DNA, in *DNA and Free Radicals: Techniques, Mechanisms and Applications*, Aruoma, O.I. and Halliwell, B., Eds., OICA International, Saint Lucia, pp. 3–26.

Duden, R. (1985) Enzymatic lipid degradation reactions in deep frozen leafy vegetables and off flavour formation, *Agrichimica*, 29, 116–121.

Dumet, D.D. and Benson, E.E. (2000) Biochemical and physical studies of cryopreserved plant germplasm, in *Cryopreservation of Tropical Plant Germplasm: Current Progress and Applications*, Engelmann, F. and Takagi, H., Eds., International Plant Genetic Resources Institute and the Japanese International Research Centre for Agricultural Sciences, Rome, pp. 43–56.

Dumet, D., Engelmann, F., Chabrillange, N., Dussert, S., and Duval, Y. (2000) Cryopreservation of oil-palm somatic embryonic cultures, in *Cryopreservation of Tropical Plant Germplasm: Current Progress and Applications,* Engelmann, F. and Takagi, H., Eds., International Plant Genetic Resources Institute and the Japanese International Research Centre for Agricultural Sciences, Rome, pp. 172–177.

Elstner, E.F. and Osswald, W. (1994) Mechanisms of oxygen activation during plant stress, *Proc. R. Soc. Edinburgh*, 102B, 131–154.

Esterbauer, H., Zollner, H., and Schaur, R.J. (1988) Hydroxyalkenals: cytotoxic products of lipid peroxidation, *ISI Atlas Sci.: Biochem.*, 311–317.

Esterbauer, H., Zollner, H., and Schaur, R.J. (1991) Chemistry and biochemistry of 4-hydroxynonenal, MDA and related aldehydes, *Free Radical Biol. Med.*, 11, 81–128.

Feierabend, J., Scaan, C., and Hertwig, B (1992) Photoinactivation of catalase occurs under both high and low-temperature stress conditions and accompanies photoinhibition of photosystem II, *Plant Physiol.*, 100, 1554–1561.

Fennema, O. and Sung, J.C. (1980) Lipoxygenase-catalyzed oxidation of linolenic acid at subfreezing temperatures, *Cryobiology*, 17, 500–507.

Fleck, R.A., Benson, E.E., Bremner, D.H., and Day, J.G. (2000) Studies of free radical-mediated cryoinjury in the unicellular alga *Euglena gracilis* using a non-destructive hydroxyl radical assay: a novel approach for developing protistan cryopreservation strategies *Free Radical Res.*, 32, 157–170.

Fleck, R.A., Day, J.G., Clarke, K.J., and Benson, E.E. (1999) Elucidation of the metabolic and structural basis for the cryopreservation recalcitrance of *Vaucheria sessilis*, Xanthophyceae, *Cryo-Letters*, 20, 271–282.

Fleck, R.A., Day, J.G., Rana, K.J., and Benson, E.E. (1997) Use of cryomicroscopy to visualise freeze-events on cryopreservation of the coenocytic alga *Vaucheria sessilis*, *Cryo-Letters*, 18, 343–354.

Floyd, R.A., West, M.S. Hogsett, W.E., and Tingey, D.T. (1989) Increased 8-hydroxyguanine content of chloroplast DNA from ozone-treated plants, *Plant Physiol.*, 91, 644–647.

Foyer, C.H. and Halliwell, B. (1976) The presence of glutathione and glutathione reductase in chloroplasts: a proposed role in ascorbic acid metabolism, *Planta*, 133, 21–25.

Fridovich, I. (1995) Superoxide radical and superoxide dismutases, *Annu. Rev. Biochem.*, 64, 97–112.

Friedberg, E.C., Walker, G.C., and Siede, W. (1995) *DNA Repair and Mutagenesis*, American Society for Microbiology Press, Washington, DC.

Fryer, M.J., Andrews, J.R., Oxborough, K., Blowers, D.A., and Baker, N.R. (1998) Relationship between CO_2 assimilation, photosynthetic electron transport and active O_2 metabolism in leaves of maize in the field during periods of low temperature, *Plant Physiol.*, 116, 571–580.

Fuller, B.J., Gower, J.D., and Green, C.J. (1988) Free radical damage in organ preservation: fact or fiction, *Cryobiology*, 25, 377–393.

Gillham, D.J. and Dodge, A.D. (1986) Hydrogen peroxide-scavenging systems within pea chloroplasts. A quantitative study, *Planta*, 167, 246–251.

Guan, L.M., Zhao, J., and Scandalios, J.G. (2000) Cis-elements and trans-factors that regulate expression of the maize Cat1 antioxidant gene in response to ABA and osmotic stress: H_2O_2 is the likely intermediary signaling molecule for the response, *Plant J.*, 22, 87–95.

Guy, C.L. (1990) Cold acclimation and freezing tolerance: roles of protein metabolism, *Annu. Rev. Plant Physiol. Mol. Biol.*, 41, 187–223.

Halliwell, B. (1982) The toxic effects of oxygen in plant tissues, in *Superoxide Dismutase*, Vol. 1, Oberley, L.W., Ed., CRC Press, Boca Raton, FL, pp. 89–124.

Halliwell, B. and Arouma, O. (1991) DNA damage by oxygen-derived species. Its mechanism and measurement in mammalian systems, *FEBS Lett.*, 281, 9–19.

Halliwell, B. and Gutteridge, J.M.C. (1999) *Free Radicals in Biology and Medicine*, 3rd edition, Oxford University Press, Oxford.

Harding, K. (1999). Stability assessments of conserved plant germplasm, in *Plant Conservation Biotechnology*, Benson, E.E., Ed., Taylor & Francis, London, pp. 7–107.

Harding, K. and Benson, E.E. (1995) Methods for the biochemical and molecular analysis of cryopreserved plant tissue cultures, in *Genetic Preservation of Plant Cells in vitro*, Grout, B.W.W., Ed., Springer, Heidelberg, pp. 113–169.

Harman, D. (1956) Aging: a theory based on free radical and radiation chemistry, *J. Gerontol.*, 11, 298–313.

Harman, D. (1988) Free radicals in ageing, *Mol. Cell. Biochem.*, 84, 155–161.

Harman, D. (1991) The ageing process: major risk factor for disease and death, *Proc. Natl. Acad. Sci. USA*, 88, 5360–5363.

Hartley, D.P., Kroll, D.J., and Patersen, D.R. (1997) Prooxidation-initiated lipid peroxidation in isolated rat hepatocytes: Detection of 4-hydroxynonenal- and malondialdehyde-protein adducts, *Chem. Res. Toxicol.*, 10, 895–905.

Heber, U. (1968) Freezing injury in relation to loss of enzyme activities and protection against freezing, *Cryobiology*, 5, 188–201.

Heckly, R.J. and Quay, J. (1983) Adventitious chemistry at reduced water activities: free radicals and poly-hydroxy agents, *Cryobiology*, 20, 613–624.

Hendry, G.A.F. (1993) Oxygen, free radical processes and seed longevity, *Seed Sci. Res.*, 3 141–153.

Hurry, V.M., Krol, M., Oquist, G., and Huner, N.P.A (1992) Effect of long-term photoinhibition on growth and photosynthesis of cold-hardened spring and winter wheat, *Planta*, 188, 369–375.

Jahnke, L.S., Hull, M.R., and Long, S.P. (1991) Chilling stress and oxygen metabolizing enzymes in *Zea mays* and *Zea diploperenius*, *Plant Cell Environ.*, 14, 97–104.

Karl, R., Matthes, G., Hackensellner, H.A., Jentzsch, K.D., and Oheme, P. (1982) Oxygen-dependent cryoin-juries and their prevention by seleno-methionine, *Cryo-Letters*, 3, 57–60.

Kaul, B.L. (1970) Studies on radioprotective role of dimethyl sulphoxide in plants, *Radiat. Bot.*, 10, 69–78.

Kingston-Smith, A.H., Harbinson, J., and Foyer, C.H. (1999) Acclimation of photosynthesis, H_2O_2 content and antioxidants in maize (*Zea mays*) grown at sub-optimal temperatures, *Plant Cell Environ.*, 22, 1071–1083.

Ku, M.S., Kano-Murakami, Y., and Matsuoka, M. (1996) Evolution and expression of C4 photosynthesis genes, *Plant Physiol.*, 111, 949–957.

Kyle, D.J., Ohad, J., and Arntzen, C.J. (1984) Membrane protein damage and repair: selective loss of a quinone-protein function in chloroplast membranes, *Proc. Natl. Acad. Sci. USA*, 81, 4070–4074.

Lane, N. (2002) *Oxygen, The Molecule that Made the World*, Oxford University Press, Oxford.

Lea, P.J. and Leegood, R.G (1999) *Plant Biochemistry and Molecular Biology*, 2nd edition, Wiley, Chichester.

Lee, E.J., Volkov, V.I., and Lee, C.H. (2001) Electron spin resonance studies of free radicals in gamma-irradiated soybean paste, *J. Agri. Food Chem.*, 49, 3457–3462.

Leprince, O., Vertucci, C.W., Hendtry, G.A.F., and Atherton, N.M. (1995) The expression of desiccation-induced damage in orthodox seeds is a function of oxygen and temperature, *Physiol. Plant.*, 94, 233–240.

Levitt, J. (1956) *The Hardiness of Plants*, Academic Press, New York.

Levitt, J. (1962) A sulfhydryl-disulfide hypothesis of frost injury and resistance in plants, *J. Theoret. Biol.*, 3, 355–391.

Levitt, J. (1966) Cryochemistry of plant tissue: protein interactions, *Cryobiology*, 3, 243–251.

Levitt, J. (1980) *Responses of Plants to Environmental Stress*, Vol. 1, 2nd edition, Academic Press, New York.

MacRae, E.A. and Ferguson, I.B. (1985) Changes in the catalase activity and hydrogen peroxide concentration in plants in response to low temperature, *Physiol. Plant.*, 65, 51–56.

Magill, W., Deighton, N., Pritchard, H.W., Benson, E.E., and Goodman, B.A. (1994) Physiological and biochemical parameters of seed storage in *Carica papaya*, *R. Soc. Edinburgh Proc.*, 102B, 439–442.

Marnet, L.J. (2000) Oxyradicals and DNA damage, *Carcinogenesis*, 21, 361–370

Matthes, G., Hackensellner, H.A., Jentzsch, K.D., and Oehme. P. (1981) Further studies on the cryoprotection supporting efficiency of selenium compounds, *Cryo-Letters*, 2, 241–245.

Maxwell, D.P., Wang, Y., and McIntosh, L. (1999) The alternative oxidase lowers mitochondrial reactive oxygen production in plants, *Proc. Natl. Acad. Sci., USA*, 96, 8271–8276.

Mazur, P., Katkov, I.I., Katkova, N., and Critser, J.K. (2000) The enhancement of the ability of mouse sperm to survive freezing and thawing by the use of high concentrations of glycerol and the presence of an *Eschericia coli* membrane preparation (Oxyrase) to lower the oxygen concentration, *Cryobiology*, 40, 187–209

McClung, C.R. (1997) Regulation of catalases in *Arabidopsis*, *Free Radical Biol. Med.*, 23, 489–496.

McKersie, B.D., Bowley, S.R., Harjanto, E., and Leprince, O. (1996) Water-deficit tolerance and field perfor-mance of transgenic alfalfa over expressing superoxide dismutase, *Plant Physiol.*, 111, 1177–1181.

McKersie, B.D., Bowley, S.R., and Jones, K.S. (1999) Winter survival of transgenic alfalfa over expressing superoxide dismutase, *Plant Physiol.*, 119, 839–847.

McKersie, B.D., Chen, Y., De Beus, M., Bowley, S.R., Bowler, C., Inzé, D., D'Halluin, K., and Botterman, J. (1993) Superoxide dismutase enhances tolerance to freezing stress in transgenic alfalfa (*Medicago sativa* L.), *Plant Physiol.*, 103, 1155–1163.

McNulty, A.K. and Cummins, W.R. (1987) The relationship between respiration and temperature in leaves of the arctic plant *Saxifraga cernua*, *Plant Cell Environ.*, 10, 319–325.

Meryman, H.T. and Williams, R.J. (1985) Basic principles of freezing injury in plant cells: natural tolerance and approaches to cryopreservation, in *Cryopreservation of Plant Cells and Organs*, Kartha, K.K., Ed., CRC Press, Boca Raton, FL, pp. 13–48.

Miller, J.S. and Cornwell, D.G. (1978) The role of cryoprotective agents as hydroxyl radicals, *Cryobiology*, 15, 585–588.

Møller, I.M. (2001) Plant mitochondria and oxidative stress: Electron transport, NADPH turnover and metabolism of reactive oxygen species, *Annu. Rev. Plant Physiol. Plant Mol. Biol.*, 52, 561–591.

Muller, K., Hardwick, S.J., Marchant, C.E., Law, N.S., Waeg, G., Esterbauer, H., Carpenter, K.L.H., and Mitchinson, M.J. (1996) Cytotoxic and chemotactic potencies of several aldehydic components of oxidized LDL for human monocyte-macrophages, *FEBS Lett.*, 388, 165–168.

Niyogi, K.K. (1999) Photoprotection revisited: genetic and molecular approaches, *Annu. Rev. Plant Physiol. Mol. Biol.*, 50, 333–359.

Nordby, H.E. and Yelenosky, G. (1985) Change in citrus lipids during freeze-thaw stress, *Phytochemistry*, 24, 1675–1679.

O'Kane, D., Gill, V., Boyd, P., and Burdon, R. (1996) Chilling, oxidative stress and antioxidant responses in *Arabidopsis thaliana* callus, *Planta*, 198, 371–377.

Okuda, T., Matsuda, Y., and Sagisaka, S. (1994) Formation of hydroxyl radicals in stems of winter wheat treated with freezing temperature, *Biosci. Biotechnol. Biochem.*, 58, 1189–1190.

Omran, R.G. (1980) Peroxide levels and activities of catalase, peroxidase and indoleacetic acid oxidase during and after chilling cucumber seedlings, *Plant Physiol.*, 65, 407–408.

Öquist, G. (1983) Effects of low temperature on photosynthesis, *Plant Cell Environ.*, 6, 281–300.

Öquist, G. and Huner, N.P.A (1993) Cold hardening-induced resistance to photoinhibition of photosynthesis in winter rye is dependent upon an increased capacity for photosynthesis, *Planta*, 89, 150–156.

Oshima H., Iida Y., Matsuda A., and Kuwabara M. (1996) Damage induced by hydroxyl radicals generated in the hydration layer of gamma-irradiated frozen aqueous solution of DNA, *J. Radiat. Res.*, 37, 199–207.

Osmond, C.B., Anderson, J.M., Ball, M.C., and Egerton, J.G. (1999) Compromising efficiency: the molecular ecology of light-resource utilization in plants, in *Physiological Plant Ecology*, Press, M.C., Scholes, J.D., and Baker, M.G., Eds., Blackwell Science, Abingdon, U.K., pp 1–24.

Öttander, C., Hundal, T., Andersson, B., Huner, N.P.A., and Öquist, G. (1993) Photosystem II reaction centres stay intact during low temperature photoinhibition, *Photosythesis Res.*, 35, 191–200.

Ouellet, F. (2002) Out of the cold: Unveiling the elements required for low temperature induction of gene expression in plants, *In vitro Cell Dev. Biol.—Plant*, 38, 396–403.

Park, J.-I., Grant, C.M., Davies, M.J., and Dawes, I.W. (1998) The Cytoplasmic Cu, Zn superoxide dismutase of *Saccharomyces cerevisiae* is required for resistance to freeze-thaw stress, *J. Biol. Chem.*, 273, 22921–22928.

Petrenko, A.Y. and Subbota, N.R. (1986) Inhibition of the activity of mitochondrial electron transport chain by low temperatures: losses of Cytochrome *c*, *Cryo-Letters*, 7, 395–402.

Pinhero, R.G., Rao, M.V., Paliyath, G., Murr, D.P., and Fletcher, R.A. (1997) Changes in activities of antioxidant enzymes and their relationship to genetic and paclobutrazol induced chilling tolerance of maize seedlings, *Plant Physiol.*, 114, 695–704.

Polge, C., Smith, A.V., and Parkes, A.S. (1949) Revival of spermatozoa after vitrification and dehydration at low temperatures, *Nature (Lond.)* 164, 666.

Prasad, T.K. (1996) Mechanisms of chilling-induced oxidative stress injury and tolerance in developing maize seedlings: changes in antioxidant system, oxidation of proteins and lipids and protease activities, *Plant J.*, 10, 1017–1026.

Prasad, T.K., Anderson, M.D, Martin, B.A., and Stewart, C.R. (1994) Evidence for chilling-induced oxidative stress in maize seedlings and a regulatory role for hydrogen peroxide, *Plant Cell*, 6, 656–674.

Rammler, D.H and Zaffaroni, A. (1967) Biological implications of Me_2SO based on a review of its chemical properties, *Ann. N.Y. Acad. Sci.*, 141, 13–23.

Riggins, J.L. and Marnet, L.J. (2001) Mutagenicity of the malondialdehyde oligomerisation products (2-(3-oxo-1-propenyl)-malondialdehyde and 2,4-dihydroxymethylene-3-(2,2-dimethoxyethylglutaraldehyde in *Salmonella*, *Mutation Res.*, 497, 153–157.

Rustin, P. and Quieroz-Claret, C. (1985) Changes in oxidative properties of *Kalanchoe blossfeldiana* leaf mitochondria during development of crassulacean acid metabolism, *Planta*, 164, 415–422.

Ruuge, E.K. and Blyumenfeld, L.A. (1965) Free radicals of ascorbic acid appearing on interaction with protein, *Biofizika*, 10, 689–692.

Senaratna, T., McKersie, B.D., and Stinson, R.H. (1985) Simulation of dehydration injury to membranes from soybean axes by free radicals, *Plant Physiol.*, 77, 472–474.

Sharma, Y.K. and Davis, K.R. (1997) The effects of ozone in antioxidant responses in plants, *Free Radical Biol. Med.*, 23, 480–488.

Singh, J. and Miller, R.W. (1984) Biophysical and ultrastructural studies of membrane alterations in plant cells during extracellular freezing: molecular mechanisms of membrane injury, in *Cryopreservation of Plant Cells and Organs*, Kartha, K.K., Ed., CRC Press, Boca Raton, FL, pp. 61–74.

Sobczyk, E.A. and Kacperska-Palacz, A. (1978) Adenine nucleotide changes during cold acclimation of winter rape plants, *Plant Physiol.*, 62, 875–878.

Sobczyk, E.A. and Marszalek, A. (1985) ATP involvement in plant tissue responses to low temperature, *Physiol. Plant.*, 63, 399–405.

Spiteller, G. (1998) Linoleic acid peroxidation—the dominant lipid peroxidation process in low density lipoprotein—And its relationship to chronic diseases, *Chem. Phys. Lipids*, 95, 105–162.

Steponkus, P. (1985) Cryobiology of isolated protoplasts: applications in plant cell cryopreservation, in *Cryopreservation of Plant Cells and Organs*, Kartha, K.K., Ed., CRC Press, Boca Raton, FL, pp. 49–61.

Stewart, J.M. and Guin, G. (1969) Chilling injury and changes in adenosine triphosphate of cotton seedlings, *Plant Physiol.*, 44, 605–608.

Strand, Å., Hurry, A., Henkes, S., Huner, N., Gustafsson, P., Gardeström, P., and Sitt, M. (1999) Acclimation of Arabidopsis leaves developing at low temperatures. Increasing cytoplasmic volume accompanies increased activities of enzymes in the Calvin cycle and in the sucrose-biosynthetic pathway, *Plant Physiol.*, 119, 1387–1397.

Streb, P. and Feierabend, J. (2000) Significance of antioxidants and electron sinks for the cold-hardening induced resistance of winter rye leaves to photo-oxidative stress, *Plant Cell Environ.*, 22, 1225–1237.

Swartz, H.M. (1971a) Effect of oxygen on freezing damage: II. Physical-chemical effects, *Cryobiology*, 8, 255–264.

Swartz, H.M. (1971b) Effect of oxygen on freezing damage: III. Modification by β-mercaptoethylamine, *Cryobiology*, 8, 243–549.

Symons, M.C.R. (1982) Radiation processes in frozen aqueous systems, *Ultramicroscopy*, 10, 97–104.

Tappell, A.L. (1966) Effects of low temperatures and freezing on enzymes and enzyme systems, in *Cryobiology*, Meryman, H.T., Ed., Academic Press, New York, pp. 163–177.

Thomas, D.J., Thomas, J.B, Prier, S.D., Nasso, N.E., and Herbert, S.K. (1999) Iron superoxide dismutase protects against chilling damage in the cyanobacterium *Synechococcus* species PCC9421, *Plant Physiol.*, 120, 275–282.

Thomashow, M.F. (1999) Plant cold tolerance: freezing tolerance genes and regulatory mechanisms, *Annu. Rev. Plant Physiol. Plant Mol. Biol.*, 50, 571–599.

Tijus, S.E., Moller, B.L., and Scheller, H.V. (1998) Photosystem I is an early target of photoinhibition in barley illuminated at chilling temperatures, *Plant Physiol.*, 116, 755–764.

Tsvetkov, T., Tsonev, L., Meranoz, N., and Minkov, I. (1985a) Functional changes in mitochondrial properties as a result of their membrane cryodestruction I. Influence of freezing and thawing on succinate-ferricyanide reductase of intact liver mitochondria, *Cryobiology*, 22, 47–54.

Tsvetkov, T., Tsonev, L., Meranoz, N., and Minkov, I. (1985b) Preservation of the inner mitochondrial membrane after freeze-thawing and freeze-drying, *Cryobiology*, 22, 303–306.

Uchida, K. and Stadtman, E.R. (1992) Modification of histidine residues in proteins by reaction with 4-hydroxynonenal, *Proc. Natl. Acad. Sci. USA*, 89, 4544–4548.

Uno, G., Storey, R., and Moore, R. (2001) *Principles of Botany*. McGraw-Hill, Boston.

Van Camp, W., Inzé, D., and Van Montagu, M. (1997) The regulation and function of tobacco superoxide dismutases, *Free Radical Biol. Med.*, 23, 515–520.

Van Kuick, F.J.G.M., Senanian, A., Handelman, G.J., and Dratz, E.A. (1987) A new role for phospholipase A_2: protection of membranes from lipid peroxidation damage, *TIBS*, 12, 31–34.

Vanlerberghe, G.C. and McIntosh, L. (1997) Alternative oxidase: from gene to function, *Annu. Rev. Plant Physiol. Mol. Biol.*, 48, 703–704.

Volk, S. and Feierabend, J. (1989) Photinactivation of catalase at low temperature and its relevance to photosynthetic and peroxide metabolism in leaves, *Plant Cell Environ.*, 12, 701–712.

Whiteley, G.S.W., Fuller B.J., and Hobbs, K.E.F. (1992) Deterioration of cold-stored tissue specimens due to lipid peroxidation: modulation of antioxidants at high sub-zero temperatures, *Cryobiology*, 29, 668–673.

Xin, Z. and Browse, J. (2000) Cold comfort farm: the acclimation of plants to freezing temperatures, *Plant Cell Environ.*, 23, 893–902.

Yoshida, S. and Niki, T. (1979) Cell membrane permeability and respiratory activity in chilling stressed callus, *Plant Cell Physiol.*, 20, 1237–1242.

Yoshida, S. and Tagawa, F. (1979) Alteration of the respiratory function in chill-sensitive callus due to low temperature stress. I. Involvement of the alternate pathway, *Plant Cell Physiol.*, 20, 1243–1250.

Yoshimura, T., Miyazakaki, T., Mochhizuki, S., Susuki, M.K., and Watanabe, M. (1992) Do ·OH radicals react with organic-substances in gamma-irradiated frozen cells of golden hamster embryos? *Radiat. Phys. Chem.*, 40, 45–48.

7 Physiology, Biochemistry, and Molecular Biology of Vertebrate Freeze Tolerance: The Wood Frog

Kenneth B. Storey and Janet M. Storey

CONTENTS

7.1   INTRODUCTION

Winter survival for hundreds of species of organisms depends on freeze tolerance, the ability to endure the conversion of a high percentage of total body water into extracellular ice. Freeze tolerance has arisen many times in phylogeny. Among animals, examples of natural freeze tolerance can be found in most invertebrate groups, with most research into the mechanisms of freeze tolerance having been done on insects and intertidal marine molluscs. A number of species of amphibians and reptiles that hibernate on land are also freeze tolerant. Studies of vertebrate freeze tolerance are not only fascinating examples of complex adaptations but, by identifying the organ-specific molecular mechanisms that support natural freezing, also provide insights into the issues that need to be addressed in the development of organ cryopreservation technology as an aid to transplant medicine.

Modern interest in vertebrate freeze tolerance was kindled by a report from Schmid (1982), who documented recovery after freezing by three species of frogs that hibernate on the forest floor: the wood frog *Rana sylvatica*, the gray tree frog *Hyla versicolor*, and the spring peeper *Pseudacris crucifer*. Subsequently, the list of freeze-tolerant vertebrates has grown to include other forest frogs (*Hyla chrysoscelis*, *Pseudacris triseriata*; Costanzo et al., 1992b; Storey and Storey, 1986a), one salamander, and several reptile species. Among the reptiles are box turtles (*Terrapene carolina*, *Terrapene ornata*), that, at ~0.5 kg in mass, are the largest known freeze-tolerant animals (Costanzo and Claussen, 1990; Costanzo et al., 1995b; Storey et al., 1993), and hatchling painted turtles (*Chrysemys picta*), for whom freezing survival aids their strategy of remaining in their natal nest over their first winter of life (Storey et al., 1988). Most freeze-tolerant vertebrates known to date are from North America, with only two Old World species known to display ecologically relevant freezing survival: the Siberian salamander *Salamandrella keyserlingi* (Berman et al., 1984) and the European common lizard *Lacerta vivipara* (Costanzo et al., 1995a; Voituron et al., 2002). Freeze-tolerant vertebrates tend to be those that spend the winter at or near the soil surface. In this habitat, substantial insulation is gained from layers of organic litter and snow so that, whereas air temperatures above the snow may fall below –20°C, buffered temperatures under the snowpack are only occasionally below –5°C. Significantly, the low-temperature limit for survival by most freeze-tolerant vertebrates is no more than about –6°C, which shows that insulation is very important for their survival in nature. Species that can winter deeper underground below the frostline, for example, toads that may dig underground or spend the winter in rodent burrows, North American salamanders that also go down natural tunnels, and snakes that winter in underground dens, show very limited or no tolerance of freezing, as do numerous species of frogs and turtles that hibernate underwater (Churchill and Storey, 1992a; Costanzo et al., 1988; Storey and Storey, 1986a, 1992; Swanson et al., 1996).

The term "ecologically relevant" has been applied to those species that use freeze tolerance as an integral part of their winter hardiness, to distinguish them from species that can endure only short-term freeze exposures (Storey and Storey, 1992). Ecologically relevant freeze tolerance has been defined as the ability to endure long-term freezing (days or weeks) at a stable maximum ice content (usually 50 to 70% of total body water frozen) at subzero temperatures that occur naturally in the hibernaculum. Animals endure the penetration of ice into all extracellular spaces (e.g., abdominal cavity, bladder, brain ventricles, eye lens, plasma) and survive the interruption of all vital functions including breathing, circulation, and nerve and muscle activity, regaining these in a coordinated manner on thawing. Like all other strategies of adaptation used by animals, freeze tolerance undoubtedly evolved from a set of preexisting physiological and biochemical capacities that are broadly present in reptiles and amphibians. For example, the general tolerance of amphibians for wide variation in body water content (Shoemaker, 1992) aids endurance of the cellular dehydration caused when body water is sequestered as extracellular ice (Storey and Storey, 1996b). Most amphibians and reptiles also have good capacities for enduring oxygen deprivation, this ability being maximized in several types of freshwater turtles (including adult *C. picta*) that can hibernate underwater for 3–4 months without breathing (Jackson, 2001). Good anoxia tolerance aids endurance of the ischemia caused by plasma freezing. Hence, it is not surprising that rudimentary abilities to endure freezing occur quite widely among amphibians and reptiles living in seasonally cold environments.

Quite a number of species can endure some freezing if the duration is short (several hours maximum), temperatures are very mild (e.g., –1° to –2°C), overall ice accumulation is low (usually <20% of total body water), and ice is largely restricted to peripheral tissues (skin and skeletal muscles; Storey and Storey, 1992). However, these animals succumb when higher ice contents are reached, probably because ice penetrates into the body core and halts heartbeat and breathing, from which nontolerant species are not able to recover. For example, Pacific tree frogs, *Hyla regilla*, survived 6–12 hours of freezing at –2°C, but only 10% of animals recovered after 24 h frozen (Croes and Thomas, 2000). Thus, their rudimentary freeze tolerance may be sufficient to let them survive through an overnight frost, but they cannot endure the prolonged freezing exposures with high body-ice contents that would be necessary for long-term winter survival. The same is true of multiple reptile species, such as wall lizards (Burke et al., 2002; Claussen et al., 1990), garter snakes (Churchill and Storey, 1992a; Costanzo et al., 1988), and several turtle species (Churchill and Storey, 1992c; Packard et al., 1999). Indeed, garter snakes typically hibernate in deep underground dens where frost does not penetrate (Macartney et al., 1989), or sometimes underwater (Costanzo, 1989). Other species show geographic or population differences in freeze tolerance. Highland populations of the European common lizard *L. vivipara* show good freezing survival—better than their lowland counterparts—and adults are more tolerant than juveniles (Costanzo et al., 1995b; Grenot et al., 2000; Voituron et al., 2002); notably, juveniles from highland populations are found at a greater vertical depth than adults in the bogs where they hibernate (Voituron et al., 2002). Juvenile painted turtles also show variation in their winter survival strategy; some populations of *C. picta* show good freeze tolerance, whereas others show extensive supercooling (to –10°C or lower), so it appears that both freeze tolerance and freeze avoidance may have roles to play among different groups (Costanzo et al., 1999c; Churchill and Storey, 1992b; Packard et al., 1999).

Freezing damage to intolerant organisms can take many forms. Chapters 1 and 2 deal extensively with freezing injuries to cells, particularly those that are relevant to cryopreservation at ultra-low temperatures. For most animals that face subzero temperatures in nature, the concern is with relatively mild and variable subzero temperatures, multiple freeze/thaw events, and a need to preserve intracellular metabolism as well as to avoid injuries associated with the growth of ice in extracellular spaces. The potential injuries from freezing, as well as the principles of protection (presented in greater detail in later sections), can be summarized as follows:

First, with only a couple of documented exceptions (Lee et al., 1993; Wharton and Ferns, 1995), intracellular freezing is lethal for animals because of the extensive damage to subcellular architecture and microcompartmentation done by crystal growth. Freeze-tolerant animals limit ice to extracellular spaces only. This is generally ensured by the laws of biophysics (the probability of a nucleation event occurring in the large, contiguous extracellular spaces of an organism being very much greater than in the small volume of cells) but is often aided by the use of extracellular ice nucleating agents.

Second, the body fluids of most organisms can supercool to temperatures substantially below their equilibrium freezing point (FP). Although extensive supercooling is key for animals that rely on the freeze-avoidance strategy for winter survival (Duman, 2001; Storey and Storey, 1989), freezing from a deeply supercooled state is highly damaging because both the instantaneous ice surge (water immediately converted to ice on nucleation) and the subsequent rate of ice accumulation are very high. For example, wall lizards survived brief freezing exposures only if the initial ice surge was 5% or less of total body water (Claussen et al., 1990). Freeze-tolerant animals ensure a slow rate of ice growth that allows sufficient time to implement physiological and biochemical adjustments by initiating ice growth at temperatures just under the FP, often employing exogenous or endogenous ice nucleating agents to stimulate ice growth.

Third, ice growth through extraorgan and extracellular spaces can cause physical damage to tissues; for example, ice expansion can burst delicate capillary walls so that on thawing, vascular integrity is compromised (Rubinsky et al., 1987). Freeze-tolerant animals manage ice growth in several ways: ice growth is promoted in large extraorgan fluid spaces (e.g., abdominal cavity, bladder, and between skin and muscle layers), a partial dehydration of organs caused by water withdrawal into extraorgan ice masses reduces the net ice growth within the vasculature and other extracellular spaces within organs, and antifreeze proteins are employed to inhibit recrystallization, the process by which small ice crystals regroup over time into larger crystals.

Fourth, extracellular ice formation sets up an immediate osmotic stress on cells. Pure water is sequestered into ice, the osmolality of the remaining extracellular fluid rises, and cells respond by losing water until a new equilibrium is reached. The net result is a large increase in intracellular osmolality and ionic strength and a large decrease in cell volume. The reduction in cell volume is of primary concern because below a critical minimum volume, irreversible damage can occur because of compression stress on the cell membrane, which causes a collapse of the bilayer structure into an amorphous gel phase, or because of rupture of the cell (occurring because of excessively rapid swelling during thawing). Freeze-tolerant animals deal with this problem using cryoprotectants: high amounts of low–molecular weight carbohydrates (glucose in wood frogs) provide colligative resistance to cell-volume reduction, and the phospholipid bilayer of membranes is stabilized by trehalose or proline (see Chapters 20 and 21). Notably, membrane stabilization by a class of small peptides named dehydrins has recently been described in freeze-tolerant plants (Thomashow, 1999; Warren et al., 2000); it will be interesting to determine whether animals have a similar system.

Fifth, extracellular freezing solidifies the blood plasma, and hence, for the duration of the freeze, all cells and organs are in an ischemic state; no oxygen or fuels can be delivered to cells, and no wastes can be removed. Although cells loose a substantial fraction of their total water and experience a strong rise in osmolality and ionic strength, most metabolic reactions in cells are not impaired except by ATP limitation and by the low metabolic rate caused by a subzero body temperature. Thus, unlike the situation for cells that are cryopreserved in liquid nitrogen, metabolic reactions continue in the cells of frozen animals, and because freezing episodes in nature could last several weeks, freeze-tolerant animals must have good anoxia tolerance (to meet all cellular energy needs from fermentative metabolism) and ischemia resistance, as well as mechanisms that suppress metabolic functions that are not needed.

Finally, ice accumulation impedes and then halts all muscle movements, including breathing and heartbeat. Nerve activity also ceases. Freeze-tolerant animals must have mechanisms that can ensure the recovery of nerve and muscle function and the coordinated reactivation of vital signs during thawing.

7.2 LIMITS AND INFLUENCES ON FREEZING SURVIVAL IN WOOD FROGS

The wood frog, *R. sylvatica*, are the primary model animal that is used for studies of vertebrate freeze tolerance, and there is, without question, far more information available on the physiology, biochemistry, and molecular biology of this species than on any other amphibian or reptile. The bulk of this chapter will deal with freeze tolerance in wood frogs, and so, despite the considerable literature on the basic parameters of freezing survival by other species of reptiles and amphibians, the following summary will be limited to wood frogs. The first studies that quantified wood frog freezing found 100 or 50% survival by juvenile wood frogs after freezing for 2 or 11 d at –4°C and 100% survival after 13 d freezing at –2.5°C by adults (Storey and Storey, 1984, 1986a). Survival was 100% for autumn-collected wood frogs frozen for up to 2 weeks at –1.5°C, but only 50% endured a 28-d freeze (Layne, 1995a; Layne et al., 1998).

Freezing survival by *R. sylvatica* is influenced by a number of variables including season, rate of freezing, temperature of freezing, time frozen, and cryoprotectant concentration. All of these variables influence the percentage of total body water that is converted to ice, and the effects of time, temperature, and season are also linked to their influence on cryoprotectant synthesis and distribution. For *R. sylvatica*, the survivable upper limit on ice content is 65 to 70% of total body water converted to extracellular ice (Costanzo et al., 1993; Layne, 1995a). The lower lethal temperature is about –5 to –6°C during autumn/winter but rises to about –3°C in the spring (summarized in Layne et al., 1998). At both seasons, these lower lethal temperature values probably represent the temperature at which the upper limit on ice content is reached, with the higher value in spring reflecting much lower cryoprotectant levels in spring animals (Storey and Storey, 1987). Autumn-collected wood frogs also endure longer freezing than do spring- or summer-collected animals. Summer wood frogs showed substantial survival only under very mild freezing conditions (24 h at −1.5°C) and did not survive either longer freezes (7 d) or deeper temperatures (24 h at −5°C) that were 100% survivable by autumn frogs (Layne, 1995a). Spring frogs showed 95% survival after 48 h frozen at −2 to −3°C but did not recover after −5 to −6°C exposure (Layne and Lee, 1987). Storey and Storey (1987) found that frogs captured on the first night of spring breeding were as tolerant of freezing as winter frogs but that tolerance decreased quickly in subsequent weeks. Loss of tolerance correlates with the reduced levels of cryoprotectant produced by spring frogs; blood levels of 15 to 60 mM in spring frogs (Layne and Lee, 1987; Storey and Storey, 1987) are substantially lower than the ~200 mM glucose in blood of winter frogs (Storey and Storey, 1984).

The rate of freezing also strongly affects postthaw survival. All spring wood frogs cooled to −2.5°C at a rate of −0.16°C/h or −0.18°C/h survived, whereas survival declined to 60 to 80% at a rate of −0.3°C/h to −1°C/h, and no frogs survived a cooling rate of −1.17°C/h (Costanzo et al., 1991a). Surviving frogs cooled at the intermediate rates required longer recovery periods after thawing, and animals showed transient neuromuscular damage. Slow rates of temperature change are normal in the insulated microhabitats of the forest floor, where frogs hibernate, and so it is not surprising that freeze survival depends on slow cooling. Slow rates of temperature change ensure a slow rate of ice formation and provide plenty of time for the implementation of cryoprotective measures (glucose synthesis, gene expression) that are triggered only when ice begins to form on the skin (Storey, 1999; Storey and Storey, 1985). Hence, if the rate of temperature decrease is too fast, the rate of ice formation becomes too high and the time available to distribute cryoprotectant and to implement other metabolic/gene responses becomes too short before ice penetration completely cuts off circulation. Indeed, laboratory freezing trials showed that glucose distribution was impaired when freezing was too fast (Costanzo et al., 1992a; Storey, 1987a). Rapid cooling also impaired the recovery of neuromuscular function after thawing (Costanzo et al., 1991a), which can be interpreted as indicating that protective measures were not fully implemented when freezing was too fast. The influence of cryoprotectant (glucose) levels on freezing survival has also been directly tested. Glucose loading markedly improved survival of adult and juvenile wood frogs; for example, 90% of glucose-loaded autumn-collected juveniles survived a 49-d freezing trial compared with none of the uninjected frogs (Costanzo et al., 1993; Layne et al., 1998). Glucose-loading elevated plasma and organ glucose content, significantly reduced ice content, lowered haemolysis and improved neuromuscular recovery after thawing (Costanzo et al., 1991b, 1993).

7.3 PHYSIOLOGY OF FREEZING AND THAWING

7.3.1 NUCLEATION AND ICE PROPAGATION

When freezing occurs just under the equilibrium FP of body fluids (about −0.5°C for wood frogs), the initial surge of ice formation and the subsequent rate of ice accumulation are minimized and the time available to implement adaptations to aid freezing survival is maximized. Hence, all freeze-tolerant animals tend to minimize their ability to supercool and to use some form of nucleation control (various freeze-tolerant plants also use nucleation control; see Chapter 5). Wood frogs have three options to initiate ice formation.

First, inoculative freezing caused by contact with environmental ice is very important and may represent the major mode of nucleation in the normally damp hibernation sites chosen by the species. Indeed, frogs chilled below their FP began to freeze within 30 seconds after coming in contact with ice crystals (Layne et al., 1990), and Costanzo et al. (1999b) showed that 98% of frogs held at −2°C froze if they were in contact with humic soil or damp leaf mold, compared with only 20% in dry containers. It is likely that once an ice nucleus catalyzes the freezing of surface moisture, that ice then penetrates the skin via pores and comes in contact with extracellular body fluids, which are then inoculated. Layne et al. (1990) detected no differences in the rate of ice propagation across samples of dorsal vs. ventral skin from *R. sylvatica*, and magnetic resonance imaging (MRI) has shown that the initial inoculation event can be on any part of the skin surface, with the freezing front then propagating inward through the animal from this point (Figure 7.1; Rubinsky et al., 1994).

Second, ice-nucleating bacteria of the *Pseudomonas* and *Enterobacter* genera have been cultured from the gut of wood frogs (Lee et al., 1995). Their presence in the gut and on the skin of wood frogs seems to be responsible for the characteristic supercooling point of –2° to –3°C displayed by frogs cooled on a dry substrate; supercooling point values for isolated skin (–3.1°C) and intestine (–3.6°C) were similar to that of intact frogs (–2.5°C), whereas other organs (liver, muscle) as well as plasma and urine cooled to –5°C or below before freezing (Layne, 1995b). Hence, if frogs cool without direct contact with external ice, then ice nucleating bacteria on skin or gut will trigger crystallization before supercooling becomes extensive. The structure and mechanism of action of bacterial ice nucleators has been well described by other authors (Fall and Wolber, 1995; see also Chapter 22 for a discussion of the principles of ice nucleation).

Third, wood frog plasma contains a proteinaceous ice nucleator that triggers freezing *in vitro* at –6° to –8°C (detected by differential scanning calorimetry) and that is susceptible to denaturation by heat and various chemical treatments (Storey et al., 1992a; Wolanczyk et al., 1990a, 1990b). If the protein stimulates nucleation *in vivo* at the same temperatures, then it is difficult to envision a role for it as a true INA. However, it is possible that the protein aids the propagation of ice through the frog's body, perhaps by modulating or regulating crystal growth through the vasculature.

Studies with freeze-tolerant insects have shown that their hemolymph may contain both ice-nucleating proteins and antifreeze proteins (Duman, 2001). The latter may seem to be an anomaly for a freeze-tolerant species, but it is proposed that their function is to stabilize ice crystal size and minimize recrystallization, the process by which small crystals regroup into larger and larger crystals that might be physically damaging to the frozen animal. Whereas a similar system could be useful for freeze-tolerant frogs, there is no evidence to date for the presence of antifreeze proteins in the blood of any freeze-tolerant vertebrate.

When wood frogs are nucleated at –2°C, the exothermic reaction of crystallization causes an immediate jump in body temperature to just under the FP (about –0.5°C). Body temperature then holds at this value for several hours while ice slowly forms at a mean rate that is typically less than 5% of total body water per hour, and then body temperature gradually drops back down to ambient (Layne and Lee, 1987). Because the rate of freezing is so slow, frogs may show nothing but a slight stiffness in some parts of their skin and limbs for the first hour or more, and maximal ice content may not be reached for 12 to 24 h. Ice propagation through the body of frogs has been monitored noninvasively using proton MRI (Rubinsky et al., 1994). Freezing begins at some peripheral point on the skin, and ice propagates inward asymmetrically through the body. Ice forms in fluid spaces such as the abdominal cavity (Figure 7.1, panel A), eye lens, and brain ventricles before the surrounding tissue freezes (Rubinsky et al., 1994). MRI also shows that heart and liver are the last organs to freeze, undoubtedly because of their high cryoprotectant content.

7.3.2 DEHYDRATION DURING FREEZING

Frogs face two kinds of dehydration stresses while frozen: evaporative water loss from the frog's body and water loss from cells and organs into extracellular and extraorgan ice masses. Net water loss from the body is a major problem for all amphibians (Shoemaker, 1992) because of their highly permeable skins, and this is the reason that most amphibians live in moist habitats. Water balance remains a problem for hibernating frogs, whether frozen or not, and indeed, evaporative water loss from frozen frogs at –2°C was just as rapid as from unfrozen frogs at 5°C (Churchill and Storey,

1993). However, wood frogs held under an insulating layer of sphagnum moss lost only 2.5% of their total body water while frozen for 7 d at –2°C, whereas without cover, water loss was 50% (Churchill and Storey, 1993). Because the lethal limit for net water loss by wood frogs is 50 to 60% of total body water, and as frozen animals cannot move to a more humid site to rehydrate themselves, it is obvious that long-term freezing survival in nature must depend on the animals having chosen a suitably humid and insulated hibernation site to begin with. In evolutionary terms, this need to hibernate in a humid or moist environment may have necessitated the original development of freeze tolerance, as frog skin is highly susceptible to nucleation by environmental ice or to contact with humic soil or organic matter (Costanzo et al., 1999c).

Within the body, freezing causes a substantial redistribution of water. Ice propagation through extraorgan fluid spaces (e.g., in the abdominal cavity, between skin and muscle) draws water out of organs, just as subsequent ice propagation through the vascular space of organs draws water out of cells. Indeed, during freezing at –2.5°C that converted 65% of total body water into ice, liver and intestine lost ~58% of their water, whereas skeletal muscles lost 23 to 36% (Lee et al., 1992). Organ dehydration occurred quite quickly and was nearly complete within the first 6 h of freezing exposure. During thawing, organ water content rises again, returning to near control values within 3 to 12 h for different organs (Lee et al., 1992).

The consequences of organ dehydration during freezing are several. On the positive side, substantial dehydration of organs minimizes the amount of ice that can grow within organ capillaries, and this reduces the potential for injury caused by ice expansion within these delicate vessels. Organ dehydration also elevates the concentration of cryoprotectant within cells. On the negative side, there is potential for metabolic damage because of the approximately threefold increase in cellular ionic strength that occurs when 65% of total body water is frozen. However, amphibians have the highest tolerance for variation in body water content and body fluid ionic strength of all vertebrates, an adaptation necessitated by their highly water-permeable skin. Thus, whereas mammals manifest nervous system dysfunction when plasma sodium rises to 30 to 60 mM above normal, amphibians can tolerate sodium at 90 to 200 mM above normal (Hillman, 1988). Desert toads such as the spadefoot toad (*Scaphiopus couchii*) display substantially higher tolerances for dehydration and high ionic strength and maintain muscle contractile performance with a lower muscle water content than do pond frogs such as the leopard frog, *Rana pipiens* (Hillman, 1982, 1988; Shoemaker, 1992). Wood frogs, which tolerate losses of 50 to 60% of total body water just like desert toads (Churchill and Storey, 1993; Shoemaker et al., 1992), undoubtedly exhibit similar tolerances for high ionic strength of body fluids. However, it is possible that extreme changes in ionic strength, or imbalances in the ionic composition of intracellular vs. extracellular fluids during melting, could be significant factors in determining recovery times for nerve and muscle functions after freezing.

7.3.3 Thawing and the Reactivation of Vital Signs

Unlike freezing, which is a directional event with ice propagating inward from the periphery to the core of the frog, MRI scans indicate that thawing occurs uniformly throughout the frog's body (Figure 7.1; Rubinsky et al., 1994). Internal organs such as heart and liver actually melt very quickly (while still surrounded by extraorgan ice) because their high cryoprotectant concentration gives them a low melting point. Rapid thawing of the heart is advantageous, for it allows the early restoration of heartbeat as the first detectable vital sign in a thawing animal (Layne and First, 1991). Indeed, Layne et al. (1989) reported that heartbeat had resumed before all ice was melted in the body. The reactivation of heartbeat allows blood flow to be reestablished very early in the recovery process and, with the subsequent recovery of breathing, allows oxygen to be delivered as soon as possible to organs that have been ischemic throughout the freeze. Along with the resumption of breathing, hind-leg retraction reflexes begin to reappear in response to pricking or pinching, followed later by increasingly complex behaviors such as righting and jumping reflexes (Layne and First, 1991; Layne and Kefauver, 1997). Complex responses can take a day or more to

Panel A. Events of freezing

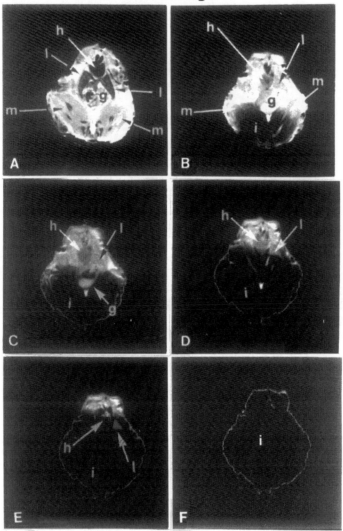

FIGURE 7.1 Proton magnetic resonance imaging showing a dorsal cross section through a wood frog at intervals during freezing and thawing. Explanation of the methodology can be found in Rubinsky et al. (1994). Data from B. Rubinsky and K.B. Storey (unpublished data). Panel A shows the events of freezing at −7°C. Timed from the initiation of freezing, images were sampled at (A) 10 min, (B) 49 min, (C) 1 h, 51 min, (D) 3 h, 14 min, (E) 3 h, 43 min, and (F) 7 h, 5 min. Tissues are heart (h), liver (l), hind-leg skeletal muscle (m), gut (g). Ice (i), which is not visible under ¹H MRI, is dark compared with unfrozen areas. Freezing in this individual began from sites on the hind legs, and early in the freeze (A), small patches of extraorgan ice can be seen dispersed over the leg muscles. Ice propagates through the leg muscles (B) and into the lower abdominal cavity, where ice surrounds the stomach, which is still unfrozen (C). Ice then moves into the upper portion of the abdomen and the lower part of the thoracic cavity, where dark margins of ice are seen around the pericardium and beneath the liver (D). With longer freezing, only one lobe of the liver remains unfrozen, as well as tissue in the neck region (E). After 7 h the frog is completely frozen (F). Panel B shows the events of thawing at 4°C for the same frog. Images were sampled at (A) 1 min, (B) 25 min, (C) 44 min, (D) 61 min, (E) 73 min, and (F) 152 min. As ice melts, the images lighten (A–C), and it can be seen that melting occurs throughout the body; hence, unlike freezing, thawing is not directional. The liver melts rapidly (B) before the dark borders of extraorgan ice surrounding it. Dark patches of abdominal ice are still present around the organs in (C) and (D) but have melted in (E). The heart has refilled with blood and resumed pumping in (E). In (F) the frog moves.

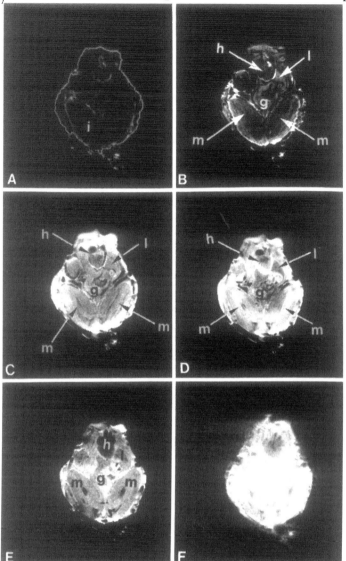

FIGURE 7.1 (continued)

reestablish after thawing, especially if freezing was prolonged. The time required for recovery of vital functions (e.g., circulation, breathing) after thawing correlates positively with the length of freezing, indicating that the longer the time frozen, the greater the metabolic disruption that must be reversed during recovery (Layne et al., 1998). Costanzo et al. (1997) also showed that freezing impairs male reproductive behaviors such as mate searching and amplexus and that even when these behaviors were restored about a day after thawing, freezing-exposed frogs competed poorly with frogs that had not been frozen. Postfreeze metabolic problems that need to be addressed include reestablishment of ionic, osmotic, and water balance in all organs, particularly excitable tissues such as nerve and muscle whose function relies on the regulated compartmentation of Na^+, K^+, and Ca^{2+}. Metabolic end products must also be cleared and interorgan communication reestablished via both blood-borne factors and nervous links.

Nerve and muscle function in most vertebrates is seriously impaired by three of the stresses associated with freezing: changes in cellular ionic strength, ischemia, and low temperature (Dalo et al., 1995; Hillman, 1988; Lutz and Reiners, 1997). Low temperature is of the least consequence to ectothermic vertebrates, as many species adapt to life at temperatures near 0°C, including polar fish that live continuously at about –2°C. Studies of motor-neuron and neuromuscular functions confirm that nerve and muscle function in *R. sylvatica* is largely unperturbed by cold exposure, with much better performance at low temperature than for tropical species and even better than cold-resistant but aquatic hibernating leopard frogs (Dalo et al., 1995; Miller and Dehlinger, 1969). Low-temperature performance may be related to selected changes to the properties of specific muscle proteins. For example, the sarco(endo)plasmic reticulum Ca^{2+}-ATPase 1 (SERCA1) from wood frog skeletal muscle shows an approximately twofold greater activity near 0°C, a significantly lower activation energy below 20°C, and altered ATP and Ca^{2+} kinetics as compared with the freeze and cold intolerant frog, *Rana clamitans* (Dode et al., 2001). Furthermore, the sequence of *R. sylvatica* SERCA1 shows seven unique amino acid substitutions, as compared with the sequences of *R. clamitans* and *Rana esculenta*, that may support improved enzyme function at very low temperatures. Notably, three of these amino acid substitutions are in the ATP-binding domain, where they may cause the altered ATP kinetics of the *R. sylvatica* enzyme.

The recovery of peripheral nerve function after thawing has been investigated in two studies. Kling et al. (1994) analyzed recovery at 5°C after 39 h freezing at –2.2°C. Isolated sciatic nerves from wood frogs were initially unresponsive to stimulation but regained excitability within 5 h, somewhat ahead of the recovery of coordinated reflexes, which reappeared at 8 h for hindlimb retraction and 14 h for the righting reflex. Compound action potentials of the nerves of freeze-treated frogs were indistinguishable from those of unfrozen frogs. Sciatic nerve lost 47% of its water during freezing, and glucose content increased eightfold. Although glucose remained high for 60 h postthaw, rehydration of the nerve occurred within 2 h of the onset of thawing, which indicates that reestablishment of normal ionic concentrations is probably crucial to the subsequent resumption of excitability. Costanzo et al. (1999a) further analyzed the effects of freezing on sciatic nerve function and ultrastructure in wood frogs. Nerves harvested from frogs given survivable freezing exposures at –2.5°C and then allowed to recover at 5°C for 14 h were responsive to electrical stimulation with no significant differences in several characteristics of action potentials (e.g., threshold stimulus, conduction velocity, amplitude, duration) between control frogs and those frozen for 30 to 50 h. Frogs also showed no differences in axonal ultrastructure after freezing. Freezing at a lower temperature, –5°C, caused minor disruptions; some electrophysiological parameters showed a tendency (not significant) for poorer responses, and evidence of minor physical damage was seen. Frogs frozen at –7.5°C did not survive, action potentials could not be induced in isolated nerves, and extensive ultrastructural damage was seen including marked shrinkage of the axon, degeneration of mitochondria, and delamination of myelin sheaths of the surrounding Schwann cells.

Muscle function appears to recover more quickly from freezing than does nerve function. Isolated gastrocnemius muscles from wood frogs thawed for 1, 2, or 24 h were all capable of generating force, although slightly less tension was generated in both twitch and tetanic contractions for the muscles from 1- or 2-h-thawed frogs compared with controls or 24-h-thawed muscles that were identical (Layne and First, 1991). In another study, Layne (1992) showed that both twitch and tetanic contraction tensions of wood frog gastrocnemius muscle depended on freezing time, being lower with longer freezes, when these parameters were assessed 3 h after the onset of thawing. After 24 h thawing, however, there were no differences in muscle contraction tension between 0-, 24-, and 48-h frozen groups. Comparable studies with leopard frogs showed that survivorship and muscle contraction properties were unaffected by a 6-h freeze at –2°C but that neither frogs nor their isolated muscles were viable after a 24-h freeze (Layne, 1992). The slower return of muscle reflex responses *in vivo*, discussed above, appears not to be a function of muscle itself but, instead, to be of the slower postthaw recovery of peripheral and central nerve functions.

Characterization of key metabolic functions of sarcolemma and sarcoplasmic reticulum from wood frog skeletal muscle provide some indications of the freeze-induced molecular events in muscle that must be reversed on thawing. Both calcium binding and oxalate-dependent calcium uptake by sarcoplasmic reticulum from muscle of frozen frogs were strongly suppressed to just 48 and 8%, respectively, of the comparable values in unfrozen controls (Hemmings and Storey, 2001). Both parameters rebounded partially after frogs were thawed (to 70 and 47% of controls, respectively). Similarly, β-adrenergic receptor binding by sarcolemma was also strongly reduced during freezing (by 88%), whereas calcium binding (a measure of calcium transport capability) was not affected by freezing but increased 2.3-fold in thawed frogs (Hemmings and Storey, 2001). β-adrenergic stimulation of muscle energy metabolism via blood-borne catecholamines is key to increasing ATP supply to meet the demands of muscle work. Suppression of this function during freezing is consistent with the cessation of muscle work and loss of muscle tone during freezing. Freeze–thaw-induced changes in parameters of Ca^{2+} metabolism (which are integral to the role of Ca^{2+} in the contraction/relaxation cycle of muscle) indicate that both freezing disruption of calcium transport and distribution in muscle cells and the need reestablish normal parameters after thawing may be critical components of the recovery process.

One of the hallmarks of metabolic recovery after stress is the repayment of an oxygen debt, reflecting the consumption of metabolic end products accumulated during the stress. Freezing is an ischemic event that cuts cells off from external supplies of oxygen for as long as blood plasma is frozen and also leads to the build-up of metabolic end products (lactate, alanine) (Storey, 1987a). Hence, an oxygen debt during thawing would be expected. Layne (2000) analyzed changes in oxygen consumption at 4, 8, and 28 h postthaw after wood frogs were frozen for 1 or 7 d at –1.5°C. Oxygen consumption after 4 h of thawing at 5°C was 25 to 30% lower than values for control, unfrozen frogs. This correlates with the low systemic recovery of frogs in the early hours of thawing, including poor postural recovery and limited responses to stimuli. However, rates of oxygen consumption were equivalent to controls at both 8 and 28 h postthaw in 1-d-frozen frogs, whereas 7-d-frozen animals showed a marked but transient increase in oxygen consumption (40% above controls) after 8 h postthaw. This peak of postthaw oxygen consumption may denote metabolic compensation for freeze-induced disturbances in cellular metabolism. Notably, cellular energetics show little disturbance after a 1-d freeze but ATP levels are reduced and glycolytic end products (lactate, alanine) accumulate in freezes lasting more than 2 d (Storey and Storey, 1986b).

7.4 CRYOPROTECTANTS: BIOCHEMISTRY, REGULATION, AND FUNCTION

7.4.1 CRYOPROTECTANTS

Most freeze-tolerant organisms produce cryoprotectants, typically of two kinds—those that act in a colligative manner to limit the extent of cell volume reduction during freezing and those that stabilize membrane structure. There has been no specific analysis of membrane cryoprotection for any freeze-tolerant vertebrate, so the reader is referred to treatments of the subject in Chapters 5 (plant freeze-tolerance) and 20 and 21 (anhydrobiosis). Colligative cryoprotection is provided by low–molecular weight carbohydrates (sugars or sugar alcohols) that are accumulated in high concentrations in cells and that provide osmotic resistance to the loss of intracellular water into extracellular ice crystals. In general, cryoprotectants are soluble in extremely high concentrations, move freely across cell membranes, are easy to synthesize from central carbohydrate reserves, are compatible solutes that do not perturb enzyme/protein function, and are excellent stabilizers of protein structure under denaturing conditions (including high protein dilution or very low water content). Notably, desiccation-resistant organisms also accumulate polyhydric alcohols (see Chapter 21), and it has been postulated, for both insect and frog systems, that cryoprotectant production by freeze-tolerant organisms grew out of preexisting metabolic responses to desiccation (Churchill and Storey, 1993; Danks, 1996; Storey and Storey, 1996b).

The blood sugar, glucose, is the common cryoprotectant among freeze-tolerant vertebrates. Glucose is the sole protectant produced by *R. sylvatica*, *P. crucifer*, and *P. triseriata*. Plasma levels typically rise to 150 to 300 mM in freeze-exposed winter animals, compared with 1 to 5 mM in unfrozen controls (Storey and Storey, 1984, 1986a). However, gray tree frogs and the Siberian salamander use glycerol as their primary cryoprotectant (Berman et al., 1984; Storey and Storey, 1986a; Layne, 1999). Glycerol levels ranging from 117 to 425 mM have been reported in *H. versicolor* blood compared with 20 to 25 mM glucose (Layne and Jones, 2001; Storey and Storey, 1986a). Interestingly, Layne and Jones (2001) reported that glycerol levels were high after cold acclimation of tree frogs and did not rise further with freezing exposure, as did glucose. Thus, glycerol production by *H. versicolor* may follow the pattern that is commonly seen in insects; that is, to accumulate polyols over the autumn before freezing (5°C is the usual trigger for synthesis), followed by maintenance of the pool over the entire winter season (Storey and Storey, 1989).

By contrast, glucose production by wood frogs and others is stimulated only when frogs begin to freeze, and the sugar is cleared again postthaw (Storey, 1987a; Storey and Storey, 1985, 1986b). The reason for this may be to limit any metabolic damage that could accrue from a long-term elevation of glucose over the winter, similar to the damage that occurs to tissues of diabetics with exposure to sustained high glucose in the 10- to 45-mM range (Kristal and Yu, 1992; Ruderman et al., 1992). Freeze-tolerant reptiles accumulate little or no cryoprotectant; blood glucose rises, but rarely to more than 20 to 30 mM, and increases in other metabolites such as glycerol, lactate, and amino acids are low (Churchill and Storey, 1992a, 1992b, 1992c; Storey et al., 1993; Voituron et al., 2002). Furthermore, reptile plasma shows no osmolality gap that would indicate the presence of unidentified low–molecular weight protectants. The lack of substantial amounts of colligative cryoprotectants in reptiles is consistent with their generally low freeze tolerance compared with amphibians.

The synthesis of glucose as a cryoprotectant by frogs is derived from the catabolism of liver glycogen reserves that are built up to high levels (~180 mg per gram wet weight in wood frogs) during summer and early autumn feeding (Storey and Storey, 1984). There is some question about whether skeletal muscle may synthesize glucose from its own reserves, but other internal organs show no depletion of their glycogen during freezing, and glucose accumulation by organs such as heart, kidney, and brain parallels the rise in plasma glucose that follows the rise in liver glucose content (Figure 7.2). Further confirmation of the liver as the source of cryoprotectant came from a comparison of organ glucose levels under conditions of fast vs. slow freezing (Storey, 1987a). After a fast freeze, which quickly cuts off circulation to other organs, glucose was much higher in liver and much lower in other organs than after slow freezing. A gradient of organ glucose levels also occurs, with frozen frogs showing highest glucose in liver and heart, lower amounts in other internal organs, and lowest levels in skin and skeletal muscle (Storey, 1987a). This gradient results because, during freezing, the ice front moves from peripheral sites inward (Figure 7.1), whereas glucose is distributed from a core organ outward.

On thawing, glucose is restored as liver glycogen, where it is available to be used again in the next freezing bout (Storey, 1987a). The process of glucose clearance is much slower than that of freeze-induced synthesis, and it can take over 1 week at 4°C for glucose to return to prefreeze levels (Storey and Storey, 1986b). Immediately postthaw, frogs are extremely hyperglycemic, with blood glucose levels that can be 200 to 300 mM. This exceeds the renal threshold for reabsorption of glucose, and glucose is copiously excreted into the urine, especially during the early hours postthaw (Layne et al., 1996). However, this important carbohydrate resource is not lost, because glucose is effectively resorbed from the frog bladder (Costanzo and Lee, 1997) and then reconverted to glycogen by the liver (Storey and Storey, 1986b) under the control of a thaw-induced activation of the enzyme glycogen synthetase (Russell and Storey, 1995).

7.4.2 REGULATION OF GLUCOSE SYNTHESIS FOR CRYOPROTECTION

The pathway of cryoprotectant synthesis by frog liver is exactly the same as that used in liver of all vertebrates for glucose production from glycogen. The three-step pathway (Scheme 7.1) is

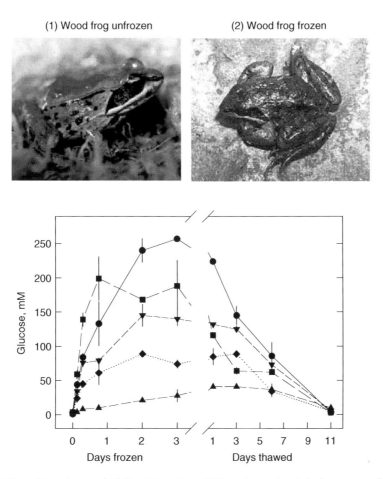

FIGURE 7.2 Effect of freezing at $-2.5°C$ and thawing at $5°C$ on glucose levels in frog organs. Data are means $\pm$ SEM, n = 3. Symbols are ●, blood; ■, liver; ▲, skeletal muscle; ▼, heart; ◆, kidney. Data compiled from Storey and Storey (1986b). Also shown are wood frogs, both unfrozen and frozen. Other images and information about wood frog life are available at www.carleton.ca/~kbstorey.

independent of a need for either ATP or reducing equivalents (e.g., NADPH), which may be one reason that glucose production in frogs can be delayed until tissues actually begin to freeze. By contrast, organisms that synthesize glycerol by its more complex biosynthetic pathway do so under nonfreezing conditions when aerobic ATP generation is not compromised.

$$\text{Glycogen} \rightarrow \text{glucose-1-phosphate} \rightarrow \text{glucose-6-phosphate} \rightarrow \text{glucose} \qquad \text{Scheme 7.1}$$

The three enzymes involved in glucose synthesis are glycogen phosphorylase (GP), phosphoglucomutase, and glucose-6-phosphatase. Cleavage of hexose phosphate units off the glycogen polymer is done by GP, and this enzyme is under strict regulatory control in liver of all vertebrates with modified regulation (discussed later) in freeze-tolerant frogs. No analysis has been made of possible regulatory controls on the other two enzymes, but notably, glucose-6-phosphate (G6P) is a substrate of other metabolic pathways in the cell, and G6P use by two of these is suppressed during freezing by inhibitory controls. The activity of glucose-6-phosphate dehydrogenase (G6PDH) that gates G6P entry into the pentose phosphate pathway is reduced by 80% in liver of frozen wood frogs as compared with control or thawed frogs (Cowan and Storey, 2001), and the activity of 6-phosphofructo-1-kinase (PFK-1) that gates G6P use by glycolysis (after its conversion to fructose-6-phosphate) is reduced by a strong freeze-induced suppression of the levels of its potent allosteric activator, fructose-2,6-bisphosphate (F2,6P$_2$; Storey, 1987a; Vazquez-Illanes and Storey, 1993). F2,6P$_2$ levels fall because freezing inhibits the activity of 6-phosphofructo-2-kinase, probably as a result of protein kinase A (PKA)-mediated phosphorylation of the enzyme (Vazquez-Illanes and Storey, 1993).

The initiation of freezing in wood frogs (detected as an exotherm caused by the heat release of crystallization) results in a near-instantaneous activation of GP in liver and a rapid rise in glucose levels. Significant increases in liver glucose occur within 2 to 4 min postexotherm, and blood glucose rises within 4 to 5 min (Storey, 1987b; Storey and Storey, 1985). GP is controlled by the interconversion of active a (phosphorylated) and inactive b (dephosphorylated) enzyme forms, and the amount of active GPa rose by twofold within 2 min postexotherm and by sevenfold within 70 min, and it continued to rise to ~13-fold higher than control activities with longer freezing (Storey, 1987b; Storey and Storey, 1989).

Freezing begins at peripheral sites on the skin, and hence, a signal transduction cascade must be involved to initiate the very rapid metabolic response by liver. The nature of the skin receptor that detects ice formation and the nature of the signal (nervous or hormonal) that is transmitted between skin and liver is still unknown, but the signal activates β-adrenergic receptors on hepatocyte plasma membranes (Hemmings and Storey, 1994); by contrast, skeletal muscle β-adrenergic receptors are not activated during freezing (Hemmings and Storey, 2001). Comparison of the effects of α- and β-adrenergic agonists and antagonists on freeze-induced glucose production in liver confirmed β-adrenergic involvement; notably, the β-adrenergic antagonist, propranolol, strongly inhibited the glycemia (Storey and Storey, 1996a). β-adrenergic signals are mediated in the cell by cyclic 3′5′adenosine monophosphate (cAMP) and PKA. The levels of cAMP levels in wood frog liver doubled within 2 min postexotherm, and the percentage of PKA present as the active catalytic subunit (PKAc) rose from 7% in control liver to 43 and 62% by 2 and 5 min postexotherm, respectively (Holden and Storey, 1996). PKAc phosphorylates and activates glycogen phosphorylase kinase, which, in turn, phosphorylates the inactive b form of GP and converts it to the active a form. PKAc also phosphorylates 6-phosphofructo-2-kinase, providing the inhibition of glycolysis that helps to channel glycogen breakdown toward glucose synthesis and export. Analysis of purified PKAc from wood frog liver indicated that differential temperature effects on the enzyme including increased substrate affinity and reduced ion inhibition at low temperature could contribute to an active role for the enzyme in the freeze-induced activation of glycogenolysis (Holden and Storey, 2000).

During fasting in healthy humans, glucose is held at about 5 mM, and during digestion of a meal, a rapid rise in insulin normally limits the transient increase in plasma glucose to 7 to 10 mM (Unger, 1991). A similar system of glucose homeostasis is present in all vertebrates, including unfrozen wood frogs. Yet, during freezing, blood glucose can rise to 150 to 300 mM. To allow this extreme hyperglycemia, one or more of the normal homeostatic controls on glucose levels must be circumvented. The full picture is incomplete, but some regulatory elements are known.

One of these elements is control over protein phosphatase-1 (PP-1), which opposes PKA action and normally intervenes to halt glycogenolysis when glucose rises, by stimulating dephosphorylation of GPa and phosphorylase kinase. PP-1 activity in liver should be suppressed during cryoprotectant synthesis, but in fact, activity rose by 2.4-fold within 20 min postexotherm (MacDonald and Storey, 1999). However, despite this, the normal consequences of PP-1 action (GP inactivation and the opposite activation of glycogen synthetase) do not occur during freezing but are instituted rapidly in liver when animals thaw (Russell and Storey, 1995; Storey and Storey, 1989). How can PP-1 be active and yet not functional? The answer seems to be because of subcellular compartmentation. In vertebrate liver, a high proportion of PP-1 catalytic subunits occur as dimers, bound 1:1 with a G subunit (glycogen binding protein) that confers glycogen-binding ability. All 3 isoforms of the catalytic subunit (α, δ, γ1) can bind to the G subunit, but the α and δ subunits are the main units associated with glycogen *in vivo* (Alessi et al., 1993). When liver samples from control (5°C) and frozen (12 h at −2.5°C) frogs were separated into cytosolic and glycogen fractions and PP-1

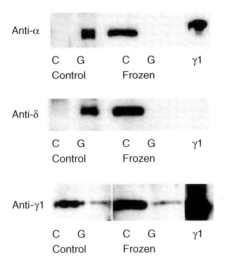

FIGURE 7.3 Protein phosphatase-1 (PP-1) isozymes in liver of control and 12-h-frozen wood frogs. PP-1 was partially purified from cytosolic (C) and glycogen (G) fractions of frog liver by microcystin-agarose affinity chromatography followed by SDS-PAGE and blotting onto PVDF membranes. Immunoblotting used antibodies to rat liver α, δ, and γ1 PP-1. Recombinant PP-1-γ1 was present in the fifth lane as a positive control. Gels show PP-1 bands at 37–39 kDA. (From MacDonald and Storey [1999]. With permission.)

isoforms were analyzed in each, the α and δ isozymes dominated in the glycogen fraction of liver from control frogs (Figure 7.3) (MacDonald and Storey, 1999). However, during freezing, nearly all the α and δ isozyme protein was translocated into the cytosolic fraction, whereas distribution of the γ1 isoform was not affected. Translocation of PP-1 to the cytosol removes the enzyme from association with glycogen and allows GP<u>a</u> to function unrestrained during freezing so that glucose levels can skyrocket. Two possible methods that could control freeze-induced translocation of PP-1 can be suggested. First, levels of the G subunit could be suppressed during freezing; indeed, low levels of the G subunit impair glycogen synthesis in insulin-dependent diabetes (Doherty et al., 1998). Alternatively, phosphorylation of the G subunit, perhaps by PKAc, could affect PP-1 binding. PKA phosphorylation of the G subunit in mammals increases the rate of PP-1 inactivation by inhibitor-1 and promotes PP-1 release from glycogen (Hubbard and Cohen, 1989). A similar situation of reduced PP-1 binding to glycogen particles also occurs in wood frog skeletal muscle during freezing (and in response to dehydration stress but not anoxia stress; MacDonald and Storey, 2002). Interestingly, this is balanced not by increased free cytsolic PP-1 but by increased amounts of PP-1 bound to myofibrils (via the M subunit). This may contribute to sustained skeletal muscle relaxation during freezing, as PP-1 reverses the phosphorylation of the myosin light chain that occurs during muscle contraction.

7.4.3 Wood Frog Insulin

Another factor that could contribute to the temporary loss of homeostatic control on glucose levels during freezing is modified hormonal control. As plasma glucose levels rise, so should insulin secretion in an attempt to halt glucose output. Indeed, plasma insulin levels did double during freezing, rising from 17 μU/mL in 5°C-acclimated frogs to 35 μU/mL in frozen frogs (Hemmings and Storey, 1996). However, insulin interaction with receptors on liver cell membranes might be blocked during freezing. Furthermore, wood frog insulin displays an anomolous structure that could limit its effectiveness (Conlon et al., 1998). Figure 7.4 shows the N-terminal sequences of insulin from four frog species compared with human insulin. One unusual feature of wood frog insulin is a two–amino acid extension (lysine-proline) to the N-terminus of the A chain. Although shared by

Insulin A-chain					
Rana sylvatica	KP GIVEQ	CCHNM	CSLYD	LENYC	S...
Rana catesbeiana	-- -----	----T	-----	-----	N...
Rana ridibunda	-- -----	----T	-----	-----	N...
Xenopus laevis	-----	---ST	--F--	--S--	N...
Homo sapiens	-----	--TSI	----Q	-----	N...

Insulin B-Chain					
Rana sylvatica	FPNQH	LCGSH	LVDAL	YMVCG	DRGFF	YSPRS...
Rana catesbeiana	----Y	-----	--E--	-----	-----	-----...
Rana ridibunda	----Y	-----	--E--	-----	E----	-----...
Xenopus laevis	LV---	-----	--E--	-L---	-----	-Y-KV...
Homo sapiens	-V---	-----	--E--	-L---	E----	-T-KT...

FIGURE 7.4 N-terminal sequences of insulin A and B chains in wood frog, bullfrog, green frog, clawed toad, and humans. (From Conlon et al. [1998]. With permission.)

other ranid frogs, this extension does not occur in any other vertebrates, and its role remains unknown. The rest of the A chain, as well as the B chain, was highly conserved in bullfrog, green frog, and clawed toad insulins as compared with the mammalian hormone, but wood frog insulin showed two anomalies. The serine residue at position A23 in wood frog insulin (A21 of human) is an asparagine in all other species, and the aspartic acid at B13 in wood frog insulin is glutamic acid in nearly all tetrapods. Both of these residues have been shown to play important roles in insulin action. A21 is key to the maintenance of the biologically active conformation via its bonding to B22/23, and B13 aids binding to the insulin receptor (Kristensen et al., 1997; Markussen et al., 1988). Both of the amino acids substitutions in wood frog insulin, as minor as they may seem, could significantly impair its function. Indeed, the only other known instance of a Glu-to-Asp substitution at B13 is in a mammal (the coypu), and this results in a low-potency insulin (Bajaj et al., 1986). Interestingly, although wood frog insulin displays novel features, its glucagon is identical with the hormone from the bullfrog and has only one amino acid substitution, as compared with the human glucagon (Conlon et al., 1998).

7.4.4 SIGNAL TRANSDUCTION: RESPONSES TO FREEZING, DEHYDRATION, AND ANOXIA

Apart from the potential for physical damage by ice, freezing places two major metabolic stresses on cells, dehydration and ischemia, and it is reasonable to postulate that various of the adaptations that defend cells during freezing are both initiated by and offer protection against one of these two stresses. This idea was first tested with respect to the regulation of cryoprotectant biosynthesis by comparing the hyperglycemic response by wood frogs to freezing vs. dehydration. Wood frogs held in dry containers (separated from an underlying layer of desiccant) at 5°C lost water at a rate of 0.5 to 1% of total body water per hour and showed a marked hyperglycemia when as little as 5% of body water was removed. Just as in freezing, glucose content increased sharply in all organs as frogs dehydrated, reaching values ranging from ninefold higher in muscle to 300-fold higher in liver of 50% dehydrated frogs, as compared with controls (Churchill and Storey, 1993). Dehydration activated liver GP and stimulated glycogenolysis exactly as seen during freezing and produced the same inhibitory block on glycolysis at the PFK reaction that serves to promote glucose export (Churchill and Storey, 1993, 1994b). Dehydration also triggered a rapid rise in liver cAMP content (by threefold at 5% body water lost), and an sharp increase in the percentage PKAc, which indicated that dehydration triggered the same β-adrenergic signal transduction pathway as freezing did (Holden and Storey, 1997). Notably, however, wood frogs given anoxia exposure at 5°C (under a nitrogen gas atmosphere) showed no activation of the PKA signal transduction pathway in liver (Holden and Storey, 1997). The hyperglycemic response to dehydration was also documented in *P. crucifer*, another freeze-tolerant frog, and a more muted response (a 24-fold increase in liver glucose) occurred in *R. pipiens*, an aquatic hibernator (Churchill and Storey, 1994a, 1995). The results of the latter study indicate that the cryoprotectant biosynthesis response to freezing grew out of a preexisting hyperglycemic response by frogs to water stress. Indeed, this is not surprising, because for cells, there would be no distinction between intracellular water lost to evaporation or that lost into extracellular ice masses.

Signal transduction via PKA clearly has a key role in the control of cryoprotectant biosynthesis in wood frogs, but we are only beginning to understand the possible roles of other signal transduction pathways in mediating responses to freezing (Storey and Storey, 2001). Both cGMP and inositol 1,4,5-trisphosphate (IP_3), the second messengers of protein kinases G and C, respectively, responded to freeze–thaw in wood frog organs (Holden and Storey, 1996). Changes in liver IP_3 levels were particularly interesting: Levels rose by 70% within 2 min postexotherm but then continued to rise to reach 11-fold higher than controls after 24 h of being frozen. By contrast, cAMP showed only a transient increase over the first hour of freezing. IP_3 levels also increased in brain during freezing. The longer response time for the rise in IP_3 and the sustained high levels of this second messenger for many hours indicates a role for PKC in events that occur with longer-term freezing, such as ischemia resistance, cell volume regulation, or gene expression. Notably, IP_3 levels also rose in wood frog liver in response to dehydration stress (Holden and Storey, 1997), perhaps indicating that PKC may be involved in mediating one or more metabolic responses to cell volume change. Interestingly, new studies of the upregulation of two novel freeze-responsive genes in wood frog liver have shown that one (*li16*) is stimulated by cGMP signals, whereas the other (*fr47*) responds to phorbol 12-myristate 13-acetate, an activator of PKC; neither responded to cAMP (McNally et al., 2002, 2003).

Mitogen-activated protein kinases (MAPKs) mediate numerous cellular responses including gene transcription, cytoskeletal organization, metabolite homeostasis, cell growth, and apoptosis in response to many different extracellular signals (Cowan and Storey, 2003; Hoeflich and Woodgett, 2001; Kyriakis, 1999). Subfamilies include the extracellular signal-regulated kinases, Jun N-terminal kinases (JNKs; also called stress-activated protein kinases), and p38. The latter is the vertebrate counterpart of yeast Hog1, which was named for its role in the high osmolarity glycerol response in cell volume regulation. An analysis of MAPK responses to freeze–thaw in wood frog organs found no evidence for the involvement of extracellular signal-regulated kinases, but both JNK and p38 responded to freeze–thaw (Greenway and Storey, 2000). JNK activities were unchanged over a 12-h freeze but showed strong increases after 90 min thawing in both liver (fivefold) and kidney (fourfold), or after 4 h thawing in heart (twofold). This indicates a role for JNK in mediating recovery responses needed during thawing. The p38 MAPK showed a widespread response to freezing in frog organs (Figure 7.5; Greenway and Storey, 2000). Within 20 min postnucleation, the amount of active, phosphorylated p38 rose by five- to sevenfold in liver and kidney, but this was reversed by 60 min. A role for p38 in one of the rapid, initial responses to freezing in these organs is therefore implicated. However, phospho-p38 content rose on a slower time course in heart peaking, at a sevenfold increase after 12 h freezing. As freezing progresses, heart undertakes an increasing work load because blood viscosity and peripheral resistance rise. Hence, changes in signal transduction in heart may be linked either to changes in heart workload or to the implementation of freeze tolerance adaptations. Parallel analyses of phospho-p38 responses to anoxia or dehydration in wood frog liver and kidney found no changes under these stresses. Interestingly, however, studies with perfused hearts of the freeze-intolerant *Rana ridibunda* showed a rapid phosphorylation of p38 in response to hyperosmotic stress, low temperature, and high perfusion pressure (Aggeli et al., 2002). In addition, JNKs were activated by high perfusion pressure and anoxia/reoxygenation in *R. ridibunda* heart (Aggeli et al., 2001a, 2001b). Hence, it is highly likely that MAPK responses to freezing address one or more of the component stresses of freezing including hyperosmolality, anoxia, and elevated workload.

7.4.5 Glucose as a Cryoprotectant and Applications to Cryopreservation

Freeze-tolerant frogs provide the good natural example of the cryopreservation of vertebrate cells, tissues, and organs. As such, an analysis of the molecular elements of natural freezing survival can

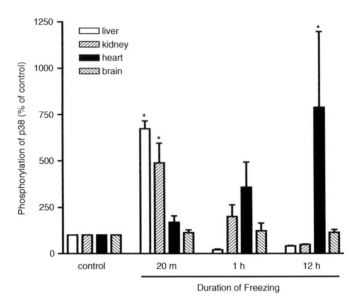

FIGURE 7.5 Effect of freezing on the amount of active phosphorylated p38 in tissues of spring wood frogs comparing controls (5°C acclimated) and frogs frozen for 20 min, 1 h (or 4 h frozen for brain only), or 12 h at −2.5°C. Phospho-p38 content was detected on Western blots, which were then scanned and quantified by densitometry. Data were standardized relative to control values and are shown as means ± SEM; * significantly different from the corresponding control value, $P < .05$. (From Greenway and Storey [2000]. With permission.)

provide novel insights that could be applied to improve the methodology for cryopreservation of mammalian and other material. A number of studies have examined the effects of freezing and cryoprotectants on wood frog cells *in vitro*. Freezing damage to isolated hepatocytes of wood frogs was assessed by measuring lactate dehydrogenase leakage into the medium after thawing. Addition of 200 m*M* glucose (similar to natural levels) to the incubation medium fully protected cells against freezing damage at −4°C, a temperature normally encountered in nature (Storey and Mommsen, 1994); this argues that the natural cryoprotectant has primary responsibility for freeze protection of wood frog cells. Higher glucose levels provided protection down to lower temperatures: 400 m*M* gave freeze protection at −8°C, and 1 *M* glucose protected cells from damage at −20°C (Storey and Mommsen, 1994). Ventricle strips from wood frogs also recovered contractility after freezing in the presence of 250 m*M* glucose but did not regain muscle activity if frozen without glucose in the bathing medium (Canty et al., 1986). Similarly, glucose and glycerol were each highly effective in preventing hemolysis of wood frog erythrocytes during freezing at −8°C, although parallel tests with human red blood cells showed that glucose was a very poor preservative for them (Costanzo and Lee, 1991). A comparison of the osmotic fragility and freezing tolerance of sperm from *R. sylvatica* vs. *R. pipiens* revealed that sperm of *R. pipiens* were more susceptible to both osmotic damage and cryoinjury at −4°C but that the sperm of both species were protected from freezing damage at −8°C by glucose or glycerol added as a protectant (Costanzo et al., 1998).

The cryoprotectant effect of glucose on wood frog cells obviously derives from the uptake of glucose in high amounts into wood frog cells. The physical effects of high glucose were documented by cryomicroscopy. Micrographs of liver slices frozen on a directional freezing stage to −7°C showed very different responses by tissue from control (5°C acclimated) frogs vs. liver slices from frogs given prior freezing exposure (24 h at −4°C to raise liver glucose to 280 m*M*; Storey et al., 1992b). Frozen liver slices from control frogs showed ice throughout an expanded vasculature with hepatocytes that were shrunken and virtually dehydrated. In the presence of the natural cryoprotectant, however, cells were much less shrunken, and ample free water remained within them. Wood frog cells take up glucose rapidly. Indeed, after a 60-min incubation in 600 m*M* glucose at 4°C, intracellular glucose concentration in wood frog hepatocytes (prepared and incubated as in Storey and Mommsen, 1994) had risen to 151 ± 35 mM, measured after cells were separated from the incubation medium by centrifugation through an oil layer (T.P. Mommsen and K.B. Storey, unpublished data). By contrast, glucose in control cells was only 2.6 ± 0.23 mM.

Uptake of glucose into vertebrate cells is via facilitated transport, and cryoprotectant uptake into wood frog cells is the result of high numbers of plasma membrane glucose transporters. Indeed, both the rate of facilitated glucose transport and the number of membrane glucose transporters were higher in wood frog than in leopard frog liver membranes, by eight- and fivefold, respectively (King et al., 1993). Glucose transporter numbers are also seasonally dependent, rising during autumn cold hardening. Plasma membrane vesicles prepared from liver of September-collected wood frogs showed sixfold higher rates of carrier-mediated glucose transport than vesicles from July-collected frogs because of an 8.5-fold higher number of glucose transporters in liver membranes of autumn frogs (King et al., 1995). High glucose transporter numbers is clearly one of the unique features needed by freeze-tolerant frogs to use glucose as a cryoprotecant, but because of this, the potential for use of the sugar in the applied cryopreservation of cells from other species is doubtful. Another factor that may contribute to effective cryoprotection by glucose in wood frog tissues is the selective inhibition of glucose catabolism at low temperatures. Studies by Brooks et al. (1999) demonstrated that whereas [U-^{14}C] glucose was taken up by wood frog erythrocytes just as effectively at 4°C as at 12°, 17°, or 23°C, it was not catabolized by cells at 4°C. Analysis by ^{13}C-NMR gave mean rates of glucose utilization of 1.27 and 0.91 mol/h per 10^{16} cells at 17° and 12°C, respectively, but the rate at 4°C, although predicted to be 0.52 mol/h per 10^{16} cells (based on a Q_{10} of 2), was not different from zero.

Freeze tolerance by wood frog embryos could provide a model for the cryopreservation of the embryos of endangered amphibian species. Wood frogs breed early in the spring (often before all the snow melts) in shallow temporary ponds, and their eggs are deposited in communal surface rafts. At this time there are still many nights with subzero air temperatures that could subject embryos to freezing. However, data from breeding ponds in Kentucky indicate that subzero exposure of eggs would be rare. Despite night air temperatures that plunged below –5° on several occasions and two instances of surface freezing, water surface temperature only dipped to about –0.5°C twice, and temperatures in communal egg masses never fell below 1°C (Frisbie et al., 2000). Furthermore, laboratory tests with wood frog eggs showed that embryos readily endured supercooling down to –2°C but that freezing survival was poor. Encapsulated embryos survived freezing for 18, 4.5, and 1 h at –0.5°, –1°, or –2°C, respectively, but survival was lower for naked embryos (those removed from their surrounding jelly) and varied with developmental stage (Frisbie et al., 2000). Hence, the data indicate that wood frog eggs have very limited freeze tolerance and are a poor model for cryopreservation.

7.5 FREEZING SURVIVAL: ISCHEMIA RESISTANCE AND ANTIOXIDANT DEFENSES

7.5.1 Ischemia Resistance

Freezing is an ischemic event. The freezing of blood plasma cuts off the delivery of oxygen and nutrients to organs, the removal of wastes, and interorgan communication via hormones and other signals. Each cell of each organ is left in isolation to survive throughout the freeze using only its own internal reserves. The subzero body temperature of a frozen organism means that metabolic rate will be very low; nonetheless, because freezing may be prolonged for many weeks, each cell must have adequate fermentative fuel reserves and a capacity to support basal metabolic needs over the long term using only the ATP generated from anaerobic pathways. Indeed, the severe effect of hypoxia or anoxia on most mammalian and human tissues and organs is one of the forces that fuels the continuing drive to develop and optimize cryopreservation technology. Both low-oxygen and low-temperature metabolic damage to mammalian tissues and organs limit the time that explants can be stored under hypothermic conditions (packed on ice; Hochachka, 1986) and spur the drive to develop techniques for preservation at ultra-low temperatures.

Freeze-tolerant frogs show a good capacity to endure long-term oxygen deprivation. Wood frogs can survive several weeks of continuous freezing (Layne et al., 1998) and readily endure 2 d of exposure to a nitrogen gas atmosphere at 5°C (Holden and Storey, 1997). Over the course of a freezing episode, wood frog organs show the typical vertebrate response to oxygen limitation: a depletion of ATP and accumulation of glycolytic end products (lactate and alanine; Storey, 1987a; Storey and Storey, 1986b). These parameters are normalized again within 3 to 11 d postthaw. Energetics during freezing have also been assessed in whole frogs using ^{31}P-NMR (Layne and Kennedy, 2002). Slow freezing of juveniles, producing high rates of postfreeze survival, were with associated stable ATP levels, a modest decline in creatine phosphate levels, a small increase in free inorganic phosphate, and a stable intracellular pH of 7.3 to 7.4. However, fast freezing disrupted all parameters and produced high mortality.

A recent study showed selected effects of freezing exposure on the maximum activities of 25 to 28 enzymes that participate in glycolysis, gluconeogenesis, the tricarboxylic acid cycle, amino acid metabolism, fatty acid metabolism, and adenylate metabolism in wood frog organs (Cowan and Storey, 2001). The ATP-limited frozen state cannot sustain normal rates of metabolic activity, and mechanisms of metabolic rate suppression are undoubtedly invoked (as occurs in response to anoxia stress in other lower vertebrates; Storey and Storey, 1990, 2004) to lower net ATP turnover and suppress most metabolic functions. Responses by wood frog kidney to freezing were especially revealing in this regard. Activities of 12 enzymes changed during freezing, with 10 of them decreasing significantly. Five that have biosynthetic functions (malic enzyme, NADP-isocitrate dehydrogenase, glyceraldehyde-3-phosphate dehydrogenase, glutamate dehydrogenase, and glutamate-pyruvate transaminase) all showed a reduction in activities of 25 to 50%, in line with a suppression of ATP-dependent biosynthesis in the frozen state (Cowan and Storey, 2001). Key enzymes involved in glucose catabolism were also targeted in kidney. The activity of hexokinase, which phosphorylates glucose for entry into central metabolic pathways, was reduced by 40% in frozen kidney, along with 43 and 40% reductions in the activities of G6PDH and 6-phosphoglu-conate dehydrogenase (6-PGDH), respectively, which gate glucose catabolism via the pentose phosphate shunt; activities of all three rebounded rapidly when frogs were thawed (Cowan and Storey, 2001). By targeting these three enzymes, catabolism of glucose as a fuel should be strictly limited in the frozen state, despite the fact that glucose is a good fermentative fuel. It is probable that endogenous glycogen, rather than colligatively active glucose, is the primary fermentative fuel for frog organs in the frozen state.

Freeze-induced suppression of enzyme activities, followed by a rebound after thawing, also occurred in liver, heart, and skeletal muscle of frogs and was the response seen in 75% of cases in which metabolic enzymes responded to freezing (Cowan and Storey, 2001). HK, G6PDH, and 6PGDH were common targets, along with multiple enzymes involved in biosynthesis. Heart showed a strong response with activities of 11 enzymes reduced during freezing (most by 50% or more); this effect may be related to the cessation of contractile work in frozen animals. Skeletal muscle showed numerous thaw-specific increases in enzyme activities, including elevated activities of three equilibrium enzymes of glycolysis and two gluconeogenic enzymes (fructose-1,6-bisphosphatase, phosphoenolpyruvate carboxykinase), which, taken together, could elevate the metabolic potential for gluconeogenic flux in muscle and the reconversion of accumulated lactate to muscle glycogen, as has been demonstrated in skeletal muscle of other frog species (Fournier and Guderley, 1993).

7.5.2 ANTIOXIDANT DEFENSES

All organisms maintain antioxidant defenses, both enzymes and metabolites (e.g., glutathione, ascorbate, thioredoxin), that deal with the continual assault to cellular metabolism by reactive oxygen species such as superoxide, hydrogen peroxide, hydroxyl radicals, and peroxynitrite. (See Chapter 6 for a detailed discussion of free radical chemistry in the context of freezing injury.) The latter two species, in particular, are highly reactive species that cause serious damage to cellular lipids, proteins, and DNA. Oxidative damage occurs in a variety of disorders including three that are of direct relevance to vertebrate freeze tolerance—ischemic heart disease, stroke, and diabetes (Ahmad, 1995; Halliwell et al., 1992). Studies in mammalian systems have linked a substantial portion of ischemic damage not to the period of oxygen deprivation itself but, rather, to the reperfusion phase, when oxygen is reintroduced. Reoxygenation results in a burst of oxyradical production that can temporarily overwhelm the antioxidant defenses of an organ and cause extensive damage to cellular macromolecules. Brain, for example, is very susceptible to oxyradical-mediated damage during reperfusion (Lipton, 1999). Freeze–thaw is an ischemia-reperfusion event that has the potential to cause oxidative damage to organs when heartbeat and breathing are restored during thawing. Furthermore, in the case of freeze-tolerant frogs that use glucose as a cryoprotectant, the potential for glucose-mediated oxidative damage is high. Reactive oxygen species are well known to play a role in tissue damage in diabetes (Kristal and Yu, 1992), and several modes of glucose-related oxidative damage are known. For example, free glucose is prone to autooxidation in the presence of transition metals (iron, copper) to form protein-reactive dicarbonyl compounds and hydrogen peroxide (leading to hydroxyl radical formation; Wolff et al., 1991).

The adaptations of antioxidant defenses that allow animals to undergo multiple cycles of freeze–thaw without oxidative injuries are of interest both for understanding the mechanisms of natural freeze tolerance and to provide insight into the principles of ischemia/reperfusion endurance that could be applied to the cryopreservation of mammalian tissues and organs. Empirical studies have shown that the inclusion of antioxidants in the perfusion medium in hypothermic preservation trials with mammalian organs improves posttransplant viability, so it is apparent that antioxidant defenses have a role to play in tissue protection under cold or frozen ischemia stresses (Bilzer et al., 1999; Southard and Belzer, 1995).

Studies with wood frogs indicate that two strategies contribute to the defense against oxidative stress associated with freeze–thaw. The first is the maintenance of high constitutive activities of antioxidant enzymes in frog organs. Activities of six antioxidant enzymes (superoxide dismutase, catalase, glutathione S-transferase, glutathione reductase, and total and Se-dependent glutathione peroxidases) were assessed in five organs of wood frogs vs. leopard frogs (Joanisse and Storey, 1996). Activities of all six enzymes were uniformly higher in liver of wood frogs, in five out of six cases by at least twofold, and except for muscle, superoxide dismutase, glutathione S-transferase, and glutathione reductase were always substantially higher in wood frog than in leopard frog organs. Wood frog organs also showed higher levels of the metabolite antioxidant, glutathione, in all organs except brain; in skeletal muscle, reduced glutathione was up to 10-fold higher in wood frogs compared with leopard frogs. Significantly, when liver and muscle enzymes were further compared with a number of other species, activities in wood frog liver proved to be high for an ectothermic vertebrate, similar to values for anoxia-tolerant turtles, and both were close to values for rat liver (Hermes-Lima et al., 2001; Storey, 1996). Antioxidant defenses in wood frog organs were also substantially better than those displayed by garter snakes that display weak freeze tolerance (Hermes-Lima and Storey, 1993). The high antioxidant defenses of wood frog organs correlated with the absence of oxidative damage to lipids over a course of freeze–thaw, as assessed by two methods (Joanisse and Storey, 1996).

The second principle of antioxidant defense in wood frogs is selective changes to the activities of some enzymes in response to freezing. For example, total glutathione peroxidase activity increased significantly (by 20 to 150%) in all five organs analyzed, and selenium-dependent glutathione peroxidase activity rose by about twofold in heart, kidney, and skeletal muscle (Joanisse and Storey, 1996). In most cases, activities had returned to near control values after 24 h thawing. This indicates that the need for enhanced defenses against peroxidative damage is greatest during freezing or immediately after thawing, a time when glucose is also very high. Selective changes to the activities of these enzymes may minimize the potential for glucose-mediated oxidative damage to macromolecules that is a significant problem associated with sustained high glucose levels in diabetes (Wolff et al., 1991).

7.6 MOLECULAR BIOLOGY OF FREEZING SURVIVAL

7.6.1 Freeze-Induced Changes in Protein Expression

The seasonal acquisition of cold or freeze tolerance in animals and plants requires multiple changes in gene expression that alter the protein make-up of cells and organs to optimize their survival. For example, over 40 cold-induced genes have been identified in the freeze-tolerant plant, *Arabidopsis thaliana* (Thomashow, 1999; see also Chapter 5). Some of these genes are involved in cold acclimation (e.g., desaturases that increase membrane fluidity), and others are involved in freeze tolerance (e.g., dehydrins, enzymes of proline biosynthesis). Numerous examples of cold-induced changes in gene and protein expression are also known in animals. Among insects, for example, activities of enzymes involved in cryoprotectant synthesis rise over the early autumn (Joanisse and Storey, 1994), and antifreeze proteins or ice nucleating proteins appear in hemolymph (Duman, 2001). Liver plasma membrane glucose transporters were 8.5-fold higher in autumn- vs. summer-collected wood frogs (Kling et al., 1994). Many protein changes occur in a preparatory manner, triggered by a critical photoperiod or temperature, in order that cold-hardiness adaptations are in place in advance of the arrival of cold weather. In other cases, changes in gene expression are directly stimulated by freezing itself, as recent studies with wood frogs have shown.

Analysis of freeze- and thaw-induced gene and protein expression by vertebrates began with studies of protein biosynthesis in wood frog organs using ^{35}S-methionine-labeling techniques to monitor protein synthesis both *in vivo* (intraperitoneal injection of ^{35}S-methionine followed by isolation of labeled proteins) and *in vitro* (isolation of tissue mRNA followed by translation in a cell-free system; Storey et al., 1997; White and Storey, 1999). Both experimental approaches used isoelectrofocusing and SDS-gel electrophoresis to analyze and compare protein-labeling patterns in control, frozen, and thawed states. Significantly, both studies revealed strong labeling of proteins of 15 to 20 kDA in size. For example, *in vitro* translation of mRNA isolated from liver of freeze-exposed frogs showed the presence of several new translation products (proteins of 45, 33.9, 21.5, 16.4, 15.8, and 14.8 kDA) as compared with controls (White and Storey, 1999). However, comparable analysis with mRNA from thawed frogs showed both no new protein peaks compared with either control or frozen profiles and the loss of several of the 16 to 22-kDA proteins that were present in frozen or control frogs. These data indicate the importance of *de novo* protein synthesis in the response to freezing by frog organs. As noted earlier, however, activities of many enzymes decreased during freezing in wood frog organs in probable response to the ATP-limitation of protein biosynthesis in the ischemic frozen state (Cowan and Storey, 2001). Hence, freeze-induced *de novo* protein synthesis should be highly selective and represent the production of proteins with key roles in freezing preservation. Although instructive in indicating that freeze-induced protein synthesis does indeed occur, the ^{35}S-protein labeling techniques described above are not conducive to easy identification of the proteins, so subsequent studies turned instead to molecular biology techniques.

7.6.2 Freeze-Induced Changes in Gene Expression

New techniques in molecular biology hold the key to tracking down the full range of gene and protein changes that support freezing survival in animals (Storey and Storey, 2001). A range of techniques are available including cDNA library screening, ddPCR, RT-PCR, and DNA array screening, each with their own advantages and disadvantages. For example, screening of cDNA libraries is highly effective for identifying genes whose transcripts are present in high copy numbers in tissues and for finding novel freeze-induced genes. DNA arrays represent the newest technology

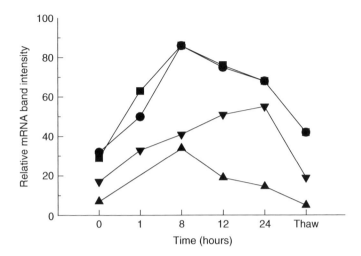

FIGURE 7.6 Freeze–thaw effects on mRNA transcript levels of freeze upregulated genes in wood frog liver. Transcript levels were quantified by Northern blots, and relative band intensities are plotted. Symbols are: ●, fibrinogen α; ■, fibrinogen γ; ▲, ATP-ADP translocase; ▼, *fr10*. Control frogs (0 h) were held at 5°C; freezing was at –2.5°C for up to 24 h, and thawed frogs were frozen for 24 h followed by 24 h back at 5°C. (Data compiled from Cai and Storey [1997a, 1997b] and Cai et al. [1997].)

and offer the ability to assess the expression of hundreds or thousands of genes simultaneously. However, the disadvantages of array screening are the inability to detect novel genes and the problem of cross-species reactivity. We have recently used 19k human cDNA chips from the Ontario Cancer Institute to screen for genes that are upregulated in wood frog heart during freezing (Storey, 2004). Overall, cross reactivity was good at 60 to 80% and the results showed over 200 putatively upregulated genes that await exploration. Among these were genes that could be involved in ischemia resistance (hypoxia inducible factor-1, adenosine A1 receptor, antioxidant enzymes), fluid dynamics (atrial natriuretic peptide receptor), and detection of protein glycation damage by high glucose (receptor for advanced glycation end products).

The first approach to analyzing freeze-induced gene expression in vertebrates involved screening of a cDNA library prepared from liver of wood frogs that were frozen for 24 h at –2.5°C. Differentially screening with ³²P-labeled single-stranded total cDNA probes from liver of control (5°C-acclimated) vs. frozen frogs revealed several freeze-responsive cDNA clones. Two were identified as the α and γ subunits of fibrinogen, the terminal protein of the clotting cascade (Cai and Storey, 1997a); another encoded the ADP/ATP translocase (*Aat*), which mediates the exchange transport of ADP and ATP across the inner mitochondrial membrane (Cai et al., 1997); and a third, named *fr10*, could not be identified, but its cDNA sequence encoded a small protein of 90 amino acids with a molecular weight of ~10 kD (Genbank accession number U44831; Cai and Storey, 1997b). The deduced amino acid sequence of the FR10 protein has an N-terminal region of 21 residues that contains ~80% hydrophobic residues and a potential export signal sequence (LALVVLVI-AISGL). This signal may also allow the exportation of FR10 from cells into the blood for a cryoprotective purpose. The predicted secondary structure of FR10 includes long sections of α helix as well as coiling structures distributed in four narrow regions and β sheet structures in the N-terminus.

Northern blots were used to gain information on the expression patterns of all four genes in wood frog liver. The genes for the two fibrinogen subunits showed coordinate expression over a time course of 24 h freezing exposure at –2.5°C followed by 24 h thawing at 5°C (Figure 7.6). The mRNA transcript levels of both subunits rose by more than threefold after 8 h freezing and remained at ~70% of this maximum after being frozen for 24 h (Cai and Storey, 1997a). When

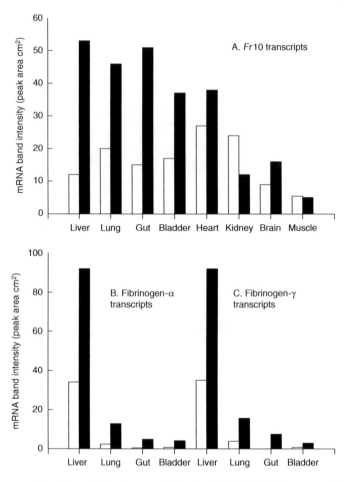

FIGURE 7.7 Organ-specific effects of freezing on the levels of *fr10* (A), fibrinogen α (B), and fibrinogen γ (C) mRNA transcripts in wood frogs. Controls (open bars) were held at 5°C; frozen frogs (solid bars) were held at −2.5°C. Total RNA was isolated from each tissue, and transcript levels were analyzed via Northern blots followed by autoradiography and densitometry. (Data compiled from Cai and Storey [1997a] and Cai et al. [1997].)

frogs were thawed, however, fibrinogen transcript levels fell and returned to near control values within 24 h. The expression pattern of *Fr10* transcripts was similar (Cai and Storey, 1997b), but *Aat* transcript levels followed a different course. After 8 hours of freezing, *Aat* transcripts had risen by 4.5-fold, but these transcripts declined again with longer freezing and were lower than controls after thawing. Immunoblotting showed that AAT protein levels lagged behind the changes in transcript levels, with the maximum increase in protein content being about twofold after 24 h of freezing (Cai et al., 1997).

Northern blots also revealed organ-specific patterns of gene upregulation. *Fr10* distribution was particularly interesting because transcripts were found in all eight organs tested, with much higher levels in tissues from frozen frogs vs. controls, except for kidney and muscle (Figure 7.7a; Cai and Storey, 1997b). This indicates that FR10 may have a universal role in the freezing protection of frog organs, and this has made it a target of continuing study in our lab. In contrast, a much narrower distribution of fibrinogen α and γ mRNA transcripts was found (Cai and Storey, 1997a). Frog liver showed the highest transcript levels in line with data that show that the synthesis of this plasma protein is liver-specific in mammals. Interestingly, low levels of fibrinogen transcripts that were upregulated during freezing were also found in lung, bladder, and gut (Figure 7.7b and 7.7c). The organ distribution of *Aat* transcripts was quite different, with elevated mRNA levels in liver, lung, and bladder during freezing; reduced levels in kidney, heart, and gut; and no change in brain and muscle. Western blots revealed that AAT protein levels followed much the same pattern (Cai et al., 1997).

Recent studies have documented the expression of two other novel, freeze-inducible genes in wood frogs, both initially detected by screening of a liver cDNA library (McNally et al., 2002, 2003). The gene *li16* encodes a protein of 115 amino acids, whereas *fr47* encodes a 390-amino acid protein (Genbank accession numbers AF175980 and AY100690). Predicted molecular masses are 12.8 and 45.7 kDA, respectively, and Western blotting using antibodies raised against the C-terminal peptides of the proteins showed crossreaction with proteins of 14 and 47 kDA. During freezing at $-2.5°C$, mRNA transcript levels of both genes rose progressively to levels that were 3.7- and 5.1-fold higher than controls for *li16* and *fr47*, respectively. Li16 and FR47 protein levels also rose, reaching 2.4- and 1.6-fold higher than control values in liver after 24 h of freezing. Both proteins continued to accumulate during the early hours of thawing, reaching peak amounts that were 2.2- to 3.5-fold higher than in 24-h-frozen liver after 2 to 4 h of thawing at $5°C$. In addition to liver, Li16 was detected in gut and heart, with particularly high levels in heart of 24-h-frozen frogs (4.4-fold above controls; McNally et al., 2002). FR47 was not found in *R. sylvatica* organs other than liver. However, it was detected in liver of other freeze-tolerant frogs (*P. crucifer, H. versicolor*) but not in freeze-intolerant species (*R. pipiens, S. couchii*).

Similar to FR10, the functions of Li16 and FR47 are not yet known, but the three novel proteins each have distinct patterns of transcript and protein accumulation during freeze–thaw, organ distribution, structural characteristics (e.g., Li16 has a transmembrane region from amino acids 1 to 21, whereas FR47 has a hydrophobic region near the C terminus), and patterns of responsiveness to dehydration and anoxia stresses that indicate that each have distinct functions in freeze tolerance. The latter parameter (response to dehydration or anoxia) is proving to be very useful for characterizing all freeze-responsive genes.

Freezing causes multiple stresses on cells and organs, including ischemia and dehydration, and one way to determine the functions of the genes/proteins that are upregulated by freezing is to assess gene responses to the individual component stresses of freezing. Analysis of mRNA transcript levels in organs from frogs given dehydration or anoxia stresses revealed that two types of responses were found. *Aat, li16,* and *fr47* transcripts in liver rose strongly under anoxic conditions but showed low (*li16, fr47*) or no (*Aat*) response to dehydration or rehydration (Cai et al., 1997; McNally et al., 2002, 2003). Western blotting also revealed a very strong increase in Li16 protein during anoxia, but surprisingly, FR47 protein did not increase under anoxia or aerobic recovery. This indicates that AAT and Li16 proteins probably have primary roles to play in anoxia/ischemia resistance, whereas FR47, although inducible by anoxia, may be regulated posttranscriptionally and have a function only during freeze-induced ischemia. In contrast, both fibrinogen and *fr10* transcripts were upregulated just as strongly during dehydration as during freezing, but transcripts were downregulated and virtually undetectable after as little as 30 min of oxygen deprivation (Cai and Storey, 1997a, 1997b). Therefore, both fibrinogen and FR10 may have roles in dealing with some aspect of water balance during freezing that could include functions in cell volume regulation or in the accommodation of extracellular ice. Interestingly, new data from plant studies has identified multiple genes that respond to both freezing and dehydration stresses (Thomashow, 1999; Warren et al., 2000), and this shows that in both plant and animal systems there is suite of genes that respond to the common problems of water balance, volume regulation, and membrane stabilization presented by both stresses. Indeed, a common response element has been found in the 5′ untranslated region of a variety of genes from plant species that display cold-hardiness, drought-resistance, or both. In *A. thaliana*, for example, the CRT/DRE (cold repeat/drought-responsive element) is present in genes that are cold-induced, and a transcription factor (CRT binding factor [CBF]) that binds to this element has been found (Thomashow, 1999; and see Chapter 5). However, the CBF gene itself responds only to freezing, whereas another transcription factor (DRE binding protein [DREB]) responds to dehydration signals. Thus, two different environmental stresses, acting via two different transcription factors, can activate the same set of freezing/dehydration responsive genes that have roles in regulating problems common to the two stresses. It is highly probable that a similar system is at work to regulate freezing- and dehydration-responsive genes in wood frog organs, and a study of the regulatory elements present in the 5′ region of freeze-responsive genes should allow us to begin to explore the signal transduction pathways operating in the nucleus to regulate freeze-specific gene expression.

Analysis of the signal transduction pathways involved in gene expression provides another way to gain insights into the function of novel freeze-responsive genes. Wood frog liver slices were incubated with second messenger analogues of protein kinase A (dibutyryl cAMP), protein kinase G (dibutyryl cGMP), or the activator of protein kinase C (phorbol 12-myristate 13-acetate) in both time-course and dose-dependency studies, and the expression patterns of *li16* and *fr47* transcripts were assessed (McNally et al., 2002, 2003). The data showed that *li16* transcript levels were elevated only in incubations that stimulated protein kinase G (by about twofold), whereas *fr47* transcript levels responded only to protein kinase C stimulation (a two- to threefold increase). These results provide a gateway for further studies of the regulation of each protein, for identification of protein role, and for linking the function of these novel proteins to other proteins that are regulated by the same signal transduction pathway. In a broader context, when taken together with the known role of protein kinase A in cryoprotectant synthesis and the freeze-stimulation of MAPK pathways, these data also illustrate an important principle: Natural freeze tolerance is a multifaceted process that involves multiple signal transduction pathways in the cell and that requires diverse signal-specific protein responses, each of which undoubtedly deals with a different type of freeze-induced stress on cells.

Whereas the roles of AAT, FR10, FR47, and Li16 in freeze tolerance are still being studied, a probable function for fibrinogen upregulation can be suggested by its known role in repairing tissue injury. The regulation and function of fibrinogen is well known in mammalian systems. This acute-phase plasma protein is synthesized by liver and secreted into the plasma; production is stimulated by stresses including infection, inflammation, and tissue injury (Huber et al., 1990). The protein has two halves, each made of three subunits (Aα, Bβ, and γ), and as the final step in the coagulation cascade, thrombin cleaves near the N-termini of the Aα and Bβ chains to release the A and B fibrinopeptides and expose sites for polymerization into the fibrin mesh of a growing blood clot. Notably, although our first study retrieved clones for just the α and γ fibrinogen subunits (Cai and Storey, 1997a), in new work we have isolated clones encoding the β and γ subunits of fibrinogen when a liver cDNA library made from glucose-loaded frogs was screened (K.B. Storey, unpublished data). Hence, coordinate expression of all three subunits appears likely as a response to freezing, dehydration, and high glucose. Freeze-stimulation of fibrinogen biosynthesis could prepare wood frogs to deal with any internal bleeding injuries that are evident on thawing, such as might result from ice damage to capillaries.

However, the question could be asked: Why is fibrinogen gene expression upregulated during freezing when its benefit is probably related to thawing? Although experimental testing is still needed, we propose that this anticipatory response occurs because the postthaw recovery of many body functions is slow. The normal mechanisms of interorgan signaling (hormone transport by blood, nervous stimulation) are cut off during freezing and may be impaired for a considerable time postthaw. Recall that it can take many hours after thawing to restore the coordinated neuro-muscular responses needed for posture and locomotion in wood frogs. If similar delays occurred in the detection and response to internal bleeding injuries during thawing, then frogs could be at high risk during the early hours postthaw. By using a freezing trigger to initiate the synthesis of clotting proteins (or other gene responses) whose probable function is linked with metabolic recovery during thawing, frogs' cells and organs are primed with the mRNA transcripts or the newly synthesized proteins that will be most needed immediately after thawing.

7.7 CONCLUSIONS

Despite all that is already known about vertebrate freeze tolerance, much more remains to be learned. One focus of continuing research in our lab is the role of gene expression in modifying the protein complement of organs to protect cells during freezing and to modify their metabolism to support long-term homeostasis in the frozen state. To that end, we are studying small proteins, such as FR10 and others, that seem to have roles in freeze-induced signaling; we are analyzing the 5′ untranslated regions of freeze upregulated genes to look for common transcription-factor binding sites that could then lead to identification of a freeze- or dehydration-specific transcription factor; and we are assessing the control of ribosomal translation from several different approaches. We are also pursuing studies of the regulation of membrane ion-motive ATPases, which, because they consume a huge fraction of total cellular ATP, must be closely controlled as a means of ischemia resistance in the frozen state and because the reinstatement of these activities after thawing is critical for the recovery of nerve and muscle function. Most metabolic and gene studies to date have focused on liver, but clearly all organs must have their own programs of freeze response and freeze protection that set them up to endure the multiple stresses imposed by freezing. The remaining challenges are exciting!

ACKNOWLEDGMENTS

We are indebted to numerous colleagues and graduate students, past and present, for their valuable contributions to the research on vertebrate freeze tolerance that is summarized here.

REFERENCES

Aggeli, I.K., Gaitanaki, C., Lazou, A., and Beis, I. (2001a) Stimulation of multiple MAPK pathways by mechanical overload in the perfused amphibian heart, *Am. J. Physiol.*, 281, R1689–R1698.

Aggeli, I.K., Gaitanaki, C., Lazou, A., and Beis, I. (2001b) Activation of multiple MAPK pathways (ERKs, JNKs, p38-MAPK) by diverse stimuli in the amphibian heart, *Mol. Cell. Biochem.*, 221, 63–69.

Aggeli, I.K., Gaitanaki, C., Lazou, A., and Beis, I. (2002) Hyperosmotic and thermal stresses activate p38-MAPK in the perfused amphibian heart, *J. Exp. Biol.*, 205, 443–454.

Ahmad, S., Ed. (1995) *Oxidative Stress and Antioxidant Defenses in Biology*, Chapman & Hall, New York.

Alessi, D.R., Street, A.J., Cohen, P., and Cohen, P.T. (1993) Inhibitor-2 functions like a chaperone to fold three expressed isoforms of mammalian protein phosphatase-1 into a conformation with the specificity and regulatory properties of the native enzyme, *Eur. J. Biochem.*, 213, 1055–1066.

Bajaj, M., Blundell, T.L., Horuk, R., Pitts, J.E., Wood, S.P., Gowan, L.K., Schwabe, C., Wollmer, A., Gliemann, J., and Gammeltoft, S. (1986) Coypu insulin: primary structure, conformation and biological properties of a hystricomorph rodent insulin, *Eur. J. Biochem.*, 238, 345–351.

Berman, D.I., Leirikh, A.N., and Mikhailova, E.I. (1984) Winter hibernation of the Siberian salamander *Hynobius keyserlingi, J. Evol. Biochem. Physiol.*, 1984, 323–327 (In Russian with English summary).

Bilzer, M., Paumgartner, G., and Gerbes, A.L. (1999) Glutathione protects the rat liver against reperfusion injury after hypothermic preservation, *Gastroenterology*, 117, 200–210.

Brooks, S.P.J., Dawson, B.A., Black, D.B., and Storey, K.B. (1999) Temperature regulation of glucose metabolism in red blood cells of the freeze-tolerant wood frog, *Cryobiology*, 39, 150–157.

Burke, R.L., Hussain, A.A., Storey, J.M., and Storey, K.B. (2002) Freeze tolerance and supercooling ability in the Italian wall lizard, *Podarcis sicula*, introduced to Long Island, NY, *Copeia*, 2002(3), 836–842.

Cai, Q. and Storey, K.B. (1997a) Freezing-induced genes in wood frog (*Rana sylvatica*): fibrinogen upregulation by freezing and dehydration, *Am. J. Physiol.*, 272, R1480–R1492.

Cai, Q. and Storey, K.B. (1997b) Upregulation of a novel gene by freezing exposure in the freeze-tolerant wood frog (*Rana sylvatica*), *Gene*, 198, 305–312.

Cai, Q., Greenway, S.C., and Storey, K.B. (1997) Differential regulation of the mitochondrial ADP/ATP translocase gene in wood frogs under freezing stress, *Biochim. Biophys. Acta*, 1343, 69–78.

Canty, A., Driedzic, W.R., and Storey, K.B. (1986) Freeze tolerance of isolated ventricle strips of the wood frog, *Rana sylvatica*, *Cryo-Letters*, 7, 81–86.

Churchill, T.A. and Storey, K.B. (1992a) Freezing survival of the garter snake *Thamnophis sirtalis*, *Can. J. Zool.*, 70, 99–105.

Churchill, T.A. and Storey, K.B. (1992b) Natural freezing survival by painted turtles *Chrysemys picta marginata* and *C. p. bellii*, *Am. J. Physiol.*, 262, R530–R537.

Churchill, T.A. and Storey, K.B. (1992c) Responses to freezing exposure by hatchling turtles *Trachemys scripta elegans*: Factors influencing the development of freeze tolerance by reptiles, *J. Exp. Biol.*, 167, 221–233.

Churchill, T.A. and Storey, K.B. (1993) Dehydration tolerance in wood frogs: a new perspective on the development of amphibian freeze tolerance, *Am. J. Physiol.*, 265, R1324–R1332.

Churchill, T.A. and Storey, K.B. (1994a) Effects of dehydration on organ metabolism in the frog *Pseudacris crucifer*: hyperglycaemic responses to dehydration mimic freezing-induced cryoprotectant production, *J. Comp. Physiol.*, B 164, 492–498.

Churchill, T.A. and Storey, K.B. (1994b) Metabolic responses to dehydration by liver of the wood frog *Rana sylvatica*, *Can. J. Zool.*, 72, 1420–1425.

Churchill, T.A. and Storey, K.B. (1995) Metabolic effects of dehydration on an aquatic frog *Rana pipiens*, *J. Exp. Biol.*, 198, 147–154.

Claussen, D.L., Townsley, M.D., and Bausch, R.G. (1990) Supercooling and freeze tolerance in the European wall lizard, *Podarcis muralis*, *J. Comp. Physiol. B*, 160, 137–143.

Conlon, J.M., Yano, K., Chartrel, N., Vaudry, H., and Storey, K.B. (1998) Freeze tolerance in the wood frog *Rana sylvatica* is associated with unusual structural features of insulin but not glucagon, *J. Mol. Endocrinol.*, 21, 153–159.

Costanzo, J.P. (1989) A physiological basis for prolonged submergence in hibernating garter snakes *Thamnophis sirtalis*: evidence for an energy-sparing adaptation, *Physiol. Zool.*, 52, 580–592.

Costanzo, J.P., Allenspach, A.L., and Lee, R.E. (1999a) Electrophysiological and ultrastructural correlates of cryoinjury in sciatic nerve of the freeze-tolerant wood frog, *Rana sylvatica*, *J. Comp. Physiol. B*, 169, 351–359.

Costanzo, J.P., Bayuk, J.M., and Lee, R.E. (1999b) Inoculative freezing by environmental ice nuclei in the freeze-tolerant wood frog, *Rana sylvatica*, *J. Exp. Zool.*, 284, 7–14.

Costanzo, J.P. and Claussen, J.P. (1990) Natural freeze tolerance in the terrestrial turtle, *Terrapene carolina*, *J. Exp. Zool.*, 254, 228–232.

Costanzo, J.P., Claussen, D.L., and Lee, R.E. (1988) Natural freeze tolerance in a reptile, *Cryo-Letters*, 9, 380–385

Costanzo, J.P., Grenot, C., and Lee, R.E. (1995a) Supercooling, ice nucleation and freeze tolerance in the European common lizard, *Lacerta vivipara*, *J. Comp. Physiol. B*, 165, 238–244.

Costanzo, J.P, Irwin, J.T., and Lee, R.E. (1997) Freezing impairment of male reproductive behaviors of the freeze-tolerant wood frog, *Rana sylvatica*, *Physiol. Zool.*, 70, 158–166.

Costanzo, J.P., Iverson, J.B., Wright, M.F., and Lee, R.E. (1995b) Cold-hardiness and overwintering strategies of hatchlings in an assemblage of northern turtles, *Ecology*, 76, 1772–1785.

Costanzo, J.P. and Lee, R.E. (1991) Freeze-thaw injury in erythrocytes of the freeze-tolerant wood frog, *Rana sylvatica*, *Am. J. Physiol.*, 261, R1346–R1350.

Costanzo, J.P. and Lee, R.E. (1997) Frogs resorb glucose from urinary bladder, *Nature*, 389, 343–344.

Costanzo, J.P., Lee, R.E., and Lortz, P.H. (1993) Glucose concentration regulates freeze tolerance in the wood frog *Rana sylvatica*, *J. Exp. Biol.*, 181, 245–255.

Costanzo, J.P., Lee, R.E., and Wright, M.F. (1991a) Effect of cooling rate on the survival of frozen wood frogs, *Rana sylvatica*, *J. Comp. Physiol. B*, 161, 225–229.

Costanzo, J.P., Lee, R.E., and Wright, M.F. (1991b) Glucose loading prevents freezing injury in rapidly cooled frogs, *Am. J. Physiol.*, 261, R1549–R1553.

Costanzo, J.P., Lee, R.E., and Wright, M.F. (1992a) Cooling rate influences cryoprotectant distribution and organ dehydration in freezing wood frogs, *J. Exp. Zool.*, 261, 373–378.

Costanzo, J.P., Litzgus, J.D., and Lee, R.E. (1999c) Behavioural responses of hatchling painted turtles (*Chrysemys picta*) and snapping turtles (*Chelydra serpentina*) at subzero temperatures, *J. Thermal Biol.*, 24, 161–166.

Costanzo, J.P., Mugnano, J.A., Wehrheim, H.M., and Lee, R.E. (1998) Osmotic and freezing tolerance in spermatozoa of freeze-tolerant and -intolerant frogs, *Am. J. Physiol.*, 275, R713–R719.

Costanzo, J.P., Wright, M.F., and Lee, R.E. (1992b) Freeze tolerance as an overwintering adaptation in Copes grey treefrog (*Hyla chrysoscelis*), *Copeia*, 1992, 565–569.

Cowan, K.J. and Storey, K.B. (2001) Freeze-thaw effects on metabolic enzymes in wood frog organs, *Cryobiology*, 43, 32–45.

Cowan, K.J. and Storey, K.B. (2003) Mitogen-activated protein kinases: New signaling pathways functioning in cellular responses to environmental stress, *J. Exp. Biol.*, 206, 1107–1115.

Croes, S.A. and Thomas, R.E. (2000) Freeze tolerance and cryoprotectant synthesis of the Pacific tree frog *Hyla regilla*, *Copeia*, 2000(3), 863–868.

Dalo, N.L., Hackman, J.C., Storey, K.B., and Davidoff, R.A. (1995) Changes in motoneuron membrane potential and reflex activity induced by sudden cooling of isolated spinal cords: differences among cold-sensitive, cold-resistant, and freeze-tolerant amphibian species, *J. Exp. Biol.*, 198, 1765–1774.

Danks, H.V. (1996) The wider integration of studies on insect cold hardiness, *Eur. J. Entomol.*, 93, 383–403.

Dode, L., van Baelen, K., Wuytack, F., and Dean, W.L. (2001) Low temperature molecular adaptation of the skeletal muscle sarco(endo)plasmic reticulum Ca^{2+}-ATPase 1 (SERCA1) in the wood frog (*Rana sylvatica*), *J. Biol. Chem.*, 276, 3911–3919.

Doherty, M.J., Cadefau, J., Stalmans, W., Bollen, M., and Cohen, P.T. (1998) Loss of the hepatic glycogen-binding subunit (GL) of protein phosphatase 1 underlies deficient glycogen synthesis in insulin-dependent diabetic rats and in adrenalectomized starved rats, *Biochem. J.*, 333, 253–257.

Duman, J.G. (2001) Antifreeze and ice nucleator proteins in terrestrial arthropods, *Annu. Rev. Physiol.*, 63, 327–357.

Fall, R. and Wolber, P.K. (1995) Biochemistry of bacterial ice nuclei, in *Biological Ice Nucleation and Its Applications*, Lee, R.E., Warren, G.J., and Gusta, L.V., Eds., American Phytopathological Society Press, St. Paul, MN, pp. 63–83.

Fournier, P.A. and Guderley, H. (1993) Glucosidic pathways of glycogen breakdown and glucose production by muscle from postexercised frogs, *Am. J. Physiol.*, 265, R1141–1147.

Frisbie, M.P., Costanzo, J.P., and Lee, R.E. (2000) Physiological and ecological aspects of low-temperature tolerance in embryos of the wood frog, *Rana sylvatica*, *Can. J. Zool.*, 78, 1032–1041.

Greenway, S.C. and Storey, K.B. (2000) Activation of mitogen-activated protein kinases during natural freezing and thawing in the wood frog, *Mol. Cell Biochem.*, 209, 29–37.

Grenot, C.J., Garcin, L., Dao, J., Herold, J.P., Fahys, B., and Tsere-Pages, H. (2000) How does the European common lizard, *Lacerta vivipara*, survive the cold of winter? *Comp. Biochem. Physiol. A*, 127, 71–80.

Halliwell, B., Gutteridge, J.M.C., and Cross, C.E. (1992) Free radicals, antioxidants and human disease: Where are we now? *J. Lab. Clin. Med.*, 119, 598–620.

Hemmings, S.J. and Storey, K.B. (1994) Alterations in hepatic adrenergic receptor status in *Rana sylvatica* in response to freezing and thawing: Implications to the freeze-induced glycaemic response, *Can. J. Physiol. Pharmacol.*, 72, 1552–1560.

Hemmings, S.J. and Storey, K.B. (1996) Characterization of γ-glutamyltranspeptidase in the liver of the frog: 3. Response to freezing and thawing in the freeze-tolerant wood frog, *Rana sylvatica*, *Cell Biochem. Funct.*, 14, 139–148.

Hemmings, S.J. and Storey, K.B. (2001) Characterization of sarcolemma and sarcoplasmic reticulum isolated from skeletal muscle of the freeze-tolerant wood frog, *Rana sylvatica*: The β(2)-adrenergic receptor and calcium transport systems in control, frozen and thawed states, *Cell Biochem. Funct.*, 19,143–152.

Hermes-Lima, M. and Storey, K.B. (1993) Anti-oxidant defenses in the tolerance of freezing and anoxia by garter snakes, *Am. J. Physiol.*, 265, R646–R652.

Hermes-Lima, M., Storey, J.M., and Storey, K.B. (2001) Antioxidant defenses and animal adaptation to oxygen availability during environmental stress, in *Cell and Molecular Responses to Stress*, Storey, K.B. and Storey, J.M., Eds., Elsevier, Amsterdam, pp. 263–287.

Hillman, S.S. (1982) The effects of *in vivo* and *in vitro* hyperosmolality on skeletal muscle performance in the amphibians *Rana pipiens* and *Scaphiopus couchii*, *Comp. Biochem. Physiol. A*, 73, 709–712.

Hillman, S.S. (1988) Dehydrational effects on brain and cerebrospinal fluid electrolytes in two amphibians, *Physiol. Zool.*, 61, 254–259.

Hochachka, P.W. (1986) Defense strategies against hypoxia and hypothermia, *Science*, 231, 234–241.

Hoeflich, K.P. and Woodgett, J.R. (2001) Mitogen-activated protein kinases and stress, in *Cell and Molecular Responses to Stress*, Vol. 2. Storey, K.B. and Storey, J.M., Eds., Elsevier, Amsterdam, pp. 175–193.

Holden, C.P. and Storey, K.B. (1996) Signal transduction, second messenger, and protein kinase responses during freezing exposures in the wood frog, *Am. J. Physiol.*, 271, R1205–R1211.

Holden, C.P. and Storey, K.B. (1997) Second messenger and cAMP-dependent protein kinase responses to dehydration and anoxia stresses in frogs, *J. Comp. Physiol. B*, 167, 305–312.

Holden, C.P. and Storey, K.B. (2000) Purification and characterization of protein kinase A catalytic subunit from liver of the freeze-tolerant wood frog: role in glycogenolysis during freezing, *Cryobiology*, 40, 323–331.

Hubbard, M.J. and Cohen, P. (1989) The glycogen-binding subunit of protein phosphatase-1G from rabbit skeletal muscle. Further characterization of its structure and glycogen-binding properties, *Eur. J. Biochem.*, 180, 457–465.

Huber, P., Laurent, M., and Dalmon, J. (1990) Human β-fibrinogen gene expression: upstream sequences involved in its tissue specific expression and its dexamethasone and interleukin 6 stimulation, *J. Biol. Chem.*, 265, 5695–5701.

Jackson, D.C. (2001) Anoxia survival and metabolic arrest in the turtle, in *Metabolic Mechanisms of Metabolic Arrest*, Storey, K.B., Ed., BIOS Scientific, Oxford, pp. 103–114.

Joanisse, D.R. and Storey, K.B. (1994) Enzyme activity profiles in an overwintering population of freeze-tolerant larvae of the gall fly *Eurosta solidaginis*, *J. Comp. Physiol. B*, 164, 247–255.

Joanisse, D.R. and Storey, K.B. (1996) Oxidative damage and antioxidants in *Rana sylvatica*, the freeze-tolerant wood frog, *Am. J. Physiol.*, 271, R545–R553.

King, P.A., Rosholt, M.N., and Storey, K.B. (1993) Adaptations of plasma membrane glucose transport facilitate cryoprotectant distribution in freeze-tolerant frogs, *Am. J. Physiol.*, 265, R1036–R1042.

King, P.A., Rosholt, M.N., and Storey, K.B. (1995) Seasonal changes in plasma membrane glucose transport in freeze-tolerant wood frogs, *Can. J. Zool.*, 73, 1–9.

Kling, K.B., Costanzo, J.P., and Lee, R.E. (1994) Post-freeze recovery of peripheral nerve function in the freeze-tolerant wood frog, *Rana sylvatica*, *J. Comp. Physiol. B*, 164, 316–320.

Kristal, B.S. and Yu, B.P. (1992) An emerging hypothesis: synergistic induction of aging by free radicals and Maillard reactions, *J. Gerontol.*, 471, B107–B114.

Kristensen, C., Kjeldsen, T., Wiberg, F.C., Schaffer, L., Hach, M., Havelund, S., Bass, J., Steiner, D.F., and Andersen, A.S. (1997) Alanine scanning mutagenesis of insulin, *J. Biol. Chem.*, 272, 12978–12983.

Kyriakis, J.M. (1999) Making the connection: coupling of stress-activated ERK/MAPK signaling modules to extracellular stimuli and biological responses, *Biochem. Soc. Symp.*, 64, 29–48.

Layne, J.R. (1992) Postfreeze survival and muscle function in the leopard frog (*Rana pipiens*) and the wood frog (*Rana sylvatica*), *J. Therm. Biol.*, 17, 121–124.

Layne, J.R. (1995a) Crystallization temperatures of frogs and their individual organs, *J. Herpetol.*, 29, 296–298.

Layne, J.R. (1995b) Seasonal variation in the cryobiology of *Rana sylvatica* from Pennsylvania, *J. Therm. Biol.*, 20, 349–353.

Layne, J.R. (1999) Freeze tolerance and cryoprotectant mobilization in the gray treefrog (*Hyla versicolor*), *J. Exp. Zool.*, 283, 221–225.

Layne, J.R. (2000) Postfreeze O_2 consumption in the wood frog (*Rana sylvatica*), *Copeia*, 2000, 879–882.

Layne, J.R., Costanzo, J.P., and Lee, R.E. (1998) Freeze duration influences postfreeze survival in the frog *Rana sylvatica*, *J. Exp. Zool.*, 280, 197–211.

Layne, J.R. and First, M.C. (1991) Resumption of physiological functions in the wood frog (*Rana sylvatica*) after freezing, *Am. J. Physiol.*, 261, R134–R137.

Layne, J.R. and Jones, A.L. (2001) Freeze tolerance in the gray treefrog: Cryoprotectant mobilization and organ dehydration, *J. Exp. Zool.*, 290, 1–5.

Layne, J.R. and Kefauver, J. (1997) Freeze tolerance and postfreeze recovery in the frog *Pseudacris crucifer*, *Copeia*, 1997, 260–264.

Layne, J.R. and Kennedy, S.D. (2002) Cellular energetics of frozen wood frogs (*Rana sylvatica*) revealed via NMR spectroscopy, *J. Thermal Biol.*, 27, 167–173.

Layne, J.R. and Lee, R.E. (1987) Freeze tolerance and the dynamics of ice formation in wood frogs (*Rana sylvatica*) from southern Ohio, *Can. J. Zool.*, 65, 2062–2065.

Layne, J.R., Lee, R.E., and Cutwa, M.M. (1996) Post-hibernation excretion of glucose in urine of the freeze-tolerant frog *Rana sylvatica*, *J. Herpetol.*, 30, 85–87.

Layne, J.E., Lee, R.E., and Heil, T.L. (1989) Freezing-induced changes in the heart rate of wood frogs (*Rana sylvatica*), *Am. J. Physiol.*, 257, R1046–R1049.

Layne, J.R., Lee, R.E., and Huang, J.L. (1990) Inoculation triggers freezing at high subzero temperatures in a freeze-tolerant frog (*Rana sylvatica*) and insect (*Eurosta solidaginis*), *Can. J. Zool.*, 68, 506–510.

Lee, M.R., Lee, R.E., Strong-Gunderson, J.M., and Minges, S.R. (1995) Isolation of ice-nucleating active bacteria from the freeze-tolerant frog, *Rana sylvatica*, *Cryobiology*, 32, 358–365.

Lee, R.E., Costanzo, J.P, Davidson, E.C., and Layne, R.E. (1992) Dynamics of body water during freezing and thawing in a freeze-tolerant frog (*Rana sylvatica*), *J. Therm. Biol.*, 17, 263–266.

Lee, R.E., McGrath, J.J., Morason, R.T., and Taddeo, R.M. (1993) Survival of intracellular freezing, lipid coalescence and osmotic fragility in fat body cells of the freeze-tolerant gall fly *Eurosta solidaginis*, *J. Insect Physiol.*, 39, 445–450.

Lipton, P. (1999) Ischemic cell death in brain neurons, *Physiol. Rev.*, 79, 1431–1568.

Lutz, P.L. and Reiners, R. (1997) Survival of energy failure in the anoxic frog brain: delayed release of glutamate, *J. Exp. Biol.*, 200, 2913–2917.

Macartney, J.M., Larsen, K.W., and Gregory, P.T. (1989) Body temperatures and movements of hibernating snakes (*Crotalus* and *Thamnophis*) and thermal gradients of natural hibernacula, *Can. J. Zool.*, 67, 108–114.

MacDonald, J.A. and Storey, K.B. (1999) Protein phosphatase responses during freezing and thawing in wood frogs: control of liver cryoprotectant metabolism, *Cryo-Letters*, 20, 297–306.

MacDonald, J.A. and Storey, K.B. (2002) Protein phosphatase type-1 from skeletal muscle enzyme of the freeze tolerant wood frog, *Comp. Biochem. Physiol. B*, 131, 27–36.

Markussen, J., Diers, I., Hougaard, P., Langkjaer, L., Norris, K., Snel, L., Sorensen, E., and Voigt, H.O. (1988) Soluble, prolonged-acting insulin derivatives, *Protein Eng.*, 2, 157–166.

McNally, J.D., Sturgeon, C.M., and Storey, K.B. (2003) Freeze induced expression of a novel gene, *fr47*, in the liver of the freeze tolerant wood frog, *Rana sylvatica*, *Biochim. Biophys. Acta*, 1625, 183–191.

McNally, J.D., Wu, S., Sturgeon, C.M., and Storey, K.B. (2002) Identification and characterization of novel freezing inducible gene, *li16*, in the wood frog, *Rana sylvatica*, *FASEB J.*, 16, 902–904.

Miller, L.K. and Dehlinger, P.J. (1969) Neuromuscular function at low temperatures in frogs from cold and warm climates, *Comp. Biochem. Physiol.*, 28, 915–921.

Packard, G.C., Packard, M.J., Lang, J.W., and Tucker, J.K. (1999) Tolerance for freezing in hatchling turtles, *J. Herpetol.*, 33, 536–543.

Rubinsky, B., Lee, C.Y., Bastacky, J., and Onik, J. (1987) The process of freezing and the mechanism of damage during hepatic cryosurgery, *Cryobiology*, 27, 85–97.

Rubinsky, B., Wong, S.T.S., Hong, J.-S., Gilbert, J., Roos, M., and Storey, K.B. (1994) ^{1}H magnetic resonance imaging of freezing and thawing in freeze-tolerant frogs, *Am. J. Physiol.*, 266, R1771–R1777.

Ruderman, N.B., Williamson, J.R., and Brownlee, M. (1992) Glucose and diabetic vascular disease, *FASEB J.*, 6, 2905–2914.

Russell, E.L. and Storey, K.B. (1995) Glycogen synthetase and the control of cryoprotectant clearance after thawing in the freeze-tolerant wood frog, *Cryo-Letters*, 16, 263–266.

Schmid, W.D. (1982) Survival of frogs in low temperature, *Science*, 215, 697–698.

Shoemaker, V.H. (1992) Exchange of water, ions, and respiratory gases in terrestrial amphibians, in *Environmental Physiology of the Amphibians*, Feder, M.E. and Burggren, W.W., Eds., University of Chicago Press, Chicago, pp. 125–159.

Southard, J.H. and Belzer, F.O. (1995) Organ preservation, *Annu. Rev. Med.*, 46, 235–247.

Storey, J.M. and Storey, K.B. (1985) Triggering of cryoprotectant synthesis by the initiation of ice nucleation in the freeze-tolerant frog, *Rana sylvatica*, *J. Comp. Physiol. B*, 156, 191–195.

Storey, J.M. and Storey, K.B. (1996a) Adrenergic, hormonal, and nervous influences on cryoprotectant synthesis by liver of the freeze-tolerant wood frog *Rana sylvatica*, *Cryobiology*, 33, 186–195.

Storey, K.B. (1987a) Organ-specific metabolism during freezing and thawing in a freeze-tolerant frog, *Am. J. Physiol.*, 253, R292–R297.

Storey, K.B. (1987b) Glycolysis and the regulation of cryoprotectant synthesis in liver of the freeze-tolerant wood frog, *J. Comp. Physiol. B*, 157, 373–380.

Storey, K.B. (1996) Oxidative stress: Animal adaptations in nature, *Brazilian J. Med. Biol. Res.*, 29, 1715–1733.

Storey, K.B. (1999) Living in the cold: freeze-induced gene responses in freeze-tolerant vertebrates, *Clin. Exp. Pharmacol. Physiol.*, 26, 57–63.

Storey, K.B. (2004) Strategies for exploration of freeze responsive gene expression: advances in vertebrate freeze tolerance, *Cryobiology*, in press.

Storey, K.B., Baust, J.G., and Wolanczyk, J.P (1992a) Biochemical modification of the plasma ice nucleating activity in a freeze-tolerant frog, *Cryobiology*, 29, 374–384.

Storey, K.B., Bischof, J., and Rubinsky, B. (1992b) Cryomicroscopic analysis of freezing in liver of the freeze-tolerant wood frog, *Am. J. Physiol.*, 263, R185–R194.

Storey, K.B., Layne, J.R., Cutwa, M.M., Churchill, T.A., and Storey, J.M. (1993) Freezing survival and metabolism of box turtles, *Terrapene carolina*, *Copeia*, 1993, 628–634.

Storey, K.B. and Mommsen, T.P. (1994) Effects of temperature and freezing on hepatocytes isolated from a freeze-tolerant frog, *Am. J. Physiol.*, 266, R1477–1482.

Storey, K.B. and Storey, J.M. (1984) Biochemical adaptation for freezing tolerance in the wood frog, *Rana sylvatica*, *J. Comp. Physiol. B*, 155, 29–36.

Storey, K.B. and Storey, J.M. (1986a) Freeze tolerance and intolerance as strategies of winter survival in terrestrially-hibernating amphibians, *Comp. Biochem. Physiol. A*, 83, 613–617.

Storey, K.B. and Storey J.M. (1986b) Freeze-tolerant frogs: Cryoprotectants and tissue metabolism during freeze/thaw cycles, *Can. J. Zool.*, 64, 49–56.

Storey, K.B. and Storey, J.M. (1987) Persistence of freeze tolerance in terrestrially hibernating frogs after spring emergence, *Copeia*, 1987, 720–726.

Storey, K.B. and Storey, J.M. (1989) Freeze tolerance and freeze avoidance in ectotherms, in *Animal Adaptation to Cold*, Wang, L.C.H., Ed., Springer, Heidelberg, pp. 51–82.

Storey, K.B. and Storey, J.M. (1990) Facultative metabolic rate depression: molecular regulation and biochemical adaptation in anaerobiosis, hibernation, and estivation, *Quart. Rev. Biol.*, 65, 145–174.

Storey, K.B. and Storey, J.M. (1992) Natural freeze tolerance in ectothermic vertebrate, *Ann. Rev. Physiol.*, 54, 619–637.

Storey, K.B. and Storey, J.M. (1996). Natural freezing survival in animals, *Ann. Rev. Ecol. Syst.*, 27, 365–386.

Storey, K.B. and Storey, J.M. (2001) Signal transduction and gene expression in the regulation of natural freezing survival, in *Cell and Molecular Responses to Stress*, Vol. 2, Storey, K.B. and Storey, J.M., Eds., Elsevier, Amsterdam, pp. 1–19.

Storey, K.B. and Storey, J.M. (2004) Metabolic arrest in animals—a molecular biology approach, *Biol. Rev.*, in press.

Storey, K.B., Storey, J.M., and Churchill, T.A. (1997) De novo protein biosynthesis responses to water stresses in wood frogs: Freeze-thaw and dehydration-rehydration, *Cryobiology*, 34, 200–213.

Storey, K.B., Storey, J.M., Brooks, S.P.J., Churchill, T.A., and Brooks, R.J. (1988) Hatchling turtles survive freezing during winter hibernation, *Proc. Natl. Acad. Sci. USA*, 85, 8350–8354.

Swanson, D.L., Graves, B.M., and Koster, K.L. (1996) Freezing tolerance/intolerance and cryoprotectant synthesis in terrestrially overwintering anurans in the Great Plains, USA, *J. Comp. Physiol. B*, 166, 110–119.

Thomashow, M.F. (1999) Plant cold acclimation: Freezing tolerance genes and regulatory mechanisms, *Annu. Rev. Plant Physiol. Mol. Biol.*, 50, 571–599.

Unger, R. (1991) Diabetic hyperglycaemia: link to impaired glucose transport in pancreatic β cells, *Science*, 251, 1200–1205.

Vazquez-Illanes, D. and Storey, K.B. (1993) 6-Phosphofructo-2-kinase and control of cryoprotectant synthesis in freeze-tolerant frogs, *Biochim. Biophys. Acta*, 1158, 29–32.

Voituron, Y., Storey, J.M., and Storey, K.B. (2002) Freezing survival, body ice content and blood composition of the freeze-tolerant European common lizard, *Lacerta vivipara*, *J. Comp. Physiol. B*, 172, 71–76.

Warren, G.J., Thorlby, G.J., and Knight, M.R. (2000) The molecular biological approach to understanding freezing-tolerance in the model plant, *Arabidopsis thaliana*, in *Cell and Molecular Responses to Stress*, Vol. 1, Storey, K.B. and Storey, J.M., Eds., Elsevier Science, Amsterdam, pp. 245–258.

Wharton, D.A. and Ferns, D.J. (1995) Survival of intracellular freezing by the Antarctic nematode *Panagrolaimus davidi*, *J. Exp. Biol.*, 198, 1381–1387.

White, D. and Storey, K.B. (1999) Freeze-induced alterations of translatable mRNA populations in wood frog organs, *Cryobiology*, 38, 353–362.

Wolanczyk, J.P., Baust, J.G., and Storey, K.B. (1990a) Seasonal ice nucleating activity in the freeze-tolerant frog, *Rana sylvatica*, *Cryo Lett.*, 11, 143–150.

Wolanczyk, J.P., Storey, K.B., and Baust, J.G. (1990b) Nucleating activity in the blood of the freeze-tolerant frog, *Rana sylvatica*, *Cryobiology*, 27, 328–335.

Wolff, S.P., Jiang, Z.Y., and Hunt, J.V. (1991) Protein glycation and oxidative stress in diabetes mellitus and aging, *Free Rad. Biol. Med.*, 10, 339–352.

Theme 3

*Freezing and Banking
of Living Resources*

8 Preservation of Fungi and Yeasts

Shu-hui Tan and Cor van Ingen

CONTENTS

8.1 INTRODUCTION

In this chapter, "life in the frozen state" will be extended to "life in the frozen and dehydrated state," as desiccation is, for microorganisms, as important and challenging a stress as freezing. Moreover, to develop strategies for long-term preservation methods under conditions in which the metabolism of microorganisms is inhibited (e.g., cryopreservation and lyophilization), it is necessary to understand the principles by which they counteract both these threats.

The processes occurring in the cell in response to osmo-, desiccation-, heat-, and to a lesser extent, cryo-stress resemble those occurring during sporulation. These processes are summarized in Figure 8.1. Sporulation is induced at the end of the exponential growth phase and during starvation, when nitrogen and carbon sources become depleted. Under these conditions, fungi and

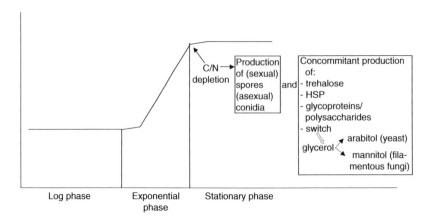

FIGURE 8.1 Processes occurring in a fungal cell during sporulation.

yeasts switch to a state of dormancy by production of (asexual) conidia or (sexual) spores. In former mentioned propagules, the metabolism is arrested by converting the cytoplasm into the glassy state by reducing the moisture content. In the dried propagules, the cellular components, especially the membranes and proteins, must be protected from denaturation. Therefore, sporulation is accompanied by the production of compounds protecting the membranes and the proteins. The dissacharide trehalose protects membranes during dehydration (see also Section 8.2.1), heat-shock proteins (HSPs) restore denatured proteins and degrade irretrievably damaged proteins (see also Section 8.2.3), and polysaccharide or glycoprotein capsules prolong dehydration during drying and accelerate rehydration (see also Section 8.2.5).

Because the cellular contents of conidia and spores must be in the glassy state, the compounds protecting the membranes and the proteins must have a high glass-transition temperature (T_g). T_g is the temperature at which the glassy state of a compound melts. For instance, in many fungi, trehalose is produced to protect the membranes because this disaccharide forms a stable glass with a relative high T_g (see also Chapter 21). Also important is the switch from the production of glycerol, which has a very low T_g, to that of arabitol in xerophilic yeasts and mannitol in filamentous fungi. Glycerol is the preferred osmolyte in hyphae and cells of growing fungi; higher polyols such as arabitol, erythritol, and mannitol become the predominant compatible solutes that accumulate in sclerotia, fruiting bodies, spores, and conidia next to trehalose because they have a significantly higher T_g than glycerol.

Interestingly, the primary function of the production and concomitant mobilization of trehalose and compatible solutes like glycerol might have been to serve as a futile cycle (Blomberg, 2000; Hottiger et al., 1987b). A futile cycle is introduced in the glycolysis under growth-inhibiting conditions. It serves to break down glucose, ATP, and reducing equivalents like $NADH_2$, whose ongoing production during growth inhibition becomes very deleterious. As it turned out, in the course of evolution, these compounds also proved to be very useful in protecting the membranes and the proteins during freezing and desiccation.

In an introductory section, the reactions of the most important classes of fungi and yeasts (Ascomycetes, Basidiomycetes, Zygomycetes, and Oomycetes) to cold-, heat-, desiccation-, and related osmo-stress will be dealt with. As mentioned before, this involves production of trehalose, (other) compatible solutes, HSPs, polymers (glyco-proteins and polysaccharides), pigments, and alterations of the membrane.

This introductory part will be followed by a section in which previously mentioned adaptations, through preconditioning, can be applied to improve survival after preservation. The final section deals with preservation (cryopreservation and lyophilization) of microorganisms.

8.2 THE STRESS RESPONSES OF FUNGI AND YEASTS

8.2.1 TREHALOSE

The disaccharide trehalose is widespread among fungi and yeasts (Thevelein, 1984) and can reach 28–30% (Feofilova, 1992) of the spore dry weight. Trehalose is produced during periods of reduced growth; for example, at the end of the exponential growth phase and during starvation, when nitrogen, carbon, P, and S sources become depleted. During germination, trehalose is metabolized. Because of the large quantities produced in conidia and spores and the rapid degradation of the compound during germination, it was generally regarded in the past to act as a reserve carbohydrate to provide energy during germination (Lillie and Pringle, 1980). At present, it is thought that trehalose is also involved in protecting the propagules against environmental-stress conditions, especially those conditions associated with desiccation and elevated environmental temperatures (Laere, 1989; Wiemken, 1990). The protection mechanisms of trehalose are various: it acts as a compatible solute, it is an excellent protector of the membranes and the proteins, and it protects the membranes during dehydration by hydrogen bonding to the phospholipid head groups (Crowe et al., 1984, 1985a, 1987, 1990). This interaction increases head group spacing, hence lowering the transition temperature of the phospholipids (Crowe et al., 1985b). By this so-called water-replacement mechanism, the transition of the membrane from the liquid crystalline to the gel is omitted. Various ways are proposed as to how trehalose should protect proteins. Timasheff and his colleagues provided evidence that the saccharides stabilize proteins during cooling because the saccharides are preferentially excluded from the surface of the proteins in aqueous solution (Arakawa and Timasheff, 1982; Back et al., 1979; Carpenter and Crowe, 1988). The saccharides repel the hydrophobic parts of the amino acid chains, thus preventing unfolding of the protein at the melting temperature. Allison et al. (1999) stated that hydrogen bonding between the saccharide and the protein in the final stages of desiccation is required to inhibit dehydration-induced protein unfolding (Allison et al., 1999). Singer and Lindquist (1998a, 1998b) suggested that the trehalose helps to suppress aggregation of denatured proteins after a heat shock by binding to partially folded stages. Here another reason for the previously mentioned rapid hydrolysis of trehalose during germination becomes obvious. Because HSPs will promote the final refolding of the denatured proteins (see Section 8.2.3), binding of trehalose to these partially denatured proteins should be temporary to not interfere with this refolding process.

8.2.2 COMPATIBLE SOLUTES

In response to dehydration and osmo-stress in general, microorganisms—in addition to excluding the stress solute—produce compatible solutes to generate the appropriate cytoplasmic solute potential so that turgidity can be maintained. A compatible solute is characterized as an osmolyte whose presence even at high concentrations does not lead to enzyme inhibition or inactivation. It can be accumulated by endogenous production or by uptake from the medium. With the exception of proline and trehalose, most compatible solutes of fungi are polyhydroxyalcohols. Besides acting as an osmolyte, these polyhydroxyalcohols, similar to the saccharides, stabilize proteins by the mechanism of preferential exclusion (Arakawa and Timasheff, 1982; Back et al., 1979; see also Section 8.2.1). For instance, exclusion of glycerol from the hydration shell around the protein minimizes protein–solvent interactions, thus preventing protein unfolding (Gekko and Timasheff, 1981).

Glycerol is the most common compatible solute produced when a salt stress is imposed on actively growing mycelium (or cells) of almost all fungi and yeasts (except the Oomycetes and the Chytridiomycetes; Hocking, 1993). Other polyols produced by these organisms are mannitol, arabitol, erythritol, D-threitol, xylitol, dulcitol, sorbitol, volemitol, ribitol, and galactitol (Blomberg and Adler, 1992; Lewis and Smith, 1967). With respect to the polyols produced, a distinction can be made according to the growth phase of the organism and the type of fungus. Glycerol is produced in growing mycelium to generate a sufficiently high cytoplasmic solute potential, as hyphal tip extension requires a positive turgor pressure relative to that of the external environment (Luard and Griffin, 1981). However, when organisms enter the stationary phase and start to sporulate, glycerol disappears and higher polyols, for example, mannitol in spores and conidia of most Ascomycetes and Basidiomycetes and arabitol, and small amounts of erythritol in xerophilic (and halophilic) yeasts (Blomberg and Adler, 1992; Brown, 1978), become the predominant compounds. For example, when the marine yeast *Debaryomyces hansenii* was grown in a medium with high salinity, glycerol was the dominant solute accumulated in the log phase, whereas arabitol was the dominant solute in the late exponential and stationary phase. These higher polyols, which are less permeable and are thus easier to retain and that have significantly higher T_g than glycerol, could be regarded as the compatible solutes for spores and conidia. In addition, they also protect their proteins against dehydration effects.

Zygomycetes produce the same polyols as the Ascomycetes and the Basidiomycetes except for mannitol (Blomberg and Adler, 1992). In addition, proline can be accumulated in this class of fungi (Luard, 1982).

Oomycetes produce few or no polyols (Luard, 1982; Pfyffer et al., 1986). Instead, proline is accumulated when these organisms are subjected to osmotic stress. It is not surprising that the Oomycetes produce compatible solutes differing from those produced by the other fungi. Although the Oomycetes are classified in the kingdom of the fungi, they are more related to the algae, most probably representing a group of pigmentless algae. Proline is also an important compatible solute produced by algae (Brown, 1978).

Apart from serving as compatible solute, polyols may have other roles in fungi. They might, like trehalose, serve as carbohydrate reserves, as translocatory compounds, as a storage for reducing power, being more reduced than saccharides (Hocking, 1993), and serve in futile cycles (Blomberg, 2000; Hottiger et al., 1987b) (see also Section 8.1).

8.2.3 HSPs

All organisms including fungi, yeasts, bacteria, and even viruses respond to sublethal elevated temperatures with the production of HSPs. These proteins help organisms to survive higher, otherwise lethal, temperatures. The proteins are classified according to their molecular weight. In general, these proteins have protease activity or interact with other proteins to facilitate their proper conformational folding, their assembly into protein complexes, and their translocation into target organelles. Because of these functions, this group of HSPs is usually referred to as molecular chaperones. Many of them are produced constitutively and play a role in the development of the organism. After a heat shock, their role is extended to prevent and repair protein aggregation damage or to promote the degradation of irretrievably damaged proteins (Feofilova, 1992; Lindquist and Craig, 1988; Plesofsky-Vig and Brambl, 1993). For example, the 104-kD HSP, induced in *Saccharomyces cerevisiae* after a heat shock and in sporulating cells of the stationary phase, promotes the final refolding of denatured proteins. In this process, trehalose suppresses aggregation by temporarily binding to the denatured protein to maintain it in the partially folded state until final refolding can take place (Singer and Lindquist, 1998a; see also Section 8.2.1).

A 98-kD HSP of *Neurospora crassa* is associated with polyribosomes, particularly the large ribosomal unit (Vassilev et al., 1992). Because ribosome production is very sensitive to heat shock, it is speculated that the related 110-kD HSP, which is induced in mammalian cells in response to elevated temperatures, serves to protect the ribosomes (Lindquist and Graig, 1988).

A 70-kD HSP of *S. cerevisiae* might assist to repair denatured proteins by binding and subsequently releasing them to disrupt aggregated or improperly folded proteins (Lindquist and Graig, 1988).

Ubiquitin, which is produced by *S. cerevisiae* after heat shock as well as constitutively, binds to other proteins, thereby marking them for ATP-dependent proteolysis. During stress it might be involved in the proteolysis of denatured proteins (Seufert and Jentsch, 1990).

8.2.4 THERMAL-HYSTERESIS PROTEINS

In addition to HSPs, thermal-hysteresis proteins are produced by some fungi. They were isolated from fruiting bodies of *Flammulina velutipes*, *Pleurotus ostreatus*, and *Coriolus versicolor*, collected in early March. Thermal-hysteresis proteins depress the freezing temperature of water by a noncolligative mechanism while not lowering the melting temperature. This phenomenon, in which a difference is created between freezing and melting temperatures, is called thermal hysteresis (Duman et al., 1993).

8.2.5 POLYSACCHARIDES AND GLYCOPROTEINS

To protect themselves against adverse environmental conditions like freezing and desiccation, fungi and yeasts also synthesize extracellular polysaccharides, mainly glucans and mannans, and glycoproteins. These polymers are excreted, often in large quantities, into the space beyond the cell wall. In liquid medium, they are loosely bound to the cell wall in the form of a capsule or mucilage. They protect the outside of conidia, spores, and fruiting bodies of mushrooms. For example, ripening of ascospores and conidia in the fruiting bodies of Ascomycetes and Coelomycetes, respectively, is often recognized by the extrusion of these propagules in a drop of slime. Likewise, encapsulation of the yeast *Cryptococcus magnus* is favored by low incubation temperatures (Golubev, 1991).

The prevalence of encapsulated organisms in the yeast soil flora of arid zones, and the increase in their relative abundance in yeast flora of drained bogs, indicates that the capsule may act as a cellular buffer system, preventing a too-rapid loss of water and providing a mechanism for efficient rehydration of the cell. Yeast flora of poor habitats, such as Arctic and Antarctic soils and the surface of tree bark, are exclusively or predominantly represented by capsule-forming species. Moreover, encapsulated yeasts showed an increased resistance to desiccation as compared to nonencapsulated ones (Golubev, 1991). Using encapsulated and nonencapsulated variants of *Cryptococcus diffluens*, Aksenov et al. (1973) observed that the presence of a large capsule prolonged dehydration during the drying of cells—but accelerated rehydration.

Extensive studies have been done on bacterial extracellular polysaccharides for industrial purposes; namely, xanthan from *Xanthomonas campestris* and dextran from *Leuconostoc mesenteroides* (Sarkar et al., 1985). In fungi, the glucan formed around the hard, insoluble sclerotia of *Sclerotium rolfsii* is isolated for commercial purposes as scleroglucan (Compere and Griffith, 1981). The mucilage, which surrounds the sclerotia, contributes to resist desiccation, temperature extremes, and irradiation. Other polysaccharides isolated from fungi and yeasts are pullulan from *Aureobasidium pullulans* (Catley and Kelly, 1966) and glomerellan from *Glomerella cingulata* (Sarkar et al., 1985).

Extracellular polysaccharides and glycoproteins of yeasts have been tested as cryoprotectants of psychrophilic yeasts. Glycoproteins secreted by the yeasts *Dipodascus australiensis* (Breierová, 1997) and *Rhodosporidium toruloides* (Breierová and Kocková-Kratochvílová, 1992) proved to be the most effective. Schizophyllan, a glucan produced by *Schizophyllum commune* (Steiner et al., 1987*)*, was tested by Berny and Hennebert (1991) for protective ability during freeze-drying.

8.2.6 PIGMENTS

Free radicals, produced during oxygen reduction, can cause damage to frozen (Fuller et al., 1988) and desiccated (Dimmick et al., 1961) organisms because their half-life is increased in dehydrated biological materials. Organisms have developed various defense mechanisms to protect themselves against free radicals and the products of lipid oxidation. A mechanism is the production of enzymes like superoxide dismutase, catalase, peroxidase, glutathione peroxidase, peroxide dismutase, malate dehydrogenase, and glutamate dehydrogenase (Breierová, 1994; Feofilova, 1994). In addition, free radicals can be scavenged by pigments such as melanin and carotenoids (Lukiewicz, 1972). Production of pigments in the cell walls of conidia and spores and in the walls of fruiting bodies is a strategy of fungi to survive desiccation and exposure to sunshine. Pycnidia and perithecia, produced on the surface of branches and twigs and fungi in tropical areas, are always melanized. Likewise, cultures of melanin- or carotene-containing fungi are more resistant to lyophilization (Tan et al., 1991b, 1994) and cryopreservation (Breierová et al., 1991).

8.2.7 MEMBRANES

In almost all organisms, including fungi and yeasts, decreasing the growth temperature results in an increased degree of unsaturated fatty acids to maintain a high membrane fluidity (Feofilova, 1994; Prasad, 1985; Wassef, 1977). In *Candida lipolytica* (Kates and Paradis, 1973), *N. crassa* (Martin et al., 1981), *Saccharomyces* ssp. (Calcott and Rose, 1982; Tanaka et al., 1985), *Rhodosporidium toruloides*, and *Mucor* sp. (Funtikova et al., 1992) adaptation to low temperature was mediated by desaturation of the fatty acids. In the Oomycetes, stress adaptation is thought to involve C20-polyunsaturation (Feofilova, 1994). In *Mucor* sp., *N. crassa*, and other fungi, the increased degree of unsaturation was accompanied by a decrease of the phosphatidyl-choline/phosphatidyl-ethanolamine ratio, as well as by a decline in the total number of these two phosphatides. Another way to increase the microviscosity of the membrane is by changing the isomer composition. For example, $\Delta 9,12$-linoleic acid dominates *Cunninghamella japonica* lipids at low temperature, whereas $\Delta 6,9$-linoleic acid is the main C18,2 isomer at 33°C (Feofilova, 1994). In this latter organism, a related phenomenon, stress-induced trehalose accumulation, is also accompanied by a change in the ratio of unsaturated to saturated lipids, with the level of the more unsaturated phospholipids, particularly phosphatidylethanolamine, increasing. Moreover, when exposed to low temperature, the degree of glycolipid unsaturation and the level of sterols were increased (Feofilova, 1992). Likewise, Calcott and Rose (1982) observed an increased survival of cryopreserved *S. cerevisiae* when the cells were enriched in sterols.

Microviscosity of the lipid bilayer can also be enhanced by altering the acyl chain length. In *Candida tropicalis* and *S. cerevisiae*, cultivation at low temperature results in production of fatty acids with shorter carbon chains (Feofilova, 1994).

8.2.8 ICE-NUCLEATING ACTIVITY

Fungi not only protect themselves against desiccation and freezing, but like some bacteria, they take advantage of freezing by acting as ice nucleators. By this mechanism the supercooling capacity of the host plant is reduced, and as a consequence, frost injury is exacerbated in the presence of ice-nucleating activity microorganisms in the phyllosphere. Lichens of the genera *Rhizoplaca*, *Xanthoparmelia*, and *Xanthuria* show ice-nucleating activity. Within the fungal kingdom, ice-nucleating activity is demonstrated for *Fusarium acuminatum* and *Fusarium avenaceum* (Pouleur et al., 1992).

8.3 PRECONDITIONING

As can be concluded from the preceding sections, microorganisms have developed many mechanisms to adapt themselves to environmental-stress conditions like desiccation and temperature and osmotic stresses. Induction of one or several of these mechanisms by preconditioning can be used to enhance freezing, thawing, and freeze-drying resistance. Actively growing microorganisms can be cryopreserved, although cells of the late exponential phase and the stationary phase survive freezing much better (Morris et al., 1988). Freeze-drying, which is a more severe stress, is much better survived by cells entering the stationary phase. Only the spores and conidia of filamentous fungi, produced at the end of the exponential growth phase, survive freeze-drying. Trehalose and compatible solutes like mannitol are produced in large quantities in these propagules. Moreover, spores and conidia have a much greater capacity to undercool and to arrest crystal growth than hyphae (Tan et al., 1991a). When studied by cryomicroscopy, ice crystals were observed to enter the hyphae at spots at which the cell walls were broken as a result of mechanical damage that occurred during preparation of the suspension. Subsequently, these crystals grew through the septal pores until all the interconnected hyphae were crystallized.

Because sporulation is induced at the end of the exponential growth phase, when carbon and nitrogen sources become limited, it is advisable to grow fungi on poor media before lyophilization. Moreover, trehalose production is induced by carbon, nitrogen, phosphorus, and sulfur limitation (Lillie and Pringle, 1980). Sporulation is also influenced by light conditions (wavelength and alternation light/dark periods); blue light and ultraviolet light especially stimulate sporulation. In addition, polysaccharide production by *S. rolfsii* was increased by growing cultures in white and blue light conditions (Miller and Liberta, 1976).

Freezing and desiccation tolerance can be enhanced by exposing fungi and yeasts to a heat shock before preservation. By shifting the temperature to 40 to 45°C, trehalose was accumulated in cells of *S. cerevisiae* (Eleutherio et al., 1993; Hottiger et al., 1987a) and *Schizosaccharomyces pombe* (Virgilio et al., 1990) and in conidia of *N. crassa* (Neves et al., 1991), and concomitantly, freezing, desiccation, and thermo-tolerance was acquired. Another effect of exposing cells to a heat shock is the production of HSPs (see also Section 8.2.3). Cryoprotection could be induced in *S. cerevisiae* by exposing the cells to 43°C because the exposure resulted in production of HSPs of 97, 85, and 70 kD (Kaul et al., 1992a). Likewise, conidia of *N. crassa* were protected against freezing injury by induction of HSPs after shifting the temperature to 45°C (Guy et al., 1986).

Increasing osmotic pressure induces the production of compatible solutes and trehalose (Volkov, 1994). Under moderate osmotic stress, glycerol is produced, making the cells more freeze resistant, whereas under severe osmotic stress and during desiccation, trehalose accumulation is increased (Hounsa et al., 1998; Lodato et al., 1999).

In addition to preexposure to high temperatures, preexposure to low temperatures enhances survival after freezing. At the International Mycological Institute's (IMI) collection of fungi in the United Kingdom, all cultures are pregrown at 5°C before cryopreservation (Smith, 1982). Moreover, preincubating cells at a low temperature increases survival after freeze-drying (Tan, unpublished results). Although both cold and heat pretreatment enhance cell tolerance to low temperatures, only heat pretreatment induces accumulation of intracellular trehalose (Diniz-Mendez et al., 1999). Preincubation at a low temperature might provide protection against freezing injury by inducing a 33-kD protein, as is the case in yeasts (Diniz-Mendes et al., 1999; Kaul et al., 1992b). Moreover, decreasing the growth temperature results in an increased degree of unsaturated fatty acids, so that high membrane fluidity can be maintained during cooling (Feofilova, 1994).

8.4 CRYOPRESERVATION OF MICROORGANISMS

Microorganisms can be preserved by various methods. The optimal preservation methods are those in which metabolism is arrested by application of low temperature (cryopreservation) or by dehydration (lyophilization). Both methods are preferred to maintenance on agar slants. This method is laborious, cultures can become infected, and repeatedly subcultured strains may degenerate.

Preservation of microorganisms is reviewed in Fennell (1960), Heckly (1978), von Arx and Schipper (1978), Smith and Onions (1980), and Kirsop and Snell (1984; see also Chapter 15). The method of cryopreservation was adopted for fungi by Hwang (Hwang, 1968; Hwang and Howells, 1968; Hwang et al., 1976) and has subsequently been introduced in most culture collections. Technical details are described by Challen and Elliot (1986) and Stalpers et al. (1987).

8.4.1 Cryoprotectants

Although survival has been obtained without the use of special protective additives, survival of most fungi and yeasts is improved by adding a cryoprotectant. Penetrating as well as nonpenetrating cryoprotectants have been tested. Most organisms were cryopreserved successfully with glycerol, but dimethyl sulfoxide (Me$_2$SO) proved to be better for the more difficult strains (Hwang and Howells, 1968; Hwang et al., 1976; Smith, 1983). Toxicity of Me$_2$SO can be reduced by the addition of glucose (Smith, 1983). The saccharides glucose (Smith, 1983) and sucrose (Ohmasa et al., 1992) solely did not provide cryoprotection, but yeast cells could be cryopreserved in 10% trehalose (Breierová, 1994; Coutinho et al., 1988). Cryoprotection by sorbitol, proline, and the nonpenetrating compounds polyvinylpyrrolidone (PVP; Ohmasa et al., 1992; Smith, 1983), hydroxyethyl starch (HES; Malik and Hoffmann, 1989), and dextrin (Ohmasa et al., 1992) was tested, but results were not superior to Me$_2$SO or glycerol. Ten percent polyethylene glycol proved to be a good cryoprotectant for Basidiomycetes (Ohmasa et al., 1992). When studying cryoprotective properties of alcohols, Tan and Stalpers (1996), obtained high survival rates with 1,2-propane-diol and ethyleneglycol. Although 1,2-butane-diol and 1,3-butane-diol are promising compounds on the basis of their high glass-forming tendency (Boutron et al., 1986) they were too toxic. Likewise, ethanol and methanol, which provide good cryoprotection to algae, were toxic, although Lewis et al. (1994) obtained good cryoprotection of yeasts by ethanol and methanol. The protective ability of the last-mentioned compounds was only obtained after rapid cooling and probably resulted from their ability to induce increased membrane permeability, leading to more rapid water equilibration during extracellular freezing. During slow cooling, both compounds acted as cryosensitizers. In the studies by Lewis et al. (1994), propan-1-ol and butan-1-ol resulted in low cell viability, but propan-2-ol showed some cryoprotection.

Analogous to viruses and bacteria, the addition of proteins like skimmed milk, serum, yeast extract, and malt extract to the penetrating component was tested for filamentous fungi (Dahmen et al., 1983; Hubalek, 1996; Smentek and Windisch, 1982). However, at present, most filamentous fungi are cryopreserved in 10% glycerol solely, whereas 5 to 10% Me$_2$SO is used when results are unsatisfactory with glycerol (Smith, 1982). Yeasts are cryopreserved in the former-mentioned cryoprotectants or in mixtures of these penetrating protectants with saccharides and nonpenetrating compounds like skimmed milk, yeast extract, peptone, serum, ficoll, and so forth (Hubalek, 1996). Breierová et al. added glycoproteins and polysaccharides, excreted by selected yeast strains, to Me$_2$SO with the purpose of cryoprotecting psychrophilic yeasts (Breierová, 1997; Breierová and Kocková-Kratochvílová, 1992; see also Section 8.2.5). In addition alpha-tocopherol protected cryopreserved yeasts against free radical damage (Breierová, 1994). To improve survival, organisms can be preincubated at 5°C before cryopreservation (Smith, 1982).

8.4.2 Cooling Rate

The cooling rate should be chosen carefully. When cells are cooled rapidly, intracellular ice crystals are produced, which is lethal. The production of intracellular ice crystals can be avoided by slow cooling. At low cooling rates, the bulk of extracellular water crystallizes, leaving a highly concentrated solution. As a result of the increased concentration of the medium, the cells dehydrate. Consequently, the cytoplasm becomes increasingly concentrated, resulting in a depression of the freezing temperature. The only cells that survive freezing are those that are dehydrated to such extent that ice crystallization is avoided completely (Morris, 1981). Unfortunately, prolonged exposure to a hypertonic solution may lead to the denaturation of proteins. Moreover, too-extensive shrinkage could result in loss of membrane material (Morris et al., 1983; Smith et al., 1986). When the shiitake fungus *Lentinus edodes* was dehydrated, deletions in the plasma and nuclear membranes, extracellular membrane-bound vesicles, and strands of membrane and cytoplasm between cell wall and membrane were observed (Roquebert and Bury, 1993). Deleterious intramembrane

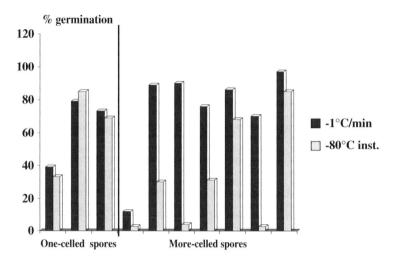

FIGURE 8.2 Survival of one- or more-celled spores after instantaneous cooling to –80°C and cooling at –1°C/min.

particle aggregation and lamellar-to-hexagonal II phase transitions have been observed in tertiary hyphae of mushrooms when inner surfaces of plasma membranes in the same hyphae were brought into direct contact as a consequence of cellular deformation caused by the formation of extracellular ice (Fujikawa, 1987, 1991). Therefore, according to the two-factor hypothesis of Mazur (1970), each cell should be cooled at the rate at which the damaging effects of both slow and fast cooling rates are minimized.

Optimal cooling rates depend on a number of parameters such as the size of the cell, permeability of the membrane, and thickness of the cell wall. During a viability check of 18,000 fungal strains, stored for up to 30 years in a lyophilized state at the Centraalbureau voor Schimmelcultures in the Netherlands, survival was good except for the fungi that produced large (length > 20 μm) or thick-walled spores. These results could be explained by the fact that the method to freeze-dry bacteria, in which drying is preceded by fast freezing (e.g., spin freeze-drying), was adopted to freeze-dry the fungal strains. When a group of 11 species of fungi producing spores of different size and wall thickness were cooled at various rates, four categories could be distinguished: organisms producing spores with one or more cells that were either thin-walled or thick-walled. Small, one-celled, thin-walled spores (diameter < 4 μm; e.g., *Trichoderma* ssp., *Penicillium* ssp., *Aspergillus* ssp.) survived instantaneous and slow cooling equally well. In contrast, survival of thin- or thick-walled spores with more cells and of one-celled thick-walled spores was significantly better when spores were cooled at –1°C/min, because these propagules dehydrated very slowly (Figure 8.2; Tan et al., 1994).

Optimal cooling rates can best be established with the aid of a cryomicroscope. In the Biotechnological Action Programme, sponsored by the Commission of the European Community, optimal cooling rates were estimated by this method for a group of 20 fungi (Smith et al., 1990).

8.4.3 VIABILITY

A comprehensive overview of viability after cryopreservation for the various species of microorganisms is presented by Hubalek (1996). When Smith studied the recovery of 3004 fungi that had been stored in liquid nitrogen at the IMI collection for up to 13 years, negative results were mainly obtained in the Chytridiomycetes, Hypochytridiomycetes, and Oomycetes (Smith, 1982). The coenocytic mycelium of these organisms, lacking cross-septa, are very sensitive to any mechanical damage sustained during preparation before preservation. The cryosensitive Oomycetes require careful optimization of the cooling rate. Slow cooling proved to be better than rapid cooling (Dahmen et al., 1983; Nishii and Nakagiri, 1991a; Xu and Huang, 1987), although –1°C/min was not the optimal cooling rate for all Oomycetes. When studied by cryomicroscopy, *Achlya ambisexualis* showed optimal recovery at –9°C/min, *Phytophthora* ssp. at rates between 3° and 11°C/min, *Pythium aphanidermatum* at –8° to –29°C/min, and *Saprolegnia parasitica* at –10°C/min (Smith et al., 1990). Within the Oomycetes, survival of the terrestic Peronosporales (*Sclerospora, Pythium, Phytophthora*) was superior to that of the aquatic Saprolegniales (e.g., *Achlya, Saprolegnia, Aphanomyces*). Best survival was obtained by *Sclerospora* ssp. followed by *Phytophthora ssp.* and *Pythium* ssp., respectively (Smith, 1982). Agar discs derived from the central part of a colony were preserved more successfully than those from the edge of the colony (Nishii and Nakagiri, 1991b). Thawing after cryopreservation at +30°C for 5 min yielded better results than thawing at +40°C for 3 min.

Downy mildew fungi—obligate plant pathogens belonging to the Peronosporales—were infective after cryopreservation of infected host tissue or (zoo) sporangia with or without cryoprotectant (Dahmen et al., 1983; Gulya et al., 1993).

Ascomycetes (Ito, 1991; Smith, 1982; Yokoyama and Ito, 1984) and yeasts (Hubalek and Kocková-Kratochvílová, 1982; Kirsop and Henry, 1984; Mikata and Banno, 1987; Smentek and Windisch, 1982) survive cryopreservation very well. When tested by Smith of the IMI collection of the United Kingdom, negative results were only obtained for the Loculoascomycetes (Smith, 1982), although Yokoyama and Ito of the Institute of Fermentation in Osaka scored high viability of all 1096 Ascomycetes tested, including the Loculoascomycetes (Yokoyama and Ito, 1984). Even the powdery mildew *Erisyphe cichoracearum* and *Podosphaera leucotricha* proved to be infective after cryopreservation (Dahmen et al., 1983). The 22,000 Ascomycetes and related hyphomycetes of the Dutch Collection of the Centraalbureau voor Schimmelcultures are stored successfully at –135°C. Failures are only encountered for some auxotrophic pathogenic strains belonging to the dermatophytes. In addition, isolates that are difficult to grow and that form small, waxy sterile colonies tend not to survive cryopreservation.

Zygomycetes (Smith, 1982; Yokoyama and Ito, 1984; Zhz and Li, 1987) and Basidiomycetes (Challen and Elliott, 1986; Ito and Nakagiri, 1996; Smith, 1982) generally show a high viability after cryopreservation. Within the Zygomycetes, the lowest viability was obtained with the Entomophthorales (*Basidiobolus, Conidiobolus*); the lowest survival of the Basidiomycetes is recorded for the ecto-mycorrhizal cultures.

Loegering et al. (1966) succeeded in cryopreserving uredospores of the Basidiomycetous obligate parasitic rust fungi *Puccinia graminis* f. sp. *tritici*.

8.5 LYOPHILIZATION

Freeze-drying of microorganisms has been reviewed by Haskins (1957), Meryman (1966), MacKenzie (1977), Heckly (1978), Jennings (1999), and Rey and May (1999). Although sterile mycelia have been dried (Croan, 2000; Tan et al., 1991a, 1991b), spores and conidia survive freeze-drying much better.

8.5.1 LYOPROTECTANTS

During freeze-drying, cells are suspended in a lyoprotectant. This lyoprotectant includes a macromolecule, which serves as a bulking agent, and a saccharide to protect the membrane. In a hydrated cell, water is hydrogen-bonded to the phospholipid head groups, yielding space for the fatty-acid acyl chains to be mobile (liquid-crystalline membrane). When the water is removed, a phase transition of the membrane to the gel phase will take place. Because the lipid components of most biological membranes are heterogeneous, they undergo phase transitions over a wide range of temperatures. As a consequence of the coexistence of solid and still fluid lipids, the proteins can no longer be anchored in the correct topography in the lipid bilayer structure. They start to drift

through the still-fluid parts of the membrane, resulting in aggregation of intramembraneous particles. In addition, during reheating, nonlamellar structures can be formed, leading to cell leakage
(Morris, 1981). Membranes are protected during freezing and drying against this transition by
saccharides (see also Section 8.2.1). The saccharides replace water by hydrogen bonding to the
phospholipid head groups (Crowe et al., 1984, 1985a, 1987, 1990). This interaction increases head
group spacing, resulting in a lower transition temperature of the phospholipids (Crowe et al., 1985b).
Disaccharides are found to be optimal, especially trehalose, which fits best within the membrane
structure. The saccharide component of the protectant can enter the cells when they are forced
through their phase transition by cooling. Then the cells become leaky and the saccharide can flow
down its concentration gradient into the cell (Leslie et al., 1995). Precaution should be taken to
avoid phase transitions during cooling as well as during rehydration. In *S. cerevisiae*, the temperature
of the dry gel-to-liquid crystal phase is lowered by trehalose from around 60° to 40°C, and therefore,
the dried cells of this yeast must be rehydrated above 40°C to avoid passing through a phase
transition during revival (Leslie et al., 1994).

Denaturation of proteins can be diminished by adding saccharides or amino acids to the
protectant (Arakawa and Timasheff, 1982; Back et al., 1979; Carpenter and Crowe, 1988). Both
types of compounds stabilize the proteins during cooling because they are preferentially excluded
(see also Section 8.2.1) from the surface of the proteins in aqueous solutions. They repel the
hydrophobic parts of the amino acid chains, thus preventing unfolding of the protein at the melting
temperature. Hydrogen bonding between saccharides and proteins in the final stages of desiccation
is required for stabilization of the dried proteins (Allison et al., 1999).

The amino acid Na-glutamate is frequently added to the lyoprotectant to dry microorganisms.
With fungi and bacteria, raising the concentration of Na-glutamate up to 5% increased survival
rates considerably. Obviously, this relative high concentration of Na-glutamate was necessary to
optimally stabilize the proteins. Moreover, in this protectant, organisms are dried more gently,
leaving the tertiary structure of the membranes and the proteins better situated.

Independent of the method of freeze-drying, denaturation of a number of the proteins after
freeze-drying can never be avoided. To remove these proteins, the addition of 10% activated charcoal
to the protectant will result in binding the often-toxic denatured proteins (Malik, 1990).

8.5.2 THE FREEZE-DRYING PROCESS

During lyophilization, both the microorganisms and the protectant are converted into a glass. In
the cooling step preceding drying, the bulk of extracellular water crystallizes, and consequently,
the cells are dehydrated. At the beginning of the primary drying phase, the dehydrated cells,
surrounded by the lyoprotectant so viscose that it is a glass (Franks, 1990), are embedded in ice
crystals. During primary drying, the ice crystals are evaporated, leaving the glass interwoven with
channels. The channels facilitate the resorption of water when the dried pellet is dissolved again.
Moreover, during the transition from the primary to the secondary drying phase and during the
secondary drying phase, water is evaporated through these channels from the glassy matrices of
the protectant and cell, making them even more viscous. Because viscosity in the protectant and
the cells increases during secondary drying, the temperature at which their glasses are stable
increases concomitantly. At the end of the secondary drying phase, when only 1 to 2% moisture
is left, glasses that are stable at room temperature are formed both in the protectant and in the cells,
and the organisms can be stored. A glass is ideal to store the dried organisms because the molecules
are immobilized, and therefore there is no chemical or enzymatic activity. In addition, the molecules
are arranged in an unordered structure, allowing them to bind to the membranes and the proteins.

The temperature applied during freeze-drying should always remain below the glass transition
temperature (T_g) curve to avoid collapse of the glass. When dried above the T_g, the cells are liquid-
dried, resulting in an enormous denaturation of the proteins. Moreover, the channel structure is lost
when the glass melts, and consequently the water vapor cannot be removed anymore. Therefore,

TABLE 8.1
Glass Transition Temperatures of
Mixtures of 5% Dextran plus 7%
Saccharide, Freeze-Dried to 2% RMC

Glucose	41
Inositol	51
Sucrose	78
Maltose	91
Trehalose	93
Lactose	94
Raffinose	112

the freeze-drying protocol should best be established with the aid of a differential scanning calorimeter (Hatley, 1990) or a freeze-drying microscope (Tan et al., 1998).

Freeze-dried organisms can be stored successfully below the T_g (Franks, 1990). Above T_g, water mobility increases, and the product consequently deteriorates. To facilitate freeze-drying and to make sure that the T_g of the dried product is above the storage temperature, constituents of the lyoprotectant should have high T_gs. Therefore, the effectiveness of the various saccharides depends, in addition to their hydrogen bonding capacity, on their T_gs (Tan et al., 1995; see also Chapters 20 and 21). When spores of fungi were freeze-dried with various types of saccharides, it was observed that the ranking order of survival rates after storage at 30°C (Figure 8.3) corresponded with the respective T_g values (Table 8.1). Pellets containing the small molecules of the monosaccharide glucose or the sugar alcohol myo-inositol showed an onset of the glass-transition curve at or below 30°C when the pellets were dried to 2% RMC. Likewise, viability of propagules dried in these protectants decreased immediately after freeze-drying or after storage at 30°C. Viability was optimal with lactose, followed by trehalose, maltose, and sucrose. Likewise, T_g values were highest for lactose and trehalose, followed by maltose and sucrose. In nature, the nonreducing disaccharides trehalose and sucrose play an important role. Trehalose, showing a relatively high T_g, is produced by biological materials (e.g., spores of fungi and yeasts, nematodes, desert plants) to make them

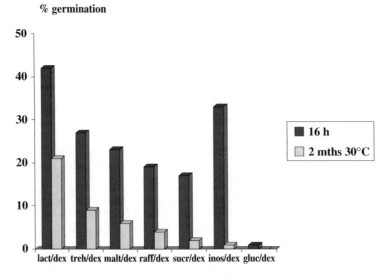

FIGURE 8.3 Survival of spores of the fungus *Artrhobotrys superba*, freeze-dried to 2% residual moisture content (RMC), protected by mixtures of 5% dextran plus 7% saccharide 16 h; 2 months 30°C.

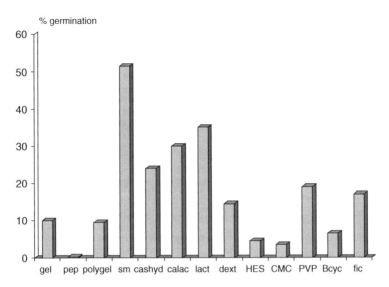

FIGURE 8.4 Survival of spores of the fungus *Arthrobotrys superba*, immediately after freeze-drying, protected by various macromolecules plus 7% trehalose, dried to 2% RMC.

drought resistant, whereas sucrose, showing a relative low T_g, is produced in plants to make them freeze resistant. Although the T_g of lactose, the saccharide in milk (the protectant routinely used to lyophilize microorganisms) equals that of trehalose, the latter compound is preferred to lactose because it is a nonreducing disaccharide, it fits better within the membrane structure, and the glass produced by trehalose is more stable than that produced by lactose, which has a strong tendency to recrystallize.

That differences in hydrogen bonding and physical stability jointly contribute to the effectively of the various saccharides is proven by the lower survival with raffinose. The glass produced by raffinose is as stable as that produced by lactose or trehalose, but survival is lower (Figure 8.3). This can be explained by the fact that disaccharides show a better hydrogen bonding capacity to the phospholipid headgroups than the to trisaccharide raffinose (Tan et al., 1995).

To increase the T_g of the lyoprotectant and to serve as bulking agent, macromolecules are added to the lyoprotectant (Tan et al., 1998). Usually when biological materials are freeze-dried, polysaccharides and proteins are applied, because they are not toxic. However, molecules like PVP also give satisfactory results.

Numerous proteins can be added to the protectant, such as algin, lactalbumin, peptone, cattle serum, bovine serum albumin, gelatin, skimmed milk, casein, casitone, and polygelin. Proteins provide a better biological protection than polysaccharides and compounds like PVP because the amino acids of the proteins might contribute to restoring the energy charge and repairing damaged proteins during revival. Moreover, by the mechanism of preferential exclusion, proteins might help to protect the tertiary structure of the cell structural proteins and enzymes. Survival immediately after freeze-drying is higher when proteins are added to the protectant (Figure 8.4). Moreover, the lag phase is increased when polysaccharides and PVP are added, and as a consequence, germination initiated more rapidly in spores dried in proteins than in spores dried in polysaccharides and PVP. Unfortunately, because of their low T_gs, some of the proteins, especially peptone, can not be dried successfully.

Polysaccharides and PVP have higher T_gs, which makes them more suitable for freeze-drying. Various polysaccharides can be applied: dextran, HES, carboxymethylstarch , ficoll, b-cyclodextrin, and so forth. A problem can occur when biological materials are freeze-dried (Tan et al., 1998). Because of the slow cooling rate and the small quantities of material included, the pellets can become so amorphous that they can scarcely be freeze-dried. Figure 8.5 shows scanning electron

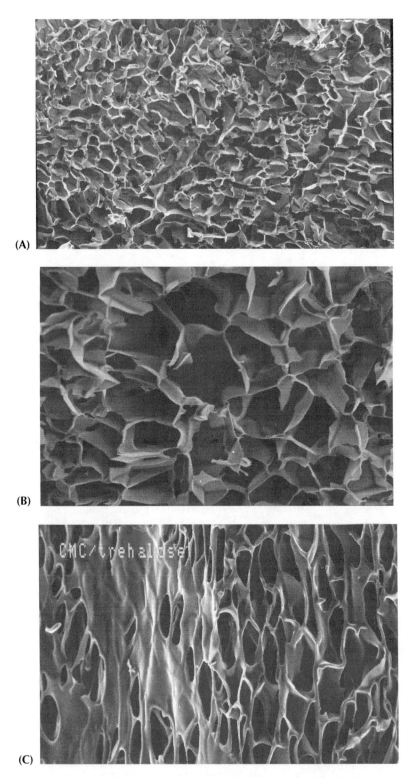

FIGURE 8.5 Scanning electron microscopy photographs of pellets of macromolecules plus 7% trehalose, freeze-dried to 2% RMC. (A) Dextran (×300) showing regular fenestrations within matrix. (B) Hydroxyethyl starch (×300), regular fenestrations. (C) Carboxymethylstarch (×300). (D) Carboxymethylstarch (×250), collapsed areas within matrix. (E) Polyvinylpyrrolidone (×300), irregular fenestrations within matrix. (F) Poly-vinylpyrrolidone (×1000) pellicular surface without pores.

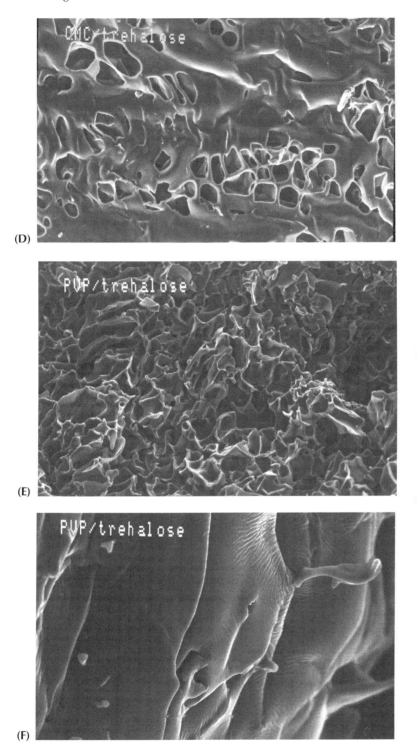

(D)

(E)

(F)

FIGURE 8.5 (continued)

microscopy photographs of dried pellets of mixtures of trehalose with dextran, HES, carboxymethylstarch, or PVP. The pellets with dextran and HES show a regular open network structure, and
they can therefore be freeze-dried successfully, although the structure of HES is a bit coarser than
that of dextran. In contrast, the structure in pellets with carboxymethylstarch is irregular, with

mazes of various size, and some parts of the pellet show a severe collapse. Pellets with PVP also show an irregular structure, and the exposed surface of the pellet is pellicular without pores, possibly indicating a phase separation that occurred during freezing, which will hamper the freeze-drying process substantially. Because annealing can be deleterious for biological materials, incipients like mannitol can be added to overcome these problems.

Although survival rates are better after cryopreservation, most sporulating fungi and yeasts survive lyophilization well, except the Oomycetes and the Basidiomycetes. Within the Ascomycetes, problems are encountered within the auxothrophic dermatophytes and the organisms producing large, thin-walled spores or conidia. Precaution should be taken that these propagules and thick-walled small, as well as large, spores and conidia are cooled slowly before dehydration (see also Section 8.4.2).

8.6 OTHER PRESERVATION METHODS

Unfortunately, membranes of some large cells like the Oomycetes, which should be cooled slowly to avoid production of intracellular ice crystals, do not stand slow cooling. For this group of recalcitrant microorganisms, two alternative preservation methods are available in which slow cooling at subzero temperatures is avoided: vitrification (Rall and Fahy, 1985; Tan and Stalpers, 1996) and storage in alginate beads (Fabre and Dereuddre, Cryo-Letters, 11, 413–426. 1990). The method of vitrification was developed to cryopreserve mammalian embryos: Alginate beads were developed to store embryogenic tissue of recalcitrant (tropical) plants. During vitrification, cells are cooled ultra rapidly before cryogenic storage. In the alginate bead technology, organisms are immobilized by encapsulation in an alginate gel, and subsequently the immobilized cells are dehydrated at room temperature before cryogenic storage. Another preservation method for freeze-sensitive microorganisms is liquid-drying. Liquid-drying involves vacuum-drying from the liquid state without freezing. The method is described in detail in Annear (1962), Ijima and Sakane (1973), and Malik (1990). The method is successfully applied for the complete collection of yeast strains of the Institute of Fermentation in Osaka and for bacteria including actinomycetes and bacteriophages. Because the residual moisture content reached with this method is 10% compared to 2% with freeze-drying, it is recommended that the dried organisms be stored at 5°C. A detailed study on compounds protecting liquid-dried cultures from mutation was performed by Sakane et al. (1983).

8.7 CONCLUSIONS

Vitrification and storage in alginate beads are promising preservation methods, but cryopreservation and lyophilization are the routinely used storage methods in culture collections. Cryopreservation is superior to lyophilization because it gives higher survival rates and is universally applicable. Moreover, lyophilization can give rise to genetic variants; for example, petite variants—respiratory deficient yeast strains lacking part of their mitochondrial DNA (Hubalek, 1996). In addition, the lag phase of lyophilized cultures is increased compared with that of cryopreserved cultures, as was proven by the fact that the production of patulin was much more retarded in a freeze-dried subculture of *Penicillium expansum* than in a cryopreserved subculture (Smith et al., 1990). Disadvantages of cryopreservation include the dependence on regular supply of liquid nitrogen or on electricity. However, the main disadvantage of cryopreservation is that organisms must be dispatched under frozen conditions or revived before transport. Therefore, cryopreservation is the preferred storage for a backup collection, whereas lyophilization is more practical when cultures are dispatched on a regular basis.

REFERENCES

Aksenov, S.I., Bab'eva, I.P., and Golubev, W.I. (1973) On the mechanism of adaptation of micro-organisms to conditions of extreme low humidity, in *Life Sciences and Space Research*, 11th edition, Sneath P.H.A., Ed., Akademie, Berlin, pp. 55–61.

Allison, S., Chang, B., Randolph, T., and Carpenter, J. (1999), Hydrogen bonding between sugar and protein is responsible for inhibition of dehydration-induced protein unfolding, *Arch. Biochem. Biophys.*, 365, 289–298.

Annear, D.I. (1962) Recoveries of bacteria after drying on cellulose fibres. A method for the routine preservation of bacteria, *Austral. J. Exp. Biol.*, 40, 1–8.

Arakawa, T. and Timasheff, S.N. (1982) Stabilization of protein structure by sugars, *Biochemistry*, 21, 6536–6544.

Back, J., Oakenfull, D., and Smith, M.B. (1979) Increased thermal stability of proteins in the presence of sugars and polyols, *Biochemistry*, 18, 5191–5196.

Berny, J.-F. and Hennebert, G. (1991) Viability and stability of yeast cells and filamentous fungus spores during freeze-drying, effects of protectants and cooling rates, *Mycologia*, 83, 805–815.

Blomberg, A. (2000) Metabolic surprises in *Saccharomyces cerevisiae* during adaptation to saline conditions, questions, some answers and a model, *FEMS Microbiol. Lett.*, 182, 1–8.

Blomberg, A. and Adler, L. (1992) Physiology of osmotolerance in fungi, *Adv. Microb. Physiol.*, 33, 145–212.

Boutron, P., Mehl, P., Kaufmann, A., and Angibaud, P. (1986) Glass-forming tendency and stability of the amorphous state in the aqueous solutions of linear polyalcohols with four carbons, *Cryobiology*, 23, 453–469.

Breierová, E. (1994) Cryoprotection of psychrophilic yeast species—by the use of additives with cryoprotective media, *Cryo-Letters*, 15, 191–197.

Breierová, E. (1997) Yeast exoglycoproteins produced under NaCl-stress conditions as efficient cryoprotective agents, *Lett. Appl. Microbiol.*, 25, 254–256.

Breierová, E. and Kocková-Kratochvílová, A. (1992) Cryoprotective effects of yeast extracellular polysaccharides and glycoproteins, *Cryobiology*, 29, 385–390.

Breierová, E., Kocková-Kratochvílová, A., Sajbidor, J., and Ladzianska, K. (1991) *Malassezia pachydermatis*, properties and storage, *Mycoses*, 34, 349–352.

Brown, A.D. (1978) Compatible solutes and extreme water stress in eukaryotic micro-organisms, *Adv. Microb. Physiol.*, 17, 181–242.

Calcott. P. and Rose H. (1982) Freeze-thaw and cold-shock resistance of *Saccharomyces cerevisiae* as affected by plasma membrane lipid composition, *J. Gen. Microbiol.*, 128, 549–555.

Carpenter, J. and Crowe, J.H. (1988) Modes of stabilization of a protein by organic solutes during desiccation, *Cryobiology*, 25, 459–470.

Catley, B. and Kelly, P. (1966) Metabolism of trehalose and pullulan during the growth cycle of *Aureobasidium pullulans*, *Biochem. Soc. Trans.*, *Edinburgh*, 3, 1079–1081.

Challen, M.P. and Elliott, T.J. (1986) Polypropylene straw ampoules for the storage of microorganisms in liquid nitrogen *J. Microbiol. Methods*, 5, 11–23.

Compere, A. and Griffith, W.L. (1981) Scleroglucan biopolymer production, properties, and economics, *Adv. Biotechnol.*, 3, 441–446.

Coutinho, C., Felix, D., and Panek, A.D. (1988) Trehalose as cryoprotectant for preservation of yeast strains, *J. Biotechnol.*, 7, 23–32.

Croan, S.C. (2000) Lyophilization of hypha-forming tropical wood-inhabiting Basidiomycotina, *Mycologia*, 92, 810–817.

Crowe, J., Carpenter, J., Crowe, L., and Anchordoguy, T.J. (1990) Are freezing and dehydration similar stress vectors? A comparison of modes of interaction of stabilizing solutes with biomolecules, *Cryobiology*, 27, 219–231.

Crowe, J.H., Crowe, L., Carpenter, J., and Wistrom, C. (1987) Stabilization of dry phospholipid bilayers and proteins by sugars, *Biochem. J.*, 242, 1–10.

Crowe, J.H., Crowe, L.M., and Chapman, D. (1984) Preservation of membranes in anhydrobiotic organisms: the role of trehalose, *Science*, 223, 701–703.

Crowe, L.M., Crowe, J.H., and Chapman, D. (1985a) Interaction of carbohydrates with dry dipalmitoylphophatidylcholine, *Arch. Biochem. Biophys.*, 236, 289–296.

Crowe, L.M., Crowe, J.H., Rudolph, A., Womersley, C., and Appel, L. (1985b) Preservation of freeze-dried liposomes by trehalose, *Arch. Biochem. Biophys.*, 242, 240–247.

Dahmen, H., Staub, T., and Schwinn, F.J. (1983) Technique for long-term preservation of phytopathogenic fungi in liquid nitrogen, *Phytopathology*, 73, 214–246.

Dimmick, R.L., Heckly, R.J., and Hollis, D.P. (1961) Free radical formation during storage of freeze-dried *Serratia marcescens*, *Nature (Lond.)*, 192, 766–777.

Diniz-Mendes, L., Bernardes, E., Araujo, P., Panek, A., and Paschoalin, V.M. (1999) Preservation of frozen yeast cells by trehalose, *Biotechnol. Bioeng.*, 65, 572–578.

Duman, J.G., Wu, D., Olsen, T., Urrutia, M., and Tursman, D. (1993) Thermal-hysteresis proteins, in *Low-Temperature Biology* Vol. 2, Steponkus, P.L., Ed., JAI Press, London, pp.131–182.

Eleutherio, E., Araujo, P., and Panek, A.D. (1993) Protective role of trehalose during heat stress in *Saccharomyces cerevisiae*, *Cryobiology*, 30, 591–596.

Fabre, J. and Dereuddre, J. (1990) Encapsulation-dehydration, a new approach to cryopreservation of *Solanum* shoot-tips, *Cryo-Letters*, 11, 413–426.

Fennell, D.I. (1960) Conservation of fungous cultures, *Bot. Rev.*, 26, 79–141.

Feofilova, E.P. (1992) Trehalose, stress, and anabiosis (review), *Mikrobiologiya*, 61, 741–755.

Feofilova, E.P. (1994) Biochemical adaptation of mycelial fungi to temperature stress, *Mikrobiologiya*, 63, 757–776.

Franks, F. (1990) Freeze drying, from empiricism to predictability, *Cryo-Letters*, 11, 93–110.

Fujikawa, S. (1987) Mechanical force by growth of extracellular ice crystals is widespread cause for slow freezing injury in teritary hyphae of mushrooms, *Cryo-Letters*, 8, 156–161.

Fujikawa, S. (1991) Lamellar to hexagonal II phase transitions in tonoplasts of mushroom hyphae caused by mechanical stress resulting from the formation of extracellular ice crystals, *Cryobiology*, 28, 191–202.

Fuller, B.J., Gower, J., and Green, C.J. (1988) Free radical damage and organ preservation, fact or fiction? *Cryobiology*, 25, 377–393.

Funtikova, N.S., Katomina, A., and Mysyakina, I.S. (1992) Composition and degree of unsaturation of the lipids of the fungus *Mucor* grown at low temperatures, *Mikrobiologiya*, 61, 793–797.

Gekko, K. and Timasheff, S.N. (1981) Mechanism of protein stabilization by glycerol, preferential hydration in glycerol-water mixtures, *Biochemistry*, 20, 4667–4676.

Golubev, V.I. (1991) Capsules, in *The Yeasts*, 2nd edition, Vol. 4, Rose, A.H. and Harrison, J.S., Eds., Academic Press, New York, pp. 175–198.

Gulya, T., Masirevic, S., and Thomas, C.E. (1993) Preservation of air-dried downy mildew sporangia in liquid nitrogen without cryoprotectants or controlled freezing, *Mycol. Res.*, 97, 240–244.

Guy, C., Plesofsky-Vig, N., and Brambl, R. (1986) Heat shock protects germinating conidiospores of *Neurospora crassa* against freezing injury, *J. Bacteriol.*, 167, 124–129.

Haskins, R.H. (1957) Factors affecting survival of lyophilized fungal spores and cells, *Can. J. Microbiol.*, 3, 477–485.

Hatley, R. (1990) The effective use of differential scanning calorimetry in the optimisation of freeze-drying processes and formulations, *Dev. Biol. Standard*, 74, 105–122.

Heckly, R.J. (1978) Preservation of microorganisms, *Adv. Appl. Microbiol.*, 24, 1–53.

Hocking, A.D. (1993) Responses of xerophilic fungi to changes in water activity, in *Stress Tolerance of Fungi*, Jennings, D.H., Ed., Marcel Dekker, New York, pp. 233–256.

Hottiger, T., Boller, T., and Wiemken, A. (1987a) Rapid changes of heat and desiccation tolerance correlated with changes of trehalose content in *Saccharomyces cerevisiae* cells subjected to temperature shifts, *FEBS Lett.*, 220, 113–115.

Hottiger, T., Schmutz, P., and Wiemken, A. (1987b) Heat-induced accumulation and futile cycling of trehalose in *Saccharomyces cerevisiae*, *J. Bacteriol.*, 169, 5518–5522.

Hounsa, C.-G., Brandt, E., Thevelein, J., Hohmann, S., and Prior, B. (1998) Role of trehalose in survival of *Saccharomyces cerevisiae* under osmotic stress, *Microbiology*, 144, 671–680.

Hubalek, Z. (1996) *Cryopreservation of Microorganisms at Ultra-Low Temperatures,* Academy of Sciences of the Czech Republic Brno, Academia, Praha.

Hubalek, Z. and Kocková-Kratochvílová, A. (1982) Long-term preservation of yeast cultures in liquid nitrogen, *Folia Microbiol.*, 27, 242–244.

Hwang, S.-W. (1968) Investigation of ultra-low temperature for fungal cultures. I. An evaluation of liquid-nitrogen storage for preservation of selected fungal cultures, *Mycologia*, 60, 613–621.

Hwang, S.-W. and Howells, A. (1968) Investigation of ultra-low temperature for fungal cultures. II. Cryoprotection afforded by glycerol and dimethyl sulfoxide to 8 selected fungal cultures, *Mycologia*, 60, 622–626.

Hwang, S.-W., Kwolek, W., and Haynes, W.C. (1976) Investigation of ultralow temperature for fungal cultures. III. Viability and growth rate of mycelial cultures following cryogenic storage, *Mycologia*, 68, 377–387.

Iijima, T. and Sakane, T. (1973) Method for preservation of bacteria and bacteriophages by drying *in vacuo*, *IFO Res. Comm.*, 6, 4–17.

Ito, T. (1991) Frozen storage of fungal cultures deposited in the IFO culture collection, *IFO Res. Comm.*, 15, 119–128.

Ito, T. and Nakagiri, A. (1996) Viability of frozen cultures of Basidiomycetes after fifteen-year storage, *Microbiol. Cult. Coll.*, 12, 67–78.

Jennings, T.A. (1999) *Lyophilization. Introduction and Basic Principles*, Interpharm Press, Denver, CO.

Kates, M. and Paradis, M. (1973) Phospholipid desaturation in *Candida lipolytica* as a function of temperature and growth, *Can. J. Biochem.*, 51, 184–197.

Kaul, S., Obuchi, K., Iwahashi, H., and Komatsu, Y. (1992a) Cryoprotection provided by heat shock treatment in *Saccharomyces cerevisiae*, *Cell. Mol. Biol.*, 38, 135–143.

Kaul, S.C., Obuchi, K., and Komatsu, Y. (1992b) Cold shock response of yeast cells, induction of 33 kDA protein and protection against freezing injury, *Cell Mol. Biol.*, 38, 553–559.

Kirsop, B. and Henry, J. (1984) Development of a miniaturized cryopreservation method for the maintenance of a wide range of yeasts, *Cryo-Letters* 5, 191–200.

Kirsop, B. and Snell, J.J.S., Eds. (1984) *Maintenance of Microorganisms*, Academic Press, London.

Laere, A. (1989) Trehalose, reserve and/or stress metabolite? *FEMS Microbiol. Rev.*, 63, 201–210.

Leslie, S.B., Israeli, E., Lighthart, B., Crowe, J.H., and Crowe, L.M. (1995) Trehalose and sucrose protect both membranes and proteins in intact bacteria during drying, *Appl. Environ. Microbiol.*, 61, 3592–3597.

Leslie, S.B., Teter, S.A., Crowe, L.M., and Crowe J.H. (1994) Trehalose lowers membrane phase transitions in dry yeast cells, *Biochim. Biophys. Acta*, 1192, 7–13.

Lewis, D.H. and Smith, D.C. (1967) Sugar alcohols (polyols) in fungi and green plants. 1. Distribution, physiology and metabolism, *New Phytol.*, 66, 143–183.

Lewis, J.G., Learmonth, R., and Watson, K. (1994) Cryoprotection of yeast by alcohols during rapid freezing, *Cryobiology*, 31, 193–198.

Lillie, S. and Pringle, J.R. (1980) Reserve carbohydrate metabolism in *Saccharomyces cerevisiae*, responses to nutrient limitation, *J. Bacteriol.*, 143, 1384–1394.

Lindquist, S. and Craig, E.A. (1988) The heat-shock proteins, *Annu. Rev. Genet.*, 22, 631–677.

Lodato, P., Segovia de Huergo, M., and Buera, M.P. (1999) Viability and thermal stability of a strain of *Saccharomyces cerevisiae* freeze-dried in different sugar and polymer matrices, *Appl. Microbiol. Biotechnol.*, 52, 215–220.

Loegering, W.Q., Harmon, D.L., and Clark, W.A. (1966) Storage of urediospores of *Puccinia graminis tritici* in liquid nitrogen, *Plant Dis. Rep.*, 50, 502–506.

Luard, E.J. (1982) Growth and accumulation of solutes by *Phytophthora cinnamomi* and lower fungi in response to changes in external osmotic potential, *J. Gen. Microbiol.*, 128, 2583–2590.

Luard, E.J. and Griffin, D.M. (1981) Effect of water potential on fungal growth and turgor, *Trans. Br. Mycol. Soc.*, 76, 33–40.

Lukiewicz, S. (1972) The biological role of melanin. I. New concepts and methodical approaches, *Folia Histochem. Cytochem.*, 10, 93–108.

MacKenzie, A.P. (1977) Comparative studies on the freeze-drying survival of various bacteria, Gram type, suspending medium, and freezing rate, *Dev. Biol. Standard*, 36, 263–277.

Malik, K.A. (1990) A simplified liquid-drying method for the preservation of microorganisms sensitive to freezing and freeze-drying, *J. Microbiol. Methods*, 12, 125–132.

Malik, K.A. and Hoffmann, P. (1989) Preservation and storage of biotechnologically important microorganisms, *Chimicaoggi*, June, 61–66.

Martin, C., Siegel, D., and Aaronson, L.R. (1981) Effects of temperature acclimation on *Neurospora* phospholipids. Fatty acid desaturation appears to be a key element in modifying phospholipid fluid properties, *Biochim. Biophys. Acta*, 665, 399–407.

Mazur, P. (1970) The freezing of biological systems. The responses of living cells to ice formation are of theoretical interest and practical concern, *Science*, 168, 939–949.

Meryman, H.T. (1966) Freeze-drying, in *Cryobiology*, Meryman, H.T., Ed., Academic Press, London, pp. 610–663.

Mikata, K. and Banno, I. (1987) Preservation of yeast cultures by freezing at –80°C. I. Viability after 2 years storage and the effects of repeated thawing-freezing, *IFO Res. Comm.*, 13, 59–68.

Miller, R.M. and Liberta, A.E. (1976) The effect of light on acid-soluble polysaccharide accumulation in *Sclerotium rolfsii* Sacc., *Can. J. Microbiol.*, 22, 967–970.

Morris, G.J., Ed. (1981) *Cryopreservation. An Introduction to Cryopreservation in Culture Collections*, Institute of Terrestrial Ecology, Cambridge.

Morris, G.J., Coulson, G.E., and Clarke, K.J. (1988) Freezing injury in *Saccharomyces cerevisiae*, the effect of growth conditions, *Cryobiology*, 25, 471–482.

Morris, G.J., Winters, L., Coulson, G., and Clarke, K.J. (1983) Effect of osmotic stress on the ultrastructure and viability of the yeast *Saccharomyces cerevisiae*, *J. Gen. Microbiol.*, 129, 2023–2034.

Neves, M., Jorge, J., Francois, J., and Terenzi H.F. (1991) Effects of heat shock on the level of trehalose and glycogen, and on the induction of thermotolerance in *Neurospora crassa*, *FEBS Lett.*, 283, 19–22.

Nishii, T. and Nakagiri, A. (1991a) Liquid nitrogen storage of oomycetous fungi, examination of cooling rates and improvement of the freezing tube case, *Bull, JFCC*, 7, 90–96.

Nishii, T. and Nakagiri, A. (1991b) Cryopreservation of oomycetous fungi in liquid nitrogen, *IFO Res. Comm.*, 15, 105–118.

Ohmasa, M., Abe, Y., Babasaki, K., Hiraide, M., and Okabe, K. (1992) Preservation of cultures of mushrooms by freezing, *Trans. Mycol. Soc. Jpn.*, 33, 467–479.

Pfyffer, G., Pfyffer, B., and Rast, D.M. (1986) The polyol pattern, chemotaxonomy, and phylogeny of the fungi, *Sydowia*, 39, 160–201.

Plesofsky-Vig, N. and Brambl, R. (1993) Heat shock proteins in fungi, in *Stress Tolerance of Fungi*, Jennings, D.H., Ed., Marcel Dekker, New York, pp. 45–68.

Pouleur, S., Richard, C., Martin, J.-G., and Antoun, H. (1992) Ice nucleation activity in *Fusarium acuminatum* and *Fusarium avenaceum*, *Appl. Environ. Microbiol.*, 58, 2960–2964.

Prasad, R. (1985) Lipids in the structure and function of yeast membrane, *Adv. Lipid Res.*, 21, 187–242.

Rall, W.F. and Fahy, G. (1985) Ice-free cryopreservation of mouse embryos by vitrification, *Nature (Lond.)*, 313, 573–575.

Rey, L. and May, J.C., Eds. (1999) *Serie Drugs and the Pharmaceutical Sciences*, Vol. 96. *Freeze-Drying/Lyophilization of Pharmaceutical and Biological Products*, Marcel Dekker, New York.

Roquebert, M.F. and Bury, E. (1993) Effect of freezing and thawing on cell membranes of *Lentinus edodes*, the shiitake mushroom, *World J. Microbiol. Biotechnol.*, 9, 641–647.

Sakane, T., Banno, I., and Iijima, T. (1983) Compounds protecting l-dried cultures from mutation, *IFO Res. Comm.*, 11, 14–24.

Sarkar, J.M., Hennebert, G., and Mayaudon, J. (1985) Optimization and characterization of an extracellular polysaccharide produced by *Glomerella cingulata*, *Biotechnol. Lett.*, 7, 631–636.

Seufert, W. and Jentsch, S. (1990) Ubiquitin-conjugating enzymes UBC4 and UBC5 mediate selective degradation of short-lived and abnormal proteins, *EMBO J.*, 9, 543–550.

Singer, M. and Lindquist, S. (1998a) Multiple effects of trehalose on protein folding *in vitro* and *in vivo*, *Mol. Cell*, 1, 639–648.

Singer, M. and Lindquist, S. (1998b) Thermotolerance in *Saccharomyces cerevisiae*, the yin and yang of trehalose, *Tibtech*, 16, 460–468.

Smentek, P. and Windisch, S. (1982) Zur Frage des Überlebens von Hefestammen unter flüssigem Stickstoff, *Zbl. Bakt. I Orig. C*, 3, 432–439.

Smith, D. (1982) Liquid nitrogen storage of fungi, *Trans. Br. Mycol. Soc.*, 79, 415–421.

Smith, D. (1983) Cryoprotectants and the cryopreservation of fungi, *Trans. Br. Mycol. Soc.*, 80, 360–363.

Smith, D., Coulson, G., and Morris, G.J. (1986) A comparative study of the morphology and viability of hyphae of *Penicillium expansum* and *Phytophthora nicotianae* during freezing and thawing, *J. Gen. Microbiol.*, 132, 2013–2021.

Smith, D. and Onions, A.H., Eds. (1980) *The Preservation and Maintenance of Living Fungi*, Commonwealth Mycological Institute, Kew, U.K.

Smith, D., Tintigner, N., Hennebert, G., Bievre, C., De Roquebert, M.F and Stalpers J.A. (1990) Improvement of preservation techniques for fungi of biotechnological importance, in *Biotechnology in the EC. II. Detailed Final Report of BAP Contractors*, Vassarotti, A. and Magnien, E., Eds., Elsevier, Paris.

Stalpers, J., Hoog, A., and Vlug, I.J. (1987) Improvement of the straw technique for the preservation of fungi in liquid nitrogen, *Mycologia*, 79, 82–89.

Steiner, W., Lafferty, R.M., Gomes, I., and Esterbauer, H. (1987) Studies on a wild strain of *Schizophyllum commune*, cellulase and xylanase production and formation of the extracellular polysaccharide schizophyllan, *Biotechnol. Bioeng.*, 30, 169–178.

Tan, C.S., van Ingen, C.W., van den Berg, C., and Stalpers, J.A. (1998) Long-term preservation of fungi, *Cryo-Letters*, 19, 15–22.

Tan, C.S., van Ingen, C.W., and Stalpers, J.A. (1991a) Freeze-drying of fungal hyphae and stability of the product, in *Genetics and Breeding of Agaricus*, Griensven, L.J.L.D., Ed., Pudoc, Wageningen, pp. 25–30.

Tan, C.S., van Ingen, C.W., and Stalpers, J.A. (1994) Microscopical observations on the influence of the cooling rate during freeze-drying of conidia, *Mycologia,* 86, 281–289.

Tan, C.S., van Ingen, C.W., Talsma, H., van Miltenburg, J.C., Steffensen, C., Vlug, I., and Stalpers, J.A. (1995) Freeze-drying of fungi, influence of composition and glass transition temperature of the protectant, *Cryobiology,* 32, 60–67.

Tan, C.S. and Stalpers, J.A. (1996) Vitrification of fungi, in *Biodiversity. International Biodiversity Seminar ECCO XIV*, Cimerman, A. and Gunde-Cimerman, N., Eds., National Institute of Chemistry, Ljubljana, Slovenia, pp. 189–191.

Tan, C.S., Stalpers, J.A., and van Ingen, C.W. (1991b) Freeze-drying of fungal hyphae, *Mycologia*, 83, 654–657.

Tanaka,Y., Shimada, S., and Saito, H.P. (1982) Freezing damage of yeast on frozen dough, in *Kobo no zoshoku to riyo*, *Reports of the 5th Kobe Godo Shinpojumu, Tokyo*, Minoda, Y., Ed., Japan Scientific Societies Press, Tokyo, pp. 211–230.

Thevelein, J.M. (1984) Regulation of trehalose mobilization in fungi, *Microbiol. Rev.*, 48, 42–59.

Vassilev, A., Plesofsky-Vig, N., and Brambl, R. (1992) Isolation, partial amino acid sequence, and cellular distribution of heat shock protein hsp98 from *Neurospora crassa*, *Biochim. Biophys. Acta*, 1156, 1–6.

Virgilio, C.D., Simmen, U., Hottiger, T., Boller, T., and Wiemken, A. (1990) Heat shock induces enzymes of trehalose metabolism, trehalose accumulation, and thermotolerance in *Schizosaccharomyces pombe*, even in the presence of cycloheximide, *FEBS Lett.*, 273, 107–110.

Volkov, V.Y. (1994) Physiological and physicochemical mechanisms of bacterial resistance to freezing and drying, *Microbiology*, 63, 1–7.

von Arx, J. A. and Schipper, M.A. (1978) The CBS fungus collection, *Adv. Appl. Microbiol.*, 24, 215–236.

Wassef, M.K. (1977) Fungal lipids, *Adv. Lipid Res.*, 15, 159–232.

Wiemken, A. (1990) Trehalose in yeast, stress protectant rather than reserve carbohydrate, *Antonie van Leeuwenhoek*, 58, 209–217.

Xu, D.-Y. and Huang, H. (1987) Storage of *Phytophthora* cultures in liquid nitrogen ultra low temperature, *Acta Mycol. Sinica*, 6, 103–109.

Yokoyama, T. and Ito, T. (1984) Long-term preservation of fungal cultures, *Jpn. J. Freezing Drying*, 30, 65–67.

Zhz, L.-Z. and Li, Z.-Q. (1987) The efficacy of liquid nitrogen freezing method for preservation of Mucorales cultures, *Acta Mycol. Sinica*, 6, 46–50.

9 Cryoconserving Algal and Plant Diversity: Historical Perspectives and Future Challenges

Erica E. Benson

CONTENTS

0-415-24700-4/04/$0.00+$1.50
© 2004 by CRC Press LLC

And now there came both mist and snow,
And it grew wondrous cold:
And ice, mast-high, came floating by,
As green as emerald

Excerpt from the
Rime of the Ancient Mariner
by Samuel T. Coleridge
(Holmes, 1996)

9.1 INTRODUCTION

Ice has had a major role in shaping the greening of our Earth; this is because it has greatly influenced the geographical distribution of plant and algal diversity. Geological and archeological records show that past plant extinctions occurred as part of natural evolutionary processes, especially during episodes of glaciation ("ice ages") or as part of natural periods of long-term climatic change. One of the coldest places on Earth, Antarctica, has a limited flora, whereas the tropical regions, not having experienced the limiting onslaughts of frequent and episodic ice ages, are "hot spots" of megadiversity.

One of the most exciting challenges faced by plant cryobiologists today is developing cryopreservation methods that can be applied to species that have never experienced cold or desiccation, much less freezing, in their normal life cycles. Cryopreserving the germplasm of tropical rain forest trees in liquid nitrogen at −196°C is very, very difficult, but it has proved possible, demonstrating exciting progress in plant cryobiological research over the last decade or so. Indeed, it appears that on occasions we have achieved the "impossible" regarding what we can plunge into cryogenic storage and hope to recover afterwards.

In contrast to medical and animal cryobiologists, plant cryoconservationists have a challenging task regarding the vast genetic diversity of the organisms they need to preserve. This is a major problem, as genotype, even within the same species, influences survival after cryogenic storage. Unraveling the molecular basis (see Chapter 5) of plant responses to the frozen state will have an increasingly important role in helping meet this challenge. Looking to the future, the tasks of contemporary plant cryoconservationists are to enable the continued stewardship of native plant diversity and to safeguard the genetic resources of our economically important food and utility crop plants and algae. Cryopreservation has a major part in this, and it is insightful to take pause and look back at the field's progression to overcome our present and future conservation challenges. This may best be undertaken not just in terms of the development of methodologies but also in terms of broader issues that relate to the international use of cryopreservation in global genebanks. This chapter will therefore integrate several historical threads. The first is the acquisition of fundamental knowledge of freezing and associated stresses, as linked to the early discovery of plant cryoprotectants. The second concerns the use of this knowledge (intercalated with that from animal and medical cryobiology) to develop cryopreservation methods for plant and algal cells, tissues, and organs. The third appraises the broader issues of international germplasm preservation and the practical need to develop complementary (to *in situ* conservation) and safe *ex situ* cryopreservation technologies for the long-term preservation of Earth's "phytodiversity." Consideration will also be given to early pioneers who have made a major impact on the field of plant and algal conservation. This historical account spans some 100 years or so and may be viewed in conjunction with the historical perspective of Leibo (see Chapter 11), which narrates the development of animal gamete cryopreservation.

Although this chapter may be of interest to plant and algal cryoconservationists, it has been written with the field newcomer in mind in the hope that it will interest cryobiologists from other disciplines. There is a friendly rivalry between the "plant," "animal," and "medical" camps of cryobiology, especially concerning "historical progress." This engenders delightfully stimulating discussions, and as a result, interdisciplinary cross-talk has supported the progression of our small, but influential, field of cryobiological research. In addition, it is imperative that we do not reinvent "cryobiological wheels." Contemporary researchers use electronic databases, and seldom do these virtual libraries travel back in time to the original and "ancient" research paper. This has ramifications: first, the work of our field's pioneers is not appropriately acknowledged and is, at best, "diluted" by secondary citation. Second, we can gain great insights into reading the original papers of our early founders. Explorations into past literature can revitalize the significance of early research findings and bring added value to newly acquired knowledge. Milestone contributions of selected, early conservationists and stress physiologists paved the way for contemporary cryobiologists, and their work will therefore be highlighted. More recently, many individuals, research groups, and organizations have made excellent progress in plant conservation, particularly regarding tropical and recalcitrant species. It is not possible to acknowledge all, but their importance in the more recent developments of the field does not go unrecognized. Therefore, although this review provides some historical account of "scientific progress" as "methods and techniques" development, its main objective is to articulate timelines of progress with respect to the present and future challenges of phytobiodiversity conservation.

9.2 WHY CRYOCONSERVE ALGAE AND PLANTS? THE CONSEQUENCES OF EXTINCTION

Plant extinctions are increasing, and conservationists consider that we are now entering the worst episode of species erosion faced in 65 million years. This is the result of human activities such as climate change, pollution, habitat erosion/destruction, and the introduction of alien species. It is the rate at which these changes are occurring that is problematic, as there is not sufficient time for the natural evolutionary processes of selection and genetic adaptation to take place.

Plants and algae are particularly vulnerable to habitat disturbance and climate change; life cycles and reproductive processes, particularly in temperate plants, are dependent on environmental cues, especially temperature and day length. Episodic environmental impacts trigger life-cycle changes, and low temperatures are very important in regulating plant dormancy, rejuvenation, senescence, and reproduction and the production of spores of algae. Flowering, seed production, and vegetative growth can be detrimentally influenced by even small changes in micro- and macroclimate. Thus, the whole of a plant's life cycle and reproductive capacity can be either directly or indirectly compromised by disturbances in environmental cues, with temperature being one of the most important. Plants and algae are the foundation for all other life forms on our planet; they are autotrophic, and through photosynthesis, they form complex sugars. A remarkable biochemical fact that endorses the importance of plant metabolism is the abundance of the CO_2 fixing enzyme, ribulose-1-5-bisphosphate carboxylase/oxygenase: It is one of the most complex and largest (480,000 daltons) enzymes ever known, and it is the most abundant protein on Earth. Plants and algae "split" H_2O during photosynthesis, and O_2 is released into the atmosphere, providing aerobes with the substrate for oxidative metabolism and respiration. In doing so, phytobiotic organisms provide the "food," "energy," and utility products for almost all other life on Earth. Increasingly, therefore, cryogenic storage (Day and McLellan, 1995), interfaced with *in vitro* conservation and biotechnological manipulations (Benson, 1999a,b; Grout, 1995), is being applied to preserve the genetic resources of crop plants, their wild relatives, forestry species (International Foundation for Science [IFS], 1998), and endangered plants (Pence, 1999). Cryopreservation has also been applied to maintain the unique biosynthetic properties of medicinal plants (Yoshimatsu et al., 2000) and their biotechnological derivatives (Benson and Hamill, 1991). The *in situ* (Maxted et al., 1997) and *ex situ* (Bajaj, 1995a, 1995b; Benson, 1999; Callow et al., 1997; Razdan and Cocking, 1997) conservation of plant and algal genetic resources thus ensures the potential for the future and sustainable exploitation of the Earth's phytobiota (Brown et al., 1989; McCormick and Cairns, 1994). One of the most important applications of plant cryopreservation is, therefore, in the support of crop plant, forestry, and agroforesty breeding programs (Bajaj, 1985a, 1985b; IFS, 1998; Ramanatha and Hodgkin, 2002).

9.3 CHARTING THE HISTORY OF PLANT CRYOPROTECTION

The development of cryopreservation protocols for the conservation of plants and algae has progressed through the use of empirical, applied, and fundamental research and, at times, serendipitous discovery. Historically, multidisciplinary, fundamental research has played an important role in the development of plant cryoprotectants. There exist few examples of other scientific disciplines in which this has occurred to the extent that it has in cryobiology, a field in which contemporary researchers engage in investigations that transcend the disciplinary boundaries of medical, plant, and animal cryobiology.

9.3.1 FOLLOWING IN THE FOOTSTEPS OF NIKOLAY MAXIMOV: DISCOVERING "CRYOBIOLOGICAL SUGARS"

Nikolay Maximov (1880–1952), a Russian scientist from St. Petersburg University, was one of the first researchers to undertake research into the effects of freezing on plants and to predict the importance of sugars and glycerol in cryoprotection. His work impinges on several threads of contemporary cryobiology, including postulating the mechanism of freezing injury. This part of the early history of plant cryobiology is often overlooked, and tribute should be given to the exceptional academic insights of early researchers. By today's standards they employed technologically simple, but well-executed, experimental systems to ascertain the effects of freezing on biological systems. Their studies, often forgotten, predate by many years the revelations of more contemporary researchers.

The discovery (see Chapter 11) of the cryoprotective properties of glycerol as applied to fowl spermatozoa is a milestone achievement in cryobiological research and is presented in the classic paper of Polge et al. (1949), but it was Maximov who first described the protective effects of glycerol in plants exposed to freezing temperatures (Maximov, 1912a, 1912b, 1912c). An excellent history of Maximov's seminal works and that of other key, early cryobiologists is presented in a paper collated by Diller (1997) and in Levitt's books (1941, 1956; and see Chapter 6).

In 1912, Maximov published (1912a, 1912b, 1912c) work from his Master's thesis, which was undertaken at St. Petersberg University. As part of his postgraduate studies, Maximov evaluated the work of Lidforss (1907), who put forward a farsighted (in terms of contemporary research) hypothesis regarding one of the most important groups of plant cryoprotectants: sugars. Working on evergreen plants, Lidforss showed that they were able to change their chemical composition, turning starch to sugar in the winter, and he proposed that the sugar was acting as a protective agent. He then went on to confirm the cryoprotective effects of sugars using *in vitro* experiments. Maximov presented the first theory of cryoprotection, which he stated as "the chemical defence of the plant against death by frost" (as reviewed by Diller, 1997 in a biographical profile), in his master's thesis. Maximov postulated that plants were killed at freezing temperatures by the accumulation of ice crystals between cells, which dehydrate and mechanically damage the cell, leading to the coagulation of the cell solutes. To test his theories, Maximov used red cabbage for his experimental system and applied glycerol, glucose, mannitol, ethyl alcohol, and methyl alcohol to tissues that had been exposed to freezing stress (within the range −5.8° to −32°C). A level of concentration-dependent cryoprotection against freezing injury (noting that noncryoprotected red cabbage released red pigments and was lethally damaged at −5° to −7°C) was afforded by each additive. Thus, Maximov identified, for the first time, the effects of a number of chemicals that would eventually become important cryoprotectants for contemporary plant cryoconservationists. These early investigations support the later studies of plant cryoprotection as reviewed by Finkle et al. (1985) and the two-factor theory of freezing injury (Meryman and Williams, 1985; and see this volume's Foreword).

Two research areas influenced the next stage of development of plant cryoprotection research. The first involved the cross-discipline interest of researchers working in medical, animal, and plant cryobiology and the finding that several chemical additives and natural compounds have cryoprotective properties. Following the studies on the protective effects of glycerol (Polge et al., 1949), Lovelock and Bishop (1959) demonstrated the cryoprotective effect of Me_2SO on both plant and animal tissues. This is one of the most potent and effective cryoprotective additives known, largely because of its highly penetrating and hence colligative properties (Finkle et al., 1985). An inventory of cryoprotective compounds, which impart freezing resistance in plant cells, was subsequently compiled by Sakai and Yoshida (1968).

The second research area pertains to studies that explored adaptive life-cycle responses in relation to temperature fluctuations, freezing, and cold hardiness (see Chapter 5). Sakai (1956, 1960) first examined the survival of winter-hardy mulberry twigs slowly frozen to different terminal temperatures, held for 16 h, and plunged into liquid nitrogen. Survival gradually increased with decreasing temperature, reaching the maximum at –30°C. Sakai proposed that there is a critical temperature range over which freezable water is extracted from cells by equilibrium freezing. On exposure to freezing, the cells are able survive because of vitrification of the remaining water. Subsequent milestone studies (Sakai, 1965, 1973; see also Chapter 10) showed that survival after freezing could be correlated to the relative cold hardiness of plants. This knowledge has been more recently applied to enhance the ability of cryopreserved plants to survive cryogenic storage by applying cold-hardening treatment before freezing (Reed, 1988).

Contemporary cryopreservation research has made good use of the fact that the sugars are "natural" plant cryoprotectants. Although trehalose, a sugar produced in naturally cold- and desiccation-tolerant organisms (Ring and Danks, 1998) has been applied in cryogenic plant cell storage (see Chapters 3 and 21), sucrose has proved to play a major role (Engelmann, 1997) in the acquisition of desiccation- and cold-tolerance in plants that are cryopreserved. Significant pioneering work on the application of sugars (sucrose and glucose) for the improvement of desiccation, and hence cryopreservation, tolerance was undertaken on somatic embryos of oil palm by Dumet et al. (1993) and on mature and immature zygotic embryos of coconut by Assy-Bah and Engelmann, (1992a, 1992b), respectively. Not surprisingly, coconut seeds are recalcitrant, and the cryopreservation of their zygotic embryos offers the best option for long-term storage. In the case of mature embryos, desiccation in a sterile airflow, with the culture on medium containing 600 g/L glucose and 15% (w/v) glycerol, and with rapid freeze/thawing achieved recovery of up to 93% (dependent on variety). Immature embryos of coconut were cryopreserved using rapid methods after a 4-h pregrowth period on glucose, glycerol, or sorbitol, with survival levels reaching 43%. Thermal analysis of oil palm embryos showed that their resistance to cryopreservation increased when treated with high loadings of sucrose and that this was caused by the formation of a vitrified rather than a frozen state. Fundamental studies undertaken by Jitsuyama et al. (2002) on embryogenic cell suspension cultures of asparagus showed that incubation in 0.8 M sucrose increased freezing tolerance. Microscopy revealed that sugar incubation induced plasmolysis in the cells as well as changes in the layering of the rough endoplasmic reticulum. Immunoblotting analysis with antidehydrin antiserum showed that a dehydrin-like protein (see Chapter 5) appeared, but only when maximal freezing tolerance was induced by exposure to sucrose. Thierry et al. (1997) evaluated the osmotic role of sucrose in the acquisition of cryogenic tolerance in carrot somatic embryos by replacing the sugar with a range of cryoprotective additives of the same osmotic pressure and concluded that the sucrose protection is to the result of osmotic as well as possible metabolic changes. To date, there are many examples, and increasingly so, of the application of sucrose as a pregrowth treatment for the cryopreservation of plant germplasm. Although the mode of action appears to be osmotic, it may also have a role in the stabilization of glasses. Clearly there is also emerging evidence (Jitsuyama et al., 2002) that the sugar has important biochemical properties that may also afford protection.

9.3.2 Traditional Cryopreservation (Cryopreservation in the Presence of Ice)

> *The ice was here, the ice was there.*
> *The ice was all around:*
> *It cracked and growled, and roared and howled,*
> *Like noises in a swound!*

> **Excerpt from the**
> ***Rime of the Ancient Mariner***
> **by Samuel T. Coleridge**
> **(Holmes, 1996)**

Ice, somewhat surprisingly, has a "protective" role in traditional cryopreservation, but its formation has to be manipulated very carefully. Key factors in traditional cryoprotection are the application of colligative cryoprotectants, controlling the rate of cooling, and manipulating the dynamics of equilibrium freezing. Extracellular ice creates a water vapor deficit between the inside and the outside of cells, and freezable water is withdrawn from the cell. Penetrating colligative cryoprotectants protect the cell from the damaging solution effects that this migration of water causes. The presentation of the two-factor (ice and dehydration; see Chapter 1) hypothesis of freezing injury in the context of plant cell cryopreservation (Meryman and Williams, 1985) and the discovery of the cryoprotective effects of glycerol (Polge et al., 1949) and Me_2SO (Lovelock and Bishop, 1959) in medical and animal cryobiology stimulated great interest in plant cryopreservation research as applied to crop plants (Bajaj, 1976, 1979). The decades of 1970–1990 enjoyed an increased application of controlled-rate cooling methods to the cryopreservation of plant cell cultures (Kartha, 1985a, 1985b; Withers, 1975, 1977, 1978, 1979, 1985). In parallel to applied studies, Steponkus (1984, 1985) used plant protoplasts as a model system for fundamental research to investigate the effects of cryoinjury on the plasma membrane. This was with a view to developing improved cryopreservation protocols. Steponkus identified the fact that deleterious alterations to the semi-permeable characteristics of membranes had injurious consequences including intracellular ice formation, loss of osmotic responsiveness during cooling, and expansion-induced lysis during warming. Steponkus (1985) put forward the theory that the latter process occurred because freezing and thawing caused large areas of deformation in the plasma membrane. He further postulated that this could be manipulated by the addition of chemical additives and by cold acclimation that would effectively increase tolerance to osmotically induced contractions.

Traditional plant cryopreservation involves the application of one or more chemical cryoprotective additives categorized as either nonpenetrating or penetrating. The former (e.g., sucrose, poly-alcohols) withdraw water osmotically from the cell, thus reducing the amount of water available for ice formation. Penetrating cryoprotectants, (usually Me_2SO) offer colligative cryoprotection and are applied in combination with controlled rate freezing. The temperature of the system is reduced slowly, usually within the range of $-0.1°$ to $-5°C/min$, to a terminal transfer temperature of approximately $-40°C$, where it is held to ensure that equilibrium freezing ensues before plunging it into liquid nitrogen. The incorporation of pregrowth additives that dehydrate the cells (mannitol, sorbitol) and pretreatment steps that simulate cold hardening, or the use of additives known to enhance stress tolerance such as proline, are also incorporated into standard protocols for both dedifferentiated cells and organized structures (Benson, 1994, 1995).

One of the most widely applicable "traditional" plant cell cryopreservation protocols developed was that of Withers and King (1980). This basic protocol has been modified (Benson, 1994, 1995) and optimized for a number of plant cell systems. The Withers and King method uses a simple controlled-rate freezing unit comprising a methanol bath that allows solvent temperature reduction to be programmed. Cells are first pretreated with mannitol, which osmotically removes water before the cryoprotection *per se*. They are then cryoproptected in a mixture of Me_2SO, glycerol, and sucrose or Me$_2$SO, glycerol, and proline, and then cooled at a rate of –1°C/min to a terminal transfer temperature of –30°C, where they are held for 30–40 min before plunging into liquid nitrogen at –196°C. Today, computer-controlled programmable freezers are used to control the cooling rate of plant, cells, tissues, and organs being cryopreserved using the traditional approach. This approach to cryopreservation has since been applied in a wide range of research sectors, one of the first being plant biotechnology.

The modest carrot (*Daucus carota*) could be considered the plant cryobiologist's guinea pig, as it has played such an important role in the early development of cryopreservation protocols (Nag and Street, 1973, 1975), which would later be applied in the biotechnology industries. Dougall and Wetherell (1974) successfully used controlled-rate freezing on 22 cell lines of wild carrot cryoprotected with 5% (v/v) Me$_2$SO and recovered the cells after 7–12 months in liquid nitrogen. Parallel studies by Nag and Street (1973, 1975) explored the application of controlled-rate freezing on carrot, sycamore, and belladonna cell cultures. Using 5% (v/v) Me$_2$SO as a cryoprotectant, they discovered that the rates of cooling (2°C/min) and thawing critically influenced cell survival after transfer to cryogenic storage and that fast thawing (120 to 80°C/min over the range –50° to –10°C) was particularly important. The cultures, once placed in liquid nitrogen, did not show a decline in survival after a 10-month storage period. Fine structural studies on carrot and sycamore were subsequently undertaken by Withers and Davey (1978), confirming the importance of freezing rate to cell survival. Slow freezing in the presence of cryoprotectants was essential to limit intracellular ice formation. Postthaw cryoinjury was assessed by Withers (1978) in carrot and sycamore cells cryoprotected with Me$_2$SO and glycerol exposed to ultra-rapid (>100°C/sec) and controlled-rate (–1 to –2°C/min) freezing, Slow cooling resulted in 70% survival in carrot; however, fine structural studies revealed heterogeneity in the survival and damage of different areas of the cell masses.

Withers and Street (1977) examined in greater detail the effects of prefreeze status of cells on the capacity to survive cryogenic storage. In the case of carrot cells, the stage of growth of the cell cycle of the suspensions in batch culture was important to survival. Highest rates occurred in cells taken from the lag or early exponential phase, and an appropriate inoculum density at the start of freezing was important to ensure recovery. Thus, only a proportion of the frozen cells survived freezing, and this was influenced by their prefreeze physiological status.

The foundation research performed on carrot in the 1970s heralded a rapid expansion in the use of *in vitro* plant cell cultures as sources of secondary metabolites including pharmacologically active compounds. It soon became apparent that cryogenic storage was an important component of cell line patenting procedures. Examples of the application of traditional cryopreservation protocols to preserve secondary product-producing cell lines include the alkaloid-producing *Catharanthus roseus* (Chen et al., 1984); *Panax ginseng* and *Dioscorea deltoidea* (Butenko et al., 1984) and *Digitalis lanata* (Diettrich et al., 1986; Seitz et al., 1983). In the decades that followed, industrially important, secondary metabolite generating plant cells were cryopreserved (Schumacher, 1999), including those produced using transgenic technologies (Benson and Hamill, 1991; Yoshimatsu et al., 2000).

Biotechnological applications of cryopreservation expanded in the 1980s–1990s to support new research programs aimed at applying genetic manipulation technologies for the improvement of world cereal crops such as rice, wheat, barley, and maize (Bajaj, 1995a,b; Benson and Lynch, 1999; Lynch et al., 1994). Cell suspension and callus cultures of cereals were produced for transformation and protoplast studies. However, the maintenance of their totipotent capacity through somatic embryogenesis was pivotal to the success of the biotechnology projects. Unfortunately, cereal cell cultures have a propensity to lose their ability to produce embryos within a relatively short time frame. Cryopreservation of totipotent cultures thus became an essential component of cereal biotechnology programs (Bajaj, 1995a). Cryogenic storage has also been used for the preservation of cell lines from the plant genome project species *Arabidopsis thaliana* (Ribeiro et al., 1996). Because of its small genome and generation time, *A. thaliana* became the subject of choice to pioneer plant genome projects. Cell suspension cultures of *A. thaliana* are used to produce mutant lines, and cryopreservation provides an excellent means of securing stable mutant collections.

Traditional cryopreservation protocols have especially proved their worth for the cryopreservation of dedifferentiated plant cultures. The carrot studies of Withers (1979) concluded by developing a method for the cryogenic storage of somatic embryos and clonal plantlets. Protocols using controlled-rate freezing have also been applied with success to organized plant structures such as meristems (Benson, 1995). Once initial capital costs are met, modern-day controlled-rate, programmable freezers are cost and labor efficient, allowing the preservation of large batches of samples in one time. This is a great advantage for large-scale germplasm repositories, which have to process hundreds of accessions. However, traditional approaches do have limitations, and as the field progressed, it soon became apparent that the preservation of difficult-to-freeze (recalcitrant) germplasm, and especially that from tropical species, required a different approach. The start of a new decade (1990s) announced the beginning of the plant "vitrification revolution," for which carrot continues to have an important role as plant biotechnology's "model system;" and in 2003, Chen and Wang reported for the first time the cryogenic storage of *D. carota* cell suspensions and protoplasts using PVS2 vitrification.

9.3.3 THE PLANT VITRIFICATION REVOLUTION

Water, water, every where,
And all the boards did shrink;
Water, water, everywhere,
Nor any drop to drink

Excerpt from the
Rime of the Ancient Mariner
by Samuel T. Coleridge
(Holmes, 1996)

To understand cryobiology, you have to understand water: how it behaves, how it interacts with cryoprotectants and cell solutes, and how it changes state on exposure to the thermal cycles of freezing and thawing (Franks, 1982, 1985; also see Chapters 1 and 2). Controlling cellular hydration and the responses of living cells to different states of water are very important determining factors for cryogenic survival. However, this becomes very complicated when, for example, dealing with the highly hydrated germplasm of a tropical plant species that has never encountered desiccation, chilling, or freezing stress; or dealing with the structurally complex, coenocytic structures of algae that have proved recalcitrant to traditional cryopreservation methods. Perhaps one factor in this problem may be that cellular heterogeneity is so great that cells possess a complex mosaic of different hydration states. Plant cells have vacuoles, and the size and distribution of these watery oases can be very different in different cell types and in cells of different ages. It can be difficult to optimize cryogenic parameters on the basis of water state, osmotic dehydration, freezing equilibrium, and colligative cryoprotection for complex and highly hydrated cells (e.g., tropical plants or algae).

As plant cryopreservation research progressed in the 1980s, it became apparent that traditional approaches could not be applied to difficult-to-preserve, desiccation- and cold-sensitive plant and algal germplasm. Importantly, understanding the behavior of water in biological systems is a complex issue requiring physical and chemical, as well as biological, know-how (Franks, 1982, 1985). A different means of protecting cells was required, and solving the problem of water turning into ice had to be the key.

Cryobiological vitrification is the process by which water forms an amorphous, metastable glass (see Chapters 10 and 22) instead of ice. It occurs when cellular viscosity reaches a critical "high" such that water molecules are unable to interact and form crystals. In the case of traditional cryopreservation, the aqueous intracellular domain becomes concentrated as water is withdrawn into the frozen extracellular milieu. On exposure to liquid nitrogen temperatures, the intracellular aqueous solution is so concentrated that the residual water forms a glass characterized by a transition temperature (T_g). In this scenario vitrification occurs, but not in the absence of ice, as extracellular freezing is required to initiate the process. In contrast, however, cryopreservation in the absence of ice can occur if the cell solute composition is manipulated such that both extra- and intracellular compartments become vitrified. This can be achieved by the application of treatments that combine osmotically dehydrating treatments, evaporative and chemical (silica gel) desiccants, and the loading of cells with high concentrations of penetrating cryoprotectants such as Me_2SO. This may be considered "ice-free" cryopreservation, although the stability of the amorphous, metastable glasses must be taken into consideration, especially on rewarming (see Chapter 22).

The plant vitrification revolution was encouraged by research ongoing in the field of mammalian cryobiology (Fahy and Fahy, 1985). In 1985, Rall and Fahy reported a new method of vitrification for cryopreserving mouse embryos. It involved the application of very high concentrations (up to 8 *M*) of cryoprotective chemicals followed by ultra-rapid cooling to $-196°C$. Taking a parallel approach through the treatment of plant suspension cultures with highly concentrated chemical additives, Langis et al. (1989) and Urugami et al. (1989) recovered plant cells and embryos from cryogenic storage, using vitrification. Several key "milestone" papers concerning plant vitrification were subsequently published in 1990. Sakai and colleagues (1990) formulated Plant Vitrification Solution 2 (PVS2) comprising 30% (w/v) glycerol, 15% (w/v) ethylene glycol, 15% (w/v) ME2SO, and 0.4 *M* sucrose. This solution has now been applied to a wide range of plant germplasm and species (see Chapter 10). Using chemical cryoprotectant additives, Langis and Steponkus (1990) published their work on the application of vitrification to rye protoplasts, and Towill (1990) also cryopreserved mint shoot-tips using chemical vitrification additives.

Fabre and Dereuddre (1990) presented a novel cryopreservation procedure for potato shoots in which plant germplasm is encapsulated in a Ca^{2+}-alginate matrix, dehydrated osmotically, and then desiccated using evaporative air drying before being directly plunged in liquid nitrogen. This method was similarly applied, with success, to pear shoot-tips (Dereuddre et al., 1990). This approach departs from the application of chemical additives to achieve vitrification and relies on concentrating solutes in the cells and the beads through osmotic dehydration (usually with sucrose) and physical or evaporative desiccation. The following year, carrot once more became a cryogenic guinea pig and was used to assess the effects of different parameters on somatic embryo survival after encapsulation. Dereuddre et al. (1991a, 1991b) studied the sucrose preculture related to the post-cryogenic survival of encapsulated and air-desiccated *D. carota* somatic embryos. In effect these are "synthetic seeds." The embryos could withstand considerable dehydration (to 16%) when precultured in the presence of 0.3 *M* sucrose, and 92% survival was achieved following direct plunging of encapsulated and dried embryos in liquid nitrogen. In a subsequent study, Dereuddre et al. (1991b) applied scanning differential calorimetry (DSC) to encapsulated/dehydrated *D. carota* somatic embryos, finding that survival after exposure to liquid nitrogen was correlated with a T_g between -50 and $-70°C$. Benson et al. (1996) applied DSC to compare the vitrification profiles of *Ribes* shoot-tips cryopreserved using the PVS2 and alginate bead encapsulation/dehydration protocol. Postcryopreservation survival of encapsulated shoots were correlated with a T_g in the region of $-66°$ to $-72°C$, confirming the earlier work of Dereuddre et al. (1991b). Survival of PVS2-treated *Ribes* shoots was associated with a T_g in the region of $-70°$ to $-76°C$. A decade after the first publication of the PVS2 and encapsulation/dehydration methods, Sakai et al. (2000) produced a "hybrid" of the two protocols in which encapsulated plant apices are precultured first with 0.3 *M* sucrose for 16 h and then simultaneously with 0.2 *M* glycerol plus 0.4 *M* sucrose for 1 h. Following silica gel desiccation for 16 h, apices were directly plunged into liquid nitrogen. Successful recovery was achieved for chrysanthemum, mint, and wasabi. To date, Sakai's and Dereuddre's vitrification protocols have been widely applied to many plant and, more recently, algal species (Benson, 1999a, 1999b; Benson et al., 2000; Engelmann and Takagi, 2000; and see Chapter 10).

Looking to the future, understanding the "glassy state" will become increasingly important because we urgently need to expand the application of vitrification protocols to desiccation-sensitive and storage-recalcitrant species (Berjak et al., 1992). To achieve this requires the use of discerning instrumentation such as DSC (Wolanczyk, 1989). Unfortunately, there still remains a paucity of information regarding our basic understanding of T_g formation, cryogenic survival, and the physical stability of cryopreserved plant germplasm held in the glassy state. Learning from the experiences of researchers from other fields (Franks, 1990, 1992; Chapter 22) should be encouraged, as this may well provide plant cryobiologists useful insights as to how we may answer some of our existing questions. Now that the scene has been set as to the early development of plant cryoprotection strategies, we will continue to chart historical progress, but this time in an international context and related to the application of cryopreservation to conserve global plant and algal diversity.

9.4 CRYOPRESERVING NOMADIC ALGAE: THE VOYAGES OF PRINGSHEIM'S CULTURES

The relatively impoverished status attributed to algae and cyanobacteria (as compared to other taxonomic groups) has led to considerable difficulties with respect to the provision of support for their conservation. Furthermore, the efforts of those who have endeavored to protect these key organisms have often been punctuated with difficulties, which at times (particularly during political unrest) necessitated the finding of safe havens for both phycologists and the precious collections held in their stewardship. The history of cryopreserved algal culture collections spans some 50 years and charts a peripatetic and difficult journey commencing in Central Europe during the early 1900s.

Prague is the historical home of both European and U.S. cryopreserved algal culture collections (see http://www.cobra.ac.uk and http://www.gwdg.de). Their establishment, as actively growing *in vitro* cultures, commenced at the start of the twentieth century in the city's prestigious Charles University. However, it was Ernest George Pringsheim (1881–1970) who founded the first stock of 49 cultures in the 1920s with his collaborating colleague Viktor Czurda (1897–1945), and their collection's details were published in 1928 (Pringsheim, 1928). Just before the Nazis occupied Czechoslovakia at the start of World War II, Pringsheim left Prague to take refuge in Cambridge, England; fortunately, he took his precious algal cultures with him (Pringsheim, 1946). In 1947, this part of the collection was taken over and expanded by E.A. George at the University of Cambridge. After the war, cultures from the original collection were subsequently used to establish the Sammlung von Algenkulturen at Göttingen when Pringsheim returned to Germany in 1953. R.C. Starr, who was Pringsheim's colleague at Cambridge, used the "Prague cultures" to initiate the USA Culture Collection of Algae at the University of Texas (Bodas et al., 1995). At the end of the war, the Czech Charles University was restored and S. Prát (who had held the cultures in safety during the war years) returned as Professor of Plant Physiology and saved the collection. In 1952 the Prague collection was taken over by the Biological Institute and subsequently by the Institute of Experimental Botany of the Czechoslovak Academy of Sciences, wherein it became the Culture Collection of Autotrophic Organisms. After several reorganizations and transfers, the collection is now housed in Trebon, in the Czech Republic, under the present-day curatorship of Dr. Jaromir Lukavsky, and cryopreservation is being used for the long-term conservation of this historically unique and precious collection.

Meanwhile, back in Cambridge in 1947, a specialist conference on the Culture Collections of Microorganisms formally established that the algal collection initiated by Pringsheim should become the United Kingdom's National Culture Collection of Algae and Protozoa and was adopted by Cambridge University. In 1970, the U.K. Government's Natural Environment Research Council took over the responsibility for the collection. In 1976, cryopreservation studies under the curatorship of John Morris (Morris, 1978) were initiated. By 1986, the algae and the Culture Collection of Algae and Protozoa were on the move again; this time, the marine cultures went north to the Scottish Marine Biological Association Laboratory at Dunstaffnage, Oban. The freshwater cultures traveled to the Windermere Laboratory of the Freshwater Biological Association in the English Lake District. This collection is presently held under the curatorship of Dr. John Day (Day, 1998), and there are more than 1500 strains of algae, 35% of which are held in cryogenic storage. The main cryoprotective additives used in the cryoprotection of algae are methanol, Me_2SO, and glycerol.

Cryopreservation is now considered the method of choice for the conservation of algae and cyanobacteria. Viability after exposure to $-196°C$ was reported for unicellular Chlorophyceae by Holm-Hansen (1963), using fast freezing and direct plunging into liquid nitrogen. Hwang and Hudock (1971) applied a slow-cooling method to *Chlamydomonas reinhardti* before cryogenic storage. Most of the strains that have been successfully cryopreserved at the Culture Collection of Algae and Protozoa were stored using a two-step protocol developed by John Morris and colleagues (Day, 1998; McLellan, 1989; Morris, 1978; Morris and Canning, 1978). The University of Texas, Austin, has developed an effective cryopreservation method (Bodas et al., 1995) for the storage of their cyanobacterial culture collection, using Me_2SO as the cryoprotectant and a simple Nalgene (Nalgene Co., Rochester, NY) freezer in which the cultures, in cryotubes, are slowly cooled to $-70°C$ before transfer to liquid nitrogen. Using this method, a wide range of cyanobacteria (e.g., including representatives from Chroococcales, Pleurocapsales, Nostocales, and Stigonematales) have been cryopreserved (Bodas et al., 1995). For reviews of the wider application of traditional two-step cryopreservation methods to algae and cyanobacteria see Day (1998) and Taylor and Fletcher (1999).

More recently, vitrification has been explored as an approach to preserve algae that are recalcitrant to traditional storage methods. Cryopreservation in the absence of ice offers great potential for complex, multicellular, and coenocytic algae, and studies of cryoinjury help develop improved conservation methods. For example, cryomicroscopical examination of *Vaucheria sessilis* (Fleck et al., 1997, 1999) demonstrated that extracellular ice crushes the organism's filaments, causing irreversible damage to cell walls and intracellular architechture. As coenocytic algae do not possess cross-walls, any ice formed intracellularly will most likely rapidly travel the length of the filaments, causing even more damage. Fleck et al. (1999) also demonstrated that extracellullar ice can disrupt organelles in *V. sessilis*, and the filamentous diatom *Fragilaria cortonensis* is lethally injured at fast cooling rates by intracellular ice, whereas slow freezing causes osmotic stress (McLellan, 1989). Vitrification may, therefore, provide an important new approach to conserve filamentous alga. Day et al. (1998a, 1998b) and Fleck (1998) report the survival of *Enteromorpha intestinalis* following cryopreservation using PVS2 (see also Chapter 10). However, the thallus fractured during thawing, indicating that glass relaxation and possibly devitrification may take place, but the filament fragments were still capable of regeneration. Rewarming treatments can lead to injurious glass relaxation events, which may best be avoided by optimizing the rate of rewarming through the use of two-step protocols (Day et al., 1998a, 2000; Fleck, 1998). Vigneron et al. (1997) applied encapsulation/dehydration to the gametophytes of the marine alga, and gametophyte survival was within the range of 25 to 75%. Hirata et al. (1996) first applied encapsulation/dehydration to microalgae, using alginate encapsulation/dehydration applied to *Dunaliella tertiolecta* that had previously proved recalcitrant to cryopreservation. Optimization of dehydration was a critical factor in ensuring survival, and the method was subsequently applied by to a wider range of microalgae and Cyanobacteria. Euglenoids are often recalcitrant to cryopreservation (Day et al., 1998a, 1998b, 1999, 2000), and cryoinjury involves flagellum loss, structural damage (Fleck, 1998), free radical–mediated oxidative stress ,and impairment of photosynthesis (Day et al., 2000; Fleck et al., 2000). Fleck (1998) and Day et al. (2000) found that PVS2 was unsuccessful in the cryopreservation of *E. gracilis* because of cryoprotectant toxicity. In comparison, encapsulation/dehydration with or without two-step cooling was applied with some success. Cryopreserved collections of algae are now spread worldwide, and storage in the frozen and vitrified state has been identified as important for their long-term storage and sustainable utilization (search http://www.biologie.uni-hamburg.de; http://www.cobra.ac.uk; Bodas et al., 1995; Cánavate and Lubian, 1995; Day et al., 1997; Day, 1998; Lee and Soldo, 1992; Taylor and Fletcher, 1999).

Concluding the journey of Pringsheim's cultures in the present provides an exciting new future for Prague's dispersed algal collection. Recently, the European Union recognized the need to use cryopreservation to secure Europe's invaluable, yet fractionated, algal collections. In 2000, the European Commission's Framework 5 Quality of Life Program for Research Infrastructures supported a new cryopreservation research project named "COBRA" (http://www.cobra.ac.uk), the aims of which are to build a collaborative research infrastructure using physical and virtual (information technology [IT]) frameworks to help develop cryopreservation methods for the conservation of algae in culture collections across Europe. The project's network embraces key European collections of algae and cyanobacteria and brings together, for the first time, Pringsheim's collections that had been scattered in the 1940s.

9.5 CROP PLANT GERMPLASM CONSERVATION: NIKOLAY VAVILOV'S LEGACY

The effects of Nikolay Ivanovich Vavilov (1887–1943), the "father" of global crop plant diversity preservation (Akeroyd, 2002; Vavilov, 1997) continue to enable and enrich the progress of contemporary cryoconservationists even to the present day. Vavilov founded the Institute of Plant Breeding in Leningrad (now St. Petersburg), and his theories pertaining to the genetics of disease resistance in plants afforded him early national acclaim in Russia. However, he is best known for his visionary international work on the centers of origin of crop plants species and their wild relatives. Most important, he was the first to recognize the need to integrate botanical and ecological field studies with applied agricultural sciences and crop plant breeding. During 1920–1930 he embarked on extensive collecting missions throughout the USSR and to over 50 other countries. Using observational field skills, he recorded and collected plants and plant germplasm from all over the world and established the concept of geographical centers of origin for crop plant diversity. Vavilov put forward (Vavilov, 1997) the elegant theory that the place of origin of a crop was the area wherein it displays the greatest level of genetic diversity. Vavilov defined eight centers of crop plant diversity, which have later been expanded to 12. Interestingly for "life in the frozen state," ice (or its absence) had a very important role in defining these "hot spots" of crop plant diversity, as they are largely found near the equator, in areas left untouched by the disturbances of the ice age. They were defined by Vavilov (Vavilov, 1997) as the Tropical Center, East Asiatic Center, Southwest Asiatic Center (containing the Caucasian Center, the Near East Center, and the Northern Indian Center), Mediterranean Center, Abyssinian Center, Central American Center (containing the mountains of Southern Mexico, the Central American Center, and the West Indian Islands), and the Andean Center. Genetic diversity is the "fuel" that drives the evolutionary process, allowing the phenotypic expression of adaptations underpinned by genetic processes, sexual reproduction, and selection pressures. Even early man cultivated plants in these "centers of diversity," and today these centers may be considered the cradles of modern agriculture. Some of them will be visited later in this chapter, providing examples as to the application of cryopreservation for the conservation of some of our most ancient and important crop plant species.

The Vavilov centers and the work that Vavilov undertook in studying the plants therein constitutes a priceless knowledge resource pertaining to the gene pools of the world's most important food crops. Vavilov must therefore be rightly acclaimed as the most influential instigator of modern plant agriculture and genetic resources conservation. He once wrote to his future wife, Yelena Barulina (Vavilov, 1997, pp. xix–xx):

> I really believe deeply in science; it is my life and the purpose of my life. I do not hesitate to give my life for the smallest bit of science.

Unfortunately, this was to become a prophetic statement, as in 1940 Vavilov was arrested under the auspices of the Stalin regime. This was at a time when his plant collection at Leningrad, in the

Institute of Plant Industry, amounted to 200,000 specimens—the largest in the world. The fact that he enjoyed such great international acclaim became his political and tragic personal undoing. Following his capture by the secret police, Vavilov eventually died in prison of starvation, in 1943. A cruel fate for a man who had the most far-reaching effect on protecting global food crop security and on engendering the need for sustainable crop plant utilization (Plucknett et al., 1987).

9.5.1 CRYOCONSERVATION AND GLOBAL NETWORKS: THE FACILITATING ROLE OF THE CONSULTATIVE GROUP ON INTERNATIONAL AGRICULTURAL RESEARCH AND THE INTERNATIONAL PLANT GENETIC RESOURCES INSTITUTE

Vavilov's work has had a major effect on crop plant conservation, as he highlighted the need to safeguard the genetic integrity of biodiversity unique to regions of crop plant origin. Vavilov's centers are of major international importance, and their protection and sustainable utilization is essential for the safekeeping the world's food crop genetic resources. The value of the genetic diversity in these areas is priceless, yet some occupy the poorest and most economically and politically unstable regions of the world. One of the most important legacies of Vavilov's centers is that they provided a progenitor model for the establishment of the international global networks of genebanks and crop plant genetic resource centers that we have today. The largest and most important resource center for developing countries is the Consultative Group on International Agricultural Research (CGIAR) and its component network organizations. CGIAR (http://www.cgiar.org) has a current mandate to

> Contribute to food security and poverty eradication in developing countries through research partnership, capacity building and policy support, promoting sustainable agricultural development based on the environmentally sound management of natural resources.

CGIAR was established at a meeting hosted by the World Bank on 19 May 1971, and the founding objective of the group was to "increase the pile of rice" in developing countries (http://www.cgiar.org). Today, the CGIAR comprises 16 independent research centers that are embraced in the overarching CGIAR umbrella. Each center is dedicated to different activities pertaining to specific activities, crops, and more recently, domestic livestock and fisheries. The protection of global biodiversity is one of the primary concerns of the CGIAR, which across its plant research centers holds one of the world's largest *ex situ* collections of plant genetic resources, comprising more than 600,000 accessions of more than 3000 crop, forage, and pasture species (http://www.cgiar.org), duplicates of which are made freely available to researchers for the conservation, improvement, and sustainable utilization of crop plants.

The second decade of CGIAR's activities (1981–1990) was defined by a research objective to increase sustainable food production in developing countries by placing an emphasis on protecting biodiversity, land, and water. Most significantly for the still fledgling field of plant cryopreservation was the fact that CGIAR placed considerable emphasis on the importance of *ex situ* genetic resource conservation. Cryopreservation was then predicted to have an essential and important role in the conservation of difficult-to-preserve crop plant germplasm (Withers, 1980; Withers and Williams, 1980). Specifically, those species that are either vegetatively propagated or that produce seeds, and those that are recalcitrant to conventional seed-banking procedures. Cryogenic storage offers an alternative approach to protecting plant diversity that cannot be conserved as orthodox seed. By virtue of the fact that many tropical plant species produce recalcitrant and difficult-to-conserve germplasm, the development of cryopreservation for tropical plants soon become a major focus of activity for many CGIAR centers. Furthermore, the need to develop cryopreservation as a complementary and parallel approach to *ex situ* field conservation was considered very important. "Growing" field banks provide an excellent conservation resource for many vegetatively propagated and cloned woody perennials. However, climate change, natural disasters, and pathogen attack place these valuable and vulnerable genetic resources at risk. Cryobanks of germplasm derived from field banks provide additional and long-term security.

In 1979, Sir Otto Frankel, a key player in promoting the cause of crop plant genetic conservation, brought to the attention of the International Genetics Federation the need to develop concerted and integrated approaches to safeguard the genetic resources of the world's crop plants. The decade that followed resulted in major advances in the field of plant cryopreservation. Under the directorship of J.T. Williams, and the expert knowledge of L.A Withers, the International Board for Plant Genetic Resource's (IBPGR's) advisory committee on *in vitro* storage (see Withers, 1980), took up the cause to develop cryopreservation as a tool to conserve the food crop germplasm (Withers and Williams, 1980). A milestone workshop on the conservation of recalcitrant material was organized and held by key conservationists and biotechnologists (Sir Otto Frankel, Dr. L.A. Withers, J.T. Williams, E.C. Cocking, and E. Roberts). Hosted by the University of Reading in September 1980, the meeting introduced cryopreservation as a key approach to solving some of the problems of recalcitrant germplasm storage. Moreover, the meeting program incorporated the expertise of two international practitioners in fundamental cryobiology: Harold Meryman and Peter Steponkus. Thus, the importance of fundamental knowledge in the development of practical cryoconservation methods was recognized from the very start of the field's applied development.

9.5.2 Cryopreserving Crop Plant Germplasm: An International Challenge

One of the most important global challenges of plant cryopreservation is the development and application of cryopreservation storage protocols to recalcitrant germplasm held in international genetic resources centers, many of which are based in developing countries (Engelmann, 1991; Engelmann and Takagi, 2000; Villalobos et al., 1991). Today, IBPGR is known as the International Plant Genetic Resources Institute (IPGRI), and to date, it has supported the development of a wide range of international cryopreservation projects in developing countries, both in and outside the CGIAR network (http://www.ipgri.cgiar.org). The early support of IBPGR in promoting international cryopreservation projects must be noted (Ashmore, 1997), as well as the efforts of key players and their host organizations (Chin, 1991, 1996a,b; Engelmann, 1991; Engelmann and Takagi, 2000; Withers, 1980; Withers and Williams, 1980). Much of the practical pioneering work targeted at the conservation of difficult-to-conserve tropical crop plant species has been undertaken by Florent Engelmann and collaborating colleagues (Engelmann, 1997), building on the formulation of simple (Engelman et al., 1994), vitrification (PVS2 and derivatives), and encapsulation/dehydration-based protocols originally devised by Sakai (see Chapter 10) and Dereuddre (Fabre and Dereuddre, 1990), respectively.

Early international crop plant cryopreservation research included a collaborative project (IPGRI/CIAT [Center Internacional de Agricultura Tropical], 1994) between IBPGR and the CGIAR's CIAT in Columbia, which pioneered the establishment and operation of a pilot *in vitro* genebanks and cryopreservation as applied to cassava (*Manihot esculenta*). Cryopreservation is now considered an important component of cassava genetic resources management at CIAT and one that in the future will be applied to maintain over 6000 clonal accessions (Escobar et al., 1995, 1997, 2000). CGIAR centers in Peru and Africa have also developed cryogenic methods for the conservation of important tuber crops such as potato, sweet potato, and yams (Golmirzaie et al., 1999, 2000; Ng et al., 1999). Similarly, IPGRI together with the International Network for the Improvement of Banana and Plantain has developed cryopreservation methodologies for *Musa* germplasm (Panis and Tien Thinh, 2001).

IPGRI has identified two main categories of plant germplasm that are particularly problematic with respect to their conservation using traditional methods. These are vegetatively propagated species, and recalcitrant seeds (Ashmore, 1997; Engelmann, 1997; Withers and Williams, 1980). Because of the importance of both problem categories, this chapter will now specifically focus on the challenges and potential solutions for the development of cryoconservation methods for recalcitrant seeds and clonally propagated plant species.

9.5.3 POTATO, AN ANCIENT CROP, HELPS CRYOPRESERVE VEGETATIVELY PROPAGATED SPECIES

Potato was one of the first major crop plants to be cryopreserved (Bajaj, 1977). It is the fourth largest food crop and historically is one of the most ancient of all our cultivated plants (Hawkes, 1990). The potato center of genetic diversity (Vavilov, 1997) is the Andes, and at the time of the Spanish Conquest it was widely spread throughout Colombia, Peru, and southern Chile. Professor Jack Hawkes (Akeroyd, 1995) played an influential role in potato conservation, and his plant collecting missions to South America brought him in contact with Nikolay Vavilov in St. Petersburg (see Section 9.5). In the 1960s Jack Hawkes, together with Otto Frankel and Erna Bennett, became expert members of a Food and Agriculture Organization of the United Nations (FAO) panel concerning plant exploration. In 1972 they were influential in the establishment of the IBPGR (see Section 9.5.2). Their work (Akeroyd, 1995) led to the creation of the regional gene banks—one of which is the CGIAR's International Potato Center in Peru (CIP), for which the cryogenic storage of potato germplasm provides an important means of *in vitro* conservation (Golmirzaie et al., 1999).

Seed storage is the usual method of choice for conserving plant germplasm, but some vegetatively propagated crops do not produce seeds, or even if they do, it may be preferable to maintain their elite genetic character through clonal propagation. Thus, one of the most important applications of plant cryopreservation is for the long-term conservation of vegetatively propagated crop and horticultural species. The potato's long cryogenic history provides an excellent example of the problems (and hopefully solutions) that plant cryobiologists have to overcome to conserve large-scale, international germplasm collections of clonally propagated plants, and especially tuber crops.

9.5.3.1 Variable Responses to Cryogenic Storage: A Conservation Challenge for Cryo-Genebanks

One of the most important challenges facing cryoconservationists is the development of storage protocols that can be applied across a wide genetic base. This has very important strategic, cost, and operational implications for the application of cryopreservation in large-scale germplasm collections such as those held by the CGIAR germplasm centers. Lack of uniformity and reproducibility in cryogenic storage responses still remains a problem, and charting the progress of potato cryopreservation can afford valuable insights as to why this may be the case.

The history of potato (*Solanum* spp.) cryopreservation starts with Bajaj (1977, 1995a, 1995b), who first recovered *Solanum tuberosum* tuber sprouts and axillary buds from liquid nitrogen storage using combinations of glycerol and sucrose as cryoprotectants. Freezing involved either ultra-rapid cryogenic exposure (in cryovials containing the shoots) or slow exposure achieved by passing the sample through liquid nitrogen vapor, resulting in maximum survival of 18%. Shoots were regenerated from cryopreserved meristems, but damage to the apical dome and the production of callus was noted for several specimens. The following year, Grout and Henshaw (1978) published a paper reporting the cryopreservation of shoot-tips from *Solanum goniocalyx*—a genotype donated by CIP, Peru. Using 10% (v/v) Me_2SO as the cryoprotectant, an ultra-rapid freezing method was devised in which the shoots were placed on the tips of hypodermic needles and plunged directly into liquid nitrogen. They were thawed in hormone-containing liquid culture medium: The shoots floated away from the needle and were first recovered under low light conditions; 20% survival was achieved, but still with some callusing. Towill (1981a, 1981b) applied controlled-rate cooling first to *Solanum etuberosum* and then to 16 different cultivars of *S. tuberosum*, following cryoprotection in 10% (v/v) Me_2SO. Slow and controlled cooling (at –0.3° to –0.4°C/min) to an intermediate transfer temperature (–30° to –40°C) was applied before plunging the samples into liquid nitrogen; importantly, ice nucleation was induced in the cryotubes at approximately 5°C. The composition of the growth regulators in the recovery medium was critical for shoot regeneration; however, survival was highly variable across the genotype range (0 to 80%) and intraexperimental variation was observed.

Lack of a uniform response to cryogenic conditions and poor poststorage shoot regeneration were identified by Towill (1981b) as major problems. In subsequent studies, solutions were afforded by using sterile, uniformly maintained *in vitro* plant cultures and by optimizing the cytokinin components of the recovery medium (Towill, 1983). Clearly, at this stage of potato's cryogenic history it became apparent that there were major limitations to cryopreserving germplasm that comprised a wide genetic base, as highly variable responses occurred even within cultivars of the same species. Furthermore, the physiological status of the donor plants and the postcryopreservation status of hormone-supplemented medium, that is, noncryogenic factors, were highly significant determinant factors for recovery. Towill (1984) explored the application of a single controlled-rate freezing and recovery protocol across a very broad genotype range (using *S. tuberosum* group representatives: Andigena, Phureja, Stenototmum, and Tuberosum). Shoot regeneration was from 0 to 100%, and Towill (1984) concluded that a single cryogenic procedure may be inadequate for all germplasm types within a species and that modification of the protocol may be necessary for both cryogenic and noncryogenic parameters.

The effects of noncryogenic factors on the ability of potato shoot-tips to survive ultra-rapid and controlled-rate freezing were examined by Benson et al. (1989; 1991) and Harding et al. (1991). Pre- and postfreeze light regimes and *in vitro* culture period were found to be critically important to survival. The importance of prestorage physiological status was later confirmed by work by Golmirzaie et al. (1999) at CIP, and lack of plant vigor was considered a bottleneck in potato cryopreservation. Importantly, CIP researchers found that when careful attention was given to maintaining *in vitro* plant vigor, average survival rate increased from 31 to 67% for 100 accessions of potato. Similarly, improvement of postthaw culture medium enhances postcryogenic storage survival by 10%.

Potato has also played a part in the "plant vitrification revolution," and together with pear (Dereuddre et al., 1990*), Solanum phureja* was one of the first species to be cryopreserved using encapsulation/dehydration (Fabre and Dereuddre, 1990). Benson et al. (1996) tested the encapsulation/dehydration protocol on six different species of potato and found that all were capable of surviving and regenerating shoots after cryogenic storage; however, highly variable responses were still observed across different genotypes and between experiments. A parallel study undertaken by Bouafia et al. (1996) refined the original encapsulation method (Fabre and Dereuddre, 1990) and applied it to a broader species range, and a 60% survival rate was obtained for each genotype tested. In this system, silica gel was used as the desiccant, and care was taken with the optimization of the water content to support the formation of stable glassy states. Hirai and Sakai (2000) applied encapsulation/vitrification and encapsulation/dehydration to shoot-tips excised from 13 cultivars of *S. tuberosum* and found that preconditioning with sucrose and glycerol improved tolerance to PVS2. Poststorage recovery was more uniform than that achieved for other previous methodologies, and no morphological abnormalities were observed. This approach certainly appears to be very promising for the conservation of potato germplasm; however, it would require testing on a broader species range. More recently, Mix-Wagner et al. (2003) reported on the application of the droplet freezing method (Kartha, 1985a, 1985b) for the long-term cryoconservation of potato apices. This study comprised an IPGRI project with the aim to apply cryogenic storage to a large scale European potato culture collection. Droplet freezing involves the protection of shoot-tips in microdroplets of cryoprotectant dispensed onto aluminium foil, which is plunged directly into liquid nitrogen. Fifty-one cultivars of potato were tested, and importantly, no major changes in survival or regeneration were observed after several years (for the period 1992–1999) of cryogenic storage. Unfortunately, major variances in recovery of viable apices were observed both among and within individual varieties. Mix-Wagner et al. (2003) suggest that this variation is a result of nonuniformity of meristem size, low experimental replication, use of different equipment, and the fact that different personnel performed the experiments in different laboratories. They suggest that the high numbers of samples must be cryopreserved to compensate for variation in responses to freezing.

9.5.3.2 Potato Pioneers' Postcryopreservation Stability Studies

Assessments of poststorage physiological, developmental, reproductive, and molecular stability are important, particularly as the application of cryopreservation in large-scale culture collections is progressed. Relative to other plant species, potato has been the subject of several long-term stability assessments, collectively offering information and methodologies, which may be provide useful insights into the assessment of stability in other cryopreserved plant systems. Ribsomal RNA genes have been applied as restriction fragment length polymorphism markers of stability in *S. tuberosum* recovered from ultra-rapid freezing, confirming molecular stability (Harding, 1991, 1997); choroplast DNA and microsatellite analysis have been similarly applied (Harding and Benson, 2000, 2001). In polyploid species such as potato, chromosome assessments provide markers of poststorage stability, and this has been confirmed for a range of different cryopreserved potato species and genotypes (Benson et al., 1996; Ward et al., 1993). Reproductive and development parameters have also been assessed using statistical and biometric analyses (Harding and Benson, 1994; Harding and Staines, 2001), indicating that there are differences in quantitative traits (height, tuber weight, petiole length) in plant recovered from cryogenic, Me_2SO, and *in vitro* treatments, suggesting that noncryogenic factors may be an important consideration; however, these differences may be the result of epigenetic rather than genetic factors. Schafer-Menuhr et al. (1997) applied flow cytometry testing, phenotypic inspections, and DNA fingerprinting to 150 different genotypes recovered from droplet freezing-cryogenic storage; all were stable with the exception of one polyploid.

9.5.3.3 Cryopreservation of Vegetatively Propagated Germplasm in International Genebanks: Identification of Critical Factors

The problem of variable responses to cryogenic and noncryogenic factors must be addressed if cryopreservation is to be applied to large numbers of accessions held in international genebanks (Reed, 2001). For large-scale germplasm repositories, however, minimizing labor costs and enhancing operational efficiency will be critically dependent on the formulation of simple, widely applicable, and economically viable storage procedures. It is important that once a storage protocol is developed it can be applied across a broad genotype range without the need for laborious optimization for each different genotype.

Potato research, now spanning 25 years, has helped identify critical factors that are very likely to influence the cryogenic responses of other major crop plant species. As well as the freezing and vitrification protocols *per se*, these are genotype, pre- and poststorage physiological status, time *in vitro*, operator, equipment, light, culture conditions, shoot-tip size, and hormone regimes before and after storage. Studies on other crop plants are now providing useful insights into the critical point factors that affect responses to cryogenic storage. CGIAR's CIP in Peru maintains the world's largest collection of sweet potato and has used experiences with potato to pave the way for the systematic determination of efficient cryopreservation methods for sweet potato (Golmirzaie et al., 2000). Similarly, CGIAR's CIAT in Columbia and the International Institute of Tropical Agriculture in Nigeria are similarly prospecting the use of rapid freezing and encapsulation for a wide range of cassava and yam cultivars (Escobar et al., 1995, 1997; Ng et al., 1999).

The United States Department of Agriculture's National Clonal Germplasm Repository, in Corvallis, Oregon, has been particularly active in considering the challenge of applying standard cryogenic procedures to the conservation of genetically diverse temperate soft fruit and nut crop germplasm held in large-scale international genebanks. Reed (2000) identified an interaction between genotypes of different *Ribes* and their responses to three cryopreservation protocols (slow cooling, PVS2 vitrification, and encapsulation/dehydration). Importantly, all genotypes responded well, producing medium to high levels of survival (approximately 50 to 80%) for at least one of the methods applied. This finding concurs with that of Mandal (2000) at the National Bureau for Plant Genetic Resources, New Delhi, India, who was able to cryopreserve yam apices from a range of different genotypes using three different methods (PVS2, encapsulation-vitrification, and encapsulation/dehydration). Thus, the way forward for the uptake of cryopreservation in large-scale international genebanks may be to develop a small number of robust and standard protocols based on a range of cryoprotective modalities (e.g., chemical vitrification, desiccation, encapsulation, dehydration, slow cooling). This will widen the possibility that one of the modalities will be applicable to diverse germplasm types that have variable sensitivities and tolerances to different types of cryoinjury and associated desiccation stress. However, noncryogenic factors such as physiological status will still need to be addressed. In the future, it will also be crucial to examine critical point factors involved in the transfer of cryopreservation technologies to personnel in different international laboratories and germplasm repositories (Reed et al., 2000, 2001). A case in point is the National Clonal Germplasm Repository, which is currently undertaking an international project with respect to cryopreservation skills technology transfer for clonally propagated germplasm in international germplasm repositories (http://www.ars-grin.gov/cor/tc/scrp.plan.html).

9.6 CRYOPRESERVING SEEDS AND EMBRYOS: A FOCUS ON RECALCITRANCE

The usual approach to plant genetic resources conservation is to place desiccated seed in conventional banks at –18 to –20°C at 5% (w/w) moisture content. Cryopreservation also offers a long-term storage solution for the conservation of orthodox seeds in large-scale germplasm repositories (Stanwood, 1985). One of the first steps toward developing a cryogenic method for seeds is to elucidate their basic storage behaviors as related to desiccation tolerance, because this will dictate the feasibility of surviving liquid nitrogen treatment.

9.6.1 CLASSIFYING SEED STORAGE BEHAVIOR

A classification of seeds on the basis of their short-, medium-, and long-term longevity was first proposed by Ewart (1908) but these categories were difficult to specify, as storage conditions can greatly influence survival. Professor Eric Roberts (1973) first defined the now-standard categories of seed storage behavior. These are orthodox seeds that are capable of surviving to less than or equal to 5% (w/w) drying and recalcitrant seeds, which are sensitive to desiccation. Later, the term "intermediate" was applied to describe seeds that behaved in between the orthodox and recalcitrant categories (Ellis et al., 1990).

With respect to ensuring the survival of orthodox, intermediate, and recalcitrant seeds in cryogenic storage, clues can be gained from studying their adaptive behavior in "nature's laboratory" (see Foreword). Villiers (1974) first noted the extended longevity of fully hydrated orthodox seeds and demonstrated (Villiers and Edgcumbe, 1975) that continuous hydration was not necessary for long-term survival, proposing that this is why annual weed seeds are able to survive for relatively long periods in the soil. Orthodox seeds frequently undergo a natural, physiological desiccation process, which renders them inactive, and metabolic activity is resumed on imbibition. This is because in nature, the germination of orthodox, temperate seeds is dependent on favorable growth conditions; they become seasonally dormant or lie dormant in the soil until conditions are suitable for their germination. Dormancy-breaking triggers are frequently water, temperature, and light.

In contrast, the natural physiological adaptations of orthodox seeds render them amenable to cryogenic storage, and often this can be achieved by simply drying seeds to a critical moisture content and plunging them directly into liquid nitrogen—no other cryoprotective strategy is required. Stanwood (1985) and Pritchard (1995) provide methodologies for the cryopreservation of orthodox seeds; practically, seeds are desiccated to a critical moisture content (MC) in large-scale drying rooms at around 12 to 15% relative humidity before they are placed in cryogenic storage.

Orthodox seeds follow a single equation with respect to the mathematical rules that can be used to predict viability (percentage) after any period of a defined storage condition (Ellis and

Roberts, 1980, 1981). This is the logarithm of any measure of longevity (e.g., time taken for 50% of a batch of seeds to lose viability) and shows an approximate linear relationship with temperature and moisture content. This enables Harrington's rule of thumb to be applied (Harrington, 1963) to estimate the effect of temperature and moisture content on orthodox seed storage responses: seed longevity (for orthodox seeds) is doubled for each 5°C lowering of temperature or each 1% decrease in moisture content. These basic rules have greatly facilitated the application of traditional seed storage methods to orthodox seeds; however, difficulties arise when considering the more complex behaviors of recalcitrant and intermediate seeds. Walters (1999) presents a useful critique for seed classification based on the practical need to quantify, in a more discerning way, the levels of desiccation tolerance of recalcitrant seeds. She states that categorizing orthodox seeds as those that can survive conventional storage protocols (at 0.02 g H_2O/dry weight at –18°C) is effective but falls short for the more complex behavior of recalcitrant seeds. The idea was thus forwarded that desiccation tolerance should be defined quantitatively as well as qualitatively for recalcitrant seeds. A study undertaken on wild rice (Zizania) embryos revealed that mature embryos could survive temperatures to –50°C, whereas the least mature embryos survived only to –18°C (Vertucci et al., 1995). These physiological and developmental differences could be assigned (using DSC) to physical properties of water at different stages of maturation. A single water activity value of 0.90 was critical for desiccation damage. Clearly these studies concur that the development status of embryos and seeds can critically influence the capacity to survive freezing and cryogenic storage. By quantifying moisture levels at which damage is observed, Walters (1999) has shown that as embryos mature, they progressively acquire tolerance to lower water potentials in discrete steps. Critical water potentials have been identified at –1.8, –5, –12, –50 and, –180 Mpa, and Walters proposes that these levels are concomitant with different levels of tolerance and damage. Thus, a hypothesis is presented that recalcitrant seeds may be categorized by the minimum water potential that they endure, which will help elucidate the limiting factors in their tolerance to desiccation stress. Such information will in turn help the optimization of desiccation treatments to ensure survival after cryogenic storage.

In contrast to orthodox seeds, many of which can survive exposure to –196°C so long as they are sufficiently dried (Stanwood, 1985), recalcitrant seeds are very difficult to conserve using cryopreservation, and as stated by Berjak et al. (1999), the quest for their cryopreservation still remains elusive.

9.6.2 FROZEN RAIN FORESTS: NATURE'S CHALLENGE

What determines life and death at low and ultra-low storage temperatures is one of the most important challenges of understanding "life in the frozen state" (see Chapter 23). Recalcitrant seeds of tropical plants are usually large, metabolically very active, and highly hydrated (e.g., imagine freezing a coconut), and they are extremely desiccation and cold sensitive. So unfortunately, not all seeds survive the desiccation and low-temperature treatments required to ensure longevity using traditional storage methods, and a major challenge pertains to the conservation of seed germplasm from tropical and equatorial, forest, and agroforestry species, especially the recalcitrant seeds of tropical rain forest trees (Chin, 1991, 1996a,b; Marzalina et al., 1999; Normah et al., 1996; Zakri et al., 1991). Most recalcitrant seeds belong to tropical timber, plantation, and fruit crops (Chin, 1988), and as they are adapted to the warm and humid tropical rain forests, they do not tolerate freezing temperatures and some fail to survive a minimum temperature of 15°C.

Farrant et al. (1988) observed that the stage of additional water required for seeds to become recalcitrant is concomitant with the onset of cell division and with increasing vacuolation. The authors present the interesting hypothesis that this set of events is similar to those that occur in imbibed orthodox seeds and that render them sensitive to desiccation. It is further proposed that as germination proceeds to a certain stage, the seeds become increasingly sensitive to desiccation and also that seeds dried rapidly just after shedding tolerate a greater degree of water loss. Thus, developmental stage is important.

One of the most important approaches for the cryopreservation of tropical seed germplasm is to excise the zygotic embryos and use this as the subject for storage. In 1986, IBPGR funded a recalcitrant seed cryopreservation project at the Universiti Pertanian, Malaysia; this was under the coordination of Professor H.F. Chin (Universiti Pertanian, Malaysia), a leading exponent for the development of cryopreservation methods as applied to tropical plants and particularly to recalcitrant seeds. The following year, Chin and Krishnapillay reported that embryos of *Artocarpus heterophyllus* could be dried to 11% MC and survive *in vitro* (see Chin, 1988). Subsequently, Pritchard and Prendergast (1986) cryopreserved *Arucaria hunstecinii* embryos and Normah et al. (1986) cryopreserved rubber seed embryos with a recovery of 20 to 69% after being desiccated to 14 to 20% MC.

Engelmann (1997) reviewed progress as to the cryopreservation of embryos and embryonic axes from a wide range of recalcitrant-seeded species: Survival ranged from 0 to 95%, and the main cryoprotective approach was evaporative desiccation.

9.6.3 INTEGRATED STUDIES MAY HELP TO SOLVE SEED RECALCITRANCE PROBLEMS

Recalcitrant tropical seeds present many additional, noncryobiological conservation problems, and increasingly it will be important to take a holistic approach to their cryoconservation. We may therefore need to intercalate knowledge of tropical forest environmental physiology and to extrapolate this to the poststorage recovery phase, which is usually undertaken *in vitro*. A challenge such as this requires information from very different fields, ranging from ecology to biotechnology.

Recalcitrant seeds usually come from two types of habitat: There are aquatic seeds, which are maintained in a fully hydrated state, and seeds from the humid tropics, which have high levels of moisture (30 to 70%) and can be killed if their MCs are reduced to 12 to 31% (Chin, 1988). Their collection and transfer from often remote areas can be problematic, as they can become physiologically unstable in transit and easily succumb to fungal contamination. Tropical seeds can often be found germinating while still in a fruit attached to the tree, and as they are exposed to high levels of pests and pathogens, they have developed elaborate phytochemical defences, particularly seeds. This is why tropical rain forests are considered "nature's medicine chests," but this is problematic, as phenolic compounds are released from injured tissues on desiccation and freezing, causing serious losses in viability through damaging autooxidation reactions. There are also other major practical problems, two of which are the prevention of fungal spoilage and seed germination before cryopreservation is in place. Phytosanitary management of forest tree seed is therefore particularly important (Sutherland et al., 2002), as contaminated seeds and embryos will interfere with recovery requiring *in vitro* manipulations.

A further conservation problem compounded by ecological factors is the fact that many tropical trees have complex life cycles and only flower and set seed on an infrequent basis, sometimes at intervals of 5 years of more. There are simply not enough seeds produced on a regular basis to provide a germplasm source for cryogenic storage. Sacrificing seed batches to assess their physiological and storage status is wasteful and compromises the size of infrequently available seed batches. Staines et al. (1999) are taking a novel approach to this problem as applied to seeds from the Malaysian tropical rain forest, through the application of Taguchi experimental design for developing cryopreservation strategies for recalcitrant seeds that are in limited supply. Taguchi-style experiments use quality-control-based statistics by minimizing variation around a target values and by therefore minimizing the number of seeds that would need to be sacrificed to gain an idea as to optimal storage parameters. In contrast to factorial experiments (ANOVAs), Taguchi methods employ signal-to-noise ratios and orthogonal arrays.

The seeds and fruits of many tropical species have complex physiologies, and their maturation and development are not so well defined, as is the case for temperate species. These factors often compromise the optimization of desiccation-sensitivity parameters required to support cryogenic storage. Berjak et al. (1992) examined the interactions between developmental stage, desiccation sensitivity, and water status in embryonic axes of *Landolphia kirkii*, a recalcitrant seed-bearing species. Axes from immature seeds were more desiccation tolerant, axes from germinating seeds were the least tolerant, and DSC profiling revealed that tolerance was associated with the level of hydration. This endorses the previous study (Walters, 1999) (that the extent and rate of desiccation influence survival) and, moreover, the fact that there is a confounding relationship between the acquisition of desiccation tolerance and intracellular responses to different modalities of drying and the maintenance of membrane integrity (Berjak et al., 1990).

In conclusion, it would therefore appear that to address the issues of seed recalcitrance requires an integrated approach, and one that not only considers the cryogenic protocol *per se* but also that fully explores other contributing factors. These factors can be broadly divided into two groups: technical issues, which related to seed availability, collection, stabilization, spoilage, and *in vitro* manipulations, and developmental/physiological factors. For example, the developmental status of recalcitrant seeds is influenced by environmental as well as physiological factors, and this can affect desiccation tolerance. Clearly, both are important, but perhaps the key to future progress will be to consider the effect and integration of cryogenic and noncryogenic critical point factors. Researchers involved in recalcitrant seed conservation may perhaps be required to become "cryoecologists" as well as cryobiologists.

9.7 CONCLUSIONS: FROM ECOSYSTEMS TO CRYOVIALS

Progress in plant cryobiology spans some 100 years and has resulted in the formulation of a wide range of cryopreservation protocols that have been successfully applied to all types of germplasm (seeds, zygotic and somatic embryos, dedifferentiated cells and callus, shoots, roots, and pollen). The taxonomic spectrum of plants that can be cryopreserved is considerable and is expanding at a rapid rate. One of the most impressive advances in plant cryopreservation has been the application of vitrification protocols for the conservation of previously difficult-to-preserve germplasm, most especially that derived from tropical species. However, there exist a number of important challenges ahead, ranging from strategic and technical issues such as how to design and operate long-term, cryopreserved plant genebanks, to more fundamental research related to understanding the "glassy state." History, however, does have an important place, and my revisiting the original thinking of Nikolay Maximov is based on the premise that reviewing the "past" helps us to understand the present, and that integrating the two will help us to devise more astute strategies for future investigations.

Maximov (1912a, 1912b) undertook research that had its roots of understanding in environmental stress physiology, the consequences of which had an effect on *in vitro* cryobiology. As Meryman (Foreword of this volume) states, "nature's laboratory" provides some of the most important clues to understanding "life in the frozen state." Taking "nature walks" with low-temperature ecologists may well provide applied cryobiologists (from all disciplines) with inspiration as to how to find solutions to their low-temperature preservation (cryogenic storage) and destruction (cryosurgery) problems.

Two scientific meetings have recently endorsed the benefits of using multidisciplinary debates to stimulate cryobiological discussions that explore ecological and applied issues of freezing stress. The first, comprising a symposium titled "Coping with Cold: Natural Solutions," was hosted by the Society of Low Temperature Biology (2000) and the University of Cambridge in September 1999. A clear output of the meeting was the great potential for cross-discipline transfer of intelligence in both ecological and applied aspects of low-temperature stress physiology (also see Chapter 3).

The Royal Society of London's (2002) Discussion Meeting, entitled "Coping With Cold: The Molecular and Structural Biology of Cold Stress Survivors," integrated generic issues of understanding freezing and cold stress at the molecular level and spanned diverse disciplines exploring low-temperature stresses in plants, animals, and bacteria in scenarios as diverse as polar lakes and industrial laboratories. The concluding remarks of this meeting are insightful. Compiled by Jeff

Bale from the University of Birmingham, United Kingdom (Royal Society of London, 2002), they distilled three modalities of "thinking about" cold stress:

- Constant cold stress: Why is it that some organisms can live almost all the year round in a continual cold-stressed state? Comparisons of the tropical and polar regions were presented
- Seasonality: How it is that organisms can respond to annual rhythms that dictate their ability to survive episodes of cold stress?
- Rapid responses to cold stress, within minutes and hours

Questions were then posed as to how many of the underlying (generic?) mechanisms and responses are the same across different groups of organisms and across different temporal ranges. The possibility of providing a common model to explore cold stress across diverse taxa was presented based on the following (modified, by the author) pathway:

cold stress signal → signal transduction → cellular response → adaptation → survival
↓
sensitivity → death

Clearly, gaining an understanding of all three modalities could provide an important framework as to how we can explore future key plant and algal cryopreservation issues. Extremely cold-tolerant organisms (e.g., polar extremophiles) may be compared with highly recalcitrant groups (tropical species) to gain an understanding of stress and acclimation responses. Historically, such studies have been applied as an aid to cryopreservation protocol development; for example, the discovery, *in vivo*, of trehalose in highly tolerant organisms and its subsequent application as a cryoprotectant in plant, animal, and medical cryobiology. Similar examples are the application of ice nucleating and antifreeze agents (see Chapter 3). Understanding seasonal tolerance, dormancy, and acclimation (in temperate species) and comparing these with the elusive life cycle and seed development processes in tropical and equatorial species may be a basis on which to study desiccation tolerance. Rapid response modes are best assigned to those treatments (rapid freezing/thawing/desiccation) that comprise cryopreservation protocols *per se*. Thus, to conclude, there exist many examples as to how the "ecosystem to cryovial" approach has been applied to great effect in developing protection strategies for contemporary cryobiologists. This experimental rationale, instigated almost a century ago by Maximov, is perhaps even more relevant and compelling today, as we can advance our cryobiological explorations using the modern "toolkits" of genome mapping and molecular and analytical technologies.

> *And through the drifts the snowy clifts*
> *Did send a dismal sheen:*
> *Nor shape of men nor beast we ken–*
> *The ice was all between*

Excerpt from the
Rime of the Ancient Mariner
by Samuel T. Coleridge
(Holmes, 1996)

ACKNOWLEDGMENTS

I acknowledge the financial support the EU Commission Grants: COBRA Project QLRT-2000-01645 and the CRYMCEPT Project QLRT-2001-01279, QLRT-2000-01645 (COBRA). The comments and discussions of Keith Harding, Barry Fuller, and Nick Lane are gratefully acknowledged.

I thank collaborating cryopreservation friends and colleagues who have contributed over the years to the development of the ideas presented in this chapter.

REFERENCES

Akeroyd, J. (1995) Jack Hawkes and his life among the potatoes, *Plant Talk*, 2, 4.

Akeroyd, J. (2002) Nicolay Ivanovich Vavilov (1887–1943) Russian pioneer of germplasm conservation, *Plant Talk*, 28, 5.

Ashmore, S. (1997) *Status Reports on the Development and Application of in vitro Techniques for the Conservation and Use of Plant Genetic Resources,* International Plant Genetic Resources Institute, Rome.

Assy-Bah, B. and Engelmann, F. (1992a) Cryopreservation of immature embryos of coconut (*Cocos nucifera* L.), *Cryo-Letters*, 13, 67–74.

Assy-Bah, B. and Engelmann, F. (1992b) Cryopreservation of mature embryos of coconut (*Cocos nucifera* L.) and subsequent regeneration of plantlets, *Cryo-Letters*, 13, 67–74.

Bajaj, Y.P.S. (1976) Regeneration of plants from cell suspensions frozen at –20, –70 and –196°C, *Physiol. Plant.*, 37, 263–268.

Bajaj, Y.P.S. (1977) Initiation of shoots and callus from potato-tuber sprouts and axillary buds frozen at –196°C, *Crop Improvement,* 4, 48–53.

Bajaj, Y.P.S (1979) Technology and prospects of cryopreservation of germplasm, *Euphytica,* 28, 267–285.

Bajaj, Y.P.S. (1995a) Cryopreservation of germplasm of cereals, in *Biotechnology in Agriculture and Forestry 32, Cryopreservation of Plant Germplasm I*, Bajaj, Y.P.S, Ed., Springer, Berlin, pp. 217–235.

Bajaj, Y.P.S. (1995b) Cryopreservation of germplasm of potato (*Solanum tuberosum* L.) and cassava (*Manihot esculenta* Crantz) in *Biotechnology in Agriculture and Forestry 32, Cryopreservation of Plant Germplasm I*, Bajaj, Y.P.S, Ed., Springer, Berlin, pp. 217–235.

Benson, E.E. (1994) Cryopreservation, in *Practical Plant Cell Culture: A Practical Approach,* Dixon, R.A., Ed., Oxford IRL Press, Oxford, pp. 147–166.

Benson, E.E. (1995) Cryopreservation of shoot-tips and meristems, in *Methods in Molecular Biology,* Vol. 20, *Cryopreservation and Freeze Drying Protocols,* McClennan, M. and Day, J.G., Eds., Humana Press, Totowa, NJ, pp. 121–132.

Benson, E.E. (1999a) Cryopreservation, in *Plant Conservation Biotechnology,* Benson, E.E., Ed., Taylor & Francis, London, pp. 83–96.

Benson, E.E. (1999b) *Plant Conservation Biotechnology,* Taylor & Francis, London.

Benson, E.E. and Hamill, J.D. (1991) Cryopreservation and post freeze molecular and biosynthetic stability in transformed roots of *Beta vulgaris* and *Nicotiana rustica*, *Plant Cell Tissue Organ Culture*, 24, 163–172.

Benson, E.E., Harding, K, and Dumet, D.J. (2000) Cryopreservation of plant cells, tissues and organs, in *The Encyclopaedia of Cell Technology*, Spier, R.E., Ed., Wiley, New York, pp. 627–635.

Benson, E.E., Harding, K., and Smith, H. (1989) The effects of pre- and post-freeze light on the recovery of cryopreserved shoot-tips of *Solanum tuberosum, Cryo-Letters,* 10, 323–344.

Benson, E.E., Harding, K, and Smith, H. (1991) The effects of *in vitro* culture period on the recovery of cryopreserved shoot-tips of *Solanum tuberosum*, *Cryo-Letters*, 12, 17–22.

Benson, E.E. and Lynch, P.T. (1999) Cryopreservation of rice cultures, in *Methods in Molecular Biology III Plant Cell Culture Protocols*, Hall, R.D., Ed., Humana Press, Totowa, NJ, pp. 83–94.

Benson, E.E., Reed, B.M., Brennan, R.M., Clacher, K.A., and Ross, D.A. (1996) Use of thermal analysis in the evaluation of cryopreservation protocols for *Ribes nigrum* L. germplasm, *Cryo-Letters*, 17, 347–362.

Benson, E.E., Wilkinson, M., Todd, A., Ekuere, U., and Lyon, J. (1996) Developmental competence and ploidy stability in plants regenerated from cryopreserved potato shoot-tips, *Cryo-Letters*, 17, 119–128.

Berjak, P., Farrant, J.M., Mycock, D.J., and Pammenter, N.W. (1990) Recalcitrant (homoiohydrous) seeds: the enigma of their desiccation-sensitivity, *Seed Sci. Technol.*, 18, 297–310.

Berjak, P., Kioko, J.I., Walker, M., Mycock, D.J., Wesley-Smith, J., Watt, P., and Pammenter, N.W. (1999) Cryopreservation and elusive goal, in *IUFRO Symposium 1998 Recalcitrant Seeds Proceedings of the Conference 12–15th October 1998*, Marzalina, M., Khoo, K.C., Jayanthi, N., Tsan, F.Y., and Krishnapillay, B., Eds., Forest Research Institute of Malaysia, Kuala Lumpur, pp. 96–109.

Berjak, P., Pammenter, N.W., and Vertucci, C. (1992) Homoiohydrous (recalcitrant) seeds: developmental status, desiccation sensitivity and the state of water in axes of *Landolphia kirkii* Dyer, *Planta*, 186, 249–261.

Bodas, K., Brennig, C., Diller, K.R., and Brand, J.J. (1995) Cryopreservation of blue-green and Eukaryotic algae in the culture collections and the University of Texas at Austin, *Cryo-Letters*, 16, 267–274.

Bouafia, S., Jelti, N., Lairy, G., Blanc, A., Bonnel, E., and Dereuddre, J. (1996) Cryopreservation of potato shoot-tips by encapsulation-dehydration, *Potato Res.*, 39, 69–78.

Brown, A.H. D., Marshal, D.R., Frankel, O.H., and Williams, J.T. (1989) *The Use of Plant Genetic Resources*, Cambridge University Press, Cambridge.

Butenko, R.G., Popov, A.S., Volkova, L.A., Chernyak, N.D., and Nosov, A.M. (1984) Recovery of cell cultures and their biosynthetic capacity after storage of *Dioscorea deltoidea* and *Panax ginseng* cells in liquid nitrogen, *Plant Sci. Lett.*, 33, 285–292.

Callow, J.A., Ford-Lloyd, B.V., and Newbury, H.J. (1997) *Biotechnology and Plant Genetic Resources, Conservation and Use. Biotechnology in Agriculture Series, No. 19,* CAB International, Wallingford, Oxon.

Cānavate, J.P. and Lubian, L.M. (1995) Some aspects of the cryopreservation of microalgae used as food for marine species, *Aquaculture*, 136, 277–290.

Chen, T.H.H., Kartha, K.K., Constabel, F., and Gusta, L.V. (1984) Freezing characteristics of cultured *Catharanthus roseus* (L.) G. Don cells in dimethyl sulfoxide and sorbitol in relation to cryopreservation, *Plant Physiol.*, 75, 720–725.

Chen, Y. and Wang, J-H. (2003) Cryopreservation of carrot (*Daucus carota* L.) cell suspensions and protoplasts by vitrification, *Cryo-Letters*, 24, 57–64.

Chin, H.F. (1988) *Recalcitrant Seeds and IBPGR Status Report*, IBPGR, Rome.

Chin, H.F. (1991) Conservation of recalcitrant seeds, in *Conservation of Plant Genetic Resources through in vitro Methods. Proceedings of an International Workshop on Tissues Culture for the Conservation of Biodiversity and Plant Genetic Resources,* 28–31 May, Kuala Lumpur, Malaysia, Zakri, A.H., Normah, M.N., Senawai, M.T., and Abdul Karim, A.G., Eds., Forest Research Institute of Malaysia Publications, Kuala Lumpur, pp. 19–28.

Chin, H.F. (1996a) Strategies for the conservation of recalcitrant species, in *In vitro Conservation of Plant Genetic Resources. Proceedings of the International Workshop on In Vitro Conservation of Plant Genetic Resources,* 4–6 July 1995, Kuala Lumpur, Malaysia, Normah, M.N., Narimah, M.K., and Clyde, M.M., Eds., Universiti Kebangsaan Malaysia Press, Kuala Lumpur, pp. 203–215.

Chin, H.F. (1996a) Strategies for the conservation of recalcitrant species, in *Dehydration and Preservation Techniques of Recalcitrant Seed*, Chin, H.F. and Krishnapillay, B., Eds., Unpublished IBPGR Report, Rome.

Day, J.G. (1998) Cryo-conservation of microalgae and cyanobacteria, *Cryo-Letters*, Suppl. 1, 7–14.

Day, J.G., Benson, E.E., and Fleck, R.A. (1999) *In vitro* culture and conservation of microalgae: Applications for aquaculture, biotechnology and environmental research, *In vitro Cell Dev. Biol. Plant*, 35, 127–136.

Day, J.G., Fleck, R.A., and Benson, E.E. (1998a) Cryopreservation of multicellular algae: problems and perspectives, *Cryo-Letters*, 19, 205–206.

Day, J.G., Fleck, R.A., and Benson, E.E. (2000) Cryopreservation recalcitrance in microalgae: novel approaches to identity and avoid cryoinjury, *J. Appl. Phycol.*, 12, 369–377.

Day, J.G and McLellan, M.R., Eds. (1995) *Cryopreservation and Freeze-Drying Protocols. Methods in Molecular Biology, Volume 38,* Humana Press, Totowa, NJ.

Day, J.G., Watanabe, M.M., Morris, G.J., Fleck, R.A., and McLellan, M.R. (1997) Long-term viability of preserved eukaryotic algae, *J. Appl. Phycol.*, 9, 121–127.

Day, J.G., Watanabe, M.M., and Turner, M.F. (1998b) *Ex situ* conservation of protistan and cyanobacterial biodiversity: A CCAP NIBS collaboration 1991–1997, *Phycol. Res.*, 46, 77–83.

Dereuddre, J., Blandin, S., and Hassen, N. (1991a) Resistance of alginate-coated somatic embryos of carrot (*Daucus carota* L.) to desiccation and freezing in liquid nitrogen. 1. Effects of preculture, *Cryo-Letters*, 12, 125–134.

Dereuddre, J., Hassen, N., Blandin, S., and Kaminski, M. (1991b) Resistance of alginate-coated somatic embryos of carrot (*Daucus carota* L.) to desiccation and freezing in liquid nitrogen. 2. Thermal analysis, *Cryo-Letters*, 12, 135–148.

Dereuddre, J., Scottez, C., Arnaud, Y., and Duron, M. (1990) Resistance of alginate-coated axillary shoot tips of plear tree (*Pyrus communis* L. cv Beurre Hardy) *in vitro* plantlets to dehydration and subsequent freezing in liquid nitrogen: effects of previous cold hardening, *CR Acad. Sci. Paris*, 310, 317–323.

Diettrich, B., Haack, U., and Luckner, M. (1986) Cryopreservation of *Digitalis lanata* cells grown *in vitro*. Precultivation and recultivation, *J. Plant Physiol.*, 126, 63–73.

Diller, K.R. (1997) Pioneers in cryobiology: Nikolay Aleksandrovich Xaximov (1890–1952), *Cryo-Letters*, 18, 81–92.

Dougall, D.K. and Wetherell, D.F. (1974) Storage of wild carrot cultures in the frozen state, *Cryobiology*, 11, 410–415.

Dumet, D., Engelmann, F., Chabrillange, N., Duval, Y., and Dereuddre, J. (1993) Importance of sucrose for the acquisition of tolerance to desiccation and cryopreservation of oil palm somatic embryos, *Cryo-Letters*, 14, 243–250.

Ellis, R.H., Hong, T.D., and Roberts, E.H. (1990) Intermediate category of seed storage behaviour I. Coffee, *J. Exp. Bot.*, 41, 1167–1174.

Ellis, R.H. and Roberts, E.H. (1980) Improved equations for the prediction of seed longevity, *Ann. Bot.*, 45, 13–30.

Ellis, R.H. and Roberts, E.H. (1981) The quantification of ageing and survival in orthodox seeds, *Seed Sci. Technol.*, 9, 373–409.

Engelmann, F. (1991) *In vitro* conservation of tropical plant germplasm—a review, *Euphytica*, 57, 227–243.

Engelmann, F. (1997) Importance of desiccation in the cryopreservation of recalcitrant seed and vegetatively propagated species, *Plant Genet. Res. Newsl.*, 112, 9–18.

Engelmann, F., Dambier, D., and Ollitrault, P. (1994) Cryopreservation of cell suspensions and embryogenic calluses of Citrus using a simplified freezing process, *Cryo-Letters*, 15, 53–58.

Engelmann, F. and Takagi, H., Eds. (2000) *Cryopreservation of Tropical Plant Germplasm: Current Research Progress and Application. JIRCAS, International Agriculture Series No. 8*, IPGRI, Rome.

Escobar, R.H., Debouck, D., and Roca, W.M. (2000) Development of cassava cryopreservation, in *Cryopreservation of Tropical Plant Germplasm: Current Research Progress and Application. JIRCAS, International Agriculture Series No. 8,* Engelmann, F. and Takagi, H., Eds., IPGRI, Rome, pp. 222–226.

Escobar, R.H., Mafla, G., and Roca, W.M. (1997) A methodology for recovering cassava from shoot-tips maintained in liquid nitrogen, *Plant Cell Rep.*, 16, 474–478.

Escobar, R.H., Roca, W.M., and Guevara, C. (1995) Cryopreservation of cassava shoot tips in liquid nitrogen, *Biotechnology Research Unit Annual Report*, Center Internacional de Agricultura Tropical, Cali, Colombia, pp. 82–84.

Ewart, A.J. (1908) On the longevity of seeds, *Proc. R. Soc. Victoria*, 21, 1–210.

Fabre, J. and Dereuddre, J. (1990) Encapsulation-dehydration a new approach to cryopreservation of *Solanum* shoot-tips, *Cryo-Letters*, 11, 413–426.

Fahy, W.F. and Fahy, G.M. (1985) Ice-free cryopreservation of mouse embryos at −196°C by vitrification, *Nature*, 313, 573–575.

Farrant, J.M., Pammenter, N.W.E., and Berjak, P. (1988) Recalcitrance: A current assessment, *Seed Sci. Technol.*, 16, 155–166.

Finkle, B.J., Zavala, M.E., Ulrich, J.M. (1985) Cryoprotective compounds in the viable freezing of plant tissues, in *Cryopreservation of Plant Cells and Organs*, Kartha, K.K., Ed., CRC Press, Boca Raton, FL, pp. 75–134.

Fleck, R.A. (1998) *The Assessment of Cell Damage and Recovery in Cryopreserved Freshwater Protists*, Ph.D. thesis, University of Abertay, Dundee, Scotland, U.K.

Fleck, R.A., Benson, E.E., Bremner, D.H., and Day, J.G. (2000) Studies of free radical-mediated cryoinjury in the unicellular green alga *Euglena gracilis* using a non-destructive hydroxyl radical assay: A new approach for developing protistan cryopreservation strategies, *Free Radical Res.*, 32, 157–170.

Fleck, R.A., Day, J.G., Clarke, K.J., and Benson, E.E. (1999) Elucidation of the metabolic and structural basis for the cryopreservation recalcitrance of *Vaucheria sessilis*, *Cryo-Letters*, 20, 271–282.

Fleck, R.A., Day, J.G., Rana, K.J., and Benson, E.E. (1997) Visualisation of cryoinjury and freeze events in the coenocytic alga *Vaucheria sessilis* using cryomicroscopy, *Cryo-Letters*, 18, 343–355.

Franks, F. (1982) *Water: A Comprehensive Treatise Vol. 7*, Plenum Press, New York.

Franks, F. (1985) *Biophysics and Biochemistry at Low Temperatures*, Cambridge University Press, Cambridge.

Franks, F. (1990) Freeze drying: from empiricism to predictability, *Cryo-Letters*, 11, 93–1990.

Franks, F. (1992) The importance of being glassy, *Cryo-Letters*, 13, 349–350.

Golmirzaie, A.M., Panta, A., and Diaz, S. (2000) Systematic determination of an adequate method for large-scale sweet potato cryopreservation at CIP, in *Cryopreservation of Tropical Plant Germplasm: Current Research Progress and Application. JIRCAS, International Agriculture Series No. 8,* Engelmann, F. and Takagi, H., Eds., IPGRI, Rome, pp. 460–464.

Golmirzaie, A.M., Panta, A., and Toledo, J. (1999) Biotechnological advances in the conservation of root and tuber crops, in *Plant Conservation Biotechnology*, Benson, E.E., Ed., Taylor & Francis, London, pp. 165–178.

Grout, B.W.W. (1995) *Genetic Preservation of Plant Cells in Vitro, Springer Laboratory Manual*, Springer, Heidelberg.

Grout, B.W.W. and Henshaw, G.G. (1978) Preservation of potato shoot-tips, *Ann. Bot.*, 42, 1227–1229.

Harding, K. (1991) Molecular stability of the ribosomal RNA genes in *Solanum tuberosum* plants recovered from slow growth and cryopreservation, *Euphytica*, 55, 141–146.

Harding, K. (1997) Stability of the ribosomal RNA genes in *Solanum tuberosum* L. plants recovered from cryopreservation, *Cryo-Letters*, 18, 217–230.

Harding, K. and Benson, E.E. (1994) A study of growth, flowering, and tuberisation in plants derived from cryopreserved potato shoot-tips: implications for *in vitro* germplasm collections, *Cryo-Letters*, 15, 59–66.

Harding, K. and Benson, E.E. (2000) Analysis of chloroplast DNA in plants regenerated from cryopreservation, *Cryo-Letters*, 21, 279–288.

Harding, K. and Benson, E.E. (2001) The use of microsatellite analysis in *Solanum tuberosum* L. *in vitro* plantlets derived from cryopreserved germplasm, *Cryo-Letters*, 22, 199–208.

Harding, K., Benson, E.E., and Smith, H. (1989) Variation in recovery of cryopreserved shoot-tips of *Solanum tuberosum* exposed to different pre- and post-freeze light regimes, *Cryo-Letters*, 10, 323–344.

Harding, K., Benson, E.E., and Smith, H. (1991) The effects of *in vitro* culture period on post-freeze survival of cryopreserved shoot-tips of *Solanum tuberosum, Cryo-Letters*, 12, 17–22.

Harding, K. and Staines, H. (2001). Biometric analysis of phenotypic characters of potato shoot-tips recovered from tissue culture, dimethyl sulphoxide treatment cryopreservation, *Cryo-Letters,* 22, 255–262.

Harrington, J.F. (1963) Practical advice and instructions on seed storage, *Proc. Int. Seed Test. Assn.,* 28, 989–994.

Hawkes, J.G. (1990) *The Potato Evolution, Biodiversity and Genetic Resources*, Bellhaven Press, London.

Hirai, D. and Sakai, A. (2000) Cryopreservation of *in vitro*–grown meristems of potato (*Solanum tuberosum* L.) by encapsulation/vitrification, in *Cryopreservation of Tropical Plant Germplasm: Current Research Progress and Application. JIRCAS, International Agriculture Series No. 8,* Engelmann, F. and Takagi, H., Eds., IPGRI, Rome, pp. 205–211.

Hirata, K., Phunchindawan, M., Kanemoto, M., Tukamoto, J., Goda S., and Miyamoto, K. (1996) Cryopreservation of microalgae using encapsulation-dehydration, *Cryo-Letters*, 17, 321–328.

Holmes, R. (1996) *Samuel Taylor Coleridge, Selected Poems*, Penguin Books, London.

Holm-Hansen, O. (1963) Viability of blue-green and green algae after freezing, *Physiol. Plant.*, 16, 530–540.

Hwang, S.W. and Huddock, G.A. (1971) Stability of *Chlamydomans reihardi* in liquid nitrogen storage, *J. Phycol.*, 7, 300–303.

International Foundation for Science. (1998) *Recent Advances in Biotechnology for Tree Conservation and Management,* International Foundation for Science, Stockholm.

International Plant Genetic Resources Institute/Center Internacional de Agricultura Tropical. (1994) *Establishment and operation of a pilot in vitro genebank. Report of a CIAT/IBPGR collaborative project using Cassava (Manihot esculenta, Crantz) as a model*, IPGRI, Rome.

Jitsuyama, Y., Suzuki, T., Harada, T., and Fujikawa, S. (2002) Sucrose incubation increases freezing tolerance of Asparagus (*Asparagus officianalis*, L.), *Cryo-Letters*, 23, 103–112.

Kartha, K.K. (1985a) *Cryopreservation of Plant Cells and Organs*, CRC Press, Boca Raton, FL.

Kartha, K.K. (1985b) Meristem culture and germplasm preservation, in *Cryopreservation of Plant Cells and Organs*, Kartha, K.K., Ed., CRC Press, Boca Raton, FL, pp. 115–134.

Langis, R., Schnable, B., Earle, E.D., and Steponkus, P. (1989) Cryopreservation of *Brassica campestris* L. cell suspensions by vitrification, *Cryo-Letters*, 10, 421–428.

Langis, R., and Steponkus, P.L. (1990) Cryopreservation of rye protoplasts by vitrification, *Plant Physiol.*, 92, 666–671.

Lee, J.J. and Soldo, A.T. (1992) *Protocols in Protozoology*, Society of Protozoologists, Kansas.

Levitt, J. (1941) *Frost Killing and Hardiness of Plants*, Burgess, Minneapolis, MN.

Levitt, J. (1956) *The Hardiness of Plants*, Academic Press, New York.

Lidforss, B. (1907) Die Wintergrune Flora, *Lunds Universitets Arsskrift (Acta Universitatis Lundensis)* 2, 13, 1–76.

Lovelock, J.E. and Bishop, M.W.H (1959) Prevention of freezing damage in living cells by dimethyl sulfoxide, *Nature (Lond.)*, 183, 1394.

Lynch, P.T., Benson, E.E., Jones, J., Cocking, E.C., and Davey, M.R. (1994) Rice cell cryopreservation: the influence of culture methods and the embryogenic potential of cell suspensions on post-thaw recovery, *Plant Sci.*, 98, 185–192.

Mandal, B.B. (2000) Cryopreservation of yam apices: A comparative study with different techniques, in *Cryopreservation of Tropical Plant Germplasm: Current Research Progress and Application. JIRCAS, International Agriculture Series No. 8*, Engelmann, F. and Takagi, H., Eds., IPGRI, Rome, pp. 233–237.

Marzalina, M, Khoo, K.C., Jayanthi, N., Tsan, F.Y., and Krishnapillay, B. (1999) IUFRO Symposium 1998 *Recalcitrant Seeds Proceedings of the Conference 12–15 October, 1998, Kuala Lumpur*, Kuala Lumpur, Malaysia Forest Research Institute of Malaysia.

Maximov, N.A. (1912a) I. Chemical protective agents of plants against freezing injury, *Berichte Deutschen Bot. Gesellschaft*, 30, 52–65.

Maximov, N.A. (1912b) II. Chemical protective agents of plants against freezing injury the protective effect of salt solutions, *Berichte Deutschen Bot. Gesellschaft*, 30, 293–305.

Maximov, N.A. (1912c) III. Chemical protective agents of plants against freezing injury concerning the nature of thr protective effect, *Berichte Deutschen Bot. Gesellschaft*, 30, 504–516.

Maxted, N., Ford-Lloyd, B.V., and Hawkes, J.G. (1997) *Plant Genetic Conservation: The in Situ Approach*, Chapman & Hall, London.

McCormick, P.V. and Cairns, J., Jr. (1994) Algae as indicators of environmental change, *J. Appl. Phycol.*, 6, 509–526.

McLellan, M.R. (1989) Cryopreservation of Diatoms, *Diatom Res.*, 4, 301–318.

Meryman, H.T. and Williams, R.J. (1985) Basic principles of freezing injury in plant cells; natural tolerance and approaches to cryopreservation, in *Cryopreservation of Plant Cells and Organs*, Kartha, K.K., Ed., CRC Press, Boca Raton, FL, pp. 75–134.

Mix-Wagner, G., Schumacher, H.M., and Cross, R.J. (2003) Recovery of potato apices after several years of storage in liquid nitrogen, *Cryo-Letters*, 243, 33–41.

Morris, G.J. (1978) Cryopreservation of 250 strains of *Chlorococcales* by the method of two-step cooling, *Bryophyte Phycol. J.*, 13, 15–24.

Morris, G.J. and Canning, C.E. (1978) The cryopreservation of *Euglena gracilis*, *J. Gen. Microbiol.*, 108, 27–31.

Nag, K.K. and Street, H.E (1973) Carrot embryogenesis from frozen cultured cells, *Nature*, 245, 270–272.

Nag, K.K. and Street, H.E. (1975) Freeze preservation of cultured plant cells II. The freezing and thawing phases, *Physiol. Plant.*, 34, 261–265.

Ng, S.Y.C, Mantel, S.H., and Ng, N.Q (1999) Biotechnology in germplasm management of cassava and yams, in *Plant Conservation Biotechnology*, Benson, E.E., Ed., Taylor & Francis, London, pp. 179–210.

Normah, M.N., Chin, H.F., and Hor, Y.L. (1986) Desiccation and cryopreservation of embryonic axes of *Hevea brasiliensis. Muell-Arg, Pertanika*, 9, 299–303.

Normah, M.N., Narimah, M.K., and Clyde, M.M., Eds. (1996) *In Vitro Conservation of Plant Genetic Resources, Proceedings of the International Workshop on in Vitro Conservation of Plant Genetic Resources*, 4–6 July 1995, Kuala Lumpur, Malaysia, Kuala Lumpur, Universiti Kebangsaan Malaysia Press.

Panis, B. and Tien Thinh, N. (2001) *Cryopreservation of Musa germplasm*, INIBAP Technical Guidelines No. 5. IPGRI, Rome.

Pence, V.C. (1999) The application of biotechnology for the conservation of endangered plants, in *Plant Conservation Biotechnology*, Benson, E.E., Ed., Taylor & Francis, London, pp. 227–250.

Plucknett, D.L., Smith, J.H., Williams, J.T., and Anishetty, N.M. (1987) *Genebanks and The World Food*, Princeton, NJ: Princeton University Press.

Polge, C., Smith, A.U., and Parkes, A.S. (1949) Revival of spermatozoa after vitrification and dehydration at low temperatures, *Nature*, 164, 666.

Pringsheim, E.G. (1928) Algenkulturen. Eine Liste der Stamme, welche auf Wunsch abegegeben wurden, *Arch. Ptistenknde*, 68, 255–258.

Pringsheim, E.G. (1946) *Pure cultures of algae*, Cambridge, Cambridge University Press.

Pritchard, H.W. (1995) Cryopreservation of seeds, in *Cryopreservation and freeze-drying protocols. Methods in Molecular Biology* Vol. 38, Day, J.G. and McLellan, M.R., Eds., Totowa, NJ: Humana Press Inc., pp. 133–144.

Pritchard, H.W. and Prendergast, F.G. (1986) Effects of desiccation and cryopreservation on the *in vitro* viability of embryos of the recalcitrant seed species *Araucaria hunstecinii* K. Schum, *Journal of Botany*, 37, 1388–1397.

Rall, W.F. and Fahy, G.M. (1985) Ice-free cryopreservation of mouse embryos at −196°C by vitrification, *Nature*, 313, 573–575.

Ramanatha V.R. and Hodgkin, T. (2002) Genetic diversity and conservation and utilization of plant genetic resources, *Plant Cell Tissue and Organ Culture*, 68, 1–19.

Razdan, M.K. and Cocking, E.C. (1997) *Conservation of Plant Genetic Resources In Vitro. Volume 1: General Aspects,* Enfield, Science Publishers Inc. USA.

Reed, B.M. (1988) Cold acclimation as a method to improved survival of cryopreserved *Rubus* meristems, *Cryo-Letters*, 9, 166–171.

Reed, B.M. (2000) Genotype consideration in temperate fruit crop cryopreservation, in *Cryopreservation of Tropical Plant Germplasm: Current Research Progress and Application. JIRCAS, International Agriculture Series No. 8,* Engelmann, F. and Takagi, H., Eds., IPGRI, Rome, pp. 200–204.

Reed, B.M. (2001) Implementing cryogenic storage of clonally propagated plants, *Cryo-Letters*, 22, 97–104.

Reed, B.M. (2002) Cryogenic storage of clonally propagated plants. *Cryo-Letters*, 22, 97–104.

Reed, B.M., Brennan, R.M., Benson, E.E. (2000) An *in vitro* method for conserving *Ribes* germplasm in international genebanks, in *Cryopreservation of Tropical Plant Germplasm: Current Research Progress and Application. JIRCAS, International Agriculture Series No. 8,* Engelmann, F. and Takagi, H., Eds., IPGRI, Rome, pp. 470–472.

Reed, B.M., Dumet, D.J., Denoma, J.M., and Benson, E.E. (2001) Validation of cryopreservation protocols for plant germplasm conservation: a pilot study using *Ribes L.*, *Biodiversity Conservation*, 10, 939–949.

Ribeiro, R.C.S., Jekkel, Z., Mulligan, B.J., Cocking, E.C., Power, J.B., Davey, M.R., and Lynch, P.T. (1996) Regeneration of fertile plants from cryopreserved cell suspensions of *Arabidopsis thaliana* (L.) Heynh, *Plant Sci.*, 115, 115–121.

Ring, R.A. and Danks, H.V. (1998) The role of trehalose in cold-hardiness and desiccation, *Cryo-Letters*, 19, 275–282.

Roberts, E.H. (1973) Predicting the storage life of seeds, *Seed Sci. Technol.*, 1, 499–514.

Royal Society of London. (2002) Coping with the cold: The molecular and structural biology of cold stress survivors, *Philos. Trans. R. Soc. Lond. B*, 2002, 357.

Sakai, A. (1956) Survival of plant tissue at super-low temperature, *Low Temp. Sci. Ser. B.*, 14, 17–23.

Sakai, A. (1960) Survival of the twig of woody plants at −196°C, *Nature,* 185, 393–394.

Sakai, A. (1965) Determining the degree of frost-hardiness in highly hardy plants, *Nature,* 206, 1064–1065.

Sakai, A. (1973) Characteristics of winter hardiness in extremely hardy twigs of woody plants, *Plant Cell Physiol.*, 14, 1–9.

Sakai, A., Kobayashi, S., and Oiyama, I. (1990) Cryopreservation of nuclear cells of navel orange (*Citrus sinensis* Osb. Var. *Brasiliensis* Tanaka) by vitrification, *Plant Cell Rep.*, 9, 30–33.

Sakai, A., Matsumoto, T., Hirai, D., and Niino, T. (2000) Newly developed encapsulation/dehydration protocol for plant cryopreservation, *Cryo-Letters*, 21, 53–62.

Sakai, A. and Yoshida, S. (1968) The role of sugar and related compounds in variations of freezing resistance, *Cryobiology*, 5, 160.

Schafer-Menuhr, A., Schumacher, H.M., and Mix-Wagner, G. (1997) Cryopreservation of potato cultivars – design of a method for routine applications in genebanks, *Acta. Hort.* 447, 477–483.

Schumacher, H.M. (1999) Cryo-conservation of industrially important plant cell cultures, in *Plant Conservation Biotechnology*, Benson, E.E., Ed., Taylor & Francis, London, pp. 125–137.

Seitz, U., Alfermann, A.W., and Renhard, E. (1983) Stability of biotransformation capacity of *Digitalis lanata* cell cultures after cryogenic storage, *Plant Cell Reports*, 2, 273–276.

Society of Low Temperature Biology. (2000) Coping with cold: natural solutions, *Cryo-Letters*, 21, 2.

Staines, H.J., Nashatul Zaimah, N.A., Marzalina, M., Benson, E.E., and Krishnapillay, B. (1999) Using Taguchi experimental design for developing cryopreservation strategies for recalcitrant seeds, in *IUFRO Symposium 1998 Recalcitrant Seeds Proceedings of the Conference 12–15ʰ October, 1998,* Marzalina, M. Khoo, K.C., Jayanthi, N., Tsan, F.Y., and Krishnapillay, B., Eds., Kuala Lumpur, Forest Research Institute of Malaysia, pp. 270–279.

Stanwood, P.C. (1985) Cryopreservation of seed germplasm for genetic conservation, in *Cryopreservation of Plant Cells and Organs,* Kartha, K.K., Ed., CRC Press, Boca Raton, FL, pp. 199–226.

Steponkus, P.L. (1984) Role of the plasma membrane in freezing injury and cold acclimation, *Annu. Rev. Plant Physiol.*, 35, 534–584.

Steponkus, P.L. (1985) Cryobiology of isolated protoplasts: applications in plant cell cryopreservation, in K.K Kartha, K.K., Ed., *Cryopreservation of Plant Cells and Organs*, CRC Press, Boca Raton, FL, pp. 50–60.

Sutherland, J.R., Diekmann, M., and Berjak, P. (2002) *Forest Trees Seed Health for Germplasm Conservation,* IPGRI Bulletin, No. 6, Rome, IPBGR.

Taylor, R. and Fletcher, R.L. (1999) Cryopreservation of eukaryotic algae: a review of methodologies, *J. Appl. Phycol.*, 10, 481–501.

Thierry, C., Tessereau, H., Florin, B., Meschine, M.-C., and Petiard, V. (1997) Role of sucrose for the acquisition of tolerance to cryopreservation of carrot somatic embryos, *Cryo-Letters*, 18, 283–292.

Towill, L.E. (1981a) *Solanum etuberosum*: A model for studying the cryobiology of shoot-tips in the tuber bearing (Solanum) species, *Plant Sci. Lett.,* 20, 315–324.

Towill, L.E. (1981b) Survival at low temperatures of shoot-tips from cultivars of *Solanum tuberosum* Group Tuberosum, *Cryo-Letters*, 2, 373–382.

Towill, L.E. (1983) Improved survival after cryogenic exposure of shoot-tips derived from *in vitro* plantlet cultures of potato, *Cryobiology,* 20, 567–573.

Towill, L.E. (1984) Survival at ultra-low temperatures of shoot-tips from cultivars of *Solanum tuberosum* group *Andigena, Phureja, Stenototmum, Tuberosum* and other tuber-bearing *Solanum* species, *Cryo-Letters*, 5, 319–326.

Towill, L.E. (1990) Cryopreservation of isolated mint shoot tips by vitrification, *Plant Cell Rep.*, 9, 178–180.

Urugami, A., Sakai, A., Nagai, M., and Takahashi, T. (1989) Survival of cultured cells and somatic embryos of *Asparagus officianalis* cryopreserved by vitrification, *Plant Cell Rep.*, 8, 418–421.

Vavilov, N.I. (1997*) Five Continents, (A book dedicated to the memory of Nicolay Ivanovich Vavilov (1887–1943),* Reznik, S. and Stapleton, P., Eds., Love, D., Trans., IPGRI, Rome.

Vertucci, C.W., Crane, J., Porter, R.A., and Oelke, E.A (1995) Survival of Zizania embryos in relation to water content, temperature and maturity status, *Seed Sci. Res.*, 5, 31–40.

Vigneron, T., Arbault, A., and Kaas, R. (1997) Cryopreservation of gametophytes of *Laminaria digitata* (l) Lameroux by encapsulation dehydration, *Cryo-Letters*, 18, 93–99.

Villalobos, V.M., Ferreira, P., and Mora, A. (1991) The use of biotechnology in the conservation of tropical germplasm, *Biotechnol. Adv.*, 9, 197–215.

Villiers, T.A. (1974) Seed ageing: Chromosome stability and extended viability of seeds stored fully imbibed, *Plant Physiol.*, 53, 875–878.

Villiers, T.A. and Edgcumbe, D.J. (1975) On the cause of seed deterioration in dry storage, *Seed Sci. Technol.*, 3, 761–764.

Walters, C. (1999) Levels of recalcitrance in seeds, *IUFRO Symposium 1998 Recalcitrant Seeds Proceedings of the Conference 12th–15th October 1998*, Marzalina, M., Khoo, K.C., Jayanthi, N., Tsan, F.Y., and Krishnapillay, B., Eds., Forest Research Institute of Malaysia, Kuala Lumpur, pp. 1–13.

Ward, A.C.W., Benson, E.E., Blackhall, N.W., Cooper-Bland, S., Powell, W., Power, J.B., and Davey, M.R. (1993) Flow cytometric assessments of ploidy stability in cryopreserved dihaploid *Solanum tuberosum* and wild *Solanum* species, *Cryo-Letters*, 14, 145–152.

Withers, L.A. (1975) Freeze-preservation of cultured cells and tissues, in *Frontiers of Plant Tissue Culture*, Thorpe, T.A., Ed., International Association of Plant Tissue Culture, University of Calgary, Calgary, Alberta, pp. 297–306.

Withers, L.A. (1977) Freeze preservation of cultures plant cells. III. The pregrowth phase, *Physiol. Plant.*, 39, 171–178.

Withers, L.A. (1978) A fine structural study of the freeze-preservation of plant tissue cultures. II. The thawed, *Protoplasma,* 94, 235–247.

Withers, L.A. (1979) Freeze preservation of somatic embryos and clonal plantlets of carrot (*Daucus carota* L.), *Plant Physiol.*, 63, 460–467.

Withers, L.A (1980) *IBPGR Report, Tissue Culture Storage for Genetic Conservation*, IBPGR, Rome.

Withers, L.A. (1985) Cryopreservation of cultured plant cells and organs, in *Cryopreservation of Plant Cells and Organs,* Kartha, K.K., Ed., CRC Press, Boca Raton, FL, pp. 243–268.

Withers, L.A. and Davey, M.R. (1978) A fine structural study of the freeze-preservation of plant tissue cultures. I. The frozen state, *Protoplasma*, 94, 207–219

Withers, L.A. and King, P. (1980) A simple freezing unit and routine cryopreservation method for plant cell cultures, *Cryo-Letters*, 1, 213–220.

Withers, L.A. and Street, H.E. (1977) Freeze preservation of cultures plant cells. III. The pregrowth phase, *Physiol. Plant.*, 39, 171–178.

Withers, L.A. and Williams, J.T. (1980) *Crop genetic resources conservation of difficult material. Proceedings of an International Workshop, University of Reading, U.K. September 1980,* International Union of Biological Sciences, International Genetic Federation, International Board for Plant Genetic Resources, IUBS Series, B-42, R. Royer Press, Paris.

Wolanczyk, J.P. (1989) Differential scanning calorimetry analysis of glass transitions, *Cryo-Letters*, 10, 75–76.

Yoshimatsu, K., Touno, K., and Shimomura, K. (2000) Cryopreservation of medicinal plant resources: retention of biosynthetic capabilities in transformed cultures, in *Cryopreservation of Tropical Plant Germplasm: Current Research Progress and Application. JIRCAS, International Agriculture Series No. 8,* Engelmann, F. and Takagi, H., Eds., IPGRI, Rome pp. 77–88.

Zakri, A.H., Normah, M.N., Senawai, M.T., and Abdul Karim, A.G. (1991) *Conservation of Plant Genetic Resources through in Vitro Methods. Proceedings of an International Workshop on Tissues Culture for the Conservation of Biodiversity and Plant Genetic Resources,* 28–31 May, Kuala Lumpur, Malaysia, Forest Research Institute of Malaysia Publications, Kuala Lumpur.

10 Plant Cryopreservation

Akira Sakai

CONTENTS

10.1 INTRODUCTION

Cryopreservation is becoming a very important tool for the long-term storage of plant genetic resources using a minimum of space and maintenance. More recently, cryopreservation is reported to offer a real hope for long-term conservation of endangered germplasm (Hargreaves et al., 1997; Touchell, 2000; Turner et al., 2000). Cryopreservation is based on the noninjurious reduction and subsequent interruption of metabolic functions of biological materials by temperature reduction to the level of LN_2 (–196°C). Availability and the development of safe, cost-effective, and reliable cryogenic protocols, and subsequent plant regeneration without genetic change, are basic requirements for plant germplasm conservation.

Conventional cryopreservation of cultured cells and tissues has been achieved by controlled slow prefreezing to about –40°C in the presence of suitable cryoprotectants before being cooled in LN_2. Almost 10 years ago, some new cryogenic, vitrification-based techniques were presented and dramatically increased the applicability to a wide range of plant materials. This chapter describes the development of cryopreservation of plant germplasm and its survival mechanism in the temperature of LN_2.

329

10.2 SURVIVAL MECHANISM OF VERY HARDY PLANTS AND CELLS COOLED TO –196°C.

10.2.1 SURVIVAL MECHANISM OF VERY HARDY TWIGS AT ULTRA-LOW TEMPERATURES

In extracellularly frozen cells, the lower the temperature, the greater the amount of ice formed at equilibrium until the temperature is reached at which all the freezable water has crystallized. Thus, equilibrium freezing is considered as effective means for freeze dehydration of hardy cells. For the purpose of testing freeze dehydration to various degrees, winter mulberry twig pieces were slowly frozen to different temperatures and held there for 16 h before immersion in LN_2. The survival gradually increased with decreasing temperature from –10° to –30°C, reaching the maximum a –30°C or below if thawed very slowly at 0°C (Sakai, 1956, 1960). The mulberry twig pieces and also twigs of *Salix koriyanagi* and *Populus sieboldii*, immersed in LN_2 after prefreezing at –30°C, remained alive at least for 1 year (Sakai, 1956, 1960). In addition, twig pieces of mulberry and twigs of *Salix sachalinensis* were not injured even when cooled from 0° to –120°C at a cooling rate of –0.5°C min^{-1}. From these results, it appears that there is a definite temperature range at which the freezable water in cells is extracted by equilibrium freezing, in this state the cells are not injured, even when exposed to an extremely low temperature. Below this temperature, the intensity of cold does not seem to exert any important effect upon these cells. From these results, the most appropriate interpretation is that completely hardy plants form aqueous glasses intracellularly during subsequent cooling to low temperatures, without crystallization. Hirsh et al. (1985) provided the evidence (freeze-etching electron micrographs) that winter-hardened poplar cortical cells resist the stresses of freezing below –28°C by glass formation of the bulk of intracellular solution during slow cooling. This implies that the formation of the solid intracellular glass prevents further water loss to extracellular ice at lower temperatures. During slow warming of the samples after slow cooling to –70°C at 3°C/min, a sudden endotherm was detected by differential scanning calorimetory (DSC) at –28°C, indicating an equilibrium glass transition (Fahy et al., 1984) without devitrification.

The prefreezing temperature required to maintain viability depends on the relative hardiness of the plant and varies from –15° to –40°C when thawed slowly (Sakai, 1965, 1973), with the hardier samples requiring less prefreezing, indicating that hardier plants significantly increase their ability to vitrify. Twigs of *Salix sachalinensis*, *Populus maximowiczii*, and *Betular platyphylla* even survived freezing to the temperature of liquid helium (about –269°C) following freezing to –20° or –30°C (Sakai, 1962; Sakai and Larcher, 1987). The treated twigs of *S. sachalinensis* rooted and grew to a tree. These results indicate that very hardy plants can tolerate freezing near the temperature of absolute zero.

10.2.2 SURVIVAL MECHANISM OF VERY HARDY CELLS COOLED TO –196°C

Cortical tissues of winter mulberry twigs can survive slow continuous freezing to –70° or –120°C. The cortical tissues, which were mounted between cover glasses with 0.05 mL water, were prefrozen at different temperatures before plunging in LN_2 (Sakai, 1966, 1985). Prefrozen cells below –20°C remained alive after rapid cooling in LN_2, regardless of warming rate (Figure 10.1).

However, in cells prefrozen above –15°C, the warming rate seriously influenced survival. In the prefrozen cells at –10°C, all cells remained alive by rapid warming in water at 30°C, but most of the cells were killed by slow warming in air at 0°C, indicating that the prefrozen cells at –10°C vitrified on rapid cooling and were killed by intracellular freezing after devitrification occurred during slow warming. The vitrified cells were considered to be killed in the temperature range between –30° and –40°C during the warming process (Sakai and Yoshida, 1967). To confirm this, tissue sections were kept in a bath at –30°C for 10 min following removal from LN_2 before being freeze-substituted with absolute alcohol at –78°C. In the electron micrographs of these cells, many ice cavities were found throughout the cells (Sakai and Otsuka, 1967).

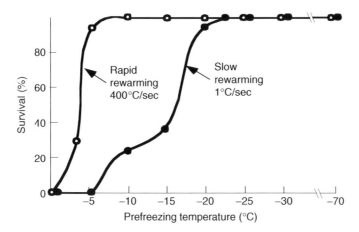

FIGURE 10.1 Effect of rewarming rate on the survival of cells immersed in LN$_2$ following prefreezing at various temperatures. Tissue sections were mounted between cover glasses with 0.05 mL water. Tissue sections were rewarmed rapidly in water at +30°C (400°C/sec) or slowly in air at 0°C (1°C/sec). (From Sakai, A. With permission.)

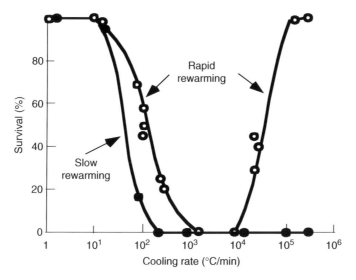

FIGURE 10.2 Survival of cortical cells from winter mulberry twigs cooled to –75°C at rates indicated on the abscissa, without prefreezing, and rewarmed either rapidly or slowly. Rewarming was done rapidly by immersion in water at +30°C and slowly in air at 0°C. (From Sakai, A. and Yoshida, S., 1967. With permission.)

The survival rates of mulberry cortical tissues as a function of cooling and rewarming rates are summarized in Figure 10.2 (Sakai and Yoshida, 1967). When cooling is carried out slowly at –1° to –10°C/min, all of the cells remained alive, even when rewarmed either rapidly or slowly. Survival rates gradually decreased to zero as the cooling rate increased from 50°C to 2000°C/min, regardless of warming rate, which indicates an increase in the number of the cells frozen intracellarly during cooling. In cells cooled at intermediate cooling rates, the intracellular water is not able to dehydrate sufficiently and rapidly enough to maintain vapor pressure equilibrium with the external ice, and so the cells become increasingly supercooled and eventually freeze intracellularly. However, at cooling rates above –10,000°C/min applied with rapid warming, the effects were reversed and survival rates increased, reaching a maximum at –2,000,000°C/min. In the case of slow warming, however, survival still remained at zero for all the rapid cooling rates used. To cool ultra rapidly, an unmounted tissue section (unicellular tissues, 1 to 2 mm wide and 2 to 3 mm long) held with forceps was directly immersed into LN$_2$ (Sakai, 1956). These ultra-rapidly cooled cells LN$_2$ remained alive when rewarmed rapidly. Using freeze-etching electron micrographs (Fujikawa, 1988), it was confirmed that these cells vitrified.

To enable very hardy cells to cool directly into LN$_2$ from ambient temperature at a practical cooling rate, hardy mulberry cortical cells were treated with different concentrations (1.0 to 4.0 *M*) of ethylene glycol (EG) for 10 min at ambient temperature before plunging in LN$_2$. During the treatment with 3 *M* EG for 10 min, the cells plasmolyzed in the beginning, and then gradually deplasmolyzed, along with the permeation of EG, into the cytosol. The cortical tissues treated with 3 *M* EG that were mounted with cover glasses with 0.05 mL EG remained alive (80% survival) after plunging in LN$_2$ when rewarmed in water at +30°C (Sakai, 1958). This procedure eliminates a freeze-dehydration step and enables hardy cells to be cooled directly in LN$_2$, provided that hardy cells are sufficiently concentrated. This procedure is a prototype of the present vitrification protocol.

10.3 DEVELOPMENT OF CRYOPRESERVATION

A prefreezing method with subsequent slow warming is widely used as the routine method for cryopreservation of very hardy plants. Successful cryopreservation of dormant vegetative buds of the apple by prefreezing to –40°C, and subsequent recovery by grafting, was first reported by Sakai and Nishiyama (1978). This protocol using apple dormant buds was refined by Stushnoff (1987) and Forsline et al. (1998). Dormant buds of the pear (Oka et al., 1991) and mulberry (Niino and Sakai, 1992) have also been successfully cryopreserved, followed by recovery of the meristems *in vitro*. Cryopreservation of dormant vegetative buds offers distinct advantages over the alternatives (Forsline et al., 1998). The cryogenic procedure is simple and reliable, with slight genotype-specific reactions to cryogenic protocols. A large number of uniform and tolerant samples are easily collected and stored in a frozen state.

The prefreezing method is also widely used as the routine method for cryopreservation to less hardy cultured cells and tissues, with the prior application of cryoprotectants. The conventional prefreezing method is depicted in Figure 10.3. Cultured cells and tissues to be cryopreserved are generally sensitive to freezing. Thus, they are previously treated with cryoprotectants (5 to 10% [v/v] Me$_2$SO and 5 to 15% [w/v] sucrose or sorbitol) so that they can survive prefreezing to –30 or –40°C before being immersed into LN$_2$ (Cyr, 2000; Kartha et al., 1988; Withers, 1985). If the frozen cells are further cooled to –50 or –70°C, their survival gradually decreases with decreasing temperature (Sugawara and Sakai, 1978). Most cultured cells are prefrozen at a rate of –0.3 to –0.5°C/min to about –40°C before plunging into LN$_2$, and they are subsequently thawed rapidly. Slow prefreezing to –40°C produces a sufficiently concentrated unfrozen fraction in the suspending solution and the cytosol to be capable of vitrifying on rapid cooling in LN$_2$. Survival of the vitrified cells depends on the fate of the glassy cytoplasm during the warming process. When warming is slow enough to permit sufficient time for crystallization to occur, then low survival was observed (Sakai et al., 1991; Yamada et al., 1991). Thus, vitrified cells following freeze-dehydration to about –40°C must be thawed rapidly in water at +30 or +40°C to produce a high level of survival, unlike the situation for cryopreservation of very hardy twigs and buds.

Almost 10 years ago, some new cryogenic procedures (refer Figure 10.3), such as vitrification (Langis et al., 1989, 1990; Sakai et al., 1990; Towill, 1990; Uragami et al., 1989) and encapsulation/dehydration (Fabre and Dreuddre, 1990) were presented. These procedures dehydrate a major part of the freezable water of specimens at nonfreezing temperatures and enable them to be cryopreserved by being directly plunged into LN$_2$ without a freeze-induced dehydration step. These alternative dehydration procedures, referred to as "vitrification-based techniques" (Engelmann, 1997), offer practical advantages compared with the classical prefreezing method. By precluding ice formation in the system, vitrification-based techniques simplified the cryogenic procedures, as they do not require the use of controlled freezers and eliminate concerns for the potentially damaging

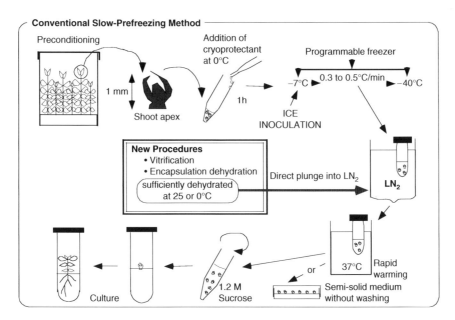

FIGURE 10.3 A scheme of conventional slow-freezing method and new procedures.

effects of intra- and extracellular crystallization. In particular, these new procedures are more appropriate for differentiated structures (apices and somatic embryos) that contain a variety of cell types, each with unique requirements for cooling rates under conditions of freeze-induced dehydration (Withers, 1979).

10.4 CLASSIFICATION OF CRYOGENIC PROCEDURE

In the vitrification-based protocols, cell dehydration is performed by exposing samples to a highly concentrated cryoprotective solution (osmotic dehydration) or by air desiccation before being plunged into LN_2. Such techniques are referred to as vitrification (complete vitrification, as both cytosol and the suspension solution vitrify), as distinct from the conventional prefreezing method (partial vitrification, where only the cytosol vitrifies). At any rate, vitrification is the only survival mechanism for maintaining the viability of hydrated cells and tissues in the temperature of LN_2 by avoiding crystallization on rapid cooling (Sakai, 1960, 1985).

Successful cryogenic procedures applying practical cooling rates by either partial or complete vitrification can be divided into four categories based on the dehydration method used before the plunge into LN_2 (Sakai, 1993, 1995):

1. Slow prefreezing (freeze-dehydration)
2. Vitrification (osmotic dehydration) with or without encapsulation
3. Encapsulation/dehydration (osmotic dehydration combined with air-drying)
4. Air-desiccation

10.5 CONCEPT OF VITRIFICATION

There are two types of liquid–solid phase transition of aqueous solution. Ice formation is the phase transition from a liquid to ice crystal. Catalysts for the water–ice phase transition are referred to as ice nuclei. There are two types of ice nucleation: homogeneous and heterogeneous. In homogeneous nucleation, the nuclei are formed spontaneously in the liquid without intervention of foreign bodies at a very low temperature, approaching –40°C. The freezing process involves two different

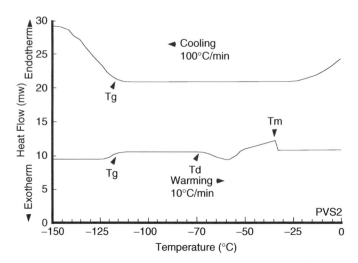

FIGURE 10.4 Differential scanning calorimetry record of a highly concentrated vitrification solution (PVS2). Tg, glass transition temperature (about −115°C); Td, initiation of freezing following devitrification (about −75°C); Tm, freezing or melting point (−37°C). (From Sakai, A. et al., 1990. With permission.)

phenomena: the formation of ice nuclei and growth of crystal units. Another phase transition is vitrification from a liquid to amorphous glass, avoiding crystallization. Water is very difficult to vitrify, because the growth rate of crystals is very high even just below the freezing point. However, highly concentrated cryoprotective solutions such as glycerol are very viscous and are easily supercooled below −70°C, which allows them to be vitrified on rapid cooling; this was first demonstrated by Rasmussen and Luyet (1970).

Evidence for vitrification requires the use of physical procedures, and one conventional method is to measure the latent heat released by the crystallization of ice during cooling and warming using DSC. As shown in Figure 10.4, the DSC profile reveals that a glycerol-based vitrification solution (Plant Vitrification Solution 2 [PVS2], 7.8 M; Sakai et al., 1990) supercooled below −100°C and finally became vitrified at about −115°C (T_g, glass transition temperature) on cooling (−100°C/min) and that a glass transition occurred during subsequent slow warming (+10°C/min) followed by exothermic devitrification (crystallization; T_d, devitrification temperature) and endothermic melting (T_m, melting temperature). Crystallization during the warming process can be prevented when the object is rewarmed rapidly in water at +30°C. If the vitrification solution is more concentrated, the value of T_g shifts to a higher temperature and T_m to a lower temperature. Finally, when T_g becomes equal to the T_m (equilibrium glass transition temperature), crystallization following devitrification is prevented and the system is virtually stable (Fahy et al., 1984).

As glass fills spaces in a tissue, glass may contribute to preventing additional tissue collapse, solute concentration, and pH alteration during dehydration. Operationally, a glass is expected to exhibit a lower water vapor pressure than the corresponding crystalline solid, thereby preventing further dehydration. As glass is exceedingly viscous and stops all chemical reactions that require molecular diffusion, its formation leads to stability over time (Burke, 1986).

10.6 VITRIFICATION PROTOCOL

10.6.1 PROTOCOL

The vitrification protocol requires the application of a highly concentrated vitrification solution, which sufficiently dehydrates cells without causing injury so that they form a stable glass along with the surrounding vitrification solution when plunged into LN_2. The author (Sakai et al., 1990)

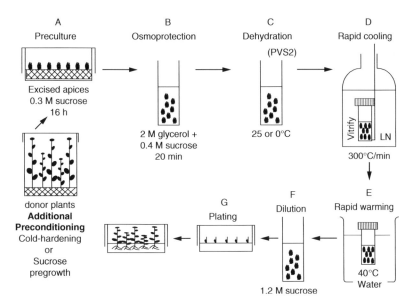

FIGURE 10.5 General view of the vitrification protocol for excised apices from *in vitro*–grown plants. (A) Preculture: 1–2 days, Murashige and Skoog medium with 0.2–0.5 *M* sucrose. (B) Osmoprotection: 20 min at +25°C, with a mixture of 2 *M* glycerol and 0.4 *M* sucrose. (C) Dehydration: 0–60 min at +25°C or 0–90 min at 0°C in PVS2. (D) Rapid cooling (at about 300°C/min): LN_2 storage, use the 2.0-mL cryotube. (E) Rapid warming: shaking 80 sec in water at +40°C. (F) Dilution: 15 min, in liquid Murashige and Skoog medium with 1.2 *M* (w/v) sucrose. (G) Plating: different media depending on species, but always with 3% (w/v) sucrose.

has developed a glycerol-based, low-toxicity vitrification solution designated PVS2. PVS2 (7.8 *M*) contains 30% (w/v) glycerol, 15% (w/v) EG, and 15% (w/v) Me_2SO in the Murashige and Skoog (MS) basal medium containing 0.4 *M* sucrose at pH 5.8 (refer to Figure 10.4). Benson et al. (1996) confirmed that DSC profiles showed no evidence for ice nucleation in PVS2 treated *ribes*-apices on cooling (at –10°C/min) and that this was reproducible for replicate samples.

The complete vitrification procedure for cryopreserving apices is shown in Figure 10.5. Vitrification is an elaborate and non-time-consuming protocol and has wide applicability. It produces higher and earlier recovery growth than the encapsulation-dehydration method (Hirai et al., 1998; Hirai and Sakai, 1999a, 1999b; Matsumoto et al., 1994, 1995b). As summarized in Table 10.1, the vitrification protocol for apices has successfully been applied to a wide range of plant materials of both temperate and tropical origins, totaling 180 or more over the last 10 years. The protocol has also been successfully established for roots and tubers, fruit trees, ornamental and medicinal plants, and plantation crops (Sakai, 1997, 2000; Thinh et al., 2000).

Successful cryopreservation by vitrification using EG-based vitrification solution and French straw (Langis and Steponkus, 1990; Langis et al., 1989) was also reported in the meristems of carnation (Langis et al., 1990) and potato (Golmirzaie and Panta, 2000; Lu and Steponkus, 1994).

In the encapsulation-vitrification protocol, samples are encapsulated in alginate beads, osmoprotected, and then dehydrated with a highly concentrated vitrification solution for 2 or 3 h before a plunge into LN_2 (Hirai et al., 1998; Hirai and Sakai, 1999a, 1999b; Matsumoto et al., 1995a; Tannoury et al., 1991).

10.6.2 THE DEHYDRATION STEP

Duration of exposure to the PVS2 must be optimized to produce high levels of recovery growth after cryopreservation by vitrification. The optimum exposure time to PVS2, and thus the

TABLE 10.1
Successful Cryopreservation (Recovery Growth > 50%)
of *in vitro*–Grown Apices Cooled to –196°C by Vitrification
Using PVS2 (from Sakai, 2000, revised)

Plant[a]		References	
Temperate Plants			
Woody Plants			
Citrus sinensis	(Nucellar cells)	Sakai et al.	1990, 1991
Citrus	(Nucellar cells, 4 spp, cvs)	Kobayashi and Sakai	1997
Grevillea scapigera	(10 clones)	Touchell	2000
Malus	(Apple, 5 spp, cvs)	Niino et al.	1992c
Mulberry	(13 spp, cvs)	Niino et al.	1992a
Persimmon	(Winter apices, 20 spp, cvs)	Matsumoto et al.	2001
Poplar		Lambardi et al.	2000
Prunus	(Cherry, 8 cvs)	Niino et al.	1997
Pyrus	(Pear, 5 cvs)	Niino et al.	1992c
Ribes nigrum		Benson et al.	1996
Tea plant		Kuranuki and Sakai	1995
Vitis	(Grape, 10 spp, cvs)	Matsumoto et al.	1998b
Herbaceous Plants			
Asparagus	(Bud cluster)	Kohmura et al.	1992
Fragaria X ananassa	(Strawberry, 4 cvs, EV)	Hirai et al.	1998
Garlic	(Post dormant bulbils, 12 cvs)	Niwata	1995
Garlic	(2 cvs)	Makowska et al.	1999
Lily	(4 spp, cvs)	Matsumoto et al.	1995b
Mint	(3 spp, EV)	Hirai & Sakai	1999a
Panax ginseng	(Hairy roots)	Yoshimatsu et al.	1996, 2000
Papaver somniferum	(Hairy roots)	Yoshimatsu et al.	2000
Solanum tuberosum	(Potato, 14cva, EV)	Hirai and Sakai	1999b
Sugar beet	(18 clones)	Vandenbussche et al.	2000
Wasabi	(Brassicaceae, 4 cvs)	Matsumoto et al.	1994
White clover	(3 spp, cvs)	Yamada et al.	1991
Tropical Plants			
Cassava		Charoensub et al.	1999
Colocasia	(Taro, 6 cvs)	Takagi et al.; Thinh	1997, 1997
Dioscorea	(Yam, 6 cvs)	Kyesmu and Takagi	2000
Musa	(Banana, 9 cvs)	Thinh et al.	1999
Cymbidium	(Orchid, 2 cvs)	Thinh and Takagi	2000
Dendrobium	(Orchid, protocorm)	Wang et al.	1998
Papaya	(Somatic embryo)	Lu and Takagi	2000
Pineapple		Thinh	1997
Pineapple		González-Arnao et al.	1998b

[a] EV: encapsulation-vitrification; cvs: cultivars

dehydration time of apices, appears to be species-specific and to increase with size of excised apices (Niino et al., 1992b). Properly dehydrated apices vitrify during rapid cooling into LN_2, preventing lethal intracellular freezing and avoiding the risks of toxic effects resulting from over-exposure to PVS2. Thus, the critical step to achieve high survival in the vitrification method is the

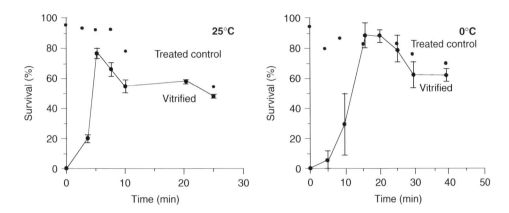

FIGURE 10.6 Effect of exposure time to PVS2 at +25° or 0°C on the shoot formation of clover apices cooled to –196°C by vitrification. Apical meristems following preculture were treated with PVS2 for different lengths of time at +25°C or 0°C and then directly plunged into LN₂ for 30 min. Approximately 20 meristems were tested for each of two replicates. Bar represents standard error. Treated control: same as before but without cooling to –196°C. No replicate in treated control. (From Yamada et al., 1991. With permission.)

dehydration step. As shown in Figure 10.6, exposure to PVS2 produced time-dependent shoot formation. In the vitrified clover apices cooled to –196°C, the shoot formation reached the highest rate (about 95%) at 5 min exposure at +25°C or at 30 min at 0°C. The rates of shoot formation of the apices dehydrated with PVS2 for 5 min at + 25°C or 15 min at 0°C without cooling in LN₂ (treated control) were nearly almost the same as those of the vitrified apices cooled to –196°C. Thus, the excursion of rapid cooling (vitrification) into LN₂ and subsequent rapid warming (devitrification) did not cause additional loss beyond that produced during dehydration process with PVS2 at nonfreezing temperatures. The same trend was observed in the vitrified tobacco cultured cells (Reinhoud, 1996) and apical meristems (Niino and Sakai, 1992; Matsumoto et al., 1994; Yamada et al., 1991). These results clearly demonstrate that the dehydration tolerance to PVS2 is sufficient for apical meristems to survive the vitrification procedure.

10.6.3 THE ACQUISITION OF DEHYDRATION TOLERANCE OF EXPLANTS

Whatever cryogenic method is applied to apical meristems excised from *in vitro*–grown plants, the induction of higher levels of dehydration tolerance is required for successful cryopreservation. Applying a high level of sugar or sugar alcohol during preculture has been reported to be very important in producing the survival of cryopreserved cells and meristems (Dereuddre et al., 1991; Niino and Sakai, 1992; Uragami et al., 1990). Reinhoud (1996) clearly demonstrated that development of osmotolerance of tobacco cells to PVS2 during preculture with 0.3 *M* mannitol solution for 1 d appeared to be a combined result of mannitol uptake and a cellular response to mild osmotic stress, caused by the preculture-induced production of Abcisic acid, proline, and certain proteins.

However, preculturing of explants with sucrose alone did not lead to a substantial increase in the survival of many apices cooled to –196°C by vitrification (Matsumoto et al., 1994, 1995a, 1995b). We found that the treatment with a mixture of 2 *M* glycerol and 0.4 *M* sucrose (termed LS solution, Nishizawa et al., 1993) for 20 min at +25°C following preculture with 0.3 *M* sucrose for 1 d was very effective in enhancing osmotolerance of Japanese horseradish (wasabi) apices to PVS2 (Matsumoto et al., 1994). The significant positive effects with LS solution were observed in the several tropical monocots such as banana, taro, and orchid apices (Thinh et al., 1999, 2000; Thinh and Takagi, 2000). More recently, the beneficial osmoprotection with LS treatment proved to be very effective in cassava (Charoensub et al., 1999), sugar beet (Vandenbussche et al., 2000), and sweet potato (Pennycooke and Towill, 2000). Thus, the LS treatment following preculture with sucrose enriched medium is a very promising step for the successful cryopreservation of apices by vitrification.

During the treatments of apical meristems with LS solution for 20 min, the meristematic cells were osmotically dehydrated and considerably plasmolyzed. However, little or no permeation of glycerol was observed in the plasmolyzed cells based on volumic change in the cytosols over 20 min. The plasmolyzed cells were subsequently subjected to PVS2 solution. Under these conditions, the plasmolyzed cells successively decreased in cytosolic volume. Thus, there seems not to be an appreciable influx of additional cryoprotectants into cytosol because of the difference in the permeability coefficient for water and solute permeation, especially in highly viscous solution (Steponkus et al., 1992). As a result, the cells of apices remain osmotically dehydrated, and increase in the cytosolic concentration required for vitrification is attained by dehydration. Matsumoto et al. (1998a) observed that in the wasabi, in longitudinal sections of apices, which were dehydrated with PVS2 for optimal time of exposure before a plunge into LN_2, meristematic cells were intensively plasmolyzed, forming contracted spherical protoplasts.

The protective effect of brief incubation with LS solution for 20 min after preculture might be caused by osmotic dehydration, resulting in the concentration of cytosolic stress-responsive solutes accumulated during preculture with 0.3 M sucrose for 16 h and in the protective effects of plasmolysis. The presence of highly concentrated cryoprotective solution in the periprotoplasmic space of plasmolyzed cells may mitigate the mechanical stress caused by successively severe dehydration (Hellergren and Li, 1981; Jitsuyama et al., 1997; Tao et al., 1983). These intracellular and extracellular protective effects may minimize the injurious membrane changes during severe dehydration, though the actual protective action mechanism is poorly understood.

10.7 ENCAPSULATION/DEHYDRATION TECHNIQUE

In the encapsulation/dehydration technique (Fabre and Dereuddre, 1990), explants are inserted into alginate beads that are osmoprotected in medium enriched with 0.75 M sucrose for 1 to 2 d before being dehydrated in the air current of a laminar airflow cabinet or with dry silica gel down to a water content of around 20% (fresh weight basis) before plunging into LN_2. In this technique, the extraction of water results in a progressive osmotic dehydration, and additional loss of water is obtained by evaporation and the subsequent increase of sucrose concentration in the beads. Thus, the sucrose molarity in the beads increases markedly during the drying process and reaches or exceeds the saturation point of the sucrose solution, resulting in glass transition during cooling to –196°C (Dereuddre et al., 1991). The dehydration technique allows much more flexibility for handling large amounts of materials, because the time schedule for all the steps is much broader than with vitrification. In addition, this technique does not use any other cryoprotectant other than sucrose. Thus, the technique has also been widely applied to apices of numerous species (Table 10.2), especially in cold-hardened or well-precultured apices treated with a progressive increase in sucrose (Niino and Sakai, 1992; Paul et al., 2000; Wu et al., 1999; Chang et al., 2000).

Our results indicate that the induction of osmotolerance with sucrose alone may be insufficient for the shoot tips to produce a high level of recovery growth (Matsumoto and Sakai, 1995). We observed that encapsulated wasabi apices treated with LS solution for 60 min produced a much higher postthaw recovery growth than those treated with 0.8 M sucrose alone for 16 h. A dramatic increase in recovery growth (from 45 to 95%) was also observed in the encapsulated dried hairy roots of horseradish treated with a mixture of 1 M glycerol and 0.5 M sucrose, compared with those treated with 0.8 M sucrose (Phunchindawan et al., 1997). More recently, Turner et al. (2000) also confirmed the positive effects of LS solution. Thus, we presented a new encapsulation/dehydration protocol using LS solution that considerably reduced the time needed for the procedure and produced higher levels of recovery growth than the conventional encapsulation/dehydration

---

TABLE 10.2
Successful Cryopreservation (Recovery > 40%) of Plants Cooled to –196°C by Encapsulation/Dehydration

Plant[a]		References	
Brassica napus	(Microembryo)	Uragami et al.	1993
Carrot	(Somatic embryo)	Dereuddre et al.	1991
Catharanthus	(Cells)	Bachiri et al.	1995
Centaurium	(Cells)	González-Benito and Perez	1997
Coffea canephora	(somatic embryo)	Hatanaka et al.	1994
Dioscorea buibifera	(Yam, apices)	Malaurie et al.	1998
Eucalyptus	(Apices)	Monod et al.	1992
Horseradish	(Hairy root)	Hirata et al.	1995
Lily	(Apices)	Matsumoto and Sakai	1995
Lolium	(5 cvs)	Chang et al.	2000
Malus	(Apple, apices)	Niino and Sakai	1992
Malus	(Apple, apices, 11 cvs)	Wu et al.	1999
Malus	(Apple, apices, 5 cvs)	Paul et al.	2000
Mint	(Apices)	Hirai and Sakai	1999a
Mulberry	(Apices)	Niino and Sakai	1992
Mulberry	(Winter bud)	Niino et al.	1992b
Poncirus trifoliata	(Citrus, apices)	González-Arnao et al.	1998a
Solanum tuberosum	(Potato, apices, 3 cvs)	Bouafia et al.	1996
Prunus	(Almond, apices)	Shatnawi et al.	1999
Pyrus	(Pear, apices)	Niino and Sakai	1992
Pyrus	(Pear, apices)	Scottez et al.	1992
Ribes nigrum	(Apices, 2 cvs)	Benson et al.	1996
Ribes	(3 spp)	Reed and Yu	1995
Sugarcane	(Apices, 5 cvs)	Paulet et al.	1993
Sugarcane	(Apices, 5 cvs)	González-Arnao et al.	1999
Tea	(Apices)	Kuranuki and Sakai	1995
Wasabi	(Brassicaceae, apices, 4 cvs)	Matsumoto and Sakai	1995
Zoysia	(5 cvs)	Chang et al.	2000

[a] Apices: excised from *in vitro*-grown plants; cvs: cultivars

---

technique (Sakai et al., 2000). We consider that the new protocol combines all the advantages of both techniques, comprising vitrification and encapsulation/dehydration.

10.8 DISCUSSION

In many temperate species and especially woody plants, cold hardening of meristem-donor plants at 0 to 5°C under an 8-h photoperiod for about 3 weeks was very effective in enhancing dehydration tolerance of the apices, followed by sucrose preculture or osmoprotection with LS (Chang et al., 2000; Lambardi et al., 2000; Niino et al., 1992b, 1992c; Paul et al., 2000; Wu et al., 1999). Thinh (1997; Thinh et al., 2000) also demonstrated pregrowth of meristem-donor plants of taro and banana cultured for 1 month on MS medium supplemented with increasing concentrations (6, 9, and 12% [w/v]) of sucrose, which could be a promising alternative to cold hardening for the tropical species (see Thinh, 1997; Thinh et al., 2000). Compared with the control plants grown on a standard concentration of 3% (v/v) of sucrose, preconditioned taro plants appeared to be shorter and more compact. Analysis of shoot parts showed that the pregrowth significantly reduced the water content but enhanced the accumulation of stress-responsive solutes (soluble sugar, free proline). The postthaw survival rates of preconditioned taro apices were significantly higher than that of the control (Thinh, 1997). Thus, cold hardening or pregrowth appears to be a promising step for preconditioning less-tolerant plants in the vitrification-based protocol.

Excised apices employed for cryopreservation must be at a physiological optimum state (Dereuddre et al., 1988) for acquisition of higher levels of dehydration tolerance and for the production of vigorous recovery growth. Thus, the quality control of apices; that is, the selection of suitable and uniform apices required for cryopreservation, appears to be a critical factor for producing a high level of recovery (average >60%). Hirai and Sakai (1999a) clearly demonstrated that about 70 nodal segments (5 mm length, with a pair of leaves) from *in vitro*–grown mint plants that were densely planted on a solidified culture media in a Petri dish (9 cm in diameter) under the light conditions produced young lateral buds about 5 days later. The apical meristems excised from the lateral buds produced very high levels (average 80%) of recovery growth by encapsulation/vitrification. The *in vitro* culture protocol also proved to be very effective (Hirai and Sakai, 1999b) for 14 cultivars of potato (average recovery 70%). This simple micropropagation protocol enabled the production of a large number of relatively homogeneous and adequate apices in terms of size, cellular composition, physiological state, and growth response, and thus increases the chances of positive and uniform responses to subsequent cryogenic treatments.

It is particularly important that cryopreserved cells, meristems, and embryos be capable of producing plants identical to the nontreated phenotypes. A callus phase before shoot formation is undesirable, as callusing potentiality increases the frequency of genetic variants. The patterns of growth of cryopreserved meristems vary considerably with cryogenic protocols and with regrowth medium (Lambardi et al., 2000; Touchell, 1996). When the vitrification-based technique is employed under well-optimized conditions, all or most of the apices remained alive, thus allowing direct, organized regrowth (Hirai and Sakai 1999a, 1999b; Matsumoto et al., 1994; Thinh et al., 1999; Yamada et al., 1991). In contrast, classical prefreezing procedures can lead to the destruction of large zones of apical domes, and callusing, or transitory callusing, is often observed before organized regrowth (Haskins and Kartha, 1980).

10.9 CONCLUSIONS

Significant progress has been made during the last ~10 years in the area of plant cryopreservation, with the development of various efficient vitrification-based protocols. An important advantage of these new techniques is their operational simplicity and efficiency.

For many vegetatively propagated species or cultivars of temperate origins, cryopreservation techniques are sufficiently advanced to envisage their immediate use for large-scale experimentation in genebanks. Research is much less advanced for tropical wild species. This is because of the comparatively limited level of research activities aiming at improving the conservation of these species. However, various technical approaches can be explored to improve the efficiency and to increase the applicability, provided that the tissue culture protocols such as apical meristem and somatic embryo culture are sufficiently operational for the species. It is hoped that new findings on critical issues such as understanding and control of dehydration tolerance will contribute significantly to the development of improved cryopreservation techniques for difficult (recalcitrant) species or cultivars.

ACKNOWLEDGMENTS

I thank Dai Hirai for processing this manuscript and Erica Benson for valuable suggestions.

REFERENCES

Bachiri, Y., Gazeau, C., Hansz, J., Morisset C., and Dereuddre, J. (1995) Successful cryopreservation of suspending cells by encapsulation-dehydration, *Plant Cell Tiss. Org. Cult.*, 43, 241–248.

Benson, E. E., Reed, B.M., Brennan, R.M., Clacher, K.A., and Ross, D.A. (1996) Use of thermal analysis in the evaluation of cryopreservation protocols for *Ribes nigrum L.* germplasm, *Cryo-Letters*, 17, 347–362.

Bouafia, S., Jelti, N., Lairy, G., Blanc, A., Bonnel, E., and Dereuddre. J. (1996) Cryopreservation of potato shoot tips by encapsulation-dehydration, *Potato Res.*, 39, 69–78.

Burke, M.J. (1986) The glassy state and survival of anhydrous biological systems, in *Membrane, Metabolism and Dry Organisms*, Leopold, A.C., Ed., Cornell University Press, Ithaca, NY, pp. 358–364.

Chang, Y., Barker, R.E., and Reed, B.M. (2000) Cold acclimation improves recovery of cryopreserved grass (*Zoysia* and *Lolium* sp.), *Cryo-Letters*, 31, 107–116.

Charoensub, R., Phansiri, S., Sakai, A., and Yongmanitchai, W. (1999) Cryopreservation of cassava *in vitro*-grown shoot tips cooled to –196°C by vitrification, *Cryo-Letters*, 20, 89–94.

Cyr, R. (2000) Cryopreservation: role in clonal preservation and germplasm conservation of conifers, in *Cryopreservation of Tropical Plant Germplasm*, Engelmann, F. and Takagi, H., Eds., JIRCAS, Tsukuba, Japan, pp. 261–268.

Dereuddre, J., Blandin, S., and Hassen, N. (1991) Resistance of alginate-coated somatic embryos of carrot (*Daucus carota* L.) to desiccation and freezing in liquid nitrogen. 1. Effect of preculture, *Cryo-Letters*, 12, 125–134.

Dereuddre, J., Fabre, J., and Basaglia, C. (1988) Resistance to freezing in liquid nitrogen of carnation (*Dianthus caryophyllus* L. var. *Kolo*) apical and axillary shoot tips excised from different aged *in vitro* plants, *Plant Cell Rep.*, 7, 170–173.

Engelmann, F. (1997) Importance of desiccation for cryopreservation of recalcitrant seed and vegetatively propagated apices, *Plant Genetic Resources Newsl.*, 112, 9–18.

Fabre, J. and Dereuddre, J. (1990) Encapsulation-dehydration: a new approach to cryopreservation of *Solanum* shoot tips, *Cryo-Letters*, 11, 413–426.

Fahy, G.M., MacFarlane, D.R., Angell, C.A., and Meryman, H.T. (1984) Vitrification as an approach to cryopreservation, *Cryobiology*, 21, 407–426.

Forsline, P.L., Towill, L.E., Wardell, W., Stuschnoff, C., Lamboy, W.F., and Ferson, J.R. (1988) Recovery and longevity of cryopreserved dormant apple buds, *J. Am. Soc. Hort. Sci.*, 123, 365–370.

Fujikawa, S. (1988) Artificial biological membrane ultrastructural changes caused by freezing, *Electron Microsc. Rev.*, 1, 113–140.

Golmirzaie, A.M. and Panta, A. (2000) Advances in potato cryopreservation at CIP, in *Cryopreservation of Tropical Plant Germplasm*, Engelmann, F. and Takagi, H., Eds., JIRCAS, Tsukuba, Japan, pp. 250–254.

González-Arnao, M.T., Engelmann, F., Urra, C., Morenza, M., and Rios, A. (1998a) Cryopreservation of citrus apices using the encapsulation dehydration technique, *Cryo-Letters*, 19, 177–182.

González-Arnao, M.T., Ravelo, M.M., Urra Villavicencio, C., Montero, M.M., and Engelmann, F., (1998b) Cryopreservation of pineapple (*Ananas comosus*) apices, *Cryo-Letters*, 19, 375–382.

González-Arnao, M.T., Urra, C., Engelmann, F., Ortiz, R., and Fe, C.D.L. (1999) Cryopreservation of encapsulated sugarcane apices: effect of storage temperature and storage duration, *Cryo-Letters*, 20, 347–352.

Gonza-Benito, M.E. and Perez, C. (1997) Cryopreservation of nodal explants of an endangered plant species (*Centaurium rigualii Esteve*) using encapsulation-dehydration method, *Biodiversity Conserv.*, 6, 583–590.

Hargreaves, C.L., Smith, D.R., Fogg, M.N., and Gorden, M.E. (1997) Conservation and recovery of Cheezemania Chalk Range an endangered New Zealand Brassicaceous plant, *Combined Int. Plant Propagation Soc.*, 47, 132–136.

Haskins, R.H. and Kartha, K.K. (1980) Freeze preservation of pea meristems: Cell survival, *Can. J. Bot.*, 58, 833–840.

Hatanaka, T., Yasuda, T., Yamaguchi, T., and Sakai, A. (1994) Direct regrowth of encapsulated somatic embryos of coffee (*Coffea canephora*) after cooling in liquid nitrogen, *Cryo-Letters*, 15, 47–52.

Hellergren, J. and Li, P.H. (1981) Survival of *Solanum tuberosum* suspension cultures to –14°C: the mode of action of proline, *Physiol. Plant.*, 52, 449–453.

Hirai, D. and Sakai, A. (1999a) Cryopreservation of *in vitro*-grown axially shoot tip meristems of mint (*Mentha spicata* L.) by encapsulation vitrification, *Plant Cell Rep.*, 19, 150–155.

Hirai, D. and Sakai, A. (1999b) Cryopreservation of *in vitro*-grown meristems of potato (*Solanum tuberosum* L.) by encapsulation-vitrification, *Potato Res.*, 42, 153–160.

Hirai, D., Shirai, K., Shirai, S., and Sakai, A. (1998) Cryopreservation of *in vitro*-grown meristems of strawberry (*Fragaria x ananassa* Duch.) by encapsulation-vitrification, *Euphytica*, 101, 109–115.

Hirata, K, Phunchindawan, M., Kanemoto, M., Miyamoto, K., and Sakai, A. (1995) Cryopreservation of shoot primordia induced from horseradish hairy root cultures by encapsulation and two-step dehydration, *Cryo-Letters*, 16, 122–127.

Hirsh, A.G., Williams, R.J., and Meryman, H. (1985) A novel method of natural cryoprotection, *Plant Physiol.*, 79, 41–56.

Jitsuyama, Y., Suzuki, T., Harada, T., and Fujikawa, S. (1997) Ultrastructural study of mechanism of increased freezing tolerance to extracellular glucose in cabbage leaf cells, *Cryo-Letters,* 18, 33–44.

Kartha, K.K, Fowke, L.C., Leung, N.L., Caswell, K.L., and Hakman, I. (1988) Induction of somatic embryos and plantlets from cryopreserved cell cultures of white spruce (*Picea glauca*), *J. Plant Physiol.*, 132, 529–539.

Kobayashi, S. and Sakai, A. (1997) Cryopreservation of *Citrus sinensis* cultured cells, in *Conservation of Plant Genetic Resources in Vitro*. Vol. 1. *General Aspects*, Razdan, M.K. and Cocking, E.C., Eds., Science Publishers, Enfield, NH, pp. 201–223.

Kohmura, H., Sakai, A., Chokyu, S., and Yakuwa, T. (1992) Cryopreservation of *in vitro*-cultured multiple bud clusters of asparagus (*Asparagus officinalis* L.) cv. Hiroshimagreen (2n=30) by the techniques of vitrification, *Plant Cell Rep.*, 11, 433–437.

Kuranuki, Y. and Sakai, A. (1995) Cryopreservation of *in vitro*-grown shoot tips of tea (*Camellia sinensis*) by vitrification, *Cryo-Letters,* 16, 345–352.

Kyesmu, P.M. and Takagi, H. (2000) Cryopreservation of shoot apices of yams (*Dioscorea* species) by vitrification, in *Cryopreservation of Tropical Plant Germplasm*, Engelmann, F. and Takagi, H., Eds., JIRCAS, Tsukuba, Japan, pp. 411–413.

Lambardi, M., Fabbri, A., and Caccavale, A. (2000) Cryopreservation of white poplar (*Populus alba* L.) by vitrification of *in vitro*-grown shoot tips, *Plant Cell Rep.*, 19, 213–218.

Langis, R., Schnabel, B., Earle, E.D., and Steponkus, P.L. (1989) Cryopreservation of *Brassica campestris* L. cell suspensions by vitrification, *Cryo-Letters*, 10, 421–428.

Langis, R., Schnabel-Preikstas, B., Earle, B.J., and Steponkus, P.L. (1990) Cryopreservation of carnation shoot tips by vitrification, *Cryobiology*, 276(69), 658–659.

Langis, R. and Steponkus, P.L. (1990) Cryopreservation of rye protoplasts by vitrification, *Plant Physiol.*, 92, 666–671.

Lu, S. and Steponkus, P.L. (1994) Cryopreservation of Solanum shoot tips by vitrification, *Cryobiology*, 31(6), 569.

Lu, T. and Takagi, H. (2000) Cryopreservation of somatic embryos of papaya (*Carica papaya* L.) by vitrification, 2000 World Congress on *in vitro* Biology, Program Issue, Society in Vitro Biology, 73.

Makowska, Z., Keller, J., and Engelmann, F. (1999) of apices isolated from garlic (*Allium sativum* L.) bulbils and cloves, *Cryo-Letters*, 20, 175–182.

Malaurie, B., Trouslot, M.F., Engelmann, F., and Chabrillange, N. (1998) Effect of pretreatment conditions on the cryopreservation *in vitro*-cultured yam (*Dioscorea alata*) *Brazo Fuerte* and *D. bulbifera* Noumea Imboro shoot apices by encapsulation-dehydration, *Cryo-Letters*, 19, 15–26.

Matsumoto, T., Mochida, K., Hamura, H., and Sakai, A. (2001) Cryopreservation of persimmon (*Diopyros kaki* Thumb.) by vitrification of dormant shoot tips, *Plant Cell Rep.*, 20, 398–402.

Matsumoto, T. and Sakai, A. (1995) An approach to enhance dehydration tolerance of alginate-coated dried meristems cooled to –196°C, *Cryo-Letters*, 16, 299–306.

Matsumoto, T., Sakai, A., and Nako, Y. (1998a) A novel preculturing for enhancing the survival of *in vitro*-grown meristems of wasabi (*Wasabia japonica*) cooled to –196°C by vitrification, *Cryo-Letters*, 19, 27–36.

Matsumoto, T., Sakai, A., and Nako, Y. (1998b) Cryopreservation of *in vitro* cultured axillary shoot tips of grape (*Vitis vinifera*) by vitrification, *J. Jpn. Soc. Hort. Sci.*, 67(1), 78.

Matsumoto, T., Sakai, A., Takahashi, C., and Yamada, K. (1995a) Cryopreservation *in vitro*-grown apical meristems of wasabi (*Wasabia japonica*) by encapsulation-vitrification method, *Cryo-Letters*, 16, 189–206.

Matsumoto, T., Sakai, A., and Yamada, K. (1994) Cryopreservation of *in vitro*-grown apical meristems of wasabi (*Wasabia japonica*) by vitrification and subsequent high plant regeneration, *Plant Cell Rep.*, 13, 442–446.

Matsumoto, T., Sakai, A., and Yamada, K. (1995b) Cryopreservation of *in vitro*-grown apical meristems of lily by vitrification, *Plant Cell Tissue Organ Cult.*, 41, 237–241.

Monod, V., Poissonnier, M., Paques, M., and Dereuddre, J. (1992) Cryopreservation of shoot tips of *in vitro* plantlets of Eucalyptus after encapsulation and air dehydration, *Cryobiology*, 29(6), 737–738.

Niino, T. and Sakai, A. (1992) Cryopreservation of alginate-coated *in vitro*-grown shoot tips of apple, pear and mulberry, *Plant Sci.*, 87, 199–206.

Niino, T., Sakai, A., Enomoto, S., Magoshi, J., and Kato, S. (1992a) Cryopreservation of *in vitro*-grown shoot tips of mulberry by vitrification, *Cryo-Letters*, 13, 303–312.

Niino, T., Sakai, A., Yakuwa, H., and Nojiri, K. (1992b) Cryopreservation of dried shoot tips of mulberry winter buds and subsesequent plant regeneration, *Cryo-Letters*, 13, 51–58.

Niino, T., Sakai, A., Yakuwa, H., and Nojiri, K. (1992c) Cryopreservation of *in vitro*-grown shoot tips of apple and pear by vitrification, *Plant Cell Tissue Organ Cult.*, 28, 261–266.

Niino, T., Tashiro, K., Suzuki, M., Oosaki, S., Magoshi, J., and Akihama, T., (1997) Cryopreservation of *in vitro*-grown shoot tips of cherry and sweet cherry by one-step vitrification, *Sci. Hort.*, 70, 155–163.

Nishizawa, S., Sakai, A., Amano, Y., and Matsuzawa, T. (1993) Cryopreservation of asparagus (*Asparagus officinalis* L.) embryogenic suspension cells and subsequent plant regeneration by the vitrification method, *Plant Sci.*, 88, 67–73.

Niwata, E. (1995) Cryopreservation of apical meristems of garlic (*Allium sativum* L.) and high subsequent plant regeneration, *Cryo-Letters*, 16, 102–107.

Oka, S., Yakuwa, H., Sato, K., and Niino, T. (1991) Survival and shoot formation *in vitro* of pear winter buds cryopreserved in liquid nitrogen, *Hort. Sci.*, 26, 65–66.

Paul, H., Daigny, G., and Sangwan-Norreel, B.S. (2000) Cryopreservation of apple (*Malus x domestica Borkh.*) shoot tips following encapsulation-dehydration or encapsulation-vitrification, *Plant Cell Rep.*, 19, 768–774.

Paulet, F., Engelmann, F., and Glaszmann, J.C. (1993) Cryopreservation of apices of *in vitro* plantlets of sugarcane (*Saccharum* sp. hybrids) using encapsulation/dehydration, *Plant Cell Rep.*, 12, 525–529.

Pennycooke, J.C. and Towill, L.E. (2000) Cryopreservation of shoot tips *in vitro* plants of sweet potato [*Ipomoea batatas* (L.) Lam.] by vitrification, *Plant Cell Rep.*, 19, 733–737.

Phunchindawan, M., Hirata, K., Sakai, A., and Miyamoto, K. (1997) Cryopreservation of encapsulated shoot primordia induced in horseradish (*Armoracia rusticana*) hairy root cultures, *Plant Cell Rep.*, 16, 469–473.

Rasmussen, D.H. and Luyet, B.J. (1970) Contribution to the establishment of the temperature-concentration curves of homogeneous nucleation in solutions of some cryoprotective agents, *Biodinamica*, 11, 33–44.

Reed, B.M. and Yu, X. (1995) Cryopreservation of *in vitro*-grown gooseberry and currant meristems, *Cryo-Letters*, 16, 131–136.

Reinhoud, P.J. (1996) Cryopreservation of tobacco suspension cells by vitrification, Ph.D. thesis, Leiden University, Institute of Molecular Plant Science.

Sakai, A. (1956) Survival of plant tissue at super-low temperature, *Low Temp. Sci., Ser. B*, 14, 17–23.

Sakai, A. (1958) Survival of plant tissue at super-low temperature. II, *Low Temp. Sci., Ser. B*, 16, 41–53.

Sakai, A. (1960) Survival of the twig of woody plants at −196°C, *Nature*, 185, 393–394.

Sakai, A. (1962) Survival of woody plants in liquid helium, *Low Temp. Sci., Ser. B.*, 20, 121–122.

Sakai, A. (1965) Determining the degree of frost-hardiness in highly hardy plants, *Nature*, 206, 1064–1065.

Sakai, A. (1966) Survival of plant tissue at super-low temperature. IV. Cell survival with rapid cooling and rewarming, *Plant Physiol.*, 41, 1050–1054.

Sakai, A. (1973) Characteristics of winter hardiness in extremely hardy twigs of woody plants, *Plant Cell Physiol.*, 14, 1–9.

Sakai, A. (1985) Cryopreservation of shoot tips of fruit trees and herbaceous plants, in *Cryopreservation of Plant Cells and Organs*, Kartha, K.K., Ed., CRC Press, Boca Raton, FL, pp. 135–158

Sakai, A. (1993) Cryogenic strategies for survival of plant cultured cells and meristems cooled to –196°C, in *Cryopreservation of Plant Genetic Resources,* JICA, Japan, pp. 5–21.

Sakai, A. (1995) Cryopreservation of germplasm of woody plants, in *Biotechnology in Agriculture and Forestry,* Vol. 32, Bajaj, Y.P.S., Ed., *Cryopreservation of Plant Germplasm,* Springer, Heidelberg, pp. 53–69.

Sakai, A. (1997) Potentially valuable cryogenic procedures for cryopreservation of cultured plant meristems, in *Conservation of Plant Genetic Resources in Vitro,* Razdan, M.K. and Cocking, E.C., Eds., Science Publishers, Enfield, NH, pp. 53–66.

Sakai, A. (2000) Development of cryopreservation techniques, in *Cryopreservation of Tropical Plant Germplasm,* Engelmann, F. and Takagi, H., Eds., JIRCAS, Tsukuba, Japan, pp. 1–7.

Sakai, A., Kobayashi, S., and Oiyama, I. (1990) Cryopreservation of nucellar cells of navel orange (*Citrus sinensis* Osb. var. brasiliensis Tanaka) by vitrification, *Plant Cell Rep.,* 9, 30–33.

Sakai, A., Kobayashi, S., and Oiyama, I. (1991) Survival by vitrification of nucellar cells of navel orange (*Citrus sinensis* Osb. var. brasiliensis Tanaka) cooled to –196°C, *J. Plant Physiol.,* 137, 465–470.

Sakai, A. and Larcher, W. (1987) *Frost Survival of Plants (Ecological Studies 62),* Springer, Heidelberg, pp. 1–321.

Sakai, A., Matsumoto, T., Hirai, D., and Niino, T. (2000) Newly developed encapsulation-dehydration protocol for plant cryopreservation, *Cryo-Letters,* 21, 53–62.

Sakai, A. and Nishiyama, N. (1978) Cryopreservation of winter vegetative buds of hardy fruit trees in liquid nitrogen, *Hort. Sci.,* 13, 225–227.

Sakai, A. and Otsuka, K. (1967) Survival of plant tissue at super-low temperature. V. An electron microscope study of ice in cortical cells cooled rapidly, *Plant Physiol.,* 42, 1680–1694.

Sakai, A. and Yoshida, S. (1967) Survival of plant tissue at super-low temperatures. VI. Effects of cooling and rewarming rates on survival, *Plant Physiol.,* 42, 1695–1701.

Scottez, C., Chevreau, E., Godard, N., Arnaud, Y., Duron, M., and Dereuddre, J. (1992) Cryopreservation of cold-acclimated shoot tips of pear *in vitro* cultures after encapsulation-dehydration, *Cryobiology,* 29, 691–700.

Shatnawi, M.A., Engelmann, F., Fratharelli, A., and Damiano, C. (1999) Cryopreservation of apices of *in vitro* plantlets of almond (*Prunus Dulcis*.Mill), *Cryo-Letters,* 20, 13–20.

Steponkus, P.L., Langis, R., and Fujikawa, S. (1992) Cryopreservation of plant tissues by vitrification, in *Advances in Low Temperature Biology,* Vol. 1, Steponkus, P.L., Ed., JAI Press, Hamptonmill, U.K., pp. 1–161.

Stushnoff, C. (1987) Cryopreservation of apple genetic resources—implications for maintenance and diversity during conservation, *Can. J. Plant Sci.,* 67, 1151–1154.

Sugawara, T. and Sakai, A. (1978) Survival of suspension-cultured sycamore cells cooled to the temperature of liquid nitrogen, *Plant Physiol.,* 54, 722–724.

Takagi, H., Thinh, N.T., Islam, O.M., Senboku, T., and Sakai, A. (1997) Cryopreservation of *in vitro*-grown shoot tips of taro (*Colocasia esculenta* (L.) Schott) by vitrification. 1. Investigation of basic conditions of the vitrification procedures, *Plant Cell Rep.,* 16, 594–599.

Tannoury, M., Ralambosoa, J., Kaminsky, M., and Dereuddre, J. (1991) Cryopreservation by vitrification of alginate-coated carnation (*Dianthus cargo* Phyllus L.) shoot tips of *in vitro* plantlets, *C.R. Acad. Sci., Paris,* t313, Serie III, 633–638.

Tao, D., Li, P.H., and Carter, J.V. (1983) Role of cell wall in freezing tolerance of cultured potato cells and their protoplasts, *Physiol. Plant.,* 58, 527–532.

Thinh, N.T. (1997) Cryopreservation of germplasm of vegetatively propagated tropical monocots by vitrification, Ph.D. thesis, Kobe University, Department of Agronomy.

Thinh, N.T. and Takagi, H. (2000) Cryopreservation of *in vitro*-grown apical meristems of terrestrial orchids (*Cymbidium* spp) by vitrification, in *Cryopreservation of Tropical Plant Germplasm,* Engelmann, F. and Takagi, H., Eds., JIRCAS, Tsukuba, Japan, pp. 441–443.

Thinh, N.T., Takagi, H., and Sakai, A. (2000) Cryopreservation of *in vitro*-grown apical meristems of some vegetatively propagated tropical monocots by vitrification, in *Cryopreservation of Tropical Plant Germplasm,* Engelmann, F. and Takagi, H., Eds., JIRCAS, Tsukuba, Japan, pp. 227–232.

Thinh, N.T., Takagi, H., and Yashima, S. (1999) Cryopreservation of *in vitro*-grown shoot tips of banana (*Musa* spp) by vitrification method, *Cryo-Letters,* 20, 163–174.

Touchell, D. (1996) Principles of cryobiology for conservation of threatened Australian plants, Ph.D. thesis, University of Western Australia, Botany Division.

Touchell, D. (2000), Conservation of threatened flora by cryopreservation of shoot apices, in *Cryopreservation of Tropical Plant Germplasm*, Engelmann, F. and Takagi, H., Eds., JIRCAS, Tsukuba, Japan, pp. 269–272.

Towill, L.E. (1990) Cryopreservation of isolated mint shoot tips by vitrification, *Plant Cell Rep.*, 9, 178–180.

Turner, S.R., Touchell, D.H., Dixson, K., and Tan, B. (2000) Cryopreservation of *Anigozanthos viridis* spp viridis and related taxa from the southwest of Western Australia, *Aust. J. Bot.*, 48, 739–744.

Uragami, A., Lucas, M.O., Ralambosoa, J., Renard, M., and Dereuddre, J. (1993) Cryopreservation of microspore embryos of oilseed rape (*Brassica napas* L.) by dehydration in air with or without alginate encapsulation, *Cryo-Letters*, 14, 83–90.

Uragami, A., Sakai, A., and Nagai, M. (1990) Cryopreservation of dried axially buds from plantlets of *Asparagus officinalis* L. *in vitro*, *Plant Cell Rep.*, 9, 328–331.

Uragami, A., Sakai, A., Nagai, M., and Takahashi, T. (1989) Survival of cultured cells and somatic embryos of *Asparagus officinalis* cryopreserved by vitrification, *Plant Cell Rep.*, 8, 418–421.

Vandenbussche, B., Weyens, G., and De Proft, M. (2000) Cryopreservation of *in vitro* sugar beet (*Beta vulgaris*. L) shoot tips by a vitrification technique, *Plant Cell Rep.*, 19, 1064–1068.

Wang, J.H., Ge, J.G., Liu, F., Bian, H.W., and Huang, C.N. (1998) Cryopreservation of seeds and protocorms of *Dendrobium candidum*, *Cryo-Letters*, 19, 123–128.

Withers, L.A. (1979) Freeze preservation of somatic embryos and clonal plantlets of carrot (*Daucus carota*), *Plant Physiol.*, 63, 460–467.

Withers, L.A. (1985) Cryopreservation of cultured plant cells and protoplasts, in *Cryopreservation of Plant Cells and Organs*, Kartha, K.K., Ed., CRC Press, Boca Raton, FL, pp. 244–265.

Wu, Y., Engelmann, F., Zhao, Y., Zhou, M., and Chen, S. (1999) Cryopreservation of apple shoot tips: importance of cryopreservation technique and of conditioning of donor plants, *Cryo-Letters*, 20, 121–130.

Yamada, T., Sakai, A., Matsumura, T., and Higuchi, S (1991) Cryopreservation of apical meristems of white clover (*Trifolium repens* L.) by vitrification, *Plant Sci.*, 78, 81–87.

Yoshimatsu, K., Touno, K., and Shimomura, K. (2000) Cryopreservation of medicinal plant resources: retention of biosynthetic capabilities in transformed cultures, in *Cryopreservation of Tropical Plant Germplasm*, Engelmann, F. and Takagi, H., Eds., JIRCAS, Tsukuba, Japan, pp. 77–88.

Yoshimatsu, K., Yamaguchi, H., and Shimomura, K. (1996) Traits of *Panax ginseng* hairy roots after cold storage and cryopreservation, *Plant Cell Rep.*, 15, 555–560.

11 The Early History of Gamete Cryobiology

Stanley P. Leibo

CONTENTS

11.1   INTRODUCTION

The effects of low temperatures on living organisms have fascinated people for centuries. Almost 320 years ago, Robert Boyle, as cited by Parkes (1957), published a monograph titled *New Experiments and Observations Touching Cold*. This was the same Robert Boyle who formulated the principle that now bears his name: that the volume of an ideal gas at constant temperature varies inversely with the pressure exerted on it. The relevance of Boyle's Law to analysis of cellular water content and low-temperature biology can hardly be exaggerated. More than 60 years ago, Lucké (1940) presented a graphical representation of the osmotic response of the eggs of marine invertebrates when they were suspended in hypertonic solutions of impermeant and permeating solutes, including shrinkage and swelling of the eggs in ethylene glycol. Even very recently, measurements relating the volume of cells subjected to hypertonic solutions are presented as Boyle-van't Hoff graphs in which the volume of a cell is plotted as a function of the reciprocal of osmotic pressure of the solution in which the cell is suspended. An example is shown in Figure 11.1, for mouse zygotes, redrawn from data described by Oda et al. (1992). This example shows that the nonosmotic volume of mouse zygotes derived from the point on the plot at which the solution is

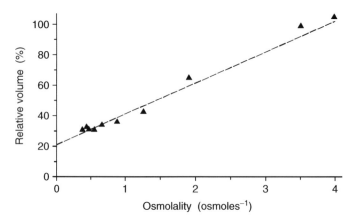

FIGURE 11.1 Boyle–van't Hoff Plot of mouse zygotes. Data of Oda et al. 1992.

infinitely concentrated, where $1/\infty = 0$, is approximately 20%. The corollary of this fact is that about 80% of the volume of mouse zygotes, like that of most mammalian oocytes and embryos, is water. It is also notable that Jacobus van't Hoff, the second name assigned to such graphs, was awarded the first Nobel Prize in Chemistry in 1901 for the "extraordinary services he has rendered by the discovery of the laws of chemical dynamics and osmotic pressure in solutions," as his Nobel citation read. Other recent examples of Boyle-van't Hoff plots as part of cryobiological analysis have been published for mouse spermatozoa (Willoughby et al., 1996), for uterine tissue (Devireddy et al., 2001), and for rhesus monkey oocytes (Songsasen et al., 2002).

In the classic monograph by Luyet and Gehenio (1940) titled *Life and Death at Low Temperatures*, the bibliography lists four treatises concerned with low temperatures that were published in the eighteenth century and another 89 publications on this same subject from the nineteenth century. Throughout the first several decades of the twentieth century, increasing numbers of observations on the effect of cold and low temperatures on viruses, microorganisms, and various species of plants and animals were made. Approximately 70 years ago, a systematic formalism began to be brought to these studies as logical analysis of the effects of low temperatures on living cells was introduced and meetings were convened to discuss observations made on the effects of cooling and freezing on various species of plants and animals. Proceedings of some of these meetings were published, so the deliberations of the participants can still be reviewed and studied today (Harris, 1954; Wolstenholme and Cameron, 1954). In 1957, a symposium titled *A Discussion on Viability of Mammalian Cells and Tissues after Freezing* was convened under the leadership of Alan S. Parkes, a British physician-scientist who directed the National Institute of Medical Research at Mill Hill in London. It was Parkes's interest in low-temperature biology that was responsible for his nurturing some of those workers in the field who were later to make major contributions to the emerging science of cryobiology, as well as to other disciplines. These researchers included James Lovelock, Audrey Smith, and Christopher Polge; they in turn fostered the investigations of Michael Ashwood-Smith, John Farrant, David Pegg, Steen Willadsen, and Ian Wilmut.

In a series of lectures and essays titled *Sex, Science and Society*, Parkes (1966) also described the rationale and background of many of the early experiments on biological effects of low temperatures. In addition, a few textbooks describing observations of the latter half of the nineteenth and first half of the twentieth century were published. These included the monograph by Luyet and Gehenio (1940); the comprehensive text authored by Smith (1961), as well as the synopsis that she edited (Smith, 1970); and the volume titled *Cryobiology*, edited by Meryman (1966). In that treatise there are important reviews written by Eizo Asahina, Jacob Levitt, Basile Luyet, Peter Mazur, Arthur Rinfret, Arthur Rowe, and Harold Meryman himself. Early investigations in the field were published in the journal founded by Luyet, titled *Biodynamica*. This unusual journal, which appeared at irregular intervals between 1934 and 1957, was devoted primarily to studies carried out in Luyet's American Foundation of Biological Research, a laboratory specifically designed to study phenomena at low temperatures. For example, the laboratory contained walk-in refrigerators at +4°C; housed within one of them was a large walk-in freezer at –20°C, and located within that freezer was yet another that operated at –40°C. The investigations conducted at American Foundation of Biological Research included microscopic observations by polarized illumination of ice formation in aqueous solutions of glycerol, albumin, and polyvinylpyrrolidone; various attempts to preserve frog and chick hearts and other tissues; and numerous studies of erythrocytes frozen in a wide variety of simple and complex solutions, supplemented with various compounds now referred to as cryoprotective additives.

The word itself, cryobiology, which now identifies this discipline, was first coined more than forty years ago by Alan Parkes. In the first issue of the journal *Cryobiology*, Parkes (1964) described the origin of the word, "Cryobiology, the study of frosty life," and gave its definition as "the study of the biological effects of low temperatures, a discipline centuries old in conception but new in its vigorous development of the last two decades." According to Parkes (1964), his immediate colleagues, James Lovelock and Audrey Smith, were opposed to such contrived words. Within a few years, however, even Smith (1970) countenanced use of the word and entitled the monograph that she edited: *Current Trends in Cryobiology*. Enigmatically, Smith dedicated this monograph "to Katie." Katie was a dog that Smith had successfully treated by transplantation of frozen-thawed bone marrow cells and that she subsequently adopted as a pet.

During the last 30 years, cryobiology has developed into a mature science, with numerous investigators conducting research all over the world, several national and international societies that convene regular meetings, and the publication of at least five journals devoted explicitly to this subject. These are *Cryobiology* and *Cell Preservation Technology*, both published in the United States; *Cryo-Letters*, published in the United Kingdom; *Problems of Cryobiology*, published by the Institute for Problems of Cryobiology and Cryomedicine, Ukraine Academy of Sciences; and *Low Temperature Medicine*, published in Japan. In addition, articles describing experiments and reviews of various aspects of cryobiology are regularly published in many other journals and textbooks. The science of cryobiology is now based on a firm foundation of a mathematical formalism and a growing mechanistic analysis of freezing injury and understanding of the actions of cryoprotectants. This maturing of the science of cryobiology has occurred as scientific publications are undergoing a profound transformation, as more and more journals are published electronically and fewer and fewer scientists and students consult and make use of bound journals collected in libraries. Furthermore, electronic databases often include no more than the last 20 years of a journal's previous publications. As a consequence, younger scientists or newcomers to a discipline may be unaware of early articles on a given subject. The purpose of this review is to summarize some of the significant observations made in cryobiology, especially those made during the first three quarters of the twentieth century. It is my view that these are the findings most likely to be "lost" or "overlooked" as investigators come to rely only on electronic retrieval of the literature of cryobiology. Now, as findings of the last 25 or so years are summarized in many recent reviews, some of which have been cited at the end of this chapter, it is less likely that they will be overlooked.

In some areas, notably assisted reproduction of humans and animals, cryopreservation of reproductive cells and gonadal tissues has become an essential adjunct to the science. Every year, more than 25 million cows are artificially inseminated with frozen-thawed bull semen (Foote, 1981; Iritani, 1980). Moreover, over the last decade, hundreds of thousands of bovine calves have been born as a result of the transfer of cryopreserved embryos into cows in which the embryos gestate (Thibier, 2002). Finally, according to tabulations prepared by the European Society for Human Reproduction and Embryology (Nygren and Andersen, 2002) and by the Society for Assisted Reproductive Technologies (2002), literally thousands of human babies have also been born by transfer of cryopreserved human embryos. The large number of humans and animals derived from cryopreserved gametes or embryos provides proof of the effect of cryobiology on the society of the twenty-first century. For these several reasons, it seems appropriate to review the early history of the cryobiology of mammalian gametes and embryos.

11.2 MEASUREMENTS OF TEMPERATURE: INVENTION OF THE THERMOMETER

Although humans had obviously always been aware of and perceived feelings of warmth or its absence, it was not until the invention of the thermometer that it became possible to assign specific values to degrees of heat and cold. According to an early description by Bolton (1900) of the invention of the thermometer, one of the first versions of a device to measure temperature was made by Galileo Galilei in the early seventeenth century. The first accurate thermometers were invented in 1714 by a German physicist, Gabriel Fahrenheit, who developed the use of mercury in glass for thermometry; these were later modified by a French physicist, René Réamur, in 1731. The thermometric scale of denoting the triple point of water (the temperature and pressure at which water exists as liquid, as ice, and as vapor) as $0.01°$ and the boiling point of water as $100°$ was originally proposed by a Swedish astronomer, Anders Celsius. Since then, measurement of temperature has become very much easier and much more accurate and can now be determined by a wide variety of instruments and of exceedingly small specimens.

11.3 CHEMISTRY OF SOLUTIONS: LIQUEFACTION OF GASES

As summarized by Smith (1961), among the events that stimulated research in low-temperature biology were advances made in chemistry and physics during the latter half of the nineteenth century. These included investigations on the properties of solutions by Jacobus van't Hoff, a chemist who was born in Rotterdam in The Netherlands in 1852, but who conducted his research as a professor at the University of Berlin. As mentioned above, van't Hoff was awarded the first Nobel Prize in Chemistry in 1901 for his studies on osmotic pressure of solutions. Other important advances during the last decades of the nineteenth century that prompted studies of low-temperature biology were the liquefaction of oxygen, hydrogen, and nitrogen by the French physicist, Louis-Paul Cailletet, in 1877; the production of liquefied air in 1895 by Linde; and the liquefaction of hydrogen by the Scottish physicist, James Dewar, in 1898. In other words, it thus became possible to use liquefied gases as refrigerants to cool specimens to extremely low temperatures.

11.4 GAMETE BIOLOGY

To understand and appreciate the cryobiology of gametes and embryos, it is necessary to consider the evolution of the science of embryology and reproductive biology. One comprehensive review of these sciences was published in 1910 in a textbook titled *Physiology of Reproduction*, edited by F.H.A. Marshall. This important textbook has now been revised and appeared as three new editions in 1922, 1960, and 1990. In his chapter on spermatozoa in the third edition, Parkes (1960) describes what was then known about the effects of cold on spermatozoa. The fourth edition of this now-classic text, edited by George Lamming, includes a comprehensive summary by Watson (1990), who provides a detailed discussion of the cryobiology of spermatozoa. Another excellent review of this subject is the two-part review by Salamon and Maxwell (1995a, 1995b). There are other excellent recent reviews of this subject; for example, in the textbook on endocrinology by Johnson and Everitt (2000), the encyclopedic treatise edited by Knobil and Neill (1994), the comprehensive text by Thibault et al. (1993), and many short summaries.

It is instructive and illuminating to begin with an early description of the science of reproductive biology. One very good early review is that of Lillie (1916), in which he summarizes "The history of the fertilization problem," beginning with the observations of Aristotle on the day-by-day development of the chick embryo. Lillie notes that there was little progress in the knowledge of reproduction from the time of ancient Greece until that of William Harvey, the English physician who discovered circulation of the blood. Lillie also notes, however, that Harvey's (1651) understanding of reproduction was seriously flawed, as Harvey "descended deeper into the slough of metaphysics, and committed himself to the fantastic idea that . . . the ovum is the product of unconscious uterine desire." According to Lillie, the improvement of the microscope and its use to study reproduction yielded the first fundamental advance in the theory of reproduction, namely the discovery of the spermatozoon by the Dutch microscopist, Anton van Leeuwenhoek, in 1677. Unfortunately, these first observations of Leeuwenhoek led to "wild flights of imagination" in which many observers claimed to have seen animacules ("the human form with naked thighs, legs, breast, both arms, the skin being pulled up higher . . . to cover the head like a cap.") contained within the single sperm cell. The next advance according to Lillie was that described in a treatise published in 1785 by the Italian physiologist and anatomist, Lazzaro Spallanzani, who discussed the then-current theories of fertilization. It was Spallanzani who first demonstrated that in frogs and toads, fertilization takes place outside the body, and who produced experimental evidence refuting the theory of spontaneous generation. It was also Spallanzani who first successfully performed artificial insemination (AI) of dogs, thus, as Lillie states, "laying the foundation for the artificial propagation of many animals." In his review, Watson (1990, p. 749) quotes the observations of Spallanzani on AI of a spaniel: "Thus did I succeed in fecundating this quadruped; and I can truly say that I never received greater pleasure upon any occasion, since I first cultivated experimental philosophy."

The next advance in understanding fertilization, according to Lillie, came from the observations of Prévost and Dumas, who studied many aspects of fertilization in frogs. They concluded that a spermatozoon penetrates each egg and becomes "the rudiment of the nervous system, and . . . furnishes all the other organs of the embryo." In a later note, Prévost (1840) described experiments that he conducted on *animalcules spermatiques*, noting that the spermatozoa would tolerate chilling to a temperature of about 8 degrees below zero without permanent loss of their motility. He also stated that he had frozen frog testes (to an unspecified temperature) and had recovered motile spermatozoa.

Although controversy regarding fertilization continued, Lillie quotes Lallemand (1841), who believed in the "union" of the egg and spermatozoon: "Each of the sexes furnishes material already organized and living.…A fluid obviously cannot transmit form and life which it does not possess.…Fertilization is the union of two living parts which mutually complete each other and develop in common." Except for those insights, the half century from 1824 to 1874 yielded relatively little advance in understanding fertilization. In 1873, however, the German zoologist, Otto Bütschli, observed within the eggs of nematodes the approach and contact of what we now know to be the pronuclei of spermatozoon and oocyte. Lillie also briefly summarized the effect of "external factors" on the process of fertilization and noted that it occurs within a definite range of temperatures; if these temperatures are exceeded, fertilization does not occur.

11.5 EFFECT OF LOW TEMPERATURES ON RABBIT SPERMATOZOA

As mentioned, it has long been recognized that temperature exerts a profound influence on biological systems. Two important publications that illustrate this are the companion papers of Hammond (1930) and of Walton (1930). Working in collaboration, although publishing in tandem, these two investigators conducted research to determine the effects of a broad range of temperatures from $0°$ to $45°C$ on rabbit spermatozoa. Hammond studied ejaculated specimens and Walton investigated those collected from the vas deferens. In the introduction to his paper, Hammond notes that if it were possible to cool spermatozoa of animals to low temperatures to reduce their metabolism without destroying their capability to fertilize oocytes, then "in these days of rapid aeroplane transport, it might be possible to move entire herds of animals around the world in the form of

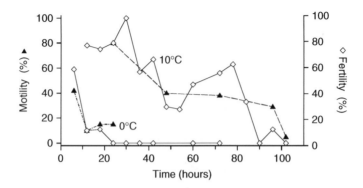

FIGURE 11.2 Effect of temperature on ejaculated rabbit spermatozoa. Data of Hammond, 1930.

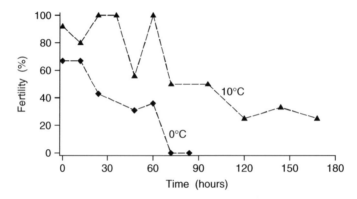

FIGURE 11.3 Effect of temperature on vas deferens rabbit spermatozoa. Data of Walton 1930.

chilled samples of semen." This prescient observation was written in 1929, only 26 years after the first powered flight by Wilbur and Orville Wright. There are several notable features of these papers. First, although they conducted their research decades before methods to perform *in vitro* fertilization would be devised, Hammond and Walton assayed the function of treated spermatozoa by using spermatozoa of one breed of rabbit to inseminate females of a second breed so that the parentage of the kits might be distinguished. In this way, they were able to assay the treated spermatozoa by examining its fertilizing capacity, as spermatozoa from the vas deferens have not yet acquired motility. Second, Hammond and Walton conducted their experiments in parallel, so that it was possible to compare the consequences of temperature on nonejaculated and ejaculated spermatozoa directly. One aspect of Walton's experiments deserves special note. He stated that as spermatozoa to be collected directly from the vas deferens would have been "protected" from oxidative effects of air, it would be preferable to prevent the cells from being exposed to air before being treated at different temperatures. As a consequence, he collected the spermatozoa from the reproductive tracts directly into degassed medium held under oil, thus preventing exposure of spermatozoa to oxygen in air.

To illustrate some of the effects of low temperatures on gametes, some of the results of the Hammond and Walton experiments have been redrawn from their tabulated data and are shown in Figure 11.2 through Figure 11.4. Hammond found that about 40 to 50% of ejaculated spermatozoa retained their motility and fertility for about 72 h at 10°C, but lost most of their function after about 90 h at that temperature. If cooled and held at 0°C, ejaculated spermatozoa seemed to lose motility and fertility within about 12 to 24 h (Figure 11.2). In contrast, Walton found that spermatozoa from the vas deferens were much more resistant than ejaculated spermatozoa (Figure 11.3). After 175 h at 10°C, such spermatozoa were able to fertilize oocytes in females that had been induced to ovulate; even after spermatozoa had been held for 60 h at 0°C, 30% of the females

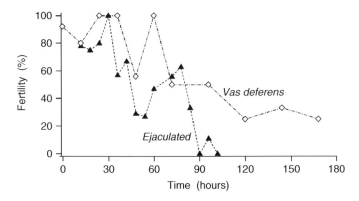

FIGURE 11.4 Pregnancy in does after insemination with rabbit spermatozoa. Data of Walton, 1930, and Hammond, 1930.

inseminated with chilled spermatozoa became pregnant and kindled offspring. In comparison, the observations of Hammond and Walton on the relative fertility of ejaculated and deferens spermatozoa held at 10°C for various times are shown in Figure 11.4. It is clear that ejaculation of spermatozoa and their exposure to seminal plasma render them much more sensitive to chilling than their nonejaculated counterparts. Since those first observations by Walton and Hammond, many other investigators have reported analogous observations of spermatozoa of various species. It is now known that exposure of ejaculated spermatozoa to the complex mixtures of hormones and other compounds exuded by the prostate and bulbourethral glands alter the composition and surface properties of spermatozoa. Apparently, these molecular alterations affect the membranes of spermatozoa so that they are less able to resist damage caused by exposure to low temperatures. One practical consequence of this difference in chilling sensitivity between ejaculated and non-ejaculated spermatozoa is that the latter are much more resistant to damage caused by freezing to low subzero temperatures. This indicates that it might be possible to cryopreserve spermatozoa collected from the testes of postmortem specimens of endangered species; in fact, this has already been achieved with the spermatozoa of a domestic dog (Marks et al., 1994) and with those of mice (Songsasen et al., 1998).

11.6 EARLY ATTEMPTS TO PRESERVE SPERMATOZOA

Many previous accounts of the beginnings of low-temperature biology often cite the fact that in 1866, an Italian military physician, P. Mantegazza, recorded the fact that human spermatozoa became immotile when cooled in snow (cited by Smith, 1961). Mantegazza proposed that, if it were possible to maintain fertility of human spermatozoa by cooling it, then it might be possible for a soldier killed in battle to father a child after his death. It is clear from such accounts from the past several centuries that people have long been interested in the possibility of controlling reproduction by use of low temperatures.

In his comprehensive review of the preservation of spermatozoa, Watson (1990) introduces the subject of AI by noting that the specific origins of this procedure are no longer traceable, but undoubtedly began no later than the first millennium. As noted above, Watson quotes the description by Spallanzani of the methods he used to inseminate a bitch with semen from a dog. He continues with a summary of the derivation during the late nineteenth and early twentieth centuries of procedures to perform AI of females of several animal species. Watson notes that by the beginning of the twentieth century, AI of horses, dogs, rabbits, guinea pigs, and humans had become commonplace. Shortly thereafter, E. Ivanoff and his Russian colleagues achieved successful AI of cattle, sheep, zebras, and Przewalskii horses. According to Watson, starting at about 1930, AI began to

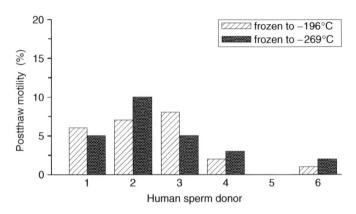

FIGURE 11.5 Postthaw survival of humans spermatozoa. Data of Shettles, 1940.

be used for commercial animal husbandry, but it was soon recognized that full exploitation of this procedure would require solution to the problems of semen preservation.

In 1938, prompted by observations that he made during experiments on freezing of the spirochetes that cause syphilis, Jahnel (1938) attempted to freeze human spermatozoa. Although the results were variable, he reported that he had succeeded in recovering motile spermatozoa. First, he placed small volumes of spermatozoa in glass tubes that were submerged into liquid nitrogen. Then, the glass tubes were placed into metal tubes that were soldered so as to make gas-tight seals. These metal tubes were then lowered into liquid helium at –269°C. After ~5 h, the tubes were warmed to –196°C, the metal tubes opened, and finally the glass tubes were warmed rapidly. When examined microscopically, although most of the spermatozoa were immotile, some of them showed a rather vivid motility pattern similar to that of fresh samples. As Jahnel discusses in his paper, one key to his investigation was the opportunity to use liquefied gases. This he did in collaboration with his colleagues, Drs. Stark and Steiner—physicists who were studying low-temperature phenomena. In the discussion of his article, Jahnel clearly appreciated the significance of being able to preserve spermatozoa at low temperatures. He posed the rhetorical question of whether such motile spermatozoa might be capable of fertilizing oocytes after having been frozen in liquid nitrogen or liquid helium.

That same year, Luyet and Hodapp (1938) published a very brief article in which they described their attempts to vitrify frog spermatozoa. They immersed spermatozoa in solutions of 1 or 2 *M* sucrose in Ringer's solution and then mounted thin films of sperm suspensions between very thin sheets of mica that were immersed in liquid air. Although ~40% of the spermatozoa became immotile in 2 *M* sucrose, Luyet and Hodapp reported that "100% of those which survived exposure to sucrose, survived vitrification in liquid air (50 experiments)." They noted that the warmed spermatozoa retained their motility as long as the controls, in one case for as long as 12 h after treatment.

Shortly thereafter, a brief study describing the survival of human spermatozoa after cooling very small volumes at high rates was published. In this investigation, Shettles (1940) cooled specimens of semen from six men by plunging small tubes containing spermatozoa directly into flasks of liquid nitrogen at –196°C or of liquid helium at –269°C. Some of the samples were warmed immediately, but others were held in the refrigerant for periods of up to 1 week. Some of Shettles's (1940) results are redrawn in Figure 11.5. Although the maximum postthaw motility was no more than 10%, the results are notable for two reasons. First, Shettles was able to achieve survival of motile human spermatozoa by rapid cooling. Second, there were significant differences in the postthaw motility of spermatozoa from different males. It is also worth noting that a very recent study has demonstrated that human spermatozoa survive and are capable of fertilizing oocytes after being cooled at very high rates in the absence of any protective compound (Nawroth et al., 2002). It is also worth mentioning that Shettles, a physician at Columbia University, would achieve

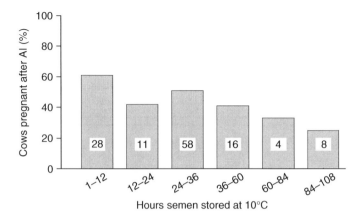

FIGURE 11.6 Storage of bull semen in fresh hens' egg yolk prepared in phosphate buffer (pH = 6.8). Data of Phillips and Lardy, 1940.

considerable notoriety some 30 years later for having attempted *in vitro* fertilization of human oocytes. Fearing political and legal consequences, the head of Shettles' academic department, Dr. Raymond Vande Wiele, terminated the procedure in which the presumptive zygotes were developing; the embryos died (Rorvik, 1974). It is also appropriate to note that Shettles (1979) reported that he had achieved nuclear transfer of diploid spermatogonia into enucleated human oocytes many years before the furor that erupted with announcements of the births of human "clones" at the end of 2002.

The same year of Shettles' experiments, Phillips and Lardy (1940) reported that a "yolk-buffer pabulum" was effective at maintaining bull spermatozoa for periods of up to 100 h when stored at 10°C. This solution, prepared from the yolk of chicken eggs, was extremely effective at preserving not only the motility of spermatozoa but also its fertility. Phillips and Lardy used semen stored at 10°C in this solution for various times of up to 180 h (~7 days) and then inseminated cows. Of 127 cows that were bred with stored semen, 38 were definitively diagnosed as being pregnant, and an additional 36 had not returned to estrus and were assumed to be pregnant. Some of their results showing the percentages of pregnant cows inseminated with semen held in their yolk-buffer pabulum are shown in Figure 11.6. This paper is notable because it is the first description of the fertility of stored spermatozoa and because it describes a solution prepared from egg yolk that is still used today as the most common diluent or "extender" used to protect spermatozoa of many species.

The intellectual and scientific heritage of both low-temperature biology and reproductive biology are illustrated by yet another experiment conducted more than 60 years ago. As he would later relate in his brief autobiographical sketch on being awarded the Pioneer Award of the International Embryo Transfer Society (Chang, 1983), Min-Chueh Chang began his doctoral investigations at Cambridge University with studies of the effects of low temperatures on ram spermatozoa. Continuing the earlier investigations of Hammond and Walton already described, Chang and Walton (1940) studied the effects on respiration of ram spermatozoa that were exposed to low temperatures from 15° to 1°C and the interaction of various cooling rates on spermatozoa. As illustrated by the results of Chang and Walton (1940) that have been redrawn in graphical form in Figure 11.7, they found that respiration was reduced to <60% of the controls when spermatozoa were cooled to 10°C or below and that the damage caused by cooling was significantly greater when spermatozoa were cooled very rapidly rather than very slowly to low temperatures. It is worth noting that 36 years later, Chang would be the senior author of papers describing the first successful cryopreservation of female gametes (Tsunoda et al., 1976; Parkening et al., 1976).

The next important observations regarding preservation of human spermatozoa were the experiments of Hoagland and Pincus (1942). Using a fine-wire bacteriological loop ~2 mm in diameter as a holder of films, they cooled specimens of rabbit, bovine, and human spermatozoa under various

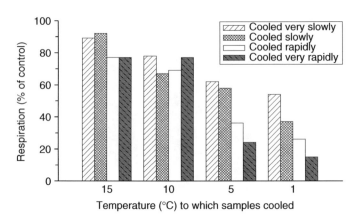

FIGURE 11.7 Effect of cooling on respiration of ram spermatozoa. Data of Chang and Walton, 1940.

conditions. In several cases, before immersing the spermatozoa in LN_2, they plasmolyzed the cells by suspending them in various sugars or other hypertonic solutions, including glycerol and butyric acid. Although sperm survival, as estimated by postthaw motility, was highly variable, in a few instances Hoagland and Pincus reported that some suspensions of human spermatozoa exhibited motility of 50% after being immersed in liquid nitrogen and then thawed. They also noted that the time after ejaculation before freezing affected the motility after thawing.

At this same time, Easley et al. (1942) began to investigate the influence of diluents, the cooling rate, and storage temperature on the motility and membrane integrity of bovine spermatozoa. They found that the osmotic pressure and pH of the diluent affected the survival of spermatozoa stored for various times at 10°C up to 72 h. By cooling samples in a slow, stepwise manner, these investigators were able to recover membrane-intact spermatozoa even after they had been stored at 0°C for as long as 216 h. They also observed motile spermatozoa in samples stored at 0°C for 24 h. They did note that semen collected from bulls in different locations seemed to display differences in their resistance to cooling and storage. That same year, Shaffner (1942) reported very preliminary observations on the freezing of fowl spermatozoa. In 1945, citing the previous observations of Jahnel (1938), Shettles (1940), and Hoagland and Pincus (1942), Parkes attempted to freeze human spermatozoa at high rates to –79°C or to –196°C. He did so by using fine capillaries with inner diameters of 0.15, 0.5, or 1 mm. When the samples were thawed rapidly by immersion in 37°C water, Parkes observed "abundant revival" of spermatozoa even after 5 h in liquid nitrogen. He then found that he could preserve sperm samples at –79°C for as long as 8 days. In his conclusion, noting that AI would no doubt become increasingly common, Parkes (1945, p. 213) stated that "elaboration of a method permitting prolonged storage and transport of the semen without affecting the genetic properties of the spermatozoa would open up remarkable possibilities."

11.7 THE "BREAKTHROUGH": DISCOVERY OF THE PROTECTIVE EFFECT OF GLYCEROL

In their review of frozen storage of ram spermatozoa, Salamon and Maxwell (1995a) state that the first published record of the use of glycerol as a cryoprotectant was for the freezing of plant cells and tissues by N.A. Maximov in 1908. Citing a paper written in Russian by A.D. Bernstein and V.W. Petropavlovsky in 1937, titled "Effect of non-electrolytes on viability of spermatozoa," Salamon and Maxwell (1995a) also state that these latter workers used 9% glycerol to successfully store spermatozoa of the rabbit, guinea pig, bull, ram, boar, and stallion, as well as of fowl and duck at –21°C. However, because of the inaccessibility of the articles cited, it has been difficult to verify those observations.

A few years later, in a brief and seldom-cited article, Jean Rostand (1946) pointed out that it would be of considerable practical interest for those interested in the study of fertilization if it were possible to preserve spermatozoa. Using spermatozoa of two species of frogs, *Rana temporaria* and *Rana esculenta*, as experimental material, Rostand found that the addition of glycerol in proportions of 10, 15, or 20% to suspensions of spermatozoa rendered the spermatozoa resistant to the effects of freezing to temperatures of –4° to –6°C. No mention was made of the fertilizing capacity of the treated spermatozoa.

In 1947, Christopher Polge was assigned the problem of developing techniques for AI of poultry, and specifically to try to preserve fowl spermatozoa. Reminiscing about the early research studies that he had conducted shortly after receiving his degree in agriculture, Polge (1968) noted that this problem had intrigued Dr. A. S. Parkes for many years. He also noted that at that time, shortly after the end of World War II, pure chemicals were hard to obtain and were very expensive. After having conducted several months of unsuccessful experiments with his colleague Audrey Smith at the National Institute of Medical Research at Mill Hill, Polge put aside this research for 6 months. He then moved to the farm laboratories at Mill Hill to be close to the animal quarters where the chickens were housed. He had the solutions and chemicals that he had used previously sent out to his new laboratory. This time, when he froze fowl spermatozoa to –79°C using what he assumed to be solutions of fructose, instead of the spermatozoa being killed by the freezing, the majority were highly motile when he thawed them. According to Polge (1968), he inseminated a large number of chickens with semen that had been frozen and thawed; of several hundred eggs that were collected from inseminated hens and incubated, he obtained one chick. Although one chick was hardly a convincing result, he attempted to repeat his observations. However, the original solution was soon exhausted. When he prepared a new solution of what he thought was fructose, no spermatozoa survived freezing. As Polge (1968) notes, this aroused suspicion and speculation as to what had happened to the original solution while stored in the media cupboard at Mill Hill. The few remaining milliliters of the "valuable vintage solution" were handed over to a chemist colleague, Dr. D. F. Elliott, for analysis. Elliott's analysis showed that the solution contained no sugars but a substantial amount of glycerol and some protein. As Polge (1968, p. 47) relates, "Nobody will ever know exactly what happened," but he and his colleagues guessed that while in storage, the label on the bottle of the original fructose solution had fallen off and had then been affixed to a bottle of Meyers solution, a mixture of glycerol and albumin in isotonic saline that was used routinely for preparing histological specimens.

These first tentative steps at devising a method to preserve spermatozoa at very low temperatures were reported in a very brief article by Polge et al. (1949). They noted that survival was negligible when fowl spermatozoa were frozen to –79°C in fructose alone, but when they were frozen in 20% glycerol, the spermatozoa exhibited full motility when thawed. Moreover, the researchers also found that they were able to remove water from the frozen specimens; when the dehydrated semen was reconstituted with water, the scientists observed a high percentage of motile spermatozoa. Polge and his colleagues (1949) also stated that similar results were obtained with propylene glycol and ethylene glycol, and they mentioned in passing that experiments with human spermatozoa suspended in a glycerol solution indicated that a much higher percentage "could be revived after vitrification."

Continuing these investigations, Smith and Polge (1950a, 1950b) reported their remarkable observations on the effects of low temperatures and freezing on spermatozoa of cattle and goats. In their *Nature* article, they stated: "Luyet believes that the preservation of life at low temperatures depends on preventing the formation of intracellular ice crystals, either by dehydrating the cells before freezing, or by ultra-rapid cooling and rewarming" (Smith and Polge, 1950b). Most cryobiologists in 2002 would undoubtedly concur with that opinion. Smith and Polge (1950a) noted that they were able to recover high percentages of motile bull and goat spermatozoa after suspending them in 15% glycerol and cooling them slowly to –79°C. They also observed that guinea pig spermatozoa suspended in 15% glycerol exhibited 75% motility after thawing, but that the acrosomes were severely damaged. Smith and Polge (1950a) found that 25% of stallion spermatozoa suspended in 5% glycerol were motile after freezing and thawing. Having tested a wide variety of compounds, they also noted that 15 or 20% ethylene glycol and propylene glycol afforded protection approaching that of glycerol.

In their companion paper, Smith and Polge (1950b) stated that removal of 90% of the water from frozen preparations of spermatozoa by vacuum distillation was "compatible" with the resumption of sperm motility by a small percentage of cells when the dry specimen was reconstituted by the addition of water. Now, some 50 years later, several groups have dehydrated spermatozoa of mice and cattle and have used intracytoplasmic sperm injection (ICSI) of oocytes to produce embryos and, ultimately, live young of mice. For example, Wakayama and Yanagimachi (1998) and their colleagues (Kusakabe et al., 2001) have subjected mouse spermatozoa to freeze-drying and then, by using ICSI, have produced live young. Furthermore, Keskintepe et al. (2002) have used a similar procedure to produce embryos by ICSI of bovine oocytes with freeze-dried bull spermatozoa. Very recently, Bhowmick et al. (2003) have produced normal mouse fetuses after ICSI of oocytes with spermatozoa dried at ambient temperature.

The most important question, of course, was whether such frozen-thawed spermatozoa had retained the ability to fertilize an oocyte. In 1951, Stewart reported that he had inseminated five cows with semen that had been frozen to –79°C and stored for various times; of the five cows, one was diagnosed pregnant and delivered a normal live bull calf. Stewart (1951) is careful to note that as these cows were being used in a metabolism experiment, "it is certain that it [the pregnant cow] could not have been served at any other time." This euphemism meant that there was no possibility that the cow might have been impregnated by an errant bull. This first trial was then repeated with a much larger sample. A total of 38 cows were artificially inseminated with spermatozoa that had been frozen to –79°C; of these, 30 became pregnant and 27 normal calves were born (Polge and Rowson, 1952a, 1952b). Subsequent experiments demonstrated that of 208 cows that were inseminated with frozen spermatozoa that had been stored in the frozen state for as long as 52 weeks, 65% became pregnant. It was later shown that even after semen had been stored for 4.5 years at –79°C, 67% of cows inseminated with such stored semen became pregnant (Polge, 1957). Within a few years, similar procedures had been used to freeze and store spermatozoa of other domestic species, such as rabbits and sheep, as well as cattle (Emmens and Blackshaw, 1950, 1955) and goats (Dauzier, 1956). One notable experiment, harking back to the observations of Walton (1930), was the demonstration that epididymal spermatozoa from a stallion could be successfully frozen, yet retain their fertilizing capability, as shown by the birth of a foal (Barker and Gandier, 1957). These authors note that this method might be used to freeze spermatozoa recovered at the time of death and preserve it for future use. Other experiments were conducted to determine the influence of various factors, such as the sperm concentration or the saccharide added to the solution, on the ultimate fertility of frozen bull spermatozoa (Martin and Emmens, 1958). The first successful cryopreservation of rabbit spermatozoa with demonstration of fertility was that by Fox (1961), who reported the birth of three litters from does inseminated with frozen semen stored for 24 h at –90°C. A few years later, Sawada and Chang (1964) inseminated 37 does with rabbit spermatozoa that had been suspended in 17.5% DMSO and frozen to –79°C; 14 of the does (38%) were found to be carrying fertilized embryos. Several years later, a very comprehensive tabulation of the international use of cryopreserved spermatozoa from a wide variety of domestic species was prepared by Iritani (1980).

11.8 INTERRUPTED RAPID COOLING OF BULL SPERMATOZOA

Pursuing the previous observations on the preservation of semen, Luyet and Keane (1955) froze duplicate samples of bull spermatozoa to various intermediate temperatures between –20° and –60°C for 5 min. Then they either thawed the samples or plunged the counterpart samples directly into liquid nitrogen and then thawed them and measured survival based on sperm motility. As

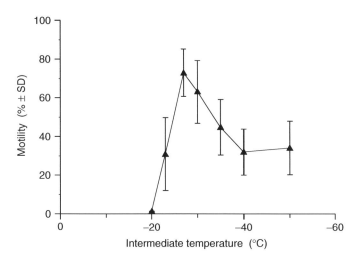

FIGURE 11.8 Postthaw motility of bull spermatozoa cooled to intermediate temperature, and then into LN_2. Data of Luyet and Keane, 1955.

shown by the results in Figure 11.8, redrawn from the tabular data of Luyet and Keane (1955), on average, samples cooled to –28°C before being plunged into LN_2 exhibited maximum survival. Samples cooled to higher or lower temperatures before being cooled to –196°C were found to have lower survival. In performing these studies, Luyet and Keane (1955) used ejaculates collected from nine bulls. Sperm survivals for different bulls varied from a high of 85% motility to a low of 30%. One of the significant aspects of these observations is that they demonstrate that cells undergo changes even when surrounded by partially frozen solutions and at temperatures as low as –25° to –30°C.

11.9   HUMAN PREGNANCIES BY AI WITH FROZEN SPERMATOZOA

Having noted the brief aside regarding human spermatozoa in the original paper by Polge et al. (1949), Sherman and Bunge (1953) treated human spermatozoa with 10% glycerol and froze them in dry ice; they found that on average, 67% of spermatozoa from five men exhibited survival of motility. Shortly thereafter, they reported that they had produced pregnancies in three women by AI with frozen spermatozoa (Bunge and Sherman, 1953). The following year, Bunge et al. (1954) described further experiments to determine the effect of four freezing methods on survival of human spermatozoa; they also provided further clinical details regarding four pregnancies produced by AI with frozen semen. Other clinicians soon began to use the procedures of Sherman and Bunge to produce pregnancies in women via AI with frozen-thawed spermatozoa (Iizuka and Sawada, 1958). As noted above, Sawada later collaborated with M.-C. Chang to produce pregnancies in rabbits, using cryopreserved spermatozoa. Sherman continued to investigate the effect of several variables on survival of frozen-thawed human spermatozoa, and he also began to attempt to freeze-dry human spermatozoa (Sherman, 1954, 1963). Among other things, Sherman found that he could successfully freeze human spermatozoa in LN_2 and that storage at –196°C was superior to that at –75°C (Sherman, 1963) in that he observed no loss of motility of spermatozoa stored for 12 months at –196°C, whereas those stored at –75°C declined. Then Perloff et al. (1964) reported four full-term pregnancies in women artificially inseminated with spermatozoa frozen in LN_2 and stored for 1 to 5.5 months; two additional pregnancies spontaneously aborted during the fourth month of gestation. In a synopsis of the status of research on sperm preservation, Sherman (1964) also discussed future clinical applications of human sperm banks and proposed research to improve such methods. Since that time, although difficult to document, undoubtedly hundreds of thousands, perhaps millions, of children have been born as a result of AI with cryopreserved spermatozoa.

11.10 CHANG'S LOW-TEMPERATURE EXPERIMENTS ON RABBIT OOCYTES AND ZYGOTES

As much as in any other discipline, the evolution of ideas regarding the preservation of gametes by cooling them to low temperatures is clearly exemplified by the investigations of M.-C. Chang. Writing in 1947, Chang stated that "We are indebted to Ivanov, Walton, and Hammond for the demonstration that mammalian spermatozoa can be kept at low temperature for some length of time without loss of fertilizing capacity." Chang had been a doctoral student at Cambridge University working under the direction of John Hammond and Arthur Walton. As described previously, his doctoral research, an investigation of the effects of low temperatures on ram spermatozoa, had been an extension of the studies of Hammond and Walton in 1930 on rabbit spermatozoa. By 1947, Chang had turned his attention to embryos. As was true for rabbit spermatozoa, Chang (1947) found that the optimum temperature to store rabbit embryos was 10°C; even after 144 h at that temperature, 24% of two cell-stage embryos cleaved in culture, whereas embryos stored at 0°, 5°, 15°, or 25°C lost their capability to develop much sooner. He also found that the rate of cooling from 25° to 10°C affected the embryos' developmental capacity, in that rapidly cooled embryos were damaged but slowly cooled embryos were not. Four litters of live young were kindled after embryos stored at 10°C for as long as ~100 h were transplanted into does. The following year, Chang (1948b) continued his studies of embryo storage by transferring a total of 1550 rabbit zygotes and cleaved embryos into recipient does that had been induced to ovulate by injection with gonadotropin. Of these transferred embryos, 28% developed into normal, live young. Of the total, 28 live young developed from zygotes stored at 10°C for 96 h, and 19 developed from zygotes that had been stored at 0°C for at least 24 h. Shortly thereafter, Chang (1950) extended these observations to rabbit blastocysts, showing that even these late-stage embryos could be stored at 10° or 0°C for at least 24 h. Although by current standards these may seem to have been modest steps, 25 years later very similar experiments were conducted with sheep and cattle embryos in an attempt to derive methods to cryopreserve them (Trounson et al., 1976a, 1976b; Wilmut et al., 1975). Pursuing the observations of Chang, Smith (1952) began experiments to try to freeze rabbit eggs.

These preliminary reports were subsequently expanded considerably by comprehensive studies of the effects of various temperatures on rabbit eggs and embryos, first of two-cell embryos (Chang, 1948a) and later on rabbit oocytes (Chang, 1952). The purpose of these very large experiments was to determine the length of time during which unfertilized eggs remained fertilizable after ovulation, and also the length of time during which early embryos remained capable of further development and implantation. In the first series, Chang (1948a) collected two-cell embryos from superovulated does and then exposed them to five temperatures (24°, 15°, 10°, 5°, or 0°C) for various times up to 168 h; he assayed survival of the treated embryos by first measuring the percentage that underwent two or three cleavages and developed into 16-cell embryos. He determined their ultimate viability by transferring treated embryos into previously mated recipients of a different breed and counting the number of live young that were born. The second series of experiments was similar in design to the first, except that Chang (1952) measured the effects of a wider range of temperatures (45°, 38°, 24°, 10°, or 0°C) on oocytes held for various times up to 120 h. These experiments were conducted many years before the derivation of methods of *in vitro* fertilization. Therefore, after treating the oocytes Chang transferred them into mated females and killed the recipients ~48 h after transfer; he then counted the number of oocytes that had been fertilized and that had developed normally into cleaved embryos. For the purposes of this review, a few examples of selected data from both series of Chang's experiments have been redrawn to illustrate his observations of the effects of low temperatures on rabbit oocytes and embryos. The results in Figure 11.9, redrawn from Chang's 1952 study, show that oocytes were rather sensitive to being held at 10° or 0°C; after ~24 h at 10°C, about 80% of the oocytes could be fertilized and develop, whereas at 0°C, only ~40% remained fertilizable. After about 72 h, only ~10% held at 10°C could be fertilized, and only very few that had been held at 0°C could be fertilized. For

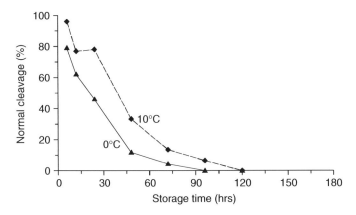

FIGURE 11.9 Effect of storage temperatures on development of rabbit oocytes. Data of Chang, 1952.

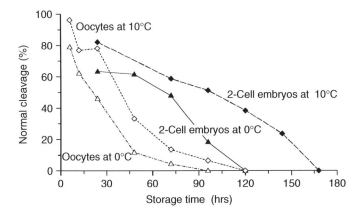

FIGURE 11.10 Effect of storage temperature on development of rabbit oocytes and two-cell embryos. Data of Chang, 1948a, 1952.

comparison, corresponding results from Chang's 1948 investigation of two-cell embryos are shown in Figure 11.10. These results clearly demonstrate that cleavage-stage rabbit embryos are much more resistant than are unfertilized oocytes. Some 50 years later, although it has been clearly established that one cause of injury to mammalian oocytes exposed to low temperatures is disassembly of the meiotic spindle, experiments are still being conducted to try to determine additional mechanisms responsible for damage to oocytes.

11.11 SHERMAN AND LIN: FREEZING AND STORAGE OF MOUSE EGGS

A few years after Chang's (1952) report on rabbit oocytes, Lin et al. (1957) used the same procedure of transferring treated oocytes into mated females of another inbred strain to determine the fertilizability of mouse oocytes. They reported that mouse oocytes contracted osmotically when suspended in 0.66 *M* glycerol, yet would develop into normal fetuses after transfer into recipients. Those investigators (Sherman and Lin, 1958a, 1958b) then found that mouse oocytes suspended in that same concentration of glycerol could be cooled to –10°C for 1 or 2 h; after being transferred into mated recipients, some of the oocytes were fertilized and 15 of 25 recipients became pregnant. Sherman and Lin then suspended mouse oocytes in 0.66 *M* glycerol, cooled them to –10°C at 2°C/min, and induced crystallization of the samples by seeding with dry ice. Even after 3.5 h at –10°C, 12% of 60 oocytes became fertilized after being transferred into mated recipients; 10 of the 14 recipients became pregnant. Sherman and Lin (1958b) also made microscopic observations of oocytes cooled on a special chamber that permitted them to observe specimens during freezing and thawing. If the oocytes underwent an abrupt "blacking out," the authors interpreted this to be evidence of intracellular ice formation; such oocytes appeared damaged. Many years later, this interpretation was verified by cryomicroscopical observations of mouse oocytes cooled over a wide range of rates (Leibo et al., 1978).

These preliminary experiments were followed by a more detailed investigation of the effects of abrupt temperature changes and storage of mouse oocytes at low temperatures. Sherman and Lin (1959) transferred a total of 1858 oocytes collected from agouti mice into 310 albino females. Whether cooled at ~12°C/min or 60°C/min to 0°C, the same percentages of oocytes survived and developed as the controls. Mouse oocytes tolerated brief exposure to 0° or –10°C, but when held at those temperatures for 6 h or longer, few survived and became fertilized. The authors concluded that unfertilized mouse eggs were very sensitive to prolonged exposure to low temperatures. Although the oocytes remained morphologically intact, they were not fertilizable after transfer.

11.12 MAZUR'S EQUATIONS

During the same time that these rather practical experiments were being conducted in an attempt to preserve mammalian gametes and embryos by cooling them to low temperatures, very basic studies were also being conducted to determine causes of injury to cells when exposed to very low subzero temperatures. Notable among these studies are the investigations of Peter Mazur. From 1957 through 1963, Mazur conducted several experiments designed to elucidate mechanisms of damage caused to microorganisms when frozen and thawed. These studies culminated in Mazur formulating a mathematical model to describe the theoretical response of cells when subjected to freezing solutions and low temperatures (Mazur, 1963). Nine years later, Mazur used his mathematical model to estimate cooling rates to be used to attempt to cryopreserve mouse embryos; this application was an important part of the experiments by Whittingham et al. (1972) to freeze mouse embryos, described below.

Over the last 40 years, Mazur's theoretical analysis has been verified by many investigators and has been found to apply with equal validity to a wide range of microbial, plant, and animal cells. Although many minor modifications have been made to his original formulation, the basic concepts remain as valid and significant as when he first formulated them. These concepts are explained and enunciated in his chapter in this volume (Chapter 1).

11.13 CRYOPRESERVATION OF MOUSE EMBRYOS

As summarized by Hafez (1969), little progress had been made to derive procedures to preserve oocytes or embryos by cooling them to low temperatures, and this is where the situation stood for several years. That year, Whittingham and Wales (1969) reported on the development of two-cell mouse embryos after they had been held at temperatures of 10°, 5°, or 0°C for various times up to 48 h. They reported that ~15% of the embryos developed into blastocysts after 24 h at 5°C and 9% developed even after 48 h. They also noted that only 3 of 40 embryos survived a 20-min exposure to 16% dimethyl sulfoxide (DMSO) and that neither DMSO nor glycerol protected the embryos against the damage caused by exposure to low temperatures. Therefore, Whittingham and Wales concluded that high concentrations either of glycerol or of DMSO damaged embryos by some undefined mechanism. For that reason, 2 years later, Whittingham (1971) selected 7.5% polyvinylpyrrolidone (PVP) as a cryoprotective solution, and he reported that 69% of 139 eight-cell mouse embryos frozen in PVP and held for 30 min at –79°C developed into blastocysts. The embryos, suspended in 0.5 mL of PVP medium in 5-mL test-tubes, were frozen at 60°C/min in a mixture of dry ice and acetone at –79°C for 5 min, held for 30 min in a flask of dry ice, and then

FIGURE 11.11 A recipient female mouse and her pups derived from cryopreserved 8-cell embryos. The embryos had been frozen in liquid nitrogen; after being thawed, they were cultured to the blastocyst stage and trasferred into the recipient foster mouse. The observations were those of Whittingham et al. (1972).

thawed at ~85°C/min to 20°C. Moreover, when 13 expanded blastocysts, obtained after freezing and thawing them as early-stage blastocysts, were transferred into two F_1 hybrid females, nine normal young were born.

Prompted by that success, Whittingham joined Peter Mazur and Stanley Leibo in April 1972 at Oak Ridge National Laboratory in the United States to continue investigations of the freezing of mammalian embryos. Basing their investigations on emerging understanding of mechanisms of freezing injury in mammalian cells (Leibo and Mazur, 1971; Leibo et al., 1970; Mazur, 1963, 1970; Mazur et al., 1969, 1972), Whittingham et al. (1972) explored the principal cryobiological factors—chiefly, suspending medium, cooling rate, final subzero temperature, and warming rate—that might influence survival of preimplantation stages of mouse embryos. They suspended embryos (zygotes, two-cell, eight-cell, and blastocysts) in solutions of 1 M DMSO and cooled replicate samples of embryos at each of nine rates to –78°C; some embryos were also frozen in 1 M glycerol. Additional embryos were first frozen to –80° or –110°C before being placed into liquid nitrogen, and some of these were then placed into liquid helium at –269°C for over 1 h. Frozen embryos were warmed at one of four rates. When cooled at the optimum rate of 0.4°C/min to –78°C and then thawed, 70% of 108 eight-cell embryos developed into blastocysts in culture. Moreover, when a total of 927 embryos that had been frozen, including some that had been frozen to –269°C (4°K), were transferred into 118 pseudopregnant recipients, 77 of the recipients became pregnant. A total of 210 near-term normal fetuses and 57 live-born pups were born. A photograph of a recipient and her foster pups derived from frozen embryos is shown in Figure 11.11; Whittingham, Leibo, and Mazur celebrated the birth of the first "frozen mice" by toasting each other with champagne that Mazur had won in an airline competition (Figure 11.12).

One month after the publication by Whittingham et al. (1972), Wilmut (1972) reported that he had also succeeded in successfully freezing eight-cell-stage and blastocyst-stage embryos of the mouse. He first tested the effects of 1.5 M DMSO, 0.6 M sucrose, and 7.5% PVP (either dialyzed or undialyzed) on development of early blastocysts that had been frozen at each of four rates (0.7°, 23°, 80°, or 690°C/min) and thawed at 360°C/min. Only in the presence of DMSO did he obtain any survival, and it was very low (1 of 18 embryos cooled at 0.7°C/min and 3 of 17 cooled at 23°C/min). Using 1.5 M DMSO as the protective solution, Wilmut cooled blastocysts at four low rates (0.07°, 0.22°, 1.2°, and 4.7°C/min) and warmed them at each of three rates (1°, 12°, and 360°C/min). He achieved 65% survival of blastocysts that had been cooled at 0.22°C/min and warmed at 12°C/min, and 25% survival of those cooled very slowly and warmed at 1°C/min. However, when he froze additional blastocysts at 0.22°C/min and warmed them at rates of 1°, 12°,

FIGURE 11.12 A photograph of Peter Mazur (left), Stanley Leibo (center), and David Whittingham (right) taken in June 1972 on the occasion of the birth of the first mammals derived from cryopreserved embryos.

or 60°C/min, he obtained respective survivals of 54, 86, and 65%. He also reported 66% survival of eight-cell embryos frozen similarly. In a recent book (Wilmut et al., 2000) describing his extraordinary research on cloning animals by nuclear transfer of adult somatic cells into enucleated oocytes (Wilmut et al., 1997), Wilmut does not cite his 1972 publication on freezing of mouse embryos. However, he does describe his production of the world's first "frozen calf" produced by transfer of a previously frozen embryo (Wilmut and Rowson, 1973). Based on those observations of Wilmut and Rowson, Willadsen et al. (1976) were able to derive an even more efficient method to cryopreserve sheep embryos.

The report by Whittingham (1971) describing survival of mouse embryos suspended in PVP and frozen at 60°C/min has now been incorporated in the literature. For example, Gunasena and Critser (1997), Gordon (1994), and Ludwig et al. (1999) cite that paper as the first report of the successful cryopreservation of mammalian embryos. However, in their article, Whittingham et al. (1972) noted that they had also frozen embryos in 7.5% PVP to –78°C at rates from 1° to 600°C/min; none of the embryos survived and the authors were unable to explain the discrepancy between their results and those previously reported by Whittingham (1971). In the introduction to his first article on embryo freezing, Wilmut (1972) also states that he was unable to replicate the preservation of mouse embryos in PVP. In a review of embryo cryopreservation written several years later, Ashwood-Smith (1986, p. 326) noted that "preservation of mouse embryos with PVP has been difficult to repeat." And Leibo and Oda (1993) attempted unsuccessfully to freeze eight-cell mouse embryos suspended in solutions of PVP alone, although they did obtain high survival of those suspended in PVP supplemented with high concentrations of ethylene glycol.

Over the last 30 years, the procedure of oocyte and embryo cryopreservation has been refined and improved and simplified. Innumerable experimental reports have been published describing major and minor alterations and modifications of the original methods. Many compounds in addition to the original ones of DMSO and glycerol have been found to act as cryoprotectants for mammalian embryos; for example, methanol, ethylene glycol, and propylene glycol, or adonitol. Embryos can now be successfully cryopreserved by being cooled slowly or rapidly or even ultra-rapidly (>10,000°C/min); they can be warmed slowly or rapidly; and they can be stored for decades, yet still develop into normal young when recovered from liquid nitrogen storage and transferred into appropriate recipients. The study of gamete and embryo cryobiology has also yielded fundamental understanding of many basic characteristics of these highly specialized cells. All of these applications have evolved from the first observations of the study of cold on biological specimens.

11.14 SUMMARIES AND REVIEWS OF GAMETE AND EMBRYO CRYOBIOLOGY

Since the two reports of the successful cryopreservation of mouse zygotes and embryos in 1972 (Whittingham et al., 1972; Wilmut, 1972), literally thousands of experimental reports and reviews have been published on the subject of preservation of reproductive cells and tissues. Live young of 22 mammalian species have been born as a result of transfer of cryopreserved embryos (Rall, 2001), and literally millions of animals, especially of domestic species, have been born as a result of AI with cryopreserved spermatozoa. Comprehensive summaries of the cryopreservation of domestic species have been published by Hasler (1992, 2001), Mapletoft (1984), and Massip and Leibo (2002). Tens if not hundreds of thousands of children have been born as a result of the practice of cryobiology applied to human gametes and embryos.

Several symposia devoted specifically to the cryobiology of mammalian gametes and embryos have been published. The first meeting, convened only 2 years after the birth of the "frozen mice" in 1972, was held at The Jackson Laboratory in the United States (Mühlbock, 1976). This was followed the next year by a Ciba Foundation Symposium in London (Elliott and Whelan, 1977) and by a third meeting that was also held in the United Kingdom (Zeilmaker, 1981). Other reviews on the cryopreservation of embryos have been published by Wood et al. (1987), Shaw et al. (2000), Rall (2001), and Rall et al. (2000), and other summaries are included in the books by Karow and Critser (1997) and by Watson and Holt (2001). Recent reviews of sperm cryopreservation have also been published (Holt, 2000a, 2000b; Leibo and Bradley, 1999; Watson, 1995); especially notable is the very comprehensive summary by Watson (1990), and other summaries are included in the books by Karow and Critser (1997), by Trounson and Gardner (1999), and by Watson and Holt (2001).

11.15 CONCLUSIONS

Almost 70 years ago, in a brief paper entitled *La vie latente de quelque Algues et Animaux inférieurs aux basses temperatures et la conservation de la vie dans l'univers* [Latent life of some algae and lower animals at low temperatures and preservation of life in the universe], Becquerel (1936) speculated about the implications of some of his investigations of low-temperature biology. He wondered what might become of life on earth when the sun becomes extinguished. He continued:

> So, on this chilled planet, wandering in the night of cosmic space, what will become of its germs of suspended life? . . . [W]ill the planet, breaking up as the result of an explosion, seed other worlds with its germ-laden debris? Should that be the case, latent life "anabiosis", truly providential for the preservation of life on Earth, would be the best means that Nature could use to give some plant and animal species a sort of immortality in the firmament.

In the twenty-first century, such musings seem not unreasonable. Some 40 years ago, Robert Ettinger (1964) proposed "The Prospect of Immortality" by suggesting that human bodies might be preserved in liquid nitrogen for extended times so as to be "revived" at some unspecified time in the future. Such a prospect seems no more achievable today than when originally proposed. In contrast, a few species of insects, amphibians, and crustaceans have been revived after having been preserved in liquid nitrogen. It is estimated that living organisms preserved at a temperature of –196°C or below will remain alive at least for millennia (DuFrain, 1976; Mazur, 1976). Therefore, one may imagine that cryopreservation of earth's flora and fauna may indeed offer the true prospect of immortality for life as we presently understand it.

ACKNOWLEDGMENT

I thank Professor Keith J. Betteridge of the University of Guelph for his translation of the 1936 article by P. Becquerel, as well as for his many thoughtful discussions of some of the concepts presented in this chapter.

REFERENCES

Ashwood-Smith, M.J. (1986) The cryopreservation of human embryos, *Hum. Reprod.*, 1, 319–332.

Barker, C.A.V. and Gandier, J.C.C. (1957). Pregnancy in a mare resulting from frozen epididymal spermatozoa, *Can. J. Comp. Med.*, 21, 47–51.

Becquerel, P. (1936). La vie latente de quelque Algues et Animaux inférieurs aux basses temperature et la conservation de la vie dans l'univers, *CR Acad. Sci.*, 202, 978–981.

Bhowmick, S., Zhu, L., McGinnis, L., Lawitts, J., Bharat, D., Nath, B.D., Toner, M., and Biggers, J. (2003) Desiccation tolerance of spermatozoa dried at ambient temperature: production of fetal mice, *Biol. Reprod.*, 68, 1779–1786.

Bolton, H.C. (1900) *Early History of the Thermometer*, Chemical Publishing, Easton, PA.

Bunge, R.G., Keettel, W.C., and Sherman, J.K. (1954) Clinical use of frozen semen. Report of four cases, *Fertil. Steril.*, 5, 520–529.

Bunge, R.G., and Sherman, J.K. (1953) Fertilizing capacity of frozen human spermatozoa, *Nature*, 172, 767–768.

Chang, M.-C. (1947) Normal development of fertilized rabbit ova stored at low temperature for several days, *Nature*, 159, 602–603.

Chang, M.-C. (1948a). The effects of low temperature on fertilized rabbit ova *in vitro,* and the normal development of ova kept at low temperature for several days, *J. Gen. Physiol.*, 31, 385–410.

Chang, M.-C. (1948b) Transplantation of fertilized rabbit ova: the effect on viability of age, *in vitro* storage period, and storage temperature, *Nature*, 161, 978–979.

Chang, M.-C. (1950) Transplantation of rabbit blastocysts at late stage: probability of normal development and viability at low temperature, *Science*, 111, 544–545.

Chang, M.-C. (1952). Fertilizability of rabbit ova and the effects of temperature *in vitro* on their subsequent fertilization and activation *in vivo, J. Exp. Zool.*, 121, 351–381.

Chang, M.-C. (1983). My work on the transplantation of mammalian eggs, *Theriogenology*, 19, 293–303.

Chang, M.-C. and Walton, A. (1940) The effects of low temperature and acclimatization on the respiratory activity and survival of ram spermatozoa, *Proc. R. Soc.* 129B, 517–527.

Dauzier, L. (1956) Quelques résultants sur l'insémination artificielle des brébis et des chèvres en France. Proceedings III International Congress of Artificial Reproduction, Cambridge. Section 3, pp. 12–14.

Devireddy, R.V., Coad, J.E., and Bischof, J.C. (2001) Microscopic and calorimetric assessment of freezing processes in uterine fibroid tumor tissue, *Cryobiology*, 42, 225–243.

DuFrain, R.J. (1976) The effects of ionizing radiation on preimplantation mouse embryos developing *in vitro,* in *Basic Aspects of Freeze-Preservation of Mouse Strains*, Mühlbock, O., Ed., Gustav Fischer, Stuttgart, pp. 73–84.

Easley, G.T., Mayer, D.T., and Bogart, R. (1942) Influence of diluters, rate of cooling, and storage temperatures on survival of bull sperm, *Am. J. Vet. Res.*, 3, 358–363.

Elliott, K. and Whelan, J., Eds. (1977) *The Freezing of Mammalian Embryos*, Ciba Foundation Symposium 52 (new series), Elsevier, Amsterdam.

Emmens, C.W. and Blackshaw, A.W. (1950) The low temperature storage of ram, bull and rabbit spermatozoa, *Aust. Vet. J.*, 26, 226–228.

Emmens, C.W. and Blackshaw, A.W. (1955) The fertility of frozen ram and bull semen, *Aust. Vet. J.*, 31, 76–79.

Ettinger, R.C.W. (1964) *The Prospect of Immortality*, Doubleday, Garden City, NY.

Foote, R. (1981) The artificial insemination industry, in *New Technologies in Animal Breeding*, Brackett, B.G., Seidel, G.E. Jr., Seidel, S.M., Eds., Academic Press, New York, pp. 13–39.

Fox, R.R. (1961) Preservation of rabbit spermatozoa: fertility results from frozen semen, *Proc. Soc. Exp. Biol. Med.*, 108, 663–665.

Gordon, I. (1994) Storage and cryopreservation of oocytes and embryos, in *Laboratory Production of Cattle Embryos*, CAB International, Wallingford, Oxfordshire, U.K.

Gunasena, K.T. and Critser, J.K. (1997) Utility of viable tissues ex vitro. Banking of reproductive cells and tissues, in *Reproductive Tissue Banking; Scientific Principles*, Karow, A.M. and Critser, J.K., Eds., Academic Press, San Diego, pp. 1–21.

Hafez, E.S.E. (1969) Storage of rabbit ova in gelled media at 10°C, *J. Reprod. Fertil.*, 2, 163–178.

Hammond, J. (1930) The effect of temperature on the survival *in vitro* of rabbit spermatozoa obtained from the vagina, *J. Exp. Biol.*, 7, 175–191.

Harris, R.J.C., Ed. (1954) *Biological Applications of Freezing and Drying*, Academic Press, New York.

Hasler, J.F. (1992) Current status and potential of embryo transfer and reproductive technology in dairy cattle, *J. Dairy Sci.*, 75, 2857–2879.

Hasler, J.F. (2001) Factors affecting frozen and fresh embryo transfer pregnancy rates in cattle, *Theriogenology*, 56, 1401–1415.

Hoagland, H. and Pincus, G. (1942) Revival of mammalian sperm after immersion in liquid nitrogen, *J. Gen. Physiol.*, 25, 337–344.

Holt, W.V. (2000a) Basic aspects of frozen storage of semen, *Anim. Reprod. Sci.*, 62, 3–22.

Holt, W.V. (2000b) Fundamental aspects of sperm cryobiology: The importance of species and individual differences, *Theriogenology*, 53, 47–58.

Iizuka, R. and Sawada, Y. (1958) Successful inseminations with frozen human semen, *Jpn. J. Fertil. Steril.*, 3, 333–337.

Iritani, A. (1980) Problems of freezing spermatozoa of different species. IX International Congress of Animal Reproduction, Madrid, pp. 115–132.

Jahnel, F. (1938) Über die Widerslendfähigkeit von menchlichen Spermatozoen gegenüber starker Kälte, *Klinische Wochenschrift*, 17, 1273–1274.

Johnson, M.H. and Everitt, B.J. (2000) *Essential Reproduction*, 5th edition, Blackwell Science, Oxford, U.K.

Karow, A.M. and Critser, J.K., Eds. (1997) *Reproductive Tissue Banking; Scientific Principles*, Academic Press, San Diego.

Keskintepe, L., Pacholczyk, G., Machnicka, A., Norris, K., Curuk, M. A., Khan, I., and Brackett, B.G. (2002) Bovine blastocyst development from oocytes injected with freeze-dried spermatozoa, *Biol. Reprod.*, 67, 409–415.

Knobil, E. and Neill, J.D., Eds. (1994) *The Physiology of Reproduction*, 2nd edition, Raven Press, New York.

Kusakabe, H., Szczygiel, M.A., Whittingham, D.G., and Yanagimachi, R. (2001) Maintenance of genetic integrity in frozen and freeze-dried mouse spermatozoa, *Proc. Natl. Acad. Sci. USA*, 98, 13501–13506.

Leibo, S.P. and Bradley, L. (1999) Comparative cryobiology of mammalian spermatozoa, in *The Male Gamete: From Basic Knowledge to Clinical Applications*, Gagnon, C., Ed., Cache River Press, Vienna, IL, pp. 501–516.

Leibo, S.P., Farrant, J., Mazur, P., Hanna, M.G., Jr., and Smith, L.H. (1970) Effects of freezing on marrow stem cell suspensions: interactions of cooling and warming rates in the presence of PVP, sucrose, or glycerol, *Cryobiology*, 6, 315–332.

Leibo, S.P. and Mazur, P. (1971) The role of cooling rates in low temperature preservation, *Cryobiology*, 8, 447–452.

Leibo, S.P., McGrath, J.J., and Cravalho, E.G. (1978) Microscopic observation of intracellular ice formation in unfertilized mouse ova as a function of cooling rate, *Cryobiology*, 15, 257–271.

Leibo, S.P. and Oda, K. (1993) High survival of mouse zygotes and embryos cooled rapidly or slowly in ethylene glycol plus polyvinylpyrrolidone, *Cryo-Letters*, 14, 133–144.

Lillie, F. (1916) The history of the fertilization problem, *Science*, 43, 39–53.

Lin, T.P., Sherman, J.K., and Willett, E.L. (1957) Survival of unfertilized mouse eggs in media containing glycerol and glycine, *J. Exp. Zool.*, 134, 275.

Lucké, B. (1940) The living cell as an osmotic system and its permeability to water, *Cold Spring Harb. Symp. Quant. Biol.*, 8, 123–132.

Ludwig, M., Al-Hasani, S., Felberbaum, R., and Diedrich, K. (1999) New aspects of cryopreservation of oocytes and embryos in assisted reproduction and future perspectives, *Hum. Reprod.*, 14, 162–185.

Luyet, B.J. and Gehenio, P.M. (1940) *Life and Death at Low Temperature*, Biodynamica, Normandy, MO.

Luyet, B.J., and Hodapp, E.L. (1938) Revival of frog's spermatozoa vitrified in liquid air, *Proc. Soc. Exp. Biol. Med.*, 39, 433–434.

Luyet, B., and Keane, J., Jr, (1955) A critical temperature range apparently characterized by sensitivity of bull semen to high freezing velocity, *Biodynamica*, 7, 281–292.

Mapletoft, R.J. (1984) Embryo transfer technology for the enhancement of animal production, *Biotechnology*, 6, 149–160.

Marks, S.L., Dupuis, J., Mickelsen, W.D., Memon, M.A., and Platz, C.C. (1994) Conception by use of postmortem epididymal semen extraction in a dog, *J. Am. Vet. Med. Assoc.*, 204, 1639–1640.

Marshall, F.H.A. (1910) *The Physiology of Reproduction*, Longmans, Green & Co., Edinburgh, Scotland.

Martin, I. and Emmens, C.W. (1958) Factors affecting the fertility and other characteristics of deep-frozen bull semen, *J. Endocrinol.*, 17, 449–455.

Massip, A. and Leibo, S.P. (2002) Embryo quality and freezing tolerance, in *Assessment of Mammalian Embryo Quality: Invasive and Non-invasive Techniques*, Van Soom, A. and Boerjan, M.L., Eds., Kluwer Academic Press, Dordrecht, pp. 121–138.

Mazur, P. (1963) Kinetics of water loss from cells at subzero temperatures and the likelihood of intracellular freezing, *J. Gen. Physiol.*, 47, 347–369.

Mazur, P. (1970) Cryobiology: the freezing of biological systems, *Science*, 168, 939–949.

Mazur, P. (1976) Freezing and low-temperature storage of living cells, in *Basic Aspects of Freeze-Preservation of Mouse Strains*, Mühlbock, O., Ed., Gustav Fischer, Stuttgart, pp. 1–12.

Mazur, P., Farrant, J., Leibo, S.P., and Chu, E.H.Y. (1969) Survival of hamster tissue-culture cells after freezing and thawing, *Cryobiology*, 6, 1–9.

Mazur, P., Leibo, S.P., and Chu, E.H. (1972) A two-factor hypothesis of freezing injury. Evidence from Chinese hamster tissue-culture cells, *Exp. Cell Res.*, 71, 345–355.

Meryman, H.T., Ed. (1966) *Cryobiology*, Academic Press, London.

Mühlbock, O., Ed. (1976) *Basic Aspects of Freeze Preservation of Mouse Strains*, Gustav Fischer, Stuttgart.

Nawroth, F., Isachenko, V., Dessole, S., Rahimi, G., Farina, M., Vargiu, N., Mallmann, P., Dattena, M., Capobianco, G., Peters, D., Orth, I., and Isachenko, E. (2002) Vitrification of human spermatozoa without cryoprotectants, *Cryo-Letters*, 23, 93–102.

Nygren, K.G. and Andersen, A.N. (2002) Assisted reproductive technology in Europe 1999. Results generated from European registers by ESHRE, *Hum. Reprod.*, 17, 3260–3274.

Oda, K., Gibbons, W.E., and Leibo, S.P. (1992) Osmotic shock of fertilized mouse ova, *J. Reprod. Fertil.*, 95, 737–747.

Parkening, A.A., Tsunoda, Y., and Chang, M.C. (1976) Effects of various low temperatures, cryoprotective agents and cooling rates on the survival, fertilizability and development of frozen-thawed mouse eggs, *J. Exp. Zool.*, 197, 369–374.

Parkes, A.S. (1945) Preservation of human spermatozoa at low temperatures, *Br. Med. J.*, 2, 212–213.

Parkes, A.S. (1957) Introductory remarks, *Proc. R. Soc.*, 147B, 424–426.

Parkes, A.S. (1960) *Marshall's Physiology of Reproduction*, 3rd edition, Longmans, London.

Parkes, A.S. (1964). Cryobiology, *Cryobiology*, 1, 3.

Parkes, A.S. (1966) *Sex, Science and Society*, Oriel Press, Newcastle upon Tyne.

Perloff, W.H., Steinberger, E., and Sherman, J.K. (1964) Conception with human spermatozoa frozen by nitrogen vapor technic, *Fertil. Steril.*, 15, 501–504.

Phillips, P.H. and Lardy, H.A. (1940) A yolk-buffer pabulum for the preservation of bull semen, *J. Dairy Sci.*, 23, 399–404.

Polge, C. (1957) Low-temperature storage of mammalian spermatozoa, *Proc. R. Soc.*, 147B, 498–508.

Polge, C. (1968) Frozen semen and the A.I. programme in Great Britain, in *Proceedings of the 2nd Technical Conference on Artifical Insemination and Reproduction*, National Association of Artificial Breeders, Chicago, IL, pp. 46–51.

Polge, C. and Rowson, L.E.A. (1952a) Fertilizing capacity of bull spermatozoa after freezing at –79°C, *Nature* 169, 626.

Polge, C. and Rowson, L.E.A. (1952b) Results with bull semen stored at –79°C, *Vet. Rec.*, 64, 851.

Polge, C., Smith, A.U., and Parkes, A.S. (1949) Revival of spermatozoa after vitrification and dehydration at low temperatures, *Nature*, 164, 666.

Prévost, M. (1840) Récherches sur les animalcules spermatiques, *CR Acad. Sci.*, 11, 907–908.

Rall, W.F. (2001) Cryopreservation of mammalian embryos, gametes, and ovarian tissues, in *Contemporary Endocrinology: Assisted Fertilization and Nuclear Transfer in Mammals*, Wolf, D.P., Zelinski-Wooten, M., Eds., Humana Press, Towata, NJ, pp. 173–187.

Rall, W.F., Schmidt, P.M., Lin, X., Brown, S.S., Ward, A.C., and Hansen, C.T. (2000) Factors affecting the efficiency of embryo cryopreservation and rederivation of rat and mouse models, *ILAR J.*, 41, 221–227.

Rorvik, D. (1974) The embryo sweepstakes, *New York Times Magazine*, September 15, pp 16ff.

Rostand, J. (1946) Glycérine et résistance du sperme aux basses températures, *CR Acad Sci.*, 222, 1524–1525.

Salamon, S. and Maxwell, W.M.C. (1995a) Frozen storage of ram semen. I. Processing, freezing, thawing and fertility after cervical insemination, *Anim. Reprod. Sci.*, 37, 185–249.

Salamon, S. and Maxwell, W.M.C. (1995b) Frozen storage of ram semen. II. Causes of low fertility after cervical insemination and methods of improvement, *Anim. Reprod. Sci.*, 38, 1–36.

Sawada, Y. and Chang, M.C. (1964) Motility and fertilizing capacity of rabbit spermatozoa after freezing in a medium containing dimethyl sulfoxide, *Fertil. Steril.*, 15, 222–229.

Shaffner, C.S. (1942). Longevity of fowl spermatozoa in frozen condition, *Science*, 96, 337.

Shaw, J.M., Oranratnachai, A., and Trounson, A.O. (2000) Cryopreservation of oocytes and embryos, in *Handbook of in Vitro Fertilization*, 2nd edition, Trounson, A.O. and Gardner, D.K., Eds., CRC Press, Boca Raton, FL, pp. 373–412.

Sherman, J.K. (1954) Freezing and freeze-drying of human spermatozoa, *Fertil. Steril.*, 5, 357–371.

Sherman, J.K. (1963) Improved methods of preservation of human spermatozoa by freezing and freeze-drying, *Fertil. Steril.*, 14, 49–64.

Sherman, J.K. (1964) Research on frozen human semen, *Fertil. Steril.*, 15, 485–499.

Sherman, J.K. and Bunge, R.G. (1953) Observations on preservation of human spermatozoa of low temperatures, *Proc. Soc. Exp. Biol. Med.*, 82, 686.

Sherman, J.K. and Lin, T.P. (1958a) Effect of glycerol and low temperature on survival of unfertilized mouse eggs, *Nature*, 181, 785–786.

Sherman, J.K. and Lin, T.P. (1958b) Survival of unfertilized mouse eggs during freezing and thawing, *Proc. Soc. Exp. Biol. Med.*, 98, 902–905.

Sherman, J.K. and Lin, T.P. (1959) Temperature shock and cold-storage of unfertilized mouse eggs, *Fertil. Steril.*, 10, 384–396.

Shettles, L.B. (1940) The respiration of human spermatozoa and their response to various gases and low temperatures, *Am. J. Physiol.*, 128, 408–415.

Shettles, L.B. (1979) Diploid nuclear replacement in mature human ova with cleavage, *Am. J. Obstet. Gynecol.*, 133, 222–225.

Smith, A.U. (1952) Behaviour of fertilized rabbit eggs exposed to glycerol and to low temperatures, *Nature*, 170, 374–375.

Smith, A.U. (1961) *Biological Effects of Freezing and Supercooling*, Williams & Wilkins, Baltimore, MD.

Smith, A.U., Ed. (1970) *Current Trends in Cryobiology*, Plenum, New York.

Smith, A.U. and Polge, C. (1950a) Storage of bull spermatozoa at low temperatures, *Vet. Rec.*, 62, 115–116.

Smith, A.U. and Polge, C. (1950b) Survival of spermatozoa at low temperatures, *Nature*, 166, 668–669.

Society for Assisted Reproductive Technologies. (2002) Assisted reproductive technology in the United States: 1998 results generated from the American Society for Reproductive Medicine/Society for Assisted Reproductive Technology Registry, *Fertil. Steril.*, 77, 18–31.

Songsasen, N., Ratterree, M.S., VandeVoort, C.A., Pegg, D.E., and Leibo, S.P. (2002) Permeability characteristics and osmotic sensitivity of rhesus monkey (*Macaca mulatta*) oocytes, *Hum. Reprod.*, 17, 1875–1884.

Songsasen, N., Tong, J., and Leibo, S. P. (1998) Birth of live mice derived by *in vitro* fertilization with spermatozoa retrieved up to 24 hours after death, *J. Exp. Zool.*, 280, 189–196.

Stewart, D.L. (1951) Storage of bull spermatozoa at low temperatures, *Vet. Rec.*, 63, 65–66.

Thibault, C., Lavasseur, M.-C., and Hunter, R.H.F., Eds. (1993) *Reproduction in Mammals and Man*, Editeur des Préparations Grands Ecoles Médecin, Paris.

Thibier, M. (2002) The IETS statistics of embryo transfers in livestock in the world for the year 1999: a new record for bovine *in vivo*-derived embryos transferred, *Embryo Trans. Newslett.*, 18, 24–28.

Trounson, A.O. and Gardner, D.K., Eds. (1999) *Handbook of in Vitro Fertilization*, 2nd edition, CRC Press, Boca Raton, FL, pp. 373–412.

Trounson, A.O., Willadsen, S.M., and Rowson, L.E. (1976a) The influence of *in-vitro* culture and cooling on the survival and development of cow embryos, *J. Reprod. Fertil.*, 47, 367–370.

Trounson, A.O., Willadsen, S.M., Rowson, L.E., and Newcomb, R. (1976b) The storage of cow eggs at room temperature and at low temperatures, *J. Reprod. Fertil.*, 46, 173–178.

Tsunoda, Y., Parkening, T.A., and Chang, M.C. (1976) *In vitro* fertilization of mouse and hamster eggs after freezing and thawing, *Experientia*, 32, 223–224.

Wakayama, T. and Yanagimachi, R. (1998) Development of normal mice from oocytes injected with freeze-dried spermatozoa, *Nat. Biotechnol.*, 16, 639–641.

Walton, A. (1930) The effect of temperature on the survival *in-vitro* of rabbit spermatozoa obtained from the vas deferens, *J. Exp. Biol.*, 7, 201–219.

Watson, P.F. (1990) Artificial insemination and the preservation of semen, in *Marshall's Physiology of Reproduction*, 4th edition, Vol. 2, Lamming, G.E., Ed., Churchill Livingstone, Edinburgh, Scotland, pp. 747–869.

Watson, P.F. (1995) Recent developments and concepts in the cryopreservation of spermatozoa and the assessment of their post-thawing function, *Reprod. Fertil. Dev.*, 7, 871–891.

Watson, P.F. and Holt, W., Eds. (2001) *Cryobanking the Genetic Resource; Wildlife Conservation for the Future?* Taylor and Francis, London.

Whittingham, D.G. (1971) Survival of mouse embryos after freezing and thawing, *Nature*, 233, 125–126.

Whittingham, D.G., Leibo, S.P., and Mazur, P. (1972) Survival of mouse embryos frozen to –196° and –269°C, *Science*, 178, 411–414.

Whittingham, D.G. and Wales, R.G. (1969) Storage of two-cell mouse embryos *in vitro*, *Austr. J. Biol. Sci.*, 22, 1065–1068.

Willadsen, S.M., Polge, C., Rowson, L.E., and Moor, R.M. (1976) Deep freezing of sheep embryos, *J. Reprod. Fertil.*, 46, 151–154.

Willoughby, C.E., Mazur, P., Peter, A.T., and Critser, J.K. (1996) Osmotic tolerance limits and properties of murine spermatozoa, *Biol. Reprod.*, 55, 715–727.

Wilmut, I. (1972) The effect of cooling rate, warming rate, cryoprotective agent and stage of development on survival of mouse embryos during freezing and thawing, *Life Sci.*, 11 (Part II), 1071–1079.

Wilmut, I., Campbell, K., and Tudge, C. (2000) *The Second Creation*, Harvard University Press, Cambridge, MA.

Wilmut, I., Polge, C., and Rowson, L.E.A. (1975) The effect on cow embryos of cooling to 20, 0 and –196°C, *J. Reprod. Fertil.*, 45, 409–411.

Wilmut, I. and Rowson, L.E.A. (1973) Experiments on the low-temperature preservation of cow embryos, *Vet. Rec.*, 92, 686–690.

Wilmut, I., Schnieke, A.E., McWhir, J., Kind, A.J., and Campbell, K.H. (1997) Viable offspring derived from fetal and adult mammalian cells, *Nature*, 385, 810–813.

Wolstenholme, G.E.W. and Cameron, M.P., Eds. (1954) *Preservation and Transplantation of Normal Tissues. Ciba Foundation Symposium*, Churchill, London.

Wood, M.J., Whittingham, D.G., and Rall, W.F. (1987) The low temperature preservation of mouse oocytes and embryos, in *Mammalian Development, A Practical Approach*, Monk, M., Ed., IRL Press, Oxford, pp. 255–280.

Zeilmaker, G.H., Ed. (1981) *Frozen Storage of Laboratory Animals*, Gustav Fischer, Stuttgart.

12 Cryobiology of Gametes and the Breeding of Domestic Animals

Alban Massip,* Stanley P. Leibo, and Elisabeth Blesbois

CONTENTS

12.1   INTRODUCTION

During the last 30 years, various innovative technologies have been devised and are now used with increasing frequency to produce animals of laboratory, domestic, and nondomestic species. These technologies include *in vitro* maturation and fertilization of oocytes; *in vitro* production of embryos; micromanipulation of embryos to produce genetically identical twins, triplets, or quadruplets; and genetic engineering by various means to alter the genome of animals, including nuclear transfer of somatic cells into enucleated oocytes, or cloning. This latter method, first reported by Wilmut

---

* Deceased.

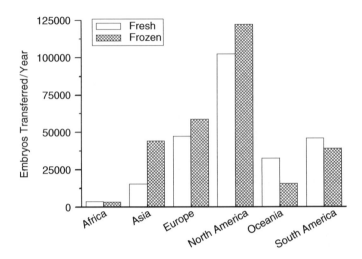

FIGURE 12.1 The numbers of bovine embryos that were transferred during the year 2000 in various parts of the world. Embryos were transferred either as fresh embryos or after having been cryopreserved. The data are those of Thibier (2001).

et al. (1997), has revolutionized the field of animal reproduction. A recent review of somatic cell cloning lists the production of live offspring of sheep, cattle, goats, pigs, and mice by this method (Brem and Kühholzer, 2002). The potential power of this method is illustrated by the fact that clones of 10 heifer calves (Wells et al., 1999), of six bull calves (Kubota et al., 2000), and of four rabbits (Chesné et al., 2002), all produced by nuclear transfer of adult somatic cells, have been reported. Such methods are being applied not only for the routine production of animals and for alteration of animal genomes but also as research tools to determine molecular mechanisms of embryonic development, and even as a means to protect endangered species. Efficient implementation of all of these methods has required cryopreservation of gametes, embryos, and other reproductive cells and tissues.

Cryobiology has thus become an integral part of methods of assisted reproduction of diverse animal species. We will examine its effect on breeding of domestic livestock species, on which the ability to preserve functionality of cells and tissues has had the greatest effect, and also on the production of domestic avian species. Evidence of the role of embryo cryopreservation on cattle breeding is shown by the synopsis of results recently tabulated by Thibier (2001) and represented graphically in Figure 12.1. Of the total of approximately 530,000 bovine embryos that were transferred throughout the world in the year 2000, more than 53% had been cryopreserved. Most important, the efficiency of calf production resulting from transfer of frozen embryos was close to that with fresh, unfrozen embryos. However, the sensitivity of gametes and embryos to cryopreservation differs according to several variables, such as the species, the stage of maturation or development, and whether the embryos were produced *in vivo* or *in vitro*. This review will consider the effect of cryobiology on various aspects of assisted reproduction.

12.2 CRYOPRESERVATION OF MAMMALIAN SPERMATOZOA AND ARTIFICIAL INSEMINATION

12.2.1 CATTLE

Artificial insemination (AI) is the oldest of the biotechnologies applied to animal breeding, and numerous reviews have been published describing the method itself and the efficiency of this technique. Historically and in terms of the number of animals bred, it has been used most extensively in cattle, first as a means to control venereal diseases and also to avoid the necessity of transporting animals from place to place for the pairing of cows and bulls. Its greater benefit, however, has been to enable genetic improvement of quantitative traits of cattle, especially of dairy breeds, made possible through intensive sire selection. In the middle of the twentieth century, AI of cattle was beginning to be applied at an expanding rate when Polge et al. (1949) first discovered that glycerol would protect certain types of cells against freezing damage and then devised a reliable and reproducible method to preserve bull spermatozoa using glycerol as a cryoprotectant (see also Chapter 11).

Used together, the techniques of AI and cryopreservation of semen have played a crucial role in securing and disseminating genetic improvements that have led to very significant increases in animal productivity during the last few decades. For example, according to Foote (1981), as a result of AI and sperm freezing, the potential number of 50 progeny per sire achievable in 1939 increased to 50,000 per sire in 1979. Furthermore, approximately one quarter the number of dairy cows are now used to produce four times the amount of milk that was produced 25 years earlier. It is not surprising, therefore, that these technologies have been described as "a breakthrough without equal in any other area of modern animal breeding" (Heap and Moor, 1995, as cited by Polge, 1998).

Freezing of sperm is now a routine procedure that is widely practiced and accepted (see also Chapter 18). Over the years, numerous reviews of this subject have been published. Notable among these is the comprehensive review of the subject of sperm preservation of various species by Watson (1990). More recent, albeit much briefer, reviews have been written by Foote and Parks (1993), Watson (1995), Leibo and Bradley (1999), and Holt (2000a, 2000b). The effect of sperm freezing on cattle breeding has been to increase the efficiency of AI and to provide safeguards for health control. More important, however, it has enabled more effective progeny-testing schemes to be implemented and has led to an extensive international trade in cryopreserved semen, resulting in genetic improvements worldwide. One disadvantage of AI with cryopreserved spermatozoa, however, is that in some countries, indigenous breeds have been lost or diluted by cross-breeding, so there is an urgent need to establish sperm banks for the purpose of genetic conservation and the maintenance of genetic diversity. One of the limitations of this approach has resulted from the overall efficiency of cattle sperm freezing, as currently practiced. On average, the freezing of semen kills up to one half of the sperm cells. Because a bull's ejaculate may contain 10 to 20 billion spermatozoa, however, until recently there seemed little need to improve the efficiency of sperm freezing (Leibo and Bradley, 1999). Now, however, intense international competition in the cattle industry means that the loss of >30% of collected sperm is no longer economically acceptable, and research on the mechanisms responsible for injury to spermatozoa is necessary. Another limitation of present methods of semen cryopreservation is that spermatozoa of different males exhibit significant differences in their sensitivity to freezing. In the case of young sires, this means that the genetics of a specific individual may be eliminated from the population, based solely on the sensitivity of his spermatozoa to freezing. Examples of such male-to-male differences are shown in Figure 12.2 and Figure 12.3. As indicated by the coefficients of variation in Figure 12.3, the postthaw motility of several ejaculates from each of five bulls was rather uniform, whereas the mean values for different bulls varied significantly. In Figure 12.3, results are shown for the membrane integrity of spermatozoa from each of nine bulls before and after freezing. Although the prefreeze values varied only slightly, from about 70 to 90%, the postthaw percentages ranged from about 10 to more than 50%. As will be discussed below, analogous male-to-male differences have been reported for several other species as well. In a very novel development, it has recently been demonstrated that spermatozoa can be preserved by freeze-drying. When such bovine spermatozoa are injected directly into oocytes, blastocysts have been produced (Keskintepe et al., 2002).

12.2.2 SHEEP AND GOATS

As with bull semen, the cryopreservation of ram spermatozoa has been the subject of innumerable investigations, and there exist many descriptions of methods that have been used to cryopreserve

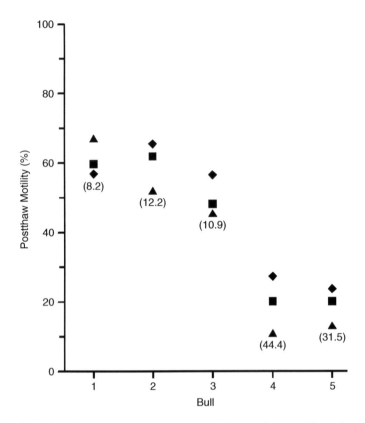

FIGURE 12.2 Survival of bovine spermatozoa, as measured by postthaw motility, of replicate ejaculates collected from five bulls. Each point represents the average motility of a single ejaculate from a given bull; the figures are the coefficients of variation. The data are unpublished observations of L. Bradley and S. P. Leibo (1998).

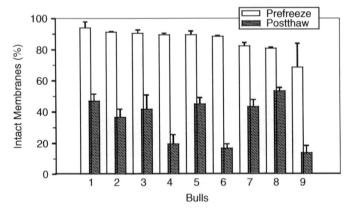

FIGURE 12.3 Survival of bovine spermatozoa, as measured by postthaw membrane integrity, of ejaculated spermatozoa from each of nine bulls. Membrane integrity was measured before and after freezing by staining replicate samples with the commercial stain, FertiLight™ (Molecular Probes, www.probes.com). The data are unpublished observations of L. Bradley and S. P. Leibo (1997).

the spermatozoa and the variables that determine their functional survival and fertility when used for AI. An extensive review of these methods and of the results achieved by insemination of ewes has been published by Salamon and Maxwell (1995). It has long been recognized that the composition

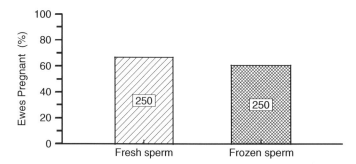

FIGURE 12.4 Pregnancies produced in ewes by artificial insemination with fresh semen or semen that had been cryopreserved and stored for 27 years. The bars show the percentage of ewes diagnosed pregnant; the figures in each bar are the number of ewes that were inseminated. The data are those of Gillan et al. (1997).

of sperm membranes influences their survival when cooled or frozen. Among the variables that have been shown to affect fertility of ram spermatozoa is the location at which the rams feed and the resultant composition of their sperm membranes (Drokin et al., 1999). A common method to cryopreserve ram spermatozoa has been to freeze it rapidly as small pellets on dry ice at –78°C and to store the frozen pellets in liquid nitrogen. In a recent large field trial, it was shown that zwitterion-buffered diluents were superior to *Tris*-citrate solutions (Molinia et al., 1996). Although seminal plasma has been reported to introduce variability among rams, its presence when added to cryopreserved ram spermatozoa after thawing can enhance pregnancy rates (Maxwell et al., 1999). The effect of synthetic macromolecules on postthaw motility of goat spermatozoa has also been studied (Kundu et al., 2002). Comparison of dextrans of various molecular weights demonstrated that 10 kDa dextran conferred maximum protection.

Although often implied, another important benefit of cryopreserved spermatozoa is their long-term stability as a function of storage time. In the early investigations of sperm freezing, attention was paid to the length of time that samples had been stored. More recently, this factor has been largely ignored. Several years ago, it was shown that bull spermatozoa that had been stored in dry ice for 4 years and in liquid nitrogen for an additional 33 years were capable of fertilizing oocytes; the resultant zygotes developed into expanded blastocysts in culture (Leibo et al., 1994). More important, in a recent comparison of the fertility resulting from AI with fresh or frozen-thawed ram spermatozoa, Gillan et al. (1997) used semen samples that had been frozen in 1968. They demonstrated that the pregnancy rate produced in ewes inseminated with 27-year-old spermatozoa was exactly the same as that in ewes inseminated with fresh spermatozoa (Figure 12.4). Another recent study examined the interactions of glycerol, pH of the medium, and an amino acid (proline) and organic osmolytes (glycine betaine) on the postthaw motility of ram spermatozoa that had been frozen as pellets (Sánchez-Partida et al., 1998). After the spermatozoa were thawed, there were significant interactions among the components of the diluents in which they had been frozen. Pregnancy rates in ewes (65 per group) were low following cervical inseminations with spermatozoa that had been frozen in any of the solutions, but they were about tenfold higher after laparoscopic inseminations.

Artificial insemination is used for the breeding of goats in only a few countries, with the exception of France. In fact, compared to the literature on cryopreservation of bovine, ovine, and equine spermatozoa, there is only a limited number of publications on freezing of caprine spermatozoa. In one recent study, the investigators compared the efficacy of skim milk and egg yolk as supplements to the diluent solution (Keskintepe et al., 1998). Using *in vitro* fertilization as an assay of sperm survival, the authors found that spermatozoa frozen in skim milk yielded far better results than those frozen in egg yolk, a finding in rather sharp contrast to that observed for spermatozoa of most other species. A review of the use and preservation of goat semen has recently been published (Leboeuf et al., 2000). In France, AI plays an important role in the production of goats

TABLE 12.1
**Comparison of the Potential Efficiency of Producing Lambs
by Natural Mating or by Artificial Insemination (AI)
with Fresh or Frozen Spermatozoa**

Natural Mating	AI, Fresh Semen	AI, Frozen Sperm
3 rams: 97 ewes	3 ejaculates ram/day;	1 L semen/ram;
	20 inseminations/ejaculate;	25,000 inseminations/ram
	1020 inseminations/cycle	
33 ewes bred/ram	1020 ewes bred/ram	25,000 ewes bred/ram
22 lambs born	500 lambs born	12,000 lambs born

and, used in conjunction with progeny testing, has led to significant improvements in milk production. Use of AI allows greater selection pressure to be exerted to identify and to select superior males. This also permits evaluation of genetically related herds and the monitoring and selection of bucks carrying major genes of importance for productivity, such as the s1 casein gene (Leboeuf et al., 1998). As shown in Table 12.1, the potential number of lambs sired by a ram each year using natural mating, by AI with fresh semen, or by AI by laparoscopy with frozen semen (using data obtained from Maxwell, 1984, and from Gordon, 1997) illustrate the potential efficiency that may be achieved by use of cryopreserved spermatozoa. In the case of goat semen, a specific problem has been the detrimental effect of seminal plasma on the viability of the spermatozoa in diluents containing egg yolk or in milk-based media.

12.2.3 HORSES

In 1957, the first specimens of equine spermatozoa were successfully frozen. Barker and Gandier (1957) froze epididymal spermatozoa to –79°C and stored them for 1 month. Of eight mares that they inseminated with thawed spermatozoa, one became pregnant and foaled. Since then, many more studies have been made of the cryobiology of stallion spermatozoa. Often, the spermatozoa have been frozen as pellets on dry ice or in relatively large volumes in ampoules. Many of these early observations are described in review articles by Graham et al. (1978) and by Amann and Pickett (1987). A very recent symposium on equine reproduction included 16 articles describing various aspects of the cryobiology of stallion spermatozoa, as well as an additional 36 articles dealing with the physiology, biochemistry, and molecular biology of the spermatozoa (Evans, 2002). Among the variables that have been investigated to determine their effect on sperm cryopreservation are the role of metal ions in seminal plasma, the effectiveness of various diluents and cryoprotectants, and the influence of reactive oxygen species or hormones on the fertility of stallion spermatozoa. As is true of several other species, important progress has recently been made in equine reproduction and genomics. It is also worth noting that very basic studies of fundamental aspects of stallion spermatozoa have been conducted to determine their role in sperm cryopreservation (Devireddy et al., 2002). Although very large numbers of mares have been successfully impregnated with cryopreserved spermatozoa, resulting in the birth of many foals, it is notable that there are significant differences among males with respect to the survival of their spermatozoa. Such differences are enumerated in the reviews by Graham et al. (1978) and by Amann and Pickett (1987) cited above.

12.2.4 PIGS

A method to cryopreserve boar spermatozoa reliably and efficaciously enough to be used commercially is still lacking (Johnson, 1985; Johnson et al., 2000). Use of cryopreserved spermatozoa for

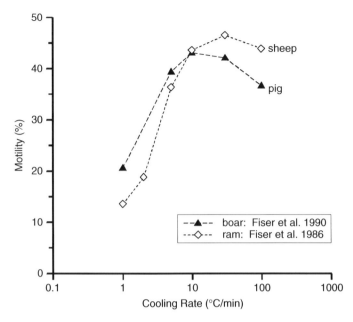

FIGURE 12.5 Survival based on postthaw motility of cryopreserved spermatozoa of boars and of rams as a function of the cooling rate. The data for boars are those of Fiser and Fairfull (1990) and data for rams are those of Fiser et al. (1986).

AI usually results in lower farrowing rates and smaller numbers of piglets per litter compared with results with fresh semen (Graham et al., 1978; Johnson et al., 2000). Such unsatisfactory results limit the commercial use of cryopreserved spermatozoa, thus impeding the agricultural effect of this technology on the swine industry.

Poor survival of cryopreserved boar spermatozoa has been attributed to various factors, including their extreme susceptibility to cold shock when abruptly cooled below 15°C and their sensitivity to osmotic shock when caused to undergo osmotic contraction or expansion (Johnson et al., 2000). Moreover, boar spermatozoa are much more sensitive to glycerol than spermatozoa of other species and are damaged simply by being suspended in freezing medium containing more than 2% glycerol (Fiser and Fairfull, 1990; Courtens et al., 1989; Wilmut and Polge, 1974). This may be a result of their extreme susceptibility to damage caused by volume changes compared with spermatozoa from other species (Gilmore et al., 1996, 1998). Furthermore, differences in sperm transport and numbers of spermatozoa required for fertilization may also contribute to species differences in fertility of cryopreserved spermatozoa (Holt, 2000a). In the last 30 years, methods to cryopreserve boar spermatozoa have been based on empirical studies of varying external factors with very limited success. Although cryopreservation methods are available, fertility of cryopreserved spermatozoa is lower than that of spermatozoa stored at +15°C (reviews in Bwanga, 1991; Johnson, 1985; Johnson et al., 2000).

In an effort to identify optimum conditions of cryopreservation, Fiser and his colleagues (Fiser and Fairfull, 1990; Fiser et al., 1993) conducted detailed studies of the interactions of cooling and warming rates and glycerol concentrations on the functional survival of boar spermatozoa. They observed maximum survival (based on measurements of motility and acrosomal integrity) when the spermatozoa were frozen in 3 or 4% glycerol (equivalent to 0.33 to 0.44 M) at a rate of ~10°C/min. Some of their results are redrawn in Figure 12.5, in which postthaw motility of boar spermatozoa in 0.44 M glycerol is shown, compared with results obtained by Fiser et al. (1986) for ram spermatozoa. It is clear that spermatozoa of both species respond in a very similar fashion to cooling rates when they are frozen. Most important, it is also evident that spermatozoa of these two species exhibit the inverted V-shaped survival curve as a function of cooling rate that has been described for many types of cells.

In the pig, there appear to be breed differences among pigs in freezing sensitivity of spermatozoa (Johnson, 1985; Thurston et al., 2002; Woelders, 1997). Cryopreserved spermatozoa from Large White boars gave higher farrowing rates than those from Landrace boars (Johnson, 1985). Cerolini et al. (2001) found that there was a large variation in the total lipid content of fresh spermatozoa from different boars. Very recently, Thurston et al. (2002) studied the survival of spermatozoa collected from 129 boars. They demonstrated that there were very significant differences among individual boars with respect to the freezing sensitivity of their spermatozoa. The group of 129 boars seemed to fall into three general categories with respect to this characteristic. Although there were no significant differences between ejaculates of a given male, multivariate pattern analysis of the viability data indicated that the boars produced spermatozoa with poor, average, or good post-thaw recovery. Using markers of amplified restriction fragment length polymorphisms, these authors analyzed spermatozoa from the boars and concluded that there is a genetic basis for variation in postthaw sperm survival among males; that is, differences in freezing sensitivity of boar sperma-tozoa may be genetically determined.

During the last decade, considerable progress has been achieved in the cryopreservation of boar semen. A method described in 1997 by Thilmant yields rates of fertility and fecundity similar to those obtained with fresh semen (79.9% vs. 80.3% and 10.1% vs. 9.8%). This method was tested *in vitro* and *in vivo* in the field with equal success (Bussière et al., 2000; Thilmant, 1999, 2001). In the future, it may be possible to use this method to establish gene banks for endangered breeds of pigs (Labroue et al., 2000). Recently, protection of boar spermatozoa from cold shock damage by 2-hydroxypropyl-β-cyclodextrin also was reported by Zeng and Terada (2001).

12.3 CRYOPRESERVATION OF DOMESTIC ANIMAL EMBRYOS AND THEIR TRANSFER

In 1972, embryos of the mouse were the first embryos to be successfully frozen that resulted in the birth of live young (Whittingham et al., 1972; see also Chapter 11). Keys to success for embryo cryopreservation in those first experiments was use of low cooling rates to avoid intracellular ice formation in cells, slow warming to prevent abrupt rehydration during thawing, and stepwise dilution to remove the cryoprotectants after the embryos were thawed. "Frosty," the first calf produced from a frozen-thawed embryo, was born in 1973 (Wilmut and Rowson, 1973). The first meeting specifically convened to discuss the freezing of embryos was convened at The Jackson Laboratory in 1975 (Mühlbock, 1976), and the first European meeting entitled "Egg Transfer in Cattle" was also held in 1975 in Cambridge. It soon became evident that effective exploitation of genetically valuable embryos would depend on the development of effective techniques for their storage at low temperature. Improvements and simplifications of the initial method were made by several authors (Leibo, 1984; Massip et al., 1984; Renard et al., 1982; Willadsen, 1977), leading to procedures that permitted direct transfer of frozen-thawed embryos of domestic species, espe-cially of cattle, into recipient females. With present methods of embryo cryopreservation and embryo transfer, the pregnancy rate with frozen-thawed bovine embryos is only slightly lower than that with fresh embryos (~60% vs. 50%).

12.3.1 CATTLE

Embryo transfer in cattle became widespread in the late 1970s and early 1980s (see reviews by Mapletoft, 1984; Seidel, 1981). Since then, cryopreservation of bovine embryos has become an absolutely integral part of the commercial practice of embryo transfer. Over the last two decades, several notable research accomplishments have become incorporated into the commercial practice of embryo transfer. Among these are the demonstration by Rall and Fahy (1985) that embryos can be successfully cryopreserved by cooling them at such high rates that the embryos themselves, as well as their suspending medium, vitrifies, or forms a glass (see also Chapter 22).

Other improvements in reproductive technologies that have been made in recent years are also of great significance. The development of effective methods for superovulation of cows to increase the number of oocytes that undergo maturation, synchronization, and embryo transfer, especially in cattle breeding, has been the most important advance. These techniques provide opportunities to increase the reproductive potential of females in much the same way as AI did for males. The successful freezing of embryos, which was achieved a little later, enables preservation of the entire genome. Embryo transfer in cattle breeding is now applied quite extensively worldwide. It has been used mainly to obtain more calves from the most valuable animals and also for international trade in genetically superior stock. An additional application has been to accelerate genetic selection through multiple ovulation and embryo transfer in so-called "MOET" breeding schemes, first described by the Canadian geneticist, Charles Smith. Embryo transfer also provides the means for development of many other technologies associated with the manipulation of eggs and embryos, including *in vitro* production of embryos, sex selection, embryo micromanipulation, nuclear transfer, and transgenesis. There are important species variations in the stages of embryonic development that tolerate cooling and freezing (Leibo et al., 1996; Polge and Willadsen, 1978). Cattle embryos are sensitive to cooling before they have reached the late morula or early blastocyst stage, whereas sheep embryos at early stages of development are more tolerant.

12.3.2 PIGS

Cryopreservation of porcine embryos at early stages of development remained a challenge for a long time because they could not be cooled below 15°C without damage (Leibo et al., 1996; Niemann, 1985; Polge, 1977). It was further suggested that their high lipid content was the major cause of their sensitivity to freezing and thawing. However, expanded and hatched blastocysts tolerate some cooling, and hatched blastocysts can even survive when subjected to controlled freezing and thawing, resulting in the birth of piglets after transfer (Hayashi et al., 1989). This decreasing sensitivity to cooling is likely caused by the metabolic utilization of intracellular lipids as the embryo develops. By the time that the porcine blastocyst hatches, it appears as if much of the intracellular lipid content has been metabolized. There are two solutions that have been used to address the problem of lipids within embryos. One novel way is to remove them. Nagashima et al. (1994) treated early-stage porcine embryos with cytochalasin, centrifuged the embryos at high force, causing the intracellular contents of the embryo to fractionate, and then removed the lipid by micromanipulation. The investigators termed this very novel procedure "delipidation." They subsequently showed that such "delipidated" embryos were much more resistant to chilling and to freezing. This is not, however, an easy procedure, nor one that could be used to treat large numbers of embryos.

An alternative that has been used with increasing success with porcine embryos is to vitrify them. It appears as if the vitrification procedure circumvents chilling injury associated with conventional cryopreservation techniques, and thus avoids damage to lipids within the embryo. To explain these differences, T.G. McEvoy (personal communication, 2001) thinks that "conventional freezing exposes the lipid component in embryos to sudden phase transitions which induce tearing of membranes and tissues." Unlike other substances, lipids do not gradually change from a fluid to a gel-like state but, rather, within a rather narrow temperature range, change rapidly and more or less *en masse* from a fluid to a gel-like phase. Other substances, for example, proteins, do not change at the same temperature, resulting in damage to lipoproteins and other structures. In those embryos in which the lipid content is high, such damage is considerable. In embryos in which the lipid content is lower, for example, in murine vs. bovine oocytes, or in bovine vs. porcine oocytes, or in delipidated vs. normal pig embryos, there is less damage initially, and consequently, postthaw repair may be possible, thereby facilitating enhanced survival.

When an aqueous solution freezes, whether intracellularly or extracellularly, ice crystals form and the volume increases as a result of the expansion that occurs as the crystalline lattice grows. In contrast, when an aqueous solution undergoes vitrification, it does not form crystalline ice, but rather the viscosity increases substantially as the liquid changes from a fluid to a solid. The process of vitrification thus avoids the potentially detrimental water/ice crystal formation that occurs in conventional freezing. It has been suggested that vitrification also may allow membranes to remain in a fluid state (Wolfe and Bryant, 1999), so it may be that the liquid phase–transition occurring in conventional cryopreservation is also avoided (see also Chapter 22).

There exist several vitrification methods, the most effective of which have been those described by Martino et al. (1996b), using electron microscope grids as a carrier of very small volumes; the Open Pulled Straw method (Vajta et al., 1997, 1998; Holm et al., 1999) and the CryoLoop method Lane et al., 1999. With these methods, because extremely small volumes of liquid are cooled at very high rates, it is likely, although still unproven by formal proof, that the samples have actually vitrified by forming a glass. The Open Pulled Straw method has been applied to pig embryos and yields satisfactory results (Berthelot et al., 2000). However achieved, vitrification seems the method of choice for pig embryos (Dobrinsky et al., 2000; Kobayashi et al., 1998; Yoshino et al., 1993; for further discussion of embryo vitrification, see Chapter 18).

12.3.3 SHEEP AND GOATS

In his early investigations, Willadsen (1977) used sheep embryos as a model for studying freezing and thawing of bovine embryos with dimethyl sulfoxide (DMSO) as the cryoprotectant. As with cattle embryos, however, ethylene glycol was soon identified as a cryoprotectant that seemed to yield high survival of embryos of several species (Songsasen et al., 1995). It has been claimed that, with sheep embryos, ethylene glycol yields high survival only with *in vivo* embryos (Tervit et al., 1994). This is inconsistent with observations of others (Songsasen et al., 1996) and with high survival exhibited after cryopreservation of *in vitro*-produced bovine embryos frozen in ethylene glycol (Hochi et al., 1996). Survival of goat embryos frozen in glycerol or DMSO does seem to increase with the stage of development (Li et al., 1990). Recently, using a vitrification mixture of 25% glycerol + 25% ethylene glycol, as described by Donnay et al. (1998), Mermillod et al. (1999) found no significant difference between fresh and vitrified *in vivo* day 7 embryos in terms of lambing rates (60% of 48 embryos vs. 50% of 50 embryos). Pregnancy rates after direct transfer and transfer after dilution were 81% and 74%, respectively, and lambing rates were 58% in both cases.

A comparison of *in vitro* and *in vivo* survival of vitrified *in vitro*-produced (IVP) goat and sheep embryos revealed significant differences between the two species (Traldi et al., 1999). With goat embryos, 60% of 177 survived and developed *in vitro*, whereas with sheep embryos, only 41% of 124 survived, a significant difference between the two species. After transfer of cryopreserved embryos, the respective birth rates were 45% of 20 goat embryos vs. 15% of 34 sheep embryos. All kids were normal, whereas the lambs tended to show excessive weight gain, dystocia, and other anomalies, or they died perinatally. These data seem to indicate that IVF sheep embryos are less able to tolerate the vitrification protocol than goat embryos. For both species, the *in vitro* survival rate after warming influenced the *in vivo* survival. Nevertheless, sheep embryos have been successfully vitrified, resulting in live birth (Baril et al., 2001).

12.3.4 HORSES

In the horse, the preferred stage of development of embryos that are to be frozen is the morula or very early blastocyst, so that embryo collection is normally performed about 6 d after ovulation. Unfortunately, timing of ovulation in the mare is much more difficult than in the cow or other domestic species. As a consequence, embryos that are collected by nonsurgical flushing of the

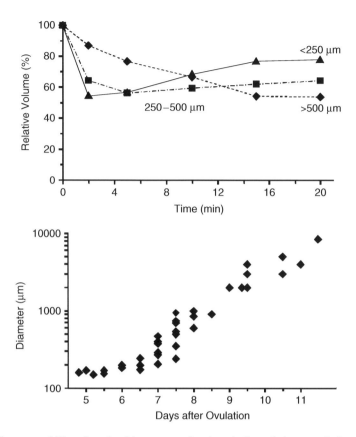

FIGURE 12.6 The permeability of equine blastocysts of various indicated sizes to ethylene glycol based on their volumetric changes. Embryos were classified as to their size and then suspended in 1.5 *M* ethylene glycol, and their volumes were measured at various times. The data are those of Pfaff et al. (1993). For comparison, the development of the equine conceptus is shown in terms of its size as a function of the days after ovulation of the oocyte from which it was conceived. The data are those of Betteridge et al. (1982).

mare's uterus may vary significantly in their stage of development. The size of the equine conceptus and its developmental stage appear to be more critical to survival after freezing and thawing than the type of cryoprotectant used (Bruyas et al., 1997; Squires et al., 1999) or the method of cryopreservation used (Oberstein et al., 2001). Beginning about day 7 after fertilization and continuing until day 16, an acellular capsule forms beneath the zona pellucida (Oriol et al., 1993a, 1993b). This capsule appears to impede loss of water from the cells of the embryo and permeation of cryoprotectants into the embryo. Pfaff et al. (1993) measured the volumetric changes exhibited by equine conceptuses of various sizes when suspended in 1.5 *M* ethylene glycol, a solute to which embryos of other species are extremely permeable. As shown by the results in Figure 12.6a, the smaller embryos (<250 μm in diameter) exhibited a typical "shrink–swell" response as a function of time. Embryos of a somewhat larger volume (250 to 500 μM) contracted but did not seem to expand, even after 20 min of exposure. In contrast, the large embryos (>500 μm in diameter) contracted very slowly and did not exhibit any increase in volume. For comparison, the diameter of equine conceptuses as a function of days after ovulation are shown in Figure 12.6b (data collected by Betteridge et al., 1982). The data indicate that as the equine conceptus develops and becomes surrounded by the capsule, it becomes increasingly impermeable to water and to solutes. However, it has been unequivocally demonstrated that embryos must undergo osmotic dehydration if they are to survive cryopreservation (Leibo et al., 1978). If the conceptus cannot undergo adequate dehydration, as occurs in most methods of embryo vitrification, then even that method may not be

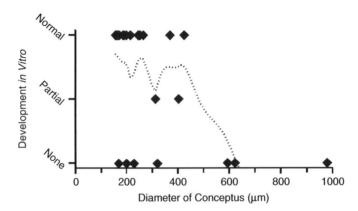

FIGURE 12.7 Survival as determined by their development *in vitro* of equine blastocysts that were vitrified at various sizes. Development was assayed in a qualitative way by assessment of the embryos as "normal," "partially developed," or "no development." The data are those of Hochi et al. (1995).

successful to vitrify late-stage equine conceptuses. Data of Hochi et al. (1995) shown in Figure 12.7 illustrate this phenomenon. The majority of smaller embryos (<500 μm) tolerate vitrification and developed in culture after vitrification, but larger embryos did not survive even after vitrification. Presumably, later-stage embryos did not dehydrate when suspended in concentrated vitrification solutions. Many biological processes appear to be unique to the horse, thus limiting the application of assisted reproductive technologies in this species. As mentioned previously for equine spermatozoa, a very recent symposium presented articles by many investigators on various aspects of the embryology and reproduction of the horse (Evans, 2002).

12.4 CRYOBIOLOGY OF MAMMALIAN OOCYTES

In vitro embryos are produced from immature oocytes collected at slaughter or by "ovum pick up," the aspiration of oocytes directly from cows using ultrasound guidance. The use of cryopreserved oocytes in bovine nuclear transfer protocols would alleviate the logistical problems associated with matching the availability of donor cells. Moreover, it would constitute a reserve of female gametes. However, few calves have yet been born after IVF of cryopreserved oocytes. Among the reasons are that, first, the meiotic spindle of the oocyte is very sensitive to chilling injury (Martino et al., 1996a; Wu et al., 1999). The oocyte is the largest cell in the mammalian body and is surrounded by granulosa cells that traverse the zona pellucida to form gap junctions with the ooplasm. Therefore, there are two types of cells—the oocyte, which is a large unique cell, and small surrounding cells—and both types must survive the freezing process. Many attempts have been conducted at different stages of maturation, with different cryoprotectants and preservation protocols, but the results were disappointing (see Niemann, 1995, for review). Some progress resulted from the application of the Open Pulled Straw method devised by Vajta et al. (1997). By reducing the diameter of a 0.25-mL straw as well as the wall thickness, they obtained a capillary tube in which oocytes were suspended in a few microliters of medium, so that the speed of cooling was very high allowing avoidance of chilling injury and toxicity of cryoprotectants. This method has proven to be very effective as used by Vajta and his colleagues. An example of some of their results is shown in Figure 12.8. Results of Le Gal et al. (2000) confirmed that very high cooling rates achieved by using a "minimum drop size" to vitrify oocytes are a key to success of this method. Other problems have been detected after ultra-rapid freezing or vitrification of mouse oocytes.

Van der Elst et al. (1998) showed that ultra-rapid freezing of mouse oocytes resulted in a reduced cell number in the inner cell mass of 5-day-old *in vitro*-cultured blastocysts. This undoubtedly impairs the developmental capacity of the blastocysts. In their studies of cryopreservation of mouse oocytes,

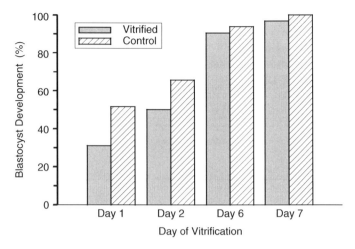

FIGURE 12.8 Survival, based on their development *in vitro* into expanded blastocysts, of bovine embryos that were vitrified at various stages of development. The respective stages of embryonic development were zygotes, two-cell, morulae and early blastocysts for embryos collected on days 1, 2, 6, or 7. For each stage, the paired bars show the development for fresh or vitrified embryos. The data are those of Vajta et al. (1998).

Wood et al. (1993) also observed a high incidence of postimplantation loss that may have been the result of an increase of chromosomal nondisjunction caused by damage during vitrification. (For further discussion of oocyte cryopreservation, see Chapter 18.)

12.5 CRYOPRESERVATION OF IVP EMBRYOS OF DOMESTIC SPECIES

An even more important research finding has been the now-routine procedure of producing bovine embryos by *in vitro* maturation and fertilization of oocytes, as summarized in the comprehensive treatise by Gordon (1994). Thousands of such IVP embryos have been transferred in many countries all over the world and have developed into live calves. Full exploitation of the capability to produce embryos *in vitro* requires the ability to cryopreserve such embryos. Although many reports have described the successful cryopreservation of IVP embryos, there are some problems, most notably the observation that IVP morulae are much more difficult to cryopreserve than their *in vivo*–derived counterparts.

To seek an explanation of the differences in freezing sensitivity between *in vitro*– and *in vivo*–derived embryos, it is useful to consider methods used to produce bovine embryos *in vitro*. These IVP embryos are produced from ovarian oocytes by either of two methods: (1) oocytes are harvested from ovaries removed from cows at slaughter; or (2) oocytes are aspirated directly from ovaries of live cows by "ovum pick-up." With both methods, the oocytes are allowed to undergo maturation for about 24 h in culture, enabling the oocytes to be fertilized *in vitro*. They are then cultured for 6 or 7 d until they have developed into morulae or blastocysts, at which point the IVP embryos are either frozen or transferred into recipients. As mentioned above, although IVP blastocysts can be cryopreserved almost as well as their *in vivo*–derived counterparts, IVP morulae have appeared to be much more sensitive than *in vivo* morulae.

In general, most references to *in vitro*–derived embryos have usually referred to embryos of cattle. Such embryos have been grown in various media that affect their metabolism and gene expression; in general, they are of somewhat poorer quality compared with their *in vivo*–derived counterparts. Nevertheless, large numbers of calves have been produced from IVP embryos (Hasler, 2001). Another manifestation of a difference between IVP and *in vivo*–derived embryos is their buoyant density. That is, the density of *in vitro*–derived embryos is lower than that of *in vivo* embryos, as reflected by their relative behavior in hypertonic, dense solutions of sucrose. The IVP embryos float in sucrose solutions, whereas *in vivo*–derived embryos sink in the same solutions. These different buoyant densities may reflect different ratios of lipids to protein within these two types of embryos (Leibo and Loskutoff, 1993; Pollard and Leibo, 1994; Pollard et al., 1993). However, for each stage of embryonic development from the oocyte to the blastocyst, it will be important to determine the effectiveness of cryopreservation when applied to new technologies such as cloning and transgenesis. Nuclear transfer is generally carried out with early embryonic cells because these have been considered to be relatively undifferentiated and the nuclei to be more capable of being reprogrammed in the oocyte cytoplasm than the nuclei of more advanced differentiated cells. The ideal donor embryos should be at precompaction stages because blastomeres can be isolated individually. Unfortunately, because these embryonic stages are chilling sensitive, they are less likely to survive unless rapid-cooling methods are applied. Recently, the effectiveness of such methods has been demonstrated (Booth et al., 1997; Vajta et al., 1997). The value of IVP embryos would be markedly increased if they could be genotyped and cryopreserved for direct transfer. Here again, it has been demonstrated by Vajta et al. (1996) that these manipulations did not decrease the overall efficiency of *in vitro* production of bovine embryos.

12.6 CRYOPRESERVATION OF SPERMATOZOA OF DOMESTIC AVIAN SPECIES

In their classic monograph summarizing low-temperature biology from 1736 to 1939, Luyet and Gehenio (1940) cite observations of Atkins (1909) and of Moran (1925) on freezing points and effects of low temperatures on hen's eggs, demonstrating that this subject has interested scientists for almost 100 years. After that, however, apparently little if any research on avian cells was conducted until Shaffner et al. (1941) showed that fertile eggs could be obtained from hens inseminated with frozen chicken semen, although no live chicks were produced. Then in 1949, Polge et al. reported their "chance observation" that glycerol protected spermatozoa against the effects of low temperatures. They found that fowl spermatozoa that were suspended in Ringer's solution containing 20% glycerol, frozen to –79°C, and then thawed rapidly exhibited motility indistinguishable from their unfrozen controls. The ramifications of that serendipitous finding cannot be exaggerated (see also Chapter 11).

Since 1949, as summarized above, innumerable studies have been performed to improve and standardize methods for the long-term preservation of spermatozoa of many animal species. These have included spermatozoa of domestic birds (see reviews by Bellagamba et al., 1993; Etches et al., 1996; Graham et al., 1984; Hammerstedt, 1995; Lake, 1986; Suraï and Wishart, 1996). These methods were first developed in the chicken and then more recently have been applied to spermatozoa of turkeys, ducks, and geese. Despite the relatively intense investment of the scientific community in research on cryopreservation of semen of avian species, however, these methods have been relatively unused in the actual breeding of domestic fowl. One of the reasons is that AI is not widely used with many domestic species of birds. Although insemination with fresh semen is extensively used in the breeding of a few species, for example, turkeys, guinea fowl and mule ducks, this method is only occasionally employed and only for selection purpose with chickens, muscovy and pekin ducks, ganders, quail, ostrich, and emu. In fact, the success of freezing procedures applicable to poultry is highly variable (Table 12.2) and depends on the species and the specific line being bred. In addition, the costs of the various stages in the preparation, storage, and use of frozen ejaculates in poultry species remain relatively high compared to the market price of day-old chicks.

Regardless of the above constraints, however, ratification of the international agreement on biodiversity in Rio de Janeiro, Brazil, in 1992 provoked new interest in development of methods of semen-freezing methods for domestic birds (see also Chapter 13). Research into cryopreservation

TABLE 12.2
Reproductive Fertility Performance with Frozen Semen in Poultry Species

	Chickens	Turkeys	Ganders
Mean fertility, % fertile eggs (min–max range)	60 (10–90)	30 (5–65)	60 (30–70)
Mean number of chicks hatched per frozen ejaculate	2	1	1.5

of avian cells, including spermatozoa, has received renewed attention so as to develop the capability of preservation of rare lines, maintenance of acceptable genetic variability in parental lines selected by primary breeders, and long-term availability of the genetic potential from exceptional animals. Recent progress in cryopreservation of avian semen, mainly of the chicken, has emphasized the demand for *ex situ* management of gene banking both in Europe and in North America, in addition to *in situ* management of these species. This is very important for domestic bird species that include a very high number of rare lines. For example, 154 rare lines of the species *Gallus gallus* exist only in France. With present methods and understanding of low-temperature biology, cryopreservation of embryos of domestic birds is not possible, undoubtedly because of the very large size and high vitellus content of fowl eggs. Cryopreservation of avian semen is, therefore, the only efficient method of *ex situ* management for domestic birds.

12.6.1 Chicken Semen

Despite early attempts to freeze chicken spermatozoa, the first methods of completely successful semen cryopreservation in chickens were published by Lake and Stewart (1978) and by Sexton (1980) more than 30 years after the report by Shaffner et al. (1941). Lake and Stewart used low cooling rates, glycerol as a cryoprotective agent, and glass ampoules to package the semen, whereas Sexton also cooled the samples slowly, but used DMSO as the cryoprotectant and straws for packaging. Both methods were later optimized by Seigneurin and Blesbois (1995) and by Van Voorst and Leenstra (1995a, 1995b), respectively. Two other methods originating from the former USSR were published by Schramm (1991) and Tselutin et al. (1995). Both groups used rapid cooling by freezing the spermatozoa as pellets either with dimethyl formamide (Schramm) or dimethyl acetamide (DMA, Tselutin) as cryoprotectants. Comparison of cryoprotectants and methods of cryopreservation under standardized conditions (Chalah et al., 1999; Tselutin et al., 1999) showed that the highest fertility rates after AI with frozen semen were obtained with semen frozen rapidly as pellets with DMA. Standardization of this method, coupled with packaging of frozen pellets in straws (to optimize identification and safety) has been chosen as a reference for gene banking of local breeds of chickens in France.

12.6.2 Turkey Semen

Several authors (Graham et al., 1982; Lake, 1986; Sexton, 1981; Tselutin et al., 1995; Zavos and Graham, 1983) have attempted to freeze turkey semen using various cryoprotectants (glycerol, DMSO, ethylene glycol, DMA), freezing the specimens either as pellets or in straws. To date, none of the above studies has yielded reproducible results of fertile spermatozoa, indicating that turkey spermatozoa are much more sensitive to damage caused by cooling/freezing procedures than chicken semen (Blanco et al., 2000). However, recent studies by Blesbois and Grasseau (2001) indicate that the variability of fertility results may be partially reduced by removing seminal plasma before freezing. Current techniques used to freeze and thaw turkey semen do not result in fertility levels comparable to those obtained in the chicken (Table 12.2), but with minor modifications, they are sufficient to allow development of sperm banks in this species.

12.6.3 Semen of Other Domestic Bird Species

Other species in which significant research efforts have been conducted to develop freezing pro-cedures for semen preservation include drakes and ganders (Tselutin et al, 1995; see review by Suraï and Wishart, 1996). As with semen of the turkey and chicken, several cryoprotectants and freezing methods have been tested. It appears from these studies that spermatozoa of ganders are fairly resistant to freezing/thawing procedures, exhibiting acceptable and reproducible rates of fertility (>60%; Lukaszewicz, 2001; Tai et al., 2001). As for drake spermatozoa, it appears as if those of muscovy ducks are more resistant to freezing than spermatozoa of pekin ducks.

It can be concluded that recent progress in freezing/thawing procedures applicable to chickens and ganders are now adequate to allow gene banking and *ex situ* management for these species. Significant progress can be expected within the next few years in the long-term storage of turkey and muscovy duck semen. However, cryopreservation of avian semen does not, *per se*, allow the long-term preservation of sex-linked genes bearing the W chromosome (only carried by females in avian species)—a justification for further research to develop techniques applicable to long-term storage of avian oocytes and embryos.

12.6.4 Cryopreservation of Embryonic Cells of Domestic Birds

In sharp contrast with the innumerable studies that have been devoted to the cryopreservation of mammalian oocytes and embryos in the last 30 years, there have been relatively few studies on the same subject performed in avian species. As indicated above, one major reason for this relative absence of interest is that, in these species, the presence of enormous quantities of lipids in the vitellus lead to highly heterogeneous (and therefore risky) freezing/thawing procedures. The main approach to reconstituting the genetic resources via the avian egg is currently the production of germ line chimera from injection of frozen-thawed embryonic cells (reviewed by Tajima, 2002). Cloning with frozen somatic cells is also being studied.

The main approaches to transferring frozen embryo cells to recipient embryos have been based on three different options depending on donor's age: transfer of blastodermal cells, transfer of circulating primordial germ cells (PGSs) and transfer of gonadal PGS. Kino et al. (1997) obtained chicken somatic cell chimeras as well as germ cell chimeras after transfer of frozen-thawed blastodermal cells into recipient embryos at stage X. Naito et al. (1994) produced germ-line chicken chimera after the transfer of frozen-thawed circulating primordial PGSs. Tajima et al. (1998) and Chang et al. (1998) obtained donor progeny from birds having developed germ-line chimeras by transferring frozen-thawed gonadal PGSs in the chicken and quail, respectively. As a consequence, recent progress in the transfer of somatic and germinal cells to recipient embryos should be considered as an important, although not final, successful step as a method for the long-term preservation of the genetics of domestic avian species. However, as with mammalian cloning, current progress in transferring somatic nuclei into oocytes may, if successful, also become increas-ingly popular in the future.

12.7 CONCLUSIONS

In this review, we have attempted to illustrate the fact that cryobiology has become an integral part of animal biotechnologies. Although much has been achieved in the last 50 years, it has in general been gained by application of empirically derived "recipes" for freezing protocols; we now need more fundamental research to understand the mechanisms of injury or success in cryopreservation to make further progress in this field. Comparative studies of embryos of domestic species, especially those that have been derived by *in vitro* production methods or those that have been genetically altered, will reveal new facets of low-temperature biology. Another increasing application in cryobiology is the establishment of banks of gametes and embryos, not only of laboratory and domestic species but also of threatened and endangered species (see also Chapter 13). Such concepts have been described and discussed by Wildt et al. (1993), Commizoli et al. (2000), Leibo and Songsasen (2002), and in a recent textbook by Watson and Holt (2001).

REFERENCES

Amann, R.P. and Pickett, B.W. 1987 Principles of cryopreservation and a review of cryopreservation of stallion spermatozoa, *Equine Vet. Sci.*, 7, 145–173.

Atkins, W.R. (1909) The osmotic pressures of the blood and eggs of birds, *Sci. Proc. Roy. Dublin Soc.*, 12, 123–130.

Baril, G., Traldi, A.L., Cognie, Y., Leboeuf, B., Beckers, J.F., and Mermillod, P. (2001) Successful direct transfer of vitrified sheep embryos, *Theriogenology*, 56, 299–305.

Barker, C.A.V. and Gandier, J.C.C. (1957) Pregnancy in a mare resulting from frozen epididymal spermatozoa, *Can. J. Comp. Med.*, 21, 47–52.

Bellagamba, F., Cerolini, S., and Cavalchini, L.G. (1993) Cryopreservation of poultry semen: A review, *World's Poult. Sci. J.*, 49, 157–166.

Berthelot, F., Martinat-Botté, F., Locatelli, A., Perreau, C., and Terqui, M. (2000) Piglets born after vitrification of embryos using the open pulled straw method, *Cryobiology*, 41, 116–124.

Betteridge, K.J., Eaglesome, M.D., Mitchell, D., Flood, P.F., and Beriault, R. (1982) Development of horse embryos up to twenty-two days after ovulation: Observations on fresh specimens, *J. Anat.*, 135 (Pt 1), 191–209.

Blanco, J.M., Gee, G., Wildt, D.E., and Donoghue, A.M. (2000) Species variation in osmotic, cryoprotectant, and cooling rate tolerance in poultry, eagle, and falcon spermatozoa, *Biol. Reprod.*, 63, 1164–1171.

Blesbois, E. and Grasseau, I. (2001) Seminal plasma affects liquid storage and cryopreservation of turkey sperm, *Cryobiology*, 43, 334.

Booth, P.J., Vajta, G., Holm, P., Greve, T., and Callesen, H. (1997) Vitrification and post-thaw *in vitro* survival of cloned bovine embryos, *Vet. Rec.*, 140, 404.

Brem, G. and Kühholzer, B. (2002) The recent history of somatic cloning in mammals, *Cloning Stem Cells*, 4, 57–63.

Bruyas, J.F., Martins-Ferreira, C., Fieni, F., and Tainturier, D. (1997) The effect of propanediol on the morphology of fresh and frozen equine embryos, *Equine Vet. J.*, 25, 80–84.

Bussière, J.F., Bertaud, G., and Guillouet, P. (2000) Conservation de la semence congelée de verrat. Résultats *in vitro* et après insemination, *Journées Rech. Porcine France*, 32, 429–432.

Bwanga, C.O. (1991) Cryopreservation of boar semen. I: A literature review, *Acta Vet. Scand.*, 32, 431–453.

Cerolini, S., Maldjian, A., Pizzi, F., and Gliozzi, T.M. (2001) Changes in sperm quality and lipid composition during cryopreservation of boar semen, *Reproduction*, 121, 395–401.

Chalah, T., Seigneurin, F., Blesbois, E., and Brillard, J.P. (1999) *In vitro* comparison of fowl sperm viability in ejaculates frozen by three different techniques and relationship with subsequent fertility *in vivo*, *Cryobiology*, 39, 185–191.

Chang, I.K., Naito, M., Kuwana, T., Mizutani, M., and Sakurai, M. (1998) Production of germline chimeric quail by transfer of gonadal primordial germ cells preserved in liquid nitrogen, *Jpn. Poult. Sci.*, 35, 321–328.

Chesné, P., Adenot, P.G., Viglietta, C., Baratte, M., Boulanger, L., and Renard, J.P. (2002) Cloned rabbits produced by nuclear transfer from adult somatic cells, *Nat. Biotechnol.*, 20, 366–369.

Comizzoli, P., Mermillod, P., and Mauget, R. (2000) Reproductive biotechnologies for endangered mammalian species, *Reprod. Nutrition Dev.*, 40, 493–504.

Courtens, J.L., Ekwall, H., Paquignon, M., and Ploen, L. (1989) Preliminary study of water and some element contents in boar spermatozoa, before, during and after freezing, *J. Reprod. Fertil.*, 87, 613–626.

Devireddy, R.V., Swanlund, D.J., Olin, T., Vincente, W., Troedsson, M.H., Bischof, J.C., and Roberts, K.P. (2002) Cryopreservation of equine sperm: Optimal cooling rates in the presence and absence of cryoprotective agents determined using differential scanning calorimetry, *Biol. Reprod.*, 66, 222–231.

Dobrinsky, J.R. (1997) Cryopreservation of pig embryos, *J. Reprod. Fertil.*, 52, 301–312.

Donnay, I., Auquier, P., Kaidi, S., Carolan, C., Lonergan, P., Mermillod, P., and Massip, A. (1998) Vitrification of *in vitro* produced bovine blastocysts: Methodological studies and developmental capacity, *Anim. Reprod. Sci.*, 52, 93–104.

Drokin, S.I., Vaisberg, T.N., Kopeika, E.F., Miteva, K.D., and Pironcheva, G.L. (1999) Effect of cryopreservation on lipids and some physiological features of spermatozoa from rams pastured in highlands and in valleys, *Cytobios*, 100, 27–36.

Etches, R.J., Clark, M.E., Toner, A., Liu, G., and Gibbins, A.M. (1996) Contributions to somatic and germline lineages of chicken blastodermal cells maintained in culture, *Mol. Reprod. Dev.*, 45, 291–298.

Evans, M.J. (2002) Equine reproduction VIII, *Theriogenology*, 58, 191–861.

Fiser, P.S. and Fairfull, R.W. (1990) Combined effect of glycerol concentration and cooling velocity on motility and acrosomal integrity of boar spermatozoa frozen in 0.5 ml straws, *Mol. Reprod. Dev.*, 25, 123–129.

Fiser, P.S., Fairfull, R.W., and Marcus, G.J. (1986) The effect of thawing velocity on survival and acrosomal integrity of ram spermatozoa frozen at optimal and suboptimal rates in straws, *Cryobiology*, 23, 141–149.

Fiser, P.S., Fairfull, R.W., Hansen, C., Panich, P.L., Shrestha, J.N., and Underhill, L. (1993) The effect of warming velocity on motility and acrosomal integrity of boar sperm as influenced by the rate of freezing and glycerol level, *Mol. Reprod. Dev.*, 34, 190–195.

Foote, R.H. (1981) The artificial insemination industry, in *New Technologies in Animal Breeding*, Brackett, B.G., Seidel, G.E., and Seidel, S.M., Eds., Academic Press, New York. pp 12–39.

Foote, R.H. and Parks, J.E. (1993). Factors affecting preservation and fertility of bull sperm: A brief review, *Reprod. Fertil. Dev.*, 5, 665–673.

Gillan, L., Evans, G., and Maxwell, W.M. (1997) Capacitation status and fertility of fresh and frozen-thawed ram spermatozoa, *Reprod. Fertil. Dev.*, 9, 481–487.

Gilmore, J.A., Du, J., Tao, J., Peter, A.T. Critser, J.K. (1996) Osmotic properties of boar spermatozoa and their relevance to cryopreservation, *J. Reprod. Fertil.*, 107, 87–95.

Gilmore, J.A., Liu, J., Peter, A.T., and Critser, J.K. (1998) Determination of plasma membrane characteristics of boar spermatozoa and their relevance to cryopreservation, *Biol. Reprod.*, 58, 28–36.

Gordon, I. (1994) *Laboratory Production of Cattle Embryos*, CAB International, Oxon, U.K.

Gordon, I. (1997) *Controlled Reproduction in Sheep and Goats*, CAB International, Oxon, U.K.

Graham, E.F., Crabo, B.G., and Pace, M.M. (1978) Current status of semen preservation in the ram, boar and stallion, *J. Anim. Sci.*, 47, 80–119.

Graham, E.F., Nelson, D.S., and Schmehl, M.K.L. (1982) Development of extender and techniques for frozen turkey semen. 1. Development, *Poult. Sci.*, 61, 550–557.

Graham, E.F., Schmehl, M.L., and Deyo, R.C.M. (1984) Cryopreservation and fertility of fish, poultry and mammalian spermatozoa, in *Proceedings, 10th Technical Conference on Artificial Insemination and Reproduction*, National Association of Animal Breeders, Columbia, MO, pp. 4–29.

Hammerstedt, R.H. (1995) Cryopreservation of poultry semen—current status and economics, in *Proceedings 1st International Symposium on the Artificial Insemination of Poultry*, Poultry Science Association, Savoy, II., pp 229–250.

Hasler, J.F. (2001) Factors affecting frozen and fresh embryo transfer pregnancy rates in cattle, *Theriogenology*, 56, 1401–1415.

Hayashi, S., Kobayashi, K., Mizuno, J., Saitoh, K., and Hirano, S. (1989). Birth of piglets from frozen embryos, *Vet. Rec.*, 125, 43–44.

Heap, R.B. and Moor, R.M. (1995) Reproductive technologies in farm animals. Ethical issues, in *Issues in Agricultural Bioethics*, Mepham, T.B., Tucker, G.A., and Wiseman, J., Eds., Nottingham University Press, Nottingham, U.K., pp. 247–268.

Hochi, S., Fujimoto, T., and Oguri, N. (1995) Large equine blastocysts are damaged by vitrification procedures, *Reprod. Fertil. Dev.*, 7, 113–117.

Hochi, S., Semple, E., and Leibo, S.P. (1996) Effect of cooling and warming rates during cryopreservation on survival of *in vitro*-produced bovine embryos, *Theriogenology*, 46, 849–858.

Holm, P., Vajta, G., Machaty, Z., Schmidt, M., Prather, R.S., Greve, T., and Callesen, H. (1999) Open Pulled Straw (OPS) vitrification of porcine blastocysts: Simple procedure yielding excellent *in vitro* survival but so far no piglets following transfer, *Cryo-Letters*, 20, 307–310.

Holt, W.V. (2000a) Basic aspects of frozen storage of semen, *Anim. Reprod. Sci.*, 62, 3–22.

Holt, W.V. (2000b) Fundamental aspects of sperm cryobiology: The importance of species and individual differences, *Theriogenology*, 53, 47–58.

Johnson, L.A. (1985) Fertility results using frozen boar spermatozoa 1970 to 1985, in *Deep Freezing Boar Semen*, Johnson, L.A. and Larsson, K., Eds., Swedish Univ. Agric. Sci., Uppsala, Sweden, pp. 199–222.

Johnson, L.A., Weitze, K.F., Fiser, P., Maxwell, W.M. (2000) Storage of boar semen, *Anim. Reprod. Sci.*, 62, 143–172.

Keskintepe, L., Pacholczyk, G., Machnicka, A., Norris, K., Curuk, M.A., Khan, I., and Brackett, B.G. (2002) Bovine blastocyst development from oocytes injected with freeze-dried spermatozoa, *Biol. Reprod.*, 67, 409–15.

Keskintepe, L., Simplicio, A.A., and Brackett, B.G. (1998) Caprine blastocyst development after *in vitro* fertilization with spermatozoa frozen in different extenders, *Theriogenology*, 49, 1265–1274.

Kino, K., Pain, B., Leibo, S., Cochran, M., Clark, M.E., and Etches, R.J. (1997) Production of chicken chimeras from injection of frozen-thawed blastodermal cells, *Poult. Sci.*, 76, 753–760.

Kobayashi, S., Takei, M., Kano, M., Tomita, M., and Leibo, S.P. (1998) Piglets produced by transfer of vitrified porcine embryos after stepwise dilution of cryoprotectants, *Cryobiology*, 36, 20–31.

Kubota, C., Yamakuchi, H., Todoroki, J., Mizoshita, K., Tabara, N., Barber, M., and Yang, X. (2000) Six cloned calves produced from adult fibroblast cells after long-term culture, *Proc. Natl. Acad. Sci. USA*, 97, 990–995.

Kundu, C.N., Chakrabarty, J., Dutta, P., Bhattacharyya, D., Ghosh, A., and Majumder, G.C. (2002) Effect of dextrans on cryopreservation of goat cauda epididymal spermatozoa using a chemically defined medium, *Reproduction*, 123, 907–913.

Labroue, F., Luquet, M., Guillouet, P., Bussière, J.F., Glodek, P., Wemheuer, W., Gandini, G., Pizzi, F., Delgado, J.V., Poto, A., and Ollivier, L. (2000) La cryoconservation des races porcines menancées de disparition. La situation en France, en Allemagne, en Italie et en Espagne, *Journées Rech. Porcine France*, 32, 419–427.

Lake, P.E. (1986) The history and future of the cryopreservation of avian germ plasma, *Poult. Sci.*, 65, 1–15.

Lake, P.E. and Stewart, J.M. (1978) Preservation of fowl semen in liquid nitrogen—An improved method, *Br. Poult. Sci.*, 19, 187–194.

Lane, M., Bavister, B.D., Lyons, E.A., and Forest, K.T. (1999) Containerless vitrification of mammalian oocytes and embryos, *Nat. Biotechnol.*, 17, 1234–1236.

Le Gal, F., De Roover, R., Verhaeghe, B., Etienne, D., and Massip, A. (2000) Development of vitrified matured cattle oocytes after thawing and culture *in vitro*, *Vet. Rec.*, 146, 469–471.

Leboeuf, B., Mandredi, E., Boue, P., Piacère, A., Brice, G., Baril, G., Broqua, C., Humblot, P., and Terqui, M. (1998) Artificial insemination of dairy goats in France, *Livestock Prod. Sci.*, 55, 193–203.

Leboeuf, B., Restall, B., and Salamon, S. (2000) Production and storage of goat semen for artificial insemination, *Anim. Reprod. Sci.*, 62, 113–141.

Leibo, S.P. (1984) A one step method for direct non surgical transfer of frozen-thawed bovine embryos, *Theriogenology*, 21, 767–790.

Leibo, S.P. and Bradley, L. (1999) Comparative cryobiology of mammalian spermatozoa, in *The Male Gamete: From Basic Science to Clinical Application*, Gagnon, C., Ed., Cache River Press, Vienna, IL, pp. 501–516.

Leibo, S.P. and Loskutoff, N.M. (1993) Cryobiology of *in vitro* derived bovine embryos, *Theriogenology*, 39, 81–94

Leibo, S.P., Martino, A., Kobayashi, S., and Pollard, J.W. (1996) Stage-dependent sensitivity of oocytes and embryos to low temperatures, *Anim. Reprod. Sci.*, 42, 45–53.

Leibo, S.P., McGrath, J.J., and Cravalho, E.G. (1978) Microscopic observation of intracellular ice formation in unfertilized mouse ova as a function of cooling rate, *Cryobiology*, 15, 257–271.

Leibo, S.P., Semple, M.E., and Kroetsch, T.G. (1994) *In vitro* fertilization of oocytes by 37-year-old cryopreserved bovine spermatozoa, *Theriogenology*, 42, 1257–1262.

Leibo, S.P. and Songsasen, N. (2002) Cryopreservation of gametes and embryos of non-domestic species, *Theriogenology*, 57, 303–326.

Li, R., Cameron, A.W.N., Batt, P.A., and Trounson, A.O. (1990) Maximum survival of frozen goat embryos is attained at the expanded, hatching and hatched blastocyst stages of development, *Reprod. Fertil. Dev.*, 2, 345–350.

Lukaszewicz, E. (2001) Effects of semen filtration and dilution rate on morphology and fertility of frozen gander spermatozoa, *Theriogenology*, 55, 1819–1829.

Luyet, B.J. and Gehenio, P.M. (1940) *Life and Death at Low Temperatures,* Biodynamica, Normandy, MO.

Mapletoft, R.J. (1984) Embryo transfer technology for the enhancement of animal production, *Biotechnology*, 6, 149–160.

Martino, A., Pollard, J.W., and Leibo, S.P. (1996a) Effect of chilling bovine oocytes on their developmental competence, *Mol. Reprod. Dev.*, 45, 503–512.

Martino, A., Songsasen, N., Leibo, S.P. (1996b) Development into blastocysts of bovine oocytes cryopreserved by ultra-rapid cooling, *Biol. Reprod.*, 54, 1059–1069.

Massip, A., Van Der Zwalmen, P., Puissant, F., Camus, M., and Leroy, F. (1984) Effects of in-vitro fertilization, culture, freezing and transfer on the ability of mouse embryos to implant and survive, *J. Reprod. Fertil.*, 71, 199–204.

Maxwell, W.M.C. (1984) Current problems and future potential of artificial insemination programmes, in *Reproduction in Sheep,* Lindsay, D.R. and Pearce, D.T., Eds., Australian Academy of Science and Australian Wool Corp., Canberra, Australia, pp. 291–298.

Maxwell, W.M., Evans, G., Mortimer, S.T., Gillan, L., Gellatly, E.S., and McPhie, C.A. (1999) Normal fertility in ewes after cervical insemination with frozen-thawed spermatozoa supplemented with seminal plasma, *Reprod. Fertil. Dev.*, 11, 123126.

Mermillod, P., Traldi, A.L., Baril, G., Beckers, J.F., Massip, A., and Cognié, Y. (1999) A vitrification method for direct transfer of sheep embryos, in *Proceedings of the 15th Annual Meeting of the AETE, Lyon,* Fondation Marcel Mérieux, Lyon, p. 212.

Molinia, F.C., Evans, G., and Maxwell, W.M. (1996) Fertility of ram spermatozoa pellet-frozen in zwitterion-buffered diluents, *Reprod. Nutrition Dev.*, 36, 21–29.

Moran, T. (1925) The effect of low temperature on hen's eggs, *Proc. Roy. Soc.,* 98B, 436–456.

Mühlbock, O., Ed. (1976) *Basic Aspects of Freeze Preservation of Mouse Strains,* Gustav Fischer, Stuttgart.

Nagashima, H., Kashiwazaki, N., Ashman, R.J., Grupen, C.G., Seamark, R.F., and Nottle, M.B. (1994). Removal of cytoplasmic lipid enhances the tolerance of porcine embryos to chilling, *Biol. Reprod.*, 51, 618–622.

Naito M, Tajima, A., Tagami, T., Yasuda, Y., and Kuwana, T. (1994) Preservation of chick primordial germ cells in liquid nitrogen and subsequent production of viable offspring, *J. Reprod. Fertil.*, 102, 321–325.

Niemann, H. (1985) Sensitivity of pig morulae to DMSO/PVP or glycerol treatment and cooling at 10°C, *Theriogenology*, 23, 213.

Niemann, H. (1995) Advances in cryopreservation of bovine oocytes and embryos derived *in vitro* and *in vivo*, in *Reproduction and Animal Breeding: Advances and Strategy,* Enne, G., Greppi, G.F., and Lauria, A., Eds., Elsevier, Milan, Italy, pp. 117–128.

Oberstein, N., O'Donovan, M.K., Bruemmer, J.E., Seidel, G.E., Jr, Carnevale, E.M., and Squires, E.L. (2001) Cryopreservation of equine embryos by open pulled straw, cryoloop, or conventional slow cooling methods, *Theriogenology*, 55, 607–613.

Oriol, J.G., Betteridge, K.J., Clarke, A.J., and Sharom, F.J. (1993a) Mucin-like glycoproteins in the equine embryonic capsule, *Mol. Reprod. Dev.*, 34, 255–265.

Oriol, J.G., Sharom, F.J., and Betteridge, K.J. (1993b) Developmentally regulated changes in the glycoproteins of the equine embryonic capsule, *J. Reprod. Fertil.*, 99, 653–664.

Pfaff, R., Seidel, G.E., Squires, E.L., and Jasko, D.J. (1993). Permeability of equine blastocysts to ethylene glycol and glycerol, *Theriogenology*, 39, 284.

Polge, C. (1977) The freezing of mammalian embryos. Prospects and possibilities, in *The Freezing of Mammalian Embryos*, Ciba Foundation, Symposium 52 (New Series), Elliot, K. and Whelan, J., Eds., Elsevier Excerpta Medica/North-Holland, Amsterdam, pp. 3–13.

Polge, C. (1998) Freezing of gametes and development of embryo technologies in farm animals, *Acta Agric. Scand. Sect. A. Anim. Sci.*, 29, 5–11.

Polge, C., Smith, A.U., and Parkes, A.S. (1949) Revival of spermatozoa after vitrification and dehydration at low temperatures, *Nature*, 164, 666.

Polge, C. and Willadsen, S.M. (1978) Freezing eggs and embryos of farm animals, *Cryobiology*, 15, 370–373.

Pollard, J.W. and Leibo, S.P. (1994) Chilling sensitivity of mammalian embryos, *Theriogenology*, 41, 101–107

Pollard, J.W., Plante, C., Songsasen, N., Leibo, S.P. 1993 Correlation of sensitivity to chilling and freezing with buoyant density of mammalian embryos, *Cryobiology*, 30, 631.

Rall, W.F. and Fahy, G. 1985 Ice-free cryopreservation of mouse embryos at –196°C by vitrification, *Nature*, 313, 573–575.

Renard, J.P., Heyman, Y., and Ozil, J.P. (1982) Congélation de l'embryon bovin: Une nouvelle méthode de décongélation pour le transfert cervical d'embryons conditionnés une seule fois en paillettes, *Ann. Méd. Vét.*, 126, 23–32.

Salamon, S. and Maxwell, W.M.C. (1995) Frozen storage of ram semen. I. Processing, freezing, thawing and fertility after cervical insemination, *Anim. Reprod. Sci.*, 37, 185–249.

Sánchez-Partida, L.G., Setchell, B.P., and Maxwell, W.M.C. (1998) Effect of compatible solutes and diluent composition on the post-thaw motility of ram sperm, *Reprod. Fertil. Dev.*, 10, 347–357.

Schramm, G.P. (1991) Eignung verschiedener gefrierschutzstoffe zur kryoprotektion von hahnensperma. Monatsh, *Veterinäermedizin*, 46, 438–440.

Seidel, G.E., Jr. (1981) Superovulation and embryo transfer in cattle, *Science*, 211, 351–358.

Seigneurin, F. and Blesbois, E. (1995) Effects of the freezing rate on viability and fertility of frozen-thawed fowl spermatozoa, *Theriogenology*, 43, 1351–1358.

Sexton, T.J. (1980) Optimal rate for cooling chicken semen from + 5°C to –196°C, *Poult. Sci.*, 59, 2765–2770.

Sexton, T.J. (1981) Development of a commercial method for freezing turkey semen. 1. Effect of pre-freeze techniques on the fertility of processed unfrozen and frozen-thawed semen, *Poult. Sci.*, 60, 1567–1573.

Shaffner, C.S., Henderson, E.W., and Card, C.G. (1941) Viability of spermatozoa of the chicken under various environmental conditions, *Poult. Sci.*, 20, 259–265.

Songsasen, N., Plante, C., Buckrell, B.C., and Leibo, S.P. (1995) *In vitro* and *in vivo* survival of cryopreserved sheep embryos, *Cryobiology*, 31, 1–14.

Songsasen, N., Walmsley, S., Pollard, J.W., Martino, A., Buckrell, B.C., and Leibo, S.P. (1996) Lambs produced from cryopreserved sheep embryos derived by *in vitro* fertilization of aspirated oocytes, *Can. J. Anim. Sci.*, 76, 465–467.

Squires, E.L., McCue, P.M., and Vanderwall, D. (1999) The current status of equine embryo transfer, *Theriogenology*, 51, 91–104.

Suraï, P. and Wishart, G.J. (1996) Poultry artificial insemination technology in the countries of the former USSR, *World's Poult. Sci. J.*, 52, 27–43.

Tai, J.L., Chen, J.C., Wu, K.C., Wang, D.S., and Tai, C. (2001) Cryopreservation of gander semen, *Br. Poult. Sci.*, 42, 384–388.

Tajima, A. (2002) Production of germ-line chimeras and their application in domestic chicken, *Avian Poult. Biol.* Review, 13, 15–30.

Tajima, A., Naito, M., Kuwana, T., and Mizutani, M. (1998) Production of germ-line chimeras by transfer of cryopreserved gonadal primordial germ cells in chickens, *J. Exp. Zool.*, 280, 265–270.

Tervit, H.R., Pugh, P.A., McGowan, L.T., Bell, A.C.S., and Wells, R.W. (1994) The freezability of sheep embryos is affected by culture system and source (*in vivo* or *in vitro* derived), *Theriogenology*, 41, 315.

Thibier, M. (2001) The animal embryo transfer industry in figures, *IETS Embryo Transfer Newsl.*, 19, 16–22.

Thilmant, P. (1997) Congélation du sperme de verrat en Paillette de 0,5 ml. Résultats sur le Terrain, *Ann. Méd. Vét.*, 141, 457–462.

Thilmant, P. (1999) Congélation du sperme de verrat. Résultats observés sur le terrain en Belgique pendant les années 1995 et 1996, *Journées Rech. Porcine France*, 31, 59–64.

Thilmant, P. (2001) Congélation du sperme de verrat en paillettes fines de 0,25 ml. Résultats observés sur le terrain, *Journées Rech. Porcine France*, 33, 151–156.

Thurston, L.M., Siggins, K., Mileham, A.J., Watson, P.F., and Holt, W.V. (2002) Identification of amplified restriction fragment length polymorphism markers linked to genes controlling boar sperm viability following cryopreservation, *Biol. Reprod.*, 66, 545–554.

Traldi, A.S., Leboeuf, B., Cognié, Y., Poulin, N., and Mermillod, P. (1999) Comparative results of *in vitro* and *in vivo* survival of vitrified *in vitro* produced goat and sheep embryos, *Theriogenology*, 51, 175 (Abstr.).

Tselutin, K., Narubina, L., Maorodina, T., and Tur, B. (1995) Cryopreservation of poultry semen, *Br. Poult. Sci.*, 36, 805–811.

Tselutin, K., Seigneurin, F., and Blesbois, E. (1999) Comparison of cryoprotectants and methods of cryopreservation of fowl spermatozoa, *Poult. Sci.*, 78, 586–590.

Vajta, G., Booth, P.J., Holm, P., Greve, T., and Callesen, H. (1997) Successful vitrification of early stage bovine *in vitro* produced embryos with the open pulled straw (OPS) method, *Cryo-Letters*, 18, 191–195.

Vajta, G., Holm, P., Greve, T., and Callesen, H. (1996) Cumulative efficiency of biopsy, vitrification and in straw dilution in a bovine *in vitro* embryo production system, *Theriogenology*, 45, 162 (Abstr.).

Vajta, G., Holm, P., Kuwayama, M., Booth, P.J., Jacobsen, H., Greve, T., and Callesen, H. (1998) Open Pulled Straw (OPS) vitrification: A new way to reduce cryoinjuries of bovine ova and embryos, *Mol. Reprod. Dev.*, 51, 53–58.

Van der Elst, J., Amerijckx, Y., and Van Steirteghem, A. (1998) Ultra-rapid freezing of mouse oocytes lowers the cell number in the inner cell mass of 5 day old in-vitro cultured blastocysts, *Hum. Reprod.*, 13, 1595–1599.

Van Voorst, A. and Leenstra, F.R. (1995a) Fertility rate of daily collected and cryopreserved fowl semen, *Poult. Sci.*, 74, 136–140.

Van Voorst, A. and Leenstra, F.R. (1995b) Effect of dialysis before storage or cryopreservation on fertilizing ability of fowl semen, *Poult. Sci.*, 74, 141–146.

Watson, P.F. (1990). Artificial insemination and the preservation of semen, in *Marshall's Physiology of Reproduction*, Lamming, G. E., Ed., Churchill Livingstone, Edinburgh, Scotland, pp. 747–869.

Watson, P.F. (1995). Recent developments and concepts in the cryopreservation of spermatozoa and the assessment of their post-thawing function, *Reprod. Fertil. Dev.*, 7, 871–891.

Watson, P.F. and Holt, W.V., Eds. (2001) *Cryobanking the Genetic Resource*, Taylor & Francis, London.

Wells, D.N., Misica, P.M., and Tervit, H.R. (1999) Production of cloned calves following nuclear transfer with cultured adult mural granulosa cells, *Biol. Reprod.*, 60, 996–1005.

Whittingham, D.G., Leibo, S.P., and Mazur, P. (1972) Survival of mouse embryos frozen to −196° and −269°C, *Science*, 178, 411–414.

Wildt, D.E., Rall, W.F., Critser, J.K., Monfort, S.L., and Seal, U.S. (1993) Genetic resource banks, *BioScience*, 47, 689–698.

Willadsen, S.M. (1977) Factors affecting the survival of sheep embryos during deep-freezing and thawing, in *The Freezing of Mammalian Embryos*, Ciba Foundation, Symposium 52 (New Series), Elliot, K. and Whalan, J., Eds., Elsevier Excerpta Medica/North-Holland, Amsterdam, pp. 175–201.

Wilmut, I. and Polge, C. (1974) The fertilizing capacity of boar semen stored in the presence of glycerol at 20, 5 and −79°C, *J. Reprod. Fertil.*, 38, 105–113.

Wilmut, I. and Rowson, L.E.A. (1973) Experiments on the low-temperature preservation of cow embryos, *Vet. Rec.*, 92, 686–690.

Wilmut, I., Schnieke, A.E., McWhir, J., Kind, A.J., and Campbell, K.H. (1997) Viable offspring derived from fetal and adult mammalian cells, *Nature*, 385, 810–813.

Woelders, H. (1997) Fundamentals and recent development in cryopreservation of bull and boar semen, *Vet. Q.*, 19, 135–138

Wolfe, J. and Bryant, G. (1999) Freezing, drying and/or vitrification of membrane-solute-water systems, *Cryobiology*, 39, 103–129.

Wood, M., Barros, C., Candy, C.J., Carroll, J., Melendez, J., and Whittingham, D.G. (1993) High rates of survival and fertilization of mouse and hamster oocytes after vitrification in dimethyl sulphoxide, *Biol. Reprod.*, 49, 489–495.

Wu, B., Tong, J., and Leibo, S.P. (1999) Effects of cooling germinal vesicle stage bovine oocytes on meiotic spindle formation following *in vitro* maturation, *Mol. Reprod. Dev.*, 54, 388–395.

Yoshino, J., Kojima, T., Shimizu, M., and Tomizuka, T. (1993). Cryopreservation of porcine blastocysts by vitrification, *Cryobiology*, 30, 413–422.

Zavos, P.M. and Graham, E.F. (1983) Effects of various degrees of supercooling and nucleation temperatures on fertility of frozen turkey semen, *Cryobiology*, 20, 553–559.

Zeng, W.X. and Terada, T. (2001) Protection of boar spermatozoa from cold shock damage by 2-hydroxypropyl-beta-cyclodextrin, *Theriogenology*, 55, 615–627.

13 Cryopreservation as a Supporting Measure in Species Conservation; "Not the Frozen Zoo!"

Amanda R. Pickard and William V. Holt

CONTENTS

13.1 INTRODUCTION

In the last years of the twentieth century, the human race became acutely aware of the effect of their activities on the world around them. Issues of habitat loss, environmental contamination, and endangered species preservation have become focal topics for members of the public throughout the developed world, and although not necessarily affording the same priority, significant awareness of these issues exists throughout the developing world. However, this awareness has come far too late to save some species that have been driven past, or to the verge of, extinction. Conservation biology has emerged as a multidisciplinary science aimed at redressing this balance across all the facets of the problem. However, issues in conservation go beyond science, and therefore conservation biology must also involve input from economists, the social sciences, and politics. Many developing countries have to balance the needs of wildlife conservation with responsibilities to sustain their human population. However, these are often the very same countries with the greatest richness of biodiversity; in these instances, conservation policies may ideally be linked with activities that attract foreign visitors and welcome investment. Viewed in this way, the value of conservation changes from a kindly and philanthropic activity to one that can provide tangible benefit to the poorest communities.

In this chapter we analyze the contribution to conservation biology that cryobiology and the reproductive sciences can make. In this context, cryobiology is mainly linked to animal breeding technologies via frozen semen, oocytes, and embryos, and is widely regarded therefore as a specialized subset of reproductive biology. However, as we intend to show, although such technologies do indeed have a potential role in conservation, it is important not to overstate it. Equally, it is important not to overlook the wider and perhaps more practical contribution that noninvasive studies and observations can make. Detailed technical issues relating to the cryobiology of gametes and embryos will not be covered in this review, as they have recently been considered by Watson and Holt (2001). In any case, with respect to conservation biology, the technicalities of cell and tissue freezing are merely one aspect of a complex multidisciplinary subject.

Germplasm cryopreservation as a means of storing sperm, eggs, and embryos, and thus contributing directly to animal breeding programs, is only one way that cryopreservation technology can contribute to conservation. Although not the focus of this chapter, it should be noted that archiving tissue samples from wild species, together with accurate genetic and geographical data about their origins, provides an invaluable source of materials for numerous studies in molecular genetics and ecology. Knowing the evolutionary origins of a species is important in developing conservation policies that affect its future.

To understand how knowledge of cryobiology and animal breeding might contribute to conservation programs, it is essential to know what those conservation programs are trying to achieve and why. In general, the objectives are to support the viability of small populations. This cannot be simply achieved by increasing the number of individuals, but by ensuring that genetic representation does not become highly skewed in favor of a few individuals, thus causing overall loss of genetic diversity. The first section of this chapter will therefore set out these background arguments before a more detailed critique of assisted reproduction and cryopreservation is presented.

13.2 REPRODUCTION AND POPULATION VIABILITY

Despite those who might argue to the contrary, reproduction has an important role to play in both *in situ* and *ex situ* conservation, simply because populations that have higher mortality rates than their reproductive output cannot survive into the future. The assertion that providing suitably protected habitat or the correct environmental stimuli is sufficient for successful procreation is, to say the least, naïve. Nevertheless, human interference is not always needed to ensure populations remain viable, and the concept that the reproductive sciences can only provide intrusive and invasive assistance shows a fundamental lack of understanding of the processes involved. Reproduction is not merely confined to conception, but the successful gestation, parturition, rearing, maturation, and subsequent reproduction of the next generation. Repeated failure of any of these processes will result in the demise of a population. However, careful observation and intervention as and when appropriate can serve to avoid this outcome. Assessing the prospective viability of a threatened population is therefore an important step in the development of conservation plans, and one that would inform any decision about the desirability of intervention and also what form such intervention might take. In many instances, the assessment might indicate the need to establish protected areas to help species recovery. In other cases, the conclusion might be that establishing a captive breeding program, with the long-term intention of reintroduction, is necessary. In both cases there may be justification for the establishment of a tissue and germplasm bank so that the extant populations could be supported by reproductive technologies should they be required in the future.

Censuses can be used to provide point-in-time information about the status of a population; for example, the relative proportion of individuals in different age classes. This information may be helpful in predicting future population dynamics, but it does not itself indicate how individuals are related or whether some individuals are contributing disproportionately to the gene pool. Long-term evaluation over a number of years, aided by the ability to identify individuals and confirm relatedness using paternity analysis, provides more useful information. Some such studies, for example, those of red deer on the Scottish Isle of Rhum (Clutton-Brock et al., 1982; Coulson et al., 1999), have provided detailed information about population responses to fluctuating environments. This type of data is useful for understanding the processes that may make species or populations more vulnerable to extinction and for initiating preventative strategies. However, the data collected from one scenario may be unsuitable for extrapolation in a generalized fashion, particularly given that relatively few populations are exclusively isolated as in the island example studied here.

Nevertheless, understanding the effect of decline and recovery on the genetic status of a population provides essential information for designing managed breeding programs. Introducing genetic material into a population using reproductive technologies relies on similar principles. Loss of genetic variability, widely known as "inbreeding," has serious consequences for reproductive performance and individual fitness at all stages of life history (Rousset, 2002; Ryan et al., 2003; Taylor, 2003). Juvenile survival is significantly depressed in several species (reviewed by Thornhill, 1983), partly because inbred individuals show increased susceptibility to diseases and parasites, and they may also show poor growth and body condition. Those that do achieve maturity also demonstrate reduced ability to successfully raise a subsequent generation (Ryan et al., 2003). When populations decline, they inevitably increase the likelihood of generating inbred individuals. By definition therefore, an endangered species is at risk from inbreeding depression and consequent further decline toward extinction. Reproductive technologies should have something to offer in such cases; transporting semen between small populations should, in principle, provide a linking mechanism that transforms several small populations into what is effectively a larger single population. Achieving this type of linkage has been likened to creating "wildlife corridors" (Bennett, 2001), a strategy that is used to connect small populations that become geographically isolated. The ability to manage genetic diversity using cryopreserved gametes is an attractive proposition to conservation biologists. This not only achieves the geographical linkage described above but also connects individuals and populations across temporal barriers. However, an important question then arises: What are the best genetic criteria on which to base a breeding program?

Knowing how to assess diversity is strongly debated among molecular and theoretical geneticists, and there is clearly no simple answer. The management of captive populations using studbooks provides a simple method for assessing relatedness. These are systems for recording the breeding histories of individuals within a population. Sophisticated computer programs are available to assist managers in the analysis of their studbook records and to generate recommended breeding strategies that will minimize inbreeding and mean kinship levels in the next generation. However, inbreeding coefficients can only be calculated from the founder individuals included in the studbook, the genetic relatedness of which may not, in fact, be known. Furthermore, this approach is unsuitable for use with most *in situ* conservation programs, where it is impossible to control, or even to record, parental identities.

A deceptively attractive alternative to the use of studbooks for the management of breeding programs is to find suitable molecular markers that indicate genetic diversity and then plan the breeding programs around them. This would have the advantage that germplasm samples could be screened in the laboratory, the marker identity noted, and the information used even if no other data were available. Several different approaches have been evaluated in an effort to establish such a system, which would then be useful for captive breeding management, and therefore for any program that involves cryopreserved germplasm.

Levels of inbreeding have been evaluated by assessing the diversity of major histocompatibility complex (MHC) class I loci between individuals within a species. This complex expresses genes involved in immune responses and disease resistance and is highly polymorphic because of its susceptibility to adaptive variation. MHC variability appears to play a major role in mate selection, with MHC-linked olfactory receptors (Penn, 2002) providing a natural mechanism to distinguish degree of relatedness in some species. It has been suggested that genetic variation at MHC loci is maintained by heterozygote advantage because heterozygotes can recognize a larger suite of pathogens than can homozygotes (Doherty and Zingernagel, 1975; Hedrick and Kim, 2000) and, therefore, have higher fitness. The high degree of diversity between individuals in this complex indicates an ability to resist exposure to novel pathogens, a major concern for endangered species. As population size declines, the degree of diversity in MHC class I reduces and ought to be detectable by molecular analysis. In theory, therefore, reduced MHC class I variability is indicative of a reduction in population variability and ability to respond to changes in the environment.

The use of MHC and other strategies, such as maximizing the occurrence of rare alleles (Hughes, 1991), for breeding management have been evaluated by theoretical modeling studies (Hedrick and Kim, 2000; Meyer and Thomson, 2001). Paradoxically, these evaluations did not show that the expected benefits were realized because they led to decreased allelic diversity in the rest of the genome (Miller and Hedrick, 1991). These strategies have recently been reviewed by Hedrick (2003).

Appropriate reproductive behavior is generally required for successful reproduction, but this is sometimes compromised by the circumstances facing endangered species. Consideration ought, therefore, to be given to how individuals interact with each other and their surroundings. Incidences of hand rearing among zoo animals can result in human imprinting and incompatible behavior toward conspecifics, precluding reproduction from occurring. The utility of interspecific embryo transfer has been questioned for some groups of species on the grounds that the newborn would become imprinted to an inappropriate species. This subtle effect has never been examined in detail, probably because too few such offspring have been available for a study. However, detailed behavioral assessments of reproductive activity, such as the social organization, mating system, life history, and seasonal patterns of reproduction of each species, are essential when deciding how populations can best be managed in the wild or how to implement captive breeding and reintroduction programs. Investing time in producing offspring by reproductive technology in subordinate females of a species that demonstrates female social dominance would be inappropriate if the dominant female were to attack and kill the resultant offspring. Similarly, attempting artificial insemination (AI) when reproduction is naturally suppressed because of seasonal breeding would be a waste of resources.

Suitable nutrition also plays a key role in successful reproduction. Translocation of animals from one natural habitat to a different, more secure location may result in nutritional imbalances that could, in the long term, adversely affect reproductive success. The Barker Hypothesis (see Barker, 2000, for a review) suggests that propensity to disease may be influenced by the environment we experience *in utero* and can be compromised by inappropriate maternal nutrition. Furthermore, embryos produced *in vitro* can result in abnormal fetal development and the production of oversize offspring, which then show a failure to thrive during further development (Khosla et al., 2001; Robinson et al., 2001). It is important not only to provide adequate nutrition in terms of energy intake but also to recognize that inappropriate quantities of key nutrients at specific stages during the reproductive processes can have significant effect on the reproductive potential of an individual and their subsequent offspring. Substituting locally available foodstuffs over the natural nutrients of choice could inadvertently be detrimental to a large number of species in the long term.

13.3 WHEN SHOULD CRYOPRESERVATION BE USED AS A SPECIES SUPPORT TOOL?

There is little evidence to indicate that cryoprotective treatments are directly responsible for inducing abnormal fetal development, although a relatively recent study (Dulioust et al., 1995) did, in fact, detect slight but significant morphological and behavioral changes in mice produced from frozen-thawed embryos. The commercial success of frozen sperm in cattle production systems, and the extensive use of both frozen semen and embryos in human clinical medicine, has not revealed the existence of major problems with cryopreserved germplasm (see also Chapters 12 and 18). Nevertheless, when considering the production of endangered species using reproductive technologies, it would be wrong to presume that treatments applied to gamete or embryos in the laboratory may not have an adverse effect on the subsequent development of the offspring.

There are some cases in which the use of cryopreservation techniques have clear advantages for conservation biology. For example, it may be possible to protect adaptations that provide the ability to occupy specific ecological niches by cryopreserving material from individuals expressing these phenotypes. Endangered Lake Victoria cichlid fish provide such an example. Dramatic declines in indigenous species have been caused by the introduction of a predatory species, the Nile Perch, for sport fishing. Moreover, the rapid proliferation of water hyacinth has choked the lake's shores and resulted in the extinction of about 60% of the cichlid species, which formerly numbered in excess of 300 (Ogutu-Ohwayo, 1997). The cichlids evolved rapidly to occupy specific niche habitats in the lake, relying on morphological adaptations to ensure they maintained an adequate food supply. Some species have developed a calcified "beak" allowing them to graze algae off rocks; others have developed molluscivorous habits and a jaw structure, which enables them to crack open shells to obtain food. When translocated to captivity for conservation purposes and kept in an artificial environment with commercially available fish food, these adaptations have been quickly lost. It is questionable, therefore, whether these species would be capable of surviving should the opportunity arise to restore future generations to their former habitat. However, it may be possible to retain some of these essential characteristics by cryopreserving representative material from wild-caught founder individuals or early captive bred generations, which can either be reintroduced at intervals during the captive breeding program or used to restore stocks into their former habitat. Unfortunately, to date there has been very limited success in obtaining and preserving Lake Victoria cichlid gametes, and therefore the effects of captive propagation cannot currently be reversed.

It is worth mentioning here that although the aquaculture industry has stimulated major advances in fish sperm cryopreservation, there are almost no comparable activities anywhere in the world that are focused on fish conservation. In addition, as fish embryo cryopreservation has still to be achieved for any species, this avenue is still completely unavailable to wildlife conservation, despite intensive research activity aimed at solving the problem (see also Chapter 14).

Although there are good reasons for developing the use of reproductive technologies for some species, in other cases it would simply be impractical. It would probably be inappropriate, for example, to consider that the development of an AI protocol for bats would have a significant role to play in the long-term conservation of species in this group, despite there being many species requiring attention. The practicalities associated with working with such cryptic species, which live in relatively large colonies, would be enormous. It would be far better to establish effective colony management strategies and habitat protection, which would support natural breeding. In contrast, recent advances in assisted reproduction have dramatically increased the success of producing African and Asian elephants in captivity. Since 1999 there have been a number of live-born elephant calves, both African and Asian, conceived by AI (Schmitt et al., 2001), and there are other ongoing pregnancies resulting from AI in the captive populations of these species. Before the successful development of AI techniques, only 27 African elephants had been born in the North American captive population since the 1800s (http://www.projectelephant.org/natural_history/ai.asp). Reproduction in Asian elephants in North America results in less than five calves born per year, which is inadequate to sustain the population (http://nationalzoo.si.edu/Publications/PressMaterials/Press Release/NZP/Kandula/Asian%20Elephant%20reproduction.doc). The success of AI with elephants is, therefore, remarkable, and its use will probably make a significant contribution to the survival of these species. To date, however, AI procedures have only been successfully performed using freshly collected semen. The ability to use cryopreserved and sex-sorted sperm would help to ensure that genetic management goals are met and that bull elephants are only born at facilities capable of housing them, as and when required.

13.4 TISSUE AND GERMPLASM BANKS FOR CONSERVATION

The storage of cryopreserved material from wild species was suggested over 20 years ago (Veprintsev and Rott, 1980), and the concept has been developed further in subsequent publications (Wildt, 1997; Wildt and Wemmer, 1999; Wildt et al., 1997). Many collections of genetic material have, in fact, been established. A survey conducted among European laboratories in 1995 identified 92 existing collections of cryopreserved tissues and cells from wild species (mammals, birds, fish, reptiles, amphibia, arthropods, and crustacae; F. Palacios, personal communication). Many more collections are in existence outside Europe, especially in countries of the former Soviet Union and the United States. Unfortunately, because these collections are managed by individual organizations, they have hitherto been unpublicized, uncoordinated, and frequently underfunded. One of the largest collections in existence is at the Zoological Society of San Diego. This currently holds samples from more than 3200 individuals representing 355 species and subspecies of mammals (http://www.sandiegozoo.com/conservation/frozen.html), with additional bird and reptile species also represented. Much of the material in this collection is in the form of fibroblast cell lines cultured for cytogenetic studies. Samples of DNA, RNA, and cDNA are available on request to legitimate researchers. However, there are as yet only limited options for using cell lines or derived genetic material to propagate endangered species. These repositories therefore provide the opportunity to examine the genetic relatedness of individuals, to undertake evolutionary studies, and to screen samples for previously undetected pathogens. These are important activities that underpin conservation programs.

The terms "genetic resource bank," "genome resource bank" (GRB), and "biological resource bank" have been coined to describe systematic collections of biological material (Holt et al., 1996b; Holt and Pickard, 1999; Wildt et al., 1997). The general intention in using these terms is to imply that although banks of biological materials include germplasm for use in animal breeding programs, they also encompass banks of somatic tissues collected for scientific research. Moreover, these collections may also contain fixed tissues, cultured cell lines, DNA, and serum samples. This level of sophistication requires a high level of curatorial organization and distinguishes the GRB or biological resource bank from the many *ad hoc* collections of materials that have accumulated almost by accident in many laboratories throughout the world.

Many issues that exercise curators of tissue banks in human clinical medicine also apply to the organization of GRBs for conservation (see also Chapter 15). For example, disease transmission risks via cryogenic storage containers have received very little consideration by centers with biological collections derived from wild species. The risks of disease transmission from one individual animal to another via semen samples are very poorly understood and need evaluating. The risks of intersample cross-contamination are even less well understood, but they have serious implications for the way in which any storage system is organized. For these reasons there is a good case for arguing against the establishment of large centrally located germplasm repositories and every reason to set up focused GRBs, located near to the extant populations that they support.

Current technologies are insufficiently effective for frozen tissue samples to be used for producing offspring. However, there is considerable interest among technologists in the prospect of using these samples to support animal breeding programs. At present, to achieve this objective, GRBs need specifically to contain germplasm material stored in such a way that it can be recovered and used for propagation. In general, this means maintaining stocks of conventionally frozen spermatozoa, embryo, and possibly oocytes, but as technology advances, the merits of keeping frozen ovarian and testicular tissues are becoming more apparent (see also Chapter 18). Mice, monkeys, and sheep have so far been generated from frozen/thawed pieces of ovary that have been replaced in a female and stimulated to ovulate (Candy et al., 2000; Gosden and Nagano, 2002; Gosden et al., 1994), whereas the principle of testicular cell freezing and transplantation has been demonstrated (Clouthier et al., 1996) and is currently being tested as a treatment for human male infertility (for review, see Schlatt and Nieschlag, 2001).

13.5 CRYOPRESERVATION, CONSERVATION, AND CLONING

Recent advances in the nuclear transfer and cloning of livestock and laboratory species have led many technologists to suggest that endangered, and even extinct, species can be propagated or recovered using these methods. In principle this is an attractive idea, especially as the raw materials, that is, sources of nuclear DNA, are already stored in the many tissue collections in existence throughout the world. Proposals to recreate the woolly mammoth and Tasmanian tiger have been published (Stone, 2001) or announced, and preliminary research is underway at Monash University aimed at developing cloning technology for the highly endangered Northern Hairy-Nosed Wombat. A case for the development of such methodology was published in a thoughtful article by Corley-Smith and Brandhorst (1999). It would seem appropriate to make a few comments here about the reality of these proposals.

As shown in Figure 13.1, the traditional reproductive technologies have been successful in a limited number of species. Success rates are generally higher with species that are linked with biomedical and agricultural research, for example, chimpanzee, marmoset monkey, and red deer, because the efforts are backed by research funds and may be the focus of attention in several different research groups. Species of no particular research or economic value tend not to be studied or supported to the same extent. There is no apparent reason why research into the cloning of endangered species should deviate from this pattern, and yet this complex technology requires a high standard of background reproductive data that simply does not exist. The difficulties of obtaining such data from endangered species are compounded by several factors. One major limitation is self-evident: There are few individuals available for study. This means that performing research on statistically meaningful numbers of animals is extremely difficult and may even be ethically unacceptable in some circumstances. Obtaining oocytes for the purposes of nuclear transfer involves not only knowing how to synchronize ovarian function accurately but also the inevitable use of anesthesia and surgical procedures. Incidentally, these considerations apply to the most mundane aspects of reproductive research; semen collection by electroejaculation for example, usually necessitates anesthesia and is therefore not without risk.

The current success rate with nuclear transfer is very low, and many embryos must be produced and transferred to host females to achieve a single pregnancy. Although the availability of females may not be a limiting factor with laboratory or agricultural species, surrogate mothers for most wild species are a rare resource, even if the species is not itself endangered. Thus, cloning a highly endangered Amur tiger would inevitably require not only an unrealistically large number of oocyte donors but a large number of recipient females as well. Given an optimistic assessment of the success rate, say 1%, how easy and how costly would it be to assemble a group of 100 female Siberian tigers to obtain a single Amur tiger offspring?

In addition to these practical questions, however, a major deficiency with the case for cloning is that it would lead directly to increased levels of inbreeding. In fact, because the offspring would be identical copies of a parent, the populations would become more highly inbred as new individuals were produced. One solution to this problem would be to clone all individuals within a small population. In fact, some endangered populations are, or have been, numerically very small indeed and have nevertheless bounced back. For example, the Mauritius kestrel population declined to two individuals but has since shown remarkable recovery to about 200 breeding pairs (Groombridge et al., 2000), and the Southern White rhinoceros declined to fewer than 50 individuals at the beginning of the twentieth century and has now reached several thousand. In such special cases, cloning may perhaps offer a realistic means of keeping sufficient individuals alive while other conservation measures are developed.

The use of cloning in this way would require highly effective supporting technologies to ensure that this extremely costly approach to conservation has the greatest chance of success, and almost by definition these technologies require development before the species has reached such a critical point. However, what about extinct species? The basic premise of our argument throughout this chapter is that cryopreservation, coupled with reproductive biology and other disciplines, offers a set of tools that augment and support conservation activities in their widest sense. It is therefore difficult to support the case that cloning technology should be used to recover extinct species, as other conservation strategies would not even exist. Moreover, although it may be theoretically possible to recreate a few individuals from extinct species, a host of controversial questions would immediately be raised. Where should the individuals be kept? Would they count as "intellectual property" on the grounds that they are a technological product? There would probably be major financial gains to be made by exhibiting such animals, but this would bear little relevance to conservation, which zoos see as an important justification for their existence. Would extinct retroviruses suddenly be released into an unsuspecting world, leading essentially to outbreaks of "new" diseases? These and other issues have been recently reviewed by Critser et al. (2003).

13.6 LIMITS AND SUCCESSES WITH CRYOPRESERVATION

Since the first successful cryopreservation of bull semen (Polge et al., 1949), efforts have been made to propagate rare and endangered species using assisted reproduction techniques. Although viable offspring have been generated in several species using cryopreserved material (see Table 13.1), the proportion of rare and endangered species represented is still incredibly low. Of the 4629 extant species of mammal (Wilson and Reeder, 1993), 1130 species are listed as critically endangered, endangered, or vulnerable (Hilton-Taylor, 2000). Therefore, in this taxonomic class alone, the use of assisted reproduction has only been successful in approximately 4% of species. For the other classes the figures are lower again. Furthermore, the number of times a procedure has been repeated successfully within a particular species is usually very low. As a consequence, propagation by assisted reproduction can be considered a "routinely" successful procedure for only a few species.

Although the results of assisted reproductive methods shown in Table 13.1 for some mammalian groups are rather poor, there are many taxonomic groups for which virtually no success at all has been achieved. Rodents provide a case study in this respect, although marsupials are another highly diverse taxonomic group containing many endangered species, and a group in which only one species, the koala, has been propagated by any form of reproductive technology (Johnston et al., 1999). It is notable that there have been many efforts to produce rodent offspring using reproductive technologies. Live offspring have been produced by transfer of frozen or vitrified embryos in laboratory mice (Wood et al., 2001, review), rats (Stein et al., 1993; Wood and Whittingham, 1981), hamsters (Lane et al., 1999; Ridha and Dukelow, 1985), and Mongolian gerbils (Mochida et al., 1999). AI has also been successful in mice (see Wood et al., 2001). Nevertheless, there are no reports in the literature describing the adaptation of these techniques to their endangered relatives. Despite there being significant interest in using GRB approaches to preserve valuable populations of laboratory rodents, which may have been generated by selective breeding or genetic manipulation, no laboratories are apparently making significant efforts to apply their knowledge to the 330 endangered species in this order. Perhaps the scale of the problem is overwhelming, but given the intensity of research effort required to achieve modest success with laboratory mice, there is almost no realistic prospect of using cryopreservation and reproductive technologies as a backup for rodent conservation. Nevertheless, successful conservation programs for rodent species have been developed, which do not use reproductive technologies but that are no less valuable for it. Conservation strategies for the water vole (*Arvicola terrestris*) in the United Kingdom have been successfully developed using knowledge of the ecology of this species and its major predator, the introduced American mink. Translocating animals where appropriate and communicating with planning authorities and developers has fostered an awareness of the plight of this species and a sense of responsibility in those who could easily contribute to its further decline (Strachan, 1998). It is to be hoped that habitat management strategies will be sufficient to prevent further threats to this species.

Figure 13.1 highlights the significant bias in technology developments for endangered species toward those species that most appeal to members of the public worldwide—carnivores, which are related to many common companion animals; primates, because of their close human ancestral association; and ungulates, which make up the majority of our agricultural species. What Figure 13.1 fails to highlight is the relatively small number of repeatable successes. For example, despite producing a Siberian tiger cub by laparoscopic AI (Donoghue et al., 1993), more than 50 further

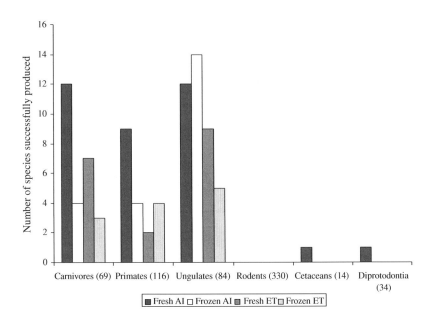

FIGURE 13.1 Taxonomic Group (Total number of species in the Red List).

attempts to repeat this procedure have failed (N. Loskutoff, personal communication). This observation is not intended to discredit those who have invested significant resources in addressing the problems associated with generating offspring using reproductive technologies, but more to emphasize the point that programs of assisted reproduction are only of value if they satisfy the "Three R's" of reproduction: All techniques should be reliable and repeatable, and the objectives of the program should be realistic.

A major limitation to success with all forms of assisted reproductive technology in wild species, whether or not the use of cryopreserved materials are involved, is knowing how to determine the correct moment for any procedure. AI depends for success on the precision with which ovulation time can be predicted or induced. If spermatozoa are delivered to the female reproductive tract too early with respect to ovulation, they will age and die before having the chance to fertilize. If they are delivered too late, the oocyte itself may have aged and subsequent embryo development may be compromised. As cryopreservation shortens the lifespan of spermatozoa, the window in which fully functional sperm and eggs actually coexist in the female reproductive tract may be as short as 6 h. Solving this problem is not easy, especially if the species in question is rare and therefore unavailable for carrying out necessary preliminary research.

Many practitioners believe that assisted reproduction techniques developed for "model" species, often a closely related nonendangered subspecies, offer the solution to this problem. Ideally the results of preliminary studies can then be translated for use in taxonomically similar species. However, this is not always necessarily the case. The application of oestrous synchronization protocols and AI in deer species can result in conception rates in excess of 65%. Similar procedures in an endangered ungulate, the Mohor gazelle (*Gazella dama mhorr*), using cryopreserved semen only result in fertilization rates of 30% (Holt et al., 1996a; Pickard et al., 2001). Close assessment of the breeding success of this species reveals that its natural fertility is similarly approximately 30%, and indicates that expectations of high success rates may be misguided. This clearly limits the efficiency of reproductive technologies as a way of reversing this species' trend toward extinction (Pickard et al., 2001). In contrast, the use of a domesticated, nonendangered model species has

TABLE 13.1
Species for which Live Offspring Have Been Produced by the Use of Assisted Reproduction Technologies

Species	Category of threat	Artificial insemination	Embryo transfer	References
Carnivora				
Silver fox (*Vulpes vulpes*)	Not listed		Fresh	Jalkanen and Lindeberg, 1998
African lion (*Panthera leo*)	Vulnerable (Asiatic subspecies critically endangered)	Fresh		Bowen et al., 1982; Reed et al., 1981
Caracal (*Felis caracal*)	Not listed		Fresh	cited by Leibo and Songsasen (2002)
Cheetah (*Acinonyx jubatus*)	Vulnerable (northwest African subspecies endangered, Asiatic subspecies critically endangered)	Fresh and frozen		Howard et al., 1992
Clouded leopard (*Neofelis nebulosa*)	Vulnerable	Fresh		Howard et al., 1996a
Indian and African wildcat (*Felis silvestris*)	Not listed (Scottish subspecies vulnerable)		Fresh and frozen	Pope et al., 1989, 1993, 2000
Leopard cat (*Felis bengalensis*)	Not listed	Fresh		Howard and Doherty, 1991
Ocelot (*Leopardus pardalis*)	Not listed (Texas subspecies endangered)	Fresh and frozen	Fresh and frozen	Moraes et al., 1997; Swanson et al., 1996, 2000
Persian leopard (*Panther pardus*)	Eight endangered or critically endangered subspecies	Fresh		Dresser et al., 1982a
Puma (*Felis concolor*)	Critically endangered	Fresh		Moore et al., 1981
Snow leopard (*Uncia uncia*)	Endangered	Fresh		Roth et al., 1997
Tiger (*Panthera tigris*)	Endangered (three critically endangered subspecies)	Fresh	Fresh	Donoghue et al., 1990, 1993; Reed et al., 1981
Tigrina (*Leopardus tigrina*)	Not listed	Fresh	Fresh and frozen	Moraes et al., 1997; Swanson et al., 2000
American black bear (*Ursus americanus*)	Not listed		Fresh	Boone et al., 1999
Black-footed ferret (*Mustela nigripes*)	Extinct in the wild (subject to a program of reintroduction)	Fresh and frozen		Carvalho et al., 1991; Howard et al., in press, 1996a
Giant panda (*Ailuropoda melanoleuca*)	Endangered	Fresh and frozen		Masui et al., 1989; Moore et al., 1984
Primates				
Baboon (*Papio cynocephalus*)	Not listed	Fresh	Fresh and frozen	Kraemer and Cruz, 1972; Kraemer et al., 1976; Pope et al., 1984
Chimpanzee (*Pan troglodytes*)	Endangered	Fresh and frozen		Gould, 1990

Species	Conservation status			References
Common marmoset (*Callithrix jacchus*)	Not listed	Fresh and frozen	Frozen	Morrell et al., 1998; Summers et al., 1987b
Gorilla (*Gorilla gorilla gorilla*)	Endangered	Fresh and frozen	Fresh	Douglass, 1981; Pope et al., 1997
Cynomolgus monkey (*Macaca fascicularis*)	Low risk	Fresh	Frozen	Balmaceda et al., 1986; Torii and Nigi, 1998
Pig-tailed macaque (*Macaca nemestrina*)	Vulnerable		Frozen (lion-tailed macaque (*M. silenus*) sperm used for IVF	Cranfield et al., 1990
Rhesus monkey (*Macaca mulata*)	Low risk	Fresh and frozen	Frozen	Sanchez-Partida et al., 2000; Wolf et al., 1989
Taiwan macaque (*Macaca cyclopis*)	Vulnerable	Fresh		Peng et al., 1973
Japanese macaque (*Macaca fuscata*)	Density dependent, subspecies endangered	Fresh		Torii and Nigi, 1998
Brown lemur (*Lemur fulvus*)	Various subspecies are low risk, vulnerable and critically endangered	Fresh		Brun et al., 1987
Ungulates				
Scimitar horned oryx (*Oryx dammah*)	Extinct in the wild (subject to a program of reintroduction)	Fresh and frozen	Fresh	Garland, 1989; Pope et al., 1991
Addax (*Addax nasomaculatus*)	Critically endangered	Frozen		Densmore et al., 1987
European mouflon (*Ovis orientalis musimon*)	Vulnerable		Frozen (transfer into domestic sheep)	Ptak et al., 2002
Blackbuck (*Antilope cervicapra*)	Vulnerable	Fresh and frozen		Holt et al., 1988
Bongo (*Tragelaphus eurycerus*)	Low risk (Eastern subspecies critically endangered)		Frozen (transfer into eland)	Dresser et al., 1984; Dresser et al., 1985
Common eland (*Tragelaphus oryx*)	Low risk		Frozen	Dresser et al., 1982b; Kramer et al., 1982
Gaur (*Bos gaurus*)	Vulnerable	Frozen (AI in domestic cow)	Fresh (transfer into domestic cow)	Hopkins et al., 1988; Johnston et al., 1994; Stover et al., 1981
Mohor gazelle (*Gazella dama mhorr*)	Endangered	Frozen		Holt et al., 1996a
Speke's gazelle (*Gazella spekei*)	Vulnerable	Fresh		Boever et al., 1980
Swamp buffalo (*Bubalus bubalis*)	Endangered	Fresh and frozen	Fresh and frozen	Numerous references available on semen freezing; for embryo freezing, see Kasiraj et al., 1993

TABLE 13.1 (continued)
Species for which Live Offspring Have Been Produced by the Use of Assisted Reproduction Technologies

Species	Category of threat	Artificial insemination	Embryo transfer	References
Proboscidae				
African elephant (*Loxodonta africana*)	Endangered	Fresh		http://www.projectelephant.org/natural_history/ai.asp; Schmitt et al., 2001
Asian elephant (*Elephas maximus*)	Endangered	Fresh		http://nationalzoo.si.edu/Publications/PressMaterials/PressRelease/NZP/Kandula/Asian%20Elephant%20reproduction.doc; Schmitt et al., 2001
Camellidae				
Bactrian camel (*Camelus bactrianus*)	Endangered	Frozen		Zhao et al., 1994
Dromedary (*Camelus dromedarius*)	Not listed	Fresh and frozen		Bravo et al., 2000
Llama (*Lama glama*)	Not listed	Fresh		Bravo et al., 2000
Alpaca (*Lama pacos*)	Not listed	Fresh and frozen	Fresh	Bravo et al., 2000; Brown, 2000; Sumar, 1996
Cervidae				
Red Deer (*Cervus elaphus*)	Various subspecies are low risk, density dependent, vulnerable, and endangered	Fresh and frozen	Fresh and frozen	Dixon et al., 1991; Fennessy et al., 1990; Haigh and Bowen, 1991; Krzywinski and Jaczewski, 1978
Wapiti (*Cervus elaphus nelsoni*)	Not listed	Frozen	Fresh	Haigh and Bowen, 1991; Wenkoff and Bringans, 1991
White-tailed deer (*Odocoileus virginianus*)	Low risk	Fresh and frozen	Fresh	Haigh, 1984; Hoekstra et al., 1984; Jacobson et al., 1989; Magyar et al., 1989; Waldhalm et al., 1989
Fallow deer (*Dama dama* subspecies)	Not listed (ssp. *Mesopotamica* endangered)	Fresh and frozen		Jabbour et al., 1993; Morrow et al., 1994; Mulley et al., 1988; Mylrea et al., 1992
Eld's deer (*Cervus eldii thiamin*)	Vulnerable	Frozen		Monfort et al., 1993

Equidae			
Przewalski's horse (*Equus przewalskii*)	Extinct in the wild	Fresh (transfer to domestic horse)	Summers et al., 1987a
Zebra (*Equus burchellii*)	Not listed (one extinct subspecies, three density-dependent subspecies)	Fresh (transfer to domestic horse)	Bennett and Foster, 1985; Summers et al., 1987a
Cetacea			
Killer whale (*Orcinus orca*)	Low risk	Fresh	http://www.seaworld.org; Robeck, 2001
Marsupials			
Koala (*Phascolarctus cinereus*)	Low risk	Fresh	Johnston et al., 1999

Note. The categories of threat have been determined from the 2000 *IUCN Red List of Threatened Species* (Hilton-Taylor, 2000), taking into account, where possible, specific subspecies differences. Those species that are not listed are not considered to be endangered.

almost certainly guaranteed the success of another species recovery program; in this case one that was set up to save the black-footed ferret (*Mustela nigripes*; Howard et al., 2003). Furthermore, this is also an example for which the application of germplasm cryopreservation technology has become crucially important.

Previously thought to be extinct in the wild, a remnant population of black-footed ferrets was discovered in 1964 in South Dakota (Biggins et al., 1997; Miller et al., 1996). Between 1972 and 1974, attempts were made to preserve this species by transferring nine individuals to captivity. However, the captive population was decimated by an outbreak of canine distemper following vaccination with a modified live virus (Carpenter et al., 1976; Hillman and Carpenter, 1983), and the remaining wild South Dakota population also disappeared. In 1981, a surprise discovery of a population in Wyoming restored hope that this species might be brought back from "extinction." Despite efforts to preserve this population in the wild, disease outbreaks repeatedly reduced numbers until, between 1985 and 1987, the remaining 18 individuals were captured by Wyoming Game and Fish Department and U.S. Fish and Wildlife Service for *ex situ* conservation (Thorne and Oakleaf, 1991). A program of captive breeding was established with a high priority to reintroduce this species into its former range. Multiple institutions were involved to avoid the possibility that a disease outbreak could destroy a population at a single establishment, as had previously occurred. However, to optimize the genetic diversity of these isolated populations, a breeding program incorporating the use of reproductive technologies, and the establishment of a GRB of frozen sperm from the most genetically valuable males was advocated.

It was recognized that the development of reproductive technologies for this species would be virtually impossible on the number of animals available for study. Therefore, the black-footed ferret conservation program was established in such a way that natural breeding was used to maintain and increase the number of individuals in the *ex situ* population. At the same time, however, steps were taken to gain increased knowledge of the basic physiology of mustelids through the study of domestic ferrets and the closely related Siberian polecat. Initially, extensive studies of the domestic ferret allowed the establishment of reliable methods for semen collection and cryopreservation. These techniques were then applied to the Siberian polecat for comparison before finally applying them to the black-footed ferret. Results were impressive; 66.7% pregnancy rates were obtained when inseminating with fresh or frozen-thawed sperm (Howard, 1999; Howard et al., 1996b). These studies have helped the black-footed ferret reintroduction program to move from strength to strength: its geographical range has been extended across several American states, and a dedicated breeding center has been established for semen banking and AI. The opportunity to carry out such detailed developments is rare when working with exotic or endangered species; however, this is an unusual but exemplary case because several conservation agencies formed a consortium with the express purpose of sharing their skills and resources.

13.7 CONCLUSIONS

The policy decisions that must be taken before investing in any program of species conservation require input from many and varied organizations. Governmental and nongovernmental organizations need to discuss relative priorities and socioeconomic factors. Ecologists, geneticists, and epidemiologists need to consider the practical aspects of the conservation plan and whether there are adequate resources to support a species protection program. The role of the reproductive biologist, and even more specifically of cryobiologists, should be to support and facilitate achieving the stated goals. However, the limitations of such a complex science must also be taken into account. Often, cryobiology techniques require specialized laboratory equipment, which may be unsuitable for use in field environments. These things should be considered during the early stages of planning a program incorporating assisted reproduction, as the ability to achieve program goals might be compromised by practical issues that have been overlooked.

Supporting technologies are also required to perform assisted reproduction in wildlife. Methods of evaluating reproductive hormone concentrations and assessing the anatomy and morphology of the reproductive tract, for example, by ultrasound, are as essential as the ability to collect semen or perform embryo transfers. The likelihood of successful AI or embryo transfer will be dramatically reduced if the endocrine control of reproduction is poorly understood or the nature of the female tract has not been explored. Frequently, more benefit will be gained by simply addressing the reasons why natural breeding is failing in a particular individual, rather than attempting to devise high-tech methods to increase reproductive output. However, suitable protocols or equipment for carrying out these assessments may not exist. There is significant merit in investing in research aimed at the development of new techniques of broad applicability and guaranteed reliability. All too often, however, priorities focus heavily on perfecting techniques that are only suitable for use in a limited number of species.

This review has concentrated mostly on the application of cryobiology techniques to the conservation of endangered mammalian species and has focused on reports available in the widely accessible literature. It should be noted, however, that many programs incorporating the use of cryobiology are currently in progress and are therefore omitted from our analysis of relative successes. Furthermore, significant efforts to reproduce nonmammalian species are being made, but to report on these is beyond the scope of this manuscript (see recent reviews by Donoghue et al., 2003 [birds]; Lance, 2003 [reptiles]; Reid and Hall, 2003 [fish]); Roth and Obringer, 2003 [amphibians]).

To be successful when applying reproductive technologies for the conservation of endangered species, the complex interaction of physiological and environmental systems needs to be taken into account. Sensationalist successes, which are often unique and cannot be readily repeated, frequently secure significant media attention and spark the public's imagination. However, their long-term benefit to conservation is questionable, and these stories do a disservice to those conservation programs aimed at finding methodical approaches to species survival. It would be far better for our limited resources be targeted toward realizing practical, achievable outcomes through cross-disciplinary collaboration and cooperation.

REFERENCES

Balmaceda, J.P., Heitman, T.O., Garcia, M.R., Pauerstein, C.J., and Pool, T.B. (1986) Embryo cryopreservation in cynomolgus monkeys, *Fertil. Steril.*, 45, 403–406.

Barker, D.J. (2000) *In utero* programming of cardiovascular disease, *Theriogenology*, 53, 555–574.

Bennett, P.M. (2001) Establishing animal germplasm resource banks for wildlife conservation: Genetic, population and evolutionary aspects, in *Cryobanking the Genetic Resource. Wildlife Conservation for the Future*? Watson, P.F. and Holt, W.V., Eds., Taylor & Francis, London, pp. 49–67.

Bennett, S.D. and Foster, W.R. (1985) Successful transfer of a zebra embryo to a domestic horse, *Equine Vet. J.*, Suppl. 3, 78–79.

Biggins, D.E., Miller, B.J., Clark, T.W., and Reading, R.P. (1997) Conservation management case studies: The Black-footed ferret, in *Principles of Conservation Biology*, Meffe, G.K. and Carroll, C.R., Eds., Sinauer Associates, Sunderland, MA, pp. 420–426.

Boever, J., Knox, D., Merilan, C., and Read, B. (1980) Estrus induction and artificial insemination with successful pregnancy in Speke's gazelle. *9th International Congress on Animal Reproduction and Artificial Insemination Madrid*, Madrid, Spain, 565–569.

Boone, W.R., Catlin, J.C., Casey, K.J., Dye, P.S., Boone, E.T., and Schuett, R.J. (1999) Live birth of a bear cub following nonsurgical embryo collection, *Theriogenology*, 51, 519–529.

Bowen, M.J., Platz, C.C., Brown, C.D., and Kraemer, D.C. (1982) Successful artificial insemination and embryo collection in the African lion (*Panthera leo*). *Proceedings of the American Association of Zoo Veterinarians Annual Conference*, 57–59.

Bravo, P.W., Skidmore, J.A., and Zhao, X.X. (2000) Reproductive aspects and storage of semen in Camelidae, *Anim. Reprod. Sci.*, 62, 173–193.

Brown, B.W. (2000) A review on reproduction in South American camelids, *Anim. Reprod. Sci.*, 58, 169–95.

Brun, B., Cranz, C., Ishak, B., Clavert, A., Hugues, F., Leclerc, M., and Rumpler, Y. (1987) Successful artificial insemination in *Lemur fulvus mayottensis*, *Folia Primatologia*, 48, 195–198.

Candy, C.J., Wood, M.J., and Whittingham, D.G. (2000) Restoration of a normal reproductive lifespan after transplantation of cryopreserved mouse ovaries, *Hum. Reprod.*, 15, 1300–1304.

Carpenter, J.W., Appel, M.J.G., Erickson, R.C., and Novilla, M.N. (1976) Fatal vaccine-induced canine distemper virus infection in black-footed ferrets, *J. Am. Vet. Med. Assoc.*, 169, 961–964.

Carvalho, C.F., Howard, J.G., Collins, L., Wemmer, C., Bush, M., and Wildt, D.E. (1991) Captive breeding of black-footed ferrets (*Mustela nigripes*) and comparative reproductive efficiency in 1-year old versus 2-year old animals, *J. Zoo Wildl. Med.*, 22, 96–106.

Clouthier, D.E., Avarbock, M.R., Maika, S.D., Hammer, R.E., and Brinster, R.L. (1996) Rat spermatogenesis in mouse testis, *Nature*, 381, 418–421.

Clutton-Brock, T.H., Guinness, F.E., and Albon, S.D. (1982) *Red Deer: Behaviour and Ecology of Two Sexes*, Chicago: University of Chicago Press.

Corley-Smith, G.E. and Brandhorst, B.P. (1999) Preservation of endangered species and populations: A role for genome banking, somatic cell cloning, and androgenesis? *Mol. Reprod. Dev.*, 53, 363–367.

Coulson, T., Albon, S.D., Pilkington, J., and Clutton-Brock, T.H. (1999) Small-scale spatial dynamics in a fluctuating ungulate population, *J. Anim. Ecol.*, 68, 658–671.

Cranfield, M.R., Berger, N.G., Kempske, S., Bavister, B.D., and Ialeggio, D.M. (1990) Successful birth of a macaque in a surrogate mother after transfer of a frozen/thawed embryo produced by *in vitro* fertilisation, *Proceedings of the American Association of Zoo Veterinarians Annual Conference*, 305–308.

Critser, J.K., Riley, L.K., and Prather, R.S. (2003) Application of nuclear transfer technology to wildlife species, in *Reproductive Science and Integrated Conservation*, Holt, W.V., Pickard, A.R., Rodger, J., and Wildt, D.E., Ed., Cambridge University Press, Cambridge, pp. 195–208.

Densmore, M.A., Bowen, M.J., Magyar, S.J., Amoss, M.S., Jr., Robinson, R.M., Harms, P.G., and Kraemer, D.C. (1987) Artificial insemination with frozen, thawed semen and pregnancy diagnosis in addax (*Addax nasomaculatus*), *Zoo Biol.*, 6, 21–29.

Dixon, T.E., Hunter, J.W., and Beatson, N.S. (1991) Pregnancies following the export of frozen red deer embryos from New Zealand to Australia, *Theriogenology*, 35, 193.

Doherty, P. and Zingernagel, R. (1975) Enhanced immunologic surveillance in mice heterozygous at the H2 complex, *Nature*, 245, 50–52.

Donoghue, A.M., Johnston, L.A., Armstrong, D.L., Simmons, L.G., and Wildt, D.E. (1993) Birth of a Siberian tiger cub (*Panthera tigris altaica*) following laparoscopic artificial insemination, *J. Zoo Wildl. Med.*, 24, 185–189.

Donoghue, A.M., Johnston, L.A., Seal, U.S., Armstrong, D.L., Tilson, R.L., Wolff, P., Petrini, K., Simmons, L.G., Gross, T., and Wildt, D.E. (1990) *In vitro* fertilization and embryo development *in vitro* and *in vivo* in the tiger (*Panthera tigris*), *Biol. Reprod.*, 43, 733–744.

Donoghue, A.M., Blanco, J.M., Gee, G.F., Kirby, Y.K., and Wildt, D.E. (2003) Reproductive technologies and challenges in avian conservation and management, in *Reproductive Science and Integrated Conservation*, Holt, W.V., Pickard, A.R., Rodger, J., and Wildt, D.E., Ed., Cambridge University Press, Cambridge, pp. 321–337.

Douglass, E.M. (1981) First gorilla born using artificial insemination, *Int. Zoo News*, 28, 9–15.

Dresser, B.L., Kramer, L., Reece, B., and Russell, P.T. (1982a) Induction of ovulation and successful artificial insemination in a Persian leopard (*Panthera pardus saxicolor*), *Zoo Biol.*, 1, 55–57.

Dresser, B.L., Kramer, L., Pope, C.E., Dahlhausen, R.D., and Blauser, C. (1982b) Super-ovulation of African eland (*Taurotragus oryx*) and interspecies ebryo transfer to Holstein Cattle, *Theriogenology*, 17, 86–86.

Dresser, B.L., Pope, C.E., Kramer, L., Kuehn, G., Dahlhausen, R.D., Maruska, E.J., Reece, B., and Thomas, W.D. (1985) Birth of bongo antelope (*Tragelaphus euryceros*) to eland antelope (*Tragelaphus oryx*) and cryopreservation of bongo embryos, *Theriogenology*, 23, 190–190.

Dresser, B.L., Pope, C.E., Kramer, L., Kuehn, G., Dahlhausen, R.D., and Thomas, W.D. (1984) Superovulation of bongo antelope (*Tragelaphus euryceros*) and interspecies embryo transfer to African eland (*Tragelaphus oryx*), *Theriogenology*, 21, 232–232.

Dulioust, E., Toyama, K., Busnel, M.-C., Moutier, R., Carlier, M., Marchland, C., Ducot, B., Roubertoux, P., and Auroux, M. (1995) Long-term effects of embryo freezing in mice, *Proc. Natl. Acad. Sci. USA*, 92, 589–593.

Fennessy, P.F., Mackintosh, C.G., and Shackell, G.H. (1990) Artificial insemination of farmed red deer (*Cervus elaphus*), *Anim. Prod.*, 51, 613–621.

Garland, P. (1989) Artificial insemination of Scimitar-horned oryx (*Oryx dammah*), *Bull. Zoo Manage.*, 27, 29–30.

Gosden, R. and Nagano, M. 2002. Preservation of fertility in nature and ART, *Reproduction*, 123, 3–11.

Gosden, R.G., Baird, D.T., Wade, J.C., and Webb, R. (1994) Restoration of fertility to oophorectomized sheep by ovarian autografts stored at –196°C, *Hum. Reprod.*, 9, 597–603.

Gould, K.G. (1990) Techniques and significance of gamete collection and storage in great apes, *J. Med. Primatol.*, 19, 537–551.

Groombridge, J.J., Jones, C.G., Bruford, M.W., and Nichols, R.A. (2000) "Ghost" alleles of the Mauritius kestrel, *Nature*, 403, 616.

Haigh, J.C. (1984) Artificial insemination of two white-tailed deer, *J. Am. Vet. Med. Assoc.*, 185, 1446–1447.

Haigh, J.C. and Bowen, G. (1991) Artificial insemination of red deer (*Cervus elaphus*) with frozen thawed wapiti semen, *J. Reprod. Fertil.*, 93, 119–123.

Hedrick, P. (2003) The major histocompatibility complex (MHC) in declining populations: An example of adaptive variation, in *Reproductive Science and Integrated Conservation,* Holt, W.V., Pickard, A.R., Rodger, J., and Wildt, D.E., Eds., Cambridge University Press, Cambridge, pp. 97–113.

Hedrick, P.W. and Kim, T.J. (2000) Genetics of complex polymorphisms: Parasites and maintenance of the major histocompatibility complex variation, in *Evolutionary Genetics: From Molecules to Morphology,* Singh, R.S. and Krimbas, C.B., Eds., Cambridge University Press, Cambridge, pp. 204–234.

Hillman, C.N. and Carpenter, J.W. (1983) Breeding biology and behavior of captive black-footed ferrets, *Mustela nigripes, Int. Zoo Yearbook*, 23, 186–191.

Hilton-Taylor, C. (compiler) (2000) *2000 IUCN Red List of Threatened Species*, IUCN, Gland, Switzerland.

Hoekstra, R.E., Magyar, S.J., Biediger, T.G., Coscarelli, K.P., Simpson, T.R., McCrady, J.D., Kraemer, D.C., and Seager, S.W.J. (1984) Artificial insemination in white tailed deer (*Odocoileus virginianus*) with frozen epididymal spermatozoa. *American Association of Zoo Veterinarians Annual Proceedings*, 174–179.

Holt, W.V., Abaigar, T., and Jabbour, H.N. (1996a) Oestrous synchronization, semen preservation and artificial insemination in the Mohor gazelle (*Gazella dama mhorr*) for the establishment of a genome resource bank programme, *Reprod. Fertil. Dev.*, 8, 1215–1222.

Holt, W.V., Bennett, P.M., Volobouev, V., and Watson, P.F. (1996b) Genetic resource banks in wildlife conservation, *J. Zool.*, 238, 531–544.

Holt, W.V., Moore, H.D.M., North, R.D., Hartman, T.D., and Hodges, J.K. (1988) Hormonal and behavioral detection of estrus in blackbuck, *Antilope cervicapra*, and successful artificial insemination with fresh and frozen semen, *J. Reprod. Fertil.*, 82, 717–725.

Holt, W.V. and Pickard, A.R. (1999) Role of reproductive technologies and genetic resource banks in animal conservation, *Rev. Reprod.*, 4, 143–150.

Hopkins, S.M., Armstrong, D.L., Hummel, S., and Junior, S. (1988) Successful cryopreservation of gaur (Bos gaurus) epididymal spermatozoa, *J. Zoo Anim. Med.*, 19, 195–201.

Howard, J.G. (1999) Assisted reproductive techniques in non-domestic carnivores, in *Zoo and Wild Animal Medicine,* Fowler, M.E. and Miller, R.E., Ed., W.B. Saunders, Philadelphia, pp. 449–457.

Howard, J.G., Byers, A.P., Brown, J.L., Barrett, S.J., Evans, M.Z., Schwartz, R.J., and Wildt, D.E. (1996a) Successful ovulation induction and laparoscopic intrauterine insemination in the clouded leopard (*Neofelis nebulosa*), *Zoo Biol.*, 15, 55–69.

Howard, J.G. and Doherty, J. (1991) *Leopard Cats Produced by Artificial Insemination*, American Association of Zoo Parks and Aquaria Communique, May, 12.

Howard, J.G., Donoghue, A.M., Barone, M.A., Goodrowe, K.L., Blumer, E.S., Snodgrass, K., Starnes, D., Tucker, M., Bush, M., and Wildt, D.E. (1992) Successful induction of ovarian activity and laparoscopic artificial insemination in the Cheetah (*Acinonyx jubatus*), *J. Zoo Wildl. Med.*, 23, 288–300.

Howard, J.G., Kwiatkowski, D.R., Williams, E.S., Atherton, R.W., Kitchin, R.M., Thorne, E.T., Bush, M., and Wildt, D.E. (1996b) Pregnancies in black-footed ferrets and Siberian polecats after laparoscopic artificial insemination with fresh and frozen-thawed semen, *J. Androl.*, Supplement P-51, 115.

Howard, J., Marinari, P.E., and Wildt, D.E. (2003) Black-footed ferret: Model for assisted reproductive technologies contributing to in situ conservation, in *Reproductive Science and Integrated Conservation,* Holt, W.V., Pickard, A.R., Rodger, J., and Wildt, D.E., Eds., Cambridge University Press, Cambridge, pp. 249–266.

Hughes, A.L. (1991) MHC polymorphism and the design of captive breeding programs, *Conserv. Biol.*, 5, 249–251.

Jabbour, H.N., Argo, C.M., Brinklow, B.R., Loudon, A.S.I., and Hooton, J. (1993) Conception rates following intrauterine insemination of European (*Dama dama dama*) fallow deer does with fresh or frozen-thawed Mesopotamian (*Dama dama mesopotamica*) fallow deer spermatozoa, *J. Zool.*, 230, 379–384.

Jacobson, H.A., Bearden, H.J., and Whitehouse, D.B. (1989) Artificial insemination trials with white-tailed deer, *J. Wildl. Manage.*, 53, 224–227.

Jalkanen, L. and Lindeberg, H. (1998) Successful embryo transfer in the silver fox (*Vulpes vulpes*), *Anim. Reprod. Sci.*, 54, 139–147.

Johnston, L.A., Parrish, J.J., Monson, R., Leibfried-Rutledge, L., Susko-Parrish, J.L., Northey, D.L., Rutledge, J.J., and Simmons, L.G. (1994) Oocyte maturation, fertilization and embryo development *in vitro* and *in vivo* in the gaur (*Bos gaurus*), *J. Reprod. Fertil.*, 100, 131–136.

Johnston, S.D., McGowan, M.R., O'Callaghan, P., Cox, R., Houlden, B., Haig, S., and Taddeo, G. (1999) Birth of Koala pouch young following artificial insemination, *Proceedings of the Australian Society of Reproductive Biology. Annual Conference.* Melbourne.

Kasiraj, R., Misra, A.K., Rao, M.M., Jaiswal, R.S., and Rangareddi, N.S. (1993) Successful culmination of pregnancy and live birth following transfer of frozen-thawed buffalo embryos, *Theriogenology*, 39, 1187–1192.

Khosla, S., Dean, W., Reik, W., and Feil, R. (2001) Epigenetic and experimental modifications in early mammalian development: Part II—Culture of preimplantation embryos and its long-term effects on gene expression and phenotype, *Hum. Reprod. Update*, 7, 419–427.

Kraemer, D. and Cruz, N.V. (1972) Breeding baboons for laboratory use, in *Breeding Primates*, Beveridge, W., Ed., Karger, Basel, pp. 42–47.

Kraemer, D.C., Moore, D.T., and Kramer, M.A. (1976) Baboon infant produced by embryo transfer, *Science*, 192, 1246–1247.

Kramer, L., Dresser, B.L., Pope, C.E., Dahlhausen, R.D., and Reed, G. (1982) Collection, transfer and cryopreservation of African eland (*Taurotragus oryx*) embryos, *American Association of Zoo Veterinarians Annual Conference*, pp. 129–131.

Krzywinski, A. and Jaczewski, Z. (1978) Observations on the artificial breeding of red deer, in *Artificial Breeding of Non-Domestic Animals*, Watson, P.F., Ed., Academic Press, London, pp. 271–287.

Lance, V.A. (2003) Reptile reproduction and endocrinology, in *Reproductive Science and Integrated Conservation*, Holt, W.V., Pickard, A.R., Rodger, J., and Wildt, D.E., Ed., Cambridge University Press, Cambridge, pp. 338–358.

Lane, M., Bavister, B.D., Lyons, E.A., and Forest, K.T. (1999) Containerless vitrification of mammalian oocytes and embryos. Adapting a proven method for flash-cooling protein crystals to the cryopreservation of live cells, *Nature Biotechnol.*, 17, 1234–1236.

Leibo, S.P. and Songsasen, N. 2002. Cryopreservation of gametes and embryos of non-domestic species, *Theriogenology*, 57, 303–326.

Magyar, S.J., Biediger, T., Hodges, C., Kraemer, D.C., and Seager, S.W.J. (1989) A method of artificial insemination in captive white-tailed deer (*Odocoileus virginianus*), *Theriogenology*, 31, 1075–1080.

Masui, M., Hiramatsu, H., Nose, N., Nakazato, R., Sagawa, Y., Tajima, H., and Saito, K. (1989) Successful artificial insemination in the giant panda (*Ailuropoda melanoleuca*) at Ueno zoo, *Zoo Biol.*, 8, 17–26.

Meyer, D. and Thomson, G. (2001) How selection shapes variation of the human major histocompatibility complex: A review, *Ann. Hum. Genet.*, 65, 1–26.

Miller, B., Reading, R.P., and Forrest, S. (1996) Prairie night: Black-footed ferrets and the recovery of an endangered species, Smithsonian Institution Press, Washington, DC.

Miller, P.S. and Hedrick, P.W. (1991) MHC polymorphism and the design of captive breeding programs: Simple solutions are not the answer, *Conserv. Biol.*, 5, 556–558.

Mochida, K., Wakayama, T., Takano, K., Noguchi, Y., Yamamoto, Y., Suzuki, O., Ogura, A., and Matsuda, J. (1999) Successful cryopreservation of Mongolian gerbil embryos by vitrification, *Theriogenology*, 51, 171.

Monfort, S.L., Asher, G.W., Wildt, D.E., Wood, T.C., Schiewe, M.C., Williamson, L.R., Bush, M., and Rall, W.F. (1993) Successful intrauterine insemination of Eld's deer (*Cervus eldii thamin*) with frozen-thawed spermatozoa, *J. Reprod. Fertil.*, 99, 459–465.

Moore, H., Bonney, R.C., and Jones, D.M. (1981) Successful induced ovulation and artificial insemination in the puma (*Felis concolor*), *Vet. Rec.*, 108, 282–283.

Moore, H.D.M., Bush, M., Celma, M., Garcia, A.-L., Hartman, T.D., Hearn, J.P., Hodges, J.K., Jones, D.M., Knight, J.A., Monsalve, L., and Wildt, D.E. (1984) Artificial insemination in the giant panda (*Ailuropoda melanoleuca*), *J. Zool.*, 203, 269–278.

Moraes, W., Morais, R.N., Moreira, N., Lacerda, O., Gomes, M.L.F., Mucciola, R.G., and Swanson, W.F. (1997) Successful artificial insemination after exogenous gonadotropin treatment in the ocelot (*Leopardus pardalis*) and tigrina (*Leopardus tigrina*), *American Association of Zoo Veterinarians Annual Conference*, 334–335.

Morrell, J.M., Nubbermeyer, R., Heistermann, M., Rosenbusch, J., Kuderling, I., Holt, W., and Hodges, J.K. (1998) Artificial insemination in *Callithrix jacchus* using fresh or cryopreserved sperm, *Anim. Reprod. Sci.*, 52, 165–174.

Morrow, C.J., Asher, G.W., Berg, D.K., Tervit, H.R., Pugh, P.A., McMillan, W.H., Beaumont, S., Hall, D.R.H., and Bell, A.C.S. (1994) Embryo transfer in fallow deer (*Dama dama*): Superovulation, embryo recovery and laparoscopic transfer of fresh and cryopreserved embryos, *Theriogenology*, 42, 579–590.

Mulley, R.C., Moore, N.W., and English, A.W. (1988) Successful uterine insemination of fallow deer with fresh and frozen semen, *Theriogenology*, 29, 1149–1153.

Mylrea, G.E., English, A.W., Mulley, R.C., and Evans, G. (1992) Artificial insemination of farmed chital deer, in *The Biology of Deer*, Brown, R.D., Ed., Springer, New York, pp. 334–337.

Ogutu-Ohwayo, R. (1997) Threats to biodiversity in Lake Victoria: The struggle between human need and biodiversity conservation, in *Workshop on Freshwater Biodiversity*, Swedish Scientific Council on Biological Diversity (Selbu), Norway.

Peng, M., Lai, Y., Yang, C., Chiang, H., New, A., and Chang, C. (1973) Reproductive parameters of the Taiwan monkey (*Macaca cyclopis*), *Primates*, 14, 203–213.

Penn, D.J. 2002. The scent of genetic compatibility: Sexual selection and the major histocompatibility complex, *Ethology*, 108, 1–21.

Pickard, A.R., Abaigar, T., Green, D.I., Holt, W.V., and Cano, M. (2001) Hormonal characterisation of the reproductive cycle and pregnancy in the female Mohor gazelle (*Gazella dama mhorr*), *Reproduction*, 122, 571–580.

Polge, C., Smith, A.U., and Parkes, A.S. (1949) Revival of spermatozoa after vitrification and dehydration at low temperatures, *Nature*, 164, 666–669.

Pope, C.E., Dresser, B.L., Chin, N.W., Liu, J.H., Loskutoff, N.M., Behnke, E.J., Brown, C., McRae, M.A., Sinoway, C.E., Campbell, M.K., Cameron, K.N., Owens, O.M., Johnson, C.A., Evans, R.R., and Cedars, M.I. (1997) Birth of a western lowland gorilla (*Gorilla gorilla gorilla*) following *in vitro* fertilization and embryo transfer, *Am. J. Primatol.*, 41, 247–260.

Pope, C.E., Gelwicks, E.J., Wachs, K.B., Maruska, E.J., and Dresser, B.L. (1989) Successful interspecies transfer of embryos from the Indian desert cat (*Felis silvestris ornata*) to the domestic cat (*Felis catus*) following *in vitro* fertilization, *Biol. Reprod.*, 40, 61.

Pope, C.E., Gelwicks, E.J., Burton, M., Reece, R., and Dresser, B.L. (1991) Nonsurgical embryo transfer in the Scimitar-horned Oryx (*Oryx dammah*)—Birth of a live offspring, *Zoo Biol.*, 10, 43–51.

Pope, C.E., Gomez, M.C., Mikota, S.K., and Dresser, B.L. (2000) Development of *in vitro* produced African wildcat (*Felis silvestris*) embryos after cryopreservation and transfer into domestic cat recipients, *Biol. Reprod.*, 62, 321.

Pope, C.E., Keller, G.L., and Dresser, B.L. (1993) *In vitro* fertilization in domestic and nondomestic cats including sequences of early nuclear events, development *in vitro*, cryopreservation and successful intraspecies and interspecies embryo transfer, *J. Reprod. Fertil.*, 47, 189–201.

Pope, C.E., Pope, V.Z., and Beck, L.R. (1984) Live birth following cryopreservation and transfer of a baboon embryo, *Fertil. Steril.*, 42, 143–145.

Ptak, G., Clinton, M., Barboni, B., Muzzeddu, M., Cappai, P., Tischner, M., and Loi, P. (2002). Preservation of the wild European mouflon: The first example of genetic management using a complete program of reproductive biotechnologies, *Biol. Reprod.*, 66, 796–801.

Reed, G., Dresser, B., Reece, B., Kramer, L., Russell, P., Pindell, K., and Berringer, P. (1981) Superovulation and artificial insemination of Bengal tigers (*Panthera tigris*) and an interspecies transfer to the African lion (*Panthera leo*), *American Association of Zoo Veterinarians Annual Conference*, 136–138.

Reid, G.M. and Hall, H. (2003) Reproduction in fishes in relation to conservation, in *Reproductive Science and Integrated Conservation*, Holt, W.V., Pickard, A.R., Rodger, J., and Wildt, D.E., Ed., Cambridge University Press, Cambridge, pp. 375–393.

Ridha, M.T. and Dukelow, W.R. (1985) The developmental potential of frozen-thawed hamster pre-implantation embryos following embryo transfer: Viability of slowly frozen embryos following slow and rapid thawing, *Anim. Reprod. Sci.*, 9, 253–259.

Robeck, T.R. (2001) Cetacean reproductive biology and ART, in *1st International Symposium on Assisted Reproductive Technology for the Conservation and Genetic Management of Wildlife*, Henry Doorly Zoo, Omaha, NE, pp. 179–188.

Robinson, J.J., McEvoy, T.G., and Ashworth, C.J. (2001) Nutrition in the expression of reproductive potential, *J. Anim. Feed Sci.*, 10, 15–27.

Roth, T.L., Armstrong, D.L., Barrie, M.T., and Wildt, D.E. (1997) Seasonal effects on ovarian responsiveness to exogenous gonadotrophins and successful artificial insemination in the snow leopard (*Uncia uncia*), *Reprod. Fertil. Dev.*, 9, 285–295.

Roth, T.L. and Obringer, A.R. (2003) Reproductive research and the worldwide amphibian extinction crisis, in *Reproductive Science and Integrated Conservation*, Holt, W.V., Pickard, A.R., Rodger, J., and Wildt, D.E., Ed., Cambridge University Press, Cambridge, pp. 359–374.

Rousset, F. 2002. Inbreeding and relatedness coefficients: What do they measure? *Heredity*, 88, 371–80.

Ryan, K.K., Lacy, R.C., and Margulis, S.W. (2003) Impacts of Inbreeding on Components of Reproductive Success, in *Reproductive Science and Integrated Conservation*, Holt, W.V., Pickard, A.R., Rodger, J., and Wildt, D.E., Ed., Cambridge University Press, Cambridge, pp. 82–96.

Sanchez-Partida, L.G., Maginnis, G., Dominko, T., Martinovich, C., McVay, B., Fanton, J., and Schatten, G. (2000) Live rhesus offspring by artificial insemination using fresh sperm and cryopreserved sperm, *Biol. Reprod.*, 63, 1092–1097.

Schlatt, S. and Nieschlag, E. (2001) Germ cell transplantation as a tool for fertility preservation of oncological patients, *Klin. Padiatrie*, 213, 250–254.

Schmitt, D.L., Hildebrandt, T.B., Hermes, R., and Goritz, F. (2001) Assisted reproductive technology in elephants, in *1st International Symposium on Assisted Reproductive Technology for the Conservation and Genetic Management of Wildlife*, Henry Doorly Zoo, Omaha, NE, pp. 22–24.

Stein, A., Fisch, B., Tadir, Y., Ovadia, J., and Kraicer, P.F. (1993) Cryopreservation of rat blastocysts: A comparative study of different cryoprotectants and freezing/thawing methods, *Cryobiology*, 30, 128–134.

Stone, R. (2001) *Mammoth: The Resurrection of an Ice Age Giant*, Perseus, Cambridge, MA.

Stover, J., Evans, J., and Dolensek, E.P. (1981) Inter-species transfer from the gaur to domestic Holstein. *American Association of Zoo Veterinarians Annual Conference*, 122–124.

Strachan, R. (1998) *The Water Vole Conservation Handbook*, Environment Agency, English Nature and the Wildlife Conservation Research Unit (WildCRU), Oxford.

Sumar, J.B. (1996) Reproduction in llamas and alpacas, *Anim. Reprod. Sci.*, 42, 405–415.

Summers, P.M., Shephard, A.M., Hodges, J.K., Kydd, J., Boyle, M.S., and Allen, W.R. (1987a) Successful transfer of the embryos of Przewalski's horses (*Equus przewalskii*) and Grant zebra (*E. burchelli*) to domestic mares (*E. caballus*), *J. Reprod. Fertil.*, 80, 13–20.

Summers, P.M., Shepard, A.M., Taylor, C.T., and Hearn, J.P. (1987b) The effects of cryopreservation and transfer on embryonic development in the common marmoset monkey, *Callithrix jacchus*, *J. Reprod. Fertil.*, 79, 241–250.

Swanson, W.F., Howard, J.G., Roth, T.L., Brown, J.L., Alvarado, T., Burton, M., Starnes, D., and Wildt, D.E. (1996) Responsiveness of ovaries to exogenous gonadotropins and laparoscopic artificial insemination with frozen-thawed spermatozoa in ocelots (*Felis pardalis*), *J. Reprod. Fertil.*, 106, 87–94.

Swanson, W.F., McRae, M.A., Callahan, P., Morais, R.N., Gomes, M.L.F., Moraes, W., Adania, C.H., and Campbell, M. (2000) *In vitro* fertilization, embryo cryopreservation, and laparoscopic embryo transfer for propagation of the endangered ocelot (*Leopardus pardalis*), in *Proceedings of the Zoological Society of London Symposium, Reproduction and Integrated Conservation Science*, Zoological Society of London.

Taylor, A.C. (2003) Assessing the consequences of inbreeding for population fitness: Past challenges and future prospects, in *Reproductive Science and Integrated Conservation*, Holt, W.V., Pickard, A.R., Rodger, J., and Wildt, D.E., Eds., Cambridge University Press, Cambridge, pp. 67–81.

Thorne, E.T. and Oakleaf, B. (1991) Species rescue for captive breeding: Black-footed ferret as an example, in *Beyond Captive Breeding: Re-Introducing Endangered Mammals to the Wild*, Gipps, J.H.W., Ed., Clarendon Press, Oxford, pp. 241–261.

Thornhill, N.W. (1983) *The Natural History of Inbreeding and Outbreeding*, University of Chicago Press, Chicago.

Torii, R. and Nigi, H. (1998) Successful artificial insemination for indoor breeding in the Japanese monkey (*Macaca fuscata*) and the Cynmologus monkey (*Macaca fascicularis*), *Primates*, 39, 399–406.

Veprintsev, B.N. and Rott, N.N. (1980) *Genome Conservation*, Academy of Sciences of the USSR, Puschino, pp. 1–49.

Waldhalm, S.J., Jacobson, H.A., Dhungel, S.K., and Bearden, H.J. (1989) Embryo transfer in the white-tailed deer—A reproductive model for endangered deer species of the world, *Theriogenology*, 31, 437–450.

Watson, P.F. and Holt, W.V. (2001) *Cryobanking the Genetic Resource: Wildlife Conservation for the Future*, Taylor & Francis, London, pp. 463.

Wenkoff, M.S. and Bringans, M.J. (1991) Embryo transfer in cervids, in *Wildlife Production: Conservation and Sustainable Development*, Renecker, L.A. and Hudson, R.J., Eds., University of Alaska, Fairbanks, pp. 461–463.

Wildt, D.E. (1997) Genome resource banking. Impact on biotic conservation and society, in *Reproductive Tissue Banking*, Karow, A.M. and Critser, J., Eds., Academic Press, New York, pp. 399–439.

Wildt, D.E., Rall, W.F., Critser, J.K., Monfort, S.L., and Seal, U.S. (1997) Genome resource banks, *Bioscience*, 47, 689–698.

Wildt, D.E. and Wemmer, C. (1999) Sex and wildlife: The role of reproductive science in conservation, *Biodiversity Conserv.*, 8, 965–976.

Wilson, D.E. and Reeder, D.M. (1993) *Mammal Species of the World*, Smithsonian Institution Press, Washington, DC, pp. 1206.

Wolf, D.P., VandeVoort, C.A., Meyer-Haas, G.R., Zelinski-Wooten, M.B., Hess, D.L., Baughman, W.L., and Stouffer, R.L. (1989) *In vitro* fertilization and embryo transfer in the rhesus monkey, *Biol. Reprod.*, 41, 335–346.

Wood, M.J., Candy, C.J., and Holt, W.V. (2001) Gamete and embryo cryopreservation in rodents, in *Cryobanking the Genetic Resource. Wildlife Conservation for the Future?* Watson, P.F. and Holt, W.V., Eds., Taylor & Francis, London, pp. 229–266.

Wood, M.J. and Whittingham, D.G. (1981) Low temperature storage of rat embryos, in *Frozen Storage of Laboratory Animals*, Zeilmaker, G.H., Ed., Gustav Fischer, Stuttgart, pp. 119–128.

Zhao, X.X., Huang, Y.M., and Chen, B.X. (1994) Artificial insemination and pregnancy diagnosis in the Bactrian camel (*Camelus bactrianus*), *J. Arid Environ.*, 26, 61–65.

14 Cryopreservation of Gametes and Embryos of Aquatic Species

Tiantian Zhang

CONTENTS

0-415-24700-4/04/$0.00+$1.50
© 2004 by CRC Press LLC

14.1 INTRODUCTION

Cryopreservation of germplasm of aquatic species offers many benefits in the fields of aquaculture, conservation, and biomedicine. It brings the possibility of preserving specific species or strains of particular interests, increasing the representation of genetically valuable animals, extending the reproductive life of a particular animal, and avoiding genetic losses through disease, catastrophe, or transfer between locations. In fish culture and farming, the successful cryopreservation of gametes, eggs, and embryos will offer new commercial possibilities, allowing the unlimited production of fry and potentially more robust and better conditioned fish as required. Further advantages include optimal utilization of hatchery facilities and facilitation of transport of stocks between hatcheries. Germplasm cryopreservation also provides a secure *ex situ* method for preserving the genomes of endangered species in diversity high enough to reconstruct stable populations when environmental conditions make it possible (see also Chapter 13). More than 65% of the European fish species are threatened (Kirchofer, 1996), and worldwide, the number of species listed as threatened or endangered grows rapidly. The extremely endangered species may be extinct before recolonization is possible, or their genetic variability may be so reduced that reconstruction of a stable population is impossible (Gilpin and Soule, 1986). Fish germplasm is also playing a significant role in human genomic studies. The relatively small size of fish genomes make them easier for sequencing, and ideal models for studies on vertebrate development and human disease. Understanding the relationship between fish and human genomes will help identify roles for human genes from fish mutations and help identify fish models for genes identified by human disease (Barbazuk et al., 2000; Brownlie et al., 1998). Successful cryopreservation of fish germplasm will be of benefit of biomedical as well as embryological and molecular biological studies.

Successful cryopreservation of reproductive materials of many aquatic species has been achieved (Table 14.1). Majority of these studies have been carried out with gametes (Barros et al., 1997; Billard and Zhang, 2001; Clark et al., 1996; Paniagua et al., 1998a; Stoss, 1983). Cryopreservation of fish gametes has been studied extensively in the last three decades, and the successful cryopreservation of the spermatozoa from many species including salmonid, cyprinids, silurids, and acipenseridae is well documented (Lahnsteiner, 2000; Magyary et al., 1996; Maisse, 1996; Rana and Gilmour, 1996; Tsvetkova et al., 1996). During the last decade, cryopreservation has became one of the most effective tools for reproduction management in fish culture and *ex situ* conservation of species. Several fish gamete cryobanks have been established in Europe, North America, and Asia for preservation and conservation of fish genetic materials. In aquaculture, cryopreserved fish spermatozoa are frequently used for artificial insemination of fish eggs when fresh spermatozoa are not available. Some studies have been carried out on cryopreservation of aquatic invertebrate semen including species from echinoderms, molluscs, polychaetes, and crustacea (Gwo, 2000b). Successful cryopreservation protocols of blastomeres have also been established for several species (Harvey, 1983; Leveroni and Maisse, 1998, 1999; Strussman et al., 1999). Limited success has been reported on the cryopreservation of eggs or embryos of aquatic species. Successful cryopreservation has been reported for certain species including the embryos of the Pacific oyster (Lin and Chao, 2000), larvae of the Eastern oyster (Paniagua et al., 1998b), embryos of the hard clam (Chao et al., 1997), larvae of the sea urchin (Naidenko and Kol'tsova, 1998), the

TABLE 14.1
Successful Cryopreservation of Reproductive Materials of Aquatic Species

Type	Reproductive Material	References
Freshwater fish		Lahnsteiner, 2000; Leveroni and Maisse, 1998,
Salmonid	Sperm, blastomere	1999; Maisse, 1996; Rana and Gilmour, 1996
Cyprinids	Sperm, blastomere	
Silurids	Sperm	
Marine fish		Bolla et al., 1987; Gwo, 2000a, b; Labbe et al.,
Clupeiformes Gadiformes	Sperm	1998; Mounib, 1978; Pullin, 1972
Perciformes	Sperm	
Pleuronectiformes	Sperm	
Tetraodontiformes	Sperm	
Echinoderms		Barros et al., 1997; Dunn and Mclachlan,
Sea urchin	Sperm, larvae	1973; Naidenko and Kol'tsova, 1998
Starfish	Sperm	
Sand dollar	Sperm	
Molluscs		Chao et al., 1997; Lin and Chao, 2000;
Oyster	Sperm, embryo, larvae	Paniagua et al., 1998a, b; Tsai and Chao, 1994
Abalone	Sperm	
Clam	Embryo	
Crustacea		Bhavanishankar and Subramoniam, 1997;
Shrimp	Sperm	Diwan and Joseph, 2000
Crab	Sperm	
Rotifer	Embryo	Toledo and Kurokura, 1990
Polychaete	Sperm, larvae	Bury and Olive, 1993; Olive and Wang, 1997

embryos of rotifer (Toledo and Kurokura, 1990), and the juveniles of marine polychaete (Olive and Wang, 1997). Despite many attempts, cryopreservation of fish embryos has not been successful (Hagedorn et al., 1996; Harvey, 1983; Zhang et al., 1989; Zhang et al., 1993). There has been virtually no published information on the cryopreservation of ovary or unfertilized eggs of fish species.

14.2   CRYOPRESERVATION OF FISH GAMETES

The major biological differences between mammalian and fish sperm and the implications on their cryopreservation are summarized as follows: first, most fish sperm do not have an acrosome that is commonly present in mammalian sperm; therefore, acrosome reaction preservation is not an issue for fish sperm cryopreservation in most of the cases. Second, most mammalian sperm have a mitochondrial sheet active enough to ensure continuous motility between ejaculation and fertilization, whereas in fish, the number of mitochondria is much less: one in trout (*Salmo tritta*; Billard, 1983), two to three in carp, and eight to 10 in turbot (*Scophthalmus maximus*; Suquet et al., 1993). Once activated, ATP depletes rapidly, and fish sperm stay motile for short period of time, ranging from 2 to 3 min in some freshwater fish to 20 to 25 min in some marine species. Third, the activation signal for fish sperm motility is osmotic change of the external environment. Fish sperm are activated when the osmolality of the seminal plasma lowers. For most fish species, the osmotic pressure of the semen fluid is within the range of 290 to 400 mosm/L. With sperm sample collection, external osmotic conditions need to be carefully controlled to avoid activation. Fourth, plasma membrane composition is influenced by the habitat temperature of a given species. Because body temperatures of mammalian species are generally higher than those of fish species, their plasma membrane lipids are less unsaturated and lipid phase transitions occur at higher temperatures. As a result, mammalian spermatozoa are more sensitive to cold shock and chilling injuries (Morris and Watson, 1984; Watson and Morris, 1987).

Cryopreservation of fish spermatozoa has been carried out on about 200 fish species (Rana and Gilmour, 1996; Tiersch and Mazik, 2000). Studies have been mainly conducted on freshwater species with economical importance, and salmonidae is the most studied family. Other freshwater groups such as cyprinids and silurids have also been well studied. Cryopreservation protocols have been developed for over 40 marine species including species from the orders of acipenseridae (Tsvetkova et al., 1996), Clupeiformes, Gadiformes, and Perciformes (Gwo, 2000a). Successful cryopreservation of fish gametes depends on many factors, and the degree of success is very often species dependent. However, generally speaking, attempts at cryopreservation of spermatozoa of marine fish species are more successful when compared with those obtained from freshwater fish. The fertilization rates obtained with cryopreserved fish sperm from marine species are comparable with those obtained with mammalian species, although sperm from freshwater fish species are generally more difficult to cryopreserve. Whereas high levels of fertilization of fresh eggs by cryopreserved sperm can be achieved in several freshwater species (Lahnsteiner et al., 1995, 1997), semen quality significantly decreases during cryopreservation, and the amount of sperm used to achieve fertilization could be 10 to 20 times greater than with fresh semen. Controlled slow cooling has been mainly used for fish spermatozoa cryopreservation. The basic steps required by this approach include collection and assessment of sperm quality, equilibration of sperm in a extender containing molar concentrations of a cryoprotective solute such as dimethyl sulphoxide (Me_2SO), freezing sperm suspension using controlled cooling to subzero temperatures, low-temperature storage at $-196°C$, and warming and thawing the sperm suspension using controlled conditions.

14.2.1 FACTORS CONSIDERED IN FISH SPERMATOZOA CRYOPRESERVATION

With controlled slow cooling, the selection and optimization of the following factors need to be considered.

14.2.1.1 Sperm Quality

When sperm is collected by stripping, milt is often contaminated by urine, which induces activation of the spermatozoa, and at least a part of their endogenous stores of ATP are exhausted (Perchec-Poupard et al., 1998). There are several ways to avoid urine contamination, including the use of a catheter introduced into the sperm duct, the clearing of the bladder before stripping, and in some cases, use of the intratesticular spermatozoa. The motility rate (percentage of motile cells) and changes with time after activation, the beat frequency of the flagellum, and the velocity are normally used as parameters to assess sperm quality. Sperm motility is normally measured either under a microscope or with computer-assisted sperm analysis. Other parameters such as ATP content have also been used.

14.2.1.2 Extender Composition

Extenders are usually based on a saline solution and mimic the osmotic pressure found in the seminal fluid. They are used to dilute the semen so all individual spermatozoa are exposed to cryoprotectant before freezing. Because the choices of salts in extenders are often not explained, in more recent studies, sugars such as sucrose or glucose were included as additives (Lahnsteiner et al., 1995), and sometimes only sugars were used in extenders (Legendre and Billard, 1980; Stoss and Refstie, 1983). Membrane stabilizers such as lipoproteins are also used in extenders. Maisse et al. (1998) reported significant increase of postthaw fertilizing capacity of rainbow trout spermatozoa with egg yolk compared that with skimmed milk, lecithin, or BSA.

14.2.1.3 Type and Concentration of Cryoprotectants

Me_2SO is the most commonly used cryoprotectant in fish sperm cryopreservation, followed by glycerol, ethylene glycol, and methanol. Other cryoprotectants used in sperm cryopreservation including propane-1,2-diol and dimethyl acetamide. The typical concentration range of cryoprotectant used is 5 to 15% (V:V%), with 10% being the most commonly used concentration.

14.2.1.4 Equilibration Period and Temperature

Spermatozoa are normally equilibrated in cryoprotectants for 10 to 15 min at 4°C. This will allow the penetration of the cryoprotectants into sperm before freezing. In some cases, when membrane permeability of sperm is considered to be high, equilibration period is kept as short as possible to avoid toxic effects of cryoprotectants and the extender. Once cryoprotectants have been added, semen is normally loaded into 0.25-, 0.5-, 1.2-, or 5-mL plastic straws or into 1- to 2-mL pellets, tubes, or vials.

14.2.1.5 Freezing Rate

After equilibration, straws or pellets are frozen in nitrogen vapor at approximately –70°C for 10 to 20 min, depending on their size, before being plunged into liquid nitrogen. The cooling rates of different straws are variable and are between 20 to 40°C/min (Mims et al., 2000; Rana and Gilmour, 1996).

14.2.1.6 Thawing and Postthaw Handling

After freezing, straws, pellets or vials are thawed in a water bath at temperatures between 0 and 80°C, with the typical temperature range being 20 to 40°C (David et al., 2000; Lahnsteiner et al., 1995; Mongkonpunya et al., 2000). Depending on the size of the straws or pellets, the thawing periods are normally between 5 and 30 seconds. Thawed sperm are immediately used for artificial insemination, as their motility and fertilizing capacity decrease with time.

14.2.1.7 Viability Assessment of Cryopreserved Sperm

A large number of criteria have been used to assess the viability of freeze-thawed sperm on the basis of motility, fertilization rate, or cell damage. Sperm motility and fertilization rate are the most commonly used criteria, whereas alteration of morphology, enzymatic activity, and membrane integrity have also been used to assess the viability of the cryopreserved sperm.

14.2.2 Cryopreservation of Sperm of Salmonid and Cyprinids Fishes

Spermatozoa have successfully been cryopreserved for 12 salmonid species including rainbow trout (*Oncorhynchus mykiss*), coho salmon (*Oncorhynchus kisutch*), atlantic salmon (*Salmo salar*), brook trout (*Salvelinus fontinalis*), brown trout (*Salmo trutta* f. fario), arctic char (*Salvelinus alpinus*), Danube salmon (*hucho hucho*), and grayling (*Thymallus thymallus*; Billard and Zhang, 2001; Erdahl and Graham, 1980; Horton and Ott, 1976; Lahnsteiner, 2000; Maisse, 1996; Stoss, 1983; Tiersch and Mazik, 2000). A typical cryopreservation protocol for salmonid species is summarized here. Mature 2- to 4-year-old male fish in breeding condition were gently massaged at the abdominal site to collect semen samples. The semen samples were then diluted in ice-cold extender containing cryoprotectants. The diluted semen was drawn into plastic straws and frozen in the vapor of liquid nitrogen for 10 to 15 min before it was plunged into liquid nitrogen. For thawing, the straws were immersed in a water bath, and thawed samples were immediately removed from the straw and poured onto the eggs (Lahnsteiner, 2000). The postthaw fertilization rate was normally between 60 to 85%, and cryopreserved spermatozoa have been routinely used in aquaculture industry for livestock handling and in the laboratories in artificial insemination for research purposes.

Cyprinids are another well-studied group. Some cyprinid fishes are amongst the most important farmed species in Europe and Asia. Successful cryopreservation protocols have been developed for several cyprinids species including common carp (*Cyprinus carpio*), rohu (*Labeo rohita*), grass carp (*Ctenopharyngodon idella*), and goldfish (*Carassius auratus*; Bercsényi et al., 1998; Magyary et al., 2000; McAndrew et al., 1995). Magyary et al. (2000) reported 95% fertilization and hatching rate for common carp using cryopreserved sperm, and these results were not significantly different from those obtained with nonfrozen controls.

14.2.3 CRYOPRESERVATION OF SPERM OF MARINE SPECIES

Since the first report of successful cryopreservation of sperm of the marine fish, the pacific herring (*Clupea palasii*; Blaxter, 1953), sperm of many marine species have been successfully cryopreserved including representatives of the orders Clupeiformes, Gadiformes, Perciformes, Pleuronectiformes, and Tetraodontiformes. The application of cryopreservation techniques is particularly important for those species with long life cycles and in the species in which male fish only mature when they are several years old. In these cases, the male brood stock is generally limited and collection of sperm is particularly difficult. Cryopreservation of sperm could reduce the number of males needed in the hatchery and minimize stress induced through stripping. Generally speaking, attempts at cryopreservation of sperm of marine species have been more successful when compared with those obtained with freshwater fish, although protocols are less consistent (Gwo, 2000). The fertilization rates of cryopreserved sperm from marine species ranged from 44 to 100% when compared with controls (Labbe et al., 1998). Fertilization rates using cryopreserved sperm from several species, including American plaice (*Pleuronectes platessoides*; Pullin, 1972), Atlantic cod (*Gadus morhua*; Mounib, 1978), and Atlantic halibut (*Hippoglossus hippoglossus*; Bolla et al., 1987), are not significantly different from those obtained with fresh sperm. A higher membrane cholestrol/phospholipid ratio has been suggested to be responsible for the higher cryopreservation tolerance of sperm from marine species (Drokin, 1993). Other factors such as number of mitochondria and energy metabolism of sperm were also identified (Labbe et al., 1998). It is possible that success of sperm cryopreservation is a species-specific multifactor effect.

14.2.4 CRYOPRESERVATION OF FISH SPERMATOZOA OF ENDANGERED SPECIES

Human activity (both directly in terms of overfishing and indirectly through environmental disturbance and damage) has resulted in a major collapse of fish populations and, even more disturbing, the final loss of some species. This loss is particularly acute in freshwater environments, which cover only 1% the planet and in which over 40% of all fish species occur. The nature of the problem is well illustrated by the case of Lake Victoria, the world's largest tropical lake (68,000 km^2), with over 300 endemic cichlid species, of which 65% have become lost or extinct during the last two decades. There is also a decline in marine fish catches worldwide. This downward trend in marine productivity is in stark contrast to the remarkable growth in world catches during most of the twentieth century—from 3 million tons in 1900 to a high of 86 million tons in 1989, when harvests peaked. Fisheries experts agree that the declines are the result of the current levels of exploitation exceeding the productive limits of many of the world's marine ecosystems. The 1990 United Nations survey of world fisheries confirmed this view, classifying nearly every commercial species it surveyed as "fully exploited, over exploited or depleted." As the number of fish species listed as threatened or endangered grows and cryopreservation of fish embryos is still problematic, cryopreservation of fish sperm is increasingly used as a tool to preserve genetic material to conserve genetic diversity. Successful cryopreservation protocols have been developed for the sperm of some endangered species such as razorback sucker, mekong giant catfish (*Pangasius hypophthalmus*), and apache trout (*Oncorhynchs apache*; David et al., 2000; Mongkonpunya et al., 1992; Tiersch, et al., 1998). Fish produced with cryopreserved sperm have been reared, and sperm cryopreservation has played an important role in the recovery programs of these species. Cryopreservation of the sperm of endangered cichlid species has also been studied.

14.2.5 ANDROGENESIS

Androgenesis is a method used for regenerating individuals with exclusively paternal inheritance of nuclear genome information. This means that the female gamete serves only as a carrier of an embryo, which is genetically originated entirely from the sperm. The three steps of induced androgenesis are the following: inactivation of the egg genome by irradiation; fertilization of the egg with intact sperm, which results in a haploid embryo; and diploidization of the embryo by either a heat or a pressure shock, which blocks the first mitotic division. As cryopreservation of fish eggs or embryos is still problematic, androgenesis is one of the alternative solutions for the preservation of rare, endangered, or near-extinct species, provided that the cryopreserved sperm of the given species and the eggs of closely related species are available. Apart from its advantages in species conservation, this technique has a very well-defined purpose in classical animal breeding, including aquaculture: Entirely homozygous individuals can be produced in just one generation, which could only be achieved after tens of generations of continuous inbreeding by natural mating.

Different methods of androgenesis have been used to regenerate fish using fresh sperm of several teleost species including rainbow trout (*Oncorhynchus mykiss*; Parsons and Thorgaard, 1984), common carp (*Cyprinus carpio*; Bongers et al., 1994), white sturgeon (*Acipenser transmountanus*; Kowtal, 1987), Siberian sturgeon (*Acipenser baeri Brandt*; Grunina and Neifakh, 1991), zebrafish (*Danio rerio*; Corley-Smith et al., 1996), and loach (*Misgurnus anguillicaudatus*; Arai et al., 1995). The androgenesis-derived broodstock of rainbow trout was established, and cloned lines were derived from androgenetic founders (Scheerer et al., 1991). Bercsényi et al. (1998) reported the hatching of goldfish from common carp eggs employing interspecific androgenesis using two cyprinid species. Androgenesis was also induced using cryopreserved fish sperm (Scheerer et al., 1991; Urbányi et al., 1999). However, the poor survival of the androgenetic offspring (typically <20% at hatching stage) is considered as the most important limiting factor of the technique for application to genetic selection programs and recovery of genotypes from cryopreserved sperm (Thorgaar and Cloud, 1993). In androgenesis studies, the following parameters have to be optimized to achieve good survival: irradiation source and dosage, incubation temperature of the fertilized eggs, and conditions for heat or pressure shock.

14.3 CRYOPRESERVATION OF FISH EGGS AND EMBRYOS

Although successful cryopreservation of fish spermatozoa has been achieved for many species, cryopreservation of fish embryos has proved to be a particularly difficult problem. Successful cryopreservation of embryos is important because the biodiversity of both paternal and maternal genomes will be preserved. In the last 20 years, attempts to cryopreserve fish eggs and embryos have been conducted on over 10 species including herring (*Clupea harengus*; Ben-Amotz and Rosenthal, 1981; Whittingham and Rosenthal, 1978), rainbow trout (*Salmo mykiss*; Haga, 1982), brown trout (*Salmo trutta*; Erdahl and Graham, 1980), brook trout (*Salvelinus fontinalis*; Zell, 1978), coho salmon (*Onorhynchus kisutch*; Stoss and Donaldson, 1983), Atlantic salmon (Salmo salar; Harvey and Ashwood-Smith, 1982; Zell, 1978), common carp (*Cyprinus carpio*; Zhang et al., 1989), medaka (*Oryzias latipes*; Onizuka et al., 1984), African catfish (*Clarias gariepinus*;

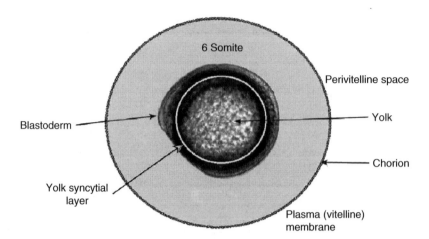

FIGURE 14.1 Image of a typical zebrafish embryo at six-somite stage, showing the structure of the embryos. The yolk syncytial layer would not be visible in this image, but its position is included for clarity. (Modified from Hagedorn et al., *Cryobiology*, 34, 251–263, 1996).

Magyary et al., 1996), turbot (*Scophthalmus maximus*; Cabrita, et al., 1999), and zebrafish (*Danio rerio*; Harvey, 1983; Zhang and Rawson, 1996a; Zhang et al., 1993). Eggs or embryos from all species have shown some degree of tolerance to subzero temperatures, but the successful cryopreservation of teleost eggs and embryos has remained elusive. There is very limited published information on the cryopreservation of ovaries or unfertilized fish eggs.

14.3.1 PROBLEMS ASSOCIATED WITH CRYOPRESERVATION OF FISH EMBRYOS

Cryopreservation of fish embryos has posed several problems associated with injuries induced during the cooling and thawing processes and four characteristics in particular have been identified as possibly being responsible for this.

14.3.1.1 The Large Size of Fish Embryos

The fertilized eggs of most fish species are greater than 1 mm in diameter. Previously, successfully cryopreserved eggs/embryos have been all considerably smaller (mammalian eggs, 70 to 150 μm; invertebrate eggs, <300 μm in diameter). This large size results in a very low surface-area-to-volume ratio. One consequence of such a low ratio is a reduction in the rate at which water and cryoprotectants can move into and out of the embryo during the steps of cryopreservation (Mazur, 1984; see also Chapters 1 and 2).

14.3.1.2 Complex Membrane Systems and Low Permeability of Fish Embryos

Fish have evolved to occupy all but a few types of natural aquatic environments ranging from those of low ionic strength (near-distilled water) freshwaters to those with salinities of 80 to 142.4‰ (Parry, 1966). Fish are osmoregulators, hyperosmotic regulators (most freshwater fish maintain body fluid concentration above that of the external environment), and hypoosmotic regulators (most marine fish maintain body fluid concentration bellow that of the external sea water). In the teleost eggs, two distinct membranes are recognized; namely, the outer chorionic membrane and the inner plasma (vitelline) membrane (Figure 14.1). Studies of the fine structure of the chorion (Hart and Donovan, 1983; Kalicharan et al., 1998; Zhang et al., 1993) of zebrafish eggs showed the chorion to be a thin envelope constructed of three distinct zones: an outer, electron-dense zone containing pore canals rich in polysaccharides (Tesoriero, 1977); a middle fibrillar zone; and an inner zone of lower electron density generally believed to be rich in proteins. Such a structure may play a role in diffusive exchange of gases as well as providing physical protection (Grierson and Neville, 1981). It also plays a role as a flexible filter for transport of some materials (Toshimori and Yasuzumi, 1976) and protects against microorganisms (Schoots et al., 1982). Studies on the chorion permeability (Hisaoka, 1958; Zhang and Rawson, 1996b) of the zebrafish eggs demonstrated it to be a porous membrane freely permeable to water, electrolytes and a range of cryoprotectants such as methanol, Me_2SO, ethylene glycol, and propane-1,2-diol.

The plasma membrane of the teleost embryos is a conspicuous and prominent region, and the organization of this membrane includes an actin-containing cytoskeleton that may function to maintain the shape of the egg and its surface specialization (Hart and Collins, 1990). The plasma membrane permits gas exchange, but it appears to be relatively impervious to most solutes (Heming and Buddington, 1988). Studies of the plasma membrane of zebrafish embryos have shown that the permeability of this membrane to water and most cryoprotectants is low (Zhang and Rawson, 1996b, 1998).

Studies on zebrafish embryos have shown that in addition to chorion and plasma membrane, a third layer develops after fertilization and the onset of embryo division by the nucleation of the cytoplasmic layer surrounding the yolk, resulting in the formation of the yolk syncytial layer (Kimmel and Law, 1986). The yolk syncytial layer has also been reported to account for the observed low cryoprotectant permeability of these embryos (Hagedorn et al., 1996). Fish embryo membrane permeability parameters are among the lowest to be reported for embryos (Hagedorn et al., 1997; Zhang and Rawson, 1996b, 1998). The plasma (vitelline) membrane and the underlying syncytial layer are believed to constitute the main permeability barrier to water and cryoprotectants (Hagedorn et al., 1996; Zhang and Rawson, 1996b). The mechanisms of osmotic and ionic regulation in fish embryos are not entirely clear (see Mazur, this volume, discussion on *Drosophila* embryos).

14.3.1.3 High Chilling Sensitivity of Fish Embryos

Stage-dependent chilling sensitivity has been reported for many species of fish embryos, including brown trout (Maddock, 1974), rainbow trout (Haga, 1982), carp (Dinnyes et al., 1998; Jaoul and Roubard, 1982; Roubaud et al., 1985), fathead minnows (Cloud et al., 1988), goldfish (Liu et al., 1993), and zebrafish (Zhang and Rawson, 1995). Most of these studies reveal that developmental stages beyond 50% epiboly are less sensitive to chilling (Figure 14.2), but that the chilling sensitivity accelerates rapidly at subzero temperatures. The reason for the extent of the stage-dependent chilling sensitivity might be related to the changes in cell and tissue types, number of cells, effectiveness of repair mechanisms, and enzymatic reactions. Although the presence of cryoprotectants somewhat mitigates against the chilling injury in fish embryos (Dinnyes et al., 1998; Zhang and Rawson, 1995), the reduction in subzero chilling injury is limited. The exact physiological process responsible for the chilling injury in fish embryos is not well understood. Mazur et al. (1992) hypothesized that the high chilling sensitivity of these embryos may result from the loss of synchrony of coupled reactions involved in embryological development. Studies involving mammalian embryos indicate that a high sensitivity to chilling injury is associated with large amounts of intraembryonic lipids (Nagashima et al., 1994; Toner et al., 1986), which are also commonly present in the yolk and cell compartments of fish embryos.

14.3.1.4 The Two-Compartment Nature of Fish Embryo with a High Yolk Content

Dechorionated fish embryos have two compartments: the blastoderm and the yolk. The shape and volume of the blastoderm (developing part of the embryo) change rapidly throughout embryo development. Because fish embryo membranes are relatively impervious to most solutes, the majority of fish embryos are dependent on endogenous yolk reserve to supply the substrates for energy production and growth. During cryopreservation, the yolk appears to act as an independent

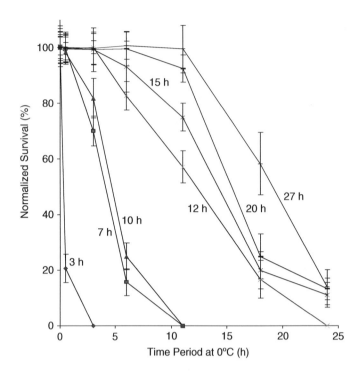

FIGURE 14.2 Survival of 3-, 7-, 10-, 12-, 15-, 20-, and 27-h stage zebrafish embryos at 0°C. Embryos were held in a low-temperature bath at 0°C for up to 24 h. Embryo survival was normalized with respect to controls at 26°C, whose survival averaged 87.3%. (From Zhang and Rawson, *Cryobiology, 32*, 239–246, 1995. With permission.)

compartment and responds osmotically in a manner analogous to the cellular cytoplasm (Hagedorn et al., 1996). The development of a single effective protocol for cryoprotectant permeation and osmotic dehydration of both the yolk and cells of the embryos may be difficult because of the distinctive nature of the two compartments (Rall, 1993). The plasma membrane of gastrula stage embryos has recently been found to have three morphologically distinct regions, being prominently ridged and folded at the surface of the blastoderm, being smooth over the syncytial layer at the vegetal pole, and having an intermediate region between the animal and vegetal pole where folding develops in advance of the expanding blastodermal disc of cells (Rawson et al., 2000; Figure 14.3). These structural differences may provide some explanation of the different osmotic properties reported for blastoderm and yolk compartments (Hagedorn et al., 1996). The ridgelike folds covering the blastoderm surface provide a larger surface-to-volume ratio, which may confer greater membrane permeability to water and solutes.

14.3.2 Factors Considered in Fish Embryo Cryopreservation

To date, two approaches have been applied to the cryopreservation of biological materials: controlled slow cooling and vitrification (see also Chapter 1 and 22). However, neither approach has yet been successfully employed to cryopreserve fish embryos. With controlled slow cooling, the selection and the optimization of the following factors need to be considered.

14.3.2.1 Slow Cooling: Embryo Developmental Stage

Fish embryos develop rapidly. Following fertilization, the embryo develops as a blastodisc at the animal pole, and the periphery of the blastodisc overgrows the yolk (epiboly), eventually enclosing it to form a gastrula. The head and eye cups are soon identifiable, the heart functions, and the trunk lengthens and separates from the yolk sac. Before hatching, the embryo becomes very active and

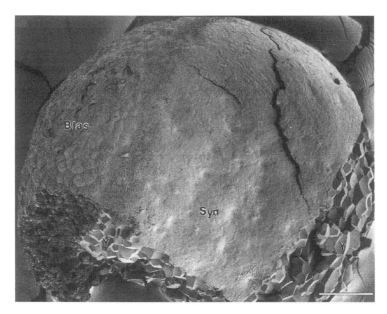

FIGURE 14.3 Field emission scanning electron microscopy (FE-SEM) view of the outer and fracture surfaces of a dechorionated gastrula stage zebrafish embryo, consisting of blastoderm (Blas) and syncytial layer (Syn) covering the yolk (Y). Bar = 100 μm (From Rawson et al., *Aquacult. Res.*, 31, 325–336, 2000. With permission.)

the chorion is softened as a result of enzymes secreted by hatching glands. During the yolk-sac period after hatching, the mouth, gut, and eyes became functional to allow the newly hatched larvae to switch from endogenous to exogenous nutrition. Cryopreservation studies have shown that intermediate embryo development stages between postgastrula and heartbeat have higher survival rates. Tail-bud stage for common carp (Zhang et al., 1989), posteyed stages for rainbow trout (Haga, 1982), and *six*-somite to heartbeat stage for zebrafish (Zhang et al., 1993) embryos showed better survival after cooling to –30°C. Factors need to be considered in choosing the appropriate embryo stage for cryopreservation including size of the embryo, membrane permeability, embryo chilling sensitivity, yolk size, and complexity of the embryos.

14.3.2.2 Slow Cooling: Cryoprotectants

The most commonly used cryoprotectants in fish egg and embryo preservation are Me_2SO, methanol, glycerol, ethanediol, and propane-1,2-diol, with typical concentration ranges of 1 to 2 *M* (Adam et al., 1995; Robertson et al., 1988; Zhang et al., 1993). Embryos are normally treated with cryoprotectants at room temperature or 0°C for 30 to 60 min in either one step or multistep additions. The selection of cryoprotectants and protocols are species related and will very much depend on their toxicity and the permeability to the embryos.

14.3.2.3 Slow Cooling: Cooling Rates

Slow-cooling rates normally result in higher survival for fish embryos as compared with high-cooling rates. The optimum cooling rates reported were in the range of 0.01 to 0.75°C/min (Harvey and Ashwood-Smith, 1982; Onizuka et al., 1984; Stoss and Donaldson, 1983; Zhang et al., 1993).

14.3.2.4 Slow Cooling: Thawing and Removal of Cryoprotectant

Optimum thawing rates reported for fish embryos vary: a slow rate (8°C/min) was optimal for common carp embryos (Zhang et al., 1989), an intermediate rate (43°C/min) was preferred for zebrafish blastoderm (Harvey, 1983), and a fast rate (300°C/min) resulted in best survival for zebrafish embryos (Zhang et al., 1993).

The use of sucrose in the diluting medium to allow controlled removal of cryoprotectant and to avoid osmotic damage was reported to improve the survival of common carp (Zhang et al., 1989) embryos after cooling, whereas no difference between postthaw handing methods on embryo survival was observed for intact zebrafish embryos (Zhang et al., 1993).

Studies on cryopreservation of fish embryos with controlled slow cooling have indicated that although embryos from all species show a certain degree of survival at subzero temperatures, they do not normally survive after cooling to –35°C. The loss of viability is mainly the result of intracellular ice formation, caused by low embryo membrane permeability to water and cryoprotectants, and embryo chilling sensitivity.

Cryopreservation of fish embryos using vitrification has so far only been performed with zebrafish embryos as a model system (Liu et al., 1998; Zhang and Rawson, 1996a). Despite differences in the methods used to produce osmotic dehydration, the factors that need to be considered with vitrification approach are very similar to those for controlled slow cooling.

14.3.2.5   Vitrification: Embryo Developmental Stage

Studies on vitrification of zebrafish embryos have shown that intermediate embryo developmental stages at 50% epiboly to prim-6 stages provided best results. There would be advantages for using early-stage embryos because their membrane systems are less complex and their membrane permeabilities appeared to be higher when compared with later-stage embryos (Zhang and Rawson, 1998). The main problems associated with these embryos are their sensitivity to chilling and cryoprotectants toxicity.

14.3.2.6   Vitrification: Stability and Toxicity of Vitrification Solutions

The formulation of the vitrification solutions for fish embryo cryopreservation must match the choice of cryoprotectants with the intrinsic permeability and toxicity properties of the embryos in question. Vitrification solutions DPP (2 M Me$_2$SO + 3 M propnane-1,2-diol + 0.5 M polyethylene glycol 400), BPP (2 M butane-2,3-diol + 3 M propane-1,2-diol + 6% polyethylene glycol 400), and 10 M methanol have been shown to be stable and have relatively low toxicities to zebrafish embryos.

14.3.2.7   Vitrification: Cooling Methods

To achieve the "glassy solid state" required, cooling rates must be sufficiently high to avoid ice crystallization in the vitrification solution/specimen. Commercially available plastic straws and electron-microscope gold grids in combination with liquid nitrogen or nitrogen slush have been applied for zebrafish embryos.

14.3.2.8   Vitrification: Thawing and Removal of Cryoprotectants

Rapid warming rates are required during thawing to avoid ice crystallization. Immersion into embryo medium at the normal culturing temperature has been used for thawing. To increase the thawing rate, the volume of vitrification solution is reduced to a minimum (<0.5 mL if plastic straws are used and <10 µL when grids are used). Immediately after thawing, the embryo is taken through a series of washing steps to remove cryoprotectants and reduce toxicity.

So far, the results obtained with the vitrification approach have shown no embryo survival, although approximately 80% of six-somite and prim-6 stage zebrafish embryos remained morphologically intact after freezing and thawing. The low membrane permeability of fish embryos to water and cryoprotectants remains the biggest problem, and approaches to overcome the problem need to be identified before effective vitrification protocols can be designed.

14.3.3 Approaches toward Successful Cryopreservation of Fish Embryos

14.3.3.1 Reduce Embryo Chilling Sensitivity

The high chilling sensitivity of early-stage fish embryos indicates that high cooling and warming rates and vitrification will be required when designing protocols for their cryopreservation. Studies have also shown that the high sensitivity of zebrafish embryos to chilling may be related to the high lipid content of the yolk and that chilling sensitivity can be reduced by reducing the amount of yolk in the embryos (Liu et al., 1999, 2001a). The survival of yolk-reduced embryos was found to be stage dependent, and later embryo developmental stages were better able to survive yolk removal. The yolk-reduced embryos showed significantly higher survival than that of chilled controls. Although embryos at these stages are highly differentiated, studies on invertebrate and insects larvae (Leopold, 1983; Lowrie, 1991) have shown that they can be successfully cryopreserved, especially when controlled slow cooling is considered.

14.3.3.2 Increase in Embryo Membrane Permeability

Identifying the best approaches to overcome the problems associated with the low membrane permeability of fish eggs/embryos is seen as vital to any further progress in their cryopreservation. Procedures for increasing embryo membrane permeability can be described as follows.

Cryoprotectants can be introduced into the yolk through deyolk procedures or microinjection. Studies on nucleation temperatures of intraembryonic water and cryoprotectant penetration in zebrafish embryos using differential scanning calorimetry (Liu et al., 2001b) showed a clear depression of intraembryonic nucleation temperatures following incorporation of cryoprotectant into the yolk through deyolk procedures. However, introducing cryoprotectants such as Me_2SO and propylene glycol into the yolk sac through microinjection did not improve the postthaw morphology of zebrafish embryo membranes (Janik et al., 2000).

Increased cryoprotectant penetration can also be achieved through permeablization of plasma membrane. Both chemical and physical methods can be used for increasing membrane permeability. Organic solvents were successfully used for the permeabilization of *Drosophila melanogaster* embryo membranes (Steponkus et al., 1990). Other chemical treatments include the use of sodium hypochlorite (Lynch et al., 1989) and enzymatic digestion using pronase (Simon et al., 1994). Recently, ultrasound was applied for the permeabilization of zebrafish embryos (Bart et al., 1999). Ionophores and electroporation have also been used to improve membrane permeability. Ideally, permeabilized fish embryos should be made sufficiently permeable to water and cryoprotectants such as methanol, Me_2SO, propylene glycol, and ethylene glycol, while retaining high viability.

14.3.4 Cryopreservation of Embryonic Cells

Although androgenesis using cryopreserved sperm has been successful with several teleost species, this technique does not overcome the lost of mitochondrial DNA, which is inherited maternally. As embryo cryopreservation is still elusive, one way to maintain the genetic diversity of both the nuclear genome and mitochondrial DNA of a fish strain is the cryopreservation of isolated blastomeres. After thawing, the cryopreserved embryonic cells could be transplanted into recipient blastulae to produce chimeras. Different methods have been successfully tested to produce fish chimeras by blastomere transplantation into blastula embryos (Hong et al., 1998; Nilsson and Cloud, 1992; Varadi and Horvath, 1997; Yamaha et al., 1997). In blastomere transplantation, dissociated embryonic cells from the donor embryo are generally injected into the deeper parts of the recipient embryo. It has been observed that the proportion of donor cells varied in the different organs of the chimeric fish, and that the contribution of the donor cells to the germ line chimeras is generally low (Yamaha et al., 1997).

Successful cryopreservation of isolated blastomere has been reported for several freshwater, estuarine and marine species including zebrafish (*Danio rerio*), rainbow trout (*Oncorhynchus mykiss*), common carp (*Cyprinus carpio*), medaka (*Oryzias latipes*), pejerrey (*Odontesthes bonariensis*), and whiting (*Sillago japonica*) (Harvey, 1983; Leveroni and Maisse, 1998, 1999; Strussmann et al., 1999). The postthaw survival rate of fish blastomeres was found to be stage dependent, and higher survival rates were obtained with older blastomeres. Me$_2$SO and propane-1,2-diol (1 to 2 *M*) have been successfully used for blastomere cryopreservation, and slow cooling was proved to be superior than fast cooling. High survival rates of 95 and 96% have been obtained for the 6C stage (72 h after fertilization) rainbow trout and for late-blastula stage common carp blastomeres respectively (Leveroni and Maisse, 1998, 1999).

14.4 CRYOPRESERVATION OF AQUATIC INVERTEBRATE

Cryopreservation of the germplasm of aquatic species has been mainly conducted with fish. However, studies have also been carried out with aquatic invertebrates including shrimp, crab, rotifer, polychaete, oyster, abalone, starfish, sand dollar, and sea urchin. Cryopreservation has been conducted with sperm as well as eggs, embryos, and larvae. Most of the studies have been carried out with oyster and shrimp. The cryoprotectants and cryopreservation procedures employed in these studies are not significantly different from those used for the cryopreservation of fish gamets and embryos. Successful semen cryopreservation has been reported for about 30 species of aquatic invertebrate (Gwo, 2000b). Limited success has been reported on the cryopreservation of eggs or embryos of aquatic species including the embryos of Pacific oyster (Lin and Chao, 2000), larvae of Eastern oyster (Paniagua et al., 1998b), embryos of hard clam (Chao et al., 1997), larvae of sea urchin (Naidenko and Kol'tsova, 1998), embryos of rotifer (Toledo and Kurokura, 1990), and juveniles of marine polychaetes (Olive and Wang, 1997).

14.4.1 CRYOPRESERVATION OF GAMETES, EMBRYOS, AND LARVAE OF OYSTER AND SHRIMP

The studies on cryopreservation of sperm, eggs, and larvae of oyster have been mainly conducted with Pacific oyster (*Crassotrea gigas*) (Bougrier and Rabenomanana, 1986; Gwo, 1995; Naidenko, 1997), other species such as *Crassotrea tulipa*, *Crassotrea iredalei*, *Saccostrea cucullata*, and species with commercial importance such as *Crassotrea virginica* (Paniagua et al., 1998a, 1998b; Yankson and Moyse, 1991; Zell et al., 1979). Successful cryopreservation of sperm from *Crassotrea gigas*, *Crassotrea virginica*, and *Crassotrea tulipa* has been reported. Me$_2$SO or a cryoprotectant mixture was used in these studies in combination with controlled-rate cooling, and fertility rates of 70 to 100% were achieved with the cryopreserved sperm (Kurokura et al., 1990; Yankson and Moyse, 1991). The majority of the studies on cryopreservation of oyster embryos and larvae were conducted in the laboratory and have yielded only a single report of oyster growth beyond the planktonoc stages (Paniagua et al., 1998b). Based on the technique developed in this study, the first commercial oyster seedstock has been recently established using cryopreserved larvae and eggs fertilized with thawed sperm (Paniagua-Chavez et al., 2000).

Successful cryopreservation of sperm from several freshwater and marine shrimp species has been reported, including *Macrobrachium rosenbergii*, *Sicyonia ingentis*, and *Penaeus vannam*; fertilization rates using cryopreserved sperm were between 43.5 and 70% (Anchordoguy et al., 1988; Chow et al., 1985; Dumont et al., 1992). Cryopreservation has been attempted on embryo and larvae of several penaeid shrimp species including *Penaeus japonicus*, *Penaeus vammamei*, *Penaeus monodon*, *Penaeus indicus*, *Penaeus orientalis*, and *Penaeus semisulcatus* (Arun and

Subramoniam, 1997; Diwan and Kandasami, 1997; Gwo and Lin, 1998; McLellan et al., 1992; Newton and Subramoniam, 1996; Zhang et al., 1992). Embryos and larvae of penaeid shrimp have been difficult to cryopreserve, and only limited success has been reported. McLellan et al. (1992) reported the recovery of shrimp embryos (*Penaeus vannamei* and *P. monodon*) from –30°C. Arun and Subramonia (1997) reported 85 to 95% survival of *P. monodon* larvae after they were recovered from final temperatures of –50 to –70°C. Gwo and Lin (1998) observed twitching movements and weak motility when these larvae were thawed from –196°C, but metamorphosis was not achieved. The problems associated with shrimp embryo cryopreservation are similar to those identified for fish embryos: Their membrane permeability to cryoprotectants is low, and the chilling sensitivity of these embryos appears to be high (Gwo and Lin, 1998; Simon et al., 1994).

14.4.2 CRYOPRESERVATION OF THE GERMPLASM OF OTHER AQUATIC INVERTEBRATES

Cryopreservation of aquatic invertebrate semen has also been carried out on sea urchin, sand dollar, starfish, abalone, polychaeta, and crab (Behlmer and Brown, 1984; Bury and Olive, 1993; Dunn and McLachlan, 1973; Jeyalectumie and Subramoniam, 1989; Matsunaga et al., 1983). Me_2SO is the most commonly used cryoprotectant for sperm cryopreservation in these species, and the typical concentration range used is 5 to 30% v/v. Controlled freezing in liquid nitrogen vapor has been widely used, and thawing is normally carried out in a water bath of 15 to 30°C. Visual estimation of sperm motility remains the major laboratory measure currently available for evaluating invertebrate sperm viability. Fertilization rates are also used for some species, such as sea urchin, in which eggs are readily available (Gwo, 2000b). Various levels of motility have been reported for cryopreserved invertebrate sperm, ranging from <5% for horseshoe crab (*Limulus polyphemus*; Behlmer and Brown, 1984) to 95% for sea urchin (*Strongylocentrous drobachiensis*; Dunn and Mclanchlan, 1973).

Cryopreservation studies on eggs, embryos, and larvae have been conducted on marine invertebrates such as blue mussel (Toledo et al., 1989), sea urchin (Asahina and Takahashi, 1979; Gakhova et al., 1988), barnacle (Gakhova et al., 1991), hard clam (Chao et al., 1997), small abalone (Lin and Chao, 2000), rotifers (Toledo et al., 1991), king scallop (Renard and Cochard, 1989), and juveniles of polychaete (Olive and Wang, 1997). Survival of most of these species has been reported after freezing to cryogenic temperatures, but only a fraction of the surviving embryos develop normally. However, a protocol allowing the mass preservation of fully differentiated nechtochaete larvae of the polychaete *Nereis virens* has been developed recently (Olive and Wang, 1997). The most successful levels of preservation and recovery were achieved using a progressive cooling of larvae equilibrated with 10% Me_2SO in seawater. Larvae were cooled at 2.5°C/min from room temperature to –35°C in 0.5-mL straws before being plunged into liquid nitrogen (Olive and Wang, 2000).

REFERENCES

Adam, M.M., Rana, K.J., and McAndrew, B.J. (1995) Effect of cryoprotectants on activity of selected enzymes in fish embryos, *Cryobiology*, 32, 92–104.

Anchordoquy, T., Crowe, J.H., Griffin, F.J., and Clark, W.H., Jr. (1988) Cryopreservation of sperm from marine shrimp *Sicyonia-ingentis*, *Cryobiology*, 25(3), 238–243.

Arai, K., Ikeno, M., and Suzuki, R. (1995) Production of androgenetic diploid loach *Misgurnus anguillicaudatus* using spermatozoa of natural tetraploids, *Aquaculture*, 137, 134–138.

Arun, R. and Subramoniam, T. (1997) Effect of freezing rates on the survival of penaeid prawn larvae: A parameter analysis, *Cryo-Letters*, 18, 359–368.

Asahina, E. and Takahashi, T. (1979) Cryopreservation of sea urchin embryos and sperm, *Dev. Growth Differentiation*, 21, 423–430.

Barbazuk, W.B., Korf, I., Kadavi, C., Heyen, J., Tate, S., Wun, E., Bedell, J.A., McPherson, J.D., and Johnson, S.L. (2000) The syntenic relationship of the zebrafish and human genomes, *Genome Res.*, 10, 1351–1358.

Barros, C., Muller, A., Wood, M.J., and Whittingham, D.G. (2001) High survival of spermatozoa and pluteus larvae of sea urchin frozen in Me$_2$SO, *Cryobiology*, 34, 341–342.

Bart, A.N., Kindschi, G., Ahmed, H., Clark, J., Young, J., and Zohar, Y. (2001) Enhanced diffusion of calcein into rainbow trout, *Oncorhynchus mykiss* larvae, using cavitation level ultrasound, *Aquaculture,* 196, 189–197.

Behlmer, S.D. and Brown, G. (1984) Viability of cryopreserved spermatozoa of the horseshoe crab, *Limulus polyphermus*, L, *Int. J. Invertebrate Reprod. Dev.*, 7, 193–199.

Ben-Amotz, A. and Rosenthal, H. (1981) Cryopreservation of marine unicellular algae and early life stage of fish for use in mariculture, *Eur. Maricult. Soc. Special Publ.*, 6, 149–162.

Bercsényi, M., Magyary, I., Urbányi, B., Orbán, L., and Horváth, L. (1998) Hatching out goldfish from commnn carp eggs: Interspecific androgenesis between two cyprinid species, *Genome*, 41, 573–579.

Bhavanishankar, S. and Subramoniam, T. (1997) Cryopreservation of spermatozoa of the edible mud crab *Scylla serrata* (Forskal), *J. Exp. Zool.*, 277, 326–336.

Billard, R. (1983) Ultrastructure of trout spermatozoa: Changes after dilution and deep-freezing, *Cell Tissue Res.*, 228, 205–218.

Billard, R. and Zhang, T. (2001) Techniques of genetic resource banking in fish, in *Cryobanking the Genetic Resource: Wildlife Conservation for the Future?* Watson, P.F and Holt, W.V., Eds., Taylor & Francis, London, 145–170.

Blaxter, J.H.S. (1953) Sperm storage and cross-fertilisation of spring and autumn spawning herring, *Nature (Lond.)*, 172, 1189–1190.

Bolla, S., Holmefjord, I., and Refsite, T. (1987) Cryogenic preservation of Atlantic halibut sperm, *Aquaculture*, 65, 371–374.

Bongers, A.B.J., Veld, E.P.C., Abo-Hashema, K., Bremmer, I.M., Eding, E.H., Komen, J., and Richter, C.J.J. (1994) Androgenesis in common carp (*Cyprinus carpio* L.) using UV irradiation in a synthetic ovarian fluid and heat shocks, *Aquaculture*, 122, 119–132.

Bougrier, S. and Rabenomanana, L.D. (1986) Cryopreservation of spermatozoa of the Japanese oyster *Crassostrea gigas*, *Aquaculture*, 58, 277–280.

Brownlie, A., Donovan, A., Pratt, S.J., Paw, B.H., Oates, A.C., Brugnara, C., Witkowska, H.E., Sassa, S., and Zon, L.I. (1998) Positional cloning of the zebrafish sauternes gene: A model for congenital sideroblastic anaemia, *Nat. Genet.*, 20, 244–250.

Bury, N.R. and Olive, P.J.W. (1993) Ultrastructural observations on membrane-changes associated with cryopreserved spermatozoa of two polychaete species and subsequent mobility induced by quinacine, *Invertebrate Reprod. Dev.*, 23, 139–150.

Cabrita, E., Chereguini, O., Luna, M.J., Polo, S., and Herráez, P. (1999) Effect of different permeability agents in the concentration of Me$_2$SO in turbot embryos, *Cryobiology*, 39, 346.

Chao, N.-H., Lin, T.-T., Chen, Y.-L., Hsu, H.-W., and Liao, I.-C. (1997) Cryopreservation of early larvea and embryos in oyster and hard clam, *Aquaculture*, 155, 31–44.

Chow, S., Taki, Y., and Ogasawara, Y. (1985) Cryopreservation of spermatophore of the fresh water shrimp *Macrobrachium rosenbergii, Biolog. Bull.* (Woods Hole), 168(3), 1985.

Clark, W.H., Jr., Chen, T.I., Pillai, M.C., Uhlinger, K., and Shoffner-McGee, J. (1996) The biology of gamete activation and fertilization in Penaeidea: Present knowledge and future directions, *Bull. Inst. Zool. Aced. Sinica Monogr.*, 16, 553–571.

Cloud, J.G., Erdahl, A.L., and Graham, E.F. (1988) Survival and continued normal development of fish embryos after incubation at reduced temperatures, *Trans. Am. Fisheries Soc.*, 117, 503–506.

Corley-Smith, G.E., Lim, C.J., and Brandhorst, B.P. (1996) Production of androgenetic zebrafish (*Danio rerio*), *Genetics*, 142, 1265–1276.

David, R.E., Wirtanen, L.J., and Ternes, M.A. (2000) Cryopreservation of the endangered Apache Trout, in *Cryopreservation in Aquatic Species*, Tiersch, T.R. and Mazik, P.M., Eds., The World Aquaculture Society, Baton Rouge, LA, pp. 104–107.

Dinnyes, A., Urbanyi, B., Baranyai, B., and Magyary, I. (1998) Chilling sensitivity of carp (*Cyprinus carpio*) embryos at different developmental stages in the presence or absence of cryoprotectants: Work in progress, *Theriogenology*, 50, 1–13.

Diwan, A.D. and Joseph, S. (2000) Cryopreservation of spermatophores of the marine shrimp *Penaeus indicus* H. Milne Edwards, *J. Aquacult. Tropics*, 15, 35–43.

Diwan, A.D. and Kandasami, K. (1997) Freezing of viable embryos and larvae of marine shrimp, Penaeus semisulcatus de haan, *Aquacult. Res.*, 28, 947–950.

Drokin, S.I. (1993) Phospholipid distribution and fatty acid composition of phosphatidylcholine and phosphatidylethanolamine in sperm of some freshwater and mariane species of fish, *Aquat. Living Res.*, 6, 49–56.

Dumont, P., Levy, P., Simon, C., Diter, A., and Francois, S. (1992) Freezing of sperm ball of the marine shrimp *Penaeus vannamei,* Abstr. Wkshp. Gamete and Embryo Storage and Cryopreservation in Aquatic Organisms, Marly Le Roi, France.

Dunn, R.S. and McLachlan, J. (1973) Cryopreservation of echinoderm sperm, *Can. J. Zool.*, 51, 666–669.

Erdahl, D.A. and Graham, E.F. (1980) Preservation of gametes of freshwater fish, in *Proc. Int. Congress Anim. Reprod. Artificial Insemin.*, 317–326.

Gakhova, E.N., Krasts, I.V., Naidenko, T., Savel, N.A., and Bessonov, B.I. (1988) Embronic development of the sea urchin after low temperature preservation, *Ontogenez*, 19, 175–180.

Gakhova, E.N., Korn, O.M., and Butsuk, S.V. (1991) Freezing of larvae of *Balanus improvisus* to –196°C, *Biol. Morya*, 4, 62–65.

Gilpin, M.E. and Soule, M.E. (1986) Minimum viable populations: Processes of species extinction, in *Conservation Biology: The Science of Scarity and Diversity*, Soule, M.E., Ed., Sinauer Associates, Sunderland, MA, pp. 19–34.

Grierson, J.P. and Neville, A.C. (1981) Helicoidal architecture of fish egg shell, *Tissue Cell*, 13, 819–830.

Grunina, A.S. and Neifakh, A.A. (1991) Induction of diploid androgenesis in the siberian sturgeon *Acipenser baeri* Brandt, *Dev. Genet.*, 22, 20–23.

Gwo, J.-C. (1995) Cryopreservation of oyster *Crassostrea gigas* embryos, *Theriogenology*, 43, 1163–1174.

Gwo, J.-C. (2000a) Cryopreservation of sperm of some marine fishes , in *Cryopreservation in Aquatic Species*, Tiersch, T.R. and Mazik, P.M., Eds., The World Aquaculture Society, Baton Rouge, LA, pp. 91–100.

Gwo, J.-C. (2000b) Cryopreservation of aquatic invertebrate seman: A review, *Aquacult. Res.*, 31, 259–271.

Gwo, J.-C., and Lin, C.H. (1998) Preliminary experiments on cryopreservation of penaeid shrimp *Penaeus japonicus* embryos, nauplii and zoea, *Theriogenology*, 49, 1289–1299.

Haga, Y. (1982) On the subzero temperature preservation of fertilised eggs of rainbow trout, *Bull. Jpn. Soc. Sci. Fisheries*, 48, 1569–1572.

Hagedorn, M., Hsu, E.W., Pilatus, U., Wildt, D., Rall, W.F., and Blackband, S.J. (1996) Magnetic resonance microscopy and spectroscopy reveal kinetics of cryoprotectant permeation in a multicompartmental biological system, *Proc. Natl. Acad. Sci. USA*, 93, 7454–7459.

Hagedorn, M., Kleinhans, F.W., Wildt, D., Rall, W.F. (1997) Chilling sensitivity and cryoprotectant permeability of dechorionated zebrafish embryos, *Brachydanio rerio*, *Cryobiology*, 34, 251–263.

Hart, N.H. and Collins, G.C. (1990) An electron-microscope and freeze-fracture study of the egg cortex of *Brachydanio rerio, Cell Tissue Res.*, 265, 317–328.

Hart, N.H. and Donovan, M. (1983) Fine structure of the chorion and site of sperm entry in the egg of *Brachydanio*, *J. Exp. Zool.*, 227, 277–296.

Harvey, B. (1983) Cooling of embryonic cells, isolated blastoderm and intact embryos of the zebrafish *Brachydanio rerio* to –196°C, *Cryobiology*, 20, 440–447.

Harvey, B. and Ashwood-Smith, M.J. (1982) Cryoprotectant penetration and supercooling in the eggs of samonid fishes, *Cryobiology*, 19, 29–40.

Heming, T.A., and Buddington, R.K. (1988) Yolk absorption in embryonic and larval fishes, in *Fish Physiology*, Hoar, W.S. and Randall, D.J., Eds., Academic Press, London, pp. 407–446.

Hong Y., Winkler C., and Schartl M., 1998. Production of medakafish chimeras from stable embryonic stem cell line, *Prod. Natl. Acad. Sci. USA*, 95, 3679–3684.

Hisaoka, K.K. (1958) Microscopic studies of the teleost chorion, *Trans. Am. Micro. Soc.*, 77, 240–243.

Horton, H.F. and Ott, A.G. (1976) Cryopreservation of fish spermatozoa and ova, *J. Fish. Res. Board Can.*, 33, 995–1000.

Janik, M., Kleinhans, F., and Hagedorn, M. (2000) Microinjection of cryoprotectants into the yolk of zebrafish embryos (Brachydanio rerio), *Biol. Reprod.*, 62, 146.

Jaoul, A., and Roubard, P. (1982) Resistance de l de carpe commune (*Cyprinus carpio L. Cypprinidae*) a des chocs thermiques chauds ou froids, *Can. J. Zool.*, 60, 3409–3419.

Jeyalectumie, C. and Subramoniam, T. (1989) Cryopreservation of spermatophores and seminal plasma of the edible crab *Scylla serrata*, *Biol. Bull.*, 177, 247–253.

Kimmel, C.B. and Law, R.D. (1986) Cell lineage of zebrafish blastomeres, *Dev. Biol.*, 108, 86–93.

Kalicharan, D., Jongebloed, W.L., Rawson, D.M., and Zhang, T.T. (1998) Variation in fixation techniques for field emission SEM and TEM of zebrafish (*Brachydanio rerio*) embryo inner and outer membranes, *J. Electr. Microsc.*, 47, 645–658.

Kirchofer, A. (1996) *Conservation of Endangered Freshwater Fish in Europe*, Birkhäuser, Basel, 341.

Kowtal, G.V. (1987) Preliminary experiments in induction of polyploidy, gynogenesis and androgenesis in the white sturgeon, *Acipenser transmontanus* Richardson, in *Proceedings World Symposium Selection, Hybridisation and Genetic Engineering in Aquaculture*, Vol. 2, Teiws, K., Ed., Heinemann, Berlin, pp. 317–324.

Kurokura, H., Namba, K., and Ishikawa, T. (1990) Lesions of spermatozoa by cryopreservation in oyster *Crassostrea gigas*, *Nippon Suisan Gakkaishi*, 56, 1803–1806.

Labbe, C., Maisse, G., and Billard, R. (1998) Fish sperm cryopreservation: What has to be improved, in *Proceedings of 33rd International Symposium on New Species for Mediterranean Aquaculture*, Alghero, April, pp. 97–110.

Lahnsteiner, F. (2000) Cryopreservation protocols for sperm of salmonid fishes, in *Cryopreservation in Aquatic Species*, Tiersch, T.R. and Mazik, P.M., Eds., The World Aquaculture Society, Baton Rouge, LA, pp. 91–100.

Lahnsteiner, F., Weismann, T., and Patzner, R.A. (1995) A uniform method for cryopreservation of semen of salmonid fishes (*Oncorhynchus mykiss, Salmo trutt* f. *fario, Salmo trutt* f. *lacustris, Coregonus* sp.), *Aquacult. Res.*, 26, 801–807.

Lahnsteiner, F., Weismann, T., and Patzner, R.A. (1997) Methanol as cryoprotectant and the suitability of 1.2ml and 5ml straws for cryopreservation of semen from salmonid fishes, *Aquacult. Res.*, 28, 471–479.

Legendre, M. and Billard, R. (1980) Cryopreservation of rainbow trout sperm by deep freezing, *Reprod. Nutr. Dév.*, 20, 1859–1868.

Leopold, R.A. (1983) Cryopreservation of third-stage larvae of *Brugia malayi* and *Dipetalonema viteae, Am. J. Trop. Med. Hyg.*, 32, 767–771.

Leveroni, S. and Maisse, G. (1998) Cryopreservation of rainbow trout blastomeres: Influence of embryo stage on post-thaw survival rate, *Cryobiology*, 36, 255–262.

Leveroni, S. and Maisse, G. (1999) Cryopreservation of carp (*Cyprinus carpio*) blastomeres, *Aquat. Living Resour.*, 12, 71–74.

Lin, T.-T. and Chao, N.-H. (2000) Cryopreservation of eggs and embryos of shellfish, in *Cryopreservation in Aquatic Species*, Tiersch, T.R. and Mazik, P.M., Eds., The World Aquaculture Society, Baton Rouge, LA, pp. 240–250.

Liu, K., Chou, T., and Lin, H. (1993) Cryosurvival of goldfish embryos after subzero freezing, *Aquat. Living Resour.*, 6, 145–153.

Liu, X.-H., Zhang, T., and Rawson, D.M. (1998) Feasibility of vitrification of zebrafish (*Danio rerio*) embryos using methanol, *Cryo-Letters*, 19, 309–318.

Liu, X.-H., Zhang, T., and Rawson, D.M. (1999) The effect of partial removal of yolk on the chilling sensitivity of zebrafish (*Danio rerio*) embryos, *Cryobiology*, 39, 236–242.

Liu, X.-H., Zhang, T., and Rawson, D.M. (2001a) Effect of cooling rate and partial removal of yolk on the chilling injury in zebrafish (*Danio rerio*) embryos, *Theriogenology*, 55, 1719–1731.

Liu, X.-H., Zhang, T., and Rawson, D.M. (2001b) Differential scanning calorimetry studies of intraembryonic freezing and cryoprotectant penetration in zebrafish (*Danio rerio*) embryos*, J. Exp. Zoology*

Lowrie, R.C. (1991) Cryopreservation of insect germplasm: Cells, tissues, and organisms, in *Insects at Low Temperatures*, Lee, R.E., Jr. and Denlinger, D.N., Eds., Chapman & Hall, New York.

Lynch, D.V., Lin, T.T., Myers, S.P., Leibo, S.P., MacIntyre, R.J. Pitt, R.E., and Steponkus, P.L. (1989) A two step method for permeabilization of Drosophila eggs, *Cryobiology*, 26, 445–452.

Maddock, B.G. (1974) A technique to prolong the incubation period of brown trout ova, *Progressive Fish-Cult.*, 36, 219–222.

Magyary, I., Dinnyes, A., Varkonyi, E., Szabo, R., and Varadi, L. (1996) Cryopreservation of fish embryos and embryonic cells, *Aquaculture*, 137, 103–108.

Magyary, I., Urbányi, B., Horváth, A., and Dinnyés, A. (2000) Cryopreservation of gametes and embryos of cyprinid fishes, in *Cryopreservation in Aquatic Species*, Tiersch, T.R. and Mazik, P.M., Eds., The World Aquaculture Society, Baton Rouge, LA, pp. 199–210.

Maisse, G. (1996) Cryopreservation of fish semen: A review, in *Refrigeration and Aquaculture Conference*, Bordeaux, 20–22 March, 1996, pp. 443–467.

Maisse, G., Pinson, A., and Loir, M. (1998) Caractérisation de l à la congélation du sperme de truite arc-en-ciel (*Salmo gairdneri*) par des critères physio-chimiques, *Aquat. Living Resour.*, 1, 45–51.

Matsunaga, H., Iwata, N., Kurokura, H., and Hirano, R. (1993) A preliminary study about cryopreservation of abalone sperm, *J. Faculty Appl. Biol. Sci. Hiroshima Univ.*, 22, 135–139.

Mazur, P. (1984) Freezing of living cells: Mechanisms and implications, *Am. J. Physiol.*, 247C, 125–142.

Mazur, P., Schneider, U., and Mahowald, A.P. (1992) Characteristics and kinetics of subzero chilling injury in *Drosophila* embryos, *Cryobiology*, 29, 39–68.

McAndrew, B.J., Rana, K.J., and Penman, D.J. (1995) Conservation and cryopreservation of genetic variation in aquatic organisms, *Recent Adv. Aquacult.*, 4, 295–337.

McLellan, M.R., MacFadzen, I.R.B., Morris, G.J., Martinez, G., and Wyban, A. (1992) Recovery of shrimp embryos (*Penaeus vannamei* and *Penaeus monodon*) from –30°C, in *Workshop on Gamete and Embryo Storage and Cryopreservation in Aquatic Organisms*, Bilard, R., Cosson, J., Fauvel, C., Loir, M., and Maisse, G., Eds., Merly le Roy, France, pp. 7.

Mims, S.D., Tsvetkova, L.I., Brown, G.G., and Gomelsky, B.I. (2000) Cryopreservation of sperm of sturgeon and paddlefish, in *Cryopreservation in Aquatic Species*, Tiersch, T.R. and Mazik, P.M., Eds., The World Aquaculture Society, Baton Rouge, LA, pp. 123–129.

Mongkonpunya, K., Pupipat, T., Pholprasith, S., Chantasut, M., Rittaporn, R., Pimolboot, S., Wiwatcharakoses, S., and Chaengkij, M. (1992) Cryopreservation of sperm of the mekong giant catfish *Pangasianodon gigas* (Chevey), in *Proceedings of a Network Meeting on Aquaculture*, National Academy Press, Washington, DC, pp. 56–60.

Mongkonpunya, K., Pupipat, T., and Tiersch, T.R. (2000) Cryopreservation of sperm of Asian catfish including the endangered mekong giant catfish, in *Cryopreservation in Aquatic Species*, Tiersch, T.R. and Mazik, P.M., Eds., The World Aquaculture Society, Baton Rouge, LA, pp. 108–116.

Morris, G.J. and Watson, P.F. (1984) Cold shock injury—A comprehansive bibliography, *Cryo-Letters*, 5, 352–372.

Mounib, M.S. (1978) Cryogenic preservation of fish and mammalian spermatozoa, *J. Reprod. Fertil.*, 53, 1–8.

Nagashima, H., Kashiwazaki, N., Ashman, R.J., Grupen, C.G., Seamark, R.F., and Nottle, M.B. (1994) Removal of cytoplasmic lipid enhances the tolerance of porcine embryos to chilling, *Biol. Reprod.*, 51, 618–622.

Naidenko, T. (1997) Cryopreservation of Crassostrea gigas oocytes, embryos and larvae using antioxidant echinochrome and an antifreeze protein AFP1, *Cryo-Letters*, 18, 375–382.

Naidenko, T. and Kol'tsova, E.A. (1998) Using antioxident echinochrome A for cryopreservation of sea urchin embryos and larvae, *Biol. Morya (Vladivostok)*, 24, 198–201.

Newton, S.S., and Subramoniam, T. (1996) Cryoprotectant toxicity in penaeid prawn embryos, *Cryobiology*, 33, 172–177.

Nilsson, E.E. and Cloud, J.G. (1992) Rainbow trout chimeras produced by injection of blastomeres into recipient blastulae, *Proc. Natl. Acad. Sci. USA*, 89, 9425–9428.

Olive, P.J.W. and Wang, W.B. (1997) Cryopreservation of *Nereis virens* (Polychaeta, Annelida) larvae: The mechanism of cryopreservation of a differentiated metazoan, *Cryobiology*, 34, 284–294.

Olive, P.J.W. and Wang, W.B. (2000) Cryopreservation of juveniles of a marine polychaete, in *Cryopreservation in Aquatic Species*, Tiersch, T.R. and Mazik, P.M., Eds., The World Aquaculture Society, Baton Rouge, LA, pp. 251–263.

Onizuka, N., Kator, K., and Egami, N. (1984) Mass cooling of embryos and fry of the fish (*Oryzias latipes*), *Cryobiology*, 21, 709–710.

Parry, G. (1966) Osmotic adaptation in fishes, *Biol. Rev. Camb. Philos. Soc.*, 41, 392–444.

Parsons, J.E. and Thorgaard, G.H. (1984) Induced androgenesis in rainbow trout, *J. Exp. Zool.*, 231, 407–412.

Paniagua, C., Buchanan, G., and Tiersch, T.R. (1998a) Effect of extender solutions and dilution on motility and fertilizing ability of eastern oyster sperm, *J. Shellfish Res.*, 17, 231–237.

Paniagua, C., Supan, J., Buchanan, J., and Tiersch, T.R. (1998b) Settlement and growth of eastern oyster produced from cryopreserved larvae, *Cryo-Letters*, 19, 283–292.

Paniagua, C.G., Buchanan, J.T., Supan, J.E., and Tiersch, T.R. (2000) Cryopreservation of sperm and larvae of the Eastern oyster, in *Cryopreservation in Aquatic Species*, Tiersch, T.R. and Mazik, P.M., Eds., The World Aquaculture Society, Baton Rouge, LA, pp. 230–239.

Perchec-Poupard, G., Paxion, C., Cosson, J., Jeulin, C. Fierville, F., and Billard, R. (1998) Initiation of carp spermatozoa motility and early ATP reduction after milt contamination by urine, *Aquaculture*, 160, 317–328.

Pullin, R.S.V. (1972) The storage of plaice (*Pleuronectes platessa*) sperm at low temperature, *Aquaculture*, 1, 273–283.

Rall, W.F. (1993) Advances in the cryopreservation of embryos and prospects for application to the conservation of salmonid fishes, in *Genetic Conservation of Salmonid Fishes*, Cloud, J.H. and Thorgarrd G.H., Eds., Plenum Press, New York, pp. 137–158.

Rana, K.J. and Gilmour, A. (1996) Cryopreservation of fish spermatozoa: Effect of cooling methods on the reproducibility of cooling rates and viability, in *Refrigeration and Aquaculture Conference*, Bordeaux, 20–22 March 1996, pp. 3–12.

Rawson, D.M., Zhang, T., Kalicharan, D., and Jongebloed, W.L. (2000) FE-SEM and TEM studies of the chorion, plasma membrane and syncytial layers of the gastrula stage embryo of the zebrafish (*Brachydanio rerio*): A consideration of the structural and functional relationship with respect to cryoprotectant penetration, *Aquacult. Res.*, 31, 325–336.

Renard, P. and Cochard, J.C. (1989) Effect of various cryoprotectants on Pacific oyster *Crassostrea giga* Thunberg, manila clam *Ruditapes philippinarum* Reeve and King scallop *Pecten maximus* (L.) embryos: Influence of the biochemical and osmotic effects, *Cryo-Letters*, 10, 169–180.

Robertson, S.M., Lawrence, A.L., Nell, W.H., Arnold, C.R., and McCarty, G. (1988) Toxicity of the cryoprotectants glycerol, dimethyl sulfoxide, ethylene glycol, methanol, sucrose, and sea salt solutions to the embryos of red drum, *Progressive Fish-Cult.*, 50, 148–154.

Roubaud, P., Chaillou, C., and Sjafei, D. (1985) Variations cycliques de la tolerance a un choc thermique froid applique au cours de la segmentation de l'embryon de la carpe commune (*Cyprinus carpio L.*), *Can. J. Zool.*, 63, 657–663.

Scheerer, P.D., Thorgaard, G.H., and Allendorf, F.W. (1991) Genetic analysis of androgenetic rainbow trout, *J. Exp. Zool.*, 260, 382–390.

Schoots, A.F.M., Stikkelbroeck, J.J.M., Bekhuis, J.F., and Denuce, J.M. (1982) Hatching in teleost fishes: Fine structure changes in the egg envelope during enzymatic breakdown *in vivo* and *in vitro*, *J. Ultra. Res.*, 80, 185–196.

Simon, C., Dumont, P., Cuende, F.X., Diter, A., and AQUACOP. (1994) Determination of suitable freezing media for cryopreservation of *Penaeus indicus* embryos, *Cryobiology*, 31, 245–253.

Steponkus, P.L., Myers, S.P., Lynch, D.V., Gardner, L., Bronshteyn, V., Leibo, S.P., Pitt, R.E., Lin, T.T., and MacIntyre, R.J. (1990) Cryopreservation of *Drosophlia melanogaster* embryos, *Nature*, 345, 170–172.

Stoss, J. (1983) Fish gamete preservation and spermatozoa physiology, in *Fish Physiology, Behavior and Fertility Control*, Vol. 9, Part B, Hoar, W.S., Randall, D.J., and Donaldson, E.M., Eds., Academic Press, Orlando, FL, pp. 305–350.

Stoss, J. and Donaldson, E.M. (1983) Studies on cryopreservation of eggs from rainbow trout (*salmo gairdneri*) and coho salmon (*oncorhynchus kisutch*), *Aquaculture*, 31, 51–65.

Stoss, J. and Refstie, T. (1983) Short-term storage and cryopreservation of milt from Atlantic salmon and sea trout, *Aquaculture*, 30, 229–236.

Strussman, C.A., Nakatsugawa, H., Takashima, F., Hasobe, M., Suzuki, T., and Takai, R. (1999) Cryopreservation of isolated fish blastomeres: Effect of cell stage, cryoprotectant concentration, and cooling rate on post-thaw survival, *Cryobiology*, 39, 252–261.

Suquet, M., Dorange, G., Omnes, M.H., Mormant, Y., Le Roux, A., Fauvel, C. (1993) Composition of seminal fluid and ultrastructure of the spermatozoa of turbot (*Scophthalmus maximus*), *J. Fish. Biol.*, 42, 509–516.

Tesoriero, J. (1977) Formation of the chorion (zona pellucida) in the teleost, *Oryzias latipes*. II polysaccharide cytochemistry of early oogenesis, *J. Histochem. Cytochem.*, 25, 1376–1380.

Thorgaard, G.H. and Cloud, J.G. (1993) Reconstitution of genetic strains of salmonids using biotechnological approaches, in *Genetic Conservation of Salmonid Fishes*, Cloud, J.H. and Thorgarrd G.H., Eds., Plenum Press, New York, pp. 189–196.

Tiersch, T.R. and Mazik, P.M. (2000) *Cryopreservation in Aquatic Species*, The World Aquaculture Society, Baton Rouge, LA.

Tiersch, T.R., Figiel, C.R., Wayman, W.R., Williamson, J.H., Carmichael, G.J., and Gorman, O.T. (1998) Cryopreservation of sperm of the endangered razorback sucker, *Trans. Am. Fisheries Soc.*, 127, 95–104.

Toledo, J.D., Kurokura, H., and Kasahara, S. (1989) Preliminary studies on the cryopreservation of the blue mussel embryos, *Nippon Suisan Gakkaishi*, 55, 1661.

Toledo, J.D., and Kurokura, H. (1990) Cryopreservation of the euryhaline rotifer *Brachionus-Plicatilis* embryos, *Aquaculture*, 91, 385–394.

Toledo, J.D., Kurokura, H., and Nakagawa, H. (1991) Cryopreservation of different strains of the euryhaline rotifer Brachionus plicatilis embryos, *Nippon Suisan Gakkaishi*, 57, 1347–1350.

Toner, M., Cravalho, E.G., Ebert, K.M., and Overstrom, E.W. (1986) Cryobiological properties of porcine embryos, *Biol. Reprod.*, 34, 98.

Toshimori, K. and Yasuzumi, F. (1976) The morphology and the function of the oocyte chorion in the teleost, *Plecoglossus altivelis*, *J. Electron Microscopy*, 25, 210.

Tsai, H.-P. and Chao, N.-H. (1994) Cryopreservation of small abalone (*Haliotis diversicolor*) sperm—Technique and its significance, *J. Fisheries Soc. Taiwan*, 21, 347–360.

Tsvetkova, L.I., Cosson, J., Linhart, O., and Billard, R. (1996) Motility and fertilizing capacity of fresh and frozen-thawed spermatozoa in sturgeons *Acipenser baeri* and *A. ruthenus*, *J. Appl. Ichthyol.*, 12, 107–112

Urbányi, B., Horváth, A., and Bercsényi, M. (1999) Androgenesis on sterlet (*Acipenser ruthenus*) using fresh and cryopreserved sperm, in *Proceedings of 6th International Symposium on Reproductive Physiology of Fish*, Bergen, Norway.

Varadi, L. and Horvath, L. (1997) Production of fish chimeras from embryonic cells, *Acta Biol. Hungarica*, 48, 95–104.

Watson, P.F. and Morris, G.J. (1987) Cold shock injury in animal cells, *Symp Soc. Exp. Biol.*, 41, 311–340.

Whittingham, D.G. and Rosenthal, H. (1978) Attempts to preserve herring embryos at subzero temperatures, *Arch. FishWiss*, 29, 75–79.

Yamaha, E., Mizuno, T., Hasebe, Y., Yamazaki, F. (1997) Chimeric fish produced by exchanging upper parts of blastoderms in goldfish blastulae, *Fish. Sci.*, 63, 514–519.

Yankson, K. and Moyse, J. (1991) Cryopreservation of the spermatozoa of Crassostrea tulipa and three other oysters, *Aquaculture*, 97, 259–267.

Zell, S.R. (1978) Cryopreservation of gametes and embryos of salmonid fishes, *Ann. Biol. Anim. Biochim. Biophys.*, 18, 1089–1099.

Zell, S.R., Bamford, M.H., and Hidu, H. (1979) Cryopreservation of spermatozoa of American oyster *Crassostrea virginica* Gmelin, *Cryobiology*, 16, 448–460.

Zhang, T.T. and Rawson, D.M. (1995) Studies on chilling sensitivity of zebrafish (*Brachydanio rerio*) embryos, *Cryobiology*, 32, 239–246.

Zhang, T.T. and Rawson, D.M. (1996a) Feasibility studies on vitrification of intact zebrafish (*Brachydanio rerio*) embryos, *Cryobiology*, 33, 1–13.

Zhang, T.T. and Rawson, D.M. (1996b) Permeability of the vitelline membrane of zebrafish (*Brachydanio rerio*) embryos to methanol and propane-1,2-diol, *Cryo-Letters*, 17, 273–280.

Zhang, T.T. and Rawson, D.M. (1998) Permeability of dechorionated 1-cell and 6-somite stage zebrafish (*Brachydanio rerio*) embryos to water and methanol, *Cryobiology*, 32, 239–246.

Zhang, T.T., Rawson, D.M., and Morris, G.J. (1993) Cryopreservation of pre-hatch embryos of zebrafish (*Brachydanio rerio*), *Aquat. Living Resour.*, 6, 145–153.

Zhang, J., Xu, W., Ren, H., Hua, Z., Zhao, L., Wang, Q., Liu, P., and Wang, Q. (1992), Toxicity and permeation of cryoprotective agents to Chinese prawn embryos, *Cryobiology*, 29, 762.

Zhang, X.S., Zhao, L., Hua, T.C., Chen, X.H., and Zhu, H.Y. (1989) A study on the cryopreservation of common carp (*Cyprinus carpio*) embryos, *Cryo-Letters*, 10, 271–278.

15 Fundamental Issues for Cell-Line Banks in Biotechnology and Regulatory Affairs

Glyn Stacey

CONTENTS

15.1 INTRODUCTION

For many years, animals and primary cell cultures from animal tissues have been used for research and for the manufacture and testing of vaccines. In the 1960s, human diploid fibroblast cell lines were developed for the manufacture of vaccines, and more recently, continuous cell lines from other species have been used in biotechnology. Notable examples include BHK cells for foot-and-mouth disease vaccine, CHO cells for production of recombinant proteins, and hybridoma cell lines for the production of monoclonal antibodies (Griffiths and Doyle, 1999; Stacey, 2000). The increased use of cell lines for the manufacture of a range of biological medicines has been driven by the need for more reproducible and safe cell substrates. In the 1950s, batches of polio vaccine were found to be contaminated with SV40 virus originating from the primary monkey kidney cells used in the manufacturing process. Although studies of vaccinees failed to show any direct adverse effects at the time, this incident in particular highlighted the need for safe and standardized cell substrates. Cell lines appeared to offer a solution but were susceptible to genetic variation and contamination if maintained by serial passage. Fortunately, mammalian cell lines are generally amenable to cryopreservation, and this made it possible to prepare reliable cell banks of homogenous aliquots that could be stored in a stable state at ultra-low temperature. Sample ampoules from the banks could then be thoroughly investigated for infectious agents before use, thus, enhancing the safety of vaccines produced from the banked cells. The benefits of standardized cell banks can be applied to all applications of cell lines, and cell-banking procedures are a vital preliminary step for any application of cell lines in which reproducibility and reliability are key issues.

Animal cell lines play an increasingly important role in the establishment of *in vitro* methods for diagnosis, prophylaxis, and treatment of human and animal diseases. They provide important substrates for biological assays of vaccine potency, vaccine efficacy, product toxicology, and detection of adventitious agents in products. In these applications, a primary driving force for use of cell lines is the international pressure to refine, reduce, and replace the use of the use of animals.

The principles of cell banking and associated safety testing and quality control are now enshrined in regulatory guidelines for the manufacture of products from animal cell substrates. They are also implicit in the development of cell lines used in processes subject to patent applications and in the provision of cell cultures from public collections for research and development. This chapter deals with key issues in the establishment, maintenance, and use of cryopreserved cell line banks.

15.2 FUNDAMENTAL ISSUES IN CELL CULTURE

Once a cell line has been established that has useful characteristics, its value is only sustained if the culture remains free of contaminating microorganisms (i.e., pure) and other cells (i.e., authentic) and shows stability of its characteristics on passage *in vitro*.

Cell culture growth media will readily support the growth of many bacteria and fungi, and thus, contamination with these organisms from the laboratory environment is often the cause of overwhelming infection and cell death. A more subtle form of contamination may occur due to mycoplasma, which are often closely associated with the cell membrane and do not produce the turbidity or colonies typical of bacterial and fungal infections, respectively. Mycoplasma are also smaller in size than most bacteria. The presence of these organisms may thus not be suspected, even when the cells are viewed by microscopy. Care must be taken to exclude environmental sources of microorganisms (e.g., via routine disinfection of waterbaths and sinks, regular laboratory cleaning regimes, exclusion of cardboard and other carriers of fungal spores) and to screen cultures for contaminants including mycoplasma (see following). The cell-doubling times of microorganisms are much shorter than for animal cells (i.e., 1 to 2 h vs. 20 to 30 h), and thus, trace contamination of cryopreserved stocks may result in overwhelming infection when thawed and recovered into culture. Therefore, wherever practicable, the use of antibiotics should be avoided, and especially when establishing cell banks for future use.

Cells may also be contaminated with viruses that can affect the characteristics of the culture or that may present a hazard to laboratory workers or patients treated with animal cell products. Viral contaminants may be introduced via growth media components of animal origin (notably bovine serum and porcine trypsin) or from the tissue of origin of the cell culture (Frommer et al., 1993; Erickson et al., 1991; Hallauer et al., 1971). Although viable virus titer may be reduced by freezing, the use of cryoprotectants to preserve cell cultures may also enhance the survival of cell-free virus and intracellular virus. In the case of cell-associated virus, snap-freezing of cells and other treatments that cause cell lysis can significantly increase the level of free viable virus in suspension on thawing. Freeze–thaw cycles are sometimes used to promote release of intracellular virus, and thus enhance virus detection.

Maintenance of cell cultures over extended periods by serial passage is a high-risk approach to provision of cells for both research and product manufacture. Even continuous cell lines that appear to be stable may show genotypic and phenotypic variation over extended periods of serial passage. Long-term passage also raises the risk of laboratory accidents, contamination with microorganisms,

TABLE 15.1
Publications Describing Cell Lines Not Matching Their Purported Origin

Reference	Cell Lines
Gartler (1967)	Breast cancer cell line cross-contamination
Culliton (1974)	HeLa cell contamination of cells worldwide
Nelson Rees et al. (1977)	Widespread cross-contamination of human breast tumor cell lines and others
Harris et al. (1981)	Putative human Hodgkins Disease cell lines cross-contaminated with nonhuman cells
De Benedetti et al. (1987)	Retraction of paper because of discovery of contaminated cells
Masters et al. (1988)	Cross-contamination of bladder cancer cell lines
van Helden et al. (1988)	Cross-contamination amongst esophageal squamous carcinoma cell lines
Chen et al. (1990)	TE671 shown to be a derived from RD cells
Drexler et al. (1993)	Cross-contamination of a leukemia cell line
Reid et al. (1995)	Cross-contamination of U937 cells
MacLeod et al. (1997)	Dami megakaryocytes found to be HEL erythroleukemia cells
Dirks et al. (1999)	ECV304 endothelial cells found to be T24 bladder cancer cells
MacLeod et al. (1999)	18% original human tumor cell lines cross-contaminated
Drexler et al. (1999)	16% human hematopoietic cell lines cross-contaminated

or cross-contamination with other cells. The latter occurrence has been well documented since the early days of cell culture, and Table 15.1 gives selected examples of reported cases of cross-contamination. Despite such frequent reports, the occurrence of cross-contaminated or mislabeled cell lines continues to be a serious scientific issue that receives insufficient attention (MacLeod et al., 1999; Stacey et al., 2000). The preparation of cryopreserved cell banks that are well characterized is central to resolving such concerns.

15.3 STANDARDIZATION

When a new cell line is established *de novo* or received in the laboratory, a viable stock of cryopreserved cells should be established at an early stage, and an initial stock of two to three ampoules will provide vital backup material in the event of accidental loss. Whether or not a cell line is likely to be used over an extended period of time, it is wise to expand cells from the initial stock of frozen cells to establish a master cell bank. It is also good practice to keep a careful note in laboratory records of early culture passages, including details of the culture medium and growth conditions. As the master cell bank will provide the reference point for all future work with a cell line, it should be well characterized and subjected to appropriate quality control tests. For future reference, it is useful to compile a record of all characterization performed on the master cell bank and other passage material. Such characterization may include cytogenetics, molecular investigations, morphology, biochemical functions, secreted products, and surface markers. Ampoules from the master bank are used to produce larger working cell banks that can be used for experimental or manufacturing purposes. The working cell bank should again be subjected to quality control. If prepared correctly, this tiered master/working bank system (Figure 15.1) can provide reproducible and reliable supplies of identical cultures over many decades.

The quality control tests that should performed as a matter of routine for all cell banks include viability (typically trypan blue dye exclusion), sterility (i.e., absence of bacteria and fungi), and testing for mycoplasma (Cord et al., 1992; Stacey and Stacey, 2000). These tests should be performed following a period of antibiotic-free culture to ensure that any contaminants that may be suppressed by antibiotics do not go undetected. Other investigations for authenticity (e.g., karyology, DNA fingerprinting, isoenzyme analysis, surface markers) and for the presence of viruses may be performed, but the exact profile of tests will depend on the type of cells involved and the intended use of the cells. For a general reference on cell banking and quality control, see Stacey

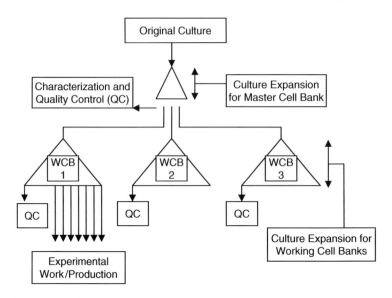

FIGURE 15.1 Preparation of master and working cell banks.

and Doyle (2000). Where cells are intended for use in the manufacture of medicines or as part of medical therapies to which patient tissues are directly exposed, a range of additional requirements are invoked that will be necessary for acceptance and licensing of any respective pharmaceutical or biological product (e.g., World Health Organization, 1998; International Conference on Harmonization, 1997). In particular, these requirements include testing for potential viral contaminants of the cells, which may have established persistent infection of a culture without any cytopathic effect. Such infections include human retroviruses (such as the case of the MT4 human leukemic cell line) and Epstein-Barr virus expressed by human B-lymphoblastoid cell lines, but a range of murine viruses (including some pathogenic for humans) have also been found in nonhuman cell lines (Nicklas et al., 1993). In addition, cell lines of mouse and hamster origin (e.g., mouse myeloma cell lines, L929, CHO) are known to express endogenous retroviruses. However, these are not considered to represent a serious health hazard, and their elimination can be demonstrated during downstream processing of cell culture products.

15.4 ESTABLISHMENT AND VALIDATION OF PROCEDURES

15.4.1 OPTIMIZATION

The preservation of mammalian cell cultures is generally assumed to be straightforward, and the most commonly used protocols involve dimethyl-sulphoxide at 5 to 10% v/v as cryoprotectant with a linear cooling rate of approximately $-1°C/min$ down to a terminal temperature of at least $-70°C$ before transfer to ultra-low-temperature storage. However, there are a number of issues in the cryopreservation of cell cultures in general that may need to be considered to ensure reliable recovery of thawed cells. When preserving an animal cell culture for the first time, it may be valuable to assess and validate the preservation method and its potential effects on the culture. A first step is to perform cryoprotection toxicity tests that will establish the optimum conditions (temperature, concentration, duration, etc.) for cryoprotection. The cooling profile may also be investigated using a cryomicroscope stage that enables direct visualization of cells as they cool and freeze. This may be particularly useful if the culture contains cells of variable morphology that may show differential survival on cooling. The optimized profiles can then be applied practically using the versatility of rate-controlled freezing equipment (see following). Subsequently, the viability and recovery of thawed cells can be investigated to establish the optimum recovery procedures (rate of warming, recovery medium, removal of cryoptoectants, etc.). The choice of a technique

TABLE 15.2
Viability Testing for Animal Cell Cultures

Method	Principle and Comments
Dye exclusion (e.g., trypan blue, Evans blue, naphthalene black)	Dyes that penetrate cells are excluded by the action of the cell membrane in viable cells; thus, cells containing no dye have functional membranes and are probably viable
	Rapid and usually easy to interpret
	Such methods overestimate viability, and apoptotic cells continue to have active membranes; thus, cells committed to this form of cell death may appear viable
Neutral red assay	The red dye is accumulated in the lysozomes of active cells and is measured by spectrophotometric analysis
	Commonly used in toxicology assays
	Relatively time-consuming, and incubation conditions should be optimized for each cell culture
3-(4,5-dimathylthiazol-2-yl)-2,5-diphenylterazolium (MTT) assay	Reduction of MTT is measured, and this is indicative of degree of biochemical activity
	Reduced metabolism in a short-term assay may be reversible and is therefore not necessarily related to cell viability
Fluorescein diacetate assay	The diacetate is split by active membrane esterases releasing flourescein, which cannot pass through the membrane, and being trapped in the cell produces flourescence in viable cells observed under ultraviolet light

for viability testing can be critical, and there are a number of methods used widely in animal cell culture (Table 15.2). Techniques such as trypan blue dye exclusion may significantly overestimate cell viability and may well assess cells committed to programmed cell death (apoptosis) as viable, as their membrane function is retained (Solis-Recendez et al., 1994). Viability assays are obviously only one indicator of the success of the preservation process, and functional assays should be employed to validate appropriate performance of recovered cells.

15.4.2 VALIDATION OF PRESERVATION PROCEDURES AND QUALIFICATION OF CELL BANKS

Homogeneity and reproducibility is a fundamental requirement of cell banks. The numbers of cells preserved in individual ampoules should be demonstrated as sufficient to regenerate cultures that faithfully reproduce the characteristics of the parental culture. Diluting out cells simply to achieve a larger number of ampoules for the bank is often a false economy, as cultures from such banks usually take much longer to achieve adequate cell density for further passage, thus delaying progress. A change of culture procedure such as adaptation to serum-free medium (see following) or scale-up of the preservation process for larger cell banks will also require careful validation.

Documentation of procedures, cultures, and reagents used are important to promote good-quality scientific work but will also be crucial to provide traceability where the cultures are used for the manufacture or testing of medicinal products. Not only will the qualification of cell banks need to be documented, but it is also important to establish the stability of critical characteristics of the culture with passage to ensure reliable performance of cells derived from cell banks. The types of documentation and how they are used to qualify the testing and validation of cell lines are illustrated in Figure 15.2.

15.4.3 PRESERVATION OF SERUM-FREE CULTURES

Where cell cultures are used in a manufacturing process (e.g., recombinant therapeutic proteins from CHO cells, viral vaccine from cell substrates, monoclonal antibodies from hybridomas),

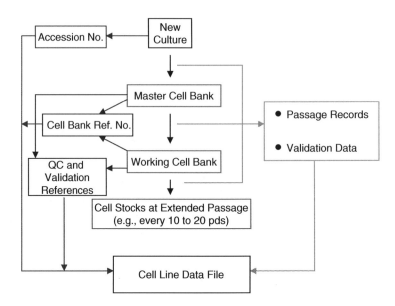

FIGURE 15.2 Documentation of cell banks and validation.

manufacturers try to avoid the use of bovine serum in the cell growth medium to simplify the downstream processing and purification and to avoid the risks associated with raw materials of animal origin. However, the transfer of cultures to serum-free growth medium generally subjects the cells to stress and may lead to permanent changes in the culture. Thus, it is important to recognize that following adaptation to serum-free growth medium a cell culture should be evaluated and characterized again to establish whether any changes have occurred that may affect performance, quality, and safety of the cell substrate. As part of this program, cell banks should be established under serum-free conditions. The removal of serum from the cryopreservation medium may significantly influence cell survival, as serum is a complex protective agent that may help to prevent cell damage during freezing and thawing. Cryoprotectants have been developed for serum-free preservation of cell lines, and some are available commercially.

For cell lines used in the manufacture of biological medicines, there are generally restrictions placed on the maximum number of population doublings permitted between master cell bank and the cells used for the production process. This is particularly important where human diploid fibroblast cultures are used that have a finite lifespan *in vitro* (Wood and Minor, 1991). Thus, in the development of new serum-free cryoprotective solutions, it may be necessary to confirm that cells recovered after thawing and on serum-free passage have not been subject to high levels of cell death, which would increase the number of population doublings at each passage and may also lead to alteration of the characteristics of the culture.

15.4.4 COOLING DEVICES

To ensure the reliability of an established cryopreservation method, it is important to have a device that will ensure a reproducible rate of cooling. This may be achieved by two-stage "passive freezing," whereby the cells are placed in a static system exposed an environment of ultra-low temperature. Some means of insulation between cells and the ultra-low-temperature environment then permits a progressive cooling profile in the cells until they reach the ambient low-temperature environment, at which point they are transferred directly to a liquid nitrogen storage vessel. Many laboratories achieve this by packing ampoules of cells insulated with paper toweling or wadding inside a small polystyrene box that is then placed in a –80°C freezer overnight. Alternative methods are also used whereby the ampoules of cells are suspended in the vapor phase of liquid nitrogen, where they gradually cool and freeze. Careful adjustment of the position of ampoules and monitoring with thermocouples can enable establishment of an effective and reliable preservation procedure. Devices for reproducible passive cooling have been developed commercially (e.g., Handi-Freeze, Taylor Wharton, UK [www.taylor-wharton.com]; Mr Frosty, Invitrogen, UK [www.invitrogen.com]), but these should be validated to ensure correct performance under local laboratory conditions before using with valuable cell stocks.

The passive freezing devices described above appear to be generally effective and have been used widely in research and routine cell culture laboratories. However, the performance of these devices may be affected by variation in the levels of liquid nitrogen in "Dewar" vessels, or by interference with cells undergoing the critical process of freezing in a –80°C freezer. Please note that although longer-term storage of cells at –80°C has been demonstrated to be successful (Sigiura et al., 1968), this should not be considered as a long-term solution in general, as in many laboratories such storage systems are open to interference when other material is retrieved, and there is a high risk of significant loss of viability even within a few months.

Enhanced control and recording of the cooling process may be required, especially in development and production programs for biological products where large cell banks comprising hundreds of ampoules are required. For such critical cryopreservation procedures, mechanical regulation of cooling enables exquisite a control of the cooling rate that can be used to optimize the cryopreservation process. In addition, the machines available commercially (e.g., Biotronics, Wantage, UK; Kryo-10, Planer Products, Sunbury, UK) provide data outputs to enable each cryopreservation run to be carefully documented, which may be important for cell banking performed under certain quality standards (see following).

15.5 STORAGE FACILITIES

15.5.1 GENERAL CONSIDERATIONS

Having invested time and resources to prepare a cryopreserved a cell bank, it is wise to ensure that the facility used to maintain the bank will provide a secure, clean, and stable environment for long-term storage. Security for stored material is assured through adoption of appropriate management systems to restrict access to authorized personnel, appropriate alarms for nitrogen storage vessels, and documented procedures for filling and maintenance of nitrogen storage. Storage management systems should include an inventory of all stored material (this is a legal requirement for infectious and recombinant materials under health and safety legislation in some countries) and routine documentation of withdrawals and entries. However, the best-designed systems will fail if they not monitored properly, and for any important stored material there should be an auditing process to ensure that maintenance and documentation are kept up to date and that any procedural changes are appropriate and recorded.

15.5.2 FACILITY SPECIFICATION AND DESIGN

Storage areas should be selected and established with a number of key criteria in mind: They should provide adequate space to maintain and hold storage vessels in a controlled area, preferably where access to vessels is restricted by use of a dedicated room or locks on vessels; to avoid infection and contamination risks (Fountain et al., 1997) areas prone to heavy environmental microbial contamination such as corridors and storage rooms with access from outdoors are undesirable and should not be used for storage of cell cultures, particularly where intended for aseptic and antibiotic-free cell culture; the storage area should be well ventilated to prevent oxygen depletion during periods of high nitrogen gas release resulting from filling procedures or accidental spills.

Liquid nitrogen storage involves significant safety issues for laboratory workers. Documented emergency procedures and use of oxygen-depletion alarms (both visible and audible) will be important features. A fixed oxygen sensor should be located below head height to detect oxygen depletion. In such systems, dual-level alarms are useful, as they can be used to trigger ventilation fans when oxygen levels are slightly depressed (e.g., 20%) thus minimizing the incidence with which a lower danger-level alarm (18%) is triggered to activate alarms and initiation of emergency procedures. It is important that if ventilation extract fans are required they should be located close to ground level, as nitrogen vapor is more dense than air.

It is important to establish whether storage will be in the liquid or vapor phase of nitrogen or whether electrical freezers (–100°C or below) are to be considered. In theory, the liquid phase of nitrogen provides the lowest and most stable storage temperature and is the method of choice for long-term storage. However, the risks of transmission of pathogenic virus should be considered, as highlighted in the past for stored bone marrow (Tedder et al., 1995). Vapor-phase storage may increase the risk of temperature cycling in stored materials, but it is generally more convenient and safer for regular access to stored material than liquid-phase storage. Furthermore, some manufacturers (e.g., CBS distributed by PhiTech Intl., Milton Keynes, U.K.) are now producing vapor-phase storage systems in which liquid nitrogen is retained in the vessel walls, which improves safety and appears to provide a temperature profile superior to standard vapor-phase systems but does not have a liquid reserve as backup. Electrical storage systems provide a very practical and maintenance-free low-temperature storage solution. However, such systems in a multiuser environment may be more prone to the effects of temperature cycling than liquid nitrogen vapor-phase storage (see long-term storage below), and this may have more serious consequences for materials stored at –135°C compared with –150°C freezers. It should also be noted that electrical freezer storage is a high-risk form of storage where power supplies may not be reliable, and even where this is not the case, liquid nitrogen or carbon dioxide backup systems will be required to cope with emergencies.

Where liquid nitrogen storage is used, the means of supply is also an important consideration in terms of safety and cost. Commercial deliveries of nitrogen are tested for gas composition at source and are generally considered to be free of microbial contamination. However, delivery trucks visit many sites, including centers storing infectious organisms, patient material, and a variety of other reagents. It is therefore possible that a supply pipe recently used, for example, to fill an open storage tank for infectious agents may go on to be used at the next location for filling tanks of patient material for transplantation. It will therefore be helpful to know what procedures are used by the nitrogen suppliers to exclude the possibility of contamination being transferred via filling devices. For critical storage systems, it may be appropriate to have on-site generation of liquid nitrogen (e.g., Cryomech [www.cryomech.com]). Although there will be significant capital cost to set up such systems, the quality and source of nitrogen can be assured and it may provide economical benefits in the long run.

Transport of nitrogen around an organization can be a hazardous process, especially where large, unwieldy, pressure vessels must be moved along public or staff corridors. Piping from the main delivery point is expensive and not efficient over long distances. Wheeled vessels with low centers of gravity or motorized "trucks" to carry or tow vessels will reduce risk significantly, but where such vessel movements are frequent, it is sensible to consider a centralization of stored material. Obviously the materials stored in a common location must be scrutinized carefully to ensure that infectious materials are physically separated in the interests of the safety of laboratory workers and subsequent cell culture products. However, centralized storage can yield economic benefits where the burden of storage maintenance can be shared and security, including out-of-hours surveillance and emergency responses, can be easier and more effective.

When setting up a new storage facility, the types of material to be stored and their physical size are obvious considerations, as is the need to separately isolate hazardous materials and those in quarantine. It is also useful to differentiate between material intended for archive storage and other material that will be accessed regularly. Material for archiving is obviously most secure when stored in vessels giving low nitrogen loss. These normally have narrow access apertures that have longer standing-times; that is, maintain acceptable storage temperatures without refilling for longer periods. For routine access, vessels with wide "bin"-type lids are most convenient and can provide a useful low-temperature working area in the top of the vessel. However, these vessels obviously loose nitrogen vapor at a much higher rate and will tend to have shorter standing-times than low-loss, narrow-necked vessels. In addition, in the upper levels of the inventory systems, "bin"-type vessels are likely to be exposed to significant temperature cycles (see Section 15.6). A decision also has to be made between use of electrical ultra-low-temperature freezers and liquid nitrogen storage vessels. Where electrical freezers are considered, the reliability of electrical supplies, provision of emergency backup (often liquid nitrogen), and need for air conditioning should be considered.

Careful consideration of the location of the storage site is also important to ensure it can be made secure (with access restricted to authorized staff) and well ventilated at all times with an appropriate alarm system (see above). Layout of the storage area and positioning of services must also be considered to ensure staff safety. For example, liquid nitrogen delivery points should not be positioned in thoroughfares, and condensation on cold delivery systems should not cause electrical hazards; that is, electrical supplies and connections should be above any transfer pipes even if insulated.

15.5.3 SAFETY OF PERSONNEL

A risk assessment should be performed in collaboration with the local safety representative (Sheeley, 1998) to establish that the proposed facility and associated procedures are appropriate and would cope with catastrophic failure and total release of liquid nitrogen. Such precautions may seem excessive, but recent cases of fatal accidents involving liquid nitrogen storage in the United Kingdom indicate the importance of careful design and risk assessment of such storage facilities. Emergency procedures should also include contingency plans for transfer of critical materials to backup storage vessels.

Liquid nitrogen presents a serious frostbite hazard, and protective gloves, masks, and aprons designed for cryogenic work should be readily available and used whenever working with storage and supply vessels. In some facilities, it may also be wise for staff to carry personal oxygen monitors. Mechanized vessel filling is usually achieved via solenoid valves. These can freeze in the open position, thus presenting a serious hazard of overfilling storage vessels. Dual solenoid valves in series reduce this risk, but a more robust system would also incorporate a compressed air–activated shut-off valve (e.g., Thames Cryogenic Ltd., Didcot, U.K.) and an in-line ice filter that can be drained periodically.

On occasion, ampoules stored in the liquid phase can explode as a result of rapid vaporization of liquid nitrogen trapped in the ampoule on warming. Serious penetration injuries can be sustained when this occurs, and this is especially hazardous when working with infectious materials. To reduce this risk, newly recovered ampoules destined for use in the laboratory may be held for a period of time in the vapor phase, and when ampoules are subsequently transferred to the laboratory, staff should continue to wear protective gloves and masks and to keep the ampoules covered until thawed. Solid carbon dioxide or "cardice," commonly used for shipment of cryopreserved cultures, carries similar risks of frostbite and asphyxiation to those of liquid nitrogen, and its use should be assessed in the same way. Emergency procedures for exposure to liquid nitrogen or cardice and low-oxygen conditions should be documented and included in staff training programs.

15.5.4 DOCUMENTATION OF STORED MATERIALS

Accurate records of stored materials are not only helpful to retrieve ampoules of cells efficiently but may also be a legal requirement for storing genetically modified, infectious, or other hazardous materials. Numerous commercial database systems are available that are specially designed for this purpose, but it is important to select a system that is flexible to the full range of user requirements. It is wise to have up-to-date, hard-copy printouts of these records and to ensure that amendments to storage records for additions or withdrawals can be made at the storage site to avoid transcriptional errors.

For critical applications, such as manufacture of biological medicines, it is important to establish reference numbering for important stages in the process of cell banking to enable accurate two-

TABLE 15.3
Traceability through Reference Numbers on Laboratory Records

Reference No.	Accession Day Book	Cell Bank Ampoules	Cell Banking Records	Quality Control Records	Media Batch Preparation Records	Sterilization Records
Accession no.	+	+	+			
Cell bank no.		+	+	+		
QC reference no.			+	+		
Media batch reference no.				+	+	+
Sterilization reference					+	+

way traceability of laboratory procedures (see Figure 15.2), which enable independent audits and tracking of reagents and cells when adverse events arise in the use of cell cultures. Key references include a cell-line accession number (assigned on each receipt of a cell culture), cell bank reference number (specific to each homogenous batch of preserved cells), and quality control and media batch references that are linked to ensure two-way traceability as illustrated in Table 15.3.

15.6 ISSUES FOR LONG-TERM STORAGE

Where cells are stored at low temperature over long periods, they may be prone to a number of potential hazards such as variation in storage temperature and contamination. Stored cell lines may be subjected to temperature cycles during routine access and maintenance of storage vessels. Intermittent warming of ampoules may be particularly marked in inventory systems in which vessels are opened frequently, and particularly where the individual storage racking must be completely withdrawn into ambient room temperature to gain access to stored vials and ampoules. The storage racking material may also be a significant factor, as good conductors of heat will promote warming cycles and temperature gradients within the storage vessel. One procedure that is guaranteed to have significant and potentially disastrous effects on viability of stored cells if not performed correctly is filling and maintenance of nitrogen storage vessels. Anecdotal cases of massive loss of important material as a result of vessel failure or breakdown in filling procedures are all too frequent given that the solutions to many of these incidents are often relatively simple.

Build-up of microbial contamination from environmental sources is known to occur (Fountain et al., 1997). Thus, long-term storage vessels will benefit from periodic cleaning at least to remove the ice sludge that accumulates at the bottom of such vessels. Careful disinfection of recovered ampoules is also important to prevent contamination of cultures, and the use of sealed ampoules or storage boxes will also help to provide protection against microbial contamination.

Natural radiation has also been considered a potential cause of loss of viability or mutation in stored cells and tissues. However, there does not appear to be any evidence for the adverse effects of long-term storage in well-maintained nitrogen vessels, even for biological systems that might be expected to be more sensitive to such effects, such as embryos (Glenister et al., 1984).

A number of straightforward procedures can be used to enhance the security of important archived material. Manual or automatic monitoring of temperature or liquid nitrogen levels can provide useful monitoring data to identify trends in the quality of maintenance over time. A system to document any filling (especially manual filling) and other maintenance procedures (e.g., visual inspection, electrical testing, vessel vacuum checks) is important and should be checked regularly and periodically audited (e.g., during safety inspections).

It is wise to store important archive material in the bottom of storage vessels or, ideally, in separate archive vessels of with long standing-times (i.e., low–nitrogen loss systems). Vessels suited to regular access with large access lids have a high nitrogen loss and, as described above, are more likely to suffer significant temperature cycling. In cases in which temperature cycling is a significant risk, it may be helpful to consider establishing "sentinel" banks of cells that are recovered periodically to determine any trends in level of viability at points in the storage vessel that may be most prone to this effect (Stacey, 1999). However, the approaches described above do not protect against catastrophic failure of vessels (usually resulting loss of the insulating vacuum). Accordingly, contingency planning is also an important aspect of the management of cryopreserved cell-line banks. A primary contingency measure is to have backup vessels available that can be brought into use rapidly should vessel failure occur (it should be borne in mind that large vessels at room temperature may take several hours to cool down sufficiently for transfers to take place). In addition, risk of damage to the storage site must be considered (e.g., failure of nitrogen filling, failure of power supply for electrical freezers, fire, sabotage), and to counter against such risk cell banks may be split to storage vessels in different locations on site and possibly also distributed to a second site. In the latter case, it is important to be sure that the storage maintenance procedures and general quality of storage are at least equivalent to those in the originating center, and periodic audits will help to assure this.

15.7 REGULATION AND QUALITY ASSURANCE OF CELL BANKS OF ANIMAL CELL SUBSTRATES

15.7.1 GUIDELINES ON CELL SUBSTRATES

A range of guidelines on best practice in cell and tissue culture have been published (Doblhof-Dier and Stacey, 2000; Freshney, 1994; Hartung et al., 2002; Stacey et al, 1998; UK Coordinating Committee for Cancer Research, 1999). Cell culture processes involved in the preparation of biological medicines are subject to more stringent guidelines from official national and international regulatory bodies including the U.S. Food and Drug Administration, the World Health Organization, and the International Conference on Harmonization (Center for Biologics Evaluation and Research, 1993a; International Conference on Harmonization, 1998; World Health Organization, 1998; European Medicines Evaluation Agency (EMEA) at http://www.emea.eu.int). The referenced guidelines were established to apply to cell substrates used in the manufacture of biological medicines. They are based on key issues relating to

- Sterility (i.e., absence of bacteria, fungi, and mycoplasma)
- Viruses and transmissible spongiform encephalopathy (TSE) contamination of cells and media
- Genetic and phenotypic stability on passage
- Tumorigenesis of cells and oncogenicity of cell DNA

Although these references deal with the substrates used in the manufacture of biological products, there is a diverse range of new cell-based therapies that will need guidelines and regulations. Regulations for aspects of gene therapy have been developed (e.g., Center for Biologics Evaluation and Research, 1993b; EMEA, 2002), and guidelines for other forms of cell therapy are under development in a number of countries, and some of these specifically refer to important issues relating to cryopreservation (U.K. Department of Health, 2002). These new regulations on cell therapy and tissue engineering introduce new issues for safety and quality of medicines that include:

- The effects of combining more than one cell type in a product
- Residual tumor cells (including tumorigenesis of stem cells)
- Influence of the scaffold materials (toxicity, adsorption, and leaching of compounds)
- Specification and validation of raw materials for clinical use

In the future, the use of embryonic stem cell lines for therapy will deliver a new set of challenges. Issues that will need to be considered include genetic imprinting, activation of endogenous viruses, and standardization of growth and differentiation. In addition, tumorigenicity studies in animal models may be difficult to assess, as teratoma formation is considered to be a positive functional indicator. In the U.K. a code of practice for the use of stem cell lines has now been established for publication by the Medical Research Council, London, U.K. (www. mrc.ac.uk).

15.7.2 QUALITY STANDARDS

Quality is the fitness for purpose of a particular product or process, and this is demonstrated in the context of a documented quality system. There are a number of formal quality standards within which the process of cell banking may be operated. The ISO9000 quality systems provide for consistent provision of a service or product that may include provision of cells from a cell bank (British Standard European Norm [BSEN] ISO9000, 1994; http://www.iso-9000-2000.com/). ISO9001 specifically addresses design, processing, and final inspection, whereas ISO9002 may be applied to research and development. However, the ISO9000 quality systems will only monitor for consistency of cell-banking procedures against in-house standards, without a requirement for external measures of quality.

Where the cells are used in the manufacture of biological medicines (e.g., recombinant proteins, antibodies, viral vaccine) or critical testing purposes such as vaccine batch control, more stringent national and international quality systems are required, including current Good Manufacturing Practices (cGMP), for which there are formal regulations published for Europe, by the World Health Organization (http://www.who.int/vaccines-documents/DocsPDF/www9651.pdf), and for some individual countries. Traditionally, cGMP has applied to the final formulation of a product, although during auditing of GMP facilities, the downstream processing will normally be expected to meet the same requirements as the final steps. Raw materials, including cell banks, used in GMP processes may also be expected to meet the requirements of cGMP. Furthermore, cGMP is also required for materials used in clinical trials.

Testing of cell banks in relation to their safety for use in manufacturing is usually performed under good labor practice (GLP) accreditation based on the guidelines established by the Organisation for Economic Cooperation and Development in 1999 (OECD, 1999). Various organizations operate worldwide to accredit such testing procedures, and these organizations are coordinated through the International Laboratory Accreditation Cooperation, established in 1996, which mediates international cooperation in this field (http://www.ilac.org/).

15.8 BIOLOGICAL RESOURCE CENTERS

Collections of microbial cultures have been existence for more than a hundred years, and the first acknowledged service collection providing cultures for industrial use was established by Kraal in 1889. Since then numerous large collections of bacteria, yeast, and fungi have been established as public service collections, and these now include a number of collections of animal and human cell lines. These public collections are coordinated internationally through organizations such as the European Culture Collection Organization (http://www.eccosite.org) and the World Federation of Culture Collections (http://wdcm.nig.ac.jp/). Culture collections and other institutions with specialist expertise in provision of cell lines are important sources of quality-controlled cell cultures that enable researchers to obtain reliable supplies of cultures. These organizations provide cell lines for various purposes, including:

- Cells representative of particular species or tissue
- Controls and reference strains for biological assays
- Seed stocks for product development (see above)

- Sources of genomic DNA carrying specific genetic lesions as controls
- Supplies of cells/DNA for genomics/proteomics research

Biological resource centers should be the first port-of-call for researchers seeking new cell lines. They provide the kind of quality-controlled and authenticated cells that will give evidence for the authenticity of the cells supplied. In addition, many resource centers also provide advice and training in culture, preservation, and quality-control techniques that will be invaluable to anyone starting out in cell culture.

15.9 OWNERSHIP OF CELL LINES AND PATENTS

Many cell lines in use for research and development have been passed between laboratories for many years and are considered to be in the public domain. Public collections, described above, maintain quality-controlled stocks of such cells for many years and have historically not claimed ownership over the cells themselves. However, in recent years, as the potential use of cell lines in the manufacture of medical products and laboratory reagents has increased, much greater attention has been directed at the ownership of cell lines in which there may be a number of interested parties.

Where the cell line has been derived from clinical procedures, the patient of origin may claim an interest in exploitation of their cells. A small number of such cases have been pursued by patients and their families, notably in relation to HeLa cells. Whereas the HeLa cell story (Gold, 1986) may be an exceptional case, it highlights the problems that may be encountered where proper informed consent has not been obtained before using patient tissues. Today it is standard practice, usually as part of local or national ethics approval, that each candidate patient is asked to sign a patient consent form that may identify the uses (this may be unrestricted use), private interests, and patent issues to which their cells can be put or waive any rights to the ownership of the cells or any subsequent discoveries or developments relating to them. Such an agreement may pass the ownership of the cells to a particular sponsor, often a research or commercial organization, and in some cases, such as embryonic stem cell lines, lead to a significant delay in the open availability of important new cell lines for research.

Public service collections (see above) have played a valuable role in establishing quality-controlled stocks of cell lines, making them available for research and development and securing their availability for later generations of scientists. This role as custodian of cell lines means that some of these resource centers now claim that they have rights to a share of any intellectual property arising from cell lines supplied by them. Most collections require that a materials transfer agreement is signed by the organization receiving cell lines and that restricts third-party distribution and requires that any commercial exploitation of the material is notified to the collection. As discussed above the originators of the cells may also have supplementary conditions on use of cells received from resource centers. Thus, when embarking on a line of research that may ultimately lead to, or otherwise assist the development of, a commercial product or process, the researchers should carefully consider the potential effect of any materials transfer agreements they sign to obtain the cells. At an early stage of project development, alternative sources of the cell line or different candidate cells can be assessed and appropriate agreements can be put in place to ensure that the development of a process or product will ultimately be commercially exploitable.

Patent applications based on a novel cell line or involving the use of a cell line as a critical part of the patent may require that samples of the cell line or a representative cell bank must be submitted to a recognized patent depository (Fritze and Weihs, 2001; for general references see Cook 1999; Crespi, 1998). Such depositories act as independent laboratories to test and hold those cultures and act as a reference point for any procedure to verify or challenge the veracity of the patent. Patents may be filed on a national basis or through the U.S. or European Patent Office. The World International Property Office is an organization that aims to provide for coordination, mutual international acceptance, and harmonization of patent procedures. Under the Budapest Treaty

(1979), certain laboratories are registered to act as International Depository Authorities (IDAs). IDAs verify the viability and purity of the deposited materials and hold them in a stable preserved state for at least 30 years. The latter commitment for these centers is a significant challenge for some cultures such as plant cell cultures, where the preservation methods now in use may not have been available to researchers until quite recently. The depository centers will have a specific list of requirements that must be met by the depositor before the patent culture accession number is formally released to permit completion of patent procedures, and it is wise to prepare a master cell bank and perform key tests such as sterility testing and mycoplasma tests before submitting samples of cells to the patent depository. Typical requirements of the IDA include:

- A minimum number of ampoules (typically 10 to 15) for storage and quality control
- Freedom from microbial contamination (primarily bacteria, fungi, and mycoplasma)
- Payment for the deposit application

Alternatively, the culture may be submitted to the IDA for its own in-house banking and testing before making the patent application. Although this may prove expensive, it carries the advantage that a culture deposited with an IDA for safekeeping may be subsequently translated to a patent deposit while retaining the original deposit date. This may be a critical advantage when the patent may need to be accepted urgently at a later date.

15.10 CONCLUSIONS

When a cell line is first established, a cryopreserved archive stock, at low passage, is essential to protect against accidental loss of the culture. For cell lines likely to be used over an extended period of time, reliable supply of cultures is achieved through the establishment of a two-tiered master and working bank system. Confirmation of viability, key characteristics, and quality control of cultures recovered from cryopreserved cells should then be performed at the earliest opportunity. To establish reliable larger-scale cell banks such as those used in manufacture of recombinant proteins and vaccines, it may be necessary to adapt and validate scale-up of the cryopreservation process to avoid loss of viability or variability between ampoules. Such banks will also be subjected to far more stringent characterization under appropriate regulatory guidelines. Cell banks stored for long-term use, including patent deposits, will require consideration of the additional important challenges, and particular attention should be paid to the maintenance and monitoring of storage conditions.

The establishment of well-characterized and quality-controlled cell banks enables enhanced reproducibility of research work and enhanced standardization of diagnostic and manufacturing processes using cell-line substrates. The ability to standardize cell culture procedures in different locations and at different times is a major contribution to high-quality research and development that is dependent on the availability of cryopreserved cell banks.

ACKNOWLEDGMENTS

I would like to acknowledge helpful comments and information on storage engineering and patent issues provided by Robin Saunders and Magnus Schoeman of NIBSC.

REFERENCES

BS EN ISO9000. (1994) *Quality Systems: Model for Quality Assurance in Design, Development, Processing, Installation and Services*, British Standards Institute, London.
Budapest Treaty Regulations. (1977) *Budapest Treaty on the International Recognition of the Deposit of Microorganisms for the Purposes of Patent Procedure*, 277 (E), World Intellectual Property Organization, Geneva, 1981.

Center for Biologics Evaluation and Research. (1993a) *Points to Consider in the Characterization of Cell Lines Used to Produce Biologicals*, Food and Drug Administration, Bethesda, MD, available at: http://www.fda.gov/cber/.

Center for Biologics Evaluation and Research. (1993b) *Application of Current Statutory Authorities to Human Somatic Cell Therapy Products and Gene Therapy*, Food and Drug Administration, Bethesda, MD, available at: http://www.fda.gov/cber/.

Chen, T.R., Dorotinsky, C., Macy, M., and Hay, R. (1989) Cell identity resolved, *Nature,* 340, 106.

Cook, T. (1999) Patent and licensing issues for biopharmaceutical products, *J. Commercial Biotechnol.,* 5, 223–230.

Cord, C., Uphoff, S.M., Gignac, M., and Drexler, H.G. (1992) Mycoplasma contamination in human leukaemic cell lines I comparison of various detection methods, *J. Immunol. Methods,* 149, 43–53.

Crespi, S. (1998) Patenting for the research scientist—Bridging the cultural divide, *Tib. Tech.,* 16, 450–455.

Culliton, B.J. (1974) HeLa cells: Contamination over the world, *Science,* 184, 1058–1059.

De Benedetti, A., Pytel, B.A., and Baglioni, C. (1987) Retraction regarding: Loss of oligoadenylate synthetase activity by production of antisense RNA results in lack of protection by interferon from viral infections, *Proc. Natl. Acad. Sci. USA,* 84, 6740.

Dirks, W.G., MacLeod, R.A., Drexler, H.G. (1999) ECV304 (endothelial) is really T24 (bladder carcinoma): Cell line cross-contamination at source, *In Vitro Cell Dev. Biol. Anim.,* 35, 558–559.

Doblhoff-Dier, O. and Stacey, G. (2000) Cell lines: Applications and Biosafety, in *Biological Safety—Principles and Practices,* 3rd edition, Fleming, D.O. and Hunt, D.L., Eds., American Society for Microbiology, Washington, DC, pp. 221–241.

Drexler, H.G., Hane, B., Hu, Z.B., Uphoff, C.C. (1993) HeLa cross-contamination of a leukemia cell line, *Leukemia,* 7, 2077–2079.

Drexler, H.G., Dirks, W.G., MacLeod, R.A. (1999) False human hematopoietic cell lines: Cross-contaminations and misinterpretations, *Leukemia,* 13, 1601–1607.

EMEA. (2002) EMEA/5843/02, *Report from the Ad hoc Meeting of CPMP Gene Therapy Expert Working Group*, EMEA, 11 February 2002.

Erickson, G.A., Bolin, S.R., and Landgraf, J.D. (1991) Viral contamination of fetal bovine serum used for tissue culture: Risks and concerns, *Dev. Biol. Stand.,* 75, 173–175.

Fritze, D. and Weihs V. (2001) Deposition of biological material for patent protection in biotechnology, *Appl. Microbiol. Biotechnol.,* 57, 443–450.

Fountain, D., Ralston, M., Higgins, N., Gorlin, J.B., Uhl. L., Wheeler, C., Antin, J.H., Churchill, W.H., and Benjamin, R.J. (1997) Liquid nitrogen freezers: A potential source of microbial contamination of hematopoietic stem cell components, *Transfusion,* 37, 585–591.

Freshney, R.I. (1994) *Culture of Animal Cells: A Manual of Basic Technique,* 3rd edition, Wiley-Liss, New York.

Frommer, W., Archer, L., Boon, B., Brunius, G., Collins, C.H., Crooy, P., Doblhoff-Dier, O., Donikian, R., Economidis, J., and Fontali, C. (1993) Safe biotechnology (5), recommendations for safe work with animal and human cell cultures concerning potential human pathogens, *Appl. Micr. Biotechnol.,* 39, 141–147.

Gartler, S.M. (1967) Genetic markers as tracers in cell culture, *Natl. Cancer Inst. Monogr.,* 26, 167–195.

Glenister, P.H., Whittingham, D.G., and Lyon, M.F. (1984) Further studies on the effect of radiation during storage of frozen 8-cell mouse embryos at −196 degrees, *C. J. Reprod. Fert.,* 70, 229–234.

Gold, M. (1986) *A Conspiracy of Cells*, State University of New York Press, Albany.

Griffiths, J.B. and Doyle, A. (2000) *Cell and Tissue Culture for Medical Research*, Wiley, Chichester.

Hallauer, C., Kronauer, G., and Siegl, G. (1971) Parvovirus contaminants of permanent human cell lines I virus isolation from 1960–1970, *Arch. Gesamte Virusforsch.,* 35, 80–90.

Harris, N.L., Gang, D.L., Quay, S.C., Poppema, S., Zamecnik, P.C., Nelson-Rees, W.A., O'Brien, S.J. (1981) Contamination of Hodgkin disease cell cultures, *Nature,* 289, 228–230.

Hartung, T., Balls, M., Bardoville, C., Blanck, O., Coecke, S., Gstraunthaler, G., and Lewis, D. (2002) Good cell culture practice, *ATLA,* 30, 407–414.

International Conference on Harmonization. (1997) *Human Medicines Evaluation Unit: ICH Topic Q5D—Quality Of Biotechnological Products: Derivation and Characterization of Cell Substrates Used for Production cf Biotechnological/Biological Products*, European Agency for the Evaluation of Medicinal Products, ICH Technical Coordination, London, available at: http://www.eudra.org/emea.eu.int.html.

MacLeod, R.A., Dirks, W.G., Reid, Y.A., Hay, R.J., Drexler, H.G. (1997) Identity of original and late passage Dami megakaryocytes with HEL erythroleukemia cells shown by combined cytogenetics and DNA fingerprinting, *Leukemia*, 11, 203–208.

MacLeod, R.A., Dirks, W.G., Matsuo, Y., Kaufmann, M., Milch, H., Drexler, H.G. (1999) Widespread intraspecies cross-contamination of human tumor cell lines arising at source, *Int. J. Cancer*, 83, 555–563.

Masters, J.R., Bedford, P., Kearney, A., Povey, S., Franks, L.M. (1988) Bladder cancer cell line cross-contamination: Identification using a locus-specific minisatellite probe, *Br. J. Cancer*, 57, 284–286.

Nelson-Rees, W.A. and Flandermeyer, R.R. (1977) Inter- and intraspecies contamination of human breast tumor cell lines HBC and BrCa5 and other cell cultures, *Science*, 195, 1343–1344.

Nicklas, W., Kraft, V., and Meyer, B. (1993) Contamination of transplantable tumours, cell lines an monoclonal antibodies with rodent viruses, *Lab. Anim. Sci.*, 43, 296–300.

Organisation for Economic Co-operation and Development (1999). OECD series on Principles of GLP and Compliance Monitoring No. 4 (revised), ENV/JM/MONO(99)20, http://www.oecd.org/ehs/ehsmono/index.htm

Reid, Y.A., McGuire, L., O'Neill, K., Macy, M., Chen, T.R., McClintock, P., Dorotinsky, C., Hay, R. (1995) Cell line cross-contamination of U-937 [correction of U-397], *J. Leukoc. Biol.*, 57, 804.

Sheeley, H. (1998) Risk assessment, in *Safety in Cell and Tissue Culture*, Stacey, G.N., Doyle, A., and Hambleton, P.J., Eds., Kluwer Academic, Dordrecht, Netherlands, pp. 173–188.

Sigiura, K., Tortolani, A.J., and Sergent, M.G. (1968) Viability of various solid and ascites tumours after prolonged storage at –78°C, *Cryobiology*, 4, 177–183.

Solis-Recendez, M., Perani, A., D'Habit, B., Stacey, G.N., and Maugras, D. (1994) Hybridoma cultures continuously undergo apoptosis and reveal a novel 100 bp fragment, *J. Biotech.*, 38, 117–127.

Stacey, G.N. (1999) Control of contamination in cell and tissue banks, *Cryo-Letters*, 20, 141–146.

Stacey, G. (2000) Cell lines used in the manufacture of biological products, in *Encyclopedia of Cell Technology*, Spier, R., Ed., Wiley Interscience, New York, pp. 79–83.

Stacey, G.N. and Doyle, A. (2000) Cell banking, in *Encyclopedia of Cell Technology*, Spier, R., Ed., Wiley Interscience, New York, pp. 293–320.

Stacey, G., Doyle, A., and Hambleton, P.H. (1998) *Safety in Cell and Tissue Culture*, Kluwer Academic Publishing, Dordrecht, Netherlands.

Stacey, G., Masters, J.R.W., Hay, R.J., Drexler, H.G., MacLeod, R.A.F., and Freshney, I.R. (2000) Cell contamination leads to inaccurate data: We must take action now, *Nature*, 403, 356.

Stacey, A. and Stacey, G.N. (2000) Routine quality control testing for cell cultures, in *Methods in Molecular Medicine, Vol. 24: Antiviral Methods and Protocols*, Kinchington, D. and Schinazi, R.F., Eds., Humana Press, Totowa, NJ, pp. 27–40.

Tedder, R.S., Zuckerman, M.A. Goldstone, A.H., Hawkins, A.E., Fielding, A., Briggs, E.M., Irwin, D., Blair, S., Gorman, A.M., and Patterson, K.G. (1995) Hepatitis B transmission from a contaminated cryo-preservation tank, *Lancet*, 346, 137–140.

U.K. Coordinating Committee for Cancer Research. (1999) *Guidelines on Cell Culture Practice*, U.K. Coordinating Committee for Cancer Research, London.

U.K. Department of Health (2002) A Code of Practice for the Production of Human-Derived Therapeutic Products. Medical Devices Agency, Hannibal House, Elephant & Castle, London, available at www.medical-devices.gov.uk

van Helden, P.D., Wiid, I.J., Albrecht, C.F., Theron, E., Thornley, A.L., Hoal-van Helden, E.G. (1988) Cross-contamination of human esophageal squamous carcinoma cell lines detected by DNA fingerprint analysis, *Cancer Res.*, 48, 5660–5662.

Wood, D.J. and Minor, P.D. (1991) Use of diploid cells in vaccine production, *Biologicals*, 18, 143–146.

World Health Organization. (1998) *WHO Expert Committee on Biological Standardization (47th Report): Requirements for the Use of Animal Cells as in Vitro Substrates for the Production of Biologicals, World Health Organization Technical Report Series 878*, World Health Organization, Geneva.

Theme 4

Medical Applications

16 Mechanisms of Injury Caused by *in Vivo* Freezing

Nathan E. Hoffmann and John C. Bischof

CONTENTS

16.1 INTRODUCTION

Cryosurgery, or tissue destruction by controlled freezing, has been investigated as a possible alternative to surgical intervention in the treatment of many diseases. This technique, which falls under the larger category of thermal therapy, has its origins in the 1800s, when advanced carcinomas of the breast and uterine cervix were treated with iced saline solutions, first by Arnott (1851). Since those early times, this technique has been used routinely to treat malignancies on the surface of the body (i.e., dermatologic tumors) and has gained some acceptance as a clinical tool for the management of internal malignancies, such as prostate and liver carcinomas. The main advantages of the technique are the potential for less invasiveness and lower morbidity compared to surgical excision. However, two main barriers to the technique gaining widespread acceptance are lack of a suitable technique to visualize the cryosurgical iceball and lack of understanding regarding the destructive process of freezing within the body. The former has been reviewed recently (Rubinsky, 2000) and will not be discussed in this chapter. The study of the destructive process of freezing is the focus of this chapter and will be divided into two main areas: understanding the thermal history

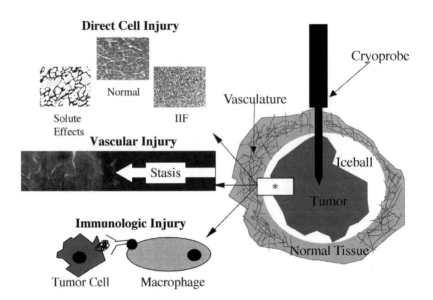

FIGURE 16.1 Overview of the main mechanisms of injury that happen as a result of cryosurgery.

that causes tissue destruction and understanding the mechanism by which freezing destroys tissue. The term "thermal history," as used in this chapter, means the time–temperature history experienced by the tissue during a thermal insult.

Many theories have been put forward as to the mechanism of tissue injury in cryosurgery (see Figure 16.1). The first, and perhaps oldest, is that of direct cellular injury. The temperature extremes from cryosurgery have been purported to cause damage to the cellular machinery and thereby lead to cell death. Cryoinjury has been proposed to cause protein damage by high solute concentrations that are created as the cell dehydrates in response to freezing, damaging the enzymatic machinery of the cell (Lovelock, 1953a, 1953b). Also, if ice crystals form with the cell, the crystals are purported to disrupt the intracellular organelles as well as the plasma membrane (Steponkus, 1984; Toner, 1993). Much work has been done with freezing of single cell suspensions to define the critical thermal thresholds and mechanisms of direct cellular injury (Jacob et al., 1985; Mazur, 1984; Roberts et al., 1997; Smith et al., 1999; Tatsutani et al., 1996; Zacarian, 1977;).

Another theory of the mechanism of freezing injury is that of freezing-stimulated immunological injury. According to this hypothesis, after cryosurgery, the immune system of the host is sensitized to the tissue destroyed by the cryosurgery. Any tissue remaining undamaged or sublethally injured by the freezing insult is destroyed by the immune system during the time after cryosurgery. Several animal models and clinical case studies indicating an immunological response to tumor tissue after cryosurgery have been reported in the literature for cryosurgery (Ablin, 1995; Lubaroff et al., 1981; Neel et al., 1973; Soanes et al., 1970b). Despite this research, this mechanism has not been conclusively proven to aid in tissue destruction after cryosurgery.

A third theory regarding the mechanism of damage is that of vascular injury (Fraser and Gill, 1967). The hypothesis is that freezing causes stasis within the vessels, particularly at the capillary level (Daum et al., 1987), and the resultant ischemia causes the tissue supplied by that vasculature to become necrotic. Cohnheim (1877) has been credited by many with first proposing the link between vascular damage and tissue necrosis. He proposed this first for freezing injury in a lecture series on general pathology. Since then, much work has been devoted to defining the events surrounding and the mechanisms causing freezing-induced vascular injury by working with models of injury *in vivo* (Kreyberg and Rotnes, 1931; Ninomiya et al., 1985; Rabb et al., 1974).

This diversity of potential mechanisms of injury indicates that the response to freezing could be a multifactorial injury process. Measuring the thermal history within the tissue during freezing is important to determining which of the mechanisms are involved in causing cryosurgical injury. For example, in an animal model of prostate cancer cryosurgery used in our laboratory, the Dunning AT-1 rat prostate tumor, vascular injury happens at much higher subzero temperatures than direct cellular injury (Hoffmann and Bischof, 2001a), and immunological injury might occur at another threshold. As a result, it is important to define all three types of injury: direct cell, vascular, and immunological. This requires looking at the underlying biophysical changes and host responses that occur as a result of freezing and at what temperatures they occur.

16.2 THE ROLE OF DIRECT CELLULAR INJURY IN CRYOSURGERY

16.2.1 THE ROLE OF DIRECT CELLULAR INJURY IN TISSUE NECROSIS

Direct cellular injury caused by freezing insult has been studied since the 1800s, when researchers observed that plant cells shrank in response to freezing (Molisch, 1897). Since then, there has been much experimental research regarding the effects of freezing on tissue. The majority of direct cell work has been on *in vitro* cells and tissues. This is because removing the cells from the host eliminates all other secondary types of injury (e.g., vascular and immunological). The evidence that freezing causes cell death is overwhelming (Bischof et al., 1997; Jacob et al., 1985; Lovelock, 1953b; Mazur, 1984; Roberts et al., 1997; Smith et al., 1999; Tatsutani et al., 1996; Zacarian, 1977), and the basis of cryosurgical research has largely been on maximizing *in vitro* cell death. Because cells experience a similar freezing event *in vivo*, this form of injury can also occur during cryosurgery. For example, Gage and Baust (1998) reviewed studies in which cells were frozen *in vivo* and removed immediately after thawing. Presumably, the other host-mediated mechanisms would not have had time to affect cell viability, yet the viability of cells from frozen tissue in these studies is lower than that in unfrozen tissue. It is therefore important to understand this type of injury to appreciate its role in overall cryosurgical destruction.

16.2.2 POTENTIAL MECHANISMS OF DIRECT CELL INJURY

Two biophysical responses that occur in cells during freezing have been linked to cell injury. At low cooling rates, as the freezing propagates extracellularly, the solute concentration outside the cell begins to rise, causing osmotic dehydration of the cells (Bischof et al., 1997). As solutes become concentrated within cells, the high concentration of solute has been hypothesized to injure the cell in several ways including damaging enzymatic machinery (Lovelock, 1953b, 1953a) and destabilizing the cell membrane (Steponkus, 1984).

The second biophysical response is intracellular ice formation (IIF), which occurs when the cooling rate is sufficiently rapid to trap water within the cell. In this case, the cell cannot osmotically equilibrate with the extracellular space. As a result, the cytoplasm cools, and ice ultimately nucleates within the cell (Mazur, 1965); these ice crystals cause injury to the organelles and membranes (Toner et al., 1990). Thus, damage resulting from solute effects happens at relatively low cooling rates, when the cells have had sufficient time to dehydrate completely, and IIF damage occurs at relatively high cooling rates, when the water is trapped inside the cells. This results in an "inverse U curve" of cell viability, with low viability at very high and very low cooling rates and high viability at cooling rates between the extremes (Mazur, 1984; see also Chapter 1). This cooling rate behavior is highly cell-type dependent, with the cooling rate that yields maximum viability (i.e., the top of the inverse U) ranging as low as 20°C/min for mouse spermatazoa (Devireddy et al., 1999) and as high as >90°C/min for rat prostate tumor (Smith et al., 1999).

Many physical and biochemical properties of cells are thought to influence the injury mechanisms discussed above, including the stage in the cell cycle (McGann and Kruuv, 1977), the state of adhesion (Hetzel et al., 1973; Kruuv, 1986), the state of membrane proteins before and after freezing (Kruuv, 1986; McGann et al., 1974), and the temperature maintained after freezing (McGann et al., 1975). These parameters, however, are difficult to control during cryosurgery and may have more influence on cryopreservation (see Chapters 1 and 2).

16.2.3 THE THERMAL HISTORY OF DIRECT CELL INJURY

Cellular injury mechanisms depend on the thermal history that a cell experiences during freezing. This thermal history is defined by four thermal parameters: cooling rate, end (or minimum) temperature, time held at the minimum temperature (hold time), and thawing rate. Variations in these thermal parameters change the mechanism by which injury to the cell occurs. The foregoing discussion of the two biophysical effects that cause direct cell injury showed that cooling rate is important for determining mechanism of injury. Variations in end temperature can also affect the mechanism of cell injury. Because extracellular solute concentrations increase with decreasing temperature during freezing, end temperature can affect how much osmotic dehydration the cell experiences and how much resultant damage occurs. End temperature can also affect IIF. Biological cells experience two different types of ice nucleation that occur at different temperature ranges: surface-catalyzed nucleation (SCN) from −5 to −20°C and volume-catalyzed nucleation from −25 to −40°C (Toner et al., 1990). Recently, data obtained in our laboratory have suggested that SCN could occur at even lower temperatures (Berrada and Bischof, 2000). Thus, variation in end temperature can affect the nucleation of ice within the cell and also affect how much IIF occurs.

Hold time is also important for both injury mechanisms that have been described. Depending on the dynamics of water movement across the membrane, increasing the hold time can allow the intracellular space to equilibrate with the extracellular space, thereby increasing intracellular solute concentration during freezing (Smith et al., 1999). Also, SCN has been shown to be both hold-time and end-temperature dependent. Thus, holding longer at subzero temperatures can increase the amount of SCN occurring within the cells, and thereby exacerbate IIF (Toner, 1993). Finally, increasing hold time is also hypothesized to allow for recrystallization, whereby smaller ice crystals fuse to form larger ice crystals to decrease surface area and minimize free energy, increasing the probability that vital intracellular machinery or the cell membrane will be disrupted by a growing ice crystal (Asahina et al., 1970). The thawing rate acts in much the same way as hold time, by increasing the amount of time spent at low subzero temperatures, increasing the probability of solute and IIF damage.

Although quantitative studies have been carried out regarding the effects of all the above parameters on cell injury, the end-temperature parameter has long been the predominate area of study in cryosurgical research. This work searches for a "lethal temperature" that is capable of completely destroying a cell or tissue. Presumably, freezing a tumor to temperatures equal to or lower than the lethal temperature would completely destroy the tumor (Gage, 1992). Lethal temperatures have been shown to be highly cell-type dependent. Some of these studies are summarized in Table 16.1. Jacob et al. (1985) found that Walker carcinoma cells had >1% viability after freezing to −35°C, whereas Zacarian (1977) found approximately 50% viability in HeLa cells at the same end temperature. Ludwin (1951) froze mouse mammary tumor tissue to −79°C and found that this tissue could grow tumors approximately half of the time. The effect of the other thermal parameters has also been studied by several laboratories, but to a lesser extent than end-temperature. Tatsutani et al. (1996) found a strong correlation between increasing cooling rate and decreasing viability in ND-1 cells, whereas McGrath et al. (1975) found little change in viability as a result of increasing cooling rate in HeLa cells. Jacob et al. (1985) also considered hold time in their study, and despite a statistically significant decrease in viability, concluded that there was very little effect of hold time on tumor cell viability.

There has been little work regarding thawing rate outside cryopreservation applications, which are usually performed under *in vitro* conditions using cryoprotectants and are thus not relevant to

TABLE 16.1
Minimum Thermal Histories Required to Induce Direct Cellular Injury

Author	Year	CR	ET	HT	TR	Cell/tissue system
McGrath et al.	1975	100°C/min	–20°C	1 min	800°C/min	HeLa S-3 cervical carcinoma
Zacarian	1977	Rapid	–35°C	1 min	N/A	HeLa cervical carcinoma
Jacob et al.	1985	1°C/min	–20°C	10 min	1°C/min	Walker mammary adenocarcinoma
Tatsutani et al.	1996	1°C/min	–30°C	5 min	37°C contact	Human prostate adenocarcinoma
Smith et al.	1999	5°C/min	–80°C	0	200°C/min	AT-1 prostate adenocarcinoma
Yang et al.	2000	~5°C/min	–10°C	0	0.5°C/min	MBT-2 bladder carcinoma
Kremer and Duffy	2000	2°C/min	–16.1°C	0	Fast	Human endometrium
Bischof et al.	2001	5°C/min	–30°C	0	200°C/min	ELT-3 uterine leiomyoma

Note For ease of comparison, the minimum thermal history that reduced viability to under 15% was entered into the table. CR, cooling rate; ET, end (or minimum) temperature; HT, time held at the minimum temperature (hold time); TR, thawing rate.

this discussion (Smith et al., 1999). A summary of thermal histories that cause a significant amount of direct cell injury can be seen in Table 16.1; see also Gage and Baust (1998) for a discussion of literature on the thermal history that causes *in vivo* cryoinjury. They present two tables, one that shows that the lethal end temperature of cryosurgery has been reported anywhere from –2 to –70°C and below, and the other of which shows the effect of multiple freeze–thaw cycles. The studies reviewed report results from tissues frozen *in vivo* that were assessed for viability both *in vivo* and *in vitro* at various points after freezing. As a result of the methodology of the experiments reviewed, the injury assessment could include injury not directly caused by the freezing insult. Because of our desire to include only thermal histories shown to cause direct cell injuries, those studies were not included.

Our laboratory has also obtained a great deal of information regarding the response of two specific model systems to variations in all thermal parameters: Dunning AT-1 rat prostate cells and ELT-3 rat uterine fibroid cells (Bischof et al., 1997, 2001; Roberts et al., 1997; Smith et al., 1999). We have found that AT-1 cells are quite resistant to cooling injury *in vitro*, having relatively high viability at low subzero temperatures—even showing survival after being plunged into liquid nitrogen. In contrast, a uterine fibroid cell line, ELT-3, has near zero viability after freezing to temperatures at or below –30°C (Bischof et al., 2001). The effects of individual thermal parameters have been investigated (Smith et al., 1999). AT-1 cells are relatively insensitive to the effects of cooling and thawing rates, and viability tends to depend on the other thermal parameters (end temperature, hold time). Similar results were found for ELT-3 cells. This information is important for drawing conclusions regarding the relative importance of the host-response injury mechanisms, which will be discussed in the following sections.

16.3 THE ROLE OF IMMUNOLOGICAL INJURY IN CRYOSURGERY

16.3.1 The Role of Immunological Injury in Tissue Necrosis

Cryosurgery was initially linked to changes in the immune system in the 1960s when, after cryosurgery of a primary prostate tumor, several investigators noted that there was a spontaneous regression of metastases (Soanes et al., 1970a). For example, a 66-year-old male with undifferentiated prostate adenocarcinoma that had lung and cervical spine metastases confirmed by x-ray received cryosurgery of a primary lesion on two separate occasions, 4 weeks apart. After the second cryosurgery, the lung metastases had disappeared from the chest x-ray and the neck pain resulting

TABLE 16.2
Experimental Results of Immune Response after Cryosurgery

Author	Year	Model System (induction)	Therapy	Immune effect
Myers et al.	1969	Mammary AC (viral)	Cryo	Positive
Neel and Ritts	1979	Mammary AC (viral) and FS (chemical)	Cryo	Positive
Lubaroff et al.	1981	R3327 prostate AC (syngeneic)	Cryo w/bCG injection	Positive
Miya et al.	1987	MRMT-1 mammary AC (chemical)	Cryo	Positive
Bayjoo et al.	1991	HSV-2-induced FS (viral)	Cryo	Positive
Misao et al.	1981	MRMT-1 mammary AC (chemical)	Cryo	Negative 1–3 weeks post, positive 7–10 weeks post
Miya et al.	1986	MRMT-1 mammary AC (chemical)	Cryo	Negative 1–3 weeks post, positive 7–10 weeks post
Eskandari et al.	1982	R3327 prostate AC (syngeneic)	Cryo	Slightly positive
Matsumura et al.	1982	MRMT-1 mammary AC (chemical)	Cryo	No effect
Hoffmann et al.	2001	AT-1 prostate AC (syngeneic)	Cryo	No effect
Muller et al.	1985	Dunn osteogenic sarcoma (syngeneic)	Cryo	Some positive and some negative effects
Hayakawa et al.	1982	1° FS (chemical)	Cryo	Negative
Yamashita et al.	1982	KMT-17 FS (chemical)	Cryo	Negative
Wing et al.	1988a	HSV-2-induced FS (viral)	Cryo	Negative
Friedman et al.	1997	MatLyLu AC (syngeneic)	Cryo of ventral prostate +/–Freund's	Negative
Roy et al.	1990	AFS Ascites FS (syngeneic)	Injection of cryodestroyed tumor	Negative

Note: The tumor type and source were described, along with the immunostimulatory therapy. Cryo, cryosurgery; AC, adenocarcinoma; FS, fibrosarcoma; bCG, bacillus Calmette-Guerin.

from the metastases in the spine had been alleviated. From this type of case history, an immuno-logical mechanism was proposed (Ablin et al., 1974a). It was theorized that during cryosurgery, the immune system of the host became sensitized to the tumor being destroyed by the cryosurgery. As the body resorbed the necrotic tissue, an active immunity was developed to the tumor tissue. Any primary tumor tissue undamaged by the cryosurgery and the metastases were destroyed by the immune system after cryosurgery. This response was termed the "cryoimmunological response."

Reviewing the literature shows that cryosurgery can either stimulate or inhibit the immune system (see Table 16.2). Several animal models and clinical case studies indicating a positive immunological response to tumor tissue after cryosurgery have been reported (Ablin et al., 1973; Lubaroff et al., 1981; Neel et al., 1973). Neel and Ritts (1979) grew chemically induced mammary sarcomas in mice and compared four different treatments (including cryosurgery) of the primary tumor as well as the effect of these treatments on a secondary tumor injected at a remote site. The cryotreated group showed the lowest incidence and smallest size of tumor after secondary challenge of the four groups. Similar data were reported from the Copenhagen rat using the syngeneic Dunning R3327 tumor cell line. After cryotreatment of small (<1 cm³) R3327 tumors in combination with an immunostimulatory therapy (bacillus Calmette-Guerin, bCG), secondary tumor challenges were rejected (Lubaroff et al., 1981).

Not all studies promote cryosurgery as a method for improving tumoricidal activity of the immune system. Hayakawa et al. (1982) found decreased resistance to a secondary tumor challenge

Mechanisms of Injury Caused by *in Vivo* Freezing

when a primary tumor was treated cryosurgically, compared with treatment by surgical excision. Other studies by this group found increased growth and metastatic rates of secondary tumors after a primary tumor had been treated with cryosurgery (Yamashita et al., 1982). It should be noted that the tumor systems used by this group were also chemically induced fibrosarcomas, similar to one of the tumors used by Neel and Ritts (1979).

The literature has shown that it is difficult to predict whether a tumor will be susceptible to immunological injury after cryosurgery. Ablin (1995) termed this subject "cryosensitivity," in which both the tumor and host are examined for their roles in generating the immune response. Table 16.2 shows some of the model systems used for investigations of the cryoimmunological effect, and the results show a spectrum of immunological consequences, from stimulation to inhibition of the immune system. The model systems studied for cryoimmunology have yielded no consistent results within tumor types. For example, cryosurgery of both adenocarcinomas (Lubaroff et al., 1981) and sarcomas (Neel and Ritts, 1979) have been shown to produce a positive immune response and to limit further tumor growth. In other studies, cryosurgery of both adenocarcinomas (Matsumura et al., 1982) and sarcomas (Yamashita et al., 1982) has had either no effect or even a negative immunological consequence, after which the tumor grows as rapidly as before the cryosurgery. Regarding syngeneic cell lines, the Dunning R3327 cell line has been shown by Lubaroff et al. (1981) to have inhibited growth after cryosurgery, but in studies by our group (Hoffmann et al., 2001), the Dunning AT-1 (a subline of R3327) cell line showed no inhibition. Attempts to correlate response with tumor type in other ways have been unsuccessful. Inclusion of other cell lines could eventually create trends, but as of now, the literature does not show any tumor type to have a predilection for being susceptible to cryoimmunological effects.

The literature indicates that the amount of antigen used to produce the antitumor immune response and the timing of the immune stimulation are critical in creating an effective antitumor immune response. The concept of "antigen excess" is that if the amount of antigen produced by cryosurgery exceeds the host's ability to remove the antigen, suppression of tumor immunity would occur through the formation of antigen–antibody complexes at the cryosurgical site (Ablin, 1995). Suppressor T-cells could be activated as well. Experimental evidence of this has been seen in other studies. Roy et al. (1990) found that injection of greater amounts of cryodestroyed tumor could actually decrease survival time in animals challenged with tumor cells. Neel and Ritts (1979) found that the growth suppression of a secondary tumor that occurred as a result of cryosurgery of a primary tumor was potentiated if, 24-h postcryosurgery, the primary tumor was excised. It is possible that removal of the primary tumor after cryosurgery allows more of the immunologically active cells to attack the secondary tumor. When the tumor was left inside the host, the amount of immune response the host generated against a second tumor could have been lower because more of the immunologically active cells were "deflected" toward the primary tumor. Studies in our laboratory (Hoffmann et al., 2001) found that using immunoadjuvants to promote an immune response to the tumor could actually increase the growth rate of the tumor, possibly because the immunoadjuvant added too much antigen for the host's immune system to absorb. With regard to the timing of immune stimulation, Roy et al. (1990) also found that challenging with live tumor cells at shorter times after injection of cryodestroyed tumor cells would decrease survival of the animal. Similarly, the work of Miya et al. (1986) has indicated that cryosurgery can suppress the immune system 1 to 3 weeks postcryosurgery and can stimulate the immune system 7 to 10 weeks postcryosurgery. Thus, timing appears critical.

From this mixed response of tumors to the immune stimulation caused by cryosurgery, investigators sought a mechanism that could explain the immune and tumor response to the surgery.

16.3.2 POTENTIAL MECHANISMS OF IMMUNOLOGICAL INJURY

There have been three main mechanisms of immune-mediated postcryosurgical injury proposed. The one that has received most research is the production of antitumor antibodies (Ablin et al.,

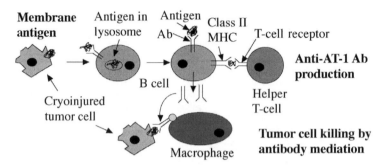

FIGURE 16.2 Antigen presentation to antibody production cascade in antibody–mediated tumor cell killing.

1974a). The hypothesis is that, as tumor cells die, they release many antigens from the proteins contained within the cell. These antigens will be of all origins, including those on the membrane of the cancer cell (see Figure 16.2). A membrane antigen will be phagocytosed and moved to the surface of an antigen-presenting cells, such as macrophages or B cells (Cotran et al., 1994a). The particular B-cell with an antibody (Ab) specific for the antigen on its surface will then be stimulated by a helper T-cell to transform into a plasma cell (the Ab-producing form of the B-cell). The Abs will then attach to the same membrane protein of other undamaged tumor cells. From there, the Ab will induce complement fixation and macrophage and neutrophil chemotaxis. Complement can produce pores in the membrane of cells, and macrophages and neutrophils can phagocytose cells as well as release enzymes and free radicals that kill the tumor cell. Macrophage-mediated cell killing is depicted in Figure 16.2. As these antibodies work to kill cells, they are often called cytotoxic antibodies (Yamashita et al., 1982).

Research in this area has investigated whether cryosurgery does indeed produce antibodies, but it has met with mixed results. Clinical cases showed that there was an overall increase in serum Ab levels, particularly IgM (Ablin et al., 1977), among other changes in serum proteins (Ablin et al., 1975). Serum from patients reacted with prostate tissue from both human and monkey prostates, indicating that Abs to prostate tissue were produced. Immunofluorescence studies indicated that these Abs were reacting to the epithelial and stromal cells of the prostate (Ablin et al., 1974a, 1974b), but the link between Ab production and cryosurgery was inconclusive. Some of these patients actually had a lower antiprostatic immune titer when compared with serum collected before surgery (Ablin et al., 1974b). In terms of animal models, cryosurgery of the normal rabbit prostate produces large immune titers to prostatic tissue (Yantorno et al., 1967). In a subsequent study, it was demonstrated that only a small percentage of the antiprostatic Abs were isoantibodies; that is, Abs that would attack the prostatic tissue in the animal that grew the Abs (Ablin et al., 1971). Presumably, it is this tissue that the Abs would have to attack for there to be an effect. Also, similar Ab levels were produced in rabbits that were injected with minced prostatic tissue (Shulman et al., 1968), and those levels could not be further increased by additional cryosurgery (Riera et al., 1968). These last studies question the specific role cryosurgery plays in developing antitumor Abs.

There have been studies in which the relationship between Ab and tumor growth was investigated both indirectly and directly. Lubaroff et al. (1979) showed that injection of tumor cells in Freund's adjuvant, an immunoadjuvant that has been shown to increase Ab production, would decrease the growth rate of a secondary tumor. Friedman et al. (1997) attempted to immunize animals to prostate cancer by performing cryosurgery on a normal prostate injected with Freund's adjuvant, assuming that the immune reaction created would cross-react with the prostate cancer cells. However, injection with adjuvant actually decreased the resistance to the tumor. Unfortunately, because neither study measured antibody levels directly, it is impossible to draw a correlation between antibody levels and tumor rejection. Yamashita et al. (1982) did look directly at the Ab-specific reaction to tumor cells. The authors showed, by enzyme-linked immunosorbent assay measurement, that cryosurgery did indeed stimulate cytotoxic Ab production. However, this did

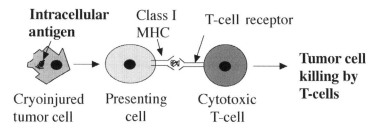

**Intracellular   Class I   T-cell receptor
antigen   MHC**

Tumor cell
killing by
T-cells

Cryoinjured   Presenting   Cytotoxic
tumor cell   cell   T-cell

FIGURE 16.3 Antigen presentation to cytotoxic T-cells in cytotoxic T-cell–mediated tumor cell killing.

not serve to protect the animal from a second tumor challenge. The animals that received cryosurgery showed less tumor rejection than those treated with surgical excision. Our group (Hoffmann et al., 2001) examined antitumor-Ab production after injection with lysed AT-1 tumor cells mixed either with PBS or Freunds. Both therapies produced an antitumor-Ab response significantly higher than both untreated tumor-bearing and non-tumor-bearing controls (as measured by serum enzyme-linked immunosorbent assay). The antitumor-Ab did not limit the growth of a second tumor. It would appear from these studies that Abs specific to tumor cells are not clearly protective from secondary tumor challenge or tumor metastasis.

A second potential method of the immune system affecting tumor growth is by cytotoxic T-cell–mediated tumor cell killing. During the immunosurveillance that goes on continuously within a host, intracellular antigens are moved to the surface of all cells by attachment to a membrane protein. If a cytotoxic T-cell recognizes that antigen, it will become activated and divide. The population of T-cells will then kill the tumor cells by secreting enzymes that mechanically disrupt the cell. However, it is important to note that this process goes on continuously, even before treatment, and that the tumor continues to grow despite this mechanism. Thus, it is hypothesized that cryosurgery somehow changes the presentation of antigen or sensitizes the T-cell by destroying many tumor cells in the vicinity. The former mechanism is depicted in Figure 16.3.

This mechanism of tumor cell killing has been shown to be the major form of malignant cell removal in the absence of treatment (Cotran et al., 1994b), yet it has not had much attention in cryosurgical research. Eskandari et al. (1982) examined T-cell activation after cryosurgery of the R3327 tumor in the Copenhagen rat. The T-cell activity was increased 2 weeks after cryosurgery when compared to preoperative levels. The activity declined from this peak but remained elevated throughout the remainder of the study. Controls (which received cryosurgery not of the tumor, but of the right thigh muscle near the tumor) did not show the initial rise of T-cell activity at 2 weeks, but activity increased at 4 weeks to the same level shown by the experimental group at 2 weeks. T-cell activity was inversely proportional to tumor growth, indicating the T-cell activity did decrease the tumor growth rate. However, it is unclear whether cryosurgery was truly the cause of T-cell activation because the reaction was also shown, albeit at a different time point, in the control group. Miya et al. (1986) examined cell proliferation in the lymph nodes after cryosurgery and found that T-cell proliferation decreased when the immune system showed other signs of being suppressed (1 to 3 weeks postcryosurgery) and increased when the immune system appeared to be stimulated (7 to 10 weeks postcryosurgery). Studies by our laboratory (Hoffmann et al., 2001) showed no difference in T-cell density (measured by immunocytochemistry) in the region of the growing tumor after cryosurgery. However, the assay used to measure T-cell response was not a functional assay. The number of T-cells present does not indicate whether these cells are activated in a tissue, and thus, it is difficult to conclude that there was no difference in T-cell activity, even though there was no difference in T-cell counts.

The final mechanism of tumor cell killing is very similar to the cytotoxic T-cell mechanism. In this mechanism, a natural killer (NK) cell destroys the tumor cells. These cells have a similar immunological mechanism as T-cells, but they do not require prior exposure to antigen for them to carry out cell destruction. Thus, under this hypothesis, cryosurgery does not sensitize the NK cell to the tumor, as it is not necessary, but rather, the cryosurgery is hypothesized to stimulate the activity of NK cells. Like the other mechanisms discussed, the research in this area has had mixed results. Bayjoo et al. (1991) showed that cryosurgery of a fibrosarcoma in rats increased NK cell activity. Wing et al. (1988a, 1988b) showed that cryosurgery decreased NK cell activity and increased immune suppressor cell activity in a fibrosarcoma of a different origin. Neither group offered any speculations as to the mechanism by which the activity of NK cells became changed. This research, at this point, does not point to cryosurgery having a specific effect on NK cell activity.

Unlike direct cell injury, where the link between cryosurgery and direct cell injury is over-whelming, the link between immunological injury and cryosurgery is at this point tenuous. As a result, linking any potential immunological changes to the thermal history experienced by the tissue appears premature at this point.

16.4 THE ROLE OF VASCULAR INJURY IN CRYOSURGERY

16.4.1 THE ROLE OF VASCULAR INJURY IN TISSUE NECROSIS

Cohnheim (1877) first hypothesized that the necrosis seen in frostbite was caused by hemostasis within the frozen tissue postthaw. Thereafter, Lewis and Love (1926) described the gross pattern of vascular response in human skin during and after freezing. After freezing specific areas of a subject's skin to $-5°C$, they observed the tissue as it warmed to room temperature. This study qualitatively showed the response of the vasculature to cooling. The previously frozen region was initially static, surrounded by hyperemia. After the tissue was completely thawed, flow returned to the previously frozen region, accompanied by edema. Rotnes and Kreyberg (1932) showed stasis that returned after the initial return of flow. By freezing small regions of rabbit ears, they observed the initial stasis and return of flow. However, in areas that were frozen for longer times (>5 sec), the blood flow did not persist in the capillaries. Twenty-four hours later, the vessels were static. On microscopic analysis, the capillaries were filled with columns of red blood cells. Rotnes and Kreyberg concluded that it was the column of blood cells that was likely occluding flow, but they did not exclude the possibility of a thrombus, which would have included fibrin in its formation.

The work that followed continued to support Cohnheim's original hypothesis. An example of this is the work done by Kreyberg and Hanssen (1950), in which they attempted to assess at what point postfreeze the tissue had been critically injured because of vascular stasis. They froze portions of mouse ears to $-78°C$ and transplanted them at various points after freezing to a subcutaneous pocket elsewhere on the mouse. For comparison, the number of ear tissue slices surviving such a procedure without being frozen was ~66%. If the tissue was transplanted 5 sec after freezing, approximately the same survival was seen (~66%). However, if the tissue was transplanted 3 h after freezing, less than 25% survived. This defined a possible time point at which the onset of irreversible vascular obstruction and resultant ischemia begins. More important, it showed that the freezing insult itself did not critically injure the tissue. It was the latent vascular effect that had a more profound effect on the tissue survival.

Further information as to the role of the vasculature in defining freezing injury was found when investigators administered vasoactive substances. Use of osmotic agents, such as inositol (Sullivan and LeBlanc, 1957) and low–molecular weight dextran (Mundth et al., 1963) to maintain plasma osmotic pressure, and therefore plasma volume, decreased tissue loss postfreeze. Several investigators (Lempke and Schumacker, 1949; Mundth et al., 1963) found that anticoagulation decreased tissue destruction. Mundth et al. (1963) also investigated the timing of anticoagulation. If the blood was heparinized for 1 h postfreezing, then the same amount of tissue destruction was seen as if the tissue was untreated. If the blood was heparinized continuously from 3 h to 5 d postfreeze, reduction in necrosis was seen. These studies further support the hypothesis that the vascular response is important, if not critical to determining tissue injury.

Not all vascular manipulations had such consistent effects. The most prominent disagreement in experimental evidence was in the area of blood flow into the region of freezing. Entin et al. (1954) found that vasodilation with Etamon chloride protected the tissue from necrosis. Rothenborg (1970), however, found that systemic vasoconstriction (from hemorrhage distant to the site of freezing) protected tissue from necrosis. Dilley et al. (1993) saw no difference between normal blood flow and complete vascular occlusion when performing cryosurgery on the liver, which was in direct contrast to earlier findings in a similar model system by Neel et al. (1971). These investigators used different thermal histories to induce freezing damage, and the effect of blood flow could have been dependent on the temperature or time at which the blood flow changes were made. This underscores the importance of thermal history in determining vascular injury.

The main limitation in the tissue-survival studies described above was that vascular injury was only investigated by indirect methods. Tissue destruction was seen as a result of vascular damage, and although the studies did attempt to remove all other variables from the possible cause of tissue death, they did not provide any data regarding the state of the vasculature. As a result, several other techniques were used to generate data regarding the state (open or static) of the vessel by direct imaging. One approach involves fluorescent vascular imaging within whole organs and limbs, which gives a macroscopic picture of injury. Lange and Boyd (1945; Lange et al., 1947) froze human and rabbit limbs and studied thrombosis and leakage using fluorescein as a contrast dye. Freezing to −78°C for 5 min created near-total stasis (signified by no fluorescence in the frozen region) for 1 h in the previously frozen region followed by increased perfusion and leakage. Stasis returned approximately 12 h postfreeze, and increasing the freezing time beyond 5 min decreased the time required for stasis to return. Stasis was found to be to the result of vascular thrombosis as assessed by histology. Similar to the findings by Mundth et al. (1963), heparin therapy maintained vessel patency (as signified by fluorescence penetrating into the frozen/thawed region) as long as the heparin was continued at least 5 d postthaw. Hurley (1957) followed this work by using mercury angiography postfreeze. The regions into which mercury was unable to penetrate became necrotic, as would be expected from vascular injury–induced ischemia.

To study the entire vasculature surrounding a frozen region, the technique of vascular casting has been used. One advantage this technique has is the ability to cast the entire vasculature that was perfused at the time of death. Scanning electron microscopy of the resulting cast allowed for great precision in estimating vascular size as well as imaging the "blind" capillaries that occurred as a result of angiogenesis or thrombosis. Hurley (1957) used vascular casting to confirm their mercury angiography results, Orita et al. (1988) used it for determining vascular damage postfreeze within the rat brain, and Daum et al. (1987, 1989, 1991) have published several papers using this technique to determine the vascular loss from freezing in rat feet. The main conclusion from these investigations was that the microvasculature (arterioles, venules, and capillaries) was preferentially destroyed in freezing injury.

The techniques of fluorescent imaging and angiography are able to determine the dynamic changes of the vasculature as a whole, but it is difficult to gain information about single vessels from these techniques. Vascular casting gives the accuracy needed, but it requires the death of the animal, providing a static picture, and cannot follow the dynamic progression of injury in a single animal. As a result, there have also been several "chambers" developed, such as the rabbit ear chamber and hamster cheek pouch, that allow direct imaging of the microvasculature with great resolution. These techniques allow accurate, dynamic measurements intravitally, but the chamber only allows for a thin "two-dimensional" vascular preparation. The thickness in the third dimension must be minimized to allow for transillumination. Furthermore, these animals have to be housed at near body temperature to prevent thermal conduction out of the chamber from causing the vasculature to regress. Thus, the advantages and limitations of each technique led to the use of all of the above techniques in defining a more complete picture of vascular injury.

The first chamber technique used for intravital viewing of the vasculature was the rabbit ear. The ear of the rabbit is very thin, and with contrast enhancement, the vasculature was easily illuminated. This was the model used by Cohnheim (1877) in framing his original hypothesis. Quintanella et al. (1947) performed an experiment similar to Rotnes and Kreyburg (1932; i.e., freezing with CO_2 snow for 30 sec) and saw stasis that began 10 min postthaw. Crimson and Fuhrman (1947) followed this work with fluorescein injection into rabbit ears frozen to approximately –60°C for 1 min. They showed that a freeze/thaw of a portion of the ear blocked fluorescein entrance into the previously frozen region and that fluorescein appeared in the edema fluid. This work was continued by Bellman and Strombeck (1960), who demonstrated thrombosis within the vasculature 3 d after freezing to –20°C using intravital techniques. Giampapa and Aufses (1981) noted that small vessels preferentially went static first after thawing, which was later confirmed by Daum et al. (1989). Finally, Lazarus and Hutto (1982) found that heparin could delay the stasis seen at 24 h postfreeze. This is similar to the findings by Mundth et al. (1963) and Lange et al. (1947) discussed above.

These studies all point to vascular injury as being critical in determining tissue damage postthaw. The fact that freezing preferentially destroys the microvasculature indicates that freezing disrupts the vasculature at its most critical site (Daum et al., 1987, 1989; Giampapa and Aufses, 1981). Without the oxygen and nutrient delivery that the microvasculature normally provides, tissue will become ischemic, and if this ischemia persists, necrosis can result. Studies by our group on AT-1 tissue grown in a dorsal skin flap chamber model of cryosurgery examined the link between stasis and necrosis after cryosurgery. Vascular stasis was produced in the chamber by freezing (Hoffmann and Bischof, 2001a), and the entire region that showed vascular stasis became necrotic at 3 d postcryosurgery. The thermal parameters required to cause this damage were much milder than those required for AT-1 cell destruction *in vitro* and were similar to temperatures that caused vascular injury in other studies. This result, combined with the results indicating that AT-1 tissue is relatively insensitive to the immunological consequences of freezing (Hoffmann et al., 2001), lends strong support to the hypothesis that vascular injury is important in determining the size of a lesion caused by freezing *in vivo*.

16.4.2 POTENTIAL MECHANISMS OF VASCULAR INJURY

While the studies discussed in the previous section do not directly show the mechanism of microvascular damage, they do point to a common source for vascular injury. Early intravital studies, such as Rotnes and Kreyberg (1932), observed leakage of large molecular dyes through the vessel wall after freezing, and several investigators observed oedema formation that could be reversed by increasing plasma osmotic pressure (Mundth et al., 1963; Sullivan and LeBlanc, 1957). Crimson and Fuhrman (1947) also showed vascular contrast dye in the edema fluid. These point to the endothelium as playing a role in vascular injury. Damage to the endothelium leads to the increased leakage of fluid and plasma proteins through the vessel wall that has been seen by many investigators (Bellman and Adams-Ray, 1956; Pollock et al., 1986). Also, removal of the endothelium from the vessel wall exposes the thrombogenic underlying connective tissue, causing the thrombus formation that has been seen postthaw in several studies (Bellman and Strombeck, 1960; Hoffmann and Bischof, 2001a; Zook et al., 1998). This could explain why anticoagulation could limit tissue injury postfreeze, as reported (Lange and Boyd, 1945; Lange et al., 1947; Mundth et al., 1963).

Histological assessment of frozen tissue at discrete time points postfreeze has shown evidence of endothelial damage as a result of freezing. An early example of this was the study by Greene (1943). He froze mouse tails for 1 min in a –62°C temperature bath and fixed and sectioned the tails at several points postfreeze. At 5 min postfreeze, he saw endothelial swelling. At 30 min postfreeze, he saw an increase in red blood cells in the vessels, and by 24 h postfreeze, there was marked inflammation and necrosis. To see the effect of thermal history on histology, he repeated the study, freezing this time in a –60°C bath for 3 min. The 5-min postfreeze time point seemed somewhat similar to the matching time point after the 1 min freeze, but by 30 min postfreeze, he saw arteriolar thrombosis and endothelial sloughing. The most interesting conclusion was that only in areas of vascular damage was there tissue necrosis. In the area he termed "destined to recover," he only saw slight vasodilation, whereas the area that would eventually be necrotic, he saw endothelial sloughing and arteriolar thrombosis.

The chamber techniques discussed in the preceding section have also suggested the importance of endothelial damage in vascular injury. Bourne et al. (1986) saw the same initial stasis and leakage in two groups of rabbit ears subjected to two different thermal histories, either –4°C for 3 min or –3°C for 4 min, for 12 h postfreeze. However, the lower freezing temperature (–4°C) produced a disruption of the endothelium. The tissue subjected to the lower freezing temperature with endothelial disruption showed a secondary stasis and a resulting tissue loss. The group subjected to the higher temperature that maintained the endothelium did not have tissue loss. Rabb et al. (1974) also reported endothelial sloughing in the region of stasis in the hamster cheek pouch. Giampapa and Aufses (1981) and Zook et al. (1998) both noted that endothelial cells were preferentially lost from the microvasculature and that it was the microvasculature that would eventually become static. Ninomiya et al. (1985), building on work by Iida and Iranpour (1972), investigated the effect of freezing on tumors grown in the hamster cheek pouch. The investigators sought to find whether the state of the endothelium could be used to predict the tissue response. They froze the tumors within the cheek pouch using a liquid nitrogen probe, and they used two markers for endothelial activity, alkaline phosphatase activity and UEA-1 lectin binding activity. They found that alkaline phosphatase and UEA-1 were both normal on the edge of the previously frozen region where the tissue showed no necrosis. There was no alkaline phosphatase or UEA-1 activity in the portion of the previously frozen region that would later become necrotic.

Several ultrastructural studies have been carried out to investigate the endothelial damage observed. Bowers et al. (1973) found that endothelial cells within vessels subject to freezing to –23°C *in vivo* lost membrane integrity. The muscle tissue surrounding the vessel containing damaged endothelium showed signs of injury, and the tissue in which the endothelium was intact showed no damage. During ultrastructural investigation, Rabb et al. (1974) showed gaps in the endothelial lining, and in some regions, the basement membrane had been obliterated by the freezing to –20°C. Beitner et al. (1989) and Zook et al. (1998) both showed similar endothelial and tissue damage to that found by Bowers et al. (1973). Rubinsky et al. (1990) found disruption of endothelia (by cryoscanning electron microscopy, which images the tissue in the frozen state) caused by cryosurgery in the liver *in vitro*. Finally, Marzella et al. (1989) did an extensive study of the timing of the endothelial events. The sequence they observed after *in vivo* freezing was: 1 min postthaw endothelial membrane disruption; 1-h postthaw vascular congestion, protein extravasation, and platelet adhesion; 6-h postthaw endothelial membrane vacuolization; and 24-h postthaw endothelial separation.

Although these studies strongly indicated that the state of the endothelium determines the state of the vasculature and surrounding tissue, they did not define how the endothelium was damaged. As a result, there have been several theorized causes for endothelial injury as a result of freezing. The first is that the formation of ice within the vasculature causes direct cellular injury to the endothelium (Rabb et al., 1974). As discussed in the section on direct cellular injury, freezing initiates several biophysical changes within the cell, the ultimate result of which can be cell death if the biophysical changes cause sufficient physiologic stress (see Figure 16.4b). The histological appearance of the endothelium and endothelial sloughing seen in chamber systems could be a result of cell death caused by direct cellular injury.

Two sources of endothelial injury occurring postthaw have been proposed, one based on free-radical propagation and the second based on neutrophil activation. These mechanisms start with the same initial steps. During the freezing process, the blood flow within the vasculature ceases. This causes ischemia to the tissue during the time the freezing process is occurring. Ischemia in

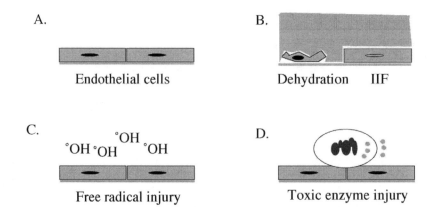

FIGURE 16.4 Mechanisms of endothelium injury: (A) uninjured endothelium, (B) direct cell injury based on biophysical changes, (C) free radical injury from postthaw hyperperfusion, (D) toxic enzyme injury from neutrophil chemotaxis postthaw (modified from Hoffmann and Bischof, 2002).

turn causes some cells to become hypoxic. When the ice thaws, blood flow is restored to the tissue (Lewis and Love, 1926). Because the hypoxic state of the cells causes them to release vasoactive factors, the vasculature dilates, and the tissue is hyperperfused. In the first postthaw mechanism, the high oxygen delivery of the hyperperfusion is theorized to cause free radical formation (Barker et al., 1987; see Figure 16.4c). It is the free radicals that are purported to cause endothelial damage by peroxidation of the lipids in the membrane. The lipid changes create pores in the membrane and destroy the endothelial cell. In support of this hypothesis, Manson et al. (1991) showed that vascular injury is reduced postthaw when the tissue is treated with superoxide dismutase, a free radical scavenger. Further support was given by Iyengar et al. (1990), who examined the use of iron chelators to inhibit free radical formation after freezing. Blood flow to the previously frozen region after thawing was much the same whether or not iron chelators were used, but the eventual necrosis was much greater in the untreated case.

The second postthaw mechanism hypothesizes that neutrophils are called to clean up the debris from the cells that died or were injured during the freeze. In the process, they adhere to the endothelium and release enzymes designed to digest the dead cells (see Figure 16.4d). However, with the large insult that comes from the tissue freezing, the neutrophils become overactivated and destroy live cells in the process. The endothelium is damaged in particular because the activated neutrophils must migrate through the endothelial lining. This would not be not surprising, as several of the freezing studies found neutrophils in the region of injured vessels (Carpenter et al., 1971; Hoffmann and Bischof, 2001a; Reite, 1965). Intravital techniques have shown neutrophil adhesion to the wall of vessels in the injury region (Zook et al., 1998). The work of Marzella et al. (1989) supported this conclusion by treating frozen tissue with a neutrophil inhibitor. This increased the patency of vessels within rabbit ears postthaw when compared to animals not treated with the inhibitor. Aliev et al. (1999) examined the aortic endothelium after freezing injury and saw significant adherence of white blood cells to the endothelium. This effect is similar to ischemia-reperfusion injury that has been described for other ischemic events, such as heart attack (Endrich et al., 1982). It should be noted that these two postthaw mechanisms of endothelial damage are not completely separate events. Neutrophil activation causes the release of free radicals, and perhaps neutrophils, as well as hyperemia, are a source of the free radical injury described above.

It is clear from this discussion that there are several theories as to how the vasculature becomes damaged after freezing, such as direct cell injury of the endothelium, free radical production, and neutrophil activation. Continued work is needed to elucidate the dominant or coupled mechanisms responsible for vascular stasis, and subsequent necrosis, postfreeze.

TABLE 16.3
Minimum Thermal Histories Required to Induce Vascular Injury

Author	Year	CR	ET	HT	TR	Tissue
Entin et al.	1954	1 °C/min	–5°C	2 h†	2°C/min	Dog leg
Mundth et al.	1963	~3°C/min	–15°C	18 min†	42°C bath	Dog leg
Arturson	1966	Bath immersion	–40°C*	5 min†	22°C bath	Dog paw
Carpenter et al.	1971	Bath immersion	–15°C*	15 min†	42°C bath	Rabbit paw
Bowers et al.	1973	2°C/min	–23.1°C	0	50°C/min	Mouse leg
Cummings and Lykke	1973	Contact w/metal plate	–20°C	30 sec†	N/A	Rat skin
LeFebvre and Folke	1975	Contact w/metal plate	–18°C	1 min†	37°C bath	Hamster cheek pouch
Rothenborg	1977	30°C/min	–30°C	0	Room temperature	Rat skin flap
Bourne et al.	1986	Contact w/metal plate	–4°C	3 min†	42°C bath	Mouse ear
Marzella et al.	1989	Bath immersion	–21°C*	1 min†	Room temperature	Rabbit ear
Daum et al.	1987	1°C/min	–30°C	0	20°C/min	Rat foot
Manson et al.	1991	Bath Immersion	–21°C*	1 min†	Room temperature	Rabbit ear
Schuder et al.	2000	7°C/min	0°C	0	5°C/min	Rat liver
Hoffmann et al.	2001	26°C/min	–16°C	0	9°C/min	AT-1 tumor in dorsal skin flap chamber

Note: (*) denotes that the bath temperature was known, but tissue temperatures were not. (†) denotes the hold time represents the total time spent cooling, rather than time spent held at a single temperature. CR, cooling rate; ET, end (or minimum) temperature; HT, time held at the minimum temperature (hold time); TR, thawing rate.

16.4.3 THE THERMAL HISTORY OF VASCULAR INJURY

The following discussion focuses on the thermal history of microvascular injury in freezing. Damage at this level is critical for determining ischemia caused by impaired gas and nutrient transport. In addition, there is evidence that the microvasculature is preferentially destroyed in freezing injury (Daum et al., 1987), and as a result, many studies have focused on finding the thermal history of microvascular injury (see Table 16.3).

In cryosurgery, the thermal history can be defined by the same thermal parameters used in the direct cellular injury studies: cooling rate, end temperature, hold time, and thawing rate. The literature shows that there is a qualitative understanding of the vascular effect of changing each thermal parameter. Decreasing end temperature increases vascular damage. This is shown most dramatically in the study by Bourne et al. (1986). In this study, freezing rabbit ears to –3°C for 4 min produced no permanent vascular damage, whereas freezing them to –4°C for 3 min produced vascular stasis and eventually tissue necrosis. It has been reported by several groups that slow thawing of the frozen region increases vascular injury and decreases tissue survival over that caused by rapid thawing (Artuson, 1966; Bellman and Adams-Ray, 1956; Entin et al., 1954; Hurley, 1957; Mundth et al., 1963; Sullivan and LeBlanc, 1957). Increasing hold time increases vascular damage, as was seen in a study by Carpenter et al. (1971). In this study, rabbit feet frozen for 18 min at –15°C showed less tissue damage and less edema (indicative of vascular damage) than rabbit feet frozen for 45 min at –15°C. Bellman and Strombeck (1960) also found that freezing to –20°C for 5 min produced more vascular stasis and thrombosis than freezing to –20°C for 2 min. Increasing cooling rate appears to increase vascular damage (Dilley et al., 1993; Salimi et al., 1986), but increased cooling rates in these studies produced lower end temperatures within the tissue. It is therefore difficult to define the effect of cooling rate by itself.

In contrast to the qualitative understanding above, developing the definition of a quantitative thermal history that causes vascular injury from the data in the literature is a difficult process. Studies that attempted to find the threshold of vascular injury are listed in Table 16.3. The thermal parameters found vary widely. For example, end temperature varies from temperatures as high as 0°C (Schuder et al., 2000) to as low as –40°C (Artuson, 1966). There are several potential reasons for this variation. As stated above, the state of the vasculature is important for determining the eventual effect of freezing. The vasculature varies from tissue to tissue and from species to species. Thus, the threshold of vascular injury may be different for each experimental model. Compounding the problem is the fact that investigators rarely use actual tissue temperatures (Mundth et al., 1963; Rabb et al., 1974). Many of these investigations used dry ice, metal blocks, air, or thermal baths of known temperature and exposed the region of interest to this medium for a set amount of time (Cummings and Lykke, 1973; Marzella et al., 1989). Thus, thermal history parameters within the tissue of interest are often unknown and are assumed to be close to the bath or rod temperature; however, the tissue temperature could be above the bath temperature because of thermal diffusion limitations. Furthermore, studies often do not report all four thermal parameters of interest. Finally, as has been discussed, after freezing, the vasculature becomes static, then reperfusion occurs, and finally a secondary, permanent stasis occurs. Depending on when the vasculature is examined, the vasculature could be in any one of the above states.

Studies in our laboratory sought to define the thermal history that causes microvascular injury. It was our goal to quantify the thermal history of tissue within a dorsal skin flap chamber and to model the thermal response of the tissue to cryosurgery (Hoffmann and Bischof, 2001b). The cryosurgical model system allowed for a gradient of thermal parameters, and it was concluded that 26°C/min to –16°C with no hold time and a 9°C/min thawing rate would destroy the vasculature of AT-1 tumor tissue grown in the dorsal skin flap. Similar results were found for normal rat hypodermis tissue. This is within the range of parameters defined in Table 16.3. We performed a similar experiment in a different model, the normal pig kidney, for comparison. In that model system, 9°C/min to –20°C with no hold time and a 12°C/min thawing rate would permanently destroy perfusion. It is interesting to note that the end temperature for damage was similar in the two model systems. The two models show very different direct cell injury characteristics, but when frozen *in vivo*, they have similar vascular injury characteristics. This could be because the site of injury (the endothelium) for the vasculature is potentially the same.

16.5 MOLECULAR ADJUVANTS FOR CRYOSURGERY

As cryosurgery continues to develop as a technique, new adjuvants are being sought to increase the destructive effect of freezing. The reason for this is during cryosurgery of the prostate and other organs such as liver, kidney, or brain, ultrasound or magnetic resonance imaging can be used to monitor the extent of the cryosurgical iceball. This iceball diameter is used at some level to predict the outcome of the procedure. The edge of the iceball is at a temperature of roughly –0.5°C, whereas thresholds for prostate cancer destruction are reported from –20 to –60°C and sometimes lower (Hoffmann and Bischof, 2002). Thus, although monitoring is useful for imaging the iceball, not all of the tissue frozen is effectively treated. In some tissues, such as liver and sometimes kidney, the iceball is allowed to progress into a margin of normal tissue beyond the tumor. However, this is more of a problem with prostate, as overfreezing into sensitive adjacent structures such as the rectum and urethra can cause complications such as rectal and urethral fistulas (Saliken et al., 2002). However, if the surgeon is too conservative and underfreezes by keeping the iceball solely within the prostate, then any cancer that (often) exists under the prostate capsule at the edge of the gland may not be effectively treated, potentially leading to recurrence of disease. Clearly, therapies that would allow freezing to minimal subzero temperature to be effective in destroying tumor tissue would greatly improve the clinical application of cryosurgery in the prostate. This would enable techniques that imaged the iceball to predict tissue destruction. This has lead to the approach

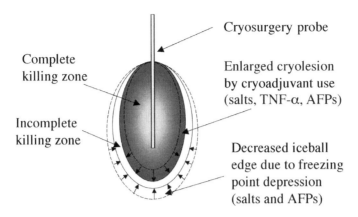

FIGURE 16.5 Overview of the potential benefit to using molecular adjuvants in cryosurgery.

outlined in Figure 16.5, where cryosurgical adjuvants that both increase freeze destruction or reduce the temperature of the iceball edge are being characterized. In this final section, several important modulators of freezing injury currently being investigated are discussed.

16.5.1 APOPTOSIS INITIATORS

Apoptosis has been recently investigated for its role in cryoinjury. The theory behind these investigations is that although a freezing insult might not destroy a cell directly or cause vascular injury leading to ischemic necrosis, it is possible that a cell with sublethal injury might undergo programmed cell death. This has been investigated extensively *in vitro*. Investigations have found evidence of apoptosis after freezing (Baust et al., 2000; Clarke et al., 1999; Hanai et al., 2001; Yang et al., 2003) or cold injury (Soloff et al., 1987), and the time and temperature ranges at which this apoptosis has occurred appear particularly significant for cryosurgery. The temperature range at which apoptosis has occurred is from as high as 6°C to as low as –10°C and perhaps lower (Clarke et al., 1999; Nagle et al., 1990; Roberts et al., 1997). The time spent at these temperatures required to create apoptosis is in the order of minutes (Nagle et al., 1990). This makes it particularly interesting for cryosurgery, because the destructive capability of cryosurgery is in question at the periphery of the iceball, where the temperatures will be close to 0°C for several minutes. Thus, use of apoptosis initiators, such as chemotherapeutic agents, could increase cell death in this peripheral zone.

Nearly all the apoptosis research in cold and freezing injury has been *in vitro*. *In vitro* apoptosis appears to depend on both the cell and microenvironment. For example, if the cells are confluent, apoptosis is more unlikely to occur than if the cells are in proliferative growth (Soloff et al., 1987). The state of cytoskeletal structures also influences apoptosis, as does the cell type (Clarke et al., 1999; Grand et al., 1995; Soloff et al., 1987). Several environmental factors appear to affect apoptosis. For example, one study showed that a sustained dose of high extracellular Ca^{2+} concentration could increase apoptosis (Perotti et al., 1990). Nagle et al. (1990) showed that rewarming the cells to 37°C would promote apoptosis compared to rewarming to 23°C. The absence of serum in the media can also lead to apoptosis (Kruman et al., 1992), as will the inclusion of chemotherapeutic agents that affect DNA synthesis (Clarke et al., 1999).

Despite these data, it is unclear what role apoptosis plays *in vivo*. There have been few results showing apoptosis after *in vivo* cryosurgery. Our laboratory attempted to show apoptosis of kidney and AT-1 tissue after cryosurgery. We saw evidence of apoptosis only after culturing the tissue *in vitro* at 37°C for 3 days. This included the classic DNA laddering (Roberts et al., 1997) and nuclear fragmentation, but it was unclear what role the culture technique vs. freezing played in this result. We have subsequently found that culturing kidney tissue at 23°C for up to 3 d can be achieved without large-scale apoptosis (Rupp et al., 2002). To confirm there was little apoptosis after freezing *in vivo*, kidney and AT-1 tissues were left *in situ* (presumably at 37°C) for 3 d after cryosurgery. Histological analysis of that tissue showed no apoptosis. It is therefore unclear what mechanism is causing the apoptosis of the kidney and AT-1 tissue during *in vitro* culture, but the mechanism appears unrelated to cryosurgery.

16.5.2 ANTIFREEZE PROTEINS

Antifreeze proteins (AFPs) are glycoproteins that are extremely efficient at inhibiting crystallization in frozen solutions. AFPs have been found in certain teleost fish from polar and north temperate regions (Ahmed et al., 1975; Ewart et al., 1992; Rubinsky et al., 1992) and in some overwintering insects (Tomchaney et al., 1982; Wu et al., 1991; see also Chapters 2, 3, 22, and 23). The proteins are also called thermal hysteresis proteins, as they noncolligatively depress the freezing temperature by up to a degree while still melting at the colligatively or osmotically predicted phase change temperature. Their presence in organisms that can survive subzero temperatures led to study of their potential use in cryopreservation. The studies have sought to use AFPs to increase the preservation of biological cells and tissues at subzero temperatures by either depressing the nucleation temperature of ice or inhibiting recrystallization (Arav et al., 1993; Carpenter and Hansen, 1992; Chao et al., 1996; Rubinsky et al., 1991a, 1991b; Wowk et al., 2000).

AFPs became of interest to the field of cryosurgery when studies attempting to preserve the viability of cells with AFPs showed increased cell damage and death with high concentrations of AFPs (Carpenter and Hansen, 1992). The mechanism for this effect appears to lie in the AFPs' changing the way ice crystals form and grow. The AFPs can change the morphology of the ice by making the crystal grow along an energetically unfavorable axis (the c-axis; Coger et al., 1994). This means that greater driving forces (hence lower subcooling) are necessary to initiate the crystal growth. However, once the crystals begin to form (if the concentration is sufficiently high), the morphology of the ice growing along this energetically unfavorable axis is changed to a spicular (from the normal dendritic) habit (Rubinsky and Devriesa, 1989). There have been several damaging cellular events shown to result from this ice crystal growth, including membrane disruption (Rubinsky and Devriesa, 1989) and increased IIF (Larese et al., 1996). It is this fact, which has been documented with increasing regularity, that may be exploited to increase the destructive capability of cryosurgery (Coger et al., 1994; Ishiguro and Rubinsky, 1994).

Data show that cooling rates as low as 1°C/min can destroy cells in the presence of the AFPs (Rubinsky and Devriesa, 1989). This is encouraging, as in cryosurgery, cooling rates at the periphery of the iceball can be 1°C/min or slower. As the destructive capability of cryosurgery is in question, particularly at the periphery of the iceball, where cooling rates are low and end temperatures relatively high, the injection of sufficient quantities of AFPs could increase the positive kill rate. This would decrease any possible tissue sparing at the boundary of the cryosurgical lesion and increase the destructive capability of cryosurgery. The accentuation of *in vivo* injury with AFPs has been demonstrated by Pham et al. (1999). After injecting a high concentration AFP/saline solution into a tumor before cryosurgery, thermal histories that did not damage control tissue (tissue that had not been injected) were shown to destroy the tumor. It appears that further investigation of AFPs as an adjuvant to cryosurgery is warranted.

16.5.3 ELEVATED CONCENTRATION SALT SOLUTIONS

The use of salt as a cryoadjuvant has received recent attention at the Cryobiology Society annual meeting (Muldrew et al., 2001). In this study, cells exposed to excess salt (3× isotonic) prior freeze thaw were shown to exhibit enhanced injury postthaw vs. control (1×) in the AT-1 cell system. Our group has also found this effect and has shown that an important part of this injury is not only traditional solute-effects injury but also induced eutectic formation, a second solid-phase transition

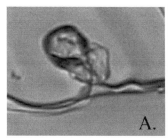 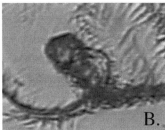

FIGURE 16.6 Cryomicroscopy images of AT-1 cell freezing in 5×PBS. (A) Cell after ice formation and some dehydration but before eutectic formation. (B) Cell after both ice and eutectic formation (modified from Han and Bischof, 2003).

that occurs after ice is in the system. A control experiment in which AT-1 cells were frozen at 5°C/min in 2× NaCl to –25°C, followed by either an induced eutectic or no eutectic, shows roughly a 50% drop in cell survival solely as a result of the effect of the eutectic (data not shown; Han and Bischof, 2003b, 2003a).

Cryomicroscopy images of AT-1 cell freezing in 5×PBS during eutectic formation are shown in Figure 16.6. Before eutectic formation, which requires significant supercooling, AT-1 cells were trapped in unfrozen fractions between ice crystals. Once eutectic formation had been initiated, it propagated along the unfrozen fraction. Eventually, the cells encountered the eutectic, which in the figure appears to actually penetrate or propagate into the cell. The control and prediction of the nucleation and crystallization of the eutectic phase both inside and outside cells may benefit from the substantial modeling framework already available for intracellular ice formation within the cryobiology literature (Toner, 1993). One possible mode of manipulation of the eutectic may be with the addition of different salts or molecules.

The eutectic can be achieved by addition (i.e., injection) of elevated concentrations of salts into a tumor system, which will then form spontaneously on cooling the iceball. Various salts have different eutectic temperatures, and thus it is possible that different thermal regimes can be preselected by use of different salts (or other eutectic forming solutions). Figure 16.7 shows the postthaw viability changes of AT-1 cell suspension in various media with end-temperature changes. The viabilities of AT-1 in culture media with either KCl or KNO3 are significantly lower than those in 1×NaCl and culture media at the same end temperature. Although these potassium salts are not likely to be used systemically, they are a tantalizing suggestion of what might be possible with the appropriate salt or eutectic-forming solution.

16.5.4 PROINFLAMMATORY CYTOKINES

Our original work in Dunning AT-1 prostate tumor vasculature convinced us that there was an important role for the endothelium in destroying the microvascular supply after cryosurgical injury in a dorsal skin-fold chamber (Hoffmann and Bischof, 2002). We have since repeated the experiments on LNCap Pro 5 human prostate cancer grown in nude mice (Chao et al., in press). We were gratified to find very similar effects with, if anything, slightly higher threshold temperatures of about –14°C in the tumor, similar to the thresholds seen in Table 16.3. The vascular nature of the injury appears to be a potentially important link among all of the tissues. As discussed previously, the role of inflammation and the endothelium are highly implicated in the progression of postcryosurgical vascular stasis and necrosis.

Of particular interest is the use of TNF-α and perhaps other cytokines as cryoadjuvants to extend the cryolesion through the induction of inflammation before freezing. TNF-α is known to promote inflammation, endothelial injury, and apoptosis (Yilmax et al., 1998) while also showing cytotoxic effects on LNCaP (Sherwood et al., 1990) and generally being injurious to tumor

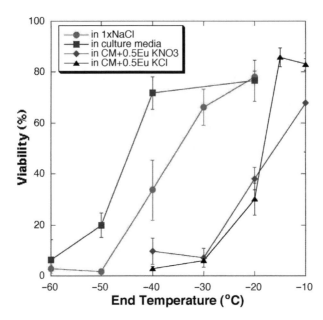

FIGURE 16.7 Freezing injury enhancement of AT-1 cell suspension by use of eutectic formation (modified from Han and Bischof, 2003).

microvasculature (Mauceri et al., 2002; Staba et al., 1998). With these benefits, TNF-α has been investigated as a therapeutic agent even in the absence of other therapies. Although systemic perfusion of TNF-α is not well tolerated, isolated perfusion can be used (Fraker and Alexander, 1994). Using it as a cryoadjuvant, one can envisage local administration by injection or perfusion followed by local freezing. In addition, we may be able to either direct cells that produce TNF-α or inject TNF-α directly to a local tissue of interest (i.e., a tumor) to increase cryosurgical (and perhaps also thermal therapeutic) efficacy.

Figure 16.8 shows the effects of a 15-min topical exposure to TNF-α (10 ng/mL) followed by a 4-h latent period before freezing or sham. Figures 16.8a and 16.8c show where LNCaP tumor was either not exposed or exposed to TNF-α, respectively, before sham cryosurgery and 3-d recovery. The vasculature is clearly patent, and the tissue histologically viable in both cases (data not shown). Figures 16.8b and 16.8d show tumors that were either not exposed or exposed to TNF-α, respectively, and that were frozen in the same way within the DSFC window, followed by 3-d recovery. The addition of TNF-α increases the tumor sensitivity to low temperatures. For example, adding TNF-α increased the thermal threshold to beyond the iceball edge (0°C) and occasionally beyond in the tumor. The end temperature at the edge of the necrosis zone was more than 10°C. Similar, but not as dramatic, results were obtained in normal nude mouse hypodermis without tumors (Chao et al., in press). These results clearly indicate the potentially important role that proinflammatory adjuvants (and others) may play in realizing the strategy of reconciling the iceball with the cryolesion, as indicated in Figure 16.5.

16.6 CONCLUSIONS

Cryosurgery has been shown to be a potent method of *in situ* tissue destruction. The precise mechanisms whereby this is achieved continue to be investigated, particularly during the conditions typical of the periphery of an iceball, where the threshold between survival and destruction is observed. Direct cell injury is clearly important in explaining freezing in any cell that was frozen; however, it does not always account for the high injury seen *in vivo* vs. *in vitro* at the conditions that occur at the periphery of an iceball. Immunological mechanisms may also play a role, but the

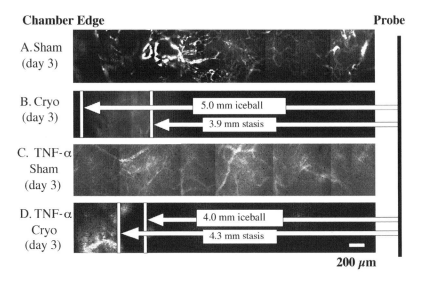

FIGURE 16.8 Fluorescent-contrast enhanced images of the vasculature in LNCaP Pro 5 tumors grown in the dorsal skin-flap chamber: (A) 3 days after sham cryosurgery without TNF-α, (B) 3 days after cryosurgery without TNF-α, (C) 3 days after sham cryosurgery with TNF-α, and (D) 3 days after cryosurgery with TNF-α (Chao et al., in press).

results of numerous studies of this type of injury remain inconclusive. Work at the microvascular level indicates that vascular injury often defines the edge of the tissue injury after freezing *in vivo*. One potentially fruitful area of investigation will be the assessment of whether "direct cell injury" to the endothelium is the major cause of microvascular injury at the periphery of a cryolesion. Finally, several molecular adjuvants to cryosurgery, including apoptosis initiators, AFPs, salt solutions, and proinflammatory cytokines, may potentiate destruction during freezing at the periphery of the iceball.

ACKNOWLEDGMENTS

This work was supported by NIH 1R29-CA75284. Thanks to Bumsoo Han for work on the salt, and Bo Chao for help on the TNF-α cryoadjuvant sections. Gifts from Candela Corp. and EndoCare to support this work are also gratefully acknowledged.

REFERENCES

Ablin, R.J. (1995) An appreciation and realization of the concept of cryoimmunology, in *Percutaneous Prostate Cryoablation*, Onik, G.M., Rubinsky, B., Watson, G., and Ablin, R.J., Eds., Quality Medical, St. Louis, MO, p. 136.

Ablin, R.J., Gonder, M.J., and Soanes, W.A. (1974a) Elution of cell-bound anti-prostatic epithelial antibodies after multiple cryotherapy of carcinoma of the prostate, *Cryobiology*, 11, 218–221.

Ablin, R.J., Gonder, M.J., and Soanes, W.A. (1974b) Immunohistologic studies of carcinoma of the prostate. 3. Elution of interepithelial antibodies from carcinomatous human prostatic tissue following cryoprostatectomy, *Oncology*, 29, 329–334.

Ablin, R.J., Gonder, M.J., and Soanes, W.A. (1975) Alterations of alpha2-globulin and the clinical response in patients with prostatic cancer following cryotherapy, *Oncology*, 32, 127–144.

Ablin, R.J., Soanes, W.A., and Gonder, M.J. (1973) Elution of *in vivo* bound antiprostatic epithelial antibodies following multiple cryotherapy of carcinoma of prostate, *Urology*, 2, 276–279.

Ablin, R.J., Soanes, W.A., and Gonder, M.J. (1977) Serum proteins in prostatic cancer. V. Alterations in immunoglobulins and clinical responsiveness following cryoprostatectomy, *Urol. Int.*, 32, 56–64.

Ablin, R.J., Witesby, E., Jagodzinski, R.V., and Soanes, W.A. (1971) Secondary immunologic response as a consequence of the *in situ* freezing of rabbit male adnexal glands tissues of reproduction, *Exp. Med. Surg.*, 29, 72–88.

Ahmed, A.I., Feeney, R.E., Osuga, D.T., and Yeh, Y. (1975) Antifreeze glycoproteins from an Antarctic fish. Quasi-elastic light scattering studies of the hydrodynamic conformations of antifreeze glycoproteins, *J. Biol. Chem.*, 250, 3344–3347.

Aliev, G., Ragazzi, E., Smith, M., Mironov, A., and Perry, G. (1999) Morphological features of regeneration of rabbit aortic endothelium after cryoinduced vascular damage, *J. Submicroscop. Cytol. Pathol.*, 31, 495–502.

Arav, A., Rubinsky, B., Fletcher, G., and Seren, E. (1993) Cryogenic protection of oocytes with antifreeze proteins, *Mol. Reprod. Dev.*, 36, 488–493.

Arnott, J. (1851) *On the Treatment of Cancer by the Regulated Application of an Anesthetic Temperature*, J. and A. Churchill, London.

Artuson, G. (1966) Capillary permeability in experimental rapid freezing with rapid and slow rewarming, *Acta. Chir. Scand.*, 131, 402–407.

Asahina, E., Shimada, K., and Hisada, Y. (1970) A stable state of frozen protoplasm with invisible intracellular ice crystals obtained by rapid cooling, *Exp. Cell Res.*, 59, 349–358.

Barker, J.H., Bartlett, R., Funk, W., Hammersen, F., and Messmer, K. (1987) The effect of superoxide dismutase on the skin microcirculation after ischemia and reperfusion, *Prog. Appl. Microcirc.*, 12, 276–281.

Baust, J.M., Van, B., and Baust, J.G. (2000) Cell viability improves following inhibition of cryopreservation-induced apoptosis, *In Vitro Cell. Dev. Biol. Anim.*, 36, 262–270.

Bayjoo, P., Rees, R.C., Goepel, J.R., and Jacob, G. (1991) Natural killer cell activity following cryosurgery of normal and tumor bearing liver in an animal model, *J. Clin. Lab. Immunol.*, 35, 129–132.

Beitner, R., Chen-Zion, M., Sofer-Bassukevitz, Y., Morgenstern, H., and Ben-Porat, H. (1989) Treatment of frostbite with the calmodulin antagonists thioridazine and trifluoperazine, *Gen. Pharmacol.*, 20, 641–646.

Bellman, S. and Adams-Ray, J. (1956) Vascular reactions after experimental cold injury, *Angiology*, 7, 339–367.

Bellman, S. and Strombeck, J.O. (1960) Transformation of the vascular system in cold injured tissue of the rabbit ear, *Angiology*, 11, 108–125.

Berrada, M.S. and Bischof, J.C. (2000) A determination of biophysical parameters related to freezing of an ELT-3 cell line, *Proc. ASME*, HTD-368/BED-47, 41–48.

Bischof, J.C., Fahssi, W.M., Smith, D.J., Nagel, T., and Swanlund, D.J. (2001) A parametric study of freezing injury in ELT-3 uterine leiomyoma tumor cells, *Hum. Reprod.*, 16, 340–348.

Bischof, J.C., Smith, D.J., Pazhayannur, P.V., Manivel, C., Hulbert, J., and Roberts, K.P. (1997) Cryosurgery of Dunning AT-1 rat prostate tumor: Thermal, biophysical and viability response at the cellular and tissue level, *Cryobiology*, 34, 42–69.

Bourne, M.H., Piepkorn, M.W., Clayton, F., and Leonard, L.G. (1986) Analysis of microvascular changes in frostbite injury, *J. Surg. Res.*, 40, 26–35.

Bowers, W.D., Hubbard, R.W., Daum, R.C., Ashbaugh, P., and Nilson, E. (1973) Ultrastructural studies of muscle cells and vascular endothelium immediately after freeze-thaw injury, *Cryobiology*, 10, 9–21.

Carpenter, H.M., Hurley, L.A., Hardenbergh, E., and Williams, R.B. (1971) Vascular injury due to cold. Affects of rapid rewarming, *Arch. Pathol.*, 92, 153–161.

Carpenter, J.F. and Hansen, T.N. (1992) Antifreeze protein modulates cell survival during cryopreservation: Mediation through influence on ice crystal growth, *Proc. Natl. Acad. Sci. USA*, 89, 8953–8957.

Chao, B. and Bischof, J. (2003) Cryosurgery of normal and LNCaP Pro 5 human prostate tumor tissue in the dorsal skin flap chamber, in *Summer Bioengineering Meeting*, Skalak, T., Ed., ASME, Key Biscayne, Florida.

Chao, H., Davies, P.L., and Carpenter, J.F. (1996) Effects of antifreeze proteins on red blood cell survival during cryopreservation, *J. Exp. Biol.*, 199, 2071–2076.

Chao, B., He, X., and Bischof, J. (in press) Pre-treatment inflammation induced by TNF-α augments cryo-surgical injury on human prostate cancer, *Cryobiology*.

Clarke, D.M., van Buskirk, R.G., and Baust, J.G. (1999) Timing dependency in cryo-chemo combination therapy: Model cell systems, *Cryobiology*, 39, 320.

Coger, R., Rubinsky, B., and Fletcher, G. (1994) Microscopic pattern of ice crystal growth in the presence of thermal hysteresis proteins, *ASME J. Offshore Mech. Arctic Eng.*, 116, 173–179.

Cohnheim, J. (1877) *Lectures on General Pathology,* Verlag von August, Hirschwald, Berlin.

Cotran, R.S., Kumar, V., and Robbins, S.L. (1994a) Diseases of immunity, in *Robbins Pathologic Basis of Disease*, Schoen, F.J., Ed., W.B. Saunders, Philadelphia, pp. 171–240.

Cotran, R.S., Kumar, V., and Robbins, S.L. (1994b) Neoplasia, in *Robbins Pathologic Basis of Disease*, Schoen, F.J., Ed., W.B. Saunders, Philadelphia, pp. 171–240.

Crimson, J.M. and Fuhrman, F.A. (1947) Studies on gangrene following cold injury V: The use of fluorescein as an indicator of local blood flow: Fluorescein tests in experimental frostbite, *J. Clin Invest.*, 26, 268–276.

Cummings, R. and Lykke, A.W. (1973) Increased vascular permeability evoked by cold injury, *Pathology*, 5, 107–116.

Daum, P.S., Bowers, W.D., Jr., Tejada, J., and Hamlet, M.P. (1987) Vascular casts demonstrate microcirculatory insufficiency in acute frostbite, *Cryobiology*, 24, 65–73.

Daum, P.S., Bowers, W.D., Jr., Tejada, J., Morehouse, D., and Hamlet, M.P. (1989) An evaluation of the ability of the peripheral vasodilator buflomedil to improve vascular patency after acute frostbite, *Cryobiology*, 26, 85–92.

Daum, P.S., Bowers, W.D., Jr., Tejada, J., Morehouse, D., and Hamlet, M.P. (1991) Cooling to heat of fusion (HOF), followed by rapid rewarming, does not reduce the integrity of microvascular corrosion casts, *Cryobiology*, 28, 294–301.

Devireddy, R.V., Swanlund, D.J., Roberts, K.P., and Bischof, J.C. (1999) Subzero water permeability parameters of mouse spermatozoa in the presence of extracellular ice and cryoprotective agents, *Biol. Reprod.*, 61, 764–775.

Dilley, A.V., Dy, D.Y., Warlters, A., Copeland, S., Gillies, A.E., Morris, R.W., Gibb, D.B., Cook, T.A., and Morris, D.L. (1993) Laboratory and animal model evaluation of the Cryotech LCS 2000 in hepatic cryotherapy, *Cryobiology*, 30, 74–85.

Endrich, B., Laprell-Moschner, C., Brendel, W., and Messmer, K. (1982) Effects of prolonged cold injury on the subcutaneous microcirculation of the hamster. I. Technique, morphology and tissue oxygenation, *Res. Exp. Med.*, 181, 49–61.

Entin, M.A., Schultz, G.A., and Baxter, H. (1954) Effect of slow and rapid warming on prolonged chilling and freezing of the legs of dogs, *Angiology*, 5, 486–499.

Eskandari, H., Ablin, R.J., and Bhatti, R.A. (1982) Immunologic responsiveness and tumor growth of the Dunning R3327 rat prostatic adenocarcinoma following cryosurgery and orchiectomy, *Ind. J. Exp. Biol.*, 20, 872–874.

Ewart, K.V., Rubinsky, B., and Fletcher, G.L. (1992) Structural and functional similarity between fish antifreeze proteins and calcium-dependent lectins, *Bioch. Biophys. Res. Comm.*, 185, 335–340.

Fraker, D.L. and Alexander, H.R. (1994) The use of tumor necrosis factor (TNF) in isolated perfusion: results and side effects. The NCI results, *Melanoma Res.*, 4, 27–29.

Fraser, J. and Gill, W. (1967) Observations on ultra-frozen tissue, *Br. J. Surg.*, 54, 770–776.

Friedman, E.J., Orth, C.R., Brewton, K.A., Ponniah, S., and Alexander, R.B. (1997) Cryosurgical ablation of the normal ventral prostate plus adjuvant does not protect Copenhagen rats from Dunning prostatic adenocarcinoma challenge, *J. Urol.*, 158, 1585–1588.

Gage, A.A. (1992) Cryosurgery in the treatment of cancer, *Surg. Gynecol. Obstet.*, 174, 73–92.

Gage, A.A. and Baust, J. (1998) Mechanisms of tissue injury in cryosurgery, *Cryobiology*, 37, 171–86.

Giampapa, V.C. and Aufses, A.H. (1981) The vascular effect of cold injury, *Cryobiology*, 19, 49–54.

Grand, R.J., Milner, A.E., Mustoe, T., Johnson, G.D., Owen, D., Grant, M.L., and Gregory, C.D. (1995) A novel protein expressed in mammalian cells undergoing apoptosis, *Exp. Cell Res.*, 218, 439–451.

Greene, R. (1943) The immediate vascular changes in true frostbite, *J. Pathol. Bact.*, 55, 259.

Han, B. and Bischof, J.C. (2003a) Direct cell injury associated with eutectic crystallization during freezing, *Cryobiology,* 48, 8–21.

Han, B. and Bischof, J. (2003b) Enhancement of cell and tissue destruction in cryosurgery by use of eutectic freezing, in *Proc. SPIE Conference BiOS2003—Thermal Treatment of Tissue: Energy Delivery and Assessment II,* Ryan, T., Ed., San Jose, CA. 4954, 106–113.

Hanai, A., Yang, W.L., and Ravikumar, T.S. (2001) Induction of apoptosis in human colon carcinoma cells HT29 by sublethal cryo-injury: Mediation by cytochrome c release, *Int. J. Cancer*, 93, 526–533.

Hayakawa, K., Yamashita, T., Suzuki, K., Tomita, K., Hosokawa, M., Kodama, T., and Kobayashi, H. (1982) Comparative immunological studies in rats following cryosurgery and surgical excision of 3-methyl-cholanthrene-induced primary autochthonous tumors, *Gann*, 73, 462–469.

Hetzel, F.W., Kruuv, J., McGann, L.E., and Frey, H.E. (1973) Exposure of mammalian cells to physical damage: effect of the state of adhesion on colony-forming potential, *Cryobiology*, 10, 206–211.

Hoffmann, N. and Bischof, J. (2002) The cryobiology of cryosurgical injury, *Urology*, 60, Suppl. 2A, 40–49.

Hoffmann, N.E. and Bischof, J. (2001a) Cryosurgery of normal and tumor tissue in the dorsal skin flap chamber II: injury response, *ASME–JBME*, 123, 310–316.

Hoffmann, N.E. and Bischof, J.C. (2001b) Cryosurgery of normal and tumor tissue in the dorsal skin flap chamber I: thermal response, *ASME–JBME*, 123, 301–309.

Hoffmann, N.E., Coad, J.E., Huot, C.S., Swanlund, D.J., and Bischof, J.C. (2001) Investigation of the mechanism and the effect of cryoimmunology in the Copenhagen rat, *Cryobiology*, 41, 59–68.

Hurley, L.A. (1957) changes associated with rapid rewarming subsequent to freezing injury, *Angiology*, 8, 19.

Iida, T. and Iranpour, B. (1972) The effect of deep freezing on hamster cheek pouch carcinoma: a preliminary report, *Oral Surg. Oral Med. Oral Pathol.*, 34, 844–849.

Ishiguro, H. and Rubinsky, B. (1994) Mechanical interactions between ice crystals and red blood cells during directional solidification, *Cryobiology*, 31, 483–500.

Iyengar, J., George, A., Russell, J.C., and Das, D.K. (1990) The effects of an iron chelator on cellular injury induced by vascular stasis caused by hypothermia, *J. Vasc. Surg.*, 12, 545–551.

Jacob, G., Kurzer, M.N., and Fuller, B.J. (1985) An assessment of tumor cell viability after *in vitro* freezing, *Cryobiology*, 22, 417–426.

Kreyberg, L. and Hanssen, O.E. (1950) Necrosis of whole mouse skin *in situ* and survival of tranplanted epithelium after freezing to 78°C and 196°C, *Scand. J. Clin. Lab. Invest.*, 2, 168–170.

Kreyberg, L. and Rotnes, P.L. (1931) La stase expérimentale, méthode pour la mettre en guidance au moyen des préparations spéciales, *C.R. Soc. Biol.*, 106, 895–897.

Kremer, C. and Duffy, S. (2000) *In vitro* studies of cryoablation of the endometrium, *Am. J. Obstet. Glyecol.*, 183, 22–27.

Kruman, I.I., Gukovskaya, A.S., Petrunyaka, V.V., Beletsky, I.P., and Trepakova, E.S. (1992) Apoptosis of murine BW 5147 thymoma cells induced by cold shock, *J. Cell. Physiol.*, 153, 112–7.

Kruuv, J. (1986) Effects of pre- and post-thaw cell-to-cell contact and trypsin on survival of freeze-thaw damaged mammalian cells, *Cryobiology*, 23, 126–133.

Lange, K. and Boyd, L.J. (1945) The functional pathology of experimental frostbite and the prevention of subsequent gangrene, *Surg. Gynecol. Obst.*, 80, 346–350.

Lange, K., Weiner, D., and Boyd, L.J. (1947) Frostbite: physiology, pathology, and therapy, *N. Engl. J. Med.*, 237, 383–389.

Larese, A., Acker, J., Muldrew, K., Yang, H., and McGann, L. (1996) Antifreeze proteins induce intracellular nucleation, *Cryo-Letters*, 17, 175–182.

Lazarus, H. M. and Hutto, W. (1982) Electric burns and frostbite: patterns of vascular injury, *J. Trauma*, 22, 581–585.

LeFebvre, J.H. and Folke, L.E. (1975) Effects of subzero temperatures on the microcirculation in the oral mucous membrane, *Microvasc. Res.*, 10, 360–372.

Lempke, R.E. and Schumacker, H.B. (1949) Studies in experimental frostbite. III: an evaluation of several methods for early treatment, *Yale J. Biol. Med.*, 21, 321–334.

Lewis, T. and Love, W.S. (1926) Vascular reactions of the skin to injury. III: some effects of freezing, of cooling, and of warming, *Heart*, 13, 27–60.

Lovelock, J.E. (1953a) The haemoloysis of human red blood cells by freezing and thawing, *Biochim. Biophys. Acta*, 10, 414–426.

Lovelock, J.E. (1953b) The mechanism of the protective action of glycerol against haemolysis by freezing and thawing, *Biochim. Biophys. Acta*, 11, 28–36.

Lubaroff, D.M., Reynolds, C.W., Canfield, L., McElligott, D., and Feldbush, T. (1981) Immunologic aspects of the prostate, *Prostate*, 2, 233–248.

Lubaroff, D.M., Reynolds, C.W., and Culp, D.A. (1979) Immunologic studies of prostatic cancer using the R3327 rat model, *Trans. Am. Assoc. Genito-Urinary Surg.*, 70, 60–65.

Ludwin, I. (1951) Survival of tumor material after freezing at 79°C, *Biodynamica*, 7, 53–55.

Manson, P.N., Jesudass, R., Marzella, L., Bulkley, G.B., Im, M.J., and Narayan, K.K. (1991) Evidence for an early free radical-mediated reperfusion injury in frostbite, *Free Radical Biol. Med.*, 10, 7–11.

Marzella, L., Jesudass, R.R., Manson, P.N., Myers, R.A., and Bulkley, G.B. (1989) Morphologic characterization of acute injury to vascular endothelium of skin after frostbite, *Plastic Reconstructive Surg.*, 83, 67–76.

Matsumura, K., Sakata, K., Saji, S., Misao, A., and Kunieda, T. (1982) Antitumor immunologic reactivity in the relatively early period after cryosurgery: experimental studies in the rat, *Cryobiology*, 19, 263–272.

Mauceri, H., Seetharam, S., Beckett, M., Lee, J., Gupta, V., Bately, S., Stack, M., Brown, C., Swedberg, K., Kufe, D., and Weichselbaum, R. (2002) Tumor production of angiostatin is enhanced after exposure to TNF-alpha, *Int. J. Cancer*, 97, 410–415.

Mazur, P. (1965) The role cell membranes in the freezing of yeast and other single cells, *Ann. N.Y. Acad. Sci.*, 125, 658–676.

Mazur, P. (1984) Freezing of living cells: Mechanisms and implications, *Am. J. Physiol.*, 143, C125–C142.

McGann, L.E. and Kruuv, J. (1977) Freeze-thaw damage in protected and unprotected synchronized mammalian cells, *Cryobiology*, 14, 503–505.

McGann, L.E., Kruuv, J., Frim, J., and Frey, H.E. (1974) Factors affecting the repair of sublethal freeze-thaw damage in mammalian cells. II. The effect of ouabain, *Cryobiology*, 11, 332–339.

McGann, L.E., Kruuv, J., Frim, J., and Frey, H.E. (1975) Factors affecting the repair of sublethal freeze-thaw damage in mammalian cells. I. Suboptimal temperature and hypoxia, *Cryobiology*, 12, 530–539.

McGrath, J.J., Cravalho, E.G., and Huggins, C.E. (1975) An experimental comparison of intracellular ice formation and freeze-thaw survival of HeLa S-3 cells, *Cryobiology*, 12, 540–550.

Misao, A., Sakata, K., Saji, S., and Kunieda, T. (1981) Late appearance of resistance to tumor rechallenge following cryosurgery. A study in an experimental mammary tumor of the rat, *Cryobiology*, 18, 386–389.

Miya, K., Saji, S., Morita, T., Niwa, H., Takao, H., Kida, H., and Sakata, K. (1986) Immunological response of regional lymph nodes after tumor cryosurgery: experimental study in rats, *Cryobiology*, 23, 290–295.

Miya, K., Saji, S., Morita, T., Niwa, H., and Sakata, K. (1987) Experimental study on mechanism of absorption of cryonecrotized tumor antigens, *Cryobiology*, 24, 135–139.

Molisch, H. (1897) *Untersuchen ueber das Erfieren der Pflanzen*, Fischer, Jena.

Muldrew, K., Liang, S., Rewcastle, J., Saliken, J., Donnelly, J., Sawchuck, S., and Sandison, G. (2001) Sodium chloride as an adjuvant for cryosurgery, *Cryobiology*, 43, 388–389.

Muller, L.-C.H., Michsche, M., Yamagata, S., and Kerschbaumer, F. (1985) Therapeutic effect of cryosurgery of marine osteosarcoma—influence on disease outcome and immune function, *Cryobiology*, 22, 77–85.

Mundth, E.D., Long, D.M., and Brown, R.B. (1963) Treatment of experimental frostbite with low molecular weight dextran, *J. Trauma*, 2, 246–257.

Myers, R.S., Hammond, W.G., and Ketcham, S. (1969) Tumor-specific transplantation immunity after cryosurgery, *J. Surg. Oncol.*, 1, 241–246.

Nagle, W.A., Soloff, B.L., Moss, A.J., Jr., and Henle, K.J. (1990) Cultured Chinese hamster cells undergo apoptosis after exposure to cold but nonfreezing temperatures, *Cryobiology*, 27, 439–451.

Neel, H.B., Ketcham, A.S., and Hammond, W.G. (1971) Ischemia potentiating cryosurgery of primate liver, *Ann. Surg.*, 174, 309–318.

Neel, H.B.D., Ketcham, A.S., and Hammond, W.G. (1973) Experimental evaluation of in situ oncocide for primary tumor therapy: comparison of tumor-specific immunity after complete excision, cryonecrosis and ligation, *Laryngoscope*, 83, 376–387.

Neel, H.B. and Ritts, R.E., Jr. (1979) Immunotherapeutic effect of tumor necrosis after cryosurgery, electro-coagulation, and ligation, *J. Surg. Oncol.*, 11, 45–52.

Ninomiya, T., Yosimura, H., and Mori, M. (1985) Identification of vascular system in experimental carcinoma for cryosurgery—histochemical observations of lectin UEA-1 and alkaline phosphatase activity in vascular endothelium, *Cryobiology*, 22, 331–335.

Orita, T., Nishizaki, T., Kamiryo, T., Harada, K., and Aoki, H. (1988) Cerebral microvascular architecture following experimental cold injury, *J. Neurosurg.*, 68, 608–612.

Perotti, M., Toddei, F., Mirabelli, F., Vairetti, M., Bellomo, G., McConkey, D.J., and Orrenius, S. (1990) Calcium-dependent DNA fragmentation in human synovial cells exposed to cold shock, *FEBS Lett.* 259, 331–334.

Pham, L., Dahiya, R., and Rubinsky, B. (1999) An *in vivo* study of antifreeze protein adjuvant cryosurgery, *Cryobiology*, 38, 169–175.

Pollock, G.A., Pegg, D.E., and Hardie, I.R. (1986) An isolated perfused rat mesentery model for direct observation of the vasculature during cryopreservation, *Cryobiology*, 23, 500–511.

Quintanella, R., Krusen, F.H., and Essex, H.E. (1947) Studies on frost-bite with special reference to treatment and the effect on minute blood vessels, *Am. J. Physiol.*, 149, 149–161.

Rabb, J.M., Renaud, M.L., Brandt, A., and Witt, C.W. (1974) Effect of freezing and thawing on the micro-circulation and capillary endothelium of the hamster cheek pouch, *Cryobiology*, 11, 508–518.

Reite, O.B. (1965) Functional qualities of small blood vessels in tissue injured by freezing and thawing, *Acta. Physiol. Scand.*, 63, 111–120.

Riera, C.M., Brandt, E.J., and Shulman, S. (1968) Studies in cryo-immunology. IV. Antibody development in rabbits after iso-immunization followed by freezing, *Immunology*, 15, 779–787.

Roberts, K.P., Smith, D., Ozturk, H., Kazem, A., Pazhayannur, P., Hulbert, J.C., and Bischof, J.C. (1997) Biochemical alterations and tissue viability in AT-1 tumor tissue after *in vitro* cryo-ablation, *Cryo-Letters*, 18, 241–250.

Rothenborg, H.W. (1970) The influence of blood flow on freezing and thawing rates of living tissues, *Cryobiology*, 6, 512–514.

Rothenborg, H.W. (1977) Cryoprotective properties of vasoconstriction, *Cryobiology*, 14, 349–361.

Rotnes, P.L. and Kreyberg, L. (1932) Eine methode zum experimentelen Nachweis von stase mittels spezieller praeparate, *Acta Path. Microbiol. Scand.*, Suppl. 11, 162–165.

Roy, A., Lahiri, S., Lahiri, P., Pal, S., Ghosh, S., and Roy, B. (1990) Immunologic and survival studies in mice immunised with cryodestroyed ascites fibrosarcoma (AFS) cells, *Ind. J. Exp. Biol.*, 28, 1026–1030.

Rubinsky, B. (2000) Cryosurgery, in *Annual Reviews of Biomedical Engineering*, Vol. 2, Yarmusch, M. and Etoner, M., Eds., Annual Reviews, Palo Alto, CA, pp. 157–188.

Rubinsky, B., Amir, A., and Devries, A. (1992) The cryoprotective effect of antifreeze glycopeptides from Antarctic fishes, *Cryobiology*, 29, 69–79.

Rubinsky, B., Arav, A., and Devries, A.L. (1991a) Cryopreservation of oocytes using directional cooling and antifreeze glycoproteins, *Cryo-Letters*, 12, 93–106.

Rubinsky, B., Arav, A., and Fletcher, G.L. (1991b) Hypothermic protection—a fundamental property of antifreeze proteins, *Biochem. Biophys. Res. Comm.*, 180, 566–571.

Rubinsky, B. and Devriesa, A.L. (1989) Effects of ice crystal habit on the viability of glycerol protected red blood cells, *Cryobiology*, 26, 580.

Rubinsky, B., Lee, C.Y., Bastacky, J., and Onik, G. (1990) The process of freezing and the mechanism of damage during hepatic cryosurgery, *Cryobiology*, 27, 85–97.

Rupp, C., Hoffmann, N., Schmidlin, F., Coad, J., and Bischof, J. (2002) Cryosurgical changes in the porcine kidney: Histologic analysis with thermal history correlation, *Cryobiology*, 45, 167–182.

Saliken, J., Donnelly, B., and Rewcastle, J.C. (2002) Evaluation and state of modern technology for prostate cryosurgery, *Urology*, 60, (Suppl. 2A), 26–33.

Salimi, Z., Wolverson, M.K., Herbold, D.R., and Vas, W. (1986) Frostbite: Experimental assessment of tissue damage using Tc-99m pyrophosphate work in progress, *Radiology*, 161, 227–231.

Schuder, G., Pistorius, G., Fehringer, M., Feifel, G., Megner, M., and Vollmer, B. (2000) Complete shutdown of microvascular perfusion upon hepatic cryothermia is critcally dependent on local tissue temperature, *Br. J. Cancer*, 82, 794–799.

Sherwood, E., Ford, T., Lee, C., and Kozlowski, J. (1990) Therapeutic efficacy of recombinant tumor necrosis factor alpha in an experimental model of human prostatic carcinoma, *J. Biol. Response Modifiers*, 9, 44–52.

Shulman, S., Brandt, E.J., and Yantorno, C. (1968) Studies in cryo-immunology. II. Tissue and species specificity of the autoantibody response and comparison with iso-immunization, *Immunology*, 14, 149–158.

Smith, D.J., Fahssi, W.M., Swanlund, D.J., and Bischof, J.C. (1999) A parametric study of freezing injury in AT-1 rat prostate tumor cells, *Cryobiology*, 39, 13–28.

Soanes, W.A., Ablin, R.J., and Gonder, M.J. (1970a) Remission of metastatic lesions following cryosurgery in prostatic cancer: immunologic considerations, *J. Urol.*, 104, 154–159.

Soanes, W.A., Gonder, M.J., and Ablin, R.J. (1970b) A possible immuno-cryothermic response in prostatic cancer, *Clin. Radiol.*, 21, 253–255.

Soloff, B.L., Nagle, W.A., Moss, A.J., Jr., Henle, K.J., and Crawford, J.T. (1987) Apoptosis induced by cold shock *in vitro* is dependent on cell growth phase, *Biochem. Biophys. Res. Comm.*, 145, 876–883.

Staba, M., Mauceri, H., Kufe, D., Hallahan, D., and Weichselbaum, R. (1998) Adenoviral TNF-alpha gene therapy and radiation damage tumor vasculature in a human malignant glioma xenograft, *Gene Ther.*, 5, 293–300.

Steponkus, P.L. (1984) Role of the plasma membrane in freezing injury and cold acclimation, *Ann. Rev. Plant Physiol.*, 35, 543–584.

Sullivan, B.J. and LeBlanc, M.F. (1957) Effect of inositol and rapid rewarming on extent of tissue damage due to cold injury, *Am. J. Phys.*, 189, 501–503.

Tatsutani, K., Rubinsky, B., Onik, G., and Dahiya, R. (1996) Effect of thermal variables on frozen human primary prostatic adenocarcinoma cells, *Urology*, 48, 441–447.

Tomchaney, A.P., Morris, J.P., Kang, S.H., and Duman, J.G. (1982) Purification, composition, and physical properties of a thermal hysteresis antifreezeprotein from larvae of the beetle, *Tenebrio molitor*, *Biochemistry*, 21, 716–721.

Toner, M. (1993) Nucleation of ice crystals inside biological cells, in *Advances in Low-Temperature Biology*, Steponkus, P., Ed., JAI Press, London, pp. 1–52.

Toner, M., Cravalho, E.G., and Karel, M. (1990) Thermodynamics and kinetics of intracellular ice formation during freezing of biological cells, *J. Appl. Phys.*, 67, 1582–1593.

Wing, M.G., Goepel, J.R., Jacob, G., Rees, R.C., and Rogers, K. (1988a) Comparison of excision versus cryosurgery of an HSV-2 induced fibrosarcoma. Survival, extent of metastatic disease and host immunocompetence following surgery, *Cancer Immunol. Immunother.*, 26, 169–175.

Wing, M.G., Rogers, K., Jacob, G., and Rees, R.C. (1988b) Characterization of suppressor cells generated following cryosurgery of normal and tumor bearing liver in an animal model, *J. Clin. Lab. Immunol.*, 35, 129–132.

Wowk, B., Leitl, E., Rasch, C.M., Mesbah-Karimi, N., Harris, S.B., and Fahy, G.M. (2000) Vitrification enhancement by synthetic ice blocking agents, *Cryobiology*, 40, 228–236.

Wu, D.W., Duman, J.G., and Xu, L. (1991) Enhancement of insect antifreeze protein activity by antibodies, *Biochim. Biophys. Acta*, 1076, 416–420.

Yamashita, T., Hayakawa, K., Hosokawa, M., Kodama, T., Inoue, N., Tomita, K., and Kobayashi, H. (1982) Enhanced tumor metastases in rats following cryosurgery of primary tumor, *Gann*, 73, 222–228.

Yang, W.H. Peng, H.H., Chang, H.C., Shen, S.Y., Wu, C.L., and Chang, H.C. (2000) An *in vitro* monitoring system for simulated thermal process in cryosurgery, *Cryobiology*, 40, 159–170.

Yang, W.L., Addona, T., Nair, D.G., Qi, L., and Ravikumar, T.S. (2003) Apoptosis induced by cryo-injury in human colorectal cancer cells is associated with mitochondrial dysfunction, *Int. J. Cancer*, 103, 360–369.

Yantorno, C., Soanes, W.A., Gonder, M.J., and Shulman, S. (1967) Studies in cryo-immunology. I. The production of antibodies to urogenital tissue in consequence of freezing treatment, *Immunology*, 12, 395–410.

Yilmax, A., Bieler, G., Bamat, J., Chaubert, P., and Lejeune, F. (1998) Evidence for the involvement of endothelial cell integrin alphaVbeta3 in the disruption of the tumor vasculature induced by TNF and IFN-gamma, *Nat. Med.*, 4, 408–414.

Zacarian, S.A. (1977) The observation of freeze-thaw cycles upon cancer cell suspensions, *J. Dermatol. Surg. Oncol.*, 3, 173–174.

Zook, N., Hussmann, J., Brown, R., Russell, R., Kucan, J., Roth, A., and Suchy, H. (1998) Microcirculatory studies of frostbite injury, *Ann. Plastic Surg.*, 40, 246–253.

17 Cryopreservation in Transfusion Medicine and Hematology

Andreas Sputtek and Rebekka Sputtek

CONTENTS

17.1 INTRODUCTION

Blood cells can be regarded as a classical field of application of low-temperature biology. Cryo-preservation methods have been developed for all different categories of blood cells (including blood stem cells) and blood corpuscles. For granulocytes, however, no clinically applicable method has been found until today. However, frozen red blood cells (also known as RBC or erythrocytes), platelets (also known as thrombocytes), lymphocytes, monocytes, and hematopoietic progenitor cells (from peripheral blood as well as from bone marrow) are currently being used for various diagnostic and clinical purposes. When Polge et al. (1949) discovered rather accidentally that glycerol was able to protect spermatozoa from damage during freezing and thawing, a rapid development of deep-freezing preservation of biological cells set in. This has led to a variety of cell-specific cryopreservation protocols. The methods differ with regard to cell concentrations, protective solutions used (cryoprotectants and their concentrations), temperature–time-histories during cooling and rewarming, and storage temperature. In addition, some of the cryoprotectants are not well tolerated in the concentrations required (e.g., dimethyl sulfoxide [Me$_2$SO] for platelets) or lead to an osmotically induced lysis of the cryoprotectant-loaded cells when transfused into an isotonic organism (e.g., glycerol for red blood cells). In these cases, a washing procedure is required after thawing before the application.

Cryopreserved erythrocytes are of advantage in case of rare blood groups, antibody problems, and civil as well as military disasters. In principle, three different methods have been established for clinical use: the Huggins (1963) technique, using glycerol in a nonionic suspension and removal of the cryoprotectant by reversible agglomeration of the RBC; the "high glycerol–slow cooling technique" according to Meryman and Hornblower (1972), which is the dominant method in the United States; and the "low glycerol–rapid cooling technique" (Krijnen et al., 1968, Rowe et al., 1968). The latter is the dominant method for the cryopreservation of RBC in Europe.

Since the first reported attempt to stop thrombocytopenic bleeding by the infusion of previously frozen thrombocytes (Klein et al., 1956), a broad variety of *in vitro* and *in vivo* studies on cryopreserved platelets have been published. Nowadays, two methods are in general use for the preparation of cryopreserved platelets: a dimethyl sulfoxide (Me$_2$SO) method (Schiffer et al., 1978) and a low-glycerol/glucose method (Dayian and Rowe, 1976; Dayian et al., 1986).

Despite of some reports appearing now and then in the literature, it is our opinion that no clinically suitable method for the preservation of granulocytes has been reported so far.

The use of cryopreserved lymphocytes and monocytes is a well established and a routine procedure for clinical laboratory testing. Most recently, there is a growing clinical interest in cryopreserved lymphocytes for the supplemental treatment of patients after blood stem cell transplantation. Usually, the lymphocytes are frozen according to methods that are more or less modifications of a technique that was first described for bone marrow by Ashwood-Smith (1961).

Today, hematopoietic progenitor cells are successfully cryopreserved using methods based on the above-mentioned Ashwood-Smith (1961) procedure. During cooling, the heat is removed either by computer-controlled and liquid nitrogen (LN$_2$)–operated machines or in mechanical (–80°C) refrigerators. Stiff et al. (1983) have demonstrated that the addition of 6% hydroxyethyl starch (HES) reduced the "original" concentration of Me$_2$SO (10%) by half. Cryopreserved autologous and homologous blood stem cells (in combination with high-dose chemotherapy or irradiation) have become a "standard" blood component for the treatment of several malignant diseases.

17.2 CRYOPRESERVATION OF ERYTHROCYTES

Examples of reviews covering the literature on cryopreserved RBC from an historical as well as cryobiological perspective are Meryman (1989), Rowe (1994, 1995, 2002), Rowe and Lenny (1982), and Sputtek and Körber (1991).

17.2.1 PROCEDURES USING THE CRYOPROTECTANT GLYCEROL

Audrey Smith (1950) was the first to report the prevention of hemolysis of RBC occurring during freezing and thawing in 15% glycerol. The problem of hemolysis occurring during the removal of the cryoprotectant was solved year later by Sloviter (1951), who removed the glycerol by dialysis. Shortly thereafter, the first transfusion of frozen RBC in a patient was performed (Mollison and Sloviter, 1951). In the next 25 years a development of different techniques for freezing, storage, and the deglycerolization process took place. In principal, three different methods using the cryo-protectant glycerol have gained clinical relevance. They differ concerning the concentration and the addition of additive, hematocrit, volume, cooling rate, storage temperature, warming rate, and removal of the additive. The data reported vary from one author to another: recovery after thawing, 90 to 99%; additional loss during removal of the cryoprotectant glycerol, 5 to 30%; time required for the deglycerolization, 0.5 to 30 h; amount of "free" hemoglobin, 50 to 250 mg/dL; leukocyte contamination, 1 to 20% of the starting value; *in vivo* survival rate after 24 h, 75 to 95%.

In 1953, Lovelock published two "milestone" papers dealing with the nature of the freeze damage in RBC and the prevention by the addition of glycerol (Lovelock, 1953a, 1953b). These classic papers related the damaging effect of freezing to the increasing electrolyte concentration in the extracellular space caused by the formation of ice rather than to a "mechanical" action of the ice itself (Lovelock, 1953a). The protective effect of glycerol was ascribed to its colligative properties; that is, to reduce the ice formation (and as a consequence the electrolyte concentration) at a given subzero temperature compared to a glycerol-free solution (Lovelock, 1953b). In 1981, Mazur et al. questioned the evidence from Lovelock's experiments (see also Chapters 1 and 2). The shrinkage of the remaining aqueous channels ("unfrozen fraction") was claimed to be the dominating damaging factor during freezing rather than the solute concentration therein. However, Pegg and Diaper (1988, 1989) have questioned the conclusions Mazur and coworkers drew from their experiments. Pegg and Diaper performed dialysis experiments that supported the more unifying (i.e., original Lovelockian) theory of freeze–thaw damage.

At the end of the 1950s, Tullis et al. (1958) and Haynes et al. (1960) developed a method suitable for a broader clinical application. They used the so-called "Cohn-ADL fractionator," one of the first continuous-flow centrifuges, for the addition of the glycerol before freezing and its removal after thawing. The additive concentration amounted to 40–50% (w/v), slow cooling (<1°C/min), and storage took place in a –80°C refrigerator.

Three years later, Huggins (1963) invented the so-called "cytoagglomeration" procedure, which was originally developed with Me_2SO as cryoprotectant, but was modified to use glycerol. For the removal of the cryoprotectant, this method uses the effect of the formation of reversible bondages between plasmatic γ-globulins and the lipoproteins of the erythrocytic membrane at pH values from 5.5 to 6.1: Lowering the ionic strength by dilution with an electrolyte-free solution (e.g., 10% glucose) leads to the precipitation of the μ-globulin fraction together with the bound RBC. The supernatant is discarded and the procedure is repeated until the cryoprotectant has been removed nearly completely. Then the agglomerated RBC are resuspended by either adding some electrolyte solution or by increasing the pH. Sumida (1974, 1993) introduced this method in Japan in the 1970s and continues to evaluate the long-term storage stability of the stored RBC.

The original high glycerol–slow cooling technique (Tullis et al., 1958) has been simplified by Valeri (1970), whereby the glycerolization was performed without the use of a centrifugal bowl and the freezing took place in the primary collection bag.

Meryman and Hornblower (1972) further simplified the addition and removal of the glycerol compared with the original Tullis procedure. They found that the RBC could be diluted after thawing with salt solutions of decreasing tonicity and washed by either continuous-flow centrifugation or by automated batch centrifugation. Today this method, with some minor modifications, is the method predominantly used in the United States and several other countries. A detailed description of this procedure can be found elsewhere (e.g., Walker, 1993). Although sickle-trait RBC can be frozen and thawed like normal RBC, a significantly higher hemolysis occurs with "standard" deglycerolization techniques. Meryman and Hornblower (1976) have modified the process for this purpose by using an extended dilution and omitting the hypertonic washing solution step.

The rejuvenation of outdated RBC with subsequent cryopreservation has been demonstrated by Valeri and Zaroulis (1972). This may be useful for unused rare blood types, autologous blood, or stocking up with group O RBC as a mainstay of disaster treatment plans.

Until recently, previously frozen RBC, once thawed, could be stored at 4°C for no longer than 24 h because of the current methods for glycerolization and deglycerolization: The opened systems were highly subject to bacterial contamination. However, a recently performed study has investigated the application of a functionally closed system using a sterile docking device and an automated system for deglycerolization (Valeri et al., 2001a, 2001b). The studies showed no risk for an increased risk of bacterial contamination. In addition, data show that cryopreserved RBC can be stored safely for more than 20 or 30 years, and perhaps indefinitely (Valeri et al., 1989, 2000). This has been taken into account in the E.U. guidelines, where "the storage may be extended to at least 10 years, if the correct storage temperature can be guaranteed" (Council of Europe, 2001).

Other investigators (Krijnen et al., 1968; Rowe 1973; Rowe et al., 1968) have approached the problem of the tedious postthaw deglycerolization by lowering the concentration of the cryoprotectant glycerol. This approach requires the use of LN_2 to generate the higher cooling rates (60 to 120°C/min) and lower storage temperatures (<–150°C) compared with the high glycerol–slow cooling technique. This "low glycerol–rapid cooling technique," as described by Rowe et al. (1968), requires 17.5% (w/v) glycerol and 3% mannitol. The method established by Krijnen et al. (1968) at the European Frozen Blood Bank in Amsterdam uses 19% (w/v) glycerol and 3% sorbitol. Freezing is complete in a couple of minutes, and the units are stored in LN_2 (–196°C) or in the liquid-vapor phase over LN_2 (<–165°C). After thawing in a 37 to 45°C water bath for several minutes, the cells are washed free of the cryoprotectant either by manual serial batch washing or by automated continuous-flow centrifugation. To avoid the osmotica stress during the removal, sodium chloride solutions of decreasing tonicity (3.5 and 0.9%, respectively), are used.

Most recently, Wagner et al. (2000) have published a paper providing evidence that a biochemical prefreeze stabilization may reduce the effect of cell density on RBC recovery after cryopreservation. The first set of experiments showed that thaw hemolysis (paralleled by changes in the fragility index) decreased with increasing hematocrit at the glycerol concentrations tested (i.e., 15%, 20%, 40% [w/v]). The overall hemolysis (and fragility index) increased with increasing hematocrit (approximately from 10 to 90%) at 15% glycerol and decreased with increasing hematocrit at 40% glycerol. In a second set of experiments, before slow freezing at glycerol concentrations as low as 15% (w/v) in a –80°C freezer was performed, the RBC were washed twice using a hypotonic solution (280 mosm/L) containing sodium chloride, glucose, glutamine, mannitol, and adenine. At a too-low glycerol concentration for slow freezing (i.e., 20%), the addition of the "biochemical stabilizing solution," which had no cryoprotective effects itself, resulted in an improvement of the thaw and overall hemolysis comparable to 40% glycerol (without the stabilizing solution). The authors concluded that the damage observed with increasing hematocrit is not an effect of the packing on the volume of the ice-free space but rather an expression of biochemical cell damage.

17.2.2 Definition, Properties, Storage, Stability, Transport, and Quality Control

Frozen RBC for transfusion are a component derived from whole blood or obtained by apheresis, cooled down preferably within 7 d after collection, using a cryoprotectant (mostly glycerol). They should be constantly maintained at –60° to –80°C if stored in an electrical freezer when a high-glycerol method is used or at –140° to –150°C in the vapor phase over liquid nitrogen, when a low-glycerol method is used (see Section 17.2.1). Before use, the cells are thawed, washed, and suspended in isotonic sodium chloride solution or in additive solutions for RBC. Reconstituted units are poor in protein, granulocytes, and platelets. The storage may be extended to at least 10 years if the correct storage temperature is guaranteed. The reconstituted product should be stored at +2° to +6°C. The storage time should be as short as possible after washing and should never exceed 24 h when an open system has been used. If transport in the frozen state is unavoidable, storage conditions should be maintained. The units should be shock- and spill-proof, surrounded with absorbing material soaked in sufficient LN_2 (a so-called "dryshipper"). The primary container should be wrapped into a leakproof secondary container not interfering with the coolant. The volume should be more than 185 mL, the "free" hemoglobin in the supernatant less than 0.2 g/U, the hemoglobin content more than 36 g/U, the hematocrit 0.65–0.75 L/L, and (in at least 75%) the residual leukocytes less than 10^8 cells/U. To avoid an *in vivo* hemolysis, the osmolarity in the supernatant should be less 340 mosm/L (Council of Europe, 2001). Refreezing of already thawed but not used RBC concentrates is possible but should only be considered in special cases.

17.2.3 INDICATIONS FOR USE AND PRECAUTIONS IN USE

Cryopreserved RBC are indicated for RBC substitution or replacement. They should only be used in special situations such as transfusion to patients with rare blood types or multiple alloantibodies, for alloimmunization purposes after at least frozen storage for 6 months to allow retesting of donors, and in some cases for autologous transfusion (Council of Europe, 2001). (We consider the following rare blood types to be candidates for cryopreservation: O_h [Bombay], Rh_{Null}, -D-, and K^0, as are cases with [multiple] antibodies to Lu^b [Lutheran], Kp^b [Rautenberg], and Co^a [Colton].) As during the preparation the component may be transferred to another bag, measures have to be taken to ensure the identification of cross-match samples and proper identification. When the processing is performed in an open system, the risk of bacterial contamination is increased (Council of Europe, 2001).

Freezing, thawing, and removal of the additive glycerol decreases plasmatic and intracellular viral contaminations, but it is not suitable to totally exclude the viral transmission. Also viral transmission from contaminated LN_2 cryopreservation tanks or contaminated units cannot be totally excluded. There are some reports in the literature (e.g., Zuckerman et al., 1995; Tedder et al.,1995), that a cross-contamination is possible. However, this requires leaky bags. For this reason, storage in the vapor phase or wrapping into a leakproof secondary container provides additional safety.

Despite a prefreeze leukodepletion, frozen/thawed, but not irradiated, RBC may lead to a Graft versus Host disease (GvHD) in immunocompromised patients. The same holds for directed donations from family donors. Although the freezing, thawing, and deglycerolization significantly decreases the number of immunocompetent T-lymphocytes, it cannot be ruled out that enough of them survive to cause the disease. For nonleucodepleted RBC that had been cryopreserved according to the high glycerol technique and subsequently deglycerolized and washed, it was shown that residual lymphocytes could not be completely eliminated and that about 60% of them had intact cell membranes (Trypan blue exclusion). Although their functional properties were significantly reduced, these cells were still considered to have some immunocompetence and recognizable immunogenicity to sensitize transfused patients (Harada et al., 1980).

17.2.4 ALTERNATIVE PROCEDURES USING EXTRACELLULAR CRYOPROTECTANTS

The use of macromolecular cryoprotectants goes back to Rinfret (1963). Water-soluble, cryoprotective macromolecules such as albumin, dextrans, modified gelatines, polyvinylpyrrolidone, polyethylene oxide, polyethylene glycol, and HES exhibit the principal advantage of not entering into the cells. This property significantly facilitates their removal after thawing. In the case of emergencies, this step could be omitted, if the additives, for example, albumin, dextrans, modified gelatines, and HES, are biodegradable and tolerated by the patient. In an investigation using dextran, Pellerin-Mendes et al. (1997) have confirmed our results obtained in a comparison of modified gelatine, dextran, and HES (Sputtek, 1995).

However, in 1967 Knorpp et al. described for the first time the successful cryopreservation of human RBC using HES and LN_2, comparing the efficacy of HES to that of polyvinylpyrrolidone. Knorpp et al. preferred the colloid HES to polyvinylpyrrolidone, as the latter is retained to a considerable extent in the recipient (as are polyethylene oxide and polyethylene glycol). Moreover, in the case of hypovolemia, albumin, dextrans, modified gelatines, and HES serve as blood volume substitutes.

The literature regarding the cryopreservation of RBC using HES until 1990 has been reviewed elsewhere (Sputtek and Körber, 1991). In the last decade, we have carried out several additional *in vitro* investigations (e.g., Sputtek et al., 1990, 1991a), experiments in dogs (e.g., Langer et al., 1993, 1994; Sputtek et al., 1991b), and an *in vivo* study including seven healthy volunteers (Sputtek et al., 1993b, 1995b). We performed the first autologous transfusion of unwashed, HES-protected frozen RBC in a 16-year-old female patient (Sputtek et al., 1995a). The RBC from three donations were frozen according to our HES procedure as described elsewhere (Sputtek et al., 1995b). After the hemoglobin concentration had dropped to 7.6 g/dL because of the blood loss during surgery, 3 units were transfused without postthaw washing. No unfavorable effects or hemoglobinuria were observed. The patient was discharged after a normal postoperative period without complications. Thomas et al. (1996) have developed a similar procedure for the freezing of RBC using HES. The major differences to our procedure are no prefreeze washing, different HES modification, lower HES concentration, higher electrolyte concentration, higher hematocrit, larger freezing bag, smaller sample thickness, smaller volume, higher viscosity, different freezing container, and uncontrolled thawing.

We have performed the first systematic clinical trial to determine the safety and tolerance of autologous, HES-cryopreserved RBC in patients (Horn et al., 1997) using our patented freezing container (Sputtek and Mingers, 1994). In this investigation the first RBC concentrate obtained from 36 patients undergoing preoperative autologous blood donation was randomly assigned to the conventional storage method (4°C, PAGGS-mannitol, group 1) or to cryopreservation with HES (weight average molecular weight: 200,000 g/mol, molar substitution: 0.5) at a final concentration of 11.5% (w/w) using LN_2 (groups 2 and 3). Before surgery, an additional 900 mL blood was drawn. Patients belonging to group 1 received the conventionally stored RBC, and those in group 2 received cryopreserved, thawed, and washed RBC. In group 3, however, no washing step before the transfusion was performed. Data were assessed after induction of anesthesia, hemodilution, and transfusion of the RBC and at regular intervals intra- and postoperatively. No significant differences between the three groups could be detected regarding hemodynamic or blood gas parameters and tissue oxygenation. No adverse reactions after transfusion of the washed and unwashed cryopreserved RBC were observed. Directly after transfusion, plasma hemoglobin levels increased twofold in group 2 and threefold in group 3, but always decreased to baseline levels within 24 h. The HES was eliminated from the plasma following first-order kinetics (Sputtek et al., 1994). The data indicate that the transfusion of one autologous unit of RBC after cryopreservation with HES is safe and well tolerated. Further investigations are deemed necessary to evaluate the effects of larger volumes of HES-cryopreserved RBC and homologous transfusions.

17.3 CRYOPRESERVATION OF THROMBOCYTES

Examples of reviews covering the literature on cryopreserved platelets from an historical as well as cryobiological perspective are Gardner (1968), Law and Meryman (1982), Meryman and Burton (1978), Rowe et al. (1980), Schiffer et al. (1985), Sputtek and Körber (1991), and Valeri (1985).

17.3.1 PROCEDURES USING THE CRYOPROTECTANTS DIMETHYL SULFOXIDE OR GLYCEROL

Djerassi et al. (1966) were the first to report on the use of 5% dimethyl sulfoxide (Me_2SO) and cooling at 1°C/min for successful cryopreservation and transfusion of human platelets. To avoid the effects related to the cryoprotectant itself (e.g., nausea, vomiting, local vasospasm, garlic-like taste and body odor), Lundberg et al. (1967) have introduced a postthaw washing step. A method used by Schiffer et al. (1978) has become the "standard" method for this purpose. A detailed description of this method can be found elsewhere (Walker, 1993). In principle, the method applies Me_2SO at a concentration of 5% (v/v), a cooling rate of approximately 10°C/min (vapor phase over LN_2 or deep freezer, storage below –120°C). Thawing takes place in a 37°C water bath within a few minutes, and removal of the cryoprotectant is performed in a washing step with autologous plasma. Dayian and Rowe (1976) described a low-glycerol/glucose procedure that does not require a washing step after thawing, because the low concentration of the cryoprotectants used. Survival of platelets frozen by this method was excellent (~65 to 70%) and the platelets exhibited a half-life of approximately 3.8 d. Transfused glycerol-frozen platelets also shortened aspirin-induced prolonged bleeding time, which indicated *in vivo* hemostatic effectiveness (Dayian et al., 1974). However, other researchers had difficulties reproducing their results (Kotelba-Witkowska and Schiffer, 1982; Redmond et al., 1983). The reason for that was probably the required relatively high cooling rate of about 30 to 40°C/min (Scheiwe et al., 1981). Later, Pert and Dayian (1979) described an inexpensive cooling technique using aluminum plates with a cardboard insulation for immersion into LN_2 instead of a controlled-rate freezer to generate the required cooling rate. It was also shown that the addition of the cryoprotective solution (CPS) plays a critical role (Armitage 1986; Kim and Baldini, 1986). In contrast to erythrocytes, thrombocytes begin to lose their function after relatively small changes in tonicity. Whereas human red cells tolerate differences of up to 1500 mosmol/L (Lovelock, 1953a), in the case of platelets, volume changes of more than 60% seem to be detrimental.

However, to judge the literature correctly, a careful distinction between numerical and functional *in vitro* recovery and *in vivo* survival after thawing/washing has to be made. In addition, numerical *in vitro* recoveries (usually in the range from 70 to 90%) may vary depending on the determination method (Brodthagen et al., 1985; Richter et al., 1983; Vecchione et al., 1981).

Functional *in vitro* recoveries also depend a lot on the parameter investigated, the method applied, and the quality of the untreated controls. Typical functional recoveries reported in the literature vary from 20 to 80%. As a consequence, a comparison of the results obtained by different investigators (using different viability assays) is difficult and often impossible. Even minor variations (e.g., platelet concentration, temperature, concentration of substances added) may strongly influence the result. Common *in vitro* methods are hypotonic stress response, ^{14}C-serotonin uptake, clot retraction, induced aggregation with various agents, thrombelastography, resonance thrombography, platelet activation by fluorescence activated cell sorting analysis, content of adenosine nucleotide/adenylate energy charge, β-thromboglobulin release, malondialdehyde formation, and morphological score by means of electron microscopy (Sputtek and Körber, 1991).

To make things even more complicated, it is not possible to predict the *in vivo* survival after transfusion or the hemostatic effectiveness from such *in vitro* results (van Imhoff et al., 1983; Zapff et al., 1988). *In vivo*, the corrected count increment 1 h after the transfusion of platelets frozen according to Schiffer's method (Schiffer et al., 1978), determined by means of the ^{51}Cr labeling method, was roughly 50% of what was measured after the transfusion of unfrozen platelets (Beaujean et al., 1979; Richter et al., 1983; Schiffer et al., 1985; van Imhoff et al., 1983; Zapff et al., 1988).

Recent investigations have demonstrated that the addition of reagents that modulate second messengers and cellular enzymes (ThromboSol) enhances *in vitro* (Vadhan-Raj et al., 1999, Lozano et al., 2000) and *in vivo* parameters of platelets frozen in 2% (v/v) Me_2SO. Time will show whether this new technique will find its way from first autologous pilot studies in healthy volunteers (Currie et al., 1999) and patients (Pedralozzi et al., 2000) into routine clinical application. Platelets are also an area of current interest, using novel freeze-drying technologies (Chapter 21), but the clinical effectiveness of these preparations remains to be established.

17.3.2 Definition, Properties, Storage, Stability, Transport, and Quality Control

Frozen platelets for transfusion are a component prepared by the freezing of platelets within 24 h of collection, using a cryoprotectant, and storing them at –8°C or below. Reconstituted cryopreserved platelets are poor in RBC and granulocytes. Two methods are in general use for the preparation of cryopreserved platelets: a dimethyl sulfoxide (Me_2SO) method and a low-glycerol/glucose method. Before use the platelets are thawed and washed (or resuspended) in autologous platelet-poor plasma or in an isotonic sodium chloride solution. Platelets in the frozen state should be constantly maintained at –80°C if stored in an electrical freezer or at –150°C if stored in the vapor phase over LN_2. If storage must be extended for more than 1 year, storage at –150°C is preferred. Thawed platelets should be used immediately after thawing. If short intermediate storage is required, they should be stored at room temperature with adequate agitation. If transport in the frozen state is necessary, storage conditions should be maintained during transportation (see Section 17.2.2. on frozen RBC). Transport of thawed platelets is limited by the short storage time. Storage conditions should be maintained during transportation. The volume should be from 50 to 200 mL, the platelet count should be higher than 40% of the prefreeze value, and the amount of residual leucocytes should be less than 2×10^5 per 6×10^8 platelets (Council of Europe, 2001).

17.3.3 Indications for Use and Precautions in Use

Thrombocytopenia (with or without bleeding) may have many reasons, but from a clinical point of view the most frequent ones are high-dose chemotherapy or whole-body irradiation in cancer patients, extended blood losses (with or without surgery), and leukemia. The "product of choice" is platelet concentrates stored at room temperature for up to 5 d.

Frozen homologous or autologous platelets should be reserved for the provision of plates that are HLA (human leukocyte antigen) or HPA (human platelet antigen) compatible, for which a compatible donor is not immediately available. Suitable HLA and HPA compatibility testing should be performed when required. The toxicity of reagents used during the processing and cryopreservation (e.g., Me$_2$SO) has to be taken into account. Nonhemolytic transfusion reactions may occur (mainly chills, fever, and urticaria). Alloimmunization, especially to the HLA and HPA series of antigens, may occur, but the risk is minimal. Viral, protozoal, and bacterial transmission as well as posttransfusion purpura are possible (Council of Europe, 2001). For additional aspects, see the last two paragraphs in Section 17.2.3. Because of the limited function (and reduced *in vivo* survival) of cryopreserved platelets (independent from the cryopreservation method applied and despite excellent numerical *in vitro* recoveries), we recommend using twice as much frozen/thawed platelets compared with fresh ones when treating bleeding in patients.

17.3.4 Use of ES as Cryoprotectant

Because of the toxicity of Me$_2$SO and the reduced quality of the thawed product (see above), many investigators have looked for alternatives using different cryoprotectants and cooling regimen. HES, which has been used for volume replacement for decades, and which had been found an effective cryoprotectant for RBC (Knorpp et al., 1967), was a promising candidate. As it is biodegradable and well tolerated by the human organism, no postthaw washing step would be required.

HES-protected cryopreserved platelets have been used by Chaudhury and Gunstone (1978) in patients for the first time. The platelets, which had been frozen in the presence of 4% (w/v) HES at 1°C/min, turned out to be hemostatically effective. Taylor (1981) performed an *in vitro* comparison of four different methods, including the HES method and the glycerol/glucose procedure. He concluded that the Me$_2$SO protocol was most effective. However, when using the HES method, optimum results can only be achieved at cooling rates from 5 to 15°C/min and when the sodium chloride concentration in the CPS is adjusted to 120 mmol/L (Sputtek et al., 1987). When comparing the HES method to Schiffer's technique (Schiffer et al., 1978), we found that both protocols were highly effective regarding the postthaw numerical platelet recovery (~90%). However, functional *in vitro* parameters as measured by resonance thrombography showed that the Me$_2$SO-protected frozen platelets were clearly inferior to the fresh controls but superior compared with the HES-protected ones (Sputtek et al., 1993a).

17.4 CRYOPRESERVATION OF GRANULOCYTES

There have been publications in the past that claim the successful cryopreservation of granulocytes (reviews: Bank, 1980; Knight, 1980). Despite reports appearing now and then in the newer literature (mostly as abstracts), it is our opinion that no clinically suitable method for the preservation of granulocytes has been found. The huge variation of the *in vitro* results shows how cumbersome the viability assays are and how unsuitable they will be to predict anything that is going to happen *in vivo*. Membrane integrity tests (e.g., trypan blue exclusion, combined vital/dead staining with fluorescein diacetate/ethidium bromide; Dankberg and Persidsky, 1976) measure only a *conditio sine qua non* (i.e., an intact cell membrane.) However, what is the meaning of these results if tests measuring typical granulocytic functions (e.g., chemotaxis, bactericidy) fail to detect any significant activity? Numerical recovery and membrane integrity are necessary, but by no means sufficient, criteria for viable cells.

Takahashi et al. (1985) have proposed some explanations as to why granulocytes are so unrewarding regarding their cryopreservation. At temperatures below $-5°C$ without the formation of ice, a significant loss of function already can be observed. This loss could be prevented by the addition of Me_2SO, whereas glycerol failed to show this effect. Because of the limited osmotic tolerance (already a twofold increase compared with isotonicity causes a significant loss of function), granulocytes are highly susceptible to the electrolyte enrichment taking place during ice formation. They also showed a limited tolerance to the hypotonic stress that may occur on thawing.

17.5 CRYOPRESERVATION OF LYMPHOCYTES AND MONOCYTES

In contrast to granulocytes, lymphocytes and monocytes (collectively known as mononuclear cells, MNC) exhibit a high osmotic tolerance, and they are rewarding candidates for cryopreservation. Lymphocytes can be cryopreserved successfully following the early descriptions in the 1960s (Ashwood-Smith, 1964; Atkins, 1962; Pegg, 1965). Useful protocols especially for monocytes have been described in the 1980s (e.g., De Boer et al., 1981; Van der Meulen et al., 1981). Most of the literature relating to the freezing of lymphocytes up to 1979 has been reviewed elsewhere (Knight, 1980).

17.5.1 CLINICAL APPLICATIONS FOR FROZEN MONONUCLEAR CELLS

Recently, there has been a growing clinical interest in fresh or cryopreserved lymphocytes for the supplemental treatment of patients after homologous blood stem cell transplantation. Donor lymphocyte infusions have shown to induce remissions in patients with chronic myelogeneous leukemia with evidence of cytogenetic or hematologic relapse after blood stem cell transplantation (Soiffer, 1997). The mechanism that underlies the response to donor lymphocytes is not clear. On the one hand, it has been discussed that effector cells recognize and react against specific cell-surface tumor antigens. On the other hand, it is speculated that donor lymphocytes may induce a secondary antitumor effect caused by an allogenic Graft versus Host reaction (GvH), characterized by an increase of cytokines and cellular mediators. The opportunities of donor lymphocyte infusions are enormous. There is already evidence that they are cytotoxic for cytomegalovirus, and Epstein-Barr virus can be transferred from donor to host, restoring antiviral immunity and diminishing post-transplant lymphoproliferative disease (Papadopoulos et al., 1994; Walter et al., 1995). The application of donor lymphocytes will be greatly facilitated by having them "on stock" in the frozen state for patients who are in danger of suffering from the above-mentioned diseases.

17.5.2 FROZEN MONONUCLEAR CELLS FOR LABORATORY TESTING

The use of cryopreserved lymphocytes and monocytes is well established as a routine procedure for clinical laboratory testing. Frozen MNC are used for various diagnostic purposes; for example, HLA typing, detection of HLA antibodies in patients on waiting lists for organ/bone marrow transplantations, and mixed lymphocyte reactions/cultures. To the best of our knowledge, in 1966, Cohen and Rowe (1968) were the first to use frozen leukocytes for the detection of cytotoxic human leukocyte antibodies as a precursor for their use in tissue typing. Frozen leukocytes are also of interest with respect to look-back procedures in transfusion medicine or diagnosis in patients. For example, no difference in isolation rates was found between fresh and frozen lymphocytes regarding the human immunodeficiency virus (Gallo et al., 1987).

The sensitivity of rat lymphocytes and macrophages to freezing procedures designed for the cryopreservation of pancreatic islets of Langerhans with Me_2SO was investigated by Taylor and Bank (1988). With respect to cell number recovery and membrane integrity, they found that a significant percentage survived freezing up to 20°C/min, whereas negligible levels were detected at 75° and 200°C/min. The functional properties, however, were drastically reduced at 20°C/min, reflected by the total inability of the lymphocytes and macrophages to respond to the mitogen by Concanavalin A (Taylor et al., 1987). It hence seems possible to deplete donor tissues of functional lymphocytes and macrophages if cooling is sufficiently rapid to destroy the MNC, thus reducing the immunogenicity of cryopreserved grafts and GvH reaction for transplantation purposes.

Moreover, lymphocytes and monocytes have served as "model cells" for studying fundamental cryobiological principles (see Körber, 1988; review in Sputtek and Körber, 1991). More recently, lymphocytes and lymphoblasts were employed to study the effect of directional solidification; that is, independent variation of the temperature gradient and ice front velocity during cooling (Beckmann et al., 1990; Hubel et al., 1992). To substantiate the visual observations of the morphological changes associated with intracellular ice formation (IIF), Körber et al. (1991) performed studies in human lymphocytes by cryomicroscopy and differential scanning calorimetry (DSC) under comparable conditions to measure the quantity of water actually transformed into the crystalline state because of the evolution of latent heat. In particular, it could be shown that the twitching type of intracellular ice formation, which is evident but difficult to observe under the cryomicroscope, can be attributed to a liquid–solid phase change within the cells as determined by DSC. The fraction of cells exhibiting intracellular ice determined as a function of the cooling rate with both methods showed a sharp demarcation zone with an increase from 0 to 100% at about the same threshold cooling rate. In contrast, the temperatures at which intracellular ice formed were found to be only weakly dependent on the cooling rate. The DSC results were hence regarded as a validation of the microscopic observations. To better understand the mechanisms of IIF, Bryant performed a DSC study using human lymphocytes in the presence of Me_2SO. Under conditions in which damage caused by IIF on the initial cooling run was 40 to 60%, the samples were studied as a function of multiple successive cooling runs. The temperature at which IIF occurred and the fraction of cell volume that underwent IIF were analyzed as functions of successive cooling runs (Bryant, 1995).

The methods for freezing mononuclear cells reported in the literature vary from one author to another (Sputtek and Körber, 1991). In general, the cell concentration ranges from 0.5×10^6 to 50×10^6/mL, the most frequently used medium is RPMI 1640 supplemented with human or fetal calf serum or plasma, and the cryoprotectant of choice is 5 to 10% Me_2SO. Cooling is performed in 1- or 2-mL vials at 1 to 2°C/min, down to a temperature of –30°C or less by means of a programmable LN_2 operated freezer, whereas thawing is usually performed in a water bath of 37°C. Numerical recoveries reported vary from 60 to 90%. We believe that cooling rates at temperatures below –40°C are not as critical as in the upper temperature region (above –40°C) and can be increased up to 10°C/min to save time. In addition, we do not think that a programmable LN_2 operated freezer is always required to generate the appropriate cooling rate: –80°C refrigerators may be suitable as long as provision is taken (e.g., by using cardboard insulations) that the cooling rates in the upper temperature region do not exceed 5°C/min. For long-term storage (e.g., months or years), however, we recommend temperatures below –123°C, which is the glass transition temperature of Me_2SO (e.g., Pegg et al., 1997). For storage of days or weeks, and up to months, –80°C freezers may also be acceptable.

Numerous technique are used for assessing postthaw viability (Sputtek and Körber, 1991); for example, staining methods for determining membrane integrity, visualization of surface receptor sites, determination of electrophoretic mobility, evaluation of surface membrane immunoglobulins, ^{3}H-thymidine uptake after stimulation with various mitogens, determination of cell subsets in a fluorescence-activated cell sorter, and mixed lymphocyte reaction or culture.

17.6 CRYOPRESERVATION OF HEMATOPOIETIC PROGENITOR CELLS

17.6.1 DEFINITION, PROPERTIES

Hematopoietic progenitor cells (HPC) are primitive pluripotent cells capable of self renewal as well as of differentiation and maturation into all hematopoietic lineages. They can be found in bone marrow, fetal liver, mononuclear cell fraction of circulating blood, and umbilical cord blood. Preparations for clinical use are intended to provide a successful engraftment of hematopoietic stem cells leading to a restoration of all types of blood cells to a normal level and function in the recipient. They can originate from the recipient (autologous) or from another individual (homologous). The size and specific gravity of HPC from different sources are similar to those of MNC in whole blood. They are characterized by their colony-forming capacities in different *in vitro* cell culture assays and by special surface antigen markers. The membrane marker CD34 is commonly used for identification, isolation, and purification (Council of Europe, 2001).

17.6.2 HISTORICAL DEVELOPMENT

Barnes and Loutit (1955) performed the first transfusion of cryopreserved homologous bone marrow stem cells in mice. Six years later, Ashwood-Smith (1961) replaced the cryoprotectant glycerol by 10% dimethyl sulfoxide (Me$_2$SO), which has remained the "golden standard" until today. In 1983, Stiff et al. showed that for bone marrow freezing, Me$_2$SO could be replaced partially by 6% HES, lowering the final Me$_2$SO concentration to 5%. Eight years earlier, HES had already been reported by Schaefer and Bayer (1975) to be an effective cryoprotectant for human and mouse bone marrow. In contrast to Me$_2$SO, HES is well tolerated by the human organism and has been used as a plasma substitute for decades. This approach helps to reduce the side-effects of Me$_2$SO, some of them being the result of its ability to induce a histamine release. Makino et al. (1991) used the combination of the two additives (together with uncontrolled cooling by means of a –80°C freezer) for the first time to freeze peripheral HPC. However, systematic clinical studies in patients comparing the two cryopreservation methods (10% Me$_2$SO/controlled cooling and 5%Me$_2$SO + 6% HES, respectively) are still missing, and very few *in vitro* investigations have been performed so far (e.g., Sputtek et al., 1997). The same holds for investigations on the storage temperatures applied in routine clinical practice (–80°C and below –120°C, respectively). In 1986, Körbling et al. performed the first treatment of a patient with a Burkitt lymphoma using frozen autologous peripheral HPC. Cryopreservation has become the method of choice when autologous peripheral HPC are needed to be preserved for weeks or longer.

HPC show a high osmotic tolerance (Law et al., 1983), and consequently numerous successful cryopreservation protocols using a broad range of cooling rates as well as various concentrations of the cryoprotectant or cryoprotectants have been reported (reviews in Sputtek and Körber, 1991; English, 2000; Hubel, 2000; Rowley, 1994, 1992a, 1992b; Slaper-Cortenbach et al., 1994; Stiff, 1995). Detailed descriptions of the procedures can be found elsewhere (e.g., Gorin, 1992; Warkentin et al., 1993). In principle, the protocols being employed for peripheral HPC are the same as those that have been used for bone marrow–derived HPC (Dicke et al., 1981). The two main differences between peripheral HPC obtained from peripheral blood by apheresis after application of hematopoietic growth factors and those obtained from bone marrow aspirates are higher total cell numbers (including granulocytes, thrombocytes, and erythrocytes) and lack of contamination with fat. The use of cytapheresis machines for further processing of bone marrow–derived HPC may lead to a considerable loss of cells during the concentration (Zingsem et al., 1992, 1993).

Other sources for human hematopoietic progenitor cells are fetal liver (Tocci et al., 1994) and umbilical cord blood (Bertoli et al., 1995; Broxmeyer et al., 1989; Gluckman, 1994; Gluckman et al., 1992; McCullough et al., 1994; Newton et al., 1993; Rubinstein et al., 1993, 1994). However, whereas successful treatment with cord blood–derived stem cells of patients up to 30 kg body weight has been reported (Rubinstein et al., 1998), a more successful treatment of adults will need an *ex vivo* expansion of the hematopoietic potential before the transplantation. All of these new technologies will require reliable cryopreservation protocols: The door is open to developing new techniques or making the existing ones more effective.

17.6.3 PREPARATION, STORAGE, TRANSPORTATION, AND THAWING

If there is any positive marker for transfusion-transmitted diseases (which may happen in the case of autologous preparations), all personnel involved in the testing, collection, processing, and storage of HPC preparations should be informed to their involvement. All containers and material that have been in direct contact with the biomaterial should then be labeled as a biohazard or disposed as hazardous waste (Council of Europe, 2001).

One serious problem regarding peripheral HPC is the danger of clotting after thawing. Although it is highly advisable to use as little anticoagulant (mostly ACD-A [acid-citrate-dextrose, formulation A] or heparin) during the apheresis procedure from the patient's/donor's point of view, this may lead to problems during the following processing and after thawing. If very low anticoagulant concentrations have been used during the apheresis, it is recommended that some ACD-A or heparin be added before freezing or after thawing into the collection bag or freeing bag, respectively. A reduction of the formation of aggregates after thawing has been reported by Stiff et al. (1983; Stiff, 1995) by using 6% HES when Me_2SO concentrations of 5% are being used. According to Stiff et al., the CPS containing 12% HES is prepared by dissolving 42 g of HES (dry substance) in 140 mL of a electrolyte carrier solution (e.g., Normosol). After sterilization and cooling, 100 mL of a solution containing 25% of human albumin and 70 mL of a solution containing 50% of Me_2SO are added.

There are several reports in the literature (e.g., Douay et al., 1982; Goldman et al., 1978) on Me_2SO toxicity at room temperature. As a consequence of these and other reports, diluted and precooled Me_2SO solutions are usually added at 0 to 4°C. In a later investigation with highly pure Me_2SO (pharmaceutical grade), however, no impairment (colony assay) within the first hour was found when 5 or 10% of Me_2SO were added at either 4 or 37°C (Rowley and Anderson, 1993). However, cytoxicity was observed at 40%. Another group has partially confirmed these observations (Branch et al., 1994): They did not find an impairment in terms of the colony-forming capacity after exposure to 10% Me_2SO for 2 h, but they did find an impairment when employing a membrane integrity test (Trypan blue exclusion). Other investigators have reported a correlation between Me_2SO and some cell surface antigens as detected by flow cytometry: The proportions of certain leukocyte subgroups and the intensity of some surface antigens were changed. Whether or not it is really required, it has become general practice to add precooled CPS containing Me_2SO under thorough mixing to an equal volume of the precooled cell suspension. However, we believe that a lot of the so-called Me_2SO toxicity reported so far is a result of the osmotic activity of Me_2SO. If the addition is performed too rapidly, the cells will shrink below a critical minimal volume. Another reason may be the heat liberated (i.e., mixing enthalpy) during the mixing of aqueous solutions/suspensions with undiluted (or not appropriately prediluted) Me_2SO, which may lead to considerable (local) overheating and subsequent protein denaturation. As a consequence of the above-mentioned effects, we recommend prediluting the Me_2SO with an electrolyte solution or autologous plasma, then cooling the cell suspension and CPS down to 0 to 4°C before adding the CPS to the cell suspension (and not the other way around). Afterward, the Me_2SO containing suspension should be frozen as soon as possible.

On occasion, after-transfusion side effects (e.g., flush, dyspnea, nausea, abdominal spasms, vomiting, diarrhea, local vasospasm, hypo- and hypertension, and cardiac or anaphylactoid reactions) have been reported (e.g., Davis et al., 1990, Kessinger et al., 1990). We believe that most of them were caused by the histamine releasing effect of the Me_2SO. (For pharmacological properties of Me_2SO, see Brayton [1986], Gerhards and Gibian [1968], Jacob and Herschler [1986], and Willhite and Katz [1984].) As a consequence, any measure taken to minimize the exposure to Me_2SO is a step in the right direction. This may either be achieved by reducing the total volume of the preparations or by decreasing the Me_2SO concentration as much as possible. Investigations concerning the LD_{50} (the dose of a toxic chemical or pathogen that will kill 50% of the test organisms to which it is given) in humans do not exist. However, from investigations in animals, it may be speculated that it is between 2.5 and 8.1 g/kg body weight (Rowley, 1992a). A washing procedure for reducing the Me_2SO content after thawing using a solution containing HES and albumin was described by Rosina et al. (1992). Beaujean et al. (1991) have described a postthaw washing method with only minimal losses concerning the nucleated cells and clonogenicity, but reports on clinical outcome (e.g., kinetics of engraftment) are missing. Cytotoxic effects of HES on HPC are not known, but there is a clinical study on the frequency of anaphylactoid reactions after the application of HES for volume replacement (Laxenaire et al., 1994). Adverse reactions were observed at a rate of 0.058% (3 out of 5231 patients investigated). The highest degree of severity was III (bronchospasm, shock) and was been observed only once.

The cell suspensions to be frozen usually contain 2 to 4% proteins, carbohydrates, and electrolytes of various compositions. It is known that proteins are able to increase the cryoprotective effect of several cryoprotectants. Moreover, thawed plasma that was frozen without any cryoprotectant was reported to contain some viable HPC (Bernvil et al., 1994). Autologous plasma, albumin, and fetal calf serum have been used frequently as protein compounds. However, the latter has become obsolete because of the variant Creutzfed-Jakob disease. To our best knowledge, no systematic investigation on the optimum protein content has been performed so far. The same holds for the electrolyte and carbohydrate composition of the carrier solution (if being used instead of autologous plasma). It is well known that some sugars (e.g., glucose, trehalose) are effective cryoprotectants. Whether this is the result of a membrane-stabilizing or vitrification-inducing effect (according to Carpenter and Crowe, 1988) or of lowering the Me_2SO "toxicity" (Clark and McCathy, 1991) remains unclear at present.

In an investigation based on 108 units transfused into patients, no dependence of the postthaw MNC recovery from the original cell concentration (average nucleated cell concentration $3.7 \pm 1.9 \times 10^8$/mL, range 0.4 to 0.8×10^8/mL; fraction of MNC 53% ± 27%, range 10 to 100%) was found. This was also not the case with regard to clonogenicity and membrane integrity. Rowley et al. (1994) concluded that even very high cell concentrations do not lead to a significant loss of the hematopoietic potential after freezing.

As already mentioned above, optimal cooling rates vary considerably from 1° to 4°C/min depending on the author. However, little attention has been paid to the temperature interval for its interpolation. We believe that the most suitable interval is from the end of the plateau phase down −30 or −40°C. Cooling is performed by means of computer-controlled, programmable apparatus (coolant: liquid nitrogen), but there are numerous studies using either simply the vapor phase over liquid nitrogen or −80°C refrigerators (e.g., Clark et al., 1991; Douay et al., 1982; Rowley, 1994). Advantages of the "uncontrolled" cooling methods are cost saving and simplicity. The major disadvantage is the missing possibility to compensate for varying sample thickness (i.e., volume if only one type of freezing bag is used). If a thermocouple is being used to measure and register the temperature–time-histories during cooling, we do not see why these methods should not be used instead of the costly and more complicated computer-controlled apparatus. However, it has to be evaluated beforehand that the "simple" cooling devices work reliably and reproducibly. The cooling process seems to have little influence on the postthaw recovery, as long as at least 5%

Me_2SO is being used (Makino et al., 1991) and the cooling rate does not exceed 5°C/min (Sputtek et al., 1997). It should be mentioned that the cool-down kinetics depend on several parameters: geometry ("undefined," cylindrical, or plate shaped), sample thickness, composition (kind and concentration of cryoprotectant or cryoprotectants), and thermal properties (heat capacity, heat conductivity) of the freezing bag and (if used) the metal envelope. That is why temperature–time-histories measured in small reference samples (e.g., tubes) may differ significantly from those measured in the corresponding product bags (Douauy et al., 1986). However, despite this bias, we did not find a significant difference with regard to cell recovery and clonogenicity (Sputtek et al., 1997), probably because of the broad range of optimal cooling rates for hematopoietic progenitor cells (i.e., 1°–5°C/min). To make things even more complicated, however, the position of the thermocouple plays also an important role (Hartmann et al., 1991; Heschel and Rau, 1993).

The temperature required for long-term storage of several is below –123°C, which is the glass transition temperature of Me_2SO (e.g., Pegg et al., 1997). For short-term storage (weeks to months), temperatures below –80°C may be sufficient. However, mechanical refrigerators are a problem in the case of an electric power breakdown. To be prepared for such a power breakdown, supplemental energy-supply systems or CO_2 cooling should be available.

Virus transmission is possible in or above liquid nitrogen (Tedder et al., 1995). However, transmission can only happen if the freezing bag in question is leaky and if the liquid nitrogen tank has already been contaminated or the infective freezing bag is leaky, too. A contact in the liquid phase is probably required, but a transmission via the vapor phase cannot be excluded. In most cases the damage has already happened before freezing (mostly because of improperly sealed freezing bags), but it becomes obvious on thawing. Freezing bags that can be filled using sterile tube-docking devices are much safer with regard to leakage than those in which manual "spiking" is being used. Sealing devices for sealing tubes work more reliably than those sealing off parts of the bag, which was the case in the report cited above (Tedder et al., 1995). Because of security reasons, possibly infected units should always be stored in the vapor phase over liquid nitrogen in separate storage tanks (or for short-term storage in –80°C freezers). A procedure for checking the viral safety of liquid nitrogen storage tanks has been described elsewhere (Shafer et al., 1976). (For general issues concerning safety and organization of cryopreserved cell banks, see Chapter 15.)

During transportation of cryopreserved HPC preparations, the temperature should remain below –120°C. Therefore, the units should be shock- and spill-proof and surrounded with absorbing material soaked in sufficient liquid nitrogen (e.g., in so-called "dryshippers"). The primary container should be wrapped into a leakproof secondary container not interfering with the cooling. The container should bear the labels reading "Do not X-ray," "Keep frozen," "Labile unique human transfusion material," and "Rush" (Council of Europe, 2001). Appropriate filling of the dryshippers before and after shipment can be checked by weighing.

Thawing usually takes place in a 37°–42°C water bath under continuous agitation immediately before transfusion. In most centers, no postthaw washing step is performed to remove the cryoprotectant and debri, because of the cell loss and the danger of clotting. In some centers, ACD-A or DNAse is added, but the risk of allergic or febrile reactions to DNAse is not known at present. However, clots are safely kept back by using 170-μm filters. The many problems outlined here may give a pessimistic viewpoint, but we can say that this is not a requiem for frozen blood (stem) cells. Their potential applications extend beyond what has become routine clinical practice today, and efforts will continue to enhance their recovery after cryopreservation.

17.7 GENERAL CONCLUSIONS

Cryopreservation methods have been developed for all different categories of blood cells (including blood stem cells) and blood corpuscles. For granulocytes, however, no clinically applicable method has been found until today. Frozen blood cells are being used for various diagnostic and clinical purposes, which has led to a variety of cell specific cryopreservation protocols. The methods differ with regard to cell concentrations, protective solutions used (cryoprotectants and their concentrations), temperature–time-histories during cooling and rewarming, and storage temperature.

Cryopreserved erythrocytes are of advantage in the case of rare blood groups, antibody problems, and civil as well as military disasters. In addition, they are recommended for immunization purposes after frozen storage to allow retesting of donors (quarantine) and could be considered in some cases for autologous transfusion.

Frozen platelets should be reserved for the provision of HLA- or HPA-compatible platelets in the case of platelet refractoryness in which a compatible donor is not immediately available. Platelets are also a focus of attention for renewed interest in freeze-drying as a storage technology (see Chapter 21), but their effectiveness in clinical practice after this treatment remains to be established.

At present, there is a growing clinical interest in cryopreserved lymphocytes for the supplemental treatment of patients after blood stem cell transplantation. Donor lymphocyte infusions have shown to induce remissions in patients with certain forms of leukemia with evidence of relapse after blood stem cell transplantation.

Cryopreserving lymphocytes and monocytes is well established as a routine procedure for clinical laboratory testing. Frozen mononuclear cells are used for various diagnostic purposes; for example, tissue typing of patients on waiting lists for organ/bone marrow transplantations.

Cryopreserved hematopoietic stem cells for clinical use are intended to provide a successful engraftment leading to a restoration of all types of blood cells to a normal level and function in the recipient. They can originate either from the recipient or from another individual and are increasingly applied successfully for the treatment of certain forms of leukemia and plasma-cell tumors.

REFERENCES

Armitage, W.J. (1986) Osmotic stress as a factor in the detrimental effect of glycerol on human platelets, *Cryobiology*, 23, 116–125.

Ashwood-Smith, M.J. (1961) Preservation of mouse marrow at –79 C with dimethyl sulphoxide, *Nature*, 190, 1204–1205.

Ashwood-Smith, M.J. (1964) Low temperature preservation of mouse lymphocytes with dimethyl sulfoxide, *Blood*, 23, 494–501.

Atkins, L. (1962) Preservation of viable leucocytes in glycerol at –80 C, *Nature*, 195, 610–611.

Bank, H. (1980) Granulocyte preservation circa 1980, *Cryobiology*, 17, 187–197.

Barnes, D.W.H. and Loutit, J.F. (1955) The radiation recovery factor: Preservation by the Polge-Smith-Parkes technique, *J. Natl. Cancer Inst.*, 15, 901–905.

Beaujean, F., Leforestier, C., and Mannoni, P. (1979) Clinical and functional studies of platelets frozen in 5% dimethyl sulfoxide, *Cryo-Letters*, 1, 98–103.

Beaujean, F., Hartmann, O., Kuentz, M., Le Forestier, C., Divine, M., and Duedari, N. (1991) A simple, efficient washing procedure for cryopreserved human hemapoietic stem cells prior to reinfusion, *Bone Marrow Transplant*, 8, 291–294.

Beckmann, J., Körber, C., Rau, G., Hubel, A., and Cravalho, E.G. (1990) Redefining cooling rate in terms of ice front velocity and thermal gradient: First evidence of relevance to freezing injury of lymphocytes, *Cryobiology*, 27, 279–287.

Bernvil, S.S., Abdulatiff, M., Al-Sedairy, S., Sasich, F., and Sheth, K. (1994) Fresh frozen plasma contains viable progenitor cells—should we irradiate? *Vox Sang.*, 67(4), 405.

Bertoli, F., Lazarri, L., Lauri, E., Corsini, C., and Sirchia, G. (1995) Cord blood-derived hematopoietic progenitor cells retain their potential for *ex vivo* expansion after cryopreservation, *Bone Marrow Transplant*, 15, 159–160.

Branch, D.R., Calderwood, S., Cecutti, M.A., Herst, R., and Solh, H. (1994) Hematopoietic progenitor cells are resistant to dimethyl sulfoxide toxicity, *Transfusion*, 34, 887–890.

Brayton, C.F. (1986) Dimethyl sulfoxide (DMSO): A review, *Cornell Vet.*, 76, 61–90.

Brodthagen, U.A., Armitage, W.J., and Parmar, N. (1985) Platelet cryopreservation with glycerol, dextran and mannitol: Recovery of 5-hydroxytryptamine uptake and hypotonic stress response, *Cryobiology*, 22, 1–9.

Broxmeyer, H.E., Douglas, G.W., Hangoc, G., Cooper, S., Bard, J., English, D., Arny, M., Thomas, L., Boyse, E.A. (1989) Human umbilical cord blood as a potential source of transplantable hematopoietic stem/progenitor cells, *Proc. Natl. Acad. Sci. USA*, 86, 3828–3832.

Bryant G. (1995) DSC measurement of cell suspensions during successive freezing runs: Implications for the mechanisms of intracellular ice formation, *Cryobiology*, 32, 114–128.

Carpenter, J.F. and Crowe, J.H. (1988) The mechanism of cryoprotection of proteins by solutes, *Cryobiology*, 25, 244–255.

Choudhury, C. and Gunstone, M.J. (1978) Freeze preservation of platelets using hydroxyethyl starch (HES): A preliminary report, *Cryobiology*, 15, 493–501.

Clark, J., Pati, A., and McCathy, D. (1991) Successful cryopreservation of human bone marrow does not require a controlled-rate freezer, *Bone Marrow Transplant*, 7(2), 121–125.

Cohen, E. and Rowe, A.W. (1965) Detection of leukoagglutinins with dimethylsulfoxide (DMSO) protected leukocytes frozen with liquid nitrogen, *Vox Sang.*, 10(5), 543–551.

Council of Europe, Ed. (2001) *Guide to the Preparation, Use and Quality Assurance of Blood Components*, Council of Europe Publishing, Strasbourg.

Currie, L.M., Lichtinger, B., Livesey, S.A., Tansey, W., Yang, D.J., and Connor, J. (1999) Enhanced circulatory parameters of human platelets cryopreserved with second-messenger effectors: An *in vivo* study of 16 volunteer platelet donors, *Br. J. Haematol.*, 105, 826–831.

Dankberg, F. and Persidsky, M.D. (1976) A test of granulocyte membrane integrity and phagocytic function, *Cryobiology*, 13, 430–432.

Davis, J.M., Rowley, S.D., Braine, H.G., Piantadosi, S., and Santos, G.W. (1990) Clinical toxicity of cryo-preserved bone marrow graft infusion, *Blood*, 75, 781–786.

Dayian, G., Reich, L.M., Mayer, K., Turc, J.M., and Rowe, A.W. (1974) Use of glycerol to preserve platelets suitable for transfusion, *Cryobiology*, 11, 563–564.

Dayian, G. and Rowe, A.W. (1976) Cryopreservation of human platelets for transfusion. A glycerol-glucose, moderate rate cooling procedure, *Cryobiology*, 13, 1–8.

Dayian, G., Harris, H.L., Vlahides, G.D., and Pert, J.H. (1986) Improved procedure for platelet freezing, *Vox Sang*, 51, 292–298.

De Boer, M., Reijneke, R., Van de Griend, R.J., Loos, J.A., and Roos, D. (1981) Large-scale purification and cryopreservation of human monocytes, *J. Immun. Meth.*, 43, 225–229.

Dicke, K.A., Vellekoop, L., Spitzer, G., Zander, A.R., Schell, F., and Verma, D.S. (1981) Autologous bone marrow transplantation in neoplasia, *Transplant Proc.*, 13, 267–269.

Djerassi, I., Farber, S., Roy, A., and Cavins, J. (1966) Preparation and *in vivo* circulation of human platelets preserved with combined dimethylsulfoxide and dextrose, *Transfusion*, 6, 572–576.

Douay, L., Gorin, N.C., David, R., Stachowiak, J., Salmon, C., Najman, A., and Duhamel, G. (1982) Study of granulocyte-macrophage progenitor (CFUc) preservation after slow freezing of bone marrow in the gas phase of liquid nitrogen, *Exp. Hematol.*, 10, 360–366.

Douay, L., Lopez, M., and Gorin, N.C. (1986) A technical bias: Differences in cooling rates prevent ampoules from being a reliable index of stem cell cryopreservation in large volumes, *Cryobiology*, 23, 296–301.

English, D. (2000) Freezing hematopoietic stem cells, *J. Hematother. Stem Cell Res.*, 9, 123–125.

Gallo, D., Kimpton, J.S., and Dailey, P.J. (1987) Comparative studies on the use of fresh and frozen peripheral blood lymphocyte specimens for isolation of human immunodeficiency virus and effects of cell lysis on isolation efficiency, *J. Clin. Microbiol.*, 25, 1291–1294.

Gardner, F.H. (1968) Platelet preservation problems, *Cryobiology*, 5, 42–48.

Gerhards, E. and Gibian, H. (1968) Stoffwechsel und Wirkung des Dimethylsulfoxids, *Naturwissenschaften*, 55, 435–438.

Gluckman, E. (1994) European organisation for cord blood banking, *Blood Cells*, 20, 601–608.

Gluckman, E., Thierry, D., Lesage, S., Traineau, R., Gerotta, J., Rabian, C., Brossard, Y., Van Nifterik, J., and Benbunan, M. (1992) Cord blood banking for human hematopoietic cell transplantation, *Prog. Clin. Biol. Res.*, 377, 591–598.

Goldman, J.M., Th'ng, K.H., Park, D.S., Spiers, A.S., Lowenthal, R.M., and Ruutu, T. (1978) Collection, cryopreservation and subsequent viability of haemopoietic stem cells intended for treatment of chronic granulocytic leukaemia in blast-cell transformation, *Br. J. Haematol.*, 40, 185–195.

Gorin, N.C. (1992) Cryopreservation and storage of stem cells, in *Bone Marrow and Stem Cell Processing—A Manual of Current Techniques*, Areman, E.M., Deeg, H.J., and Sacher, R.A., Eds., Davis, Philadelphia, pp. 292–362.

Harada, M., Yoshimoto, R., Ishino, C., and Hattori, K. (1980) Immunocompetence of the residual lymphocytes in frozen blood, *Cryobiology*, 17, 100–107.

Hartmann, U., Nunner, B., Körber, C., and Rau G. (1991) Where should the cooling rate be determined in an extended freezing sample? *Cryobiology*, 28, 115–130.

Haynes, L.L., Tullis, J.L., Pyle, H.M., Wallach, S., and Sproul, M.T. (1960) Clinical use of glycerolized frozen blood, *J. Am. Med. Assoc.*, 173, 1657.

Heschel, I. and Rau, G. (1993) Numerische Berechnungen zum Erstarrungsverlauf in wäßrigen Salzlösungsproben mit Bezug zur Kryobiologie, *Chem. Ing. Technol.*, 65, 63–66.

Horn, P., Sputtek, A., Standl, T., Rudolf, B., Kühnl, P., and Schulte am Esch, J. (1997) Transfusion of autologous, hydroxyethyl starch cryopreserved red blood cells in patients, *Anaesth. Analg.*, 85, 739–745.

Hubel, A., Cravalho, E.G., Nunner, B., and Körber, C. (1992) Survival of directionally solidified B-lymphoblast under various crystal growth conditions, *Cryobiology*, 29, 514–531.

Hubel, A. (2000) Parameters of cell freezing: Implications for the cryopreservation of stem cells, *Transfus. Med. Rev.*, 11, 224–233.

Huggins, C.E. (1963) Preservation of blood by freezing with dimethyl sulfoxide and its removal by dilution and erythrocyte agglomeration, *Vox Sang*, 8, 99–100.

Jacob, S.W. and Herschler, R. (1986) Pharmacology of DMSO, *Cryobiology*, 23, 14–27.

Kessinger, A., Schmit-Pokorny, K., Smith, D., and Armitage, J. (1990) Cryopreservation and infusion of autologous peripheral blood stem cells, *Bone Marrow Transplant*, 5, 25–27.

Kim, B.K. and Baldini, M.G. (1986) Glycerol stress and platelet integrity, *Cryobiology*, 23, 209–213.

Klein, E., Toch, R., Farber, S., Freeman, G., and Fiorentino, R. (1956) Hemostasis in thrombocytopenic bleeding following infusion of stored, frozen platelets, *Blood*, 11, 693–698.

Knight, S.C. (1980) Preservation of leucocytes, in *Low Temperature Preservation in Medicine and Biology*, Ashwood-Smith, M.J. and Farrant, J., Eds., Pitman Medical, Tunbridge Wells, UK, pp. 121–138.

Knorpp, C.T., Merchant, W.R., Gikas, P.W., Spencer, H.H., and Thompson, N.W. (1967) Hydroxyethyl starch: Extracellular cryophylactic agent for erythrocytes, *Science*, 57, 1312–1313.

Körber, C. (1988) Phenomena at the advancing ice-liquid interface: Solutes, particles and biological cells, *Quart. Rev. Biophys.*, 21, 229–298.

Körber, C., Englich, S., and Rau, G. (1991) Intracellular ice formation: Cryomocroscopical observation and calorimetric measurement, *J. Microsc.*, 191, 313–325.

Körbling, M., Dörken, B., Ho, A.D., Pezzutto, A., Hunstein, W., and Fliedner, T.M. (1986) Autologous transplantation of blood derived haemopoietic stem cells after myeloablative therapy in a patient with Burkitt's lymphoma, *Blood*, 67, 529–532.

Kotelba-Witkowska, B. and Schiffer, C.A. (1982) Cryopreservation of platelet concentrates using glycerolglucose, *Transfusion*, 22, 121–124.

Krijnen, H.W., Kuivenhoven, A.C.J., and De Wit, J.J.F.M. (1968) The preservation of blood cells in the frozen state. Experiences and current methods in the Netherlands, *Cryobiology*, 5, 136–143.

Langer, R., Düpre, H.J., Kron, W., Sputtek, A., Steigerwald, R., Trenkel, K., Rau, G., and Henrich, H.A. (1993) Untersuchung zur funktionellen Verträglichkeit autologer, mit Hydroxyäthylstärke kryokonservierter Erythrozyten beim Hund, in *Klinische Mikrozirkulation und Hämorheologie*, Landgraf, H., Jung, F., and Ehrly, A.M., Eds., Blackwell Wissenschaft, Berlin, pp. 233–239.

Langer, R., Albrecht, R., Hempel, K., Krug, S., Sputtek, A., Steigerwald, R., Trenkel, K., and Henrich, H.A. (1994) Charakterisierung der 24-h-Überlebensrate und Lebensdauer von mittels Hydroxyethylstärke kryokonservierten Erythrozyten nach autologer Transfusion im Hund, *Infus. Ther. Transfus. Med.*, 21, 393–400.

Law, P. and Meryman, H.T. (1982) Cryopreservation of platelets: Current status, *Plasma Ther. Transfus. Technol.*, 3, 317–326.

Law, P., Alsop, P., Dooley, D.C., and Meryman, H.T. (1983) Studies of cell separation: A comparison of the osmotic response of human lymphocytes and granulocyte-monocyte progenitor cells, *Cryobiology*, 20, 644–651.

Laxenaire, M.C., Charpentier, C., and Feldman, L. (1994) Réactions anaphylactoides aux substituts colloidaux du plasma: Incidence, facteurs de risque, méchanismes, *Ann. Fr. Anesth. Réanim.*, 13, 301–310.

Lovelock, J.E. (1953a) The hemolysis of human red blood cells by freezing and thawing, *Biochim. Biophys. Acta*, 10, 414–426.

Lovelock, J.E. (1953b) The mechanism of protective action of glycerol against haemolysis by freezing and thawing, *Biochim. Biophys. Acta*, 11, 28–36.

Lozano, M., Escolar, G., Mazzara, R., Connor, J., White, J.G., DeLecea, C., and Ordinas, A. (2000) Effects of the addition of second-messenger effectors to platelet concentrates separated from whole-blood donations and stored at 4 degrees C or –80 degrees C, *Transfusion*, 40, 527–534.

Lundberg, A., Yankee R.A., Henderson E.S., and Pert, J.H. (1967) Clinical effectiveness of blood platelets preserved by freezing, *Transfusion*, 7, 380–381.

Makino, S., Harada, M., Akashi, K., Taniguchi, S., Shibuya, T., Inaba, S., and Niho, Y. (1991) A simplified method for cryopreservation of peripheral blood stem cells at –80 degrees C without rate-controlled freezing, *Bone Marrow Transplant*, 8, 239–244.

Mazur, P., Rall, W.F., Rigopoulos, N. (1981) Relative contributions of the fraction of unfrozen water and of salt concentration to the survival of slowly frozen human erythrocytes, *Biophys. J.*, 36, 653–675.

McCullough, J., Clay, M.E., Fautsch, S., Noreen, H., Segall, M., Perry, E., and Stroncek, D. (1994) Proposed policies and procedures for establishment of a cord blood bank, *Blood Cells*, 20, 609–626.

Meryman, H.T. (1989) Frozen red cells, *Transfus Med. Rev.*, 3,121–127.

Meryman, H.T. and Burton J.L. (1978) Cryopreservation of platelets, *Prog. Clin. Biol. Res.*, 28, 153–165.

Meryman, H.T. and Hornblower, M. (1972) A method for freezing and washing red blood cells using a high glycerol concentration, *Transfusion*, 12, 145–156.

Meryman, H.T. and Hornblower, M. (1976) Freezing and deglycerolizing sickle-trait red blood cells, *Transfusion*, 16, 627–632.

Mollison, P.L., Sloviter, H.A. (1951) Successful transfusion of previously frozen human red cells, *Lancet II*, 261, 862–864.

Newton, I., Charbord, P., Schaal, J.P., and Hervé, P. (1993) Toward cord blood banking: Density-separation and cryopreservation of cord blood progenitors, *Exp. Hematol.*, 21, 671–674.

Papadopoulos, E.B., Ladanyi, M., Emanuel, D., Mackinnon, S., Boulad, F., Carabasi, M.H., Gulati, S.C., Castro-Malspina, H., Childs, B.H., Gillio, A.P., Small, T.N., Young, J.W., Kernan, N.A., and O'Reilly, R.J., (1994) Infusion of donor leukocytes to treat Epstein-Barr virus-associated lymphoproliferative disorders after allogeneic bone marrow transplantation, *N. Engl. J. Med.*, 330, 1185–1191.

Pedralozzi, P., Noris, P., Perotti, C., Schiavo, R., Ponchio, L., Beletti, S., Da Prada, G.A., Balduini, C.L., Salvaneschi, L., Robustelli-Della-Cuna, G., and Siena, S. (2000) Transfusion of platelet concentrates cryopreserved with ThromboSol plus low-dose dimethylsulfoxide in patients with severe thrombocytopenia: A pilot study, *Br. J. Haematol.*, 108, 653–659.

Pegg, P.J. (1965) The preservation of leucocytes for cytogenetic and cytochemical studies, *Br. J. Haematol.*, 11, 586–591.

Pegg, D.E. and Diaper, M.P. (1988) On the mechanism of injury to slowly frozen erythrocytes, *Biophys. J.*, 54, 471–488.

Pegg, D.E. and Diaper, M.P. (1989) The unfrozen fractionhypothesis of freezing injury to human erythrocytes: A critical examination of the evidence, *Cryobiology*, 26, 30–43.

Pegg, D.E., Wusteman, M.C., and Boylan, S. (1997) Fractures in elastic arteries, *Cryobiology*, 34, 183–192.

Pellerin-Mendes, C., Million, L., Marchand-Arvier, M., Labrude P., and Vigneron C. (1997) Study on the protective effect of trehalose and dextran during freezing of human red blood cells in liquid nitrogen, *Cryobiology*, 35, 173–186.

Pert, J.H. and Dayian, G. (1979) Statically controlled cooling rate device, *Cryobiology*, 16, 90–96.

Polge, C., Smith, A.U., and Parkes, A.S. (1949) Revival of spermatozoa after vitrification and dehydratation at low temperatures, *Nature*, 164, 666.

Redmond, J., Bolin, R.B., and Cheney, B.A. (1983) Glycerol-glucose cryopreservation of platelets. *In vivo* and *in vitro* observations, *Transfusion*, 23, 213–214.

Richter, E., Fitzner, E., Freund, B., Hackensellner, H.A., Matthes, G., Schürer, M., and Ziemer, S. (1983) Eine Methode zur Kryokonservierung von Thrombozyten, *Folia Haematol. (Leipzig)*, 110, 725–730.

Rinfret, A.P. (1963) Some aspects of preservation of blood by rapid freeze-thaw procedures, *Fed. Proc.*, 22, 94–101.

Rosina, O., Zhou, S., Tabrizi, D., and Scigliano, E. (1992) Thawing and dilution of bone marrow, in *Bone Marrow and Stem Cell Processing—A Manual of Current Techniques*, Areman, E.M., Deeg, H.J., and Sacher, R.A., Eds., Davis, Philadelphia, pp. 340–341.

Rowe, A.W. (1973) Preservation of blood by the low glycerol-rapid freeze process, in *Red Cell Freezing*, American Association of Blood Banks, Ed., Washington, DC, pp. 55–71.

Rowe, A.W. (1994) Cryopreservation of red blood cells, *Vox Sang.*, 67, 201–206.

Rowe, A.W. (1995) Cryopreservation of blood—An historical perspective, *Infus. Ther. Transfus. Med.*, 22, 36–40.

Rowe, A.W. (2002) Cryopreservation of red cells by freezing and vitrification—Some recollections and some predictions, *Infus. Ther. Transfus. Med.*, 29, 25–30.

Rowe, A.W., Eyster, E., and Kellner, A. (1968) Liquid nitrogen preservation of red blood cells for transfusion: A low glycerol-rapid freeze procedure, *Cryobiology*, 5, 119–128.

Rowe, A.W. and Lenny, L.L. (1982) Red blood cells, in *Organ Preservation for Transplantation*, Karow, A.M. and Pegg, D.E., Eds., Marcel Dekker, New York, pp. 285–312.

Rowe, A.W., Lenny, L.L., and Mannoni, P. (1980) Cryopreservation of red cells and platelets, in *Low Temperature Preservation in Medicine and Biology*, Ashwood-Smith, M.J. and Farrant, J., Eds., Pitman Medical, Tunbridge Wells, UK, pp. 85–120.

Rowley, S.D. (1992a) Hematopoietic stem cell cryopreservation: A review of current techniques, *J. Hematother.*, 1, 233–250.

Rowley, S.D. (1992b) Techniques of bone marrow and stem cell cryopreservation and storage, in *Marrow Transplantation: Practical and Technical Aspects of Stem Cell Reconstitution*, Sacher, R.A. and AuBuchon, J.P., Eds., American Association of Blood Banks, Bethesda, MD, pp. 105–128.

Rowley, S.D. (1994) Secondary processing, cryopreservation, and reinfusion of the collected product, in *Practical Considerations of Apheresis in Peripheral Blood Stem Cell Transplantation*, Kessinger, A. and McMannis, J.D., Eds., Cobe BCT, Lakewood, CO, pp. 53–62.

Rowley, S.D. and Anderson, G.L. (1993) Effect of DMSO exposure without cryopreservation on haematopoietic progenitor cells, *Bone Marrow Transplant*, 11, 389–393.

Rowley, S.D., Bensinger, W.I., Gooley, T.A., and Buckner, C.D. (1994) Effect of cell concentration on bone marrow and peripheral blood stem cell cryopreservation, *Blood*, 83, 2731–2736.

Rubinstein, P., Rosenfield, R.E., Adamson, J.W., and Stevens, C.E. (1993) Stored placental blood for unrelated bone marrow reconstitution, *Blood*, 81, 942–949.

Rubinstein, P., Taylor, P.E., Scaradavou, A., Adamson, J.W., Migliaccio, G., Emanuel, D., Berkowitz, R.L., Alvarez, F., and Stevens, C.E. (1994) Unrelated blood for bone marrow reconstitution: Organization of the placental blood program, *Blood Cells*, 20, 587–600.

Rubinstein, P., Carrier, C., Scaradavou, A., Kurtzberg, J., Adamson, J., Migliaccio A.R., Berkowitz R.L., Cabbad, N.L., Taylor, P.E., Rosenfield, R.E., Stevens, C.E. (1998) Outcomes among 562 recipients of placental-blood transplants from unrelated donors, *N. Engl. J. Med.*, 22, 1565–1577.

Schaefer, U.W. and Beyer, J.H. (1975) Die protektive Wirkung von Hydroxyäthylstärke bei der Kryokonservierung von Knochenmark der Maus und des Menschen, *Anaesthesist.*, 24, 505–506.

Scheiwe, M.W., Breidenbach, B., Tarkkanen, P., and Stürner, K.H. (1981) Cryopreservation of human platelets with glycerol-glucose suspended in a large fluid volume: Influence of the cooling rate, *Cryo-Letters*, 2, 247–256.

Schiffer, C.A., Aisner, J., and Wiernik, P.H. (1978) Frozen autologous platelet transfusion for patients with leukemia, *N. Engl. J. Med.*, 299, 7–12.

Schiffer, C.A., Aisner, J., and Wiernik, P.H. (1978) Frozen autologous platelet transfusion for patients with leukaemia, *N. Engl. J. Med.*, 299, 7–12.

Schiffer, C.A., Aisner, J., and Dutcher, J.P. (1985) Platelet cryopreservation using dimethyl sulfoxide, *Ann. NY Acad. Sci.*, 459, 161–169.

Shafer, T.W., Everett, J., Silver, G.H., and Came, P.E. (1976) Biohazards: Virus contaminated liquid nitrogen, *Science*, 19, 24–25.

Slaper-Cortenbach, I.C., Wijngaarden-du-Bois, M.J., de Vries-van Rossen, A., and Zadurian, A. (1994) Conditions for optimal cryopreservation of peripheral stem cells, *Prog. Clin. Biol. Res.*, 389, 649–656.

Sloviter, H.A. (1951) Recovery of human red blood cells after freezing, *Lancet I*, 261, 823.

Smith, A.U. (1950) Prevention of haemolysis during freezing and thawing of red blood-cells, *Lancet II*, 259, 910–911.

Soiffer, R.J. (1997) Donor lymphocyte infusions: The door is open, *J. Clin. Oncol.*, 15, 416–417.

Sputtek, A. (1995) Cryopreservation of human red blood cells: A comparison of the cryoprotectants gelatine, dextran, and hydroxyethyl starch, *Cryo-Letters*, 16, 58.

Sputtek, A., Barthel, B., and Rau, G. (1993a) Cryopreservation of human platelets: An *in vitro* comparison of a 4% HES method with the standard 5% DMSO method, *Cryobiology*, 30, 656–657.

Sputtek, A., Brohm, A., Classen, I., Hartmann, U., Körber, C., Scheiwe, M.W., and Rau, G. (1987) Cryopreservation of human platelets with hydroxyethyl starch in a one-step procedure, *Cryo-Letters*, 8, 216–229.

Sputtek, A., Jetter, S., Hummel, K., and Kühnl, P. (1997) Cryopreservation of peripheral blood progenitor cells: Characteristics of suitable techniques, *Beitr. Infus. Ther. Transfus. Med*, 34, 79–83.

Sputtek, A. and Körber, C. (1991) Cryopreservation of red blood cells, platelets, lymphocytes, and stem cells, in Fuller, B.J. and Grout, B.W.W., Eds., *Clinical Applications of Cryobiology*, CRC Press, Boca Raton, FL, pp. 95–147.

Sputtek, A., Körber, C., and Rau, G. (1990) Cryopreservation of human erythrocytes with hydroxyethylated starches under variation of starch modification and concentration, electrolyte content, hematocrit, and cooling rate, *Cryobiology*, 27, 667–668.

Sputtek, A., Körber, C., and Rau G. (1991a) Tieftemperaturkonservierung menschlicher roter Blutkörperchen mit dem Kryoprotektiv Hydroxyethylstärke—Bedeutung der HES-Modifikation, *Z. Klin. Med.*, 46, 1567–1570.

Sputtek, A. and Mingers, B. (1994) Freezing Container, European Patent 0,786,981, Priority 17.10.1994.

Sputtek, A., Roling, C., Singbartl, G., Schleinzer, W., and Rau, G. (1993b) First autologous retransfusion of human red blood cells (450 ml) cryopreserved in the presence of 11.5% HES without postthaw washing, *Cryobiology*, 30, 657.

Sputtek, A., Schmid, H., Henrich, H.A., Körber, C., and Rau, G. (1991b) Cryopreservation of dog red blood cells (DRBC) with hydroxyethyl starch (HES) for autologous retransfusion as an animal study, *Cryobiology*, 28, 546.

Sputtek, A., Sensendorf, H., Fründ, V., Birmanns, H., and Kühnl, P. (1995a) Ein Fallbericht zur Kryokonservierung von Erythrozyten mit Hydroxyethylstärke, in Mempel, W., Schwarzfischer, G., and Mempel, C., Eds., *Eigenbluttransfusion heute*, Sympomed, Munich, pp. 206–211.

Sputtek, A., Singbartl, G., Langer, R., Schleinzer, W., Henrich, H.A., and Kühnl, P. (1995b) Cryopreservation of red blood cells with the non-penetrating cryoprotectant hydroxyethyl starch, *Cryo-Letters*, 16, 283–288.

Sputtek, A., Warnken, U.H., Langer, R., Henrich, H.A., and Kühnl, P. (1994) Hydroxyethyl starch plasma concentration profiles after administration of HES protected frozen erythrocytes, *Cryobiology*, 31, 584.

Stiff, P.J. (1995) Cryopreservation of hematopoietic stem cells, in *Marrow and Stem Cell Processing for Transplantation*, Lasky, L.C. and Warkentin P.I., Eds., American Association of Blood Banks, Bethesda, MD, pp. 69–82.

Stiff, P.J., Murgo, J.A., Zaroulis, C.G., DeRisi, M.F., and Clarkson, B.D. (1983) Unfractionated human marrow cell cryopreservation using dimethylsulfoxide and hydroxyethyl starch, *Cryobiology*, 20, 17–24.

Sumida, S. (1974) *Transfusion of Blood Preserved by Freezing*, Thieme, Stuttgart.

Sumida, S. (1993) Cryopreservation of human blood cells, in *Recent Advances in Cryomedicine and Cryobiology*, Chinese Association of Refrigeration, Ed., International Academic Publishers, Beijing, pp. 3–15.

Takahashi, T., Hammett, M., Cho, M.S. (1985) Multifaceted freezing injury in human polymorphonuclear cells at high subfreezing temperatures, *Cryobiology*, 22, 215–236.

Taylor, M.A. (1981) Cryopreservation of platelets: An *in-vitro* comparison of four methods, *J. Clin. Pathol.*, 34, 71–75.

Taylor, M.J. and Bank, H.L. (1988) Function of lymphocytes and macrophages after cryopreservation by procedures for pancreatic islets: Potential for reducing tissue immunogenicity, *Cryobiology*, 25, 1–17.

Taylor, M.J., Bank, H.L., and Benton, M.J. (1987) Selective destruction of leucocytes by freezing as a potential means of modulating tissue immunogenicity: Membrane integrity of lymphocytes and macrophages, *Cryobiology*, 24, 91–102.

Tedder, R.S., Zuckerman, M.A., Goldstone, A.H., Hawkins, A.E., Fielding, A., Briggs, E.M., Irwin, D., Blair, S., Gorman, A.M., Patterson, K.G., Linch, D.C., Heptonstall, J., and Brink, N.S. (1995) Hepatitis B transmission from contaminated cryopreservation tank, *Lancet*, 346, 137–140.

Thomas, M.J.G., Perry, E.S., Nash, S.H., and Bell, S.H. (1996) A method for the cryopreservation of red blood cells using hydroxyethyl starch a cryoprotectant, *Transfus. Sci.*, 17, 385–396.

Tocci, A., Rezzoug, F., Aitouche, A., and Touraine, J.L. (1994) Comparison of fresh, cryopreserved and cultured haematopoietic stem cells from fetal liver, *Bone Marrow Transplant,* 13, 641–648.

Tullis, J.L., Kethel, M.M., Pyle, H.M., Pennell, R.B., Gibson, J.G., Tinch, R.J., and Driscoll, S.G. (1958) Studies on the in-vivo survival of glycerolized and frozen human red blood cells, *J. Am. Med. Assoc.*, 168, 399.

Vadhan-Raj, S., Currie, L.M., Bueso-Ramos, C., Livesey, S.A., and Connor, J. (1999) Enhanced retention of *in vitro* functional activity of platelets from recombinant human thrombopoietin-treated patients following long-term cryopreservation with a platelet-preserving solution (ThromboSol) and 2% DMSO, *Br. J. Haematol.*, 104, 403–411.

Valeri, C.R. (1970) Recent advances in techniques for freezing red cells, *CRC Crit. Rev. Clin. Lab. Sci.,* 1, 381–425.

Valeri, C.R. (1985) Cryopreservation of human platelets and bone marrow and peripheral blood totipotential mononuclear stem cells, *Ann. NY Acad. Sci.*, 459, 353–366.

Valeri, C.R., Pivacek, L.E., Gray, A.D., Cassidy, G.P., Leavy, M.E., Dennis, R.C., Melaragno, A.J., Niehoff, J., Yeston, N., Emerson, C.P., and Altschule, M.D. (1989) The safety and therapeutic effectiveness of human red cells stored at –80°C for as long as 21 years, *Transfusion*, 29, 429–437.

Valeri, C.R., Ragno, G., Pivacek, L.E., Cassidy, G.P., Srey, R., Hansson-Wicher, M., and Leavy, M.E. (2000) An experiment with glycerol frozen red cells stored at –80°C for up to 37 years, *Vox Sang*, 79, 168–174.

Valeri, C.R., Ragno, G., Pivacek, L.E., and O'Neill, E.M. (2001a) *In vivo* survival of apheresis RBCs, frozen 40-percent (wt/vol) glycerol, deglycerolized in the ACP 215, and stored at 4°C in AS-3 for up to 21 days, *Transfusion*, 41, 928–932.

Valeri, C.R., Ragno, G., Pivacek, L.E., Srey, R., Hess, J.R., Lippert, L.E., Metille, F., Fahie, R., O'Neill, E.M., and Szymanski, I.O. (2001b) A multicenter of *in vitro* and *in vivo* values in human RBCs frozen with 40-percent (wt/vol) glycerol and stored after deglycerolization for 15 days at 4°C in AS-3: Assessment of RBC processing in the ACP 215, *Transfusion*, 41, 933–939.

Valeri, C.R. and Zaroulis, C.G. (1972) Rejuvenation and freezing of outdated stored human red cells, *N. Engl. J. Med.*, 287, 1307–1313.

Van der Meulen, F.W., Reiss, M., Stricker, E.A.M., van Elven, E., von dem Borne, A.E.G.K. (1981) Cryopreservation of human monocytes, *Cryobiology*, 18, 337–343.

van Imhoff, G.W., Arnaud, F., Postmus, P.E., Mulder, N.H., Das, P.C., and Smit-Sibinga, C.T. (1983) Autologous cryopreserved platelets and prophylaxis of bleeding in autologous bone marrow transplantation, *Blut*, 47, 203–209.

Vecchione, J.J., Chomicz, S.M., Emerson, C.P., and Valeri, C.R. (1981) Enumeration of previously frozen platelets using the Coulter counter, phase microscopy and the Technicon optical system, *Transfusion*, 21, 511–516.

Wagner, C.T., Burnett, M.B., Livesey, S.A., and Connor J. (2000) Red blood cell stabilization reduces effect of cell density on recovery following cryopreservation, *Cryobiology*, 41, 178–194.

Walker, R.H., Ed. (1993) *Technical Manual*, 11th edition, American Association of Blood Banks, Bethesda, MD.

Walter, E.A., Greenberg P.D., Gilbert, M.J., Finch, R.J., Watanabe, R.J., Thomas, E.D., and Riddell, S.R. (1995) Reconstitution of cellular immunity against cytomegalvirus in recipients of allogeneic bone marrow by transfer of T-cell clones from the donor, *N. Engl. J. Med.*, 333, 1038–1044.

Warkentin, P.I., Jackson, J.D., and Kessinger, A. (1993) Cryopreservation and infusion of peripheral blood stem cells, in *Peripheral Blood Stem Cell Autografts*, Wunder, E.W. and Henon, R.R., Eds., Springer, Berlin, pp. 199–205.

Willhite, C.C. and Katz, P.I. (1984) Toxicology updates, dimethyl sulfoxide, *J. Appl. Toxicol.*, 4, 155–160.

Zapff, G., Scheibe, I., Toursel, W., Dekowski, D., and Richter, H. (1988) Ergebnisse (*in vitro* und *in vivo*) der Kryokonservierung von Human-Thrombozyten und Glycerol, *Folia Haematol. (Leipzig)*, 115, 755–767.

Zingsem, J., Zeiler, T., Zimmermann, R., Weisbach, V., Mitschulat, H., Schmid, H., Beyer, J., Siegert, W., and Eckstein, V. (1993) Automated processing of human bone marrow grafts for transplantation, *Vox Sang* 65, 293–299.

Zingsem, J., Zeiler, T., Zimmermann, R., Weisbach, V., Mitschulat, H., Siegert, W., and Eckstein, V. (1992) Bone marrow processing with the Fresenius AS 104: Initial results, *J. Hemother.*, 1, 273–278.

Zuckerman, M.A., Hawkins, A.E., Briggs, M., Waite, J., Balfe, P., Thom, B., Gilson, R.J., and Tedder, R.S. (1995) Investigation of hepatitis B virus transmission in a health care setting: Application of direct sequence analysis, *J. Infect. Dis.*, 172(4), 1080–1083.

18 Cryopreservation of Human Gametes and Embryos

Barry Fuller, Sharon J. Paynter, and Paul Watson

CONTENTS

18.1 INTRODUCTION

The cryopreservation of reproductive cells and embryos has, in many ways, reflected the development of the science of cryobiology as a whole. Many of the methods and analyses used to study cryobiology in mammalian cells were developed from work on oocytes and embryos (see Chapter 11). The groundbreaking work on cryopreservation of animal semen (Polge et al., 1949) focused attention on the fundamental principles which underscore the survival of cells in the frozen state and fueled the application of cryopreservation in animal reproductive technologies over the last three decades (see Chapter 12). The parallel developments of techniques for *in vitro* manipulation of human reproductive cells for the treatment of infertility, over the same time period, have ensured that interest in their cryopreservation has remained high. It is fair to say that the ability to store these cells in the frozen state has greatly facilitated the progression of clinical assisted reproduction from esoteric research into mainstream patient treatment. Across the spectrum of patient treatments now available, cryopreservation has been attempted for most of the developmental stages of both male and female reproductive cells, ranging from the immature gametes residing in ovarian or testicular tissues through to mature oocytes and spermatozoa and various stages of pre-implantation embryos. However, each variant of the cohort of reproductive cells required to be frozen has introduced their own problems in achieving high-yield recoveries from the frozen state, and it has been realized that many aspects of the particular physiology of the cells will dictate how they respond to cryopreservation. More recent studies have been aimed at establishing the fundamental cryobiological properties of human reproductive cells, to improve safety and stability in the frozen state and recovery of high yields of fully functional cells, and these studies will be discussed in this chapter. Better success has been achieved with some reproductive stages (such as early-cleavage embryos) than with others (such as the mature oocyte), and we are beginning to understand why this happens, but even in the best cases, improvements are essential to achieve both a fundamental appreciation of the underlying mechanisms and an enhanced clinical outcome.

18.2 THE NATURE OF THE PROBLEM

The established principles of cell injury or survival during the freezing process have been set out in detail in previous chapters (see Chapters 1 and 2). However, there are additional complicating factors resulting from the specialized nature of human reproductive cells that affect survival after cryopreservation. These relate, in the main, to the fact that both male and female cells are designed to deliver the haploid set of chromosomes at the point of fertilization, in the correct sequence of events, to achieve normal embryo production and eventual normal birth. In both male and female, the mature gametes develop from immature precursor cells in their respective sites (testicular tissue or ovary). It is beyond the scope of the current chapter to describe in detail these developmental processes, and they can be found elsewhere (Edwards and Brody, 1995; Jones, 1997). However, it is worth pointing out some overriding principles that affect cryopreservation of the various stages.

On achieving maturation, both male and female gametes have acquired highly specialized structural components essential to normal fertilization (e.g., the sperm-tail contractile machinery, or the oocyte protective coat of the zona pellucida) that may respond to the freezing process in ways different from that of basic cell structures such as the plasma membrane. Also, the haploid state exists with chromosomes devoid of a surrounding nuclear membrane. Earlier developmental stages of the gametes exist as a less complex cell population, which superficially may be more attractive to consider for cryopreservation. However, in the human patient, the question then becomes how to access these primary-stage cells, and how to achieve their essential maturation after cryopreservation. In many ways, in the development of clinical reproductive technologies, progress in cryopreservation has gone hand in hand with that for *in vitro* maturation by cell culture methods. The comparative complexities of *in vivo* and *in vitro* maturation (of both male and female gametes) have been described in an excellent review by Leibfried-Rutledge et al. (1997).

18.3 CRYOPRESERVATION OF THE FEMALE GAMETE: OVARIAN TISSUE TO OOCYTE

At birth, the ovaries contain the lifetime complement of primary oocytes, which have developed from primordial germ cells during fetal development. The oocytes are arrested in the prophase stage of meiosis 1 and are surrounded by a squamous, single-layered epithelium to form the primordial follicles, which can be identified microscopically in the cortex of the tissue. With the onset of reproductive maturity, individual oocytes enter the maturation cycle, during which the oocyte increases in diameter, proceeds through the second meiotic division, and interacts with the increasingly complex follicular cell layers, which develop into granulosa and thecal cells. The whole process is under a multitiered control resulting from both systemic and local hormonal production. The complex kinetic interaction of these components ensures that in a typical monthly cycle, only one follicle reaches full maturation and a single oocyte is released during ovulation. It is thus obvious that, to achieve successful cryopreservation of ovarian tissue, it is essential to maintain the functional status of the whole mixture of different cell types.

18.3.1 HISTORICAL REVIEW OF OVARIAN TISSUE CRYOPRESERVATION

Much of the early work on ovarian tissue cryopreservation was performed in animal studies, and these set the groundwork for investigations into human ovarian tissue during the last 5 years. In fact, a rather comprehensive study on fundamental research on ovarian tissue cryopreservation was conducted as long ago as the 1950s in London, under the direction of Sir Alan Parkes (Parkes, 1957, 1958; Parkes and Smith, 1953, 1954), on animal tissues. Many of the observations are pertinent to the current interest in clinical application, and thus will be described in some detail. This was the same team that had established the cryoprotective properties of glycerol for sperm (Polge et al., 1949), and they applied the same technology to ovarian tissue. Before this, culturing or implanting rat ovaries that had been frozen as small 1-mm^3 pieces in liquid air without added cryoprotective additive (CPA) had failed to detect any functional activity in the thawed tissues (Payne and Meyer, 1942). The earliest positive results were obtained when glycerol was applied as CPA (up to levels of 15%) for cryopreservation of rabbit granulosa cells (Smith, 1952). A slow-cooling protocol was used, as was an end storage temperature of −79°C (liquid nitrogen was not readily available at that time). A series of investigations was then undertaken on rat ovarian tissue pieces exposed to 15% glycerol-saline for 1 to 2 h at room temperature before cooling at a slow rate using the Lovelock apparatus (Polge and Lovelock, 1952) to −79°C, as it was established early on that use of slow-cooling rates was more successful than of rapid rates (Parkes and Smith, 1953, 1954). In these early studies, the method of assessment of long-term function of the cryopreserved grafts was microscopical investigation of the tissue after explant and vaginal cornification in animals that had been previously oophorectomized—an indicator that estrogen production and reproductive cycling had been reestablished by the grafted tissue. These early studies established many important principles concerning ovarian tissue cryopreservation, including the facts that the tissues could be damaged by exposure to the CPA alone without freezing, particularly if the exposure time was prolonged; that serum was an important secondary additive to the CPA mixture that improved recovery; and that autografts (i.e., tissue that had been cryopreserved and returned to the same individual after oophorectomy) were much more successful than allografts (tissue from a separate donor animal grafted after cryopreservation into the oophorectomized recipient). The latter observation was made at a time when the immunology of tissue transplantation was just starting to be understood, and it has remained an important consideration when moving toward a clinical application of ovarian tissue cryopreservation.

Working in the same group, Deansly (1954) studied cryopreservation of bisected rat ovaries using the glycerol–slow cooling protocol. Histological evaluation of the tissue immediately after thawing demonstrated that healthy follicles were predominantly to be found in the exterior, whereas those deeper in the tissue showed extensive damage. This was an early indication that permeation of the CPA into all levels of the tissue was a necessary prerequisite for survival and that attention would need to be paid to achieve sufficient CPA equilibration throughout the tissue. Nevertheless, these thawed tissues were capable of reestablishing hormonal cycling in the oophorectomized recipients. Further papers by Green et al. (1956), Deansly and Parkes (1956), and Parkes (1957, 1958) comprehensively reviewed the status of ovarian tissue cryopreservation at that time. The problems of achieving CPA permeation throughout a large tissue specimen was solved in a pragmatic fashion by moving toward smaller samples, and for the rat, quarter ovaries were used. Glycerol at a concentration of 15% was found to be more protective than at 10 or 5%, as long as exposure times were maintained at 1 to 2 h at room temperature. The slow-cooling protocol was retained, and rapid rewarming by plunging the samples into a water bath at 40°C was the most effective procedure. Other cryoprotective agents (diethylene glycol, methanol, monoethyl ether, and carbawax) were all found to be inferior to glycerol. Some success was achieved using a two-stage cooling protocol that was not closely controlled but that had a holding period at –20°C.

A landmark paper was published by Parrott in 1960, in which a more detailed investigation was made into the establishment of fertility in oophorectomized mice grafted with tissue cryopreserved using slow cooling with 12% glycerol serum. Her criteria of success were the histological assessment of the numbers of surviving oocytes in grafts taken at autopsy, the proportion of recipients that became fertile, and the timescale of graft function *in vivo*. Some live births were recorded in animals receiving cryopreserved grafts, but the numbers of oocytes surviving in the tissues were low, and the reproductive lifespan of the grafted recipients was curtailed compared with those receiving unfrozen tissues.

It was some years before attempts were made to transfer the technology of ovarian tissue cryopreservation to human tissue. In 1976, Stahler et al. reported studies on ovaries obtained during the follicular phase from oophorectomy patients aged between 30 and 41 years. Intact ovaries were flushed with ice-cold heparinized medium via the ovarian artery, and CPA solution was introduced via the same route. The ovaries were cooled slowly to –18°C and then rapidly rewarmed by arterial perfusion with a warm, colloid-supplemented diluent. Measurements of oxygen and glucose consumption by the ovarian tissues showed that some degree of protection was obtained under these conditions, but further deep cooling to liquid nitrogen temperatures proved lethal.

Over the next decades, additional progress was made in cryopreservation of animal ovarian tissues (Candy et al., 1995; Harp et al., 1994), including a move toward using dimethyl sulfoxide (Me$_2$SO) as CPA of choice. A significant stimulus of interest in potential clinical application followed impressive studies in large-animal (sheep) ovarian tissue cryopreservation by Gosden et al. (1994). These workers successfully demonstrated restoration of fertility and the birth of a lamb after autografting ovarian tissue recovered from –196°C into the ovarian bursa of an oophorectomized recipient. The cryopreservation protocol was now based on Me$_2$SO as CPA, and the treated tissues were reduced in volume to thin slices of ovarian cortex. The slices were cooled at –2°C/min

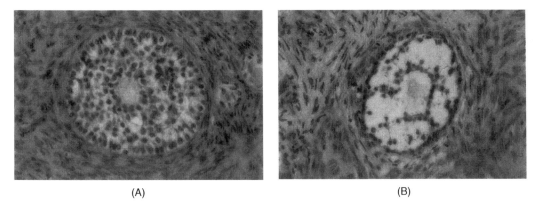

(A)   (B)

FIGURE 18.1 Bovine preantral follicles in tissue slices after cryopreservation using slow cooling with Me₂SO as cryoprotectant. Some follicles presented a normal morphology with a central oocyte surrounded by evenly distributed granulosa cells (A). In others, the oocytes were still centrally located, but there was evidence of destruction and displacement of the granulosa cell mass (B).

to −7°C, at which point ice crystallization was induced by manual seeding. A cooling rate of −0.3°C/min was used to an end temperature of −40°C before increasing the rate to from −10° to −140°C before plunging to −196°C. Samples were allowed to thaw initially at room temperature before transfer to a warm water bath to complete the process. Histological investigations revealed that follicles at all stages of development could be detected in cryopreserved grafts up to 9 months after operation. However, these authors commented on an observation that has been consistently reported since the earliest studies on ovarian tissue cryopreservation—if the grafts were examined after only 1 week, all developing follicles were degenerating and only primordial follicles could be detected. Thus, cryopreservation could only maintain viable immature follicles, which then needed a supportive physiological environment to mature and grow to preantral and antral stages.

18.3.2 APPLICATION OF CRYOPRESERVATION TO HUMAN OVARIAN TISSUES

Cryopreservation and subsequent autografting has been suggested as an ideal solution to the problem of restoring fertility in female patients who must undergo ablative chemo- or radiotherapy during cancer therapy (Gosden and Aubard, 1996; Hovatta et al., 1997). However, the technique must still be viewed as an experimental therapy until improvements and consistency of results can be achieved. Most of the recent studies have employed cryopreservation regimes based around those used for successful human embryo storage or ovarian tissue recovery in animal experiments. Both Me₂SO and 1,2 propane diol were investigated by Hovatta et al. (1996), who used human ovarian cortical tissue fragments (0.3 to 2 mm diameter). CPA concentrations were 1.5 M, and slow cooling (−0.3°C/min to −30°C, then −50° to −150°C) was used, with subsequent rapid warming. Histological appearance of primordial or primary follicles immediately after thawing was not different to that in fresh cortical tissue. The same group went on to perform *in vitro* culture of the thawed tissues for up to 21 d (Hovatta et al., 1997). Two thirds of the population of early follicular stages were viable on morphological grounds after 10 to 15 d in culture, and there were clear indications of follicular growth as the proportion of secondary follicles increased. As discussed above, the process of follicle development and oocyte maturation are interdependent and require concerted interaction between follicular cell layers and the oocyte. There is evidence from animal studies that cryopreservation of ovarian fragments (using slow cooling with Me₂SO) may lead to disruption of theca or granulosa cells without overt destruction of the oocyte (Paynter et al., 1999a; see Figure 18.1). A similar effect in human tissues has recently been reported after both slow and rapid cooling, where survival of follicular oocytes was high but overall follicle structure was compromised (Seibzehnrubl et al., 2000).

The issues surrounding establishment of basic parameters of human ovarian tissue cryobiology have begun to be addressed in recent years. The underlying problem of achieving sufficient CPA permeation by surface diffusion into the entire tissue, composed as it is of densely distributed cell/matrix components, has required the development of a sensitive chemical assay. Proton nuclear magnetic resonance spectroscopy provides one way to investigate chemical composition, and it was employed in studies of the penetration of Me_2SO into human ovarian tissue pieces at 4°C by Thomas et al. (1997). These results demonstrated that, using exposure times up to 20 min, only approximately 50% of the theoretically maximum volume of tissue water had reached equilibration. A similar incomplete permeation of human tissue slices by Me_2SO at low temperatures was reported by Newton et al. (1998), whereas increasing the exposure temperature to 37°C provided a more complete equilibration with 1.5 M Me_2SO. The highest degree of equilibration was reported for 1,2 propane diol at 37°C, but it remains to be established whether chemical toxicity of the CPA will follow from using higher exposure temperatures.

Propane diol was used by Gook et al. (1999) in human ovarian cortical slices where equilibration of CPA was achieved at room temperature, and sucrose (0.1 M) was used as a secondary CPA to achieve a degree of tissue dehydration before freezing. In addition, the traditional slow cooling (–0.3° to –30°C) was compared with faster rates, achieved by exposing the vials to either liquid nitrogen vapor or directly plunging them into liquid nitrogen. After rapid warming and dilution of the CPA, histological evaluations were performed. The highest survival of morphologically normal follicles that contained oocytes was found when the CPA-dehydration step was extended to 90 min at room temperature and the traditional slow cooling method was used. This longer CPA exposure time would certainly increase the percentage of tissue water equilibrated with the propane diol, and it fit with the observations on kinetics of CPA uptake discussed above.

The impressive results from histological studies on cryopreserved human ovarian tissue have required confirmation using a more biologically relevant assay that could provide information about potential folliculogensis and oocyte maturation. Unfortunately, current techniques of *in vitro* culture have not been adequately sophisticated to support ovarian tissues for sufficient periods to permit this (it takes in the region of 3 months to achieve the development of human primordial follicles to the level of the hormone-dependent growth phase). This has been circumvented by the use of a xenografting technique, whereby human cortical fragments were implanted under the skin of immunodeficient mice. Using this model, Weissman et al. (1999) have been able to demonstrate follicle growth and antral follicle development under the control of exogenously delivered gonadotropin. It has been previously suggested that one of the reasons why small immature follicles are the only ones that survive transplantation in ovarian cortical slices (either fresh or cryopreserved) might be the degree of hypoxia encountered in the grafted tissue until a new blood supply becomes established, but this remains to be thoroughly investigated. Another way to assess follicle function after recovery of cryopreserved ovarian tissue is to investigate growth of isolated follicles, and the strengths and weaknesses of this approach will be discussed later.

Using the mouse xenograft model, Nisolle et al. (2000) showed that cryopreserved human ovarian tissue implanted subcutaneously or intraperitoneally for 24 d showed clear immuno-histochemical evidence of the development of a new blood supply, whereas the numbers of primordial and primary follicles were equal to those in fresh tissue grafts. Using the propane diol–sucrose method with slow cooling, Gook et al. (2001) made further progress by showing growth of antral follicles in human ovarian tissue implanted under the kidney capsule of immuno-deficient mice (this site has been recognized for some time as providing good early blood supply to grafted tissues). When grafts were examined after less than 20 weeks, only early stages of follicle development were seen, but in longer-term grafts, antral follicles were seen in 75% of cases and the structure of the developing oocyte within the follicular cavities was normal. Again, no difference was detected between grafts from fresh or cryopreserved tissues.

The summation of all these experimental studies has led in the recent past to the establishment of the first clinical applications of ovarian tissue cryopreservation. In one case, orthotopic reimplantation of cryopreserved strips of ovarian cortex was performed in a patient undergoing ablative chemo-therapy for Hodgkin's lymphoma (Radford et al., 2001). One ovary had been removed before the start of the chemotherapy, and cortical slices had been cryopreserved using Me_2SO and slow cooling. After 19 months, thawed tissues were reimplanted, and evidence of ovarian hormonal function, absent before grafting, was seen over the next 9 months. Under similar circumstances in a second study, ovarian hormonal function and ovulation were reported during 4 months after reimplanting cryopreserved tissue to the ovarian fossa (Oktay et al., 2001). Again, the slow-cooling technique with propane diol as CPA was used.

These reports herald an undoubted upswing in activity in the field of ovarian tissue cryopreservation. Many questions remain to be answered as the technique becomes a more routine clinical practice, including the true reproductive potential of oocytes reaching maturity and ovulation in the cryopreserved tissues and the functional lifespan of such grafts. New ethical questions may be posed, such as the possibility of using donor ovarian tissue to replenish hormonal functions in patients with ovarian failure. Alternative cryopreservation techniques may be found to be more successful—for example, vitrification methods may be applicable in the clinical situation (these have already been investigated in limited animal experiments; Sugimoto et al., 1996, 2000; see Chapter 12), but little information is available so far for human tissue.

18.3.3 CRYOPRESERVATION OF OVARIAN TISSUE AS A SOURCE OF OOCYTES

Although reimplantation of cryopreserved ovarian tissue is intuitively the most logical use of this resource, a parallel interest has developed in the concept of using banked tissues as a source of immature oocytes that could then be matured *in vitro* for fertility treatment, either for the same patient or as a source of donor oocytes. A question also remains over the reliability of tissue-grafting methods in the clinic, although these problems do seem to be soluble if the early successes (Oktay et al., 2001; Radford et al., 2001) can be repeated. However, isolation and *in vitro* culture of immature oocytes from ovarian tissues trades one difficulty (expertise in tissue grafting) for another (long-term *in vitro* culture to allow follicle growth and oocyte maturation). From the discussions above, it will be recalled that long periods of time (in the human, several months) are needed to achieve full development of primordial follicles and associated oocytes. Over the last decade sophisticated culture methods have been devised that allow isolated immature follicles from mouse ovarian tissue to grow and produce oocytes capable of fertilization (Spears et al., 1994) and live birth (Eppig and O'Brien, 1996). It has proved possible to isolate immature follicles from mouse ovarian tissue slices cryopreserved with Me_2SO, propane diol, or glycerol to and place them in maturation culture (Newton and Illingworth, 2001). Glycerol was an ineffective CPA, but the other agents provided sufficient protection during slow cooling to permit isolation of viable preantral follicles, which yielded mature oocytes after culture.

Human ovarian tissue is more dense and fibrous than that of small mammals. Nevertheless, Oktay et al. (1997) studied enzymatic digestion of human cortical tissue slices after cryopreservation using propane diol and slow cooling. It was shown that similar numbers of primordial follicles could be isolated from either fresh or cryopreserved tissues. Vital staining methods indicated that the majority of the follicles were undamaged by cryopreservation, but future work will need to demonstrate that their growth potential has been retained after thawing.

It is also possible to envisage isolation of immature follicles before cryopreservation. In this way, the isolated follicles may, as small structures with relatively low numbers of cells, respond to the biophysical changes during freezing in a way more in common with that of multicellular embryos, rather than as part of a complex three-dimensional tissue structure in cortical slices. The thawed immature follicles would still require a prolonged period of maturation culture to yield a mature oocyte, but it has been known for some time, from studies in mice, that isolated primary follicles could be cryopreserved using Me_2SO and slow cooling and be matured to yield mature oocytes capable of fertilization and live birth (Carroll et al., 1990; see also Chapter 11). In this case, maturation was achieved by *in vivo* placement of the follicles in a gel structure under the kidney capsule of a host animal, rather than by *in vitro* culture. For the future, follicle cryopreservation is another area of reproductive medicine in which advances in cryobiology will need to be matched by progress in *in vitro* maturation techniques. Primordial follicles are the most numerous follicles in ovarian tissue and can be recovered by digestion or teasing apart of ovarian tissue, and thus, cryopreservation of these would be an attractive option, but much progress needs to be made both in their cryopreservation and maturation before it will be a useful clinical therapy (Oktay et al., 1998). For later-stage follicles, a recent study of note was that undertaken on mouse ovarian tissues, and in this, Sztein et al. (2000) recovered cryopreserved tissues, and immediately on thawing, harvested oocytes from antral follicles. They were able to show that these oocytes had survived and were capable of maturation and fertilization. The study goes some way to addressing the question of why, in cryopreserved cortical slices, only the small primary follicles appear to survive the freezing process and can be identified early postgrafting. Sztein's study indicates that it is not a result of freeze–thaw damage of the follicular oocyte, but must relate to other tissue factors that ultimately lead to necrosis of the antral stages.

The relative explosion of interest in ovarian tissue cryopreservation and allied clinical technologies have increased options for treatment of infertility or associated hormonal pathologies and have raised complex ethical issues about how the advances should be introduced. There is also an ongoing debate about the safety of reintroducing cryopreserved tissues, taken from patients exhibiting overt malignant disease, after successful anticancer therapy—Will the grafted tissues harbor cells that can lead to secondary malignancies? These issues are beyond the scope of this chapter, but current opinions can be found in several recent papers and reviews (Kim et al., 2001a, 2001b; Shaw et al., 1999).

18.4 CRYOPRESERVATION OF HUMAN OOCYTES

With the development of *in vitro* fertilization (IVF) techniques, it is possible to artificially stimulate women to produce several mature oocytes per cycle rather than the single oocyte usually produced. These oocytes can be fertilized *in vitro*, but current U.K. legislation stipulates that a maximum of three, but preferably two, embryos be replaced in the uterus. Excess embryos can be cryopreserved; such a technique now forms a routine part of the service offered by many fertility clinics, as will be discussed later. However, embryo cryopreservation raises ethical dilemmas such as ownership and fate of the embryos, and indeed, embryo storage is banned or strictly limited in some European countries. Storage of unfertilized gametes raises fewer ethical concerns. Oocyte storage would also allow women without a partner who object to using donor sperm the possibility of future pregnancy. This issue is particularly relevant to young female cancer patients whose fertility is likely to be undermined by cancer therapies (Apperley and Reddy, 1995). Storage of oocytes would also make donation of oocytes more practicable, avoiding the need to synchronize donor and recipient cycles.

18.4.1 FEATURES OF THE HUMAN OOCYTE OF SIGNIFICANCE IN CRYOPRESERVATION

Oocytes retrieved following hormonal stimulation are mature, fertilizable metaphase II oocytes or are, at most, a few hours of culture away from potentially becoming so. Hence they do not require the extensive periods of culture *in vitro* that would be necessary with oocytes contained within immature follicles.

18.4.1.1 Size

The diameter of mature oocytes is around 130 μm, and as such, they are the largest human cell type. Their size has important consequences for successful cryopreservation. Large single cells have a low surface-area-to-volume ratio, and hence they are less efficient at taking up CPAs and

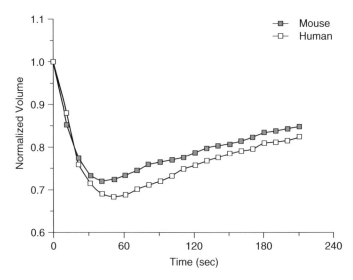

FIGURE 18.2 Comparison of osmotic response of oocytes on exposure to 1.5 mol/L 1,2 propane diol at 23°C. Each point represents a mean of 10 oocytes. Each oocyte was held in position, within a 5-μL droplet of phosphate buffered medium, using negative pressure generated through a holding pipette. Each oocyte was flushed with 1 mL of cryoprotectant solution and the osmotic response recorded on videotape. Oocyte volume was calculated from diameter measurements taken from freeze-frame images.

at losing water. The overall effect is that oocytes are more likely to retain water during freezing and thus be damaged by the formation and growth of intracellular ice. In addition, the integrity of oocytes can be seriously undermined even before freezing begins if addition and dilution of cryoprotectant results in excessive osmotic stress.

18.4.1.2 Membrane permeability

The permeability of a cell to water and solutes determines the extent of the shrink–swell response on exposure to a given CPA. The more permeable the cell to the permeating solute, the less extensive are the cell volume fluctuations during addition or removal of CPA. Water permeability also determines the extent of cellular dehydration during freezing. For cells to survive freezing, they must be sufficiently dehydrated, or contain sufficient CPA, that intracellular ice formation does not occur. It is possible to measure the shrink–swell response of oocytes on exposure to a given concentration of cryoprotectant and, from these measurements, to calculate theoretical values for water permeability and cryoprotectant permeability of the cell. Such data have been collected, and permeability coefficients calculated, for murine oocytes in the presence of the CPAs Me_2SO, 1,2 propane diol (Paynter et al., 1997b), ethylene glycol, and glycerol (Paynter et al., 1999c, 1999d). Murine oocytes are often used as model systems for designing cryopreservation protocols for human oocytes. However, murine oocytes are considerably smaller (with a diameter of 80 μm) than human oocytes and hence could be expected to react differently in the presence of cryoprotectant. Both murine and human oocytes have a low permeability to water compared with other cell types. However, human oocytes tend to shrink more than murine oocytes (Figure 18.2) and exhibit higher values per unit volume under similar conditions for permeabilities to both water and cryoprotectant (Paynter et al., 1999b, 2001). Interestingly, human oocytes are also more permeable to propane-1,2-diol than to Me_2SO, whereas for murine oocytes, the response is almost identical for the two CPAs (Paynter et al., 1997b). Permeability characteristics also change during maturation of the oocyte, with immature germinal-vesicle-stage oocytes reacting in a different manner than mature oocytes of the same species (Le Gal et al., 1994; Younis et al., 1996).

Clearly, to optimize survival following cryopreservation, the protocols used need to be designed specifically for the particular cell type to be stored, rather than simply adopting protocols that have been successfully used for other systems. Over the last two decades, the majority of studies involving cryopreservation of human oocytes have applied the cryopreservation techniques found to be successful for the storage of human embryos. More recently, methods have begun to be proposed from information derived from experimental studies on both human and animal oocytes.

18.4.1.3 Cytoskeleton

The mature human oocyte contains condensed chromosomes arranged on a microtubular spindle, the appropriate organization of which is essential for the correct alignment and segregation of chromosomes. Microtubules and microfilaments within the cell are also involved in such events as spindle rotation, polar body formation, and pronuclear formation, which are all crucial steps in the process of meiosis. Studies using murine oocytes revealed the spindle to be sensitive to cooling (Pickering and Johnson, 1987; Sathananthan et al., 1992) although capable of repair if incubated at 37°C for sufficient time. However, the spindle of the human oocyte has shown irreversible disruption following exposure to room temperature for 10 or 30 min (Pickering et al., 1990). Less than half of the oocytes displayed normal spindle morphology after incubation at 37°C for 4 h. Dispersal of chromosomes was evident after a 30-min incubation at room temperature. Human oocytes contain less foci of pericentriolar material, which form sites of tubulin organization, than do murine oocytes, which may explain their reduced ability to undergo spindle reformation. Similar disruption of the spindle was reported by Almeida and Bolton (1995), although returning oocytes to 37°C for 1 or 4 h after a 2-min incubation at room temperature did allow restoration of normal spindle structure. The authors reported the presence of normal spindle structure with dispersed chromosomes as well as abnormal spindle structure with compact chromosomes. However, the oocytes used in the study were those that had failed to fertilize 18 to 20 h postinsemination and hence may have been compromised. A study using freshly collected immature human oocytes that were matured *in vitro* to the metaphase II stage looked at the effect of cooling to 0°C for 1 to 10 min (Zenzes et al., 2001). After 1 min at 0°C, spindle damage was negligible, but in oocytes cooled for 2 to 3 min the spindle was shortened. At 4 to 9 min, disruption was increasingly severe, and by 10 min, the spindles had completely disappeared. Despite this depolymerization of the microtubular spindle at 0°C, the chromosomes did not become dispersed. The chromosomes remained anchored even in the absence of the spindle. The authors concluded that the microtubules associated with the kinetochores of chromosomes react differently to chilling than do the microtubules that pass among the chromosomes through the metaphase plate.

Exposure to cryoprotectants has also been shown to result in disruption of the microtubular spindle (Johnson and Pickering, 1987), but the effect has been reduced by performing exposure at 4°C rather than at 37°C in human oocytes exposed to Me_2SO (Pickering et al., 1991). Cooling to 0°C for 20 or 60 min has been shown to result in disorganized spindles in human oocytes in the presence or absence of Me_2SO (Sathananthan et al., 1988), although widespread scattering of the chromosomes was not observed. Human oocytes exposed to 1,2-propanediol, either with or without freezing, displayed no stray chromosomes despite absence or abnormality of the spindle (Gook et al., 1993).

One way of avoiding any danger of disruption to the sensitive microtubular spindle is to cryopreserve immature germinal-vesicle oocytes, at which stage the spindle is not present and the chromosomes are decondensed. Immature oocytes are often collected during conventional IVF, or oocyte collection techniques can be modified to allow retrieval of fully grown germinal-vesicle oocytes (Trounson et al., 1994). Immature oocytes can also be collected without hormonal stimulation, conferring an additional advantage of this technique on women in danger of ovarian hyperstimulation syndrome. A live birth has been reported following cryopreservation of a human immature oocyte after hormonal stimulation (Tucker et al., 1998a,1998b), and a pregnancy has been reported from oocytes retrieved from isolated, unstimulated ovarian tissue (Wu et al., 2001). However, there is evidence of spindle abnormality and abnormal chromosome arrangements in cryopreserved immature oocytes from stimulated and unstimulated ovaries following maturation (Baka et al., 1995; Park et al., 1997). In general, development of embryos resulting from cryopreserved immature oocytes is impaired (Toth et al., 1994a; Son et al., 1996). Results are generally better with oocytes from patients receiving ovarian stimulation than with those from unstimulated ovaries (Toth et al., 1994a, 1994b; Salha et al., 2001). This may be a reflection of the quality of the oocytes, with greater variability being evident in unstimulated ovaries. For all immature oocytes, it seems likely that maturation has been undermined such that subsequent embryo development is affected. The cumulus cells, which surround the oocyte prior to fertilization, provide nourishment via gap-junction communication during the growth phase and are also likely to be necessary for the final stages of maturation. Cumulus cells are often lost during cryopreservation (Cooper et al., 1998; Goud et al., 2000), and hence cryopreservation protocols need to be designed to allow survival of these two different cell types as well as preserving communication pathways between the two.

18.4.1.4   Zona pellucida

The zona pellucida is a glycoprotein coat that surrounds the oocyte. In the mature oocyte, cortical granules lie around the periphery of the oocyte, which if caused to release their contents, induce changes, known as "zona hardening," in the zona pellucida that prevent the penetration of spermatozoa into the oocyte. During the course of fertilization the zona is rendered impenetrable to spermatozoa once one spermatozoon has entered. However, if the cortical granules are damaged before fertilization, such that their contents are released, fertilization is effectively blocked. Conversely, the zona pellucida may be damaged or the cortical granules caused to migrate to the center of the oocyte, resulting in multiple sperm entry. Cryopreservation has been shown to induce cracks in the zona pellucida of human oocytes (Sathananthan et al., 1987). Ultrastructural evidence of premature cortical granule release has been found in oocytes exposed to 1.5 mol/L Me$_2$SO or 1,2-propanediol at room temperature. Conversely, in another study, cortical granules were identified and their abundance reported after thawing and dilution of 1.5 mol/L 1,2-propanediol in 15 human oocytes (Gook et al., 1993). Whether or not zona hardening occurs following cryopreservation has less significance now that intracytoplasmic sperm injection (ICSI) can be used to bypass the zona pellucida and inject a single spermatozoon. Studies comparing conventional IVF with ICSI-fertilized oocytes that have been cryopreserved showed comparable or better fertilization and cleavage with ICSI (Gook et al., 1995b; Kazem et al., 1995). However, zona hardening may have consequences for the ability of the embryo to hatch from the zona pellucida before implantation in the uterus.

18.4.2   EARLY CLINICAL EXPERIENCE IN CRYOPRESERVATION OF HUMAN OOCYTES

Despite the potential problems associated with the cryopreservation of oocytes, several live births have resulted from frozen human eggs (for a review, see Paynter, 2000). The first birth was reported in 1986. Fifty oocytes were cryopreserved by slow cooling in 1.5 mol/L Me$_2$SO. Of these, 38 were intact following thawing, at least 75% were fertilized when exposed to spermatozoa, and embryo transfers were performed in seven patients. Two pregnancies culminated in one twin (Chen, 1986) and one singleton (Chen, 1988) birth. A further birth quickly followed (van Uem et al., 1987) using a similar technique, but only one quarter of the oocytes survived the freeze/thaw. Poor results then followed in studies using a variety of techniques (Al Hasani et al., 1987; Diedrich et al., 1987; Hunter et al., 1991, 1995; Pensis et al., 1989; Todorow et al., 1989). The poor results of these studies may well reflect the quality of the oocytes cryopreserved. Often the best-quality oocytes were inseminated as fresh oocytes, and none of the studies used as high a quality oocyte as those selected for the original study by Chen.

In 1993, considerably better survival of human oocytes (64%) was reported following cryo-preservation by a slow-freeze, rapid-thaw method in the presence of 1,2-propanediol and sucrose (Gook et al., 1993). The method was one that had been successfully applied to the cryopreservation of human embryos (Lassalle et al., 1985). The karyotypes of four human oocytes cryopreserved using the same technique were found to be normal following fertilization (Gook et al., 1994), although a later study showed 27% of thawed oocytes to be parthenogenetically activated, whereas none of the control untreated oocytes, nor the oocytes exposed to CPA without freezing, were activated (Gook et al., 1995a). A number of live births have now been reported using this cryo-preservation method and fertilization by ICSI (Polak de Fried et al., 1998; Porcu et al., 1997; Tucker et al., 1998a, 1998b,) with the percentage of live births from thawed oocytes varying from 1 to 10. The largest study performed to date included 1769 frozen oocytes, with 1502 having been thawed (Porcu et al., 2000). The survival rate was 54%, with 58% fertilization and 91% cleavage. Sixteen pregnancies resulted in 11 births (seven singleton and two twin deliveries). All patients included in the study were under 38 years of age, had tubal infertility, had no previous IVF failure, and had at least 10 retrieved oocytes, all of which were cryopreserved.

Recently, modifications of the 1,2-propanediol/sucrose slow-freeze, rapid-thaw protocol have been tried. Increasing the temperature of exposure to CPA has shown some improvements. The temperature of exposure to the cryoprotectant before freezing has been increased from room temperature to 37°C with a concomitantly reduced exposure time (Yang et al., 1998). The births of two sets of twins and four ongoing pregnancies have since been reported (Yang et al., 1999). Optimizing the time of exposure to CPA such that sufficient CPA enters the cell but the time of exposure to suboptimal temperatures is minimized has also been beneficial. An increase in survival of oocytes exposed to 1.5 mol/L propane diol for 10 min (65%) rather than 15 min (53%) at room temperature before cryopreservation has been documented (Fabbri et al., 2000). However, in a later study (Fabbri et al., 2001), it was shown that exposure at room temperature to 1.5 mol/L propane diol plus 0.2 mol/L sucrose for 10.5 to 15 min gave greater survival post cryopreservation than exposure for shorter times (5.5 to 10 min, or 1 to 5 min) (70, 56, and 55%, respectively; Fabbri et al., 2001). More significantly, these authors reported higher survival when the sucrose concentration was doubled to 0.2 mol/L (60%), and it was even higher (80%) when the sucrose concentration was 0.3 mol/L. The increased survival may be a result of greater dehydration of the oocyte leading to a lower incidence of intracellular ice formation.

A number of studies have investigated whether cryopreservation of oocytes with attached cumulus cells affords any protection during freezing. One study reported better results with cumulus intact than with denuded oocytes (Imoedemhe and Sigue, 1992), whereas Mandelbaum et al. (1988) and a large study performed by Fabbri et al. (2001) suggested that there was no difference between the two.

Replacement of sodium with choline in the cryoprotectant vehicle solution has been found beneficial for mature murine oocytes (Stachecki et al., 1998) and for immature and *in vitro* matured human oocytes (Goud et al., 2000) after slow cooling in 1,2-propanediol. Another area in which an improvement in survival has been seen is in increasing the temperature at which ice formation was induced before cooling. An increase from –8° to –6° to –4.5°C improved survival of immature and failed-to-fertilized human oocytes from 32 to 56 to 95%, respectively (Trad et al., 1999).

Success has also been reported recently following cryopreservation by a completely different technique—that of vitrification (see also Chapter 22). These studies have used ethylene glycol, a cryoprotectant found to be well tolerated by oocytes at the high concentrations required for vitrification. A live birth has been achieved from a batch of 17 oocytes vitrified in ethylene glycol plus sucrose in open pulled straws (Kuleshova et al., 1999), whereas two births and an ongoing pregnancy have been achieved using the same CPAs but vitrified on electron microscope (EM) grids (Yoon et al., 2000). The rapid cooling rates used to achieve vitrification mean that the oocytes quickly traverse the potentially damaging temperature range, whereas both EM grids and open pulled straws allow faster cooling rates than conventional straws, and hence, less time for ice crystal formation. Vitrification has also been applied to immature oocytes (Chung et al., 2000; Wu et al.,

2001) and to oocytes matured *in vitro* (Chen et al., 2000). In the latter case, survival rates were high (65 to 100%) depending on the length of exposure to CPA and the dilution techniques. No blastocysts were found in culture, although blastocyst formation for control oocytes was also low (8%).

Cryopreserved oocytes have been used in conjunction with cryopreserved spermatozoa, as well as with epididimal and testicular sperm to produce live births (Porcu et al., 2000). However, a pregnancy using cryopreserved testicular sperm and thawed oocytes resulted in abortion and an XXY karyotype that was probably the result of fertilization of a digynic egg (i.e., one in which the second polar body or its chromosomes had been retained; Chia et al., 2000). The oocytes had been cultured for 12 h before cryopreservation, which may have contributed to the result. Also, an abnormal three-pronuclear zygote resulted after ICSI of fresh eggs from the same patient, indicating that the quality of the eggs was a major cause of triploidy resulting from the frozen oocyte. Another study looking at chromosomal abnormalities in embryos used probes for chromosomes 13, 18, 21 X and Y, and found no increase in chromosomal abnormalities between fresh and cryopreserved oocytes (Cobo et al., 2001).

18.4.3 CURRENT STATUS OF OOCYTE CRYOPRESERVATION

The recent live births are very encouraging, but results from cryopreserved human oocytes remain variable. Although fertilization, cleavage, and implantation rates approach those of fresh oocytes in some studies (Cobo et al., 2001; Porcu et al., 2000), the survival of the oocytes is still a problem. One of the main determining factors in survival is oocyte quality. It is understandable that the poorest-quality oocytes are the ones donated for research, but this unfortunately affects the results. Because studies seem to indicate that cumulus cell attachment to mature oocytes confers no advantage during cryopreservation, it may be best to denude oocytes so that assessment of their quality and maturity can be performed before freezing.

A major breakthrough in the use of cryopreserved oocytes has been the application of ICSI to achieve fertilization. However, once fertilized, the embryo still has to divide and retain a normal complement of chromosomes. Despite reports of spindle disruption, scattering of chromosomes is less common, but few studies have looked at chromosomal abnormalities in resulting embryos. Encouragingly, no abnormalities have been reported in the births resulting from cryopreserved oocytes.

Success, in terms of live births, has now been achieved using a variety of cryopreservation techniques. However, the overall success rates are still poor and need to be improved, particularly in view of the fact that, even with hormonal stimulation, relatively few oocytes (usually <20) are retrieved. This number is likely to be less for cancer patients (Posada et al., 2001). New protocols, such as the replacement of sodium chloride with choline to obviate salt-induced damage during slow cooling (see also Chapter 1) are being tested in a clinical setting (Quintans et al., 2002), but they remain to be fully evaluated. Despite this, cryopreservation of mature oocytes is currently the best option for preserving fertility for patients unable to preserve embryos. Before embarking on such a program, patients should be informed of the risks involved with oocyte cryopreservation and of the currently poor success rates. Clearly, further work is needed to improve survival rates.

18.5 CRYOPRESERVATION OF THE MALE GAMETE

18.5.1 HISTORICAL REVIEW

The first reported success of human sperm cryopreservation (Bunge et al., 1954) came at a time when sperm freezing was in its relative infancy. Glycerol had recently been discovered as a suitable cryoprotectant for spermatozoa, and successes had been reported in cattle and in several other species as diverse as fish and fowl (see also Chapter 11). Nevertheless, it was a bold step to attempt to produce human babies at that stage. Throughout the next few years there were a few pioneers

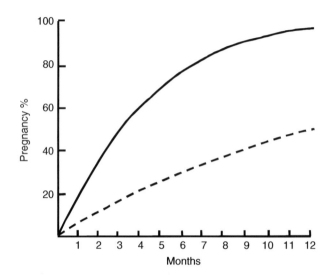

FIGURE 18.3 Cumulative theoretical pregnancy with donor insemination using fresh and frozen semen calculated from relative fecundity studies using the patient as her own control and life table analysis methods. Redrawn from Richter et al. (1984). Fresh semen, solid line; frozen semen, dotted line.

who championed the cryostorage of human spermatozoa, but there was little general enthusiasm because of the difficulty of storage of glass vials in dry ice (solid CO_2). It was only the availability of liquid nitrogen and the advent of the plastic straw that generated more enthusiasm for cryopreservation. In the 1980s, several studies showed that cryopreserved spermatozoa were less successful at achieving a pregnancy. Indeed, using the patient as her own control, Richter et al. (1984) found that frozen semen was three times less likely to achieve a pregnancy, which translated into a theoretical 12-month cumulative pregnancy rate of 93% for fresh semen and 45% for cryopreserved semen (Figure 18.3). A substantial reduction in fertility is still generally acknowledged.

The most important stimulus for cryopreservation of spermatozoa was the realization that HIV could be transmitted via semen (Stewart et al., 1985). It is now mandatory that semen samples for donor insemination be cryopreserved for a minimum of 6 months while the donor is tested before semen collection and 6 months later for absence of seroconversion.

Following on from cattle artificial insemination (AI) and the animal studies that showed that cryopreserved spermatozoa were more likely to achieve a pregnancy if they were inseminated deeper in the female tract, a number of studies have shown that intrauterine insemination (IUI) was preferable with cryopreserved human semen. However, because of the high concentration of Prostaglandin E in human semen, there is a need to wash the cells free of seminal plasma before insemination to avoid painful uterine contractions following the process. A recent meta-analysis confirmed the advantage of IUI (O'Brien and Vandekerckhove, 2002).

Alongside these developments the treatment of human infertility has embraced IVF, a technique that allows relatively few spermatozoa to achieve fertilization. The control of the interaction of spermatozoa and oocyte has resulted in many subfertile couples being enabled to conceive a child. This itself becomes a driving force for sperm cryopreservation, as many who were incapable of conceiving with natural fertilization or AI are now interested in storing sperm for IVF attempts.

Perhaps the most far-reaching development has been the widespread use of intra-cytoplasmic sperm injection (ICSI). This has meant that semen samples that are oligospermic can now be considered for cryopreservation, opening up the possibility of fathering a child for individuals who are effectively infertile by natural means. Moreover, the survival of cancer patients is increasing very dramatically, and there is considerable interest in cryopreservation of semen to retain the possibility of future paternity when they must undergo radiation treatment that will render them infertile. ICSI also opens the possibility that immature cells even as primitive as spermatids can be used to create a pregnancy, although it is not yet a widespread procedure. The procedure has encouraged attempts to recover and cryopreserve germ cells from infertile individuals even if they can only be recovered in very small numbers.

18.5.2 FEATURES OF THE SPERM OF SIGNIFICANCE IN CRYOPRESERVATION

The spermatozoon is a highly differentiated cell with a single ultimate purpose. To achieve this, it must display a number of unique activities given the appropriate environmental stimuli, and it must interact physically and metabolically with cellular structures of the female tract. Thus, it must display motility and be capable of hyperactivated motility given the right capacitating environment. It must undergo capacitating changes in the female tract, possibly binding to the isthmus cells of the Fallopian tube before being unbound at ovulation. It must reach a state of responsiveness to the oocyte and its vestments in the ampulla of the Fallopian tube with receptor proteins exposed on the cell surface. It must bind to the zona surface and be capable of the acrosome reaction on a progesterone or zona glycoprotein stimulus, and it must penetrate the zona and fuse with the oolemma. It must activate the oocyte and achieve fertilization and embryonic development, implantation, and fetal growth to term. Faults in any of these vital stages result in failure of conception or pregnancy.

Cryopreservation interferes with several cell organelles including the plasma membrane, acrosome, mitochondria, and nucleus, and can thus have an effect on many aspects of cell function. The phospholipid bilayer of the plasma membrane is subjected to phase changes as cells are cooled and rewarmed. This can result in phase separation with clustering of functional protein, and there is evidence that the changes are not all simply reversed on rewarming. Permeability changes also occur with ion imbalance, with a potential effect on intracellular signaling. Cooled spermatozoa display capacitation-like changes as a result of cryopreservation and will prematurely undergo an acrosome reaction. The acrosome may be damaged during cryopreservation and be incapable of a physiological acrosome reaction. Mitochondria show signs of damage in cryopreserved semen, indicating that their energy production may be compromised. Motility is often depressed, although the quality of motility displayed even in a normal ejaculate is quite variable. Finally, recent evidence indicates that there is an increased nuclear alteration as a result of cryopreservation, which may affect subsequent chromosomal function. All these changes add up to a proportion of lethal cell injuries, reducing the number of live cells present, and to further sub-lethal injuries compromising the fertilization potential of the live subpopulation. Although IVF and ICSI remove many of the functional requirements of the mature spermatozoon, cryopreserved spermatozoa are inevitably less likely to achieve fertilization.

18.5.3 CRYOPRESERVATION OF IMMATURE FORMS

In contrast to the oocyte, the difficulties encountered with preserving mature spermatozoa were not the reason for searches to find alternative, more primitive, cells to store. Instead, it is the problems created by the infertile individual that has driven this research. Oligozoospermia and obstructive aspermia, resulting in insufficient or no spermatozoa in the ejaculate to cryopreserve in conventional ways, led to trials to recover spermatozoa from more proximal positions in the reproductive tract. Attempts have then been made to store these cells to build up sufficient numbers for IVF or ICSI. There can be no doubt that the advent of ICSI as a common procedure in the treatment of infertility has revolutionized the approach to treatment of conditions such as oligospermia and azoospermia. No longer is it necessary to preserve spermatozoa in six-figure numbers; even individual spermatozoa are considered worthy of cryopreservation (Borini et al., 2000). Thus, methods that permit recovery of even a few competent germ cells now provide a stimulus for the development of cryopreservation techniques. Storage of these cells until a stimulated female cycle has produced suitable oocytes for fertilization avoids wasted attempts, with their attendant health risks, and provides opportunity to improve the chances of conception by stockpiling samples of germ cells before attempted fertilization.

In the testis, the germ cells go through a complex series of divisions as they undergo meiosis from the primitive diploid spermatogonium through the primary and secondary spermatocyte stages to the spermatid. This latter, now haploid, cell undergoes a remarkable restructuring and elongation to form the familiar spermatozoon with head and tail components. Recovery of primitive cells with a view to *in vitro* maturation is currently under investigation; harvesting of spermatids as immature round cells or as differentiating forms provides a source of potentially fertile male gametes without further maturation (although it is not permissible to transfer embryos created using round cells in the United Kingdom at present). Testicular and epididymal spermatozoa are recovered by a range of puncture and aspiration or biopsy techniques for use in ICSI procedures, these spermatozoa can be cryopreserved. Providing the haploid nucleus is present, the developmental status is of apparently relatively little significance, but concerns have been expressed regarding early spermatids lacking either appropriate genomic imprinting or oocyte activating factors (Tesarik et al., 1998).

18.5.3.1 Testicular Germ Cell Cryopreservation

Studies in rodents indicate that spermatogonia can be harvested and cryopreserved and subsequently are capable of colonizing the spermatogenic tubule and reestablishing spermatogenesis (Avarbock et al., 1996). The expectation is that such technology might permit young men with cancer requiring radiation therapy that causes sterility, to recover their fertility and subsequently be able to impregnate their partners. Initial trials with human spermatogonial cells have not been encouraging (Reis et al., 2000), but cryopreservation of these primitive cells is relatively easy (Brook et al., 2001).

Postcryopreservation maturation of *in vitro*–cultured germ cells was limited to the first 24 h of culture of testicular biopsy samples, perhaps related to the DNA injury sustained by Sertoli cells during cryopreservation (Tesarik et al., 2000), raising the distant possibility of *in vitro* production of spermatozoa.

Spermatids can be recovered by testicular sperm aspiration or extraction (biopsy). In a comparison of developmental stages of cryopreserved germ cells, Trombetta et al. (2000) achieved 64% fertilization rate in IVF with testicular spermatozoa and a lower rate with cryopreserved round spermatids. Results with elongated spermatids were no different from those with spermatozoa, but the number of spermatids involved was very small. Nevertheless, cryopreservation does not totally destroy fertilizing ability. Cryopreservation significantly reduced the fertilization rate of human spermatids microinjected into hamster oocytes, and more primitive stages of spermatid were less fertile (Aslam and Fishel, 1999).

By far the most effort is directed toward cryopreservation of testicular spermatozoa, partly because it is currently permitted to transfer an embryo derived from these germ cells. Recent clinical studies obtained very successful pregnancy results by ICSI with cryopreserved testicular spermatozoa and found no adverse consequences of cryopreservation (Habermann et al., 2000; Huang et al., 2000). These results were confirmed in an experimental study showing that testicular spermatozoa survived cryopreservation as well as vasal spermatozoa, although their motility was understandably much poorer (Bachtell et al., 1999). Injury to plasma and acrosomal membranes was detected by EM in testicular spermatozoa and late spermatid stages (Nogueira et al., 1999); this was similar to injury seen in cryopreserved ejaculated spermatozoa. In earlier stages, particularly, the primary spermatocytes were vulnerable. This technology is being considered for those aspermic men who have a chromosomal abnormality. Successful testicular sperm cryopreservation was recently reported in a 15-year-old individual with Klinefelter's syndrome, offering him the potential of future biological paternity (Damani et al., 2001).

Many people recognize the importance of confining the few cells for cryopreservation into empty zonae to aid retrieval (Borini et al., 2000); a standard TEST(Tes/Tris)-egg yolk–glycerol buffer was used. Others have used a Hepes-buffered human tubal fluid with glycerol as the

TABLE 18.1
Donor Insemination Clinical Pregnancy and Live Birthrates per treatment cycle August 1, 1991 to March 31, 1999 (Data Include GIFT Using Donor Gametes and Intrauterine Insemination)

Reporting period	Number of treatment cycles	Clinical pregnancy rate per treatment cycle (%)	Live birthrate per treatment cycle (%)
08/01/91 to 03/31/92[a]	16,299	6.6	5.0
04/01/92 to 03/31/93	25,623	6.9	5.4
04/01/93 to 03/31/94	23,869	8.6	7.0
04/01/94 to 03/31/95	20,604	9.7	7.9
04/01/95 to 03/31/96	16,874	11.2	9.3
04/01/96 to 03/31/97	14,333	11.6	9.6
04/01/97 to 03/31/98	12,753	11.6	9.6
04/01/98 to 03/31/99	11,035	12.1	9.9

Note: All percentages are expressed per treatment cycle. Taken from *Ninth Annual Report,* Human Fertilisation and Embryology Authority (2000), with permission.

[a] Data for 8 months only.

cryopreservative; no advantage was seen for culture for 3 d before cryopreservation (Liu et al., 2000). A significant gain in survival of germ cells was observed when the testicular tissue was shredded before cryopreservation, presumably allowing better access of the cryoprotectant to the cell membranes (Crabbe et al., 1999).

18.5.3.2 Epididymal Spermatozoa

From animal studies it was expected that spermatozoa recovered from the caudal epididymis could be cryopreserved. Unlike that in other mammals, the human epididymis seems to be less clearly segmented, and thus spermatozoa from quite high in the caput epididymis can be capable of fertilization. More proximal spermatozoa from the caput or corpus epididymis can be recovered and preserved as well as, or better than, ejaculated spermatozoa. In two recent studies, cryopreservation of epididymal spermatozoa did not adversely influence the outcome of ICSI procedures (Patrizio, 2000; Tournaye et al., 1999). However, a much smaller study found that cryopreservation significantly reduced ICSI outcome (Shibahara et al., 1999). Although not all authors would agree, there is considerable evidence that good fertility results can be expected from ICSI using cryopreserved epididymal spermatozoa.

18.5.4 THE CURRENT STATUS OF MATURE (EJACULATED) SPERM PRESERVATION

The fertility results in the United Kingdom following AI with donor cryopreserved semen have almost doubled over the past 8 years, indicating an improvement in techniques (Table 18.1); interestingly, in the same period the number of treatment cycles has declined by over 50% (Human Fertilisation and Embryology Authority, 2000). In a survey of results over the past 18 years, Botchan et al. (2001) found an average fertility of 12.6% per cycle and a 12-month cumulative rate of 75% with no change over time. Centers differ considerably in their success rates because of variations in patient group and clinical conditions. Age of women patients is a most important determinant of success; because the average age has increased, Botchan et al. (2001) concluded that this is the most likely cause of failure to increase fertility rates with improved methods. Alternatively, it may be that their sample of earlier studies was already at the high end of the range.

Of particular interest at the present is the prospect of successful preservation of semen from individuals with compromised fertility. Among this group, the most pressing are those diagnosed with cancer whose treatment will render them infertile; with the increased expectation of life given with these cancer treatments, this possibility becomes important. Two recent studies obtained similar results; namely, that semen quality in these cancer patients was generally poorer than that of normal donors, but that cryodamage was similar to that in normal population (Agarwa, 2000; Hallak et al., 1999); neither study found any relationship with the stage of the cancer. The relatively poor quality of the cryopreserved semen need not be a disincentive because ICSI now offers good prospects.

18.5.4.1 Novel Diluents and Cryopreservation Methods

There has been very little novel development in suitable buffers to cryopreserve human semen. Recent studies have compared two common diluents, HPSM (Mahadevan and Trounson, 1983) and TEST-egg yolk buffer (TEST-EY), and concluded that TEST-EY was significantly the better medium in preserving acrosomal morphology (Stanic et al., 2000) and motility (Hammadeh et al., 2001).

The benefits of seminal plasma as a component of the cryopreservation medium are being rediscovered. Seminal plasma offers a slight but consistent advantage for both normal and low-quality semen. Moreover, a definite advantage to recovery of optimal motility and DNA integrity was also seen in the presence of seminal plasma (Donnelly et al., 2001a).

The use of swim-up selection of normal motile spermatozoa improved the overall quality and resulted in better motility, proportion of intact acrosomes, and spermatozoa able to undergo the acrosome reaction in response to Ca^{2+} ionophore postfreeze (Esteves et al., 2000). The continual search for methods to improve the recognized poorer quality of oligospermic samples has led to the discovery of some possible beneficial additives. Amann et al. (1999b) reported that, for more than 30% of samples, the addition of a synthetic peptide fragment of prosaposin (FertPlus®, BioPore, Inc, State College, PA, U.S.A.) to cryopreserved semen improved the binding of spermatozoa in an *in vitro* assay for zona binding ability; they propose that this is a benefit in those individuals who have subfertility related to zona binding deficiency. Briton-Jones et al. (2001) claim that platelet-activating factor improved the motility of cryopreserved oligospermic samples, aiding the identification of suitable live spermatozoa for ICSI. Wolf et al. (2001) have devised a preparative method for cryopreservation that obviates the need for postthaw preparation for IUI; they show that this method yields as good or better pregnancy rates per cycle than conventionally frozen samples.

18.5.4.2 Permeation Concepts

It has long been a hope that studies of permeability of the sperm plasma membrane to water and cryoprotectants at various temperatures would allow the cryopreservation process to be modeled and a protocol for optimum survival devised. Previous results have indicated optimum cooling at vary fast rates, well exceeding those that were known empirically to be appropriate (Curry et al., 1994). This discrepancy now appears to be receding, as recent studies indicate that the earlier results were inaccurate in that the assumptions were erroneous. Gilmore et al. (2000) presented data, collected in the presence of several different cryoprotectants, that indicate that the increased activation energy (E_a) of hydraulic conductivity at low temperatures in the presence of cryopro-tectant is the main reason why previous studies overestimated optimum cooling rates; the method was based on the volume change measured by means of a Coulter particle counter of cells exposed to water and cryoprotectant solutions. Devireddy et al. (2000) used a cell shape–independent differential scanning calorimeter to estimate volume change throughout the freezing process and concluded that the cooling rates estimated from their model accord with known empirical rates. There is, therefore, a realistic possibility that the cryopreservation process can be modeled in such a way to design the optimum protocol for maximal survival.

The standard approach in these studies is to assume a linear cooling rate. This has been criticized in a recent study as being inappropriate, as it takes no account of the water fluxes at different temperatures (Morris et al., 1999). In this study, the authors found that curvilinear cooling rates giving a chosen nonlinear concentration profile achieved almost complete recovery of all cells motile before freezing. Using a cryomicroscope and freeze substitution, they demonstrated that these optimal cooling protocols resulted in spermatozoa that were neither severely dehydrated nor contained any visible intracellular ice.

18.5.4.3 Membrane Studies

Giraud et al. (2000) demonstrated that cryopreservation resulted in membranes that were less pliable. They found that the higher the membrane fluidity was before freezing, the better was the response of spermatozoa to cryopreservation. In relation to plasma membrane fluidity, James et al. (1999) have shown that fluidity varies between regions of the sperm cell, being most fluid over the acrosome. Cryopreservation resulted in a reduction in lateral lipid movement, indicating a decreasing fluidity. Lipid analyses suggested that the lipids most likely to be targeted by the changes are sphingomyelin and phosphatidylcholine composed of unsaturated fatty acids, especially docosahexanoic acid. After cryopreservation there was an increase in lysophospholipids and ceramide, indicating the activation of phospholipase A2 and sphingomyelinase (Schiller et al., 2000).

The presence of lysophospholipids is also known to be associated with the development of reactive oxygen species, which have been shown to result from cryopreservation (Mazilli et al., 1995). However, not all studies confirm this view, as poor survival seems to be associated with a reduction in reactive oxygen species production (Duru et al., 2001; Wang et al., 1997). Lipid peroxidation, a consequence of the activity of reactive oxygen species, was increased by dilution and cryopreservation, although this was not simply associated with the freeze–thaw process but, rather, with exposure to the cryoprotectant diluent (Schuffner et al., 2001). Translocation of phosphatidylserine to the outer leaflet of the plasma membrane was also a consistent observation in cryopreserved spermatozoa (Duru et al., 2001; Schuffner et al., 2001). These lipid studies all indicate profound changes to the plasma membrane as a result of cryopreservation, which may have serious consequences for fertilizing potential.

18.5.4.4 Nuclear Damage and Cryopreservation

In recent years, as techniques have become available to examine the condition of the nucleus, attention has been devoted to consideration of the effect of cryopreservation (Donnelly et al., 2001a). Duru et al. (2001) could find no evidence of DNA deterioration using terminal deoxynucleotidyl transferase-mediated dUTP nick end labeling assay, although they did detect phospholipid translocation. In contrast, using a flow cytometric acridine orange-labeled chromatin assay, Spano et al. (1999) found that the proportion of cells with no detectable chromatin deterioration was consistently lower after cryopreservation, a conclusion supported by an aniline blue staining method to detect chromatin decondensation (Hammadeh et al., 1999). Further confusion is created by the observation that sperm DNA studied by single cell gel electrophoresis (comet) assay was unaffected in semen from normal individuals, but that of infertile individuals was significantly damaged by cryopreservation (Donnelly et al., 2001b).

At present, one may conclude that there are aspects of sperm nuclear structure that are affected by cryopreservation, and these may be detectable or not depending on the particular assay methods used. There is no clear evidence as yet that they contribute to the lower fertility seen with cryopreserved semen.

18.5.4.5 Capacitation-Like Changes

One aspect of semen cryopreservation that has received recognition in the past few years is the increased readiness to undergo the acrosome reaction (Critser et al., 1987; Drobnis et al., 1993) caused by a capacitation-like change to the plasma membrane (Watson, 1995). Our own evidence from pig spermatozoa is that the induced change is primarily an increased permeability to calcium ions that then has attendant intracellular consequences resembling but not identical with capacitation (Green and Watson, 2001). The loss of selective barrier function of the plasma membrane is most likely a correlate of the known lipid changes induced by temperature reduction.

18.5.4.6 Vitrification

Vitrification has been considered the obvious step to avoid the rigors of ice crystal formation and dissolution. However, the response of spermatozoa to large concentrations of cryoprotectants has been a significant problem to realizing this goal. Nonetheless, Nawroth et al. (2002) have shown that it is conceptually possible; they have frozen human spermatozoa in liquid nitrogen slush (~10,000°C/min) without cryoprotectants and have achieved motile and fertile cells in IVF experiments.

18.5.4.7 Advances in Sperm Assessment

Over the last few years there has been a move to develop more *in vitro* tests of sperm function to assess the action of cryopreservation. Those used regularly continue to be motility assessment by either subjective or CASA methodology, the hyposmotic swelling test, which indicates functional plasma membranes, or viability by fluorescent markers such as propidium iodide, ethidium homodimer, or SYBR14.

Membrane function postcryopreservation may well be impaired, although the cells remain viable. Glander and Schaller (1999) showed that Annexin V conjugated to a fluorescent marker would distinguish cells with increased phosphatidyl serine in the outer leaflet of the plasma membrane, although still excluding propidium iodide. This change is recognized as an early stage of apoptosis. After cryopreservation, a number of spermatozoa fell into the category of Annexin V positive, propidium iodide negative, and the proportion correlated with a change in motility pattern. This test may well provide a more sensitive assay of normal sperm membrane function postcryopreservation.

Tests associated with sperm in the oviduct assess functions related to fertilizing ability. The proportion of spermatozoa attaching to monolayers derived from bovine oviductal epithelium proved to have the sensitivity to distinguish poorer binding ability of frozen-thawed spermatozoa compared with fresh (Ellington et al., 1999). More recently, other tests have been proposed, such as binding to a substrate derived from hen's eggs that reflects zona binding ability (Amann et al., 1999a). This test was able to distinguish between fresh and frozen spermatozoa, but it has yet to be correlated with fertilizing ability. A sperm penetration assay, in which cryopreserved spermatozoa are exposed to oocytes *in vitro* and numbers penetrating over time are counted, was found to be potentially useful in screening fertilizing ability at donor insemination in 11 fertile donors; the authors concluded that it might enable screening out of poor-quality samples (Navarrete et al., 2000).

18.5.4.8 Conclusions

With substantial loss of function in many spermatozoa in the ejaculate as a result of cryopreservation, there is a continual hope for a technical breakthrough. So far, many aspects of methodology have been improved to optimize the situation. ICSI has reduced the need for so many functional spermatozoa. Vitrification, to avoid the consequences of ice crystal formation and dissolution, shows some future promise, but perhaps the most important discovery recently published (Kusakabe et al., 2001) is that lyophilized mouse spermatozoa retain the ability to fertilize when microinjected into oocytes. (Discussion on freeze-drying in mammalian cells can be found in Chapters 20 and 21.) The cells were stored at ambient temperature for many days. The possibility of avoiding frozen storage altogether, albeit combined with ICSI, offers alternative, cheaper ways to store germplasm in the future.

18.6 EMBRYO CRYOPRESERVATION

18.6.1 The Nature of the Embryo

The process of fertilization brings about profound changes that have a significant effect on the success of cryopreservation. At a prosaic level, the combination of changes following sperm fusion, the reformation of a nuclear envelope, and onward development of two pronuclei, enhances survival after thawing in a fashion that is not completely understood. There is evidence that membrane permeability to CPA is increased in the one-cell zygote (Jackowski et al., 1980), whereas the other structural changes may go some way to avoiding chilling damage—for example, the meiotic spindle predominant in oocytes is now no longer an issue. Fertilization is also a self-defining step—whereas it is difficult to predict the functional normality of a particular oocyte on morphological grounds within a cohort of aspirated cells, fertilization by normal fusion processes is to some degree selective for normal cells. This is not necessarily true if fertilization by ICSI is undertaken, but this will be discussed later. Fertilization also raises important ethical and logistical questions about the fate of the resulting embryos: If there are numbers in excess of those required for immediate transfer, then cryopreservation offers a potential option and safeguard against failure of the first treatment cycle.

The progressive development of the fertilized oocyte proceeds by a series of divisions at intervals of 16 to 24 h in the early stages, leading to preimplantation embryos with increasing blastomere numbers and a parallel reduction in size (Figure 18.4). Eventually, blastocysts are formed with an inner fluid space or blastocoel and 64 blastomeres on days 5 to 6 of culture. Finally, the blastocyst hatches from the surrounding remnant of zona pellucida as an essential step toward implantation in the uterine wall (Figure 18.4). More detailed explanations can be found in other texts (Brinsden, 1999; Edwards and Brody, 1995). The stage of embryo development selected for cryopreservation depends on a complex mixture of clinical and pragmatic factors—progressive cell divisions provide indicators that particular embryos have the potential to yield a successful outcome, but culture of human embryos to the blastocyst stage is difficult and time-consuming (Alper et al., 2001), whereas good pregnancy rates have been achieved in some centers when cryopreserving early-stage embryos (Nikolettos and Al-Hasani, 2000; Testart et al., 1986).

18.6.2 The History of Clinical Embryo Cryopreservation

As with other cryobiological developments in fertility treatment, many of the techniques have been transferred from studies in animals, particularly those in mice. Early studies on mouse embryos at the eight-cell stage demonstrated that is was possible to achieve success using Me_2SO (1.5 mol/L) as CPA as long as cooling rates were slow (−0.3°C/min to −80°C before transfer to liquid nitrogen) and warming rates were also slow (Whittingham et al., 1972; see also Chapter 11). The protocol was based on the assumption that slow cooling rates would be required to achieve dehydration during cooling and thus avoid intracellular ice formation—a viewpoint largely supported by subsequent work (Leibo et al., 1974). It was also established that addition, and particularly washing out, of the Me_2SO should be performed in a stepwise manner to avoid osmotic damage. The earliest attempts to cryopreserve human embryos used a similar protocol, with some success (Trounson and Mohr, 1983). Other studies used slightly modified procedures, in which slow cooling was interrupted at −40°C before rapid transfer to liquid nitrogen and thereafter rapid warming. It gradually became apparent that such protocols were equally successful with a range of embryo stages (Gelety and Surrey, 1993).

A major step forward was made when it was demonstrated that 1,2-propanediol was an effective cryoprotectant for human pronuclear (single-cell) embryos (Testart et al., 1986). The medium also

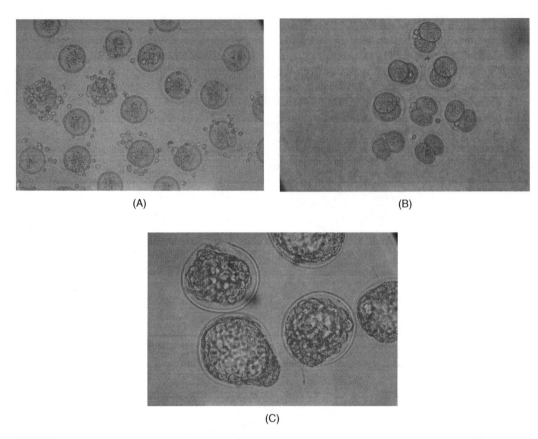

FIGURE 18.4 A comparison of the cellular changes with embryo development in the mouse. The mature unfertilized oocytes are seen as large cells filling the zona pellucida, after removal of the cumulus cells by hyaluronidase treatment (A); original magnification ×200. After fertilization and culture for 1 d, the two-cell embryos present two equal blastomeres of half size and the much smaller, extruded polar bodies (B); original magnification ×200. By day 5 of culture, the embryos have developed to blastocysts, with many small blastomeres surrounding the fluid-filled cavity (blastoceol). Some of the blastocysts show evidence of "hatching" or breaking through the zona pellucida (C); original magnification ×320.

contained sucrose (0.1 mol/L) to act as an osmotic buffer, and slow cooling was interrupted at –30°C, with subsequent rapid rewarming. This has generally become accepted in IVF centers across the world as the method of choice for pronuclear embryos (Garrisi and Navot, 1992; Byrd, 2002).

There were fewer earlier studies on cryopreservation of late-stage human embryos, in part because few centers were prepared to undertake culture to this stage in a clinical setting. Propanediol was not found to be a useful CPA for slow cooling of late-stage embryos (Friedler et al., 1988). In contrast, glycerol (at concentrations of 8 to 10% v/v) was used most successfully with slow cooling (Fehilly et al., 1985), and again stepwise addition and removal of the CPA were essential to avoid osmotic stress to the embryos.

18.6.3 Recent Developments in Human Embryo Banking

18.6.3.1 The Debate about the Developmental Stage for Embryo Cryopreservation

Mixed in with the biophysical changes that affect survival of different-stage preimplantation embryos are other, more philosophical, factors. One reason why single-cell pronuclear embryos have been chosen for cryopreservation in some countries is an ethical issue—in countries such as Germany, it is outside the law to culture more than three embryos in a treatment cycle, whereas single-cell fertilized oocytes are not accorded the definition of embryo, and in this situation, cryopreservation of any excess pronuclear stages is of obvious advantage (Nikolettos and Al-Hasani, 2000). After thawing, pronuclear oocytes can be cultured for a relatively short period (overnight) and allow selection for transfer of only those which have cleaved, a type of selective viability assay. However, other studies have reported that embryos cryopreserved and thawed at the four-cell stage have better implantation rates than thawed two-cell embryos, even if subsequently cultured on to four cells, indicating that longer periods of *in vitro* culture may not be beneficial for thawed embryos or that embryos of different growth rates have different potential outcome (Edgar et al., 2000). For later-stage embryos, it may be possible to achieve further embryonic development in culture (although at a reduced rate) in thawed embryos in which some of the individual blastomeres have been damaged (van den Abbeel and van Steirteghem, 2000), whereas for single-cell pronuclear embryos, cryosurvival will be an "all or nothing" event. In those groups skilled in blastocyst culture techniques, cryopreservation at the blastocyst stage can result in good rates of implantation and ongoing pregnancies (Kaufman et al., 1995). More recently, the value of cryopreservation at the late compact stage (morula, day 4 embryo) has been discussed (Tao et al., 2001) on the basis that this combines the advantage of longer culture to select viable embryos and that of being safer to undertake the "assisted hatching" *in vitro* at this stage (because the cell mass is still compacted away from the zona pellucida that is to be breached; see Section 6.3.5).

18.6.3.2 Practical Embryo Cryopreservation—Slow or Rapid Cooling?

The slow-cooling protocols described above are now in routine use in infertility clinics. However, there are documented limitations to the current methods. Damage to the zona pellucida (ranging from small splits to complete destruction) may result from the biophysical changes and has been noted in clinical embryo cryopreservation programs and correlated with poor outcome (van Steirteghem et al., 1987). Improvements can be gained by modifying the cryopreservation protocols to stabilize the transition between different states during freezing or thawing by inclusion of polymers (Dumoulin et al., 1994) or by modifying the type of container used for cryopreservation (van den Abbeel and van Steirteghem, 2000). The propanediol/sucrose protocol is widespread in its application, but some centers continue to use slow cooling with Me_2SO with better results in selected clinical trials (van der Elst et al., 1995). Other practical considerations include the length of time required for slow cooling methods (several hours), which dictates the number of freezing runs that can be completed in normal working hours. Nevertheless, traditional embryo cryopreservation can be seen as a highly positive contribution to overall patient treatment.

The explosion of interest in the use of vitrification for cryoconservation of animal embryos (Paynter et al., 1997a) and human oocytes (see above) is beginning to affect clinical embryo storage. Feichtinger et al. (1991) reported normal live births following ultra-rapid cooling of early stage embryos, and this has been confirmed by other recent studies in which CPA concentrations high enough to achieve vitrification have been used (Mukaida et al., 1998; for discussions on vitrification, see Chapters 10 and 22) The effectiveness of "ultra-rapid" cooling (in which intermediate–high CPA concentrations have been used but are still insufficient to prevent all formation of ice crystals) in human embryo cryopreservation remains in question. In studies on animal embryos, some groups have achieved high success rates (Shaw et al., 1991), whereas others reported that slow-cooling protocols still gave better success than ultra-rapid cooling in comparative studies (Liu et al., 1993). This same group undertook a randomized clinical trial, which again showed that slow cooling was more effective than ultra-rapid methods (van den Abbeel et al., 1997). The problems may lie in the many small interactive conditions (times and temperatures of embryo exposure to CPA, related osmotic and chemical toxicity, method of transfer to containers, types of containers, and more) that need to combine optimally to achieve successful outcomes from ultra-rapid methods but that are difficult to reproduce on a daily basis in the clinic. As more has been understood about these factors, success with vitrification techniques has improved. The study by Mukaida et al. (1998) used 40% ethylene glycol, 30% Ficoll, and 0.5 M sucrose to vitrify multicellular human embryos, and a successful twin delivery was reported. The successful application of vitrification to clinical blastocyst-stage embryos was described by Choi et al. (2000), with some pregnancies to term. Following this, a case report of successful pregnancy after blastocyst vitrification using a mixture of ethylene glycol and Me_2SO has been made (Yokota et al., 2000). There is still much to learn about the reproducibility, efficacy, and long-term stability of embryo vitrification, but if these are eventually elucidated, the technology may be seen as a way to simplify and speed up the cryobanking procedure.

18.6.3.3 The Place of Embryo Cryopreservation in Infertility Treatment

As pointed out already, embryo cryopreservation is a routine practice within assisted reproduction technologies, with many healthy deliveries. However, it has been realized that it is difficult to quantify the exact contribution of embryo cryopreservation across centers because of differences in reporting, in patient populations, and indeed in populations of embryos selected for fresh transfer or for cryopreservation. One way of assessing the effect of embryo cryopreservation has been to investigate patient-specific pregnancy as a way of identifying those additional patients pregnant from use of cryopreserved embryos (Schnorr et al., 2000) who did not achieve pregnancies from the same fresh transfer or from previous attempts. Recipient factors affected the outcome, but in the best-case cohort of patients, transfer of cryopreserved embryos was able to raise pregnancy rates from 40 (after fresh transfer only) to 57% after use of cryopreserved resources, which is obviously a significant increase. Of paramount importance is the health of children born from transfer of cryopreserved embryos. Suttcliffe (2000) reviewed the current evidence and concluded that these children had an overall satisfactory well-being, but cautioned that consistent and continuing follow-up is required.

18.6.3.4 Current Clinical Perspectives of Embryo Cryobanking

The benefits of using cryopreservation to bank embryos in excess of those required for the current treatment cycle, and thus enhance the chances of overcoming an unsuccessful outcome or additional deliveries at a later date, are self-evident. However, there are also more subtle reasons for using cryopreservation. One of these is to deal with ovarian hyperstimulation, which is a condition experienced by some patients in response to the hormonal treatments and that carries significant risks of morbidity and mortality. The condition can be relieved by canceling the treatment cycle, but this can be costly and frustrating for the patient. In such circumstances, some centers have promoted cryopreservation of all embryos obtained during the original terminated cycle and then offering frozen embryo transfer at a later date (Amso et al., 1990), when a more sparing use of hormonal therapy can avoid the complications. The outcome in terms of pregnancy rates can be as good as, or in some cases, better than, those in matched groups of patients receiving fresh embryo transfer (Queenan, 2000).

As infertility treatments have become more successful, the question of the number of embryos that should be transferred has been raised. In early reports, multiple-embryo transfers were undertaken to enhance the likelihood of achieving pregnancy, but the incidence of multiple births (triplets or more) forced a reappraisal. Some centers are now advocating single-embryo transfer in selected groups of patients and where good-quality embryos have been produced. In such situations, cryopreservation of the remaining embryos for later transfer if pregnancy is not achieved has been recommended. Such a policy has been found to enhance eventual cumulative pregnancy rates by approximately one third (Tiitinen et al., 2001).

18.6.4.5 Cryopreservation of Embryos after Micromanipulation

The increased sophistication of manipulations employed in infertility treatment has led to the provision of embryos in which the zona pellucida has been breached for a variety of reasons. There is a growing use of ICSI to achieve fertilization, whereas issues of prenatal diagnosis, in which individual blastomeres may be removed for genetic analysis before cryoconservation of the embryo, is another potential reason. Alternatively, the zona may be mechanically/chemically disrupted (zona "drilling" or "slitting") to try and enhance implantation. In animal experiments, both slow cooling with propanediol (Sandalinas et al., 1994) and rapid cooling with high concentrations of Me_2SO (Wilton et al., 1989) have been successfully used in embryos in which the zonae have been breached. In human embryos, breaching the zona pellucida by a variety of methods was found to result in lower survival after cooling with the propane diol protocol (Ciotti et al., 2000), even if blastomeres were not removed from the embryo. These results confirm earlier reports where in the mouse, the larger the breach in the zonae, the greater the degree of damage (Thompson et al., 1995). In contrast, successful pregnancies have been reported using human embryos subjected to blastomere biopsy after thawing (Magli et al., 2000).

In comparison with the size of breach needed for embryo biopsy, needle puncture of the zona pellucida to achieve fertilization by ICSI is quite small. Van den Abbeel et al. (2000; van den Abbeel and van Steiteghem, 2000) reported that ICSI embryos cryopreserved at early cleavage stages (two to eight cells) showed morphological survival at rates similar to cryopreserved embryos achieved by conventional IVF, using a slow-cooling protocol and Me_2SO as CPA. However, this study commented on a trend toward slightly increased pregnancy loss from the cryopreserved ICSI embryos; whether this resulted from the ICSI procedure coupled with cryopreservation or from other factors such as male-factor poor-semen quality (which was the reason for many of the ICSI cycles) remains to be clarified. Other groups (Hu et al., 1999) found that ICSI embryos could be cryopreserved and achieve similar pregnancy outcomes as IVF embryos. Thus, in the clinical situation, cryopreservation of micromanipulated embryos will need considerable and ongoing further evaluation.

18.6.4 CRYOPRESERVATION OF EMBRYONIC STEM CELLS

One area of embryo manipulation in which interest is expected to increase over the next decade is that of production of embryonic stem cells. These are the pluripotent cells of the inner cell mass in blastocysts, which have the potential for indefinite *in vitro* propagation and differentiation into a variety of different somatic tissues, with far-reaching applications in medicine and biotechnology. Thus, an effective cryopreservation protocol for these important cells will be an essential development. Reubinoff et al. (2001) have reported one of the first such studies, comparing slow cooling using Me_2SO and vitrification in mixtures of ethylene glycol and Me_2SO. Recovery after slow cooling was relatively low, but vitrification yielded improved results. Alterations in the background, unstimulated, levels of differentiation of the stem cells after vitrification were highlighted as one area in which further investigations will be required.

18.7 SAFETY ISSUES IN CRYOBANKING HUMAN REPRODUCTIVE MATERIALS

As with banking of all biomaterials (see Chapter 15), there are recognized issues concerning the low-temperature storage of human reproductive tissues (Clarke, 1999). Unwitting recent viral contamination of samples when stored in liquid nitrogen in proximity to samples from another patient in a bone marrow cryobank has highlighted the issue for reproductive clinics (Tomlinson and Sakkas, 2000). There are measures, such as storage of samples only in the vapor phase, manufacture of vials or straws of materials that will not shatter at ultra-low temperatures, or use of a sterile sleeve to enclose straws or vials, which can reduce the risk of cross-contamination and that are likely to be seen as essential good laboratory practice.

18.8 DISCUSSION

Cryopreservation techniques remain one of the key pieces in the overall picture of clincial infertility treament in the twenty-first century. An improved understanding of cryobiological principles is rapidly being transferred into practice; for example, vitrification methods are being used for blastocyst preservation (Reed et al., 2002; Son et al., 2002), whilst there is a widening appreciation of the value of cryopreservation for ovarian tissues or immature gametes (Gosden, 2002; Gosden et al., 2002). However, there still remain many areas in which our full appreciation of fundamental low-temperature principles has not yet been achieved and that need to be addressed before consistently high success rates can be guaranteed. There are also equally important areas of ethics and education about the use of cryopreservation in reproductive medicine in general, the role of cryobiology in therapuetic cloning, and the moral position on reproductive cloning, which form part of an ongoing debate in society at large. In the 50 years since the initial freezing of bull spermatozoa, the application of cryobiology and society's expectations have expanded at an amazing pace; the next 50 years are likely to be equally amazing from a cryobiologist's standpoint.

REFERENCES

Agarwa, A. (2000), Semen banking in patients with cancer: 20-year experience, *Int. J. Androl.,* 23, 16–19.

Al-Hasani, S., Diedrich, K., van der Ven, H., Reinecke, A., Hartje, M., and Krebs, D. (1987) Cryopreservation of human oocytes, *Hum. Reprod.,* 2, 695–700.

Almeida, P.A. and Bolton, V.N. (1995) The effect of temperature fluctuations on the cytoskeletal organisation and chromosomal constitution of the human oocyte, *Zygote,* 3, 357–365.

Alper, M., Brinsden P., Fischer, R., and Wirkland, M. (2001) To blastocyst or not to blastocyst? That is the question, *Hum. Reprod.,* 16, 617–619.

Amann, R.P., Shabanowitz, R.B., Huszar, G., and Broder, S.J. (1999a) *In vitro* sperm-binding assay to distinguish differences in populations of human sperm or damage to sperm resulting from cryopreservation, *J. Androl.,* 20, 648–54.

Amann, R.P., Shabanowitz, R.B., Huszar, G., and Broder, S.J. (1999b). Increased *in vitro* binding of fresh and frozen-thawed human sperm exposed to a synthetic peptide *J. Androl.,* 20, 655–60.

Amso, N.N., Ahuja, K.K., Morris, N., and Shaw R.W. (1990) The management of predicted ovarian hyperstimulation involving gonadotropin-releasing hormone analog with elective cryopreservation of all pre-embryos, *Fertil. Steril.,* 53, 1087–1090.

Apperley, J.F. and Reddy, N. (1995) Mechanism and management of treatment-related gonadal failure in recipients of high dose chemoradiotherapy, *Blood Rev.,* 9, 93–116.

Aslam, I. and Fishel, S. (1999) Evaluation of the fertilization potential of freshly isolated, *in-vitro* cultured and cryopreserved human spermatids by injection into hamster oocytes, *Hum. Reprod.,* 14, 1528–1533.

Avarbock, M.R., Brinster, C.J., and Brinster, R.L. (1996) Reconstitution of spermatogenesis from frozen spermatogonial stem cells, *Nat. Med.,* 2, 693–696.

Bachtell, N.E., Conaghan, J., and Turek, P.J. (1999) The relative viability of human spermatozoa from the vas deferens, epididymis and testis before and after cryopreservation, *Hum. Reprod.,* 14, 3048–3051.

Baka, S.G., Toth, T.L., Veeck, L.L., Jones, H.W., Jr., Muasher, S.L., and Lanzendorf, S.E. (1995) Evaluation of the spindle apparatus of in-vitro matured human oocytes following cryopreservation, *Hum. Reprod.,* 7, 1816–1820.

Borini, A., Sereni, E., Bonu, E., and Flamigni, C. (2000) Freezing a few testicular spermatozoa retrieved by TESA, *Mol. Cell Endocrinol.,* 169, 27–32.

Botchan, A., Hauser, R., Gamzu, R., Yogev, L., Paz, G., and Yavetz, H. (2001). Results of 6139 artificial insemination cycles with donor spermatozoa, *Hum. Reprod.,* 16, 2298–304.

Brinsden, P., Ed. (1999) *A Textbook of in vitro Fertilisation and Assisted Reproduction: The Bourn Hall Guide to Clinical and Laboratory Practice,* 2nd edition, CRC Press, Boca Raton, FL.

Briton-Jones, C., Yeung, Q.S., Tjer, G.C., Chiu, T.T., Cheung, L.P., Yim, S.F., Lok, I.H., and Haines, C. (2001) The effects of follicular fluid and platelet-activating factor on motion characteristics of poor-quality cryopreserved human sperm, *J. Assist. Reprod. Genet.,* 18, 165–170.

Brook, P.F., Radford, J.A., Shalet, S.M., Joyce, A.D., and Gosden, R.G. (2001). Isolation of germ cells from human testicular tissue for low temperature storage and autotransplantation, *Fertil. Steril.,* 75, 269–274.

Bunge, R.G., Keettel, W., and Sherman, J. (1954) Clinical use of frozen semen, report of 4 cases, *Fertil. Steril.,* 5, 520–529.

Byrd, W. (2002) Cryopreservation, thawing and transfer of human embryos, *Semin. Reprod. Med.,* 20, 37–43.

Candy, C., Wood, M., and Whittingham, D. (1995) Follicular development in cryopreserved marmoset ovarian tissues after transplantation, *Hum. Reprod.,* 10, 2334–2338.

Carroll, J., Whittingham, D., Wood, M., Telfer, E., and Gosden, R. (1990) Extra-ovarian production of mature oocytes from frozen primordial follicles, *J. Reprod. Fertil.,* 90, 321–327.

Chen, C. (1986) Pregnancy after human oocyte cryopreservation, *Lancet,* I, 884–886.

Chen, C. (1988) Pregnancies after human oocyte cryopreservation, *Ann. NY Acad. Sci.,* 541, 541–549.

Chen, S., Lien, Y., Chao, K., Lu, H., Ho, H., and Yang, Y. (2000) Cryopreservation of mature human oocytes by vitrification with ethylene glycol in straws, *Fertil. Steril.,* 74, 804–808.

Chia, C.M., Chan, W.B., Quah, E., and Cheng, L.C. (2000) Triploid pregnancy after ICSI of frozen testicular spermatozoa into cryopreserved human oocytes, *Hum. Reprod.,* 15, 1962–1964.

Choi, D., Chung, H., Lim, J., Ko, J., Yoon, T., and Cha, K. (2000) Pregnancy and delivery of healthy infants developed from vitrified blastocysts in an IVF-ET program, *Fertil. Steril.,* 74, 838–899.

Chung, H.M., Hong, S.W., Lim, J.M., Lee, S.H., Cha, W.T., Ko, J.J., Han, S.Y., Choi, D.H., and Cha, K.Y. (2000) *In vitro* blastocyst formation of human oocytes obtained from unstimulated and stimulated cycles after vitrification at various maturational stages, *Fertil. Steril.,* 73, 545–551.

Ciotti, P., Lagalla, C., Ricco, A., Fabbri, R., Forabosco, A., and Porcu, E. (2000) Micromanipulation of cryopreserved embryos and cryopreservation of micromanipulated embryos, *Mol. Cell Endocrinol.,* 169, 63–67.

Clarke, G.N. (1999) Sperm cryopreservation: Is there a significant risk of cross-contamination? *Hum. Reprod.,* 14, 2941–2943.

Cobo, A., Rubio, C., Gerli, S., Ruiz, A., Pellicer, A., and Remohi, J. (2001) Use of fluorescence *in situ* hybridization to assess the chromosomal status of embryos obtained from cryopreserved oocytes, *Fertil. Steril.,* 75, 354–360.

Cooper, A., Paynter, S.J., Fuller, B.J., and Shaw, R.W. (1998) Differential effects of cryopreservation on nuclear or cytoplasmic maturation *in vitro* in immature mouse oocytes from stimulated ovaries, *Hum. Reprod.,* 13, 971–978.

Crabbe, E., Verheyen, G., Tournaye, H., and van Steirteghem, A. (1999) Freezing of testicular tissue as a minced suspension preserves sperm quality better than whole-biopsy freezing when glycerol is used as cryoprotectant, *Int. J. Androl.,* 22, 43–48.

Critser, J.K., Arneson, B.W., Aaker, D.V, Huse-Benda, A.R., and Ball, G.D. (1987) Cryopreservation of human spermatozoa. II. Postthaw chronology of motility and of zona-free hamster ova penetration, *Fertil. Steril.,* 47, 980–984.

Curry, M.R., Millar, J., and Watson, P. (1994) Calculated optimal cooling rates for ram and human sperm cryopreservation fail to conform with empirical results, *Biol. Reprod.,* 51, 1014–1021.

Damani, M.N., Mittal, R., and Oates, R.D. (2001) Testicular tissue extraction in a young male with 47,XXY Klinefelters syndrome: Potential strategy for preservation of fertility, *Fertil. Steril.,* 76, 1054–1056.

Deansly, R. (1954) Immature rat ovaries grafted after freezing and thawing, *J. Endocrinol.,* 11, 197–200.

Deansly, R. and Parkes, A.S. (1956) Delayed development of grafts from frozen ovarian tissue, *J. Endocrinol.,* 14, 35–36.

Devireddy, R.V., Swanlund, D.J., Roberts, K.P., Pryor, J.L., and Bischof, J.C. (2000) The effect of extracellular ice and cryoprotective agents on the water permeability parameters of human sperm plasma membrane during freezing, *Hum. Reprod.,* 15, 1125–1135.

Diedrich, K., Al Hasani, S., van der Ven, H., and Krebs, D. (1987) Successful *in vitro* fertilization of frozen-thawed rabbit and human oocytes, *Abstracts 5th World Congress on IVF and ET*, 562–570.

Donnelly, E.T., McClure, N., and Lewis, S.E. (2001a) Cryopreservation of human semen and prepared sperm: Effects on motility parameters and DNA integrity, *Fertil. Steril.*, 76, 892–900.

Donnelly, E.T., Steele, E.K., McClure, N. and Lewis, S.E. (2001b) Assessment of DNA integrity and morphology of ejaculated spermatozoa from fertile and infertile men before and after cryopreservation, *Hum. Reprod.*, 16, 1191–1199.

Drobnis, E.Z., Clisham, P.R., Brazil, C.K., Wisner, L.W., Zhong, C.Q., and Overstreet, J.W. (1993) Detection of altered acrosomal physiology of cryopreserved human spermatozoa after sperm residence in the female reproductive tract, *J. Reprod. Fertil.*, 99, 159–165.

Dumoulin, J.C., Bergers-Janssen, J.M., Pieters, M.H., Enginsu, M.E., Geraedts, J.P., and Evers, J. (1994) The protective effects of polymers in the cryopreservation of human and mouse zonae pellucidae and embryos, *Fertil. Steril.*, 62, 793–798.

Duru, N.K., Morshedi, M.S., Schuffner, A., and Oehninger S (2001) Cryopreservation—Thawing of fractionated human spermatozoa is associated with membrane phosphatidylserine externalization and not DNA fragmentation, *J. Androl.*, 22, 646–651.

Edgar, D.H., Bourne, H., Jericho, H., and McBain, J.C. (2000) The developmental potential of cryopreserved human embryos, *Mol. Cell Endocrinol.*, 169, 69–72.

Edwards, R. and Brody, S. (1995) *Principles of Assisted Reproduction*, W.B. Saunders, London.

Ellington, J.E., Broemeling, L.D., Broder, S.J., Jones, A.E., Choker, D.A., and Wright, R.W. (1999) Comparison of fresh and cryopreserved human sperm attachment to bovine oviduct (uterine tube) epithelial cells *in vitro*, *J. Androl.*, 20, 492–499.

Eppig, J.J. and O'Brien, M.J. (1996) Development *in vitro* of mouse oocytes from primordial follicles, *Biol. Reprod.*, 54, 197–207.

Esteves, S.C., Sharma, R.K., Thomas, A.J., Jr. and Agarwal, A. (2000) Improvement in motion characteristics and acrosome status in cryopreserved human spermatozoa by swim-up processing before freezing, *Hum. Reprod.*, 15, 2173–2179.

Fabbri, R., Porcu, E., Marsella, T., Primavera, M.R., Rochetta, G., Ciotti, P.M., Magrini, O., Seracchioli, R., Venturoli, S., and Flamigni, C. (2000) Technical aspects of oocyte cryopreservation, *Mol. Cell. Endocrinol.*, 169, 39–42.

Fabbri, R., Porcu, E., Marsella, T., Rochetta, G., Venturoli, S., and Flamigni, C. (2001) Human oocyte cryopreservation: New perspectives regarding oocyte survival, *Hum. Reprod.*, 15, 411–416.

Fehilly, C., Cohen, J., Simons, R., Fishel, S., and Edwards, R. (1985) Cryopreservation of cleaving embryos and expanded blastocysts in the human: A comparitive study, *Fertil. Steril.*, 44, 638–644.

Feichtinger, W., Hochfellner, C., and Ferstl, U. (1991) Clinical experience with ultra-rapid freezing of embryos, *Hum. Reprod.*, 6, 735–736.

Friedler, S., Giudice, L.C., and Lamb, E.J. (1988) Cryopreservation of embryos and ova, *Fertil. Steril.*, 49, 743–764.

Garrisi, G.J., and Navot, D. (1992) Cryopreservation of semen, oocytes, and embryos, *Curr. Opin. Obstet. Gynecol.*, 4, 726–731.

Gelety, T. and Surrey, E. (1993) Cryopreservation of embryos and oocytes: An update, *Curr. Opin. Obstet. Gynaecol.*, 5, 606–614.

Gilmore, J.A., Liu, J., Woods, E.J., Peter, A.T., and Critser, J.K. (2000) Cryoprotective agent and temperature effects on human sperm membrane permeabilities: Convergence of theoretical and empirical approaches for optimal cryopreservation methods, *Hum. Reprod.*, 15, 335–343.

Giraud, M.N., Motta, C., Boucher, D., and Grizard, G. (2000) Membrane fluidity predicts the outcome of cryopreservation of human spermatozoa, *Hum. Reprod.*, 15, 2160–2164.

Glander, H.J. and Schaller, J. (1999) Binding of annexin V to plasma membranes of human spermatozoa: A rapid assay for detection of membrane changes after cryostorage, *Mol Hum. Reprod.*, 5, 109–115.

Gook, D.A., Osborn, S.M., Bourne, H., and Johnston, W.I.H. (1994) Fertilisation of human oocytes following cryopreservation: Normal karyotypes and absence of stray chromosomes, *Hum. Reprod.*, 9, 684–691.

Gook, D.A., Osborn, S.M., and Johnston, W.I.H. (1993) Cryopreservation of mouse and human oocytes using 1,2-propanediol and the configuration of the meiotic spindle, *Hum. Reprod.*, 8, 1101–1109.

Gook, D.A., Osborn, S.M., and Johnston, W.I.H. (1995a) Parthenogenetic activation of human oocytes following cryopreservation using 1,2-propanediol, *Hum. Reprod.*, 10, 654–658.

Gook, D.A., Schiewe, M.C., Osborn, S.M., Asch, R.H., Jansen, R.P.S., and Johnston, W.I.H. (1995b) Intracytoplasmic sperm injection and embryo development of human oocytes cryopreserved using 1,2-propanediol, *Hum. Reprod.,* 10, 2637–2641.

Gook, D., Edgar, D., and Stern, C. (1999) Effect of cooling rate and dehydration regimen on the histological appearance of human ovarian cortex following cryopreservation in 1,2-propanediol, *Hum. Reprod.,* 14, 2061–2068.

Gook, D., McCully, B., Edgar, D., and McBain, J. (2001) Development of antral follicles in human cryopreserved ovarian tissue following xenografting, *Hum. Reprod.,* 16, 417–422.

Gosden, R.G. (2002) Gonadal tissue cryopreservation and transplantation, *Reprod. Biomed. Online,* 4, 64–67.

Gosden, R.G. and Aubard, Y. (1996) *Transplantation of Ovarian and Testicular Tissue, Medical Intelligence Unit,* R. Landes Press, Austin, TX.

Gosden, R.G., Baird, D.T., Wade, J.C., and Webb, R. (1994) Restoration of fertility to oophorectomized sheep by ovarian autografts stored at −196°C, *Hum. Reprod.,* 9, 597–603.

Gosden, R.G., Mulan, J. Picton, H., Yin, H., and Tan, S. (2002) Current perspectives on primordial follicle cryopreservation and culture for reproductive medicine, *Hum. Reprod. Update,* 8, 105–110.

Goud, A., Goud, P., Qian, C., van der Elst, J., van Maele, G., and Dhont, M. (2000) Cryopreservation of human germinal vesicle stage and *in vitro* matured MII oocytes: Influence of cryopreservation media on the survival, fertilization, and early cleavage divisions, *Fertil. Steril.,* 74, 487–494.

Green, C. and Watson, P. (2001) Comparison of the capacitation-like state of cooled boar spermatozoa with true capacitation, *Reproduction,* 122, 889–898.

Green, S.H., Smith, A.U., and Zuckerman, S. (1956) The numbers of oocytes in ovarian autografts after freezing and thawing, *J. Endocrinol.,* 13, 330–334.

Habermann, H., Seo, R., Cieslak, J., Niederberger, C., Prins, G.S., and Ross, L. (2000) *In vitro* fertilization outcomes after intracytoplasmic sperm injection with fresh or frozen-thawed testicular spermatozoa, *Fertil. Steril.,* 73, 955–960.

Hallak, J., Kolettis, P.N., Sekhon, V.S., Thomas, A.J., Jr., and Agarwal, A. (1999) Sperm cryopreservation in patients with testicular cancer, *Urology,* 54, 894–899.

Hammadeh, M.E., Askari, A.S., Georg, T., Rosenbaum, P., and Schmidt, W. (1999) Effect of freeze-thawing procedure on chromatin stability, morphological alteration and membrane integrity of human spermatozoa in fertile and subfertile men, *Int. J. Androl.,* 22, 155–162.

Hammadeh, M.E., Georg, T., Rosenbaum, P., and Schmidt, W. (2001) Association between freezing agent and acrosome damage of human spermatozoa from subnormal and normal semen, *Andrologia,* 33, 331–336.

Harp, R., Leibach, J., Black, J., Kehldahl, C., and Karow, A. (1994) Cryopreservation of murine ovarian tissue, *Cryobiology,* 31, 336–343.

Hovatta, O., Silye, R., Abir, R., Krausz, T., and Winston, R.M.L. (1997) Extracellular matrix improves survival of both stored and fresh human primordial and primary ovarian follicles in long-term culture, *Hum. Reprod.,* 12, 1032–1036.

Hovatta, O., Silye, R., Krausz, T., Abir, R., Margara, R., Trew, G., Lass, A., and Winston, R.M.L. (1996) Cryopreservation of human ovarian tissue using dimethylsulphoxide and propanediol-sucrose as cryoprotectants, *Hum. Reprod.,* 11, 1268–1272.

Hu, Y., Maxson, W.S., Hoffman, D.I., Ory, S.J., and Eager, S. (1999) A comparison of post-thaw results between cryopreserved embryos derived from intracytoplasmic sperm injection and those from conventional IVF, *Fertil. Steril.,* 72, 1045–1048.

Huang, F.J., Chang, S.Y., Tsai, M.Y., Kung, F.T., Lin, Y.C., Wu, J.F., and Lu, Y.J. (2000) Clinical implications of intracytoplasmic sperm injection using cryopreserved testicular spermatozoa from men with azoospermia, *J. Reprod. Med.,* 45, 310–316.

Human Fertilisation and Embryology Authority. (2000) *Donor Insemination Data, Ninth Annual Report,* Human Fertilisation and Embryology Authority, London.

Hunter, J.E., Bernard, A.G., Fuller B.J., Amso, N., and Shaw, R.W. (1991) Fertilization and development of the human oocyte following exposure to cryoprotectants, low temperatures and cryopreservation: A comparison of two techniques, *Hum. Reprod.,* 6, 1460–1465.

Hunter, J.E., Fuller, B.J., Bernard, A., Jackson, A., and Shaw, R.W. (1995) Vitrification of human oocytes following minimal exposure to cryoprotectants: Initial studies on fertilization and embryonic development, *Hum. Reprod.,* 10, 1184–1188.

Imoedemhe, D.G. and Sigue, A.B. (1992) Survival of human oocytes cryopreserved with or without the cumulus in 1,2-propanediol, *J. Assist. Reprod. Gen.*, 9, 323–327.

Jackowski, S., Liebo, S., and Mazur, P. (1980) Glycerol permeabilities of fertilized and unfertilized mouse ova, *J. Exp. Zool.*, 212, 329–341.

James, P.S., Wolfe, C.A., Mackie, A., Ladha, S., Prentice, A., and Jones, R. (1999) Lipid dynamics in the plasma membrane of fresh and cryopreserved human spermatozoa, *Hum. Reprod.*, 14, 1827–1832.

Johnson, M.H. and Pickering, S.J. (1987) The effect of dimethylsulphoxide on the microtubular system of the mouse oocyte, *Development*, 100, 313–324.

Jones, R.E. (1997). *Human Reproductive Biology*, Academic Press, New York.

Kazem, R., Thompson, L.A., Srikantharajah, A., Laing, M.A., Hamilton, M.P.R., and Templeton, A. (1995) Cryopreservation of human oocytes and fertilization by two techniques: *In vitro* fertilization and intracytoplasmic sperm injection, *Hum. Reprod.*, 10, 2650–2654.

Kaufman, R., Menezo, Y., Hazout, A., Nicollet, B., DuMont, M., and Servy E. (1995) Co-cultured blastocyst cryopreservation: Experience of more than 500 transfer cycles, *Fertil. Steril.*, 64, 1125–1129.

Kim, S., Battaglia, D., and Soules M. (2001a) The future of human ovarian cryopreservation and transplantation: Fertility and beyond, *Fertil. Steril.*, 75, 1049–1056.

Kim, S., Radford, J., Harris, M., Varley, J., Rutherford, A., Lieberman, B., Shalet, S., and Gosden, R.G. (2001b) Ovarian tissue harvested from lymphoma patients to preserve fertility may be safe for autotransplantation, *Hum. Reprod.*, 16, 2056–2060.

Kuleshova, L., Gianaroli, L., Magli, C., Ferraretti, A., and Trounson, A. (1999) Birth following vitrification of a small number of human oocytes, *Hum. Reprod.*, 14, 3077–3079.

Kusakabe, H., Szczygiel, M.A., Whittingham, D.G., and Yanagimachi, R. (2001) Maintenance of genetic integrity in frozen and freeze-dried mouse spermatozoa, *Proc. Natl. Acad. Sci. USA*, 98, 13501–13506.

Lassalle, B., Testart, J., and Renard, J.P. (1985) Human embryo features that influence the success of cryopreservation with the use of 1,2-propanediol, *Fertil. Steril.*, 44, 645–657.

Le Gal, F., Gasqui, P., and Renard, J.P. (1994) Differential osmotic behaviour of mammalian oocytes before and after maturation: A quantitative analysis using goat oocytes as a model, *Cryobiology*, 31, 154–170.

Leibfried-Rutledge, M., Dominko, T., Critser, E., and Critser, J. (1997) Tissue maturation *in vivo* and *in viro*: Gamete and early embryo ontogeny, in *Reproductive Tissue Banking*, Karow, A.M., and Crister, J.K., Eds., Academic Press, New York, pp. 23–138.

Leibo, S., Mazur, P., and Jackowski, S. (1974) Factors affecting the survival of mouse embryos during freezing and thawing, *Exp. Cell Res.*, 89, 79–88.

Liu, J., van den Abbeel, E., and van Steirteghem, A. (1993) Assessment of ultrarapid and slow freezing procedures for 1-cell and 4-cell mouse embryos, *Hum. Reprod.*, 8, 1115–1119.

Liu, J., Zheng, X.Z., Baramki, T.A., Compton, G., Yazigi, R.A., and Katz, E. (2000) Cryopreservation of a small number of fresh human testicular spermatozoa and testicular spermatozoa cultured *in vitro* for 3 days in an empty zona pellucida, *J. Androl.*, 21, 409–413.

Magli, C., Gianaroli, L., Fortinini, D., Ferraretti, A., and Munne, S. (2000) Impact of blastomere biopsy and cryopreservation techniques on human embryo viability, *Hum. Reprod.*, 14, 770–773.

Mahadevan, M. and Trounson, A.O. (1983) Effect of cryoprotective media and dilution methods on the preservation of human spermatozoa, *Andrologia*, 15, 355–66.

Mandelbaum, J., Junca, A.M., Tibi, C., Plachot, M., Alnot, M.O., Rim, H., Salat-Baroux, J., and Cohen, J. (1988) Cryopreservation of immature and mature hamster and human oocytes, in *In Vitro Fertilization and Other Assisted Reproduction*, Jones, Jr., H.W. and Shrade, C., Eds., *Ann. NY Acad. Sci.*, 541, 550–561.

Mazzilli, F., Rossi, T., Sabatini, L., Pulcinelli, F.M., Rapone, S., Dondero, F., and Gazzaniga, P.P. (1995) Human sperm cryopreservation and reactive oxygen species (ROS) production, *Acta Eur. Fertil.*, 26, 145–148.

Morris, G.J., Acton, E., and Avery, S. (1999) A novel approach to sperm cryopreservation, *Hum. Reprod.*, 14, 1013–1021.

Mukaida, T., Wada, S., Takahashi, K., Pedro, P.B., An, T.Z., and Kasai, M. (1998) Vitrification of human embryos based on the assessment of suitable conditions for 8-cell mouse embryos, *Hum. Reprod.*, 13, 2874–2879.

Navarrete, T., Johnson, A., Mixon, B., and Wolf, D. (2000) The relationship between fertility potential measurements on cryobanked semen and fecundity of sperm donors, *Hum. Reprod.*, 15, 344–350.

Nawroth, F., Isachenko, V., Dessole, S., Rahimi, G., Farina, M., Vargiu, N., Mallman, P., Dattena, M., Capobianco, G., Perts, D., Orth, I., and Isachenko, E. (2002) Vitrification of human spermatozoa without cryoprotectants, *Cryo-Letters,* 23, 93–102.

Newton, H., Fisher, J., Arnold, J.R., Pegg, D.E., Faddy, M.J., and Gosden, R.G. (1998) Permeation of human ovarian tissue with cryoprotective agents in preparation for cryopreservation, *Hum. Reprod.,* 13, 376–380.

Newton, H. and Illingworth P. (2001) *In vitro* growth of murine pre-antral follicles after isolation from cryopreserved ovarian tissue, *Hum. Reprod.,* 16, 423–429.

Nikolettos, N., and Al-Hasani, S. (2000) Frozen pronuclear oocytes: Advantages for the patient, *Mol. Cell Endocrinol.,* 169, 55–62.

Nisolle, M., Casanas-Roux, F., Qu, J., Motta, P., and Donnez, J. (2000) Histologic and ultrastructural evaluation of fresh and frozen-thawed human ovarian xenografts in nude mice, *Fertil. Steril.,* 74, 122–129.

Nogueira, D., Bourgain, C., Verheyen, G., and van Steirteghem, A.C. (1999) Light and electron microscopic analysis of human testicular spermatozoa and spermatids from frozen and thawed testicular biopsies, *Hum. Reprod.,* 14, 2041–2049.

O'Brien, P. and Vandekerckhove, P. (2000) Intra-uterine versus cervical insemination of donor sperm for subfertility, *Cochrane Database Syst. Rev.,* Issue 2, CD000317.

Oktay, K., Aydin, B.A., and Karlikaya, G. (2001) A technique for laparoscopic transplantation of frozen-banked ovarian tissue, *Fertil. Steril.,* 75, 1212–1216.

Oktay, K., Nugent, D., Newton, H., Salha, O., Chatterjee, P., and Gosden, R.G. (1997) Isolation and characterization of primordial follicles from fresh and cryopreserved human ovarian tissue, *Fertil. Steril.,* 67, 481–486.

Oktay, K., Newton, H., Mullan, J., and Gosden, R.G. (1998) Development of human primordial follicles to antral stages in SCID/*hpg* mice stimulated with follicle stimulating hormone, *Hum. Reprod.,* 13, 1133–1138.

Park, S., Son, W., Lee, S., Lee, K., Ko, J., and Cha, K. (1997) Chromosome and spindle configurations of human oocytes matured *in vitro* after cryopreservation at the germinal vesicle stage, *Fertil. Steril.,* 68, 920–926.

Parkes, A.S. (1957) Viability of ovarian tissue after freezing, *Proc. R. Soc. Lond. Ser. B,* 140, 520–528.

Parkes, A.S. (1958) Factors affecting the survival of frozen ovarian tissue, *J. Endocrinol.,* 17, 337–340.

Parkes, A.S. and Smith, A.U. (1953) Regeneration of rat ovarian tissue grafted after exposure to low temperatures, *Proc. Roy. Soc. Lond. Ser. B.,* 140, 455–470.

Parkes, A.S. and Smith, A.U. (1954) Preservation of ovarian tissue at –79°C for transplantation, *Acta Endocrinol.,* 17, 313–320.

Parrott, D.M. (1960) The fertility of mice with orthotopic ovarian grafts derived from frozen tissue, *J. Reprod. Fertil.,* 1, 230–241.

Patrizio, P. (2000) Cryopreservation of epididymal sperm, *Mol. Cell Endocrinol.,* 169, 11–14.

Payne, M. and Meyer, R. (1942) Endocrine function of ovarian tissue after growth or storage *in vitro, Proc. Soc. Exp. Biol. NY,* 51, 188–189.

Paynter, S.J. (2000) Current status of the cryopreservation of human unfertilized oocytes, *Hum. Reprod. Update,* 6, 449–456.

Paynter, S.J., Cooper, A., Fuller, B.J., and Shaw, R.W. (1999a) Cryopreservation of bovine ovarian tissue: Structural normality of follicles after thawing and culture *in vitro, Cryobiology,* 38, 301–309.

Paynter, S.J., Cooper, A., Gregory, L., Fuller B.J., and Shaw, R.W. (1999b) Permeability characteristics of human oocytes in the presence of the cryoprotectant dimethylsulphoxide, *Hum. Reprod.,* 14, 2338–2342.

Paynter, S.J., Cooper, A., Thomas, N., and Fuller, B.J. (1997a) Cryopreservation of multicellular embryos and reproductive tissues, in *Reproductive Tissue Banking,* Karow, A. and Critser, J., Eds., Academic Press, San Diego, pp. 359–397.

Paynter, S.J., Fuller, B.J., and Shaw, R.W. (1997b) Temperature dependence of mature mouse oocyte membrane permeabilities in the presence of cryoprotectant, *Cryobiology,* 34, 122–130.

Paynter, S.J., Fuller, B.J., and Shaw, R.W. (1999c) Temperature dependence of Kedem-Katchalsky membrane transport coefficients for mature mouse oocytes in the presence of ethylene glycol, *Cryobiology,* 39, 169–176.

Paynter, S.J., McGrath, J.J., Fuller, B.J., and Shaw, R.W. (1999d) A method for differentiating nonunique estimates of membrane transport properties: Mature mouse oocytes exposed to glycerol, *Cryobiology*, 39, 205–214.

Paynter, S.J., O'Neil, L., Fuller, B.J., and Shaw, R.W. (2001) Membrane permeability of human oocytes in the presence of the cryoprotectant propane-1,2-diol, *Fertil. Steril.*, 75, 532–538.

Pensis, M., Loumaye, E., and Psalti, I. (1989) Screening of conditions for rapid freezing of human oocytes: Preliminary study toward their cryopreservation, *Fertil. Steril.*, 52, 787–794.

Pickering, S.J., Braude, P.R., and Johnson, M.H. (1991) Cryoprotection of human oocytes: Inappropriate exposure to DMSO reduces fertilization rates, *Hum. Reprod.*, 6, 142–143.

Pickering, S.J., Braude, P.R., Johnson, M.H., Cant, A., and Currie, J. (1990) Transient cooling to room temperature can cause irreversible disruption of the meiotic spindle of the human oocyte, *Fertil. Steril.*, 54, 102–108.

Pickering, S.J. and Johnson, M.H. (1987) The influence of cooling on the organization of the meiotic spindle of the mouse oocyte, *Hum. Reprod.*, 2, 207–216.

Polak de Fried, E., Notrica, J., Rubinstein, M., Marazzi, A., and Gonzalez M. (1998) Pregnancy after human donor oocyte cryopreservation and thawing in association with intracytoplasmic sperm injection in a patient with ovarian failure, *Fertil. Steril.*, 69, 555–557.

Polge, C. and Lovelock, J. (1952) Preservation of bull sperm at –79°C, *Vet. Rec.*, 64, 396–397.

Polge, C., Smith, A.U., and Parkes, A. (1949) Revival of spematozoa after vitrification and dehydration, *Nat. (Lond.)* 164, 666.

Porcu, E., Fabbri, R., Damiano, G., Giunchi, S., Fratto, R., Ciotti, P.M., Venturoli, S., and Flamigni, C. (2000) Clinical experience and applications of oocyte cryopreservation, *Mol. Cell Endocrinol.*, 169, 33–37.

Porcu, E., Fabbri, R., Seracchioli, R., Ciotti, P.M., Magrini, O., and Flamigni, C. (1997) Birth of a healthy female after intracytoplasmic sperm injection of cryopreserved human oocytes, *Fertil. Steril.*, 68, 724–726.

Posada, M.N., Kolp, L., and Garcia, J.E. (2001) Fertility options for female cancer patients: Facts and fiction, *Fertil. Steril.*, 75, 647–653.

Queenan, J. (2000) Embryo freezing to prevent ovarian hyperstimulation syndrome, *Mol. Cell Endocrinol.*, 169, 79–84.

Quintans, C., Donaldson, M., Bertolino, M., and Pasqualina, R. (2002) Birth of two babies using oocytes that were cryopreserved in a choline-based freezing medium, *Hum. Reprod.*, 17, 3149–3152.

Radford, J., Lieberman, B., Brison, D., Smith, A., Critchlow, J., Russell, A., Clayton, J., Harris, M., Gosden, R., and Shalet, S. (2001) Orthotopic reimplantation of cryopreserved ovarian cortical strips after high-dose chemotherapy for Hodgkin lymphoma, *Lancet*, 357, 1172–1175.

Reis, M.M., Tsai, M.C., Schlegel, P.N., Feliciano, M., Raffaelli, R., Rosenwaks, Z., and Palermo, G.D. (2000) Xenogeneic transplantation of human spermatogonia, *Zygote*, 8, 97–105.

Reed, M., Lane, M., Gardner, D, Jensen, N., and Thompson, J. (2002) Vitrification of human blatsocysts using the cryoloop method: Successful clinical application and birth of offspring, *J. Assist. Reprod. Genet.*, 19, 304–306.

Reubinoff, B., Pera, M., Vajta, G., and Trounson, A. (2001) Effective cryopreservation of human embryonic stem cells by the open pulled straw vitrification method, *Hum. Reprod.*, 16, 2187–2194.

Richter, M.A., Haning, R.V., and Shapiro, S.S. (1984) Artificial donor insemination: Fresh versus frozen semen; the patient as her own control, *Fertil. Steril.*, 41, 277–280.

Salha, O., Picton, H., Balen, A., and Rutherford, A. (2001) Human oocyte cryopreservation, *Hosp. Med.*, 62, 18–24.

Sandalinas, M., Grossman, M., Egozcu, J., and Santalo, J. (1994) Blastomere disaggregation of cryopreserved two-cell zona free mouse embryos, *Cryo-Letters*, 15, 343–352.

Sathananthan, A.H., Henry, A., Trounson, A., Freeman, L., and Brady T. (1988) The effects of cooling human oocytes, *Hum. Reprod.*, 3, 968–977.

Sathananthan, A.H., Kirby, C., Trounson, A., Philipatos, D., and Shaw, J. (1992) The effects of cooling mouse oocytes, *J. Assist. Reprod. Genet.*, 9, 139–148.

Sathananthan, A.H., Trounson, A., and Freeman, L. (1987) Morphology and fertilizability of frozen human oocytes, *Gamete Res.*, 16, 343–354.

Schiller, J., Arnhold, J., Glander, H.J., and Arnold, K. (2000) Lipid analysis of human spermatozoa and seminal plasma by MALDI-TOF mass spectrometry and NMR spectroscopy—Effects of freezing and thawing, *Chem. Phys. Lipids*, 106, 145–156.

Schnorr, J., Muasher, S., and Jones, H. (2000) Evaluation of the clinical efficacy of embryo cryopreservation, *Mol. Cell Endocrinol.,* 169, 85–90.

Schuffner, A., Morshedi, M., and Oehninger, S. (2001) Cryopreservation of fractionated, highly motile human spermatozoa: Effect on membrane phosphatidylserine externalization and lipid peroxidation, *Hum. Reprod.,* 16, 2148–2153.

Shaw, J.M., Diotallevi, L., and Trounson, A. (1991) A simple rapid 4.5 M dimethyl-sulfoxide freezing technique for the cryopreservation of one-cell to blastocyst stage preimplantation mouse embryos, *Reprod. Fertil. Dev.,* 3, 621–626.

Shaw, J., Oranratnachai, A., and Trounson, A.O. (1999) Fundamental cryobiology of mammalian oocytes and ovarian tissue, *Theriogenology,* 53, 59–72.

Shibahara, H., Hamada, Y., Hasegawa, A., Toji, H., Shigeta, M., Yoshimoto, T., Shima, H., and Koyama, K. (1999) Correlation between the motility of frozen-thawed epididymal spermatozoa and the outcome of intracytoplasmic sperm injection, *Int. J. Androl.,* 22, 324–328.

Seibzehnrubl, E., Kohl, J., Dittrich, R., and Wildt, L. (2000) Freezing of human ovarian tissue—Not the oocytes but the granulosa is the problem, *Mol. Cell Endocrinol.,* 169, 109–111.

Smith, A.U. (1952) Culture of ovarian granulosa cells after cooling to very low temperatures, *Exp. Cell Res.,* 3, 574–583.

Son, W., Park, S., Lee, K., Lee, W., Ko, J., Yoon, T., and Cha K. (1996) Effects of 1,2-propanediol and freezing-thawing on the *in vitro* developmental capacity of human immature oocytes, *Fertil. Steril.,* 66, 995–999.

Son, W., Yoon, S., Park, S., Yoon, H., Lee, W., and Lim, J. (2002) Ongoing twin pregnancy after vitrification of blastocysts by *in vitro* matured oocytes retreived from a woman with polycystic ovary syndrome: Case Report, *Hum. Reprod.,* 17, 2963–2966.

Spano, M., Cordelli, E., Leter, G., Lombardo, F., Lenzi, A., and Gandini, L. (1999) Nuclear chromatin variations in human spermatozoa undergoing swim-up and cryopreservation evaluated by the flow cytometric sperm chromatin structure assay, *Mol. Hum. Reprod.,* 5, 29–37.

Spears, N., Boland, N., Murray, A., and Gosden, R.G. (1994) Mouse oocytes derived from *in vitro* grown primary ovarian follicles are fertile, *Hum. Reprod.,* 9, 527–532.

Stachecki, J.J., Cohen, J., and Willadsen, S.M. (1998) Cryopreservation of unfertilized mouse oocytes: The effect of replacing sodium with choline in the freezing medium, *Cryobiology,* 37, 346–354.

Stahler E., Sturm, G., Spottling, L., Daume, E., and Bucholz, R. (1976) Investigations into the cryopreservation of human ovaries by means of a cryoprotectant, *Arch. Gynokol.,* 226, 339–344.

Stanic, P., Tandara, M., Sonicki, Z., Simunic, V., Radakovic, B., and Suchanek, E. (2000) Comparison of protective media and freezing techniques for cryopreservation of human semen, *Eur. J. Obstet. Gynecol. Reprod. Biol.,* 91, 65–70.

Stewart, G., Tyler, J., Cunningham, A.L., Barr, J.A., Driscoll, G.L., Gold, J., and Lamont, B.J. (1985) Transmission of human T-cell lymphotrophic virus type III (HTLV-III) by artificial insemnination by donor, *Lancet,* 2, 581–585.

Sugimoto, H., Miyamoto, H., Kabasawa, T., and Manabe, N. (1996) Follicle survival in neonatal rat ovaries cryopreserved by vitrification, *Cryo-Letters,* 17, 93–98.

Sugimoto, M., Maeda, S., Manabe, N., and Miyamoto, H. (2000) Development of infantile rat ovaries after cryopreservation by vitrification, *Theriogenology,* 15, 1093–1103.

Sutcliffe, A. (2000) Follow up of children conceived from cryopreserved embryos, *Mol. Cell Endocrinol.,* 169, 91–94.

Sztein, J., O, M., Farley, J., Mobraaten, L., and Eppig, J. (2000) Rescue of oocytes from antral follicles of cryopreserved mouse ovaries: Competence to undergo maturation, embrygenesis and development to term, *Hum. Reprod.,* 15, 567–571.

Tao, J., Tamis, R., and Fink, K. (2001) Cryopreservation of mouse embryos at morula/compact stage, *J. Assist. Reprod. Genet.,* 18, 235–243.

Tesarik, J., Sousa, M., Greco, E., and Mendoza, C. (1998) Spermatids as gametes: Indications and limitations, *Hum. Reprod.,* 13, 89–107.

Tesarik, J., Mendoza, C., Anniballo, R., and Greco, E. (2000) In-vitro differentiation of germ cells from frozen testicular biopsy specimens, *Hum. Reprod.,* 15, 1713–1716.

Testart, J., Lassalle, E., Belaisch-Allart, J., Hazout, A., Forman, R., Rainhorn, J., and Frydman, R. (1986) High pregnancy rate after early human embryo freezing, *Fertil. Steril.,* 46, 268–272.

Thomas, N., Busza, A., Cooper, A., Paynter, S., Fuller, B., and Shaw, R. (1997) Measurement of permeating levels of cryoprotectant during ovarian tissue cryopreservation using ^{1}H NMR spectroscopy in human and porcine ovaries, *Cryo-Letters,* 18, 179–184.

Thompson, L., Srikantharajah, A., Hamilton, M., and Templeton, A. (1995) A comparison of the effects of different biopsy strategies on the post-thaw survival of 8-cell stage mouse embryos: Implications for preimplantation diagnosis, *Hum. Reprod.,* 10, 659–663.

Tiitinen, A., Halttunen, M., Harkki, P., Vuoristo, P., Hyden-Granskog, C. (2001) Elective single embryo transfer: The value of cryopreservation, *Hum. Reprod.,* 16, 1140–1144.

Todorow, S., Seibzehnrubl, E., Koch, R., Wildt, L., and Lang, N. (1989) Comparative results on survival of human and animal eggs using different cryoprotectants and freeze-thawing regimens II Human, *Hum. Reprod.,* 4, 812–816.

Tomlinson, M. and Sakkas, D. (2000) Is a review of standard procedures for cryopreservation needed? Safe and effective cryopreservation—should sperm banks and fertility centres move toward storage in nitrogen vapour? *Hum. Reprod.,* 15, 2460–2463.

Toth, T.L., Baka, S.G., Veeck, L.L. Jones, H.W., Muasher, S., and Lanzendorf, S.E. (1994a) Fertilization and *in vitro* development of cryopreserved human prophase I oocytes, *Fertil. Steril.,* 61, 891–894.

Toth, T.L., Lanzendorf, S.E., Sandow, B.A., Veeck, L.L., Hassen, W.A., Hansen, K., and Hodsgen, G.D. (1994b) Cryopreservation of human prophase oocytes collected from unstimulated follicles, *Fertil. Steril.,* 61, 1077–1982.

Tournaye, H., Merdad, T., Silber, S., Joris, H., Verheyen, G., Devroey, P., and van Steirteghem, A. (1999) No differences in outcome after intracytoplasmic sperm injection with fresh or with frozen-thawed epididymal spermatozoa, *Hum. Reprod.,* 14, 90–95.

Trad, F.S., Toner, M., and Biggers, J.D. (1999) Effects of cryoprotectants and ice-seeding temperature on intracellular freezing and survival of human oocytes, *Hum. Reprod.,* 14, 1569–1577.

Trombetta, C., Liguori, G., Gianaroli, L., Magli, M.C., Selman, H.A., Colpi, G., Belgrano, E., Vitali, G., and Ferraretti, A.P. (2000) Testicular sperm extraction combined with cryopreservation of testicular tissue in the treatment of azoospermia, *Urol. Int.,* 65, 15–20.

Trounson, A. and Mohr, L. (1983) Human pregnancy following cryopreservation. Thawing and transfer of an eight-cell embryo, *Nature,* 305, 707–709.

Trounson, A.O., Pushett, D., MacLellan, L.J., Lewis, I., and Gardener, D.K. (1994) Current status of IVM/IVF and embryo culture in humans and farm animals, *Theriogenology,* 41, 57–66.

Tucker, M.J., Morton, P.C., Wright, G., Sweitzer, C.L., and Massey, J.B. (1998a) Clinical application of human egg cryopreservation, *Hum. Reprod.,* 3156–3159.

Tucker, M.J., Wright, G., Morton, P.C., and Massey, J.B. (1998b) Birth after cryopreservation of immature oocytes with subsequent *in vitro* maturation, *Fertil. Steril.,* 70, 578–579.

van den Abbeel, E., Camus, M., Joris, H., and van Steirteghem, A. (2000) Embryo freezing after intracyto-plasmic sperm injection, *Mol. Cell Endocrinol.,* 169, 49–54.

van den Abbeel, E., Camus, M., van Waesberghe, L., Devroey, P., and van Steirteghem, A. (1997) A randomized comparison of the cryopreservation of one-cell human embryos with a slow controlled-rate cooling procedure or a rapid cooling procedure by direct plunging into liquid nitrogen, *Hum. Reprod.,* 12, 1554–1560.

van den Abbeel, E. and van Steirteghem, A. (2000) Zona pellucida damage to human embryos after cryo-preservation and the consequences for their blastomere survival and *in-vitro* viability, *Hum. Reprod.,* 15, 373–378.

van der Elst, J., Camus, M., van den Abbeel, E., Maes, R., Devroey, P., and van Steirteghem, A.C. (1995) Prospective randomized study on the cryopreservation of human embryos with dimethylsulfoxide or 1,2-propanediol protocols, *Fertil. Steril.,* 63, 92–100.

van Steirteghem, A., van den Abbeel, E., Camus, M., van Waesberghe, L., Braeckmans, P., Khan, I., Nijs, M., Smitz, J., Staessen, C., Wisanto, A., and de Vroey, P. (1987) Cryopreservation of human embryos obtained after gamete intra-fallopian transfer and/or *in vitro* fertilisation, *Hum. Reprod.,* 7, 593–598.

van Uem, J.F.H.M., Siebzehnrubl, E.R., Schuh, B., Koch, R., Trotnow, S., and Lang, N. (1987) Birth after cryopreservation of unfertilized oocytes, *Lancet,* 2, 752–753.

Wang, A.W., Zhang, H., Ikemoto, I., Anderson, D.J., and Loughlin, K.R. (1997) Reactive oxygen species generation by seminal cells during cryopreservation, *Urology,* 49, 921–925.

Watson, P.F. (1995) Recent developments and concepts in the cryopreservationof spermatozoa and the assessment of their post-thawing function, *Reprod. Fert. Dev.*, 7, 871–891.

Weissman, A., Gotlieb, L., Coglan, T., Juriscova, A., Greenblatt, E., and Caspar, R. (1999) Preliminary experience with subcutaneous human ovarian cortex transplantation in the NOD-SCID mouse, *Biol. Reprod.*, 60, 1179–1182.

Whittingham, D., Leibo, S., and Mazur, P. (1972) Survival of mouse embryos frozen to –196°C and –269°C, *Science,* 178, 411–414.

Wilton, L., Shaw, J., and Trounson, A. (1989) Successful single-cell biopsy and cryopreservation of preimplantation mouse embryos, *Fertil. Steril.,* 51, 513–517.

Wolf, D.P., Patton, P.E., Burry, K.A., and Kaplan, P.F. (2001) Intrauterine insemination-ready versus conventional semen cryopreservation for donor insemination: A comparison of retrospective results and a prospective, randomized trial, *Fertil. Steril.,* 76, 181–185.

Wu, J., Zhang L., and Wang X. (2001) *In vitro* maturation, fertilisation and embryo development after ultra-rapid freezing of immature human oocytes, *Reproduction,* 121, 389–393.

Yang, D.S., Blohm, P.L., Cramer, L., Nguyen, K., Zhao, Y.L., and Winslow, K.L. (1999) A successful human oocyte cryopreservation regime: Survival, implantation and pregnancy rates are comparable to that of cryopreserved embryos generated from sibling oocytes, *Fertil. Steril.,* 72, S86.

Yang, D.S., Blohm, P.L., Winslow, K.L., and Cramer, L. (1998) A twin pregnancy after microinjection of human cryopreserved oocyte with a specially developed oocyte cryopreservation regime, *Fertil. Steril.,* 70, S239.

Yokota, Y., Sato, S., Yokota, M., Ishikawa, Y., Makita, M., Asada, T., and Araki, Y. (2000) Successful pregnancy following blastocyst vitrification: Case report, *Hum. Reprod.,* 15, 1802–1803.

Yoon, T.K., Chung, H.M., Lim, J.M., Han, S.Y., Ko, J.J., and Cha, K.Y. (2000) Pregnancy and delivery of healthy infants developed from vitrified oocytes in a stimulated *in vitro* fertilization-embryo transfer program, *Fertil. Steril.,* 74, 180–181.

Younis, A.I., Toner, M., Albertini, D.F., and Biggers, J.D. (1996) Cryobiology of non-human primate oocytes, *Hum. Reprod.,* 11, 156–165.

Zenzes, M.T., Bielecki, R., Casper, R.F., and Leibo, S.P. (2001) Effects of chilling to 0°C on the morphology of meiotic spindles in human metaphase II oocytes, *Fertil. Steril.,* 75, 769–777.

19 The Scientific Basis for Tissue Banking

Monica Wusteman and Charles J. Hunt

CONTENTS

19.1 INTRODUCTION

Clinical demand for human tissue grafts continues to grow both in volume and in the range of tissues required for transplantation. At present, however, banking is limited to tissues that either do not require live cells for optimum function (e.g., bone, tendon) or that are sufficiently simple in structure to survive rudimentary freezing techniques. These methods frequently consist of either uncontrolled freezing in a –80°C freezer or use protocols adapted from those developed for use with suspensions of cells of a type often unconnected with the tissues themselves. In fact, it can be said that, with little exception, the introduction of new tissue products has progressed with little *a priori* research or understanding of the effects of freezing on tissue structure and function and in ignorance of a growing body of research into the cryobiology of freezing injury in multicellular systems.

The overriding objective of tissue banking is to preserve the tissue in such a state as to enable it to perform its function to a clinically acceptable level on transplantation. The presence of living cells is only required for some tissues. If live cells are required, methods have to be devised by which the cellular function, or viability, of the stored tissue can be assessed. These methods are required to optimize methods of preservation and for the purposes of quality control. Briefly, to be of value, the assay chosen should reflect some active function of the tissue *in vivo* (e.g., muscle contraction, active membrane ion transport, hormone release) rather than a single process (e.g., respiratory activity), which may not reflect damage to other cellular functions. Aspects of structural integrity (e.g., exclusion of vital stains by the cell membrane) are also very limited in usefulness for anything other than preliminary experimental observations. The more complex the process identified by the assay, the more likely it is to be capable of revealing incipient cell or tissue damage. Viability assays based on the ability of the stored cells to divide (Wusteman and Pegg, 2001) will therefore provide a rigorous test of cell function *in vivo*. Discussion of the definition of tissue viability, and the nature of appropriate assays to determine it, are beyond the scope of this review. A symposium on the subject was published in 1989 that addressed the problem of identifying suitable assays for a whole range of banked cells and tissues (Pegg, 1989; Southard, 1989). Although this symposium took place more than 14 years ago, the requirements have not changed and the problems addressed still remain today.

19.2 BANKING OF TISSUES WHERE CELLULAR VIABILITY IS NOT A PREREQUISITE

19.2.1 BONE AND TENDON

It has been estimated that some 150,000 bone allografts are used currently in the United States each year, making allograft bone by far the most implanted tissue (Tomford and Mankin, 1999). Tendons (usually Achilles and patella) are also banked and used most often for anterior and posterior cruciate ligament repair. Experimental studies have confirmed an immune reaction to allograft bone, the clinical significance of which is not clear. However, both freezing and freeze-drying have been shown to reduced immunogenicity (Stevenson, 1999). The methods used to process and store bone allografts and tendon—freezing (in an uncontrolled manner, in the absence of a cryoprotectant with storage at –80°C) and freeze-drying, as well as the common methods of sterilization (gamma irradiation, ethylene oxide)—render the cells within the graft nonviable. This has consequences for the ultimate performance of the graft (Boyce et al., 1999).

Allograft bone, processed by the methods routinely used in the bone bank, is not osteogenic, as living cells do not reside in the graft after processing. Both cortical bone, used to provide structural support, and morcellized cancellous bone, used for impaction grafting in hip revision surgery and elsewhere to fill bone defects, are osteoconductive but not osteoinductive, as the bone matrix remains mineralized. The graft provides a three-dimensional trellis, permitting vascular ingrowth and penetration of osteoprogenitor cells from the host site. Demineralized bone produced by methods that involve acid treatments is both osteoconductive and osteoinductive: Bone morphogenic proteins stimulate recruitment and differentiation of pluripotent cells from the host bed. Although small amounts of cancellous allograft are usually completely remodeled and incorporated into the host site, large cortical allografts are incorporated generally only at the junction with the host site, and the remaining bone may never be remodeled, with consequent weakening of the graft over time.

Methods to improve osteogenesis by cryopreservation may not be beneficial. One recent study has indicated that the use of cryoprotectants may be detrimental to the long-term integrity of the implant (Wohl et al., 1998).

19.2.2 AMNIOTIC MEMBRANE

Both frozen and glycerolized amniotic membrane have been used as a temporary biological dressing in the treatment of burns and in reconstructive surgery (Chang and Yang, 1994; Maral et al., 1999).

However, its ineffectiveness in reducing fluid loss caused by a lack of a multilayer epithelium has seen it largely superseded in Western Europe and the USA by allogeneic skin grafts (see following). Its use in the treatment of a wide variety of ocular surface defects, particularly in conjunction with limbal stem cell transplantation, is growing (Sippel et al., 2001). Optimal preservation and storage conditions are unknown. At present, the grafts are frozen in a 50% glycerol, balanced salts solution according to the protocol outlined by Kim and Tseng (1995). The grafts frozen by this protocol are nonviable (Adds et al., 2001), and their effectiveness in stimulating reepithelialization of the ocular surface appears, at least in part, to be the result of the preservation of the amniotic subepithelial basement membrane (Koizumi et al., 2000).

19.3 BANKING OF TISSUES WHERE CELLULAR VIABILITY IS OR IS PRESUMED TO BE AN ESSENTIAL REQUIREMENT FOR OPTIMUM FUNCTION

19.3.1 CORNEAS

Corneal grafting is one of the most frequent and successful of all allograft procedures. It is carried out for a whole range of conditions in which the cornea has been damaged by either trauma or disease. The immune response generated by this tissue is only significant in a minority of clinical cases, and over 90% of procedures are successful in restoring sight. As a consequence, the banking of corneas is a relatively well-regulated and well-organized process both in Europe and the United States. Collection and transplant data can be found at http://www.restoresight.org (United States), http://www.eeba.net (Europe), and http://www.uktransplant.org.uk (United Kingdom).

The cornea is a relatively simple but highly specialized tissue consisting of an outer epithelial and an inner endothelial layer separated by the stroma, which occupies 90% of the thickness of the tissue (~550 μm for human cornea). The stroma consists of sheets of collagenous fibrils, or lamellae, embedded in a ground substance of glycosaminoglycans. Scattered amongst the lamellae are flattened modified fibroblasts, the keratocytes. The endothelium, which sits on the structureless Descemets' membrane, maintains the degree of stromal hydration. Acting not only as a physical barrier to the passive diffusion of water and solutes but also as an active metabolic bicarbonate pump driven by Na+/K+ ATPase, the endothelium maintains the transparency of the cornea, which is essential for normal vision. Damage to the endothelium leads to stromal swelling and corneal opacity. A corneal graft, therefore, needs an intact and functional endothelial layer. The human corneal endothelium does not possess any measurable mitotic activity, and damage is repaired by cell enlargement rather than division. Moreover, endothelial cells continue to be lost from grafts after transplantation at a faster rate than that associated with ageing (Bourne et al., 1994). Maintenance of endothelial cell density during storage is therefore vital.

Storage of the intact eye within a moist chamber at 4°C limits preservation to less than 48 h. Alternatively, storage of excised corneas either at 4°C or by organ culture at 31 to 37°C in media adapted to optimize endothelial integrity can extend storage time substantially (Chu, 2000; Moller Pedersen et al., 2001). At present, 4°C storage, the method used by eye banks in the United States, permits storage for up to 2 weeks (Chu, 2000), whereas organ culture, the technique used by the majority of eye banks in Europe, allows storage for 1 month or even longer (Armitage and Easty, 1997; Ehlers et al., 1999). This method also permits increased microbiological safety compared with hypothermic storage. Storage at 4°C has been shown to cause disruption of the f-actin cytoskeleton and apical junctional complex, which may contribute to corneal swelling during storage (Hsu et al., 1999). Organ culture, in contrast, maintains the integrity of tight junctions and the actin cytoskeleton in both endothelial and epithelial cell layers (Crewe and Armitage, 2001). Nevertheless, for the majority of U.S. eye banks, the advantages of organ culture are considered to be outweighed by what are believed to be greater technical demands and costs. To date, there have not been any differences reported between the two techniques with respect to clinical outcome (Frueh and Böhnke, 2000; Rijneveld et al., 1992).

The role of apoptosis in cell loss during storage has recently attracted some attention. It has been reported in both the epithelium and endothelium of organ-cultured corneas (Crewe and Armitage, 2001) as well as in corneas stored at 4°C (Komuro et al., 1999). Results for organ-cultured endothelium, however, remain contradictory (Albon et al., 2000). Further studies are needed to determine the role that this mechanism plays in cell survival during storage.

For almost a decade, the medium of choice for 4°C storage of corneas in the U.S. has been Optisol™ (Bausch and Lomb Surgical, San Dimas, CA). This is a HEPES buffered solution based on TC 199 tissue culture medium, which is supplemented with antibiotics and with dextran and chondroitin sulphate (Chu, 2000). A modification of this has become available (Chen and Chen, 1994). The most significant change has been the supplementation with the ketone body, β-hydroxybutyrate, to reduce lactate formation and maintain ATP levels and metabolic activity during cold storage. The limited trials carried out to date with Chen medium have not shown any benefit in either indices of preservation (Nelson et al., 2000; Yap et al., 2001) or in clinical outcome (Bourne et al., 2001).

Cryopreservation is currently the only technique that would permit the longer periods of storage necessary to make optimal use of the limited number of available grafts and to allow for tissue typing. Early reports in the 1960s of successful transplantation of cryopreserved grafts were soon followed by laboratory and clinical studies showing an unacceptable degree of endothelial injury following freezing (Taylor, 1986). Therefore, at present, cryopreserved corneas are used only rarely, and then only in emergencies when fresh tissue is unavailable (Brunette et al., 2001). Despite a substantial body of experimental work carried out over recent years aimed at avoidance of freezing injury in corneas (see Taylor, 1986; Wusteman et al., 1997; Bourne et al., 1999), the most successful method for cryopreservation (cooling at 1°C/min in 20% w/w dimethyl sulphoxide, added and removed in stages to limit osmotic stress, and rapid thawing; see Wusteman et al., 1997) does not retain sufficient endothelial integrity to justify its clinical use. Freeze substitution of corneas cryopreserved with dimethyl sulphoxide as detailed above has revealed the presence of ice throughout the tissue (see Figure 19.1), although not in the zone immediately under the endothelium, which remained physically intact and attached. The reason for the irregular pattern of ice crystal size throughout the stroma is not understood but is likely to be related, at least in part, to the anterior–posterior variation in stromal hydration. The consequences for graft function of this disruption of the stroma by ice have not been evaluated. It is reasonable to assume that repair to the stroma can occur through the action of surviving keratocytes, although infiltration of keratocytes from host tissue is also a possibility (Jakobiec et al., 1981). Again, little research has been carried out into the optimum conditions for keratocyte preservation, which may differ significantly from those of the endothelium (Cheng et al., 1996; Borderie et al., 1998).

At present, the avoidance of ice altogether by vitrification is being investigated as a means of corneal cryopreservation and some promising results have been achieved. Corneal endothelium can withstand exposure to vitrifiable concentrations of propane-1,2-diol (Rich and Armitage, 1991) and shows some signs of recovered function after vitrification in this medium (Armitage et al., 2002). This approach, whereby a single vitrifying solute is used in preference to complex mixtures of solutes, is relatively novel and is also being pursued in other studies on larger tissues and organs (Wusteman et al., 2002). The use of electromagnetic fields for the rewarming of bulky tissues is also applicable to small tissues like the cornea for uniform rewarming of the tissue at rates sufficiently rapid to avoid devitrification (Armitage et al., 2002). The results using this approach indicate that long-term storage of corneas at low temperatures may yet be a possibility.

19.3.2 VASCULAR GRAFTS

In recent years there has been an increased demand for small-diameter vascular grafts for use in cardiac bypass and lower limb salvage procedures when the preferred material, autologous saphenous vein or

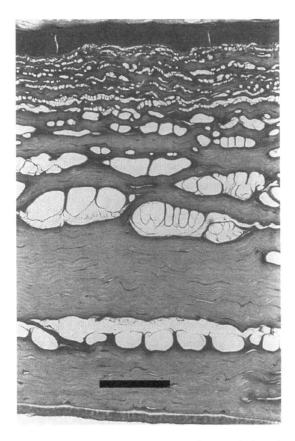

FIGURE 19.1 Ice in a cryopreserved cornea. A rabbit cornea, freeze substituted at –80°C after cryopreservation in 2 *M* dimethyl sulphoxide with cooling at 1°C/min to –196°C in air (without bulk medium). Note uneven pattern of ice formation (white spaces) with crystal size increasing from the epithelium toward the inner endothelial layer, which is intact and fully attached. Bar = 50 μM. (Micrograph prepared by C.J. Hunt.)

the internal mammary (internal thoracic) artery, is not available. Cryopreservation is the only prospect for their long-term storage. To date, the use of these grafts is limited to mainland European centers, from which clinical results are becoming available. Blood vessel allografts are therefore the most complex tissue currently being banked. Exceptionally, there is also a substantial body of data from experimental animal research into the nature of freezing injury in vascular tissue, on its prevention, and on the function of these cryopreserved grafts after transplantation. This has been studied extensively using both *in vitro* and *in vivo* techniques for viability testing. *In vitro* techniques (smooth muscle contractility, endothelial-dependant smooth muscle relaxation), though providing no insight into the degree of function that the tissue can exhibit after transplantation, have been of substantial value in the optimization of the cryopreservation processes in studies with animal tissue. Using these techniques, it has been possible to optimize a number of the parameters in the cryopreservation process, such as extraction and handling of the tissue, choice of cryoprotectant and its concentration, vehicle solution, cooling and rewarming protocols, and so forth (for review, see Wusteman et al., 2000), to maintain optimum postthaw function and to avoid the development of the fractures that otherwise frequently develop during thawing in cryopreserved vascular conduits (Hunt et al., 1994; Pegg et al., 1997; Wassenaar et al., 1995). Thus, it is now possible to devise a clinically practical method for the banking of vascular grafts for clinical transplantation (Wusteman et al., 2000). Studies with human blood vessels have confirmed that the presence of serum in the cryopreservation medium is unnecessary (Müller-Schweinitzer et al., 1997; Schilling et al., 1995), a fact that has important implications for the avoidance of disease transmission in the banking of vascular grafts. Recently, significant success has been achieved in attempts to cryopreserve small sections of blood vessel by an ice-free method, or vitrification. Early graft results are encouraging and the details of the approach are discussed in Chapter 22. However, it remains to be seen whether this method would offer any advantage over the current conventional approach to the banking of vascular conduits for clinical transplantation.

In vitro testing of arteries and veins from several species including human confirm that, even when controlled cryopreservation techniques are used, significant loss of smooth muscle contractility occurs following cryopreservation. The extent of the loss depends on species, vessel, and the agonist used to elicit the contraction, but it can be up to 50%. The degree of retention of endothelium-dependant vasodilatory responses also varies (see Wusteman et al., 2000). Such effects may or may not be reversible. Pharmacological studies in human internal mammary artery (IMA) indicate that activation of protein kinase C and a resulting increase in calcium channel influx could be responsible for some of these changes in pharmacological responsiveness of vessels after cryopreservation (Müller-Schweinitzer et al., 1998, 2000). Such an understanding of the molecular mechanisms involved could lead to the development of therapeutic interventions to improve posttransplantation graft function (Müller-Schweinitzer et al., 1998). Good preservation of mechanical properties of frozen vessels has been reported (Adham et al., 1996; Rosset et al., 1996).

Cryoinjury is only one of the many factors that will determine the patency of cryopreserved allografts. The anatomical origin of the graft is also of primary importance because it will determine both physical characteristics (diameter, medial thickness, compliance) and the pharmacological receptor profile of the smooth muscle and of the endothelium, the integrity of which plays such a vital role in the regulation of vascular tone (Dzimiri et al., 1996). For example, when saphenous vein is grafted into the arterial circulation, major morphological changes occur in the graft, leading to thickening and eventual fibrosis of the arterial wall (Davies and Hagen, 1994). This, together with changes in endothelial responsiveness and smooth muscle contractility, has been implicated in the inferior long-term patency of fresh veins compared with arterial grafts (Loop et al., 1986). Internal mammary artery has proven to be the most effective source of autograft to date.

All living vascularized allografts, whether fresh or cryopreserved, will inevitably stimulate an immunological response. Clinical results are complicated by factors such as variation in the clinical status of the recipients, drug regime, surgical methods, and so forth, and therefore they give limited insight in to the causes of graft failure. Animal experiments allow these factors to be controlled. A comparison of the fate of allograft and autograft veins in a variety of experimental models indicates a major role for immunological factors in the failure of vessel allografts (Neves et al., 1997; Miller et al., 1993; Vischjager et al., 1996; Wusteman et al., 2000). Little is known about the interaction of cryopreservation with the systems involved in the immune response in any tissue. These are likely to be complex, and therefore, any meaningful experimental investigation into these interactions is going to be very difficult to devise.

Cryopreserved allogenic saphenous veins, and aortic allografts, are commercially available in the United States from CryoLife Cardiovascular. The cryopreservation process resembles the method described earlier in several important aspects and produces a comparable degree of post-thaw function. Early clinical experience with these vessels in both coronary and peripheral grafting procedures appeared promising (Brockbank et al., 1992), but other centers have since reported unsatisfactory clinical results in both coronary bypass and limb salvage procedures, particularly after long-term follow up (Harris et al., 2001). Comparisons with the results with prosthetic grafts are ambiguous, as is the role of rejection in the clinical outcome. Some centers have report a high incidence of signs of rejection in failed grafts, whereas others have found signs in as little as 30% of cases. Elsewhere, matching for ABO blood group and immunosuppression has also failed to improve graft survival rates. For a review of the data on clinical transplantation of cryopreserved veins, see Wusteman et al. (2000).

Cryopreserved arterial allografts are currently being evaluated in a range of vascular reconstructive procedures in a number of European centers. In general, the clinical experience to date is more promising, although the small number of patients in many of the studies limits their interpretation. In one study of 100 patients receiving large, cryopreserved arteries in reconstructive surgery for the treatment of a number of conditions, the cryopreserved arterial grafts were found to provide a better outcome than prosthetic grafts in nonimmunosuppressed patients with vascular infections (Goffin et al., 1998). The authors indicate that either the reduced immunogenicity of the cryopreserved grafts or their reduced resistance to bacterial infection was responsible for their relative success. The maintenance of patency is more difficult for small-diameter vascular grafts, in particular those needed for below-the-knee revascularization for limb salvage. Early clinical results in patients with severe distal arterial occlusive disease indicate that cryopreserved allografts are a valuable alternative to prosthetic grafts when autologous material is not available. These grafts appear to be particularly effective in immunosuppressed patients and when ABO matching is carried out. (Gournier et al., 1995; da Gama et al., 1994). One application of cryopreserved vessels in which rejection is probably less relevant is the use of cryopreserved conduits for vascular access for hemodialysis in patients in whom access has been lost because of infection. Early results with these grafts give promising results, in particular with avoidance of infection (Matsuura et al., 2000).

The present failure rate of cryopreserved allograft vessels appears to be immunological rather than cryobiological in origin, and in most cases arteries appear to be preferable to veins. However, more controlled data are urgently required before any firm conclusions can be drawn. If current trials confirm these initial conclusions, and the surgical community requires them, there will be a need for the tissue banking community to provide these tissues on a regular basis.

19.3.3 ARTICULAR CARTILAGE AND OSTEOCHONDRAL ALLOGRAFTS

Joint defects requiring cartilage repair are increasingly common conditions, and considerable effort is now being directed toward the improvement of methods of grafting this tissue. The clinical importance of the immune response to cartilage allografts is not understood, but there is no evidence yet to indicate that it plays a major role in graft failure (Czitrom et al., 1986), whereas allograft meniscal transplantation is a procedure that has already been shown to have significant short-term clinical benefits (Rodeo, 2001). A bank of allograft cartilage, or tissue-engineered equivalents, would therefore be of tremendous benefit for a range of orthopedic procedures. The use of cryopreserved cartilage is, however, not routine, because there is not yet a method for cryopreservation that maintains adequate cell viability throughout the tissue. Cartilage is a very atypical tissue, with no blood vessels or nerve supply. The cells in cartilage (chondrocytes) only occupy 5% of its total volume, but they are responsible for the maintenance of the extensive surrounding matrix, which comprises a highly complex network of collagen fibrils, associated proteoglycans, and other non-collagenous proteins. Because these give the tissue its unique mechanical properties, and little or no cell division is seen in mature cartilage, it is assumed that viable cells are necessary for optimum graft function. Support for this assumption is to be found in results of experimental transplantation in which the long-term function of transplanted cartilage has been was found to be proportional to number of living chondrocytes in the graft (Malinin et al., 1985; Schachar et al., 1999; Tomford et al., 1992).

Attempts have been made to cryopreserve cartilage for almost 50 years. Chondrocytes, isolated from the matrix by enzyme digestion, survive cryopreservation by a number of different procedures (Smith, 1965; Schachar et al., 1989; Tomford et al., 1984), but the same techniques produce unsatisfactory cell survival in the intact tissue (Schachar and McGann, 1986). This problem may be related to differences in cell morphology, matrix composition, or the confinement of the chondrocytes inside rigid capsules, or lacunae. For example, it has been calculated that cells frozen to $-50°C$ in 10% Me_2SO will shrink by approximately 50% in volume (Pegg et al., 1987). The results of freeze-substitution studies at $-80°C$ with cartilage after cryopreservation in 10% Me_2SO indicate that chondrocytes appear to remain tethered to parts of the capsule during freeze-induced shrinkage, thereby producing additional stresses on the cells (Figure 19.2). Furthermore, excessive swelling

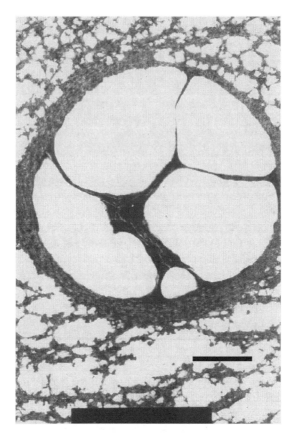

FIGURE 19.2 Cryopreserved ovine cartilage. Freeze-substituted chondrocyte and surrounding matrix follow-ing freezing to –80°C in 10% dimethyl sulphoxide. Note shrunken cell, which remains tethered to the surrounding capsule. The large spaces were formerly occupied by ice. Bar = 2 μM. (Micrograph prepared by C.J. Hunt.)

during removal of the cryoprotectant after preservation could also induce damaging hydrostatic pressures within the cell. Either—or both—of these factors may well be responsible for the excessive damage to chondrocytes found after cryopreservation of intact cartilage. Strategies to avoid exces-sive excursions in cell volume during addition and removal of cryoprotectant have already been devised for other cells (Wusteman and Pegg, 2001). If applied to cartilage, such an approach may produce a substantial improvement in survival of the chondrocytes in intact cartilage.

Experiments with animal tissue have established that Me_2SO is the least toxic cryoprotectant for chondrocytes, both in suspension and in intact tissue. (Schachar and McGann, 1986; Yang and Zhang, 1991). Chondrocytes in intact cartilage respond to variations in cooling rate in the same manner as other cell types, with an optimum cooling rate of between 1° and 4°C/min. This permits maximum dehydration while at the same time minimizing the time and temperature of exposure to hypertonic conditions. Several studies have shown what appear to be regional differences in survival of chondrocytes in intact cartilage, the cells of the intermediate layer being more susceptible to crypreservation injury than the cells of the superficial or deep zones (Muldrew et al., 1994; Ohlendorf et al., 1996). Permeation of the Me_2SO through the matrix is too rapid to provide an explanation for this phenomenon (Pegg, 1998), but regional differences in water, collagen, and proteoglycan content (Mow et al., 1992) or regional differences in chondrocyte susceptibility to freezing injury may be responsible. Certainly, chondrocytes differ in their ultrastructural appearance in the three zones (Tavakol et al., 1993). It has also been proposed that the damage to the intermediate layer of cryopreserved cartilage is the result of high solute concentrations caused by the advance of planar ice fronts from the surfaces of the tissue (Muldrew et al., 2000), but this hypothesis has not yet been tested in other studies. There is the possibility that cell damage is secondary to damage to the pericellular matrix, which controls chondrocyte metabolic activity. The degenerative changes of osteoarthritis are initiated by changes in the interaction between the cell and the pericellular matrix, leading to matrix degradation and eventually to the appearance of chondrocyte clusters resulting from chondrocyte proliferation. Similar clusters have been observed in cryopreserved ovine cartilage 12 months after transplantation (Muldrew et al., 2001).

Most experimental studies rely on *in vitro* assessments of functionality. Clearly, transplantation is the ultimate surgical test. However, the technical problems associated with whole-tissue allografting (mismatches in graft size, surgical and immunological variables, postoperative immobilization) are numerous and lead to difficulties in interpretation of the results of animal studies (Hurtig et al., 1998; Schachar et al., 1992). The additional problem of wide variability of ovine articular cartilage has been noted when metabolic indicators of cell recovery are used (D.E. Pegg, personal communication). Similar problems associated with experimental studies are well illustrated in a recent transplantation study with osteochondral dowels carried out in crossbred ewes (Muldrew et al., 2001).

Not withstanding the poor cryopreservation of chondrocytes *in vivo*, however, the banking and implantation of osteochondral allografts has been carried out for many years for the repair of larger defects, which involve more extensive damage to the articular surface, particularly those to the subchondral bone (Gross et al., 1983; Aho et al., 1997). Massive allografts, both with and without an articular surface, have been shown to be effective in the repair of large skeletal defects produced during the resection of a range of bone and connective tissue tumors as well as other conditions affecting the skeleton (Mankin et al., 1996).

The success of "fresh" allografts for osteochondral transplantation has been well documented (Aubin et al., 2001). Chondrocyte viability is maintained in these grafts during the storage period (generally no longer than 72 h at 4°C) and persists for many years postimplantation. Two recent follow-up studies of transplanted osteochondral shell allografts demonstrate that the repair of even larger articular surface defects is possible (Chu et àl., 1999; Salai et al., 1997). However, the use of "fresh" allografts poses a number of problems: coordination of donor and recipient operations, the nonelective nature of the surgery, and the risk of disease transmission. Although the transplantation of articular cartilage allografts does not appear to elicit a damaging immunogenic response, so long as the articular cartilage remains intact, transplantation of large osteochondral grafts, as with other bone grafts, invokes an immune response that is variable and that is believed to be responsible for major complications such as nonunion at host–donor junction, allograft fracture, and high infection rates (Mankin et al., 1998). Both freezing and freeze-drying appear to reduce the immune response invoked by implantation of allograft bone (Friedlaender et al., 1976). Unfortunately, data from clinical studies, which include reliable outcome measures, are sparse.

Given the large size of these grafts, methods used within the tissue bank for cryopreservation are relatively crude, involving exposure of the articular surface to Me_2SO followed by uncontrolled slow cooling (Mankin et al., 1998; Tomford, 1983). Despite this, cryopreserved osteochondral allografts have been shown in a number of studies to provide an effective alternative to metallic prostheses or amputation. Clinical, radiographic, and other functional criteria reflect success rates of 60 to 90% (Friedlaender et al., 1999). In the largest study to date, 450 patients receiving a cryopreserved osteoarticular allograft were followed intermittently over a period of 26 years (Mankin et al., 1998). The functional capacity of the body part requiring grafting and disease recurrence was evaluated. Seventy percent of those who were followed for 2 or more years showed no evidence of disease, pain, or major disability. In this and other studies, chondrocytes have not been reported to survive in any numbers (Campanacci et al., 1999), but the effects that a nonviable articular surface has on the long-term function of the joint is unclear. Though these studies report osteoarthritic changes and degeneration of the articular surface in many of the patients receiving allografts, only a small number were reported as requiring resurfacing or joint replacement. Nevertheless, the need for improved chondrocyte viability *in vivo* is held to be one of the goals for improved clinical efficacy of this procedure.

19.3.4 HEART VALVES

Allograft aortic and pulmonary valves have been used clinically since the early 1960s for both left and right ventricular outflow tract reconstruction (Goffin et al., 2000), as well as in the repair or replacement of damaged mitral and tricuspid valves (Angell et al., 1968). More recently, replacement or repair of these latter valves has been reported using cryopreserved mitral valves (Acar, 1997; Miyagishima et al., 2000; Reardon and Oury, 1998).

Allograft cardiac valves offer an alternative to mechanical and bioprosthetic valves and are considered the valve of choice where long-term postoperative anticoagulation therapy is contraindicated. Because allograft valves are also more resistant to bacterial growth than other valve prostheses (Staab, 1998), this also holds true in the case of valve replacement for endocarditis of either a native or a prosthetic valve.

Initially, valves were harvested under sterile conditions and implanted either immediately or following a brief period of storage in tissue culture medium at +4°C. Such valves have shown good durability in at least one long-term follow-up study (Yacoub et al., 1995). As the clinical use of allograft heart valves increased, the need for some method of storage became necessary to permit routine surgery. Because sterile harvesting had already been abandoned, an effective sterilization method was also required. Early methods (using ethylene oxide, β-propiolactone, gamma radiation) and attempts at preservation and storage by flash freezing or freeze-drying were found to damage the valve matrix and biomechanical properties. Such treatments led to a high incidence of early failure after implantation (Brock, 1968; Missen and Roberts, 1970) and, in a separate study, calcification of a substantial proportion of those still functioning after 3 years (Moore et al., 1975).

These studies also reported the loss of the cells from such valves. As a consequence, the belief grew that viability of aortic valve fibroblasts and endothelium was a prerequisite for long-term function of transplanted valves and that this was related to the methods used for sterilization and storage. Viable fibroblasts were thought to be required within the matrix of the valve to continue to synthesize collagen and glycosaminoglycans and to remodel and repair the implanted valve (Angell et al., 1973, 1989). This led to the use of serum-supplemented tissue culture medium for 4°C storage, often for some weeks. Such storage conditions were incapable of maintaining cell viability for this length of time, and cryopreservation techniques were reintroduced (O'Brien et al., 1987). Further clinical studies with cryopreserved valves have demonstrated improved long-term durability compared with antibiotic-treated 4°C-stored valves (Kirklin et al., 1993; O'Brien et al., 2001; Tweddell et al., 2000; Yankah et al., 1996). On the basis of such studies, most heart valve banks and clinicians worldwide have adopted the antibiotic-treated, cryopreserved valve as the allograft of choice.

Unlike vascular grafts, current methods for the cryopreservation of heart valves do not rest on a substantial body of experimental data but have been derived generally from protocols used for the preservation of single-cell suspensions. In most tissue banks, valves are cooled slowly in 10% dimethyl sulphoxide and stored in the vapor phase above liquid nitrogen. Though there has been inconclusive debate over the years on conditions for optimum storage and transportation (Feng et al., 1996), on the effect of cryopreservation on early calcification (Brockbank et al., 2000; Yankah et al., 1995), and on the immunological status of the implanted cryopreserved valve (Ketheesan et al., 1996), maintenance of cellular viability has been an *a priori* assumption. Yet the very need for cellular preservation to provide long-term durability of the graft is still open to question.

Viability is as poorly defined a concept with heart valves as it is with other tissues. Many different assays have been applied to heart valve tissue *ex vivo*, including incorporation of radioactive proline, electron microscopic and histologic studies, vital dye staining, and *in situ* hybridization techniques.

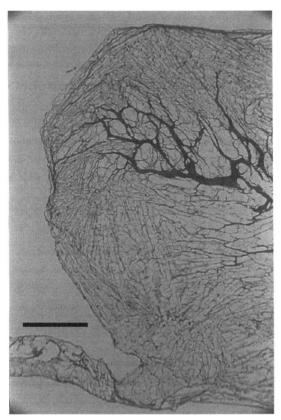

FIGURE 19.3 Cryopreserved heart valve. Freeze-substituted human heart valve leaflet cooled in 10% DMSO at 1°C/min, freeze substituted at –80°C. Note extensive ice (white spaces) throughout. Bar = 200 μM. (Micrograph prepared by C.J. Hunt.)

For the unimplanted, cryopreserved heart valve, successful preservation of both endothelium and leaflet fibroblasts has been reported in a number of studies (Crescenzo et al., 1993; Gall et al., 1998; Niwaya et al., 1995), though by no means in all (Armiger, 1995). On the basis of such assays, long-term storage at –180°C would appear more effective in comparison to storage at –80°C (Brockbank et al., 1992; Feng et al., 1996), though care is required when rewarming from this temperature to avoid the cracking phenomenon also seen with other vascular tissue (Hunt et al., 1994; Wassenaar et al., 1995). There is little evidence that short periods of storage at –80°C, such as those required for transportation, are detrimental to the valve (Brockbank et al., 1992). Freeze-substitution studies (Figure 19.3) indicate that ice formation occurs throughout the valve. Quantitative assessment indicated that periods of storage of up to 1 month at –80°C did not affect the amount or distribution of ice in the valve compared with storage at –180°C (Hunt et al., 1991).

Viability as a prerequisite for valve durability has not gone unquestioned. As early as 1977, Wheatley and McGregor concluded from a comparative study of viable and nonviable canine pulmonary allografts that viability was disadvantageous, as it increased the possibility of early tissue degradation (Wheatley and McGregor, 1977). Accelerated calcifying degeneration of allografts in young children is well documented (Yankah et al., 1995), and though the effect of immunological factors on the durability of cryopreserved allografts is still uncertain (Dignan et al., 2000), at least one group has suggested the avoidance of allografts with a high degree of cellular viability in an effort to reduce early valve failure in children (Bodnar and Ross, 1991). Although general cellular preservation in the unimplanted cryopreserved valves is good (particularly when considering leaflet fibroblasts), explanted valves tend to show a much reduced and variable degree of cellularity largely independent of the period of implantation. Though the valves are largely devoid of endothelium, the presence of fibroblasts within otherwise acellular leaflets has often been reported (Lupinetti et al., 1993; O'Brien et al., 1987). The origin of such cells has been hotly debated. Donor cells were identified in early studies in which there were sex mismatches between donor an recipient (Gonzalez-Lavin et al., 1990; O'Brien et al., 1988), but more recently, DNA *in situ* hybridization on 22 explanted allograft valves with a donor sex mismatch found that donor cells survived in the graft no longer than several weeks (Hazekamp et al., 2003). Further studies of human and animal explanted valves have shown that the residual fibroblast population residing in the valve leaflet is of both donor and recipient origin (Braun et al., 1997; Koolbergen et al., 1998).

It has been argued that, even if isolated cells persist, whether they are of donor or recipient origin, this will not materially contribute to matrix remodeling. Schoen, in a study comparing explanted cryopreserved valves with valves from transplanted hearts, concluded that the well-preserved collagen fibers and matrix glycoproteins found in the acellular leaflets of long-term-implanted cryopreserved valves was the structural basis for their continued function (Schoen et al., 1995). Goffin has reported similar findings (Goffin et al., 1994).

In a recent study, sheep aortic valves were subjected to novel sterilization and preservation methods that devitalized the tissue grafts, which were then compared with cryopreserved valves (Aidulis et al., 2002; Farrington et al., 2002; Neves et al., 2002). Mechanical, microscopic, and functional assays, as well as implantation studies, all indicated that provided the tissue matrix was well preserved, the presence of leaflet fibroblasts did not materially improve the performance of the valve. These and others studies argue that within a short time following implantation, the cryopreserved valve becomes essentially acellular, and thus no growth or remodeling of the extra-cellular matrix takes place. Durability of cryopreserved valves must be attributed mainly to the preservation of the collagen matrix and leaflet ground substance, with cellular viability acting as an index for the integrity of such structures.

19.3.5 SKIN

Skin, the largest organ in the body, is a two-layered structure comprising an outer stratified epithelial layer of keratinocytes and an inner matrix, the dermis. The epidermis provides a barrier function and is capable of rapid regeneration. The underlying dermis, composed of collagen, elastin, and glycosaminoglycans laid down by dermal fibroblasts, provides strength and elasticity. This complex, highly organized layer has some regenerative powers after injury, but the inelastic scar tissue eventually formed results in functional impairment. Skin defects requiring grafting arise from a variety of insults including chronic ulceration, acute injury, and burns. The driving force for the clinical use of allograft skin has been its use in the treatment of burns. In the case of full-thickness burns, allograft skin restores barrier function (reducing fluid loss and infection), helps to reduce pain and heat loss, and mitigates the hypermetabolic stress response to thermal injury. Skin grafts may be autologous or allogeneic. They may be partial or split-thickness grafts (i.e., epidermis and some dermis) or full-thickness grafts (i.e., epidermis, dermis, and subcutaneous tissue), or they may be composed solely from autologous or allogeneic keratinocytes grown in culture and applied directly to the wound bed or onto allogeneic dermis or bioengineered substrates to form composite grafts (Balasubramani et al., 2001; Wood and Harris, 1995).

In the case of full-thickness burns, split-thickness autologous grafts provide permanent wound closure and restoration of the barrier function as well as providing a dermal component that will reduce or prevent scarring. However, in the case of extensive burns, there is often insufficient body surface remaining to provide sufficient material for autologous grafting. Allograft split-thickness skin is rejected within a few weeks and so provides only temporary wound closure that allows the maximum use of limited autograft donor sites and permitting reharvesting of healed sites and culture of autologous keratinocyte sheets (Hickerson et al., 1994).

19.3.6 PRESERVATION OF SPLIT-THICKNESS SKIN GRAFTS

The maintenance of both cell viability and structural integrity is viewed as the key to successful engraftment and revascularization of allograft skin (Kagan, 1998). "Fresh" allogeneic skin is therefore generally the preferred biological dressing over cryopreserved allograft because of its perceived superior cellular viability. Unused "fresh" autologous skin is often stored at 4°C in saline or nutrient media and may used by the clinician after anything up to 7 d storage. Studies using a variety of viability assays have reported survival of keratinocytes reducing at variable rates for up to 2 weeks, depending on storage conditions (Robb et al., 2001; Sterne et al., 2000). Common practice dictates that unused allograft skin stored at 4°C for longer than about a week is subsequently cryopreserved; this is a practice that was recently advocated for autologous skin as well (Sheridan et al., 1998). Thus, the perceived superiority of fresh skin over cryopreserved may in part be a result of this practice of preserving suboptimally viable grafts. One recent study has concluded that cryopreserved skin is no more or less "viable" than skin stored at 4°C for 72 h (Bravo et al., 2000). In another recent study, paired comparison between fresh and cryopreserved allograft indicated that there was no significant difference in graft performance with up to 5 years storage at vapor-phase nitrogen temperatures (Ben-Basset et al., 2001). Thus, the perception that any graft labeled as "fresh allograft" can provide a highly viable biological dressing superior to that of cryopreserved tissue should be viewed with caution.

In most cryopreservation studies, the focus of attention has been maintenance of cellular viability. Over the years, a large number of cryopreservation protocols have been developed largely empirically and have been advocated for improving viability and the quality of banked skin. These have been the subject of a number of comprehensive reviews (Baxter et al., 1985). Where studies have been carried out using prognostic indicators of engraftment (Kearney et al., 1990; Villalba et al., 1996), dimethyl sulphoxide was found to be marginally the better cryoprotectant (cf. glycerol, ethanediol, or propanediol), and slow cooling (–1°C/min) was shown to be preferable to rapid cooling (by direct immersion in liquid nitrogen). This parallels the experience in most tissue banks. Recently, very rapid cooling has been advocated as a means of improving the recovery of split-thickness skin (Zeiger et al., 1996). Design of optimum cryopreservation protocols must await more information on cell permeability and the kinetics of cryoprotectant diffusion into skin (Aggarwal et al., 1988; Zeiger et al., 1997), as well as on studies on reducing postthaw deterioration (Kearney, 1998a).

As with other tissues, cellular viability of allograft skin as a prerequisite for efficacy remains an area of controversy. In addition to cryopreserved skin, nonviable (freeze-dried and glycerolized skin) has been used to treat burns (de Backere, 1994). Viable grafts stimulate revascularization of the wound bed, and dermal elements may become incorporated into the wound site (so call graft "take"). The viable epidermis maintains a barrier function, which is lost or reduced, in nonviable grafts. Immunogenicity has been reported to be reduced in grafts cryopreserved at intermediate cooling rates (Ingham et al., 1993) but not at cooling rates currently used for the banking of skin (Tomita et al., 1998). Both freeze-drying and glycerolization have been reported to reduce antigenicity clinically, though in the case of glycerol this has not been supported by *in vitro* studies (Hettich et al., 1994). The clinical situation is clouded by the use of allograft in combination with meshed autograft, cultured autologous keratinocyte sheets, and other composite grafting techniques (Valencia et al., 2000).

It has been speculated (Greenleaf and Livesey, 1999) that the characteristics attributed to viability may be surrogate indicators of the preservation of the extracellular matrix of the graft. Thus, conventional methods that attempt to preserve cells paradoxically may do so at the expense of the dermal matrix (damaged through ice formation and solute denaturation of proteins), accelerating the immune response through a nonspecific inflammatory reaction to the damaged matrix. As such, the desirable attributes of "fresh" skin may relate more to an undamaged matrix than to cellular viability. The successful use of glycerolized skin has been cited as evidence for this point of view (Richters et al., 1996).

19.3.7 PRESERVATION OF CULTURED KERATINOCYTES AND ALLOGENEIC DERMIS

The use of allogeneic cultured keratinocyte sheets as an alternative to allogeneic split skin is another treatment available for use in partial and full-thickness burns (Valencia et al., 2000). Such grafts have been shown to improve wound healing in a number of studies. Cultured keratinocytes are not rejected; nevertheless, they do not persist on the wound bed but are progressively replaced by recipient epithelium, and as such, the grafts act as a biological dressing. The use of keratinocyte sheets is problematic, as the sheets are fragile and the grafts lack long-term stability because of the absence of a dermal component. This can be overcome by the use of allogeneic dermis in conjunction with the cultured keratinocyte sheets (Wood and Harris, 1995). The dermal component may be supplied either by grafting allogeneic split-thickness skin, permitting the graft to "take" and subsequently removing the epithelial layers by dermabrasion, or through the application of cryopreserved or glycerolized dermis from which the immunogenic epithelium has been removed (McKay et al., 1994). Recently, decellularized, freeze-dried allogeneic dermis has become commercially available (Greenleaf and Livesey, 1999). Though cryopreservation studies of cultured keratinocytes have been reported, the conditions for optimal preservation of keratinocyte sheets attached to dermis or dermal equivalents are still being determined (Harringer et al., 1997).

19.4 ETHICAL AND SAFETY CONSIDERATIONS

As the clinical demand for banked tissues grows, so does the need for effective regulation of tissue banking activities. The ethical demands of donor, recipient, and society have all to be met as well as the provision of safeguards to minimize the transmission of disease from donor to recipient. Discussion of these issues is well beyond the scope of this chapter, and the reader is referred to the following articles: Womack and Gray (2000), Kearney (1998b), and Warwick et al. (1996), and the following Web sites: The Medical Research Council (http://www.mrc.ac.uk), Royal College of Pathologists (http://www.rcpath.org), Joint UK Blood Transfusion Services/National Institute for Biological Standards and Control (http://www.transfusionguidelines.org.uk), American Association of Tissue Banks (http://www.aatb.org), British Association for Tissue Banking (http://www.batb.org.uk), The UK Department of Health (http://www.doh.gov.uk), and the American Food and Drug Administration (http://www.fda.gov) for further information.

19.5 FUTURE PROSPECTS FOR TISSUE BANKING

19.5.1 PANCREATIC ISLET CELL TRANSPLANTATION

One prospect for the expansion of tissue banking in the near future is in the use of pancreatic islet cell transplantation for the treatment of diabetes mellitus. This has been an alternative to whole-pancreas transplantation for the past decade or more (Hering et al., 1988). To date, despite the demonstration of the successful return to normoglycemia in both large and small animal models, its clinical potential remains largely unrealized (White et al., 2001). A recent review of results reported to the International Islet Transplant Registry between 1990 and 1998 found that only 12% of those receiving islet allografts for type I diabetes achieved insulin independence for more than 7 d, with only 8% remaining independent at 1 year (White et al., 2001). The reasons for this lack of success are varied and well beyond the scope of this chapter, but they are the subject of two recent reviews (Shapiro et al., 2001; White et al., 2001). However, recent advances in the clinical protocol for islet transplantation have led to an apparent dramatic improvement in insulin independence (Shapiro et al., 2000). Using this protocol, a cohort of patients experiencing type I diabetes underwent allografting with "fresh" islets. All were reported insulin independent after a mean follow-up period of 12 months. Further data on 12 patients reported recently by this same Edmonton group have shown insulin independence ranging out to almost 2 years (Ryan et al., 2001). This has led to a considerable renewal of interest in islet transplantation and has focused attention once again on the need for effective methods of preservation.

Among the major changes introduced in the Edmonton protocol was the use of "double-donor" transplantation to provide a sufficient islet cell mass to ensure that the majority of recipients remained insulin-free. An obvious solution to the logistical problems associated with finding multiple donors would be the provision of banks of cryopreserved islets. The provision of a sufficient islet mass for long-term insulin independence is not the only advantage: It would also provide for the matching of human lymphocyte antigen (HLA)-typed islets between donor and recipient (Rajotte et al., 1990), ease logistical problems, permit proper quality control and sterility testing (Bretzel et al., 1994), and may, as a number of studies have indicated, improve the purity of the islets and modulate immunogenicity (Catrall et al., 1993; Evans et al., 1987; Taylor et al. 1992).

Many studies have been undertaken to define optimum conditions for cryopreservation (Piemonti et al., 1999), but the protocol that has so far gained widest acceptance is that devised by Rajotte and coworkers, which has been used to successfully preserve both animal and human islets (Rajotte, 1999). However, the development of cryopreservation procedures for islets has been largely empirical, and the assays used to assess viability are disparate. Furthermore, the cryopreservation protocols so far developed are not easily translated into the tissue bank, nor have they been successful in achieving insulin independence in the limited clinical experience of transplanting cryopreserved islets alone (Hering et al., 1995).

Recently, the biophysical parameters necessary to define optimum cryoprotectant addition and elution protocols have been described (Woods et al., 1999a, 1999b; Zeiger et al., 1999). When an understanding has been reached about other important criteria, such as the probability of intracellular ice formation and its relation to cooling rate, it is probable that an effective cryopreservation protocol will be developed. This should eventually permit the pancreatic islet to be added to the list of tissues banked for clinical use.

19.5.2 OTHER FUTURE PROSPECTS FOR EXPANSION IN TISSUE BANKING

As the understanding of the nature of cryoinjury and its avoidance in tissues increases, so will the availability of a wider range of allograft tissues for transplantation. The advances with the use of vitrification as an alternative method of storing blood vessels and corneas may well increase the range of bankable tissues, particularly in the direction of large vascularized tissues and organs. In the longer term, however, the increased understanding of the molecular mechanisms involved in cell differentiation and tissue development will inevitably make engineered tissue constructs available as an alternative to allograft tissue. This development will ultimately have a profound effect on tissue banking. Although large elements of tissue processing and storage will no longer be required, the safe procurement and handling of tissues, and the extraction and storage of the component cells, will still provide a crucial role for tissue banks. Tissue engineered constructs will be designed to mimic the natural tissue as closely as possible and will also require preservation to maintain operational stocks for distribution and use. The role of cryobiology in the field of tissue repair and transplantation will therefore remain a crucial one for the foreseeable future.

REFERENCES

Acar, C. (1997) Mitral valve homograft, *Adv. Card. Surg.,* 9, 1–13.

Adds, P.J., Hunt, C.J., and Dart, J.K.G. (2001) Amniotic membrane grafts, fresh or frozen? A clinical and *in vitro* comparison, *Br. J. Ophthalmol.,* 85, 905–907.

Adham, M., Gournier, J.P., de la Roche, E., Ducerf, C., Bauileux, J., et al. (1996) Mechanical characteristics of fresh and frozen human descending thoracic aorta, *J. Surg. Res.,* 64, 32–34.

Aggarwal, S.J., Diller, K.R., and Baxter, C.R. (1988) Hydraulic permeability and activation energy of human keratinocytes at subzero temperatures, *Cryobiology,* 25, 203–211.

Aho, A.J., Eskola, J., Ekfors, T., Manner, I., Kouri, T., et al. (1997) Immune response and clinical outcome of massive human osteoarticular allografts, *Clin. Orthop.,* 346, 196–206.

Aidulis, D., Pegg, D.E., Hunt, C.J., Goffin, Y.A., Yanderkelen et al. (2002) Processing of ovine cardiac valve allografts: 1. Effects of preservation method on structure and mechanical properties, *Cell Tissue Banking,* 3, 79–89.

Albon, J., Tullo, A.B., Aktar, S., and Boulton, M.E. (2000) Apoptosis in the endothelium of human corneas for transplantation, *Invest. Ophthalmol. Vis. Sci.,* 41, 2887–2893.

Angell, W.W., Iben, A.B., and Shumway, N.E. (1968) Fresh aortic homografts for multiple valve replacement, *Arch. Surg,* 97, 826–830.

Angell, W.W., Lanerolle, P.D.E., and Shumway, N.E. (1973) Valve replacement: Present status of homograft valves, *Prog. Cardiovasc. Dis.,* 15, 589–622.

Angell, W.W., Oury, J.H., Lamberti, J.J., and Koziol, J. (1989) Durability of the viable aortic allograft, *J. Thorac. Cardiovasc. Surg.,* 98, 48–56.

Armiger, L.C. (1995) Viability studies of human valves prepared for use as allografts, *Ann. Thorac. Surg.,* 60, S118–S121.

Armitage, W.J. and Easty, D.L. (1997) Factors influencing the suitability of organ-cultured corneas for transplantation, *Invest. Ophthalmol. Vis. Sci.,* 38, 16–24.

Armitage, W.J., Hall, S.C., and Routledge, C.R. (2002) Recovery of endothelial function after vitrification of cornea at –110°C, *Invest. Ophthalmol. Vis. Sci.,* 43, 2160–2164.

Aubin, P.P., Cheah, H.K., Davis, A.M., and Gross, A.E. (2001) Long-term follow-up of fresh femoral osteochondral allografts for posttraumatic knee defects, *Clin. Orthop.,* 391, S318–S327.

Balasubramani, M., Kumar,T.R., and Babu, M. (2001) Skin substitutes: A review, *Burns,* 27, 534–544.

Baxter, C., Aggarwal, S., and Diller, K.R. (1985) Cryopreservation of skin, *Transplant Proc.,* 17, 112–120.

Ben-Basset, H., Chaouat, M., Segal, N., Zumai, E., Wexler, M.R., and Eldad, A. (2001) How long can cryopreserved skin be stored to maintain adequate graft performance, *Burns,* 27, 425–431.

Bodnar, E. and Ross, D.N. (1991) Valvular homografts, in *Replacement Cardiac Valves,* Bodnar, E. and Frater, R.W.M., Eds., Pergamon Press, New York, pp. 287–306.

Borderie, V.M., Lopez, M., Lombert, A., Carvajal-Gonzalez, S., Cywiner, C., and Laroche, L. (1998) Cryopreservation and culture of human corneal keratocytes, *Invest. Ophthalmol. Vis. Sci.,* 39, 1511–1519.

Bourne, W.M., Hodge, D.O., and Nelson, L.R. (1994) Corneal endothelium 5 years after transplantation, *Am. J. Ophthalmol.,* 118, 185–196.

Bourne, W.M., Nelson, L.R., and Hodge, D.O. (1999) Comparison of three methods for human corneal cryopreservation that utilize dimethyl sulphoxide, *Cryobiology,* 39, 47–57.

Bourne, W.M., Nelson, L.R., Maguire, L.J., Baratz, K.H., and Hodge, D.O. (2001) Comparison of Chen medium and Optisol GS for human corneal preservation at 4°C, *Cornea,* 20, 683–686.

Boyce, T., Edwards, J., and Scarborough, N. (1999) Allograft bone. The influence of processing on safety and performance, *Orthop. Clin. North Am.,* 30, 571–581.

Braun, J., Hazekamp, M.G., Koolbergen, D.R., Sugihara, H., Goffin, Y.A., et al. (1997) Identification of host and donor cells in porcine homograft heart valve explants by fluorescence *in situ* hybridization, *J. Pathol.,* 183, 99–104.

Bravo, D., Ridgley, T.H., Gibran, N., Strong, D.M., and Newman-Gage, H. (2000) Effect of storage and preservation methods on viability in transplantable human skin allografts, *Burns,* 26, 367–378.

Bretzel, R.G., Alejandro, R., Herin, B.J., van Suylichem, P.T.R., and Ricordi, C. (1994) Clinical islet transplantation: Guidelines for islet quality control, *Transplant Proc.,* 26, 388–992.

Brock, R.C. (1968) Long-term degenerative changes in aortic segment homografts, with particular reference to calcification, *Thorax,* 23, 249–255.

Brockbank, K.G., Carpenter, J.F., and Dawson, P.E. (1992) Effects of storage temperature on viable bioprosthetic heart valves, *Cryobiology,* 29, 537–542.

Brockbank, K.G., Lightfoot, F.G., Song, Y.C, and Taylor, M.J. (2000) Interstitial ice formation in cryopreserved homografts: A possible cause of tissue deterioration and calcification *in vivo, J. Heart Valve Dis.,* 9, 200–206.

Brockbank, K.G.M., McNally, R.T., and Walsh, K.A. (1992) Cryopreserved vein transplantation, *J. Cardiac Surg.,* 7, 170–177.

Brunette, I., Le Francois, M., Tremblay, M.C., and Guertin, M.C. (2001) Corneal transplant tolerance of cryopreservation, *Cornea,* 20, 590–596.

Campanacci, D.A., Caldora, P., Beltrami, G., Gluckert, B., and Capanna R. (1999) Osteoarticular allografts and megaprostheses reconstruction in bone tumour surgery, in *Advances in Tissue Banking*, Vol. 3, Phillips, G.O., Strong, D.M., von Versen, R., and Nather, A., Eds., World Scientific, Singapore, pp. 67–85.

Catrall, M.S., Warnock, G.L., Kneteman, N.M., Halloran, P.F., and Rajotte, R.V. (1993) The effect of cryo-preservation on the survival and MHC antigen expression of murine islet allografts, *Transplantation*, 55, 159–163.

Chang, C.-J. and Yang, J.-Y. (1994) Frozen preservation of human amnion and its use as a burn wound dressing, *Chang Gung Med. J.*, 17, 316–324.

Chen, C.H. and Chen, S.C. (1994) The efficacy of non-lactate-generating metabolites as substrates for maintaining donor tissues, *Transplantation*, 57, 1778–1785.

Cheng, H.C., Armitage, W.J., Yagoubi, M.I., and Easty, D.L. (1996) Viability of keratocytes in epikeratophakia lenticules, *Brit. J. Ophthalmol.*, 80, 367–372.

Chu, C.R., Convery, F.R., Akeson, W.H., Meyers, M., and Amiel, D. (1999) Articular cartilage transplantation. Clinical results in the knee, *Clin. Orthop.*, 360, 159–168.

Chu, W. (2000) The past twenty-five years in eye banking cornea, *Cornea*, 19, 754–765.

Crescenzo, D.G., Hilbert, S.L., Messier, R.H., Jr., Domkowski, P.W., Barrick, M.K., et al. (1993) Human cryopreserved homografts: Electron microscopic analysis of cellular injury, *Ann. Thorac. Surg.*, 55, 25–31.

Crewe, J.M. and Armitage, W.J. (2001) Integrity of epithelium and endothelium in organ cultured human corneas, *Invest. Ophthalmol. Vis. Sci.*, 42, 1757–1761.

Czitrom, A.A., Langer, F., McKnee, N.H., and Bone, A.E. (1986) Bone and cartilage allotransplantation. A review of 14 years of research and clinical studies, *Clin. Orthop.*, 208, 141–145.

da Gama, A.D., Sarmento, C., Viera, T., and do Carmo, G.X. (1994) The use of arterial allografts for vascular reconstructions in patients receiving immunosuppression for organ transplantation, *J. Vasc. Surg.*, 20, 271–278.

Davies, M.G. and Hagen, P.-O. (1994) Structural and functional consequences of by-pass grafting with autologous vein, *Cryobiology*, 31, 63–70.

de Backere, A.C.J. (1994) Euro skin bank: Large scale skin-banking in Europe based on glycerol-preservation of donor skin, *Burns*, 20, S4–S9.

Dignan, R., O'Brien, N., Hogan, P., Passage, J., Stephens, F., et al. (2000) Influence of HLA matching and associated factors on aortic valve homograft function, *J. Heart Valve Dis.*, 9, 504–511.

Dzimiri, N., Chester, A.H., Allen, S.P., Duran, C., and Yacoub, M.H. (1996) Vascular reactivity of arterial coronary artery bypass grafts-implications for their performance, *Clin. Cardiol.*, 19, 165–171.

Ehlers, H., Ehlers, N., and Hjortdal, O. (1999) Corneal transplantation with donor tissue kept in organ culture for 7 weeks, *Acta Ophthalmol. Scand.*, 77, 277–278.

Evans, M.G., Rajotte, R.V., Warnock, G.L., and Procyshen, A.W. (1987) Cryopreservation purifies canine pancreatic fragments, *Transplant Proc.*, 19, 3471–3477.

Farrington, M., Wreghitt, T., Matthews, I., Scarr, D., Sutehall, G. et al. (2002) Processing of ovine cardiac valve allografts: 2. Effects of antimicrobial treatment on sterility, structure, and mechanical properties, *Cell Tissue Banking*, 3, 91–103.

Feng, X.J., Van Hove, C.E., Walter, P.J., and Herman, A.G. (1996) Effects of storage temperature and fetal calf serum on the endothelium of porcine aortic valves, *J. Thorac. Cardiovasc. Surg.*, 111, 218–230.

Friedlaender, G.E., Strong, D.M., Tomford, W.W., and Mankin, H.J. (1999) Long-term follow-up of patients with osteochondral allografts. A correlation between immunologic responses and clinical outcome, *Orthop. Clin. North Am.*, 30, 583–588.

Friedlander, G.E., Strong, D.M., and Sell, K.W. (1976) Studies on the antigenicity of bone: I. Freeze-dried and deep-frozen bone allografts in rabbits, *J. Bone Joint Surg.*, 58A, 854–858.

Frueh, B.B., and Böhnke, M. (2000) Prospective randomised clinical evaluation of Optisol vs organ culture corneal storage media, *Arch. Ophthalmol.*, 118, 757–760.

Gall, K.L., Smith, S.E., Willmette, C.A., and O'Brien, M.F. (1998) Allograft heart valve viability and valve-processing variables, *Ann. Thorac. Surg.*, 65, 1032–1038.

Goffin, Y.A., Van Hoeck, B., Jashari, R., Soots, G., and Kalmar, P. (2000) Banking of cryopreserved heart valves in Europe: Assessment of a 10-year operation in the European Homograft Bank (EHB), *J. Heart Valve Dis.*, 9, 207–214.

Goffin, Y.A., Henriques de Gouveia, R., Szombathelyi, T., Toussaint, H., et al. (1994) European Homograft Bank heart valve explants: A five year pathology study with reference to unimplanted valves, *Proceedings of the 4th Symposium of the European Homograft Bank*, European Homograft Bank, Brussels, pp. 55–60.

Goffin, Y.A., Grandmougin, D., Wozniak, G., Keppenne, V., Nevelsteen, A., et al. (1998) Banking and distribution of large cryopreserved arterial homografts in Brussels: Assessment of 4 years of activity by the European Homograft Bank (EHB) with reference to implantation results in reconstruction of infected infrarenal arterial prostheses and mycotic aneurisms, *Vascular Surg.*, 32, 19–32.

Gonzalez-Lavin, L., Spotnitz, A.J., Mackenzie, J.W., Gu, J., Gadi, I.K., et al. (1990) Homograft valve durability: Host or donor influence? *Heart Vessels*, 5, 102–106.

Gournier, J.-P., Favre, J.-P., Gay, J.-L., and Barral, X. (1995) Cryopreserved arterial allografts for limb salvage in the absence of suitable saphenous vein: Two year results in 20 cases, *Ann. Vasc. Surg.*, 9, S7–S14.

Greenleaf, G. and Livesey, S.A. (1999) Expanding the use of allograft skin: The properties and use of the acellular dermal matrix, in *Advances in Tissue Banking*, Vol. 3, Phillips, G.O., Strong, D.M., von Versen, R., and Nather, A., Eds., World Scientific, Singapore, pp. 167–184.

Gross, A.E., McKee, N.H., Pritzker, K.P., and Langer, F. (1983) Reconstruction of skeletal deficits at the knee: A comprehensive osteochondral transplant program, *Clin. Orthop.*, 174, 96–106.

Harringer, M.D., Supp, A.P., Swope, V.B., and Boyce, S.T. (1997) Reduced engraftment and wound closure of cryopreserved cultured skin substitutes grafted to athymic mice, *Cryobiology*, 35, 132–142.

Harris, L., O'Brien-Irr, M., and Ricotta, J.J. (2001) Long-term assessment of cryopreserved vein bypass grafting success, *J. Vasc. Surg.*, 33, 528–532.

Hazekamp, M.G., Koolbergen, D.R., Braun, J., Bruijn, J.A., Cornelisse, C.J., et al. (2003) Cell viability: Cell origins and fates following transplantation of cryopreserved allografts, in *Cardiac Reconstruction with Allograft Tissue*, 2nd edition, Hopkins, R.A., Ed., Springer Verlag, New York.

Hering, B.J., Schultz, A.O., Geier, C., Bretzel, R.G., and Federlin, K. (1995) *Newsletter 6, Int. Islet Transplant Registry*, 5, 18.

Hering, B.J., Bretzel, R.G., and Federlin, K. (1988) Current status of clinical islet transplantation, *Horm. Met. Res.*, 20, 537–545.

Hettich, R., Ghofrani, A., and Hafemann, B. (1994) The immunogenicity of glycerol-preserved donor skin, *Burns*, 20, S71–S76.

Hickerson, W.L., Compton, C., Fletchall, S., and Smith, L.R. (1994) Cultured epidermal autografts and allodermis combination for permanent burn wound coverage, *Burns*, 20, S52–S56.

Hsu, J.K.W., Cavanagh, H.D., Jester, J.V., Ma, L., and Petroll, W.M. (1999) Changes in corneal endothelial apical junctional protein organization after corneal cold storage, *Cornea*, 18, 712–720.

Hunt, C.J., Song, Y.C., Bateson, E.A.J., and Pegg, D.E. (1994) Fractures in cryopreserved arteries, *Cryobiology*, 31, 506–515.

Hunt, C.J., Hayes, A.R., van Hoeck, B., and Goffin, Y. (1991) Ice formation in cryopreserved heart valves: A freeze substitution study with process monitoring using a new printing thermometer, *Cryobiology*, 28, 565.

Hurtig, M.B., Novak, K., McPherson, R., McFadden, S., McGann, L.E., Muldrew, K., and Schachar, N.S. (1998) Ostechondral dowel transplantation for repair of focal defects in the knee: An outcome study using an ovine model, *Vet. Surg.*, 27, 5–16.

Ingham, E., Matthews, J.B., Kearney, J.N., and Gowland, G. (1993) The effect of variation of cryopreservation protocols on the immunogenicity of allogeneic skin grafts, *Cryobiology*, 30, 443–458.

Jakobiec, F.A., Koch, P., Iwamoto, T., Harrison, W., and Troutman, R (1981) Keratophakia and keratomileusis: Comparison of pathologic features in penetrating keratoplasty specimens, *Ophthalmology*, 88, 1251–1259.

Kagan, R.J. (1998) Human skin banking: Past present and future, in *Advances in Tissue Banking*, Vol. 3, Phillips, G.O., Strong, D.M., von Versen, R., and Nather, A., Eds., World Scientific, Singapore, pp. 297–321.

Kearney, J.N., Wheldon, L.A., and Gowland, G. (1990) Effects of cryobiological variables on the survival of skin using a defined murine model, *Cryobiology*, 27, 164–170.

Kearney, J.N. (1998a) Evaluation of proteinase inhibitors and free radical inhibitors/scavengers in reducing post-thaw viability loss of cryopreserved skin, *Burns*, 24, 507–512.

Kearney, J.N. (1998b) Quality issues in skin banking: A review, *Burns*, 24, 299–305.

Ketheesan, N., Kearney, J.N., and Ingham, E. (1996) The effect of cryopreservation on the immunogenicity of allogeneic cardiac valves, *Cryobiology*, 33, 41–53.

Kim, J.C. and Tseng, S.C.G. (1995) Transplantation of preserved human amniotic membrane for surface reconstruction in severely damaged rabbit corneas, *Cornea*, 14, 473–484.

Kirklin, J.K., Smith, D., Novick, W., Naftel, D.C., Kirklin, J.W., et al. (1993) Long-term function of cryopreserved aortic homografts. A ten-year study, *J. Thorac. Cardiovasc. Surg.*, 106, 154–165.

Koizumi, N., Fullwood, N.J., Bairaktaris, G., Inatomi, T., Kinoshita, S., and Quantock, A.J. (2000) Cultivation of corneal epithelial cells on intact and denuded human amniotic membrane, *Invest. Ophthalmol. Vis. Sci.*, 41, 2506–2513.

Komuro, A., Hodge, D.O., Gores, G.J., and Bourne, W.M. (1999) Cell death during corneal storage at 4°C, *Invest. Ophthalmol. Vis. Sci.*, 40, 2827–2832.

Koolbergen, D.R., Hazekamp, M.G., Kurvers, M., de Heer, E., Cornelisse, C.J., et al. (1998) Tissue chimerism in human cryopreserved homograft valve explants demonstrated by *in situ* hybridization, *Ann. Thorac. Surg.*, 66, S225–S232.

Loop, F.D., Lytle, B.W., Cosgrove, D.M., Stewart, R.W., Goormastic, M., et al. (1986) Influence of the internal mammary graft on 10 year survival and other cardiac events, *N. Engl. J. Med.*, 314, 1–6.

Lupinetti, F.M., Tsai, T.T., Kneebone, J.M., and Bove, E.L. (1993) Effect of cryopreservation on the presence of endothelial cells on human valve allografts, *J. Thorac. Cardiovasc. Surg.*, 106, 912–917.

Malinin, T.I., Wagner, J.L., Pitta, J.C., and Lo, H. (1985) Hypothermic storage and cryopreservation of cartilage, *Clin. Orthop. Rel. Res.*, 197, 15.

Mankin, H.J., Gebhardt, M.C., and Tomford, W.W. (1998) Long-term results of allograft replacement in the management of bone tumors: A retrospection, in *Advances in Tissue Banking*, Vol. 3, Phillips, G.O., Strong, D.M., von Versen, R., and Nather, A., Eds., World Scientific, Singapore, pp. 209–239.

Mankin, H.J., Gebhardt, M.C., Jennings, L.C., Springfield, D.S., and Tomford, W.W. (1996) Long-term results of allograft replacement in the management of bone tumors, *Clin. Orthop.*, 324, 86–97.

Maral, T., Borman, H., Arslan, H., Demirhan, B., Akinbingol, G., and Haberal, M. (1999) Effectiveness of human amnion preserved long-term in glycerol as a temporary biological dressing, *Burns*, 25, 625–635.

Matsuura, J.H., Johansen, K.J., Rosenthal, D., Clark, M.D., Clarke, K.A., et al. (2000) Cryopreserved femoral vein grafts for difficult hemodialysis access, *Ann. Vasc. Surg.*, 14, 50–55.

McKay, I., Woodward, B., Wood, K., Navsaria, H.A., Hoekstra, H., and Green, C. (1994) Reconstruction of human skin from glycerol-preserved allodermis and cultured keratinocyte sheets, *Burns*, 20, S19–S22.

Miller, V.M., Bergman, R.T., Gloviczki, P., and Brockbank, K.G. (1993) Cryopreserved venous allografts: Effects of immunosuppression and anti-platelet therapy on patency and function, *J. Vasc. Surg.*, 18, 216–226.

Missen, G.A. and Roberts, C.I. (1970) Calcification and cusp-rupture in human aortic-valve homografts sterilised by ethylene oxide and freeze-dried, *Lancet*, 2, 962–964.

Miyagishima, R.T., Brumwell, M.L, Jamieson, W.R., and Munt, B.I. (2000) Tricuspid valve replacement using a cryopreserved mitral homograft. Surgical technique and initial results, *J. Heart Valve Dis.*, 9, 805–809.

Moller-Pedersen, T., Hartmann, U., Moller, H.J., Ehlers, N., and Engelmann, K. (2001) Evaluation of potential organ culture media for eye banking using human donor corneas, *Br. J. Ophthalmol.*, 85, 1075–1079.

Moore, C.H., Martelli, V., Al-Janabi, N., and Ross, D.N. (1975) Analysis of homograft valve failure in 311 patients followed up to 10 years, *Ann. Thorac. Surg.*, 20, 274–281.

Mow, V.C., Ratcliffe, A., and Poole, R.A. (1992) Cartilage and diarthroidal joints as paradigms for hierarchical materials and structures, *Biomaterials*, 13, 67–97.

Muldrew, K., Hurtig, M., Novak, K., Schachar, N., and McGann, L.E. (1994) Localization of freezing injury in articular cartilage, *Cryobiology*, 31, 31–38.

Muldrew, K., Novak, K., Studholme, C., Wohl, G., Zernicke, R., Schachar, N.S., and McGann, L.E. (2001) Transplantation of articular cartilage following a step-cooling cryopreservation protocol, *Cryobiology*, 43, 260–267.

Muldrew, K., Novak, K., Yang, H., Zernicke, R., Schachar, N., and McGann, L.E. (2000) Cryobiology of articular cartilage: Ice morphology and recovery of chondrocytes, *Cryobiology*, 40, 102–109.

Müller-Schweinitzer, E., Mihatsch, M.J., Schilling, M., and Haefeli, W.E. (1997) Functional recovery of human mesenteric and coronary arteries after cryopreservation at –196°C in a serum free medium, *J. Vasc. Surg.*, 25, 743–750.

Müller-Schweinitzer, E., Brett, W., Zerkowski, H.R., and Haefili, W.E. (2000) The mechanism of cryoinjury: *In vitro* studies on human internal mammary arteries, *Br. J. Pharmacol.*, 130, 636–640.

Müller-Schweinitzer, E., Stultz, P., Striffler, H., and Haefili, W.E. (1998) Functional activity and transmembrane signalling systems for cryopreservation of human internal mammary arteries, *J. Vasc. Surg.*, 27, 528–537.

Nelson, L.R., Hodge, D.O., and Bourne, W.M. (2000) *In vitro* comparison of Chen medium and Optisol GS medium for human corneal storage, *Cornea*, 19, 782–787.

Neves, J.P., Gulbenkian, S., Ramos, T., Martins, A.P., Caldas, M., et al. (1997) Mechanisms underlying degeneration of cryopreserved vascular homografts, *J. Thorac. Cardiovasc. Res.*, 113, 1014–1021.

Neves, J., Abecassis, M., Santiago, T., Ramos, T., Melo, J. et al. (2002) Processing of ovine cardiac valve allografts: 3. Implantation following antimicrobial treatment and preservation, *Cell Tissue Banking*, 3, 105–119.

Niwaya, K., Sakaguchi, H., Kawachi, K., and Kitamura, S. (1995) Effect of warm ischemia and cryopreservation on cell viability of human allograft valves, *Ann. Thorac. Surg.*, 60, S114–S117.

O'Brien, M.F., Harrocks, S., Stafford, E.G., Gardner, M.A., Pohlner, P.G., et al. (2001) The homograft aortic valve: A 29-year, 99.3% follow up of 1,022 valve replacements, *J. Heart Valve Dis.*, 10, 334–344.

O'Brien, M.F., Johnston, N., Stafford, G., Gardner, M., Pohlner, P., et al. (1988) A study of the cells in the explanted viable cryopreserved allograft valve, *J. Card. Surg.*, 3, 279–287.

O'Brien, M.F., Stafford, E.G., Gardner, M.A., Pohlner, P.G., and McGiffin, D.C. (1987) A comparison of aortic valve replacement with viable cryopreserved and fresh allograft valves, with a note on chromosomal studies, *J. Thorac. Cardiovasc. Surg.*, 94, 812–823.

Ohlendorf, C., Tomford, W.W., and Mankin, H.J. (1996) Chondrocyte survival in cryopreserved osteochondral articular cartilage, *J. Orthop. Res.*, 14, 413–416.

Pegg, D.E (1998) Cryobiology of cells *in situ*: Experiments with ovine articular cartilage, *Cryobiology*, 37, 381–382.

Pegg, D.E. (1989) Viability assays for preserved cells, tissues and organs, *Cryobiology*, 26, 212–231.

Pegg, D.E., Hunt, C.J., and Fong, L.P. (1987) Osmotic properties of rabbit corneal endothelium and their relevance to cryopreservation, *Cell Biophys.*, 10, 169–189.

Pegg, D.E., Wusteman, M.C., and Boylan, S. (1997) Fractures in cryopreserved elastic arteries, *Cryobiology*, 34, 183–192.

Piemonti, L., Bertuzzi, F., Nano, R., Leone, B.E., Socci, C., et al. (1999) Effects of cryopreservation on *in vitro* and *in vivo* long-term function of human islets, *Transplantation*, 68, 655–662.

Rajotte, R.V. (1999) Islet cryopreservation protocols, *Ann. NY Acad. Sci.*, 875, 2000–2007.

Rajotte, R.V., Evans, M.G., Warnock, G.L., and Kneteman, N.M. (1990) Islet cryopreservation, *Horm. Metab. Res. Suppl.*, 25, 72–81.

Reardon, M.J. and Oury, J.H. (1998) Evolving experience with cryopreserved mitral valve allografts, *Curr. Opin. Cardiol.*, 13, 85–90.

Rich, S.J., and Armitage, W.J. (1991) Corneal tolerance of vitrifiable concentrations of propane 1,2, diol, *Cryobiology*, 28, 159–170.

Richters, C.D., Hoekstra, M.J., Van Baare, J., Du Pont, J.S., and Kamperdijk, E.W.A. (1996) Morphology of glycerol-preserved human cadaver skin, *Burns*, 22, 113–116.

Rijneveld, W.J., Beekhuis, H., van Rij, G., Rinkel-van Driel, B., and Pels, E. (1992) Clinical comparison of grafts stored in McCarey-Kaufmann medium at 4°C in corneal organ culture at 31°C, *Arch. Ophthalmol.*, 110, 203–205.

Robb, E.C., Bechmann, N., Plessinger, R.T., Boyce, S.T., Warden, G.D., and Kagan, R.J. (2001) Storage media and temperature maintain normal anatomy of cadaveric human skin for transplantation to full-thickness skin wounds, *J. Burn Care Rehabil.*, 22, 393–396.

Rodeo, S.A. (2001) Meniscal allografts—where do we stand? *Am. J. Sports Med.*, 29, 246–261.

Rosset, E., Friggi, A., Novakovitch, G., Rolland, P.-H., et al. (1996) Effects of cryopreservation on the viscoelastic properties of human arteries, *Ann. Vasc. Surg.*, 10, 262–272.

Ryan, E.A., Lakey, J.R., Rajotte, R.V., Korbutt, G.S., Kin, T., et al. (2001) Clinical outcomes and insulin secretion after islet transplantation with the Edmonton protocol, *Diabetes*, 50, 710–719.

Salai, M., Ganel, A., and Horoszowski, H. (1997) Fresh osteochondral allografts at the knee joint: Good functional results in a follow-up study of more than 15 years, *Arch. Orthop. Trauma Surg.*, 116, 423–425.

Schachar, N., McAllister, D., Stevenson, M., Novak, K., and McGann, L.E. (1992) Metabolic and biochemical status of articular cartilage following cryopreservation and transplantation: A rabbit model, *J. Orthop. Res.*, 10, 603–609.

Schachar, N., Nagao, M., Matsuyama, T., McAllister, D., and Ishii, S. (1989) Cryopreserved articular chondrocytes grow in culture, maintain cartilage phenotype and synthesize matrix components, *J. Orthop. Res.*, 7, 344–351.

Schachar, N.S. and McGann, L.E. (1986) Investigations of low temperature storage of articular cartilage for transplantation, *Clin. Orthop.*, 208, 146–150.

Schachar, N.S., Novak, K., Hurtig, M., Muldrew, K., McPherson, R., Wohl, G., Zernicke, R.F., and McGann, L.E. (1999) Transplantation of cryopreserved osteochondral dowel grafts for repair of focal articular defects in an ovine model, *J. Orthop. Res.*, 17, 909–920.

Schilling, A., Glusa, E., and Müller-Schweinitzer, E. (1995) Nature of the vehicle solution for cryopreservation of human peripheral veins: Preservation of reactivity to pharmacological stimuli, *Cryobiology*, 32, 109–113.

Schoen, F.J., Mitchell, R.N., and Jonas, R.A. (1995) Pathological considerations in cryopreserved allograft heart valves, *J. Heart Valve Dis.*, 4, S72–S76.

Shapiro, A.M., Lakey, J.R., Ryan, E.A., Korbutt, G.S., Toth, E., et al. (2000) Islet transplantation in seven patients with type I diabetes mellitus using a glucocorticoid-free immunosuppressive regimen, *N. Engl. J. Med.*, 343, 230–238.

Shapiro, A.M.J., Ryan, E.A., and Lakey, J.R.T. (2001) Pancreatic islet transplantation in the treatment of diabetes mellitus, *Best Pract. Res. Clin. Endocrinol. Metab.*, 15, 241–264.

Sheriden, R., Mahe, J., and Walters, P. (1998) Autologous skin banking, *Burns*, 24, 46–48.

Sippel, K.C., Ma, J.J.K., and Foster, C.S. (2001) Amniotic membrane surgery, *Curr. Op. Ophthalmol.*, 12, 269–281.

Smith, A.U. (1965) Survival of frozen chondrocytes isolated from the cartilage of adult animals, *Nature*, 205, 782–784.

Southard, J.H. (1989) Viability assays in organ preservation, *Cryobiology*, 26, 232–238.

Staab, M.E., Nishimura, R.A., Dearani, J.A., and Orszulak, T.A. (1998) Aortic valve homografts in adults: A clinical perspective, *Mayo Clin. Proc.*, 73, 231–238.

Sterne, G.D., Titley, O.G., and Christie, J.L. (2000) A qualitative histological assessment of various storage conditions on short term preservation of human split skin grafts, *Br. J. Plast. Surg.*, 53, 331–336.

Stevenson, S. (1999) Biology of bone grafts, *Orthop. Clin. North Am.*, 30, 543–552.

Tavakol, K., Miller, R.G., Bazett-Jones, D.P., Hwang, W.S., McGann, L.E., and Schachar, N.S. (1993) Ultra-structural changes of articular chondrocytes associated with freeze-thawing, *J. Orthop. Res.*, 11, 1–9.

Taylor, M.J. (1986) Clinical cryobiology of tissues: Preservation of corneas, *Cryobiology*, 23, 323–353.

Taylor, M.J., Foreman, J., Biwata, Y., and Tsukikawa, S. (1992) Prolongation of islet allograft survival is facilitated by storage condition using cryopreservation involving fast cooling and/or tissue culture, *Transplant Proc.*, 24, 2860–2862.

Tomford, W.W. (1983) Cryopreservation of articular cartilage, *Osteochondral Allografts*, in Friedlaender, G.E., Mankin, H.J., and Sell, K.W., Eds., Little Brown, Boston, pp. 215–218.

Tomford, W.W., Fredericks, G.R., and Mankin, H.T. (1984) Studies on the cryopreservation of articular cartilage chondrocytes, *J. Bone Joint Surg.*, 66, 253–259.

Tomford, W.W. and Mankin, H.J. (1999) Bone banking. Update on methods and materials, *Orthop. Clin. North Am.*, 30, 565–570.

Tomford, W.W., Springfield, D.S., and Mankin, H.J. (1992) Fresh and frozen articular cartilage allografts, *Orthopaedics*, 15, 1183–1188.

Tomita, Y., Zhang, M., Yoshikawa, M., Uchinda, T., Nomoto, K., and Yasui, H. (1998) Lack of effect of cryopreservation on the class I and class II antigenicities of skin allografts, *Transplant Proc.*, 30, 60–62.

Tweddell, J.S., Pelech, A.N., Frommelt, P.C., Mussatto, K.A., Wyman, J.D., et al. (2000) Factors affecting longevity of homograft valves used in right ventricular outflow tract reconstruction for congenital heart disease, *Circulation*, 102, 130–135.

Valencia, I.C., Falabella, A.F., and Eaglestein, W.H. (2000) Skin grafting, *Dermatol. Clin.*, 18, 521–532.

Villalba, R., Benitez, J., de No-Lowis, E., Rioja, L.F., and Gómez-Villagrán, J.L. (1996) Cryopreservation of human skin with propane-1,2-diol, *Cryobiology*, 33, 525–529.

Vischjager, M., Van Gulik, T.M., Van Marle, J., Pfaffendorf, M., and Jacobs, M.J. (1996) Function of cryo-preserved artery allografts under immunosuppressive protection form cyclosporin A, *J. Vasc. Surg.*, 24, 876–882.

Warwick, R.M., Eastlund, T., and Fehily, D. (1996) Role of the blood transfusion service in tissue banking, *Vox Sang*, 71, 71–77.

Wassenaar, C., Wijsmuller, E.G., Van Herwerden, L.A., Aghai, Z., Van Tricht, C., et al. (1995) Cracks in cryopreserved aortic allografts and rapid thawing, *Ann. Thorac. Surg.*, 60, S165–S167.

Wheatley, D.J., and McGregor, C.G.A. (1977) Influence of viability on canine allograft heart valve structure and function, *Ann. Thorac. Surg.*, 60, 422–427.

White, S.A., James, R.F.L., Swift, S.M., Kimber, R.M., and Nicholson, M.L. (2001) Human islet cell transplantation—future prospects, *Diabetic Med.*, 18, 78–103.

Wohl, G., Goplen, G., Ford, J., Novak, K., Hurtig, M., et al. (1998) Mechanical integrity of subchondral bone in osteochondral autografts and allografts, *Can. J. Surg.*, 41, 228–233.

Womack, C. and Gray, N.M. (2000) Human research tissue banks in the UK National Health Service: Laws, ethics, controls and constraints, *Br. J. Biomed. Sci.*, 57, 250–253.

Wood, E.J. and Harris, I.R. (1995) Reconstructed human skin: Transplant, graft or biological dressing? *Essays Biochem*, 29, 65–85.

Woods, E.J., Liu, J., Zeiger, M.A, Lakey, J.R., and Critser, J.K. (1999a) The effects of microencapsulation on pancreatic islet osmotically induced volumetric response, *Cell Transplant*, 8, 699–708.

Woods, E.J., Liu, J., Zeiger, M.A., Lakey, J.R., and Critser, J.K. (1999b) Water and cryoprotectant permeability characteristics of isolated human and canine pancreatic islets, *Cell Transplant*, 8, 549–559.

Wusteman, M.C., Pegg, D.E., Robinson, M.P., Wang, L.-H., and Fitch, P. (2001) Vitrification media: Toxicity, permeability and dielectric properties, *Cryobiology*, 44, 24–37.

Wusteman, M.C. and Pegg, D.E. (2001) Differences in the requirements for cryopreservation of porcine aortic smooth muscle and endothelial cells, *Tissue Eng.*, 7, 507–518.

Wusteman, M.C., Pegg, D.E., and Warwick, R.M. (2000) The banking of arterial allografts in the United Kingdom. A technical and clinical review, *Cell Tissue Banking*, 1, 295–301.

Wusteman, M.C., Boylan, S., and Pegg, D.E. (1997) Cryopreservation of rabbit corneas in dimethyl sulphoxide, *Invest. Ophthalmol. Vis. Sci.*, 38, 1934–1943.

Yacoub, M., Rasmi, N.R., Sundt, T.M., Lund, O., Boyland, E., et al. (1995) Fourteen-year experience with homovital homografts for aortic valve replacement, *J. Thorac. Cardiovasc. Surg.*, 110, 186–194.

Yang, X.-H. and Zhang, Z.-H. (1991) Effects of DMSO and glycerol on ^{35}S incorporation of articular cartilage, *Cryo-Letters*, 12, 52–58.

Yankah, A.C., Alexi-Meskhishvili, V., Weng, Y., Schorn, K., Lange, P.E., et al. (1995) Accelerated degeneration of allografts in the first two years of life, *Ann. Thorac. Surg.*, 60, S71–S76.

Yankah, A.C., Weng, Y., Hofmeister, J., Alexi-Meskhishvili, V., Siniawski, H., et al. (1996) Freehand subcoronary aortic valve and aortic root replacement with cryopreserved homografts: Intermediate term results, *J. Heart Valve Dis.*, 5, 498–504.

Yap, C., Wong, A.M.F., Naor, J., and Rootman, D.S. (2001) Corneal temperature reversal after storage in Chen medium compared with Optisol GS, *Cornea*, 20, 501–504.

Zeiger, M.A.J., Tredget, E.E., and McGann, L.E. (1996) Mechanisms of cryoinjury and cryoprotection in split thickness skin, *Cryobiology*, 33, 376–389.

Zeiger, M.A.J., Tredget, E.E., Sykes, B.D., and McGann, L.E. (1997) Injury and protection in split-thickness skin after very rapid cooling and warming, *Cryobiology*, 35, 53–69.

Zeiger, M.A., Woods, E.J., Lakey, J.R., Liu, J., and Critser, J.K. (1999) Osmotic tolerance limits of canine pancreatic islets, *Cell Transplant*, 8, 277–284.

20 Engineering Desiccation Tolerance in Mammalian Cells: Tools and Techniques

Jason P. Acker, Tani Chen, Alex Fowler, and Mehmet Toner

CONTENTS

20.1 INTRODUCTION

Advances in tissue engineering, cell transplantation, and genetic technologies have created intense interest in the use of living cells as therapeutic tools in clinical care. The field of tissue engineering has already developed several commercial products used to temporarily or permanently replace organ functions, with the promise of many more to come. Cell transplantation is being used for many clinical indications to replace or enhance certain organ functions. Gene therapy is based on

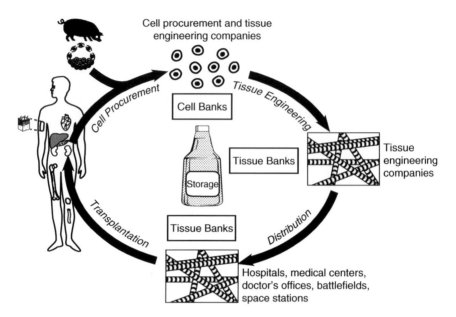

FIGURE 20.1 Long-term storage of cells is central to the successful application of cellular therapies in clinical medicine. Effective preservation techniques ensure that a readily available supply of biological material is available at the various steps in the development and transplantation of engineered cells and tissues.

the overexpression or downregulation of one or several cellular processes to provide enhanced functions to cells for various therapeutic end goals, such as wound healing, and tissue repair. Recent advances in stem cell biology further fuel the excitement for these burgeoning approaches by providing a potentially unlimited source of cells for reparative medicine.

As several living cell–based therapies are approaching clinical utility, emphasis must now be placed on the fundamental and practical issues associated with the translation of these new technologies from bench-to-bedside (Langer and Vacanti, 1993). Some of the important translational technologies include cell isolation, cell and tissue culture and differentiation, scale-up of bioreactors, advances in biomaterials and scaffolds, development of long-term storage strategies, and implementation of safety and regulatory policies. Although there have been significant advances both at the fundamental and practical levels for most of these technologies, the tools and understanding needed for the storage of living cells and complex tissue constructs lags significantly behind.

20.2 APPROACHES TO LONG-TERM STORAGE

Long-term storage of living cells or tissues is needed to provide a readily available supply of cells and engineered tissue constructs to end-users at medical centers, hospitals, clinics, and physician offices (Figure 20.1). Effective preservation procedures are required at various steps in the production of engineered cells and tissues, including screening of source cells, cell banking, inventory control, quality control, product distribution, and tissue banking (Karlsson and Toner, 2000). Cryopreservation is the established modality for long-term storage of living cellular systems and can be subdivided into two approaches based on overall methodology: standard freeze-thawing and vitrification. Although both approaches use cryogenic temperatures (typically below -80°C), there are important fundamental differences (Karlsson and Toner, 1996; Song et al., 2000), which are discussed in Chapters 7 and 20.

A more recent approach for cryopreservation involves the use of low concentrations of intracellular sugars to stabilize cells. Trehalose is loaded into cells at about 0.1 to 0.2 M using several different approaches. In the absence of conventional cryoprotectants such as dimethyl sulfoxide

(DMSO), results using intracellular trehalose show that fibroblasts, keratinocytes, and oocytes survive cryopreservation (Eroglu et al., 2000, 2002). The major advantage of nontoxic sugars is the potential to infuse freeze-thawed cells directly into patients without the cumbersome steps involved in the removal of traditional cryoprotectants.

Another recent development involves increasing the desiccation tolerance of mammalian cells and the development of novel strategies to store living cells at ambient temperatures. In nature, many organisms can survive in a dry, glassy state for extended times in a phenomenon called "anhydrobiosis" (Crowe et al., 1998; Potts, 1994). Anhydrobiosis is found in a variety of organisms including plant seeds, bacteria, yeast, brine shrimp, fungi and fungal spores, and cysts of certain crustaceans. Extensive studies using these organisms revealed that there is a series of complex molecular and physiological adaptations that permit these organisms to survive drying stress. Among these adaptations is the accumulation of internal sugars or mixtures of sugars, such as trehalose, sucrose, and raffinose. Sugars are believed to play a major role in the stabilization of membranes, proteins, and other key cellular structures in the dry state. The mechanism of sugar protection is an active area of research that includes the role of the glassy state in long-term stabilization of living cells and the interaction of sugars with biological molecules and supramolecular structures to afford stabilization. This is covered in detail in other chapters.

In desiccated (or dry) storage, living cells are put into a stasis state at ambient temperatures by dehydrating the cells using either freeze-drying or convective drying techniques. This approach is based on removing water to achieve a glassy (i.e., amorphous or vitrified) state in and around cells at ambient temperatures. The glassy state is known to have an exceedingly high viscosity ($>10^{12}$ Pa) that may inhibit the chemical, biological, and physical processes that lead to cell deterioration (Buitink et al., 1998, 2000; Sun and Leopold, 1997).

Dry storage provides a long-term preservation strategy that alleviates many of the problems associated with other preservation technologies. First, dried storage is the only preservation method that permits the ambient-temperature, long-term storage of biological molecules. This simplifies the distribution of the therapeutic product and reduces the need for stringent storage requirements at the manufacturing site and end user's facility. Second, as the process of drying removes a substantial proportion of the sample water, the resultant product is much smaller and lighter than conventionally preserved products. This can significantly increase the storage capacity of a facility, with a resulting increase in inventory levels. Third, the general acceptance of drying as a suitable method for the manufacturing of therapeutic products by regulatory authorities (Franks, 1998) makes this technology an appealing option for biopreservation. Finally, the significantly lower concentrations of stabilizers used in dry storage compared to standard freeze–thaw or vitrification protocols reduces the need for the careful and skilled removal of these agents before injection, transfusion, or transplantation.

Interest in engineering desiccation tolerance in mammalian cells is growing because the ability to store mammalian cells in the dry state has many advantages over the current low-temperature-based methods used for long-term storage. Recent work has focused on the role that intracellular sugars have in protecting cells from the effects of excessive dehydration (Bieganski et al., 1998; Chen et al., 2001; Garcia de Castro and Tunnacliffe, 2000; Guo et al., 2000; Puhlev et al., 2001; Wolkers et al., 2001). In the process of examining the effects of intracellular sugars, a number of technologies have been developed to incorporate protectants into cells, to control the drying and storage of cells, and to rehydrate the cells following drying. In this chapter we will discuss current techniques and technologies that are being developed for the desiccation and storage of mammalian cells in the dry state. More details of this new approach to living cell storage and mechanisms of sugar stabilization are covered in Chapter 5.

20.3 LOADING OF PROTECTANTS INTO MAMMALIAN CELLS

There is strong evidence that sugars need to be present on both sides of the membrane to afford maximum protection against the damaging effects of freezing and dehydration (Chen et al., 2001;

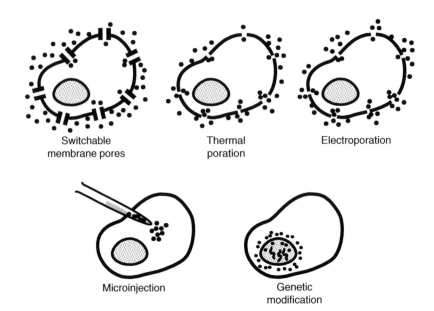

FIGURE 20.2 Schematic representation of the approaches currently being developed to overcome the impermeability of cell membranes to sugars and other compounds.

Crowe et al., 1985; Eleutherio et al., 1993; Eroglu et al., 2000, 2002; Womersley et al., 1986). Among the key impediments to using sugars such as trehalose during the desiccation of mammalian cells has been the impermeability of the plasma membrane to these molecules. A number of approaches have been used to overcome the impermeability of cell membranes to sugars and other compounds (Figure 20.2). Many of these techniques have been recently used to load protectants into mammalian cells to improve their desiccation tolerance. These methods are discussed below.

20.3.1 Switchable Membrane Pores

One approach that has been recently used to load trehalose and other sugars into mammalian cells involves the use of a genetically engineered mutant of *Staphylococcus aureus* α-hemolysin to reversibly permeabilize the plasma membrane (Bayley, 1997; Russo et al. 1997). Monomers of this protein, on introduction into lipid bilayers, form uniform 2-nm-diameter transmembrane pores that can be closed using divalent cations. Previous studies on the self-assembly of α-hemolysin protein (Walker et al., 1994) and on the structure of α-hemolysin (Song et al., 1996) have resulted in a detailed understanding of the mechanism by which this protein functions to porate mammalian cells. As a monomer, this 293–amino acid polypeptide spontaneously binds to the plasma membrane and forms a heptameric prepore complex with other bound monomers (Bayley, 1997). Membrane insertion follows oligomerization and results in the formation of a functional pore (Walker and Bayley, 1995). Using genetic engineering to replace five sequential native residues with histidines allows the heptameric pore to be toggled between an open and closed state by the removal or addition of micromolar concentrations of divalent cations (Bayley, 1997; Russo et al., 1997; Walker et al., 1994).

Genetically engineered α-hemolysin has been used in the development of techniques for the cryopreservation of mammalian cells (Eroglu et al., 2000). Eroglu and coworkers have shown that the introduction of low concentrations of trehalose (0.2 *M*) using H5 α-hemolysin resulted in a significant improvement in the postthaw recovery of rapidly frozen fibroblasts and keratinocytes (Eroglu et al., 2000). Critical to the survival of the porated and frozen cells was the ability to selectively close the pores following the loading of intracellular trehalose (Figure 20.3).

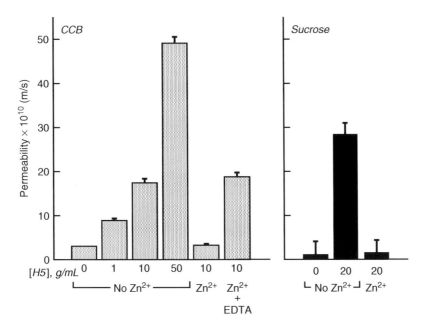

FIGURE 20.3 The permeability of mammalian cells to small molecules can be modulated using the switchable membrane pore H5 α-hemolysin. The addition of Zn^{2+} blocks the pore and prevents the uptake of carboxy-calcein blue and sucrose in NIH/3T3 fibroblasts. Permeability values for and sucrose were obtained by applying a simple transport model to data from fluorescence intensity and radiolabeled sucrose uptake assays, respectively. Figure reproduced from Russo et al. (1997).

   Using α-hemolysin to permeabilize mammalian cells has permitted the detailed examination of the effect that intracellular trehalose has in protecting mammalian cells from dessication injury (Chen et al., 2001). Studies have shown a beneficial effect of intracellular trehalose on the membrane integrity of dried fibroblasts (Chen et al., 2001). Fibroblasts were loaded with increasing concentrations of trehalose, using the H5 α-hemolysin pore, and then dried using natural convection. Examination of the postrehydration membrane integrity revealed not only that intracellular trehalose was necessary for the stabilization of the plasma membrane during drying and rehydration but that there was a minimal amount of intracellular trehalose required for protection ($0.2\ M$). Furthermore, the postrehydration membrane integrity was found to be a function of the residual moisture content of the cells following drying, with a threshold for survival at around 15%. Using this information, Chen and coworkers were able to recover a high percentage (>75%) of cells with intact membranes following dehydration and storage for up to 90 d at –20°C.

   Quantitative measurement of the intracellular trehalose concentration using gas chromatography/mass spectroscopy (GC-MS) has further demonstrated the advantage of using switchable membrane pore technology for the controlled loading of protectants (Acker et al., 2003). Using the GC-MS method, Acker and coworkers have shown that the switchable characteristics of H5 α-hemolysin provide an excellent means to regulate the permeabilization of cells and control the accumulation and removal of intracellular trehalose. By coupling GC-MS with the H5 α-hemolysin technology, verification of the loading of the high concentrations of intracellular trehalose necessary for engineering desiccation tolerance in mammalian cells was demonstrated.

   Whereas the α-hemolysin equipped with a metal-actuated switch has been exceedingly beneficial in reversible, controlled permeabilization and loading of sugars into mammalian cells, recent advances in engineering membrane pores indicates that it may be possible to further improve this technology. Using genetic engineering to modify the structure of α-hemolysin, novel mutants of this membrane pore have been developed that are highly selective, thus permitting the stochastic sensing of single molecules (Bayley and Cremer, 2001). Following this trend, it may become possible to engineer specific membrane pores that are more selective to sugars, resulting in much greater control over the movement of these molecules across the plasma membrane. As further advances are made in understanding the specific requirements for mammalian dessication tolerance, concurrent improvements in the engineering of switchable pores will ensure that this technology remains a powerful method to overcome the permeability barrier of mammalian cells.

20.3.2 THERMAL PORATION

When phospholipid bilayers and biological membranes are cooled, they undergo a liquid crystalline-to-gel phase transition at a defined temperature. This thermotropic phase transition has been shown to result in a transient increase in the permeability of the plasma membrane (Crowe et al., 1989). Also referred to as thermal shock, this phenomenon has been widely studied as it relates to the transient permeabilization of cells during cryopreservation, which often results in the colloidal osmotic lysis of sensitive cells (Lovelock, 1955; Daw et al., 1973). Using the thermotropic phase transition, Beattie and coworkers where able to load trehalose into pancreatic islet cells and show that a combination of DMSO and trehalose was effective at maintaining β-cell function following cryopreservation (Beattie et al., 1997). Recent efforts to permeabilize human fibroblasts to trehalose before the desiccation of these cells resulted in a measureable uptake of the disaccharide (Puhlev et al., 2001).

In another example of thermal poration, human platelets were successfully freeze-dried following permeabilization and loading of intracellular trehalose (Wolkers et al., 2001). Contrary to the standard treatment, whereby cells are permeabilized by cooling to 0° or 4°C, trehalose uptake in platelets was shown to occur following incubation at 37°C for extended times in trehalose-containing solutions. It has been suggested that a thermotropic transition in platelet membranes occurs at this physiologic temperature (Wolkers et al., 2001). On stimulation with thrombin, collagen, and ADP, rehydrated trehalose-loaded platelets responded by forming clots in a manner similar to untreated controls. Efforts to scale-up the freeze-drying process to permit the preparation of clinically relevant numbers of platelets are underway (Wolkers et al., 2002).

Using the thermotropic phase change represents a unique approach to loading sugars and other nonpermeable chemicals into mammalian cells. However, further work is required to determine the suitability of using this technique for the desiccation of mammalian cells. As it has been shown that not all mammalian cells undergo a distinct thermotropic phase transition (Drobnis et al., 1993; Watson, 1981), using this loading procedure may not be appropriate for certain cell types. In addition, the size range and specificity of the membrane pores to a variety of desiccation-important molecules needs to be resolved. Although further work is required to characterize and optimize this loading technique, recent successes using thermal poration to load sugars into mammalian cells have demonstrated the utility of this permeabilization strategy.

20.3.3 GENETIC MODIFICATION

Genetically engineering mammalian cells to express foreign genes has been proposed as a method of overcoming the permeability barrier of the plasma membrane to protectants. Initially developing this method as a technique to introduce intracellular sugars into *Escherichia coli* (Billi et al. 2000; Kaasen et al., 1994), researchers have recently transfected mammalian cells with the genes responsible for the coding of sucrose-6-phosphate synthase (Potts, 2000) and trehalose-6-phosphate synthase (Garcia de Castro and Tunnacliffe, 2000; Guo et al., 2000; Lao et al., 2001; Puhlev et al., 2001). These gene products, when expressed in conjunction with a respective phosphatase, function to convert uridine diphosphate glucose (UDP-Glc) into sucrose and trehalose. Accumulation of millimolar concentrations of intracellular sucrose (Billi et al., 2000) and trehalose (Garcia de Castro and Tunnacliffe, 2000) have been reported using this genetic engineering approach.

Although expression of intracellular trehalose in mammalian cells has been demonstrated, it is unclear whether desiccation tolerance is conferred by the intracellular concentrations being achieved using genetic engineering (Garcia de Castro et al., 2000; Levine, 2000). It has been reported that 0.3 to 0.5 pg of trehalose per cell is sufficient to maintain human primary skin fibroblasts in a dried state (Guo et al., 2000). This is in direct conflict with the work of Garcia de Castro and Tunnacliffe, who demonstrated that genetically engineered mouse fibroblasts, which accumulated 40 pg/cell trehalose, were unable to survive complete desiccation (Garcia de Castro and Tunnacliffe, 2000). Recent work by Levine and coworkers further complicates the issue, as they have concluded that the concentrations of intracellular trehalose produced using genetic expression are insufficient for the stable storage of mammalian cells in a desiccated state (Puhlev et al., 2001). As studies of other biological systems have indicated that a minimum concentration of ~0.1 M trehalose may be needed for protection during drying (Cerrutti et al., 2000; Chen et al., 2001; Israeli et al., 1993; Tunnacliffe et al., 2001), further improvements in the efficiency of the transfection and expression of bacterial sucrose and trehalose synthase genes in mammalian cells are necessary before this can become an effective technique to promote desiccation tolerance.

20.3.4 MICROINJECTION

Recent efforts to overcome the permeability barrier of oocytes (Eroglu et al., 2002; Kubisch et al., 1995) and embryos (Janik et al., 2000) to cryoprotectants have involved the microinjection of these molecules directly into the cell. Microinjection is the process whereby micrometer-size glass micropipettes are used to inject molecules directly into individual cells. This technique is well suited for the loading of high concentrations of large molecules into individual cells. By carefully controlling the injection conditions, accurate loading of precise amounts of sugars into human oocytes has been demonstrated (Eroglu et al., 2002), which permits the careful examination of the effect of intracellular sugar concentration on cell survival. The microinjection of small amounts of intracellular trehalose (0.15 M), in the absence of any other cryoprotectant, was shown to afford significant protection to cryopreserved human oocytes (Eroglu et al., 2002). Although microinjection is inherently a very tedious procedure that is not well suited for use on large populations of cells, for specific applications it is a useful method by which the permeability barrier of cells can be overcome.

20.3.5 ELECTROPORATION

Applying a transmembrane electric potential can alter the permeability of cell membranes. Electroporation (also known as electropermeabilization, -injection, or -transfection) is based on the temporary increase in membrane permeability caused by the reversible electric breakdown of the plasma membrane on application of external high-intensity field pulses of very short duration (pulsed electric field). When an electric pulse is applied to a plasma membrane, it has been proposed that formation of hydrophilic pores occurs when the dipoles of the phospholipid molecules are reoriented within the electric field (Tsong, 1987). Local Joule heating caused by the movement of ions across the membrane further enhances the formation of pores resulting from thermal phase transitions. This technique has been used successfully for gene transfer (Wong and Neumann, 1982; Kirchmaier, 2001), cell fusion (Senda et al., 1979), and the loading of molecules intracellularly (Kinosita and Tsong, 1978; Tsong, 1987; Mir, 2001).

As membrane pore formation is a highly dynamic process, with the pore size and transient stability of the pores being strongly dependent on characteristics of the cell, the shape characteristics of the pulse, and the composition of the intra- and extracellular media (Tsong, 1991; Djuzenova et al., 1996), developing optimal conditions for the electroporation and loading of high concentrations of large molecules can be difficult. New techniques to study the structure and dynamics of pore formation, using flow cytometry (Bartoletti et al., 1989; Prausnitz et al., 1993) and fluorescent microscopy (Sowers and Lieber, 1986), have allowed researchers to examine molecular transport in detail and to deduce the effectiveness of different electroporation parameters on the uptake of extracellular molecules. As this work develops, a better understanding of the intricate relationships between the many variables involved in electroporation will allow one to more effectively permeabilize and load protectants into mammalian cells.

Although electroporation has been successfully used to load extracellular agents into cells, much work remains to be done to understand the biological effects of the pulsed electric fields. Preventing irreversible cell damage such as membrane bleb formation (Gass and Chernomordik, 1990), disruptions in biochemical pathways (Vernhes et al., 1999), DNA denaturation (Lips and Kaina, 2001), and cell lysis (Kinosita and Tsong, 1977) will all be required before electroporation will find widespread use in the loading of protectants into mammalian cells.

20.3.6 INTRACELLULAR PROTECTANTS—SUMMARY

Enhancing the desiccation tolerance of mammalian cells by engineering the accumulation of intracellular sugars is one method that is currently being rigorously explored. The techniques discussed above use physical, chemical, or molecular methods to promote the accumulation of intracellular sugars by overcoming the inherent impermeability of mammalian cells to these molecules. Although each method suffers from one or more practical limitations, each technique has been successfully shown to enhance the intracellular concentration of sugars. These initial successes would strongly suggest that continued efforts to characterize and understand these technologies would improve the utility of these strategies for the intracellular accumulation of protectants.

As the role that intracellular sugars will play in the final engineering of desiccation tolerance in mammalian cells is still uncertain (Oliver et al., 2001; Tunnacliffe et al., 2001), additional strategies are being investigated. For example, other protective mechanisms used by natural systems that undergo seasonal exposure to environmental stresses include downregulation of metabolism to enter a hypometabolic state (diapause), scavenging of reactive oxygen species, and the intracellular accumulation of proteins and amphiphilic solutes (reviewed in Oliver et al., 2001). Researchers are studying these adaptations with the aim of metabolically and genetically engineering mammalian cells to more effectively mimic the complex response of these natural systems to desiccation. As access to the intracellular environment will likely be a necessary element in this process, expanding and improving current techniques for the expression or permeation of endogenous protectants into mammalian cells will be critical.

20.4 DRYING TECHNIQUES

Sugars have been shown to stabilize biomolecules through the formation of stable amorphous glassy states during drying (Crowe et al., 1992; Franks, 1999; Hirsh, 1987). The formation of a glass caused by the removal of water can be achieved through two general paths (Figure 20.4). The simplest path constitutes drying at ambient temperature (Figure 20.4, path AD). In contrast, freeze-drying or lyophilization consists of three distinct phases: freezing, where water is separated from the solution as pure water (ice) phase, with the concomitant concentration of solutes in the residual liquid (freeze concentration) as shown by path AB; the removal of the ice (usually 80% of the water in the original solution) by sublimation (primary drying), as depicted by path BC; and finally, the removal of most of the residual unfrozen water from the freeze-concentrated sample by diffusion/desorption from the sample (secondary drying) as in path CD.

As the formation of a glass caused by the removal of water can be achieved through at least two distinct paths, there exist a number of ways by which biological systems can be stabilized in a desiccated state using amorphous glasses. As these processes are analogous to conditions observed in nature, where anhydrobiotic organisms can experience air- and flow-drying conditions as well as drying at subzero temperatures in Artic regions, there has been much interest in their application

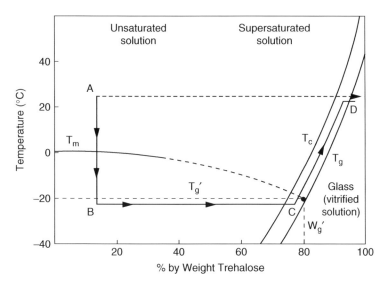

FIGURE 20.4 Representative supplemented phase diagram for a trehalose/water system. Path AD indicates removal of water from a 15% trehalose solution at ambient temperatures (A) and the formation of an amorphous glassy state (D). Path AöBöCöD depicts a hypothetical optimal freeze-drying pathway.

to biopreservation. However, during desiccation, extreme stresses are placed on cells, such as liquid-to-glass transitions or hyperosmolarity resulting from supersaturation or precipitation. These stresses, and how cells can survive these stresses, are a function of the drying conditions as well as the internal composition of the cells. Understanding these drying processes is vital when designing methods to engineer dessication tolerance in cells and tissues.

20.4.1 Drying at Ambient Temperature

Natural convection drying is one of the simplest methods to dry cells and tissues. The sample of cells or tissue is placed into a closed environment, or is just left out in the open. If the air or other gas in the environment of the sample has a lower water vapor pressure than that at the sample surface, then drying will occur. If the air is completely quiescent, then the drying will occur by simple diffusion through the surrounding gas; but in practice this does not occur. As long as gravity is present, a natural circulation of the environmental gases will occur as a result of the drying itself. The nature of this circulation is complex because there are two competing effects that influence the flow. The increase in water molecules in the gas near the sample surface usually decreases the density of the gas because water molecules have a lower molecular weight than nitrogen, which is the most commonly used gas for drying environments as well as the principal component of air. The evaporation of water from the surface, however, also leads to surface cooling, which causes an increase in the density of the gas near the sample. Depending on the drying rate, temperature, and local humidity, either of these effects can be dominant; but they are virtually never in perfect balance, so a gas flow will result. This naturally occurring gas flow, natural convection, will increase the drying rate relative to that of pure diffusion through still air, but the drying rate will still tend to be relatively slow compared to other drying techniques. Drying a small sample of aqueous sugar solution to a low moisture content using natural convection takes hours or even days, depending on sample geometry and temperature and, most important, on the humidity of the drying environment. In nature, many organisms enter the anhydrobiotic state through a process of slow water loss, as rapid drying will kill these organisms (Aguilera and Karel, 1997). Exploration of natural convection drying, therefore, is important insofar as it mimics nature.

The major concern when using natural convection drying is that it is not currently possible to predict the final moisture content in the sample that can be achieved even when drying for extremely long times. As mentioned earlier, drying will only occur if the water vapor pressure at the sample surface is greater than that in the environment. This is true for all drying protocols. One impediment to predicting final moisture contents and drying rates for the drying of aqueous sugar solutions is that the water vapor pressure as a function of concentration has not been determined for many sugars. Although there are idealizations such as Raoult's law that can be used to predict vapor pressure at high moisture contents, the vapor pressure at low moisture contents needs to be determined empirically. Whereas some of this data has been collected in the food processing industry for fructose and sucrose (Norrish, 1966; Venegas and Marinos-Kouris, 1992), the data are incomplete and do not exist at all for many of the proposed protectants for mammalian cells.

The inability to predict water vapor pressure for high-concentration sugar solutions is an impediment to understanding and predicting drying in all protocols, but it is of particular concern in natural convection drying because, in general, the drying environment has some humidity. This is obviously true for experiments that rely on room conditions, but it is also true for closed environments because even if those closed environments start out completely dry, the evaporation of water from the sample itself quickly raises the environmental humidity. For a room-temperature cubic chamber with a 10-cm side length, the internal environment will reach 100% relative humidity after 23 mg of water have evaporated. After that, even pure water will no longer dry. Because one does not know the water vapor pressure at the surface of a concentrated sugar solution, one cannot know how much drying will occur before no more drying is possible. In the above-mentioned chamber, it would certainly be less than 23 mg, as the presence of sugar will reduce the water vapor pressure at the surface. This difficulty can be partially overcome by the addition of desiccants or saturated salt solutions (Winston and Bates, 1960; Young, 1967) to the closed environment. Using these, the relative humidity can be held at known levels within the chamber. This will decrease the effect that evaporation from the sample has on the internal humidity of the chamber. The final moisture of the sample, even after extremely long drying times, cannot be predicted without more complete vapor pressure data for concentrated sugar solutions.

More rapid drying can be accomplished using forced convection. This technique relies on gas flow forced over the sample to cause drying. The gases used are typically either desiccated air or nitrogen gas. The rate of drying depends on the temperature of the sample and gas, the velocity of the forced gas, and the geometry of the chamber. When using desiccated air or nitrogen with very low humidity, one can be sure the sample will ultimately become extremely dry, as the moisture from the sample is carried downstream and out of the system. Lack of knowledge about vapor pressure at the surface of highly concentrated solutions still makes modeling and predicting the drying rates difficult, as the solution becomes very low in moisture; but the sample will continue to dry as long as the gas is dry and there is water left in the sample.

The difficulty with forced convection drying is that it usually results in nonuniform drying rates within a chamber. Drying rates at the entrance of the flow chamber will be greater than drying rates at the back. Many simple chamber designs will result in flow recirculation zones where the same gas passes repeatedly over the sample and becomes progressively more humid, rather than being purged out of the chamber so that new, dry gas can pass over the sample. Increasing the gas flow rates and shortening the channel length can minimize these problems. However, high gas flow rates are not practical for all samples, such as in the case of cells suspended in liquid droplets. These samples may be simply blown away. Design of a good forced-convection drying environment requires a fairly detailed understanding of fluid mechanics and mass transfer. In general, nonuniformities in the drying rate will exist inside the chamber, and this consequently makes it difficult to dry biological samples.

More uniform rapid drying can be achieved using vacuum drying of liquid. When the vacuum is pulled, the liquid will begin to evaporate very rapidly, and the sample will cool because of the fact that the phase change from liquid to gas requires energy. Liquid droplets can actually freeze during a fast evaporative process if there is no additional source of energy provided. In general, samples dried in this manner are placed in good thermal contact with a large warm surface, or active heating is provided. The drying rate depends strongly on the temperature of the sample. The sample temperature is strongly influenced by radiative heat transfer, as there is very little gas to conduct heat to the sample. If the radiative transfer to different areas of the chamber is uniform, then the drying rate will be uniform. Vacuum drying works very well for drying small samples, provided freezing can be avoided. For larger samples, vacuum drying requires some method for removing the water vapor as drying progresses. Although a standard vacuum pump will maintain fairly low overall pressures within the chamber, that pressure will be made up entirely of water vapor as drying progresses. The small water vapor pressure in the "evacuated" chamber can be enough to inhibit further drying. In freeze-drying, the water vapor is removed by collection on the condenser; but in simple vacuum drying, the water vapor is not removed. A system for purging the water vapor could be incorporated by repeated introduction of gas into the chamber or by adding a cold condenser. For simple vacuum systems, vacuum drying will provide rapid uniform drying only for small sample sizes.

20.4.2 FREEZE-DRYING

Freeze-drying, or lyophilization, is commonly used to dry biological compounds such as proteins or drugs (Franks, 1998; Gatlin and Nail, 1994). Although freeze-drying has been extensively studied, less work has been done to preserve mammalian cells and tissues by this technique. The past success with freezing-drying bacteria and other simple organisms (Conrad et al., 2000; Israeli et al., 1993; Leslie et al., 1995) and the recent report of the successful freezing-drying of platelets (Wolkers et al., 2001) have resulted in a great interest in extending these studies to mammalian cells.

Freeze-drying is usually performed on samples that undergo vitrification. Disaccharides such as trehalose or sucrose can be used as additives, as they have been shown to have both cryoprotective and lyoprotective capabilities (Crowe et al., 1990; Franks, 1999). Freeze-drying generally consists of three steps: freezing, primary drying, and secondary drying. The freezing of cells and tissues is a complicated process that involves careful optimization of the freezing regime. Although research has been performed on freezing cells under various conditions, less work has been performed on the freezing of tissue sections in a uniform and controllable manner. Complicating the freezing step is that many additives and cryoprotective agents often used during the freezing process are reactive or toxic to cells at room temperature, preventing their use in dried storage.

During primary drying, the temperature of the shelf containing the samples is kept low, usually $-5°$ to $-40°C$, under high vacuum (typically <100 mTorr). Under these conditions, ice within the sample sublimates directly into the vapor phase. The vapor condenses on a condenser, which is held at a temperature lower than the shelf temperature, typically $-80°C$. During the primary drying, the frozen sample is typically held at a constant temperature that is below the glass phase transition temperature (T_g') of the freeze-concentrated solution W_g'. Primary drying occurs until the glass transition is reached, at which point all of the ice within the sample has been sublimated away. Residual amounts of water remaining in the glass phase (so-called "bound" water) are removed during secondary drying.

In the secondary drying phase, the temperature is raised in a controlled way slightly above the glass transition so that the water in the freeze-concentrated solution gains mobility and thus diffuses from the bulk to the surface, where it is removed by desorption. The rise in temperature alters the phase equilibrium, and less water is present in the glassy state (see Figure 20.4). Thus, additional amounts of water can be removed from the sample. When done correctly, the secondary drying essentially "climbs" the glass transition curve. At a temperature called the collapse temperature, T_c, the freeze-dried preparations lose the open texture and the ability to be rapidly rehydrated because the sample exhibits observable deformation. Therefore, the secondary drying phase must be carefully performed at a temperature below T_c. The secondary drying requires controlled increases of temperature to effectively remove water from the glassy matrix.

Freeze-drying is a very slow technique that can take up to several days to complete. The drying rate is a function of the shelf temperature and the vacuum pressure of the freeze-drying chamber.

However, as with other drying techniques, the drying rate for tissues will also depend on the size and surface area of the tissue. The interior of large tissue sections will dry more slowly than the surface.

Freeze-drying puts two major stresses on the cells: freezing and desiccation stresses (Crowe et al., 1990). Careful design of the experimental protocol is necessary to minimize both of these stresses. Freeze-drying avoids problems with supersaturation and crystallization and removes water very slowly while the cells are immobilized. In addition, a large amount of research has already been conducted into freeze-drying, and many excipients, bulking agents, antioxidants, and other preservatives have already been developed to protect proteins and other molecules of biological interest (Arakawa et al., 2001; Franks, 1998, 1999). Much of this knowledge will undoubtedly prove useful in designing techniques to preserve cells and tissues.

20.4.3 Drying Summary

In conclusion, several different drying techniques have recently been developed that show promise in the preservation of mammalian cells and tissues during drying. Understanding and modeling of the physical processes that occur during drying will be vital in controlling drying rates and final moisture contents. Careful design of the drying protocols is needed to minimize the stresses experienced by the cells.

20.5 STORAGE AND BIOLOGICAL GLASSES

During long-term dry storage, cells and tissues are typically preserved in a vitrified state. Degradation reactions are significantly slower in a glassy state than in aqueous solution, and labile molecules such as proteins can retain their native conformations during storage; a fact that is being used to preserve proteins in the pharmaceutical industries. However, cells and tissues are much more complicated than individual proteins, and additional issues may arise during the storage of these materials.

20.5.1 Storage

The storage conditions chosen will affect survival in the dried state. Although it is common practice to use storage conditions below the T_g, this does not necessarily mean that all molecular motion stops. In fact, many studies have shown that molecular motion and chemical reactions can still occur well below the T_g (Hancock et al., 1995; Shamblin and Zografi, 1998; Sun and Leopold, 1997). For example, ion transport has been reported within glassy polymer substrates (Imrie et al., 1999). Significant degradation can occur on the timescale of weeks, even at 30°C below the T_g (Hancock et al., 1995). Likewise, studies of liposomes have shown that leakage can occur at temperatures below the T_g (Sun et al., 1996). The nature of these reactions is not understood. Reactions in the glassy state typically occur on a timescale of days to months. They may be the result of the diffusion of compounds through the glassy matrix or of chemical reactions between adjacent molecules. Because of these possible reactions, temperatures significantly lower than the T_g of the matrix may be required for successful long-term storage, once desiccation and vitrification of the cells has occurred.

Studies of pharmaceutical compounds preserved in glassy states have shown that molecular motion occurs within the glassy state and that temperatures at least 50°C below the T_g may be required for long-term preservation (Hancock et al., 1995; Shamblin et al., 1999). Similarly, recent work examining the effect of storage temperature on the stability of the plasma membrane of desiccated mammalian cells has shown that long-term preservation of intact cells requires a storage temperature of at least 30 to 50°C below the T_g (Chen et al., 2001). Degradation of the plasma membrane was shown to occur, despite preservation within a glassy matrix. Although more experiments are needed to elucidate the mechanism by which damage in the glassy state occurs, it would appear that storage of mammalian cells just below the T_g might not be enough for long-term preservation of cells in the dried state.

The relative humidity plays an important factor in the long-term state of the dried sample. Equilibrium moisture contents can generally be predicted from moisture isotherm data, with higher moisture contents occurring at higher relative humidities (Saleki-Gerhardt and Zografi, 1994). However, vitrified samples in storage can continue to dry, as water leaves the sample through surface diffusion-type processes. The drying rate of the sample is generally controlled by the relative humidity of the ambient gas. Over time, even tissue sections that have been uniformly dried will still develop moisture gradients. Higher relative humidities during storage will slow down the rate of water loss, with a relative humidity approaching 100% required to completely stop water loss. Detailed studies of the equilibrium vapor pressure for concentrated solutions of sugars and other proposed protectants need to be carried out to determine proper storage conditions.

20.5.2 BIOLOGICAL GLASS FORMATION

It is not clear whether the cells themselves vitrify once the external matrix has vitrified, and how this vitrification actually occurs. The overall glass transition of the cell is affected by all of the proteins and lipids within the cell. The transition itself may involve hundreds of reversible thermal events caused by changes in lipids, DNA, or protein structures (Lepock et al., 1993). In some cases, the cell may not vitrify at a single defined T_g; heterogeneous mixtures such as block copolymers and starch often show this behavior (Wunderlich, 1990; Sikora et al., 1999). Glass transitions have been studied extensively in plant cells, such as pollen or seeds, which often contain very high concentrations of sugars such as sucrose or raffinose. Although similar to pure sugar solutions, differences in glass formation in biological systems have been observed (Sun and Leopold, 1997; Leopold et al., 1994).

As a typical cell is a highly compartmentalized structure, it is not clear whether all parts of the cell vitrify or not, and how that affects long-term survival. Furthermore, it is not clear how cells tolerate partial vitrification. Enclosed organelles, such as mitochondria or the nucleus, may not have the same concentrations of sugars as the cytosol, and therefore may not vitrify under the same conditions. Organisms that produce sugars do not distribute them evenly (McBride and Ensign, 1990; Tettero et al., 1994), but concentrate them in the cytoplasm (Bruni and Leopold, 1992; Buitink et al., 1998). Furthermore, water transport across the lipid membranes of the organelles, as well as the plasma membrane, may be rate-limiting and will thus affect how vitrification occurs. The viscosity of the cytoplasm is significantly higher than that of water, even in fully hydrated cells (Pollack, 2001), and this may impede molecular motion or alter glass formation. Intracellular sugars will have correspondingly more interactions with proteins and lipids within the cell. It is not clear how these differences affect vitrification of the intracellular space, and further studies will be needed to understand the vitrification process on an organelle level.

In conclusion, the formation of biological glasses is an area in which further study is needed, as the vitrification processes within cells and tissues are not well understood. Although techniques such as differential scanning calorimetry, electron paramagnetic/spin resonance, and dynamic mechanical analysis have been used to study biological glass formation, these have typically been used to measure bulk properties on the cellular level in plants (Buitink et al., 1998; Miller et al., 1997; Williams, 1994). Extending these studies to other cells and tissues, as well as down to the organelle level, is a significant challenge.

20.6 IMPORTANCE OF REHYDRATION

Regardless of the method used to dry mammalian cells, rehydration is a critical step in their successful recovery. Just as improper drying and storage conditions can result in cell death, neglecting the severe physico-chemical changes that occur during rehydration can have lethal consequences. Understanding the rate at which these rehydration-induced changes occur will greatly aid in improving the survival of dried cells.

Although very little work has focused on the understanding the effects of rehydration on the survival of mammalian cells, studies with bacteria and yeast systems have demonstrated that the rehydration conditions can significantly affect viability. For example, survival rates of *Saccharomyces cerevisiae* have been shown to improve when cells are slowly rehydrated at relatively high temperatures (Becker and Rapoport, 1987; Poirier et al., 1999; Van Stevenick and Ledeboer, 1974). Membrane destabilization and solute leakage following rehydration have been proposed to occur because of a desiccation-induced increase in the lipid gel-to-liquid crystalline phase transition temperature (T_m; Crowe and Crowe, 1982; Crowe et al., 1992). Increasing the rehydration temperature, or adding agents to depress T_m have been shown to significantly improve the survival of yeast and bacteria following desiccation (Leslie et al., 1994, 1995; Poirier et al., 1999). Slow rehydration of yeast, bacteria, and platelets by placement in atmospheres of low moisture before complete rehydration in media has also been shown to improve cell survival (Marechal and Gervais, 1994; Poirier et al., 1997, 1999; Wolkers et al., 2001).

In addition to controlling the rate and temperature of rehydration, improved cell survival has been shown to occur when efforts are made to minimize oxidative damage to cells. As a result, the lethal effects of storage and rehydration under vacuum (Becker and Rapoport, 1987; Puhlev et al., 2001), in nitrogen environments (Matsuo et al., 1995; Poirier et al., 1999), and in the absence of light (Puhlev et al., 2001; Seel et al., 1991) have been examined.

In the few studies that have focused on the desiccation of nucleated mammalian cells, rapid one-step rehydration has been the dominant technique used (Chen et al., 2001; Garcia de Castro and Tunnacliffe, 2000; Guo et al., 2000; Wakayama and Yanagimachi, 1998). The knowledge gained working with bacteria and yeast indicates that uncontrolled rapid addition of media to desiccated mammalian cells is unlikely to be the best method for maximizing cell survival. A detailed examination of the effects of rehydration on the survival of desiccated mammalian cells is clearly warranted.

20.7 OUTLOOK

With the advancements being made in tissue bioengineering and genetic technologies, the clinical demand for effective long-term storage methods for cells and tissues will continue to increase. Desiccation of mammalian cells is an attractive storage strategy that has the potential to alleviate some of the practical constraints of using cryopreservation. However, as with all new technologies, there are a number of challenges that must be overcome before the practical application of desiccation engineering will become a reality.

The challenges facing the development and implementation of drying technologies for mammalian systems are similar to those that were faced in the field of cryopreservation in the late 1950s. Although the discovery of the effects of glycerol and dimethyl sulfoxide on cell survival following freezing created the promise of an effective method for the long-term storage of cells, there was little knowledge of the mechanism of action and the sites of damage during freezing. This hindered the development and widespread application of the technology. Similarly, the introduction of intracellular sugars has been shown to improve the survival of cells following desiccation; however, little is known about the mechanisms of damage, nor is there a consensus on the protective effects of sugars. The diversity of potential pathways for cell damage during the incorporation of intracellular sugars, drying, and rehydration necessitates the careful examination of the biochemical and biophysical effects of drying and of the stabilizing effects of sugars in the cellular microenvironment.

The critical need for effective preservation technology in cell and tissue engineering will be the impetus for accelerated efforts toward solving some of the fundamental problems surrounding the use of dessication engineering. Although the remaining scientific and technological challenges are formidable, significant progress has been made through interdisciplinary research activities, and prospects for the future are bright.

ACKNOWLEDGMENTS

We are grateful to Drs. Sankha Bhowmick and Gloria Elliott for their critical review of this chapter and to the many technicians who made valuable contributions to this work. Research from the authors' laboratories was partially supported by the National Institutes of Health (DK46270), Defense Advanced Research Projects Agency (N00173_01_1 G011) and the Shriners Hospitals for Children. Dr. Jason Acker was supported by a Post-Doctoral Fellowship from the Canadian Institutes of Health Research.

REFERENCES

Acker, J.P., Lu, X.-M., Young, V., Cheley, S., Bayley, H., Fowler, A., and Toner, M. (2003) Measurement of trehalose loading of mammalian cells porated with a metal-actuated switchable pore, *Biotechnol. Bioeng.*, 82, 525–532.

Aguilera, J.M., and Karel, M. (1997) Preservation of biological materials under desiccation, *Crit. Rev. Food Sci. Nutr.*, 37, 287–309.

Arakawa, T., Prestrelski, S.J., Kenney, W.C., and Carpenter, J.F. (2001) Factors affecting short-term and long-term stabilities of proteins, *Adv. Drug. Deliv. Res.*, 46, 307–326.

Bartoletti, D.C. Harrison, G.I., and Weaver, J.C. (1989) The number of molecules taken up by electroporated cells: Quantitative determination, *FEBS Lett.*, 256, 4–10.

Bayley, H. (1997) Building doors into cells, *Sci. Am.*, 277, 62–67.

Bayley, H. and Cremer, P.S. (2001) Stochastic sensors inspired by biology, *Nature*, 413, 226–230.

Beattie, G.M., Crowe, J.H., Lopez, A.D., Cirulli, V., Ricordi, C., and Hayek, A. (1997) Trehalose: A cryoprotectant that enhances recovery and preserves function of human pancreatic islets after long-term storage, *Diabetes*, 46, 519–523.

Becker, M.J. and Rapoport, A.I. (1987) Conservation of yeasts by dehydration, *Adv. Biochem. Eng. Biotechnol.*, 35, 127–171.

Bieganski, R.M., Fowler, A., Morgan, J.R., and Toner, M. (1998) Stabilization of active recombinant retroviruses in an amorphous dry state with trehalose, *Biotechnol. Prog.* 14, 615–620.

Billi, D., Wright, D.J., Helm, R.F., Potts, M., and Crowe, J.H. (2000) Engineering desiccation tolerance in *Escherichia coli*, *Appl. Environ. Microbiol.* 66, 1680–1684.

Bruni, F. and Leopold, A.C. (1992) Cytoplasmic glass formation in maize embryos, *Seed Sci. Res.*, 2, 251–253.

Buitink, J., Claessens, M.A.E., Hemminga, M.A., and Hoekstra, F. (1998) Influence of water content and temperature and molecular mobility and intracellular glasses in seeds and pollen, *Plant Physiol.*, 118, 531–541.

Buitink, J., Hemminga, M.A., and Hoekstra, F.A. (2000) Is there a role for oligosaccharides in seed longevity? An assessment of intracellular glass stability, *Plant Physiol.*, 122, 1217–1224.

Cerrutti, P., Segovia de Huergo, M., Galvagno, M., Schebor, C., and del Pilar Buera, M. (2000) Commercial baker yeast stability as affected by intracellular content of trehalose, dehydration procedure and the physical properties of external matrices, *Appl. Microbiol. Biotechnol.*, 54, 575–580.

Chen, T., Acker, J.P., Eroglu, A., Cheley, S., Bayley, H., Fowler, A., and Toner, M. (2001) Beneficial effect of intracellular trehalose on the membrane integrity of dried mammalian cells, *Cryobiology*, 43, 168–181.

Conrad, P.B., Miller, D.P., Cielenski, P.R., and de Pablo, J.J. (2000) Stabilization and preservation of *Lactobacillus acidophilus* in saccharide matrices, *Cryobiology*, 41, 17–24.

Crowe, J.H., Carpenter, J.F., and Crowe, L.M. (1998) The role of vitrification in anhydrobiosis, *Ann. Rev. Physiol.* 60, 73–103.

Crowe, J.H., Carpenter, J.F., Crowe, L.M., and Anchordoguy, T.J. (1990) Are freezing and dehydration similar stress vectors? A comparison of modes of interaction of stabilizing solutes with biomolecules, *Cryobiology*, 27, 219–231.

Crowe, J.H. and Crowe, L.M. (1982) Induction of anhydrobiosis: Membrane changes during drying, *Cryobiology*, 19, 317–328.

Crowe, J.H., Crowe, L.M., and Hoeskstra, F.A. (1989) Phase transitions and permeability changes in dry membranes during rehydration, *J. Bioenerg. Biomembr.*, 21, 77–91.

Crowe L.M., Crowe J.H., Rudolph, A., Womersley, C., and Appel, L. (1985) Preservation of freeze-dried liposomes by trehalose, *Arch. Biochem. Biophys.*, 242, 240–247.

Crowe, J.H., Hoekstra, F.A., Crowe, L.M. (1992) Anhydrobiosis, *Annu. Rev. Physiol.*, 54, 579–599.

Daw, A., Farrant, J., and Morris, G.J. (1973) Membrane leakage of solutes after thermal shock or freezing, *Cryobiology*, 10, 126–133.

Djuzenova, C.S., Zimmermann, U., Frank, H., Sukhorukov, V.L., Richter, E., and Fuhr, G. (1996) Effect of medium conductivity and composition on the uptake of propidium iodide into electropermeabilized myeloma cells, *Biochim. Biophys. Acta*, 1284, 143–152.

Drobnis, E.Z., Crowe, L.M., Berger, T., Anchordoguy, T.J., Overstreet, J.W., and Crowe, J.H. (1993) Cold shock damage is due to lipid phase transitions in cell membranes: A demonstration using sperm as a model, *J. Exp. Zool.*, 265, 432–437.

Eleutherio, E.C.A., de Araujo, P.S., and Panek, A.D. (1993) Role of the trehalose carrier in dehydration resistance of *Saccharomyces cerevisiae*, *Biochim. Biophys. Acta*, 1156, 263–266.

Eroglu, A., Russo, M.J., Bieganski, R., Fowler, A., Cheley, S., Bayley, H., and Toner, M. (2000) Intracellular trehalose improves the survival of cryopreserved mammalian cells, *Nat. Biotechnol.*, 18, 163–167.

Eroglu, A., Toner, M., and Toth, T.L. (2002) Beneficial effect of microinjected trehalose on the cryosurvival of human oocytes, *Fertil. Steril.*, 77, 152–158.

Franks, F. (1998) Freeze-drying of bioproducts: Putting principles into practice, *Eur. J. Pharm. Biopharm.*, 45, 221–229.

Franks, F. (1999) Thermomechanical properties of amorphous saccharides: Their role in enhancing pharmaceutical product stability, *Biotechnol. Genet. Eng. Rev.*, 16, 281–292.

Garcia de Castro, A., Lapinski, J., and Tunnacliffe, A. (2000) Anhydrobiotic engineering, *Nat. Biotechnol.*, 18, 473.

Garcia de Castro, A., and Tunnacliffe, A. (2000) Intracellular trehalose improves osmotolerance but not desiccation tolerance in mammalian cells, *FEBS Lett.*, 487, 199–202.

Gass, G.V. and Chernomordik, L.V. (1990) Reversible large-scale deformations in the membranes of electrically-treated cells: Electroinduced bleb formation, *Biochim. Biophys. Acta*, 1023, 1–11.

Gatlin, L.A. and Nail, S.L. (1994) Protein purification process engineering. Freeze drying: A practical overview, *Bioprocess Technol.*, 18, 317–367.

Guo, N., Puhlev, I., Brown, D.R., Mansbridge, J., and Levine, F. (2000) Trehalose expression confers desiccation tolerance on human cells, *Nat. Biotechnol.*, 18, 168–171.

Hancock, B.C., Shamblin, S.L., and Zografi, G. (1995) Molecular motion of amorphous pharmaceutical solids below their glass transition temperatures, *Pharm. Res.*, 12, 799–806.

Hirsh, A.G. (1987) Vitrification in plants as a natural form of cryoprotection, *Cryobiology*, 24, 214–228.

Imrie, C.T., Ingram, M.D., and McHattie, G.S. (1999) Ion transport in glassy polymer electrodes, *J. Phys. Chem. B*, 103, 4132–4138.

Israeli, E., Shaffer, B.T., and Lightheart, B., (1993) Protection of freeze-dried *Escherichia coli* by trehalose upon exposure to environmental conditions, *Cryobiology*, 30, 519–523.

Janik, M., Kleinhans, F.W., and Hagedorn, M. (2000) Overcoming a permeability barrier by microinjecting cryoprotectants into Zebrafish embryos (*Brachydanio rerio*), *Cryobiology*, 41, 25–34.

Kaasen, I., McDougall, J., and Strøm, A.R. (1994) Analysis of the otsBA operon for osmoregulatory trehalose synthesis in *Escherichia coli* and homology of the OtsA and OtsB proteins to the yeast trehalose-6-phosphate synthase/phosphatase complex, *Gene*, 145, 9–15.

Karlsson, J.O.M. and Toner, M. (1996) Long-term storage of tissues by cryopreservation, *Biomaterials*, 17, 243–256.

Karlsson, J.O.M. and Toner, M. (2000) Cryopreservation: Foundations and applications in tissue engineering, in *Principles of Tissue Engineering*, 2nd edition, Lanza, R., Langer, R., and Vacanti, J., Eds., Academic Press, New York, pp. 293–307.

Kinosita, K. and Tsong, Y.T. (1977) Hemolysis of human erythrocytes by a transient electric field, *Proc. Natl. Acad. Sci. USA*, 74, 1923–1927.

Kinosita, K. and Tsong, T.Y. (1978) Survival of sucrose-loaded erythrocytes in circulation, *Nature*, 272, 258–260.

Kirchmaier, A.L. (2001) Introduction of plasmid vectors into cells via electroporation, *Methods Mol. Biol.,* 174, 137–145.

Kubisch, H.M., Hernandez-Ledezma, J.J., Larson, M.A., and Sikes, J.D. (1995) Expression of two transgenes in *in vitro* matured and fertilized bovine zygotes after DNA microinjection, *J. Reprod. Fertil.,* 104, 133–139.

Langer, R. and Vacanti, J.P. (1993) Tissue engineering: The development of functional substitutes for damaged tissue, *Science,* 260, 920–926.

Lao, G., Polayes, D., Xia., J.L., Bloom, F.R., Levine, F., and Mansbridge, J. (2001) Overexpression of trehalose synthase and accumulation of intracellular trehalose in 293H and 293FtetR:Hyg cells, *Cryobiology,* 43, 106–113.

Leopold, A.C., Sun, W.Q., and Bernal-Lug, I. (1994) The glassy state in seeds: Analysis and function, *Seed Sci. Res.,* 4, 267–274.

Lepock, J.R., Frey, H.E., and Ritchie, K.P. (1993) Protein denaturation in intact hepatocytes and isolated cellular organelles during heat shock, *J. Cell Biol.,* 122, 1267–1276.

Leslie, S.B., Teter, S.A., Crowe, L.M., and Crowe, J.H. (1994) Trehalose lowers membrane phase transitions in dry yeast cells, *Biochim. Biophys. Acta,* 1192, 7–13.

Leslie, S.B., Israeli, E., Lighthart, B., Crowe, J.H., and Crowe, L.M. (1995) Trehalose and sucrose protect both membranes and proteins in intact bacteria during drying, *Appl. Environ. Microbiol.,* 61, 3592–3597.

Levine, F. (2000) *Nat. Biotechnol.,* 18, 473.

Lips, J. and Kaina, B. (2001) DNA double-strand breaks trigger apoptosis in p53-deficient fibroblasts, *Carcinogenesis,* 22, 579–585.

Lovelock, J.E. (1955) Haemolysis by thermal shock, *Br. J. Haematol.,* 1, 117–129.

Marechal, P.A. and Gervais, P. (1994) Yeast viability related to water potential variation: Influence of the transient phase, *Appl. Microbiol. Biotech.,* 42, 617–622.

Matsuo, S., Toyokuni, S., Osaka, M., Hamazaki, S., and Sugiyama, T. (1995) Degradation of DNA in dried tissues by atmospheric oxygen, *Biochem. Biophys. Res. Commun.,* 208, 1021–1027.

McBride, M. J., and Ensign, J.C. (1990) Regulation of trehalose metabolism by *Streptococcus griseus* spores, *J. Bacteriol.,* 172, 3637–3643.

Miller, D.P., de Pablo, J.J., and Corti, H. (1997) Thermophysical properties of trehalose and its concentrated aqueous solutions, *Pharm. Res.,* 14, 579–590.

Mir, L.M. (2001) Therapeutic perspectives of *in vivo* cell electropermeabilization, *Bioelectrochem. Bioenerg.,* 53, 1–10.

Norrish, R.S. (1966) An equation for the activity coefficients and equilibrium relative humidities of water in confectionary syrups, *J. Food Technol.,* 1, 25–39.

Oliver, A.E., Leprince, O., Wolkers, W.F., Hincha, D.K., Heyer, A.G., and Crowe, J.H. (2001) Non-disaccharide-based mechanisms of protection during drying, *Cryobiology,* 43, 151–167.

Poirier, I., Marechal, P.A., and Gervais, P. (1997) Effects of the kinetics of water potential variation on bacteria viability, *J. Appl. Microbiol.,* 82, 101–106.

Poirier, I., Maréchal, P.-A., Richard, S., and Gervais, P. (1999) *Saccharomyces cerevisiae* viability is strongly dependent on rehydration kinetics and the temperature of dried cells, *J. Appl. Microbiol.,* 86, 87–92.

Pollack, G.H. (2001) *Cells, Gels, and the Engines of Life: A New, Unifying Approach to Cell Function,* Ebner and Sons, Seattle, WA.

Potts, M. (1994) Desiccation tolerance of prokaryotes, *Microbiol. Rev.,* 58, 755–805.

Potts, M. (2000) Metabolic engineering of desiccation tolerance, *Cryobiology,* 41, 325.

Prausnitz, M.R., Lau, B.S., Milano, C.D., Conner, S., Langer, R., and Weaver, J.C. (1993) A quantitative study of electroporation showing a plateau in net molecular transport, *Biophys. J.,* 65, 414–422.

Puhlev, I., Guo, N., Brown, D.R., and Levine, F. (2001) Desiccation tolerance in human cells, *Cryobiology,* 42, 207–217.

Russo, M.J., Bayley, H., and Toner, M. (1997) Reversible permeabilization of plasma membranes with an engineered switchable pore, *Nat. Biotechnol.,* 15, 278–282.

Saleki-Grehardt, A., and Zografi, G. (1994) Non-isothermal and isothermal crystallization of sucrose from the amorphous state, *Pharm. Res.,* 11, 1166–1173.

Seel, W., Hendry, G., Atherton, N., and Lee, J. (1991) Radical formation and accumulation *in vivo*, in desiccation tolerant and intolerant mosses, *Free Radic. Res. Commun.,* 15, 133–141.

Senda, M., Takeda, J., Shunnosuke, A., and Nakamura, T. (1979) Induction of cell fusion of plant protoplasts by electrical stimulation, *Plant Cell Physiol.,* 20, 1441–1443.

Shamblin, S.L., and Zografi, G. (1998) Enthalpy relaxation in binary amorphous mixtures containing sucrose, *Pharm. Res.* 15, 1828–1834.

Shamblin, S.L., Tang, X., Chang, L., Hancock, B.C., and Pikal, M.J. (1999) Characterization of the time scales of molecular motion in pharmaceutically important glasses, *J. Phys. Chem. B,* 103, 4113–4121.

Sikora, M., Mazurkiewicz, J., Tomasik, P., and Pielichowski, K. (1999) Rheological properties of some starch-water-sugar systems, *Int. J. Food Sci. Technol.,* 34, 371–383.

Song, L., Hobaugh, M.R., Shustak, C., Cheley, S., Bayley, H., and Gouaux, J.E. (1996) Structure of Staphylococcal α-hemolysin, a heptameric transmembrane pore, *Science,* 274, 1859–1866.

Song, Y.C., Kheirabadi, B.S., Lightfoot, F.G., Brockbank, K.G.M., and Taylor, M.J. (2000) Vitreous cryopreservation maintains the function of vascular grafts, *Nat. Biotechnol.,* 18, 296–299.

Sowers, A.E. and Lieber, M.R. (1986) Electropore diameter, lifetimes, numbers and locations in individual erythrocyte ghosts, *FEBS Lett.,* 205, 179–184.

Sun, W.Q., and Leopold, A.C. (1997) Cytoplasmic vitrification and survival of anhydrobiotic organisms, *Comp. Biochem. Physiol. A,* 117, 327–333.

Sun, W.Q., Leopold, A.C., Crowe, L.M., and Crowe, J.H. (1996) Stability of dry liposomes in sugar glasses, *Biophys. J.,* 70, 1769–1776.

Tettero, F.A.A., Bomal, C., Hoekstra, F.A., and Karssen, C.M. (1994) Effect of abscisic acid and slow drying on soluble carbohydrate content in developing embryoids of carrot (*Daucus carota L.*) and alfalfa (*Medicago sativa L.*), *Seed Sci. Res.,* 4, 203–210.

Tsong, T.Y. (1987) Electric modification of membrane permeability for drug loading into living cells, *Methods Enzymol.,* 149, 248–259.

Tsong, Y.T. (1991) Electroporation of cell membranes, *Biophys. J.,* 60, 297–306.

Tunnacliffe, A., Garcia de Castro, A., and Manzanera, M. (2001) Anhydrobiotic engineering of bacterial and mammalian cells: Is intracellular trehalose sufficient? *Cryobiology,* 43, 124–132.

Van Stevenick, J. and Ledeboer, A.M. (1974) Phase transitions in the yeast cell membrane. The influence of temperature on the reconstitution of active dry yeast, *Biochim. Biophys. Acta,* 352, 64–70.

Venegas, G.K. and Marinos-Kouris, D. (1992) Use of the Wilson equation for the prediction of the sorptional equilibrium of sugar-based foodstuffs, *Fluid Phase Equilibria,* 78, 191–207.

Vernhes, M.C., Cabanes P.A., and Teissie, J. (1999) Chinese hamster ovary cells sensitivity to localized electrical stresses, *Bioelectrochem. Bioenerg.,* 48, 17–25.

Wakayaman, T. and Yanagimachi, R. (1998) Development of normal mice from oocytes injected with freeze-dried spermatozoa, *Nat. Biotechnol.,* 16, 639–641.

Walker, B. and Bayley, H. (1995) Key residues for membrane binding, oligomerization, and pore forming activity of Staphylococcal α-hemolysin identified by cysteine scanning mutagenesis and targeted chemical modification, *J. Biol. Chem.,* 270, 23065–23071.

Walker, B., Kasianowicz, J., Krishnasastry, M., and Bayley, H. (1994) A pore-forming protein with a metal-actuated switch, *Protein Eng.,* 7, 655–662.

Watson, P.F. (1981) The effects of cold shock on sperm cell membranes, in *Effects of Low Temperatures on Biological Membranes*, Morris, G.J. and Clark, A., Eds., Academic Press, New York, pp. 189–218.

Williams, R.J. (1994) Methods for determination of glass transitions in seeds, *Ann. Bot.,* 74, 525–530.

Winston, P.W. and Bates, D.H. (1960) Saturated solutions for the control of humidity in biological research, *Ecology,* 41, 232–237.

Wolkers, W.F., Tablin, F., and Crowe, J.H. (2002) From anhydrobiosis to freeze-drying of eukaryotic cells, *Comp. Biochem. Physiol. A.*

Wolkers, W.F., Walker, N.J., Tablin, F., and Crowe, J.H. (2001) Human platelets loaded with trehalose survive freeze-drying, *Cryobiology,* 42, 79–87.

Womersley, C., Uster, P.S., Rudolph, A.S., and Crowe, J.S. (1986) Inhibition of dehydration-induced fusion between liposomal membranes by carbohydrates as measured by fluorescence energy transfer, *Cryobiology,* 23, 245–255.

Wong, T.-K. and Neumann, E. (1982) Electric field mediated gene transfer, *Biochem. Biophys. Res. Commun.,* 107, 584–587.

Wunderlich, B. (1990) *Thermal Analysis*. Academic Press, Boston.

Young, J.F. (1967) Humidity control in the laboratory using salt solutions—A review, *J. Appl. Chem.,* 17, 241–245.

21 Stabilization of Cells during Freeze-Drying: The Trehalose Myth

John H. Crowe, Lois M. Crowe, Fern Tablin, Willem Wolkers, Ann E. Oliver, and Nelly M. Tsvetkova

CONTENTS

21.1 INTRODUCTION

Stabilization of cells by freeze-drying has been one of the major goals of cryobiology for decades, but success has been limited mainly to prokaryotes (e.g., Conrad et al., 2000; Leslie et al., 1994, 1995; Malik et al., 1996). The first reports of preserving mammalian cells in a dry state — red blood cells (Goodrich et al., 1992; Sowemimo-Coker et al., 1993) — soon led to controversy (Franks, 1996; Spieles et al., 1996), as a result of which a consensus developed that preservation of mammalian cells by freeze-drying was probably not possible. However, with the development of new technologies and the application of molecular techniques (Eroglu et al., 2000; Guo et al., 2000; Wolkers et al., 2001), this field has taken on a new life.

We will review progress toward stabilizing mammalian cells in the freeze-dried state and suggest fertile lines for further research in this regard. Before doing so, however, we provide some comments on the nature of the stress induced by freeze-drying and distinguish it carefully from that induced by freezing. Because freezing clearly involves removal of water from the cell, there is at least a superficial resemblance between the stresses induced by freezing and freeze-drying. We have previously suggested that the resemblance is indeed only superficial (Crowe et al., 1990), a proposition that led to considerable controversy in the literature over the past decade, beginning with a paper given the same title as the following section (Crowe et al., 1990).

21.2 ARE FREEZING AND DEHYDRATION SIMILAR STRESS VECTORS?

21.2.1 SPECIFICITY OF PROTECTIVE EFFECTS OF SOLUTES DURING FREEZING

Addition of solutes stabilizes both proteins and membranes during freezing, but cryoprotection is not limited to a single class of compounds. Instead, cryoprotection can be achieved by a diversity of molecules, including sugars, polyalcohols, certain salts, and even proteins. (We point out, however, the interesting observations of Sei et al. [2002] on effects of trehalose and sucrose on ice crystal growth; those workers reported that trehalose is about twice as effective as sucrose in inhibiting the crystal growth.) We believe that this lack of specificity is best explained by the preferential exclusion mechanism for protein stabilization in solution, first proposed by Timasheff and his colleagues (see Xie and Timasheff, 1997, for references). This proposal, which had its roots in the experimental observation that solutes that stabilize proteins in solution are preferentially excluded from the domain of the protein, has been developed into an elegant thermodynamic theory: Exclusion of the solute from the domain of the protein is clearly unfavorable thermodynamically, as it results in a decrease of the entropy of the system. If the protein were to unfold, exposing more surface area (and excluding even more of the solute), the already unfavorable situation becomes more unfavorable. As a result, Timasheff and colleagues argue, the protein does not unfold; indeed, it is driven towards its native conformation. Conversely, if a solute such as urea binds to the protein, unfolding would be favored thermodynamically.

Timasheff and colleagues have shown that the effects described above are actually derived from a balance between the destabilizing effects of solute binding and the stabilizing effects of solute exclusion. For instance, all salts are electrostatically attracted to polar domains in the protein and will bind there. When this binding effect is offset by exclusion, however, the protein will tend to be stabilized. Thus, the nature of the solute and the balance between binding and exclusion will have a profound effect on stability of the protein.

The preferential exclusion mechanism has been applied directly to cryoprotection of proteins, and it appears that it adequately explains cryoprotection without the necessity of resorting to more complex mechanisms. Indeed, in every case a solute that stabilizes a protein in solution has been found to be preferentially excluded from the domain of the protein (Crowe et al., 1990). Such solutes are invariably good cryoprotectants.

21.2.2 Specificity of Protective Effects of Solutes during Drying

When proteins or membranes are dried, they undergo profound damage resulting from the close approach of surfaces that are normally separated by bulk water and from the removal of the water of hydration near polar surfaces, a matter that will also be considered in the next section. For instance, phosphofrucokinase is a tetrameric enzyme that dissociates into its component parts during drying. When this enzyme was dried progressively, it was damaged when only a relatively small part of the water was removed (reviewed in Crowe et al., 1990), but when a solute — proline — known to stabilize proteins in solution was added, the enzyme survived to a much lower water content. We suggest that this stabilization by proline can be explained by the preferential exclusion mechanism, a mechanism that clearly cannot operate when all bulk water is removed. As a result, at the lowest water contents, enzyme activity was lost in the presence of proline. By contrast, when the enzyme was dried with any one of a number of disaccharides, activity was preserved. Similar results have been obtained for the specificity of solutes for preservation of membranes during freezing and dehydration.

We have proposed elsewhere that, unlike the mechanism involved in stabilization during freezing—preferential exclusion of the solute—stabilization during extreme dehydration requires direct interaction between the solute and biomolecule. In other words, we have proposed that the stabilizing solute binds directly to polar residues in a dry protein or membrane, thus satisfying the hydrogen bonding requirements. A summary of the evidence concerning this idea, which has come to be known as the water replacement hypothesis, is beyond the scope of the present essay, but for a discussion of the water replacement hypothesis as well as an alternative theory, see Crowe et al. (1998), Oliver et al. (2002), and Koster (2000) and references therein. Nevertheless, the clear observation that there is a high specificity for stabilization of membranes and proteins during drying that is lacking for stabilization during freezing indicates that freezing and drying are two fundamentally different stress vectors, requiring different mechanisms for stabilization.

21.2.3 The Matter of Freezable and Nonfreezable Water

It is well known that isolated proteins (Miura et al., 1995), membranes (Crowe et al., 1994), carbohydrate polymers (Cornillon et al., 1995), and even intact cells (Buitink et al., 1996; Pritchard et al., 1995) contain a fraction of water that does not freeze readily. The amount of unfrozen water is not large—in the region of 0.2 to 0.3 g water per gram dry weight—and represents only about 5% of the intracellular water in a typical living cell. However, its removal has profound consequences; for instance, the phase transition temperature of membrane phospholipids rises about 80°C as that water is progressively removed. We have suggested previously that this is a major damaging event.

The status of this water and its physical properties has been hotly debated, with some workers questioning whether it is nonfreezable. Nevertheless, there is abundant evidence in the literature that water near surfaces, including abiotic ones (e.g., Sklari et al., 2001), or water trapped in micropores of a wide variety of substances, including even dirt, does not freeze readily and is commonly referred to as nonfreezable water.

The question of whether this water is indeed nonfreezable is of some importance in deciding whether the mechanism of freezing damage is similar to that induced by dehydration. In a comprehensive, carefully done study on this matter, Bronstyn and Steponkus (1993) showed that a substantial fraction of the "nonfreezable" water can indeed be frozen, under defined conditions and depending on the thermal history. In multilamellar vesicles of dipalmitoylphosphatidyl choline (DPPC) with an initial water concentration between 0.13 and 0.18 g H_2O per gram dry weight (Bronstyn and Steponkus presented these values as percentages; we have converted them to the present format for uniformity), ice formation occurred only after homogeneous nucleation at temperatures below –40°C. In suspensions containing less than 0.13 g/g, ice formation during cooling was undetectable by differential scanning calorimetry (DSC). However, an endotherm resulting from ice melting during warming was observed in suspensions containing 0.12 g/g or more. In suspensions containing less than 0.12 g/g, an endotherm corresponding to the melting of ice was not observed during warming. Despite extended incubation of the sample at low temperatures and repeated thermal cycling, this fraction appeared to remain unfrozen. Thus, about half the "nonfreezable" water apparently will freeze under the experimental conditions used by these investigators.

However, a word of caution is necessary: The method used, DSC, is an indirect measurement of the unfrozen water fraction. The way this is done is to measure the enthalpy of melting of ice in the sample, and from that enthalpy the amount of frozen water is calculated. This value is subtracted from the total water content to obtain a measurement of the unfrozen fraction. The only problem with this approach is that there is no easy way of assigning unambiguously the measured enthalpy to melting of ice alone. It is conceivable in a mixture of lipids and water that there may be a significant enthalpic contribution from the lipid component (for instance, because of a conformational change accompanying ice melting). This contribution need not be large to introduce a significant error in the calculation.

There is very little information available on this point, but Hsieh and Wu (1996) showed, using nuclear magnetic resonance, that no more than one or two of the 10 to 12 water molecules that appear to be associated in some way with the headgroups freeze, even down to $-70°C$. Somewhat surprisingly, about half this water is apparently a clathrate-like complex around the choline, with the remainder apparently hydrogen bonded to the phosphate. Because the measurement was made on ^2H_2O deuterons, they may be somewhat less questionable than those done with DSC. At any rate, the unfrozen water fraction is probably somewhere between the values presented by Bronstyn and Steponkus (1993) and those of Hsieh and Wu (1996). Let us now consider what this means for freeze-dried liposomes.

Careful estimates of the water contents of freeze-dried liposomes, made by using ^3H_2O as a marker (Crowe et al., 1987), show that the residual water content is at most 0.01 g water per gram dry weight. (Because of tritium exchange between the ^3H_2O and polar residues in the phospholipids, the estimate represents an upper limit and is likely to be an overestimate.) Thus, the unfrozen water content of frozen liposomes is at least an order of magnitude greater than the total water content obtained with freeze-dried liposomes (depending on whose values for unfrozen water one accepts). Furthermore, inspection of the hydration-dependent phase diagram for DPPC shows that most of the effect of dehydration on physical properties occurs during removal of this final fraction of the water. In fact, direct measurement of effects of freezing on phase behavior (Sanderson et al., 1993) showed that freezing increased T_m by only about 3.5°C, whereas complete dehydration increases T_m by as much as 80°C.

Clearly, the matter of nonfreezable water and membranes has not been satisfactorily resolved. However, we believe that the available evidence leads to the conclusion that freezing and dehydration remove significantly different amounts of water—water that is critical to the physical behavior of biomolecules. Taken with the markedly different solute requirements for preservation during freezing and freeze-drying, we again suggest that freezing and freeze-drying are fundamentally different stress vectors, as we proposed more than a decade ago.

21.3 TREHALOSE AND FREEZE-DRYING INTACT CELLS

21.3.1 TREHALOSE PRODUCTION AND STRESS

Trehalose is accumulated at high concentrations—as much as 20% of the dry weight—by many organisms capable of surviving complete dehydration, spread across many major taxa in all kingdoms (reviewed in Crowe et al., 1998). For example, baker's yeast cells, which have been the subject of the most intensive investigation, do not survive drying in log phase of growth and do not contain significant amounts of trehalose, but in stationary phase they accumulate the sugar and may then be dried successfully (reviewed in Argüelles, 2000). Until relatively recent years, trehalose was thought to be a storage sugar in these cells, and the correlation between survival in the dry state and its presence was believed to be related to repair functions during rehydration, providing a ready energy source. Evidence is accumulating that trehalose production may be a universal stress response in yeasts; it can even prevent damage from environmental insults such as ethanol production during fermentation (Gimeno-Alcañiz, 1999; Lucero et al., 2000; Mansure et al., 1994, 1997; Sharma, 1997). In fact, overproduction of trehalose in some yeast strains caused by increasing synthesis (Kaasen et al., 1994; Soto et al., 1999) and decreasing degradation (Reinders et al., 1999) is being used to increase ethanol-tolerance and thus to boost industrial ethanol production. Trehalose production may be more widespread as a stress response than had been previously appreciated; the analogue of trehalose in higher plants has been thought to be sucrose (reviewed in Hoekstra et al., 1997), but several higher plants have recently been reported to produce trehalose in response to drought stress, either by the plant itself or by a symbiont (Farías-Rodríguez et al., 1998; Ghasempour et al., 1998; Goddijn and Van Dun, 1997; Iturriaga et al., 2000).

21.3.2 TREHALOSE AND BIOSTABILITY

In the early 1980s we established that molecular assemblages such as membranes and proteins can be stabilized in the dry state in the presence of trehalose. When comparisons were made with other sugars, trehalose appeared to be clearly superior (reviewed in Crowe et al., 1987, 2001).

Since that time, an astonishing array of applications for trehalose has been reported (summarized in Table 21.1). Some studies indicated that it might even be efficacious in treatment of dry eye syndrome (Matsuo, 2001) or dry skin (Norcia, 2000) in humans. According to one group, trehalose inhibits bone resorption in ovariectomized mice (Nishizaki et al., 2000), apparently by suppressing osteoclast differentiation (Yoshizane et al., 2000). Another group reported that trehalose inhibits senescence in cut flowers (Otsubo and Iwaya-Inoue, 2000). Thus, a myth has grown up about trehalose and its properties, as a result of which it is being applied to a myriad of biological and clinical problems. We revisit this myth and ask whether trehalose really has any of the special properties to which it has been linked.

At least half the applications for trehalose listed in Table 21.1 deal with fully hydrated cells, and thus the solution properties of trehalose are particularly relevant. Because we are concerned principally with dry cells here, we point out only in passing that a considerable body of evidence is developing on solution properties of this molecule (Ballone et al., 2000; Batta and Kövér, 1999; Branca et al., 1999; Conrad and De Pablo, 1999; Kacuráková and Mathlouthi, 1996; Liu et al., 1997; Luzardo et al., 2000; Magazu et al., 1998; Miller et al., 1997, 1999; Poveda et al., 1997; Sola-Penna and Meyer-Fernandes, 1998). Among the most intriguing findings is that the hydrated radius of trehalose is anomalously large—at least 2.5 times that of the other sugars tested (Sola-Penna and Meyer-Fernandes, 1998). This would seem to be in good agreement with the report of Lin and Timasheff (1996) that, unlike other sugars, trehalose is totally excluded from the hydration shell of the proteins studied. This effect would, in turn, presumably maximize the stabilization of proteins by the preferential exclusion mechanism (Timasheff, 2000; Xie and Timasheff, 1997), a possibility that warrants further investigation. In fact, it seems possible that many of the properties reported for trehalose for stabilization of biomaterials in bulk water or during freezing (cf. Table 21.1) might be related to this apparent anomaly. Faraone et al. (2001) produced data from quasielastic neutron scattering of trehalose in aqueous solution that seem consistent with this anomalous behavior.

21.3.3 THE TREHALOSE MYTH: IS TREHALOSE "SPECIAL" FOR STABILIZING DRY BIOMATERIALS?

The proposition that trehalose has special properties has its roots in a lengthy series of studies. In the first, *Sarcoplasmic reticulum*, isolated from lobster muscle (reviewed in Crowe et al., 1987),

TABLE 21.1
Some Novel Applications for Trehalose

Application	Treatment	References
Enzymes and other proteins	f, ad, fd, sd	1–8
Vaccines and antibodies	f, fd, ad	9–12
Nanoparticles	fd, sd	13–15
Membranes	fd, f	16–19
Liposomes	fd	20–24
DNA and DNA–lipid complexes	fd	25–27
Bacteria and yeasts	fd	28–32
Nucleated mammalian cells	ad, f	33–37
Mammalian blood cells	f, fd	38–42
Mammalian organs	hs	43–48

Note: In most of these studies other sugars or polymers were tested as well as trehalose. This is by no means a complete list, but represents only a sampling of what is being done in this field. f, freezing; ad, air drying; fd, freeze-drying; sp, spray dried; hs, hypothermic storage.

[1]Adler and Lee (1999), [2]Baptista et al. (2000), [3]Cardona et al. (1999), [4]Carninci et al. (1998), [5]Heller et al. (1999), [6]Kreilgaard et al. (1999), [7]Murray and Liang (1999), [8]Uritani et al. (1995), [9]Arya (2000), [10]Dráber et al. (1995), [11]Esteves et al. (2000), [12]Worrall et al. (2000), [13]Cavalli et al. (1997), [14]De Jaeghere et al. (1999), [15]Schwarz and Mehnert (1997), [16]Felix et al. (1999), [17]Reshkin et al. (1988), [18] Sampedro et al. (2001), [19]Crowe et al. (1987), [20]Crowe and Crowe (1992), [21]Kim et al. (1999), [22]Martorell et al. (1999), [23]van Winden et al. (1997), [24]Harrigan et al. (1990), [25]Anchordoguy et al. (1997), [26]Li et al. (2000), [27] Yoshinaga et al. (1997), [28]Conrad et al. (2000), [29]Leslie et al. (1994), [30]Leslie et al. (1995), [31] Lucero et al. (2000), [32]Malik and Lang (1996), [33]Beattie et al. (1997), [34]DeCastro and Tunnacliffe (2000), [35]Eroglu et al. (2000), [36]Guo et al. (2000), [37]Matsuo (2000), [38]Boutron and Peyridieu (1994), [39]Goodrich et al. (1992), [40]Pellerin-Mendes et al. (1997), [41]Sowemimo-Coker et al. (1993), [42]Wolkers et al. (2001), [43]Bando et al. (1994a), [44]Bando et al. (1994b), [45]Hirata et al. (1993), [46]Kitahara et al. (1996), [47]Kitahara et al. (1998), [48]Yokomise et al. (1995).

served as the first model. When these membranes were dried without trehalose, massive fusion was observed, and the ability to transport calcium was lost. When the membranes were dried with trehalose, however, both the morphology and biological activity were maintained intact. Trehalose was by far the most effective sugar tested. That is not to say that other sugars did not also stabilize the SR; indeed, sucrose, for example, worked about as well as trehalose, but much higher concentrations of the sugar were required. Some years later, however, we obtained evidence that these SR membranes have a mechanism for translocating trehalose across the bilayer. When we measured the space accessible to trehalose compared with that accessible to sucrose, we found that trehalose has access to the aqueous interior of the vesicles, whereas sucrose at comparable concentrations does not. We still do not know the mechanism by which trehalose crosses the bilayer, but it seems likely that a carrier for this sugar is present. At any rate, the fact that trehalose penetrates into the interior of the vesicles may explain its superior properties in preserving these membranes in particular. We suggest that other sugars such as sucrose might preserve the membranes at concentrations similar to those seen with trehalose if they had access to the aqueous interior.

The second study involved enzymes. Phosphofructokinase is a tetrameric enzyme that is readily denatured by freezing or drying, but when it is dried in the presence of trehalose, secondary and higher-order structure are preserved, and enzyme activity is almost perfectly retained (reviewed in Crowe et al., 1987, 1998). In the initial studies, trehalose appeared to be unique in preserving this enzyme. Subsequently, however, we discovered that the trehalose preparation used contained a contaminant of trace amounts of Zn^{++}, used in preparation of the sugar. When similar amounts of Zn^{++} were added to other disaccharides, they became as effective as trehalose.

The third study was on liposomes. In the initial studies, from the mid-1980s (Crowe and Crowe, 1992; Crowe et al., 1997a, b, 1998), liposomes were prepared from a lipid with low T_m, palmitoyl-oleoylphosphatidyl choline (POPC). A fluorescent marker, carboxyfluorescein, was trapped in the aqueous interior. When the liposomes were freeze-dried with trehalose and rehydrated, the vesicles were seen to be intact, and nearly 100% of the carboxyfluorescein was retained. It quickly emerged that stabilization of POPC liposomes, and other vesicles prepared from low–melting point lipids, had two requirements: inhibition of fusion between the dry vesicles, and depression of T_m in the dry state. In the hydrated state, T_m for POPC is about $-1°C$ and rises to about $+70°C$ when it is dried without trehalose. In the presence of trehalose, T_m is depressed in the dry state to $-20°C$. Thus, the lipid is maintained in liquid crystalline phase in the dry state, and phase transitions are not seen during rehydration. The significance of this phase transition during rehydration is that when phospholipids pass through such transitions, the bilayer becomes transiently leaky. (The physical basis for this leakiness has recently been investigated in some detail by Hays et al. [2001].) Thus, the leakage that normally accompanies this transition must be avoided if the contents of membrane vesicles and whole cells are to be retained.

These effects were reported first for trehalose (reviewed in Crowe and Crowe, 1992). When we compared the effects of other sugars and polymers on the preservation, we found that, with vesicles made from lipids with low T_m, trehalose appeared to be significantly superior to the best of the additives tested. Oligosaccharides larger than trisaccharides did not work at all (Crowe and Crowe, 1992). Other sugars, particularly disaccharides, did provide good stabilization of POPC vesicles in the dry state, but much higher concentrations than trehalose were required, at least according to initial reports. However, as freeze-drying technology improved, the differences between disaccharides tended to disappear, and the myth eventually got modified to encompass disaccharides in general. Nevertheless, the observation that trehalose was significantly more effective at low concentrations under suboptimal conditions for freeze-drying requires explanation, which we provide later.

Liposomes with saturated acyl chains were also studied. DPPC is a lipid with saturated acyl chains, and thus an elevated T_m ($41°C$). When it is dried without trehalose, T_m rises to about $110°C$; with trehalose present, T_m rises to about $65°C$ (reviewed in Crowe et al., 1998). Thus, DPPC is in gel phase at all stages of the freeze-drying and rehydration process, and one would expect that inhibition of fusion might be sufficient for the stabilization. In other words, any inert solute that would separate the vesicles in the dry state and thus prevent aggregation and fusion should stabilize the dry vesicles. That appears to be the case; a high–molecular weight (450,000) hydroxyethyl starch (HES) has no effect on T_m in dry DPPC, but it preserves the vesicles nevertheless. With scanning electron microscopy, the dry vesicles are seen to be embedded in a matrix of HES (Crowe et al., 2001), with no change in diameter from the freshly prepared vesicles. By contrast, DPPC vesicles dried without HES showed massive fusion (Crowe et al., 1997a, b) and leaked all their contents when rehydrated.

The final study involved the effects of polymers and monosaccharides on liposomes. Polymers like HES alone will not stabilize dry liposomes with low T_m, but a combination of a low–molecular weight sugar and HES can be effective. Even glucose and HES are effective (Crowe et al., 1997a, b). Here is the apparent mechanism: glucose depresses T_m in the dry lipid but has little effect on inhibiting fusion, except at extremely high concentrations. In contrast, the polymer has no effect on the phase transition, but inhibits fusion. Thus, the combination of the two meets both requirements, whereas neither alone does so (Crowe et al., 1997a, b). A glycan isolated from the desiccation-tolerant alga *Nostoc* apparently has a similar role in conjunction with oligosaccharides (Hill et al., 1997). Recent results from Hincha et al. (Hincha et al., 2000) have shown that certain polymers from desiccation-tolerant higher plants will by themselves both inhibit fusion and reduce T_m in dry phospholipids such as egg phosphatidyl choline (PC). The mechanism behind this effect is still unclear, but there is some evidence that the fructans insert between the polar headgroups (Vereyken et al., 2001), much as trehalose is thought to do (Crowe et al., 1998).

There is little about trehalose and its effects on preserving dry biomolecules that is consistent with "special properties." Once an understanding of the physical requirements for preservation was achieved, it became apparent that many routes can lead to the same end. Similar observations on the stability of proteins dried with sugars and polymers have been made by Carpenter and his group, with similar conclusions (e.g., Allison et al., 2000; Anchordoguy et al., 2001; Heller et al., 1999).

21.4 REVIVAL OF THE MYTH: TREHALOSE WORKS UNDER SUBOPTIMAL CONDITIONS

We implied above that trehalose works well for freeze-drying liposomes under conditions that are less than optimal. The same applies for storage under conditions that would normally degrade the biomaterial. Leslie et al. (1995) reported that bacteria freeze-dried in the presence of trehalose showed remarkably high survival immediately after freeze-drying. Furthermore, we found that the bacteria freeze-dried with trehalose retained a high viability even after long exposure to moist air. By contrast, when the bacteria were freeze-dried with sucrose, they showed lower initial survival, and when they were exposed to moist air, viability deceased rapidly. Along the same lines, Conrad et al. (2000) showed that addition of borate ions to trehalose dramatically improved survival of freeze-dried bacteria. It is not clear how borate imparts these effects, but it may affect the stability of the glass. More recently, Esteves et al. (2000) reported that when immunoconjugates were freeze-dried with trehalose or other disaccharides, all the sugars provided reasonable levels of preservation. However, when the dry samples were stored at high relative humidities and temperatures, those dried with trehalose were stable for much longer than those dried with other sugars. This finding is of some considerable significance, as such immunoconjugates, vaccines, antisera, and the like are being shipped as freeze-dried preparations to areas such as the Amazon, where they would be exposed to high temperatures and humidities as soon as they are exposed to air. That proposition is already in practice for stabilizing viral vaccines (Worall et al., 2000).

21.4.1 DOES NONENZYMATIC BROWNING CONTRIBUTE TO INSTABILITY?

The Maillard (browning) reaction between reducing sugars and proteins in the dry state has often been invoked as a major source of damage (e.g., Li et al., 1996), and the fact that both sucrose and trehalose are nonreducing sugars may explain at least partly why they are the natural products accumulated by anhydrobiotic organisms. However, the glycosidic bonds linking the monomers in sucrose and trehalose have very different susceptibilities to hydrolysis (reviewed in O'Brien, 1996). When O'Brien (1996) incubated a freeze-dried model system with sucrose, trehalose, and glucose at water activity 0.33 and pH 2.5, the rate of browning seen with sucrose approached that of glucose—as much as 2000 times faster than that with trehalose. Using somewhat different techniques, Schebor et al. (1999) showed considerable hydrolysis of sucrose stored in the presence of amino-containing compounds, whereas such hydrolysis was minimal in samples containing trehalose. In the samples with sucrose, considerable browning resulted from the sucrose hydrolysis. Thus, under less than optimal conditions for storage, it is clear that trehalose is preferred.

21.4.2 DO GLASS TRANSITIONS EXPLAIN STABILITY?

Using liposomes as a model, we attempted to find a mechanism for long-term stability in the presence of trehalose. As with the bacteria and immunoconjugates, the dry liposomes exposed to increased relative humidity rapidly leaked their contents when they were dried with sucrose, but not when they were dried with trehalose (Crowe et al., 1996; Sun et al., 1996). The liposomes underwent extensive fusion in the moist air when dried with sucrose, but not with trehalose.

Examination of the state diagram for trehalose provides a possible explanation for this effect (see Chen et al., 2000, for an excellent review and extended state diagram). T_g for trehalose is much higher than that for sucrose, a finding first reported by Green and Angell (1989). As a result, one would expect that addition of small amounts of water to sucrose by adsorption in moist air would decrease T_g to below the storage temperature, while at the same water content T_g for trehalose would be above the storage temperature. This proved to be the case. Furthermore, we point out again that degradation does proceed in samples below T_g, albeit at a slower rate. With trehalose, a sample at 20°C would be nearly 100°C below T_g. By contrast, one dried with sucrose would be only about 45°C below T_g. Under these conditions, one would expect the sample dried with sucrose to be degraded more rapidly. Along the same lines, Schebor et al. (2000) reported that significant losses of cellular integrity were seen in dry baker's yeast stored below T_g. Interestingly, Nagase et al. (2002) recently reported the existence of an amorphous phase with intriguing physical properties that the authors suggested might be important in stabilization.

Aldous et al. (1995) suggested an additional interesting property of trehalose, which we were able to confirm. They proposed that since the crystalline structure of trehalose is a dihydrate, some of the sugar might, during adsorption of water vapor, be converted to the crystalline dihydrate, thus sparing the remaining trehalose from contact with the water. Experimental observation showed that this suggestion is correct: With addition of small amounts of water, the crystalline dihydrate immediately appeared, and T_g for the remaining glassy sugar remained unexpectedly high (Crowe et al., 1996). Iglesias et al. (1997) produced a sorption isotherm for vitrified trehalose that seems consist with this observation. More recently, Yoshii et al. (2000) showed that trehalose readily converts between the crystalline and amorphous forms under certain conditions, so this process might be reversible.

We stress, however, that the elevated T_g seen in trehalose is not anomalous. Indeed, trehalose lies at the end of a continuum of sugars that show increasing T_g (Crowe et al., 1996), although the basis for this effect is not understood.

Is trehalose special? Under ideal conditions for drying and storage, no, it is not. However, under suboptimal conditions, it provides stability when other sugars do not. This is not to say that trehalose is the magic bullet that will be preferred over all other excipients; indeed, as we described above, combinations of polymers and other sugars may work just as well and may even be preferable in some circumstances. Nevertheless, it is still the preferred first excipient to be tested.

21.5 CAN WE USE WHAT WE HAVE LEARNED FROM MODEL SYSTEMS TO PRESERVE INTACT CELLS IN THE DRY STATE?

Clearly, trehalose must be introduced into the cytoplasm of a cell if it is to be effective at stabilizing intracellular proteins and membranes during dehydration. Current efforts are centered around this fundamental problem, as summarized below.

21.5.1 TREHALOSE BIOSYNTHESIS

Because trehalose biosynthesis is a simple two-step process, it seemed amenable to genetic engineering; two enzymes are involved, the substrates for which are normal metabolites in virtually all cells, so substrate availability would not seem to be a problem. The genes were cloned some time ago, from bacteria, designated otsA and B (83,84.150); yeasts, designated tps 1 and 2 (e.g., Bell et al., 1992; Kaasen et al., 1994; Serrano et al., 1999); and higher plants, also designated tps 1 and 2 (Blázquez et al., 1998; Zentella et al., 1999). The first transfections into higher-plant cells showed reasonably high levels of expression and led to improved drought tolerance (Garg et al., 2002; Pilon-Smits et al., 1998). Subsequent investigations showed that inhibition of trehalase activity increased trehalose production (Goddijn et al., 1997; Müller et al., 1995). Even more recently, Sode et al. (2001) have synthesized a trehalose derivative that is a potent inhibitor of trehalase; it may be even more promising as a mechanism for increasing trehalose production. Similar transfections of the genes for sucrose synthesis into *Escherichia coli* also led to accumulation of sucrose and improved resistance to dehydration damage (Billi et al., 2000).

The complete operon-containing gene complex for trehalose synthesis has recently been cloned (Kong et al., 2001) and subjected to directed evolution, in an effort at improving trehalose production. The results look very promising, with as much as a fivefold increase in production over controls. Along the same lines, Seo et al. (2000) produced a fusion protein encompassing ots A and B. The logic here is to get the enzymes in close proximity, which Seo et al. suggest will increase trehalose production. Bloom, Levine, and colleagues have taken up this approach in their work on transfection of mammalian cells, as described below.

Expression of the enzymes for trehalose synthesis in mammalian cells was only recently achieved. Guo et al. (2000) engineered the genes for trehalose synthesis into mammalian cells in an adenovirus vector. With multiple infections, the trehalose biosynthesis increased. The cells, which were then dried to a level at which free water was no longer detectable, retained viability for 3 to 5 days. It seemed reasonable to suspect that the level of viability might be improved by altering the storage or drying conditions (Gordon et al., 2001). One of the difficulties in such studies has been in obtaining expression of the genes at high levels to produce large amounts of trehalose.

De Castro and Tunnacliffe (2000) reported they did not obtain any viable cells after a similar transfection. Although it may be true that the transfected cells do not survive complete drying, it is also possible that differences in the methods for drying may account for the widely differing results. Chen et al. (2002), Gordon et al. (2001), and Tunnacliffe et al. (2001) have reported further studies aimed at resolving this major discrepancy.

There is an alternative pathway to trehalose synthesis involving conversion of maltose to trehalose, first reported by Panek and her colleagues (Paschoalin et al., 1986). This pathway was more recently investigated in some detail by several laboratories, including cloning the genes involved (Gueguen et al., 2001; Kato et al., 1996; Nishimoto et al., 1995) and investigations of the regulatory pathways (Matthijs et al., 2000). There is at present no clear advantage to transfections with these genes compared with the trehalose phosphate synthase and phosphatase genes, at least for mammalian cells, but we nevertheless point out the existence of this pathway. It has been proposed as a method for industrial production of trehalose (Yoshida et al., 1998). Still other known pathways are summarized by De Smet et al. (2000).

21.5.2 TREHALOSE TRANSPORT

Panek and her colleagues have shown that a trehalose transporter is required in yeasts to transport trehalose out of the cell during drying (Cuber et al., 1997; Eleutherio et al., 1993). Because stabilization requires that trehalose be on both sides of the membrane (reviewed in Crowe et al., 1998), this is a way of meeting that requirement. The gene for this transporter has now been cloned (Han et al., 1995) and expressed at a high level in yeasts (Stambuk et al., 1998), so there is no obvious reason why it could not be incorporated into the cassettes already in use for trehalose synthesis (Guo et al., 2000; Lao et al., 2001; Tunnacliffe et al. 2001), a treatment that might well improve viability. It also provides an obvious route for introducing trehalose into the cell. In the yeast from which the gene was cloned, the transporter is an active H^+ symporter (Crowe et al., 1991), but it may also act as a weak passive carrier in the absence of a H^+ gradient across the membrane (Araujo et al., 1991), so the prospects of using it as a passive carrier in mammalian cells seems reasonable. A related transporter from bacteria has been characterized extensively, and a crystal structure was recently produced (Diez et al., 2001)

21.5.3 A PORE FOR PERMEATION

Eroglu et al. (2000) engineered an elegant pore-forming hemolytic protein, α-hemolysin, so that the pore could be switched on and off. By substituting a number of residues with histidines, the pore could be regulated; by adding μM quantities of Zn^{++} the pore could be closed; and by removing the Zn^{++}, the pore could be reopened. Because the pore protein spontaneously inserts into membranes,

Eroglu et al. found that they could simply incubate the cells in its presence, add trehalose in the absence of Zn^{++}, and introduce the sugar via the pore. Using this procedure, they were able to obtain very high rates of survival of two lines of mammalian cells in the frozen state. However, they more recently reported that when the cells were dried, survival was disappointingly low, although the cells were protected from damage relative to the controls (Chen et al., 2001).

21.5.4 USE OF PHASE TRANSITIONS

Beattie et al. (1997) discovered that the insulin-producing cells from mammalian pancreas have a membrane lipid phase transition well above the freezing point. Because membranes are known to become transiently leaky during the phase transition, Beattie et al. used that leakiness to introduce trehalose into the cells, which were then successfully frozen, and kept frozen for extended periods. The thawed cells were transplanted into rats, where they were found to remain viable for many months. These findings are being developed into a commercial product, which, it is proposed, will provide a stable, transplantable device for treatment of diabetes.

21.6 FREEZE-DRYING HUMAN BLOOD PLATELETS

Human blood platelets have a blood bank storage lifetime of 5 d, after which they are discarded. They cannot be chilled below room temperature without activating them, and thus are stored at temperatures above 20°C. Even with such a restricted shelf-life, bacterial contamination frequently occurs, and septic shock resulting from transfusion of contaminated platelets continues to be a serious clinical problem (Chernoff and Snyder, 1992; Moroff et al., 1994).

We have set out to find means for improving the storage of platelets, including hypothermic storage (Tablin et al., 1996) and freeze-drying (Wolkers et al., 2001). Cryopreservation of platelets with Me_2SO (Bock et al., 1995; Valeri et al., 1974) has had some success, but the thawed platelets must be subjected to extensive washing to remove the potentially harmful Me_2SO, which is toxic at physiological temperatures (Arakawa et al., 1990). During the washing steps, the platelets are frequently activated and are of little clinical value. Thus, stabilization of platelets in the presence of excipients that also are injectable is desirable.

Platelets would seem to be good candidates for stabilization by freeze-drying; they are cell fragments, with very little intracellular compartmentation, so one might expect the requirements for stabilization during freeze-drying to be far less stringent than with nucleated cells. Indeed, that appears to be the case

21.6.1 PHYSIOLOGICAL EFFECTS OF CHILLING ON PLATELETS

Platelets are exquisitely sensitive to chilling; when they are cooled below 20°C, these cell fragments markedly change shape from discoid, resting cells to spherical cells, with multiple filopodia (White and Krivit, 1967; Zucker and Borelli, 1954). Changes in cell shape are accompanied by a net increase in filamentous actin (Winokur and Hartwig, 1995) and a depolymerization of the platelet microtubule coil (White et al., 1985). In addition, incubation of platelets at 4°C leads to a sharp rise in intracellular calcium (Oliver et al., 1995; Winokur and Hartwig, 1995). Platelets stored in the cold for greater than 24 h also undergo fusion and secretion of their dense, lysosomal, and alpha granules (Bode, 1999), a process that mimics physiological activation (Stenberg et al., 1985), after which they have minimal clinical value.

21.6.2 LIPID PHASE TRANSITIONS AND CHILLING DAMAGE TO PLATELETS

The chain of events that leads to activation at low temperatures begins with passage of platelet membrane lipids through a phospholipid phase transition between 10° and 20°C (Tablin et al., 1996; Crowe et al., 1999). This main phospholipid transition is seen in the major membrane components, the dense tubular system (an analog of sarcoplasmic reticulum from muscle; it serves as a membrane delimited Ca^{++} store), and in the plasma membrane (Tsvetkova et al., 1999). Passage through this transition is highly correlated with shape changes during chilling (Tablin et al., 1996), but the transition *per se* is only part of the story; the shape changes seen during the phase transition are completely reversible for up to about 24 h in the cold, after which they become irreversible, and the platelets become physiologically activated. Tsvetkova and her colleagues (2000) obtained evidence concerning events during long-term storage. They showed that storage of equine platelets at 4°C for greater than 24 h resulted in the presence of multiple phase transitions, presumably because of lateral phase separations. More recently, we provided direct evidence that lateral phase separation of membrane lipids and proteins are responsible for this long-term storage damage (Tablin et al., 2001).

21.6.3 PLATELETS ALSO HAVE A HIGH-TEMPERATURE LIPID PHASE TRANSITION

More detailed analysis of membrane phase transitions in intact, resting human platelets, using Fourier transform infrared spectroscopy (FTIR) and DSC, showed two membrane phase transitions, a main transition between 10° and 20°C and a second one between 30° and 40°C. We have previously demonstrated that this lower phase transition can be assigned to membrane phospholipids (Crowe et al., 1999; Tablin et al., 1996), and more recently we have shown that the higher transition is probably the result of a liquid ordered (L_o)-to-liquid crystalline transition for sphingomyelin (Gousset et al., 2002). This upper transition is caused by the presence of liquid ordered (L_o) domains, also known as "rafts" (Brown and London, 1998), which are composed of mixtures of sphingomyelin and cholesterol (Gousset et al., 2002). Rafts isolated from the platelets gave predominantly the upper transition, with the lower one greatly reduced, indicating that the upper transition seen in the intact platelets is caused by the rafts. This high-temperature transition was of considerable interest as a marker for platelet integrity, particularly as rafts are widely thought to be signaling platforms, into which receptors and accessory proteins are thought to be organized (Brown and London, 1998). However, it assumed unexpected importance when we attempted to introduce trehalose into platelets.

21.6.4 INTRODUCTION OF TREHALOSE INTO PLATELETS

There is an obvious problem with the proposal that we freeze-dry platelets with trehalose: The sugar must be introduced into the cell for this to work. The methods of molecular biology clearly are inappropriate, as the biosynthetic apparatus is lacking in these enucleated cell fragments. The elegant trehalose pore developed by Toner's group (Chen et al., 2000; Eroglu et al., 2000) might provide a way of introducing the trehalose, but platelets treated in such a way could not be used clinically.

The remaining major possibility is to use the phase transitions themselves. Membranes are known to become transiently leaky during phase transitions (see Hays et al., 200) for evidence concerning the mechanism of leakage), thus offering an opportunity to introduce the trehalose. It would clearly be advantageous to be able to load trehalose into cells using the cell's own biological functions without having to permeabilize the membrane. Beattie et al. (1997) previously demonstrated that Me_2SO could be used to incorporate trehalose into pancreatic islet cells for stabilization during freezing. However, Me_2SO has been shown to be damaging to membranes at physiological temperatures (Arakawa et al., 1990), and its presence would require extensive washing, a treatment that can easily result in activation.

We first tried to introduce trehalose by chilling through the 10 to 20°C transition (a procedure that works with some cells), but it failed with platelets. However, when the platelets were heated to temperatures in excess of 30°C (in the range of the transition assigned to the rafts), rapid uptake of trehalose was seen. The uptake appears to be by an endocytotic or pinocytotic pathway, which is only active above 30°C (Wolkers et al., 2001). Studies on endocytosis in other model cell systems have shown that endocytosis is associated with raft domains (Rodal et al., 1999). We speculate that local perturbations in membrane fluidity in the range of the sphinogomyelin raft phase transition facilitate endocytosis. Making use of this high-temperature transition, we can incorporate significant concentrations of trehalose (20 mM or more, depending on the extracellular concentration and time of incubation) into the platelets (Wolkers et al., 2001).

21.6.5 Is the Trehalose Contained in Endocytotic Vesicles?

Two lines of evidence indicate that it gets released into the cytosol (but we do not yet understand the mechanism by which this happens): first, when we used a fluorescent dye, lucifer yellow, which is roughly the same size as trehalose, as a marker, the dye was seen to accumulate inside the platelets, first as punctate endocytotic vesicles, but then within a short time as a diffuse fluorescence seen throughout the cytosol. Second, we measured the amount of trehalose in the cells and calculated the concentration, assuming it was homogeneously distributed in the cytosol. We then counted the number of endocytotic vesicles in a cell, calculated the volume encompassed by those vesicles, and calculated the concentration of trehalose in the vesicles, assuming it is all in that compartment. The result is that the concentration required in the vesicles exceeded the extracellular concentration by several-fold, which seems highly unlikely, if not impossible. We conclude that the trehalose, like the lucifer yellow, escapes the vesicles and is free in the cytosol (Wolkers et al., 2001).

21.6.6 Successful Freeze-Drying of Trehalose-Loaded Cells

We have recently reported successful freeze-drying of platelets, with a detailed discussion of the procedure, which results in survival exceeding 90% (Wolkers et al., 2001). We have only a few new observations to add here about further progress: first, the dry platelets are stable for at least 8 months when stored at room temperature, under vacuum. During that time, we have seen no loss of platelets. Second, the freeze-dried, rehydrated cells respond to normal platelet agonists including thrombin, ADP, collagen, and ristocetin. Third, studies on the morphology of the trehalose-loaded, freeze-dried, and rehydrated platelets, using scanning electron microscopy, show that they are affected by the drying, but are morphologically similar to fresh platelets. When they were dried without trehalose, most of the platelets disintegrated during the rehydration event, but of the small number that were left, most had fused with adjacent cells, forming an insoluble clump. Fourth, we have extended the freeze-drying to mouse and pig platelets as animal models for *in vivo* testing—studies that are currently in progress.

21.6.7 Is the Phase-Separation Model Correct for Freeze-dried Platelets?

FTIR analysis indicates that the phase-separation model is correct for freeze-dried platelets. These platelets have phase transitions that are virtually identical to those of fresh platelets, with transitions of 10 to 20°C and 28 to 40°C. Similar analysis of cells freeze-dried in the absence of trehalose show multiple transitions, indicative of lateral phase separation of membrane lipids. The available evidence is consistent with this viewpoint, but these studies are still in progress.

21.7 SUMMARY AND CONCLUSIONS

We believe that freezing and the extensive dehydration seen following freeze-drying are distinctly different stress vectors, requiring different stabilizing molecules. Under ideal conditions for drying and storage, trehalose is probably no more effective than other oligosaccharides in this regard. Nevertheless, under suboptimal conditions it can be very effective and is thus still a preferred excipient. Furthermore, there is an emerging consensus that trehalose is safe for use as an injectable excipient (Richards et al., 2002). Nevertheless, despite the promise of trehalose in the stabilization process, there is growing evidence that additional modifications to the cellular milieu will probably be required if we are to achieve a stable, freeze-dried mammalian cell. However, most workers in this field have already concluded that the best approach is to start with trehalose, for the reasons presented here.

ACKNOWLEDGMENTS

This work was supported by grants HL57810 and HL98171 from NIH, 98171 from ONR, and N66001-00-C-8048 from DARPA. We gratefully acknowledge many useful discussions with our colleagues Fred Bloom, Malcolm Potts, Richard Helm, John Battista, and James Clegg.

REFERENCES

Adler, M. and Lee, G. (1999) Stability and surface activity of lactate dehydrogenase in spray dried trehalose, *J. Pharm. Sci.,* 88, 199–208.

Aldous, B.J., Auffret, A.D., and Franks, F. (1995) The crystallisation of hydrates from amorphous carbohydrates, *Cryo-Letters,* 16, 181–186.

Allison, S.D., Manning, M.C., Randolph, T.W., Middleton, K., Davis, A., and Carpenter, J.F. (2000) Optimization of storage stability of lyophilized actin using combinations of disaccharides and dextran, *J. Pharm. Sci.,* 89, 199–214.

Anchordoquy, T.J., Carpenter, J.F., and Kroll, D.J. (1997) Maintenance of transfection rates and physical characterization of lipid/DNA complexes after freeze-drying and rehydration, *Arch. Biochem. Biophys.,* 348, 199–206.

Anchordoquy, T.J., Izutsu, K.-I., Randolph, T.W., and Carpenter, J.F. (2001) Maintenance of quaternary structure in the frozen state stabilizes lactate dehydrogenase during freeze-drying, *Arch. Biochem. Biophys.,* 390, 35–41.

Arakawa, T., Carpenter, J.F., Kita, Y.A., and Crowe, J.H. (1990) The basis for toxicity of certain cryoprotectants: An hypothesis, *Cryobiology,* 27, 401–415.

Araujo, P.S. de, Panek, A.C., Crowe, J.H., Crowe, L.M., and Panek, A.D. (1991) Trehalose-transporting membrane vesicles from yeast, *Biochem. Int.,* 24, 731–737.

Argüelles, J.C. (2000) Physiological roles of trehalose in bacteria and yeasts: A comparative analysis, *Arch. Microbiol.,* 174, 217–224.

Arya, S.C. (2000) Stabilization of vaccines: To be or not to be, *Vaccine,* 19, 595–597.

Ballone, P., Marchi, M., Branca, C., and Magazú, S. (2000) Structural and vibrational properties of trehalose: A density functional study, *J. Phys. Chem. B,* 104, 6313–6317.

Bando, T., Liu, C.J., Kosaka, S., Yokomise, H., Inui, K., Yagi, K., Hitomi, S., and Wada, H. (1994a) Twenty-hour canine lung preservation using newly developed solutions containing trehalose, *Transplant. Proc.,* 26, 871–872.

Bando, T., Kosaka, S., Liu, C., Hirai, T., Hirata, T., Yokomise, H., Yagi, K., Inui, K., Hitomi, S., and Wada, H. (1994b) Effects of newly developed solutions containing trehalose on twenty-hour canine lung preservation, *J. Thorac Card. Surg.,* 108, 92–98.

Baptista, R.P., Cabral, J.M.S., and Melo, E.P. (2000) Trehalose delays the reversible but not the irreversible thermal denaturation of cutinase, *Biotechnol. Bioeng.,* 70, 699–703.

Batta, G. and Kövér, K.E. (1999) Heteronuclear coupling constants of hydroxyl protons in a water solution of oligosaccharides: Trehalose and sucrose, *Carbo. Res.,* 320, 267–272.

Beattie, G.M., Crowe, J.H., Lopez, A.D., Cirulli, V., Ricordi, C., and Hayek, A. (1997) Trehalose: A cryoprotectant that enhances recovery and preserves function of human pancreatic islets after long-term storage, *Diabetes,* 46, 519–523.

Bell, W., Klaassen, P., Ohnacker, M., Boller, T., Herweijer, M., Schoppink, P., Van der Zee, P., and Wiemken, A. (1992) Characterization of the 56-kDa subunit of yeast trehalose-6-phosphate synthase and cloning of its gene reveal its identity with the product of *CIF1*, a regulator of carbon catabolite inactivation, *Eur. J. Biochem.,* 209, 951–959.

Billi, D., Wright, D.J., Helm, R.F., Prickett, T., Potts, M., and Crowe, J.H. (2000) Engineering desiccation tolerance in Escherichia coli, *App. Env. Micro.,* 66, 1680–1684.

Blázquez, M.A., Santos, E., Flores, C.L., Martínez-Zapater, J.M., Salinas, J., and Gancedo, C. (1998) Isolation and molecular characterization of the *Arabidopsis TPS1* gene, encoding trehalose-6-phosphate synthase, *Plant J.,* 13, 685–689.

Bock, M., Schleuning, M., Heim, M.W., and Wempel, W. (1995) Cryopreservation of human platelets with dimethyl-sulfoxide: Changes in biochemistry and cell function, *Transfusion,* 35, 921–924.

Bode, A.P. (1990) Platelet activation may explain the storage lesion in platelet concentrations, *Blood Cells,* 16, 109–126.

Boutron, P. and Peyridieu, J.-F. (1994) Reduction in toxicity for red blood cells in buffered solutions containing high concentrations of 2,3-butanediol by trehalose, sucrose, sorbitol, or mannitol, *Cryobiology,* 31, 367–373.

Branca, C., Magazù, S., Maisano, G., and Migliardo, P. (1999) Anomalous cryoprotective effectiveness of trehalose: Raman scattering evidences, *J. Chem. Phys.,* 111, 281–287.

Bronshteyn, V.L., and Steponkus, P.L. 1993. Calorimetric studies of freeze-induced dehydration of phospholipids, *Biophys. J.,* 65, 1853–1865.

Brown, D. and London, E. (1998) Functions of lipid rafts in biological membranes, *Ann. Rev. Cell Dev. Biol.,* 14, 111–136.

Buitink, J., Walters-Vertucci, C., Hoekstra, F.A., and Leprince, O. (1996) Calorimetric properties of dehydrating pollen—analysis of a desiccation-tolerant and an intolerant species, *Plant Physiol.,* 111, 235–242.

Cardona, 0.S., Schebor, C., Buera, M.P., Karel, M., and Chirife, J. (1997) Thermal stability of invertase in reduced-moisture amorphous matrices in relation to glassy state and trehalose crystallization, *J. Food Sci.,* 62, 105–112.

Carninci, P., Nishiyama, Y., Westover, A., Itoh, M., Nagaoka, S., Sasaki, N., Okazaki, Y., Muramatsu, M., and Hayashizaki, Y. (1998) Thermostabilization and thermoactivation of thermolabile enzymes by trehalose and its application for the synthesis of full length cDNA, *Proc. Nat. Acad. Sci. USA,* 95, 520–524.

Cavalli, R., Caputo, O., Carlotti, M.E., Trotta, M., Scarnecchia, C., and Gasco, M.R. (1997) Sterilization and freeze-drying of drug-free and drug-loaded solid lipid nanoparticles, *Int. J. Pharm.,* 148, 47–54 (1997).

Chen, T., Acker, J.P., Eroglu, A., Cheley, S., Bayley, H., Fowler, A., and Toner, M. (2001) Beneficial effect of intracellular trehalose on the membrane integrity of dried mammalian cells, *Cryobiology,* 43(2), 168–81.

Chen, T.N., Fowler, A., and Toner, M. (2000) Literature review: Supplemented phase diagram of the trehalose-water binary mixture, *Cryobiology,* 40, 277–282.

Chernoff, A. and Snyder, E.L. (1992) The cellular and molecular basis of the platelet storage lesions: A symposium summary, *Transfusion,* 32, 386–390.

Conrad, B. and De Pablo, J.J. (1999) Computer simulation of the cryoprotectant disaccharide α,α-trehalose in aqueous solution, *J. Phys. Chem. A,* 103, 4049–4055.

Conrad, P.B., Miller, D.P., Cielenski, P.R., and De Pablo, J.J. (2000) Stabilization and preservation of *Lactobacillus acidophilus* in saccharide matrices, *Cryobiology,* 41, 17–24.

Cornillon, P., Andrieu, J., Dupla, J.C., and Laurent, M. (1995) Use of nuclear magnetic-resonance to model thermophysical properties of frozen and unfrozen model food gels, *J. Food Eng.,* 25, 1–19.

Crowe, J.H., Carpenter, J.F., Crowe, L.M., and Anchordoguy, T.J. (1990) Are freezing and dehydration similar stress vectors? A comparison of modes of interaction of stabilizing solutes with biomolecules, *Cryobiology,* 27, 219–231.

Crowe, J.H., Carpenter, J.F., and Crowe, L.M. (1998) The role of vitrification in anhydrobiosis, *Ann. Rev. Physiol.,* 6, 73–103.

Crowe, J.H. and Crowe, L.M. (1992) Preservation of liposomes by freeze drying, in *Liposome Technology,* 2nd edition, Gregoriadis, G., Ed., CRC Press, Boca Raton, FL, pp. 229–251.

Crowe, J.H., Crowe, L.M., Carpenter, J.F., Aurell Wistrom, C. (1987) Stabilization of dry phospholipid bilayers and proteins by sugars, *Biochem. J.,* 242, 1–10.

Crowe, J.H., Hoekstra, F.A., Nguyen, K.H.N., and Crowe, L.M. (1997a) Is vitrification involved in depression of the phase transition temperature in dry phospholipids? *Biochim. Biophys. Acta,* 1280, 187–196.

Crowe, J.H., Oliver, A.E., Hoekstra, F.A., and Crowe, L.M. (1997b) Stabilization of dry membranes by mixtures of hydroxyethyl starch and glucose: The role of vitrification, *Cryobiology,* 3, 20–30.

Crowe, J.H., Panek, A.D., Crowe, L.M., Panek, A.C., and Soares de Araujo, P. (1991) Trehalose transport in yeast cells, *Biochem. Int.,* 24, 721–730.

Crowe, J.H., Spargo, B.J., and Crowe, L.M. (1987) Preservation of dry liposomes does not require retention of residual water, *Proc. Nat. Acad. Sci., USA,* 84, 1537–1540.

Crowe, J.H., Tablin, F., Tsvetkova, N., Oliver, A.E., Walker, N., and Crowe, L.M. (1999) Are lipid phase transitions responsible for chilling damage in platelets? *Cryobiology,* 38, 180–191.

Crowe, L.M., Spargo, B.J., Ioneda, T., Beaman, B.L., and Crowe, J.H. (1994) Interaction of cord factor (α,α-trehalose-6,6-dimycolate) with phospholipids, *Biochim. Biophys. Acta,* 1194, 53–60.

Crowe, L.M., Reid, D.S., and Crowe, J.H. (1996) Is trehalose special for preserving dry biomaterials? *Biophys. J.,* 71, 2087–2093.

Cuber, R., Eleutherio, E.C.A., Pereira, M.D., and Panek, A.D. (1997) The role of the trehalose transporter during germination, *Biochim. Biophys.,* 1330, 165–171.

De Castro, A.G. and Tunnacliffe, A. (2000) Intracellular trehalose improves osmotolerance but not desiccation tolerance in mammalian cells, *FEBS Lett.,* 487, 199–202.

De Jaeghere, E., Allémann, J.C., Leroux, W., Stevels, J., Feijen, E., Doelker, R., Gurny, R. (1999) Formulation and lyoprotection of poly(lactic acid-*co*-ethylene oxide) nanoparticles: Influence on physical stability and *in vitro* cell uptake, *Pharm. Res.,* 16, 859–866.

De Smet, K.A.L, Weston, A., Brown, I.N., Young, D.B., Robertson, B.D. (2000) Three pathways for trehalose biosynthesis in mycobacteria, *Microbiology,* 146, 199–208.

Diez, J., Diederichs, K., Greller, G., Horlacher, R., Boos, W., and Welte, W. (2001) The crystal structure of a liganded trehalose/maltose-binding protein from the hyperthermophilic Archaeon *Thermococcus litoralis* at 1.85Å, *J. Mol. Biol.,* 305, 905–915.

Dráber, P., Dráberová, E., and Nováková, M. (1995) Stability of monoclonal IgM antibodies freeze-dried in the presence of trehalose, *J. Immunol. Meth.,* 181, 37–43.

Eleutherio, E.C.A., De Araujo, P.S., and Panek, A.D. (1993) Role of the trehalose carrier in dehydration resistance of *Saccharomyces cerevisiae, Biochim. Biophys. Acta,* 1156, 263–266.

Eroglu, A., Russo, M.J., Bieganski, R., Fowler, A., Cheley, S., Bayley, H., and Toner, M. (2000) Intracellular trehalose improves the survival of cryopreserved mammalian cells, *Nat. Biotechnol.,* 18,163–167.

Esteves, M.I., Quintilio, W., Sato, R.A., Raw, I., De Araujo, P.S., and Da Costa, M.H.B. (2000) Stabilisation of immunoconjugates by trehalose, *Biotechnol. Lett.,* 22, 417–420.

Faraone, A., Magaza, S., Lechner, R.E., Longeville, L., Maisano, G., Majolino, D., Migliardo, P., and Wanderlingh, U. (2001) Quasi-elastic neutron scattering from trehalose aqueous solutions, *J. Chem. Phys.,* 115, 3282–3286.

Farías-Rodríguez, R., Mellor, R.B., Arias, C., and Peña-Cabriales, J.J. (1998) The accumulation of trehalose in nodules of several cultivars of common bean (*Phaseolus vulgaris*) and its correlation with resistance to drought stress, *Physiol. Plant.,* 102, 353–359.

Felix, C.C., Moreira, M.S., Oliveira, M., Sola-Pena, J., Meyer-Fernandes, R., Scofane, H.M., and Ferreira-Pereira, A. (1999) Protection against thermal denaturation by trehalose on the plasma membrane ATPase from yeast's synergistic effect between trehalose and the phospholipid environment, *Eur. J. Biochem.,* 266, 660–664.

Franks, F. (1996) Freeze-dried blood: Reality or confidence trick? *Cryo-Letters,* 17, 1.

Garg, A.K., Kim, J-K, Owens, T.G., Ranwala, A.P., Choi, Y.D., Kochian, L.V., and Wu, R.J. (2002) Trehalose accumulation in rice plants confers high tolerance levels to different abiotic stresses, *Proc. Natl. Acad. Sci. USA,* 99, 15898–15903.

Ghasempour, H.R., Gaff, D.F., Williams, R.P.W., and Gianello, R.D. (1998) Contents of sugars in leaves of drying desiccation tolerant flowering plants, particularly grasses, *Plant Growth Regul.,* 24, 185–191.

Gimeno-Alcañiz, J.V., Pérez-Ortín, J.E., and Matallana, E. (1999) Differential pattern of trehalose accumulation in wine yeast strains during the microvinification process, *Biotechnol. Lett.,* 21, 271–274.

Goddijn, J.M. and Van Dun, K. (1999) Trehalose metabolism in plants, *Trends Plant Sci.,* 4, 315–319.

Goddijn, O.J.M., Verwoerd, T.C., Voogd, E., Krutwagen, P.W.H.H., De Graaf, P.T.H.M., Poels, J., Van Dun, K., Ponstein, A.S., Damm, B., and Pen, J. (1997) Inhibition of trehalase activity enhances trehalose accumulation in transgenic plants, *Plant Physiol.,* 113, 181–190.

Goodrich, R.P., Sowemimo-Coker, S.O., Zerez, C.R., and Tanaka, K.R. (1992) Preservation of metabolic activity in lyophilized human erythrocytes, *Proc. Natl. Acad. Sci., USA,* 89, 967–971.

Gordon, S.L., Oppenheimer, S.R., Mackay, A.L., Brunnabend, J., Pihle, I., and Levine, F. (2001) Recovery of human mesenchymal stem cells following dehydration and rehydration, *Cryobiology,* 43(2), 182–187.

Gousset, K., Wolkers, W.F., Tsvetkova, N.M., Oliver, A.E., Field, C.L., Walker, N.J., Crowe, J.H., and Tablin, F. (2002) Evidence for a physiological role for membrane rafts in human platelets, *J. Cell Physiol.,* 190, 117–128.

Green, J.L. and Angell, C.A. (1989) Phase relations and vitrification in saccharide-water solutions and the trehalose anomaly, *J. Phys. Chem.,* 93, 2880–2882.

Gueguen, Y., Rolland, J.L., Schroeck, S., Flament, D., Defretin, S., Saniez, M.H., and Dietrich, J. (2001) Characterization of the maltooligosyl trehalose synthase from the thermophilic archaeon *Sulfolobus acidocaldarius, FEMS Microbiol. Lett.,* 194, 201–206.

Guo, N., Puhlev, I., Brown, D.R., Mansbridge, J., and Levine, F. (2000) Trehalose expression confers desiccation tolerance on human cells, *Nat. Biotechnol.,* 18, 168–171.

Han, E.-K., Cotty, F., Sottas, C., Jiang, H., and Michels, C.A. (1995) Characterization of AGT1 encoding a general alpha-glucoside transporter from Saccharomyces, *Mol. Microbiol.,* 17, 1093–1107.

Harrigan, P.R., Madden, T.D., and Cullis, P.R. (1990) Protection of liposomes during dehydration or freezing, *Chem. Phys. Lipids,* 52, 39–149.

Hays, L.M., Crowe, J.H., Wolkers, W., and Rudenko, S. (2001) Factors affecting leakage of trapped solutes from phospholipid vesicles during thermotropic phase transitions, *Cryobiology,* 42, 88–102.

Heller, M.C., Carpenter, J.F., and Randolph, T.W. (1999) Protein formulation and lyophilization cycle design: Prevention of damage due to freeze-concentration induced phase separation, *Biotech. Bioeng,.* 63, 166–174.

Hill, D.R., Keenan, T.W., Helm, R.F., Potts, M., Crowe, L.M., and Crowe, J.H. (1997) Extracellular polysaccharide of Nostoc commune (Cyanobacteria) inhibits fusion of membrane vesicles during desiccation, *J. Appl. Phycol.,* 9, 237–248.

Hincha, D.K., Hellwege, E.M., Meyer, A.G., and Crowe, J.H. (2000) Plant fructans stabilize phosphatidyl-choline liposomes during freeze-drying, *Eur. J. Biochem.,* 267, 535–540.

Hirata, T., Yokomise, H., Fukuse, T., Muro, K., Ono, N., Inui, K., Hitomi, S., and Wada, H. (1993) Successful 12-hour lung preservation with trehalose, *Transplant. Proc.,* 25, 1597–1598.

Hoekstra, F.A., Wolkers, W.F., Buitink, J., Golovina, E.A., Crowe, J.H., and Crowe, L.M. (1997) Membrane stabilization in the dry state, *Comp. Biochem. Physiol. [A],* 117A, 335–341.

Hsieh, C.H. and Wu, W.G. (1996) Structure and dynamics of primary hydration shell of phosphatidylcholine bilayers at subzero temperatures, *Biophys. J.,* 71, 3278–3287.

Iglesias, H.A., Chirife, J., and Buera, M.P. (1997) Adsorption isotherm of amorphous trehalose, *J. Sci. Food Agric.,* 75, 183–186

Iturriaga, G., Gaff, D.F., and Zentella, R. (2000) New desiccation-tolerant plants, including a grass, in the central highlands of Mexico, accumulate trehalose, *Aust. J. Bot.,* 48, 153–158.

Kaasen, I., McDougall, J., and Strom, A.R. (1994) Analysis of the *otsBA* operon for osmoregulatory trehalose synthesis in *Escherichia coli* and homology of the OtsA and OtsB proteins to the yeast trehalose-6-phosphate synthase/phosphatase complex, *Gene,* 145, 9–15.

Kacuráková, M. and Mathlouthi, M. (1996) FTIR and laser-Raman spectra of oligosaccharides in water: Characterization of the glycosidic bond, *Carbo. Res.,* 284, 145–157.

Kato, M., Miura, Y., Kettoku, M., and Shindo, K. (1996) Purification and characterization of new trehalose-producing enzymes isolated from the hyperthermophilic archae, *Sulfolobus solfataricus* KM1, *Biosci. Biotechnol. Biochem.,* 60, 546–550.

Kim, C.K., Chung, H.S., Lee, M.K., Choi, L.M., and Kim, M.H. (1999) Development of dried liposomes containing galactosidase for the digestion of lactose in milk, *Int. J. Pharm.,* 183, 185–193.

Kitahara, A.K., Suzuki, Y., Zhan, C.W., Wada, H., and Nishimura, Y. (1996) Preservation of skin free-flap using trehalose, *J. Surg. Res.,* 62, 130–134.

Kitahara, A.K., Suzuki, Y., Zhan, C.W., Wada, H., and Nishimura, Y. (1998) Evaluation of new improved solution containing trehalose in free skin flap storage, *Br. J. Plast. Surg.,* 51, 118–121.

Kong, X.D., Liu, Y., Gou, X.J., Zhang, H.Y., Wang, X.P., and Zhang, J. (2001) Directed evolution of operon of trehalose-6-phosphate synthase/phosphatase from *Escherichia coli, Biochem. Biophys. Res. Comm.,* 280, 396–400.

Koster, K.L., Lei, Y.P., Anderson, M., Martin, S., and Bryant, G. (2000) Effects of vitrified and nonvitrified sugars on phosphatidylcholine fluid-to-gel phase transitions, *Biophys. J.,* 78, 1932–1946.

Kreilgaard, L., Frokjaer, S., Flink, J.M., Randolph, T.W., and Carpenter, J.F. (1999) Effects of additives on the stability of *Humicola lanuginosa* lipase during freeze-drying and storage in the dried solid, *J. Pharm. Sci.,* 88, 281–290.

Lao, G., Polayes, D., Xia, J.L., and Bloom, F.R. (2001) Overexpression of trehalose synthase and accumulation of intracellular trehalose in 293H and 293FTetR:Hyg cells, *Cryobiology,* 43(2), 106–113.

Leslie, S.B., Teter, S.A., Crowe, L.M., and Crowe, J.H. (1994) Trehalose suppresses phase transitions in dry yeast, *Biochim. Biophys. Acta,* 1192, 7–13.

Leslie, S.B., Israeli, E., Lighthart, B., Crowe, J.H., and Crowe, L.M. (1995) Trehalose and sucrose protect both membranes and proteins in intact bacteria during drying, *Econ. Env. Microbiol.,* 61, 3592–3597.

Li, B., Li, S., Tan, Y.D., Stolz, D.B., Watkins, S.C., Block, L.H., and Huang, L. (2000) Lyophilization of cationic lipid-protamine-DNA (LPD) complexes, *J. Pharm. Sci.,* 89, 355–364.

Li, S., Patapoff, T.W., Overcashier, D., Hsu, C., Nguyen, T.H., and Borchardt, R.T. (1996) Effects of reducing sugars on the chemical stability of human relaxin in the lyophilized state, *J. Pharm. Sci.,* 85, 873–877.

Lin, T.Y. and Timasheff, S.N. (1996) On the role of surface tension in the stabilization of globular proteins, *Protein Sci.,* 5, 372–381.

Liu, Q., Schmidt, R.K., Teo, B., Karplus, P.A., and Brady, J.W. (1997) Molecular dynamics studies of the hydration of α,α-trehalose, *J. Am. Chem. Soc.,* 119, 7851–7862.

Lucero, P., Penalver, E., Moreno, E., and Lagunas, R. (2000) Internal trehalose protects endocytosis from inhibition by ethanol in *Saccharomyces cerevisiae, Appl. Environ. Microbiol.,* 66, 4456–4461.

Luzardo, M.D., Amalfa, F., Nuñez, A.M., Díaz, S., De Lopez, A.C.B., and Disalvo, E.A. (2000) Effect of trehalose and sucrose on the hydration and dipole potential of lipid bilayers, *Biophys. J.,* 78, 2452–2458.

Magazu, S., Maisano, G., Middendorf, H.D., Migliardo, P., Musolino, A.M., and Villari, V. (1998) trehalose-water solutions. II. Influence of hydrogen bond connectivity on transport properties, *J. Phys. Chem. B,* 102, 2060–2063.

Malik, K.A. and Lang, E. (1996) Successful preservation of *Campylobacteraceae* and related bacteria by liquid-drying under anaerobic conditions, *J. Microbiol. Methods,* 25, 37–42.

Mansure, J.J., Panek, A.D., Crowe, L.M., and Crowe, J.H. (1994) Trehalose inhibits ethanol effects on intact yeast cells and liposomes, *Biochim. Biophys. Acta Bio-Membr.,* 1191, 309–316.

Mansure, J.J., Souza, R.C., and Panek, A.D. (1997) Trehalose metabolism in *Saccharomyces cerevisiae* during alcoholic fermentation, *Biotechnol. Lett.,* 19, 1201–1203.

Martorell, D., Siebert, S.T.A., and Durst, R.A. (1999) Liposome dehydration on nitrocellulose and its application in a biotin immunoassay, *Anal. Biochem.,* 271, 177–185.

Matsuo, T. (2001) Trehalose protects corneal epithelial cells from death by drying, *Br. J. Ophthalmol.,* 85, 610–612.

Matthijs, S., Koedam, N., Cornelis, P., and De Greve, H. (2000) The trehalose operon of *Pseudomonas fluorescens* ATCC 17400, *Res. Microbiol.,* 151, 845–851.

Miller, D.P., De Pablo, J.J., and Corti, H.R. (1999) Viscosity and glass transition temperature of aqueous mixtures of trehalose with borax and sodium chloride, *J. Phys. Chem. B,* 103, 10243–10249.

Miller, D.P., De Pablo, J.J., and Corti, H. (1997) Thermophysical properties of trehalose and its concentrated aqueous solutions, *Pharm. Res.,* 14, 578–590.

Miura, N., Hayashi, Y., Shinyashiki, N., and Mashimo, S. (1995) Observation of unfreezable water in aqueous-solution of globule protein by microwave dielectric measurement, *Biopolymers,* 36, 9–16.

Moroff, G., Holme, S., George, V.M., and Heaton, W.A. (1994) Effect on platelet properties of exposure to temperatures below 20°C for short periods of storage at 20 to 24°C, *Transfusion,* 34, 317–321.

Müller, J., Boller, T., and Wiemken, A. (1995) Effects of validamycin A, a potent trehalase inhibitor, and phytohormones on trehalose metabolism in roots and root nodules of soybean and cowpea, *Planta,* 197, 362–368.

Murray, B.S. and Liang, H.J. (1999) Enhancement of the foaming properties of protein dried in the presence of trehalose, *J. Agric. Food Chem.,* 47, 4984–4991.

Nagase, H., Endo, T., Ueda, H., Nakagaki, M. (2002) An anhydrous polymorphic form of trehalose, *Carbo. Res.,* 337, 167–173.

Nishimoto, T., Nakano, M., Ikegami, S., Chaen, H., Fukuda, S., Sugimoto, T., Kurimoto, M., and Tsujisaka, Y. (1995) Existence of a novel enzyme converting maltose into trehalose, *Biosci. Biotechnol. Biochem.,* 59, 2189–2190.

Nishizaki, Y., Yoshizane, C., Toshimori, Y., Arai, N., Akamatsu, S., Hanaya, T., Arai, S., Ikeda, M., and Kurimoto, M. (2000) Disaccharide-trehalose inhibits bone resorption in ovariectomized mice, *Nutr. Res.,* 20, 653–664.

Norcia, M.A. (2000) Compositions and methods for wound management, *Off. Gaz. U. S. Patent Trademark Office* 1232, 424–448.

O'Brien, J. (1996) Stability of trehalose, sucrose and glucose to nonenzymatic browning in model systems, *J. Food Sci.,* 61, 679–682.

Oliver, A.E., Hincha, D.K., and Crowe, J.H. (2002) Are sugars sufficient for preservation of anhydrobiotes? *Comp. Biochem. Physiol.,* 131A, 515–525

Otsubo, M. and Iwaya-Inoue, M. (2000) Trehalose delays senescence in cut gladiolus spikes, *Hort Sci.,* 35, 1107–1110.

Paschoalin, V.M.F., Costa-Carvalho, V.L.A., and Panek, A.D. (1986) Further evidence for the alternative pathway of trehalose synthesis linked to maltose utilization in *Saccharomyces, Curr. Genet.,* 10, 725–732.

Pellerin-Mendes, C., Million, L., Marchand-Arvier, M., Labrude, P.P., and Vigneron, C. (1997) *In vitro* study of the protective effect of trehalose and dextran during freezing of human red blood cells in liquid nitrogen, *Cryobiology,* 35, 173–186.

Pilon-Smits, E.A.H., Terry, N., Sears, T., Kim, H., Zayed, A., Hwang, S., Van Dun, K., Voogd, E., Verwoerd, T.C., Krutwagen, R.W.H.H., and Goddijn, O.J.M. (1998) Trehalose-producing transgenic tobacco plants show improved growth performance under drought stress, *J. Plant Physiol.,* 152, 525–532.

Poveda, A., Vicent, C., Penadés, S., and Jiménez-Barbero, J. (1997) NMR experiments for the detection of NOEs and scalar coupling constants between equivalent protons in trehalose-containing molecules, *Carbohydr. Res.,* 301, 5–10.

Pritchard, H.W., Tompsett, P.B., Manger, K., and Smidt, W.J. (1995) The effect of moisture-content on the low-temperature responses of *Araucaria hunsteinii* seed and embryos, *Ann. Bot.,* 76, 79–88.

Reinders, A., Romano, I., Wiemken, A., and De Virgilio, C. (1999) The thermophilic yeast *Hansenula polymorpha* does not require trehalose synthesis for growth at high temperatures but does for normal acquisition of thermotolerance, *J. Bacteriol.,* 181, 4665–4668.

Reshkin, S.J., Cassano G., Womersley, C., and Ahearn, G.A. (1988) Preservation of glucose transport and enzyme activity in fish intestinal brush border and baolateral membrane vesicles, *J. Exp. Biol.,* 140, 123–135.

Richards, A.B.,,Krakowka, S., Dexter, L.B., Schmid, H., Wolterbeek, A.P.M., Waalkens-Berendsen, D.H., Shigoyuki, A., and Kurimoto, M. (2002) Trehalose: A review of properties, history of use and human tolerance, and results of multiple safety studies, *Food Chem. Toxicol.,* 40, 871–898.

Rodal, S.K., Skretting, G., Garred, O., Vilhardt, F., van Deurs, B., and Sandvig, K. (1999) Extraction of cholesterol with methyl-beta-cyclodextrin perturbs formation of clathrin-coated endocytic vesicles, *Mol. Biol. Cell,* 10, 961–974.

Sampedro, J.G., Cortés, P., Muñoz-Clares, R.A., Fernández, A., and Uribe, S. (2001) Thermal inactivation of the plasma membrane H^+-ATPase from *Kluyveromyces lactis.* Protection by trehalose, *Biochim. Biophys. Acta,* 1544, 64–73.

Sanderson, P.W., Williams, W.P, Cunningham, B.A., Wolfe, D.H., and Lis, L.J. (1993) The effect of ice on membrane lipid phase behavior, *Biochim. Biophys. Acta,* 1148, 278–284.

Schebor, C., Galvagno, M., del Pilar Buera, M., and Chirife, J. (2000) Glass transition temperatures and fermentative activity of heat-treated commercial active dry yeasts, *Biotechnol. Prog.,* 16, 163–168.

Schebor, C., Burin, L., del Pilar Buera, H., and Chirife, J. (1999) Stability to hydrolysis and browning of trehalose, sucrose and raffinose in low-moisture systems in relation to their use as protectants of dry biomaterials, *Lebensm.-Wiss. u.-Technol.,* 32, 481–485.

Schwarz, C. and Mehnert, W. (1997) Freeze-drying of drug-free and drug-loaded solid lipid nanoparticles (SLN), *Int. J. Pharm.,* 157, 171–179.

Sei, S., Gonda, T., and Arima, Y. (2002) Growth rate and morphology of ice crystals growing in a solution of trehalose and water, *J. Crystal Growth,* 240, 218–229.

Seo, H.S., Koo, Y., Lim, J.Y., Song, J.T., Kim, C.H., Kim, J.K., Lee, J.S., and Choi, Y.D. (2000) Characterization of a bifunctional enzyme fusion of trehalose-6-phosphate synthetase and trehalose-6-phosphate phosphatase of *Escherichia coli*, *Appl. Environ. Microbiol.*, 66, 2484–2490.

Serrano, R., Culianz-Macia, F.A., and Moreno, V. (1999) Genetic engineering of salt and drought tolerance with yeast regulatory genes, *Sci. Hort.*, 78, 261–269.

Sharma, S.C. (1997) A possible role of trehalose in osmotolerance and ethanol tolerance in *Saccharomyces cerevisiae*, *FEMS Microbiol. Lett.*, 152, 11–15.

Sklari, S., Rahiala, H., Stathopoulos, V., Rosenholm, J., and Pomonis, P. (2001) The influence of surface acid density on the freezing behavior of water confined in mesoporous MCM-41 solids, *Microporous Mesoporous Materials*, 49, 1–13.

Sode, K., Akaike, E., Sugiura, H., and Tsugawa, W. (2001) Enzymatic synthesis of a novel trehalose derivative, 3,3-diketotrehalose, and its potential application as the trehalase enzyme inhibitor, *FEBS Lett.*, 489, 42–45.

Sola-Penna, M. and Meyer-Fernandes, J.R. (1998) Stabilization against thermal inactivation promoted by sugars on enzyme structure and function: Why is trehalose more effective than other sugars? *Arch. Biochem. Biophys.*, 360, 10–14.

Soto, T., Fernández, J., Vicente-Soler, J. Cansado, J., and Gacto, M. (1999) Accumulation of trehalose by overexpression of *tps1*, coding for trehalose-6-phosphate synthase, causes increased resistance to multiple stresses in the fission yeast *Schizosaccharomyces pombe*, *Appl. Environ. Microbiol.*, 65, 2020–2024.

Sowemimo-Coker, S.O., Goodrich, R.P., Zerez, C.R., and Tanaka, K.R. (1993) Refrigerated storage of lyophilized and rehydrated, lyophilized human red cells, *Transfusion*, 33, 322–329.

Stenberg, P.E., McEver, R.P., Shuman, M.A., Jacques, Y.V., and Bainton, D.F. (1985) A platelet alpha-granule membrane protein (GMP-140) is expressed on the plasma membrane after activation, *J. Cell Biol.*, 101, 880–886.

Spieles, G., Heschel, I., and Rau, G. (1996) An attempt to recover viable human red blood cells after freeze-drying, *Cryo-Letters*, 17, 43–52.

Stambuk, B.U., Panek, A.D., Crowe, J.H., Crowe, L.M., and De Araujo, P. (1998) Expression of high-affinity trehalose-H+ symport in *Saccharomyces cerevisiae*, *Biochim. Biophys. Acta*, 1379, 118–128.

Sun, W.Q., Leopold, A.C., Crowe, L.M., and Crowe, J.H. (1996) Stability of dry liposomes in sugar glasses, *Biophys. J.*, 70, 1769–1776.

Tablin, F., Wolkers, W.F., Walker, N.J., Oliver, A.E., Tsvetkova, N.M., Crowe, L.M., and Crowe, J.H. (2001) Membrane reorganization during chilling: Implications for long-term stabilization of platelets, *Cryobiology*, 43, 114–23.

Tablin, F., Oliver, A.E., Walker, N.J., Crowe, L.M., and Crowe, J.H. (1996) The membrane phase transition of intact human platelets—correlation with cold-induced activation, *J. Cell Physiol.*, 168, 305–331.

Timasheff, S. N. (2000) Control of protein stability and reactions by weakly interacting cosolvents: The simplicity of the complicated, in *Advances in Protein Chemistry: Linkage Thermodynamics of Macromolecular Interactions*, Di Cera, E., Ed., Academic Press, San Diego, CA, pp. 192–234.

Tsvetkova, N.M., Crowe, J.H., Walker, N.J., Crowe, L.M., Oliver, A.E., Wolkers, W.F., and Tablin, F. (1999) Physical properties of membrane fractions isolated from human platelets: Implications for chilling induced platelet activation, *Mol. Mem. Biol.*, 16, 265–272.

Tsvetkova, N.M., Walker, N.J., Crowe, J.H., Field, C.L., Shi, Y., and Tablin, F. (2000) Lipid phase separation correlates with activation in platelets during chilling, *Mol. Mem. Biol.*, 17, 209–218.

Tunnacliffe, A., García de Castro, A., and Manzanera, M. (2001) Anhydrobiotic engineering of bacterial and mammalian cells: Is intracellular trehalose sufficient? *Cryobiology*, 43(2), 124–132.

Uritani, M., Takai, M., and Yoshinaga, K. (1995) Protective effect of disaccharides on restriction endonucleases during drying under vacuum, *J. Biochem. (Tokyo)*, 117, 774–779.

Valeri, C.R., Feingold, H., and Marchionni, L.D. (1974) A simple method for freezing human platelets using 6% dimethylsulfoxide and storage at –80°C, *Blood*, 43, 131–136.

van Winden, E.C.A., Zhang, W., and Crommelin, D.J.A. (1997) Effect of freezing rate on the stability of liposomes during freeze-drying and rehydration, *Pharm. Res.*, 14, 1151–1160.

Vereyken, I.J., Chupin, V., Demel, R.A., Smeekens, S.C.M., and De Kruijff, B. (2001) Fructans insert between the headgroups of phospholipids, *Biochim. Biophys. Acta*, 1510, 307–320.

White, J.G. and Krivit, W. (1967) An ultrastructural basis for shape changes induced in platelets by chilling, *Blood*, 30, 625–635.

White, J.G., Krumwiede, M., and Sauk, J.J. (1985) Microtubule reassembly in surface-activated platelets, *Blood,* 65, 1494–1503.

Winokur, T. and Hartwig, J.H. (1995) Mechanism of shape change in chilled human platelets, *Blood,* 85, 349–360.

Wolkers, W.F., Walker, N.J., Tablin, F., and Crowe, J.H. (2001) Human platelets loaded with trehalose survive freeze-drying, *Cryobiology,* 42, 79–87.

Worrall, E.E., Litamoi, J.K., Seck, B.M., and Ayelet, G. (2000) Xerovac: An ultra rapid method for the dehydration and preservation of live attenuated Rinderpest and Peste des Petits ruminants vaccines, *Vaccine,* 19, 834–839.

Xie, G. and Timasheff, S.N. (1997) The thermodynamic mechanism of protein stabilization by trehalose, *Biophys. Chem.,* 64, 25–43.

Yokomise, H., Inui, K., Wada, H., Hasegawa, S., Ohno, N., and Hitomi, S. (1995) Reliable cryopreservation of trachea for one month in a new trehalose solution, *J. Thorac. Cardiovasc. Surg.,* 110, 382–385.

Yoshida, N., Nakamura, M., and Horikoshi, K. (1998) Production of trehalose by a dual enzyme system of immobilized maltose phosphorylase and trehalose phosphorylase, *Enzyme Microb. Technol.,* 22, 71–75.

Yoshii, H., Furuta, T., Kudo, J., and Linko, P. (2000) Crystal transformation from anhydrous α-maltose to hydrous β-maltose and from anhydrous trehalose to hydrous trehalose, *Biosci. Biotechnol. Biochem.,* 64, 1147–1152.

Yoshinaga, K., Yoshioka, H., Kurosaki, H., Hirasawa, M., Uritani, M., and Hasegawa, K. (1997) Protection by trehalose of DNA from radiation damage, *Biosci. Biotechnol. Biochem.,* 61, 160–161.

Yoshizane, C., Arai, N., Arai, C., Yamamoto, M., Nishizaki, Y., Hanaya, T., Arai, S., Ikeda, M., and Kurimoto, M. (2000) Trehalose suppresses osteoclast differentiation in ovariectomized mice: Correlation with decreased *in vitro* interleukin-6 production by bone marrow cells, *Nutr. Res.,* 20, 1485–1491.

Zentella, R., Mascorro-Gallardo, J.O., Van Dijck, P., Folch-Mallol, J., Bonini, B., Van Vaeck, C., Gaxiola, R., Covarrubias, A.A., Nieto-Sotelo, J., Thevelein, J.M., and Iturriaga, G. (1999) A *Selaginella lepidophylla* trehalose-6-phosphate synthase complements growth and stress-tolerance defects in a yeast tps1 mutant, *Plant Physiol.,* 119, 1473–1482.

Zucker, M. and Borelli, J. (1954) Reversible alterations in platelet morphology produced by anticoagulants and cold, *Blood,* 9, 602–208.

22 Vitrification in Tissue Preservation: New Developments

Michael J. Taylor, Ying C. Song, and Kelvin G.M. Brockbank

CONTENTS

0-415-24700-4/04/$0.00+$1.50
© 2004 by CRC Press LLC

22.1 INTRODUCTION

Cryopreservation has been notably effective for banking and shipping of isolated cells, but much less so for more complex, integrated multicellular systems. We now have a broad understanding of the mechanisms that injure cells during freezing and thawing, and techniques have been developed that limit or prevent this injury, so that very low temperatures can now be used to preserve, virtually indefinitely, many cell-types with very high recovery rates (Karlsson and Toner, 1996; Taylor, 1984; Mazur, 1984a). These techniques are all aimed at preventing intracellular freezing and minimizing the damaging changes that occur in the remaining liquid phase as a consequence of the separation of water to form ice. In tissues and organs, however, it is not sufficient to maintain cellular viability—it is also important to maintain the integrity of the extracellular structure on both a micro- and macroscale. Techniques that are effective for the cryopreservation of cell suspensions do not always maintain this integrity because highly organized multicellular tissues present a special set of problems; foremost among these is the effect of extracellular ice formation, which disrupts the tissue architecture (Hunt et al., 1982; Karlsson and Toner, 1996; Pegg, 1987; Pegg et al., 1979; Taylor and Pegg, 1982). However, extracellular ice cannot explain all the problems encountered in moving from single cells to tissues—for example, the detachment of the corneal endothelium has a different, and as yet unknown, cause (Taylor, 1986), but prevention of this detachment is crucial for corneal viability. The effects of cryopreservation on the mechanisms of cellular adhesion have not yet been studied widely. To date, there have been very few studies focused on the effects of cryopreservation on anchorage-dependent cells, and in general, these studies have shown that postthaw recovery of adherent cells is lower than comparable cells frozen and thawed as cell suspensions (Ohno, 1994; Hetzel et al., 1973).

As outlined throughout this book, the science of cryobiology has defined many factors that must be optimized for cells to be stored for lengthy periods at low temperatures. Survival of cells through the rigors of freezing and thawing in cryopreservation procedures is only attained by using appropriate cryoprotective agents, and in general, these techniques are applicable to isolated cells in suspension or small aggregates of cells in simple tissues. More complex tissues and organs that have a defined architecture are not easily preserved using conventional cryopreservation techniques; this is principally because of the deleterious effects of ice formation on organized multicellular structures (Karlsson and Toner, 1996; Pegg, 1987; Pegg et al., 1979). The application of cryobiological principles to engineered tissue constructs is likely to be fraught with problems similar to those identified during the course of trying to extrapolate the successes of freezing cell suspensions to organized tissues and organs. Avoidance of ice damage has therefore become the principal focus in research to develop effective storage techniques for multicellular tissues and organs. In this chapter we review some of the underlying principles of the various approaches to "ice-free" cryopreservation, with particular emphasis on vitrification, which has recently been shown to provide a practical solution for the cryopreservation of complex tissues that can not be adequately preserved by freezing/thawing methods.

The aim of this chapter is to deal with "ice-free" cryopreservation of tissues, which at face value might be considered to be a departure from the main theme of a book titled *Life in the Frozen State*. However, frozen systems usually embody a vitreous component, and our objective here is to review the approach to cryopreservation that aims to achieve vitrification at the outset in an attempt to circumvent the hazards of water crystallization as ice formation. The vitreous state is essentially a solidified, amorphous liquid state obtained by specific conditions of cooling and solute concentration that inhibit ice nucleation and growth. During cooling, molecular motions within the liquid are slowed and eventually arrested with extreme viscosity; the "arrested liquid" state is known as a glass. It is the conversion of a liquid into a glass that is called vitrification (derived from *vitri*, the Greek word for glass). The specifics of this process are addressed more completely below.

22.2 BACKGROUND AND HISTORICAL PERSPECTIVE

Readers of this book will be aware that the foundations of modern-day cryopreservation were laid in the middle of the last century following the milestone discovery of the cryoprotective effects of glycerol by Polge et al. (1949; see Chapters by 11 and18). The practical successes for cryopreservation of a wide variety of cells that ensued from these empirical studies led to considerable fundamental work that has defined a number of mechanistic principles underlying the cryopreservation of cells. Critically important mechanisms include the fundamental importance of the total quantity of ice and its location in relation to the cells, the toxicity of cryoprotectants and the temperature dependence of that toxicity, and the extent of osmotically induced changes in cell volume. Because an understanding of these factors is the foundation of cryobiology, a detailed discussion of these mechanistic principles can be found in Chapters 1 and 2, as well as in numerous other literature reviews on the subject.

Following the advent of modern-day cryobiology in the 1950s, it was confidently anticipated that cryopreservation techniques for tissues and organs would quickly ensue from the notable successes with cells in suspension. Unfortunately, this transition proved to be very difficult, and it was quickly realized that there were significant additional hurdles to be overcome in the freezing of mammalian tissues and organs (Smith, 1961). Only recently has it been possible to demonstrate successful cryopreservation of tissues using the vitrification approach that avoids some of the destructive events inherent in freezing; this approach is the focus of this chapter. In the intervening years, both empirical and systematic studies have contributed to a greater understanding of the mechanisms of cryoinjury in multicellular systems; these mechanisms have in turn mandated consideration of vitrification as the most realistic approach to circumvent the problems of which extracellular ice is undoubtedly a primary concern. Nevertheless, avoidance of ice for optimum preservation is not a new idea, as even before the ground-breaking discovery of Polge et al. (1949), Luyet had concluded that ice formation is not compatible with the survival of living systems and ought to be avoided if possible (Luyet, 1937; Luyet and Gehenio, 1940). At that time, the idea of vitrifying biological systems at low temperatures was born, but it was conceptually confined to cooling small living systems at extremely high rates to achieve a vitreous state and avoid ice crystallization (Luyet, 1937; Luyet and Gehenio, 1940). These constraints have been minimized to some extent in the modern era of cryobiology with the introduction of cryoprotective solutes that include the ability to promote vitrification during cooling as one of their inherent properties.

It is important to appreciate that vitrification and freezing (water crystallization) are not mutually exclusive processes and that the crystalline phase and vitreous phase can, and often do, coexist within a system. In fact, during conventional cryopreservation involving controlled freezing of cells, a part of the system vitrifies. This occurs because during freezing, the concentration of solutes in the unfrozen phase increases progressively until the point is reached at which the residual solution is sufficiently concentrated to vitrify in the presence of ice. Conventional cryopreservation techniques are optimized by designing protocols that avoid intracellular freezing. Under these cooling conditions, the cell contents actually vitrify because of the combined processes of dehydration, cooling, and the promotion of vitrification by intracellular macromolecules. In this context, the use of the term "vitrification" in the title of the seminal paper by Polge et al. (1949) was justified, as the spermatozoa would be vitrified in the presence of extracellular ice during cryopreservation with glycerol. However, the term is now used more generally to refer to a process in which attempts are made to vitrify the whole system to avoid any ice formation (Armitage and Rich, 1990; Fahy, 1988, 1989; Fahy et al., 1984; Pegg and Diaper, 1990; Rall, 1987).

The conditions necessary to achieve this objective are discussed more completely late. It will suffice in this section on the historical perspective to record that the current interest in vitrification was spurred by two independent approaches to achieving cryopreservation in the absence of damaging ice. The first has been referred to as the "equilibrium approach," in that it seeks to preequilibrate the biological specimen with a sufficiently high concentration of cryoprotectants before cooling such that freezing is prevented irrespective of the cooling rate (Elford, 1970; Elford and Walter, 1972a, 1972b; Farrant, 1965; Taylor et al., 1978). The second approach is called the "nonequilibrium approach," as the initial concentrations of solutes in the system are not of themselves sufficient to prevent ice nucleation and growth. This approach is achieved by using rapid cooling with or without the application of increased pressures to promote vitrification (Fahy, 1988, 1998; Fahy et al., 1984; MacFarlane, 1987; MacFarlane et al., 1992). In practice, because of constraints of the maximum tolerated limits of cryoprotective additive (CPA) toxicity, this leads to a metastable condition that is highly dependent on cooling and warming conditions if ice nucleation and crystal growth are to be avoided. Over the last 20 years, interest in this second approach to vitrification has in large part been kindled by the resolve of Fahy et al. to employ this as the only feasible way to achieve the cryopreservation of whole organs (Fahy, 1989; Fahy et al., 1984). Although this milestone has still to be achieved, the basic studies underpinning the research in this area have contributed to a better understanding of the ground rules that will ultimately lead to successful cryopreservation of organs. In the meantime, we have pursued this approach for the cryopreservation of complex tissues that may be regarded as intermediate between cell suspensions and whole organs with considerable recent success, as outlined below. Ten years ago, Pegg and Diaper reviewed the basic principles of freezing vs. vitrification as markedly different approaches to cryopreservation (Pegg and Diaper, 1990). Their review of the status of this technology was summed up with the reminder that at that time, "no system that is susceptible to damage by extracellular ice has yet been successfully vitrified, and all those systems that have been preserved by vitrification (early embryos, monocytes, and pancreatic islets) can equally well be preserved by conventional freeze-preservation methods" (Pegg and Diaper, 1990, p. 68). In such systems a vitrification method is often preferred because of practical benefits of operational simplicity, avoiding the need for expensive cooling equipment. Thus, in 1990 the challenges of vitrifying complex tissues remained formidable but approaches toward ice-free cryopreservation were still regarded as the way forward. Ten years later, this barrier has now been removed and we will review here the developments that have led to the successful vitrification of several tissues, two of which, blood vessels and articular cartilage, were previously refractory to cryopreservation with a high degree of functional survival.

22.3 EVIDENCE THAT EXTRACELLULAR ICE IS HARMFUL

Over the years since the early discoveries of cryoprotectants, attempts to extrapolate from the successful freeze-preservation of a wide variety of cell types to multicellular tissues and organs have been fraught with frustrating failures. We now recognize that the cryopreservation of complex tissues imposes a set of additional problems over and above the known mechanisms of cryoinjury that apply to single cells in suspension. These have been discussed in a number of reviews and will not be recounted here (Karlsson and Toner, 1996; Mazur, 1984b; Pegg et al., 1979; Taylor, 1984). Most important, it is now generally accepted that extracellular ice formation presents a major hazard for cryopreservation of multicellular tissues. As we have stated above, the predominance of this as a primary mechanism of cryoinjury in complex tissues has led to a focus on the development of low-temperature preservation techniques that avoid ice crystallization and *ipso facto* circumvent the associated problems. The evidence for the damaging role of ice in tissue cryopreservation has been outlined in a series of prior publications (Hunt et al., 1982; Jacobsen et al., 1984; Pegg, 1987; Pegg et al., 1979; Taylor, 1984; Taylor and Pegg, 1982). We will review briefly some of the early studies that provided clear evidence for a definitive role of ice *per se* as a principal hazard during cryopreservation of smooth muscle tissue. This is summarized here as a prelude to a review of the approaches to ice-free cryopreservation methods.

Smooth muscle is a good model tissue, intermediate between cells and organs, that requires for its function not only the survival of a high proportion of its cells but also the structural integrity

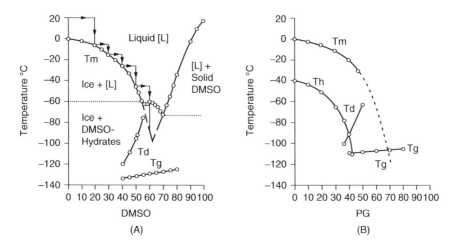

FIGURE 22.1 Supplemented binary phase diagrams for aqueous mixtures of Me$_2$SO (A) and Propane-1,2-diol (B) showing the principal events and phase changes associated with cooling and heating. A supplemented phase diagram combines nonequilibrium data on a conventional equilibrium phase diagram and serves to depict the important transitions inherent in cooling and warming aqueous solutions of cryoprotective solutes. Details are described elsewhere (Fahy, 1998; MacFarlane, 1987; Rasmussen and MacKenzie, 1968; Taylor, 1987) and in the text. T_m, equilibrium melting point curve (liquidus curve); T_h, homogeneous nucleation curve; T_d, devitrification curve; T_g, glass transition curve. The stepped line above the Me$_2$SO–H$_2$O liquidus T_m curve represents a scheme for incremental equilibration of a tissue with sufficient cryoprotective additive such that the system does not freeze during cooling. (See text for details.)

of the tissue so that the contractile response of the cells can remain coordinated. Experiments were designed to discover whether ice formation as such was damaging (Taylor and Pegg, 1982). Strips of muscle were cooled to –21°C in 2.56 *M* dimethyl sulfoxide (Me$_2$SO), held at that temperature overnight, and then thawed; then the Me$_2$SO was removed in excess Krebs's solution at 37°C and contractile response to histamine was measured. As illustrated in Figure 22.1, phase diagram data show that medium containing 20% (2.56 *M*) Me$_2$SO freezes at –8°C and concentrates 1.7 times to 34% (4.35 *M*) Me$_2$SO at –21°C. Another group of muscles was therefore equilibrated with 35% (4.49 *M*) Me$_2$SO containing 1.75 times the normal salt concentration and then cooled to and held at –21°C, unfrozen, for the same length of time. It was found that the unfrozen samples gave 72 to 76% recovery whether they were cooled at 2°C min^{-1} or 0.3°C min^{-1}, whereas the frozen samples gave only 21% ± 4% recovery if cooled at 2°C min^{-1} or 53% ± 7% if cooled at 0.3°C min^{-1}. Comparison of the frozen and unfrozen groups cooled at 2°C min^{-1} shows that ice damaged the tissue ($P < .001$), and comparison of the groups cooled at 0.3°C min^{-1} leads to the same conclusion ($P < .05$). In each case, the cooling rate, final temperature, concentration of Me$_2$SO, and concentration of other solutes was the same: Clearly, solute concentration was relatively innocuous, but ice formation was harmful. However, there was a striking difference between the recovery of tissue frozen at 0.3°C min^{-1} and tissue frozen at 2°C min^{-1} (Figure 22.2A). In the absence of freezing, there was no significant effect of cooling rate, but frozen tissue was more severely damaged when cooled at 2°C min^{-1} than were frozen muscles cooled at 0.3°C min^{-1}. A similar effect of cooling rate was found when muscles were frozen to –60°C (Figure 22.2A).

It seemed likely that, as with cells, the slower cooling rate would have allowed more complete dehydration of the tissue during cooling, and hence less ice in the tissue as a whole. Thus, it was proposed that a reduction in the quantity of extracellular ice could explain the difference in survival. This notion was subsequently confirmed by freeze substitution studies carried out for groups of muscle treated similar to the function studies. As illustrated in Figure 22.2C and D, a noticeable difference was detected in the patterns of ice formation between the two groups. Cooling at 2°C

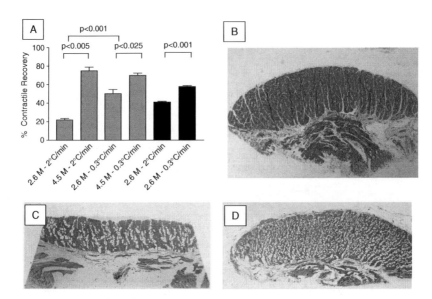

FIGURE 22.2 Structure and function of smooth muscle tissue after cooling to either –21°C, or –60°C in either the frozen or unfrozen state. (A) Histograms of postwarming contractility (mean ± SEM) normalized to the control responses derived before cooling at either –21°C (grey bars) or –60°C (black bars) under the conditions indicated. (B) Light micrograph of a section of nonfrozen control taenia coli smooth muscle showing the normal configuration of the muscle blocks that are the effector units of muscle contraction. (C) and (D) Light micrographs of freeze-substituted taenia coli smooth muscle depicting the location of ice domains after cooling to –21° at either 0.3°C/min (C), or 2°C/min (D). Details have been published elsewhere (Taylor and Pegg, 1982; Hunt et al., 1982) and are described in the text.

min^{-1} (Figure 22.2D) caused a random distribution of ice throughout the muscle tissue, whereas cooling at 0.3°C min^{-1} produced ice cavities that were predominantly in the extracellular matrix separating the muscle bundles and at the periphery of the bundles (Figure 22.2C). Because the muscle bundles are the effector units of muscle contraction, the greater disruption by ice of the muscle fasciae after cooling at the faster rate can explain the functional differences produced by varying the cooling rate (Hunt et al., 1982). The unequivocal conclusion from studies of this type was that the amount and location of extracellular ice has a dramatic effect on the postthaw function of complex tissues and organs. As a result, it is generally thought that cryopreservation of multicellular tissues and organs will mandate that the amount of ice in the system is limited, restricted to harmless sites, or preferably, that ice crystallization is prevented altogether.

22.4 APPROACHES TO ICE-FREE CRYOPRESERVATION

22.4.1 EQUILIBRIUM APPROACH

This approach to the avoidance of freezing during subzero cryopreservation was first proposed by Farrant in 1965. He suggested that if 60% of the cell water was replaced by a cryoprotective solute such as Me$_2$SO, freezing would be prevented at temperatures as low as –70°C; hence, any damage associated with the formation of ice crystals and the simultaneous rise in the concentration of solutes would be avoided (Farrant, 1965). Actually, Farrant's focus in developing this approach was principally on attenuating problems relating to increased solutes and, notably, electrolytes. The significance of the concomitant benefits of avoiding ice was appreciated more recently. This technique was explored using smooth muscle and involved the progressive mutual exchange of tissue water with Me$_2$SO during cooling and the gradual removal of the CPA during rewarming to minimize the known toxic effects of the CPA additive. The stepwise addition and removal of CPA such that the system remained above the equilibrium freezing point at each stage is best appreciated by reference to a phase diagram as illustrated in Figure 22.1A. (Elford, 1970; Farrant, 1965). Although ice crystallization and solute concentration (other than that of the CPA itself) could play no part in cryoinjury sustained by tissue during storage, other factors such as the pH and the anionic composition of the medium in which the tissue was immersed were found to have a profound effect on survival (Elford and Walter, 1972a; Taylor, 1982; Taylor et al., 1978). Freezing could be avoided completely, irrespective of the cooling rate, provided the tissue was fully equilibrated with CPA at each stage. This approach, which was tested using smooth muscle cooled and kept unfrozen at –79°C, failed to provide adequate contractile function until adequate steps were also taken to optimize the ionic composition of the CPA medium (Taylor, 1982; Elford and Walter, 1972a, 1972b). Nevertheless, this equilibrium approach has not been pursued, presumably because the technique demands lengthy periods of exposure to toxic solutes to ensure equilibration and because permeation studies have shown that adequate exchange of tissue water with CPAs at subzero temperatures may not be achievable either in a practical timescale or without exceeding the tolerable limits of solute toxicity in the tissue (Elford, 1970; Elford and Walter, 1972b).

22.4.2 NONEQUILIBRIUM APPROACH—VITRIFICATION

We will now turn our attention to the nonequilibrium approach to the avoidance of ice by focusing on the details of vitrification as it is currently practiced. The basic principles of this approach have been reviewed in great detail by others (Armitage and Rich, 1990; Fahy, 1988, 1989; Fahy et al., 1984; MacFarlane et al., 1992; Pegg and Diaper, 1990; Rall, 1987; see also Chapter 10). Thus, it will only be necessary here to outline the salient points that will enable an understanding of the practical applications to be discussed in this chapter.

Vitrification is the solidification of a liquid without crystallization. This state is achieved in systems that are sufficiently concentrated or that are cooled sufficiently rapidly that the increase in viscosity inhibits molecular rearrangement into a crystalline pattern. As cooling progresses, viscosity increases to the point at which translational molecular motion is essentially halted and the solution becomes a glass. The resultant solid retains the random molecular arrangement of a liquid but has the mechanical properties of a solid (MacFarlane, 1987; MacFarlane et al., 1992). We advisedly use the term "essentially halted" in referring to the attainment of molecular stasis during vitrification because there is a kinetic component to the process. In practical terms, the glass is a liquid that is too cold or viscous to flow, and although it is metastable in a strict thermodynamic sense, it is regarded as possessing pseudostability on the timescale of practical interest for biological preservation.

When considering the physico-chemical and biophysical responses of biological systems to low temperatures, it is important to be aware that events rarely take place under true equilibrium conditions. For example, the phase-change phenomena depicted schematically by the phase boundaries (e.g., T_m in Figures 22.1A and 22.1B) would hold true only for a simple binary system of CPA–H_2O in which supercooling was avoided by ensuring that nucleation occurred at the equilibrium freezing/melting point (T_m) during cooling at a sufficiently slow rate to prevent appreciable temperature gradients. Many interdependent factors determine whether an aqueous system, such as a biological system, approaches the thermodynamic state of lowest free energy during cooling. Metastability is thus often unavoidable, especially in concentrated systems. Many of these nonequilibrium states are, however, sufficiently reproducible and permanent to have been described as pseudoequilibrium states, and conversion of such metastable thermodynamic states to more stable forms may be subject to large kinetic barriers. The prevalence of so-called "unfreezable" or "bound" water in the vicinity of macromolecules is a prime example, where the expected path of thermodynamic stabilization by way of crystallization is prevented by large kinetic restraints (Franks, 1982a, 1982b; Taylor, 1987). A clear understanding of the occurrence and effects of metastable states during the cooling of compartmentalized living systems is complicated by the interaction of thermodynamic and kinetic factors, which are difficult to separate. Moreover, these complexities are compounded when such systems are cooled rapidly to low subzero temperatures. Nevertheless, some basic principles have been established with the aid of model systems such as aqueous solutions of cryoprotective solutes and other macromolecules that interact with water by hydrogen bonding (Franks, 1977; MacKenzie, 1977). Such studies have permitted some qualitative interpretation of the nonequilibrium phase behavior of the fluids in cells and tissues during cooling and warming. To this end, phase diagram data such as those depicted in Figure 22.1 have proved to be a useful tool in understanding the physico-chemical relationship between temperature, concentration, and change of phase. A detailed discussion of the role and interpretation of solid–liquid state diagrams in relation to low-temperature biology has been given in a previous review (Taylor, 1987). In particular, supplemented phase diagrams that combine nonequilibrium data on conventional equilibrium phase diagrams serve to depict the important transitions inherent in cooling and warming aqueous solutions of cryoprotective solutes. Reference to these major transitions, illustrated in Figure 22.1, can serve to explain and summarize the principles of achieving vitrification as follows.

The equilibrium freezing curve, labeled T_m, is often described as the liquidus curve and represents the points at which a solution having a particular concentration will freeze (or melt) under equilibrium conditions of temperature change. Hence this curve represents the phase-change boundary for the two-component solution as a function of temperature. Cooling a solution below the liquidus curve will result in ice formation if the conditions are favorable for nucleation, with the result that the remaining liquid phase becomes more concentrated in the solute as defined by the curve. A discussion of the details of nucleation is beyond the scope of this chapter but has been the subject of excellent reviews published in recent years (Fahy, 1998; Mehl, 1996; see also Chapters 1 and 2). In practice, it is well known that freezing is rarely initiated at the liquidus point. Inherently, solutions tend to undercool to varying degrees before nucleation and ice crystal growth proceed at a significant rate. In pure water at temperatures above –38.5°C, ice formation is catalyzed by surfaces, usually particulate impurities, that act as seeds for crystal growth. This is the process of heterogeneous nucleation. If the process is avoided, pure samples of water will self-nucleate at –38.5°C, known as the homogeneous nucleation temperature (T_h; Taylor, 1987). As shown in Figure 22.1B, the temperatures of both heterogeneous and homogeneous nucleation are progressively lowered by increasing concentration of dissolved solutes.

The phase diagram for propane diol (Figure 22.1B) shows that in the region of 0 to 35% freezing will occur at some point 5 to 20°C below T_m, invariably by heterogeneous nucleation. At sufficiently high concentrations and low temperatures, the kinetics of the process become so slow that T_h is difficult to detect and any nucleated crystals that form in the region of T_h remain microscopic. As temperature is lowered further, molecular motion is slowed to the point at which translational and rotational molecular motion is essentially halted and the system is trapped in a high-energy state that resembles a liquid-like configuration, or a vitreous glass (MacFarlane et al., 1992). This glass transition (T_g) is associated with a marked change in physical properties such as specific heat and refractive index and certain mechanical properties such that T_g can be clearly identified. Determination of the transition temperatures that provide data for the construction of supplemented phase diagrams is usually derived from thermograms generated using differential scanning calorimetry or the related technique, differential thermal analysis. The kinetic nature of these transitions means that T_g has to be defined with reference to a particular set of experimental conditions. For example, changing the cooling rate means that the thermal events would occur over a different range of temperatures (MacFarlane et al., 1992; Moynihan et al., 1976).

Reference to Figure 22.1B shows that in the region of 35 to 40% propane diol, it is possible to cool samples through the T_h curve without apparent freezing and to form what have been referred to as doubly unstable glasses (Angell et al., 1981). This term reflects the fact that the vitreous system almost certainly contains ice nuclei and that if warming is not sufficiently rapid, further nucleation and crystallization will occur (devitrification), as signified by the curve T_d. Hence, during cooling, the sample attains the glassy state, but it invariably contains ice nuclei—the growth of which is arrested along with all other molecular motions in the sample. However, on rewarming, crystallization can be detected, either visibly or by an exothermic event in a thermogram, reflecting the growth of ice by devitrification and recrystallization. The phenomenon of crystallization on warming a glassy sample to temperatures in the vicinity of T_g is often referred to as devitrification of a doubly unstable glass, as it is unstable with respect to both the liquid and solid states (Angell et al., 1981; MacFarlane, 1986). Hence, the process by which a metastable glass, or supercooled liquid obtained by heating the glass above its T_g, forms the stable crystalline phase is generally referred to as devitrification (MacFarlane, 1986). The growth of small existing ice crystals into larger more stable ice forms occurs by recrystallization (Forsyth and MacFarlane, 1986), and the important distinction between devitrification and recrystallization has been discussed in a previous review (Taylor, 1987).

In the higher concentration range of 41 to 50% for propane diol (Figure 22.1B), the T_h curve meets T_g, and in this region it is possible to slowly cool even bulk liquids directly to T_g without experiencing any detectable freezing events. However, it is noteworthy that in this region of the supplemented phase diagram, devitrification still occurs and has been taken as evidence for significant heterogeneous nucleation during cooling. Nevertheless, the amount of ice formed under these conditions is extremely small, and further growth can be prevented by using moderate warming rates, thereby avoiding devitrification. It can also be seen in Figure 22.1B that at concentrations above 50%, devitrification ceases to be detectable even at low warming rates and the system can be regarded as stable, as nucleation is prevented. The intersection of the melting curve and the glass transformation curve at T_g' indicates the minimum concentration of propane diol in aqueous solution that will vitrify irrespective of cooling rate. The concentration at which a glass transition occurs varies according to the nature of the solute. It appears that those systems with the strongest solute–solvent hydrogen bonding provide the best suppression of ice nucleation and promote vitrification (MacFarlane and Forsyth, 1990; MacFarlane et al., 1992). In extreme cases, when appropriate concentrations of cryoprotectant solutions are maintained in the amorphous state, even during slow cooling and warming rates, the biological component should in principle be protected (in the absence of cold shock or osmotic stresses befire cooling) because there would be no phase transition during cooling and warming and the injuries associated with the coexistence of two phases would be avoided.

22.4.2.1 Stability of the Amorphous State

If the physical phenomena outlined above for the attainment of a vitreous state are to have practical value for the cryopreservation of tissues, then the stability of the amorphous state is of paramount importance. Boutron and his colleagues have made comprehensive studies of the stability of the amorphous state for a variety of potentially important cryoprotective mixtures with a view to identifying the most useful compounds for improved cryopreservation without freezing (Boutron et al., 1986). The stability of the amorphous state has been defined empirically in terms of the critical heating rate, V_{cr}, above which there is insufficient time for a vitreous sample to crystallize, even to a limited extent, before T_m is reached. The smaller the value of V_{cr}, the more stable the amorphous state. The dependence of T_d on the rate of warming can be measured, and the difference $T_m - T_d$, corresponding to a given warming rate, has been used to define the stability of the amorphous state (Boutron and Kaufmann, 1979; Boutron et al., 1986). The warming rate for which $T_m - T_d$ is zero is defined as the critical heating rate, V_{cr}, for which the supercooled mixture does not devitrify or recrystallize. On the basis of these considerations, it has been shown that the stability of the wholly amorphous state of aqueous solutions of 1,2-propane diol and its glass-forming tendency are much greater, for the same water contents, than for all other solutions of commonly used cryoprotectants, including glycerol, Me$_2$SO, and ethylene glycol. Butane-2,3-diol is the only new cryoprotectant to have emerged in recent years with comparable, or slightly better, physical characteristics for vitrification than any of the aforementioned cryoprotective solutes (Boutron,

1990). Nevertheless, solutions of polyalcoholic CPAs, such as propane diol and butane diol, that show the most promise in terms of cooling rates and concentrations necessary for vitrification also required unrealistically high heating rates to avoid devitrification. Moreover, principally because of isomeric impurities that crystallize a hydrate at reduced temperatures, 2,3-butanediol has proved to have an unanticipated biological toxicity at concentrations below that necessary for vitrification (Hunt et al., 1991; Mejean and Pegg, 1991; Mehl and Boutron, 1988; Taylor and Foreman, 1991). This disappointing development led to attempts to use lower concentrations of butane-2,3-diol by adding polymers such as polyethylene glycol to promote the vitreous state (Sutton, 1992), but to our knowledge, successful vitrification of a biological system using such mixtures has not been reported.

Despite developments to devise solutions that would vitrify at practically attainable cooling rates for sizeable biological tissues, achieving the corresponding critical warming rate necessary to avoid devitrification remains a critical challenge. Conceptually, elevated pressures (MacFarlane and Angell, 1981), electromagnetic heating (Robinson and Pegg, 1999; Ruggera and Fahy, 1990; Marsland et al., 1987), and the use of antifreeze molecules (DeVries, 1983) have been proposed as ways to tackle the problem. Nevertheless, each of these approaches presents a set of new problems that must be overcome if practical solutions are to be realized. Control of ice crystal growth using appropriate natural or synthetic molecules is a promising area of research that we discuss in more detail below.

24.4.2.2 Material Properties and Cracking

On a macroscopic scale, instability of the amorphous state can be manifest as fracturing or cracking, with devastating consequences for a biological tissue encased within the glassy matrix (Pegg et al., 1997; Rall and Meyer, 1989). The formation of cracks during the vitrification of glycerol solutions was reported by Kroener and Luyet (1966), and more recently, fracturing has been recognized as a hazard during cryopreservation of a variety of tissues (Pegg et al., 1997; Wassenaar et al., 1995; Wolfinbarger et al., 1991). In addition to the anticipated and observed mechanical destruction of tissues by fracturing, it has been reported that fractures provide an interface for nucleation that can initiate devitrification (Williams, 1989).

In his review of the physical properties of vitreous aqueous systems, MacFarlane (1987) emphasized that information on the material properties of vitreous aqueous solutions does not exist. Material properties such as thermal conductivity and fracture strength of aqueous solutions in the glassy state have many similarities with their inorganic analogues that exist at normal temperatures, for example, window glass and ceramics, but studies of these properties in the context of cryobiology have not yet been made extensively. Nevertheless, some information on thermomechanical stresses in frozen systems has begun to emerge in recent years, paving the way for comparable studies to be undertaken in fully or partially vitrified tissues, as we outline in the following section.

24.4.2.3 Thermomechanical Stress during Cryopreservation of Tissues

Mechanical stress in a material is related to pressure, and it is the force per unit area that either pulls the material apart (tensile stress) or presses it together (compressive stress). The magnitude of stress is related to the deformation of the material, where deformation, or strain, is defined as the change in geometric size relative to the initial size. When a material is at its original length, with no force acting on it, the stress is zero. This stress increases as the material is stretched (strained) while being maintained at a constant temperature. The rate at which stress increases with increasing strain depends on the material; this material stiffness (elastic modulus) is much greater, for example, in steel than in rubber. Some materials exhibit a more complex relation between stress and strain. For some materials that have been stretched and then held at the stretched length, the stress decreases with time; this is called stress relaxation. Biological materials often exhibit such time-dependent behavior. The rate of stress relaxation in materials generally tends to be lower when the material is held at lower temperatures. Extensive testing is often necessary to determine how the stress depends on straining, time, and temperature.

Changes of temperature produce another independent effect. Any material that is unrestrained will undergo a change in size (thermal strain) when subjected to a change in temperature. Materials in general, and tissues as they are cryopreserved in particular, shrink when they are brought from physiological temperature down to lower temperatures. Extensive testing is necessary to determine how the thermal strain (shrinkage) of a material depends on the temperature and possibly on the rate at which the temperature is changed. Under these unrestrained conditions, when thermal expansion or contraction is free to occur, the stress remains zero; that is, no forces act on the material.

As tissues are cryopreserved, they are externally free to shrink. However, in practice it is impossible to cool a tissue, of realistic size, uniformly; the outside surface decreases in temperature more rapidly than the inside. The outside of the tissue is forced to shrink less and the inside to shrink more. The level of stresses that must arise to accommodate the differential shrinkage is dependent on the stiffness and relaxation of the material. If these stresses are too severe, they have the potential to produce fractures. There are methods of predicting the stresses that arise because of nonuniform changes in temperature. These methods combine mathematical analysis with data that capture both the thermal strain caused by uniform temperature changes (material unrestrained at zero stress) and the time-dependent response of stress to strain as a function of temperature.

The development of stresses in biological tissues during freezing began being investigated after Rubinsky et al. (1980) proposed that mechanical destruction to cell membranes could arise from thermal expansion during freezing (known also as "thermal stress" or "thermomechanical stress"). Calculations of stress in freezing biological tissues have been carried out by Rabin et al. showing that thermal stress can easily reach the yield strength of the frozen tissue, resulting in plastic deformations or fractures (Rabin and Steif, 1998, 2000; Rabin and Podbilewicz, 2000). The driving mechanism of thermal stress is the constrained contraction of the frozen or vitrified tissue. It is commonly assumed that thermal expansion of frozen biological tissues is similar to that of pure water ice crystals (Rabin and Podbilewicz, 2000; Rubinsky et al., 1980), and Rabin et al. (1998) have confirmed this experimentally. Moreover, their studies provided some preliminary insight with regard to the effect of the presence of cryoprotectants on the thermal expansion. Results of pilot expansion tests of rabbit muscle permeated by the cryoprotectants Me_2SO and glycerol solutions, and pig liver perfused with Me_2SO solution, indicated that the cryoprotectants dramatically reduced the thermal expansion at higher temperatures and created a maximum value of thermal expansion within the temperature range of $-70°$ to $-100°C$. A significant effect of the Me_2SO concentration on the thermal expansion of pig liver was demonstrated, and it appears that the thermal expansion decreases with the increase in Me_2SO concentration. A rapid change in thermal strain was observed in the lobe suspected of attaining the highest concentration of Me_2SO, which could be related to a change in physical properties associated with a glass transition. A more complete understanding of the effect of cryoprotectants on thermal expansion during cooling will require further detailed study in fully equilibrated tissues and with an experimental device designed specifically for vitrified specimens (Y. Rabin, personal communication).

Further insight into the mechanical properties of frozen soft biological tissues has also been provided by measuring the response of frozen liver, kidney, and brain to externally applied compressive stresses (Rabin et al., 1996, 1997). The mechanical properties under study in this work were the compressive strength and the elastic modulus. A new load chamber for measuring the stress–strain relationship of frozen biological tissues in the cryogenic temperature range was designed and constructed to enable such measurements. It was found that the stiffness of the frozen tissues is of the same order of magnitude as that of sea ice and that the yield strength of frozen tissues is up to one order of magnitude higher than that of sea ice; sea ice data are widely available in the literature.

A unique response of frozen biological tissues to compression was observed. We found elastic behavior up to rather small strains, on the order of 0.005, and a sawtooth pattern of stress thereafter, featuring a series of sudden stress drops followed by a linear return to a roughly constant upper level of stress. It was suggested that the stress drops are associated with the formation of micro cracks, which steadily accumulate until final failure. The highly heterogeneous nature of this material may allow such cracks to appear, but not to propagate. Complete unloading leaves the material with a permanent plastic strain; continued microcracking only resumed when the stress was returned to the previous level at which microcracking occurred. Hence, it appears that the mechanical response of frozen tissues can usefully be idealized by elastic perfectly plastic models (Rabin et al., 1997). It was argued that the relationship between the thermal expansion coefficient and the strength-to-stiffness ratio is the dominant factor for fractures to occur, as it represents the relationship between the driven source and the consequential mechanical response in the frozen material.

Rubinsky et al. (1980) were the first to suggest a model for prediction of thermal stress in the context of cryobiology. Unfortunately, predictions of mechanical stresses were inconsistent with one important observation of tissue destruction; namely, in cryobiological applications, severe fractures often form at the early stages of thawing and not, as commonly expected, during cooling. This phenomenon has recently been observed in cryopreserved blood vessels (Pegg et al., 1996). Comparable observations in the context of cryosurgical applications have also been reported (Rabin and Steif, 2000). This inconsistency prompted Rabin et al. to reexamine the assumptions underlying the models of freezing tissues presented to date, as it is now thought that behavior at the freezing front has not been properly modeled heretofore (Y. Rabin, personal communication). Specifically, the deviatoric stress should be zero at an advancing freezing front. Parenthetically, the deviatoric stress is the total stress minus the hydrostatic pressure; excessive deviatoric stress is known to be linked with the likelihood of fracture formation in the theory of solid mechanics. Any volume-preserving strain that occurs while the material is still in the liquid state cannot contribute to the deviatoric stress. Therefore, material that has just solidified at an expanding freezing front must start with zero deviatoric stress.

This new approach for thermal stress modeling of freezing tissues has been investigated recently by Rabin and Steif (Rabin and Steif, 1998, 2000). Typical cryopreservation procedures were analyzed by simulating an inward freezing of a sphere (Rabin and Steif, 1998), and a typical cryosurgical protocol was analyzed by simulating an outward freezing of a sphere (Rabin and Steif, 2000). In both cases, closed-form solutions were obtained, and it was shown that simulation results qualitatively follow experimental data. It was shown that for cryopreservation involving crystallization, the attendant potential for tissue destruction are unavoidable regardless of how slowly the freezing is carried out, provided there is a substantial expansion associated with phase transition. It was noted that the phase transition temperature may significantly decrease during the cryopreservation process, because of the elevated hydrostatic pressure in the unfrozen region. This line of research has not yet been expanded for the case of high cryoprotectant concentration and of the very high cooling rate applicable to vitrification.

In summary, attention to the thermomechanical aspects of cryopreservation in recent years has identified the key physical phenomena that contribute to mechanical stress and fracture formation. These include the thermal expansion, which is the driving mechanism of the process; the stress–strain relationship, which represents the behavior of the material under mechanical load; and the strength of the material, a mechanical stress threshold above which tissue destruction or fractures will occur. These phenomena are affected by many factors such as the temperature, the cooling rate, the warming rate, and the cryoprotectant type and concentration. It is anticipated that systematic study of these parameters and models will generate a more complete understanding of the conditions necessary to avoid thermomechanical stresses during cryopreservation. In the meantime, practical experience has shown that fracturing can generally be avoided by cooling and warming slowly below the T_g. Studies in both frozen (Pegg et al., 1997) and vitrified (Song et al., 2000b; Taylor et al., 1999) blood vessels have shown that relatively slow warming to –100°C, at which temperature the vitreous material has softened, reduces the thermomechanical stresses and avoids macroscopic fractures.

22.4.3 ALTERNATIVE STRATEGIES FOR ACHIEVING VITREOUS CRYOPRESERVATION

It is clear from the foregoing discussion that the principal objective for achieving "ice-free" cryopreservation is to limit the nucleation and growth of ice during cooling and warming. Although there is now a considerable understanding of the physical processes involved in achieving this objective, achievement in bulky biological samples remains a challenge principally because of the constraints of heat transfer. Hence, successful application of this approach to the cryopreservation of tissues calls for innovative ways to either limit ice nucleation and growth or restrict ice formation to harmless sites within the tissue, as we have outlined above. It is worth emphasizing that ice *per se* is not always synonymous with cryoinjury, provided that its growth is constrained within appropriate limits (Mazur, 1988).

In view of this, it has been suggested that paradoxically, the promotion of ice nucleation might provide an alternative approach to controlling ice damage. The basic premise is that the intentional induction of very large numbers of nucleation events in the extracellular space would be less destructive than an equivalent total amount of ice organized as sizeable ice crystals. Conceptually, it might be feasible to induce nucleation on a massive scale in the region of a single temperature and thereby avoid the consequences of large ice crystal formation, provided of course, that recrystallization is also prevented. MacFarlane's group (Forsyth and MacFarlane, 1986) demonstrated that cooling conditions can be manipulated to produce an extremely high number density of small crystallites that, even after growth is complete, remain sufficiently small (<0.5 μm) to cause insignificant, observable scattering of ordinary light. For example, a 39% by weight solution of propane diol was annealed (held at constant temperature) near T_g for 2 h and remained visibly transparent. This experiment was attempted on the basis of previous observations that the nucleation rate increases with decreasing temperature in aqueous solutions such as these, whereas the growth rate, being a function of the transport properties, falls continuously as temperature is lowered. Thus, growth of the individual crystallites will be slow and nucleation of any new crystallites can continue such that the low-temperature annealing produces a sample containing a much higher density of ice nuclei than a comparable sample would contain during a simple cool/warm cycle. The concept is to anneal a vitrified tissue near T_g to provide for maximum nucleation and thereby prevent further growth during warming, as most of the freezable water would have already been consumed in the formation of the nuclei. It was also noted in these studies that the concentrated solutions (39%) did not appear to recrystallize (Forsyth and MacFarlane, 1986). By inference, ice crystallites smaller than the resolution of light might be expected to be innocuous to the integrity of tissues in which they form. However, this assumption remains to be validated by direct experimentation in cryopreserved tissues.

Another approach to produce maximally nucleated samples is to spike the solutions with special agents that promote nucleation. Products such as freeze-dried *Pseudomonas syringae* (commercially available as Snowmax for the snow-making industry) are potent ice nucleators that have been used experimentally to minimize supercooling in frozen samples (Fahy, 1998). It has been proposed that combining efficient ice nucleators with an abundance of antifreeze compounds, such as antifreeze peptides, might be a way to facilitate the formation and stabilization of myriads of supposedly harmless ice nuclei (Fahy, 1998; see also Chapter 5 for a discussion of similar principles in plants). This concept also awaits specific investigation as a practical approach to vitreous cryopreservation of tissues.

22.4.4 ICE-GROWTH INHIBITORS

Through evolution, nature has produced several families of proteins that help animals (e.g., fish and insects) and plants survive cold climates (see Chapters 3, 5, and 7). These proteins are known collectively as antifreeze proteins (AFPs). AFPs have the ability to modify ice structure, the fluid properties of solutions, and the response of organisms to harsh environments. The natural AFPs found in polar fish and certain terrestrial insects are believed to adsorb to ice by lattice-matching (Davies and Hew, 1990) or by dipolar interactions along certain axes (Knight and Duman, 1986). These molecules actually bind to the forming face of ice nuclei. By default, when temperature is lowered sufficiently, growth occurs preferentially in the *c*-axis direction (perpendicular to the basal plane) in a series of steps. This abnormal growth mode produces long ice needles, or spicules, that are much more destructive to cells and tissues than normal ice (Mugnano et al., 1995). Regardless, these molecules confer a survival advantage on certain animals. These observations led to the hypothesis that naturally occurring antifreeze molecules might be improved on by synthesis of molecules that will bind either to other ice nuclei domains or on stable ice crystals.

Discovery of new ice-inhibiting cryoprotectants for use in either classical cryopreservation or in molecular ice control techniques and vitrification has become an important focus of the research program of Organ Recovery Systems, (Des Plaines, IL and Charleston, SC). Chou (1992) mentioned an intention to specifically design ice crystal growth inhibitors. However, his interest was confined to minor modifications of existing naturally occurring AFPs and did not include preparation of *de novo* synthetic nonprotein antifreeze molecules. Historically, serendipity has been responsible for most discoveries of cryoprotectants. A major focus of our research has been the intentional design of synthetic ice blockers, which will combine with conventional cryoprotectants, and possibly naturally occurring antifreeze compounds, to minimize ice nucleation and growth during deep subzero cooling and subsequent warming. Two proprietary synthetic ice blockers have already demonstrated exceptional ice-blocking capabilities in our preliminary studies, as described below.

A complementary approach has recently been published by Wowk et al. (2000), who described the enhancement of vitrification solutions using synthetic polyvinyl alcohol. The mechanism of action of this compound is not clear, but based on visual observations and calorimetry, Wowk et al. suggest that polyvinyl alcohol blocks ice primarily by inhibition of heterogeneous nucleation. Such compounds might therefore be classified as antinucleating agents, as opposed to "ice blockers" that bind in some way to an ice nucleus and prevent or slow its growth into a damaging ice crystal.

Ice blockers are compounds that interact directly with ice nuclei or crystals to modify their structure or rate of growth. Examples of naturally occurring compounds include the antifreeze peptides and glycoproteins. Examples of synthetic compounds include the cyclohexanetriols and cyclohexanediols. Their properties are distinct from those of other cryoprotective solutes that lower the freezing point of solutions on a colligative basis. The latter are independent of chemical nature, whereas the former are highly dependent on chemical structure.

22.4.5 PRELIMINARY PHYSICAL STUDIES ON THE EFFECT OF SYNTHETIC ICE BLOCKING MOLECULES

We have used molecular modeling techniques to identify molecular conformations that might complement the atomic spacing of hydrogen-bonding sites on the prism face of an ice crystal. Hypothetically, these structures might be expected to hinder the growth of ice by lattice-matching with available sites on the basal plane surface of an ice crystal, as illustrated in Figure 22.3. Such considerations revealed that 1,3,5 cyclohexanetriol or its -diol derivatives possess the required bond angles and distances to conform with this hypothesis and were selected as lead compounds in preliminary physical studies to determine their efficacy in controlling ice growth (Fahy, 2001). A variety of related molecular structures were tested, but 1,3 cyclohexanediol (1,3 CHD) and 1,4 cyclohexanediol (1,4 CHD) were found to demonstrate significant ice-blocking capability and proved to be more soluble than 1,3,5 cyclohexanetriol, which was impractical to use at concentrations greater than 3% (0.2 *M*).

Tests were conducted with concentrations up to 0.5 *M* (6%) of the new agents added to one of our preferred cryoprotectant mixtures. This solution (designated V49) is a slightly diluted version of the VS55 baseline vitrification solution comprising Me$_2$SO (2.75 *M*), 1,2-propane diol (2.0 *M*),

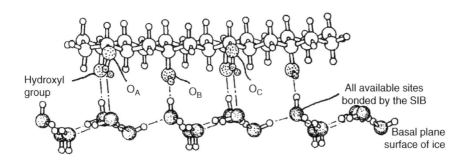

FIGURE 22.3 Molecular-modeling representation of the orientation of a synthetic ice blocker (SIB) with the basal plane of ice to demonstrate lattice matching. The model illustrates a remarkable coincidence between the spacing of strategically located hydroxyls on the SIB backbone and the 4.5 and 7.4 Å spacing of forward-projecting oxygen atoms of ice.

TABLE 22.1
Bulk Phase Ice Crystallization Measurements from Image Analysis

Cryoprotectant Solution	Number of Ice Crystals	Total Area Occupied by Ice (%)
V49 (7.5M CPAs)	Indefinite	100
VS55 (8.4M CPAs)	50 ± 6	1.3 ± 0.1
V49 + NaCl (6%)	609 ± 104	22.2 ± 1
V49 + Sucrose (6%)	5 ± 2	99.8 ± 0.1
V49 + 1,3 CHD(6%)	173 ± 76	2.3 ± 0.7
V49 + 1,4CHD (6%)	107 ± 66	1.7 ± 0.5
DP6 (6.0M CPAs)	2300 ± 385	39 ± 8
DP6 + 1,3 CHD (6%)	0 ± 0	0 ± 0
DP6 + 1,4CHD (6%)	6 ± 3	0.1 ± 0.1

Note: CPA, cryoprotective additive; 1,3 CHD, 1,3 cyclohexanediol; 1,4 CHD, 1,4 cyclohexanediol.

and formamide (2.75 M) and would not be expected to vitrify at low cooling rates under 1 atm pressure (Fahy et al., 1995). When cooled to temperatures below −34°C at slow rates (<3°C/min), V49 freezes with extensive ice crystallization throughout the sample (see Table 22.1). Cooling tests with V49 were performed in the presence of single synthetic ice blocking (SIB) compounds (6% w/v) or alternative control solutes of the same concentration known to have high colligative activity. The purpose was to identify specific ice-blocking activity compared with the more general colligative freezing-point depression function of additive solutes such as sodium chloride and sucrose. The data in Table 22.1 show that the presence of these SIBs caused a dramatic reduction of ice crystal formation and growth in bulk samples (75 mL) cooled to −100°C under slow cooling conditions.

Combinations of V49 and colligative solutes (e.g., NaCl and sucrose) yielded ice crystals in more than 20% of the volume (nearly 100% in the case of sucrose). By contrast, the SIBs were effective in reducing ice formation and growth such that the total ice volume in the bulk samples was less than 2%. In the case of our newer vitrification medium (DP6), which is a modification of VS55 that omits formamide (demonstrated to be a toxic component for several cell types; Campbell et al., 1999) and contains only 6.0 total CPA solutes (3 mols/l each of Me_2SO and Propanediol), the addition of the SIBs was effective in reducing the amount of ice from approximately 40% to negligible levels (data summarized in Table 22.1).

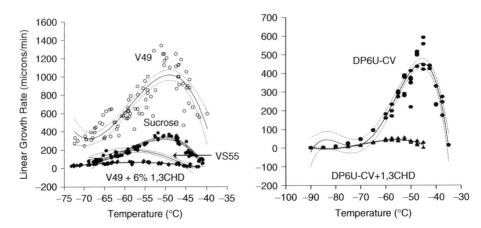

FIGURE 22.4 Kinetics of linear ice crystallization growth in vitrification solutions as a function of temperature. Solid lines represent the third-order regression curves fitted to the data with 95% confidence limits shown as dotted lines. DP6U-CV is DP6 solution prepared in Unisol (Organ Recovery Systems, Des Plaines, IL) cryoprotectant vehicle solution (Taylor et al., 2001; see text for details).

In additional experiments, the most effective ice-blocking compounds were evaluated for their ability to affect the kinetics of ice growth using cryomicroscopy. Figure 22.4 shows the linear ice growth rate measurements in both V49 and DP6 solutions alone and in solutions containing 6% sucrose or 1,3 CHD. Similar to the results in bulk solutions, the SIB molecule was more effective at slowing the rate of ice crystal growth in both solutions compared with sucrose at equal concentrations. 1,3 CHD has been shown to be highly effective for reducing the rate of growth to a very low level over the entire temperature range below the freezing point of V49 or DP6. Importantly, this combination of solutes proved more effective at controlling ice crystal growth than the baseline VS55 vitrification solution.

In comparable experiments using AFPs and antifreeze glycoproteins (AFGP), there was no statistical reduction in ice crystal growth rate by addition of either AFP III or AFGP to the V49 solution. This indicates that the small synthetic compounds exert their effect by a different mode of action to the much larger AFPs. Traditionally, the physical properties of AFP in solution have been quantitatively evaluated by nanoliter osmometry to determine the degree of thermal hysteresis. This is the difference between the freezing and melting points of a solution containing so-called "thermal hysteresis proteins" and typically amounts to a fraction of a degree depending on concentration (Barrett, 2001; Clarke et al., 2002; Duman et al., 1993; Ewart et al., 1998; Tyshenko et al., 1997). Using this technique, we have undertaken some preliminary experiments to examine the effect of our synthetic compounds on the thermal hysteresis (TH) of AFGP. Figure 22.5 illustrates the thermal hysteresis for AFGP as a function of concentration measured using a nanoliter osmometer (Clifton, Hartford, NY). Addition of 0.5 M 1,4-CHD (60 mg/mL) to AFGP had no significant effect, but in preliminary experiments using a similar concentration of 1,3-CHD, TH of the glycoprotein appears to be increased by ~35% at every protein concentration increment (Figure 22.5B).

Another highly significant finding was the changes seen in the ice crystal shapes as they formed in the presence of AFGP and 1,3-CHD. It is well documented that when an AFP solution is cooled to subzero temperature, the protein binds to the prism surface of hexagonal ice crystals and limits the growth along the a-axis (basal plane), but it does not affect the growth of the crystal along the c-axis (perpendicular to the basal plane). The result is the formation of needle-shaped ice spicules that are far more damaging to the frozen tissues and cells than normal hexagonal ice crystals (Mugnano et al., 1995). Addition of 1,3-CHD to the AFGP solution not only retarded the growth of ice crystals (at very low temperatures), it also changed the shapes of the forming ice crystals to

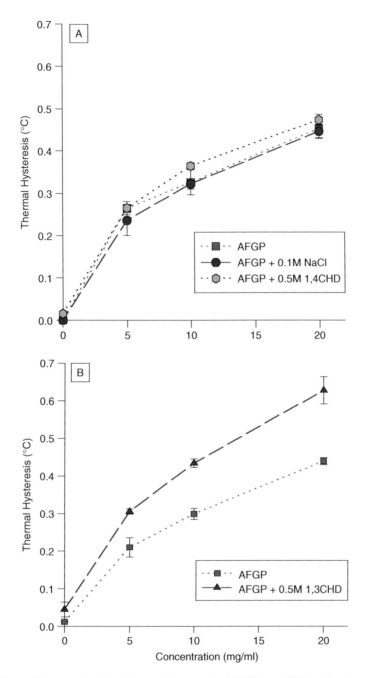

FIGURE 22.5 Thermal hysteresis of antifreeze glycoprotein (AFGP) and SIB molecules. Addition of 1,4 cyclohexanediol (0.5 M) or NaCl (0.1 M) to AFGP did not affect thermal hysteresis of the protein (A). A similar concentration of 1,3 cyclohexanediol potentiated thermal hysteresis of AFGP by 35% at various protein concentrations (B).

hexagonal, rectangular, or trapezoid forms as detected in the nanoliter osmometer. These observations could have a significant effect on designing new preservation media for freezing and storage of biological samples.

22.5 SYNOPSIS OF VITRIFICATION COMPARED WITH FREEZING, AND THE ADVANTAGES OF VITREOUS CRYOPRESERVATION

Conventional cryopreservation techniques, which require the substitution of up to 30% of cell water by a cryoprotective compound permit storage of many types of cells at deep subzero temperatures (typically $<-100°C$). When the rate of cooling is low enough, ice forms exclusively outside the cells and the external osmolality rises, dehydrating the cells. In fact, the ice is external to the system that it is desired to conserve, namely the cell, and the concentrated cell contents eventually solidify as an amorphous glass; that is, the cells vitrify. If cooling is too rapid to permit dehydration, and the cell contents actually freeze, the cell is invariably destroyed. It is noteworthy that this result shows that cells can tolerate the vitreous state. It has now been established beyond any doubt that the principal problem in attempting to cryopreserve tissues and organs is that ice forms within the system that it is desired to preserve, albeit outside the cells, and destroys both structure and function (Pegg, 1989; Taylor, 1984; Taylor and Pegg, 1982). It is clear that more than cell survival is needed in tissue preservation; complete structural integrity is vital. We have shown that some tissues and organs are severely damaged by extracellular ice and a mechanism that is adequate to account for the effect of extracellular ice in vascularized tissues – the rupture of capillaries by accumulating ice – has been demonstrated (Pegg, 1987; Pollock et al., 1986; Rubinsky and Pegg, 1988).

22.5.1 AVOIDANCE OF ICE

If a sufficiently high concentration of CPA could be used, the formation of ice would be avoided completely. The rates of cooling and warming are then unimportant because there is no driving force for transmembrane water movement and no ice to recrystallize during warming. The concentration of CPA necessary to avoid freezing is very high (typically ~60%) and "compatibility" (the absence of deleterious effects of the solute itself) is the essential problem such that the concentration of solute required is unattainable at suprazero temperatures. By taking advantage of the temperature dependence of most toxic actions, it is possible to increase the concentration progressively as the temperature is reduced. It was shown some years ago that by using this approach to increase the concentration of Me_2SO in a stepwise manner to remain above the equilibrium freezing point during cooling, it was possible to recover smooth muscle tissue with a high degree of stimulated contractile function (Taylor et al., 1978; Elford and Walter, 1972b).

More recently, an alternative approach has been explored, based on dynamic features, to reduce the amount of ice by selecting sufficiently high cooling rates to prevent ice nucleation. This approach produces a metastable state that is at risk of devitrifying (recrystallization) during warming, and ice formation during warming is just as injurious as during cooling. Nevertheless, vitrification procedures by this technique have been developed and shown to provide effective preservation for a number of cells, including monocytes, ova and, early embryos and pancreatic islets (Bodziony et al., 1994; Fahy, 1988; Jutte et al., 1987a, 1987b; Rall, 1987; Rall and Fahy, 1985; Takahashi et al., 1986). Vitrification refers to the physical process by which a concentrated solution of CPAs solidifies during cooling without crystallization. The solid, called a glass, retains the normal molecular and ionic distributions of the liquid state and is therefore usually considered to be an extremely viscous supercooled liquid. The difference between conventional cryopreservation and vitrification lies not in the occurrence of vitrification in only the latter method, but in the means by which vitrification is produced—by extracellular freezing and progressive cell dehydration during cooling in conventional preservation, and by achieving a vitrifiable system at the outset in vitrification (Pegg and Diaper, 1990).

When materials are vitrified, no ice forms, even at cryogenic temperatures. The formation of ice is prevented by the presence of high concentrations of chemicals that interact strongly with water and, therefore, prevent water molecules from interacting to form ice. It has been shown that depressing the homogeneous nucleation temperature until it equals T_g permits vitrification of macroscopic biological systems. Prevention of freezing means that the water in a tissue remains liquid during cooling. As cooling proceeds, however, the molecular motions in the liquid permeating the tissue decrease. Eventually, an "arrested liquid" state known as a glass is achieved. A glass is a liquid that is too cold or viscous to flow. A vitrified liquid is essentially a liquid in molecular stasis. Vitrification does not have any of the biologically damaging effects associated with freezing because no appreciable degradation occurs over time in living matter trapped within a vitreous matrix. Vitrification is potentially applicable to all biological systems.

22.5.2 Advantages of the Vitrification Approach

Cryopreservation by the complete vitrification of the tissue suspension offers several important advantages compared with procedures that allow or require crystallization of the suspension. First, complete vitrification eliminates concerns for the known damaging effects of intra- and extracellular crystallization. Second, tissues cryopreserved by vitrification are exposed to less-concentrated solutions of CPAs for shorter periods of time. For example, during a typical cryopreservation protocol involving slow freezing to $-40°$ or $-70°C$, cells are exposed to solutions whose concentration increases gradually to 21.5 and 37.6 osmolal, respectively. In contrast, cells dehydrated in vitrification solutions are exposed for much shorter periods of time to less than 18 osmolal solution, although the temperature of exposure is higher (Rall, 1987). Third, unlike conventional procedures that employ freezing, vitrification does not require controlled cooling and warming at optimum rates—cooling and warming need only be rapid enough to prevent crystallization, and this can generally be achieved without the need for specialist equipment. It is widely anticipated, therefore, that for many integrated multicellular tissues, vitrification may offer the only feasible means of achieving cryopreservation without ice damage, and for some tissues such as pancreatic islets that appear to partially withstand cryopreservation by freezing, vitrification offers a number of practical advantages that will be attractive in tissue engineering, as indeed they have been for embryo banking (Rall, 1987). On this basis we have committed to pursuing vitrification techniques for complex tissues, as we discuss in the remainder of this chapter.

22.6 APPLICATION TO VIABLE TISSUES

In a recent editorial article in the journal *Science* (Kaiser, 2002), "New Prospects for Putting Organs on Ice" were discussed with the focus on the need for ice-free cryopreservation methods. The consensus opinion is that viable tissues such as blood vessels, corneas, and cartilage that have proved refractory to cryopreservation using conventional freezing methods can only be successfully preserved with an adequate degree of poststorage function if steps are taken to prevent or limit the amount of ice that forms during the cooling and warming. Some of the recent work that has contributed to this consensus will be summarized here.

Kaiser's 2002 editorial article in *Science* accurately summarizes the state-of-the-art of tissue and organ cryopreservation in that year. Clearly, ice-free approaches have taken preeminence over the futile prospects of fine-tuning conventional freezing techniques to yield adequate methods for cryopreserving structured tissues and organs. The historical background to the ideals of a vitrification approach has already been outlined earlier in this chapter, and much of the current-day impetus for pursuing this approach is credited to Fahy's dedication over two decades to attempt vitreous cryopreservation of kidneys (Fahy, 1989; Fahy and Hirsch, 1982; Fahy et al., 1990). Although his ultimate objective remains elusive, the basic science that he and his collaborators have generated along the way has contributed significantly to the recent advances in this field. The successful application of vitrification as an alternative method for cryopreservation of embryos (Paynter et al., 1997; Rall, 1987; Rall and Fahy, 1985) and pancreatic islets (Bodziony et al., 1994; Jutte et al., 1987a, 1987b) provided enthusiasm for the prospect of applying these techniques to other tissues such as corneas, blood vessels, and cartilage that cannot be adequately preserved using freezing methods.

22.6.1 CORNEAS

Armitage et al. (Armitage and Rich, 1990; Armitage et al., 2002; Hall and Armitage, 1999; Rich and Armitage, 1991) focused principally on the formulation of the cryoprotectant solutions that would promote vitrification of rabbit corneas. Their work in this area culminated in a demonstration that corneas equilibrated in a solution containing 50% (v/v) propane-1,2-diol, 0.25 M sucrose, 6% (w/v) polyethylene glycol, and 2.5% (w/v) chondroitin sulfate could be vitrified, as judged by the absence of visible ice, when cooled to −110°C at approximately 6°C/min. Nevertheless, under these conditions devitrification was reported to be a persistent problem, indicating that the cooling and warming conditions have not yet been sufficiently optimized to avoid ice crystallization altogether.

Similar problems were encountered by Bourne and Nelson (1994) during their attempts to vitrify human corneas using a solution containing 3.1 M Me$_2$SO, 3.1 M formamide, and 2.2 M 1,2-propane diol in a corneal storage solution with 2.5% (w/v) chondroitin sulfate. It was demonstrated in these studies that increasing the time of exposure of each step in the cryoprotectant addition protocol from 10 to 25 min at 0°C permitted sufficient CPA permeation into the tissue to avoid any detectable ice in the corneas during cooling at ~10°C/min. However, this prolonged CPA-loading protocol led to unacceptable damage to the corneal endothelium, with the conclusion that further advances for cryopreservation of corneas using a vitrification technique will require careful optimization of the loading and unloading protocols for the cocktails known to vitrify at practical cooling rates (Bourne and Nelson, 1994). It is clear, therefore, that significant advances have been made toward ice-free cryopreservation of corneas, giving encouragement that this remains a reasonable approach for the development of improved methods of cryopreservation.

22.6.2 VASCULAR GRAFTS

In recent years, an increasing and sometimes urgent need has developed for prosthetic blood vessels for arterial bypass surgery. These prostheses are required for graft replacements in redo procedures and when autologous vessels are not available. In the present era of arterial replacement, at least 345,000 to 485,000 autologous coronary grafts (either arteries or veins; American Heart Association, 1996; British Cardiac Society, 1991) and over 200,000 autogenous vein grafts (Callow, 1983) into peripheral arteries are performed each year. A recent marketing report indicated that at least 300,000 coronary artery bypass procedures are performed annually in the United States, involving in excess of 1 million vascular grafts (Frost and Sullivan, 1997). Many of these patients do not have autologous veins suitable for grafts because of preexisting vascular disease, vein stripping, or use in prior vascular procedures. It has been estimated that as many as 30% of the patients who require arterial bypass procedures will have saphenous veins unsuitable for use in vascular reconstruction (Edwards et al., 1966).

Cryopreserved allogeneic veins are being used clinically (Brockbank et al., 1992; McNally et al., 1992). However, *in vivo* studies using these grafts in both animal models and patients have demonstrated poor long-term patency rates (Müller-Schweinitzer et al., 1998; Stanke et al., 1998; Almassi et al., 1996). These grafts also demonstrate reduced endothelial cell functions and impaired smooth muscle contractility after cryopreservation (Brockbank, 1994). In light of this, we have entertained the hypothesis that prevention of ice formation in blood vessels by an alternative cryopreservation approach, vitrification, will optimize cell functions and minimize extracellular matrix damage, resulting in more effective, durable grafts. We have recently completed studies on the vitrification of both veins and arteries (described in the next section).

22.6.2.1 Vitrification of Veins

A study was designed to evaluate a vitrification approach to storing a vascular tissue model (rabbit jugular vein) compared with a standard commercial method employing slow cooling with dimethyl

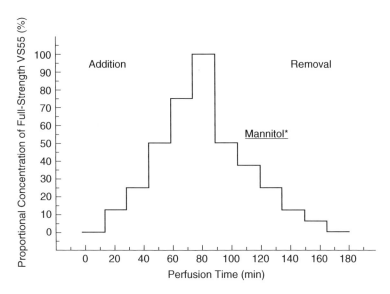

FIGURE 22.6 Incremental steps for the addition and removal of VS55 solution in the baseline vitrification process. During dilution the EuroCollins vehicle solution was supplemented with 300 mM mannitol as an osmotic buffer.

sulfoxide (Me$_2$SO) and chondroitin sulfate as the cryoprotective agents (Brockbank, 1994; Brockbank et al., 1992; McNally et al., 1992). This method was developed during several years of optimization studies by Brockbank and his colleagues for CryoLife, Inc., and remains the most widely used method of cryopreservation for clinical vascular grafts (Brockbank et al., 1992; McNally et al., 1992) because of CryoLife's dominance in the U.S. marketplace.

Following Brockbank's method for conventional cryopreservation involving freezing, the tissue was initially immersed for 20 min in HEPES-buffered Dulbecco's Modified Eagle's Medium (DMEM) containing 1 M Me$_2$SO, 2.5% chondroitin sulfate, and 10% fetal calf serum at 4°C. Samples were then cooled at a controlled rate of 1.0°C/min to –80°C and finally transferred to liquid nitrogen for storage. Thawing was accomplished by immersing the containers in a water bath controlled at 37°C until all ice had visibly disappeared, whereupon the containers were transferred to an ice-bath for elution of the CPA. This was achieved in sequential steps in which the tissue samples were transferred to DMEM containing 0.5 M, 0.25 M, and finally 0 M mannitol as an osmotic buffer.

A method for vitrification of vein rings and segments was developed in which a baseline vitrification medium (designated VS55 to reflect that it comprises 55% [w/v] total cryoprotective solutes, but previously designated as VS41A by its originators [Mehl, 1993]) was used to replace at least 50% of the tissue water with a combination of CPAs. The VS55 solution consisted of 3.1 M Me$_2$SO, 3.1 M formamide, and 2.2 M 1,2-propane diol in EuroCollins solution (Fahy, 1988; Rall and Fahy, 1985); the full strength mixture was added, and removed in stepwise manner as outlined in Figure 22.6. This protocol was introduced as a baseline technique, and no attempt to optimize the method was attempted at this stage (Song et al., 2000b; Taylor et al., 1999). In the vitrification experiments with vein rings, the tissue was immersed in vitrification solutions in glass vials at each step. Experiments with vein segments were carried out using a perfusion technique as follows.

The external jugular vein was perfused *in situ* to remove blood from lumen. A 4- to 5-cm length of vein was cannulated *in situ* at its distal end, and perfusion was performed for addition and removal of vitrification solution in these isolated veins. The perfusion system consisted of a reservoir (a 60-cc syringe) connected to the cannula with three-way stopcock. The reservoir was adjusted to

TABLE 22.2
Maximal Physiological Responses of Rabbit Jugular Veins

Drug	Control (g)	Frozen (g)	%	Control (g)	Vitrified (g)	%
Histamine	2.08 ± 0.15	0.52 ±0.05*	25.0	1.78 ±0.19	1.55 ±0.27*	87.1
Bradykinin	1.70 ±0.18	0.50 ±0.05*	29.0	1.75 ±0.18	1.49 ±0.15*	85.1
Angiotensin II	0.94 ±0.09	0.17 ±0.04*	18.0	0.58 ±0.06	0.49 ±0.09*	84.5
Norepinephrine	0.87 ±0.19	0.13 ±0.03*	14.9	0.99 ±0.12	0.83 ±0.14*	83.8

Note: Data are expressed in grams of maximal tension generated. Values are means ± S.E.M. **%,** percentage of corresponding fresh controls; frozen, vein rings cryopreserved with 1.0 M Me_2SO (n = 28–32); vitrified, vein rings vitrified with vitrification solution (n = 26); control, fresh vein rings (control for cryopreserved vein rings, n = 12–16; control for vitrified vein rings, n = 11–15).

* $P < .0001$ vs. fresh controls.

provide physiologic pressure by adjusting its height to provide a hydrostatic pressure of 80 to 100 mmHg at the cannula. The vein was placed in a Petri dish containing vitrification solution precooled to 4°C, and the dish was placed on ice during the perfusion process. Vitrification solution was added in six steps and removed in seven steps as shown in Figure 22.6.

Vitrification was achieved by cooling the samples rapidly (43° ± 2°C/min) to –100°C, followed by slow cooling (3° ± 0.2°C/min) to –135°C, whereupon they were transferred to a –135°C freezer for at least 24 h. Rewarming was accomplished in two stages: initially, samples were warmed slowly to –100°C (30° ± 2°C/min) and then warmed rapidly (225° ± 15°C/min) to melting, whereupon the vitrification solution was eluted in a stepwise manner as shown in Figure 22.6. Finally, all preserved samples were returned to physiological DMEM medium in preparation for viability testing or transplantation (Song et al., 2000b; Taylor et al., 1999).

In vitro function of fresh control tissue from each rabbit or preserved rings was assessed using a physiological organ-bath technique (Song et al., 1994, 1995). Each vein ring segment was mounted between two stainless steel wire hooks suspended in a custom organ bath (Radnoti, Monrovia, CA) containing 5 mL Krebs'-Henseleit solution, which was gassed continuously with 95% O_2/5% CO_2 at 37°C. One hook was fixed to the base of the organ chamber and the other was connected to a force transducer (Myograph F-60, Narco Bio-Systems, Houston, TX). Isometric contractile tensions were measured by adding a variety of agonists and antagonists to the tissue in the organ baths and recording the changes in developed tension relative to baseline values. The panel of drugs used in this study included histamine, bradykinin, angiotensin II, norepinephrine, and sodium nitroprusside. Because baseline responses can vary between different freshly isolated veins, the experimental design included paired controls for each preserved vein by testing the contractile responses of fresh untreated sample rings from each jugular vein harvested for the preservation studies.

The maximal contraction of fresh, frozen, and vitrified vein rings in response to the panel of agonists is shown in Table 22.2 (The data has been published in graphical form elsewhere; Song et al., 2000b). It can be seen that the maximum contractions achieved by the vitrified blood vessel rings in response to all four agonists were greater than 80% of fresh matched controls. In marked contrast, the maximum contraction index for frozen rings was less than 30% of fresh matched controls. Smooth muscle relaxation tests using sodium nitroprusside as the agonist drug showed that vitrified veins produced maximum relaxation of the precontraction. This response was similar to that of fresh control veins. In contrast, cryopreserved veins reached only 66% relaxation ($P < .01$). Moreover, the dose response curves showed that the vitrified vessels demonstrated similar, if not slightly enhanced, drug sensitivities compared with untreated controls, whereas frozen vein rings exhibited decreased drug sensitivities (Song et al., 2000b).

22.6.2.2 *In Vivo* Studies

22.6.2.2.1 *Graft Patency*

Autologous vein implantation studies showed that the patency of fresh and vitrified rabbit jugular veins was not significantly different after 2 and 4 weeks, with both groups exhibiting short-term patency rates of ~90% (Song et al., 2000a).

22.6.2.2.2 *Graft Histology*

Graft rupture, aneurysm, thrombosis, or inflammatory infiltration was not noted in any of the patent grafts. Vitrification had not altered the pathophysiological cascade of events that occur when a vein graft is inserted into the arterial system. The vitrification process did not appear to induce any adverse effects locally or systemically *in vivo*. Morphological studies at both the light and ultrastructural level confirmed that vitrification had preserved endothelial cell and smooth muscle cell integrity posttransplantation. Details of these studies have been published elsewhere (Song et al., 2000a).

The demand for cryopreserved allogeneic veins is growing despite the well-documented immune response to these grafts and the low clinical patency rates. Between 1985 and 1992, approximately 3000 cryopreserved allogeneic vein segments were used for arterial bypass (McNally et al., 1992); however, the allograft veins cryopreserved using conventional cryopreservation methods produced less satisfactory results. Walker et al. (1993) reported that the cumulative survival rate was 14% and the cumulative secondary patency rate was 37% at 18 months. The grafts demonstrated reduced endothelial cell functions and impaired smooth muscle contractility after cryopreservation *in vitro* and poor long-term patency rates *in vivo* (Field et al., 1969; Gelbish et al., 1986; Jackson and Abel, 1972; Sellke et al., 1991; Showlater et al., 1989; Stephen et al., 1978; Tice and Zerbino, 1972). Our recent studies using an autologous animal model clearly demonstrate a significant benefit of vitrification for preservation of graft function.

22.6.2.3 Effects of Storage Temperature and Duration on the Stability of Vitrified Blood Vessels

As discussed above, a major concern regarding vitrified tissue storage has been glass stability at very low storage temperatures. Because of this, it has proved advisable for vitrified tissue samples to be stored a few degrees below the solution glass transition point (T'_g; $-123°C$ for the VS55 solution used as a baseline solution in our studies). We further tested the hypothesis that vitrified materials would be stable at less than $-160°C$ ($\pm10°C$) compared to $-130°C$. Rabbit jugular veins were vitrified using our standard vitrification protocol. The stability of the glass (the absence of ice crystallization) during storage and on rewarming was verified by visual inspection. Cell viability was assessed using the smooth muscle physiology method. There were five groups in this study: group 1, fresh control; group 2, vitrified veins stored at $-130°C$ for 4 weeks; group 3, vitrified veins stored at $-130°C$ for 4 months; group 4, vitrified veins stored at the temperature less than $-160°C$ (vapor-phase liquid nitrogen) for 4 weeks; group 5, vitrified veins stored at the temperature less than $-160°C$ for 4 months. The results showed that there was no ice formation in the vitrified samples during storage at either $-130°C$ or less than $-160°C$. *In vitro* function of fresh control tissue from each rabbit or preserved rings was assessed using a panel of drugs as before. The maximum contractions achieved by the vitrified blood vessel rings stored at either $-130°C$ or less than $-160°C$ for either 4 weeks or 4 months were similar to veins stored for 24 h. The responses of stored samples were not significantly different to fresh controls for three of the four agonists. Smooth muscle relaxation tests using sodium nitroprusside (endothelium-independent) as the agonist drug showed that vitrified veins produced maximum relaxation of the precontraction. This response was similar to that of fresh control veins. Endothelium-dependent smooth muscle relaxation was tested using acetylcholine. No deterioration was observed at $-160°C$ or over time. Dose response curves showed that the stored vessels demonstrated similar drug sensitivities compared with untreated fresh controls. This study demonstrated that the glass stability of vitrified samples can be retained during vapor-phase liquid nitrogen storage and that the cell viability is maintained for at least 4 months.

22.6.2.4 Vitrification of Arteries

Additional studies focused on arteries to determine whether the methods effective for veins also work for arteries. There is an extensive literature indicating that autologous internal mammary, gastroeploic, and radial arteries are superior in terms of patency to autologous venous grafts. The autologous internal mammary artery has superior long-term patency when it is employed with a pedicle so that its vasa vasoral blood supply is uninterrupted (Barner et al., 1985; Loo et al., 1986; Zeff et al., 1988). Autologous arterial conduits used as free grafts, including internal mammaries and gastroeploic, radial, and inferior epigastric arteries, have similar or better patency rates than autologous saphenous veins (Acar et al., 1992; Loop et al., 1986; Mills and Everson, 1989; Puig et al., 1990).

A study was therefore undertaken to evaluate the feasibility of vitrification for arterial graft preservation of small-caliber arteries (<6 mm internal diameter) for clinical implantation. For this, our baseline vitrification method developed for veins was applied to arterial segments. However, there have not been any studies of small-diameter cryopreserved or fresh allogeneic arteries to compare with the studies of cryopreserved allogeneic saphenous veins for either coronary or peripheral artery bypass. Larger diameter allogeneic arteries may have been used as arterio-venous shunts for dialysis or in association with pulmonary or aortic allograft heart valves. Justification of research on small-diameter artery preservation based on the need for peripheral and coronary grafts is inappropriate and misleading, as allogeneic internal mammary arteries, gastroeploic arteries, and redial arteries are not being used clinically.

During rewarming of vitrified samples, a transparent glassy vitrification solution could be visualized at the early stage of warming, providing visible evidence that, at least by visual criteria, vitrification of the samples had been achieved. As with the prior vein study, cryosubstitution was used to confirm there was no detectable ice in the vitrified arteries, but extensive ice cavities were prevalent throughout the arterial specimens cryopreserved by the traditional method of freezing.

After rewarming and removal of cryoprotective agents, vessel function was evaluated using an *in vitro* contractility test. The maximum contractions achieved by the vitrified blood vessel rings in response to norepinephrine and phenylephrine were similar to those of fresh controls. In marked contrast, the maximum contraction index for frozen rings was less than 30% of fresh control arteries. In addition, vitrified arteries did not show significantly different sensitivity to the agonists compared to controls, whereas frozen arteries were significantly less sensitive to the agonists compared to control arteries ($P < .05$; Figure 22.7). Smooth muscle relaxation tests using sodium nitroprusside (endothelium independent) showed that vitrified and fresh control arteries relaxed to a similar degree. In contrast, frozen arteries reached only 9% relaxation ($P < .001$). Endothelium-dependent smooth muscle relaxation was tested using the calcium ionophore A23107. Although the maximum response in vitrified samples was compromised when compared with fresh controls (37 vs. 68%), the mean response was superior to the frozen vessels, which achieved only 28% relaxation. As with the veins, it was clear that prevention of extracellular ice formation improved vascular smooth muscle function.

Our current study outlined above demonstrated that vitrification is superior to conventional cryopreservation methods in preservation of rabbit carotid arteries. The VS55 solution preserved vascular smooth muscle function, but endothelial functions were not significantly better than in conventionally preserved arteries. VS55 was originally designed for vitrification of kidney slices (Mehl, 1993), and its composition may not be optimal for preservation of arterial endothelial function. Earlier studies had demonstrated significant improvements in endothelial function of vitrified veins compared with conventionally frozen veins (see previous section). Modification of the formulation of the vitrification cocktail and process are being investigated to improve arterial functions.

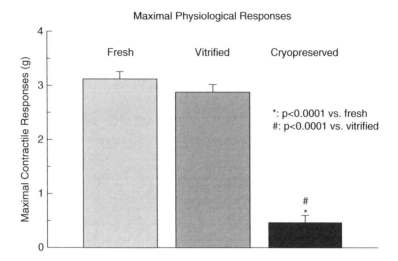

FIGURE 22.7 Maximal contractile responses of fresh, vitrified, and cryopreserved rabbit carotid arteries to noradrenaline. Arteries were either vitrified with VS55 solution or cryopreserved at a controlled rate (1°C/min) with 15% Me$_2$SO. All values are means $\pm$ SEM (n = 8). The P values were calculated using an unpaired Student's t-test.

22.6.3 Articular Cartilage

Fresh osteochondral allografts have proven to be effective and functional for transplantation. However, the limited availability of fresh allograft tissues necessitates the use of osteoarticular allograft banking for long-term storage (Bakay et al., 1998; Marco et al., 1992; Malinin et al., 1985; Ohlendorf et al., 1996). Although cryopreservation by means of freezing is currently a preferred method for storing tissue until needed, conventional protocols result in death of 80 to 100% of the chondrocytes, along with damage to the extracellular matrix as a result of ice formation. These detrimental effects are the main obstacles preventing successful clinical outcome of osteochondral allografts (Ohlendorf et al., 1996; Stone et al., 1998; Tomford et al., 1984) and commercial success of tissue-engineered cartilage constructs. Allogeneic cartilage is generally considered immunologically privileged because of an absence of both blood vessels and a lymphatic system. Recipient immunosurveillance cells do not, therefore, come into contact with the graft's chondrocytes, and immunotherapy is not required for graft function. As a consequence, the search continues for a better preservation method.

Isolated chondrocytes can be preserved using conventional cryopreservation methods involving freezing; however, chondrocytes embedded in their natural matrix are extremely difficult to preserve. Studies using a variety of animal articular cartilage models (Ohlendorf et al., 1996; Marco et al., 1992; Muldrew et al., 1994; Wu et al., 1998) and human cartilage biopsies (Stone et al., 1998) have revealed no more than 20% chondrocyte viability following conventional cryopreservation procedures employing either Me$_2$SO or glycerol as cryoprotectants. Ohlendorf et al. (1996) used a bovine articular cartilage, osteochondral plug model to study their clinical cryopreservation protocol. This protocol employed slow-rate cooling and 8% Me$_2$SO as the cryoprotectant. They observed loss of viability in all chondrocytes except those in the most superficial layer at the articular surface. Muldrew et al. (1994) previously investigated chondrocyte survival in a similar sheep model. These researchers observed cells surviving postcryopreservation close to the articular surface and deep at the bone/cartilage interface. The middle layer was devoid of viable cells. More recently, Muldrew et al. (2001) demonstrated improved results using a step-cooling cryopreservation protocol, but cell survival posttransplantation was poor, and again there was significant loss of cells in the midportion of the graft. The reason for lack of cell survival deeper than the superficial layers

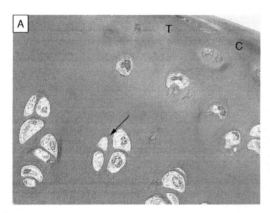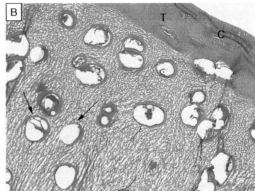

FIGURE 22.8 Light microscopy of vitrified and frozen cryosubstituted articular cartilage. (A) Vitrified cartilage. The matrix shows little evidence of ice; the tangential layer (T) reveals an elliptical chondrocyte (C) within its lacuna. The deeper chondrocytes (arrow) show some shrinkage. (B) Frozen cartilage. The chondrocytes within the ice-filled lacunae (arrows) appear totally disrupted; the deep matrix has considerable ice formation (white spaces). The chondrocytes (C) within the superficial and tangential layer (T) appear to have less ice. (100×).

of articular cartilage is most likely multifactorial and related principally to heat and mass transfer considerations (Karlsson and Toner, 1996). Surface cells freeze and thaw more rapidly than cells located deep within the matrix. This phenomenon could result in a greater opportunity for ice to form, both within cells and in the extracellular matrix, deeper within the articular cartilage. Furthermore, typically employed concentrations of Me_2SO (8 to 20%) may not penetrate adequately to limit intracellular ice formation.

In view of the consistently poor outcome of classical cryopreservation of articular cartilage reported by several groups, we hypothesized that vitreous cryopreservation would provide improved recovery. This hypothesis was tested by using the baseline vitrification method that had proved to be effective for the improved cryopreservation of vascular grafts (described above; Song et al., 2000a, 2000b). Refrigerated cartilage was used as a positive control for viability.

Cryosubstitution studies of frozen and vitrified articular cartilage samples revealed negligible ice in the vitrified specimens (Figure 22.8A) and extensive ice formation, both in the extracellular matrix and deeper lacunae, in frozen specimens (Figure 22.8B) by light microscopy. Some cell shrinkage was observed in the lacunae of vitrified specimens (Figure 22.8A), which was most likely related to high concentrations of cryoprotectants.

Viability was assessed using two methods: fluorescent microscopy using a live/dead stain, and a metabolic assay employing alamar blue. The fluorescence studies allowed observation at 30 to 50 μm deep into the tissue, and the metabolic assay gave an overview of cell viability for the entire implant. Figure 22.9 is representative of the three experiments performed in a preliminary study. Figure 22.9A depicts a quantitative estimate of the relative fluorescent intensity readings using calcein (viable) stain and indicates vitrified tissue had approximately 80% of the viability of fresh controls, whereas the frozen cryopreserved tissue was less than 13% viable. Moreover, the oxidation-reduction indicator, alamar blue assay indicated that the metabolic activity of vitrified samples was approximately 85% of fresh samples (Figure 22.9B) (Song et al., 2004b).

These studies combine to demonstrate that the vitrification process results in ice-free preservation of rabbit articular cartilage tissue and that between 80 and 85% of the cells were alive following rewarming. Frozen samples contained ice within the cells and the matrix, with the exception of the articular surface, where some viable cells were observed. For an *in vivo* assessment of survival and function of osteochondral allografts, tissue samples were randomly assigned to the three treatment groups: fresh, vitrified, and frozen. The O'Driscoll Score (O'Driscoll et al., 1985)

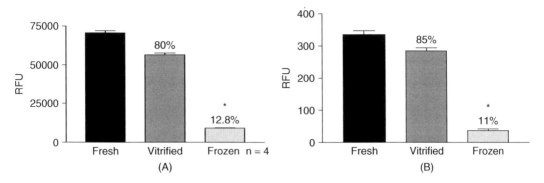

FIGURE 22.9 Viability assessments of articular cartilage samples using two fluorescence assays. (A) shows the mean (± sem) of four experiments using Calcein AM (measures intracellular esterase activity). (B) shows the mean (±sem) of ten samples of articular cartilage assayed using alamarBlue. Viability is expressed as relative fluorescence units (RFU) relative to the dry weight of each articular cartilage sample. *p<0.05 by one-way ANOVA, using the Kruskal-Wallis Test, Dunns Post-test.

is a composite of multiple factors that allows comparisons of overall graft survival between the three treatment groups. Analysis of variance showed that there was a significant treatment effect ($P = .037$), and a significant time effect ($P = 0.031$). The difference between the fresh and the vitrified samples was not significant ($P > 0.05$), but the differences between the fresh and the frozen ($P = 0.05$) and the vitrified and the frozen were significant ($P < 0.01$) at the 95% level of confidence.

The histology of fresh (4°C preservation) and vitrified explants was essentially the same, although the cells were less well organized within the explants when compared with unoperated control cartilage (Figure 22.10). The frozen cryopreserved explants were devoid of chondrocytes, and only fibroblast-like cells were present (Figure 22.10D) (Song et al., 2004a).

These data make a strong case for cryopreservation of cartilage by vitrification. The vitrification process protected the cartilage from damage associated with ice and resulted in 80 to 85% retention of cell viability. Moreover, fresh refrigerated and vitrified cartilage plugs performed similarly *in vivo* and were statistically superior to frozen plugs.

22.6.4 THE EMERGING "80/20" RULE

In the process of collating the information on tissue cryopreservation for this chapter, an interesting observation was made that warrants brief mention in summary of this section on the vitrification of tissues. Table 22.3 summarizes the viability data contrasting the outcome of vitrification with conventional cryopreservation involving freezing. Interestingly, the functional survival of vitrified tissues was approximately 80% or higher, whereas the frozen counterparts yielded less than 20% survival. This marked contrast (~80% vs. 20 to 30%) appears consistent irrespective of the nature of the tissue or the method of assay. This phenomenon serves to emphasize that avoidance of large amounts of ice in organized tissues can improve the outcome of cryopreservation from a meager 20% survival to a respectable 80%. It will be interesting in future studies to see whether the magnitude of this difference is also manifest in other tissues and whether the maximum recovery can be further improved by optimization of the vitrification protocols.

In addition, a better understanding of molecular signaling pathways involved in cell injury caused by cryopreservation is required. It is likely that both inhibition of apoptotic cell death and stabilization of cell membranes during cryopreservation may have significant benefits. The idea that there is a preservation threshold for the cryopreservation of viable tissues is depicted schematically in Figure 22.11. This schematic was constructed from the data given in Table 22.3, which shows for a variety of tissues that survival after cryopreservation could be increased from ~20% to ~80% by employing ice-free techniques. Althogh this represents a marked improvement that

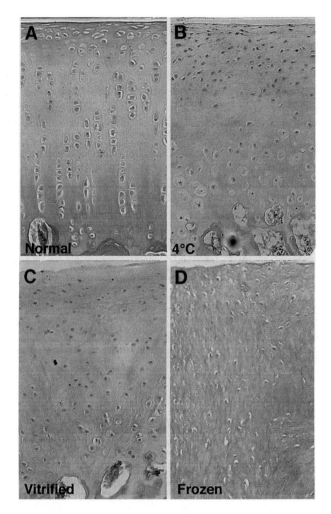

FIGURE 22.10 Histological comparisons of representative 12-week explants from three preservation procedures and control cartilage. (A) *Normal nonoperated rabbit articular cartilage:* In the deeper zone, large
cubical chondrocytes lie in vertical columns. In the superficial perichondrial zone, cells are smaller, flat and
lie parallel to the joint surface. (B) *4°C preservation:* Composed of hyaline cartilage-like tissue in the deeper
zone and fibrous cartilage-like tissue in the superficial perichondrial zone. (C) *Cryopreserved by vitrification:*
Similar to B in composition. (D) *Cryopreserved by Freezing:* The tissue is devoid of chondrocytes and most
cells are representative of fibrous connective tissue.

elevates the preservation technique from one that yields inferior survival using freezing to one that
provides respectable recovery using vitrification, it nonetheless represents a threshold that provides
room for further improvement. The ultimate goal is 100% survival, and this might be approached
by further optimization of ice-free technologies or by attention to the "molecular paradigm" that
has recently been purported to be a way forward to remove the so-called preservation cap (Baust,
2002; Baust et al., 2001, 2002).

Moreover, as mentioned in section 22.4.1, the nature of the vehicle solution used to expose
cells and tissues to cryoprotectants at low temperatures has been shown to affect the outcome of
cryopreservation (Elford and Walter, 1972a, 1972b; Taylor, 1982; Taylor et al., 1978) and has
recently become the focus of additional research aimed at optimization and attenuation of the
cryopreservation cap (Baicu and Taylor, 2002; Baust, 2002; Baust et al., 2001, 2002; Taylor, 2002;
Taylor et al., 2001).

TABLE 22.3
Contrasting Survival of Tissues after Cryopreservation in Either the Frozen or Ice-Free State

Model Tissue	Assay	Survival Outcome (%)		Reference
		Frozen	Ice-Free	
Taenia coli smooth muscle (Guinea pig)	Histamine-induced contractility	21	78	Taylor and Pegg, 1983
Jugular vein (Rabbit)	Various contractile agonists	6–22	84–87	Taylor et al., 1999 Song et al., 2000a
Carotid artery (Rabbit)	Various contractile agonists	<30	>80	Song et al., 2001b
Articular cartilage (Rabbit)	Esterase activity (Calcein AM)	13	80	Song et al., 2001a Brockbank et al., 2002
	Metabolic activity (alamarBlue)	11	85	Song et al., 2004a, b

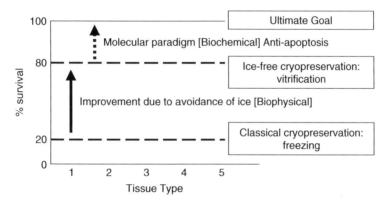

FIGURE 22.11 Schematic diagram to illustrate the paradigms that are considered to influence the outcome of the cryopreservation of multicellular tissues.

22.7 FUTURE DEVELOPMENTS IN RELATION TO TISSUE BANKING AND TISSUE ENGINEERING FOR TRANSPLANTATION

22.7.1 THE COMMERCIAL OPPORTUNITY FOR CRYOPRESERVED TISSUES

Traditional tissue and cell storage cryopreservation methods have well-recognized technical problems that include tissue cracking, matrix disruption, and posttransplantation apoptosis and calcification. Freeze-drying of tissues is still nowhere near a commercial process. The potential for U.S. economic benefits of widely available tissue products was made very clear in the request for proposals from the National Institute of Standards and Technology, Advanced Technology Program, in 1997 which specifically identified "defining and designing conditions for long-term tissue and cell storage that will make products globally available in varying environmental conditions" (http://www.atp.nist.gov). The bottom line is that it was estimated that tissue engineering may address diseases and disorders that account for about one half of the existing U.S. health care costs, which in 1995 dollars had exceeded $1 trillion. We anticipate that this technology will first be applied to research products followed by orthopedic clinical products, such as tissue-engineered cartilage and bone constructs. Cardiovascular products using this technology, including allogeneic human tissues, will follow close behind the orthopedic products. The potential markets are enormous. In 1996, a Frost and Sullivan report indicated that the heart valve replacement and skin

TABLE 22.4
Potential U.S. Organ and Tissue Markets

Structure	Procedures/Year
Skin	4,750,000
Cartilage	1,132,100
Blood Vessels	1,100,000
Pancreas	728,000
Kidney	600,000
Breast	261,000
Liver	155,000
Tendon and Ligament	123,000
Intestine	100,000
Ureter and Urethra	81,900
Heart Valves	65,000
Bladder	57,200

Sources: American Heart Association, 1996; Langer and Vacanti, 1993.

repair product markets alone have maximum potential market sizes of $225 and $5945 million, respectively. Revenues in the total market are anticipated to continue growth at double-digit rates (Frost and Sullivan, 1996). The ultimate potential annual U.S. market size for organ and tissue transplants projected by Drs. Vacanti and Langer (Langer and Vacanti, 1993), assuming unlimited supply of transplants for all potential applications, is indicated in Table 22.4. Many scientific advances in the fields of tissue engineering and xenotransplantation, however, are required for this potential to materialize. The numbers are also predicated on transplantation being such a safe procedure that it would be considered appropriate therapy for a wide range of organ and tissue diseases. Tissue engineering may eventually address diseases and disorders that account for approximately half of all existing health care costs (Langer and Vacanti, 1993; Nerem and Sambanis, 1995; Wilkerson Group, 1992). This market is expected to be affected by major technological changes, and most products will contain highly specialized cell components that will require effective transport solutions and devices to enable product distribution and increase shelf-life.

Research efforts in bioengineering of tissues and organs are driven by the shortfall in allogeneic tissues and organs. It is anticipated by both the scientific and business communities that with new discoveries in tissue engineering, xenotransplantation, and the development of new immunosuppressive therapies, many more diseases will be treated by replacement of defective components (Table 22.4). Hence, the need for reliable methods of storage and shipping will become increasingly important as these developments emerge.

22.7.2 REMAINING HURDLES FOR COMMERCIALIZATION OF VITRIFICATION PROCEDURES

Vitrification is a relatively well-understood physical process. There are, however, a number of significant challenges for commercial deployment of this new technology. Vitrification approaches to preservation have some of the limitations associated with conventional freezing approaches. First, both approaches require low-temperature storage and transportation conditions. Neither can be stored above their T_g for long without significant risk of product damage caused by inherent instabilities leading to ice formation and growth (see Chapters 20 and 21). Both approaches also use CPAs with their attendant problems, discussed below, and require competent technical support during rewarming and CPA elution phases before product use. Therefore, it is possible to employ vitrified products in highly controlled environments, such as a commercial manufacturing facility or an operating theatre, but not in a doctor's outpatient office, or in third-world environments.

22.7.2.1 Challenges Relating to the Use of Cryoprotectants

The high concentrations of CPAs necessary to facilitate vitrification are potentially toxic because the cells may be exposed to these high concentrations at higher temperatures than in freezing methods of cryopreservation. (This point is discussed more fully in Section 22.5.) Cryoprotectants can kill cells by direct chemical toxicity, or indirectly by osmotically induced stresses during suboptimal addition or removal. On completion of warming, the cells should not be exposed to temperatures above 0°C for more than a few minutes before the glass-forming cryoprotectants are removed. Chemical toxicity of CPAs is invariably temperature dependent, but in some cases even subzero exposure can be detrimental. (Chilling injury is a thermal event and not related to CPAs, which may actually protect against thermal shock.) There may be issues concerning safety of some CPAs in conjunction with medical products. Formamide, one of the components of VS55, is a known mutagen. Alternatives to formamide with fewer safety risks and potentially easier clinically acceptance are being tested. In addition, a better understanding of molecular signaling pathways involved in cell injury caused by vitrification is required. It is likely that both inhibition of apoptotic cell death (Baust, 2002; Baust et al., 2000, 2001) and stabilization of cell membranes during vitrification may have significant benefits.

Since 1986, computer-operated organ perfusion devices have been employed to introduce and remove cryoprotectants from animal organs (Fahy, 1994). The primary organs studied to date have been the rabbit kidney and rat liver. At the present time, it is possible to introduce into these organs CPA solutions that are capable of vitrification at ambient pressure, by employing subzero perfusion temperatures for both addition and removal. Use of subzero perfusion temperatures helps to overcome cryoprotectant cytotoxicity. Kidneys, which have been perfused with vitrification solution, cooled to subzero temperatures (but not vitrified), and then perfused to remove the vitrification chemicals, have been reimplanted. These kidneys were shown to function *in vivo* and in some cases supported life for several months (Fahy et al., 1995). This, and our own demonstration of the tolerance of vascular grafts and articular cartilage to vitrifiable concentrations of CPA cocktails, shows that the constraints of CPA toxicity and fluxes can be overcome by selection and optimization of appropriate conditions. This leaves heat transfer issues as a primary hurdle for scaling up the successes in model tissue samples to larger specimens of clinical dimensions.

22.7.2.2 Challenges Relating to Thermomechanical Stresses

As outlined in this chapter, techniques have recently been developed to preserve native tissues such as blood vessels and articular cartilage in an essentially ice-free condition. However, although vitreous cryopreservation has been demonstrated to provide superior preservation compared with conventional freezing methods in these small model systems, cryopreservation of large tissue samples continues to be hampered by thermomechanical constraints. These include problems arising from the limits of heat and mass transfer in bulky systems and damage induced by mechanical stresses including fractures. Such fractures are attributed to stresses that can arise because of the nonuniform cooling of larger tissues. In fact, the higher cooling rates that facilitate vitrification will typically lead to higher mechanical stresses. The competing needs of vitrification and minimization of mechanical damage demand a greater understanding of both vitrification and stress development. Although mechanical stress has long been recognized as an important mechanism of tissue destruction, it has received very little attention in the context of cryobiology. It is our opinion that in some circumstances, even a single major fracture may prevent the tissue from recovery, or effective use after cryogenic storage. Reduction of mechanical stress, and thereby prevention of fracture formation, is a necessary integral condition for successful cryopreservation of large specimens.

We have already indicated that heating constraints impose severe limitations on the ability to rewarm vitrified specimens without danger of devitrification. Hence a major hurdle for deployment of vitrification methods is the development of effective rapid-warming techniques for larger specimens (>10 mL volume) to prevent devitrification and ice growth by recrystallization; devitrification is the freezing of a formerly vitrified solution. To prevent devitrification with current technology, the vitrified material must be warmed uniformly at up to 300°C/min so that ice does not have the opportunity to form in significant quantities. To achieve this warming rate, materials must be warmed in 20 sec or less. Many simple structures, such as single-cell suspensions or cell aggregates (with which vitrification has been successful) are small enough that rapid-warming rates can be achieved using immersion in warm fluids.

The same is true for the small model tissues for which vitrification has been successful. Unfortunately, tissue that has been engineered and allogeneic organs and tissues are generally of a much larger volume; thus, more sophisticated techniques of warming are required. Microwave warming has been attempted but has never been successful because of the uneven warming of specimens and problems with thermal runaway, which results in heat-denatured tissues. Ruggera and Fahy (1990) reported success in warming test solutions at rates of up to about 200°C/min using a novel technology based on electromagnetic techniques (wire length resonance radio frequency warming) developed and owned jointly by the American Red Cross and the U.S. Food and Drug Administration. Unfortunately, unpublished results indicate that this method is also problematic because of the uneven warming of specimens and problems associated with thermal runaway. Pegg et al. are credited with the most comprehensive studies of this approach to rapid heating of cryopreserved specimens. During a period spanning more than a decade, they have systematically developed a new device for dielectric heating to achieve uniform and high rates of temperature change (Marsland et al., 1987; Robinson and Pegg, 1999). At this point in time, the focus has been on the electromechanical developments of the technique, which has been reported to yield maximum warming rates of more than 10°C/sec (600°C/min). Using Pegg et al.'s device, frozen samples may be warmed from –65°C to room temperature in less than 30 sec, with final spatial differences of less than 20°C. Application of this technology to the survival of cells and tissues has not yet been reported, so it is still not possible to say whether this approach will provide an answer to the problem of rapid heating of cryopreserved tissues with adequate retention of cell viability and tissue function.

As we have mentioned already, an alternative, or even adjunctive, approach involves the use of molecular ice-control techniques to prevent the damaging growth of ice crystals during cooling or warming. The objective is to identify molecules that specifically interact with ice nuclei, resulting in either prevention of ice nucleus development or modification of ice crystal phenotype. These natural or synthetic molecules promise to have benefits in both freezing and vitrification preservation protocols either by rendering ice crystals less damaging or by permitting reduction of CPA concentrations, respectively.

There is now unequivocal evidence that ice formation within the extracellular matrix of multicellular tissues is the principal event that limits the survival of cryopreserved tissues using conventional freezing techniques. This mode of injury can be circumvented using vitreous cryopreservation, which in the case of relatively small tissue specimens has recently been shown to markedly improve functional outcome. However, this ice-free method of cryopreservation is not easily applied to larger bulk samples of clinically relevant dimensions because of the effects of nonuniform cooling and rewarming, which are the source of thermomechanical stresses in the vitrified tissue samples. Moreover, these additional stresses in cryopreservation have not heretofore been studied in sufficient detail to affect the rational design of improved methods of preservation. It is our opinion that effective methods of cryopreservation of tissues and engineered tissue constructs will only be realized using methods of preservation that not only minimize the effects of ice formation (via vitrification) but also avoid thermomechanical stresses in the cryopreserved samples during cooling, storage, and rewarming.

Finally, it should be understood that vitrification is a technology that must be deployed with extreme care and diligence. Errors in technique are common and can lead to failure to obtain satisfactory results. A large part of our current research program is definition of the limits for each step in the vitrification process including reduction of CPAs to the lowest possible concentrations, long-term storage conditions and duration, and simplification of the CPA addition and removal steps. Such efforts will produce vitrification procedures that can be deployed with less risk of tissue quality being compromised.

ACKNOWLEDGMENTS

We gratefully acknowledge the invaluable contributions of our colleagues who have contributed significantly to the work described in this chapter. Specific thanks go to Bijan Khirabadi and John Walsh for their work on the physical measurements of ice crystal growth kinetics and vitrification solution formulations; Fred Lightfoot for the cryosubstitution and histology studies; and Elizabeth Greene, Janet Boggs, Zhen Zhen Chen, and Chaoyang Li for excellent technical support. We also thank Yoed Rabin for his insightful discussions on the thermomechanical properties of biological tissues at low temperatures. This work was supported in part by grants from the U.S. Department of Commerce—National Institutes of Standards and Technology (Cooperative Agreement 97-07-0039), and the National Institutes of Health (1R43-AR47273-01).

REFERENCES

Acar, C., Jebara, V., and Portoghese, M. (1992) Revival of the radial artery for coronary bypass grafting, *Ann. Thoracic Surg.*, 54, 652–660.

Almassi, G.H., Farahbakhsh, B., Wooldridge, T., Rusch, N.J., and Olinger, G.N. (1996) Endothelium and vascular smooth muscle function in internal mammary artery after cryopreservation, *J. Surg. Res.*, 60, 355–360.

American Heart Association, (1996) *Heart and Stroke Facts: Statistical Supplement*, www.americanheart.org.

Angell, C.A., Sara, E.J., Donnella, J., and MacFarlane, D.R. (1981) Homogeneous nucleation and glass transition temperatures in solutions of Li salts in D_2O and H_2O: A doubly unstable glass region, *J. Phys. Chem.*, 85, 1461–1464.

Armitage, W.J., Hall, S.C., and Routledge, C. (2002) Recovery of endothelial function after vitrification of cornea at −110°C, *Invest. Ophthalmol. Vis. Sci.*, 43, 2160–2164.

Armitage, W.J. and Rich, S.J. (1990) Vitrification of organized tissues, *Cryobiology*, 27, 483–491.

Baicu, S. and Taylor, M.J. (2002) Acid-base buffering in organ preservation solutions as a function of temperature: New parameters for comparing buffer capacity and efficiency, *Cryobiology*, 45, 33–48.

Bakay, A., Csonge, L., Papp, G., and Fekete, L. (1998) Osteochondral resurfacing of the knee joint with allograft. Clinical analysis of 33 cases, *Int. Orthop.*, 22, 277–281.

Barner, H.B., Standeven, J.W., and Resse, J. (1985) Twelve–year experience with internal mammary artery for coronary artery bypass, *J. Thoracic Cardiovasc. Surg.*, 90, 668–675.

Barrett, J. (2001) Thermal hysteresis proteins, *Int. J. Biochem. Cell Biol.*, 33, 105–117.

Baust, J.M. (2002) Molecular mechanisms of cellular demise associated with cryopreservation failure, *Cell Preservation Technol.*, 1, 17–32.

Baust, J.M., van Buskirk, R.G., and Baust, J.G. (2000) Cell viability improves following inhibition of cryopreservation-induced apoptosis, *In Vitro Cell Dev. Biol.*, 36, 262–270.

Baust, J.M., Vogel, M.J., van Buskirk, R.G., and Baust, J.G. (2001) A molecular basis of cryopreservation failure and its modulation to improve cell survival, *Cell Transplant.*, 10, 561–571.

Baust, J.M., Van Buskirk, R., and Baust, J.G. (2002) Modulation of the cryopreservation cap: Elevated survival with reduced dimethyl sulfoxide concentration, *Cryobiology*, 45(2), 97–108.

Bodziony, J., Schmitt, P., and Feifel, G. (1994) In vitro function, morphology, and viability of cryopreserved rat pancreatic islets: Comparison of vitrification and six cryopreservation protocols, *Transplant. Proc.*, 26, 833–834.

Bourne, W.M. and Nelson, L.R. (1994) Human corneal studies with a vitrification solution containing dimethyl sulfoxide, formamide, and 1,2-propane diol, *Cryobiology*, 31, 52–530.

Boutron, P. (1990) Levo- and dextro-2,3butanediol and their racemic mixture: Very efficient solutes for vitrification, *Cryobiology*, 27, 55–69.

Boutron, P. and Kaufmann, A. (1979) Stability of the amorphous state in the system water–1,2-Propanediol, *Cryobiology*, 16, 557–568.

Boutron, P., Mehl, P., Kaufmann, A., and Angibaud, P. (1986) Glass-forming tendency and stability of the amorphous state in the aqueous solutions of linear polyalcohols with four carbons, *Cryobiology*, 23, 453–469.

British Cardiac Society (1991) Report of a working party of the British Cardiac Society: Coronary Angioplasty in the United Kingdom, *Br. Heart J.*, 66, 325–331.

Brockbank, K.G.M. (1994) Effects of cryopreservation upon vein function *in vivo*, *Cryobiology*, 31, 71–81.

Brockbank, K.G.M., McNally, R.T., and Walsh, K.A. (1992) Cryopreserved vein transplantation, *J. Card. Surg.*, 7, 170–176.

Callow, A.D. (1983) Historical overview of experimental and clinical development of vascular grafts, in *Biologic and Synthetic Vascular Prosthesis*, Stanley J., Ed., Grune and Stratton, New York.

Campbell, L.H., Rutledge, R.N., Taylor, M.J., and Brockbank, K.G.M. (1999) Evaluation of the relative cytotoxicity of the components of a vitrification solution in a variety of cardiovascular cells, *Cryobiology*, 39, 362–362.

Chou, K.-C. (1992) Energy-optimized structure of antifreeze protein and its binding mechanism, *J. Mol. Biol.*, 223, 509–517.

Clarke, C.J., Buckley, S.L., and Lindner, N. (2002) Ice structuring proteins—A new name for antifreeze proteins, *Cryo-Letters*, 23, 89–92.

Davies, P.L. and Hew, C.L. (1990) Biochemistry of fish antifreeze proteins, *FASEB J.*, 4, 2460–2468.

DeVries, A.L. (1983) Antifreeze peptides and glycopeptides in cold-water fishes, *Ann. Rev. Physiol.*, 45, 245–260.

Duman, J.G., Wu, D.W., Olsen, T.M., Urrutia, M., and Tursman, D. (1993) Thermal-hysteresis proteins, in *Advances in Low-Temperature Biology*, Vol. 2, Steponkus, P.L., Ed., JAI Press, Greenwich, CT, pp. 131–182.

Edwards, W.S., Holdefer, W.F., and Motashemi, M. (1966) The importance of proper caliber of lumen in femoral popliteal artery reconstruction, *Surg. Gynecol. Obstet.*, 122, 37.

Elford, B.C. (1970) Diffusion and distribution of dimethyl sulphoxide in the isolated guinea-pig taenia coli, *J. Physiol.*, 209, 187–208.

Elford, B.C. and Walter, C.A. (1972a) Effects of electrolyte composition and pH on the structure and function of smooth muscle cooled to –79°C in unfrozen media, *Cryobiology*, 9, 82–100.

Elford, B.C. and Walter, C.A. (1972b) Preservation of structure and function of smooth muscle cooled to –79°C in unfrozen aqueous media, *Nat. New Biol.*, 236, 58–60.

Ewart, K.V., Li, Z., Yang, D.S.C., Fletcher, G.L., and Hew, C.L. (1998) The ice-binding site of atlantic herring antifreeze protein corresponds to the carbohydrate-binding site of C-Type lectins, *Biochemistry*, 37, 4080–4085.

Fahy, G.M. (1988) Vitrification, in *Low Temperature Biotechnology: Emerging Applications and Engineering Contributions*, McGrath, J.J. and Diller, K.R. Eds., The American Society of Mechanical Engineers, New York, pp. 113–146.

Fahy, G.M. (1989) Vitrification as an approach to organ cryopreservation: Past, present, and future, in *Cryopreservation and Low Temperature Biology in Blood Transfusion*, Smit Sibinga, C.T., Das, P.C., and Meryman, H.T., Eds., Kluwer Academic Publishers, Dordrecht, pp. 255–268.

Fahy, G.M. (1994) Organ perfusion equipment for the introduction and removal of cryoprotectants, *Biomed. Instrumentation Technol.*, 28(2), 87–100.

Fahy, G.M. (1998) The role of nucleation in cryopreservation, in *Biological Ice Nucleation and Its Applications*, Lee, R.E., Jr., Warren, G.J., and Gusta, L.V., Eds., APS Press, St. Paul, MN, pp. 315–336.

Fahy, G.M. (2001), Process for Preparing Novel Ice-Controlling Molecules, U.S. Patent 6,303,388B1.

Fahy, G.M., DaMouta, C., Tsonev, L., Khirabadi, B.J., Mehl, P., and Meryman, H.T. (1995) Cellular injury associated with organ cryopreservation: Chemical toxicity and cooling injury, in *Cell Biology of Trauma*, Lemasters, J.J. and Oliver, C., Eds., CRC Press, Boca Raton, FL, pp. 333–356.

Fahy, G.M. and Hirsch, A. (1982) Prospects for organ preservation by vitrification, in *Organ Preservation: Basic and Applied Aspects*, Pegg, D.E., Jacobsen, I.A., and Halasz, N.A., Eds., MTP Press, Lancaster, PA, pp. 399–404.

Fahy, G.M., MacFarlane, D.R., Angell, C.A., and Meryman, H.T. (1984) Vitrification as an approach to cryopreservation, *Cryobiology*, 21, 407–426.

Fahy, G.M., Saur, J., and Williams, R.J. (1990) Physical problems with the vitrification of large biological systems, *Cryobiology*, 27, 492–510.

Farrant, J. (1965) Mechanism of cell damage during freezing and thawing and its prevention, *Nature*, 205, 1284–1287.

Field, P., Matar, A., and Agrama, H. (1969) An assessment of allograft veins for arterial grafting, *Circulation*, 3, 39–40.

Forsyth, M. and MacFarlane, D.R. (1986) Recrystallization revisited, *Cryo-Letters*, 7, 367–378.

Franks, F. (1977) Solution and conformational effects in aqueous solutions of biopolymer analogues, *Philos. Trans. R. Soc. Lond. B*, 278, 33–57.

Franks, F. (1982b) Physiological water stress, in *Biophysics of Water*, Franks, F. and Mathias, S., Eds., Wiley, Chichester, pp. 279–294.

Franks, F. (1982a) The properties of aqueous solutions at sub-zero temperatures, in *Water: A Comprehensive Treatise*, Vol. 7, Franks F., Ed., Plenum, New York, pp. 215–338.

Frost and Sullivan (1996), *U.S. Organ Transplant and Related Product Markets,* Market Report, Mountain View, CA.

Frost and Sullivan (1997), *World Cell Therapy Markets*, Market Report, Mountain View, CA.

Gelbish, J., Jacobowitz, I.J., and Rose, D.M. (1986) Cryopreserved homologous saphenous vein: Early and late patency in coronary artery bypass surgical procedures, *Ann. Thor. Surg.*, 42, 70–73.

Hall, S.C. and Armitage, W.J. (1999) Vitrification of rabbit corneas, *Cryobiology* 39, 310–310.

Hetzel, F.W., Kruuv, J., McGann, L.E., and Frey, H.E. (1973) Exposure of mammalian cells to physical damage: Effect of the state of adhesion on colony-forming potential, *Cryobiology*, 10, 206–211.

Hunt, C.J., Taylor, M.J., and Chapman, D. (1991) Exposure of rabbit corneas to multimolar concentrations of butane-2,3-diol: A transmission and scanning electron microscope study of the endothelium, *Cryobiology*, 28, 561–561.

Hunt, C.J., Taylor, M.J., and Pegg, D.E. (1982) Freeze-substitution and isothermal freeze fixation studies to elucidate the pattern of ice formation on smooth muscle at 252K (–21 C), *J. Microscopy*, 125, 177–186.

Jackson, D.R. and Abel, D.W. (1972) The homologous saphenous vein in arterial reconstruction, *Vasc. Surg.*, 6, 85–92.

Jacobsen, I.A., Pegg, D.E., Starklint, H., Chemnitz, J., Hunt, C.J., Barfort, P., and Diaper, M.P. (1984) Effect of cooling rate and warming rate on glycerolized rabbit kidneys, *Cryobiology*, 21, 637–653.

Jutte, N.H.P.M., Heyse, P., Jansen, H.G., Bruining, G.J., and Zeilmaker, G.H. (1987a) Vitrification of human Islets of Langerhans, *Cryobiology*, 24, 403–411.

Jutte, N.H.P.M., Heyse, P., Jansen, H.G., Bruining, G.J., and Zeilmaker, G H. (1987b) Vitrification of mouse Islets of Langerhans: Comparison with a more conventional freezing method, *Cryobiology*, 24, 292–302.

Kaiser, J. (2002) New prospects of putting organs on ice, *Science*, 295, 1015.

Karlsson, J.O.M. and Toner, M. (1996) Long-term storage of tissues by cryopreservation: Critical issues, *Biomaterials*, 17, 243–256.

Knight, C.A. and Duman, J.G. (1986) Inhibition of recrystallization of ice by insect thermal hysteresis proteins: A possible cryoprotective role, *Cryobiology*, 23, 256–262.

Kroener, C. and Luyet, B. (1966) Formation of cracks during the vitrification of glycerol solutions and disappearance of the cracks during rewarming, *Biodynamica*, 10, 47–52.

Langer, R. and Vacanti, J.P. (1993) Tissue engineering, *Science*, 260, 920–926.

Loo, F.D., Lytle, B.W., and Cosgrove, D.M. (1986) Influence of the internal mammary artery graft on 10-year survival and other cardiac events, *N. Engl. J. Med.*, 314, 1–6.

Loop, F.D., Lytle, B.W., and Cosgrove, D.M. (1986) Free (aorto-coronary) internal mammary artery graft: Late results, *J. Thor. Cardiovasc. Surg.*, 92, 827–831.

Luyet, B.J. (1937) The vitrification of organic colloids and of protoplasm, *Biodynamica*, 1, 1–14.

Luyet, B.J. and Gehenio, P.M. (1940) *Life and Death at Low Temperatures*, Biodynamica, Normandy, MO.

MacFarlane, D.R. (1986) Devitrification in glass-forming aqueous solutions, *Cryobiology*, 23, 230–244.

MacFarlane, D.R. (1987) Physical aspects of vitrification in aqueous solutions, *Cryobiology*, 24, 181–195.

MacFarlane, D.R. and Angell, C.A. (1981) Homogenous nucleation and glass formation in cryoprotective systems at high pressures, *Cryo-Letters*, 2, 353–358.

MacFarlane, D.R. and Forsyth, M. (1990) Recent insights on the role of cryoprotective agents in vitrification, *Cryobiology*, 27, 345–358.

MacFarlane, D.R., Forsyth, M., and Barton, C.A. (1992) Vitrification and devitrification in cryopreservation, in *Advances in Low Temperature Biology*, Vol. 1, Steponkus, P.L., Ed., JAI Press, Greenwich, CT, pp. 221–278.

MacKenzie, A.P. (1977) Non-equilibrium freezing behaviour of aqueous systems, *Philos. Trans. R. Soc. Lond. B*, 278, 167–189.

Malinin, T.I., Martinez, O.V., and Brown, M.D. (1985) Banking of massive osteoarticular and intercalary bone allografts, *Clin. Orthopaed. Related Res.*, 197, 44–57.

Marco, F., Leon, C., Lopez-Oliva, F., et al. (1992) Intact articular cartilage cryopreservation. *In vivo* evaluation, *Clin. Orthopaed. Related Res.*, 283, 11–20.

Marsland, T.P., Evans, S., and Pegg, D.E. (1987) Dielectic measurements for the design of an electromagnetic rewarming system, *Cryobiology*, 24, 311–323.

Mazur, P. (1984a) Freezing of living cells: Mechanisms and implications, *Am. J. Physiol.*, 247, C125–C142.

Mazur, P. (1984b) Fundamental cryobiology and the preservation of organs by freezing, in *Organ Preservation for Transplantation*, 2nd ed., Karow, A. Jr. and Pegg D.E., Eds., Marcel Dekker, New York, pp. 143–175.

Mazur, P. (1988) Stopping biological time: The freezing of living cells, *Ann. NY Acad. Sci.*, 541, 514–531.

McNally, R.T., McCaa, C., Brockbank, K.G.M., Heacox, A.E., and Bank, H.L. (1992), Method for Cryopreserving Blood Vessels, U.S. Patent 5,145,769.

McNally, R.T., Walsh, K., and Richardson, W. (1992) Early clinical evaluation of cryopreserved allograft vein, *Cryobiology*, 29, 702 (Abstract).

Mehl, P. (1993) Nucleation and crystal growth in a vitrification solution tested for organ cryopreservation by vitrification, *Cryobiology*, 30, 509–518.

Mehl, P. (1996) Crystallization and vitrification in aqueous glass-forming solutions, in Steponkus, P.L., Ed., *Advances in Low Temperature Biology*, Vol. 3, JAI Press, Greenwich, CT, pp. 185–255.

Mehl, P. and Boutron, P. (1988) Cryoprotection of red blood cells by 1,3-butanediol and 2,3- butanediol, *Cryobiology*, 25, 44–54.

Mejean, A. and Pegg, D.E. (1991) Development of a vehicle solution for the introduction and removal of butane-2,3-diol in rabbit kidneys, *Cryobiology*, 28, 518–518.

Mills, N.L. and Everson, C.T. (1989) Right gastroepiploic artery: A third conduit for coronary artery bypass, *Ann. Thor. Surg.*, 47, 706–711.

Moynihan, C.T., Easteal, A.J., DeBolt, M.A., and Tucker, J. (1976) Dependence of the fictive temperature of glass on cooling rate, *J. Am. Chem. Soc.*, 59, 12–26.

Mugnano, J.A., Wang, T., Layne, J.R., Jr., DeVries, A.L., and Lee, R.E., Jr. (1995) Antifreeze glycoproteins promote intracellular freezing of rat cardiomyocytes at high subzero temperatures, *Am. J. Physiol.*, 269, R474–R479.

Muldrew, K., Hurtig, M., Schachar, N., and McGann, L.E. (1994) Localization of freezing injury in articular cartilage, *Cryobiology*, 31, 31–38.

Muldrew, K., Novak, K., Studholme, C., Wohl, G., Zernicke, R., Schachar, N.S., and McGann, L.E. (2001) Transplantation of articular cartilage following a step-cooling cryopreserved protocol, *Cryobiology*, 43, 260–267.

Müller-Schweinitzer, E., Stulz, P., Striffeler, H., and Haefeli, W. (1998) Functional activity and transmembrane signaling mechanisms after cryopreservation of human internal mammary arteries, *J. Vasc. Surg.*, 27, 528–537.

Nerem, R.M. and Sambanis, A. (1995) Tissue engineering: From biology to biological substitute, *Tissue Eng.*, 1, 3–13.

O'Driscoll, S.W., Salter, R.B., and Keeley, F.W. (1985) A method for quantitative analysis of ratios of types I and II collagen in small samples of articular cartilage, *Anal. Biochem.*, 145, 277–285.

Ohlendorf, C., Tomford, W.W., and Mankin, H.J. (1996) Chondrocyte survival in cryopreserved osteochondral articular cartilage, *J. Orthop. Res.*, 14, 413–416.

Ohno, T. (1994) A simple method for *in situ* freezing of anchorage-dependent cells, in *Cell and Tissue Culture: Laboratory Procedures*, Doyle, A., Griffiths, J.B., and Newell, D.G., Eds., Wiley, Chichester.

Paynter, S., Cooper, A., Thomas, N., and Fuller, B.J. (1997) Cryopreservation of multicellular embryos and reproductive tissues, in *Reproductive Tissue Banking: Scientific Principles*, Karow, A.M. and Critser, J.K., Eds., Academic Press, San Diego, CA, pp. 359–397.

Pegg, D.E. (1987) Ice Crystals in Tissues and Organs, in *The Biophysics of Organ Preservation*, Pegg, D.E. and Karow, A.M., Jr., Eds., Plenum, New York, pp. 117–140.

Pegg, D.E. (1989) The nature of cryobiological problems, *Low Temp. Biotechnol.*, 10, 3–21.

Pegg, D.E. and Diaper, M.P. (1990) Freezing versus vitrification: Basic principles, in *Cryopreservation and Low Temperature Biology in Blood Transfusion*, Vol. 24, Smit Sibinga, C.T., Das, P.C., and Meryman, H.T., Eds., Kluwer Academic Publishers, Dordrecht, pp. 55–69.

Pegg, D.E., Jacobsen, I.A., Armitage, W.J., and Taylor, M.J. (1979) Mechanisms of cryoinjury in organs, in *Organ Preservation II*, Pegg, D.E. and Jacobsen, I.A., Eds., Churchill Livingstone, Edinburgh, pp. 132–146.

Pegg, D.E., Wusteman, M.C., and Boylan, S. (1996) Fractures in cryopreserved elastic arteries: Mechanism and prevention, *Cryobiology*, 33, 658–659.

Pegg, D.E., Wusteman, M.C., and Boylan, S. (1997) Fractures in cryopreserved elastic arteries, *Cryobiology*, 34, 183–192.

Polge, C., Smith, A.Y., and Parkes, A.S. (1949) Revival of spermatozoa after vitrification and de-hydration at low temperatures, *Nature*, 164, 666.

Pollock, G.A., Pegg, D.E., and Hardie, I.R. (1986) An isolated perfused rat mesentery model for direct observation of the vasculature during cryopreservation, *Cryobiology*, 23, 500–511.

Puig, L.B., Ciongolli, W., and Cividanes, G.L. (1990) Inferior epigastric artery as a free graft for myocardial revascularization, *J. Thoracic Cardiovasc. Surg.*, 99, 251–255.

Rabin, Y., Olson, P., Taylor, M.J., Steif, P.S., Julian, T.B., and Wolmark, N. (1997) Gross damage accumulation in frozen rabbit liver due to mechanical stress at cryogenic temperatures, *Cryobiology*, 34, 394–405.

Rabin, Y. and Podbilewicz, B. (2000) Temperature-controlled microscopy for 4-D imaging of living cells: Apparatus, thermal analysis, and temperature dependency of embryonic elongation in *C. elegans*, *J. Microscopy*, 199, 214–223.

Rabin, Y. and Steif, P.S. (1998) Thermal stresses in a freezing sphere and its application to cryobiology, *ASME J. Appl. Mech.*, 65, 328–333.

Rabin, Y. and Steif, P.S. (2000) Thermal stress modeling in cryosurgery, *Int. J. Solids Struct.*, 37, 2363–2375.

Rabin, Y., Steif, P.S., Taylor, M.J., Julian, T.B., and Wolmark, N. (1996) An experimental study of the mechanical response of frozen biological tissues at cryogenic temperatures, *Cryobiology*, 33, 472–482.

Rabin, Y., Taylor, M.J., and Wolmark, N. (1998) Thermal expansion measurements of frozen biological tissues at cryogenic temperatures. *J. Biomech. Eng.*, 120, 259–266.

Rall, W.F. (1987) Factors affecting the survival of mouse embryos cryopreserved by vitrification, *Cryobiology*, 24, 387–402.

Rall, W.F. and Fahy, G.M. (1985) Ice-free cryopreservation of mouse embryos at −196°C by vitrification, *Nature*, 313, 573–575.

Rall, W.F. and Meyer, T.K. (1989) Zona fracture damage and its avoidance during the cryopreservation of mammalian embryos, *Theriogenology*, 31, 683–692.

Rasmussen, D. and MacKenzie, A.P. (1968) Phase diagram for the system water-dimethylsulphoxide, *Nature*, 220, 1315–1317.

Rich, S.J. and Armitage, W.J. (1991) Corneal tolerance of vitrifiable concentrations of propane-1,2-diol, *Cryobiology*, 28, 159–170.

Robinson, M.P. and Pegg, D.E. (1999) Rapid electromagnetic warming of cells and tissues, *IEEE Trans. Biomed. Eng.*, 46, 1413–1425.

Rubinsky, B., Cravalho, E.G., and Mikic, B. (1980) Thermal stress in frozen organs, *Cryobiology*, 17, 66–73.

Rubinsky, B. and Pegg, D.E. (1988) A mathematical model for the freezing process in biological tissue, *Proc. R. Soc. Lond. B*, 234, 343–358.

Ruggera, P.S. and Fahy, G.M. (1990) Rapid and uniform electromagnetic heating of aqueous cryoprotectant solutions from cryogenic temperatures, *Cryobiology*, 27, 465–478.

Sellke, F.W., Stanford, W., and Rossi, N.P. (1991) Failure of cryopreserved saphenous vein allografts following coronary artery bypass surgery, *J. Cardiovasc. Surg.*, 32, 820–823.

Showlater, D., Durham, S., and Sheppeck, R. (1989) Cryopreserved venous homografts as vascular conduits in canine carotid arteries, *Surgery*, 106, 652–659.

Smith, A.U. (1961) *Biological Effects of Freezing and Supercooling,* Edward Arnold, London.

Song, Y.C., An, Y.H., Kang, Q.K., Li, C., Boggs, J.M., Chen, Z., Taylor, M.J., and Brockbank, K.G.M. (2004a) Vitreous preservation of articular cartilage grafts, *J. Invest. Surg.*, 17, 1–6.

Song, Y.C., Hagen, P.O., Lightfoot, F.G., Taylor, M.J., Smith, A.C., and Brockbank, K.G.M. (2000a) *In vivo* evaluation of the effects of a new ice-free cryopreservation process on autologous vascular grafts, *J. Invest. Surg.*, 13, 279–288.

Song, Y.C., Hunt, C.J., and Pegg, D.E. (1994) Cryopreservation of the common carotid artery of the rabbit, *Cryobiology*, 31317–31329.

Song, Y.C., Khirabadi, B.S., Lightfoot, F.G., Brockbank, K.G.M., and Taylor, M.J. (2000b) Vitreous cryopreservation maintains the function of vascular grafts, *Nat. Biotechnol.*, 18, 296–299.

Song, Y.C., Lightfoot, F.G., Chen, Z., Taylor, M.J., and Brockbank, K.G.M. (2004b) Vitreous preservation of rabbit articular cartilage, *Cell Preserv. Technol.*, 2(1), 67–74.

Song, Y.C., Lightfoot, F.G., Li, C., Boggs, J.M., Taylor, M.J., and An, Y.H. (2001a) Successful cryopreservation of articular cartilage by vitrification, *Cryobiology* 43, 353.

Song, Y.C., Pegg, D.E., and Hunt, C.J. (1995) Cryopreservation of common carotid artery of the rabbit: Optimization of dimethyl sulfoxide concentration and cooling rate, *Cryobiology*, 32, 405–421.

Song, Y.C., Taylor, M.J., and Brockbank, K.G.M. (2001b) Ice-free cryopreservation of arterial grafts, *Cryobiology*, 41, 370.

Stanke, F., Riebel, D., Carmine, S., Cracowski, J.-L., Caron, F., Magne, J.-L., and Egelhoffer, H. (1998) Functional assessment of human femoral arteries after cryopreservation, *J. Vascular Surg.*, 28, 273–283.

Stephen, M., Sheil, A.G.R., and Wong, J. (1978) Allograft vein arterial bypass, *Arch. Surg.*, 113, 591–593.

Stone, B.B., Defranzo, B.E., Dicesare, C., Rapko, S.M., Brockbank, K.G.M., Wolfrum, J.M., Wrenn, C.A., and Grossman, J.D. (1998) Cryopreservation of human articular cartilage for autologous chondrocyte transplantation, *Cryobiology*, 37, 445–446.

Sutton, R.L. (1992) Critical cooling rates for aqueous cryoprotectants in the presence of sugars and polysaccharides, *Cryobiology*, 29, 585–598.

Takahashi, T., Hirsh, A.G., Erbe, E.F., Bross, J.B., Steere, R.L., and Williams, R.J. (1986) Vitrification of human monocytes, *Cryobiology*, 23, 103–115.

Taylor, M.J. (1982) The role of pH* and buffer capacity in the recovery of function of smooth muscle cooled to −13°C in unfrozen media, *Cryobiology*, 19, 585–601.

Taylor, M.J. (1984) Sub-zero preservation and the prospect of long-term storage of multi-cellular tissues and organs, in *Transplantation Immunology: Clinical and Experimental*, Calne, R.Y., Ed., Oxford University Press, Oxford, pp. 360–390.

Taylor, M.J. (1986) Clinical cryobiology of tissues: Preservation of corneas, *Cryobiology*, 23, 323–353.

Taylor, M.J. (1987) Physico-chemical principles in low temperature biology, in *The Effects of Low Temperatures on Biological Systems*, Grout, B.W.W. and Morris, G.J., Eds., Edward Arnold, London, pp. 3–71.

Taylor, M.J. (2002) System for Organ and Tissue Preservation and Hypothermic Blood Substitution. U.S. Patent 6,492,103.

Taylor, M.J., Campbell, L.H., Rutledge, R.N., and Brockbank, K.G.M. (2001) Comparison of Unisol with EuroCollins solution as a vehicle solution for cryoprotectants, *Transplant. Proc.*, 33, 677–679.

Taylor, M.J. and Foreman, J. (1991) Tolerance of isolated pancreatic islets to butane-2,3-diol at 0°C, *Cryobiology*, 28, 566–567.

Taylor, M.J. and Pegg, D.E. (1983) The effect of ice formation on the function of smooth muscle tissue following storage at −21°C and −60°C, *Cryobiology*, 20, 36–40.

Taylor, M.J., Song, Y.C., Khirabadi, B.S., Lightfoot, F.G., and Brockbank, K.G.M. (1999) Vitrification fulfills its promise as an approach to reducing freeze-induced injury in a multi-cellular tissue, *Adv. Heat Mass Transfer Biotechnol.*, 44, 93–102.

Taylor, M.J., Walter, C.A., and Elford, B.C. (1978) The pH-dependent recovery of smooth muscle from storage at −13°C in unfrozen media, *Cryobiology*, 15, 452–460.

Tice, D.A. and Zerbino, V.R. (1972) Clinical experience with preserved human allografts for vascular reconstruction, *Surgery*, 72, 260–267.

Tomford, W.W., Fredericks, G.R., and Mankin, H.J. (1984) Studies on cryopreservation of articular cartilage chondrocytes, *J. Bone Joint Surg. Am.*, 66, 253–259.

Tyshenko, M.G., Doucet, D., Davies, P.L., and Walker, V.K. (1997) The antifreeze potential of the spruce budworm thermal hysteresis protein, *Nat. Biotechnol.*, 15, 887–890.

Walker, P.J., Mitchell, R.S., McFadden, P.M., James, D.R., and Mehigan, J.T. (1993) Early experience with cryopreserved saphenous vein allografts as a conduit for complex limb-salvage procedures, *J. Vascular Surg.*, 18, 561–569.

Wassenaar, C., Wijsmuller, E.G., Van Herwerden, L.A., Aghai, Z., Van Tricht, C., and Bos, E. (1995) Cracks in cryopreserved aortic allografts and rapid thawing, *Ann. Thoracic Surg.*, 60, S165–S167.

Wilkerson Group, Inc. (1992), *Research on Market Potential for Tissue Engineering.*

Williams, R.J. (1989) Four modes of nucleation in viscous solutions, *Cryobiology*, 26, 568–568.

Wolfinbarger, L., Jr., Adam, M., Lange, P., and Hu, J.F. (1991) Microfractures in cryopreserved heart valves: Valve submersion in liquid nitrogen revisited, *Appl. Cryogenic Technol.*, 10, 227–233.

Wowk, B., Leitl, E., Rasch, C.M., Mesbah-Karimi, N., Harris, S.B., and Fahy, G.M. (2000) Vitrification enhancement by synthetic ice blocking agents, *Cryobiology*, 40, 228–236.

Wu, F.J., Davisson, T.H., and Pegg, D.E. (1998) Preservation of tissue-engineered articular cartilage, *Cryobiology*, 37, 410 (Abstract).

Zeff, R.H., Kongathahworn, C., and Iannone, L.A. (1988) Internal mammary artery versus saphenous vein graft to the left anterior descending coronary artery: Prospective randomized study with 10-year follow-up, *Ann. Thoracic Surg.*, 45, 451–454.

Theme 5

The Future of Cryobiology

23 The Future of Cryobiology

Nick Lane

CONTENTS

23.1 PROGRESS AND STASIS: 50 YEARS OF CRYOBIOLOGY

Few scientific problems have proved as intractable as cryopreservation. In his foreword to this book, Harry Meryman recalls the astonishing shifts in scientific paradigms since the 1950s. If we go back further, to the early 1940s, the differences become even more radical. When the Father of cryobiology, Basil Luyet, published his seminal work *Life and Death at Low Temperatures* in 1940, Avery had yet to prove that genes are composed of nucleic acids. Even scientists of the calibre of J.B.S. Haldane wrote scornfully that DNA did not have the variability of structure necessary to encode genes and argued that histones were more likely candidates. Yet at this time, Luyet had a strikingly contemporary view of vitrification—still the holy grail of cryobiology. Compare Luyet and Gehenio writing in 1940 with Taylor in Chapter 22 in this volume:

> Good vitrification is not injurious, there being no molecular disturbance, while an incomplete vitrification or devitrification and, a fortiori, crystallization, are injurious to the extent that they disrupt the living structure. (Luyet and Gehenio, 1940)

> A vitrified liquid is essentially a liquid in molecular stasis. Vitrification does not have any of the biologically damaging effects associated with freezing because no degradation occurs over time in living matter trapped within a vitreous matrix. Vitrification is potentially applicable to all biological systems. (Taylor et al., 2004)

Conceptually, then, little has changed in our understanding of vitrification for more than 60 years. Yet these passages also serve to highlight the tremendous distance that cryobiology has actually covered since the 1940s. Luyet and Gehenio conceded failure to vitrify anything more complex than moss, listing what must have been a frustrating succession of failures to vitrify cell suspensions, as well as plant and animal tissues, including leaves and muscle fibres. In contrast, in their chapter in this volume, Taylor et al. report success in vitrifying complex tissues, if not yet organs, including corneas, blood vessels, and articular cartilage.

It is not just vitrification that has leapt ahead. Chris Polge's felicitous discovery of the cryoprotective effects of glycerol in 1949 (which, incidentally, he referred to as vitrification) sparked a revolution in the science of cryobiology—indeed, some claim it was the beginning of scientific cryobiology. Since then, there have been tremendous developments in both the fundamental understanding and empirical practice of cryobiology. To give a single practical example, cited in the Chapter 18, fertility results following artificial insemination with donor cryopreserved semen have almost doubled in the United Kingdom in the last 8 years, whereas the number of treatment cycles has declined by over 50%. Similar far-reaching advances have transformed the prospects in almost all fields of cryobiology, from the apparently successful freeze-drying of platelets with trehalose (see Chapter 21), to the vitrification of meristems from recalcitrant tropical plants, such as bananas (see Chapter 10).

Perhaps the most remarkable fact about these achievements is that they are based, at least philosophically, on protocols and techniques pioneered in the 1950s by Polge, Audrey Smith, and James Lovelock. Not only are their early methodologies still in use today, but they are still regularly discussed. Several chapters in this book (1, 2, and 17) contain passages in which the validity of Lovelock's conclusions is debated with vigor, if not firm consensus. It is hard to think of many other fields of science in which the significance of experiments carried out 50 years ago is still in dispute today (though quantum physics is one good example, lest readers confound protracted problems with intellectual stagnation). If cryobiology seems to be running to stand still (or better, standing still while giving the appearance of running), the reason is that most practical advances have been built on incremental refinements in methodology. Thus, even though the triumphs might make the pioneers green with envy, remarkably few modern cryobiological methods would take them by surprise. The difficulty is that cryobiology has been straitjacketed by its need to conform to the intractable laws of biophysics. For all its successes, cryobiology has been stuck in a rut.

23.2 LIMITATIONS OF THE BIOPHYSICAL APPROACH TO CRYOBIOLOGY

The incremental refinements in practical methodology over the last 50 years have been based on fundamental advances in our understanding of the biophysics of cryobiology, pioneered by Peter Mazur, Akira Sakai, and others. These biophysical principles are detailed in the chapters by Mazur, Muldrew et al., Taylor et al., and Sakai in this book.

In essence, the survival of cells and simple tissues when subjected to cryogenic temperatures describes an "inverted U" according to the cooling rate. Very fast cooling rates supercool the intracellular environment below the homogeneous nucleation point (about –40°C for cytoplasm), typically causing lethal intracellular freezing. Conversely, very slow cooling (or "equilibrium" cooling) results in the osmotic dehydration of cells as ice crystallizes in the extracellular spaces. The osmotic stresses alone may be severe enough to cause lethal injury, for example, by "salting" proteins into solution and permuting biochemical reactions, whereas changes in membrane permeability to Na^+ may lead to swelling and rupture, as originally argued by Lovelock. Extracellular ice is likely to "seed" nucleation within cells and to directly injure cell membranes. Cells compressed together in the dehydrated state by an advancing ice front are also liable to interact in ways that are not easily reversible; for example, by the fusion of membranes. Even if ice formation is avoided by vitrification, the rewarming of metastable glasses can be critically rate dependent, which poses the problems of devitrification and recrystallization. Although the precise mechanisms of injury in particular circumstances are often unproved, the outcome is clear: poor survival, especially of larger tissues, which are also stressed by purely mechanical forces.

Many of the triumphs of cryobiology over the last four decades stem from manipulating these forces, using controlled freezing and rewarming protocols, along with cryoprotective agents (CPAs). CPAs bring their own problems. At high concentrations, most are toxic in their own right. Exaggerating this toxicity, the equilibration of CPAs between the intracellular and extracellular compartments is slow, especially at subzero temperatures, in comparison with the osmotic movements of water (which may be orders of magnitude faster, even in the absence of aquaporins). The final distribution of CPAs, and the osmotic stresses generated, therefore depends on the permeability of the cell membranes and the temperature. Nonpermeating CPAs have the potential to exacerbate intracellular dehydration without stabilizing intracellular proteins or membranes, whereas permeating CPAs may not be cleared from cells quickly enough on rewarming, leading to swelling and possibly rupture (assuming that osmotic equilibration is rarely perfect, even when cells are subject to "equilibrium" cooling). This means that successful cryopreservation depends largely on the permeability of cells to water and CPAs, and on their sensitivity to the toxicity of high concentrations of CPAs. Sometimes these factors can be balanced to give an optimal protocol, sometimes not. Peter Mazur concludes his chapter with a beautifully concise summary of the practical limitations of the traditional biophysical approach to cryopreservation:

> In some cases, difficulties in cryopreservation may stem from the complex concatenation of conflicting variables. Thus, cooling rates that are low enough to avoid intracellular ice formation may be so slow as to induce damage from solution effects or chilling. The use of higher CPA concentrations to minimize solution effects may introduce toxicity or exacerbate osmotic damage. Toxicity may be reduced at lower temperatures, but lower temperatures slow the permeation and further exacerbate osmotic damage. Damage from external ice can be prevented by vitrification, but the induction of the vitrified state requires high CPA concentrations that exacerbate both toxicity and osmotic problems. These incompatibilities may not be challenges to our understanding but in some cases they remain challenges to achieving successful cryopreservation. (Mazur 2004)

The problem today is that applying the basic principles of biophysics simply cannot solve many of the remaining challenges in cryobiology. The fact is that some cells or tissues deal with physical stresses better than others. Cells that are at once osmotically intolerant and sensitive to high concentrations of CPAs will fare badly under virtually any conventional cryopreservation protocol. Success is likely to be partial at best, and dependent on an empirical testing of different cooling and warming rates, as well as on the concentrations and toxicities of various CPAs. This is, in fact, exactly how some of the more cumbersome and difficult cryopreservation protocols have been developed over the last decades, but as the remaining challenges become steadily more refractory, we can predict that the successes will become correspondingly rare—especially in complex tissues in which different cell types have exacting and diverse requirements. The failure to vitrify large or complex mammalian tissues and organs using CPAs alone illustrates the intractable difficulties involved.

23.3   NATURE'S LABORATORY

There is another way out, and herein lies the probable future of cryobiology: a future that is closely tied in with the spectacular advances in molecular biology and genomics of the last decade. We are today on the verge of an explosion in our understanding of genetic adaptation. As advocated by Harry Meryman in the foreword, we can learn from nature's own laboratory: a marvelous arena with hundreds of specimens in each group—bacteria, algae, fungi, plants, invertebrates, fish, and amphibians—and millions of years run-time. The future of cryobiology is surely the dovetailing of formal biophysics with the study of life's adaptations to similar problems. In effect, if some cells and tissues are refractory to cryopreservation protocols, the trick is not necessarily to change the protocol but to change (precondition) the cells themselves. If the mountain will not come to Mohammed, then Mohammed must come to the mountain.

Even so, a number of failures show the difficulties involved in trying to apply the "tricks" of nature to the problems of cryobiology. Just as the incremental advances in conventional cryobiology have been achieved by the application of a mathematical formalism, so too we must in future apply, at the very least, a philosophical formalism to the remaining challenges. That is to say, we need to think carefully about the selection pressures involved in evolution and the extent to which they coincide with the biophysical pressures of cryogenic freezing. It is not necessarily true that "nature knows best," when nature has never been called on to adapt to –196°C. In this final chapter, then, I will explore some of the most likely avenues of the future. The discussion is inspired by broad themes discussed in the foregoing chapters, and is not intended (or referenced) as a review; rather, it is a "preview" of future possible worlds.

A good example of enthusiasm running before formal analysis is the genetic engineering of tomatoes to express fish antifreeze proteins (AFPs), in the hope of protecting them against frost injury. This was first accomplished in the early 1990s, but proved a disappointment: The engineered tomatoes duly expressed the fish AFPs but were, if anything, more vulnerable to frost than normal tomatoes. Today we know the reason why: At high concentrations, fish AFPs alter the crystalline habit of ice, from dendritic to spicular, and sharp spicules of ice are more destructive of tissue structure than dendritic crystals. This outcome is worth dwelling on for a moment, as it highlights the uneasy relationship between the two Janus faces of science: the fundamental and the applied.

From a fundamental point of view, a detailed understanding of the mechanism of AFPs, apart from being fascinating in its own right, feeds back into applied science in unexpected ways. In this case, fish AFPs are now being used to potentiate damage in cryosurgery, as discussed in Chapter 3 by Elster and Benson, and Chapter 16 by Hoffmann and Bischoff. Moreover, close scrutiny of the mechanisms of spicule formation has underpinned the development of synthetic ice blockers (SIBs), such as 1,3-cyclohexanediol, discussed in the chapter by Taylor et al. By retarding ice formation in general, and by altering the crystalline habit of any ice that does form to hexagonal, rectangular, or trapezoid shapes, SIBs have the power to revolutionize vitrification. If they really can limit the devitrification and recrystallization of metastable glasses, SIBs may overcome the decades-old challenge of successfully rewarming vitrified organs (and, potentially, recalcitrant hydrated tropical seeds such as coconuts, as discussed in the chapter by Benson).

From a strictly "applied" point of view, however, the engineering of tomatoes to express fish AFPs betrays a tendency toward an empirical "wishful thinking" that falls well short of the philosophical formalism discussed above. Any serious progress in resolving the future practical challenges in cryobiology will need to take a more structured approach. Empirical "shots in the dark" are no longer enough. Parallel fields can sometimes cloud the issue. The burgeoning interest in astrobiology, for example—the pursuit of life in space and on planets like Mars—has reinvigorated polar research, especially into microbial adaptations to extreme conditions, including extreme cold. As discussed in the chapters by Elster and Benson, and by Ponder et al., astrobiology shares common ground with cryobiology, but should not be conflated with it. The adaptations of life to cold take two broad forms: freeze tolerance and freeze avoidance. The latter is of profound importance to polar and permafrost ecology. Many permafrost bacteria, for example, avoid freezing and show signs of limited metabolism. In terms of the prospects of finding life actually living on Mars or Europa (rather than anabiotic in space), the study of metabolic adaptations and freeze avoidance is of far greater relevance than "mere" long-term preservation.

From the practical perspective, though, freeze avoidance at high subzero temperatures is a very different matter than freeze avoidance or freeze tolerance at cryogenic temperatures, where cellular metabolism certainly stops. (In both Antarctic sea ice and permafrost bacteria, metabolic activity and protein synthesis are undetectable below about –20°C.) This is the trouble with fish AFPs: Their evolutionary "purpose" is to prevent fish from freezing without the need for high concentrations of osmolytes. This they do very effectively, within a restricted temperature range, by inducing a thermal hysteresis of 1 to 2°C (a lowering of the freezing point without affecting the melting point). Playing around with thermal hysteresis, however, is a dangerous game that can have catastrophic consequences, such as the formation of spicular ice if the temperature supercools below the hysteresis point. In ecological terms, this is unlikely to happen at sea: Fish live in a stable, if icy, environment. Looking to the evolution of fishes for an answer to the very different problems of freeze tolerance or avoidance on land (where temperatures are far more variable and extreme), or worse, at cryogenic temperatures, is sloppy thinking.

23.4 THE CRUCIBLE OF EVOLUTION

Evolution is a grand experimental crucible, but it differs from human experimental research in two crucial respects: first, evolution is "applied" to a degree that would make purists blench, and second, it is inherently multivariate. By "applied" in this case I mean that life evolves to cope with very particular conditions, but only when this type of adaptation, rather than another, is advantageous. The lack of an adaptation may or may not be evidence that it cannot happen. Is the absence of frozen penguins in the Antarctic evidence that large, warm-blooded animals cannot be frozen, or merely that frozen penguins would make an easy meal for marauding seals? In North America, smaller animals, such as frogs and turtles, do indeed freeze. Does the fact that they are smaller and cold-blooded reflect ecological, physiological, or biophysical constraints on freezing tolerance (or all three)?

There are other problems with the use of adaptations as a window on future cryopreservation. Inherent in the very word "adaptation" is a stable environment; obviously no organism can adapt to an asteroid impact, but given a stable environment, life's adaptations are often astonishingly precise. Conversely, the same adaptations are useless if the parameters are suddenly shifted. As discussed in the chapter by Kenneth and Janet Storey, wood frogs (*Rana sylvatica*) can survive freezing for days or even weeks at –4°C, but they cannot survive much longer at this temperature, or at all below about –6°C. Similarly, as discussed by Elster and Benson, the Antarctic alga *Zygnema* can endure repeated overnight exposures to temperatures of about –4°C and still maintain photosynthetic capacity during the day; but if exposed to more prolonged periods of freezing, or to lower temperatures, the cells leak solutes, lose their photosynthetic capacity, and die. The alga is in fact adapted to the Antarctic summer, when nocturnal temperatures rarely fall below –4°C. Wood frogs manage to restrict the temperature to within narrow limits by insulating themselves with snow, leaves, and moss, by which means they also contrive to limit evaporative water loss. In other words, both wood frogs and Antarctic algae have adapted with precision to expected average conditions, which they may help to regulate through their behavior, and natural selection has never been called on to protect them against lower temperatures. One of the most fascinating and difficult questions for cryobiology in future will be the extent to which adaptations to freezing at high subzero temperatures can help precondition against subsequent cryogenic storage. Certainly it is beyond our current competence to cryogenically preserve a frog already frozen at –4°C. We will need to learn from the spirit of the frog's adaptations rather than the letter.

The wood frog is also a good example of the multivariate nature of evolution. As outlined by the Storeys, the expression of a large number of genes is either upregulated or downregulated in response to freezing. The function of several of these genes is still unknown, but those that have been identified do give some sense of the range of adaptations. Broadly speaking, the adaptive response falls into three categories: cryoprotection, suppression of metabolism, and stress response. Glucose is the main colligative CPA, and it probably also helps to stabilize cell proteins during the controlled dehydration of organs. Ice-nucleating agents (INAs) are found in the plasma and extracellular spaces, and presumably help direct extracellular ice formation and reduce osmotic stress. Central metabolism is selectively suppressed, especially ion-motive ATPase activity, muscle energy catabolism, and general biosynthetic pathways, with only a limited fermentative energy production to cover basal metabolism and specific biosynthetic pathways. Of these, the mitochondrial inner membrane ADP/ATP translocase is upregulated severalfold in response to freezing, but how this affects basal metabolism is as yet unknown.

The stress response encompasses both biochemical and biomechanical adaptations, notably a rise in the activity of several antioxidant enzymes including glutathione peroxidase, and indeed a tenfold rise in glutathione levels in some organs, such as the brain. The most striking biomechanical adaptation is the production of fibrinogen peptides, which presumably aid in blood clotting in damaged capillaries following rewarming. The main point, however, is that the frog's cryoprotective strategy is exquisitely fine-tuned, both temporally and spatially. Within minutes, freezing triggers a carefully modulated chain of events, controlled by various signal-transduction pathways and transcription factors. Different genes are transcribed at different rates and in different organs. A loss of synchronization would in all likelihood undermine success. For example, fibrinogen is transcribed during the freezing period, rather than on rewarming—an energy-consuming task in difficult times. Presumably it is vital, literally, to block internal bleeding immediately on thawing, rather than potentially hours afterwards.

23.5   PARALLELS IN PLANTS

Plants orchestrate their responses to freezing in an equally complex manner, as discussed in the chapter by Roger Pearce. At least 60 plant genes are upregulated or downregulated in response to freezing, though not necessarily all at once in the same species (and indeed not necessarily all in response to freezing: Some may be adaptations to other aspects of the winter environment, such as wind and waterlogging). As in freeze-tolerant animals, the identity and function of many of these genes is still unknown, but the parallels in function of the known gene products are striking. Plant adaptations fall within the same three broad categories of cryoprotection, metabolic suppression, and stress response. In terms of cryoprotection, the classes of adaptation are again similar: raised concentrations of colligative cryoprotectants (such as raffinose, sucrose, or fructans) and nucleation of ice in extracellular compartments with INAs. In addition, some plants express AFPs, which (with the odd exception of the carrot) are evolutionarily distinct from fish and insect AFPs. Given my earlier remarks on fish AFPs, it is interesting to note that plant AFPs exert little thermal hysteresis, usually a few tenths of a degree, but are strong inhibitors of recrystallization on rewarming. They also tend to be present at lower concentrations than in animals, consistent with the lower thermodynamic driving force for recrystallization. Thus, in conjunction with INAs, the purpose of the AFPs seems to be to direct and restrict recrystallization, rather than to prevent ice formation at all, as is the case in fish and some insects.

Perhaps most important of all are the various plant adaptations to dehydration and desiccation. Intracellular desiccation, caused by the growth of extracellular ice, is the main cause of death in overwintering cereals and other crops. Plant adaptations to dehydration stress are orchestrated by plant hormones (such as abscisic acid) and transcription factors (such as drought-response element binding-factors, or DREBs). Several signal transduction pathways run in parallel or even converge. Overall, these raise the levels of proline, amino acids, and a number of late-embryogenesis-abundant proteins such as the dehydrins, which were first described in desiccating seeds. The exact mechanism of dehydrins is still unknown, but they are usually supposed to interact with hydrophilic groups on lipids and proteins to stabilize their structure during dehydration. Another interesting possibility is that they stabilize the glassy state in dehydrated tissues at freezing temperatures by raising the glass transition temperature (T_g). In seeds, this would reduce the likelihood of repeated cycles of vitrification and devitrification at freezing temperatures (which are associated with free radical–mediated decomposition). However, there is a danger here, as they also appear to lower the T_g at high water content (presumably preventing immature seeds from entering the glassy state, and again protecting against damaging cycles of vitrification and devitrification). Thus, at high water content, dehydrins seem to inhibit entry to glassy state, and at low water content, they stabilize the glassy state. If this is true, we can predict that the dehydrins should actually be detrimental to the cryopreservation of samples with high water content, as they would tend to destabilize the glassy state, especially at higher subzero temperatures (from –60° to –20°C). Again, we need to think carefully about what dehydrins evolved to do. In applying them to cryopreservation protocols, there is no substitute for a deeper fundamental understanding of their mode of action and evolutionary purpose.

The metabolic response of plants to freezing depends on the broader strategy adopted: innate dormancy (as in woody species, which show the greatest known tolerance of freezing) or potential activity (as in plants that continue to grow during milder spells). In the latter case, metabolic rate is initially greatly suppressed by cold, but then partially recovers, with some degree of oxidative phosphorylation, photosynthesis, and protein synthesis taking place. Such changes are accompanied by other adaptations that facilitate metabolic activity, such as shifts in the spectrum of membrane lipids toward greater unsaturation, to maintain the fluid state. For example, active thylakoid membranes of chloroplasts consistently show a high level of lipid (e.g., phosphatidylglycerol) unsaturation in response to chill (low positive temperatures). In extreme cold, and therefore cryogenic preservation, such changes are probably counterproductive, as unsaturated lipid membranes are likely to lose their normal lamellar structure when dehydrated, and these lipids are also more vulnerable to lipid peroxidation on rewarming. Thus, adaptations favoring winter activity may prove detrimental to successful cryopreservation.

One factor that does seem to be common to both metabolically active and dormant (freeze-tolerant) species is the upregulation of antioxidant enzymes and other stress proteins. A number of antioxidant enzymes are expressed at higher levels in cold-acclimated plants, such as freeze-tolerant wheat, including ascorbate peroxidase, glutathione reductase, and catalase. However, freezing inactivates enzymes such as superoxide dismutase, so high levels of simple antioxidants such as ascorbate and reduced glutathione may play a more important role at deep-freezing temperatures. Again, there is a striking parallel with the wood frog, in which reduced glutathione levels show the most marked rise. As noted in the chapter by Benson and Bremner, many colligative cryoprotectants, including glycerol, sucrose, and Me_2SO, are also excellent free radical scavengers, and may exert a part of their protective effect through this mechanism.

23.6 FIRST STEPS TO THE "NEW" CRYOBIOLOGY

Taken together, the main difficulty in basing the "new" cryobiology directly on nature's laboratory is that evolution selects for phenotypes adapted to specific conditions. Put another way, evolution selects blindly for multigene combinations, gambling on average conditions, whereas the scientific method prescribes the variation of single parameters, while maintaining other parameters at constant levels. It becomes very difficult to interpret the results of experiments in which multiple variables are changed simultaneously. This may be necessary, nonetheless, as individual changes can easily be counterproductive in isolation. For example, the seeding of small extracellular ice crystals with INAs to restrict supercooling prevents the sudden and inequitable osmotic stresses that accompany uncontrolled rapid freezing, whereas dehydration helps to promote vitrification of the intracellular environment. These changes, however, may be far from helpful if there are no compensating defenses against the mechanical injury caused by small extracellular ice crystals (e.g., raised fibrinogen levels in the frog), no defenses against recrystallization (for example AFPs), or no intracellular accumulation of compatible solutes to protect against osmotic stress, membrane fusion, and the precipitation of proteins. Most conventional cryopreservation protocols consider only dehydration and colligative cryoprotection, although INAs and AFPs are beginning to make an appearance in vitrification protocols. With a few exceptions, the other broad issues—the stress response and metabolic suppression—are yet to be incorporated in a systematic way into cryopreservation protocols.

This last consideration returns us to the question of which changes that are adaptive at high subzero temperatures will also protect cells and tissues at cryogenic temperatures. Clearly, colligative cryoprotection is a must. The similarity in the choice of CPAs in bacteria, protists, fungi, algae, plants, and animals, compared with cryopreservation protocols, is striking. Chris Polge's accidental discovery of glycerol has been equally "accidentally" replicated in nature hundreds of times. Interestingly, most species use only one or two colligative CPAs, such as glucose in frogs, sucrose in many plants, and trehalose in many bacteria, yeasts, and fungi (see chapter by Tan and van Ingen). Why this should be the case has never been satisfactorily explained, but the precise choice does matter. A CPA that works well in one species may work badly in another. Thus glucose is an excellent CPA for frog erythrocytes but a poor CPA for human erythrocytes. The problem with glucose is presumably its reactivity with proteins in the Maillard (browning) reaction, as happens in people with diabetes. How wood frogs deal with such high levels of glucose is a fascinating question, with manifest medical applications, and the Storeys discuss a number of secondary adaptations to insulin structure and glucose transport that may have a bearing on the matter.

There is a broader point here—the evolutionary and energetic cost of multiple adaptations. The concentrations of CPA required to protect cells against dehydration, so as to enable vitrification, are likely to be toxic in their own right. Combining a number of CPAs at subtoxic levels is a possible solution to the problem (as in most vitrification protocols), but from a biological point of view, this requires potentially costly controls over the concentration and toxicity of individual components. Each must be homeostatically regulated within tightly defined limits. An alternative possibility is to guard against the toxicity of a single CPA, such as glucose, as in the wood frog. Here the evolutionary rationale is to precondition (in good time) against that CPA, to mitigate its toxic effects. Thus, the wood frog reacts within minutes to ice formation, but its tolerance of high glucose levels varies seasonally. Although such preconditioning involves quite a few concerted adaptations, it is easier to envisage these evolving from an existing stress response—to dehydration, say—than to imagine separate adaptations to subtoxic levels of numerous CPAs. How far we might be able to simulate preconditioning in organisms that (unlike frogs) have not had the benefit of millions of years of evolutionary honing to dehydration is an open question, but it certainly makes sense to amplify natural responses where possible, rather than to superimpose an "alien" protocol.

23.7 PRECONDITIONING DEHYDRATION TOLERANCE

Dehydration tolerance is critical in plants. As discussed in the chapter by Sakai, the twigs of extremely hardy plants such as *Salix sachalinensis*, *Populus maximowiczii*, and *Betular platyphylla* can even survive freezing to nearly absolute zero in liquid helium (at −269°C), if they are first equilibrium-cooled down to about −30°C. The procedure partially freeze-dries the twigs, with much of the freezable water in the cells extracted by equilibrium freezing. When plunged into liquid nitrogen, the remaining intracellular solution vitrifies, preventing further water loss to extracellular ice at lower temperatures, and so avoiding lethal dehydration damage. The stability of such glassy matrices depends on the T_g, which in turn varies with the concentration of the intracellular solution—in other words, the degree of dehydration. This depends on time, as well as the tolerance of cells to dehydration. If the cooling rate is too quick, or is held only a little below zero, then cells do not dehydrate sufficiently, and the intracellular solution may devitrify on rewarming, leading to intracellular ice formation and recrystallization. However, if the prefreezing phase is too protracted, or continues much below −30°C, then the intracellular environment desiccates too far and the cells are injured by dehydration. The "safe zone" varies according to the desiccation tolerance of the plant (and, as we have seen, can be modified by proteins such as dehydrins).

The vitrification protocols pioneered by Akira Sakai replace the prefreezing stage with either solute dehydration or air desiccation, to prevent the formation of ice in both the intracellular and extracellular compartments. Again, tolerance of dehydration is critical. Freeze-drying, solute dehydration, and air-desiccation all require either an innate or an acquired dehydration tolerance. Extremely hardy species have an innate dehydration tolerance, but, excitingly, it seems that even the meristems of tropical monocots, such as bananas and orchids, can acquire dehydration tolerance. By taking advantage of the plant's own stress response, the procedure is remarkably simple, if time-consuming. Thus, preconditioning of meristem donor plants can be achieved simply by supplementing their growth medium with rising concentrations of sucrose over a period of a month before vitrification. Similar protocols, and others such as transient heat shock (at about +45°C), also precondition fungi and yeasts, enhancing freeze-tolerance and survival (see chapter by Tan and van Ingen).

Although cause for celebration, there are greater difficulties in applying such elegant and simple methods to mammalian tissues. For a start, animal cells do not have a cell wall. The extent to which plant cell walls protect against osmotic dehydration is uncertain, but vitrified plant cells are considerably plasmolyzed (their cytoplasm contracts away from the cell wall). This probably protects in two ways, neither of which applies to animal cells: first, the CPA solution fills the periplasmic space, protecting membranes and preventing collapse of cell volume; and second, the cell walls preserve tissue structure and integrity, and prevent fusion of dehydrated cells. However, even though there is no equivalent to cell walls in mammalian cells, the microscopic structure of mammalian tissues is bulwarked by collagen scaffolding, and the absence of cell walls has not prevented the successful vitrification, using SIBs, of quite complex tissues, as discussed earlier (or see the chapter by Taylor et al.). Indeed, the cryobanking of many mammalian tissues does not even require cell survival, but rather the preservation of structural integrity, which acts as a framework for the regrowth of cells from the perimeter (see the chapter by Wusteman and Hunt). Of course, this approach only applies to small grafts with a periphery that is contiguous with the adjoining tissues, from which the appropriate cells can grow, and not to whole organs (although cryopreserved cells can be "seeded" into an organs, as in hepatocyte or islet cell transplants, which repopulate parts of damaged or dysfunctional organs).

A second problem may pose more of an obstacle to the successful vitrification of mammalian cells in organs: the lack of an appropriate stress response to cold. Even in plants, this can be a problem. The strongest protection against freezing injury in plants is produced by upregulating the DREB transcription factors. However, DREBs regulate the downstream transcription of many genes, each of which has an incremental effect on frost tolerance. The difficulty is that DREBs are also found in tropical or temperate plants. In these cases, they seem to control the transcription of a different or overlapping suite of genes in response to other stresses. Rice, for example, expresses DREBs but has no tolerance to cold below +15°C. In mammals and other warm-blooded animals, the problem is even more acute: Whereas mammalian cells are perfectly capable of mounting stress responses, they are extremely sensitive to changes in temperature and do not produce any systematic stress response to chilling below about +32°C. Worse, mammalian homeostasis regulates salt balance strictly, and mammals are never subject to the kind of salt stress that is a major selective pressure for amphibians like frogs. For these reasons there has been little interest in preconditioning mammalian cells to freezing.

Even so, such preconditioning might still be possible. Not all mammalian cells are so molly-coddled. For example, the cells of the renal medulla can tolerate fluctuating concentrations of salts in the molar range as part of the urine-concentrating countercurrent multiplication system. They do so by accumulating compatible solutes, including sorbitol, myoinositol, glycine betaine, and taurine, via transcription factors acting on osmotic response elements. The system appears to be an ancient throwback to similar mechanisms in bacteria and protists. Interestingly, the renal medulla cells are not the only mammalian cells that can respond in this way to osmotic stress. There is some evidence that glucose triggers an osmotic response in vulnerable tissues (such as the micro-circulation) in people with type II diabetes. Other cells show similar responses in cell culture, implying that the osmotic response elements are still functional in many mammalian cells. Whether preconditioning with glucose or salts, or genetic manipulation of the osmotic response elements, might improve CPA loading and tolerance before vitrification is a fascinating question for the future. Certainly, harnessing endogenous osmotic response elements to the task of CPA loading has the potential to overcome many of the difficulties encountered in the engineering of desiccation tolerance in mammalian cells (such as transfection of trehalose-synthase genes, or membrane poration), discussed in the chapter by Acker et al.

23.8 STABILITY OF GLASSES

Assuming that these methods do ultimately enable the stabilization and vitrification of dehydrated organs, one question looms: How stable are metastable glasses? Is there an upper limit to the temperature at which they can be stored, or might it be possible to desiccate to glass and store at room temperature? Predictions about the future have a knack of returning to haunt those reckless enough to make them, but it would be unfair to readers to address the future of cryobiology without venturing an opinion. As befits one of the editors of *Life in the Frozen State*, I hazard the view that desiccation followed by storage at room temperature will prove less fruitful than cryogenic vitrification. Quite apart from the hazards associated with the "collapse temperature" (the temperature at which freeze-dried preparations exhibit observable deformation and cannot easily be rehydrated) the most interesting reasons relate to the stability of glasses below T_g. As discussed by Acker et al., it is not true to say that "molecular motion stops" below T_g. On the contrary, although molecular motion and chemical reactions are much slower in the glassy state than in aqueous solution, membrane degradation can take place in a matter of weeks, even at 50°C below T_g.

The basis of this degeneration is little known, but it may relate to oxidative reactions, especially if the glass is stored not far below T_g and the temperature or moisture content fluctuates (which promotes cycles of vitrification and devitrification and accompanying free radical reactions). Certainly, oxidative damage is known to take place in viable desiccated seeds stored at –20°C. The degree of damage is influenced by the moisture content and the prestorage status of seeds, such as age and lipid composition. Cell survival improves when steps are taken to minimize oxidative damage; for example, by lowering the moisture content, or by storing under vacuum, under nitrogen, or in the absence of light. In this context, Crowe et al. suggest one reason why trehalose seems to be superior to other nonreducing sugars like sucrose, at least under non-ideal conditions: The glycosidic bond between the sugar monomers in trehalose is far less vulnerable to hydrolysis than that in sucrose. As a result of such hydrolysis, the rate of Maillard browning in samples stored with sucrose approaches that of glucose and is up to 2000 times faster than with trehalose. Even with trehalose, however, it is safer to assume that oxidative degradation will continue to some extent at ambient temperatures. For example, lyophilization of some yeasts with trehalose gives rise to genetic variants, such as the petite, respiratory-deficient variants that lack part of their mitochondrial DNA (see chapter by Tan and van Ingen). For long-term storage, it may be necessary to preserve at temperatures well below T_g.

The extent to which oxidative damage can occur at cryogenic temperatures, in either the frozen or the vitrified state, is an open question discussed in the chapter by Benson and Bremner. Gamma irradiation of cells stored in liquid nitrogen produces a detectable hydroxyl-radical signature on electron paramagnetic resonance spectroscopy, but the lack of decay implies that the radicals are "trapped" in the matrix and do not react (at least until rewarmed). Interestingly, the hydroxyl-radical signal does decay at –162°C, which presumably means that hydroxyl-radical reactions can take place in the vapor phase of liquid nitrogen, especially if the storage facility is regularly accessed (see Chapter 15). Similar reactions may take place in the permafrost, which is exposed to ultraviolet and gamma radiations. Even deeply buried permafrost bacteria express catalase, possibly to protect against gamma radiation from radioactive potassium-40 in the surrounding rocks; see the chapter by Ponder et al. Polar bacteria appear to repair DNA at temperatures as low as –17°C, as judged by the incorporation of radio-labeled thymidine, presumably implying that it is damaged at these temperatures, perhaps by oxidative reactions. Whether some degree of DNA damage or epigenetic instability (such as changes in DNA methylation) can be caused by oxidative reactions at cryogenic temperatures is unknown (and seems unlikely), but the stability of genes and genomes over the whole course of cryogenic storage is an issue of growing interest, recently christened "cryobionomics" by Keith Harding. The apparent genetic variability of freeze tolerance between individuals implies that selection for cryogenically stable genotypes might take place before, during, or after storage. Such selection could reflect natural variations in resistance to oxidative stress rather than other, perhaps ultimately more desirable, phenotypic traits related to wild-type vigor.

Nonetheless, if suitable precautions are taken (such as shielding against radiation or light, and safeguarding against large temperature fluctuations in the vapor phase, to prevent degenerative cycles of vitrification and devitrification), it seems unlikely that hydroxyl radicals would be generated by oxidative reactions during cryogenic storage. There is little detectable activity of oxidative enzymes, such as lipoxygenase, below about $-20°C$. At lower temperatures, the concentration of reactants by ice can promote nonenzymatic oxidative reactions and lipid peroxidation. Data from the frozen food industry give an indication of how far such reactions can proceed: In frozen parsley, for example, nearly 20% of phospholipids are lost within a month of deep freezing at $-32°C$, and remarkably, 70% are lost within 3 d at $-18°C$. Thus, cryogenic temperatures may be required to inhibit oxidative reactions altogether.

23.9 STABILIZING ENERGY-TRANSDUCING MEMBRANES

Regardless of how many oxidative reactions take place during cryogenic storage, the most serious threat posed by oxidative injury is postponed until rewarming. This damage is a sequel of the spatial and organizational disruption of energy-transducing membranes during cryopreservation. The vulnerability of mitochondrial membranes was demonstrated by Tappell and others during the 1960s, when they showed that ATP synthesis is uncoupled by cycles of freezing and thawing. The chloroplast membranes are similarly (if not more) vulnerable in plants. Damage to the photosynthetic and respiratory redox proteins (which are very densely packed, accounting for 60% of membrane composition), and to the highly unsaturated lipid bilayer in which they are embedded, is likely to be manifest only on rewarming, when electron transport begins again. Now, oxygen is freely available as an electron acceptor, but the integrity of the electron transport chains is compromised. Such conditions are practically guaranteed to generate superoxide radicals, hydrogen peroxide, and hydroxyl radicals. Apart from the ensuing failure to restore primary energy metabolism when most critically needed, mitochondrial and chloroplast free radical production distorts cellular redox signaling, and if unrestrained, it may promote apoptosis. Release of cytochrome c from damaged mitochondria is redox dependent, and a critical component of the apoptosis execution machinery.

Such vicious circles are well known in the context of ischemia-reperfusion injury in transplanted organs, and although cryogenically preserved tissues are not ischemic (because metabolism has ceased altogether), membrane disruption is likely to provoke a similar outcome. That this really is the case is illustrated not only by studies of cryopreservation (e.g., many studies of oocyte cryopreservation show mitochondrial degeneration) but also by research into the mechanisms of tissue destruction following cryosurgery. For many years a "Cinderella" subject, modern cryosurgery is coming into its own as a sophisticated molecular science with direct relevance to cryopreservation as well as surgery. As discussed in the chapter by Hoffmann and Bischoff, free radical–mediated mechanisms underpin both microvascular and immune injury, which in turn produce indirect cryosurgical damage well beyond the confines of the ice-ball itself. Although the intention of cryosurgery is to maximize damage, it seems probable that the only reason such injury is not routinely detected following suboptimal cryopreservation is that it is not routinely looked for. In this context, Taylor et al. speculate that the "80% ceiling" of success in the vitrification of mammalian tissues may be a consequence of apoptotic cell death and membrane instability. The situation is even more extreme in conventional cryopreservation. For example, the number of live chondrocytes in cryopreserved articular cartilage (40% survival immediately after thawing) is "dramatically lower" 3 months after transplantation, a loss that Muldrew et al. ascribe to apoptosis. The question is, Can we learn from nature's laboratory how to limit the disruption of the energy-transducing mitochondrial and chloroplast membranes?

Plants provide a clue. In photosynthesis, several alternative pathways of cyclic, noncyclic, and pseudocyclic photophosphorylation coexist, and it seems likely that their joint operation ensures redox poise under extremely variable ambient conditions. Similar, albeit less sophisticated, controls are found in mitochondria. Indeed, the existence of independent platoons of mitochondrial and chloroplast genes may be a testament to the importance of rapid genetic responses to environmental changes, to maintain redox poise. Regarding freeze-tolerance, the most interesting adaptation of plant mitochondria is the rapid expression of an alternative oxidase in response to cold, which bifurcates the respiratory chain. The alternative oxidase generates little energy, as it bypasses complexes II, III, and IV, but it does have two important effects: first, it quickly reduces excess oxygen to water, and second, it leaves the respiratory complexes in a relatively oxidized state. Thus, in response to the sudden onset of cold, the mitochondria are effectively "parked" in an oxidized state, with few respiratory electrons kicking around to cause trouble and with low levels of oxygen available to react on rewarming. At the same time, the pool of reduced antioxidants, in particular glutathione and ascorbate, is built up. These antioxidants are concentrated in the mitochondria and chloroplasts. Whatever else happens there deserves closer scrutiny. For example, sugars such as trehalose have been shown to stabilize the respiratory complexes (such as ATPase activity) following freeze–thaw of isolated mitochondria, but whether sugars—or proline and dehydrins—do concentrate in the organelles of freeze-tolerant plants is unknown. Clearly we need to learn a lot more about the subcellular compartmentalization of freezing tolerance.

Given that animal cells do not benefit from the plant's alternative oxidase pathway, is it possible to "park" animal mitochondria in a stable, relatively oxidized state before freezing? It is certainly feasible. Presumably this happens naturally in starvation. Many freeze-tolerant insects survive the polar winters by voiding their guts. Such behavior is usually interpreted in terms of the ice-nucleating properties of feces—removing the feces removes the most potent ice nucleators—but the lack of sustenance may provide the additional benefit of "parking" respiratory chains in a relatively oxidized state. Certainly, as discussed in the chapter by Elster and Benson, the larvae of the Arctic diptera *Heleomyza borealis* enter a preconditioned hypometabolic state before the onset of winter, and this dormant state appears to be its main protective strategy. It is conceivable that we could stimulate similar changes in mammalian cells. For example, calorie restriction in mammals promotes major changes in gene expression, which regulate mitochondrial function and suppress free radical leakage. It would be interesting to see whether calorie restriction, perhaps for a period of a few weeks or months before freezing, could improve mitochondrial integrity during and after cryo-preservation.

The other beneficial effect of expressing an alternative oxidase in plants is the removal of excess oxygen. This may be less easy to manipulate in mammals, as oxygen delivery to cells is very tightly regulated. However, there are some signs that elimination of oxygen is desirable. Mazur's group have shown that the motility of mouse sperm is nearly doubled after cryopreservation with an *Escherichia coli* membrane preparation called Oxyrase, which reduces the oxygen tension to less than 3% of atmospheric pressure. Although such preparations are hardly applicable to whole tissues, it is already known that the outcome of vitrification is enhanced by storage under nitrogen, as discussed earlier. Thus, experimental manipulations to reduce the oxygen tension in stored tissues may well be a profitable avenue in future.

23.10 THE FUTURE OF CRYOBIOLOGY

So where does all this leave the future of cryobiology? We have much to learn from Mother Nature, but if we consider the spirit rather than the letter of evolutionary adaptations, and how to dovetail these adaptations with the formal biophysics of cryobiology, the years ahead look rich with possibilities. The fields that seem especially rewarding relate to the stability of glasses, the preconditioning of dehydration tolerance, and the preservation of metabolic integrity during and after cryogenic storage. If each of these aspects can be incorporated into cryopreservation protocols, I see no reason why the future challenges of cryobiology should not be successfully overcome. Above all, it is time to cast off the old biomechanical view of cells as passive osmomoters and to embrace the dynamism of the new biology.

ACKNOWLEDGMENTS

I would like to thank Professor Barry Fuller, Dr. Erica Benson, Dr. Ana Hidalgo, Dr. Keith Harding, and Dr. David Bremner for stimulating discussions on the content of this chapter.

Index

A

Aat gene, 265
ABA response elements, 192
ABC-transporter proteins, 162
Achlya ambisexualis, 286
Acinonyx jubatus, 402
Acipenser baeri Brandt, 421
Acipenser transmountanus, 421
Actinobacteria, permafrost, 155
Active life zones, of polar regions, 120–121
Adaptation
    membrane, 159
    to permafrost conditions, 158–163
    strategies in polar regions, 125–128, 130–132
Addax nasomaculatus, 403
Adonitol, 364
ADS1/ADS2 genes, 182
Aerophytic communities, in polar regions, 121–122
AFGP gene, 137
A-hemolysin, 590–591
Alaskozetes antarcticus, 134
Alcohol, polyvinyl, 616
Alcohol dehydrogenase, 182
Alcohols, in fungi and yeasts, 284
Algae
    permafrost, 155–157
    in polar environment, 111–150. *See also* Polar
        environment
    rationale for cryopreservation, 301–302
Alginate bead storage, of fungi and yeasts, 292
Alternative oxidase, 656
Alternative oxidase (AOX) pathway, 217–218, 219
Amniotic membrane, tissue banking, 542–543
Anabaena, permafrost, 156
Anabiosis (dormancy), 163–164
Andreaea regularis/gainii, 134
Angstrom-sized membrane pores, 32–34
Anhydrobiosis, 565
Anopheles eggs, ice nucleation in, 31–32
Antifreeze glycoproteins (AFGPs), 618–619
Antifreeze proteins (AFPs), 131
    applications, 136–138, 138
    in cryosurgery, 472
    extracellular ice and, 80
    fish in tomatoes, 648
    naturally occurring, 616
    in permafrost conditions, 160
    in red blood cells, 46–47
    synthetic, 616–619
    in vitrification, 618–619
    vitrification and, 615–616
Antigen excess concept, 461

Antilope cervicapra, 403
Antioxidants, 219–225
    ascorbate, 223–225
    catalases, 221–222
    glutathione, 223–225
    H_2O_2 removal, 221–222
    peroxidases, 221–222
    superoxide dismutase, 220–221
    vitamin E, 223–225
    in wood frog survival, 262–264
Antitumor antibodies, 461–463
Aparticulate domains, 176
Apoplastic proteins, 185
Apoptosis initiators, 471–472
Aquapore theory, of nucleation, 32–34
Aquatic species, gamete cryopreservation, 415–435. *See
    also* Fish
Aqueous vacancies, in articular cartilage, 92–93
Arabidopsis, 190
Arabidopsis thaliana, 182, 184–185, 186, 267, 305
Archaea, permafrost, 157
Arctic Collembola, 130
Arctic vs. Antarctic environments, 112–115. *See also* Polar
    environment
Arrhenius relation, 24–26
Arteries. *See also* Vascular grafts
    vitrification, 626–627
Arthrobacter, permafrost, 154, 155
Articular cartilage
    ice growth in, 91–101
    tissue banking, 547–549
    vitrification, 627–629
Artificial insemination, 518
    cattle, 372–373
    early history, 359
    horses, 376
    pigs, 376–378
    sheep and goats, 373–376
Artocarpus heterophyllus, 318
Arucaria hunstecinii, 318
Arvicola terrestris, 400
Ascomycetes, 280, 286, 292. *See also* Fungi
Ascorbate, 223–225
Asparagus, 336
Aspergillus spp., 285
A-tocopherol (vitamin E), 223–225
Aureobacterium, permafrost, 155
Aureobasidum pullulans, 281
Avian embryonic cells, 386
Avian spermatozoa, 384–386
Avoidance, as adaptive strategy, 125–126